Norwegian Petroleum Society (NPF), Special Publication No. 1

Structural and Tectonic Modelling and its Application to Petroleum Geology

Proceedings of Norwegian Petroleum Society Workshop,
18–20 October 1989, Stavanger, Norway

Norwegian Petroleum Society (NPF), Special Publication No. 1

Structural and Tectonic Modelling and its Application to Petroleum Geology

Proceedings of Norwegian Petroleum Society Workshop, 18–20 October 1989, Stavanger, Norway

Edited by

R.M. Larsen
Statoil, P.O. Box 300, 4001 Stavanger, Norway

H. Brekke
Norwegian Petroleum Directorate, P.O. Box 600, 4001 Stavanger, Norway

B.T. Larsen
Statoil, P.O. Box 300, 4001 Stavanger, Norway

and

E. Talleraas
A/S Norske Shell, P.O. Box 40, 4056 Tananger, Norway

ELSEVIER
Amsterdam – London – New York – Tokyo 1992

ELSEVIER SCIENCE PUBLISHERS B.V.
Sara Burgerhartstraat 25
P.O. Box 211, 1000 AE Amsterdam, The Netherlands

ISBN 0-444-886079

This book is printed on acid-free paper.

Printed in The Netherlands

Preface

A conference entitled "Structural and Tectonic Modelling and Its Application to Petroleum Geology" was arranged by the Norwegian Petroleum Society in Stavanger, Norway, and took place on the 18th to 20th of October, 1989. At the conference, more than 80 papers and posters were presented. Thirty-six of these papers are contained in this volume.

Structural geology and tectonic models have implications on all scales and at several phases during the exploration for and exploitation of hydrocarbons. Large-scale models define the basic concepts of regional deformation, which influences subsidence history and, hence, basin modelling and maturation studies. Semi-regional and local structural studies are necessary to understand fault geometry, timing, migration paths and hydrocarbon traps. Fractures, from composite faults to simple joints, can act as seals and enhance reservoir communication.

A complete structural analysis includes interpretation of seismic data, geodynamic modelling, geometric analysis and studies of fracture systems down to micro-scale. Analyses should construct realistic three-dimensional models in order to unravel structural development in both time and space. The aim of this international conference was to present examples of structural and tectonic models at all scales. Emphasis was put on models for prediction and their applications and limitations in petroleum geology.

The conference was organized in three parts: each topic having a keynote speaker. The three parts were *Regional Tectonics and Basin History*, *The Application of Tectonic and Structural Aanalysis to Hydrocarbon Accumulation* and *Structural Geology on Reservoir and Field Scale*. In addition, two special sessions were arranged; "Tertiary Uplift/Barents Sea" and "Fault Processes and Geometries".

The meeting was initiated by the Geology and Geophysics Committee of the Norwegian Petroleum Society. The organization committee consisted of A. Andresen, H. Brekke, R.H. Gabrielsen, S. Hanslien, B.T. Larsen, R.M. Larsen, E. Talleraas, E. Holter and K. Haugnæss from NPF.

The Norwegian Petroleum Society has decided to introduce a special publication series of which this is the first. This volume, therefore, is the Norwegian Petroleum Society, Special Publication No. 1. It is recommended that this volume is referred to in one of the following forms, as appropriate:

Larsen, R.M., Brekke, H., Larsen, B.T. and Talleraas, E. (Editors), 1992. *Structural and Tectonic Modelling and its Application in Petroleum Geology.* Norwegian Petroleum Society, Special Publication No. 1. Amsterdam, Elsevier.

Jensen, L.N. and Sørensen, K. 1992. Tectonic framework and kalokinesis of the Nordkapp Basin, western Barents Sa. In: R.M. Larsen, H. Brekke, B.T. Larsen and E. Talleraas (Editors), *Structural and Tectonic Modelling and its Application in Petroleum Geology.* Norwegian Petroleum Society, Special Publication No. 1. Amsterdam, Elsevier, pp. 109–120.

The symposium in Stavanger was the ninth in a generally biennial series of meetings organized by the Norwegian Petroleum Society (Norsk Petroleumsforening, NPF) comprising:

- Jurassic Northern North Sea Symposium, Stavanger (1975)
- Mesozoic Northern North Sea Symposium, Oslo (1977)
- Norwegian Sea Symposium, Tromsø (1979)
- The Sedimentation of the North Sea reservoir rocks, Geilo (1980)
- Norwegian Symposium on Exploration, Bergen (1981)
- North European Margin Symposium, Trondheim (1983)
- Habitat of Hydrocarbons Symposium, Stavanger (1985)
- Correlation in Hydrocarbon Exploration, Bergen (1988)

Information on the first five symposia and their publication can be obtained from: NPF, P.O. Box 1897 – Vika, 0124 Oslo, Norway.

The proceedings of the next three symposia, plus those of the "Petroleum Geochemistry in the Exploration of the Norwegian Continental Shelf" and an "atlas" of hydrocarbon discoveries, containing a full description of 35 of Norway's major oil and gas fields and finds, "Geology of the Norwegian Oil and Gas Fields", have been published by and can be obtained from Graham and Trotman Ltd., Stirling House, 66 Wilton Road, London SW1V 1DE, England.

The committee would like to acknowledge the administrative support of T.H. Hult, Statoil.

The Editorial Committe
September 1991

ACKNOWLEDGEMENTS

The editing committee wish to thank the following reviewers for their thoughtful comments and suggestions.

Andresen, A.
Beach, A.
Dangerfield, J.
Dengo, C.A.
Ehrenberg, S.
Eldholm, O.
Faleide, J.I.
Forth, Ph.
Færseth, R.B.
Gabrielsen, R.H.
Gowers, M.B.
Graham, R.H.
Jensen, L.N.
Knipe, R.
Koestler, A.G.
Larsen, B.T.
Larsen, R.M.
Lippard, S.
Michelsen, O.
Milnes, A.G.
Myhre, A.M.
Nash, D.F.
Needham, C.
Nelson, E.B.
Nøttvedt, A.
Olaussen, S.
Olsen, T.
Oprinsen, F.
Pallesen, S.
Roberts, A.M.
Sales, J.K.
Skarpnes, O.
Skogseid, J.
Sørensen, S.
Thon, A.
Tørudbakken, B.
Vejbæk, O.V.
Welbon, A.I.
Wiggins, B.D.
Yielding, G.
Young, R.

List of Contributors

A. ANDRESEN	*University of Oslo, Department of Geology, P.O. Box 1047, Blindern, 0316 Oslo 3, Norway*
N. BARTON	*Norges Geoteknisk Institutt, P.O. Box 40 Tåsen, 0801 Oslo 1, Norway*
A. BEACH	*Alastair Beach Associates, 11 Royal Exchange Square, Glasgow G1 3AJ, Scotland, U.K.*
D. BERING	*Geology Institute, University of Bergen, Bergen, Norway* Present address: *Norwegian Petroleum Directorate, P.O. Box 600, N-4001 Stavanger*
K. BJØRLYKKE	*Department of Geology, University of Oslo, P.O. Box 1047 Blindern, N-0316 Oslo 3, Norway*
A. BROWN	*BP Norway Limited U.A., P.O. Box 197, N-4033 Forus, Norway*
R.W.H. BUTLER	*Department of Earth Sciences, The University, Leeds, LS2 9JT, U.K.*
B.P.R. CALINE	*Koninklijke/Shell Exploratie en Produktie Laboratorium, Volmerlaan 6, 2288 GD Rijswijk ZH, The Netherlands*
J. DANGERFIELD	*Phillips Petroleum Company Norway, Utenlandsk Aksjeselskap, P.O. Box 220, N-4056 Tananger, Norway*
R.L. DART	*Branch of Geologic Risk Assessment, U.S. Geological Survey, Box 25046, MS 966, Denver Federal Center, Denver, CO 80225, U.S.A.*
M. DeBUYL	*Western Geophysical, 10001 Richmond Avenue, Houston, TX 77252, U.S.A.*
C.A. DENGO	*Exxon Exploration Company, P.O. Box 4279, Houston, TX 77210-4279, U.S.A.*
J.B. DUNHAM	*Unocal International Division, P.O. Box 7600, Los Angeles, CA 90051, U.S.A.*
J.E. ILIFFE	*Department of Geological Sciences, University of South Carolina, Columbia, SC 29208, U.S.A.*
H. FARRELL	*Phillips Petroleum Company Norway, Utenlandsk Aksjeselskap, P.O. Box 220, N-4056 Tananger, Norway*
P.S. FEATHERSTONE	*Koninklijke/Shell Exploratie en Produktie Laboratorium, Volmerlaan 6, Rijswijk, The Netherlands*
W. FJELDSKAAR	*Rogaland Research Institute, P.O. Box 2503, Ullandhaug, N-4004 Stavanger, Norway*
R.H. GABRIELSEN	*Geological Institute, Department A, University of Bergen, Allégaten 41, 5007 Bergen, Norway* Present address: *Norsk Hydro A/S, Forskningssenteret, N-5020 Bergen, Norway*
K. GERDES	*BP Exploration, Britannic House, Moore Lane, London EC2Y 9BU, U.K.* Present address: *BP Norway Limited U.A., P.O. Box 197, N-4033 Forus, Norway*
P. HAREMO	*University of Oslo, Department of Geology, P.O. Box 1047, Blindern, 0316 Oslo 3, Norway* Present address: *Norsk Hydro A/S, Drammensveiem 264, P.O. Box 200, N-1321 Stabekk, Norway*

L. HERMANS	*Koninklijke/Shell Exploratie en Produktie Laboratorium, Volmerlaan 6, 2288 GD Rijswijk, The Netherlands*
R.G. HICKMAN	*Unocal Science & Technology Division, 5460 East La Palma Avenue, P.O. Box 68076, Anaheim, CA 92817-8076, U.S.A.*
J. HURST	*BP Exploration, London, U.K.*
R. JEANS	*BP Exploration, London, U.K.*
L.N. JENSEN	*Statoil A/S, P.O. Box 300, N-4001 Stavanger, Norway*
A. KJEMPERUD	*READ Production Geology Services A/S, P.O. Box 145, N-1300 Sandvika, Norway* Present address: *PETEC, P.O. Box 63, N-3478 Nærnes, Norway*
O.S. KLØVJAN	*Norsk Hydro A/S, Harstad, P.O. Box 31, 9401 Harstad, Norway*
I. KNIGHT	*Statoil A/S, P.O. Box 300, 4001 Stavanger, Norway* Present address: *Phillips Petroleum Company Norway, Utenlandsk Aksjeselskap, P.O. Box 220, N-4056 Tananger, Norway*
R.J. KNIPE	*Department of Earth Sciences, The University of Leeds, Leeds, LS2 9JT, U.K.*
A.G. KOESTLER	*Geo-Recon A/S, Bernhard Herres vei 3, 0376 Oslo 3, Norway*
J.A. KORSTGÅRD	*Geologisk Institut, Aarhus Universitet, DK-8000 Århus C, Denmark*
V.B. LARSEN	*Statoil A/S, Stavanger, Norway*
S.E. LAUBACH	*Bureau of Economic Geology, University of Texas at Austin, University Station Box X, Austin, TX 78713-7508, U.S.A.*
F.K. LEHNER	*Koninklijke/Shell Exploratie en Produktie Laboratorium, Volmerlaan 6, Rijswijk, The Netherlands*
I. LERCHE	*Department of Geological Sciences University of South Carolina, Columbia, SC 29208, U.S.A.*
S. LIPPARD	*Continental Shelf and Petroleum Technology Research (IKU), N-7034 Trondheim, Norway* Present address: *Department of Geology and Mineral Resources Engineering, Norwegian Institute of Technology (NTH), N-7034 Trondheim, Norway*
G. LIU	*Continental Shelf and Petroleum Technology Research (IKU), N-7034 Trondheim, Norway*
S.E. LIVERA	*Koninklijke/Shell Exploratie en Produktie Laboratorium, Volmerlaan 6, 2288 GD Rijswijk ZH, The Netherlands*
A. MAKURAT	*Norges Geoteknisk Institutt, P.O. Box 40 Tåsen, 0801 Oslo 1, Norway*
O. MICHELSEN	*Geological Survey of Denmark, Copenhagen, Denmark* Present address: *Department of Geology, University of Aarhus, DK-8000 Århus C, Denmark*
A.G. MILNES	*Department of Earth Sciences, ETH, Zürich, Switzerland* Present address: *Geological Institute, Department A, University of Bergen, Allegt 41, N-5007 Bergen, Norway*
A.W. MITCHELL	*BP Norway Limited U.A., P.O. Box 197, N-4033 Forus, Norway*
T.E. MOGENSEN	*Aarhus University, Århus, Denmark*
K. MONSEN	*Norges Geoteknisk Institutt, P.O. Box 40 Tåsen, 0801 Oslo 1, Norway*
H. MORIZOT	*Elf Aquitaine Norge A/S, Stavanger, Norway*
K. NAKAYAMA	*Department of Geological Sciences University of South Carolina, Columbia, SC 29208, U.S.A.*
I.R. NILSSEN	*BP Norway Limited U.A., P.O. Box 197, N-4033 Forus, Norway*

B. NYLAND	*Norwegian Petroleum Directorate, P.O. Box 600, N-4001 Stavanger, Norway*
N.E. ODLING	*Continental Shelf and Petroleum Technology Research Institute (IKU), NHåkon Magnusson's gt. 1B, P.O. Box 1883, Jarlesletta, 7001 Trondheim* Present address: *Bergen Environmental Centre, IBM, Thormøhlensgt. 55, Bergen, N-5008, Norway*
T. PEDERSEN	*Department of Geology, University of Oslo, Oslo, Norway*
A. RASMUSSEN	*Norsk Hydro A/S, Harstad, P.O. Box 31, 9401 Harstad, Norway*
R.P. RATTEY	*BP Exploration, Glasgow, U.K.*
S.J. REYNOLDS	*Arizona Geological Survey, Tucson, AZ 85719, U.S.A.*
F. RIIS	*Norwegian Petroleum Directorate, P.O. Box 600, N-4001 Stavanger, Norway*
A. ROBERTS	*Badley Ashton and Associates Limited, Winceby House, Winceby, Horncastle, Lincolnshire, LN9 6PB, U.K.*
K.G. RØSSLAND	*Esso Norge A/S, P.O. Box 60, N-4033 Forus, Norway* Present address: *Statoil A/S, UND-LS, ST-FH D1, P.O. Box 300, N-4001 Stavanger, Norway*
M.G. ROWAN	*Alastair Beach Associates, 11 Royal Exchange Square, Glasgow G1 3AJ, Scotland, U.K.*
W. SASSI	*Koninklijke/Shell Exploratie en Produktie Laboratorium, Volmerlaan 6, 2288 GD Rijswijk ZH, The Netherlands* Present address: *Institut Géologie-Géochimie, Institut Français du Pétrole, 14 Avenue de Bois Préau, B.P. 311, 92506 Rueil-Malmaison Cedex, France*
W.J. SCHMIDT	*Conoco Norway Inc., Norway* Present address: *Conoco, 600 N. Dairy Ashford, Houston, TX 77079, U.S.A.*
J. SKAGEN	*Statoil A/S, P.O. Box 40, N-9401 Harstad*
O. SKARPNES	*Statoil A/S, P.O. Box 40, N-9401 Harstad* Present address: *Statoil A/S, U&P INT, DT-FH B3, P.O. Box 300, N-4001 Stavanger, Norway*
J. SKOGSEID	*Department of Geology, University of Oslo, P.O. Box 1047 Blindern, 0316 Oslo 3, Norway*
S. SKOTTHEIM	*Elf Aquitaine Norge A/S, Stavanger, Norway*
K. SØRENSEN	*Statoil A/S, St. Annæ Plads 13, DK-1298 København K, Denmark*
S. SØRENSEN	*Elf Aquitaine Norge A/S, Stavanger, Norway* Present address: *Norex A/S, Luragaarden, Standnes N-4300, Norway*
I.J. STEWART	*BP Norway Limited U.A., P.O. Box 197, N-4033 Forus, Norway* Present address: *BP Exploration Inc., P.O. Box 4587, Houston, TX 77210, U.S.A.*
T. STØLAN	*Norsk Hydro A/S, P.O. Box 31, N-9401 Harstad, Norway*
A. STORLI	*Statoil A/S, Bergen, Norway*
P.T. SVELA	*BP Norway Limited U.A., P.O. Box 197, N-4033 Forus, Norway*
E. SVERDRUP	*Saga Petroleum A/S, Kjørbovn. 16, P.O. Box 490, N-1301 Sandvika, Norway*
H.S. SWOLFS	*U.S. Geological Survey, Box 25046, MS 966, Denver Federal Center, Denver, CO 80225, U.S.A.*
R.O. THOMSEN	*Geologisk Institut, Aarhus Universitet, DK-8000 Århus C, Denmark, and Department of Geological Sciences, University of South Carolina, Columbia, SC 29208, U.S.A.* Present address: *Saga Petroleum A/S, P.O. Box 490, N-1301 Sandvika, Norway*
A.D. van KUYK	*Koninklijke/Shell Exploratie en Produktie Laboratorium, Volmerlaan 6, Rijswijk, The Netherlands*

I.R. VANN	*BP Exploration, London, U.K.*
O.V. VEJBÆK	*Geological Survey of Denmark, Thoravej 8, DK-2400 København, Denmark* Present address: *Amoco Production Company, Houston, TX 77253-3092, U.S.A.*
B.C. VENDEVILLE	*Bureau of Economic Geology, University of Texas at Austin, Austin, TX 78713-7508, U.S.A.*
T. VORREN	*University of Tromsø, P.O. Box 3085, N-9001 Tromsø*
A.I. WELBON	*Department of Geology, University of Oslo, P.O. Box 1047 Blindern, 0316 Oslo 3, Norway* Present address: *Department of Geological Sciences, Queens University, Union Street, Kingston, Ont. K7L 3N6, Canada*
G. YIELDING	*Badley Ashton and Associates Limited, Winceby House, Winceby, Horncastle, Lincolnshire, LN9 6PB, U.K.*
R. YOUNG	*BP Exploration, London, U.K.* Present address: *BP Norway Limited U.A., Forusbeen 35, P.O. Box 197, N-4033 Forus, Norway*

Contents

Structural geology on reservoir and field scale

Geodynamic modelling of the Danish Central Trough

O.V. Vejbæk

A geodynamic model is applied to the post-Middle Jurassic tectonic development of the northern and middle Danish Central Trough. The model uses stretching and phase transformations as subsidence mechanisms. It uses a definition of the lower boundary below the lithosphere in terms of heat flow that contrasts with other models.

Following the Middle Jurassic uplift and erosion of major parts of the Central Trough, an important stretching phase caused rapid subsidence, centred in the Tail End Graben area. The subsidence was controlled by the boundary fault separating the Ringkøbing Fyn High and the Tail End Graben-Salt Dome Province. Local initial stretching factors are calculated to be up to 2. Due to an asymmetric model, the relatively large stretching factors are interpreted as having affected mainly the crust locally, leaving the lithospheric mantle almost unaffected by stretching, as indicated by the assumed presence of a detachment at the deep crustal continuation of the master fault, and by backstripping. As the tectonic activity continued, faulting spread to the west to the Feda Graben, the Gertrud Graben and the Heno Plateau.

The modelling shows that heat flow reached its peak in the deepest parts of the Tail End Graben area during the main Middle to Late Jurassic rifting phase. This was later reversed leading to a concentration of relatively high heat flow in relatively high areas. This is explained by a more shallow location of anomalous heat in the lithosphere and, therefore, a shorter decay constant in the Graben areas, than in the relatively high areas to the west.

It is shown that stretching, thermal subsidence and the metamorphic mechanism incorporated in the model cannot account for the Cenozoic portion of the subsidence pattern. It is suggested that a gabbro–eclogite phase transition may explain the Cenozoic subsidence, with only negligible effect on surface heat flow.

Introduction

The purpose of this paper is to present a quantitative analysis of the Middle Jurassic to Cenozoic subsidence pattern of the Danish Central Trough. This is carried out on the basis of the geodynamic model proposed by Vejbæk (1989).

The purpose of modelling the evolution of a particular basin is to increase the understanding of the interplay between the tectonism, as interpreted from seismic data, and the responsible lithospheric processes. It is also to constrain heat flow variation through time used for modelling maturation of organic matter. Geodynamic modelling has its main origin in the work of McKenzie (1978), which was later applied to the North Sea area (e.g., Sclater and Christie, 1980; Dewey, 1982; Barton and Wood, 1984). The basic concept in this hypothesis is that sedimentary basins are formed by stretching alone. In the North Sea, this has been questioned, because industrial reflection seismic data show too little faulting to explain crustal thinning by stretching (e.g., Ziegler, 1982, p. 90). A recent regional North Sea quantitative subsidence analysis study by Thorne and Watts (1989) has also challenged the stretching hypothesis. They found that additional mechanisms must be active such as gabbro–eclogite phase transition, deep crustal dewatering and anomalous asthenospheric convection. This is especially true in the central parts of the middle and southern North Sea, where a linear to accelerating tectonic subsidence during the Cretaceous, Tertiary and Quaternary is observed. Thermal subsidence alone can thus be rejected for that period.

The concept that metamorphism may play an important role in the formation of sedimentary basins has also been proposed in the form of a one-dimensional model (Vejbæk, 1989). This model differs from the original concept of McKenzie (1978) by including heat production in the lithosphere, phase transformations and, most importantly, by having a heat flow condition below the lithosphere. The lower boundary of the fixed frame used for the calculations is at a fixed depth equal to or below the base of the lithosphere and differs from the frequently used temperature definition. The heat flow condition accounts for a longer duration of anomalous heat flow as required by data (Vejbæk, 1989, 1990). The model

Structural and Tectonic Modelling and its Application to Petroleum Geology edited by R.M. Larsen, H. Brekke, B.T. Larsen and E. Talleraas. NPF Special Publication 1, pp. 1–17. Elsevier, Amsterdam.

handles thermal perturbations induced by one or several phases of stretching, variations in asthenospheric heat flow or sill intrusions.

General geology of the Danish Central Trough

The main structural units of the Danish Central Trough, as defined by Andersen et al. (1982) and Gowers and Sæbøe (1985) and revised by Møller (1986) and Vejbæk (1986), are summarized in Fig. 1. In the western part of the region, the Late Jurassic Grensen Nose and the Outer Rough Basin separate the Inge High and the Mads High from the Mid North Sea High, located in the adjacent British sector (Gowers and Sæbøe, 1985). The Heno Plateau represents an intermediate block in which relatively thin Upper Jurassic deposits are unconformably overlain by Lower and Upper Cretaceous sediments. The Feda Graben contains up to 2.5 km of Upper Jurassic shales (Møller, 1986). It is characterized by continuous sedimentation from Late Jurassic to Late Cretaceous times. This trough is separated from the Gertrud Graben by the narrow Gert Ridge and it extends northwards into Norwegian waters. The Gertrud Graben probably has a subsidence history similar to that of the Feda Graben, but it contains considerably thinner Upper Jurassic deposits.

The Gertrud Graben and the Heno Plateau are separated from the Tail End Graben by the Arne-Elin Graben. The latter has a flower-type structural geometry and is suggested to have been formed by Late Jurassic dip-slip faulting with significant NNW–SSE left-lateral strike-slip components (Møller, 1986; Vejbæk, 1986). The Tail End Graben came into existence during the Permian(?) to Triassic and rapid subsidence during the Late Jurassic resulted in the accumulation of more than 4 km of shale-dominated sediments.

It has been suggested that strike-slip movements along a conjugate set of WNW–ESE right-lateral and NNW–SSE left-lateral fault systems also influenced the rapid Late Jurassic subsidence of the pull-apart basins of the Feda and Tail End Grabens (Gowers and Sæbøe, 1985). However, faulting was probably dominated by dip-slip. The Mandal High and the associated Søgne Basin represent an intermediate fault block between the Ringkøbing Fyn High and the Tail End Graben. The eastern margin of the Tail End Graben is formed by a major basement block of the stable Ringkøbing Fyn High.

The Tail End Graben is bounded to the south by the Salt Dome Province. Here the structural framework becomes strongly overprinted by successive diapirism of the Zechstein salt, initiated during the Late Triassic. During the Cretaceous all of the structural highs became submerged and the individual subbasins coalesced into a regionally subsiding basin.

Subsidence history

The Danish Central Trough probably started to subside in the Early Permian (e.g., Ziegler, 1975; Michelsen, 1982; Gowers and Sæbøe, 1985). The extent of Permian deposits below the central part of the Danish Central Trough is limited. Only the northeastern part of the Danish Central Trough in the southern Søgne Basin and the Salt Dome Basin, show Zechstein salt structures. The only connection between the northern and southern Zechstein basins in the Central Trough area seems to have been through the Tail End Graben and the Søgne Basin. However, the deeper parts of the Tail End Graben, which is mainly filled with Mesozoic sediments, cannot be resolved seismically. Hence, it remains uncertain whether or not there has actually been any connection between the northern and southern Zechstein basins through the Central Trough. If a connection existed between the two basins, it probably was of a limited extent, since no salt structures can be detected in the central part of the region.

Triassic sedimentation in the Central Trough is indicated by more than 1400 m of Triassic sediments in the U-1 well (Jacobsen, 1982). This is a minimum thickness, since the well did not penetrate the entire Triassic succession, and the top of the section is mildly truncated by the Fjerritslev Formation. Where the overlying Lower Jurassic Fjerritslev Formation is present, the unconformity is weak and the youngest Triassic belongs to the Lower Rhaetian. Where the Fjerritslev Formation has been removed during the Middle Jurassic uplift, the Triassic may be more severely truncated or even completely removed, as is seen in the western part of the Central Trough area. How much of the Central Trough was actively subsiding as a graben feature is uncertain due to these later erosion events. Data from the V-1 well suggest that the Ringkøbing Fyn High was covered by sediments before the Middle Jurassic uplift (Møller, 1986). Similarly, the absence of Triassic sediments in wells such as P-1 may indicate Middle Jurassic erosion rather than a primary distribution of Triassic sediments. The present distribution of Triassic sediments is not well known due to limited well control and uncertain seismic correlation across major structural features. Triassic sediments are likely to be present in the entire Salt Dome Basin, Tail End Graben and Søgne Basin. To the west, the Triassic is thin or absent due to erosion, and Middle and Upper Jurassic sediments rest with angular unconformity on Palaeozoic strata. Possible Triassic sediments in the

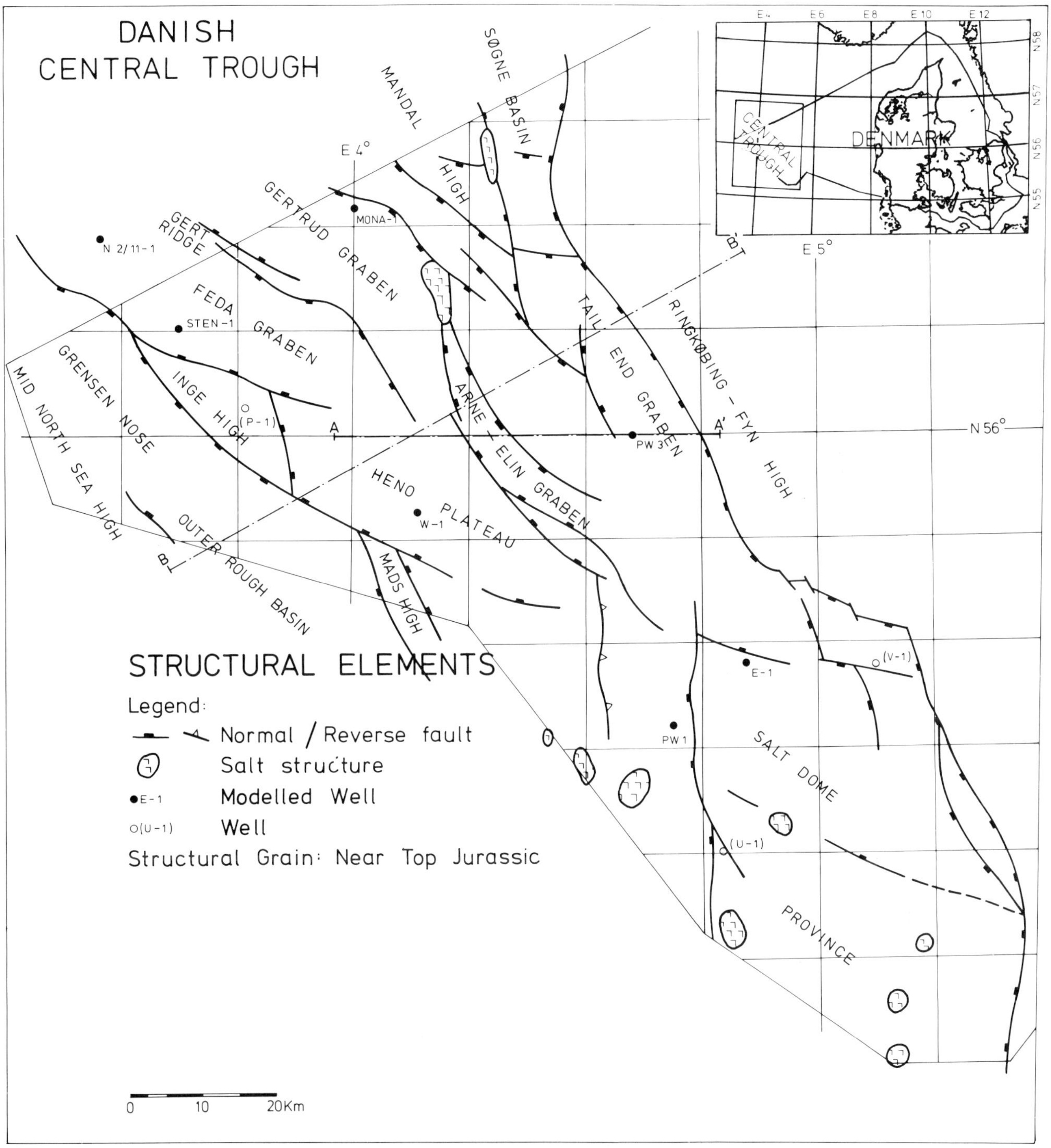

Fig. 1. Structural outline of the Danish Central Trough with location of the wells and profiles used in the modelling (based on Møller, 1987). *A-A′* and *B-B′* are profiles shown in Figs. 8 and 10, respectively.

Feda Graben are probably truncated to the south.

A minor unconformity separates the Triassic from the Lower Jurassic in the Salt Dome Province (Andersen et al., 1982). However, this minor event probably represents an episode in a decreasingly active tectonic environment. This may also be indicated by reconstructions of tectonic subsidence elsewhere in the North Sea (Sclater and Christie, 1980; Wood and Barton, 1983). Hence, the Early Jurassic was characterized by tectonic stability, with rather uniform and possibly widespread deposition of shales in a much-expanded basin. This situation was reversed in the Middle Jurassic, when renewed tectonic activity caused updoming and erosion of the central part of the North Sea. This was accompanied by marked narrowing of depositional areas to fault-bound basins and deposition of mainly coarse clastics (Eynon, 1981). As noted above, this tectonism resulted in extensive erosion to the west, whereas very little material was removed in the vicinity of the Ringkøbing Fyn High. During this period the general post-Triassic structural framework of the Central Trough evolved. Since the magnitude of the erosion is hard to assess with sufficient precision for quantitative modelling, the ensuing modelling attempt begins with the post-Middle Jurassic.

During the Late Jurassic, the rate of tectonic subsidence increased locally as wrenching gave rise to considerable lateral variations in subsidence rates. This initiated the major Upper Jurassic depocentres such as the Feda and Tail End Grabens (Gowers and Sæbøe, 1985; Møller, 1986). During the Middle to Late Jurassic and Early Cretaceous, the depositional area gradually expanded towards the west.

The structural framework of the Lower Cretaceous is very similar to the Upper Jurassic, with extensional block faulting continuing under the influence of wrenching. The importance of the WNW–ESE orientated right-lateral wrench system suggested by Gowers and Sæbøe (1985) gradually ceased, but the conjugated left-lateral NNW–SSE system continued to be active during the Early Cretaceous (Vejbæk, 1986). The rate of subsidence decreased markedly and the Lower Cretaceous deposits became separated more into local depocentres, partly as a result of decreased sediment supply and partly due to the accentuation of ridges and structural highs separating the subbasins. At this time, the Tail End Graben was separated from the Søgne Basin due to uplift of the Mandal High Complex. Also, the connection between the Tail End Graben and the Salt Dome Province became restricted due to the initiation of the low local east-west trending Pollerne Ridge.

Block faulting gradually ceased from Hauterivian and onwards and gave way to regional subsidence, interrupted by periods of inversion (Vejbæk and Andersen, 1987). Inversion produced strong reverse faulting. Throws exceed 600 m locally in conjunction with mild dextral strike-slip faulting along reactivated NNW–SSE striking fault systems. During the latest Cretaceous times the inversion was dominated by flexing rather than faulting.

Inversion ceased in the Tertiary and regional subsidence turned the North Sea Basin into the large north-west European sedimentary basin seen today. Deposition rates changed gradually from moderate to high through the Tertiary and Quaternary with shifting depocentres, but with very little variation in total thickness. The total thickness of the Cenozoic within the Danish Central Trough varies from about 2 km in the southern part to about 3 km in the northern part (Day et al., 1981; Vejbæk and Andersen, 1987).

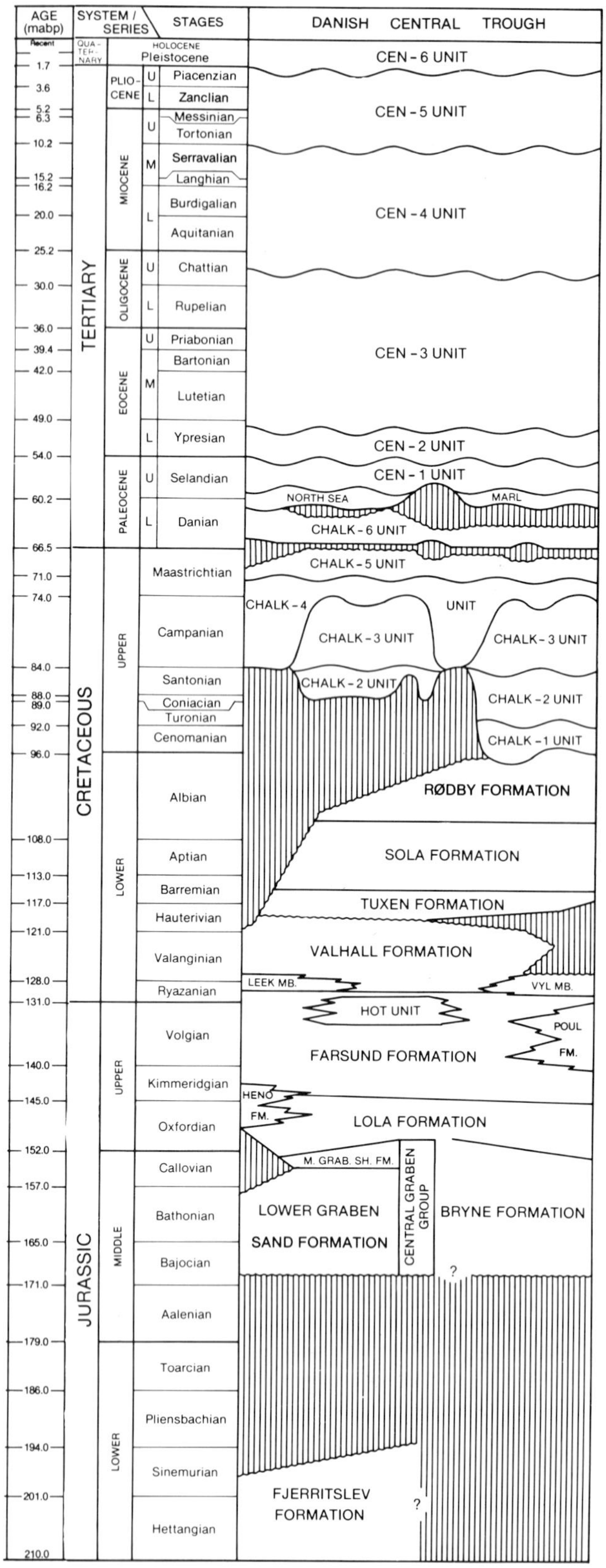

Fig. 2. General stratigraphy of the post Triassic sediments of the Danish Central Trough. Absolute ages are from Haq et al. (1987). The figure is modified from Michelsen et al. (1981) and Jensen et al. (1986).

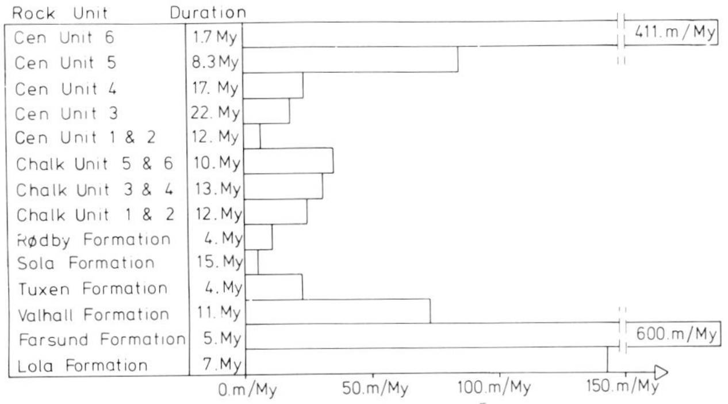

Fig. 3. Estimated maximum rates of deposition for major rock units. Thicknesses are not decompacted. The durations noted are based on the ages cited in Fig. 2.

The average rates of deposition for the major stratigraphic units (Fig. 2) are shown in Fig. 3. Periods of relatively vigorous tectonic activity are indicated by the high depositional rates, as water depths are assumed to have remained below a few hundred metres. High deposition rates are recorded for the Lola and Farsund Formations and the youngest part of the Cenozoic. The idea of McKenzie (1978), that short episodic events of rapid subsidence are caused by rifting and account for the bulk of the sediments, and that they are separated by longer periods of thermal subsidence, seems to be justified (Fig. 3). However, as will be discussed later, the Late Cenozoic subsidence is not a rifting period, although subsidence rates compare with those recorded by deposition of the syn-rift Late Jurassic Farsund Formation.

The base of the modelled succession is taken as the Middle Jurassic, because the amount of preceding erosion is considered ambiguous, thus rendering quantitative reconstruction of tectonic subsidence unreliable.

Geodynamic modelling

Modelling is divided into a study of the subsidence history and an assessment of what may have caused it. The subsidence is suggested to be controlled by three major factors: crustal thinning, phase transformations and the heat budget of the lithosphere. Modifications of subsidence patterns by variations in flexural rigidity are disregarded. Quantification is done mainly by fitting the one-dimensional backstripped tectonic burial graphs to tectonic subsidence curves. The stretching factors are also constrained by evidence seen in industrial seismic data. In contrast to the original McKenzie (1978) model, in which the crust and immediately subjacent lithospheric mantle are stretched by the same factor, an asymmetric model is used here, in which the

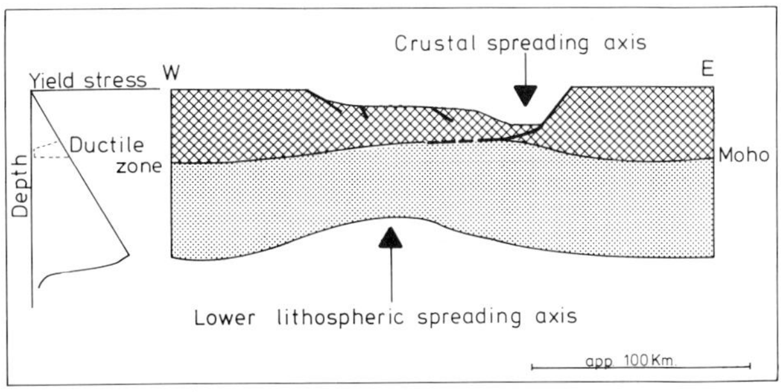

Fig. 4. Tentative model to unify the modelling results (modified from Braun and Beaumont, 1989). A detachment level develops as temperature increases during stretching (dotted line in yield stress diagram). This may cause stretching to be transferred to a pre-existing weakness zone, represented here by the main fault of the Tail End Graben. Lithospheric stretching factors will be low at that location, as opposed to the spreading axis area.

two layers are stretched differentially (cf. Coward, 1986). Although the regionally averaged stretching factor should be the same for both layers, the axis of maximum stretching offsets considerably from one layer to another. The two zones of maximum extension are accommodated by a detachment zone in the deep crust (Fig. 4). The two main steps in the exercise, backstripping and the modelling itself, are briefly summarized below. Detailed discussions may be found in Yükler et al. (1978) and Vejbæk (1989, 1990) with regard to backstripping and subsidence modelling, respectively.

Backstripping

Backstripping was performed using the procedure described by Yükler et al. (1978) which, among other things, accounts for the effects of overpressuring on the burial history. It is a finite difference method based on the principles of conservation of mass, momentum and energy. Compaction is considered to be governed by fluid flow influenced by thermo-mechanical changes. The reason for using this method, instead of the more simple backstripping procedure based on exponential decrease of porosity with depth, is that substantial overpressuring is seen in the Central Trough and the Viking Graben (e.g., Chiarelli and Duffaud, 1980; Thorne and Watts, 1989). It causes significant breakdown of the simple empirical porosity–depth relationship, that would cause serious errors if it were not considered. According to the Yükler (1978) model, overpressuring occurs mainly as a consequence and during periods of high deposition rates. For instance, if high deposition rates during stretching were not considered they might lead to overpressuring and thus underestimated initial subsidence.

The backstripping is followed by tectonic backstripping. Tectonic backstripping is referred to here as a separate procedure whereby the effects of sediment loading are subtracted to yield the tectonic subsidence under the assumption of isostatic equilibrium. It is represented as waterloaded basin subsidence (modified from Steckler and Watts, 1978; Vejbæk, 1990) :

$$Y = \frac{Z\rho_{\mathrm{m}}(1-\alpha T_{\mathrm{i}})-M}{\rho_{\mathrm{m}}(1-\alpha T_{\mathrm{i}})-\rho_{\mathrm{w}}} + W_{\mathrm{d}} - \Delta S_{\mathrm{l}}\frac{\rho_{\mathrm{m}}(1-\alpha T_{\mathrm{i}})}{\rho_{\mathrm{m}}(1-\alpha T_{\mathrm{i}})-\rho_{\mathrm{w}}}$$

where Y is water loaded basement elevation, ρ_{m} is mantle density, ρ_{c} is average crustal density, ρ_{w} is water density, T_{i} is the mantle temperature at the base of the lithosphere, α is the thermal expansion coefficient, Z is decompacted sediment thickness, W_{d} is palaeo-water depth and ΔS_{l} is palaeo-sealevel. The total mass of the sediments (M) is given by:

$$M = \left[\rho_{\mathrm{w}} \int_0^Z \phi(z')\,\mathrm{d}z' + \int_0^Z \rho_{\mathrm{s}}(z')[1-\phi(z')]\,\mathrm{d}z'\right]$$

where $\phi(z')$ is the porosity–depth function and $\rho_{\mathrm{s}}(z')$ is the density–depth function of the sediment matrix.

The geodynamic model

The model that was used for geodynamic modelling (Vejbæk, 1989) differs from the original concept of McKenzie (1978) mainly by including heat production in the lithosphere, allowing subsidence to be enhanced by phase transformations and by supplying a flow condition at the lower boundary below the lithosphere. The heat flow definition at the base of the model frame differs from the frequently used temperature definition. The model handles thermal perturbations caused by one or several phases of stretching, variations in asthenospheric heat flow and sill intrusions.

Three heat sources are assumed to govern the steady-state conditions prior to stretching: in the crust, in the lithospheric mantle and a heat flow from the asthenospheric into the lithosphere. By including these sources, it is possible to fit modelled surface heat flow to observed surface heat flow and to match present lithospheric thickness, as well as honour constraints imposed by burial history. The latter causes a fourth thermal contribution, in the form of a relatively high transient component in present surface heat flow.

Here the traditional temperature definition at the base of the lithosphere is replaced by a heat flow condition at a fixed depth. This is because the former leads to large variations in asthenospheric heat flow (Vejbæk, 1989, 1990). The heat flow condition used also accounts for a longer duration of raised transient surface heat flow that is better able to honour the data and may also explain the prominence of the low velocity layer of the asthenospheric mantle below Phanerozoic tectonic provinces (Calcagnile and Scarpa, 1985). In the present study, the asthenospheric heat flow has been allowed to vary slightly in order to give minor adjustments to the rate of decay of the thermal anomaly responsible for thermal subsidence.

The phase transformation mechanism included in the model is very similar to those proposed by Falvey and Middleton (1981) and Sandiford and Powell (1986). It assumes that stretching causes a release of volatiles in the crust that increase its density, provided the volatiles escape. The theoretical pressure–temperature path that rocks in the vicinity of the Moho follow during stretching and subsequent cooling is shown in Fig. 5. During stretching, the rocks first experience a pronounced pressure drop without any significant fall in temperature. During thermal subsidence temperatures slowly decrease. A slight pressure increase owing to sediment loading occurs, but is not included in the model. These strong pressure/temperature changes are assumed to create possibilities for a granulite-type metamorphism exemplified by the serpentine–forsterite transition. This transition may not be a typical granulite facies process, but is chosen because densities honour the data and fit very well into the simple two layer description of the system. The traditional eclogite/gabbro system (represented by the eclogite–garnet granulite transition) is unable to contribute to subsidence during stretching, because a volume of rock

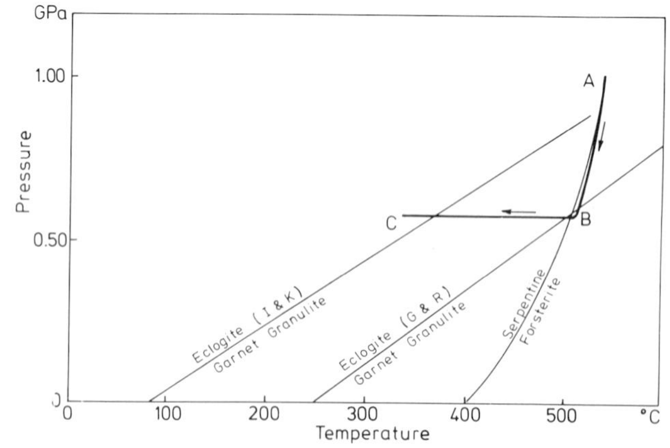

Fig. 5. Pressure–temperature diagram, showing the theoretical P/T time development in the vicinity of the Moho. One GPa corresponds roughly to 30 km. Pressure decreases represent periods of stretching (A to B), temperature decreases indicate periods with decay of the thermal anomaly (B to C). The former period is relatively fast, the latter is relatively slow. The garnet-granulite to eclogite (relevant for the gabbro to eclogite transition) and the serpentine to forsterite phase boundaries (Kitahara et al., 1966) representing amphibolite to granulite transformations are also shown. $G\,\&\,R$ = Green and Ringwood (1971); $I\,\&\,K$ = Ito and Kennedy (1971).

will move from the high density eclogite field to the low density garnet granulite field during stretching (Fig. 5). However, this phase transition may become important long after stretching. This may be of particular importance if gabbroic rocks are intruded during, and as a consequence of, stretching.

Application

The evaluation of the post-Middle Jurassic subsidence history of the Danish Central Trough is based on analysis of seven wells (Fig. 1). Two of these (PW1 and PW3) are pseudo-wells, constructed with the aid of seismic data and using rock types, porosities, etc. derived by extrapolation from nearby wells. The location of the seven wells are chosen as representative of the central part of each subbasin in order to minimize effects relating to higher orders of dimension and, therefore, cannot be monitored by the one-dimensional model. These effects include lateral heat flow (Cochran, 1983; Steckler, 1986; Moretti and Chenet, 1987). Data pertinent for the backstripping, such as palaeo-water depths, rock types, thermal and hydraulic conductivities, etc., are given by Jensen (1990).

In order to ensure comparability and mapability of the modelling results, the starting conditions for each simulation are the same (Table 1). Present-day crustal thicknesses follow those of Barton and Wood (1984). According to their interpretations, based on refraction seismic data, the crustal thickness excluding sediments may be less than 10 km in central areas of the adjacent Norwegian Central Trough.

The heat production parameters for the crust and lithospheric mantle and the average asthenospheric heat flow are fitted to data in the E-1 well to yield present-day heat flow values (Madsen, 1975; Jensen, 1990). Madsen (1975) estimated the heat flow to be in the order of 59 mW/m^2 for the Danish Central Trough area, but this value was not corrected for palaeoclimatic effects and may thus be around 10 mW/m^2 higher (e.g., Balling, 1979). It is important to note that the temperatures of the sediments within the Central Trough are unlikely to be in steady-state equilibrium with regard to basement heat flow. This is due to effects from compaction and fluid flow caused by the high deposition rates in the Quaternary (Fig. 3), which are likely to perturb the temperature profile. Results may, therefore, not be exactly comparable to present heat flow estimates obtained from wells, even if the modelling is correct with respect to heat flow at top basement level. However, it is suggested that values between 60 and 70 mW/m^2 are acceptable.

TABLE 1

Pertinent parameters common for all of the modelled wells. The crustal thicknesses etc. refer to the initial conditions before the post-Middle Jurassic stretching commenced

Parameter	Symbol	Value
Crust	C_z	35.542 km
Lithosphere	L_z	11.019 km
Heat production crust	A_{cz}	6.13×10^{-7} W/m^3
Heat production lithosphere	A_{lz}	4.00×10^{-8} W/m^3
Asthenopheric heat flow	Q_{ast}	0.0325 W/m^2
Mantle density	ρ_m	3330 kg/m^3
Crustal density	ρ_c	2800 kg/m^3
Water density	ρ_w	1000 kg/m^3
Expansion coefficient	α	3.28×10^{-5} K^{-1}
Thermal conductivity	K	3.24 W/m/K
Thermal diffusivity	κ	8.0×10^{-7} m^2/s^{-1}

The stretching history as obtained from the simulations is summarized in Fig. 6. The tectonic activity began in the Callovian along the boundary faults of the Tail End Graben and Salt Dome Province trend, where large stretching factors are applied. Initial stretching factors are modelled to be 2.1, 1.6 and 1.4 for the PW3, the PW1 and the E-1 locations, respectively. These are crustal stretching factors, whereas the immediately subjacent lithospheric mantle is thinned only by factors of 1.15 for the PW1 and PW3 locations and 1.03 for the E-1 location. The difference in stretching between crust and lithospheric mantle is necessary in order to match

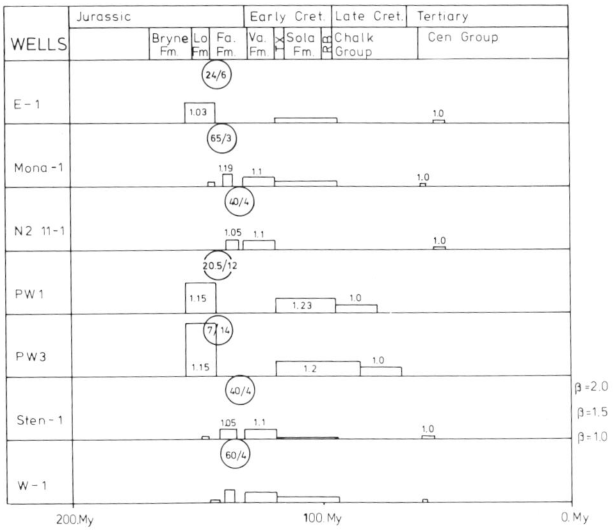

Fig. 6. Diagram showing the timing, amount and the duration of stretching events. All stretching factors are as shown for the Sten-1 well. When the lithospheric mantle has not been stretched by an equal amount as the crust, the mantle stretching factor is indicated in the block. The encircled values indicate the introduction of volcanic sills simulated by a raised temperature of 1400°C. The first number indicates the depth to the top, the second the thickness of the sill in kilometers. *Lo. Fm.* = Lola Formation; *Fa. Fm.* = Farsund Formation; *Va. Fm.* = Valhall Formation; *TX* = Tuxen Formation; *RB* = Rødby Formation

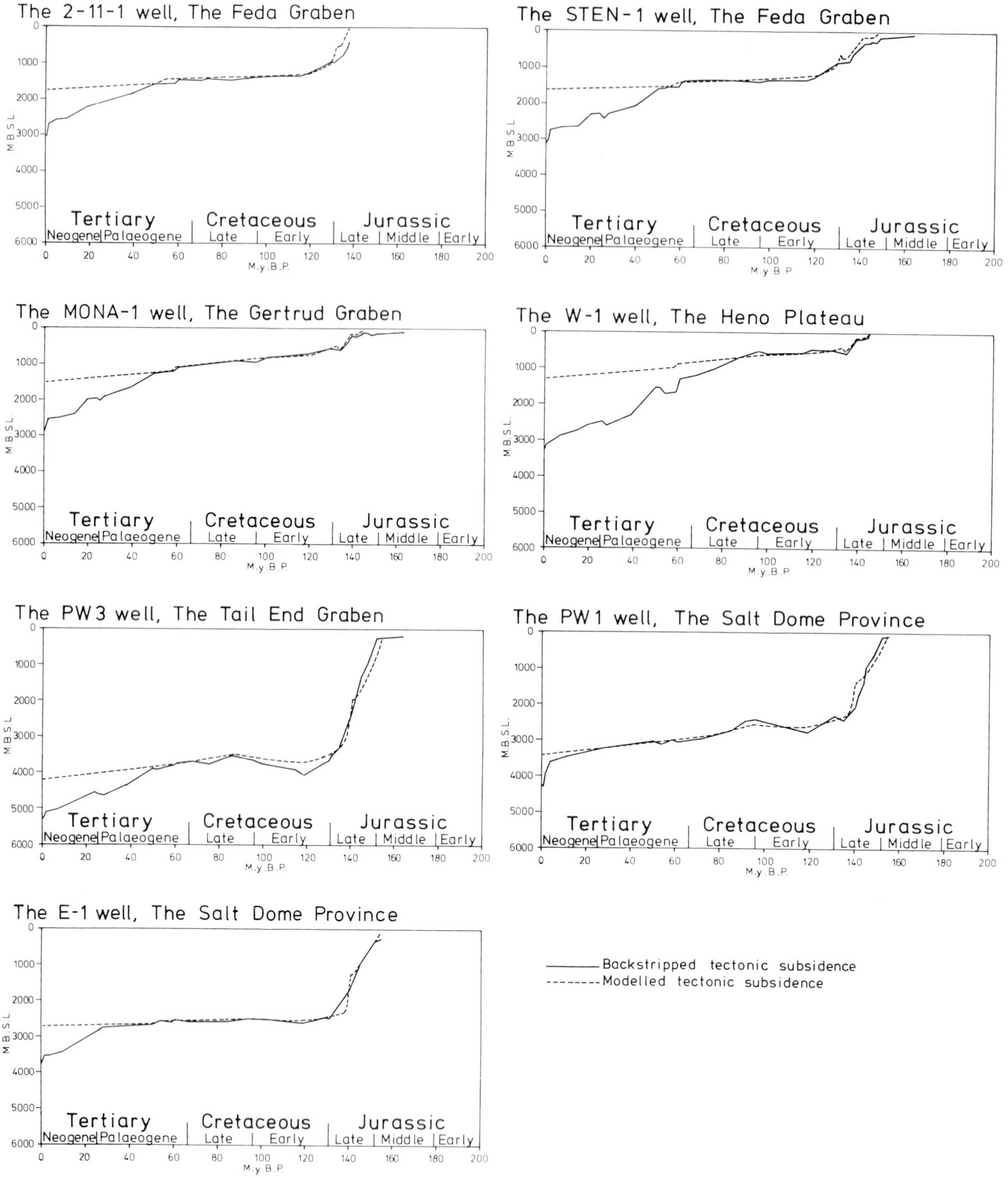

Fig. 7. Backstripping and the best fitting modelled tectonic subsidence curves of the seven wells (Fig. 1). All curves represent water-loaded basement movements where effects of sediment loading have been subtracted. The actual sediment accumulation is about twice as large. The general mis-match between the Cenozoic portion of the modelled and the backstripped curves is intentional, because no mechanism in the model can provide a satisfactory explanation of that part of the backstripped subsidence curve. See text for discussion.

the backstripped curves, which display remarkably little thermal subsidence following stretching. This is particularly evident in the case of the E-1 well (Fig. 7).

Although stretching clearly has occurred as seen in seismic data (Fig. 8), the tectonism has given rise to surprisingly little anomalous heat build-up in the lithosphere during the stretching. The small stretch-

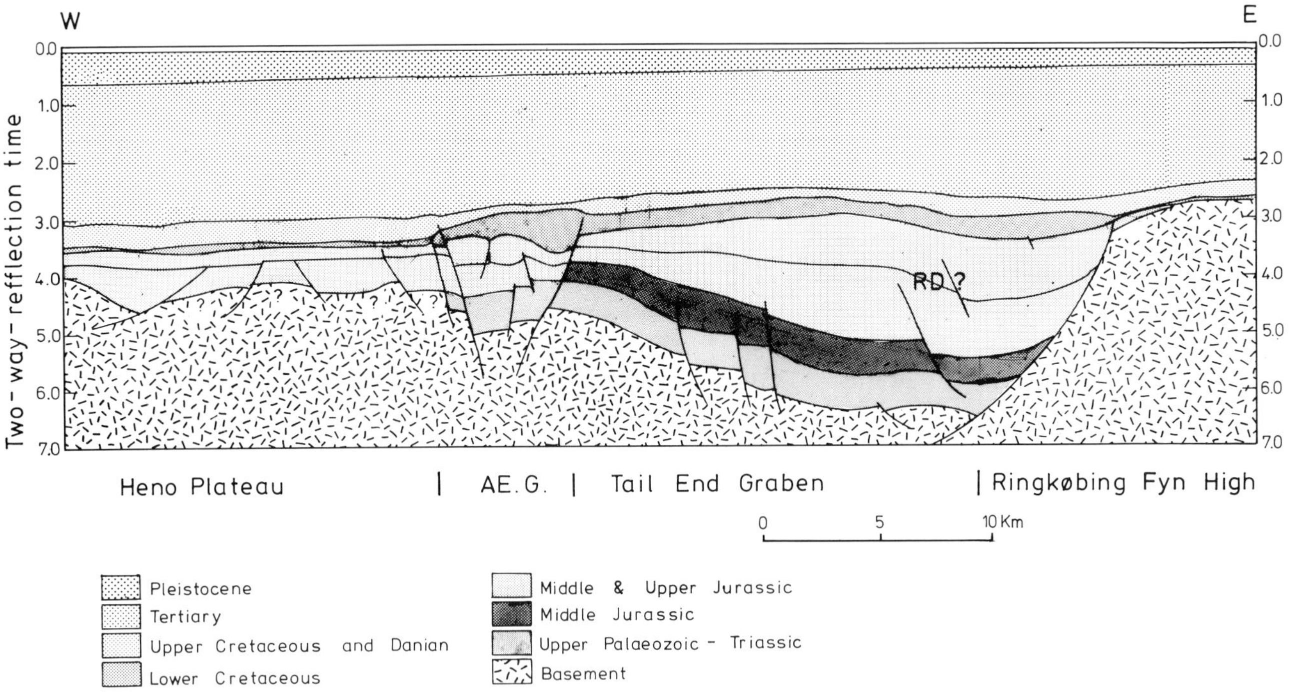

Fig. 8. Profile across the Tail End Graben (Fig. 1). Note the zone with an increased dip in the central to eastern part of the graben (*RD?*), which may be equivalent to the reverse drag of Groshong (1989). The divergence of reflectors in the Upper Jurassic succession into the graben reflects the syn-depositional block-rotation. *AE. G* = Arne-Elin Graben.

ing factors for the lithospheric mantle accounts for this and requires the presence of a detachment level in the vicinity of the Moho (cf. Coward, 1986). The apparent block rotation of the Tail End Graben along the boundary fault to the Ringkøbing Fyn High may suggest a deep detachment for the fault, possibly via a listric fault geometry (Fig. 8). The fault-controlled localized Late Jurassic subsidence is also reflected in the isopach map for the Upper Jurassic, which shows thicknesses in excess of 3.5 km adjacent to the boundary fault between the Tail End Graben and the Ringkøbing Fyn High (Fig. 9).

The tectonic activity gradually spread to the west during the Late Oxfordian, culminating around the Kimmeridgian–Volgian transition. This is simulated as several stretching events with maximum factors of 1.25 at the Mona-1 and W-1 locations and 1.2 at the N2/11-1 and the Sten-1 locations. Although stretching in these locations (except for the W-1 location) also is concentrated in the crust, the differentiation is less pronounced with factors of 1.1 to 1.2 for the lithospheric mantle. The migration of extension westward is also evidenced by the progressively younger sediments on top of the basal Middle Jurassic unconformity westwards indicating later onset of subsidence (Johannesen, 1988).

In addition to the initial stretching, simulation of intrusive activity is required to enable a better fit to the initial part of the thermal subsidence curves following main rifting. The temperature increase modelled thereby also enhances modelled phase transformations. This activity is simulated as sill intrusions with a temperature of 1400°C (Fig. 6). The most important locations for intrusions were under the Upper Jurassic depocentres at the PW3, PW1 and E-1 locations, where density increases are seen caused by the intrusion simulation equivalent to thinning factors of 3, 1.4 and 1.5, respectively. Subsidence originating from phase changes modelled for the N2/11-1, the Sten-1 and the Mona-1 locations only equals thinning factors of 1.05, 1.05 and 1.02, respectively, but has no effect at the W-1 location.

The relatively few stretching events needed at the PW1, PW3 and E-1 locations compared to the other locations may be caused by the fact that they are located in the main Upper Jurassic depocentres. As the later events at the other locations are relatively small, they may have been masked completely by compaction at the PW1, PW3 and E-1 locations, thus preventing their detection in the subsidence curves. However, in the Hauterivian and the Late Cretaceous interval, stretching of up to 1.3 is applied at the PW1 and PW3 locations, whereas extension ceased in all other areas. These areas continued to subside, whereas compressional fault activity affected most of the other locations (cf. Vejbæk and Andersen, 1987).

The modelled curves fit well for the pre-Cenozoic. However, the Cenozoic portion of the curves fits only

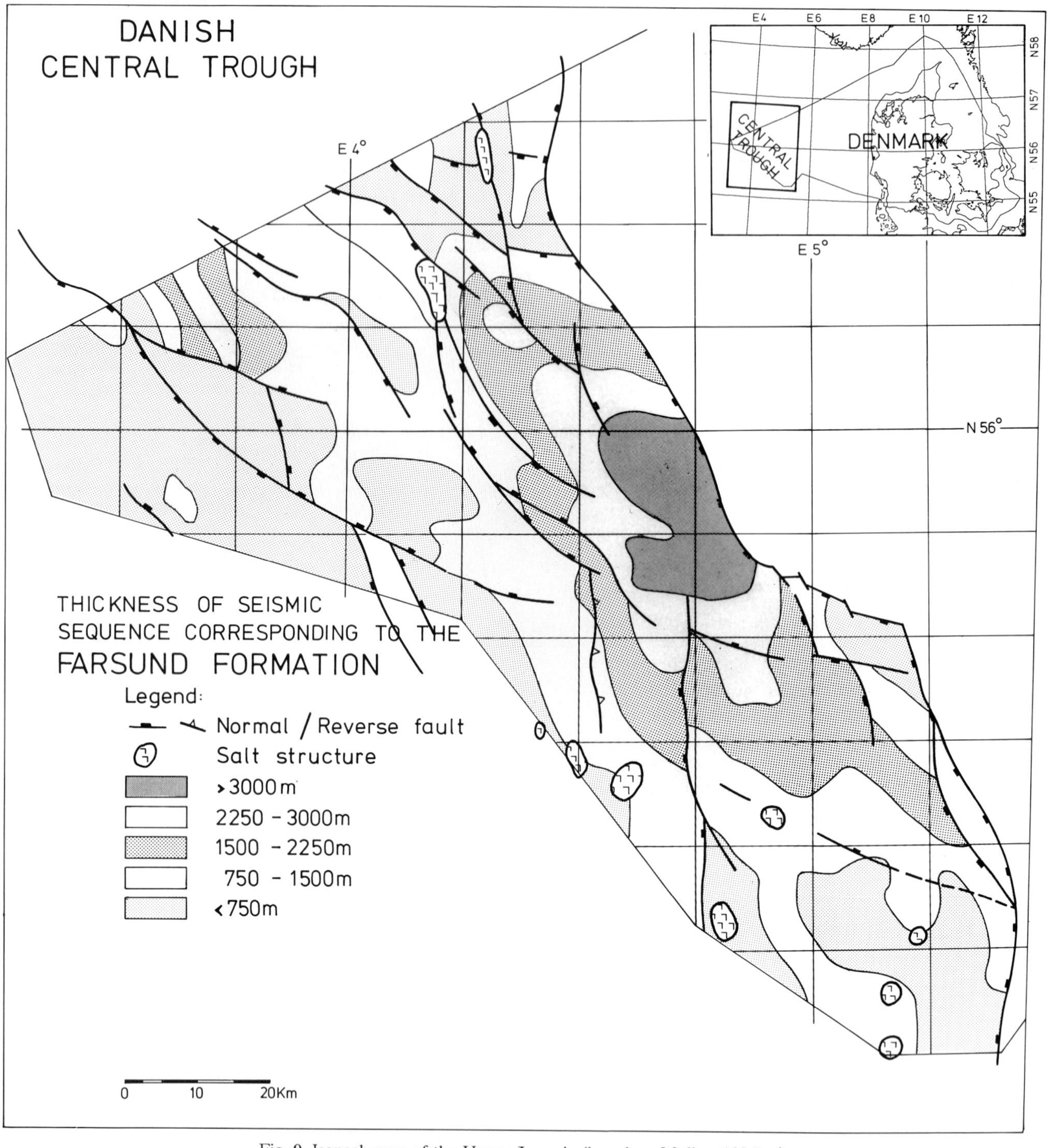

Fig. 9. Isopach map of the Upper Jurassic (based on Møller, 1986).

for the Palaeogene of the PW1 well (Fig. 7). Thus, the Cenozoic period is characterized by increasing subsidence rates despite the absence of faulting (Figs. 3 and 8), as has also been pointed out by Thorne and Watts (1989). As thermal subsidence is expected to decrease with time corresponding to the decay of the underlying thermal anomaly, this precludes the possibility that the Cenozoic was governed only by thermal subsidence following the preceding stretching. Stretching cannot be postulated on the basis of the seismic data (Fig. 8). The North Sea region formed a regionally subsiding basin almost devoid of any faults during the Cenozoic (e.g., Ziegler, 1982, pp. 89). None of the mechanisms accounted for in the geodynamic model can explain this subsidence pattern. Consequently, no attempt is made to fit these parts of the curves. However, as stretching probably is the most efficient way of perturbing the heat flow, the anomalous Cenozoic subsidence pattern probably had affected the heat flow

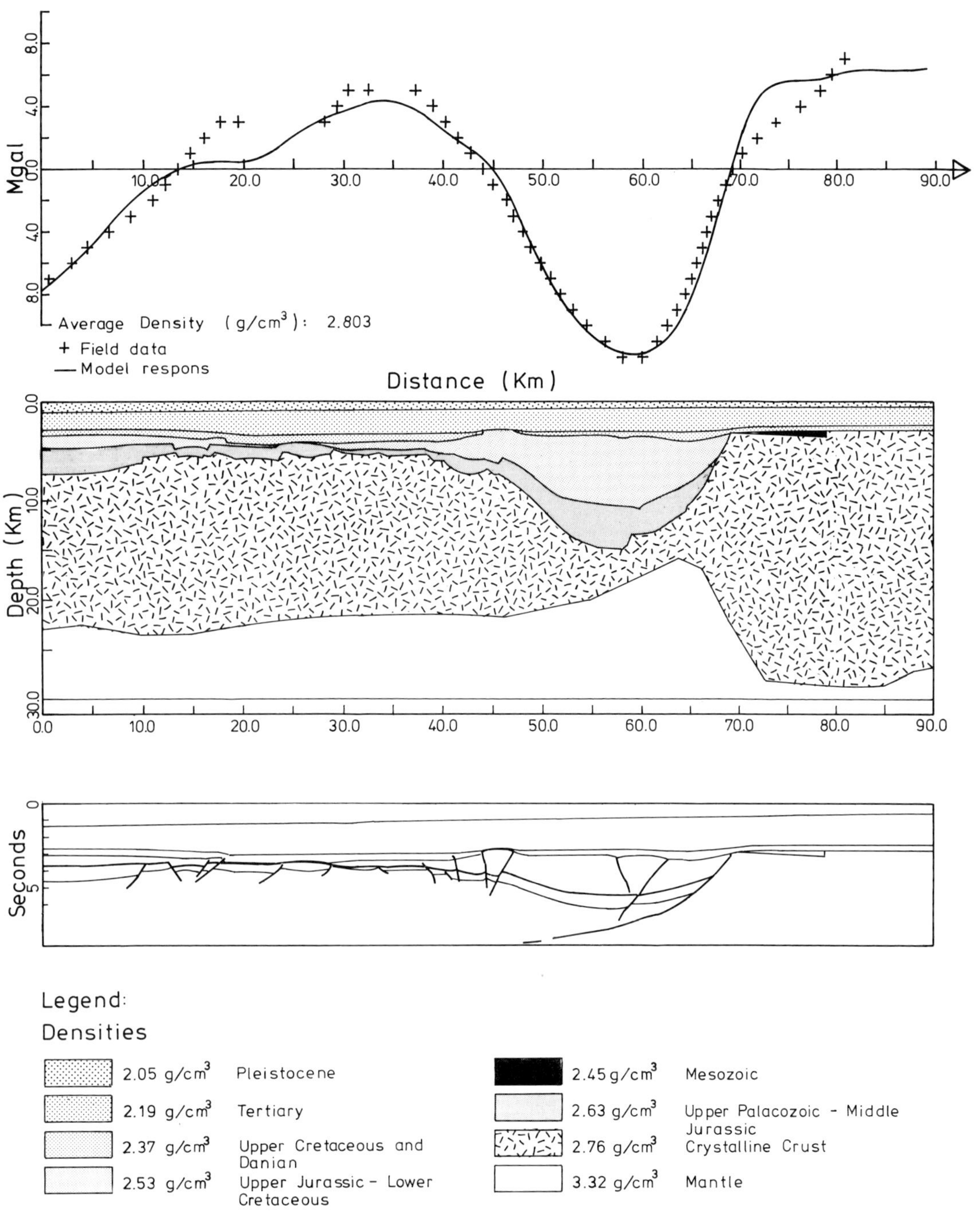

Fig. 10. Tentative gravity modelling across the Tail End Graben using the two-dimensional foreward modelling scheme of Talwani et al. (1959) (Fig. 1). Gravity data are from Michelsen (1982). Sediment densities are based on Uldall and Balling (1989).

only slightly, although the mechanism remains an enigma.

Total stretching, as indicated by the modelling has a maximum of 3.15 at PW3, decreasing from there, to 2.4 at PW1, 1.6 at E-1 and Sten-1, 1.8 at Mona-1 and W-1 and 1.5 at N2/11-1. Enhancement of subsidence by simulation of phase transformations is only significant at PW3, PW1 and E-1 as mentioned above. These factors have resulted in crustal thicknesses excluding sediments of only 4 km at PW3, 9 km at PW1, 15 km at E-1, 20 km at Mona-1 and W-1, 21 km at Sten-1 and 22 km at N2/11-1. These thicknesses are generally somewhat larger than the 10–15 km of Barton and Wood (1984), except for PW3, which may be regarded as a local extreme. The discrepancy between Barton and Wood's values and the values in the other wells is suggested to reflect the present inability to model the Cenozoic subsidence. The rather large crustal thickness variations seems to be in agreement with gravity data (Fig. 10). Mapping of L_g-wave attenuation transmitted across the Central Trough area has also been suggested to

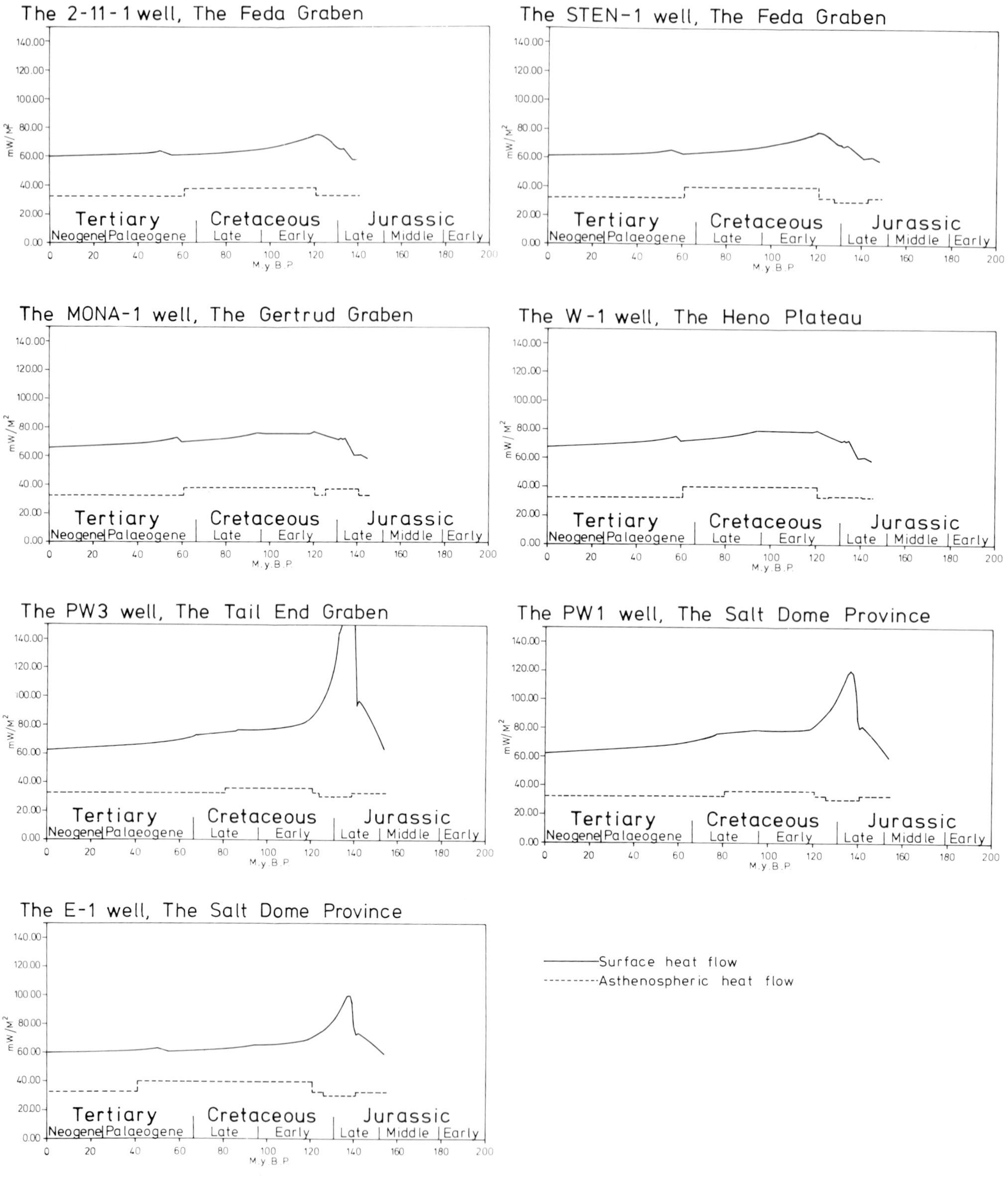

Fig. 11. Calculated heat flow for the seven wells (Fig. 1). Compare with Fig. 7.

be indicative of large horizontal gradients in crustal thickness (Kennett et al., 1985).

Heat flow development

The calculated heat flow history is shown as plots of heat flow versus time (Fig. 11) and is summarized in contoured heat flow maps (Fig. 12) depicting the calculated situation at 131, 96, 60 and 0 million years. Machine contouring is used to avoid bias from ill-defined structural constraints on heat flow patterns. Map A (Fig. 12) depicts heat flow near the beginning of the Ryazanian (131 million years). The Tail End Graben area still had a relatively high heat flow, although activity had abated according to

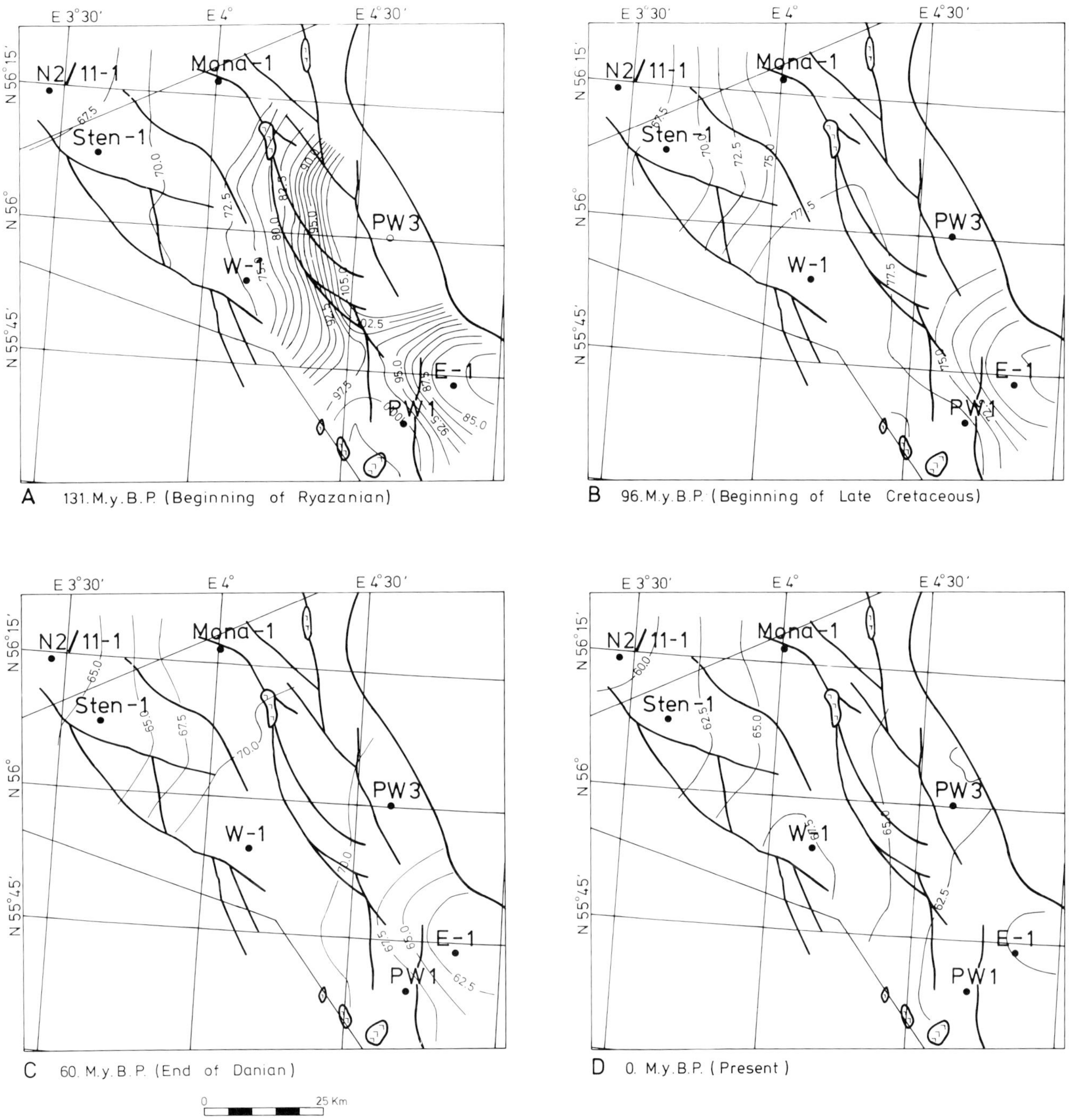

Fig. 12. Maps showing calculated heat flow in latest Jurassic (131 million years) end of Early Cretaceous (96 million years) end of Danian (Early Paleocene, 60 million years) and at present. The rifting activity is reflected in the heat flow maximum in the Tail End Graben on the first map. Subsequently, the heat flow decayed. At present the highest heat flow is found in the Heno Plateau area.

the calculation. The differences are probably overestimated, because an anomaly of this short duration (less than 5 million years, Fig. 11) is likely to be subdued by the effects of the sediment cover and lateral heat flow, not included in the model (cf. Lucazeau and Douaran, 1985; Steckler, 1986; Moretti and Chenet, 1987). The strong gradient at the PW3 location should cause significant lateral heat transfer and some error in the estimate of magnitude and duration of heat flow.

Apart from the southern part of the area, which also has relatively high values, the northern and the western parts of the area display heat flow values that are above steady-state conditions. Even though this situation represents maximum heat flow at E-1, it is below that of the nearby PW1 and PW3 locations, apparently because of the pronounced difference between crustal and lithospheric stretching at E-1. If most of the stretching is in the crust, only a small thermal anomaly will result and it will decay faster than when the whole lithosphere is stretched. This correlates with the end of the Hot Shale deposition in the Farsund Formation, but effects on the maturation patterns should be minimal, as no overburden had accumulated to blanket in the heat.

Map B shows the beginning of the Late Cretaceous, when faulting had decreased. However, at this time heat flow was close to its peak at W-1, although it never reached values as high as in other places. At this time, the present-day heat flow pattern started to develop and maps C and D have similar trends, but lower values. However, some areas cooled faster than others. The area at W-1 and Mona-1 cooled only slightly during that period, whereas the PW3 location especially, and to some extent the PW1 location, cooled much more. The E-1 location seems to have remained fairly cold through the whole 131 million-year period. This is explained by asymmetric extension, that causes low anomalous heat flow and different cooling trends relative to other locations. This pattern is due to both the amount and depth of anomalous heat. Map D reflects the present remaining heat anomaly and differences in crustal thickness, because little anomalous heat is left and the steady-state heat production potential in the model is dependent on the crustal thickness.

The present lithospheric thicknesses calculated, as defined by the depth to the 1330°C isotherm, reflect the amount of transient heat left in the lithosphere and are consequently roughly inversely correlated with present surface heat flow. In decreasing order thicknesses are 95.2 km at N2/11-1, 87.7 km at Sten-1, 86.3 km at E-1, 81.5 km at PW1, 81 km at Mona-1 and 78 km at PW3. A thickness of 77.4 km at W-1 coincides with the present maximum surface heat flow. A significant transient component is still present in the surface heat flow, as the approximate ultimate steady-state heat flow is estimated to vary between 43 mW/m^2 at PW3 and 50 mW/m^2 at N2/11-1 (cf. Fig. 12). This variation is dependent on stretching factors alone, since conservation of heat producing elements is assumed during phase transformations.

Discussion

The notion that stretching in the crust locally may exceed stretching in the lithospheric mantle is based on the conspicuous absence of thermal post-rift subsidence, best seen in the case of the E-1 well. This differential stretching, accommodated by a detachment level, may be supported by theoretical studies of tensile strength profiles of the lithosphere (Fig. 13). According to Lynch and Morgan (1987), the rock composition change with depth in the litho-

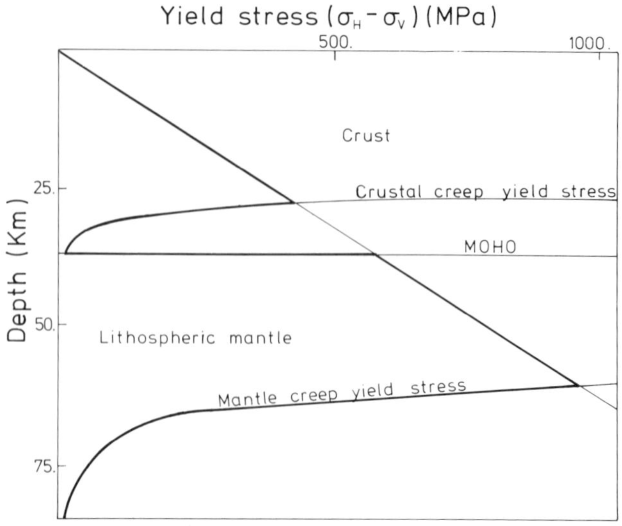

Fig. 13. Theoretical yield stress diagram for the lithosphere based on Lynch and Morgan, 1987). The diagram shows the calculated distribution of the horizontal deviatoric stress needed to cause non-elastic extension. The presence and size of the yield stress minima are strongly dependent on ambient temperatures and, thus, heat flow along with other things such as strain rate.

sphere may give rise to yield stress minima, because the temperature at which ductility increases drastically (i.e., the rocks approach their melting point) is different for different rock types. A yield stress minimum may occur, if the lithosphere is heated, and disappear again as it cools. The resulting decoupling level (i.e., at the base of the crust) makes the lithosphere mechanically discontinuous and renders uniform stretching unlikely. As syn-rift subsidence at the Upper Jurassic depocentre in the Tail End Graben clearly was accompanied by major faulting along the boundary fault to the Ringkøbing Fyn High (Figs. 8 and 9), it is likely that movement along a possible detachment level was intimately connected to movements along the fault. It is difficult to ascertain whether the main fault is listric or planar, owing to absence of direct fault plane images in the seismic data and to uncertainties in velocities for the necessary depth conversion. However, the apparent block rotation of the Tail End Graben along the boundary fault to the Ringkøbing Fyn High may suggest a deep detachment. Analysis of recent earthquake data shows that active faults are all approximately planer and evidence for detachments in a ductile lower crust is also sometimes present (Jackson, 1987). Jackson (1987) suggests that rotation in the brittle upper crust is mainly due to domino-type fault plane rotation by continuing extension. However, whether the fault plane is rotating or is curved, both may be accommodated by a ductile lower crust.

According to Groshong (1989) the depth to a detachment assuming a planar fault geometry may

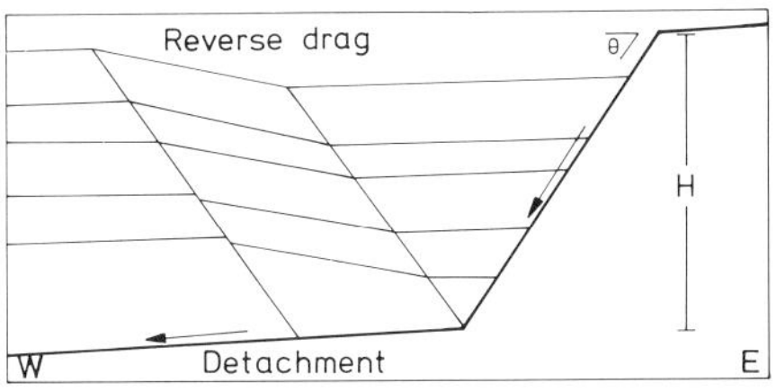

Fig. 14. Sketch showing the concept of reverse drag that may be used to determine the depth of a detachment level (based on Groshong, 1989). The abrupt change in the fault plane dip requires an equally abrupt change in the ductility. Compare with Fig. 8.

be estimated from the distance between the reverse drag section in the hanging wall and the main fault and the dip of the fault (Fig. 14). This model seems to be applicable to the Tail End Graben profile, as a reverse drag section seems to be present in the hanging wall (Fig. 8). Unfortunately, the exact geometry of the fault is ill-constrained, for reasons stated above. However, the depth to a possible detachment level below the Tail End Graben is likely to be located within the first second of two-way reflection time below the top of the basement, corresponding to at least 2 km. The calculated very thin crust below the Tail End Graben may, therefore, be supported by this interpretation of the fault and seems to honour gravity data (Fig. 10).

The conclusive tentative model for the middle to northern Danish Central Trough may be similar to the sketch on Fig. 4, which is based on theoretical studies by Braun and Beaumont (1989). They showed, by application of a two-dimensional finite element model, that extension in the lithospheric mantle may be transferred laterally within about 100 km to a pre-existing weakness zone in the crust, provided a detachment level is also developed. This results in different stretching factors for the lithospheric mantle as compared to the crust at points along a line perpendicular to strike (Fig. 4), although the regional stretching values are the same. If the boundary fault between the Tail End Graben and the Ringkøbing Fyn High corresponds to this pre-existing crustal weakness zone, then this model may explain the large localized stretching factors applied to the crust as well as the small stretching factors applied to the lithospheric mantle at PW3, PW1 and E-1.

A major problem is the Cenozoic subsidence, which cannot be explained by the geodynamic model. The problem is particularly conspicuous when comparison is made between the deposition rates of the Upper Jurassic Farsund Formation and the Quaternary Cen Unit 6 (Fig. 3), which are comparable. The former is a syn-rift deposit associated with strong faulting, the latter with no faulting. Faulting ceased in earliest Tertiary, apart from local faulting at salt diapirs. As stretching, therefore, can be ruled out as a cause for Cenozoic subsidence, it is suggested that some metamorphic process leading to a density increase and additional subsidence is active. Because it cannot be induced by stretching, the mechanism must be different from that incorporated in this study. The mechanism of the present study is mainly sensitive to pressure drops and temperature increases, which take place during convective uplift. However, the mechanism causing Cenozoic subsidence is suggested to be a self-supporting interaction between sediment loading and a pressure-induced phase transformation. A thermal drive for the transformation does not seem to be a viable mechanism, because the immediate response, as seen in the case of the Quaternary subsidence and deposition, cannot possibly be thermally driven as a consequence of the insulating effect of sediments. It takes a very long time before the insulating effect is felt at depths where possible metamorphic reactions occur. It is suggested that the mechanism is driven mainly by pressure, because this will provide an immediate response to an extra load, as may be the case with the Quaternary loading. An obvious mechanism may be the well-known eclogite/garnet-granulite system, which is pressure-sensitive (e.g., O'Connell and Wasserburg, 1972; Haxby et al., 1976). As seen in Fig. 5, an extra load may cause portions of the lower crust to enter the eclogite field and thus lead to a density increase. However, the lithosphere should be close to thermal steady-state to allow the pressure mechanism to be active (Fig. 5), so some cooling is required to make the system sensitive to loading. To incorporate the eclogite/garnet-granulite system in the model, a much larger complexity of the equations is required as sediments have to be directly included in the calculations. This is regarded as extremely difficult with the semi-analytical approach used in the theory of the present model and, therefore, has not been attempted.

Conclusion

The post-Middle Jurassic tectonic development started with major faulting along the boundary fault between the Tail End Graben and the Ringkøbing Fyn High, thus giving rise to the well-defined depocentres of the Farsund Formation. Southwards towards the Salt Dome Province this depocentre divides, because stretching was accommodated by two major faults. This early faulting was much stronger than the later activity. The resulting heat flow also was at its peak during this period, with a maximum in the Tail End Graben.

The tectonism gradually spread to the west during the Upper Jurassic, leading to a later maximum heat flow in this region, but with a lower level than in the Tail End Graben. On the basis of the tectonic history, it is concluded that much of the variation in subsidence patterns in the Danish Central Trough is generated by differential stretching between crust and lithospheric mantle. However, during the Cretaceous, stretching gradually abated and gave way to regional subsidence. This subsidence could not be modelled, as the suggested pressure-induced density increases from gabbro-eclogite metamorphism is not represented in the model of the present study.

Acknowledgements

The STATOIL Group and the Danish Research Academy are thanked for financial support of the research project upon which this paper is based. The working team of the Danish Modelling Group is also thanked for making their studies of the subsidence history using the Yükler model available for the geodynamic studies of this paper. Comments and suggestions of Claus Andersen, Finn Surlyk, John Sales (who acted as referee) and one anonymous reviewer have been helpful and are gratefully acknowledged. Finally I thank Kaj Jensen-Jørgensen (Geokon) for drafting the illustrations.

References

Andersen, C., Olsen, J.C., Michelsen, O. and Nygaard, E., 1982. Structural outline and development. In: O. Michelsen (Editor), Geology of the Danish Central Graben. Danm. Geol. Unders., Ser. B, 8: 9–26.

Balling, N.P., 1979. Subsurface temperatures and heat flow estimates in Denmark. In: V. Cermak and L. Rybach (Editors), Terrestrial Heat Flow in Europe. Springer-Verlag, Berlin, pp. 161–171.

Barton, P. and Wood, R., 1984. Tectonic evolution of the North Sea Basin: crustal stretching and subsidence. Geophys. J. R. Astron. Soc., 79: 987–1022.

Braun, J. and Beaumont, C., 1989. Dynamical models of the role of crustal shear zones in asymmetric continental extension. Earth Planet. Sci. Lett., 93: 405–423.

Calcagnile, G. and Scarpa, R., 1985. Deep structures of the European Mediterranean area from seismological data. Tectonophysics, 118: 93–111.

Chiarelli, A. and Duffaud, F., 1980. Pressure origin and distribution in Jurassic of Viking Basin (United Kingdom-Norway). Am. Assoc. Pet. Geol., Bull., 64: 1245–1266.

Cochran, J.R., 1983. Effects of finite rifting times on the development of sedimentary basins, Earth Planet. Sci. Lett., 66: 289–302.

Coward, M.P., 1986. Heterogeneous stretching, simple shear and basin development. Earth Planet. Sci. Lett., 80: 325–336.

Day, G.A., Cooper, B.A., Andersen, C., Burgers. W.F.J., Rønnevik, H.C. and Schøneich, H., 1981. Regional seismic structure maps of the North Sea. In: L.V. Illing and G.D. Hobson (Editors), Petroleum Geology of the Continental Shelf of North-West Europe. Institute of Petroleum/Heyden, London, pp. 76–84.

Dewey, J.F., 1982. Plate tectonics and the evolution of the British Isles. J. Geol. Soc. London, 139: 371–412.

Eynon, G., 1981. Basin development and sedimentation in the Middle Jurassic of the northern North Sea. In: L.V. Illing and G.D. Hobson (Editors), Petroleum Geology of the Continental Shelf of North-West Europe. Institute of Petroleum/Heyden, London, pp. 196–204.

Falvey, D.A. and Middleton, M.F., 1981. Passive continental margins: evidence for a pre-breakup deep crustal metamorphic subsidence mechanism. Proc. 26th Int. Geology Congr., Geology of Continental Margins Symp., Paris, 7–17 July, 1980, Oceanol. Acta, 4: 103–114.

Gowers M.B. and Sæbøe, A., 1985. On the structural evolution of the Central Trough in the Norwegian and Danish sectors of the North Sea. Mar. Pet. Geol., 2: 298–318.

Green, D.H. and Ringwood, A.E., 1972. A comparison of recent experimental data on the gabbro–garnet granulite–eclogite transition, J. Geol., 80: 277–288.

Groshong, R.H., Jr., 1989. Half-graben structures: balanced models of extensional fault-bend folds. Geol. Soc. Am. Bull., 101: 96–101.

Haq, B.U., Hardenbol, J. and Vail, P.R., 1987. Chronology of fluctuating sea levels since the Triassic (250 million years ago to present). Science, 235: 1156–1167.

Haxby, W.F., Turcotte, D.L. and Bird, J.M., 1976. Thermal and mechanical evolution of the Michigan Basin. Tectonophysics, 36: 57–75.

Ito, K. and Kennedy, G.C., 1971. An experimental study of the gabbro–garnet granulite–eclogite transition. In: J.G. Heacock (Editor), Am. Geophys. Union, Geophys. Monogr., 14, pp. 303–314.

Jackson, J.A., 1987. Active normal faulting and crustal extension. In: M.P. Coward, J.F. Dewey and P.L. Hancock (Editors), Continental Extensional Tectonics. Geol. Soc. London, Spec. Publ., 28: 3–17.

Jacobsen, F., 1982. Triassic. In: O. Michelsen (Editor), Geology of the Danish Central Graben. Danm. Geol. Unders., Ser. B, 8: 32–36.

Jensen, P.K., 1990. Analysis of the temperature field around salt diapirs. Geothermics, 19(3): 273–283

Johannesen, P., 1988. Upper Jurassic sandstones Part 1. Depositional environments and basin development, CENBAS Sub-project B. Reservoir rocks and depositional models. Danm. Geol. Unders., Internal Rep.

Kennett, B.L.N., Gregersen, S., Mykkeltveit, S. and Newmark, R., 1985. Mapping of crustal heterogeneity in the North Sea basin via the propagation of L_g-waves. Geophys. J. R. Astron. Soc., 83: 299–306.

Kitahara, S. Takenouchi, S. and Kennedy, G.C., 1966. Phase relations in the system MgO–SiO_2–H_2O at high temperatures and pressures, Am. J. Sci., 264: 223–233.

Lucazeau, F. and Le Douaran, S., 1985. The blanketing effect of sediments in basins formed by extension: a numerical model. Application to the Gulf of Lion and Viking graben. Earth Planet. Sci. Lett., 74: 92–102.

Lynch, H.D. and Morgan, P., 1987. The tensile strength of the lithosphere and the localization of extension. In: M.P. Coward, J.F. Dewey and P.L. Hancock (Editors), Continental Extensional Tectonics. Geol. Soc. London, Spec. Publ., 28: 53–65.

Madsen, L., 1975. Approximate geothermal gradients in Denmark and the Danish North Sea Sector. Danm. Geol. Unders., Årbog, 1974: 5–16.

McKenzie, D.P., 1978. Some remarks on the development of sedimentary basins. Earth Planet. Sci. Lett., 40: 25–32.

Michelsen, O. (Editor), 1982. Geology of the Danish Central Graben. Danm. Geol. Unders., Ser. B, 8, 133 pp.

Moretti, I. and Chenet, P.Y., 1987. The evolution of the Suez rift. A combination of stretching and secondary convection, Tectonophysics, 133: 229–234.

Møller, J.J. 1986. Seismic structural mapping of the Middle and Upper Jurassic in the Danish Central Trough. Danm. Geol. Unders., Ser. A, 13, 37 pp.

O'Connell, R.J. and Wasserburg, J.G., 1972. Dynamics of submergence and uplift of a sedimentary basin underlain by a phase-change boundary. Rev. Geophys. Space Phys., 10: 335–368.

Sandiford, M. and Powell, R., 1986. Deep crustal metamorphism during crustal extension. Modern and ancient examples. Earth Planet. Sci. Lett., 79: 151–158.

Sclater, J.G. and Christie, P.A.F., 1980. Continental stretching. An explanation of the post-mid-Cretaceous subsidence of the Central North Sea basin. J. Geophys. Res., 85: 3711–3739.

Steckler, M.S., 1986. Uplift and extension of the Gulf of Suez, indications of induced mantle convection. Nature, 317: 135–139.

Steckler, M.S. and Watts, A.B., 1978. Subsidence of the Atlantic type continental margin off New York. Earth Planet. Sci. Lett., 41: 1–13.

Talwani, M., Worzel, J.L. and Landisman, M., 1959. Rapid gravity computations for two-dimensional bodies with application to the Mendocino Submarine Fracture Zone. J. Geophys. Res., 64: 49–60.

Thorne, J.A. and Watts, A.B., 1989. Quantitative analysis of North Sea subsidence. Am. Assoc. Pet. Geol., Bull., 73: 88–116.

Uldall, A. and Balling N., 1989. Basin analysis of the Danish Central Graben. CENBAS Sub-project F: Petrophysics, Department of Geology, Geophysical Laboratory, Aarhus University, 95 pp.

Vejbæk, O.V., 1986. Seismic stratigraphy of the Lower Cretaceous in the Danish Central Trough. Danm. Geol. Unders., Ser. A, 11, 46 pp.

Vejbæk, O.V. and Andersen, C., 1987. Cretaceous–Early Tertiary inversion tectonism in the Danish Central Trough. Tectonophysics, 137: 221–238.

Vejbæk, O.V., 1989. Effects of asthenospheric heat flow in basin modelling exemplified with the Danish Basin. Earth Planet. Sci. Lett., 95: 97–114.

Vejbæk, O.V., 1990. Geodynamic modelling of the Danish Basin. In: B. Pinet and C. Bois (Editors), The Potential of Deep Seismic Profiling for Hydrocarbon Exploration. Institut Français du Pétrole, 1989, Arles, Edition Technip, 10 pp.

Wood, R. and Barton P., 1983. Crustal thinning and subsidence in the North Sea. Nature, 302: 134–136.

Yükler, M.A., Cornford, C. and Welte, D., 1978. One-dimensional model to simulate geologic, hydrodynamic and thermodynamic development of a sedimentary basin. Geol. Rundsch., 67: 960–979.

Ziegler, P.A., 1975. Geologic evolution of North Sea and its tectonic framework. Am. Assoc. Pet. Geol., Bull., 59: 1073–1097.

Ziegler, P.A., 1982. Geological Atlas of Western and Central Europe. Elsevier, Amsterdam, 130 pp. and encl.

O.V. VEJBÆK *Geological Survey of Denmark, Thoravej 8, DK-2400 København, Denmark*
Present address: *Amoco Production Company, Houston, TX 77253-3092, USA*

A tectonostratigraphic analysis of the southeast Norwegian North Sea Basin

S. Sørensen, H. Morizot and S. Skottheim

In the South East North Sea several wells have been drilled. Oil has been found in four of them. In order to better understand the hydrocarbon distribution in the area, Elf Aquitaine Norge A/S made a regional study of the Egersund Basin. In this paper the regional aspect with respect to structural history, depositional history and stratigraphy will be discussed.

The study includes a biostratigraphic update of 26 wells in general. The regional seismic study included the interpretation of 17300 km of seismic.

The area between 57°N and 59°N and 3°E and 5°E can be divided structurally into a mainly salt induced basin to the south of 58°N, Egersund Basin sensu latio and a more fault controlled basin to the north of 58°N, Åsta Graben.

The Late Jurassic period was dominated by an overall transgressive phase in the lower part and a regressive phase in the upper part.

A compression during the transition from Early to Late Cretaceous has been observed and has a certain bearing on the Chalk distribution of the area.

The seismic interpretation further allowed us to define the shift of basin axis through time and to define the importance of the different faults through time and their geographical position. This, together with seismic stratigraphy, better age control and seismic facies analysis increased our confidence as regards the mapping of the sediment distribution and facies shift through time.

This presentation concludes with a set of paleogeographic maps constructed from well data, seismic facies analysis and the structural mapping.

Introduction

The study was performed in order to evaluate the regional geological evolution of the area between 57°N and 59°N and 3°20′E and 5°40′E. This area includes several structural provinces, such as the Åsta Graben, Egersund Basin, Fiskebank Basin and part of the Norwegian Danish Basin (Fig. 1). During the study and in the text we have applied the common term "South East North Sea" for these structural provinces, and it is further abbreviated to the "SENS study". The study includes a biostratigraphical update of several wells in addition to seismic interpretation, geological modelling and reinterpretation. The methodology is a combination of different geological and geophysical disciplines applied in order to better appreciate the sediment distribution in relationship to the tectonic evolution through time.

Data base

In this study, lines from seismic surveys with acquisition years ranging from 1977 to 1987 were used. A total of 17 300 km of seismic data were interpreted. The sources were Maxipulse, Airgun or Watergun and the coverage varied from 24-fold to 60-fold. The size of the seismic grid ranged from 5 × 5 km regionally to 1 × 1 km locally. A total of 27 wells were used in the seismic calibration.

Basement

In the study area three wells (8/3-1, 10/5-1 and 18/11-1) reached metamorphic basement (Fig. 2).

Well 8/3-1 is located in the western part of the study area on the southernmost part of the Sele High. The basement rocks in this well comprise schistose rocks of Silurian age (411 ± 4 Ma) (Myhre, 1975). Well 10/5-1 is located on the Lista Nose structure. It stopped in granites of indicated Pre-Cambrian age. Well 18/11-1 reached a weathered and slightly folded quartzitic basement.

Paleozoic

During the transition from Caledonian to Variscan orogenesis, the stress pattern seems to have rotated

Structural and Tectonic Modelling and its Application to Petroleum Geology edited by R.M. Larsen, H. Brekke, B.T. Larsen and E. Talleraas. NPF Special Publication 1, pp. 19–42. Elsevier, Amsterdam.

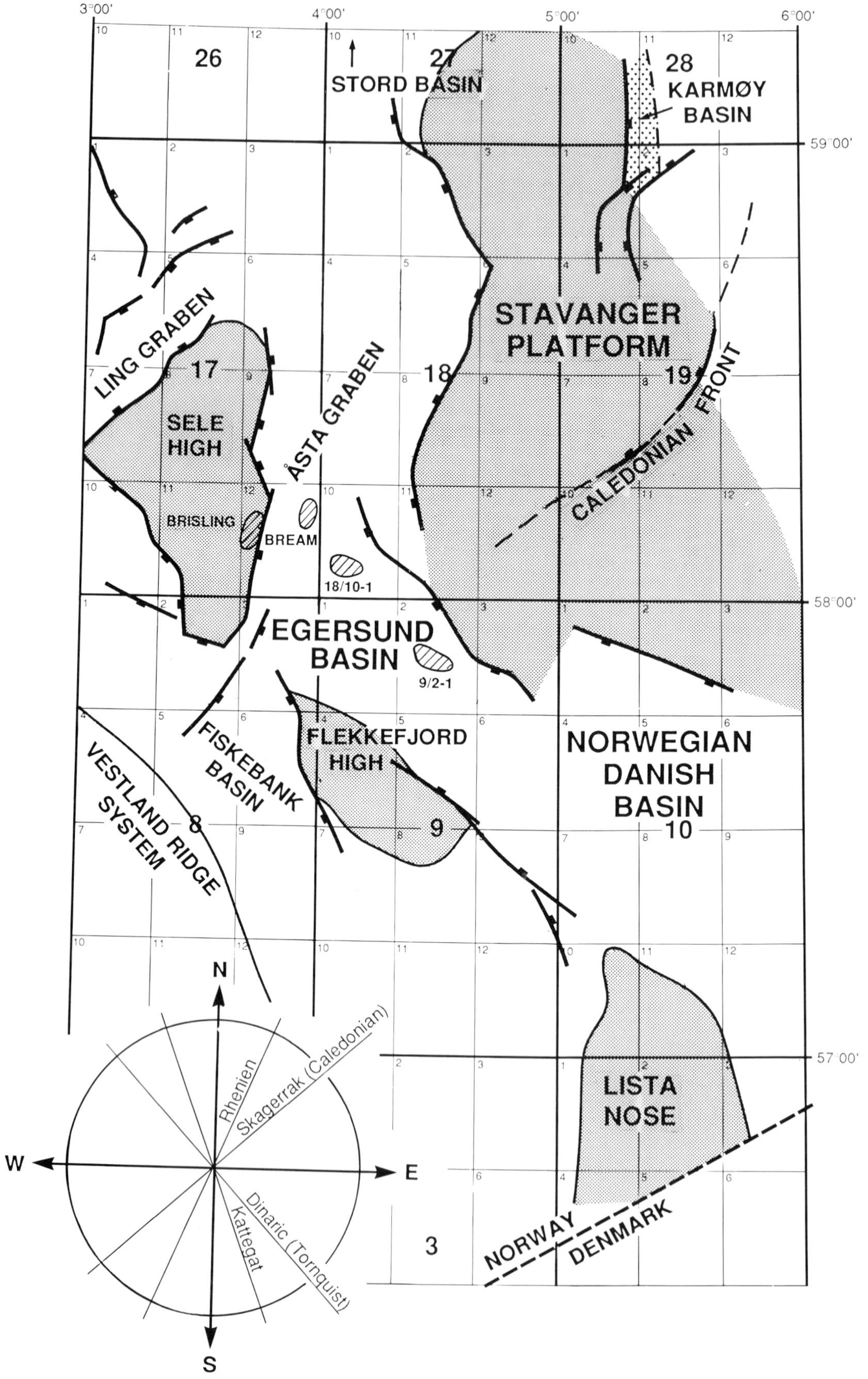

Fig. 1. Principal structural elements — SENS (South East North Sea).

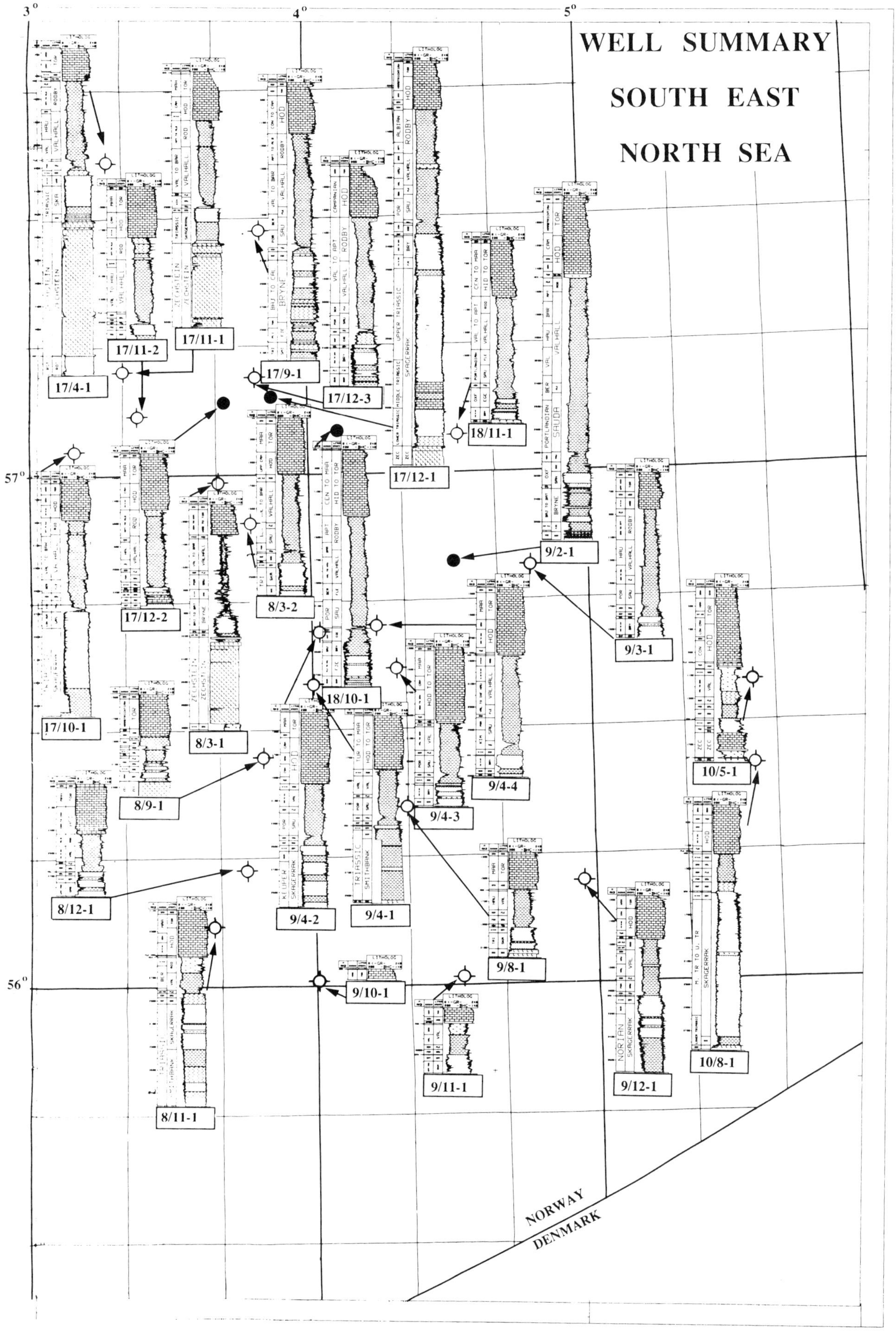
WELL SUMMARY
SOUTH EAST
NORTH SEA
3°
4°
5°
57°
56°
17/4-1
17/11-2
17/11-1
17/9-1
17/12-3
17/12-1
18/11-1
9/2-1
9/3-1
17/10-1
17/12-2
8/3-1
8/3-2
18/10-1
9/4-4
9/4-3
10/5-1
8/9-1
8/12-1
9/4-2
9/4-1
9/8-1
8/11-1
9/10-1
9/11-1
9/12-1
10/8-1
NORWAY
DENMARK

TABLE 1

Structural trends through time, with respect to geographical position: south-east North Sea

Time map	Area				
	56°30′–57°N	57°–57°30′N	57°30′58°N	58°–58°30′N	58°30′–59°N
Basement	D R	D R	S D (R)	S D	S D
Pre-Zechstein	D K	D (K)	D (S)	S R (K)	S K R (S)
Top Triassic	D	K D (R)	D K (R)	S R (KD)	D K R (S)
Top Middle Jurassic	D	K D (R)	D K	S R	D K R
Top Kimmeridgian	D (K)	D(K)	D (K)	K D R	K D
Base Cretaceous Unconformity	D	D K (R)	D (KR)	K D R (S)	R K D (S).
Base Chalk	D (K)	D K	K D	K S R	S

D = N130–150°E, Dinaric (Tornquist Trend); K = N150–170°E, Kattegat; R = N10–30°E, Rhenian (Oslo Trend); S = N35–55°E, Skagerrak (Caledonian Trend).
(X) = Minor influence when trend between brackets.

clockwise from a NW-SE direction of compressional stress in Silurian times to a more N–S direction of compressional stress in Eifelian-Givetian (Middle Devonian times). This clockwise rotation of the stress field seems to have continued into a NE–SW stress in Early Carboniferous times (Ziegler, 1988). While the NW–SE compressional direction represents the Caledonian orogeny, the N–S to NE–SW compressional direction represents the Variscan orogeny. During the transition from Carboniferous to Permian times, the stress system changed from a predominantly compressional to a predominantly extensional system.

The fault pattern at basement level is dominated by three trends (Table 1). The area south of 58°N is dominated by the 130°–150°N trend (Dinaric, Table 1) initiated in Early Paleozoic times, which, as we shall see later, was reactivated during several later periods.

The second trend in the south, oriented 10°–30°N, might be associated with an extension phase in Devonian-Permian times. This orientation is similar to the main trends of the Oslo Graben. The third trend 150°–170°N is analogous to the orientation of the northern part of the Horn Graben. This fault trend was initiated prior to mid Carboniferous times. The NW–SE trends of 130°–150°N and 150°–170°N are the main axes for the salt structures created in Early Triassic times.

The area north of 58°N is dominated by the 35°–55°N (Skagerrak or Caledonian) trend and the 130°–150°N (Dinaric or Tornquist trend) at basement level, indicating the Early Paleozoic initiation of these trends. On the Pre-Zechstein map (Fig. 3) the importance of the 130°–150°N trend diminishes north of 58°N, and the NW–SE trending fault orientation turns to 150°–170°N between 58°N and 59°N.

The Carboniferous and Rotliegendes sediment distribution is assumed to follow the same controlling fault trends. However, the significance of the different faults is rather uncertain, in the sense that the throws on the faults are unknown at this time.

Devonian

The oldest sediments in the area might be of Devonian age. This age has been suggested for the lowermost section penetrated in well 17/12-2. Its association with the sandstones in the Hornelen Basin in Western Norway has been suggested (Jacobsen, 1982).

However, a Rotliegendes age for this sandstone is favoured by these authors, based on regional interpretation.

Permian

Rotliegendes

Clastic deposits of Rotliegendes age are likely to be widely present in the SENS area, as indicated by their seismic character. At present only two wells (10/5-1 and 17/12-2) have reached the Rotliegendes series (Fig. 2).

Well 17/12-2 is located on the Sele High. As discussed in the paragraph on the Devonian, the age dating is uncertain. However, based on the regional context it is believed that the rocks belong to the Rotliegendes suite. In well 10/5-1 only 7 m of postulated Rotliegendes siltstones were penetrated before the well bottomed in Pre-Cambrian granite.

A northern Permian Basin probably existed over parts of or the whole SENS area, with connections both to the southeast (Norwegian Danish Basin) and maybe to the north into Stord Basin (Ziegler, 1988, Sørensen and Martinsen, 1987) (Fig. 1).

The Rotliegendes rocks of Early Permian age were deposited in an arid climate with desert conditions. The arid conditions led to the deposition of aeolian

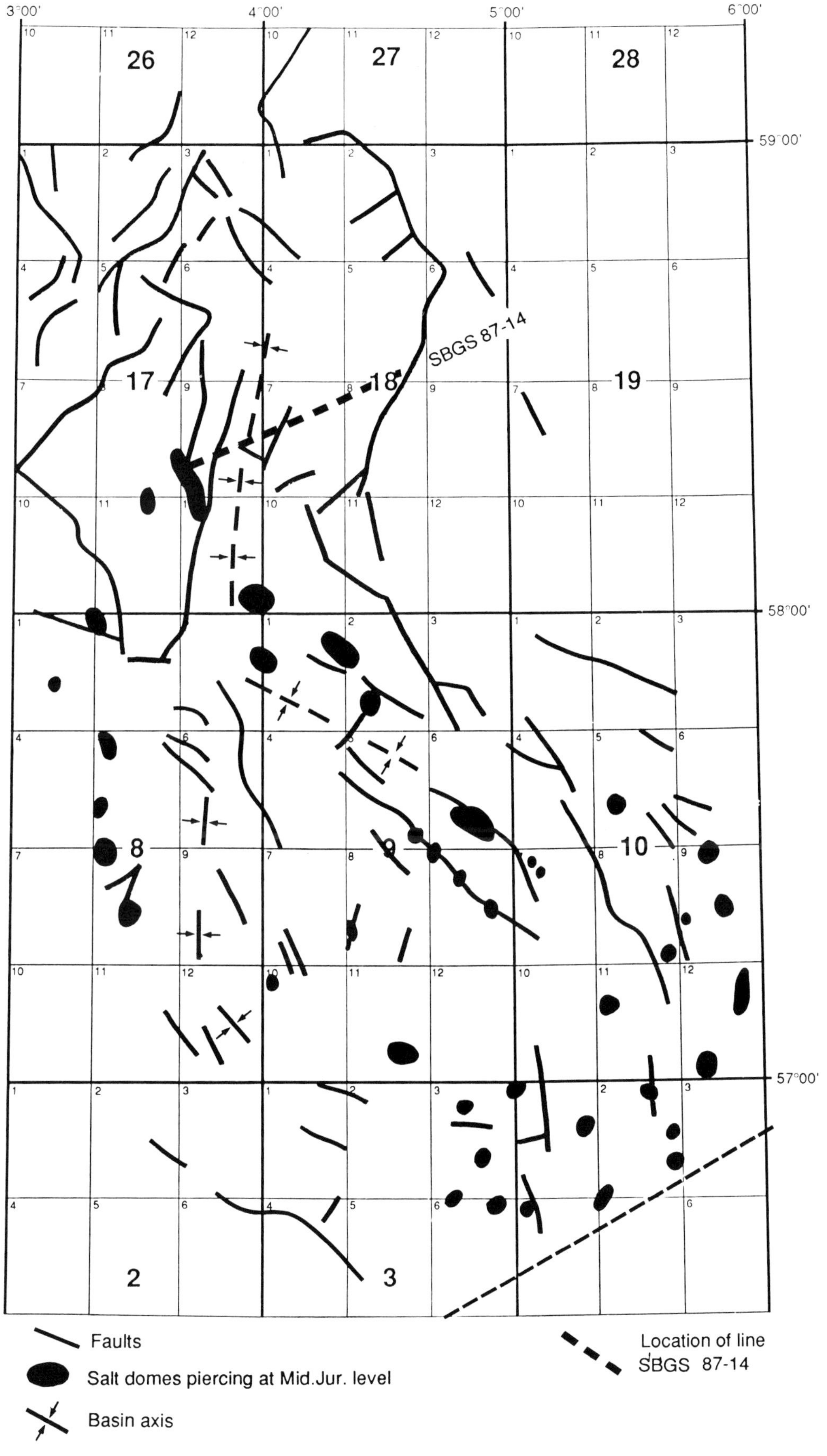

Fig. 3. Fault orientation at Top Pre-Zechstein level.

sands and sabkhas in structurally low areas. On the Sele High (well 17/12-2) the continental deposits are mainly arkosic.

During this period the Åsta Graben and Egersund Basin already existed as sedimentary basins. The structural framework was initiated with a central basin flanked by the somewhat structurally elevated Stavanger Platform to the east and the Sele High to the west. The boundary fault seen at Base Zechstein level along the eastern side of the Sele High was in existence and governed the assumed alluvial fan deposits on the downthrown side of the Sele High (Fig. 3). A drainage area from the east is expected along the gently westerly dipping margins of the Stavanger Platform.

Rotliegendes deposits are known to have fair to good reservoir characteristics, depending on the degree of diagenesis. The facies range from alluvial, continental and fluvial to aeolian sandstones. Accurately mapping this sequence is a problem. In places, it can be seismically observed as a transparent (white) sequence below the pre-Zechstein marker, whilst in other areas it cannot be mapped.

Kupferschiefer equivalent

This sequence is supposed to be widespread in the SENS area and blankets the Rotliegendes deposits, being the beginning of the "marine" transgression at the end of Rotliegendes time. Generally, these deposits are dominated by organic-rich shales, which can be of excellent source rock quality, but in restricted quantity.

The Kupferschiefer is only drilled in the structurally elevated areas, such as the Lista Nose (well 10/5-1) and on the southern side of Ling Graben (17/4-1), whilst it has been eroded or not deposited on the Sele High. In the structurally deeper parts of the basin the sequence has not been reached, but in this interpretation it is assumed to be present.

Zechstein

For the Zechstein deposits it seems that the fault activity was of minor importance during deposition, i.e., that little synsedimentary faulting took place. The area south of 58°N is linked directly to the Norwegian-Danish Basin, in what is termed the Northern Permian Basin (Ziegler, 1988). To the north of 58°N, the Zechstein deposits seem locally to be fault controlled. The deposits are thin on the eastern side of the basin and do not extend beyond the eastern boundary fault (Fig. 4), indicating that this fault probably existed prior and during the Zechstein deposition. On the western side the sediments cross the eastern boundary faults of the Sele High. However, the sediment thickness is reduced and a clear change in facies occurs (halite-carbonate). Between the two fault patterns there was a fairly flat lying or slightly westward tilted basin dominated by halite deposits. Generally, the Late Permian experienced a gentle subsidence in the central parts of the present basins. The subsidence axis was probably located in a N–S orientation east of the Sele High in the north with the axis having a more NW–SE orientation to the south of 58°N.

Ten of the wells in the SENS study area reached or penetrated the Zechstein interval. These wells represent all the different structural provinces. Of the ten wells, eight are in a definite salt facies (8/3-1, 8/9-1, 9/4-1, 9/8-1, 10/8-1, 17/4-1, 17/11-1 and 17/12-1), whilst two are in a marginal marine facies (10/5-1 and 17/12-2) (Fig. 2). The evaporitic sequences (salts) are divided into four cycles for the North Sea (Rhys, 1974), however some authors (Ormaasen et al., 1980) claim that only three of the four cycles are present in the Northern Permian Basin.

Zechstein deposits are believed to be restricted to the west of the Stavanger Platform. If this is the case, it has yet to be resolved whether the Stavanger Platform was an area of non-deposition or an erosional feature during Zechstein times.

On the structural highs flanking the basin, we find the marginal marine platform sediment, as in well 10/5-1. The other well with Zechstein platform sediments is 17/12-2, located on the Sele High. This well is located at a structurally higher position than the 17/11-1 which encountered salt deposits.

The general thickness of the salt deposits in the SENS area is in the order of 1000–1500 m. As no northern limit of the deposits has been defined in the SENS area, the Zechstein evaporitic series should be expected in the Stord Basin.

Triassic

Triassic deposits have been penetrated in several wells. The thickest series encountered so far is in well 17/12-1 where more than 1500 m of Triassic rocks was penetrated. Even on the structural highs, thick Triassic series have been penetrated (e.g., well 17/11-1) (Fig. 2).

According to Vollset and Doré (1984), no revision of the lithostratigraphy has been carried out. However, from the seismic data, it seems that the Triassic sequence can be divided into three sequences. The lowermost part of the Triassic series comprises conformable seismic reflectors on top of the underlying (often Permian) sequences. The top of this series is cut by an unconformity which we believe might be the Hardegsen Unconformity.

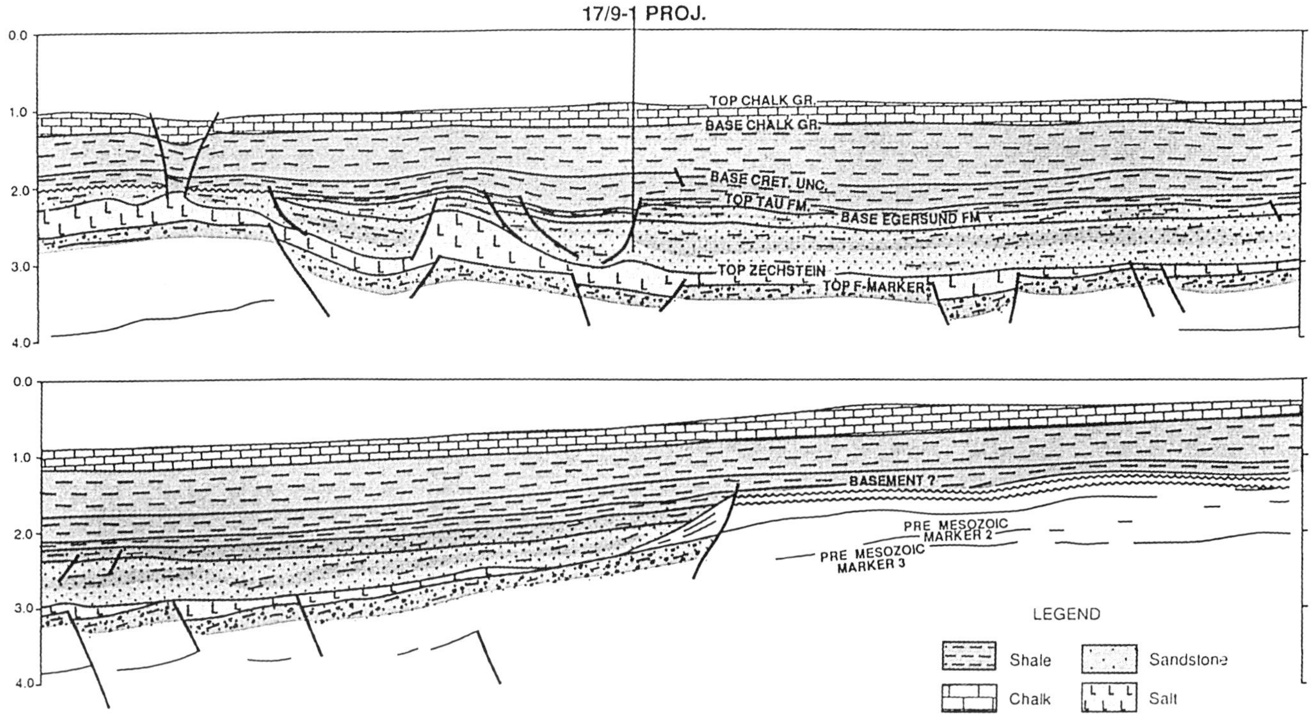

Fig. 4. Geoseismic section line SBGS 87-14.

The middle part of the Triassic series contains a fairly uniform series and the top of it is often marked by a distinct seismic marker. The uppermost Triassic series is normally cut by an unconformity.

The Triassic period was affected by a global sea level drop at the end of the Paleozoic era, changing the overall depositional environment from the marine Zechstein deposits to the dominantly continental Triassic deposits. The evolution of the Triassic in the area is related to the activity along the Tornquist Teisseyre Line, seen as the development of the Horda-Egersund-North Danish system of Triassic half-grabens, sub-paralleling the southwest coast of Norway and the west coast of Sweden (Ziegler, 1988).

The study area can be partly separated into two tectonic domains during Triassic times, one severely affected by salt movement south of 58°N (Egersund Basin) and one controlled by the westerly dipping basement to the north of 58°N (Åsta Graben) (Fig. 5).

For the whole Triassic period the basin axes seem to be located on the east side of the Sele High going southwards where they split into several axes located between the active growing salt walls. A minor axis west of the Sele High in Block 17/10 and parts of 17/11.

The distribution of Early Triassic sediments generally seems to be limited to the east by the Stavanger Platform (Fig. 4). The Middle Triassic has a somewhat more widespread distribution. During the Middle Triassic a marine transgression occurred from the south and southeast, epitomized by the distribution of the Muschelkalk. How far north the transgression can be traced is debatable, but the limestone stringers of probable Anisian age in well 17/12-1 have been indicated to be of marine origin (Ziegler, 1988). However, Jacobsen et al. (1982) claim that these limestones are sabkha deposits. Deposits of lacustrine origin are found in wells 17/12-2 and 17/12-3.

From in-house geochemical analysis of the Triassic interval in well 17/11-2, a deltaic to lacustrine origin of the shaly sediments has been postulated (Fig. 6).

Following these observations and the structural trends, we believe it likely that a fairly large lake existed periodically located on the eastern side of the Sele High, with sabkha deposits as seen in well 17/12-1 and somewhat more fluvial deposits on the structurally elevated sections, like the Sele High (well 17/11-2). The position of this lake is linked to the development of a half-graben east of the Sele High, the Åsta Graben.

During the Late Triassic times the climate changed from arid to humid principally due to the northward plate movement of the area. In addition, there was a strong rejuvenation of the tectonic regime, with dike intrusions onshore (Sunnhordland) dated at 220 Ma (Færseth et al., 1975) and volcanic activity offshore (well 17/9-1).

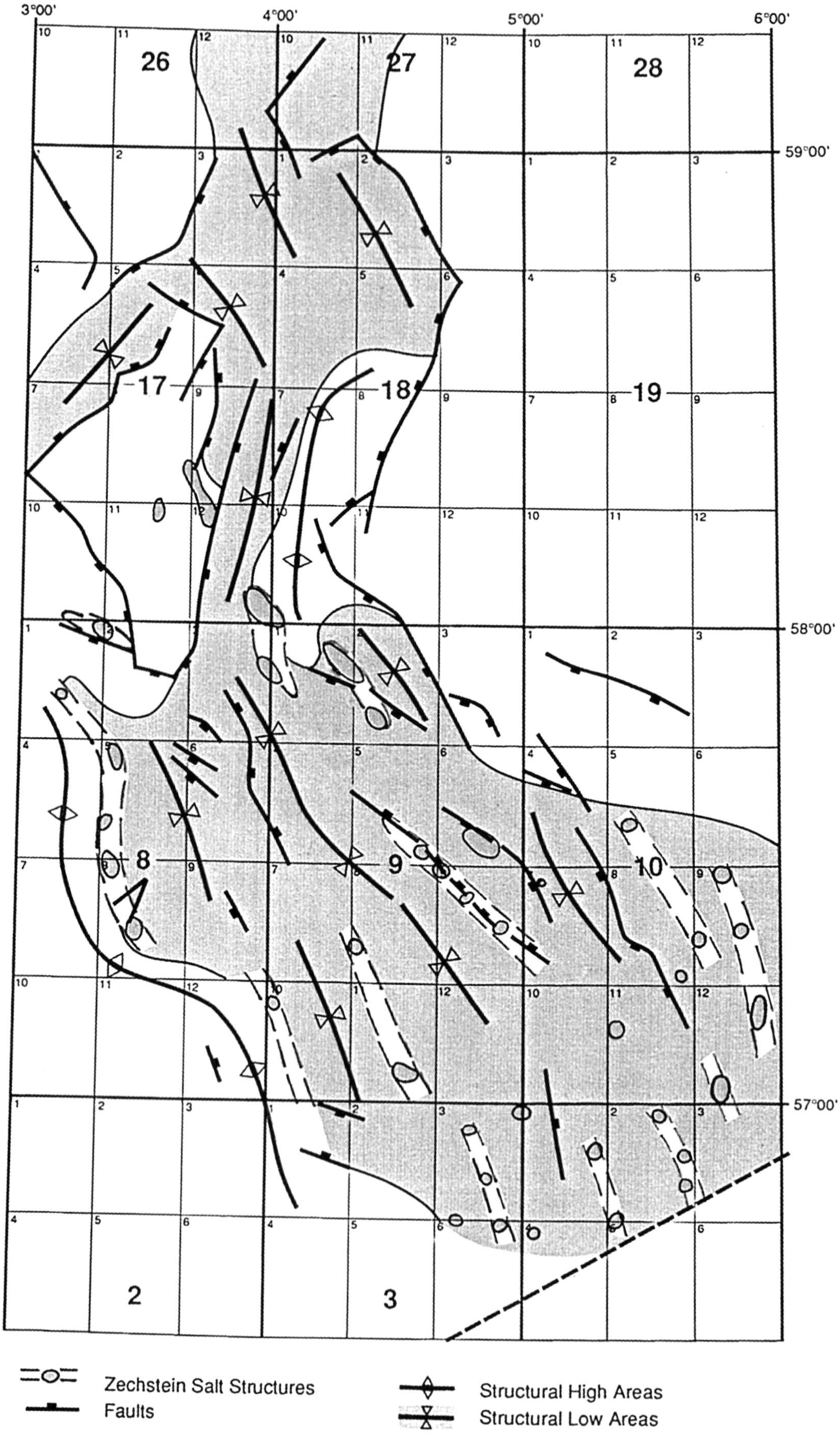

Fig. 5. Triassic basin configuration.

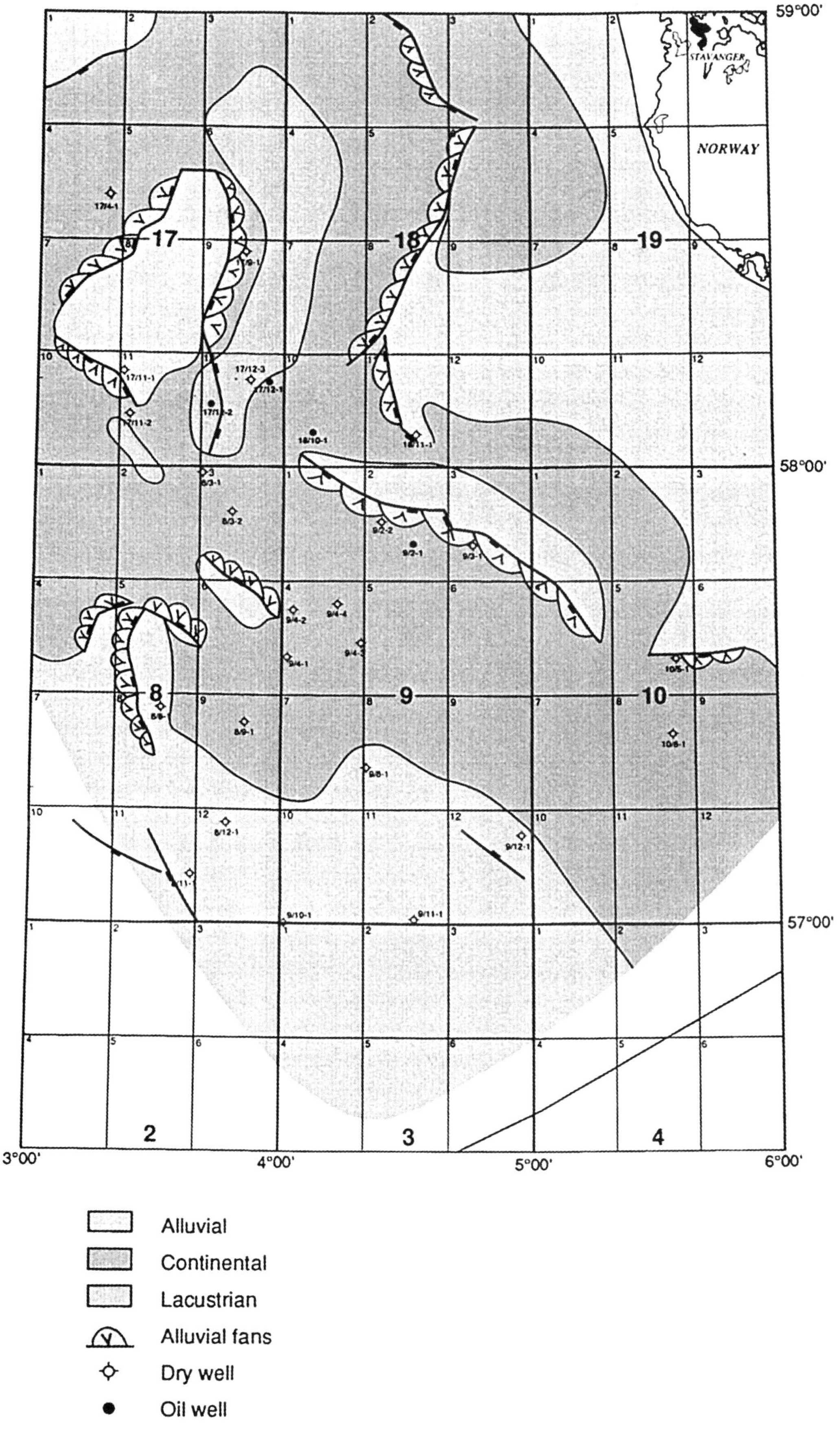

Fig. 6. Triassic paleogeography.

The clastic input from the Fennoscandian hinterland was probably significant. The increased chemical weathering of the Pre-Cambrian feldspar-rich rocks to the east due to the climate change led to an increased development of clay. This may partly explain the increased shale content in the Upper Triassic (pre-Rhaetian) series in the SENS area.

In the very Late Triassic time (Rhaetian) the sea level rose and the rate of subsidence in the SENS area decreased as one entered the tectonically uneventful Early Jurassic period. The Late Triassic transgression led to the deposition of basal transgressive sands termed the Gassum Formation of Rhaetian age.

Jurassic

Early Jurassic

The tectonic activity during the Early Jurassic was probably fairly limited. The period as such was characterized by a transgressive sequence (Vail and Todd, 1981). The thickest Lower Jurassic series is 165 m thick (17/9-1) representing a time span of some 20 Ma (Fig. 2).

The Early Jurassic is termed the Fjerritslev Formation and it is dominated by shales, indicative of shallow marine deposits. In this study, we have included the sandy facies encountered in wells 17/12-1 and 17/12-3, representing near coast to deltaic deposits in this formation and not in the Gassum Formation. These sandy facies attributed to the Fjerritslev Formation are known in the Danish sector.

In Early Jurassic times the evolutions of the Egersund Basin and Åsta Graben were slightly different (Fig. 6). A fluvial system with drainage from the Stavanger Platform probably had its front in the geographical centre (north–south) of block 17/12. A marine connection existed between the northern part and the southern part of the SENS area. In the northern part, Fjerritslev shales where deposited in an anoxic environment indicating restricted bottom water circulation (Fig. 7).

The sandy deposits in well 18/10-1 are indicative of a fluvial to deltaic environment. In well 9/2-1 to the south of the threshold the Fjerritslev Formation was deposited in a shallow marine shelf environment.

The main basin axis was on the east side of the Sele High, extending from 59°N southwards to 58°10′N along a fairly narrow band between 3°20′E and 4°E. The shaly Fjerritslev Formation can be traced on seismic data into the Stord Basin. To the south (south of 58°N) the main Early Jurassic Basin axis has a slight shift in orientation from N–S (north of 58°N) to NNW–SSE (south of 58°N). The main axes are located in Blocks 9/2, 9/5, 9/6 and 9/9 (Fig. 8). Southeast of this, ther seems to be a new threshold before the Early Jurassic basin axis extends into the Norwegian Danish Basin.

The extension of the Early Jurassic deposits is debatable, but based on their seismic facies character, they can be traced almost to the eastern edge of the SENS area in the northeast (Blocks 18/5 and 18/6) and also into the area named Fiskebank Basin (Blocks; 8/3, 8/6 and 8/9).

Middle Jurassic

The Middle Jurassic deposits range in age from Aalenian through Callovian. However, from a sequence point of view the Callovian deposits belong to the Upper Jurassic transgressive mega cycle. For the SENS study area, the Middle Jurassic can be split into two main cycles.

(a) The Aalenian through Bathonian generally fluvial deltaic deposits, with some marine influx from the south. During periods of the Middle Jurassic a marine connection might have existed to the north along the main basin axis. Several of the wells near the axis contain marginal marine to nearshore sediments in the Middle Jurassic series (i.e., 17/9-1, 17/12-1, 17/12-3 and 18/10-1).

(b) The Callovian deposits dominated by marginal marine deposits (sands). During the Middle Jurassic the SENS area experienced a dextral movement along the Tornquist-Teisseyre Line (Fig. 8). Activity along this trend around 58°N continued westward to the southeastern edge of the Sele High. This dextral movement led to the reactivation of the normal fault flanking the western side of the Åsta Graben and Egersund Basin (the eastern boundary fault of the Sele High) with the result that the Åsta Graben to the north of approximately 58°N shows a gentle tilt to the west at pre-Zechstein level. In the area where the Tornquist-Teisseyre Line links the N–S normal faults (Block 17/12), a thick post-Early Jurassic basin developed as seen on the isopach map Top Triassic to Top Middle Jurassic (Fig. 9).

Furthermore, the area to the west of the study area (Montrose-Forties High) was a centre of domal uplift, where the relief might have reached 2000–3000 m (Ziegler, 1988). This uplifted area acted as a sediment source also in a southeast direction. Generally, the whole North Sea underwent severe tectonic activity involving the up-doming with associated volcanic activity, normal faulting (perpendicular to wrench movements), wrenching and subsidence (Ziegler, 1988). To the south of 58°N, in Quadrants 9 and 10, increased tectonic activity can be observed. In addition to the effect of wrench movement and normal faulting, this southern part of the SENS area was affected by halokinetic movements.

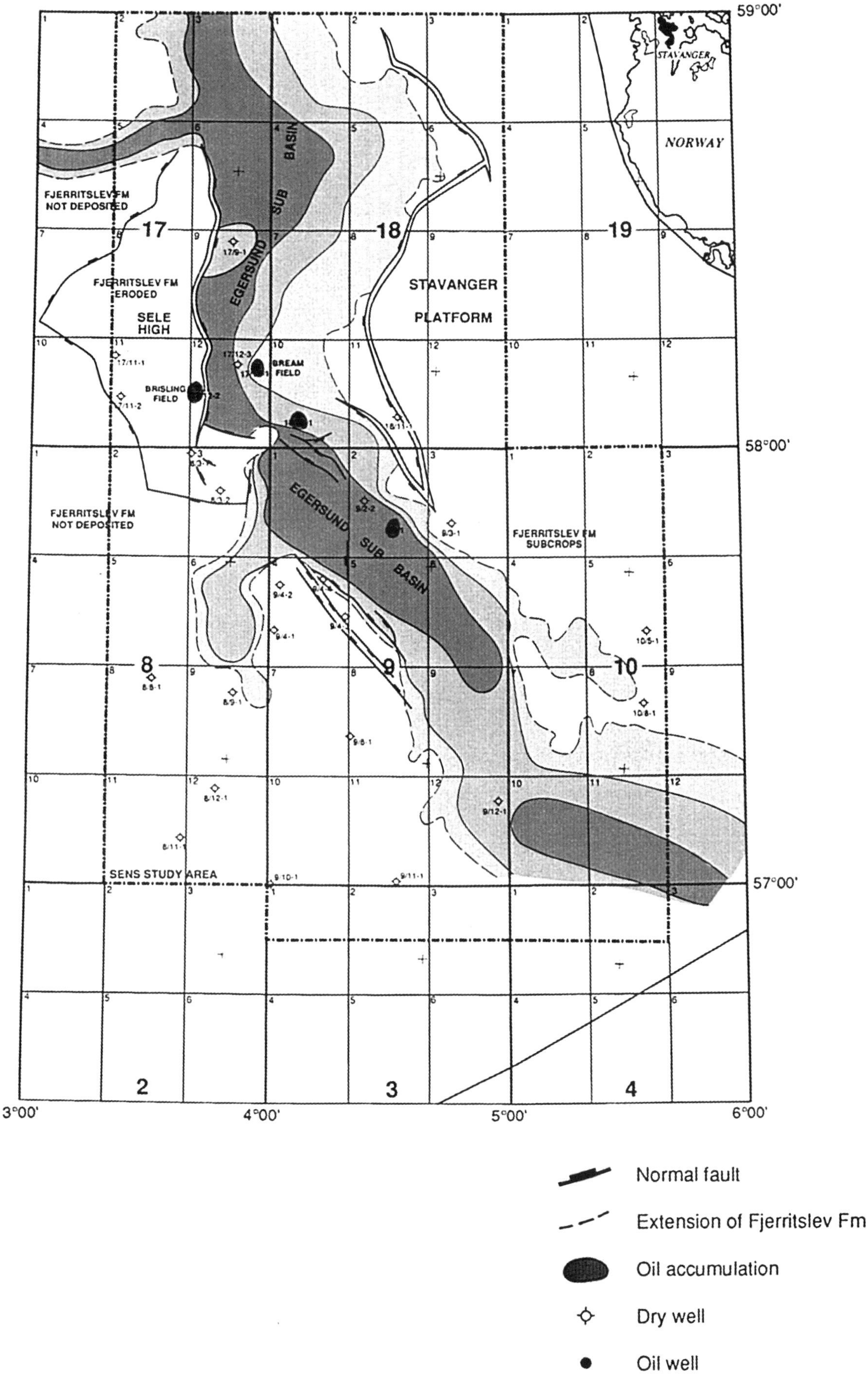

Fig. 7. Distribution of the Early Jurassic, Fjerritslev Formation.

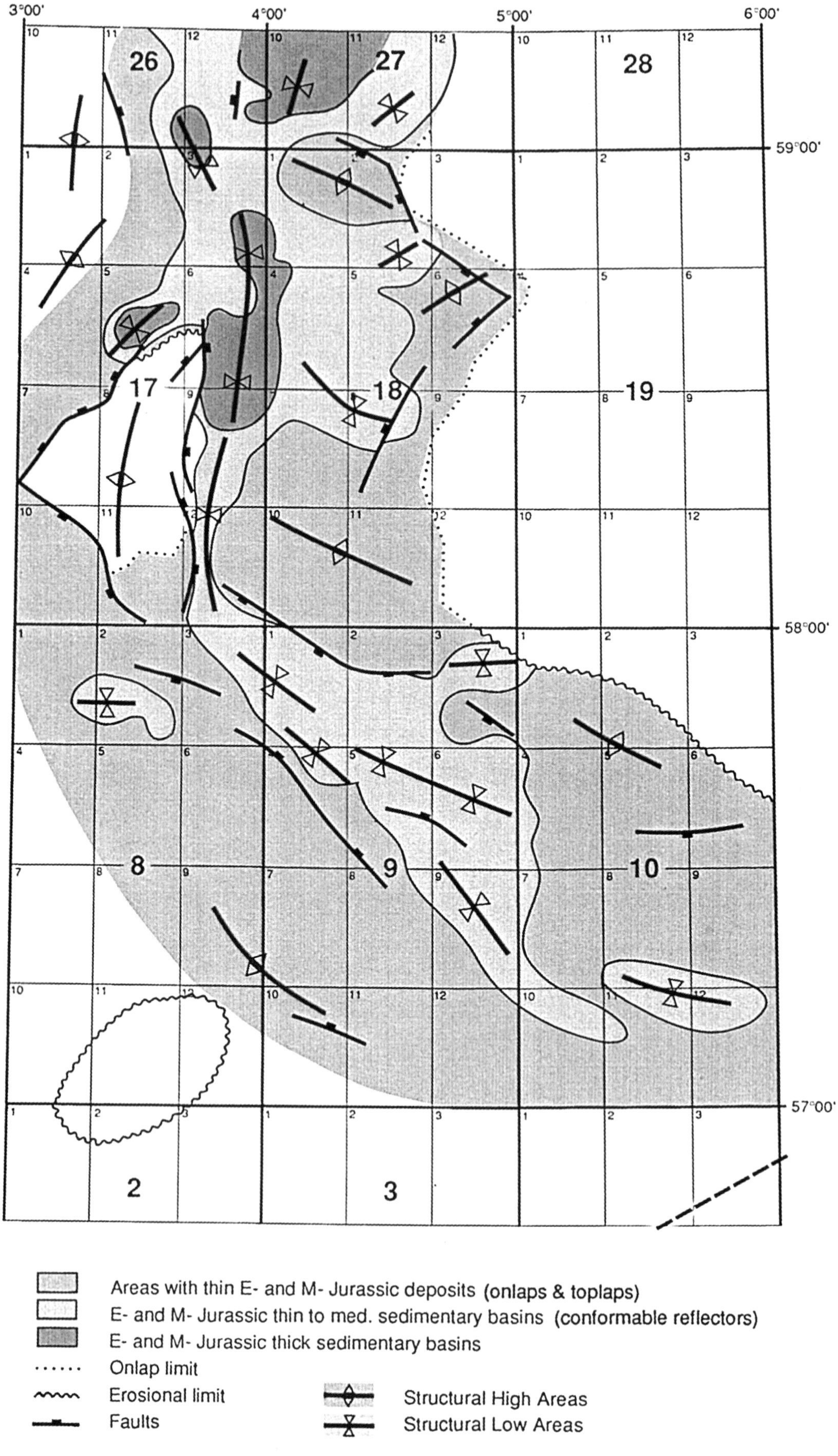

Fig. 8. Early and Middle Jurassic basin configuration.

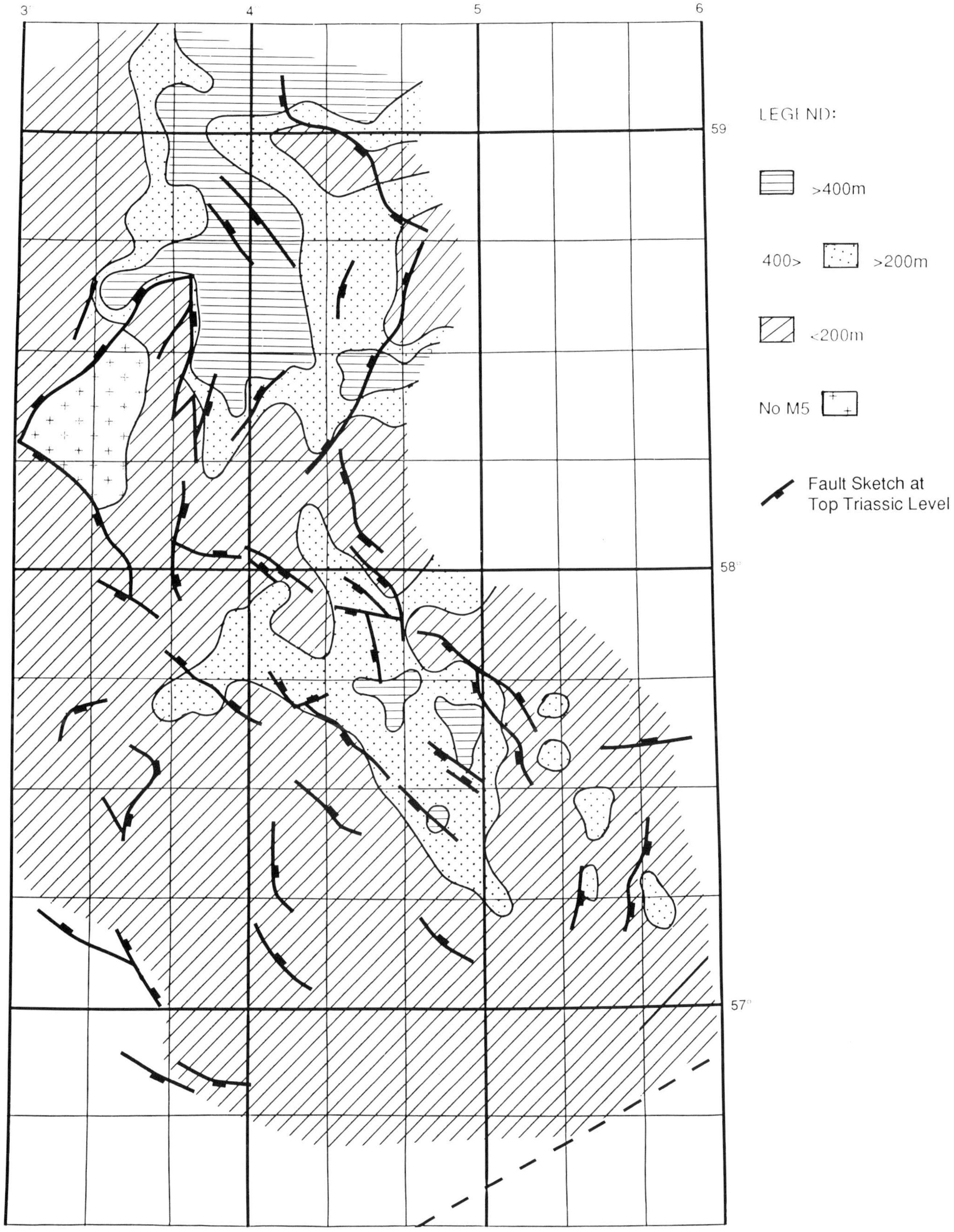

Fig. 9. Isopach map between Top Triassic and Top Middle Jurassic (metres).

At the Middle Jurassic level the main fault orientation between 56°30′N and 58°N is 130°–150°N with a slight change towards 150°–170°N in the SW-corner of the study area (Table 1). The third trend of less importance is the 10°–30°N.

In the area between 58°N and 59°N, the western boundary fault orientated 10°–30°N has the largest throw, whilst that in the eastern Egersund boundary fault orientated 10°–30°N is very limited. In the northeastern corner of the study area the throw is almost zero.

The Middle Jurassic Bryne Formation is widespread in the study area, as can be documented by the 27 wells involved in the study, of which only four (17/11-1, 17/10-1, 17/11-2 and 8/11-1) did not penetrate Bryne Formation. Whether the formations were eroded or not deposited in these wells is not clear. All these wells have encountered older stratigraphic horizons. But it appears as if it has been eroded, particularly in areas on the Sele High (17/10-1, 17/11-1 and 17/11-2).

The Bryne Formation was deposited in fluvial deltaic system, with basinal connections to the north (Stord Basin) and to the south (Norwegian Danish Basin). Marine incursions are observed as far north as the northern limit of Quadrant 9 (Fig. 10). The deposits are dominantly sands, silts and shales, with some volcanics and coal beds. The volcanic material is basically found in well 17/9-1, where the Middle Jurassic series is more than 500 m thick.

In Callovian times a global transgression started (Vail and Todd, 1981), which also affected the SENS area, represented by the Late Middle Jurassic Sandnes Formation, which is a marginal marine sandstone of limited distribution and thickness. The Sandnes Formation ranges in age from Callovian to Middle Oxfordian (Vollset and Doré, 1984). A marine connection to the north is believed to have existed frequently during Callovian times (Fig. 11). In the study area only 50% of the wells encountered the Sandnes Formation. This is probably due to non-deposition of this formation on the structurally elevated areas. Wells 17/9-1 and 18/11-1 are believed to have been located in a more basinal position, hence no marginal marine sandstone is encountered in the wells. We prefer to assign these Callovian age deposits to the Egersund Formation.

The Middle Jurassic sand deposits are the main reservoir rocks in the Egersund Basin and the Åsta Graben. Wells 9/2-1 and 17/12-2 both tested oil from the Sandnes Formation, whilst wells 17/12-1 and 18/10-1 tested oil from the Bryne Formation.

Late Jurassic

The second major extensional event in the Jurassic occurred at the beginning of Late Jurassic (approximately 160 Ma — Oxfordian). This event is well observed in the Central and Viking Grabens, but seems to be of fairly minor significance in the SENS area, its effects being some reactivation with a minor vertical throw along the existing faults. According to Ziegler (1988), some of the eastern parts of the Egersund Basin were affected by compressional foreland deformation, but the main parts of the Horda-Åsta half-grabens became inactive. To the west of the SENS area the Vestland Ridge system probably formed an archipelago during Oxfordian (Fig. 12). This archipelago might have been a local source for more coarse clastic sediments in the westernmost part of the area, which is partly reflected in the lithology of the formations and the kerogen composition of the Oxfordian to Kimmeridgian deposits. Rocks of a more (deep) marine origin in the central parts of the basin gradually become more marginal marine deposits to the west. In the northeastern corner of the basin (Blocks 18/2, 18/6 and 18/9) deposits of Oxfordian age are fairly thick indicating a fairly strong subsidence and perhaps significant fluvial drainage from the hinterland to the east. In the northwestern corner the marine Egersund Formation is found in the Ling Graben (17/4-1), indicating a marine connection at least as far west as this. It is, however, likely that the Ling Graben was only an embayment connected to the southeast at that time.

The old threshold, more or less orientated east to west at approximately 58°N, was still in existence, and the thickness of Oxfordian sediments are reduced in this area. To the southeast of 58°N the deposits decrease in thickness. In the western part, on the southern part of Sele High, some local Oxfordian basins occur (Blocks 17/10 and 17/11).

During Late Jurassic Kimmeridgian to Volgian times, the area had a more direct contact to the west (Central Graben), with the eastern platform edge between 4°E and 4°20′E and between 58°N and 59°N (Fig. 13). The isopach map between Base Cretaceous and Middle Jurassic clearly shows the effect of the active faulting at the Middle Jurassic level in terms of sediment thickness, as a result of faulting and subsidence (Figs. 9 and 14).

At the Intra Upper Jurassic level (Top Tau Formation) the number of faults affecting this horizon is reduced compared with the map of the Top Middle Jurassic level. Again, the overall dominating trends are 130°–150°N and 150°–170°N with a 35°–55°N influence being the northwestern boundary of the Sele High (Table 1). During Kimmeridgian times, the

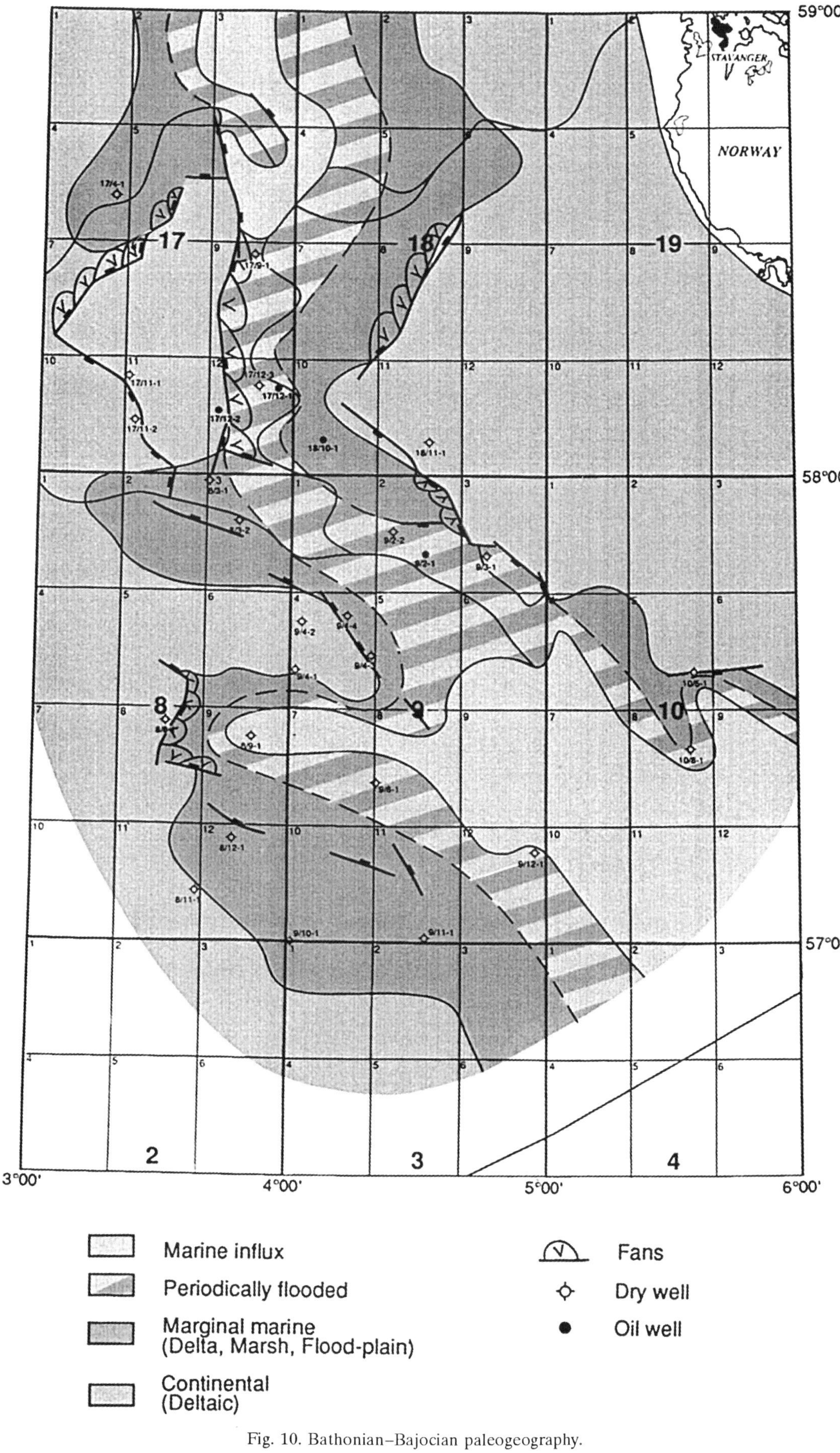

Fig. 10. Bathonian–Bajocian paleogeography.

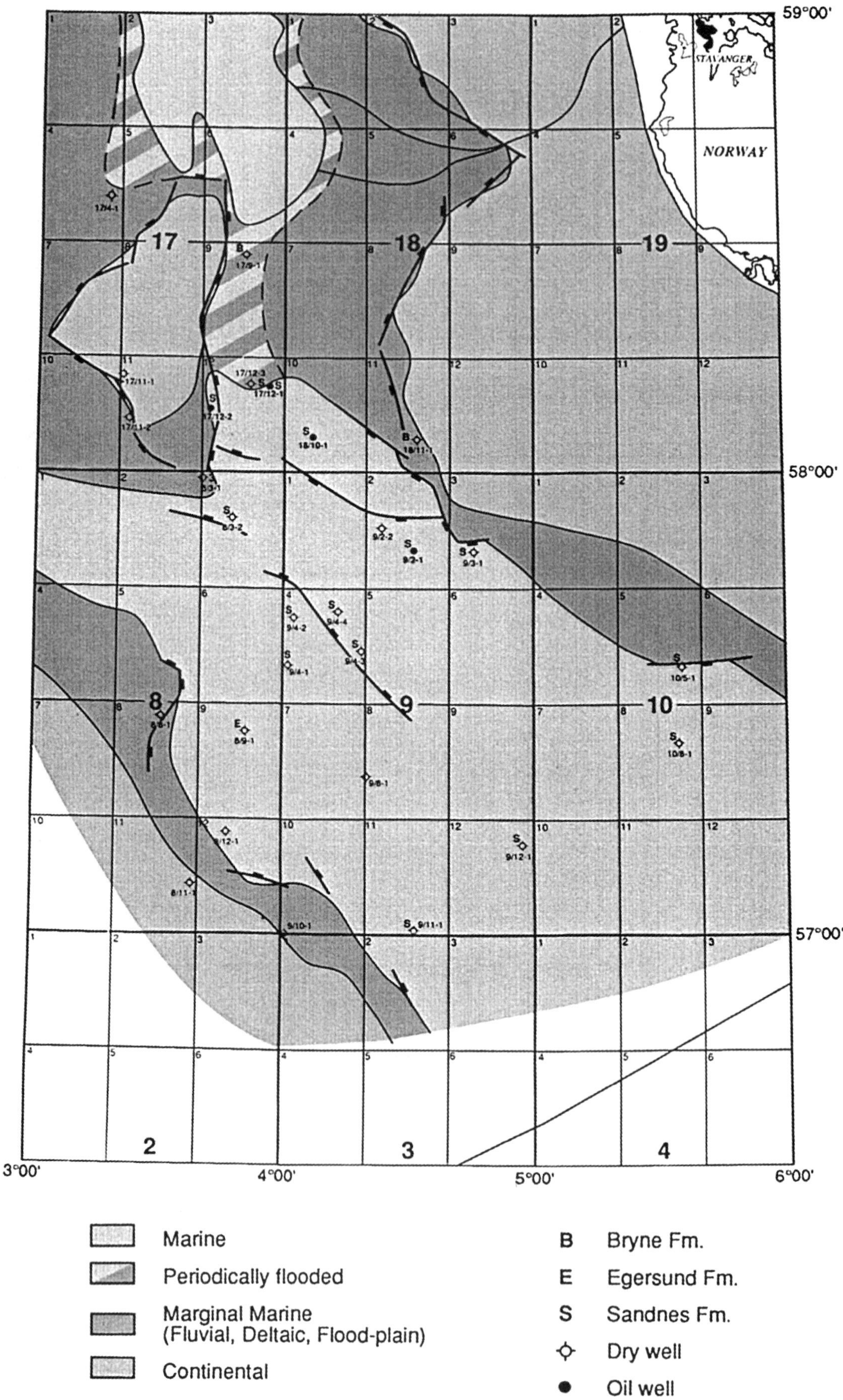

Fig. 11. Callovian paleogeography.

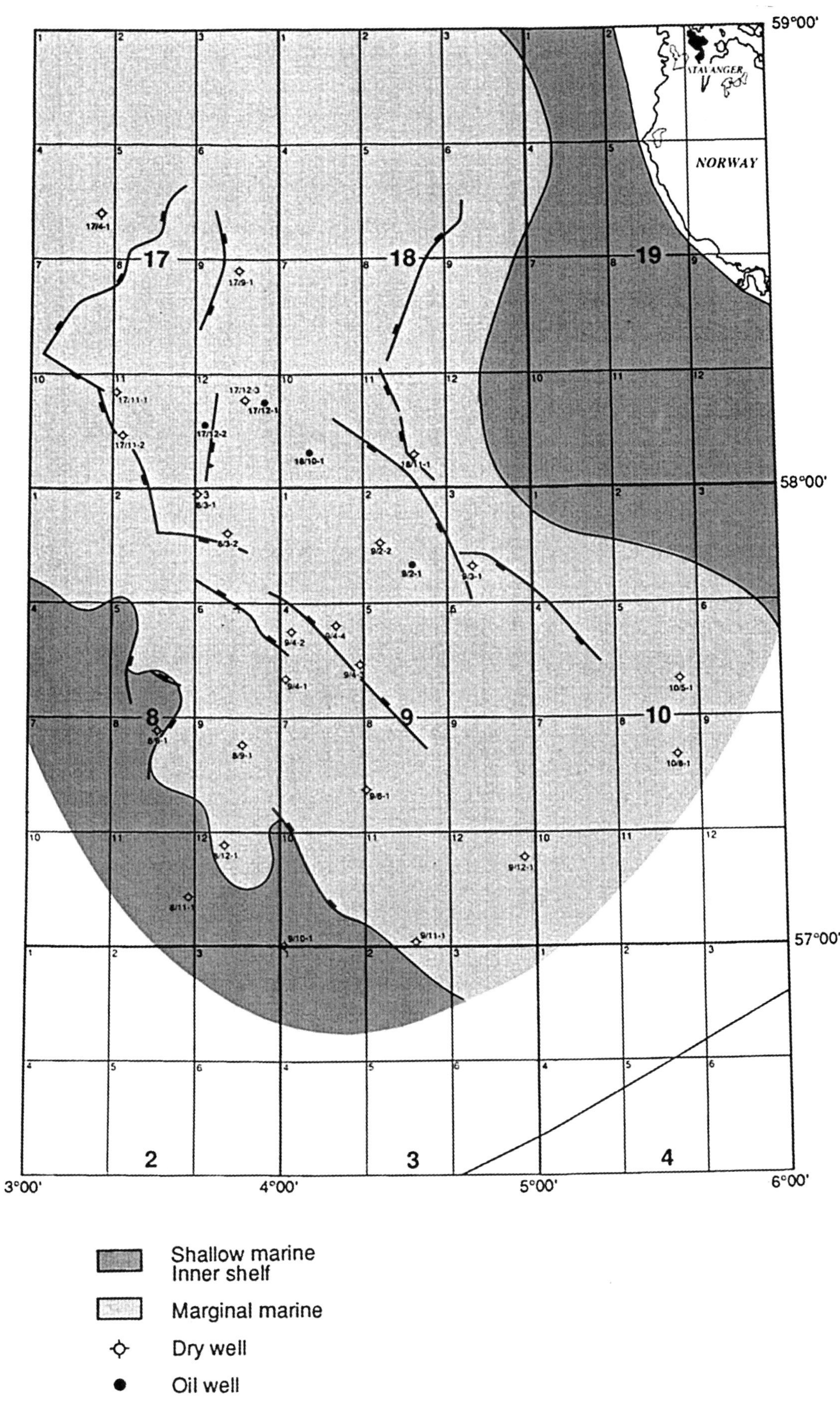

Fig. 12. Oxfordian paleogeography.

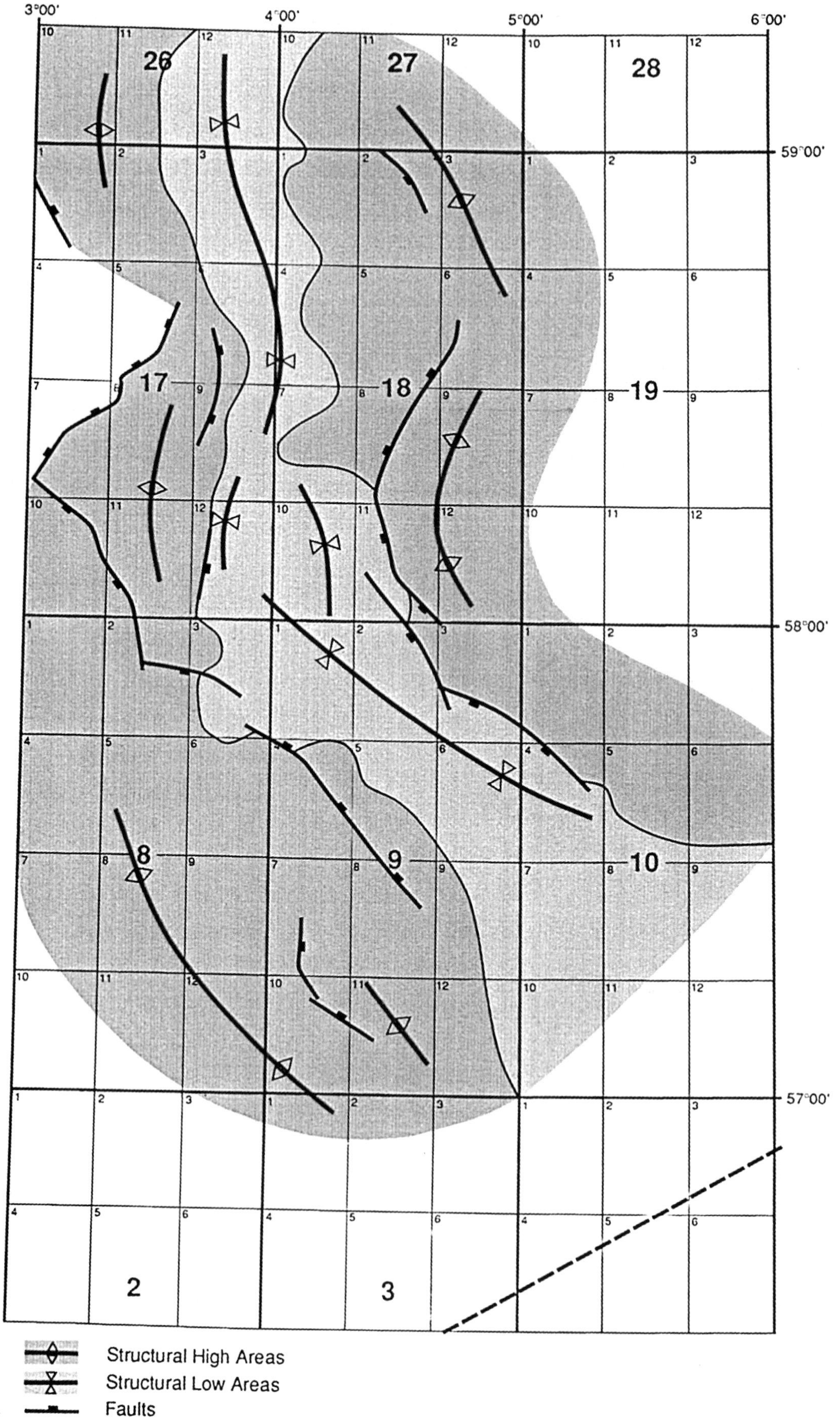

Fig. 13. Late Jurassic basin configuration.

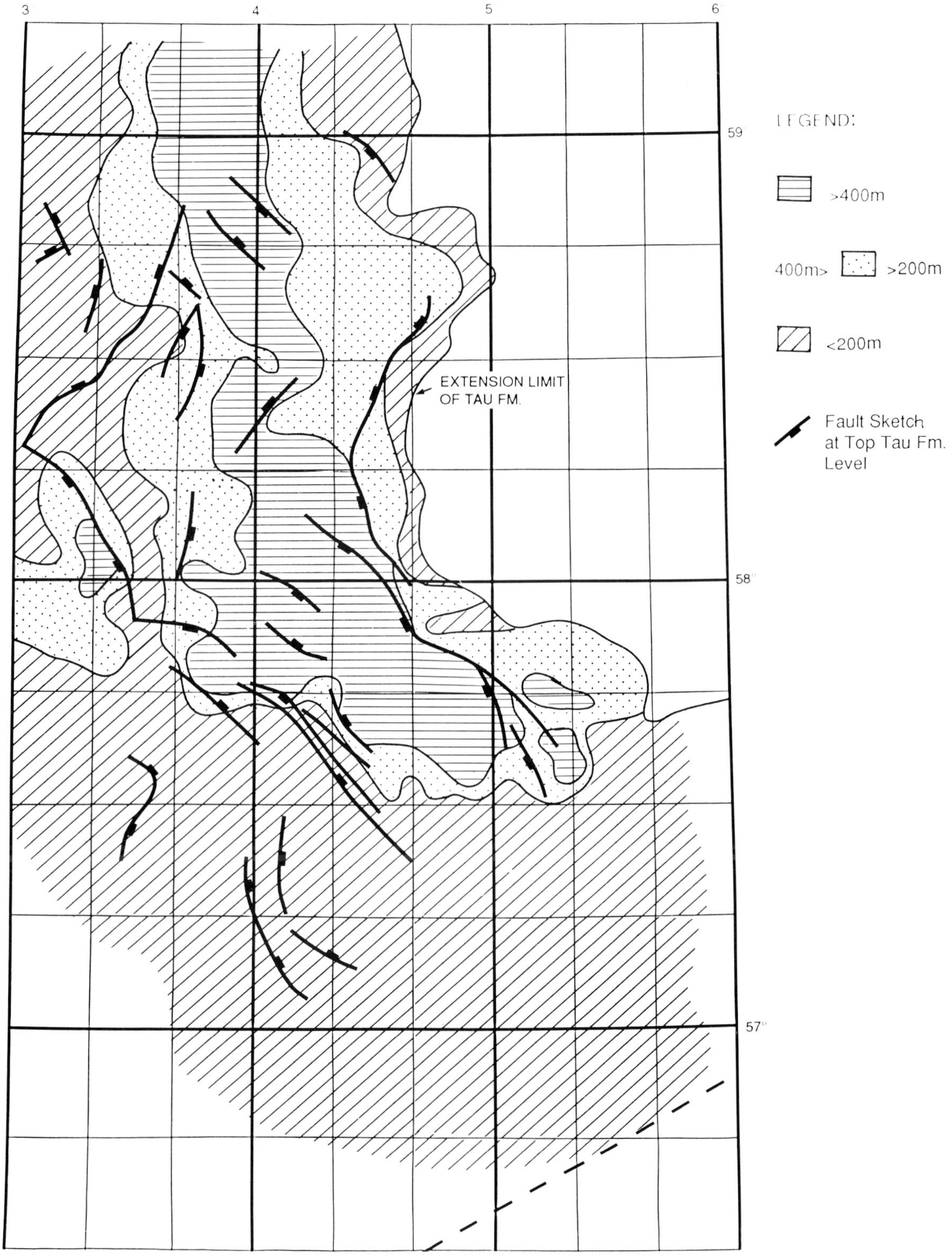

Fig. 14. Isopach map between Top Middle Jurassic and Top Tau (metres).

sea level rose with inundation of the Sele High and probably covered most of the Vestland Ridge. The existence of local islands, however, is likely (Figs. 13 and 15). The Egersund Basin and the Åsta Graben continued to subside, this being epitomized by the significant increase in sediment thicknesses. On the Stavanger Platform, the Kimmeridgian transgression was extensive, probably covering most of the area. The Kimmeridgian (Tau Formation) can be observed as an onlapping sequence. At this time a marine connection existed to the north (Stord Basin) and to the west (Viking Graben and Central Graben). This Late Jurassic ocean was connected to the Tethys through the Polish-Danish trough, and not through

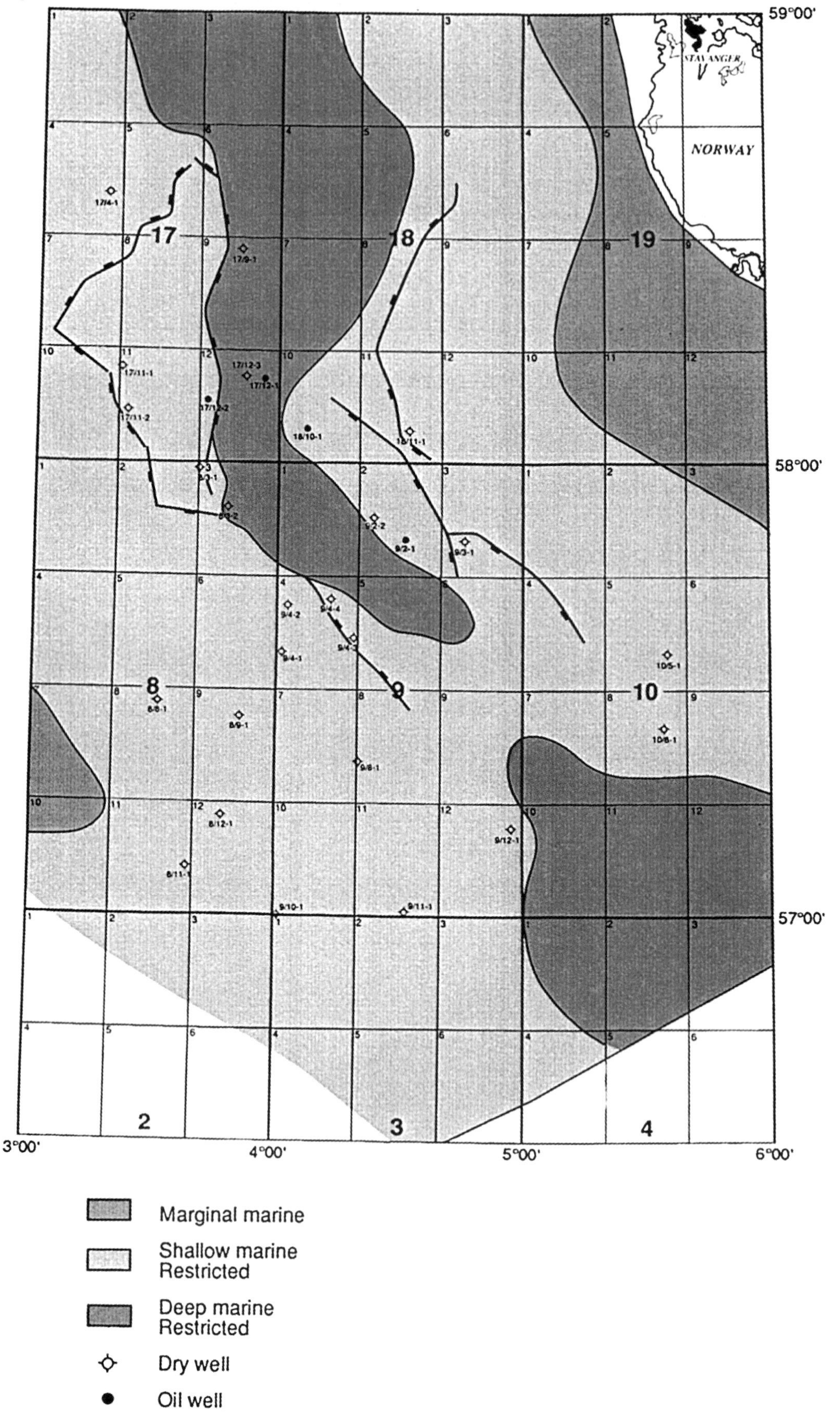

Fig. 15. Kimmeridgian paleogeography.

the Böhemian seaway, which came to an end at this time (Ziegler, 1988).

Due to differential subsidence in the SENS area during Kimmeridgian times, the area was separated into several sub-basins. This led to very limited bottom water circulation and, hence, anoxic bottom water conditions were developed in the basins. On the margins of the basin, water circulation was better due to shallow marine longshore currents. The effect being reduced preservation of the organic material and a less favourable kerogen composition in terms of source rock.

During Volgian times, the sea level dropped somewhat, once more exposing parts of the Vestland Ridge. To the east, this is seen as a regression on the Stavanger Platform with the prograding sequences of the Sauda Formation. These prograding sequences indicate the westernmost extension of the Late Jurassic shelf (Fig. 16). To the west of this limit, at approximately 4°20′E (north of 58°N), the depositional environment was open marine, well oxygenated, probably bounded westwards by some islands located along the Vestland Ridge.

At this level, it becomes obvious that the Egersund Basin and Åsta Graben ceased to exist as separate structural features (sensu stricto) and should be regarded as a part of the overall Late Jurassic North Sea basin.

In the area there is no clear-cut evidence of an unconformity between the Late Jurassic and Early Cretaceous, and the Early Cretaceous series should be regarded as a part of the transgressive megasequence initiated in early Callovian times.

The Late Jurassic deposits range in age from Base Oxfordian through Volgian. Lithostratigraphically speaking, they consist of the Egersund, Tau and Sauda Formations.

The Egersund Formation. The formation ranges in age from Base Oxfordian to Base Volgian. It was deposited in an open marine, generally low energy basinal environment throughout most of the area. It is encountered in most of the wells in the study area, except for the ones on the Sele High where the formation probably was not deposited (Fig. 12).

The Tau Formation. The Tau Formation ranges in age from Base Kimmeridgian to Middle Volgian. It was deposited in an anaerobic marine environment with high organic productivity and restricted bottom water circulation. It also represents the maximum of the Upper Jurassic transgressive mega-cycle in the Egersund Basin (Fig. 15).

The Sauda Formation. The type well in the Sauda Formation is 9/4-3. It ranges in age from Late Kimmeridgian to Top Ryazanian. It shows a downlap character on the Tau Formation in the northeastern part of the study area. It was deposited in an open marine, generally low energy basinal environment (Fig. 16).

Cretaceous

Early Cretaceous

During the Early Cretaceous, the marine transgressive phase continued with a fairly large subsidence in the area. At the Top Early Cretaceous level, the fault trends can be separated south and north of 58°N. The southern part being dominated by 130°–150°N and 150°–170°N with a minor 10°–30°N component (Table 1), whereas north of 58°N, the main trends are 10°–30°N and 35°–55°N.

The Early Cretaceous deposits are lithostratigraphically divided into the Flekkefjord Formation, Valhall Formation and Rødby Formation. The Flekkefjord Formation of Ryazanian age being a part of the Late Jurassic mega-cycle.

The Flekkefjord and Valhall Formations. The Flekkefjord Formation ranges in age from Late Volgian through Ryazanian. It was deposited in a marine, low energy, basinal environment.

Near the base of the Valanginian there was a significant drop in sea level, followed by an overall transgressive sequence ending in Mid Barremian time. The lithostratigraphic unit representative of this time span is the Valhall Formation. The Valhall Formation reached extensive thicknesses, as observed even on the former structural highs (wells 17/11-1, 17/11-2 and 17/12-2). A local depocentre has been postulated to have developed in Block 17/10 during this time. It has further been speculated that the Sele High became inverted (Jacobsen, 1982), although we find no evidence of inversion of the Sele High during Valanginian through Barremian times. However, inversion along the eastern side of the Egersund Basin and Åsta Graben probably was initiated during this time. During the deposition of the Valhall Formation the basin subsidence kept pace with the rate of sedimentation.

The Rødby Formation. The Rødby Formation ranges in age from Barremian to Base Cenomanian. The lower part was deposited in a marine environment in an overall regressive sequence (Jacobsen, 1982). From Mid to Late Albian times, the area had a shift of depositional system from a general regressive phase to a Late Albian-Cenomanian transgressive

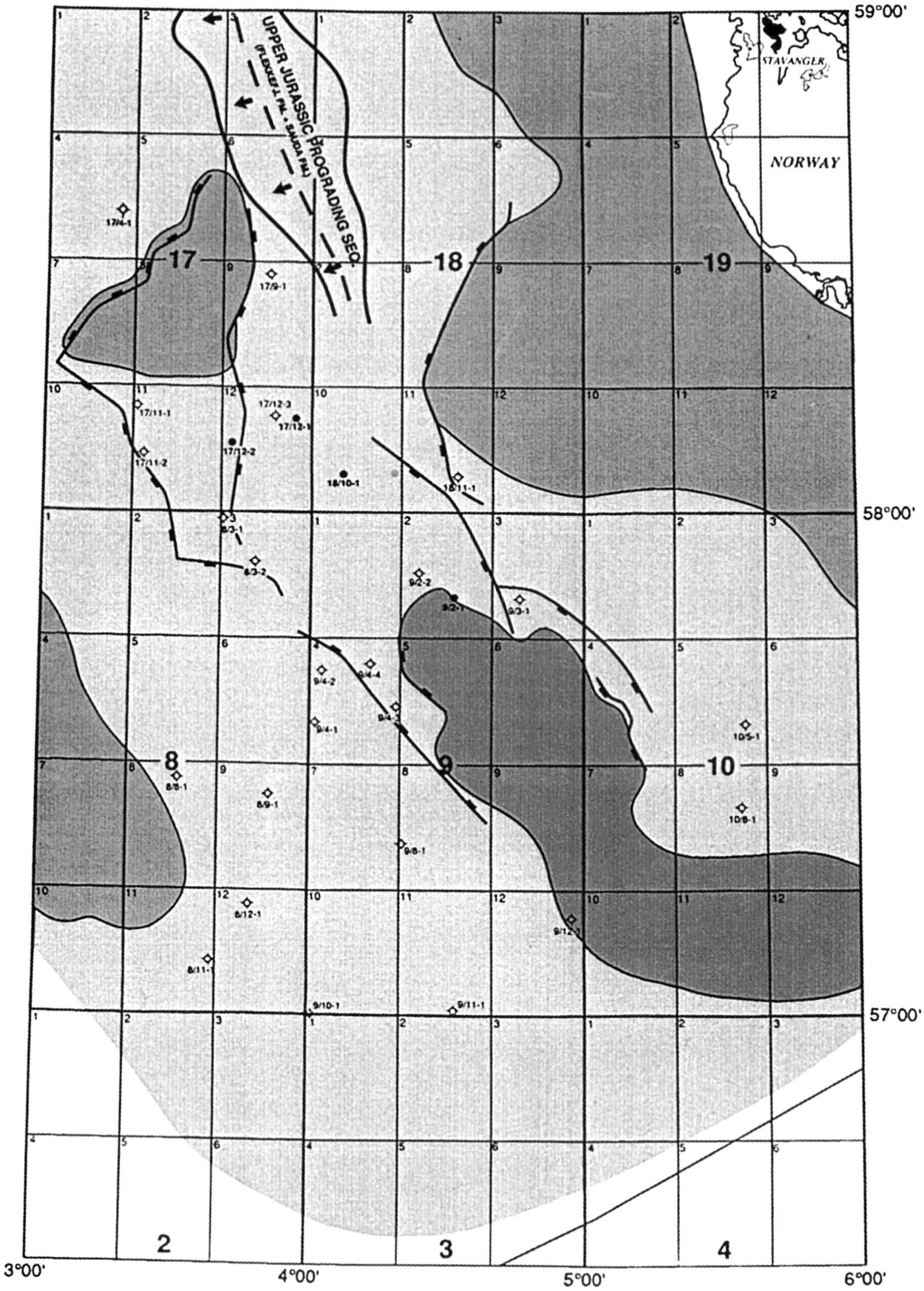

Marginal marine

Shallow marine Shelf

Deep marine

Dry well

Oil well

Fig. 16. Volgian paleogeography.

mega-cycle (Vail and Todd, 1981). The formation is fairly thin and is eroded or not deposited on the flank of the basin.

During the transition from Early to Late Cretaceous, compression starts to have an effect in the study area, with inversion accomplished by salt deformation (i.e., reactivation of the old salt structures). The compression phase was affecting the area periodically, until Early Eocene times.

Late Cretaceous

At the onset of the Late Cretaceous, the acceleration of the global sea floor spreading rate induced a major regional transgression inundating the whole North Sea area. The Late Cretaceous deposits can be divided geographically into two zones along a NE–SW line north of Quadrant 25. The southern part being the chalk province to which the SENS area belongs, and the northern part being the clastic province.

The southern province is lithologically dominated by chalk, chalky limestone and limestone deposits. The depositional environment for the carbonate province was one of warm sea with water depths in the range of 300–500 m (Sørensen et al., 1986). The chalk group has been divided into five formations ranging in age from Cenomanian through Danian. The chalk in the Egersund Basin ranges in thickness from 60 m to almost 1000 m.

The oldest formation, the Hidra Formation, ranges in age from Cenomanian to Turonian. This formation generally has a "high" clay content (with respect to pure chalk) and as such is not regarded as a potential reservoir rock.

The Blodøks Formation (Isaksen and Tonstad, 1989) of Turonian age is a thin shaly formation. It seems that the drop in sea level in Turonian times might have led to erosion of the Blodøks and Hidra Formation on the somewhat structurally elevated areas like the Sele High. Neither of the two formations are present in the wells in the area (17/11-1, 17/11-2 and 17/12-2).

The Hod Formation, (Coniacian to Late Campanian) was deposited in a calcareous province with fairly limited tectonic activity. The boundary between Hod and Tor Formations is often seen as a hardground. The boundary between Tor (Maastrichtian) and Ekofisk (Danian) Formations is also seen as a hardground in places. Both the Tor and Ekofisk Formations are represented by both allochthonous and autochthonous chalks. The allochthonous chalks are found locally, and represent deposition due either to faulting activity along pre-existing faults, or to local halokinetic movements. These allochthonous chalks are known to be the best reservoir chalks. However, in the area there is no proof of the existence of either of these types of deposit, but they might be present in the southern part of the basin along the old faults and salt structures, and along the eastern Egersund Basin and Åsta Graben basin margin faults, which are inverted features formed during Late Cretaceous to Early Tertiary times.

Tertiary

For the whole of the Tertiary period, the area was tectonically associated to the North Sea basin which experienced thermal subsidence. Tertiary sediments are found all over the SENS area, but they gradually thin out towards the Fennoscandian shield.

During the Paleocene, the eastern Egersund Basin and Åsta Graben were uplifted (Jacobsen, 1982), leading to the condensed sequences of Lista and Sele Formations.

The boundary between Paleocene and Eocene is represented by the volcanic ash-rich Balder Formation. This formation is clay dominated with a high content of tuffaceous material. In the SENS area, the formation is assigned to the lowermost Eocene.

During Eocene times, sands were shed into the southeastern part of the Egersund Basin. The sand deposits have been termed lithostratigraphically the Fiskebank Formation (Deegan and Scull, 1977). The Late Eocene and Early Oligocene deposits are dominantly shaly. However, in the Early Chattian time a significant drop in sea level occured with the development of Chattian sands along the eastern margins of the SENS area. This drop in sea level could be associated with the uplift of the Fennoscandian Shield to the east. One might infer that this uplift has continued to the present times and is associated either with the non-deposition or the erosion of some of the Tertiary deposits. During the Late Oligocene and Miocene times the subsidence continued with the deposition of significant thicknesses of shales. However, all these Upper Tertiary series thin towards the east.

Conclusions

From Permian to Mesozoic times, the study area can be divided into several structural provinces. Both the Sele High and the Stavanger Platform are old structural highs. The major structural basins can be divided in two: the Egersund Basin, which was heavily salt affected from Mid Triassic times, and the Åsta Graben, which actually is a half-graben. At the end of Mesozoic the study area ceased to exist as an independent tectonic province and became part of the North Sea as a whole. The basinal part of the study area comprises sediments ranging in age from at least Early Permian to recent.

Acknowledgements

We would like to thank Elf Aquitaine Norge A/S for granting permission to publish this paper. Further, we would like to thank the drafting department for the figures and Karin Nielsen who kindly typed the manuscript. In addition, we would like to thank our colleagues in the Exploration Division for fruitful discussions during this study.

References

Deegan, C.E. and Scull, B.J., 1977. A proposed standard lithostratigraphic nomenclature for the Central and Northern North Sea. Institute for Geological Sciences, London, Rep. 77/25.

Færseth, R.B., Macintyre, R.M. and Naterstad, J., 1975. Mesozoic alkaline dykes in the Sunnhordland region, western Norway: ages, geochemistry and regional significance. Lithos 9, 331–345.

Isaksen, D. and Tonstad, K. (Editors), 1989. A revised Cretaceous and Tertiary lithostratigraphic nomenclature for the Norwegian North Sea. Norw. Pet. Dir., Bull., 5, 59 pp.

Jacobsen, H.K., Hamar G.P., Ormaasen, D.E. and Skarpnes, O., 1980. Triassic facies in the North Sea Reservoir Rocks. Norwegian Petroleum Society, Article XV.

Jacobsen, V.W., 1982. Sedimentation and paleogeography. In: R.C. Olsen and I.F. Strass (Editors), 1982. The Norwegian Danish Basin. Norwegian Petroleum Directorate, NPD Paper No. 31, pp. 54–66.

Myhre, L., 1975. Lithology, Well 8/3-1. Norwegian Petroleum Directorate, NPD Paper No. 1, 20 pp.

Ormaasen, D.E., Hamar, G.P., Jacobsen, K.H. and Skarpnes, O., 1980. Permo-Triassic correlations in the North Sea area north of the Central Highs. In: The Sedimentation of the North Sea Reservoir Rocks. NPF Meet. Geilo, 11–14 May, 1980. Institute of Petroleum, Heyden & Son, London.

Rhys, G.H. (Editor), 1974. A proposed standard lithostratigraphic nomenclature for the Southern North Sea and an outline structural nomenclature for the whole of the (UK) North Sea. A report of the joint Oil Industry–Institute of Geological Nomenclature. Rep. Inst. Inst. Geol. Sei. 74/8: 14 pp.

Sørensen, S., Jones, M., Hardman, R.F.P., Leutz, W.K. and Schwarz, P.H., 1986. Reservoir characteristics of high and low productivity chalks from the Central North Sea. In: Habitat of Hydrocarbons on the Norwegian Continental Shelf. Norwegian Petroleum Society, Graham and Trotman, London, pp. 91–110.

Sørensen, S. and Martinsen, B., 1987. A paleogeographic reconstruction of the Rotliegendes Deposits in the Northeastern North Sea. In: J. Brook and K. Glennie (Editors), Petroleum Geology of North-West Europe. Graham and Trotman, London, pp. 497–508.

Vail, P.R. and Todd, R., 1981. Northern North Sea Jurassic Unconformity, chronostratigraphy and sea level from seismic stratigraphy. In: L.V. Illing and G.D. Hobsen (Editors), Petroleum Geology of the Continental Shelf of North-West Europe. Norwegian Petroleum Society, Graham and Trotman, London, pp. 216–235.

Vollset, J. and Doré, A.G., 1984. A revised Triassic and Jurassic lithostratigraphic nomenclature for the Norwegian North Sea. Norw. Pet. Dir., Bull., 3.

Ziegler, P.A., 1988. Evolution of the Arctic–North Atlantic and the Western Tethys. Am. Assoc. Pet. Geol., Mem., 43, 197 pp.

S. SØRENSEN *Elf Aquitaine Norge A/S, Stavanger, Norway*
Present address: *Norex A/S, Luragaarden, Standnes N-4300, Norway*

H. MORIZOT *Elf Aquitaine Norge A/S, Stavanger, Norway*

S. SKOTTHEIM *Elf Aquitaine Norge A/S, Stavanger, Norway*

Tectonic modelling of the northern North Sea using program RIFT

Stephen Lippard and Guojiang Liu

A one-dimensional pure shear lithospheric extension model has been applied to modelling the subsidence of the northern Viking Graben using a depth converted and backstripped regional seismic section along 61°N. Early rifting events are poorly constrained by the seismic data, but the mid- to late Jurassic rifting event is clearly recognised and gives rise to block faulting and the regional Base Cretaceous unconformity. Post-rifting thermal subsidence following this event resulted in the deposition of a largely unfaulted Cretaceous and Cenozoic sequence of up to 5 km thick. A single phase, uniform stretching model can account for the subsidence in the graben if paleobathymetries of up to 650 m occurred in the Cretaceous and early Tertiary. The β factors for the mid- to late Jurassic rifting event in the graben centre are modelled as 1.7 to 1.8 assuming that prior to this event the crust was already thinned by earlier rifting to about 25 km. On the margins of the rift, most notably on the northern Horda Platform, the subsidence curves depart from those predicted by the uniform stretching model. In particular, initial subsidences are too small to account for the total subsidence. In this case, a non-uniform stretching model, with $\beta_{\mathrm{mantle}} > \beta_{\mathrm{crust}}$, can be applied. This suggests that the zone of mantle extension is distributed over a wider area than the crustal extension, which is more restricted to the graben centre. A heterogeneous stretching model, with both laterally and vertically varying β factors, is suggested. By interpolating the top of the basement beneath the graben from published data, an attempt has been made to model the effects of earlier rifting phases. Two scenarios; late Permian–early Triassic and late Triassic rifting, have been tested. In both cases, the early rift phase can be modelled by β factors of 1.3–1.5. This gives a total β value for all rifting phases of 2.7 in the graben centre, which is in agreement with the present-day crustal thickness. This differs substantially from published results which seem to have underestimated the total β factors in the area.

Introduction

The northern part of the North Sea between 60° and 62°N, encompassing the northern part of the Viking Graben and the flanking Tampen Spur and Horda Platform areas, has been the subject of considerable study since the discovery of major oil and gas fields in the area in the 1970s. In early descriptions of the geology, the dominantly extensional nature of the structures was recognised (Blair, 1975; Bowen, 1975; Halstead, 1975). The first complete geological cross-section across the area was presented by Ziegler (1982), who showed tilted fault blocks containing Jurassic and Triassic sediments on the flanks of a central trough. The fault blocks are buried beneath an infill of Cretaceous and Cenozoic sediments up to 5 km thick in the centre of the basin. Ziegler (1982) suggested that, although the main rift phase occurred in the Jurassic, the presence of thick Triassic sediments suggested an earlier rift phase. He also noted that the amount of extension that could be measured on the faults is relatively small and suggested that not all of the crustal thinning could be accounted for by crustal stretching. He proposed instead a model of sub-crustal erosion with thinning of the lower crust by a "thermally active mantle plume". Later work in the Viking and Central Grabens and other parts of the North Sea rift system showed, however, that the McKenzie model of pure-shear crustal extension can be successfully applied (Christie and Sclater, 1980; Wood and Barton, 1983; Barton and Wood, 1984; Giltner, 1987; Hellinger et al., 1989).

Beach (1985, 1986) and Beach et al. (1987) interpreted a deep reflection seismic profile across the northern Viking Graben and suggested a simple shear detachment model (Wernicke, 1985) to account for the asymmetry of the structures. The model implies that the upper and lower crustal extensions are decoupled by a major crustal shear zone dipping eastwards from the western flank under the Viking Graben and passing through the Moho beneath the Horda Platform.

Beach et al. (1987) presented a generalised subsidence curve for the Viking Graben which showed three distinct periods of increased subsidence in the Triassic, late Jurassic and late Cretaceous. These

Structural and Tectonic Modelling and its Application to Petroleum Geology edited by R.M. Larsen, H. Brekke, B.T. Larsen and E. Talleraas. NPF Special Publication 1, pp. 43–54. Elsevier, Amsterdam.

events were related to three rift phases with β factors of 1.5, 1.8–2.2 and 1.1, respectively, giving a total β factor for all three phases of over 3.0. Giltner (1987) applied the McKenzie model to the North Viking Graben using Ziegler's cross-sections as a basis. He obtained tectonic subsidences in the basin centre that suggested a total β factor of about 1.8 consisting of Triassic (β 1.5) and late Jurassic (β 1.2) rifting phases. Vially (1988) made a regional study of subsidence in the northern North Sea. He also distinguished the Permo-Triassic and late Jurassic rifting phases, but suggested that they were offset from one another with the axis of earlier rifting lying to the east of the Viking Graben. Vially shows a maximum β factor for the Permo-Triassic rifting beneath the Horda Platform of 1.4 and for the late Jurassic rifting a maximum β factor of about 1.5 in the centre of the Viking Graben.

Badley et al. (1988) calculated β factors for the mid-Jurassic to early Cretaceous rifting phase in the northern Viking Graben. They noted that the β factors obtained for the post-rift subsidences (1.14–1.49) were greater than those for the syn-rift subsidences (1.03–1.15) and proposed a depth-dependent extension model to account for this. They also suggested that the symmetry of the syn- and post-rift sequences is incompatible with the Wernicke-type model proposed by Beach (1986). In an earlier paper, Badley et al. (1984) published part of a seismic line from the eastern part of the Horda Platform showing deep-lying tilted fault blocks which they related to the earlier Permo-Triassic rifting phase postulated by Ziegler (1982). Badley et al. (1988) suggested that both rift phases were formed by more or less orthogonal east–west extension and that there is no evidence of major strike-slip or compressional phases of deformation.

Klemperer (1988) interpreted the NSDP-84 deep seismic profiles in the northern North Sea, one of which (NSDP-84-1) crosses the north Viking Graben at about 61°N approximately coincident with the line interpreted by Beach (1986). Klemperer (1988) interpreted the reflection Moho on most of the profile and provided confirmatory evidence for crustal thinning beneath the graben. Crustal thicknesses were estimated to vary from about 20 km beneath the graben to about 30 km beneath the Horda Platform. In contrast to Gibbs' (1987) interpretation of the same data, Klemperer (1988) found no evidence for the presence of continuous crustal shear zones, but emphasised the continuity of flat-lying lower crustal events. Dipping events occur beneath the Moho in the upper mantle, but these appear to show a symmetrical distribution about the graben. Klemperer (1988) and White et al. (1989) indeed suggest that the deep seismic data lend most support to a pure-shear rather than simple-shear model of basin development.

Both Hollinger and Klemperer (1988) and Fichler and Hospers (1991) have studied the NSPD-84 data and, combined with gravity modelling, have made maps of depth to Moho in the Viking Graben. They show that the reflection Moho and the gravity Moho are more or less coincident except locally under the centre of the graben. Fichler and Hospers (in press) calculate β factors of more than 3.0 in the deepest part of the graben between 60° and 61° where the crustal thickness, between the base of the sediments and the Moho, is interpreted to be only 10 km. They assumed a pre-stretching crustal thickness of 31 km.

Seismic interpretation

In the present study a detailed interpretation has been made of a regional seismic line NNST-84-08, made available by Nopec a.s. The line trends E–W just to the north of lat. 61° and extends from the Øygarden Fault in the east across the northern part of the Horda Platform ("Sogn spur") and the Viking Graben. On the west flank of the graben it crosses part of the Gullfaks structure on the Tampen Spur before crossing the Norwegian/U.K. sector boundary near to the Statfjord field (Fig. 1). The seismic line continues further east in the U.K. sector, but for the purpose of this study we have used only the Norwegian part.

The interpretation was tied to well data, notably wells 34/10-8 and 35/8-2 which lie on the line. It was depth converted using interval velocity data corrected by a time/depth correlation from the wells. The depth converted cross-section is shown on Fig. 2.

The Base Cretaceous unconformity is the easily recognised reflector truncating tilted fault blocks below and with onlapping Cretaceous reflections above. Over the highs this event represents a considerable time gap, perhaps as much as 100 m.y., but, where the sequence is more complete in the deepest part of the troughs, it is assigned an age of 131 Ma (Ma = million years before present, m.y. = million years). The time scale of Haq et al. (1987) was used for dating.

The Top Brent reflector (Top Bathonian 156 Ma) has been tied to the well data and interpreted along the whole section. There are numerous faults at this level and the interpretation is most problematic in the deepest part of the graben. Deeper events (Top Statfjord, Top Lomvi) can be interpreted, but these are parallel to the Top Brent and the interpretation in the graben centre is even less confident. The Top

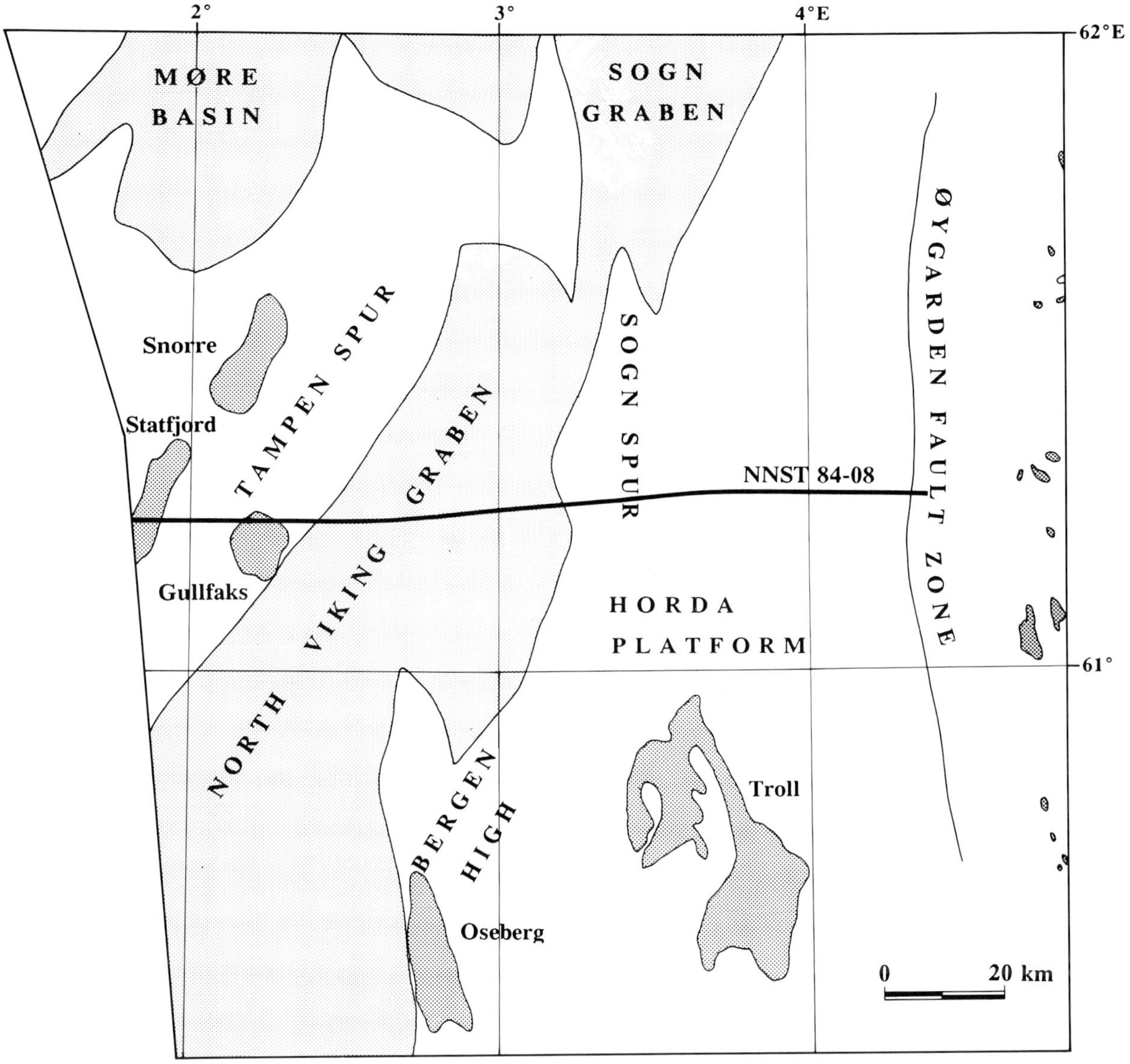

Fig. 1. Location of the area in the northern North Sea. Seismic line NNST-84-08, major structural features and important oil and gas fields are shown.

Brent reflector is taken to be the top of the pre-rift sequence with respect to the late Jurassic rifting phase (Badley et al. 1988).

The post-rift sequence, containing several intra-Cretaceous reflectors — Intra Albian (100 Ma), Top Cenomanian (92 Ma) and Top Turonian (89 Ma) — and the Base Tertiary (65 Ma), Top Palaeocene (54 Ma) and Intra Eocene (49 Ma) events, is tied to the wells and fairly well constrained by the interpretation.

The base and top of the Miocene Utsira Formation are dated as 25 Ma and 5 Ma, respectively, following Rundberg and Smalley (1989). The Base Quaternary (2 Ma) forms an erosion surface which truncates older Cenozoic strata towards the east so that Quaternary sediments rest on Palaeocene strata at the eastern end of the section. The Øygarden Fault juxtaposes basement rocks against the Quaternary sediments and was active during Cenozoic time (Rundberg, 1989).

Subsidence analysis

An important method of estimating crustal and lithospheric extension is an analysis of basin subsidence history (van Hinte, 1978; Bessis, 1986). In order to do this, the subsidence must be corrected for compaction due to the progressive loading of the sediments. The isostatic effects of sediment loading are removed by backstripping to give the "tectonic" subsidence due to water loading alone (Steckler and Watts, 1978). The subsidence history at any number of selected points along the seismic line can be obtained by decompaction and backstripping using

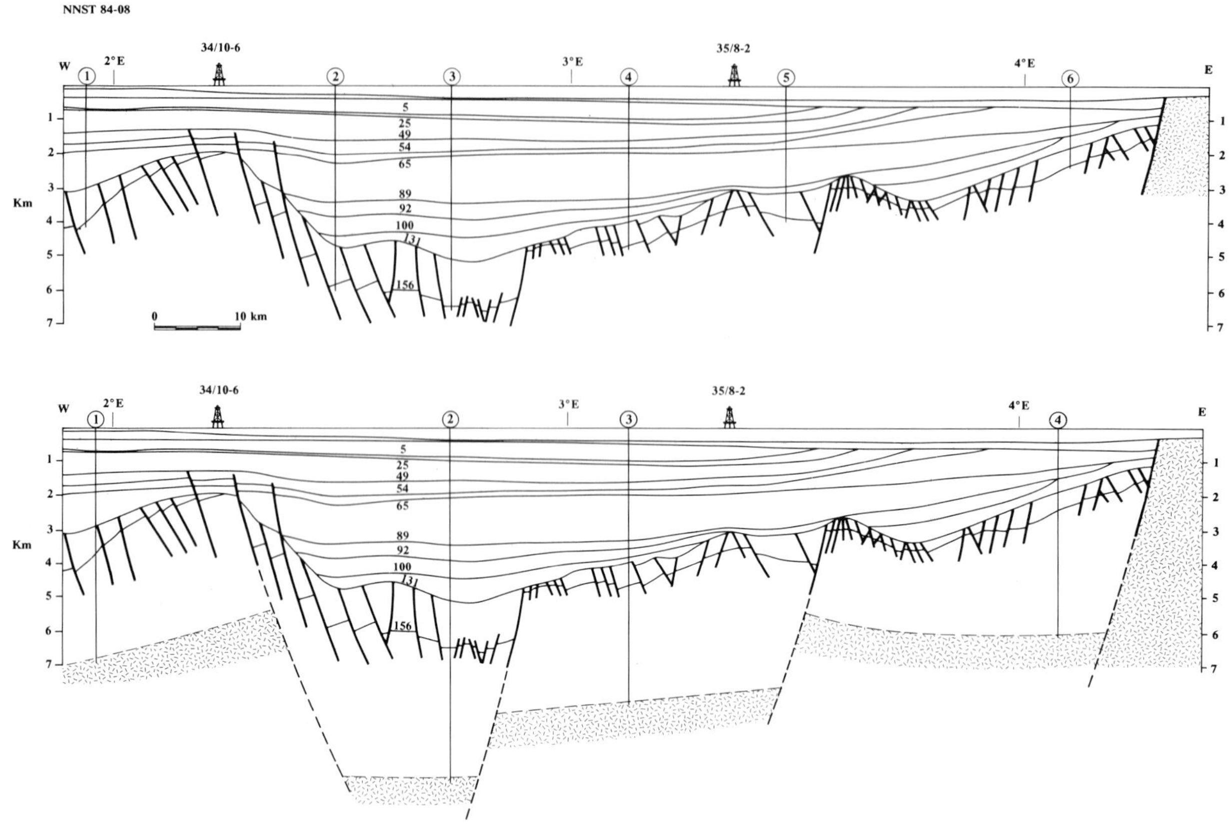

Fig. 2. Depth-converted geoseismic line NNST-84-08 with top basement interpolated below. Ages of reflectors in million years (Ma) — see Table 1 for details. Points *1–6* (above) and *1–4* (below) refer to modelling locations.

TABLE 1

Parameters used in backstripping

Horizon	Age (Ma)	Lithology	C	f_o
Top Basement		ss/sh	0.39	0.56
Top Brent	156	sh	0.51	0.62
Base Cretaceous	131	sh (lst)	0.46	0.60
Intra Albian	100	sh	0.51	0.62
Top Cenomanian	92	sh	0.51	0.62
Top Turonian	89	sh	0.51	0.62
Base Tertiary	65	sh (tuff)	0.43	0.59
Top Paleocene	54	sh	0.51	0.62
Intra Eocene	49	sh	0.51	0.62
Top Oligocene	25	ss	0.31	0.52
Top Miocene	5	ss/sh	0.39	0.56
Base Quaternary	2	ss/sh	0.39	0.56

Lithology: ss = sand; sh = shale/mudstone; lst = marl/limestone. C = compaction factor (10^{-5} cm^{-1}); f_o = surface porosity. Sediment grain density = 2.7 g/cm^3.

the methods first applied to the North Sea basin by Sclater and Christie (1980). Compaction coefficients and initial porosities were applied according to the lithologies of each unit using the end member values of Sclater and Christie (1980) (Table 1).

Past water depths or paleobathymetry are a part of the basin's subsidence history and can be the result of the paleotopography of the basin and eustatic sea level variations. Since mid-Jurassic times eustatic sea levels have generally been above those of the present day, with maximum levels of 100–300? m reached in the mid-late Cretaceous (Haq et al., 1987). Paleobathymetry is usually given by paleontological data or can be estimated from seismic stratigraphy (Vail et al., 1977). Later in this paper we will attempt to derive some information about the paleobathymetry from the subsidence curves.

The backstripping method we have assumed to be local, or Airy isostasy. Barton and Wood (1984) and Hellinger et al. (1989) have addressed the problem of local versus regional isostatic compensation by modelling gravity data and have concluded that the elastic thickness of the lithosphere in the North Sea is small and that, as a first approximation, flexural effects can be ignored.

Tectonic modelling

Program RIFT (Odling, this volume) is based on a McKenzie (1978) model of pure shear extension of the lithosphere. Finite rate extension (Jarvis and McKenzie, 1980) and multiple rift phases are incorporated in the model by calculating the effects of extension on any pre-existing initial temperature profile. Non-uniform extension (Royden and Keen, 1980; Hellinger and Sclater, 1983; Rowley and Sahagian, 1986) can be included by dividing the lithosphere into a number of layers each of which can be stretched by a different amount. Between each rifting phase the crustal thickness is calculated by filling the basin with sediment to a specified water depth. In this way it is possible to use the crustal thickness data as a control on the validity of the β factors obtained by subsidence analysis.

In the modelling of the northern Viking graben profile we have initially modelled the mid-Jurassic to present-day subsidence patterns as a single rifting phase which began at 156 Ma, lasted 25 m.y. and was followed by 131 m.y. of post-rift subsidence. The effect of earlier rifting phases has been taken into account by varying the initial crustal thicknesses across the graben until a best fit was obtained with the present crustal thickness. In the second stage we attempted to model the earlier rifting phase more accurately by testing two two-phase rifting models in which the timing of the first rift phase differs.

One-phase rifting model

The mid Jurassic to present-day (156–0 Ma) subsidence history was modelled for six selected points along the seismic line. The points were chosen to lie in the deepest parts of the troughs or in the centres of tilted fault blocks in order to avoid the effects of footwall uplift or hanging wall subsidence due to fault block tilting (Fig. 3) or flexural effects (Marsden et al., 1989). The intra-basin highs; for example, the two-well locations (Fig. 2), give anomalous subsidence patterns with thin or absent syn-rift sequences and are difficult to model. Two subsidence curves; the decompacted sediment-loaded subsidence ("total subsidence") and the backstripped water-loaded subsidence ("tectonic subsidence"), were obtained for each point (Fig. 4).

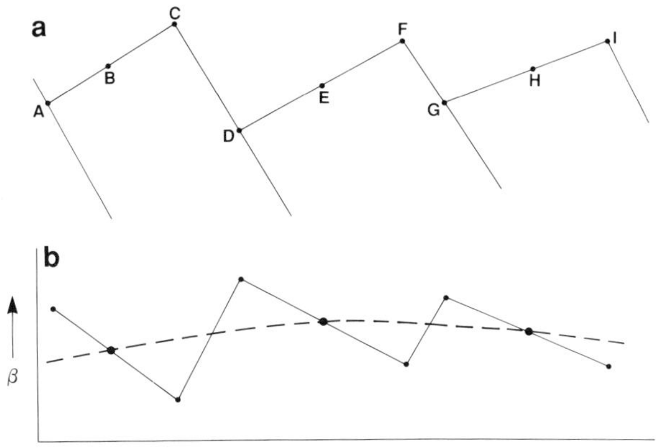

Fig. 3. (a) Tilted fault blocks showing modelling locations *A–I*. (b) The irregular β factor profile (solid line) is obtained using all modelling locations. The dashed line shows a smoothed β factor profile obtained using only the mid-points of the tilted fault blocks (*B, D, H*).

TABLE 2

One-phase (mid to late Jurassic) rifting results

	1	2	3	4	5	6
β_c	1.4	1.75	1.8	1.45	1.35	1.10
β_m	1.4	1.75	1.8	2.0	1.7	1.6
C_i	27.5	25	25	27.5	27.5	30
C_p	25.1	21.5	21.2	24.3	24.6	27.5
T_a	0.94	1.18	1.25	0.68	0.61	0.56
T_{ip}	0.92	1.16	1.29	0.61	0.60	0.51
T_{ta}	1.74	2.45	2.56	1.96	1.64	1.52
T_{tp}	1.82	2.47	2.53	1.99	1.68	1.46

1 to 6 are the modelled points on the seismic line (Fig. 2). β_c = crustal β factor; β_m = mantle β factor; C_i = initial crustal thickness; C_p = predicted present crustal thickness (including sediments); T_{ia} and T_{ip} = actual and predicted syn-rift tectonic subsidences; T_{ta} and T_{tp} = actual and predicted total tectonic subsidences.

Theoretical tectonic subsidence curves derived from the RIFT program were fitted to each of the actual tectonic subsidence curves (Fig. 4). The mid-late Jurassic rifting event is assumed to have lasted 25 m.y. and the tectonic subsidence for this period is fitted as closely as possible to the theoretical syn-rift subsidence. The initial crustal thicknesses were varied from 27.5 to 25 km according to their position in the graben (Table 2). The initial lithosphere thickness was fixed at 125 km following McKenzie (1978). The total subsidence was obtained by modelling to a point in the mid-Cenozoic (about 30 Ma). This is the point at which there was a change from deep to shallow water conditions in the area (Rundberg and Smalley, 1989). For points 1–3, the data fit a uniform stretching model ($\beta_{crust} = \beta_{mantle}$) with β factors of 1.4 on the west flank and 1.75 and 1.8 in the graben centre. For points 4–6 further east on the Horda Platform, the curves cannot be fitted to a uniform stretching model mainly because the syn-rift subsidences are too small compared with the total subsidences. However, a non-uniform stretching model ($\beta_{mantle} > \beta_{crust}$) can be used with β_{mantle} ranging from 1.7 to 2.0 and β_{crust} from 1.45 to 1.1 (Table 2).

It should be noted that in all cases the actual subsidence curves deviate substantially from the theoretical curves, most notably in the Cretaceous and early Tertiary (130–30 Ma), where the actual curves lie above the theoretical ones. The discrepancies are shown by the shaded areas on Fig. 4. An explanation for this is the extent of the water depths

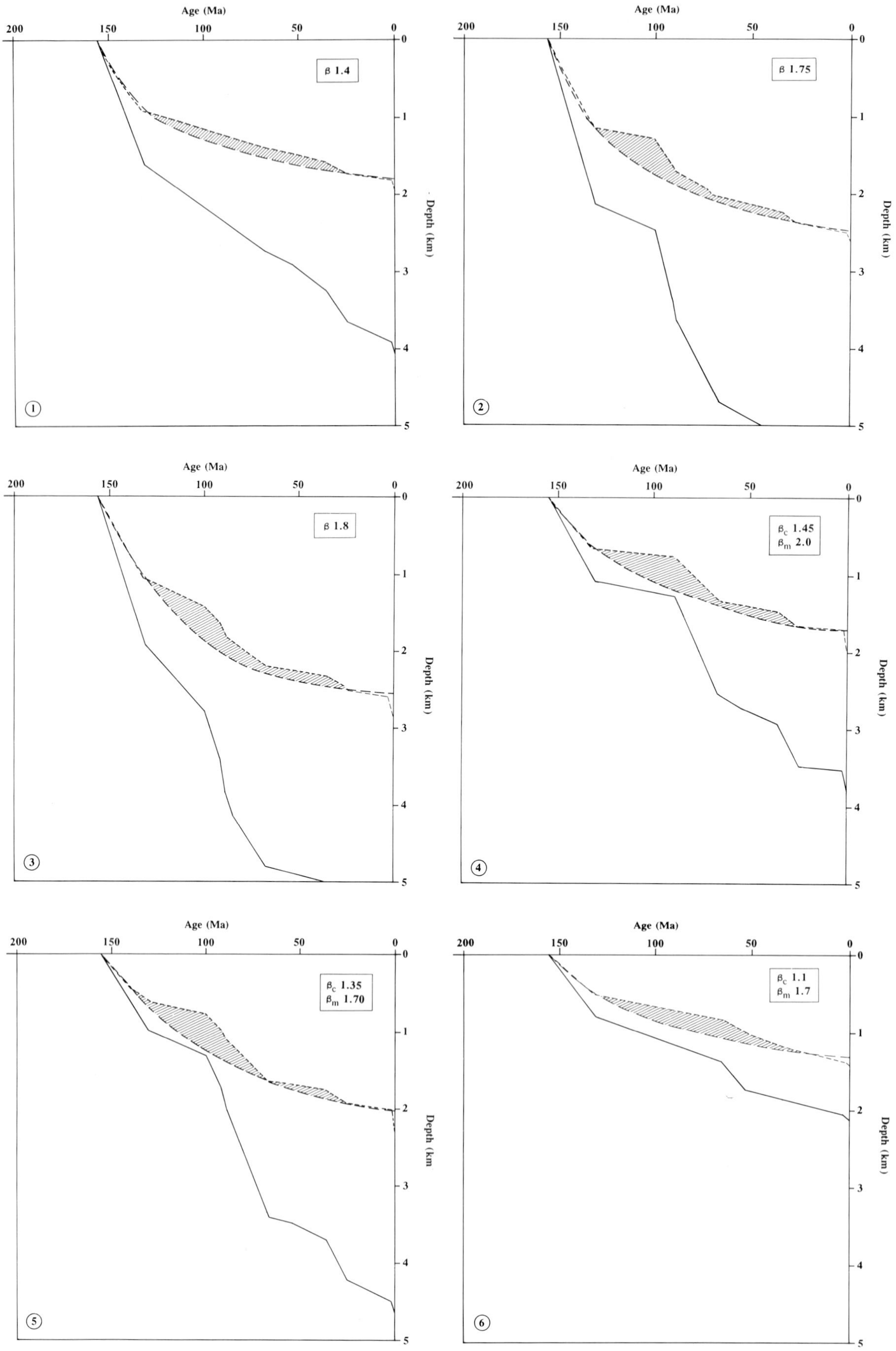
Age (Ma)
200 150 100 50 0
Depth (km)
0 1 2 3 4 5
β 1.4
β 1.75
β 1.8
β_c 1.45
β_m 2.0
β_c 1.35
β_m 1.70
β_c 1.1
β_m 1.7
1
2
3
4
5
6

TABLE 3

Paleobathymetries from subsidence analysis

	1	2	3	4	5	6
100 Ma	350	650	580	650	320	400
89 Ma	450	500	600	560	–	–
65 Ma	300	210	250	150	210	–

1 to 6 are the modelled points on the seismic line (Fig. 2). 100 Ma, 89 Ma and 65 Ma refer to Near Top Albian, Top Turonian and Base Tertiary. The predicted paleobathymetries are given in metres.

during the Cretaceous and early Tertiary. This is in agreement with paleobathymetric evidence (Nelson and Lamy, 1987) for deep water conditions (as much as 1000 m?) in this period in the Viking Graben. The onlap of the post-rift Cretaceous sequences onto the irregular fault blocks is evidence of a paleotopography of several hundred metres at this time, although the effects of compaction and differential subsidence must also be taken into account. The small amount of sediment deposited in the early Cretaceous suggests that the basin was sediment-starved at this time (Bertram and Milton, 1988). From the differences between the theoretical and actual subsidence curves, it is possible to make some estimates of water depths (Table 3). These results suggest that water depths reached a maximum of at least 650 m in the mid-Cretaceous (100 Ma). Note that no eustatic sea level correction (+100 to 300 m?) has been made.

Two-phase rifting models

The seismic data do not allow the Top Basement to be identified with any confidence. A depth to basement map was produced by Hospers and Ediriweera (1988) by analysis of well data and modelling of the gravity and magnetic data. This has been used to fill in the top of the basement for modelling the effects of earlier rifting phases on the mid-late Jurassic rifting phase.

The timing of the earlier rifting event in the northern Viking Graben is uncertain. Ziegler (1982) proposed a phase of late Permian–early Triassic rifting, whereas Morton et al. (1987) and Gabrielsen et al. (1989) have recognised late Triassic rifting events in the area. Lervik et al. (1989) mention that in well 30/6-1 on the Horda Platform, Triassic rests directly on the basement, suggesting that the rifting commenced during the Triassic. They conclude from a regional study that faulting occurred throughout the Triassic, but that details of the timing are as yet unknown. Because of these uncertainties, we tested two models; Permo-Triassic (260–235 Ma) and late Triassic (225–200 Ma) rifting, followed by mid-late Jurassic rifting. The Triassic sediments were assumed to be an equal mixture of sand and shale for the decompaction calculations. The theoretical subsidence curves for four modelled points (shown on the lower section on Fig. 2) are shown in Fig. 5. In each case it can be seen that the β factor for the late Triassic rifting has to be slightly larger (0.05) than for Permo-Triassic rifting to obtain the same amount of subsidence at the onset of the mid-late Jurassic event. Table 4 shows that for Permo-Triassic rifting the β factors range from 1.3 on the flanks to 1.45 in the graben centre, for late Triassic rifting the corresponding figures are 1.35 and 1.5. It is interesting to note that the β factors for the earlier rifting events are more even across the Viking Graben than for the mid-late Jurassic event, although the largest values still occur in the graben centre. This supports the symmetrical distribution of the early and late rifting phases as proposed by Ziegler (1982) and Badley et al. (1988). In order to obtain the best fit for the subsidence data, it was necessary to vary the initial crustal thickness from 28 km in the centre of the graben to 30–32 km on the flanks (Table 4). This implies that the crust beneath the graben was already thinned prior to

Fig. 4. Subsidence curves for points 1–6 (upper section on Fig. 2). Solid line, total decompacted subsidence (sediment-loaded); short-dash line, tectonic subsidence (water-loaded); long-dash line, theoretical modelled tectonic subsidence from program RIFT (see Table 2 for details). β factors given in the top right-hand corner (β_c, crust and β_m, mantle for two-layer stretching). Shaded areas represent misfits between actual and theoretical tectonic subsidence curves (see text and Table 3 for further explanation).

TABLE 4

Two-phase rift rifting models

	1	2	3	4
(a) Permo-Triassic followed by mid-late Jurassic rifting				
β_{P-T}	1.3	1.45	1.45	1.35
C_i	30	28	30	32
C_{156}	28.0	25.3	27.2	29.6
C_p	25.7	22.6	24.8	27.6
(b) Late Triassic followed by mid-late Jurassic rifting				
β_{LT}	1.35	1.5	1.5	1.4
C_i	30	28	30	32
C_{156}	26.9	24.3	26.1	28.5
C_p	25.6	22.4	24.6	27.4

1 to 4 are the modelled points on the seismic line (Fig. 2). β_{P-T} = β factor for Permo-Triassic (260–235 Ma) rifting; β_{LT} = β factor for late Triassic (225–200 Ma) rifting; C_i = initial crustal thickness; C_{156} = predicted crustal thickness at 156 Ma (prior to mid-late Jurassic rifting); C_p = predicted present day crustal thickness (including sediments).

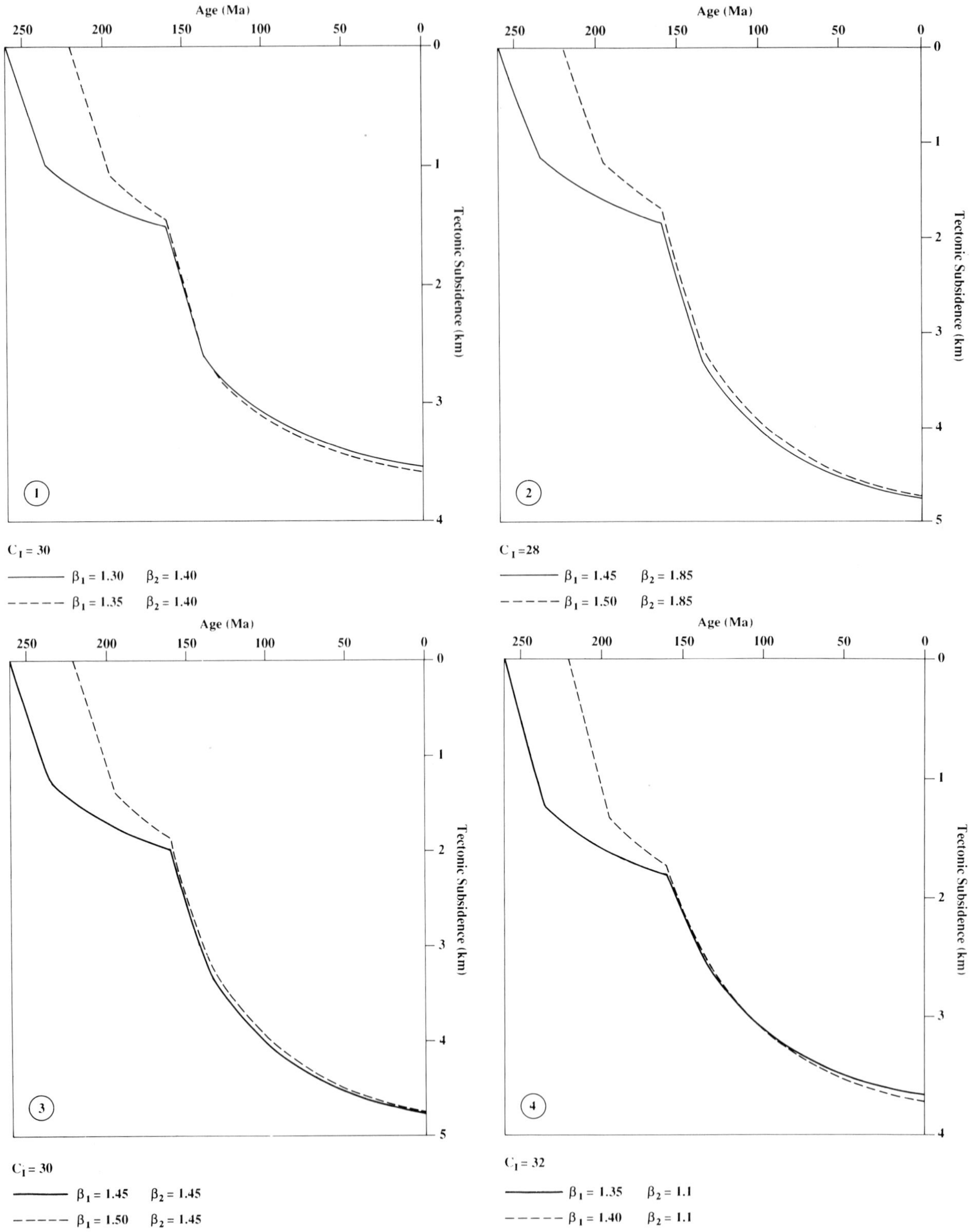

Fig. 5. Theoretical tectonic subsidence curves modelled by program RIFT for points *1–4* (lower section on Fig. 2). Solid line, scenario 1 (Permo-Triassic rifting followed by mid-late Jurassic rifting). Dashed line, scenario 2 (late Triassic rifting followed by mid-late Jurassic rifting). C_I, initial crustal thickness; β_1, β factor for first rift phase; β_2, β factor for second rift phase (see Table 4 for further details).

the first rifting event. This suggests earlier Devonian or, perhaps, Carboniferous rifting although there is no evidence from the seismic data to support this. It is possible that the overthickened crust following the Caledonian orogeny extended and thinned by gravitational collapse, a model that has been used for the Devonian basins of western Norway (Norton, 1986; Seranne and Seguret, 1987).

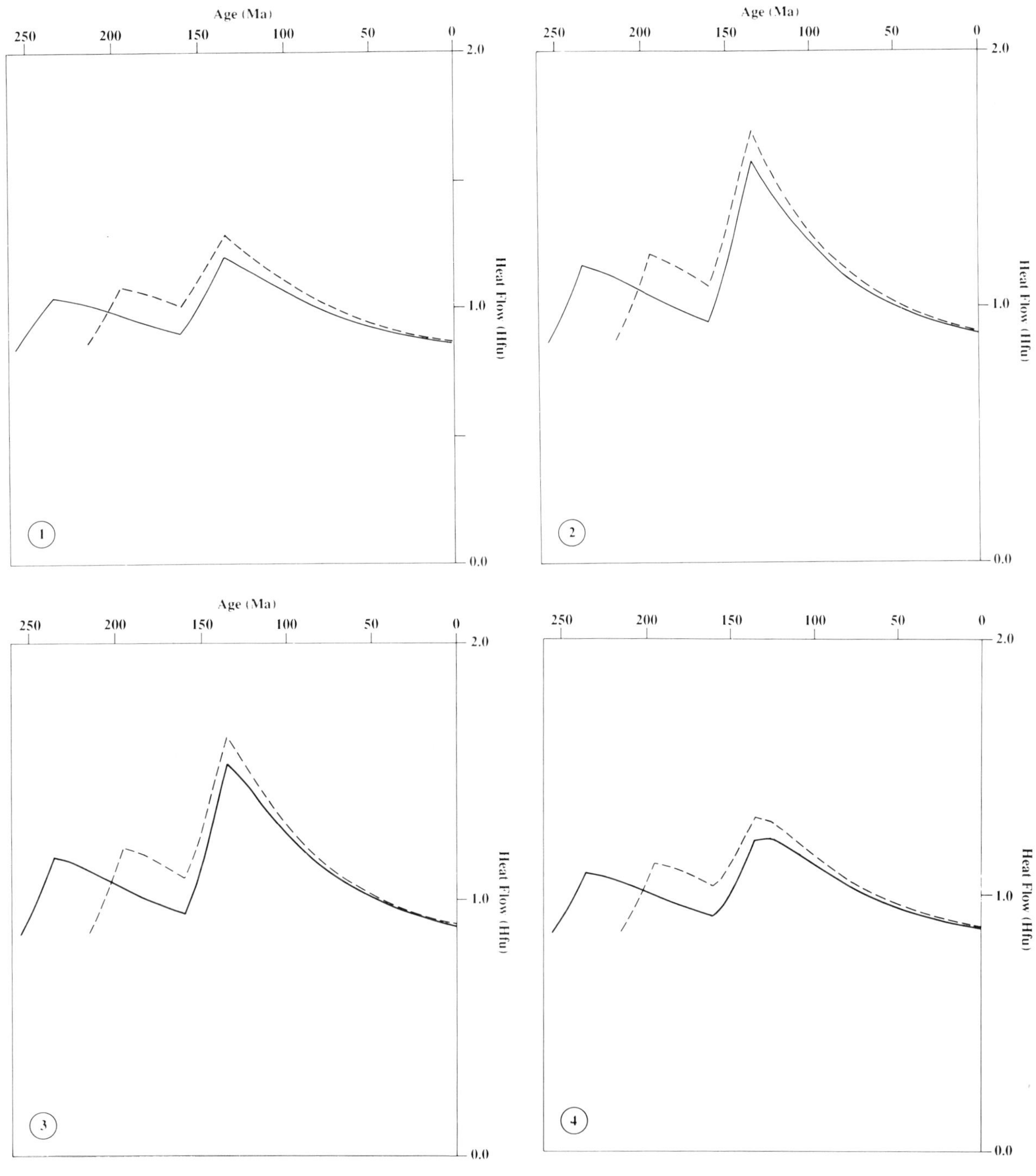

Fig. 6. Theoretical basement heat flow modelled by program RIFT for points 1–4. Scenario 1, solid line; scenario 2, dashed line.

Figure 6 shows the theoretical basement heat flow evolution predicted by the two-phase rifting models. The effect of the late Triassic event is to increase the heat flow during the mid-late Jurassic rifting event more than in the case of the Permo-Triassic event. Note that, in both cases, the heat flow anomaly caused by the rifting events has decayed to background value by the present day.

Discussion

Single and two-phase rifting models have been applied to the subsidence history of the northern Viking Graben. The results are summarised and compared with the result of others in Table 5. It can be seen that the present results differ substantially from those of Giltner (1987) and Vially (1988). Giltner (1987) modelled two-phase rifting in much the same way as we have done. He modelled a first Triassic rift phase lasting for 35 Ma from 248 Ma to 213 Ma

TABLE 5

Comparison of β factors obtained in the present work and published ones

	Tampen Spur	Viking Graben	Horda Platform
(a) Giltner (1987)			
β_{P-T}	1.2	1.5	1.2
β_{J-K}	1.2	1.2	<1.2
β_{TOT}	1.4	1.8	<1.4
(b) Vially (1988)			
β_{P-T}	1.15	1.2	1.4
β_{J-K}	1.2	1.5	1.2
β_{TOT}	1.4	1.8	1.7
(c) Badley et al. (1988)			
β_{J-K}	1.05/1.38 [a]	1.15/1.49 [a]	1.03/1.14 [a]
(d) This work			
β_{P-T}	1.35	1.5	1.4
β_{J-K}	1.4	1.8	1.4/1.85 [b]
β_{TOT}	1.9	2.7	2.0/2.6 [b]

β_{P-T} = β factor Permo-Triassic rifting; β_{J-K} = β factor Jurassic–Cretaceous rifting; $\beta_{TOT} = \beta_{P-T} \times \beta_{K-J}$.

[a] First value from initial syn-rift subsidence; second from total subsidence.

[b] First value for crustal extension, second for mantle extension.

and a second Jurassic–Cretaceous phase from 169 Ma to 98 Ma (although he admits that most of the tectonic activity had ceased by 130 Ma). The initial crustal thickness was taken as 35 km. The difference with our results can be largely accounted for by this large initial crustal thickness which results in the need for lower β factors to obtain the same tectonic subsidence. However, it is difficult to see how Giltner arrived at lower β factors for the Jurassic–Cretaceous event than for Triassic in the centre of the graben. Vially (1988) presented maps of total subsidence, tectonic subsidence and subsidence rates for various time periods and maps of β factor distribution for Permo-Triassic and "Cimmerian" rifting phases. As noted previously, he showed the axis of Permo-Triassic rifting underlying the Horda Platform and offset from the later rifting event in the Viking Graben proper. This is supported by Lervik et al. (1989) who stated that, far from the Triassic rifting being a precursor to the later rifting, the two may have little in common. Vially's total β factor of about 1.8 for the centre of the Viking Graben at 61°N is the same as Giltner's, but the values given to the early and late phases are reversed. Vially gives no information about the parameters (i.e., initial crustal thickness, etc.) used in the modelling.

Fichler-Fettig (1989) and Fichler and Hospers (1991) show that the present-day crustal thickness, between top basement and the Moho, varies from 20 km on the Horda Platform to less than 14 km and probably close to 12 km thick in the graben centre at 61°N. Their β factor map shows values of 1.5–2.0 on the flanks and 2.5–3.0 in the graben centre (based on a uniform initial crustal thickness of 31 km).

We have shown that the mid-late Jurassic rifting event can be accounted for by a uniform rifting model in the centre of the graben, but that a non-uniform model, with more stretching in the mantle part of the lithosphere than the crust, can be applied to the Horda Platform. The deep seismic data (Gibbs, 1987; Klemperer, 1988) provide evidence for detachment zones in the upper mantle beneath the area, which may indicate that the crust and mantle are partly decoupled. Badley et al. (1988) also proposed that two-layer extension may account for the discrepancies between initial and total subsidences in the area. White and McKenzie (1988) proposed that a non-uniform stretching model with mantle extension extending over a wider area than crustal extension could account for the onlap of post-rift sequences onto the rift margins. The model assumes overall equal amounts of stretching in the crust and mantle so that the crustal stretching is greater than the mantle in the graben centre. Another mechanism by which the syn-rift subsidence can be reduced is lateral transfer of heat from the graben centre to the flanks (Cochran, 1983). This is not taken into account by the McKenzie model which assumes vertical heat transfer. In large extensional basins such as the North Sea vertical heat transfer seems, however, to be a reasonable assumption. The present results are thus similar to the model of White and McKenzie (1988), but suggest more irregular lateral and vertical variations in β factors and perhaps conform most closely to a heterogeneous stretching model of the type proposed by Coward (1986).

Conclusions

Analysis of backstripped subsidence curves using a pure-shear model of lithospheric extension has shown that the amounts of extension for the mid- to late Jurassic rifting event in the northern Viking Graben at 61°N vary from 1.75–1.8 in the graben centre to 1.35–1.45 on the flanking Tampen Spur and Horda Platform areas. Whereas uniform stretching ($\beta_{crust}=\beta_{mantle}$) fits the data in the graben centre, non-uniform stretching ($\beta_{crust} < \beta_{mantle}$) provides a better fit on the flanks.

The non-conformity of the subsidence curves with the theoretical ones for the Cretaceous and early part of the Tertiary can be accounted for by relatively large paleobathymetries, probably as much as 500–600 m in the mid-Cretaceous. This agrees with published biostratigraphic data and seismic stratigraphy.

The effects of an earlier rifting event are taken into account in the one-phase modelling by varying the initial crustal thickness using present-day crustal thickness as a control. Two-phase rifting models have been carried out by reconstructing the depth to top of the basement beneath the rift using published data. Two scenarios; late Permian–early Triassic and late Triassic rifting, were tested. It is found that late Triassic rifting requires slightly larger β factors to obtain the required amount of subsidence prior to the onset of the mid-late Jurassic event and has a greater influence on the heat flow during the latter event. The basement heat flow anomaly associated with all rifting events should have decayed by the present day.

The β factors for Triassic rifting are fairly uniform across the section suggesting a broad zone of fairly even extension contrasting with the later mid-late Jurassic event the effects of which are strongly concentrated in the Viking Graben.

The effect of both early rifting phases is an increase in the predicted basement heat flow during the mid-late Jurassic event, but this does not significantly affect the β factors calculated by the one-phase model.

The total β factor obtained from both rifting events is about 2.5 for the centre of the graben. This is in better agreement with the present-day crustal thickness than previously published results.

Acknowledgement

We thank IKU and Nopec a.s for granting permission to publish this paper.

References

Badley, M.E., Egeberg, T. and Nipen, O., 1984. Development of rift basins illustrated by the structural evolution of the Oseberg feature, Block 30/6, offshore Norway. J. Geol. Soc. London, 141 639–649.

Badley, M.E., Price, J.D., Rambech Dahl, C. and Agdestein, T., 1988. The structural evolution of the northern Viking Graben and its bearing upon extensional models of basin formation. J. Geol. Soc. London, 145: 455–472.

Barton, P. and Wood, R., 1984. Tectonic evolution of the North Sea basin; crustal stretching and subsidence. Geophys. J. R. Astron. Soc., 79: 987–1022.

Beach, A., 1985. Some comments on sedimentary basin development in the northern North Sea. Scott. J. Geol., 21: 493–512.

Beach, A., 1986. A deep seismic reflection profile across the northern North Sea. Nature, 323: 53–55.

Beach, A., Bird, T. and Gibbs, A.D., 1987. Extensional tectonics and crustal structure: deep seismic reflection data from the northern North Sea Viking Graben. In: M.P. Coward, J.F. Dewey and P.L. Hancock (Editors), Continental Extension Tectonics. Geol. Soc. London, Spec .Publ., 28: pp. 467–476.

Bertram, G.T. and Milton, N.J., 1988. Reconstructing basin evolution from sedimentary thickness: the importance of paleobathymetric control, with reference to the North Sea. Basin Res., 1: 247–259.

Bessis, F., 1986. Some remarks on the study of subsidence of sedimentary basins. Mar. Pet. Geol., 3: 37–63.

Blair, D.G., 1975. Structural styles in North Sea oil and gas fields. In: A.W. Woodland (Editor), Petroleum and the Continental Shelf of North West Europe, Vol. 1. Geology. Applied Science Publishers, London, pp. 327–338.

Bowen, W., 1975. The Brent oil field. In: A.W. Woodland (Editor), Petroleum and the Continental Shelf of North West Europe, Vol. 1. Geology. Applied Science Publishers, London, 353–361.

Christie, P.A.F. and Sclater, J.G., 1980. An extensional origin for the Buchan and Witchground graben in the North Sea. Nature, 283: 730–733.

Cochran, J.R., 1983. Effect of finite rifting times on the development of a sedimentary basin. Earth Planet. Sci. Lett., 66: 289–302.

Coward, M.P., 1986. Heterogeneous stretching, simple shear and basin development. Earth Planet. Sci. Lett., 80: 325–336.

Fichler, C. and Hospers, J., 1991. Gravity modelling in the Viking Graben area, North Sea. In: D.J. Blundell and A.D. Gibbs (Editors), Tectonic Evolution of the North Sea Rifts. Clarendon Press, Oxford, pp. 71–80

Fichler-Fettig, C., 1989. The North Sea Viking Graben from 3-D gravity modelling and deep seismic profiling. Terra Abstracts, 1: 36.

Gabrielsen, R., Færseth, R.B., Idil, S. and Steel, R.J., 1989. Architectural styles of basin fill in the North Viking Graben. Terra Abstracts, 1: 34.

Gibbs, A.D., 1987. Deep seismic profiles in the northern North Sea. In: J. Brooks and K.W. Glennie (Editors), Petroleum Geology of North-West Europe. Graham and Trotman, Lonmdon, pp. 1025–1028.

Giltner, J.P., 1987. Application of extensional models to the Northern Viking Graben. Nor. Geol. Tidsskr., 67: 339–352.

Halstead, P.H., 1975. Northern North Sea faulting. In: K.G. Finstad and R.C. Selley (Editors), Jurassic Northern North Sea Symposium. Norw. Pet. Soc., NP5, pp. 1–38.

Haq, B.U., Hardenbol, J. and Vail, P.R., 1987. Chronology of fluctuating sea levels since the Triassic (250 million years ago to present). Science, 235: 1156–1166.

Hellinger, S.J. and Sclater, J.G., 1983. Some comments on two-layer extensional models for the evolution of sedimentary basins. J. Geophys. Res., 88: 8251–8269.

Hellinger, S.J., Sclater, J.G. and Giltner, J., 1989. Mid-Jurassic through mid-Cretaceous extension in the Central Graben of the North Sea — Part 1. Estimates from subsidence. Basin Res., 1: 191–200.

Hollinger, K. and Klemperer, S.L., 1988. A comparison of the Moho interpreted from gravity data and from deep seismic reflection data in the northern North Sea. Geophys. J., 97: 247–250.

Hospers, J. and Ediriweera, K.K., 1988. Mapping of the top of the crystalline continental crust in the Viking Graben area, North Sea. Nor. Geol. Unders., Spec. Publ., 3: 21–28.

Jarvis, G.T. and McKenzie, D.P., 1980. Sedimentary basin formation with finite extension rates. Earth Planet. Sci. Lett., 48: 42–52.

Klemperer, S.L., 1988. Crustal thinning and the nature of extension in the northern North Sea from deep seismic reflection profiling. Tectonics, 7: 803–821.

Lervik, K.S., Spencer, A.M. and Warrington, G., 1989. Outline of Triassic stratigraphy and structure in the central and northern North Sea. In: J.D. Collinson (Editor), Correlation in Hydrocarbon Exploration, Norwegian Petroleum Society, Graham and Trotman, London, pp. 173–189.

Marsden, G., Kusznir, N., Roberts, A. and Yielding, G., 1989. Numerical modelling of basin formation in the northern North Sea. Terra Abstracts, 1: 34.

McKenzie, D.P., 1978. Some remarks on the development of sedimentary basins. Earth Planet. Sci. Lett. 40: 25–32.

Morton, N., Smith, R.M., Golden, M. and James, A.V., 1987. Comparative stratigraphic study of the Triassic–Jurassic sedimentation and basin evolution in the northern North Sea and northwest of the British Isles. In: J. Brooks and K.W. Glennie (Editors), Petroleum Geology of North-West Europe. Graham and Trotman, London, pp. 697–710.

Nelson, P.H.H. and Lamy, J.M., 1987. The Møre-West Shetlands area: a review. In: J. Brooks and K.W. Glennie (Editors), Petroleum Geology of North-West Europe. Graham and Trotman, London, pp. 775–784.

Norton, M.G., 1986. Late Caledonide extension in Western Norway; a response to extreme crustal thickening. Tectonics, 5: 195–204.

Odling, N.E., 1992. RIFT, a model of sedimentary basin evolution by finite rate, non-uniform, pure shear extension of the lithosphere. In: R.M. Larsen, H. Brekke, B.T. Larsen and E. Talleraas (Editors), Structural and Tectonic Modelling and its Application to Petroleum Geology. Norwegian Petroleum Society (NPF), Special Publication 1. Elsevier, Amsterdam, pp. 457–467 (this volume).

Rowley, D.B. and Sahagian, D., 1986. Depth-dependent stretching: a different approach. Geology, 14: 32–35.

Royden, L. and Keen, C.E., 1980. Rifting processes and thermal evolution of the continental margin of Eastern Canada determined from subsidence curves. Earth Planet. Sci. Lett., 51: 343–361.

Rundberg, Y., 1989. Tertiary sedimentary history and basin evolution of the Norwegian North Sea between 60–62°N — an integrated approach. Dr.-Ing. Thesis, University of Trondheim, NTH, 292 pp.

Rundberg, Y. and Smalley, P.C., 1989. High resolution dating of Cenozoic sediments from the Northern North Sea using $^{87}Sr/^{86}Sr$ stratigraphy. Am. Assoc. Pet. Geol., Bull., 73: 298–308.

Sclater, J.G. and Christie, P.A.F., 1980. Continental stretching: an explanation of the post-mid-Cretaceous subsidence of the North Sea basin. J. Geophys. Res., 85: 3711–3739.

Seranne, M. and Seguret, M., 1987. The Devonian basins of western Norway, tectonics and kinematics of an extending crust. In: M.P. Coward, J.F. Dewey and P. Hancock (Editors), Continental Extensional Tectonics. Geol. Soc. London, Spec. Publ., 28: 537–548.

Steckler, M.S. and Watts, A.B., 1978. Subsidence of the Atlantic-type continental margin off New York. Earth Planet. Sci. Lett., 41: 1–13.

Vail, P.R., Mitchum, R.M. and Thompson, S., 1977. Seismic stratigraphy and global changes of sea level, Part 3. Relative changes of sea level from coastal onlap. In: C.E. Payton (Editor), Seismic Stratigraphy — Application to Hydrocarbon Exploration. Am. Assoc. Pet. Geol., Mem. 26: 63–81.

van Hinte, J.E., 1978. Geohistory analysis — application of micropaleontology to exploration geology. Am. Assoc. Pet. Geol., Bull., 62: 201–222.

Vially, R., 1988. La subsidence dans le Viking Graben (mer du nord septentrionale). Rev. Inst. Fr. Pétrole, 43: 629–646.

Wernicke, B., 1985. Uniform sense simple shear of the continental lithosphere. Can. J. Earth Sci., 22: 108–125.

White, N. and McKenzie, D., 1988. Formation of the "steer's head" geometry of sedimentary basins by differential stretching of the crust and mantle. Geology, 16: 250–253.

White, N. and the BIRPS Group, 1989. Nature and magnitude of extension in the North Sea from regional seismic reflection profiles. Terra Abstracts, 1: 33.

Wood, R. and Barton, P., 1983. Crustal thinning and subsidence in the North Sea. Nature, 302: 134–136.

Ziegler, P.A., 1982. Geological Atlas of Western and Central Europe. Elsevier, Amsterdam, 130 pp.

S. LIPPARD *Continental Shelf and Petroleum Technology Research (IKU), N-7034 Trondheim, Norway*
Present address: *Department of Geology and Mineral Resources Engineering, Norwegian Institute of Technology (NTH), N-7034 Trondheim, Norway*

G. LIU *Continental Shelf and Petroleum Technology Research (IKU), N-7034 Trondheim, Norway*

Vøring Basin: subsidence and tectonic evolution

J. Skogseid, T. Pedersen and V.B. Larsen

The post-Caledonian history of the Vøring Basin has been examined by tectonic subsidence and a uniform lithosphere extension model. The study is based on depth converted seismic reflection lines integrated with seismic velocities and well data. Tectonic subsidence calculated for the pre-Cretaceous, Cretaceous and Cenozoic time intervals shows that the main episode of late Jurassic–early Cretaceous extension was centred in the eastern Vøring Basin, but affected the entire basin province, whereas the early Cenozoic rifting, associated with high magmatic activity and continental separation, was restricted to the central and western parts of the basin. NE–SW trending zones of weakness, probably inherited from the Caledonian crustal configuration, are inferred throughout the evolution of the basin. In some regions high extension rates during the early Cenozoic caused only minor faulting. This is attributed to ductile deformation/mobilization of Cretaceous shales and to intrusions. Although the uniform extension modelling approach has relevance in a regional setting, we suggest that differential crustal thinning, lateral heatflow and magmatic underplating have to be considered when analysing selected local basin features. The importance of previous tectonic episodes is demonstrated in the Bodø High region. Here the pre-Mesozoic structural configuration has a large rotated basement block and a deep detachment plane which controls the later pattern of crustal deformation and subsidence. Source rocks for petroleum hydrocarbons include the Kimmeridgian Clay formation and possibly early Cretaceous and Paleocene sediments believed to have been deposited under conditions of restricted water circulation in the western Vøring Basin. A potential reservoir unit may occur in redeposited sediments derived from a 250 km wide Paleocene–earliest Eocene landmass along the line of continental separation.

Introduction

Scientific and commercial investigations of the Vøring continental margin off mid-Norway (Fig. 1) document a post-Caledonian history of progressive, predominantly extensional tectonics leading towards continental breakup and formation of the Norwegian-Greenland Sea at about 57 Ma. The regional sedimentary Vøring Basin, which occupies the central part of the margin, provides important information about the pre-Cenozoic intracontinental basin evolution between Norway and Greenland, the sequence of events during breakup and the post-opening development of the passive continental margin.

Although the margin has been explored for two decades, it has only recently been possible to perform quantitative modelling of the margin in space and time. This is largely due to the acquisition of a database of multichannel seismic (MCS) lines covering the entire margin (Bøen et al., 1984; Skogseid and Eldholm, 1989). In addition, borehole data from commercial drilling on the continental shelf (Spencer et al., 1984; 1986; Dalland and Worsley, 1988) and scientific drilling at the Vøring Plateau (Talwani et al., 1976; Eldholm et al., 1987; 1989) tied to the seismic lines have established a stratigraphic framework. Finally, the first-order configuration of the deeper parts of the basin and the crust has been obtained by digital sonobuoy and expanded spread profiling (ESP) experiments (Mutter et al., 1984; Eldholm and Mutter, 1986; Planke et al., 1991).

Based on these data this study investigates the history of vertical movements at the margin focusing on the Vøring Basin. In particular, we attempt to resolve and quantify the different tectonic episodes by means of their tectonic subsidence. Furthermore, we discuss the basin evolution and aspects related to the temperature history of the sediments and their hydrocarbon potential.

Geological setting

The Vøring Margin comprises three main geological provinces: the Trøndelag Platform, the Vøring Basin and the Vøring Marginal High (Fig. 2). The area's main structures and structural units, as defined at the base Cretaceous level, are described and given formal names by Gabrielsen et al. (1984) and Blystad et al. (in press).

The Trøndelag Platform is a 150 km wide area between the Norwegian mainland and the Vøring

Structural and Tectonic Modelling and its Application to Petroleum Geology edited by R.M. Larsen, H. Brekke, B.T. Larsen and E. Talleraas. NPF Special Publication 1, pp. 55–82. Elsevier, Amsterdam.

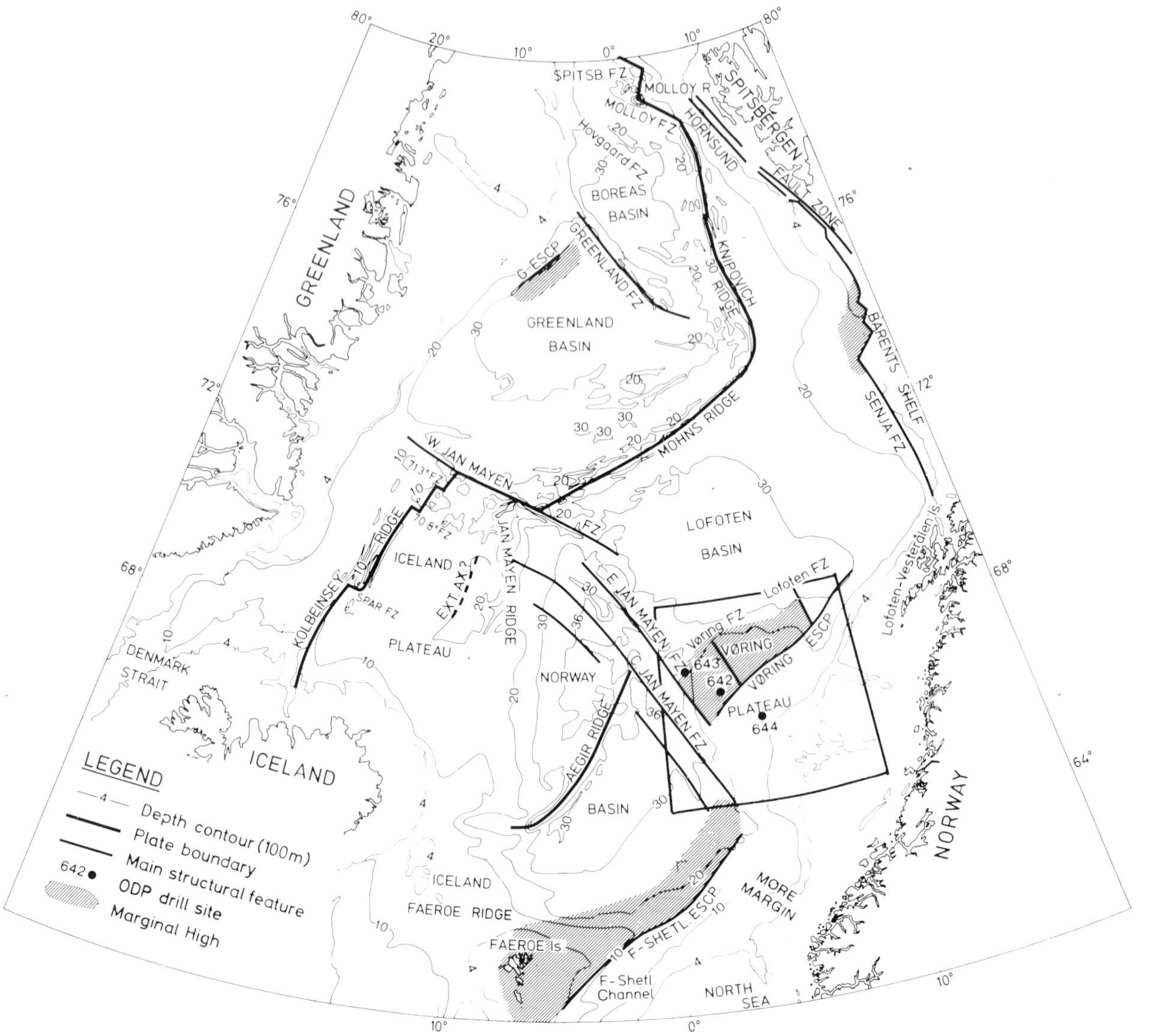

Fig. 1. Regional physiography and structural features of the ocean basins and continental margins in the Norwegian-Greenland Sea (Eldholm et al., 1990). The box shows the study area.

Basin. Although representing a relatively stable area during late Mesozoic times, it was previously divided into several structural units.

The Vøring Basin is a large sedimentary basin province with grabens, sub-basins and structural highs. The eastern basin province includes the deep Træna Basin and Rås Subbasin, flanked by the Halten and Dønna terraces adjacent to the elevated Nordland Ridge in the east, and the Fulla Ridge in the west. The Fulla Ridge forms a prominent ridge across the entire Vøring Basin separating the eastern basin province from the Fenris and Hel Grabens west of the ridge. The northern part of the Fulla Ridge includes a synform, the Nyk Syncline. In addition to the base Cretaceous structures, separate structures also exist at deeper stratigraphical levels, exemplified by the Bodø High, recognized by a major gravity anomaly between the Træna Basin and the northernmost Fulla Ridge (Planke et al., 1991).

The Vøring Basin is terminated to the west by the Vøring Escarpment: the eastern boundary of the Vøring Marginal High. The formation of this high, during early Cenozoic breakup, was associated with massive extrusive and intrusive magmatic activity. A sill/flow complex in a 10–40 km wide area east of the Vøring Escarpment inhibits seismic resolution to deeper levels (Fig. 2; flow limit basalts).

The seismic resolution in the Vøring Basin deteriorates below the base Cretaceous horizon, although a series of deeper reflectors are observed in most of the study area. A minimum of 3 km of pre-Cretaceous sediments is inferred regionally (Bøen et al., 1984; Skogseid and Eldholm, 1989), while as much as 10 km may exist locally (Eldholm and Mutter, 1986; Planke et al., 1991). Thus, the pre-Cretaceous sediment thicknesses in the Vøring Basin

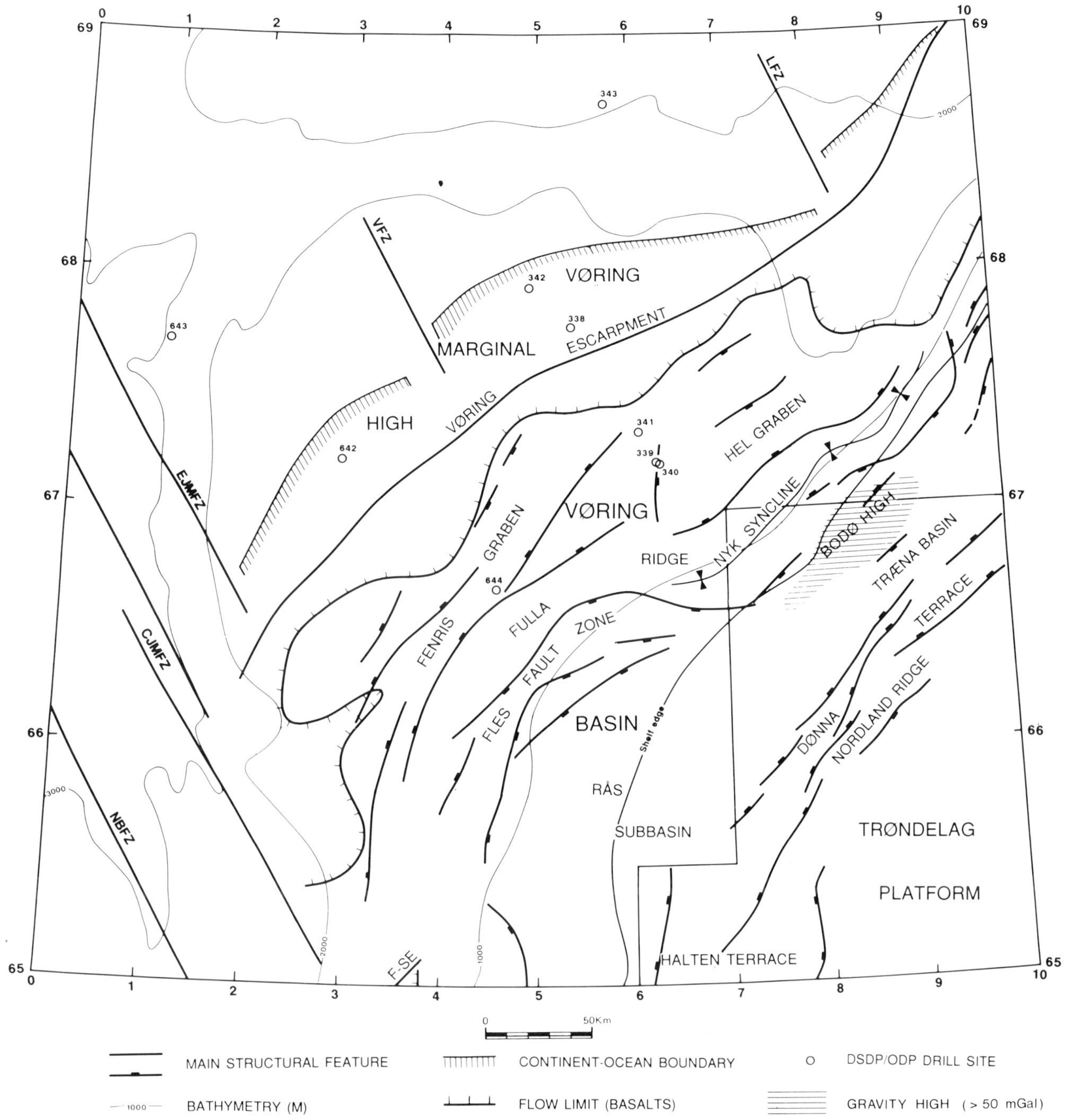

Fig. 2. Main structural features in the Vøring Basin and adjacent areas based on Skogseid and Eldholm (1989) and Blystad et al. (in press). F-SE, Faeroe–Shetland Escarpment; COB, Continent-ocean boundary; NBFZ: Norway Basin fracture zone; CJMFZ, central Jan Mayen fracture zone; EJMFZ, East Jan Mayen fracture zone; LFZ, Lofoten fracture zone; VFZ: Vøring fracture zone.

are comparable to those underlying the Trøndelag Platform as well as the East Greenland sedimentary sequences. These regions document several Paleozoic and Mesozoic tectonic episodes prior to the late Jurassic (Bukovics et al., 1984; Surlyk et al., 1984; Larsen, 1987; Ziegler, 1988; Graham and Robinson, 1989; Larsen et al., 1989; Manby and Hambrey, 1989).

The present structural division between the Trøndelag Platform and the Vøring Basin was established during a late Jurassic–early Cretaceous extensional episode and subsequent Cretaceous subsidence. This episode represented a late stage in a tectonic evolution dominated by extension following the end of the Caledonian Orogeny. Seismic information recording the pre-Cretaceous events exists mainly from the Trøndelag Platform, whereas onshore geology documents large-scale extensional deformation dating back about 400 Ma. Based on these data, information from gravity anomalies

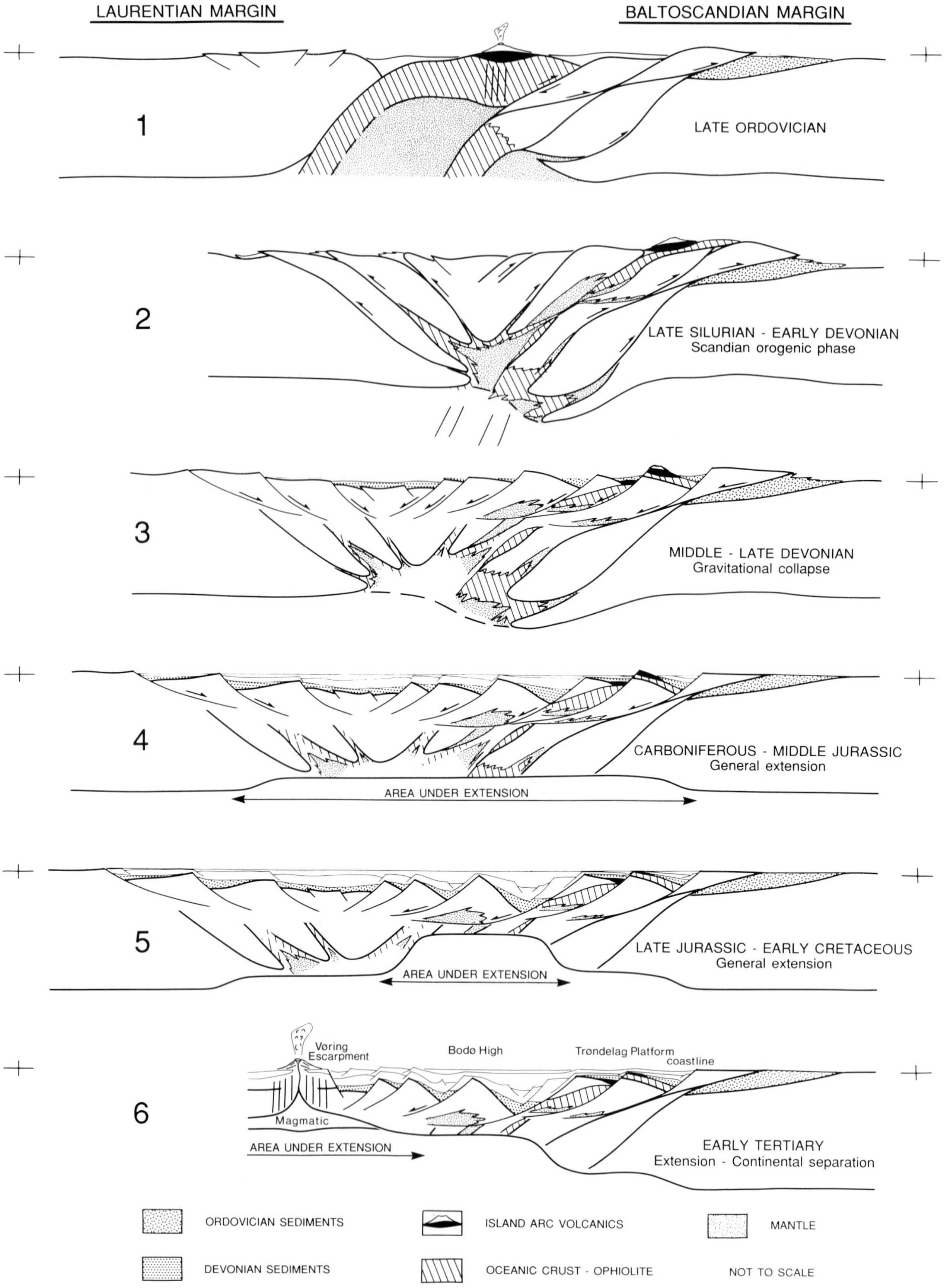

Fig. 3. Schematic diagrams illustrating an evolutionary model of the margin off mid-Norway from late Ordovician to early Tertiary times.

(Talwani and Eldholm, 1972; Planke et al., 1991) and the subsidence modelling of this study, we summarize the tectonic evolution of this region since late Ordovician times (Fig. 3, stages 1–6):

(1) Continental convergence between the Laurentian and Baltoscandian margins, referred to as the Caledonian Orogeny, has been suggested to have occurred in two main phases; the Finnmarkian, lasting until late Ordovician (stage 1), and the Scandian that caused the final closure of the Iapetus Ocean during the late Silurian–early Devonian (stage 2) (Dallmeyer, 1988). Obducted island arcs along the orogen suggest a phase of ocean-ocean collision which may argue for the existence of an early subduction zone to the west. The Scandian phase caused reactivation of former thrust zones, obduction of oceanic crust and mantle materials in front of the Laurentian continental crust, and finally continent-continent collision with large-scale lithospheric thickening (Austrheim, 1987; Jamtveit, 1987; Bucher-Nurminen, 1991).

(2) The thick isostatically unbalanced lithosphere responded by gravitational collapse and extensional crustal thinning (Dewey, 1988; England and Houseman, 1988) during the Devonian, mainly by reactivation of the original thrusts by normal slip (stage 3). This process is documented onshore where Devonian half-grabens formed and filled with deposits of Old Red sandstone (Surlyk et al., 1984; Hossack, 1986; McClay et al., 1986; Norton, 1986; Ziegler, 1988). Similar offshore structures are described by Gowers and Lunde (1984), Duindam and Van Hoorn (1987), Nelson and Lamy (1987) and Graham and Robinson (1989).

(3) Between Devonian and late Jurassic times several extensional episodes occurred (stage 4): late Carboniferous–early Permian regional rifting and basement-involved block faulting, mid-to-late Triassic basement-involved block faulting and, finally, late Triassic–early Jurassic growth faulting with detachment in Triassic evaporites (Surlyk et al., 1984; B.T. Larsen, pers. comm.).

(4) Late Jurassic–early Cretaceous extension led to major fault activity including reactivation of older fault zones which generally created slightly rotated fault blocks and caused subsequent subsidence along major rift systems (stage 5) (Rønnevik et al., 1983; Bøen et al., 1984; Hinz et al., 1984; Mutter, 1984; Skogseid and Eldholm, 1989). Along the Nordland Ridge, the Halten Terrace and the eastern flank of the Vøring Basin, the base Cretaceous reflector is an erosional unconformity which truncates late Jurassic fault blocks (Skogseid and Eldholm, 1989). Brekke and Riis (1987) found that these regions remained relatively elevated until middle Cretaceous times when the entire Trøndelag Platform and Halten Terrace were tilted westward causing a contemporaneous transgression. The base Cretaceous reflector is faulted along the Fulla Ridge and in the vicinity of the Fenris and Hel Grabens, revealing early Cretaceous subsidence apparently without erosion. Consequently the base Cretaceous reflector may define a diachron, implying that the central and western parts of the Vøring Basin developed in a later stage of the late Jurassic–early Cretaceous tectonic episode or, more likely, experienced a more rapid subsidence.

(5) In the early Cenozoic, renewed extension centred in the western Vøring Basin led to breakup and continental separation between Norway and Greenland in the earliest Eocene (stage 6). Associated with continental breakup, the region experienced syn-tectonic listric faulting, uplift and erosion, and large-scale magmatic activity and volcanism (Eldholm et al., 1989; Skogseid and Eldholm, 1989). Post-tectonic Cenozoic subsidence was restricted to the 150–200 km wide region of the Vøring Basin east of the Vøring Escarpment.

(6) During Neogene time the Norwegian mainland and adjacent regions were uplifted and eroded. This process produced huge amounts of Plio-Pleistocene sediments which were transported westward and deposited as the shelf edge prograded to its present position.

Tectonic subsidence

Modelling procedure

Vertical movements, corrected for sediment compaction and loading as well as paleo water depth, provide important information for understanding the tectonic evolution of sedimentary basins. To quantify this tectonic subsidence (negative values representing uplift), we use the "backstripping" procedure (Steckler and Watts, 1978; Sclater and Christie, 1980). Here, the one-dimensional tectonic subsidence, Y, is:

$$Y = H_s \frac{\rho_m - \rho_s}{\rho_m - \rho_w} + W_d \qquad (1)$$

where H_s is the decompacted sediment thickness, W_d is the paleo water depth, and ρ_m, ρ_s and ρ_w are the densities of mantle, sediments and water, respectively. Equation 1 assumes an Airy compensation mechanism and eustatic sea level changes are ignored.

From eqn. 1 and observational data the "observed" tectonic subsidence can be calculated as a function of time for a given point within the basin. Furthermore, it is possible to derive first-order estimates of lithosphere evolution as a function of time and space by comparing the "observed" tectonic subsidence at different locations with synthetic subsidence curves from geodynamic models. We use the McKenzie (1978) model with instantaneous and uniform lithosphere extension, as formulated by Le Pichon and Sibuet (1981).

The initial fault controlled subsidence, S_i, is:

$$S_i = (C_1 - W)\left(1 - \frac{1}{\beta}\right) \quad (2)$$

where C_1 is the "hot" mantle geoid depth, W is the pre-rift water depth (after sediment unloading) and β is the lithosphere stretching factor. The lithosphere returns to thermal equilibrium an infinitely long time after extension with a total tectonic subsidence, S_{tot}:

$$S_{tot} = (C_2 - W)\left(1 - \frac{1}{\beta}\right) \quad (3)$$

where C_2 is the "cold" mantle geoid depth. The thermal subsidence, S_{the}, approaches its final value with a time constant, τ:

$$S_{the} = (S_{tot} - S_i)(1 - e^{-t/\tau}) \quad (4)$$

where t is time.

Note that the "hot" mantle geoid depth is equivalent to the depth of a mid-ocean ridge without an oceanic crust, and the "cold" value is the corresponding depth after cooling of the asthenospheric anomaly (Turcotte et al., 1977; Celerier, 1986). S_{tot} is the theoretical equivalent of the observational "total tectonic subsidence", or TTS, of Sawyer (1985).

The rationale for using this approach is threefold. First, extension is documented to play an important role in the evolution of the Vøring Basin. Second, the above extension model contains a minimum of parameters. These are: the "hot" and "cold" mantle geoids, the lithospheric thermal time constant and stretching factor. This is important as it minimizes the number of necessary a priori known crust and mantle lithosphere properties. Third, the temporal resolution of subsidence in the Vøring Basin at the present time does not support the use of more complex extension models, in particular since it is difficult to separate syn- and post-tectonic subsidence components. We do believe, however, that this approach allows us to make first-order estimates of vertical movements and lithosphere extension during the Vøring Basin history.

To quantify the spatial (two-dimensional) basin evolution, the tectonic subsidence and β factors were calculated at about 750 locations forming a 15 × 15 km grid in the Vøring Basin. The grid allows us to present both subsidence profiles and maps. Five stratigraphic horizons, with ages of 0, 65, 144, 200 and 380 Ma, were used in the calculations. Other parameters are shown in Table 1. There are two reasons for incorporating the 200 and 380 Ma horizons in the modelling in spite of their relatively poor definition. Neglecting sediments below base Cretaceous may lead to: (1) overestimation of the tectonic subsidence, or "pseudosubsidence" (Roll, 1974), due to compaction of the pre-Cretaceous sediments (Fig. 4), and (2) underestimation of tectonic subsidence, due to an erroneous pre-rift water depth (sediment unloaded basement depth) in the expression for both initial and total tectonic subsidence (eqns. 2 and 3) (Fig. 4). These effects will generally not cancel one another out despite their opposite signs.

TABLE 1

Lithospheric parameters used in this study

C_1 = 3.6 km	Hot mantle geoid
C_2 = 7.8 km	Cold mantle geoid
L = 125 km	Lithosphere thickness
τ = 65 Ma	Thermal time constant
ρ_m = 3.3 g cm^{-3}	Mantle density
ρ_w = 1.03 g cm^{-3}	Water density

The modelling recognizes the two well-documented late Jurassic-early Cretaceous and early Cenozoic rift episodes. We incorporated the Jurassic–Cretaceous rifting in the entire Vøring Basin, whereas the early Cenozoic episode only affected the western and central Vøring Basin (Pedersen and Skogseid, 1989; Skogseid and Eldholm, 1989).

Data

Our study is based mainly on the seismic interpretation of Skogseid and Eldholm (1987; 1989) (Fig. 5) which outlines the lateral distribution of sedimentary units and their structural setting. However, these data resolve only the "shallow" structure of the crust, generally down to the base Cretaceous level. To evaluate better the deeper basin and the crustal configuration we have integrated velocity-depth curves interpreted from seismic refraction data (Mutter et al., 1984; Eldholm and Mutter, 1986; Planke et al., 1991). Assuming a 6.0–6.5 km/s velocity for the crystalline basement and using the SEASAT free-air gravity field as a guideline, we have inferred a 380 Ma basement level in the basin. The velocity-depth information has also been used to determine whether seismic reflection

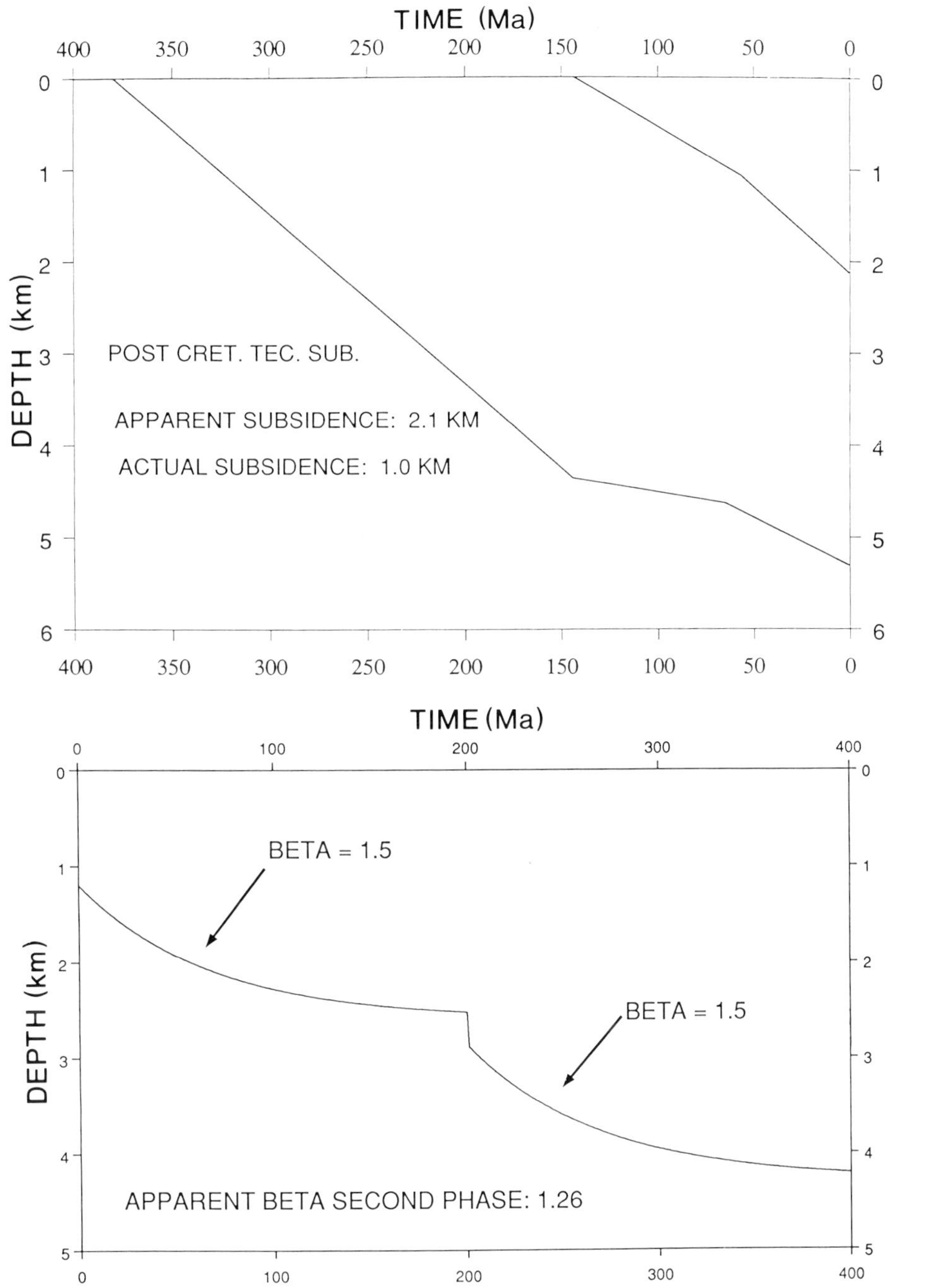

Fig. 4. Top: An example of pseudosubsidence, the difference between apparent and actual subsidence due to compaction of underlying sediments. The apparent tectonic subsidence of 2.1 km (upper curve) is reduced here to 1.0 km when compaction of pre-Cretaceous sediments is accounted for (lower curve). Bottom: A two-phase uniform extension model with β values of 1.5 is erroneously interpreted as a two phase model with β values of 1.5 and 1.26 if the pre-rift sediments unloaded water depth of 2500 m at 200 Ma is ignored and a zero depth is assumed.

interpretations are consistent within each subregion (Table 2) and to depth convert the seismic reflection profiles. To demonstrate regional trends and anomalies in the crust we have constructed crustal profiles across the margin based on seismic reflection, refraction and gravity data. The profiles provide a tool to compare deep velocity anomalies to shallow structures, and to correlate these with the gravity field (Fig. 6).

Lithology and paleo water depth data are necessary in order to calculate tectonic subsidence. Lithologies were obtained from commercial wells west of the Nordland Ridge and on the Halten Terrace, and from the DSDP and ODP wells on the Vøring Plateau (Fig. 2). Extrapolation into the Vøring Basin was guided by general relationships between lithology, seismic reflectivity and velocity. The sediments were separated into five main sequences,

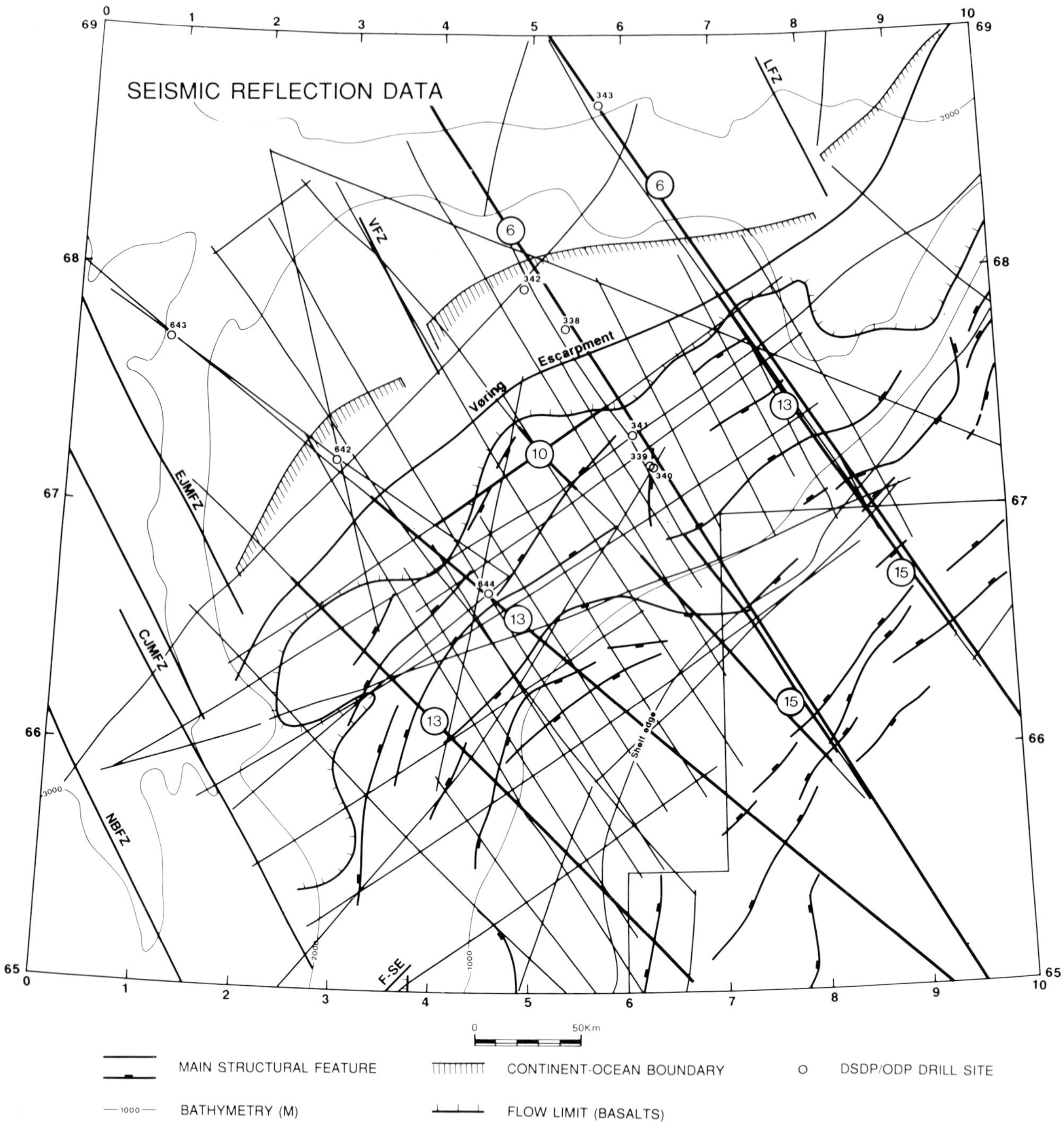

Fig. 5. MCS profiles in the Vøring Basin and Vøring Marginal High. The location of profiles presented in this paper are showindicated by bold lines with figure number annotated. For abbreviations see Fig. 2.

to which we ascribed three different lithologies (Table 3). Paleo water depth is difficult to determine, both from well and seismic data. However, because the Jurassic–Cretaceous and early Cenozoic tectonic episodes are bounded by erosional unconformities, we have assumed an averaged shallow water depth at these times.

Results

The subsidence modellng is exemplified by a profile across the main structural elements in the Vøring Basin (Fig. 7). The total tectonic subsidence postdating the base Cretaceous level increases westward. There is also a general increase in Cenozoic tectonic subsidence from east to west, reflecting increased tectonic activity as the zone of final continental separation between Greenland and Norway is approached (Skogseid and Eldholm, 1987). Moreover, initial uplift is observed both in the Fenris Graben and at the western Fulla Ridge. The best-fitting β factors, estimated from the tectonic subsidence curves after correction for remaining thermal anomaly from

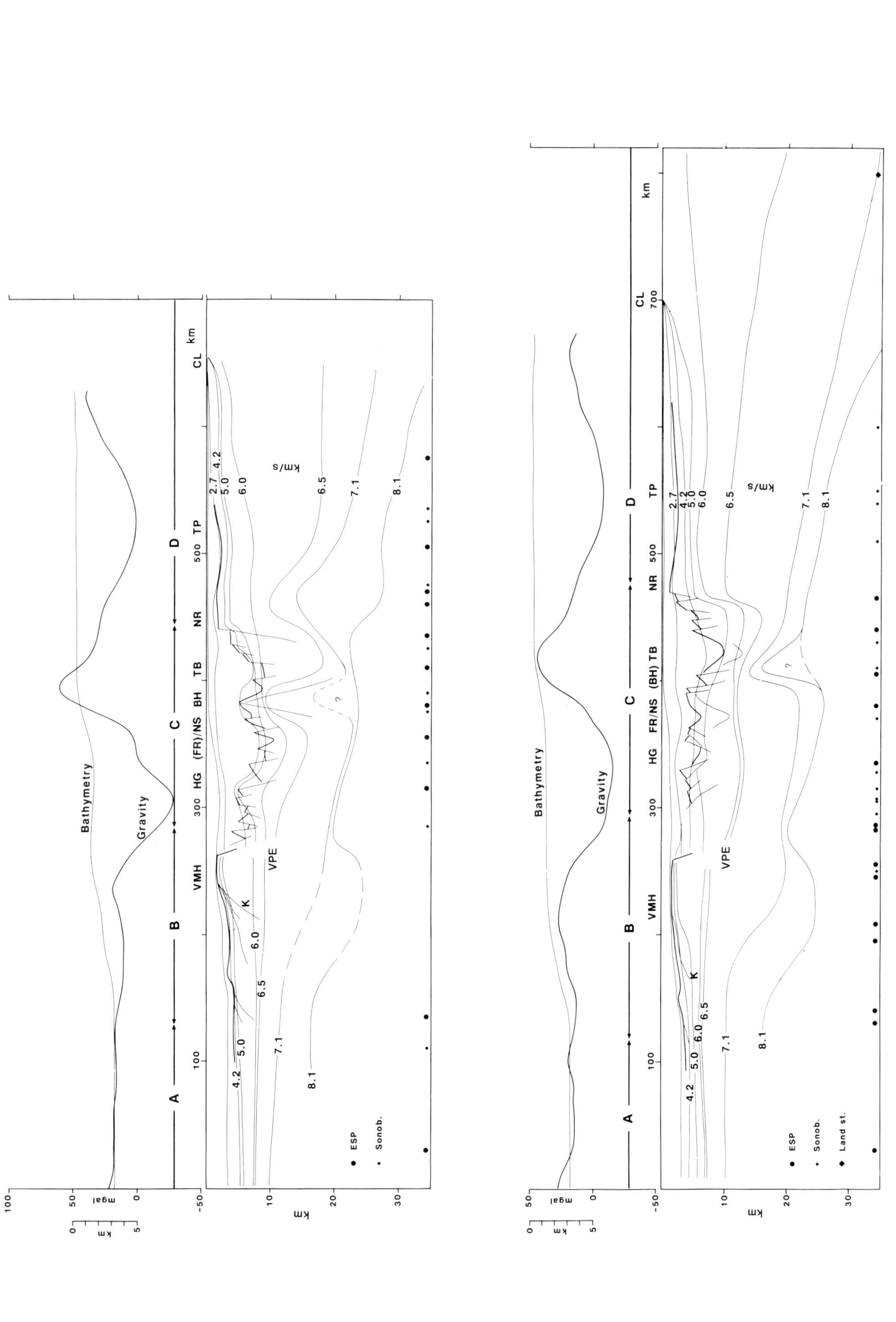

Fig. 6. Vøring Margin crustal profiles including isovelocity contours, the faulted base Cretaceous horizon (bold lines), and other prominent deep reflectors (thin lines). Zone *A* = normal oceanic crust; zone *B* = sub-aerially formed oceanic and transitional crust; zone *C* = Vøring Basin; and zone *D* = Trøndelag Platform region. *CL* = coast line; *BH* = Bodø High; *FR* = Fulla Ridge; *HG* = Hel Graben; *NR* = Nordland Ridge; *NS* = Nyk Syncline; *TB* = Træna Basin; *TP* = Trøndelag Platform; *VMH* = Vøring Marginal High; *VE* = Vøring Escarpment. From Planke et al. (1991). Location in Fig. 5 (upper profile is northern profile).

TABLE 2

Mean seismic refraction velocities (km/s) at identified seismic horizon levels

Horizon	VB	Sub-province						Wells
		1	2	3	4	5	6	
O′	^{27}2.3	^{1}2.3	^{1}2.3	^{6}2.3	^{5}2.2	^{8}2.0	^{3}2.0	^{4}2.5
A	^{44}2.3		^{3}2.4	^{7}2.4	^{11}2.4	^{13}2.1	^{4}2.2	^{4}2.2
A′	^{28}2.6			^{2}2.5	^{7}2.6	^{13}2.5	^{1}1.9	^{4}2.2
TP	^{32}2.8		^{1}2.5	^{4}2.5	^{11}2.7	^{5}3.4	^{5}2.5	^{3}2.2
C	^{44}3.0	^{1}2.4	^{3}2.5	^{7}2.4	^{11}3.1	^{10}3.4	^{7}2.6	^{4}2.4
D′	^{7}4.1		^{2}4.0	^{5}3.9				^{4}3.0
D	^{31}4.3	^{1}2.8	^{3}4.0	^{4}4.5	^{11}4.4	^{5}4.3	^{3}3.7	^{3}3.2
E	^{12}5.1		^{1}5.5	^{4}4.9	^{6}5.2		^{1}4.7	^{3}4.1
BM	^{4}5.9			^{1}5.3	^{2}6.5		^{1}5.1	
LCR	^{4}8.0				^{1}8.0	^{2}8.1	^{1}8.0	

Abbreviations: VB: Vøring Basin, 1: Nordland Ridge and Trøndelag Platform, 2: Halten and Dønna terraces, 3: Træna Basin and Rås Subbasin, 4: Fulla Ridge, 5: Fenris Graben, 6: Hel Graben. O′: base Pliocene, A: lower Miocene, A′: middle Oligocene, TP: earliest Eocene tuff marker, C: base Tertiary, D′: middle Cretaceous, D: base Cretaceous, E: approximate base Jurassic, BM: Basement, LCR: Lower Crustal Reflector. Indexes show number of velocity values used to compute averages.

TABLE 3

Stratigraphical division and parameters of the exponential relation between porosity and depth for various normal pressure North Sea lithologies

Sequence	Lithology	f_o	C (km^{-1})	ρ_{sg} (g cm^{-3})	Cond$_{sg}$ (Wm^{-1} K^{-1})
Pliocene, Pleistocene and Paleocene	sandy shale	0.58	0.41	2.70	3.3
Miocene, Oligocene, Eocene and Cretaceous	shale	0.63	0.51	2.72	2.0
Pre-Cretaceous	shaley sand	0.53	0.34	2.68	3.0

Sediment grain densities (ρ_{sg}) and conductivities (Cond$_{sg}$) are for the same lithologies (from Sclater and Christie, 1980, and Zoth and Haenel, 1988). Fractional porosity $f = f_o\, e^{-Cz}$ where z is depth in kilometer, C is a decay constant, and f_o is surface porosity.

earlier rift phases, are summarized in Table 4. Note that the pre-Cretaceous sediments were used only to estimate the water-loaded basement depth (W_d) in eqns. 2 and 3 for the Jurassic–Cretaceous and early Cenozoic stretching.

Cenozoic. Cenozoic tectonic subsidence indicates that a broad region of the western Vøring Basin was affected by extension prior to continental separation (Fig. 8). The crustal cross-section in Fig. 9 illustrates several important observations related to the early Cenozoic rifting.

TABLE 4

Stretching factors for individual tectonic episodes assuming uniform extension (McKenzie, 1978)

Area	Pre-Cretaceous β_1	Cretaceous β_2	Cenozoic β_3
Nordland Ridge	–	1.2	–
Dønna Terrace	–	1.3	–
Rås Subbasin	–	1.7	1.4[a]
Fulla Ridge	–	1.1	1.9
Fenris Graben	–	1.3	2.2

[a] Western basin.

(1) The western Vøring Basin experienced regional uplift, listric faulting and the formation of a well-defined rift unconformity of early Tertiary age (Fig. 10).

(2) The pre-Cenozoic Fenris Graben (Fig. 10), and possibly also the Hel Graben underwent additional local inversion. In the seismic data the inverted areas are generally observed in apparent association with a well-defined lower crustal reflector at typical depths between 10 and 14 km.

(3) Areas affected by large-scale erosion associated with both the regional uplift and the local inversion (Fig. 11a) are probable provenance areas for the deposition of the thick Paleocene sequence towards the east (Figs. 9 and 11b). Assuming a conformable pre-erosional Cretaceous sediment sequence, the amount of erosion was estimated by adding layer thicknesses truncated by the base Tertiary rift unconformity.

(4) Extensive magmatic activity is represented by the inferred large magmatic body underplating the continent–ocean transitional region, sills in the Cretaceous sedimentary sequence, and massive emplacement of lavas during breakup (Fig. 9).

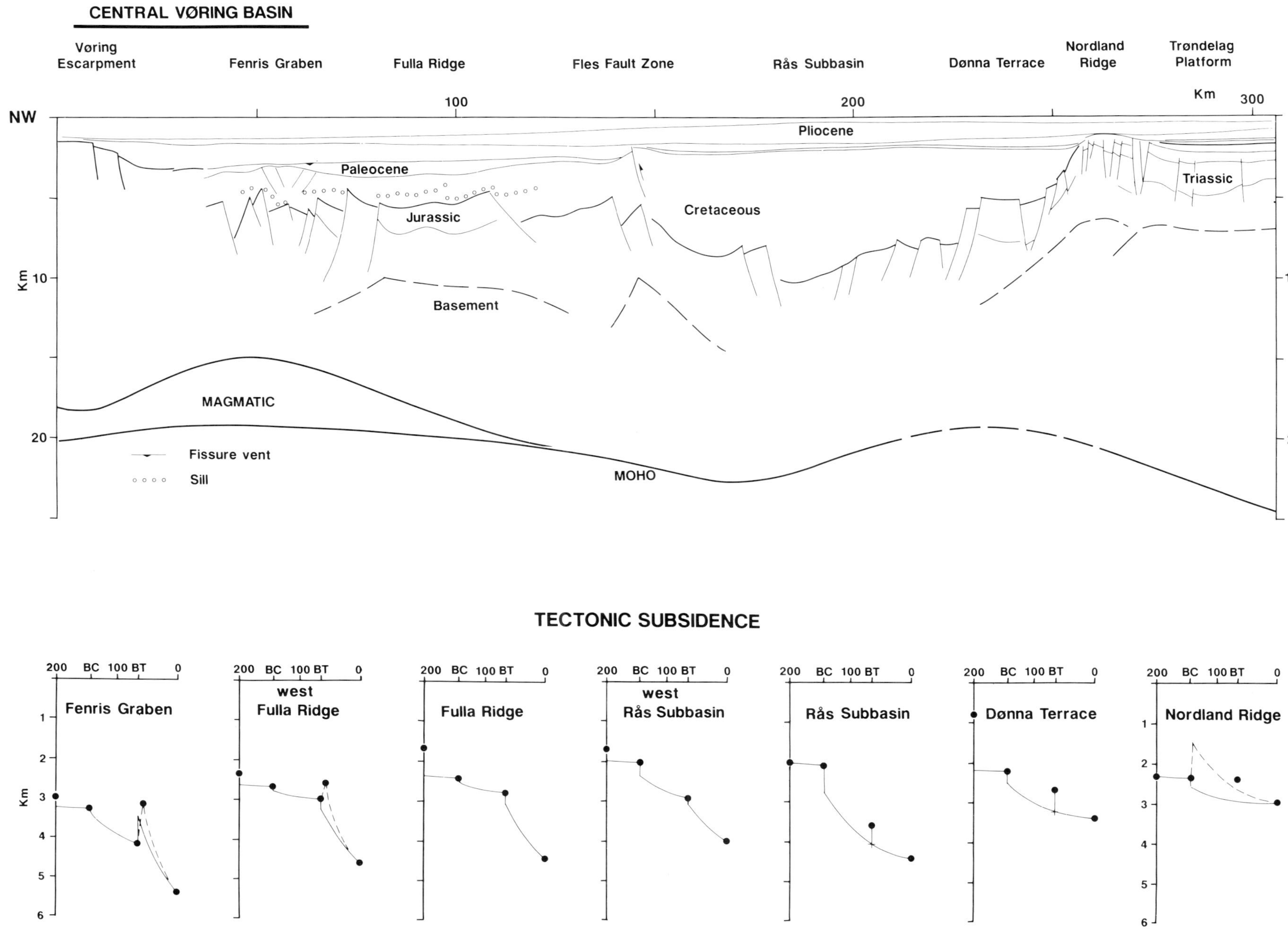

Fig. 7. Tectonic subsidence of selected stratigraphic levels computed for different sub-provinces in the Vøring Basin (black dots). Between the dots subsidence is inferred according to a uniform lithospheric extension model (solid line) with adjustments made to fit observation points (dashed lines). *BC* = base Cretaceous; *BT* = base Tertiary. Location in Figs. 8 and 12.

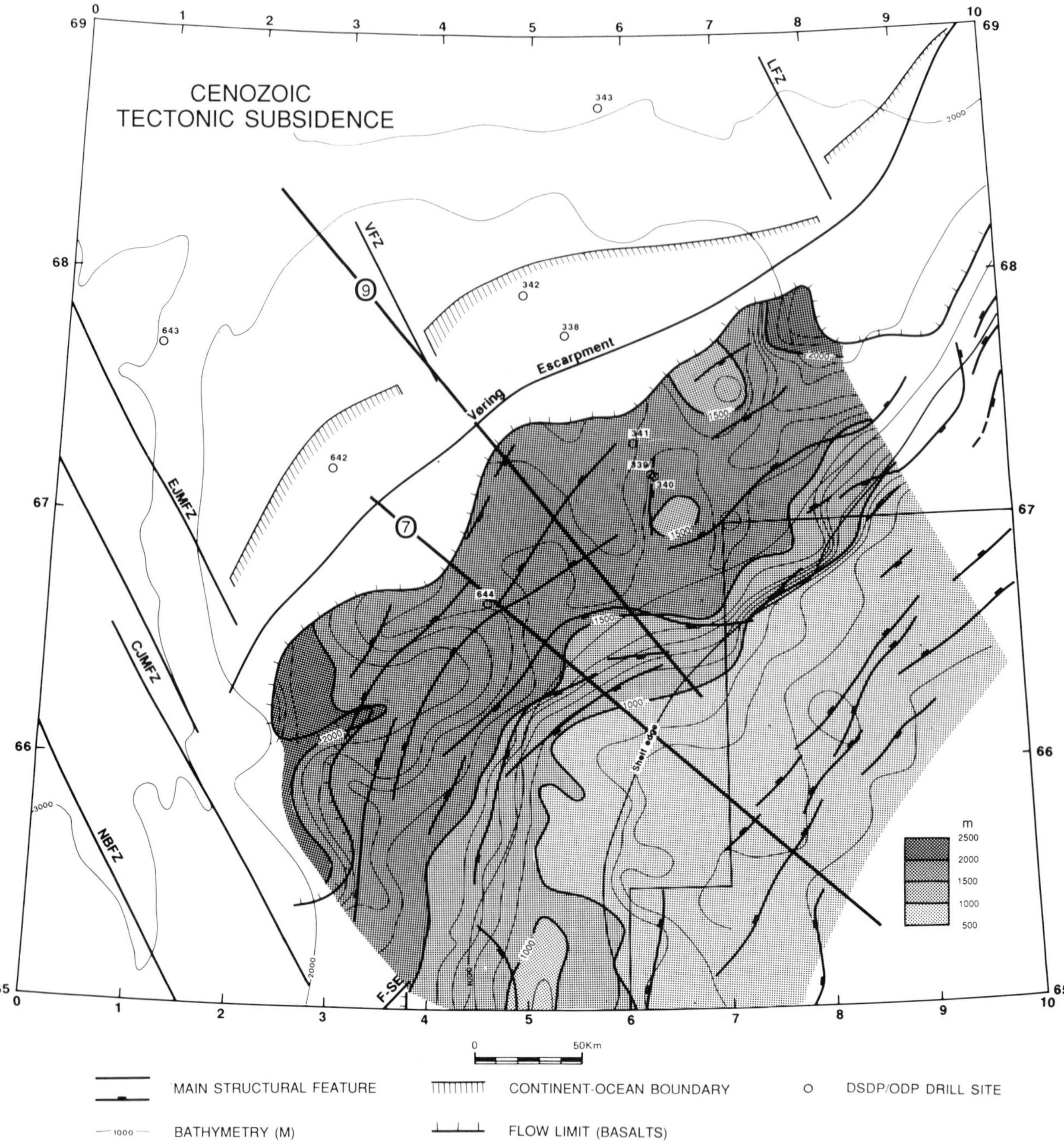

Fig. 8. Cenozoic tectonic subsidence. Contour interval 100 m. Location of profiles in Figs. 7 and 9 shown with figure numbers annotated. For abbreviations see Fig. 2.

Cretaceous. Late Jurassic–early Cretaceous rifting created the main structural subdivision of the Vøring Margin and is represented by Cretaceous tectonic subsidence (Fig. 12). The main area of extension and subsequent subsidence occurred in the Rås Subbasin and Træna Basin, whereas the Trøndelag Platform/Nordland Ridge experienced rift flank uplift (Figs. 7 and 12). The Fulla Ridge and Fenris Graben regions experienced relatively minor Cretaceous subsidence (Fig. 7).

A comparison of the Cretaceous and Cenozoic subsidence patterns (Figs. 8 and 12) demonstrates a major westward shift in the location of the rift zones. This trend has also been observed by Bukovics and Ziegler (1985) and Graham and Robinson (1989).

Specifically, the Cretaceous subsidence patterns reveal (Figs. 12 and 13):

(1) The zone of maximum subsidence narrows towards the northeast, and north of 67°N it shifts to the west.

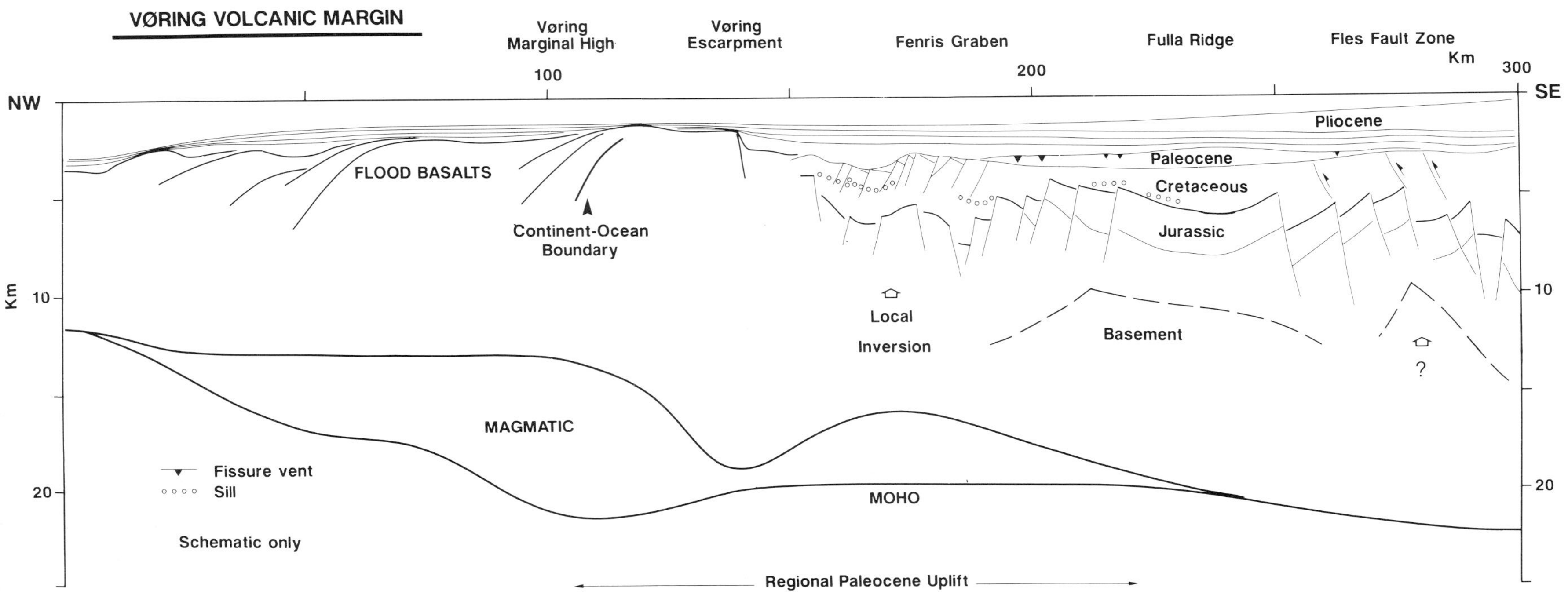

Fig. 9. Schematic crustal profile across the western Vøring Basin and transition between continental and oceanic crust. The profile shows relative distribution of different magmatic/volcanic and structural features forming the Vøring Volcanic Margin. Approximate location in Fig. 8.

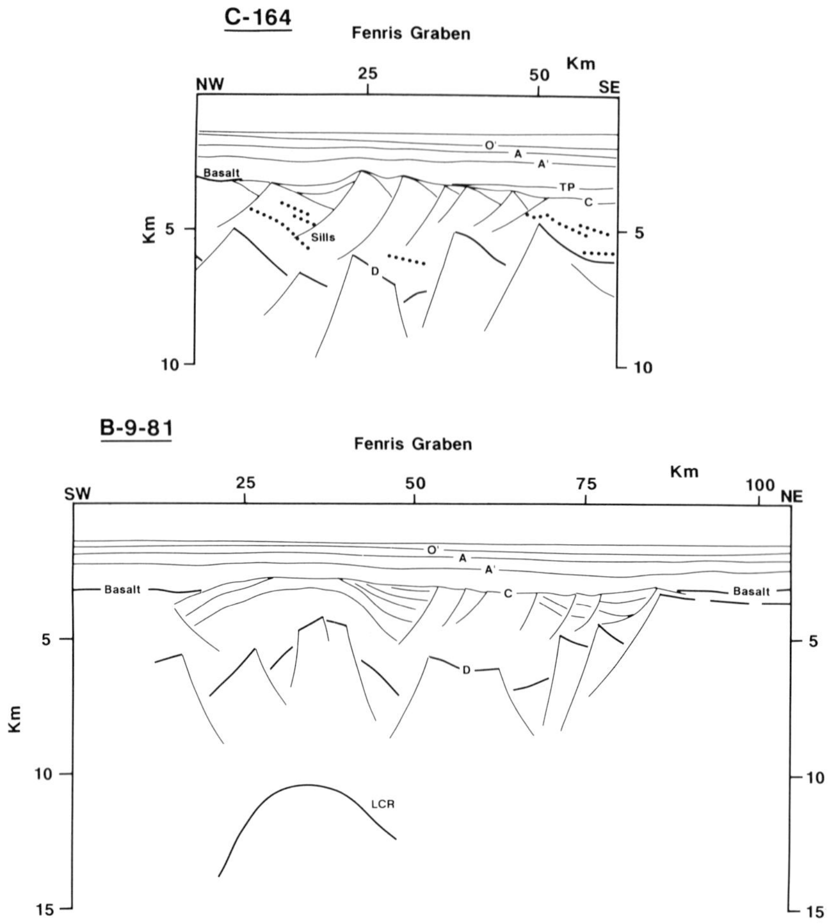

Fig. 10. Top: line drawing of profile C-164 illustrating listric faults truncated by the base Tertiary rift unconformity, *C*. Bottom: profile B-9-81 documenting Paleocene local inversion in the Fenris Graben, with earliest Eocene flood basalts onlapping the inverted region. The uplift is apparently associated with a prominent lower crustal reflector, *LCR*. *O'* = base Pliocene; *A* = lower Miocene; *A'* = middle Oligocene; *TP* = earliest Eocene tuff marker; *C* = base Tertiary; *D* = base Cretaceous. Location in Fig. 5.

(2) The Fenris and Hel Grabens subsided less than the eastern basin province.

(3) There is an apparent overall relationship between late Jurassic–early Cretaceous faulting and the underlying basement topography. In particular, the Nyk Syncline may have been formed passively due to differential extension and subsidence between the deeper basement structures (Fig. 13).

(4) An apparent asymmetry exists between maximum basin subsidence and Moho shallowing. This is particularly evident in the central Vøring Basin profile (Fig 13), but also apparent in the northern Vøring Basin profile. We point out, however, that the Moho is still poorly defined.

Pre-Cretaceous. We are not able to resolve the various pre-Cretaceous tectonic episodes within the Vøring Basin. Nevertheless, the incorporation of a basement surface, albeit poorly constrained, permits construction of a pre-Cretaceous subsidence map (Fig. 14). This shows a pattern of fairly uniform subsidence with the exception of local NE–SW trending troughs and ridges. The fact that the map represents a time span of about 240 Ma, encompassing several tectonic episodes, may explain the much wider region of subsidence compared to those for the Cretaceous and Cenozoic. This implies that pre-Cretaceous extension took place regionally, or shifted between separate rift zones.

The pre-Cretaceous subsidence map is considered poorly constrained, except for those regions where there is reasonable basement control, for example in the vicinity of the Bodø High (Fig. 6). Here ESP velocities and the positive gravity anomaly indicate a high-density lower crustal body (Talwani and Eldholm, 1972; Planke et al., 1991), whereas the reflection profiles identify a horizon which correlates with the 6.0–6.5 km/s velocity contours and dips deep into the Træna Basin (Fig. 15). We believe that the Bodø High represents a large rotated fault block, once adjacent to the Nordland Ridge, that formed above a "shallow detachment" during the Devonian collapse of the Caledonides (Fig. 3). The following observations are supporting evidence:

(1) The block contains a deep core with high seismic velocity and density.

(2) The base of the rotated block appears to "rest" on the Moho discontinuity.

(3) The regions on each side of the Bodø High have experienced differential subsidence with respect to each other and to the rotated block.

(4) The Bodø High gravity anomaly is part of a regional elongated system of gravity belts between Lofoten and Scotland that follow the trend of the Caledonian orogenic belt (Planke et al., 1991).

Evaluation of observational data

We have used seismic refraction velocities averaged for each subregion within the basin to evaluate the overall and local velocity-depth relationships (Table 2). The large differences in velocities along the stratigraphic boundaries are related to the amount of overburden, whereas velocities as a function of depth below seafloor are fairly consistent throughout the Vøring Basin (Table 5). Average velocities for each sub-region have been used in the depth conversion assuming constant velocity gradients. We have observed that the refraction velocities generally are slightly higher than corresponding well and interval velocities derived during processing of MCS data. This may cause an overestimation of depths in the deepest basins.

Only Neogene well control exists in the Vøring Basin proper. Basin subsidence and lithospheric extension are therefore obtained from an interpreted seismic data set. However, interpretation of seismic data in areas with poor well control is often ambiguous.

The reliability of the history of vertical movements is critically dependent on the correct definition of stratigraphic boundaries. In particular, identification of the base Cretaceous horizon over the northern Fulla Ridge may need reinterpretation in view of a denser grid of new better-quality data. The base Cretaceous horizon (reflector D) of Skogseid and Eldholm (1989) could be interpreted as early Tertiary sills, placing the base Cretaceous unconformity at a deeper level. Although we agree that sills exists at these stratigraphic levels, the question is whether the sills lie within the thick Cretaceous sequence or were emplaced near the base Cretaceous level. We chose the latter interpretation for two reasons. First, the Cretaceous sediments generally exhibit poorly developed stratification compared with the pre-Cretaceous and Cenozoic sediments. At the Fulla Ridge, there are thick, generally well-stratified sediments below reflector D, in contrast to the Cretaceous sediments in the adjacent basins. Second, our interpretation of reflector D is supported by analysis of velocity-depth curves from a great number of seismic refraction profiles in the different Vøring Basin sub-provinces (Table 2; Fig. 16). These show typical velocity gradient changes associated with the base Tertiary and the base Cretaceous horizons. Eldholm and Mutter (1986) have suggested that the stratigraphically lower gradient change takes place just below reflector D. This is in agreement with wells at the Halten Terrace which document that the velocity increase occurs within the upper Jurassic sequence rather than at the Jurassic-Cretaceous boundary. Thus, the velocities appear to favour the interpretation of Skogseid and Eldholm (1989) that the Fulla Ridge experienced non-deposition and/or erosion during the early Cretaceous. The early Tertiary sills may have been preferentially emplaced between Jurassic sands and Cretaceous marine shales which are of different competences. It may be argued, however, that sills or composite sills may increase the velocity although this effect is not observed elsewhere along the margin (Eldholm and Mutter, 1986).

The parameters in the backstripping calculations may introduce errors in the tectonic subsidence. Both the decompacted sediment thicknesses and the sediment densities depend on the lithology. In addition, the use of proper compaction formulae, present porosities and grain densities contain uncertainties. Because we have used parameters that essentially are "North Sea" averages (Tables 1 and 3), the uncertainties for each lithology may be large. However, if the lithologic parameters are fairly uniform across the basin, the relative subsidence pattern may be representative, even though the absolute values are poorly constrained.

Water depth changes affect subsidence estimates linearly, and lack of precise paleo water depth indicators will cause errors. Present water depths across the Vøring Basin range from 360 to 1400 m, showing that sedimentation has not kept pace with subsidence during Cenozoic time. Nevertheless, the erosional unconformities that formed during rifting indicate fairly shallow waters. We cannot exclude local water depths in the range of 0–500 m, or even more, within this large basin region. In this context we would like to draw attention to the Cenozoic history of the Vøring Basin and the adjacent Nordland Ridge region. During this period the maximum subsidence was in the western Vøring Basin. In the basin modelling we have assumed that the early Cenozoic extension did not affect areas east of the central Rås Subbasin (Figs. 7 and 8). Consequently, the Rås Subbasin has not undergone major extension since Jurassic–Cretaceous rifting. The model curve in Fig. 7 does not pass through the base Tertiary "obser-

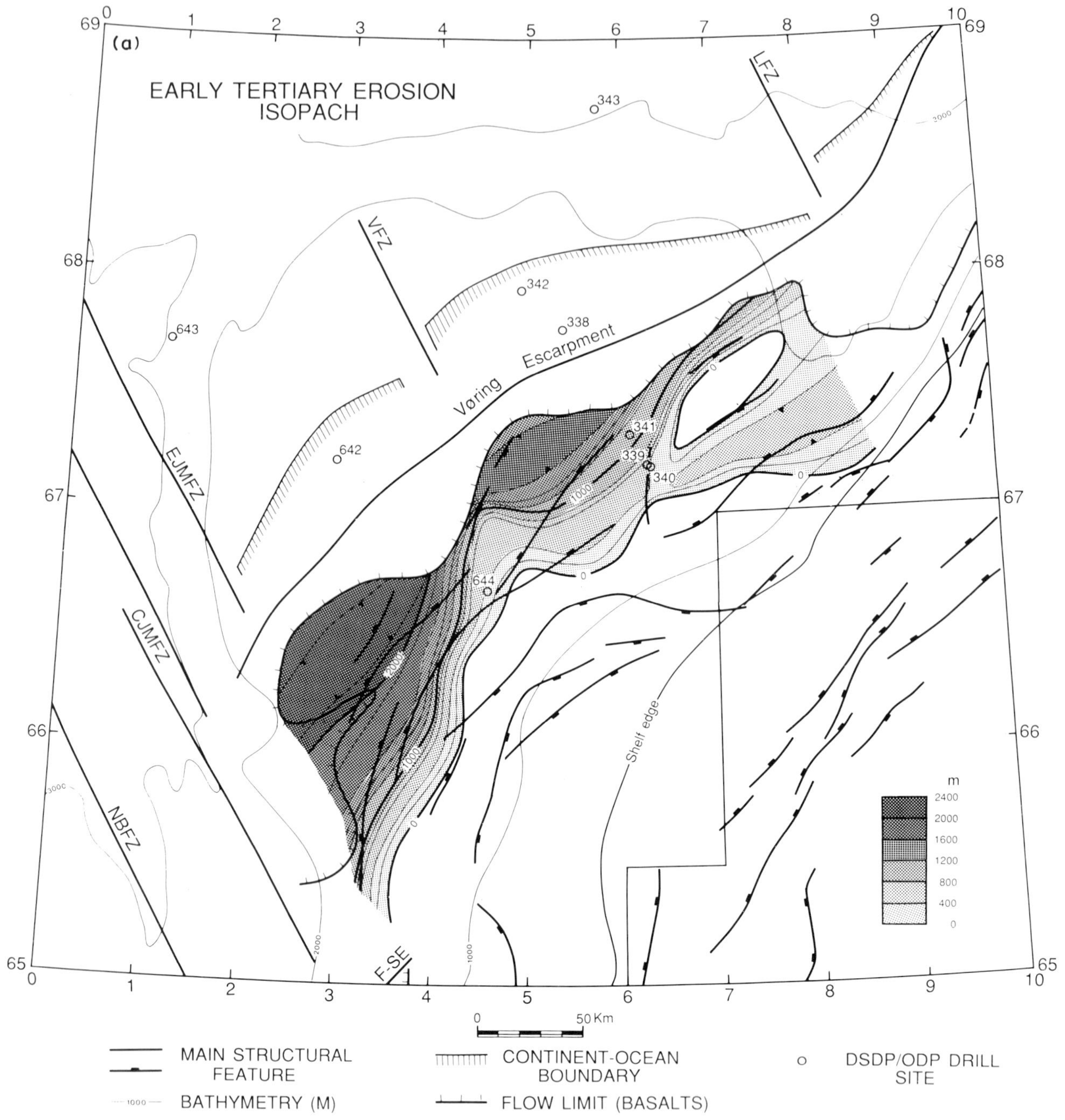

Fig. 11. (a) Contours of estimated amounts of Paleocene erosion. Contour interval 200 m. For abbreviations see Fig. 2.

TABLE 5

Mean seismic refraction velocities (km s^{-1}) at depth (s) below seafloor

Depth	VB	Sub-province:					
		1	2	3	4	5	6
1	$^{58}2.4$	$^{13}2.6$	$^{6}2.5$	$^{15}2.4$	$^{12}2.4$	$^{14}2.1$	$^{11}2.5$
2	$^{51}3.1$	$^{13}3.2$	$^{4}2.8$	$^{14}2.7$	$^{12}3.0$	$^{14}3.3$	$^{7}3.5$
3	$^{42}4.1$	$^{10}5.2$	$^{4}4.1$	$^{14}3.8$	$^{11}4.1$	$^{10}4.3$	$^{3}4.4$
4	$^{26}4.7$		$^{4}4.6$	$^{10}4.4$	$^{7}4.7$	$^{3}4.6$	$^{2}5.0$
5	$^{13}5.2$		$^{1}5.5$	$^{5}4.8$	$^{4}5.4$	$^{2}5.3$	$^{1}5.2$

Abbreviations: VB: Vøring Basin, 1: Nordland Ridge and Trøndelag Platform, 2: Halten and Dønna terraces, 3: Træna Basin and Rås Subbasin, 4: Fulla Ridge, 5: Fenris Graben, 7: Hel Graben. Indexes show number of velocity measurements used to compute averages. Velocities higher than 6.0 km s^{-1} are assumed to represent crystalline basement and are not shown.

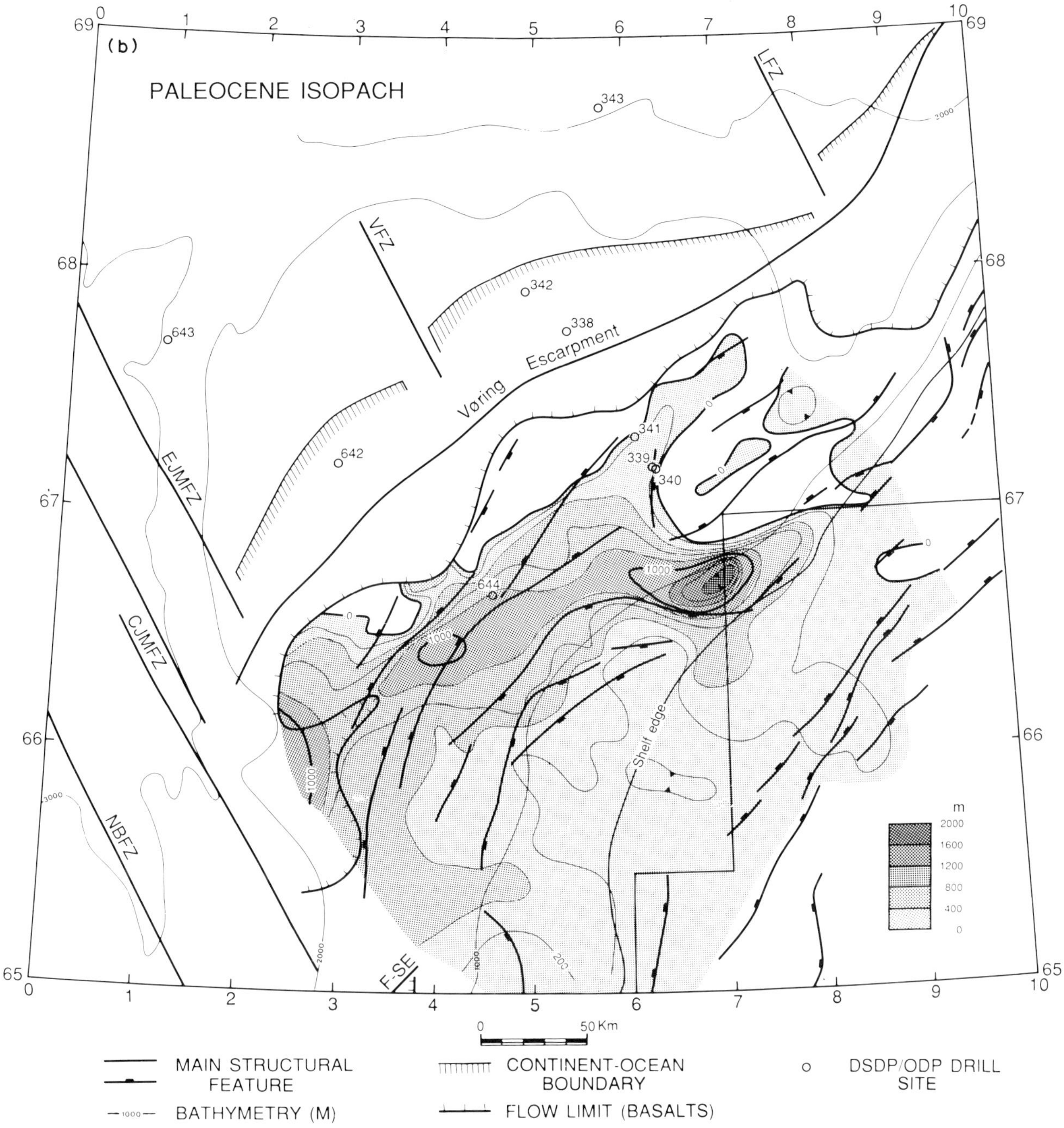

Fig. 11 (continued). (b) Isopach map of Paleocene sediments. Contour interval 200 m. For abbreviations see Fig. 2.

vation" points for either the Rås Subbasin or the Dønna Terrace and Nordland Ridge. If the difference between observation and model is due entirely to paleo water depth, a depth of about 500 m is inferred during the early Tertiary.

The Nordland Ridge model curve is incorrect because there is evidence of syn-rift uplift during the late Jurassic–early Cretaceous extension. Yielding and Roberts (this volume) have ascribed this to "footwall" uplifting along the Vøring Basin boundary fault and have calculated effective uplift of maximally 1500 m. This is not considered in the simple uniform extension model applying Airy isostasy, but has been approximated by the dotted line in Fig. 7.

In view of the gravity field, Airy isostasy seems reasonable (Talwani and Eldholm, 1972), although the condition of Airy isostasy may not have prevailed in the past. Considering an elastic plate model, structures with short wave-lengths will be most affected by deviations from Airy isostasy. Thus, features such as the Nordland Ridge, the Fenris Graben and the Bodø High region may have experienced a flexural-

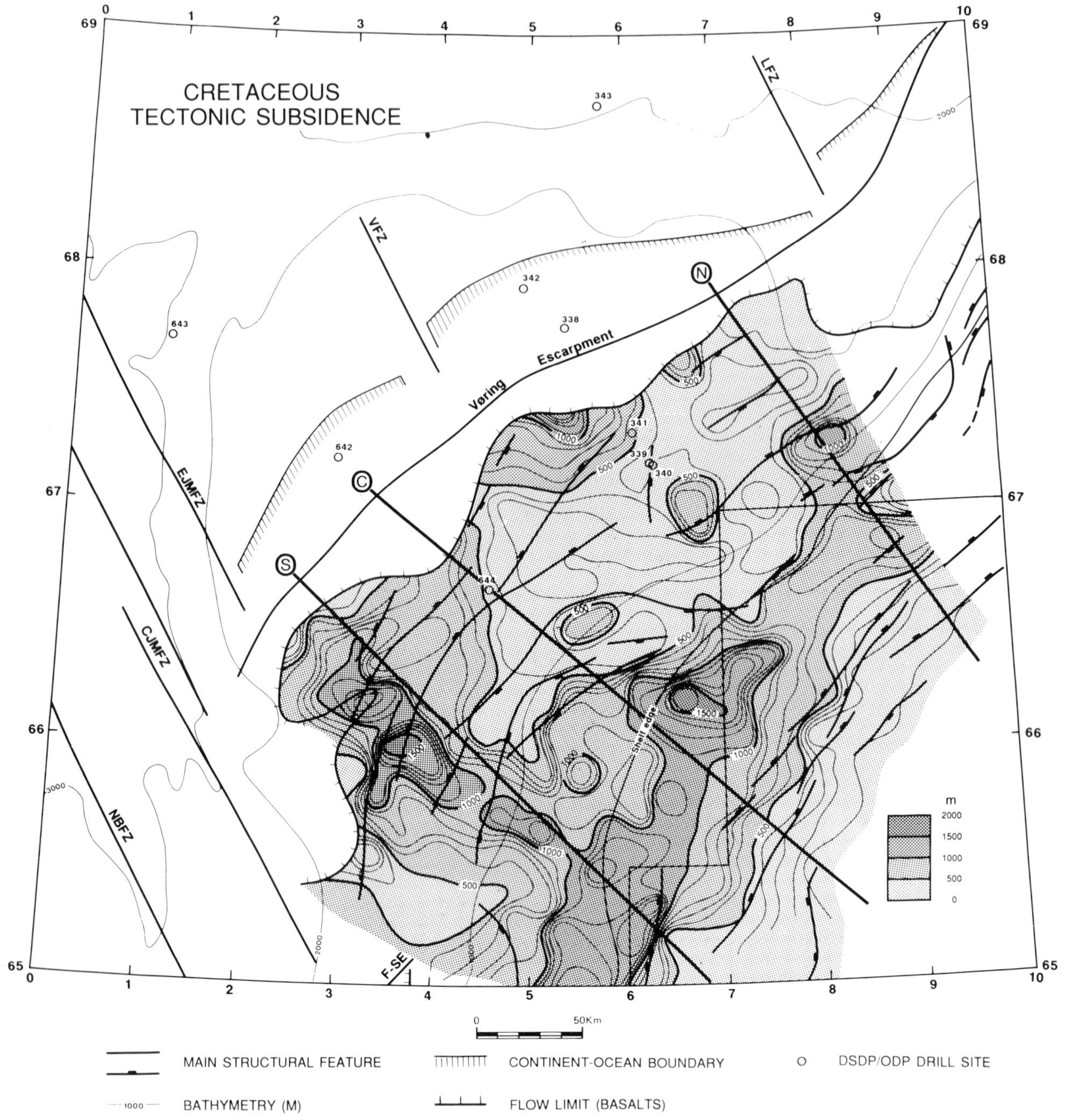

Fig. 12. Cretaceous tectonic subsidence. Contour interval 100 m. Location of N, northern, C, central and S, southern Vøring Basin profiles in Fig. 13. The profile in Fig. 7 corresponds to profile C. For abbreviations see Fig. 2.

type basement response. A finite flexural rigidity model of the lithosphere could increase the observed tectonic subsidence in the Fenris Graben by approximately 10%, maximally 150–200 m, and contribute to the assumed Jurassic–Cretaceous uplift at the Nordland Ridge by 100–300 m.

Ignoring eustatic sea level changes may affect the absolute values of tectonic subsidence, but, by definition, will not influence its geographical pattern. The maximum error here probably relates to the base Tertiary level where subsidence may have been overestimated by 180–300 m. This estimate is obtained by multiplying an assumed sea level rise of 120–200 m (Watts and Thorne, 1984; Haq et al., 1987) with an isostatic loading factor of 1.5 (Steckler and Watts, 1978).

Aspects of basin development

By applying relatively simple models of lithospheric extension and basin subsidence, we have

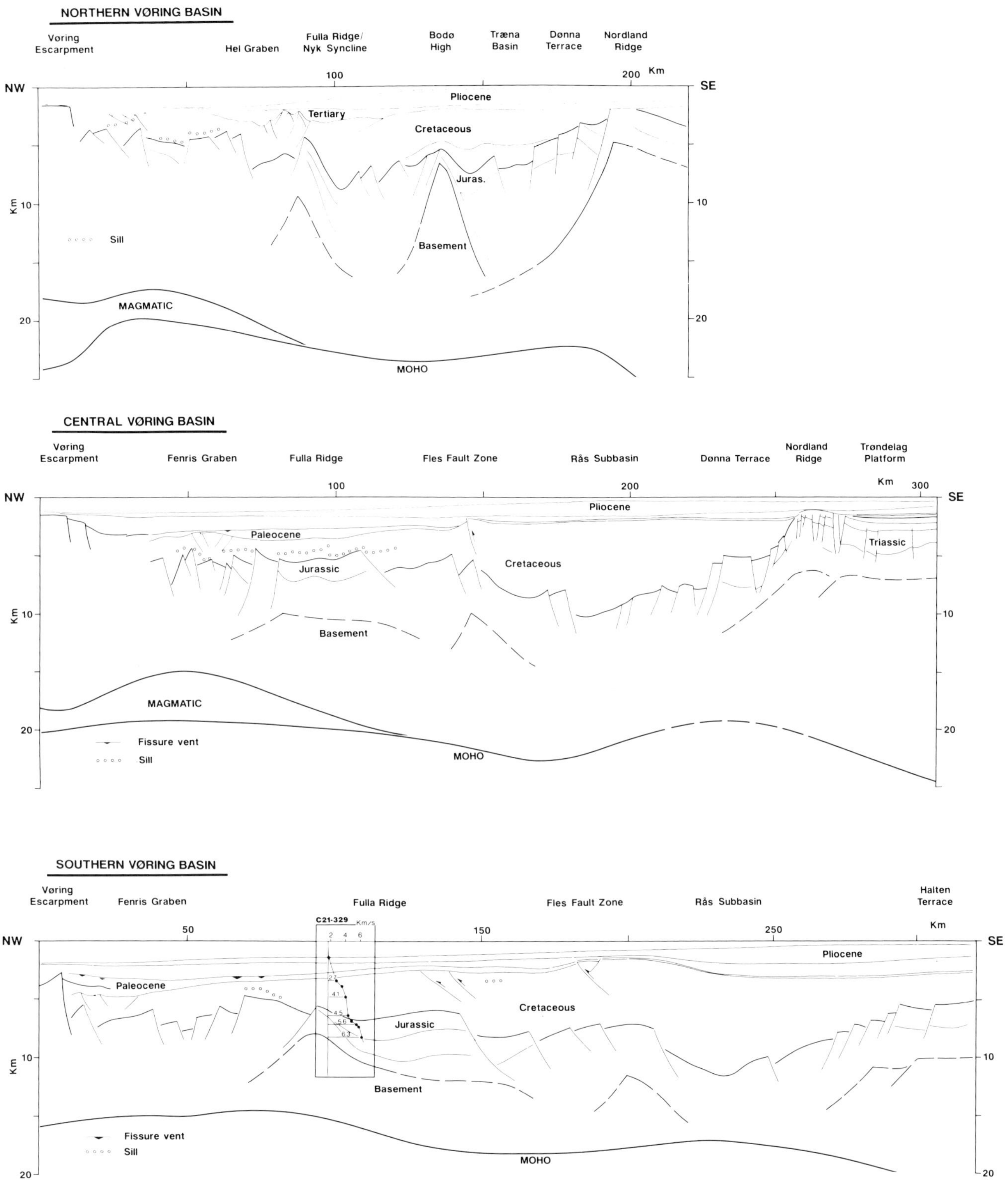

Fig. 13. Crustal profiles across the Vøring Basin. The southern profile includes the velocity-depth curve from sonobuoy C21-329 (Eldholm and Mutter, 1986) showing "shallow" basement velocities. Location in Figs. 5 and 12.

described the regional aspects of the basin evolution based on a two-phase rift history: the late Jurassic–early Cretaceous and early Cenozoic rift episodes. Several pre-Cretaceous rift events are probable, but cannot be resolved. Furthermore, we have indicated that the regional modelling approach does not explain some local features in the Vøring Basin which we address below.

Pre-Cenozoic basin

The Bodø High region is the only "window" into the pre-Cretaceous part of the basin. The large Bodø

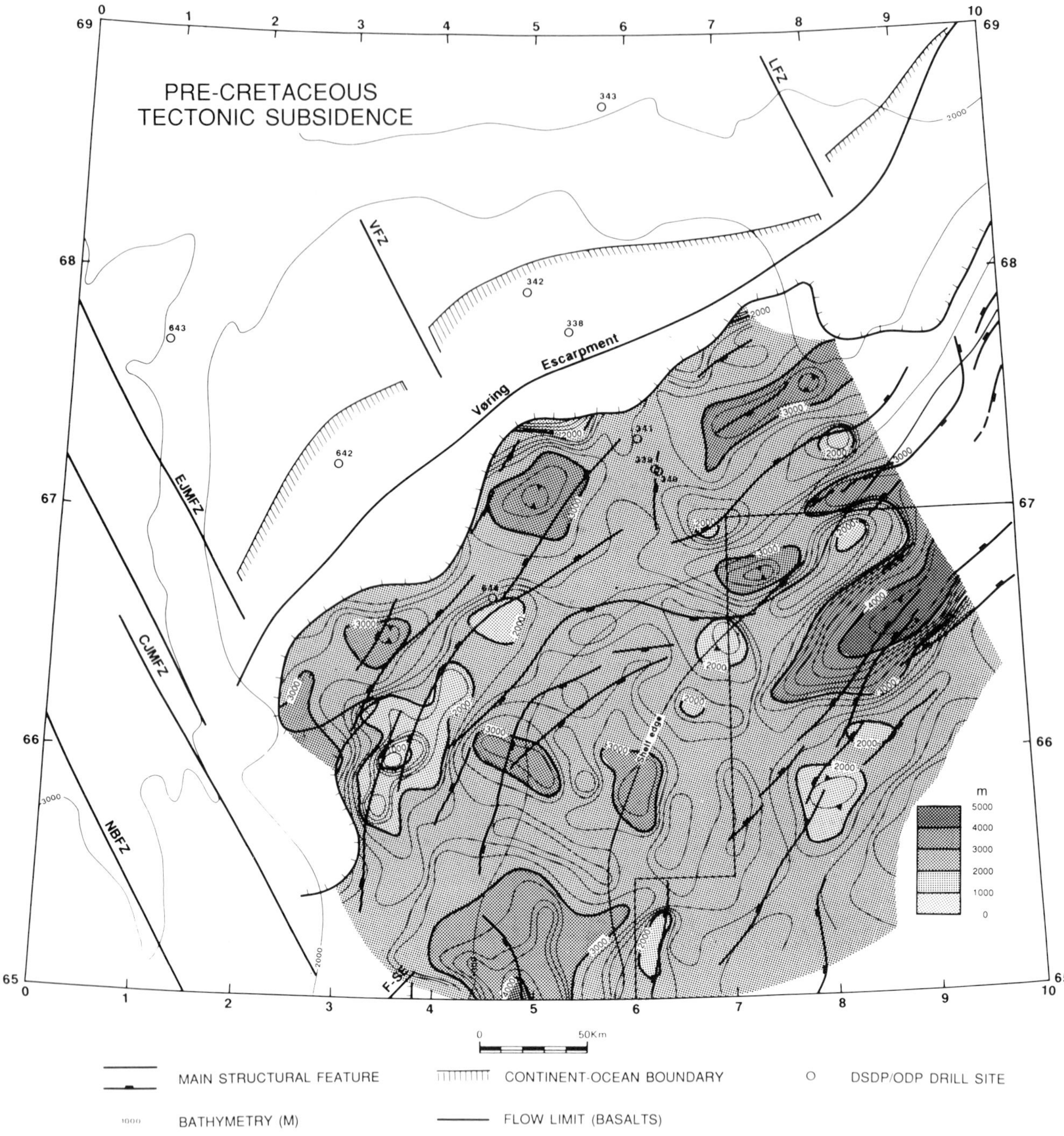

Fig. 14. Pre-Cretaceous tectonic subsidence. Contour interval 200 m. For abbreviations see Fig. 2.

High block of crystalline basement (Fig. 15) was formed during Scandian compression and the following Devonian collapse of the orogen (Fig. 3). It appears, however, that the block has been laterally displaced without experiencing large-scale internal deformation during the subsequent extensional episodes. Speculatively we relate this to successive reactivation of the bounding listric faults and associated formation of large half-grabens on either side of the main basement blocks. The lower crust, on the other hand, reacted to extension by ductile deformation and thinning, consequently bringing the mantle up towards the base of the observable basement block. This is schematically illustrated in Fig. 3, whereas Figs. 6, 13 and 15 show the present configuration where the Bodø High basement block appears to be bounded at depth by the hanging wall fault plane, directly above Moho. During the respective episodes of extension this configuration caused differential syn-tectonic vertical movements, which may have been expressed in actual uplift of the block proper.

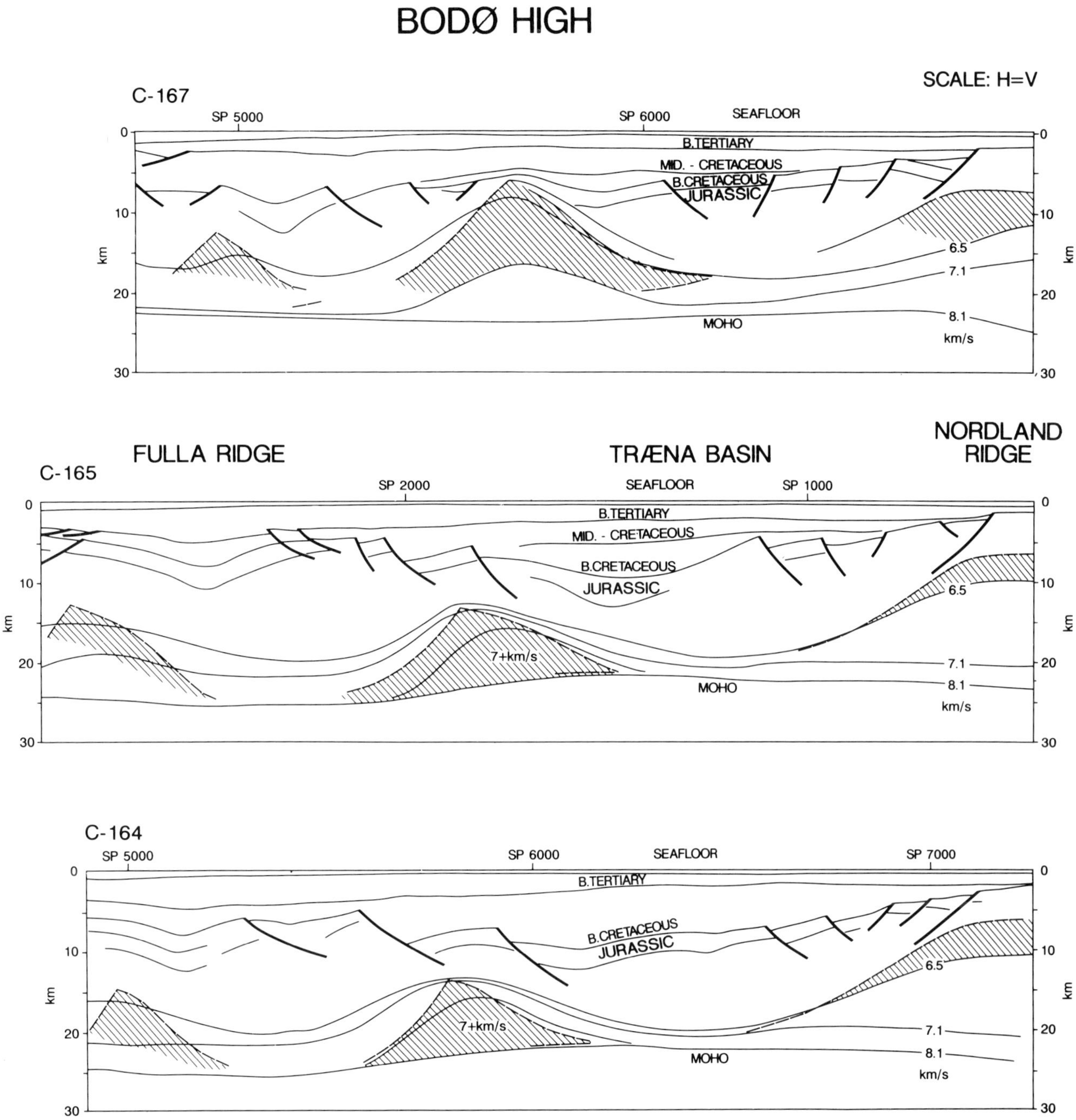

Fig. 15. Crustal sections and isovelocities across the Bodø High region. The faulted base Cretaceous horizon (bold lines) and other prominent deep reflectors (thin lines) are modified from Skogseid and Eldholm (1989). Shading indicates crystalline basement underneath the Nordland Ridge and in the rotated fault blocks discussed in the text. There is no vertical exaggeration. Locations in Fig. 5.

The Bodø High lies at the southern extension of a positive gravity anomaly belt which trends along the Lofoten shelf subparallel to a more prominent anomaly over the Lofoten Islands (Talwani and Eldholm, 1972). These gravity anomalies are also characterized by shallow Moho (Goldschmidt-Rokita et al., 1988) and shallow basement (Chroston and Brooks, 1989), most likely composed of rotated basement blocks (Mokhtari et al., 1989). We speculate that similar gravity anomaly belts which exists along the entire mid-Norwegian margin, although offset by apparent transfer zones, overlie a series of crystalline basement blocks similar to the Bodø High. However, the most prominent gravity anomalies are produced only where the basement blocks are associated with high-density rocks, probably mantle material and/or eclogite zones (Talwani and Eldholm, 1972).

The basement configuration has implications for the lateral extent of Cretaceous basin subsidence. If the Bodø High experienced uplift during Cre-

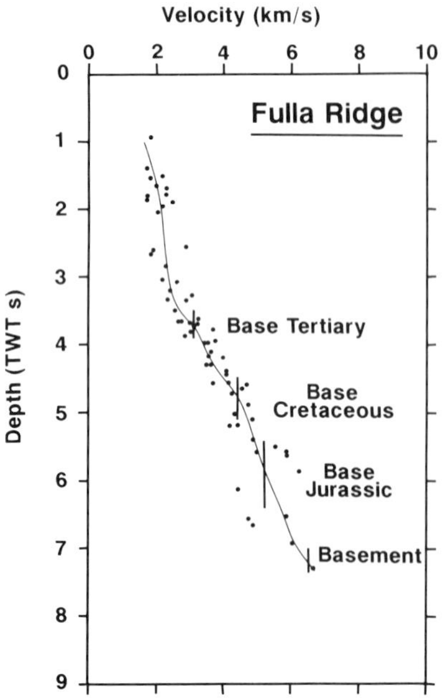

Fig. 16. Relationship of stratigraphy and refraction velocities along the Fulla Ridge. Vertical bars indicate average velocities (Table 1) and range in depth of the respective horizons.

taceous rifting and, subsequently, less and delayed subsidence compared with the adjacent basins, the tectonic subsidence map (Fig. 12) does not accurately represent the area of lithospheric deformation. In this case Cretaceous rifting becomes more uniform along strike, and the apparent westward shift of the area of maximum subsidence north of 67°N becomes less pronounced.

We observe that the late Jurassic–early Cretaceous extension generally formed collapse grabens within older half-graben basins (Figs. 13 and 15). This relation between Jurassic–Cretaceous faults and the underlying basement topography shows that the deep structural setting influenced the location of later faulting. The old basement-involved fault zones were locally reactivated, but Jurassic-Cretaceous faults are generally basement-detached.

The asymmetry observed between the location of maximum basin subsidence and Moho shallowing along the central Vøring Basin profile (Fig. 13) may be explained in light of the early Cretaceous erosion and rift flank uplift described above. If the Moho relief also reflects the region of maximum mantle-lithosphere thinning, the main thermal anomaly was probably located under the east flank of the basin in the Halten and Dønna terraces and the Nordland Ridge area. In this case this area would have experience syn-tectonic uplift and associated erosion rather than subsidence. Brekke and Riis (1987) have suggested that these areas were peneplaned during the early Cretaceous and related the uplift to regional sinistral shear movements along a Great Glen Fault trend. Drawn from our observations, we put forward an alternative explanation where the uplift, or relative lack of subsidence, is caused by differential lithospheric extension accompanied by footwall uplift along the major normal faults (Yielding and Roberts, this volume), and that shear movements (Gabrielsen and Robinson, 1984; Gowers and Lunde, 1984; Caselli, 1987; Grønlie and Roberts, 1987) represent a second-order effect.

Cenozoic basin

The magmatic surge during breakup at 57.5 Ma is manifested by a large volume of extrusive rocks including the wedges of seaward dipping reflectors that partly compose the Vøring Marginal High. The high is underlain by a thickened, high-velocity, lower crustal body (Fig. 6).

This "underplated" body is believed to be a result of decompressional melting as the asthenosphere rose passively beneath the stretched and thinned lithosphere (White et al., 1987). Many, often conflicting, evolutionary models have been advanced to account for the Vøring volcanic margin, and the drilling of ODP Site 642 (Fig. 2) recently has provided key boundary conditions. The volcanic margin evolution and its relationship to the North Atlantic Volcanic Province (Morton and Parson, 1988) have been discussed in detail by Eldholm et al. (1989). Here, we restrict ourselves to the thermomechanical modelling of Pedersen and Skogseid (1989). Assuming instantaneous rifting, they estimated the thermal anomaly needed to produce the observed amount of melt and regional uplift across the Vøring Margin. By calculating the amount of partial melting, they concluded that an asthenospheric thermal anomaly of about +50°C is necessary.

Seismic data in the Vøring Basin (Skogseid and Eldholm, 1989) and the Møre Basin (Myhre et al., 1989) document that the construction of the marginal highs was preceded by intrusive activity which progressively moved up-section and was focused towards the line of continental breakup. The time between initial rift-related faulting and final breakup may have been as long as 17 Ma if the observed faults are dated as old as Campanian/Maastrichtian as proposed by Skogseid et al. (in press) In the models we have placed the initiation of Cenozoic rifting at 65 Ma, the Cretaceous-Tertiary boundary.

This episode is characterized by high amounts of extension, with β values ranging from 1.4 in the western Rås Subbasin to 2.2 in the Fenris Graben (Table 4). Extension of this magnitude is not reflected in the observed faults. However, rift-related faulting

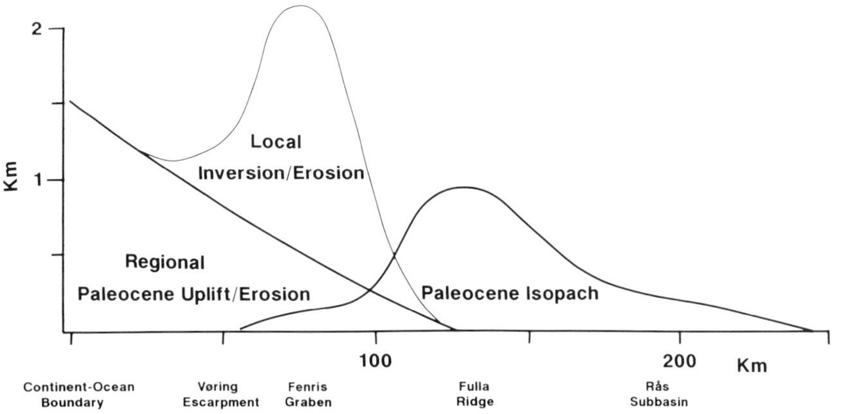

Fig. 17. Diagram showing the general relationship between early Tertiary uplift/erosion and Paleocene redeposition.

is normally associated with syn-tectonic subsidence. But Fig. 7 shows that this subsidence is almost at a minimum during the early Cenozoic rifting. Because extension took place as gradual stretching associated with intrusions and heating over a period of 17 Ma, we question whether relatively unconsolidated Cretaceous shales would have deformed in a manner resolvable as faults in seismic data, instead undergoing ductile deformation. The Cretaceous sequence generally shows poor internal stratification.

The β values derived from subsidence only (Table 4) may be reduced by taking into account the intrusive activity expressed as dikes in the lower crust and sills in the sediments (Royden et al., 1980). However, a reduction in the stretching estimates is not supported by the 4–6 km crustal thickness following early Cenozoic rifting. This implies relatively high rates of extension and thinning.

With reference to eqn. 2, which describes the initial "fault-controlled" subsidence due to extension, the modelling in the Fenris Graben yields an initial uplift component consistent with the local inversion within the graben as observed on seismic data (Fig. 10). This local inversion, which is superposed on the regional uplift (Figs. 11a and 17), is caused in part by the large pre-Cenozoic depth to basement in the graben (cf. eqn. 2), equivalent to a much reduced pre-rift crustal thickness (McKenzie, 1978).

In addition the high magmatic production would have affected both the initial and total tectonic subsidence/uplift (Foucher et al., 1982). To examine the influence of magmatism on the early Cenozoic β estimates in the Fenris Graben we isostatically corrected the total tectonic subsidence (Fig. 7) for the presence of the magmatic underplating. This gives an increase, ΔS:

$$\Delta S = H_{\rm mag} \frac{\rho_{\rm m} - \rho_{\rm mag}}{\rho_{\rm m} - \rho_{\rm w}} \quad (5)$$

where $H_{\rm mag}$ is the magma thickness, and $\rho_{\rm mag}$, $\rho_{\rm m}$ and $\rho_{\rm w}$ are the densities of magma, mantle and water, respectively. Using the values in Table 1 for mantle and water, 3–4 km for $H_{\rm mag}$ and 2.8–2.9 g cm^{-3} for $\rho_{\rm mag}$ (McKenzie, 1984), ΔS is 500–900 m with a corresponding increase in β.

In addition to magmatic effects, lateral heat flow will dominate the heat budget for narrow basins such as the Fenris and Hel Grabens (Alvarez et al., 1984). This has two important consequences: the total subsidence will be approached much faster than indicated in eqn. 4, and the lateral heat flow will cause syn-rift thermal uplift of the flanks. A first-order estimate of the rapid cooling effect was obtained by approximating $S_{\rm tot}$ in eqn. 3 based on the present total subsidence. Adding this correction to ΔS calculated above, we obtain an early Cenozoic β factor of 1.9–2.4. A flank uplift at the adjacent Fulla Ridge of about 300–400 m is inferred (Alvarez et al., 1984) and in good agreement with observations.

We believe these calculations made at selected locations (Fig. 7) provide first-order estimates of effects due to rapid cooling and magmatic underplating along the entire length of these grabens because the Cenozoic subsidence, graben width and underplated "magma thickness" vary little along strike. The local inversion of the Fenris Graben, due to the underplating and "infill" of the Moho relief, probably will not be reversed during the later subsidence as the underplated magma represents an actual growth of the crustal thickness.

Apparent differential vertical movements in the early Tertiary are also observed along the western flank of the Fulla Ridge/Nyk Syncline (Fig. 13). These movements are related to similar processes of differential crustal thinning as discussed at the Bodø High. The basement block beneath the western Fulla Ridge has probably experienced syn-tectonic uplift or differential subsidence in the early Tertiary, and formation of the Nyk Syncline may be due, in part, to these relative vertical movements. Similarly, we observe a correlation between an inverted Cenozoic structure along the Fles Fault Zone (Fig. 13) and deeper basement structures. This inverted area, associated with reverse faults and termed the Helland–Hansen Arch (Blystad et al., in press.), however, is marginally within the area that experienced early Cenozoic extension and it may be difficult to document syn-tectonic uplift.

Petroleum potential

Hydrocarbon exploration in parts of the eastern Vøring Basin will provide valuable information when assessing the hydrocarbon potential in the western Vøring Basin. However, the impact of the early Cenozoic rifting and associated magmatic activity must also be considered.

The main source rock for oil in the North Sea and offshore mid-Norway is the Kimmeridgian hot-shale (Faber and Stahl, 1984; Baird, 1986). These organic-rich sediments are also found onshore East Greenland and have been correlated with the North Sea sequences both stratigraphically and geochemically. This indicates deposition under similar environmental conditions throughout the North Atlantic realm (Requejo et al., 1989). In addition to this source, we speculate that potential source rocks may have been deposited within basins with restricted water circulation conditions during two other periods. During the early Cretaceous, favourable conditions for source rock deposition may have existed in the Fenris and Hel Grabens, which were separated from the eastern basins by the Fulla Ridge–Bodø High complex. During the Paleocene and earliest Eocene a land area along the central axis of the developing Norwegian-Greenland Sea may have created conditions for restricted basins along the present margin (Eldholm et al., 1989). This was a period dominated by a hot, humid climate and high rates of biogenic production (Froget et al., 1989). In particular, the Paleocene–early Eocene Vøring Basin may have been closed towards the north, as the margin approached the uplifted Lofoten–Vesterålen area.

From seismic data we infer the existence of potential reservoir rocks deposited in two different stratigraphic intervals and tectonic settings. One is the product of the major erosional event caused by early Cenozoic extension and uplift. These sediments, derived from the area west of the Fulla Ridge, were transported eastward and deposited in the Vøring Basin where they reach a maximum thickness of about 1500 m (Fig. 17). Using the Paleocene and Eocene sands derived from the East Shetland Platform and deposited in the Southern Viking Graben (Lovell, 1984; Conort, 1986), or even more likely the sands within the thick (3000 m) Paleocene sequence in the Flett Basin (Hitchen and Ritchie, 1987) as analogs, we infer sands within the Vøring Basin Paleocene sequence. During the late Paleogene and Neogene the western Vøring Basin subsided rapidly and marine shales provide a suitable cap rock. Sand bodies may also exist along both sides of the Fulla Ridge and locally along the Bodø High. These highs seem to have received little to no sediments throughout the early Cretaceous, because they experienced local rift-related uplift and, thus, were prone to erosion. These sequences may be genetically correlated with the Cenomanian and Turonian sand-prone deposits drilled west of the Nordland Ridge (Hastings, 1986). Cretaceous marine shales interfingering with the sand lobes could form suitable cap rocks for these possible reservoirs.

The main uncertainty in evaluating the area's hydrocarbon potential are related to quantifying the amount of extension and the associated thermal influence. Too many parameters are unknown making it impossible to determine accurate temperatures and degrees of maturity. In particular we have poor control on the actual distribution of different lithologies. Thus, conductivity may exceed acceptable ranges for one or more units, which could alter temperatures dramatically. However, we tentatively infer that the potential reservoirs are promising, as shown in Fig. 18.

Summary and conclusions

Although uniform extension modelling is a simplification of the actual geodynamic processes, it provides a method for quantifying lithospheric extension and amounts of subsidence. Because many critical parameters for basin modelling, particularly thermal modelling, remain unknown in the Vøring Basin, our results are primarily presented as tectonic subsidence maps.

The tectonic evolution of the Vøring Basin comprises a series of predominantly extensional episodes since Devonian time, which governed the initial crustal and lithospheric configuration in the basin modelling. The two latest main episodes, the late Jurassic–early Cretaceous and early Cenozoic rifting, have had a dominant influence on the regional structural and subsidence pattern in the study area, whereas older structures are only exposed locally. However, the regional configuration of the crystalline basement surface can be derived from seismic refraction data.

β values for Jurassic–Cretaceous extension range from 1.1 to 1.7 along a basin transect. Maximum extension occurred in the Rås Subbasin and the

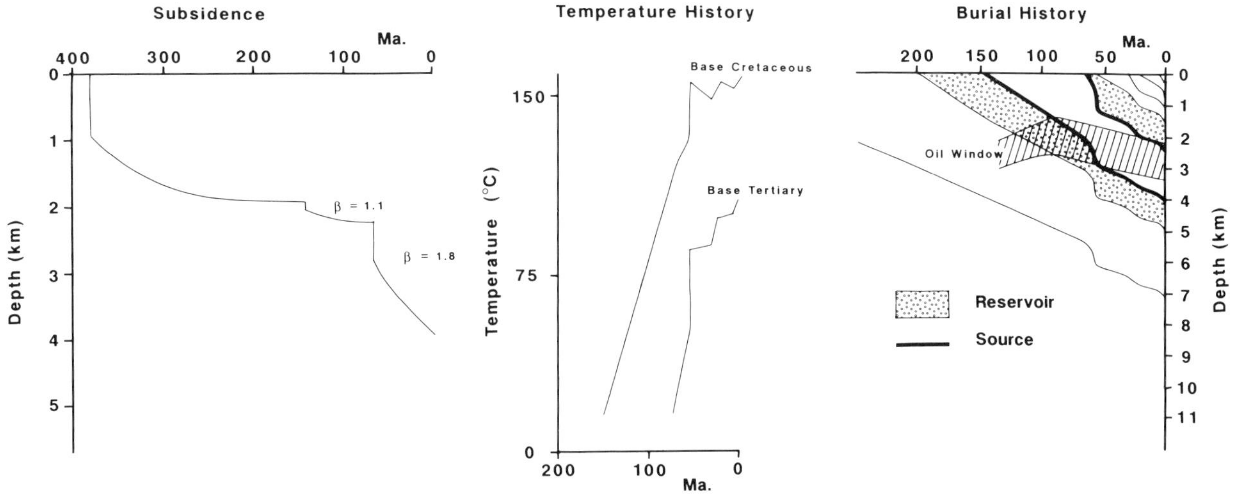

Fig. 18. Correlation between tectonic subsidence, temperature and burial history for an assumed well at the western Fulla Ridge. Lithology and conductivity parameters are given in Table 3.

Træna Basin, whereas the Fenris and Hel Grabens experienced minimal extension on the order of 1.1–1.3.

The locus of maximum early Cenozoic stretching shifted westward with respect to that of the Jurassic–Cretaceous episode. Estimates of β values range from 1.4 in the western Rås Subbasin to 1.9–2.5 in the Fenris Graben. This Cenozoic extension lasted for at least 10 Ma before it finally caused continental separation in the earliest Eocene. During the syn-rift phase the western Vøring Basin was uplifted, and an area about 250 km wide (Fig. 17) may have been exposed to erosion. Also associated with rifting was high magmatic activity which caused accretion of thick magmatic bodies at the base of the crust, emplacement of sills in Cretaceous sediments and regional fissure eruptions. During breakup a surge of extrusive volcanism formed large lava wedges represented as seaward dipping reflectors.

Local features in the Vøring Basin show subsidence patterns that deviate from a uniform extension model. Among these are:

(1) The Bodø High region which represents a "window" into the pre-Cretaceous part of the basin. Here the pre-Jurassic structural setting caused apparent differential crustal thinning during the main extensional episodes. The lithospheric thinning caused probable syn-tectonic uplift of the Bodø High block followed by post-tectonic differential subsidence.

(2) Offset between Moho shallowing and maximum basin subsidence may explain rift-flank uplift during Jurassic–Cretaceous rifting, particularly in the eastern Vøring Basin, the Halten and Dønna terraces and Nordland Ridge. This development suggests that simple shear models of basin subsidence should be considered.

(3) Early Cenozoic magmatic activity caused both subsidence and thermal anomalies not inherent in uniform extension modelling. In particular the high amounts of extension in the western Vøring Basin do not correlate with observed extensional faults. This is accounted for in part by assuming ductile deformation of the Cretaceous shales and by introducing extension by magmatic intrusion documented by the dikes and sills in the western Vøring Basin region. Moreover, the effects of underplating will increase the amount of initial uplift and decrease total subsidence.

(4) Finally, lateral heatflow might be important in the development of the "narrow" Fenris and Hel Grabens. The high amount of extension, β=1.9–2.5, in the Fenris Graben resulted in a crystalline crust only 4–6 km thick. This suggests that these regions were close to the stage of seafloor spreading in the late Paleocene. The narrow rift configuration caused considerable lateral heatflow and rapid cooling, which may have strengthened the lithosphere and subsequently lead to continental separation further west. Lateral heatflow is also indicated by uplift of the western Fulla Ridge and the relatively rapid thermal subsidence during Cenozoic time.

These results, which provide an initial quantitative description of the Mesozoic and Cenozoic evolution of the Vøring Basin, impact assessment of the petroleum potential of the basin. Although lithology and conductivity data are required in order to resolve the temperature history and allow reliable maturity estimates, the most prospective formations appear to be possible Cretaceous sands along the Fulla Ridge and Bodø High and the thick Paleocene sequence in the western Vøring Basin.

Acknowledgments

This study has benefited from the drilling results and, in particular, subsequent seismic studies presented in the Leg 104 Scientific Result Volume (Eldholm, Thiede, Taylor et al., 1989). We also acknowledge data made available by Statoil A.S. We thank Steinar Thor Gudlaugsson and Sverre Planke for stimulating discussions and are particularly grateful to Olav Eldholm for initiating the project and for providing advice and critical review. We also thank two anonymous reviewers for their critical and constructive reviews and for providing editorial expertise. This study was supported by grants from the Norwegian Research Council for Science and the Humanities, Statoil A.S. and VISTA.

References

Alvarez, F., Virieux, J. and Le Pichon, X., 1984. Thermal consequences of lithospheric extension over continental margins: the initial stretching phase. Geophys. J. Astron. Soc., 78: 389–411.

Austrheim, H., 1987. Eclogitization of lower crustal granulites by fluid migration through shear zones. Earth Planet. Sci. Lett., 81: 221–232.

Baird, R.A., 1986. Maturation and source rock evaluation of Kimmeridge Clay, Norwegian North Sea. Am. Assoc. Pet. Geol., Bull., 70: 1–11.

Blystad, P., Færseth, R.B., Larsen, B.T., Skogseid, J. and Tørudbakken, B., in press. Structural elements of the Norwegian continental shelf, Part 2. The Norwegian Sea. Norw. Pet. Dir., Bull., 6

Brekke, H. and Riis, F., 1987. Tectonics and basin evolution of the Norwegian Shelf between 62°N and 72°N. Nor. Geol. Tidsskr., 67: 295–322.

Bucher-Nurminen, K., 1991. Mantle fragments in the Scandinavian Caledonides. Tectonophysics, 190: 173–192

Bukovics, C. and Ziegler P.A. 1985. Tectonic development of the Mid-Norway continental margin. Mar. Pet. Geol., 2: 2–22.

Bukovics, C., Shaw, N.D., Cartier, E.G. and Ziegler, P.A., 1984. Structure and development of the Mid-Norway continental margin. In: A.M. Spencer et al. (Editors), Petroleum Geology of the North European Margin. Graham and Trotman, London, pp. 407–423.

Bøen, F., Eggen, S. and Vollset, J., 1984. Structures and basins of the margin from 62–69°N and their development. In: A.M. Spencer et al. (Editors), Petroleum Geology of the North European Margin. Graham and Trotman, London, pp. 3–28.

Caselli, F., 1987. Oblique-slip tectonics, mid-Norway shelf. In: J. Brooks and K.W. Glennie (Editors), Petroleum geology of North-West Europe. Graham and Trotman, London, pp. 1049–1063.

Celerier, B., 1986. Models for the evolution of the Carolina Trough and their limitations. Ph.D. Thesis, Massachusetts Institute of Technology, Cambridge, Mass., 206 pp.

Chroston, P.N. and Brooks, S.G., 1989. Lower crustal seismic velocities from Lofoten-Vesterålen, North Norway. Tectonophysics, 157: 251–269.

Conort, A., 1986. Habitat of Tertiary hydrocarbons, South Viking Graben. In: A.M. Spencer et al. (Editors), Habitat of Hydrocarbons on the Norwegian Continental Shelf. Graham and Trotman, London, pp. 159–170.

Dalland, A. and Worsley, D., 1988. A lithostratigraphic scheme for the Mesozoic and Cenozoic succession offshore Norway north of 62°N. Norw. Pet. Dir., Bull., 4 (ISBN 82–7275-241-9), 67 pp.

Dallmeyer, R.D., 1988. Polyphase tectonothermal evolution of the Scandinavian Caledonides. In: A.L. Harris and D.J. Fettes (Editors), The Caledonian–Appalachian Orogen. Geol. Soc. London, Spec. Publ., 38: 365–379.

Dewey, J.F., 1988. Extensional collapse of orogens. Tectonics, 7: 1123–1139.

Duindam, P. and Van Hoorn, B., 1987. Structural evolution of the western Shetland continental margin. In: J. Brooks and K.W. Glennie (Editors), Petroleum Geology of North-West Europe. Graham and Trotman, London, pp. 765–773.

Eldholm, O. and Mutter, J.C. 1986. Basin structure of the Norwegian Margin from analysis of digitally recorded sonobuoys. J. Geophys. Res., 91: 3763–3783.

Eldholm, O., Thiede, J., Taylor, E. et al. 1987. Proceedings of the Ocean Drilling Program, Initial Rep. 104, (Pt. A), 783 pp.

Eldholm, O., Thiede, J., Taylor, E., 1989. Evolution of the Vøring volcanic margin. In: O. Eldholm, J. Thiede, E. Taylor et al. (Editors), Proc. ODP, Sci. Results, 104, College Station, Texas, (Ocean Drilling Program), pp. 1033–1065.

Eldholm, O., Thiede, J., Taylor, E. et al. 1989. Proc. ODP, Sci. Results, 104, College Station, Texas, (Ocean Drilling Program), 1141 pp.

Eldholm, O., Skogseid, J., Sundvor, E. and Myhre, A.M., 1990. The Norwegian-Greenland Sea. In: A. Grantz, G.L. Johnson and J. Sweeney (Editors), The Arctic Ocean Region, Vol. L, The Geology of North America. Geological Society of America, Boulder, Colo., pp. 351–364.

England, P.C. and Houseman, G.A., 1988. The mechanics of the Tibetan Plateau. Philos. Trans. R. Soc. London, Ser. A, 326: 301–320.

Faber, E. and Stahl, W., 1984. Geochemical surface exploration for hydrocarbons in the North Sea. Am. Assoc. Pet. Geol., Bull., 68: 363–386.

Foucher, J.-P., Le Pichon, X. and Sibuet, J.-C., 1982. The ocean-continent transition in the uniform lithospheric stretching model: role of partial melting in the mantle. Philos. Trans. R. Soc. London, Ser. A, 305: 27–43.

Froget, C., Desprairies, A., Latouch, C. and Miallet, N., 1989. Paleoenvironmental significance of Cenozoic clay deposits from the Norwegian Sea: ODP Leg 104. In: O. Eldholm, J. Thiede, E. Tayloret al. (Editors), Proc. ODP, Sci. Results, 104, College Station, Texas, (Ocean Drilling Program), pp. 41–60.

Gabrielsen, R.H. and Robinson C., 1984. Tectonic inhomogeneities of the Kristiansund-Bodø Fault complex, off shore mid-Norway. In: A.M. Spencer et al. (Editors), Petroleum Geology of the North European Margin. Graham and Trotman, London, pp. 397–406.

Gabrielsen, R.H., Færseth, R., Hamar, G. and Rønnevik, H.C., 1984. Nomenclature of the main structural features on the Norwegian Continental Shelf north of the 62nd parallel. In: A.M. Spencer et al. (Editors), Petroleum Geology of the North European Margin. Graham and Trotman, London, pp. 41–60.

Goldschmidt-Rokita, A., Sellevoll, M.A., Hirschleber, H.B. and Avedik, F., 1988. Results of two seismic refraction profiles off Lofoten, Northern Norway. Nor. Geol. Under., Spec. Publ., 3: 49–57.

Gowers, M.B. and Lunde, G., 1984. The geological history of Trænabanken. In: A.M. Spencer et al. (Editors), Petroleum Geology of the North European Margin. Graham and Trotman, London, pp. 237–252.

Graham, R.H. and Robinson, N.D., 1989. A regional structural study of the Norwegian Continental Shelf between 62°50′ and 67°. Paper presented at the Norwegian Petroleum Society Workshop, 18–20 October, 1989, Stavanger.

Grønlie, A. and Roberts, D., 1987. Dextral strike-slip duplexes of Mesozoic age along the Hitra-Snåsa and Verran faults, Møre-Trøndelag Fault Zone, Central Norway. NGU-Rep., 87, 139 pp.

Hastings, D.S., 1986. Cretaceous stratigraphy and reservoir potential, mid Norway continental shelf. In: A.M. Spencer et al. (Editors), Habitat of Hydrocarbons on the Norwegian Continental Shelf. Graham and Trotman, London, pp. 287–298.

Haq, B.U., Hardenbol, J. and Vail, P.R., 1987. Chronology of fluctuating sea levels since the Triassic. Science, Reprint Ser., 235: 1156–1167.

Hinz, K., Dostman, H.J. and Hanisch, J. 1984. Structural elements of the Norwegian Sea continental margin, Geol. Jahrb., A75: 193–211.

Hitchen, K. and Ritchie, J.D., 1987. Geological review of the West Shetland area. In: J. Brooks and K.W. Glennie (Editors), Petroleum Geology of North-West Europe. Graham and Trotman, London, pp. 737–749.

Hossack, J.R., 1986. The geometry of listric faults in the Devonian basins of Sunnfjord, W. Norway. J. Geol. Soc. London, 141: 629–637.

Jamtveit, B., 1987. Metamorphic evolution of the Eiksunddal eclogite complex, Western Norway, and some tectonic implications. Contrib. Mineral. Petrol., 95: 82–99.

Larsen, H.C., Armstrong, G., Marcussen, C., Moore, S. and Stemmerik, L., 1989. Deep seismic data from the Jameson Land Basin, East Greenland. Abstracts, IFP/ILP /IUGS Exploration Research Conference, Arles. Terra Abstracts, 1.

Larsen, V.B., 1987. A synthesis of tectonically-related stratigraphy in the North Atlantic-Arctic region from Aalenian to Cenomanian time. Nor. Geol. Tidsskr., 67: 281–193.

Le Pichon, X. and Sibuet, J.-C., 1981. Passive margins: a model of formation. J. Geophys. Res., 86: 3708–3720.

Lovell, J.P.B., 1984. Cenozoic. In: K.W. Glennie (Editor), Introduction to the Petroleum Geology of the North Sea. Blackwell, Oxford, pp. 151–169.

Manby, G.M. and Hambrey, M.J., 1989. The structural setting of the Late Proterozoic tillites of East Greenland. In: R.A. Gayer (Editor), The Caledonide Geology of Scandinavia. Graham and Trotman, London, pp. 299–312.

McClay, K.R., Norton, M.G., Coney, P. and Davis, G.H., 1986. Collapse of the Caledonian orogen and the old red sandstone. Nature, 323: 147–149.

McKenzie, D., 1978. Some remarks on the development of sedimentary basins. Earth Planet. Sci. Lett., 74: 25–32.

McKenzie, D. 1984. A possible mechanism for epiorogenic uplift. Nature, 307: 616–618.

Mokhtari, M., Pegrum, R.M. and Sellevoll, M.A., 1989. A geophysical study of the Norwegian continental margin between 67° and 69°N. Seismological Observatory, Univ. of Bergen, Seismo-Series.

Morton, A.C. and Parson, L.M., 1988. Early Tertiary volcanism and the opening of the NE Atlantic. Geol. Soc. London, Spec. Publ., 39, 477 pp.

Mutter, J.C., 1984. Cenozoic and late Mesozoic stratigraphy and subsidence history of the Norwegian margin. Bull. Geol. Soc. Am., 95: 1135–1149.

Mutter, J.C., Talwani, M. and Stoffa, P.L., 1984. Evidence for a thick oceanic crust adjacent to the Norwegian margin. J. Geophys. Res., 89: 483–502.

Myhre, A.M., Skogseid, J., Eldholm, O., Faleide, J.I., Gudlaugsson, S.T., Planke, S. and Vågnes, E., 1989. Norwegian Continental margin: structural and stratigraphical styles. 28th Int. Geol. Congr., Washington, D.C. (abstract).

Nelson, P.H.H. and Lamy, J.M., 1987. The Møre/West Shetland area: a review. In: J. Brooks and K.W. Glennie (Editors), Petroleum Geology of North-West Europe. Graham and Trotman, London, pp. 775–784.

Norton, M.G., 1986. Late Caledonian extension in western Norway: A response to extreme crustal thickening. Tectonics, 5: 195–204.

Pedersen, T. and Skogseid, J., 1989. Vøring Plateau volcanic margin: extension, melting and uplift. In: O. Eldholm, J. Thiede, E. Taylor et al. (Editors), Proc. ODP, Sci. Results, 104, College Station, Texas, (Ocean Drilling Program), pp. 985–991.

Planke, S., Skogseid, J. and Eldholm, O., 1991. Crustal structure off Norway, 62°–70°N. Tectonophysics, 189: 91–107.

Requejo, A.G., Hollywood, J. and Halpern, H.I., 1989. Recognition and source correlation of migrated hydrocarbons in upper Jurassic Hareelv Formation, Jameson Land, East Greenland. Am. Assoc. Pet. Geol., Bull., 73: 1065–1088.

Roll, A., 1974. Langfristige Reduktion der Mächtigkeit von Sedimentgesteinen und ihre Auswirkung — eine Übersicht. Geol. Jahrb., Ser. A, 14, 76 pp.

Royden, L., Sclater, J.G. and von Herzen, R.P., 1980. Continental margin subsidence and heatflow: important parameters in formation of petroleum hydrocarbons. Am. Assoc. Pet. Geol., Bull., 64: 173–187.

Rønnevik, H.C., Eggen, S. and Vollset, J., 1983. Exploration of the Norwegian Shelf. In: J. Brooks (Editor), Petroleum Geochemistry and Exploration of Europe. Blackwell, Oxford, pp. 71–94.

Sawyer, D.S., 1985. Total tectonic subsidence: a parameter for distinguishing crust type at the U.S. Atlantic continental margin. J. Geophys. Res., 90: 7751–7769.

Sclater, J.G. and Christie, P.A.F., 1980. Continental stretching: an explanation of the post-mid-Cretaceous subsidence in the central North Sea basin. J. Geophys. Res., 85: 3711–3739.

Skogseid, J. and Eldholm, O., 1987. Early Cenozoic crust at the Norwegian continental margin and the conjugate Jan Mayen Ridge. J. Geophys. Res., 92: 11471–11491.

Skogseid, J. and Eldholm, O. 1989. Vøring Continental Margin: seismic interpretation, stratigraphy and vertical movements. In: O. Eldholm, J. Thiede, E. Taylor et al. (Editors), Proc. ODP, Sci. Results, 104, College Station, Texas, (Ocean Drilling Program), pp. 993–1030.

Skogseid, J., Pedersen, T., Eldholm, O. and Larsen, B.T., in press. Tectonism and magnatism during during NE Atlantic continental break-up: The Vøring Margin. In: B.C. Storey, T. Alablaster and R.J. Plankhurst (Editors), Magmatism and the Causes of Continental Break-up. Geol. Soc. London, Spec. Publ.

Spencer, A.M. et al. (Editors), 1984. Petroleum Geology of the North European Margin. Graham and Trotman, London, 436 pp.

Spencer A.M. et al. (Editors), 1986. Habitat of Hydrocarbons on the Norwegian Continental Shelf. Graham and Trotman, London, 354 pp.

Steckler, M.S. and Watts, A.B., 1978. The Gulf of Lion: subsidence of a young continental margin. Nature, 287: 425–429.

Surlyk, F., Piasecki, S., Rolle, F., Stemmerik, L., Thomsen, E. and Wrang, P., 1984. The Permian basin of East Greenland. In: A.M. Spencer et al. (Editors), Petroleum Geology of the North European Margin. Graham and Trotman, London, pp. 303–315.

Talwani, M. and Eldholm, O., 1972. The continental margin off Norway. A geophysical study. Geol. Soc. Am. Bull., 83: 3375–3608.

Talwani, M., Udintsev, G. et al. 1976. Initial Reports of the DSDP, Vol. 38. U.S. Government Printing Office, Washington, D.C., 1256 pp.

Turcotte, D.L., Haxby, W.F. and Ockendon, J.R., 1977. Lithospheric instabilities. In: M. Talwani and Pitman W.C. III (Editors), Island Arcs, Deep Sea Trenches and Back-arc Basins. Am. Geophys. Union, Maurice Ewing Ser., 1: 63–69.

Watts, A.B. and Thorne, J.A., 1984. Tectonics, global changes in sea-level and their relationship to stratigraphic sequences at the U.S. Atlantic continental margin. Mar. Pet. Geol., 1: 319–339.

White, R.S., Spence, G.D., Fowler, S.R., McKenzie, D.P., Westbrook, G.K. and Bowen, A.N., 1987. Magmatism at rifted continental margins. Nature, 330: 439–444.

Yielding, G. and Roberts, A., 1992. Footwall uplift during normal faulting — implications for structural geometries in the North Sea. In: R.M. Larsen, H. Brekke, B.T. Larsen and E. Talleraas (Editors), Structural and Tectonic Modelling and its Application to Petroleum Geology. Norwegian Petroleum Society (NPF), Special Publication 1. Elsevier, Amsterdam, pp. 289–304 (this volume).

Ziegler, P.A., 1988. Evolution of the Arctic-North Atlantic and the western Tethys. Am. Assoc. Pet. Geol., Mem., 43: 198 pp.

Zoth, G. and Haenel, R., 1988. Thermal conductivity. In: R. Haenel, L. Rybach and L. Stegena (Editors), Handbook of Terrestrial Heat-flow Density Determination. Kluwer Academic Publisher, Dordrecht, pp. 449–466.

J. SKOGSEID *Department of Geology, University of Oslo, P.O. Box 1047, Blindern, 0316 Oslo 3, Norway*
T. PEDERSEN *Department of Geology, University of Oslo, Oslo, Norway*
V.B. LARSEN *Statoil A/S, Stavanger, Norway*

The orientation of minor fault plane striae and the associated deviatoric stress tensor as a key to the fault geometry in part of the Møre–Trøndelag Fault Zone, on-shore central Norway

D. Bering

The Verran Fault and the associated Hitra Fault are prominent regional features within the northern part of the Møre–Trøndelag Fault Zone. On the northern side of the Beitstadfjord the main Verran Fault Zone consists of at least three main parallel faults. These structures, together with smaller interconnecting faults, are considered to form part of an intricately linked system. The faults are clearly post-Caledonian in age with a poly-phasal brittle deformation history which can be seen from the pervasive brecciation on all scales together with hydrothermal alteration of the rock products. Measurements of the orientation of slickensides and other kinematic indicators on minor polished fault planes establish a system of slip-vectors with an oblique-slip normal to pure dip-slip normal main component. The slip-vectors together with the orientation of the minor slip planes can be used to compute the deviatoric stress tensor responsible for the deformation. It is concluded that the fault pattern along the main Verran Fault can best be interpreted as a set of dip-slip normal faults which are a response to pure crustal extension. The minor interconnecting faults between the major zones are considered to represent transfer faults linking the motions on the main faults.

Introduction

The purpose of this paper is to demonstrate the importance and usefulness of establishing the paleostress responsible for the deformations when discussing the geometry and movements along major fault systems which lack mappable marker horizons. The technique is applied to the Verran Fault Zone which is part of the Møre–Trøndelag Fault Zone (Fig. 1). The study of this major fault zone is the subject of ongoing work and a more detailed description of the field relationship, an assessment of various methods of paleostress estimation and a discussion of the different stress systems active along the zone, will be the subject of future papers.

The Møre–Trøndelag Fault Zone (Gabrielsen and Ramberg, 1979; Gabrielsen et al., 1984) was defined by satellite imagery analysis and consists of strong lineaments following the Caledonian structural grain from Ålesund on the coast of Norway through Møre and Trøndelag (Fig. 1). Based on the similarity of the geometry and orientation, these authors linked the lineaments with established faults on land and with the development of fault structures off-shore.

On the basis of detailed analyses of Landsat TM image data, Rindstad and Grønlie (1986) and Grønlie and Roberts (1989) established an interesting pattern of anastomosing major and minor faults in the Verran area north of the Beitstadfjord in the northern part of Trøndelag (Fig. 1). This pattern resembles the "strike-slip duplex" structures of Woodcock and Fischer, 1986, and, together with the field observations of strong cataclasis and the development of thick breccias along the majority of the faults, led the authors to postulate that major dextral lateral movements are responsible for this structure. This would imply that at least part of the Møre–Trøndelag Fault Zone, with its off-shore extension, had a strong component of strike-slip movements.

However, the bedrock consists of ductile polydeformed ortho- and paragneisses lacking typical marker horizons, and the relative motions along the different faults are thus difficult to establish. This paper is a part of a major work in which orientation data for minor fault planes with striae are used to classify the individual slip and to establish the stress system(s) responsible for the deformation and, thus, possibly provide arguments for the fault geometry and kinematics operating on a larger scale.

Structural and Tectonic Modelling and its Application to Petroleum Geology edited by R.M. Larsen, H. Brekke, B.T. Larsen and E. Talleraas. NPF Special Publication 1, pp. 83–90. Elsevier, Amsterdam.

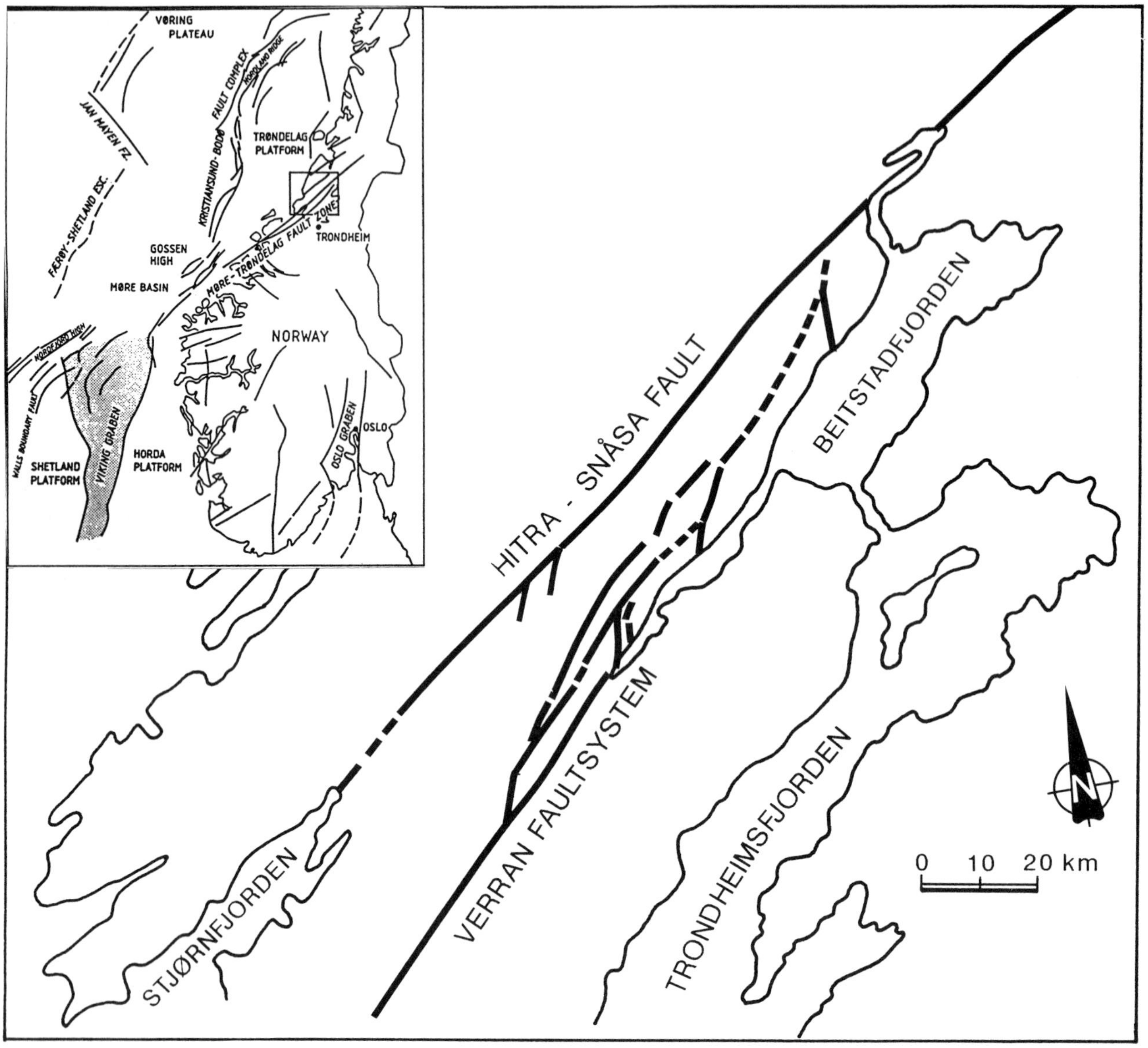

Fig. 1. The Møre–Trøndelag Fault Zone in relation to the main structural elements off-shore Norway. The main map shows the structural features north of the Trondheimsfjord discussed in the text. Note the three almost parallel faults with the minor interconnecting faults resembling strike-slip duplexes (see text for discussion).

The method

The fault products along the major and minor faults in the Verran System are characterized by polyphasal brecciation, hydrothermal alteration and local prehnite and stilbite veining (Grønlie and Roberts, 1989). It is evident from the field that the bedrock has undergone more than one phase of brittle deformation. One earlier phase is marked by quartz-filled extensional veining and rather coherent fault products with quartz-epidote matrix.

The structures considered in this paper cut the earlier ones, and are considered to be of the second (or third) phase in the brittle deformation history. They consist of numerous small discrete fault planes in dm and metre scales. They are characterized by polished hard greenish surfaces developed in coherent fine-grained roch powder. The planes show striae developed as rather coarse grooves and locally occurrence of kinematic indicators, such as comb structures, tool marks etc. (Hancock and Barke, 1987), are observed. The similarity of the slip planes and their structural style are believed to reflect the same deformational event.

The stilbite veining serves as a further time constraint as it is sometimes cut and polished by the slip planes: in other situations the slip planes are cut by small extensional veins filled with stilbite.

After development of the slip planes, the major fault breccias show evidence of one (or possibly two)

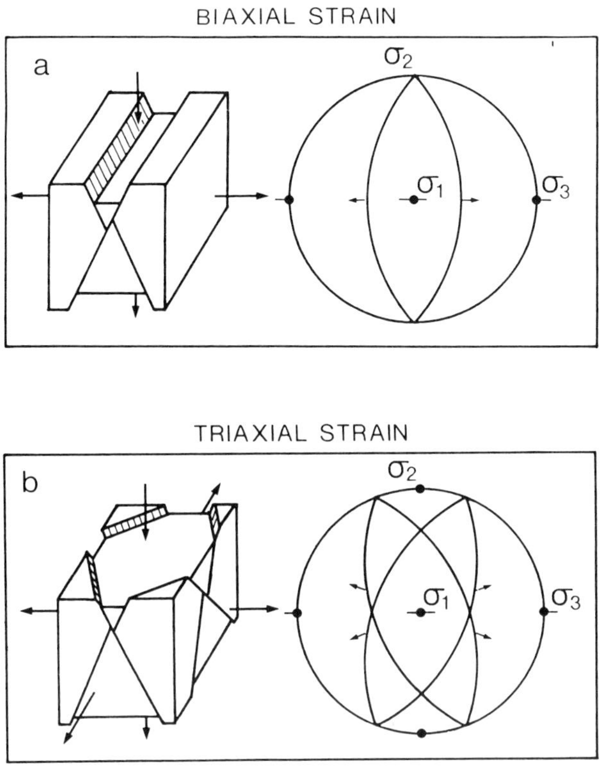

Fig. 2. The three principal compressive stress axes in relation to the orientation of conjugate faults under (a) biaxial, and (b) triaxial strain condition. Modified from Hancock, 1985.

later phase of brittle deformation. This deformation is marked by development of irregular kink bands in the dm and metre scales and produced a rather loose and incoherent late brecciae.

The latest deformation in the area is not just confined to the fault breccias, but is also observed in the relatively undisturbed bedrock. The structures developed are mostly small fractures and fissures, but some fault planes with slickensides show evidence of movements. Both in their orientation and the associated mineralization, these structures are evidently separated, both in time and association, from the brittle deformation structures related to the development of the thick fault breccias, and will not be analysed further in this contest. The deformational events subject to analysis in this paper are, thus, considered to be somewhere in the middle of the brittle deformation history of the major fault zones; not the earliest and not the latest, but they do reflect an event which had a great influence in the development of the thick breccias along the major faults as seen on a regional scale.

The movement along the regional faults is thus accommodated (at least in part) by development of the minor slip planes. The orientation and type of movement along one particular small fault may be just a local accommodation, not necessarily reflecting the movement on a larger scale. However, taken together at different localities, the minor fault planes must in some way reflect the overall stress-strain relationship for that particular deformation on a larger scale. Classification of the individual planes into normal, reverse, dextral and sinistral slip planes, and computation of the deviatoric stress tensor responsible for their development, will therefore contribute to the discussion of the sense of movements and the kinematics of the faults on a regional scale.

From the work of Anderson (1942) it is known that the orientation of faults is geometrically linked to the orientation of the principal axes of the stress ellipsoid responsible for the deformation. In a volume of rock deformed under plane strain condition (Fig. 2a), one conjugate fault system will develop. These faults will be oriented in such a way that the intersection line between them will define σ_2, σ_1 will bisect the acute angle, and σ_3 will bisect the obtuse angle ($\sigma_1 > \sigma_2 > \sigma_3$ being the three principal directions in the stress ellipsoid responsible for the deformation).

Under triaxial strain conditions, the situation is more complicated (Fig. 2b). Two systems of conjugate faults must develop to accommodate the strain. σ_2 will in this situation bisect the acute angle between the lines of intersection of the two systems. Likewise, σ_3 will bisect the larger angle and σ_1 will be oriented normal to the σ_2–σ_3 plane (Reches and Dieterich, 1983).

Most natural faults are oblique-slip and do not frequently occur in conjugate systems, and the symmetry criterion above is seldom met. Based on the Bott (1959) analysis, many methods of stress estimation assume that the slip along a fault occurs in the direction of the resolved shear stress (e.g., Carey and Bruinier, 1974; Angelier, 1979; 1984; Gephart and Forsyth, 1984; Reches, 1987; and others), and several inversion techniques have been developed. The one used here is the direct inversion method of Angelier et al. (1984), which minimizes a function that depends on both the slip-shear angle and the shear.

Application of paleostress estimates to the Verran fault zone

In the Fosen area, northwest of the Trondheimsfjord, two main fault systems are observed — the Hitra Fault and the Verran Fault (see Figs. 1 and 3). The Hitra Fault is considered to be an older system than the Verran Fault, mainly on the basis of deformation products (see also Grønlie and Roberts, 1989). The fault rock along this zone displays older elements of ductile-semibrittle deformation of the minerals quartz and feldspar and almost always have

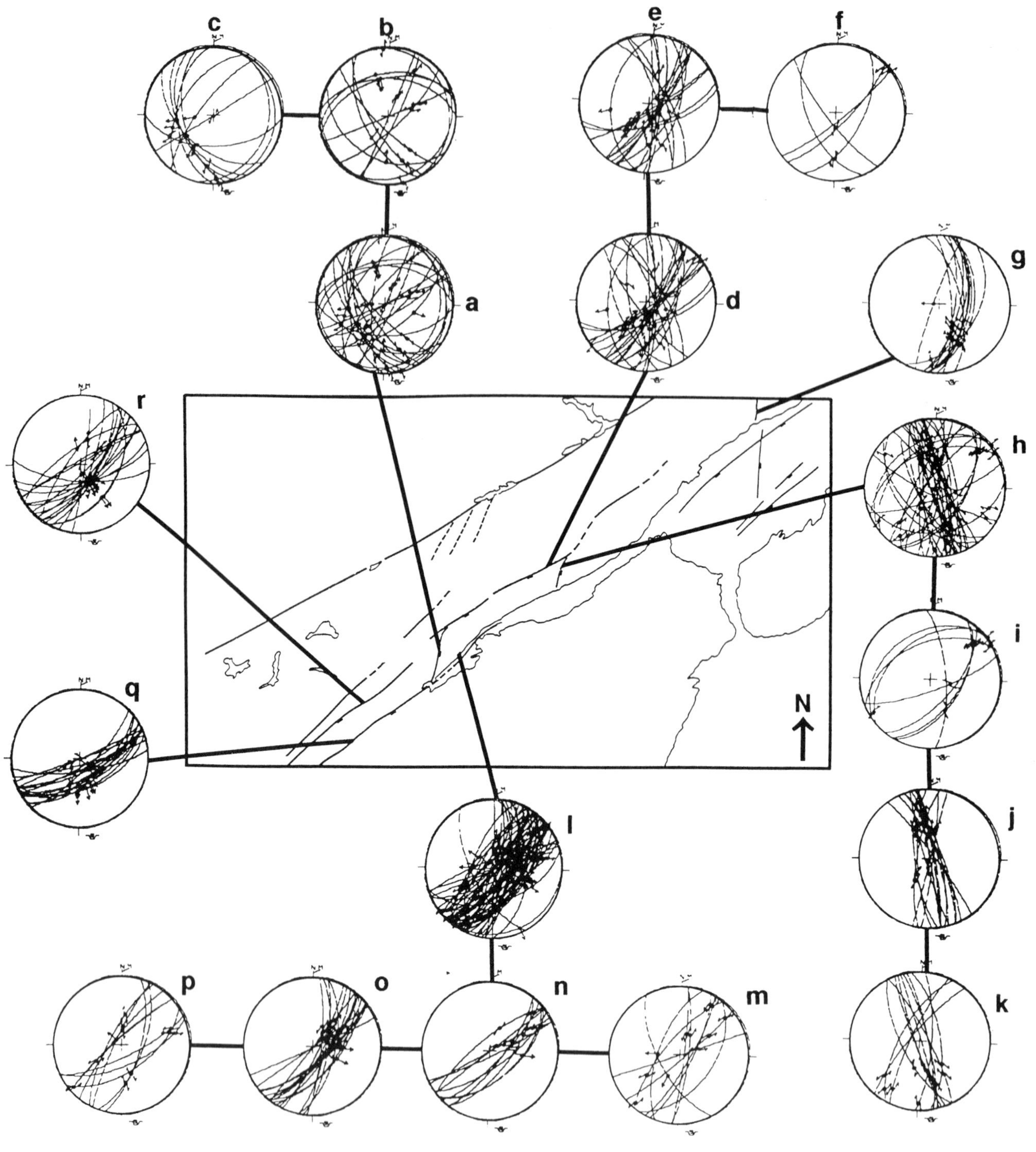

Fig. 3. Map of the Verran area showing the location of the Hitra and Verran Fault. The Hitra Fault displays Riedel-like shears (upper central) and the Verran Fault System consists of three almost nearly parallel faults with minor interconnecting faults. Eaqal-area (lower hemisphere) projection of minor faults with their associated striae from the main localities a, d, g, h, l, q and r are shown together with the different populations after separation. See text for discussion.

a fine-grained cataclastic matrix. Very little veining is observed, except for localities where the fault has been reactivated. Analyses of the stress system responsible for the Hitra Fault are not presented here. It should be mentioned, however, that the TM image displays some Riedel-like shears in the central part of the fault, indicating that the intermediate stress axis could have had a near vertical orientation at some point in the history of the fault, thus indicating a strike-slip movement.

The Verran Fault system consists of three almost parallel major faults (Fig. 3), with at least three minor interconnecting faults between the two larger ones. In the Beitstadfjord, seismic profiling by NGU

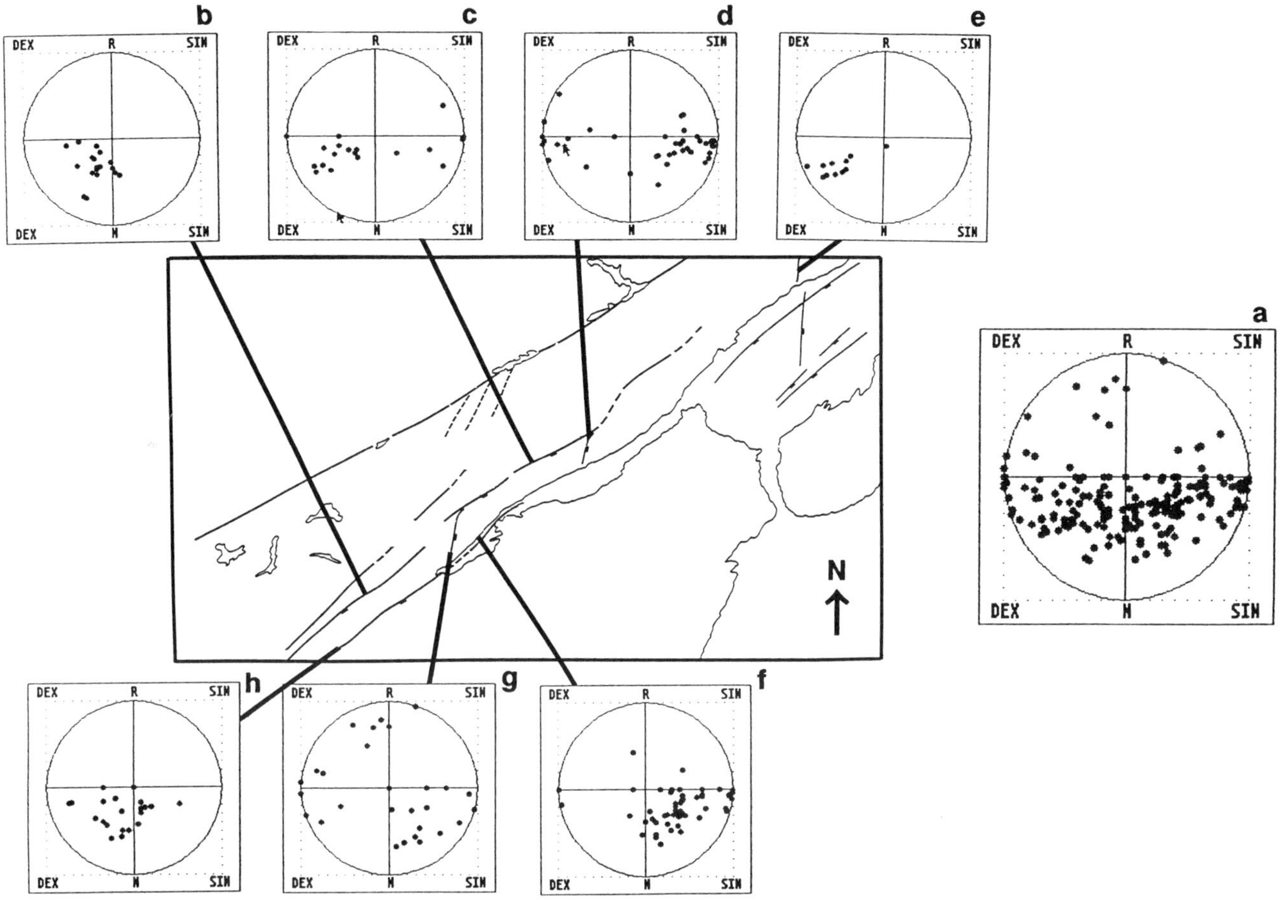

Fig. 4. Classification of the minor fault planes into dip-slip normal or reverse, sinistral or dextral strike-slip faults (from Angelier, 1984). In (a) faults with a high degree of certainty regarding the relative movements from the whole area are plotted, and diagrams b–f show the areal distribution of the different types. The majority of the faults show oblique-slip normal movements although some pure strike-slip faults are recognized, particulary in the central part of the main zone and in the interconnecting faults.

(Bøe, 1989) shows a series of faults developed in what is considered to be rocks of Middle Jurassic age. These faults follow the trend to both the major and minor ones on land (Fig. 3).

All the faults are marked in the field by strong brecciation and development of numerous minor slip planes. Although the slip planes show a wide variety in orientation, the majority of the smaller faults are oriented almost parallel to the larger structures. (Fig. 3, stereograms a, d, g, h, l, q and r). Where fault populations show conjugate systems, the intersection lines of these faults are parallel with the larger structure (i.e., stereograms j, k, m and p in Fig. 3). There is then a strong indication of a link between the development of the major and minor structures.

Based on the measurements of the attitude of the minor faults and the direction of the slip vector observed from the striations, all the faults have been classified as normal, reverse, dextral and sinistral relative movements. By using the principles drawn up by Angelier et al., 1984, a good picture of fault movements is given by computing the distribution of transverse and lateral movements, expressed as ratios relative to the net slip. The components are computed as simple trigometric functions of the dip of the fault and the pitch of the striation. As the sum of the three vectors that represent vertical, horizontal transverse and horizontal lateral components of slip equals the net slip vector, all the points in the diagram fall within a circle (Fig. 4)

In Fig. 4a faults from the whole area are plotted. Only those faults with a very high degree of certainty regarding the relative movements are used. Although faults with all types of relative motion are found, oblique-slip normal faults are in the majority. Of the 206 faults plotted, 82% show a normal-slip component. The oblique-slip and pure strike-slip faults show a near equal distribution of sinistral and dextral components.

Looking at the areal distribution of the various faults (Fig. 4b–h), some general pattern can be seen. The central part of the southern main fault (the Verran Fault proper), Fig. 4h, contributes most to the population of oblique-slip faults with a sinistral component. In two of the interconnecting faults, i.e., Fig. 4g and d, faults with a larger degree of "obliqueness" are found, and some classify as pure strike-slip faults with both sinistral and dextral relative movements.

From the stereograms in Fig. 3, it is evident that more than one population exists in most of the main localities chosen for this analysis (i.e., Fig. 3a, d, h, l). Inversion of these unmodified data sets will obviously give a solution for the deviatoric stress tensor which has no geological meaning. One must first try to separate the data sets into homogeneous sub-sets. The separation technique adopted here is firstly based on the small amount of field evidence of cross-cutting relationships that exists in a few localities. Subsequently, the whole population from one locality is inverted and the mean stress tensor is computed. This tensor is then used, for each fault plane, to compute the theoretical slip vector. The angular difference between this and the actual slip direction (from the striations), called the misfit angle, is then a measure of how good the tensor solution is for that particular fault. Faults with high misfit angles are then eliminated from the population and a new stress tensor is computed. By repeatedly eliminating faults, and accepting a certain amount of misfit (10–15 degrees in this analysis), one obtains a population of faults with one homogenous solution. The faults which were eliminated are then used for computing a second tensor, with the same process of eliminating faults with high misfit angles as in the first case and the process is repeated. After computing a series of tensors in this way, one must check whether the faults could belong to more than one solution, indicating that one or more of the tensors are of little geological meaning. This process is very important, particularly in cases where field evidence is scarce or lacking.

The results of this separation procedure are shown in Fig. 3, where the sub-sets of fault slip data are plotted. The resulting deviatoric stress tensors are represented in Fig. 5a–n by plotting the orientations of the principal axes in the different stress ellipsoids. The stereograms also show the directions of the horizontal extension and compression.

Most of the data sets from the major NE–SW striking faults give solutions with a near vertical orientation of the maximum principal axis, σ_1, (i.e., Fig. 5a, c, i, j, k, l, m, n). The three solutions i, j and k from the same locality have almost the same orientations and probably represent the same tectonic event. One solution from one locality (d) has the intermediate principal axis, σ_2, in the vertical position, indicating strike-slip movements.

Looking at the minor interconnecting faults, one gets a more complicated picture. Although some solutions show pure extension (a and e, and to some extent f and g), one solution has the minimum principal axis, σ_3, near the vertical (Fig. 5b), showing that a pure compressional event existed locally. Solution h gives an intermediate position of the principal axes with both extension and compression in the horizontal plane, i.e., strike-slip faulting.

The horizontal extension directions from localities in the major NE–SW striking faults (Fig. 5l–n) have the same NW–SE direction. Solution c gives a uniaxial stress ellipsoid and an extension direction cannot be inferred. The central part of the southern fault, together with this general extension direction, also shows an almost E–W to ENE–WSW direction, indicating that the region as a whole departs from a plane strain condition (see Fig. 2 and discussion above).

The paleostress pattern from the whole region then gives a picture of a general stress regime with the maximum principal compressive stress, σ_1, oriented almost vertical. This stress regime will then generally favour development of normal, more or less dip-slip faults. The departure from this general stress state found locally, is believed to represent local stress, variable both in time and space, induced by movements along the major faults. This conclusion is also supported by the fact that the minor interconnecting faults show the largest degree of departure from the regional stress state. It is believed that the overall fault geometry in the context of this stress pattern is that of a large system with parallel major faults with dip-slip movements. The interconnecting faults in this picture will represent transfer faults linking the relative motions on the bigger structures.

In a large fault system like the Møre–Trøndelag Fault Zone, with its long history, it is likely that different stress systems have been active at different times and at different places, and it is not believed that the analysis presented above represent other than a part of the total deformation history. It cannot, then, be ruled out that lateral relative movements have been active at some time in the region, although no field evidence to support this has been found in the Verran Fault System. On the basis of the data and analysis presented above, it is believed that the mappable fault pattern now met within the Verran area is the response to pure crustal extension.

Conclusions

The orientation of minor fault planes along with their associated striae in fault zones can be a valuable tool in the discussion of the kinematics of large

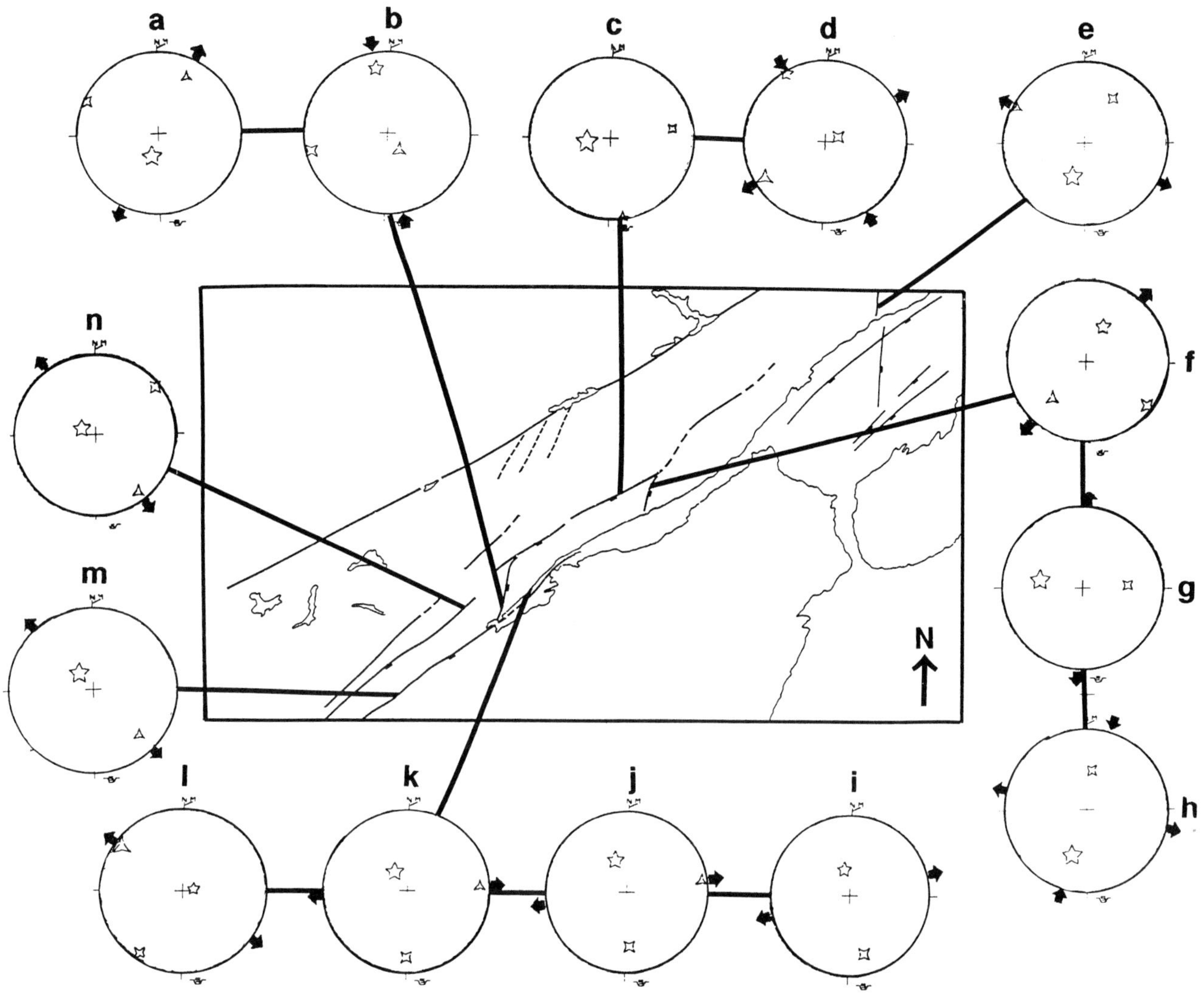

Fig. 5. The results after computation of the deviatoric stress tensors (a–n) from the fault populations in Fig. 3, shown here as an orientation of the principal stress axes (symbols: five-, four- and three-legged stars mark the position of σ_1, σ_2 and σ_3 axes, respectively, in the stress ellipsoids. Also shown (arrows) are the direction of the horizontal extension (compression) directions.

faults. This is particularly important in areas lacking mappable marker horizons and where the fault pattern can be interpreted on a large scale both as strike-slip or dip-slip movement.

In the case of the Verran area, the analysis above points to the conclusion that the major and minor faults in the area are of dip-slip normal type, and the fault pattern on the map scale is not evidence for major strike-slip movements along the Møre–Trøndelag Fault Zone.

Acknowlegdement

The author wishes to thank the Norwegian oil companies Norsk Hydro A/S and Statoil for financing the work. The cooperation with B.T. Larsen and R. Færseth is especially acknowledged.

References

Anderson, E.M., 1942. The Dynamics of Faulting. Olivier and Boyd, Edinburgh, 1st ed.

Angelier, J., 1979. Determination of the mean principal directions of stresses for a given fault population. Tectonophysics, 56: T17–T26.

Angelier, J., 1984. Tectonic analysis of fault slip data sets. J. Geophys. Res., 98(B7): 5835–5848.

Bott, M., 1959. The mechanisms of oblique faulting. Geol. Mag., 96.

Bøe, R., 1989. NGF Vintermøte, Geolognytt Nr. 22 (abstract).

Carey, E. and Brunier, B., 1974. Analyse théorique et numerique d'un modèle mécanique élémentaire appliqué à l'étude d'une population de failles. C.R. Acad. Sci. Paris, 279D: 891–894.

Gabrielsen, R.H. and Ramberg, I.B., 1979. Fracture patterns in Norway from Landsat imagery: results and potential use. Proc. Norwegian Sea Symp. Tromsø, 23: 1–28.

Gabrielsen, R.H., Færseth, R., Hamar, G. and Rønnevik, H., 1984. Nomenclature of the main structural features of the Norwegian Continental Shelf north of the 62nd parallel. Pet.

Geol. N. Eur. Margin, pp. 41–60.
Gephart, J.W. and Forsyth, D.W., 1984. An improved method for determining the regional stress tensor using earthquake focal mechanism data: application to the San Fernando earthquake sequence. J. Geophys. Res., 89: 9305–9320.
Grønlie, A. and Roberts, D., 1989. Resurgent strike-slip duplex development along the Hitra–Snåsa and Verran Faults, Møre–Trøndelag Fault Zone, Central Norway. J. Struct. Geol., 11: 295–305.
Hancock, P.L., 1985. Brittle microtectonics: principles and practices. J. Struct. Geol., 7: 437–457.
Hancock, P.L. and Barka, A.A., 1987. Kinematic indicators on active normal faults in Western Turkey. J. Struct. Geol., 9: 573–584.
Reches, Z. and Dieterich, J.H., 1983. Faulting of rocks in three-dimensional strain fields. I. Failure of rocks in polyaxial, servo-control experiments. Tectonophysics, 95: 111–132.
Reches, Z., 1987. Determination of the tectonic stress tensor from slip along faults that obey the Coloumb yield condition. Tectonics, 6: 849–861.
Woodcock, N.H. and Fischer, M., 1986. Strike-slip duplexes. J. Struct. Geol., 8: 725–735.

D. BERING *Geology Institute, University of Bergen, Bergen, Norway*
Present address: *Norwegian Petroleum Directorate, P.O. Box 600, N-4001 Stavanger*

Extensional tectonic history of the western Barents Sea

C.A. Dengo and K.G. R癤ssland

Extensional tectonics was the dominant mode of deformation in the western Barents Sea beginning in Late Devonian–Early Carboniferous time and culminating with the break-up between Greenland and Norway in Paleocene time. This paper describes extensional tectonic elements that have been recently mapped, places them within a regional tectonic framework and, discusses the significance that this extensional deformation may have to petroleum exploration in the Barents Sea. The oldest tectonic event that can be mapped regionally in the western Barents Sea occurred in Late Devonian–Early Carboniferous time in response to the initial rifting between Greenland and Norway. This event established a fundamental basement architecture of half-grabens and intervening highs that influenced the location of younger basins and hydrocarbon traps, the deposition of source, reservoir and seal facies and the timing of hydrocarbon maturation and migration. Major basins that formed at that time are the Tromsø, Bjørnøya, Hammerfest and Nordkapp basins, as well as a newly identified basin named the Dia Graben. The Nordkapp and Bjørnøya basins are each sub-divided into two major half-grabens.

Salt deposition in these basins marks the end of rifting and the beginning of thermal subsidence, in Late Carboniferous time. During Permian through Late Triassic, a time of relative tectonic quiescence, some basement-involved normal faults were reactivated, possibly in response to the Uralian Orogeny and loading by westward prograding clastic sediments. Fault reactivation is apparent along the eastern margin of the Dia Graben which provided a structural control for the position of the Early-Mid Triassic shelf edges. Basement-detached normal faults developed in association with salt withdrawal, mostly in the Dia Graben and Nordkapp basin.

In Late Jurassic–Early Cretaceous times there was renewed crustal extension in the Barents Sea. However, only those basins west of the Loppa High underwent rapid tectonic subsidence. The Cretaceous Tromsø, Bjørnøya and south-western Hammerfest basins, and perhaps the Kvaløy and Harstad basins are interpreted as pull-apart basins that formed in response to oblique rifting between the Barents Sea and Greenland. Basement faults bounding the Troms–Finnmark Platform and southern Nordkapp Basin also were reactivated. Basement-detached normal faulting developed east of the Loppa High and west of the southern Nordkapp Basin. There was additional salt withdrawal in the Nordkapp basins, Dia Graben and possibly the Tromsø Basin.

Northward-propagating rifting led to sea floor spreading in Late Paleocene time and the development of a passive margin along the western Barents Sea. Most of the observed deformation is west of the Senja Ridge; however, there was continued movement along basement-involved normal faults bounding the Harstad, Tromsø, Bjørnøya and southern Nordkapp basins, accompanied by additional salt withdrawal. Neogene contractional deformation, manifested primarily as basin inversion, is apparent along most of the passive margin, west of Loppa High.

Introduction

The western Barents Sea is located on a passive continental margin that developed as the North Atlantic and Arctic oceans opened in response to the break-up between Greenland and Norway in Late Paleocene time. Multiple pre-break-up tectonic events are recorded in the Barents Sea. The earliest is the Caledonian Orogeny, related to the collision and suturing of Laurentia and Baltica (and intervening microplates) in Late Silurian–Early Devonian time (see review by Smith, 1988). Subsequent tectonic events were characterized by crustal extension that began with the initial rifting between Greenland and Norway in Late Devonian–Early Carboniferous time, continued with rifting in Late Jurassic–Early Cretaceous time and culminated with sea floor spreading in Late Paleocene time.

This paper discusses the extensional deformation that occurred in the western Barents Sea, focusing primarily on the Late Devonian–Early Carboniferous deformation. Mapping at this structural level has permitted identification of tectonic elements in addition to those already described (Rønnevik and Jacobsen, 1984; Faleide et al., 1984; Berglund et al., 1986). During the Late Devonian–Early Carboniferous, asymmetric crustal extension imparted to the basement of the western Barents Sea a fundamental architecture characterized by half-grabens and intervening highs. Recognizing the importance of this rifting event is critical to understanding the petroleum geology of the western Barents Sea because it es-

Structural and Tectonic Modelling and its Application to Petroleum Geology edited by R.M. Larsen, H. Brekke, B.T. Larsen and E. Talleraas. NPF Special Publication 1, pp. 91–107. Elsevier, Amsterdam.

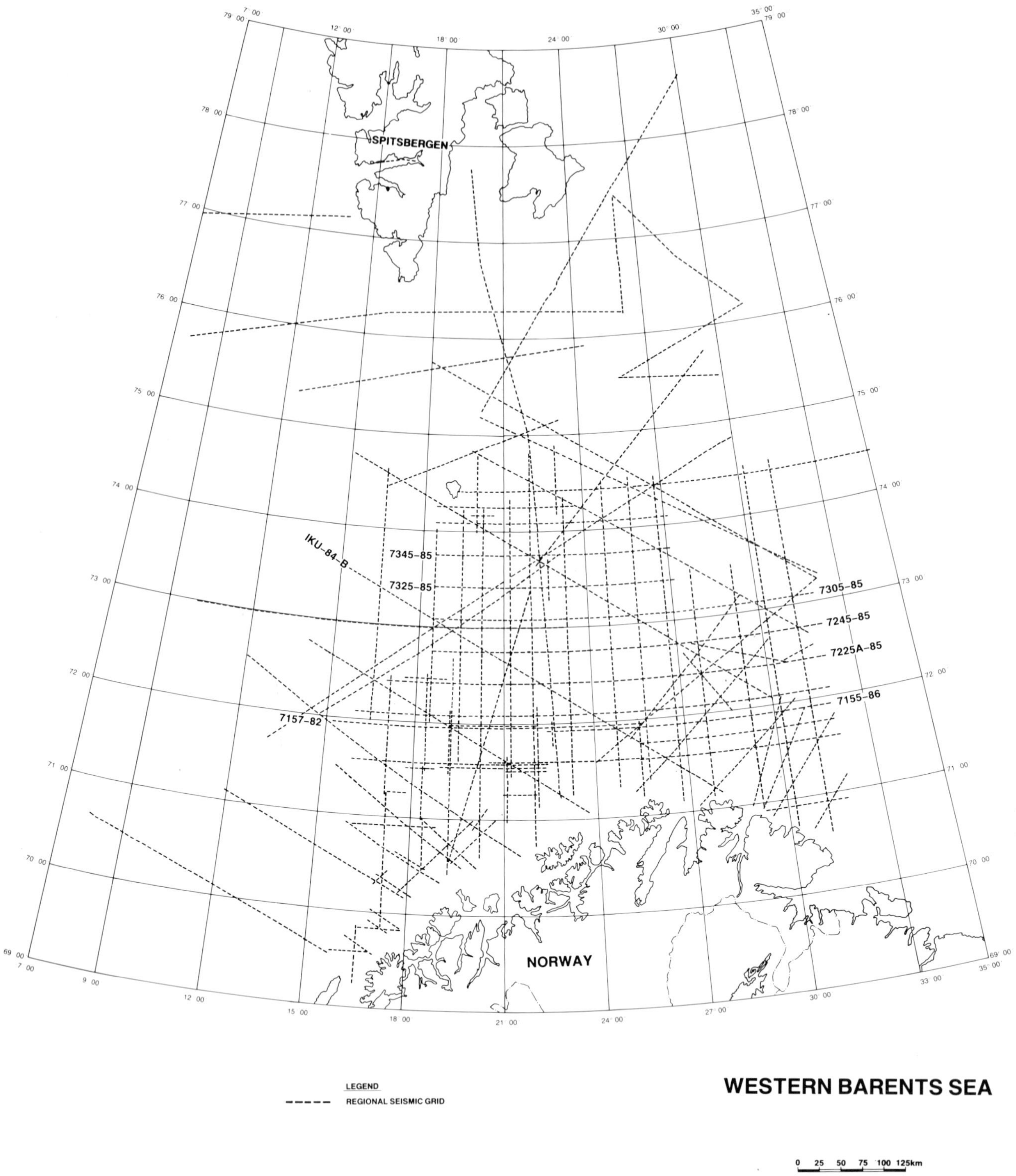

Fig. 1. Location map of the western Barents Sea. The study area is located between the northern coast of Norway and 75°N and between 15° and 35°E. The regional seismic grid used for this study is indicated and numbered lines represent geoseismic lines referenced in the text.

tablished the tectonic framework that influenced (1) the location of younger basins and structures, (2) the distribution of source, reservoir and seal facies and (3) hydrocarbon maturation and migration. Some of these aspects are discussed at the end of this paper.

The area of study is located between longitudes 13°00′ and 33°00′ E and between latitudes 74°30′ and 70°30′ N (Fig. 1). This area covers only a portion of the entire Barents Shelf; however, it is the area that has been opened for hydrocarbon exploration and, consequently, it is where the geophysical data

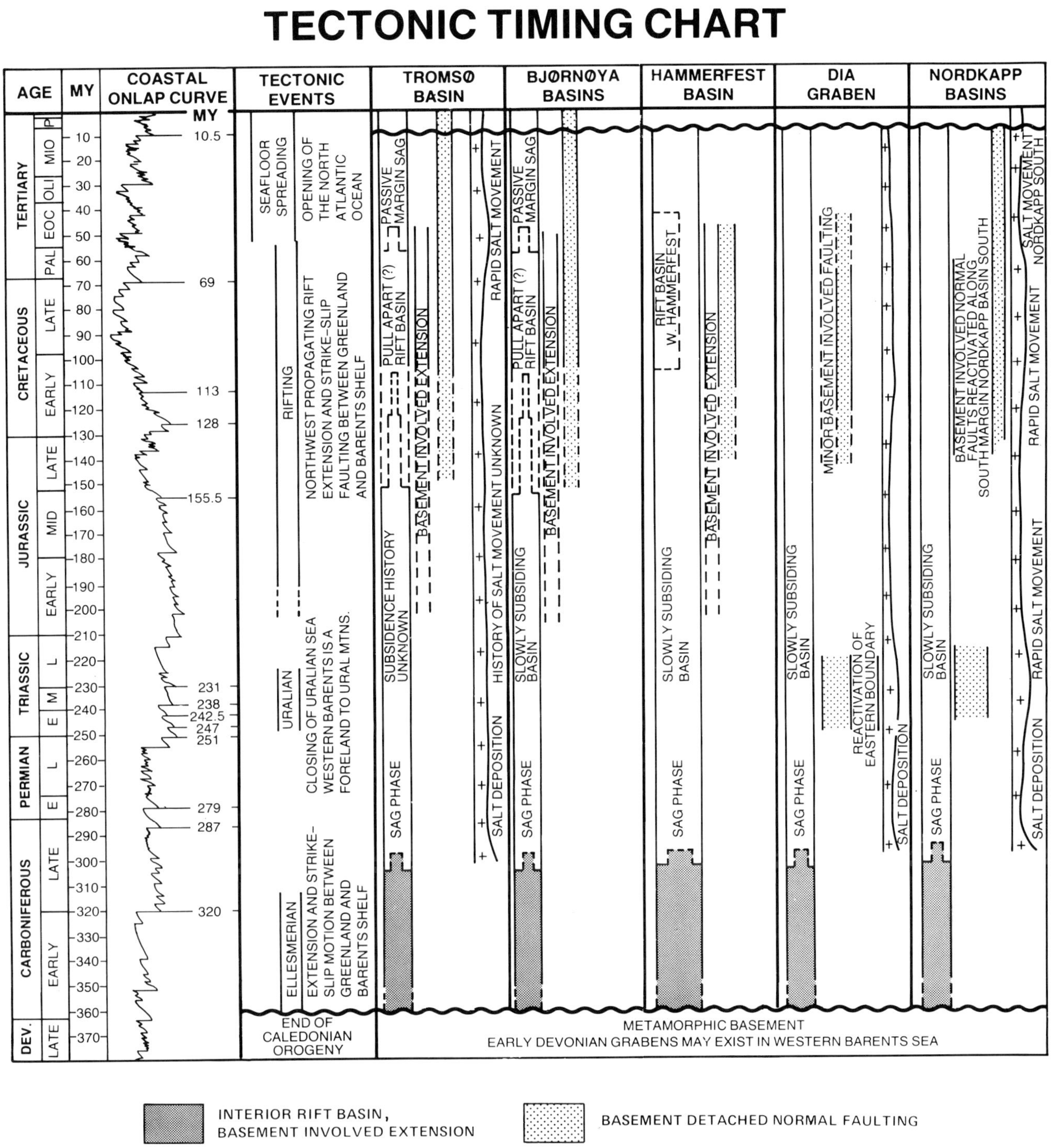

Fig. 2. Summary chart showing timing of major tectonic events and resultant styles of deformation in the western Barents Sea.

coverage is best. Regional syntheses of the entire Barents Shelf and its tectonic relationship to adjacent Svalbard, Franz Josef Land, the Soviet Arctic and Greenland have recently been published in a volume on the geological evolution of the Barents Shelf (Harland and Dowdeswell, 1988).

Three regional paleo-tectonic maps were made to represent the major periods of crustal extension, the Late Devonian–Early Carboniferous, Late Jurassic–Early Cretaceous and Tertiary. Tectonic features were mapped using the regional seismic grid shown in Fig. 1. An infill seismic grid and gravity data were used to compliment the regional grid. Key stratigraphic horizons were dated using available well data and the seismic horizons were correlated regionally using concepts of sequence stratigraphy. A summary of the tectonic events and associated deformational styles for the major basins described in this paper is presented in Fig. 2.

Pre-rift deformation

The Caledonian Orogeny, which occurred in Early Paleozoic time with the closure of the Iapetus Ocean and the suturing of Greenland to Norway and Spitsbergen, formed the metamorphic basement of the Barents Sea. Caledonian structures have a north-east trend in northern Norway (Sturt et al., 1978; Townsend, 1987), a north-west trend in Spitsbergen (see Harland, 1985) and most likely extend across the Barents Sea (Faleide et al., 1984). The trend of Caledonian structures is remarkably coincident with the trend of many of the younger extensional basins, suggesting that Caledonian structures had fundamental control in localizing the position of major north-east and north-west trending basin-bounding normal faults. The influence that pre-existing thrust faults have in the localization of younger extensional faults has been documented in many rift basins world-wide, particularly where deep seismic reflection data are available (e.g., Nelson et al., 1985; Cheadle et al., 1987; Ratcliffe and Burton, 1985; Swanson, 1986). Chadwick (1985) and Lake and Karner (1987) proposed that reactivation of Caledonian and Variscan-age thrust faults as normal faults controlled the development of several basins in southern Britain and adjacent offshore.

Following the closure of the Iapetus Ocean, in Late Silurian–Early Devonian time, erosion of the Caledonian Orogen led to the deposition of thick continental clastic sediments, collectively termed the Old Red Sandstone (see Roberts and Sturt, 1980). In Spitsbergen, these rocks are found largely in a half-graben that was interpreted to have formed in response to divergent strike-slip movement on the paleo-Hornsund and Billefjorden faults (Harland et al., 1974; Harland, 1985). Arguments in favour and against major strike-slip movements in Late Devonian time have been proposed (Lamar et al., 1986; Harland et al., 1988). Not withstanding, in the western Barents Sea there are indications of deep extensional structures that may be of the same age as those in Spitsbergen. These structures are poorly imaged on seismic data and to date have not been mapped.

Early Devonian–Early Carboniferous interior rift phase

The oldest sequence of rocks that can be mapped regionally in the western Barents Sea is composed of Upper Devonian to Lower Carboniferous rift-fill alluvial fan and floodplain clastic sediments mixed with carbonates and evaporites (Faleide et al., 1984; Steel and Worsley, 1984). These were deposited in half-grabens that formed during the initial phase of crustal extension between Greenland and Norway (Rønnevik, 1981). Within the study area, these half-grabens constitute the Tromsø, Bjørnøya, Hammerfest and Nordkapp basins and the Dia Graben.* Basement-involved normal faulting was the dominant style of deformation (Fig. 2).

Rønnevik and Jacobsen (1984) and Faleide et al. (1984) used the tectonic reconstructions performed by Harland (1969) and Ziegler (1978) to propose that the Late Devonian–Early Carboniferous crustal extension in the western Barents Sea resulted from the combined effects of sinistral strike-slip faulting in the western part of the Barents Sea and a conjugate dextral strike-slip fault in the central Barents Sea. The interpretations presented here support and build on the concept that the Late Devonian–Early Carboniferous deformation is important in the Barents Sea. However, they differ from those of previous workers in that we propose that the deformational kinematics is dominantly dip-slip normal faulting with little demonstrable strike-slip deformation.

Tectonic elements

The major tectonic elements that were active In Late Devonian–Early Carboniferous time are shown on the paleo-tectonic map (Fig. 3) and on six parallel, east-trending, regional geoseismic lines (Figs. 4–9). The structures represented on the paleo-tectonic map are at deeper structural levels than the Upper Carboniferous horizon interpreted on the geoseismic lines (Figs. 4–9). Three regional structural trends are observed on the paleo-tectonic map (Fig. 3): (1) north-west trending structures in Finnmark East (south-east of Nordkapp), (2) north-east trending structures that dominate the fabric and which extend across most of the study area from the Norwegian coast to about 74°N and (3) north-north-west trending structures that extend from 74°N to Spitsbergen.

The principal features of the Late Devonian–Early Carboniferous deformation are discussed below. Although the Kvaløy Basin is shown on the paleo-tectonic map it is not possible at this time to determine whether or not it is sited over a Late Devonian–Early Carboniferous predecessor.

The Troms–Finnmark and Berlevåg Faults. Two major faults separate the less deformed Troms–

*At the time of this writing this article, there is not yet an officially approved nomenclature for some tectonic elements described here. The Norwegian Petroleum Directorate has proposed that the names Maud Basin and Nordsel High be used for elements described here, i.e., the Dia Graben and the Saiika High, respectively.

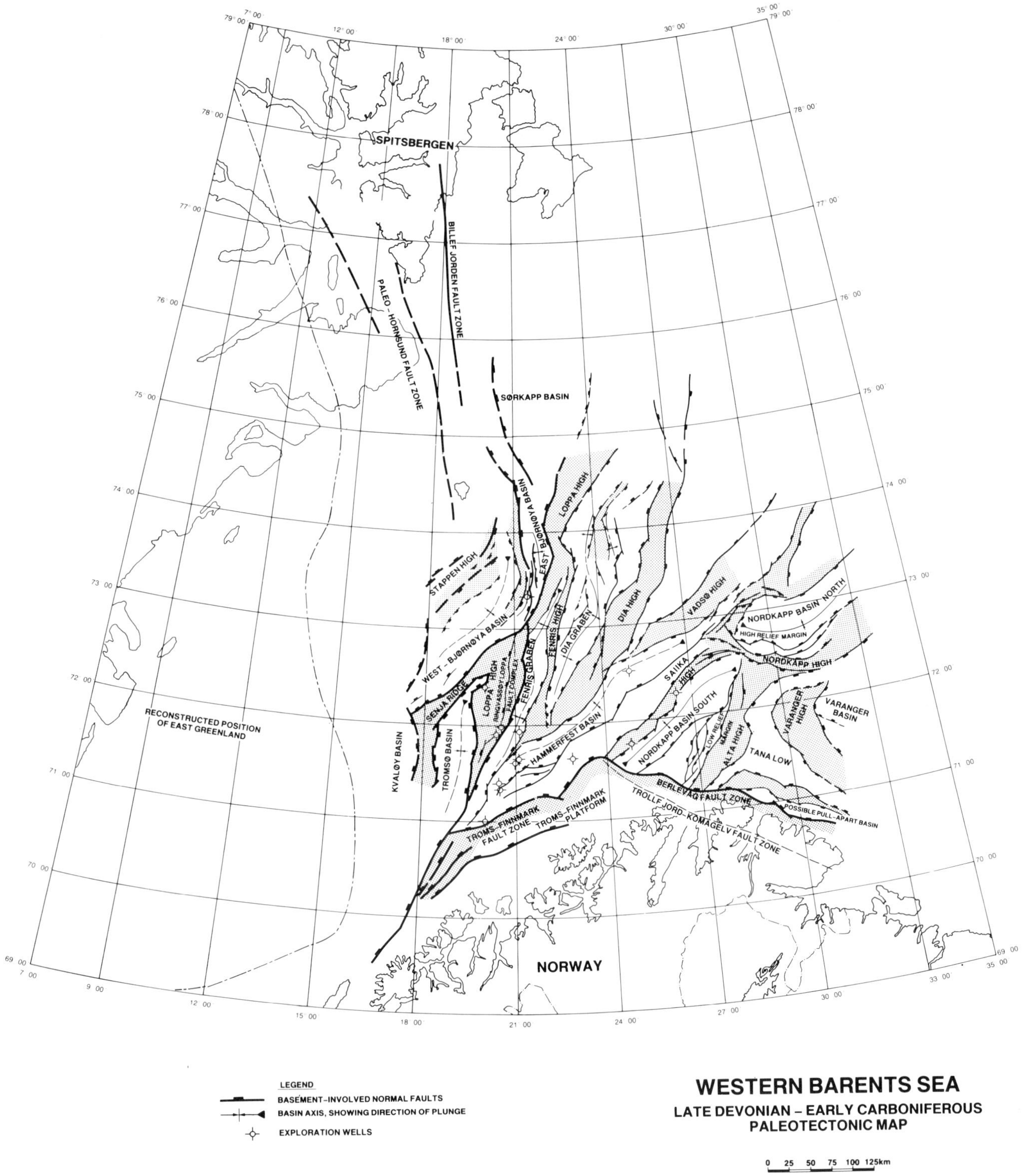

Fig. 3. Late Devonian–Early Carboniferous paleo-tectonic map. Areas highlighted in grey represent basement highs relative to adjacent basins.

Finnmark Platform from the much more deformed Barents Shelf. West of 24°E, the Troms–Finnmark Fault zone separates the Platform from the Hammerfest and Tromsø basins. East of 24°E, the structural boundary is a fault located to the north of the Trollfjord–Komagelv fault, named here the Berlevåg Fault. The Berlevåg Fault is the southern boundary of the Nordkapp Basin. It has a consistent down-to-the-north offset, yet it appears to have a more complicated history than just one of dip-slip motion. Displacement along the basin-bounding normal faults at the

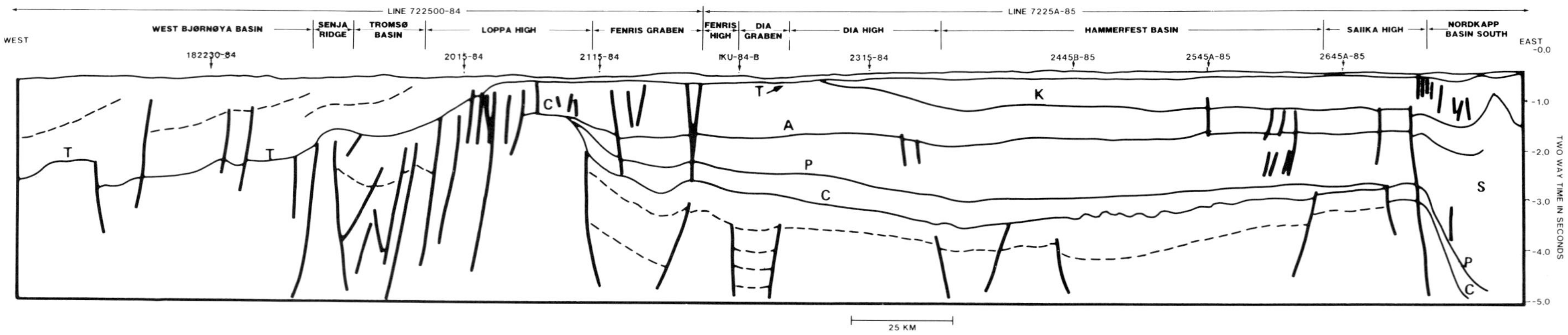

Fig. 4. Geoseismic lines 722500-84 and 7225A-85. Horizons shown represent base Tertiary (*T*), base Cretaceous unconformity (*K*), Mid Triassic (Anisian, *A*), top Permian (*P*), and top Carboniferous (*C*). Dashed lines are form lines shown only to represent configuration of stratal patterns. *S* represents salt. Lines have been squeezed and are vertically exaggerated five times. For the locations of lines see Fig. 1.

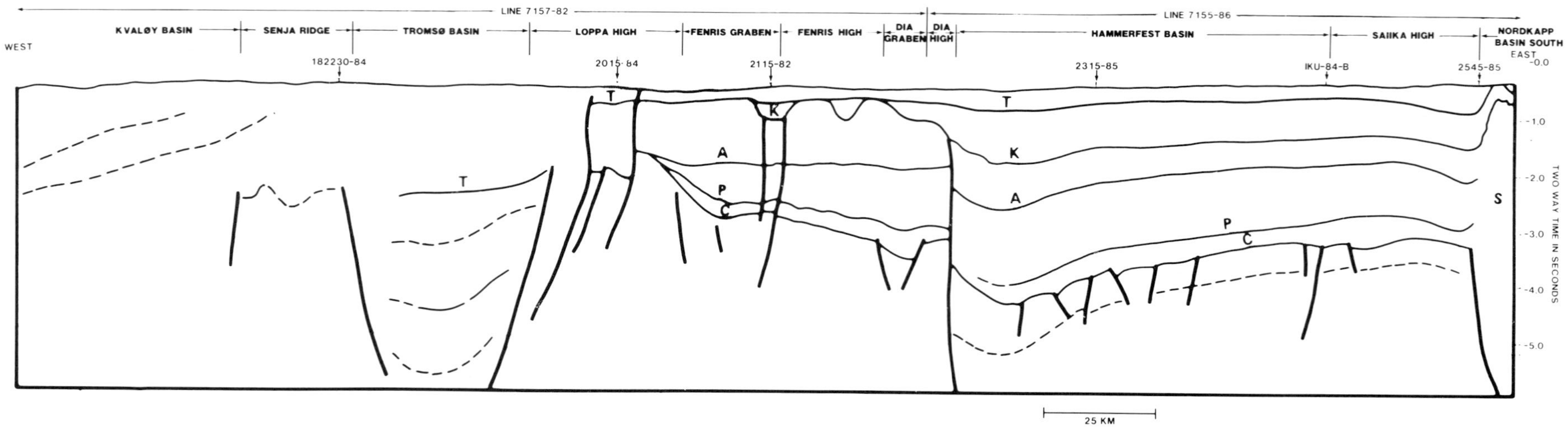

Fig. 5. Geoseismic lines 7157-82 and 7155-86. Lines have been squeezed and are vertically exaggerated five times. A legend for the horizons shown can be found in Fig. 4. For the location of lines see Fig. 1.

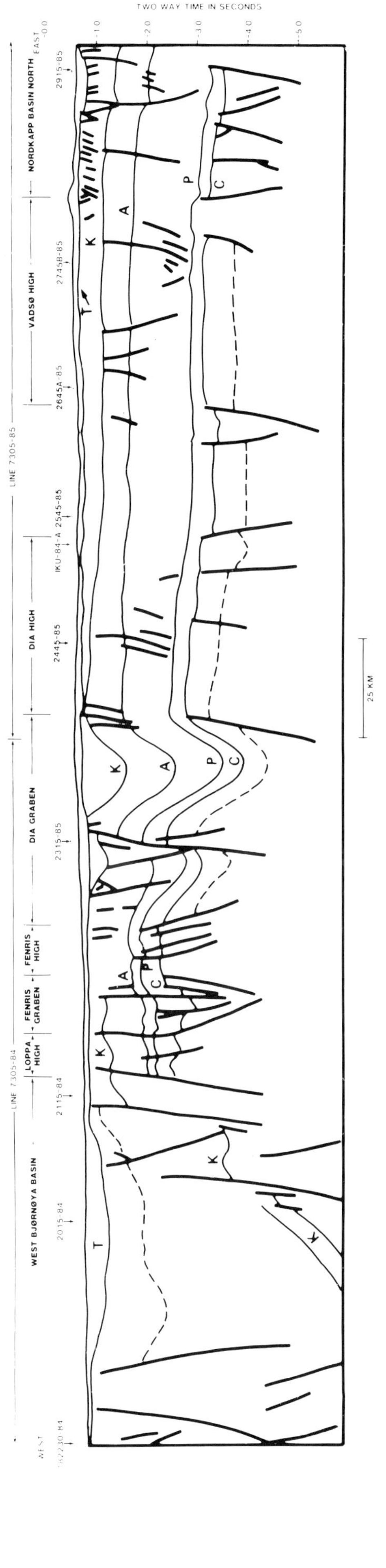

Fig. 6. Geoseismic lines 7305-84 and 7305-85. Lines have been squeezed and are vertically exaggerated five times. A legend for the horizons shown can be found in Fig. 4. For the location of lines see Fig. 1.

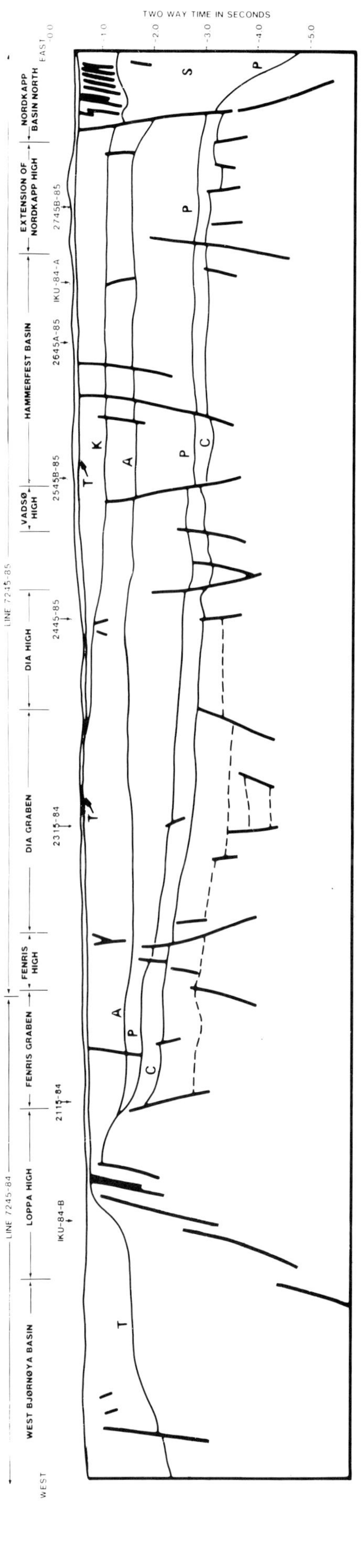

Fig. 7. Geoseismic lines 7245-84 and 7245-85. Lines have been squeezed and are vertically exaggerated five times. A legend for the horizons shown can be found in Fig. 4. For the location of lines see Fig. 1.

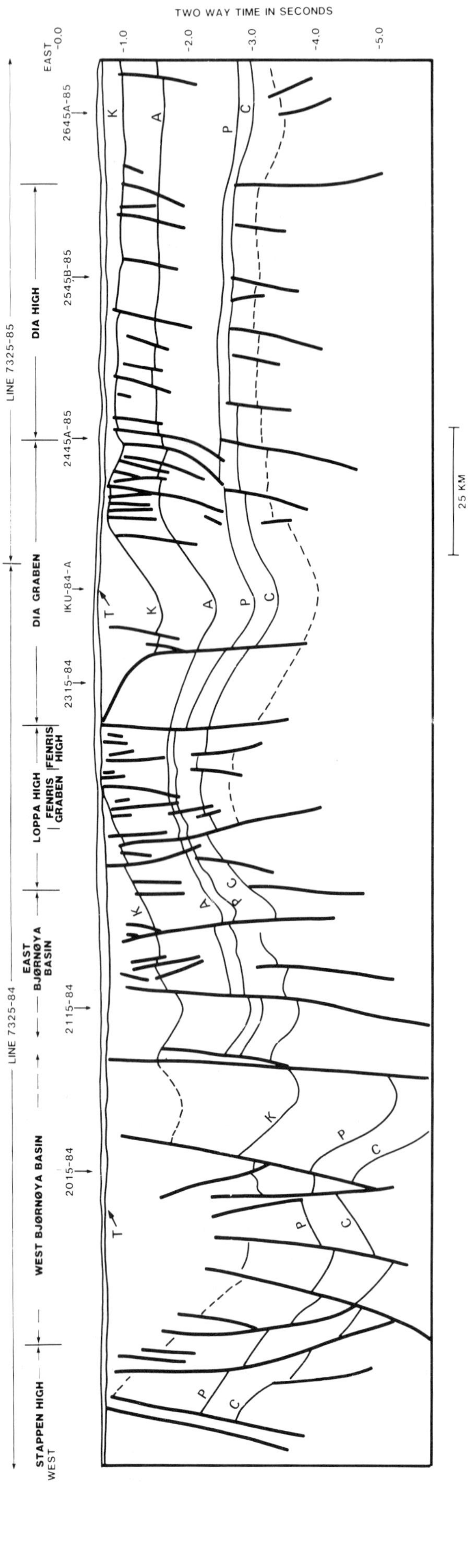

Fig. 8. Geoseismic lines 7325-84 and 7325-85. A legend for the horizons shown can be found in Fig. 4. Lines have been squeezed and are vertically exaggerated five times. For the location of lines see Fig. 1.

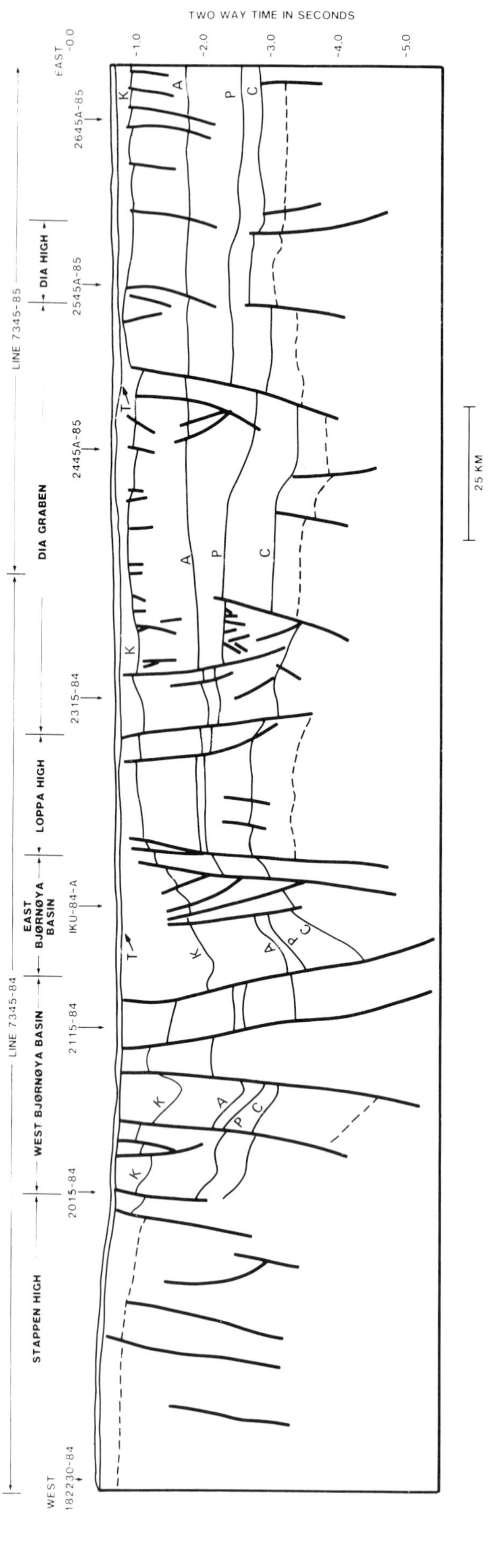

Fig. 9. Geoseismic lines 7345-84 and 7345-85. Lines have been squeezed and are vertically exaggerated five times. A legend for the horizons shown can be found in Fig. 4. For the location of lines see Fig. 1.

southern end of Nordkapp must be transferred to the Berlevåg Fault as strike-slip motion. Between 28°–31°E there is a small elongated basin that is bounded to the south by the Berlevåg Fault and to the north by a basement-high. This basin may be a pull-apart basin that opened in response to strike-slip motion along the Berlevåg Fault. No evidence was found to indicate a north-westward extension of either the Berlevåg or Trollfjord–Komagelv fault zones into the Hammerfest Basin, west of 24°E.

Finnmark East area. The Finnmark East area, located between the Nordkapp and Varanger basins and the Troms–Finnmark Platform (Fig. 3), is unique because it is the only area where north-west trending faults are dominant.

North-west and north-east trending faults intersect to form the broad Tana low (Fig. 3). Flanking this low are two basement highs (Alta and Varanger) that are the basement flanks of the Nordkapp and Varanger basins, respectively.

Nordkapp Basins. The Nordkapp Basin is divided into two adjacent, opposing half-grabens: Nordkapp Basin South and Nordkapp Basin North. The two half-grabens are resolved on the free-air gravity map as distinct lows (Fig. 10). Nordkapp North is the deeper of the two basins and is separated form Nordkapp South by a prominent interbasinal ridge (Nordkapp High) that trends north-west, oblique to the basinal axes. Both basins are flanked by basement highs and have high-relief margins and an opposing low-relief margin. The Saiika High forms the high-relief, north-west margin of Nordkapp South and the Alta High forms the low-relief, south-east margin of this basin. The Saiika and Nordkapp highs flank the Nordkapp Basin North. A complex pattern of multidirectional fault trends marks the intersection of the Nordkapp High with the graben-flanking highs. The asymmetry of Nordkapp has previously been recognized, particularly for Nordkapp Basin South (Rønnevik et al., 1982; Faleide et al., 1984; Rønnevik and Jacobsen, 1984).

Hammerfest Basin. In Late Devonian–Early Carboniferous time the Hammerfest Basin was a north-east trending half-graben with the high-relief margin on the north-west flank. The basin is flanked by the Dia High to the north-west (Figs. 4 and 5), the Vadsø High to the north (Figs. 6 and 7) and the Saiika High to the south-east. The south-west boundary of Hammerfest is a complex fault zone that displaces the Tromsø Basin down-to-the-west relative to the Hammerfest Basin.

Loppa–Nordkapp Platform area. A significant change has been made to previously published maps in the area traditionally called the Loppa–Nordkapp Platform. This area has been divided into a series of subparallel, north-east trending half-grabens and intervening highs. The Late Devonian–Early Carboniferous elements now recognized and displayed on Figs. 3–9 are: (1) Dia High, (2) Dia Graben, (3) Fenris High, (4) Fenris Graben and (5) Loppa High.

The Loppa High is mapped to extend north of 73°05′ N. At this latitude the Loppa High is faulted, with the northern segment of the block being structurally lower than the southern segment. The adjacent Fenris High continues northward, unfaulted. At 73°35′ N the Fenris and Loppa Highs merge and become a single basement block extending north as the eastern flank of the Sørkapp Basin. The north-east trending Senja Ridge merges with the Loppa High forming a single basement block at 72°40′ N, an interpretation that is also supported by the gravity data (Fig. 10). The Ringvassøy–Loppa Fault Complex, which was active in Late Permian time, may have been active during Early Carboniferous time.

The Fenris and Dia Highs form a single basement block south of latitude 72°N, north of which they bifurcate into two distinct basement highs (Figs. 3–7). The Dia High trends north-east as one of the most prominent features of the study area. The Dia Graben, flanked by the Dia, Fenris and Loppa Highs is a narrow graben at about 72°N and widens to the north (Figs. 3–9). A small interbasinal ridge is located within the Dia Graben towards its northward extension (Fig. 3).

Bjørnøya and Tromsø Basins and Stappen High. The Bjørnøya Basin has been subdivided into two half-grabens, East Bjørnøya and West Bjørnøya, the latter being the larger and deeper of the two basins (Figs. 3, 8 and 9). The eastern margins of both basins is formed by the faults flanking the Senja Ridge and its continuation as the Loppa High. The north-western margin of the West Bjørnøya Basin is the Stappen High. This margin is characterized by a series of down-to-the-south-east normal faults. A narrow intrabasinal ridge, located at about 21°E, separates East from West Bjørnøya basins (Figs. 3, 8 and 9).

The Tromsø Basin has a very thick Cretaceous fill that obscures the deeper structure (Fig. 5). The basin-bounding faults show Early Carboniferous offset; however, the presence of salt in the Tromsø Basin, which is of Late Carboniferous age elsewhere in the western Barents Sea, is indirect evidence that Tromsø was an actively subsiding basin at that time.

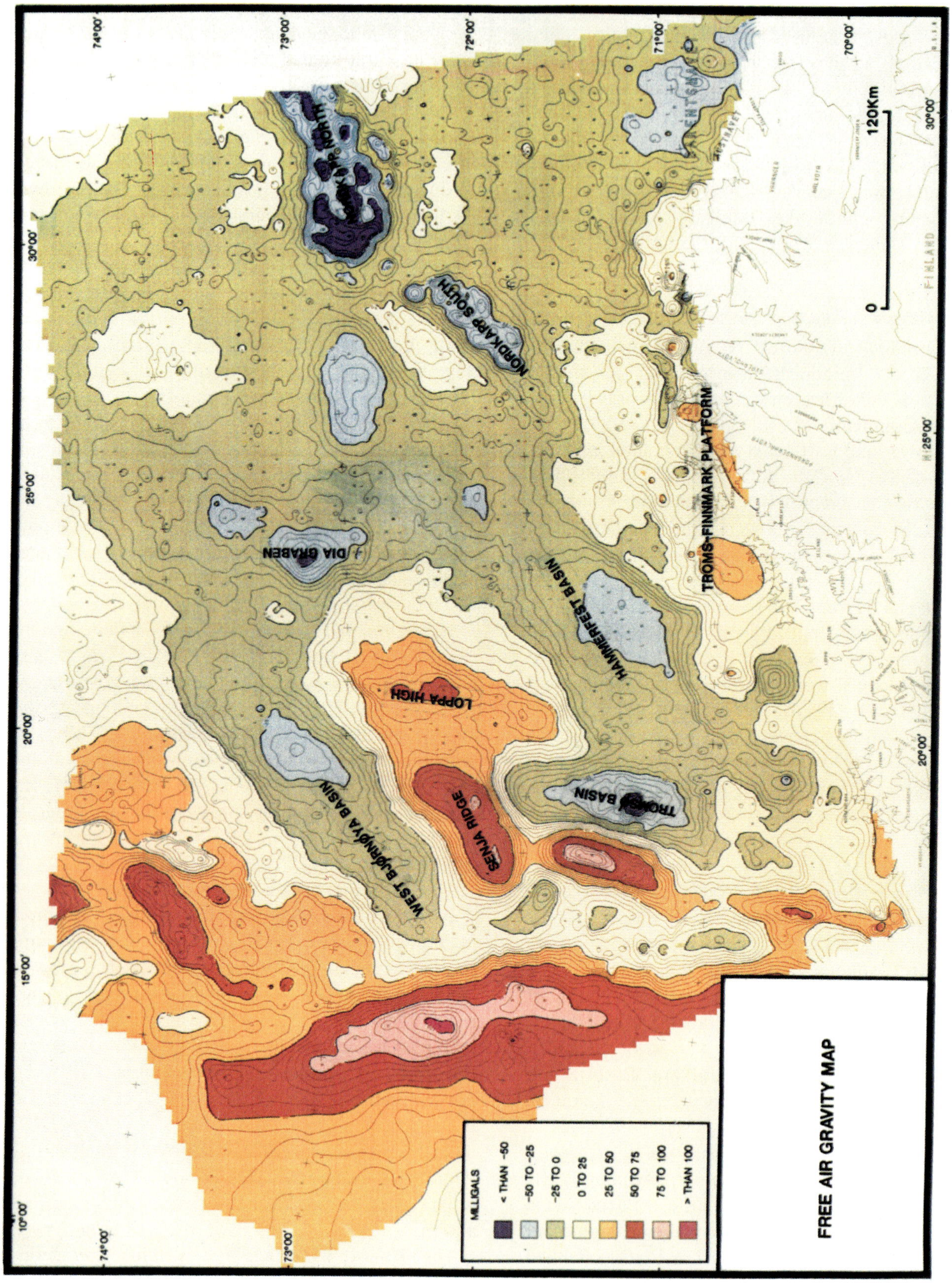

Fig. 10. GECO free-air gravity map.

Late Carboniferous–Jurassic sag phase

In the Late Carboniferous active crustal extension ceased, leading to a transition into a slowly subsiding interior sag (Fig. 2). Sedimentation changed from continental to shallow-marine siliciclastic deposition and ultimately to carbonate and evaporite deposition (Rønnevik and Jacobsen, 1984). Salt deposition is interpreted to have occurred in the Dia Graben, Tromsø, Nordkapp, West Bjørnøya and Hammerfest basins (Fig. 2). Slow subsidence continued through most of the Permian. In Late Permian times the western portion of the Loppa High was faulted down to the west along the Ringvassøy Loppa Fault Complex. The Loppa High was subsequently rotated down to the east.

Closing of the Uralian Sea during the Late Permian through Early Triassic placed the Barents Shelf in a distal foreland position to the Ural Mountains. Great volumes of sediment were deposited in Finnmark East, the Nordkapp Basins and across the Dia Graben and Loppa High. Sediment loading had a two-fold effect. First, there was reactivation of some basement-involved normal faults, most notably the eastern margin of the Dia Graben, which provided a structural control for the position of the Early-Mid-Triassic shelf edges (Figs. 2, 3 and 8). Secondly, there was increased withdrawal of salt, primarily in the Dia Graben and the Nordkapp basins, accompanied by the local development of basement-detached normal faults.

Late Jurassic–Early Cretaceous rift phase

Renewed crustal extension between Greenland and Norway initiated a continued and long-term uplift of the Loppa High. This episode of rifting was most intense during Late Jurassic–Early Cretaceous (Barremian) time (Fig. 2). Major structures that were active at this time are shown on the paleo-tectonic map in Fig. 11.

Regionally, most of the deformation occurred west of the Loppa High. A thick, Hauterivian–Albian section, sourced primarily from the west, was deposited in the rapidly subsiding Tromsø, Bjørnøya and Kvaløy Basins. These basins can be interpreted to have formed as pull-apart basins in response to oblique rifting between the Barents Sea and Greenland (Fig. 11).

The Tunheim Terrace developed on the Loppa High, characterized by a relatively wide zone of tilted fault blocks underlain by detached normal faults. A prominent set of faults with multi-directional trends extends along the western boundary of the Loppa High from the Tromsø Basin, across the Senja Ridge, to the West Bjørnøya Basin. This is the position of the Cretaceous hinge line, which for the most part is controlled by reactivation of Late Devonian–Early Carboniferous basin-bounding normal faults (Figs. 2, 4–9 and 11; compare Figs. 3 and 11). East of the Loppa High, prominent basement faults that were reactivated are the Troms–Finnmark fault, the north-west portion of the Berlevåg Fault and the southern bounding faults of Nordkapp Basin South.

The north-east structural trends inherited from the Late Devonian–Early Carboniferous deformation predominate across the western Barents Sea. Two exceptions to this are noted in the Hammerfest Basin. In the central Hammerfest, at about 21°40′E, north-west trending faults are superimposed on the pre-existing north-east trending fabric (Fig. 11). These faults are found along strike of the Berlevåg Fault. West-trending faults are also superimposed on earlier structural trends along the north-west boundary of the Hammerfest Basin, forming the boundary between it and the Loppa High (Fig. 11). This fault system continues to the north-east, but the displacement is transferred to a drape flexure located above the Late Devonian–Early Carboniferous margin of the basin.

Detached normal faults and low-amplitude folds, east of Loppa High, were partially localized over deeper, pre-existing Late Devonian–Early Carboniferous faults (Figs. 4–9). The folds are either flexures over normal fault blocks, or low-side roll-over structures (extensional fault-bend folds; Groshong, 1989). Salt-cored folds are also present, but no effort was made to differentiate these on the map.

Two possible zones of strike-slip faulting have been identified, each having minor displacements (Fig. 11). A north-east trending divergent(?) strike-slip fault is located along the eastern margin of the Fenris Graben. Displacement along this fault appears to be minimal, yet it was sufficient to form a small depression along its strike. The southern boundary of the Vadsø High was also reactivated, possibly as a convergent(?) strike-slip fault.

Tertiary passive margin phase

Tectonic activity during the Tertiary is mostly related to the progressive northward opening of the North Atlantic and Arctic Oceans in Late Paleocene time (magnetic anomaly 24; Figs. 2 and 12). Opening began near the Vøring Plateau escarpment and propagated northwards along the Senja fracture zone. By Early to mid-Miocene times the Knipovich and Mohns Ridges were established as spreading centres (Fig. 12). An excellent review of the Cenozoic evolution of this margin has been presented by

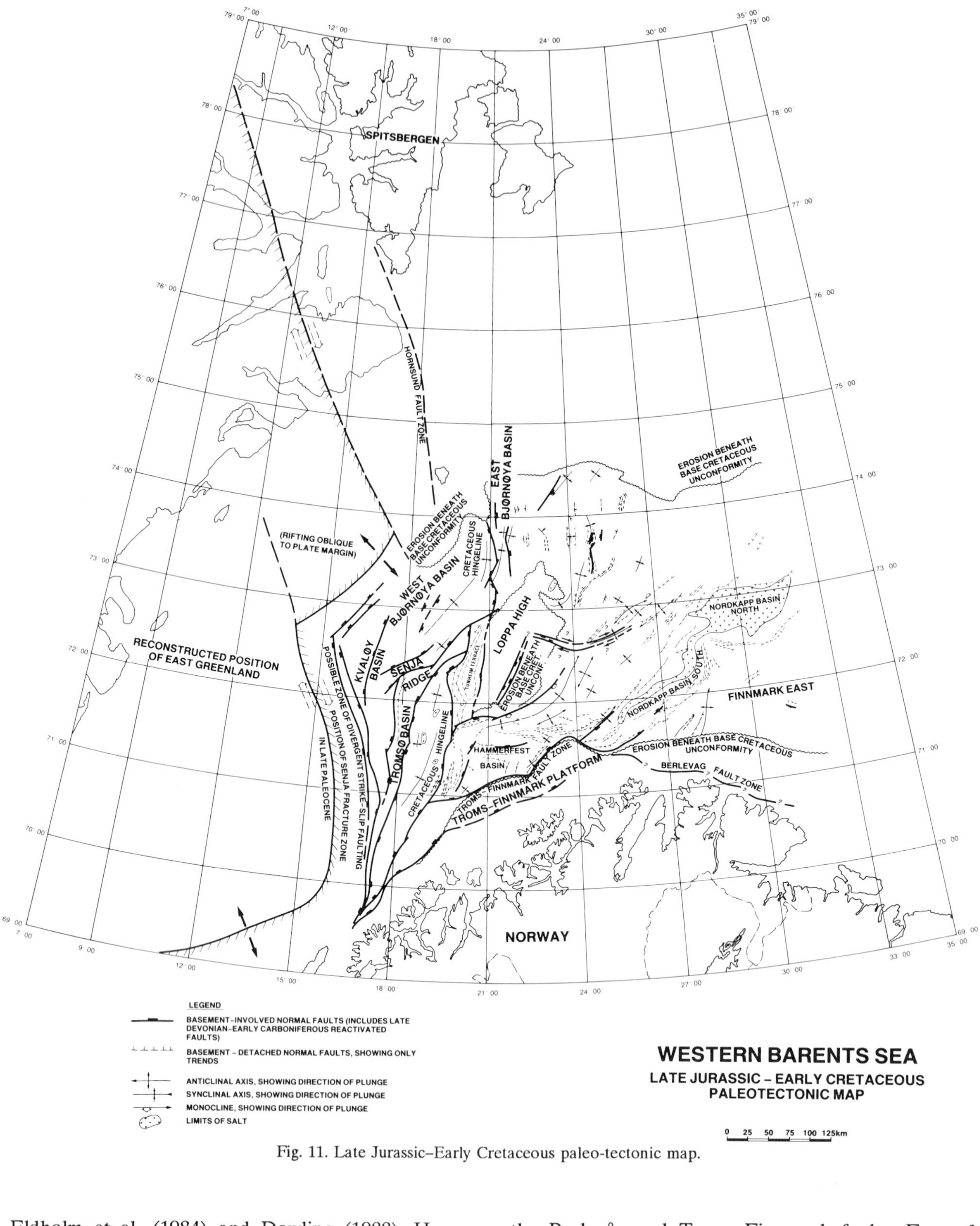

Fig. 11. Late Jurassic–Early Cretaceous paleo-tectonic map.

Eldholm et al. (1984) and Dowling (1988). Here, we note only that most of the deformation occurred west of the Loppa High and Senja Ridge, along reactivated, pre-existing faults. Active basement-involved normal faults are the basin-bounding faults of the Harstad, Kvaløy, Tromsø, Bjørnøya and Nordkapp Basin South. Some movement is also recorded along the Berlevåg and Troms–Finnmark faults. East of the Loppa High the region was relatively stable. There was deformation related to salt movement and basement-detached normal faulting overlying older basement-involved faults.

Subsequent to the initiation of sea floor spreading (Early Eocene), the western margin of the Barents

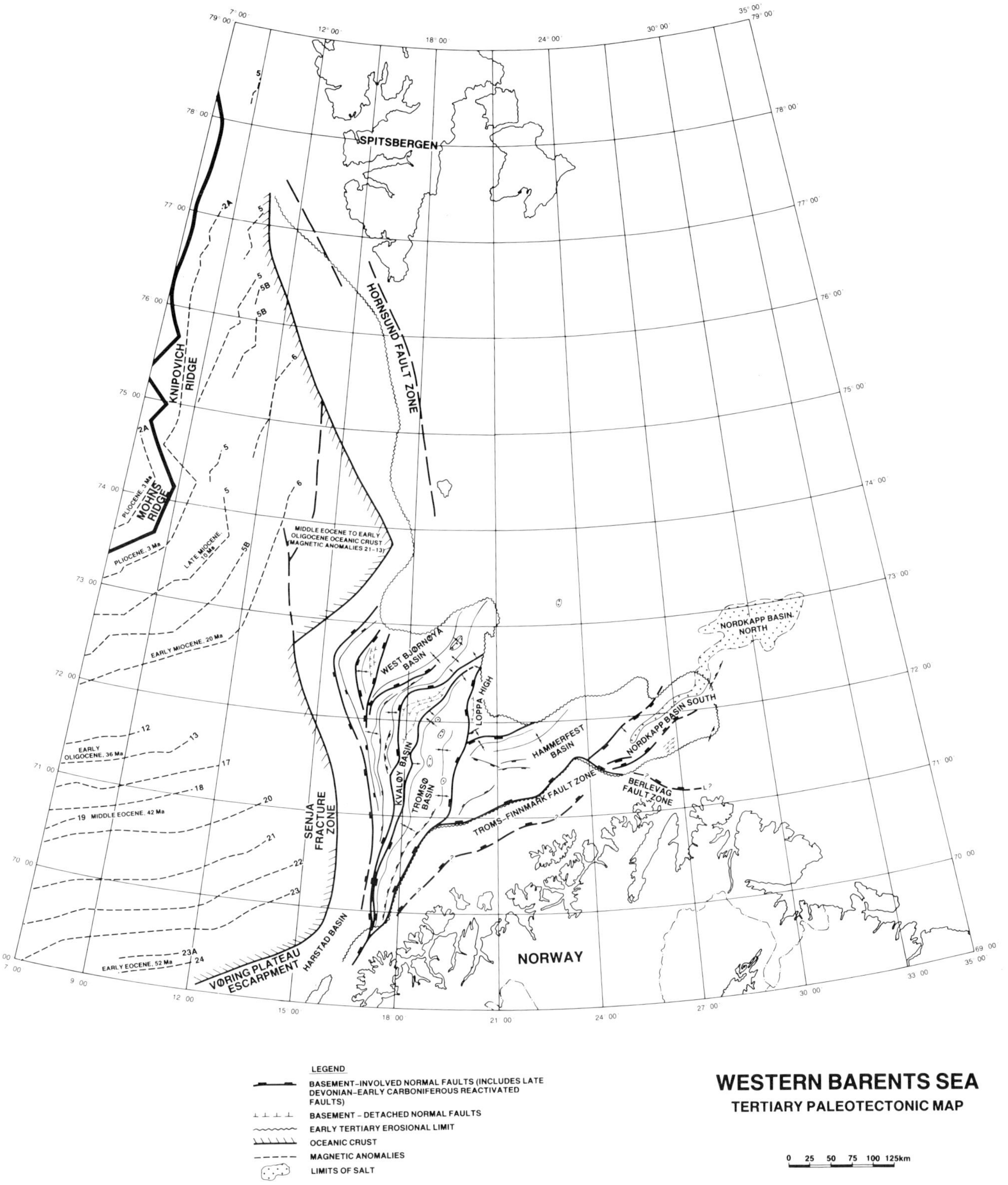

Fig. 12. Tertiary paleo-tectonic map.

Sea developed as a sheared margin (Myhre et al., 1982) with as much as 550 km of dextral strike-slip displacement on the Hornsund Fault. Compressive deformation associated with this strike-slip faulting led to the development of the Spitsbergen orogenic belt, and to the south led to compressive structures in Bjørnøya Island and the Bjørnøya West region. To the east there is little evidence of compressive deformation in the Barents Sea. From mid-Miocene time to the present the western Barents Sea has been regionally uplifted.

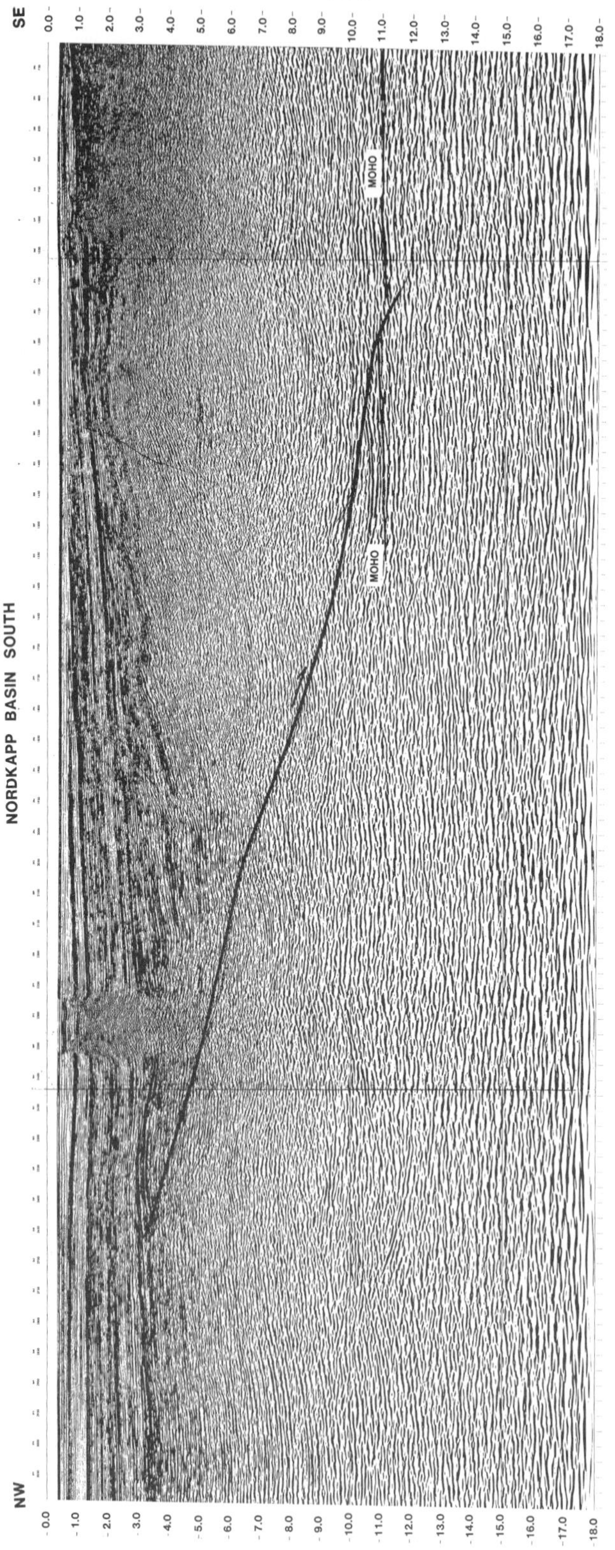

Fig. 13. Portion of the deep seismic reflection line, IKU-B, across Nordkapp Basin South. A low-angle normal fault is interpreted to extend beneath the entire basin to the base of the crust. The line has been squeezed and is vertically exaggerated five times. For the location of the seismic line see Fig. 1.

Discussion

The principal conclusions derived from this study are that: (1) extensional tectonics has been the dominant mode of deformation in the western Barents Sea since Late Devonian–Early Carboniferous time with little demonstrable large-scale strike-slip deformation, (2) a fundamental basement architecture of half-grabens and intervening highs formed in Late Devonian–Early Carboniferous time and these controlled the location of younger basins, and (3) deformation progressed westwards from the time of initial extension until the time of break-up in the Late Paleocene. The recognition in the last several years that lithospheric extension leads to the development of asymmetrical basins (half-grabens) characterized by tilted fault-blocks (e.g., Wernicke and Burchfiel, 1982; Jackson et al., 1982; Gibbs, 1984; Rosendahl et al., 1986) provides a framework with which to view the extensional deformation of the western Barents Sea. Equally important, the concepts of asymmetrical extensional structuring have a direct application in hydrocarbon exploration because they can be used to improve models of heat flow and maturation and to predict trap geometry and reservoir distribution.

The western Barents Sea extensional fabric shares similarities with extensional areas elsewhere. Regionally, the Late Devonian–Early Carboniferous paleotectonic map shows a fabric of subparallel ridges and basins that are similar in scale and arrangement to the Basin and Range Province of the western United States (see Myers and Hamilton, 1964; Wernicke and Burchfiel, 1982). Low-angle detachments may have a significant control on the extensional geometries observed in the western Barents Sea, as indicated by the deep seismic reflection profiles available. For example, in the case of the Nordkapp Basin South, it is possible to interpret a low-angle normal fault extending beneath the entire basin to lower crustal levels (Fig. 13). Additional studies may reveal that regionally extensive basement highs (for example the Loppa High) may be analogous to the uplifted footwall block associated with low-angle crustal detachment systems, as described for parts of the Basin and Range Province (Wernicke and Axen, 1988). It is important to recognize, however, that the asymmetry observed in the Barents Sea does not have to represent a process that is unique to lithospheric extensional deformation everywhere, but may represent the control imposed by the pre-existing Caledonian structural fabric.

Locally in the western Barents Sea, half-grabens are arranged with alternating polarity across interbasinal ridges (e.g., Nordkapp Basins across the Nordkapp High and the Tromsø and West Bjørnøya Basins across the Senja Ridge). The arrangement of these half-grabens and highs is similar to that observed in the East African Rift (Rosendahl et al., 1986). Although the types of structural trap developed in these basins should be typical of graben settings (Harding, 1984), the mode of linkage of the basement-involved normal faults can modify and or enhance trap closure capacity, as illustrated recently for the North Viking Graben (Scott and Rosendahl, 1989). This is particularly evident where the interbasinal highs intersect the graben-flanking highs, for example the intersection of the Nordkapp and Saiika Highs. Cross-cutting (multi-directional) trends can form traps that otherwise would not be expected within a full-graben.

Basin asymmetry will also have an important influence on the distribution of the rift-fill clastic sediments. Different types of facies relationship will develop as a function of margin relief and slope, fault linkage patterns, drainage directions, etc. (Leeder and Gawthrope, 1987). Thus, predictive models for reservoir and seal distribution can be achieved with knowledge of the basin architecture. This was demonstrated by Gerdes et al. (this volume) who describe a case for the Barents Sea where the structural analysis of the margin of a half-graben, in conjunction with other data, was used to construct depositional models for different seismic sequences and from these to determine areas favoured for reservoir deposition.

Additional studies are required in order to further the understanding of the deep crustal structure of the Barents Sea and the mechanism(s) by which lithospheric extension has occurred — whether this has been dominantly by lithospheric-scale pure shear (McKenzie, 1978), simple shear (Wernicke, 1985) or some other mechanism. To understand the entire rift evolution fully it will also be necessary to take into account the structure of the complimentary margin of north-east Greenland.

Identification of the mechanism(s) of lithospheric extension during the various rift phases may help in understanding problems such as the cause of the Tertiary uplift and erosion in the western Barents Sea (Fig. 12). Some possible causes for this uplift are discussed by Nyland et al., Kjemperud and Fjeldskaar and Riis and Fjeldskaar (all in this volume). Although glacial rebound accounts for some of the uplift, other processes, operative at the lithospheric scale and probably directly related to the mechanism of lithospheric extension, must account for the remaining uplift. Thermal subsidence and the time-integrated anomalous heat flow will differ depending on the mode of crustal extension; for example, calculations for a low-angle detachment model indicate that they will be about one-half the values calculated for a model of pure shear extension (Voorhoeve and Houseman, 1988). Consequently, accurate maturation modelling of source rock intervals, hydrocarbon type and yield predictions, and construction of thermal subsidence curves for geohistory analysis depend directly on the extensional model assumed. It is important to continue our studies of the deep crustal structure of the western Barents Sea.

Acknowledgments

We thank Esso Norge a.s., Exxon Company International and Exxon Production Research Company for permission to publish this work. We also acknowledge the Norwegian Petroleum Directorate, GECO and IKU for permitting us to publish the geoseismic sections, the gravity map and the seismic line IKU-B, respectively. This study benefited from the contributions made by many of our colleagues at Exxon, in particular T. Nardin, B. Wiggins, D. Bergslien, O. Hostad, S. Ballestad, J. Demarest and S. Greenlee. The manuscript was improved by reviews made by T. Hauge and L. Fairchild from Exxon Production Research Co., I. Stewart from BP Norway and an anonymous reviewer.

References

Berglund, T., Auguston, J., Faerseth, R., Gjelberg, J. and Ramberg-Moe, H., 1986. The evolution of the Hammerfest Basin. In: Habitat of Hydrocarbons on the Norwegian Continental Shelf. Norw. Pet. Soc., Graham and Trotman, London, pp. 319–338.

Chadwick, R.A., 1985, Upper Jurassic to Cretaceous sedimentation and subsidence. In: A. Whittaker (Editor), Atlas of Onshore Sedimentary Basins in England and Wales: Post-Carboniferous Tectonics and Stratigraphy. Blackie, Glasgow, pp. 49–52.

Cheadle, M.J., McGeary, S., Warner, M.R. and Matthews, D.H., 1987. Extensional structures on the western U.K. continental shelf: a review of evidence from deep seismic profiling. In: M.P. Coward, J.F. Dewey and P.L. Hancock (Editors), Continental Extensional Tectonics. Geol. Soc. London Spec. Publ., 28: pp. 445–465.

Dowling, L.M., 1988. Cenozoic evolution of the western margin of the Barents Shelf. In: W.B. Harland and E.K. Dowdeswell (Editors), Geological Evolution of the Barents Shelf Region. Graham and Trotman, London, pp. 157–169.

Eldholm, O., Sundvor, E., Myhre, A.M. and Faleide, J.I., 1984. Cenozoic evolution of the margin off Norway and Svalbard: In: A.M. Spencer et al. (Editors), Petroleum Geology of the North European Margin. Graham and Trotman, London, pp. 3–18.

Faleide, J.I., Gudlaugsson, St.T. and Jacquart, G., 1984. Evolution of the western Barents Sea. Mar. Pet. Geol., 1: 123–149.

Gerdes, K., Hurst, J. and Jeans, R., 1992. Seismostratigraphic study of a synrift megasequence, from the Barents Sea, northern Norway. In: R.M. Larsen, H. Brekke, B.T. Larsen and E. Talleraas (Editors), Structural and Tectonic Modelling and its Application to Petroleum Geology. Norwegian Petroleum Society (NPF), Special Publication 1. Elsevier, Amsterdam, pp. 133–152 (this volume).

Gibbs, A.D., 1984. Structural evolution of extensional basin margins. J. Geol. Soc. London, 141: 609–620.

Groshong, R.H., 1989. Half-graben structures: balanced models of extensional fault bend folds. Geol. Soc. Am. Bull., 101: 96–195.

Harding, T.P., 1984. Graben hydrocarbon occurrences and structural style: Am. Assoc. Pet. Geol. Bull., 68: 333–362.

Harland, W.B., 1969. Contribution of Spitsbergen to understanding of tectonic evolution of North Atlantic Region. Am. Assoc. Pet. Geol. Mem., 12: 817–851.

Harland, W.B., 1985. Caledonide Svalbard. In: D.G. Gee and B.A. Stuat (Editors), The Caledonide Orogen — Scandinavia and Related Areas. John Wiley and Sons, Chichester, pp. 999–1016.

Harland, W.B. and Dowdeswell, E.K. (Editors), 1988. Geological Evolution of the Barents Shelf Region. Graham and Trotman, London, 176 pp.

Harland, W.B., Cutbill, J.L., Friend, P.F., Gobbett, D.J., Holliday, D.W., Maton, P.I., Parker, J.R. and Wallis, R.H., 1974. Billefjorden Fault Zone, Spitsbergen. Nor. Polarinst. Skr., 161, 72 pp.

Harland, W.B., Perkins, P.J. and Smith, M.P., 1988. Cambrian through Devonian stratigraphy and tectonic development of the western Barents Shelf. In: W.B. Harland and E.K. Dowdeswell (Editors), Geological Evolution of the Barents Shelf Region. Graham and Trotman, London, pp. 73–88.

Jackson, J.A., King, G. and Vita-Finzi, C., 1982. The neotectonics of the Aegean: an alternative view. Earth Planet. Sci. Lett., 61: 303–318.

Kjemperud, A. and Fjeldskaar, W., 1992. Pleistocene glacial isostacy — implications for petroleum geology. In: R.M. Larsen, H. Brekke, B.T. Larsen and E. Talleraas (Editors), Structural and Tectonic Modelling and its Application to Petroleum Geology. Norwegian Petroleum Society (NPF), Special Publication 1. Elsevier, Amsterdam, pp. 187–195 (this volume).

Lamar, D.L., Reed, W.E. and Douglass, D.N., 1986. Billefjorden fault zone, Spitsbergen. Is it part of a major Late Devonian transform? Geol. Soc. Am. Bull., 97: 1083–1088.

Lake, S.D. and Karner, G.D., 1987. The structure and evolution of the Wessex basin, southern England: an example of inversion tectonics. Tectonophysics, 137: 347–378.

Leeder, M.R. and Gawthrope, R.L., 1987. Sedimentary models for extensional tilt-block/half-graben basins. In: M.P. Coward, J.F. Dewey and P.L. Hancock (Editors), Continental Extensional Tectonics. Geol. Soc. Londen Spec. Publ., 28: pp. 139–152.

McKenzie, D.P., 1978. Some remarks on the development of sedimentary basins. Earth Planet. Sci. Lett., 40: 25–32.

Myers, W.B. and Hamilton, W., 1964. Deformation associated with the Hebgen Lake earthquake of August 17, 1959. U.S. Geol. Surv. Prof. Pap., 435: 55–98.

Myhre, A.M., Eldholm, O. and Sundvor, E., 1982. The margin between Senja and Spitsbergen fracture zones: implications from plate tectonics. Tectonophysics, 89: 1–32.

Nelson, K.D., Arnow, J.A., McBride, J.H., Willemin, J.H., Huang, J., Zheng, L., Oliver, J.E., Brown, L.D. and Kaufman, S., 1985. New COCORP profiling in the southeastern United States, Part I: Late Paleozoic suture and Mesozoic rift basin. Geology, 13: 714–717.

Nyland, B., Jensen, L.N., Skagen, J., Skarpness, O. and Vorren, T., 1992. Tertiary uplift and erosion in the Barents Sea; magnitude, timing and consequences. In: R.M. Larsen, H. Brekke, B.T. Larsen and E. Talleraas (Editors), Structural and Tectonic Modelling and its Application to Petroleum Geology. Norwegian Petroleum Society (NPF), Special Publication 1. Elsevier, Amsterdam, pp. 153–162 (this volume).

Ratcliffe, N.M. and Burton, W.C., 1985. Fault reactivation models for the origin of the Newark basin and studies related to eastern U.S. seismicity. In: G.P. Robinson and A.J. Froelich (Editors), Proc. Sec. U.S.G.S. Workshop on the Early Mesozoic Basins of the Eastern United States. U.S. Geol. Surv. Circ., 946: pp. 36–45.

Riis, F. and Fjeldskaar, W., 1992. On the magnitude of the Late Tertiary and Quaternary erosion and its significance for the uplift of Scandinavia and the Barents Sea. In: R.M. Larsen, H. Brekke, B.T. Larsen and E. Talleraas (Editors), Structural and Tectonic Modelling and its Application to Petroleum Geology. Norwegian Petroleum Society (NPF), Special Publication 1. Elsevier, Amsterdam, pp. 163–185 (this volume).

Roberts, D. and Sturt, B.A., 1980. Caledonian deformation in Norway. J. Geol. Soc. London, 137: 241–150.

Rosendahl, B.R., Reynolds, D.J., Lorben, P.M., Burgess, C.F., McGill, J., Scott, D., Lambiase, J.J. and Derksen, S.J., 1986. Structural expressions of rifting: lessons from Lake Tanganyika, Africa: sedimentation in the African Rifts. Geol. Soc. London, Spec. Publ., 25: 29–43.

Rønnevik, H.C., 1981. Geology of the Barents Sea. In: Petroleum Geology of the Continental Shelf of North West Europe. Institute of Petroleum, London, pp. 395–406.

Rønnevik, H.C., Beskow, B. and Jacobsen, H.P., 1982. Structural and stratigraphic evolution of the Barents Sea. ONS-82, Norw. Pet. Soc., pp. 1–28

Rønnevik, H.C. and Jacobsen, H.P., 1984. Structural highs and basins in the western Barents Sea. In: Petroleum Geology of the North European Margin, Norwegian Petroleum Society, Graham and Trotman, London, pp. 19–32.

Scott, D.L. and Rosendahl, B.R., 1989. North Viking Graben: an East African perspective. Am. Assoc. Pet. Geol. Bull., 73: 155–165.

Smith, M.P., 1988. Cambrian through Devonian Panarctic tectonic events. In: W.B. Harland and E.K. Dowdeswell (Editors), Geological Evolution of the Barents Shelf Region. Graham and Trotman, London, pp. 7–17.

Steel, R.J. and Worsley, D., 1984. Svalbard's post-Caledonian strata — an atlas of sedimentational patterns and paleogeographic evolution. In: A.M. Spencer et al. (Editors), Petroleum Geology of the North European Margin. Nor. Pet. Soc., Graham and Trotman, London, pp. 109–135.

Sturt, B.A., Pringle, I.R. and Ramsay, D.M., 1978. The Finnmarkian phase of the Caledonian Orogeny. J. Geol. Soc. London, 135: 597–610.

Swanson, M.T., 1986. Pre-existing fault control for Mesozoic basin formation in eastern North America. Geology, 14: 419–422.

Townsend, C., 1987. Thrust transport directions and thrust sheet restoration in the Caledonides of Finnmark, north Norway. J. Struct. Geol., 9: 345–352.

Wernicke, B. and Burchfiel, B.C., 1982. Modes of extensional tectonics: J. Struct. Geol., 4: 105–115.

Wernicke, B., 1985. Uniform-sense normal simple shear of the continental lithosphere. Can. J. Earth Sci., 22: 108–125.

Wernicke, B. and Axen, G.J., 1988. On the role of isostacy in the evolution of normal fault systems. Geology, 16: 848–851.

Voorhoeve, H. and Houseman, G., 1988. The thermal evolution of lithosphere extending on a low-angle detachment zone. Basin Res., 1: 1–9.

Ziegler, P.A., 1978. North western Europe: tectonics and basin development. Geol. Mijnbouw, 57: 589–626.

C.A. DENGO — *Exxon Exploration Company, P.O. Box 4279, Houston, TX 77210-4279, U.S.A.*

K.G. RØSSLAND — *Esso Norge A/S, P.O. Box 60, N-4033 Forus, Norway*

Present address: *Statoil, UND-LS, ST-FH D1, P.O. Box 300, N-4001 Stavanger, Norway*

Tectonic framework and halokinesis of the Nordkapp Basin, Barents Sea

L.N. Jensen and K. Sørensen

The Nordkapp Basin (NKB) is a deep Paleozoic basin on the Barents Shelf, initiated in a regional rift episode of Late Devonian to Carboniferous age. The basin contains numerous major salt structures, but is surrounded by flat-lying platforms with only minor Post-Paleozoic tectonism.

The Trollfjord–Komagelv Fault Zone (TKFZ) is a fundamental strike-slip zone in northern Finnmark, with a late reactivation in Devonian to Visean time. The late dextral strike-slip is reported to be less than 5 km. We propose that the contemporaneous rifting in the Barents Sea and the movements on the TKFZ were related. The rifting occurred north of the TKFZ, with the Nordkapp Basin as the most prominent rift basin. The extension needed to create these basins was accommodated by dextral strike-slip reactivation of the TKFZ.

A Late Carboniferous to Asselian age is inferred for the salt in the Nordkapp Basin. In Permian times both the platforms and the basin were covered by a carbonate sequence of regional extent, followed in the Early Triassic by a thick sequence of fine-grained clastic sediments. The even thickness of these two sequences demonstrates that no salt movements took place in the Nordkapp Basin at that time.

The salt structures in the NKB are walls and diapirs, while salt pillows are only seen on the basin margins. The normal (Trusheim) model for halokinesis, comprising a pillow stage with primary rim synclines followed by a diapiric phase with secondary rim synclines, cannot be applied to the Nordkapp Basin. Instead, the whole basin evolved as secondary rim synclines with closely spaced diapirs in Triassic times. Based on volume calculations of the secondary rim synclines and the salt we estimate an original maximum salt thickness of more than 2 km in the southwestern part of the basin, and more than 4 km in the northern part. Apparently, such abnormal salt sequences give rise to an unusual halokinetic development which deviates from the Trusheim model, and an alternative model is proposed.

Introduction

Since the opening of the Barents Sea for hydrocarbon exploration the area has attracted much interest and, although drilling results in the last 2 years have been disappointing, there are still major basins in the area that have scarcely been explored. The Nordkapp Basin is one of these (Fig. 1).

Two exploration wells have been drilled on the western flank of the Nordkapp Basin, finding subcommercial quantities of gas and oil (Fig. 2). A third well has been drilled in the central segment of the basin, near the western margin (7228/2-1, Fig. 2). The oil and gas found in all three wells indicates that hydrocarbons have been generated and migrated in the Nordkapp Basin and that seal failure may be the major problem in the area.

The Barents Sea covers an intracratonic basin with Devonian to Quaternary sediments over a basement of Caledonian and pre-Caledonian age. The geological history and tectonic subdivision of the southern Barents Sea have been described in a number of recent publications (Faleide et al., 1984; Gabrielsen et al., 1984, 1990; Rønnevik and Jacobsen, 1984) and only a brief summary focusing on the Nordkapp Basin will be given here.

In Late Devonian to Early Carboniferous time the Barents Sea, including Bjørnøya and Svalbard, was affected by a regional rift episode, and a number of N–S to NE–SW trending basins formed. The most prominent of these, the Nordkapp Basin, is 350 km long and 30–80 km wide, with a general NE–SW trend and a dog-leg shape. The basin is defined at base Cretaceous level (Fig. 1; Gabrielsen et al., 1990). At the Paleozoic level, the southwestern part of the basin is a half-graben, while the central part is a graben (Figs. 2 and 3). Shallow flat-lying platforms, with only minor post-Paleozoic tectonism, surround the Nordkapp Basin, and the basin is characterized by numerous major salt structures. There are no publications solely concerning the basin, but a number of recent abstracts (Lind, 1987; Jensen and Sørensen,

Structural and Tectonic Modelling and its Application to Petroleum Geology edited by R.M. Larsen, H. Brekke, B.T. Larsen and E. Talleraas. NPF Special Publication 1, pp. 109–120. Elsevier, Amsterdam.

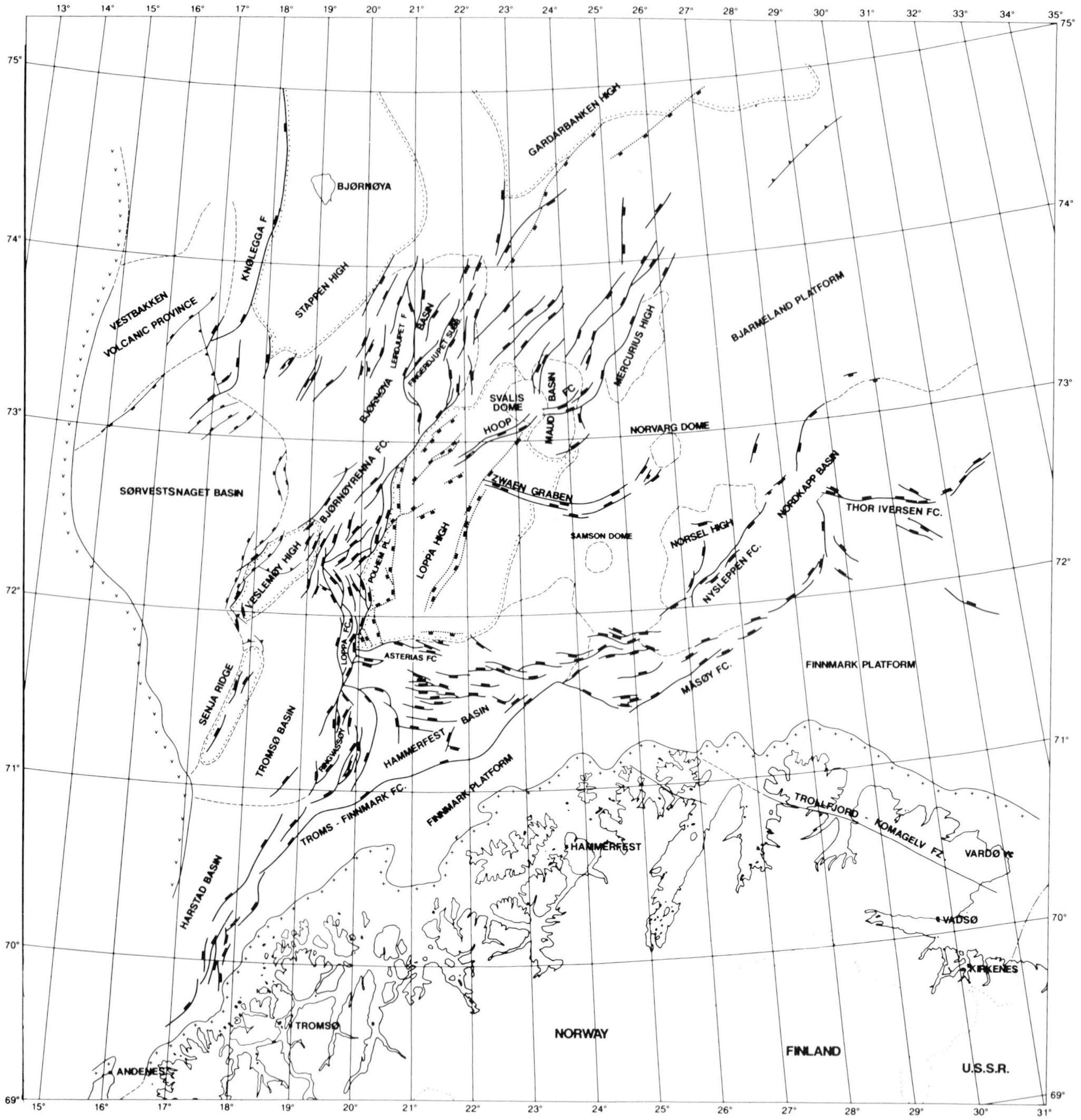

Fig. 1. The main structural elements in the Barents Sea (from Gabrielsen et al., 1990).

1988; Jensen and Broks, 1988) and theses (Krokan, 1988; Bergendahl, 1989) have focussed on the basin history.

The Trollfjord–Komagelv Fault Zone (TKFZ) is a fundamental fault zone in northern Finnmark (Fig. 1) which forms part of the tectonic framework for the Nordkapp Basin. The Proterozoic to Devonian history of the TKFZ is one of large-scale dextral strike-slip movements (Kjøde et al., 1978) followed by a minor reactivation in Devonian to Visean time. The amount and sense of the strike-slip movements have recently been debated (Rice et al., 1989), and the late movements may be transtensional with a normal fault component (cf. Gabrielsen and Færseth, 1989, p. 61).

Jenyon (1988, p. 316) has mentioned the possibility of salt diapirs developing without a preceding pillow phase and primary rim synclines. In this paper we document the existence of such diapirs and show that the whole Nordkapp Basin developed in this way in Triassic time. The salt in the Nordkapp Basin is inferred to be of Late Carboniferous to Asselian age.

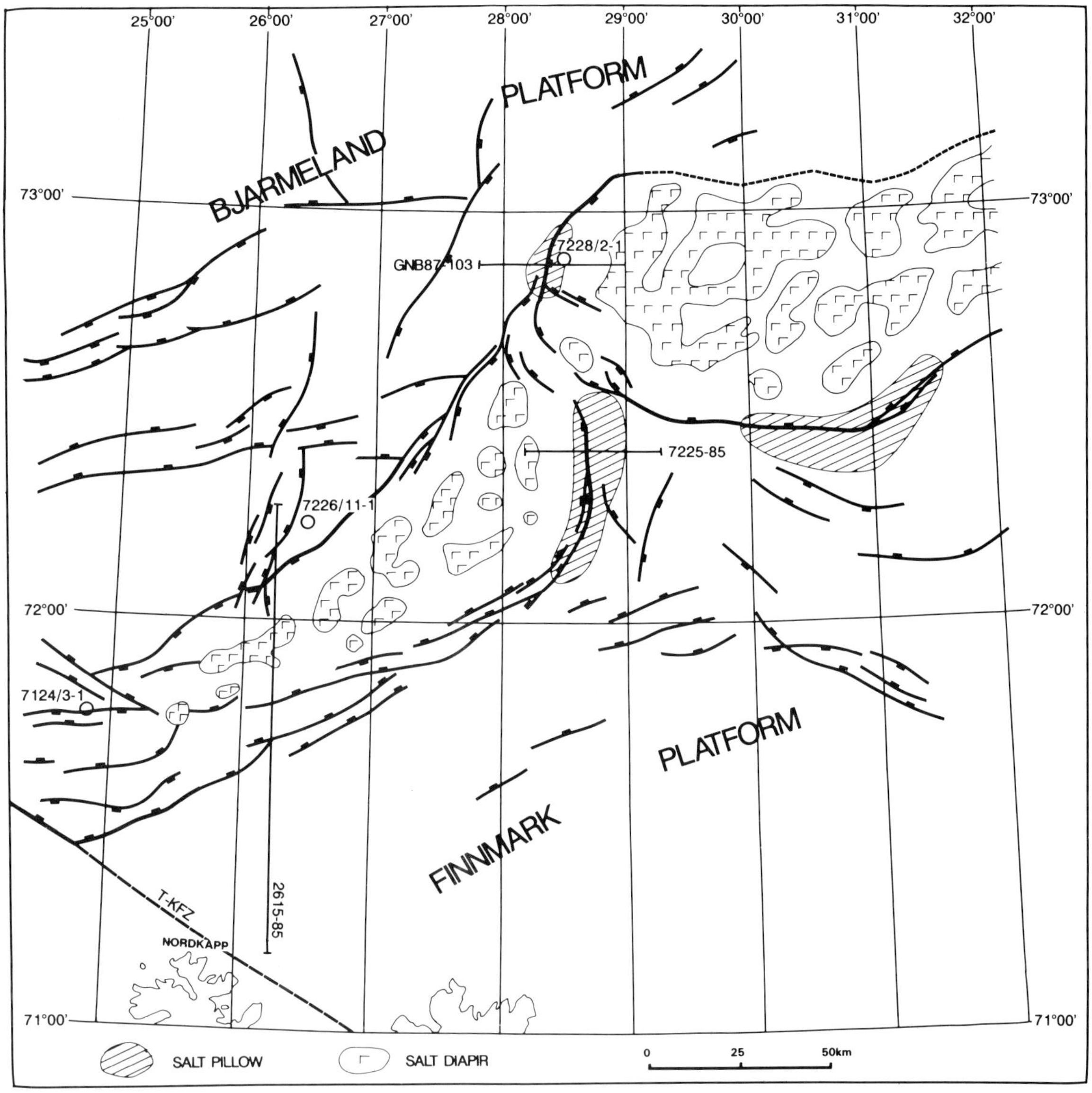

Fig. 2. The main structural elements of the southwestern and central part of the Nordkapp Basin and adjacent platforms. The locations of wells and seismic lines used in this study are shown.

Halokinesis started in Late Scythian time and was reactivated in Tertiary time. The salt structures in the Nordkapp Basin are walls, irregular massifs and diapirs, all piercing to Quaternary level. Salt pillows are only seen on the basin margin. The normal model for halokinesis (Trusheim, 1957), comprising a pillow stage with primary rim synclines followed by a diapiric phase with secondary rim synclines, cannot be applied to the Nordkapp Basin. Instead, in Triassic time, the whole basin evolved as a secondary rim syncline with closely spaced diapirs.

The aim of this paper is to present and discuss two aspects of the tectonic evolution of the Nordkapp Basin: a model for the relation between Late Paleozoic rifting and late movements on the Trollfjord–Komagelv Fault Zone and one for the unusual halokinesis in the basin.

Late movements on the Trollfjord–Komagelv Fault Zone and rifting in the Nordkapp Basin

Rifting in the Barents Sea

A Late Paleozoic rift episode affected the North Atlantic region including the Barents Sea (Hazeldine, 1984). In Svalbard this episode is well documented, with Devonian and Carboniferous sedimentation in grabens and half-grabens (Steel and Worsley, 1984; Worsley and Aga, 1986). On Bjørnøya this episode started in Late Devonian (Fammenian) time with sedimentation in a half-graben east of a major fault (Gjelberg, 1981). The syntectonic sedimentary development is similar in both places, starting with non-marine coal bearing sediments, followed by more marine influenced sequences and ending with

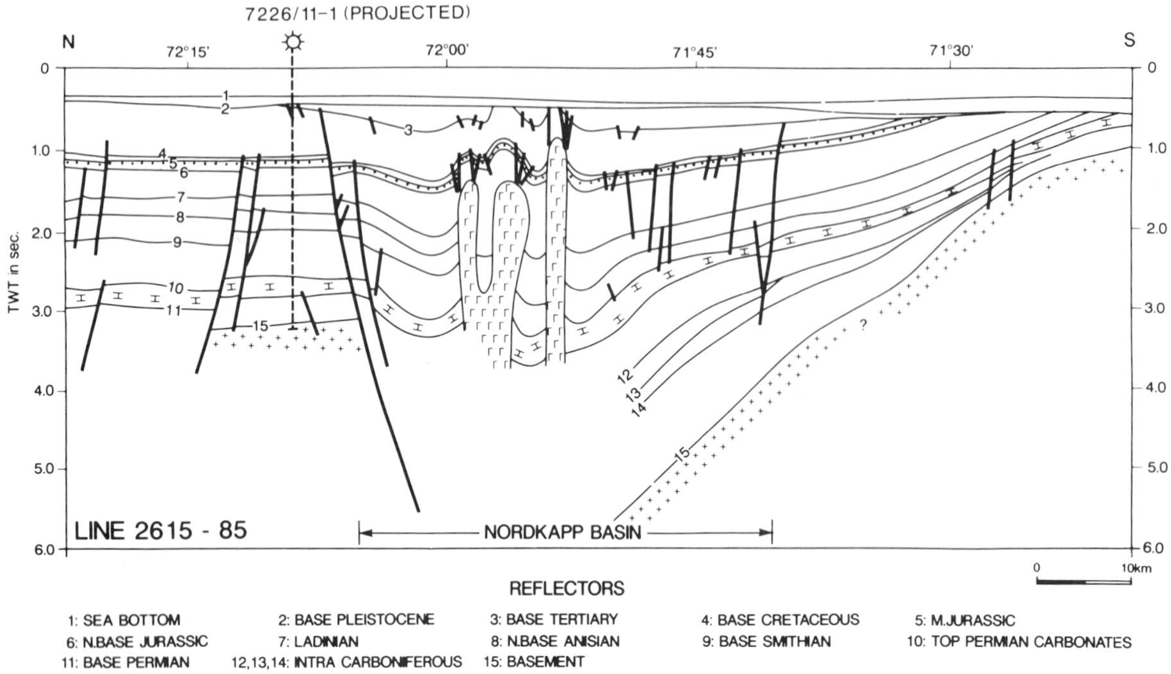

Fig. 3. Profile across the southwestern segment of the Nordkapp Basin. Reflectors are tied to the nearby well, 7226/11-1. The location of the profile is shown in Fig. 2.

carbonate and evaporite dominated sedimentation in the Late Carboniferous.

When hydrocarbon exploration in the Barents Sea started, a similar development was expected there (Rønnevik and Jacobsen, 1984; Faleide et al., 1984) and this has now been confirmed by seismic mapping and exploration drilling. The Late Paleozoic rift basins in the southern Barents Sea are mainly half-grabens, but grabens are also known. The Nordkapp Basin is the most prominent of these Late Paleozoic rift basins in the Barents Sea, and the southwestern segment of the basin is used as an example (Figs. 2 and 3).

The Norsel High (Fig. 1) is a paleo high, with Permian and Carboniferous carbonate and evaporite strata resting on a crystalline basement (Fig. 3). Southeast of the Norsel High, the Nordkapp Basin is developed as a deep (12–14 km) half-graben with more than 6 km of pre-Permian sediments (Fig. 3). The Permian carbonate sequence between reflector 10 and 11 (Fig. 3) is used as a reference level. A deep reflector is seen on this line (2615-85, Fig. 3), and its position at top basement level has been confirmed by shallow drilling near the Finnmark coast. The sequences between the basement and the Permian carbonates are of Late Devonian to Permian age and we anticipate a stratigraphic development similar to that on Svalbard and Bjørnøya.

The known Late Paleozoic basins in the Barents Sea and their major boundary faults have a general N–S to NE–SW trend (Fig. 4), and they all contain evaporites. The Paleozoic origin of the Harstad Basin (Fig. 1) is uncertain, but we assume that the basin formed part of a Paleozoic trend through the Tromsø Basin and Bjørnøya Basin. The salt in the Tromsø Basin is believed to be of Late Carboniferous to Asselian age as in the other Late Paleozoic basins. The minor basin on the southwestern margin of the Loppa High (Fig. 4) has been confirmed by three exploration wells, and the stratigraphic development in this basin is similar to that on Bjørnøya.

Gabrielsen (1984) discussed the importance of the Troms–Finnmark Fault Zone (Fig. 1), but we are not sure about the Late Paleozoic activity or the existence of marked Late Paleozoic basins to the north of this fault. The fault zone is therefore indicated by a dotted line on Fig. 4.

The on-shore position of the Trollfjord–Komagelv Fault Zone is clear, and has been mapped in detail (Lippard and Roberts, 1987). Caledonian and pre-Caledonian activity on the fault zone is believed to be large-scale strike-slip movements (Kjøde et al., 1978). In this paper we will only focus on the post-Caledonian reactivation of the zone. Based on age relations between dyke intrusions and activity on the TKFZ, a Devonian to Visean reactivation with (?dextral) strike-slip of less than 5 km is reported (Kjøde et al., 1978; Gabrielsen and Færseth, 1989).

The offshore continuation of the TKFZ is less clear, and has been discussed in a number of recent publications. Our interpretation is in agreement with that of Gabrielsen and Færseth (1989, their fig. 1)

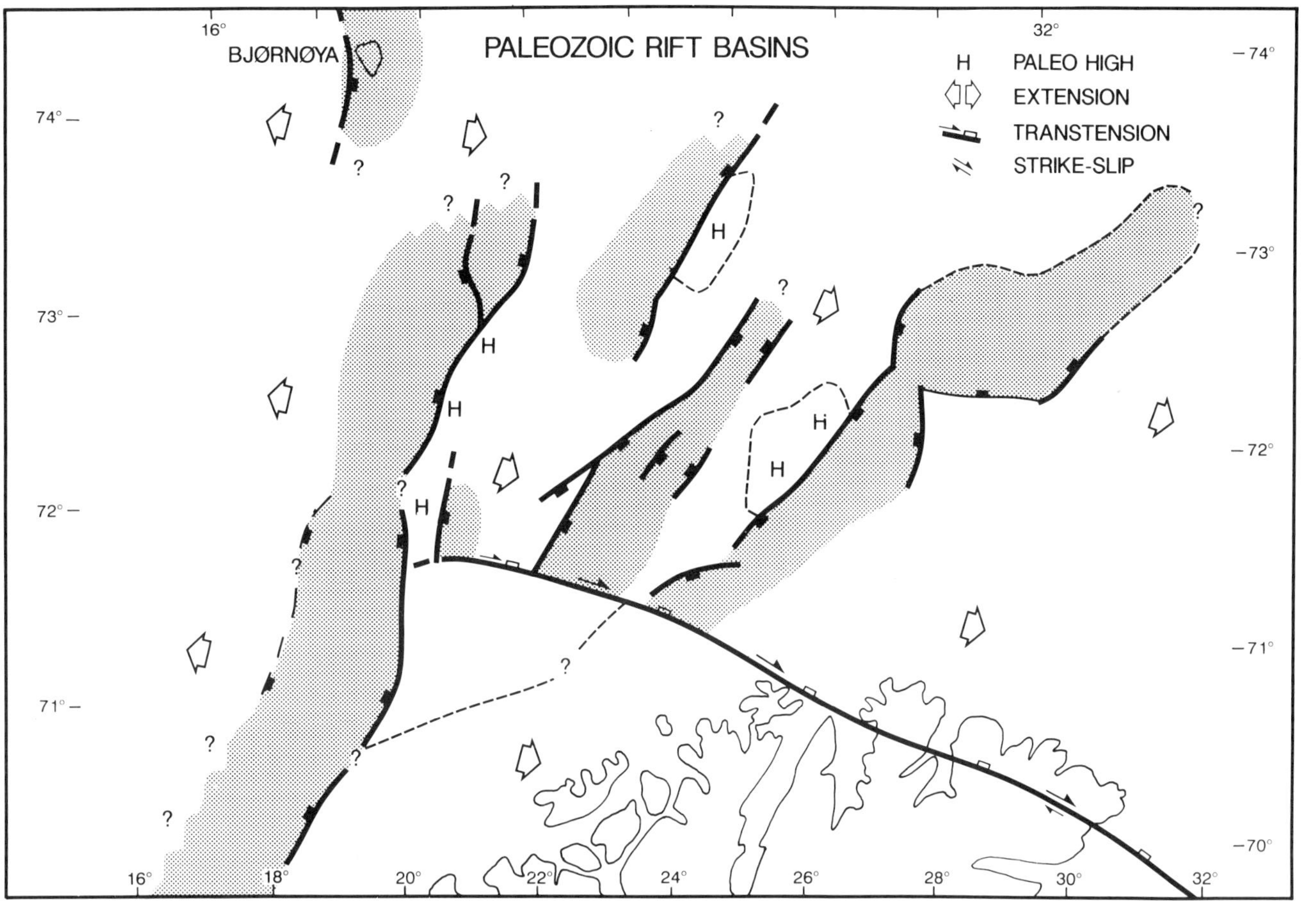

Fig. 4. Paleozoic basins in the southwestern Barents sea. The proposed model is illustrated by arrows indicating areas of regional extension in late Devonian and Carboniferous time. This rifting was accommodated by dextral strike-slip or transtension on the Trollfjord–Komagelv Fault Zone.

and is shown in a simplified form in Fig. 4. The offshore part of the fault zone is discontinuous and obscure in places, and the zone seems to die out to the northwest near the Loppa High.

Rifting related to the Trollfjord–Komagelv Fault Zone

Figure 4 shows our model for the northwestern termination of the Trollfjord–Komagelv Fault Zone and its relation to the Late Paleozoic rift basins in the southwestern Barents Sea. The rifting needed to create the Nordkapp Basin and other basins north of the TKFZ was accommodated as dextral strike-slip on the fault zone. This strike-slip decreases to the northwest and the fault zone dies out in an area of regional extension west of the Loppa High and the Hammerfest Basin (Fig. 4). This extensional termination of a major intracontinental strike-slip zone could be the reason why the offshore part of the zone is so difficult to define.

The inspiration for this model is taken from Reading (1980, his fig. 3d), who described the termination of major intracontinental strike-slip zones, and Harding et al. (1985, their fig. 22f) who described a tear fault accommodating differential extension. An analogous example is the Garlock Fault in California, as described by Davis and Burchfiel (1973, their fig. 4).

It is important to stress that the tectonic model (Fig. 4) is partly based on published data for the Trollfjord–Komagelv Fault Zone, and that these data are under debate (cf. Rice et al., 1989). The model proposed will also work if the late movements on the TKFZ turn out to be larger (dextral shear sense) or if they have a transtensional component. In fact, the model will work in all cases where the late movements on the TKFZ accommodate differential extension north and south of the zone.

Geologic history and halokinesis of the Nordkapp Basin

Regional stratigraphic information and shallow drilling indicate that the Nordkapp Basin was initiated in Late Devonian time. We expect the Late Devonian and Early Carboniferous sedimentation in the basin to be dominated by clastic sequences, followed by carbonate and evaporite sedimentation in Late Carboniferous time. The geographical extent of the Late Paleozoic rift basin was larger than that of the Nordkapp Basin as defined at the base Cretaceous level. The most important difference is in the southwestern part, where the Paleozoic basin extends further to the SW and terminates against the offshore part of the Trollfjord–Komagelv Fault Zone (Fig. 4).

Age of the salt in the Nordkapp Basin

Using the Early Permian regional carbonate sequence as a reference level, it is possible to demonstrate that the salt in the Nordkapp Basin must be of Late Carboniferous to Asselian age (Figs. 3 and 5). Around the central part of the basin, a thin salt sequence of similar age is extending onto the platforms 30–60 km beyond the basin margins. We assume that the lateral equivalent of the salt in the basin consists of interbedded sequences of carbonates and anhydrite, as found in wells 7124/3-1 and 7226/11-1.

South of well 7124/3-1 (Fig. 2), seismic lines indicate the existence of deep salt pillows near the basin boundary fault and the offshore part of the TKFZ soles out in these salt pillows. In an early interpretation of the Nordkapp Basin, Faleide et al. (1984, their fig. 13) indicated the existence of two salt layers. Based on our work, we see no need for two salt layers to explain the features seen in the Nordkapp Basin, and it is hard to imagine that an early salt layer could have survived the pre-salt tectonic episode seen in Fig. 5. This tectonic episode is anticipated to be of Early Carboniferous age.

The amount of salt in the Nordkapp Basin

In order to estimate the original salt thickness in the southwestern and central part of the Nordkapp Basin, volume calculations have been carried out. Two methods were used; direct calculation of the present salt volume and indirect calculation based on excess volume in the secondary rim synclines. These methods are described by Seni and Jackson (1983) and Sørensen (1986).

Due to Late Tertiary erosion all diapirs in the Nordkapp Basin pierce to Base Quaternary level, and have their maximum horizontal cross-sectional area at a level somewhere between Upper Triassic and Base Quaternary. The depth to the base of the diapirs is listed using three different depth estimates (A,B and C in Table 1). As a first approximation, the present salt volume was calculated assuming vertical sides for the diapirs (total volume, Table 1). This is an overestimation, because some diapirs have a spherical top and some are clearly mushroom shaped. However, some salt may still exist in the source layer and some may have been lost due to dissolution, extrusion and Late Tertiary erosion. Assuming an overestimation of 25% or 50% and using the three different depth estimates, we end up with a total salt volume in the Nordkapp Basin of 22 500 to 30 000 km^3 (Table 1). Of this salt volume, approximately 25% is in the southwestern segment of the basin and 75% in the central segment west of 32°E (Table 1).

The areas of the Nordkapp Basin at Middle Jurassic and Base Permian levels have been calculated (Table 1). If we redistribute the above volumes of salt in the basin using the area at Base Permian level, the original average salt thickness is 1400–1800 m in the southwestern segment and 2300–3100 m in the central segment of the basin (Table 1). The base of the salt is not horizontal. It is clearly faulted in the central part of the basin, and forms a half-graben in the southwestern part (Fig. 3). An original maximum salt thickness of 2000 m in the SW part of the basin and 4000–5000 m in the central part (cf. Jensen and Sørensen, 1988) is therefore reasonable. This has also been confirmed by two-dimensional calculations of the excess rim syncline volume.

Bergendahl (1989) has confirmed these estimates for the SW part of the Nordkapp Basin, using both methods for volume calculations. Based on gravimetric modelling, Krokan (1988) also indicated an original maximum salt thickness of 4000–5000 m in the central part of the basin, but he gave no salt volume estimate.

TABLE 1

Areal content, salt volume and original salt thickness in the Nordkapp Basin west of 32°E

Areal content (km^2)

Stratigraphic level:	Middle Jurrasic	Base Permian
Southwestern segment	5507	4244
Central segment	8271	7120
Total (W of 32°E)	13778	11364

Salt volume (km^3)

Base salt depth estimate:	A	B	C
Southwestern segment	8700	9200	9700
Central segment	25100	26400	27700
Total (W of 32°E)	33800	35600	37400
25% overestimated	27700	28500	30000
50% overestimated	22500	23700	25000

Original salt thickness[a] (m)

Southwestern segment:	1400–1800 m
Central segment:	2300–3100 m

[a] Thickness range for depth estimate A to C and 25% to 50% overestimated salt volume.

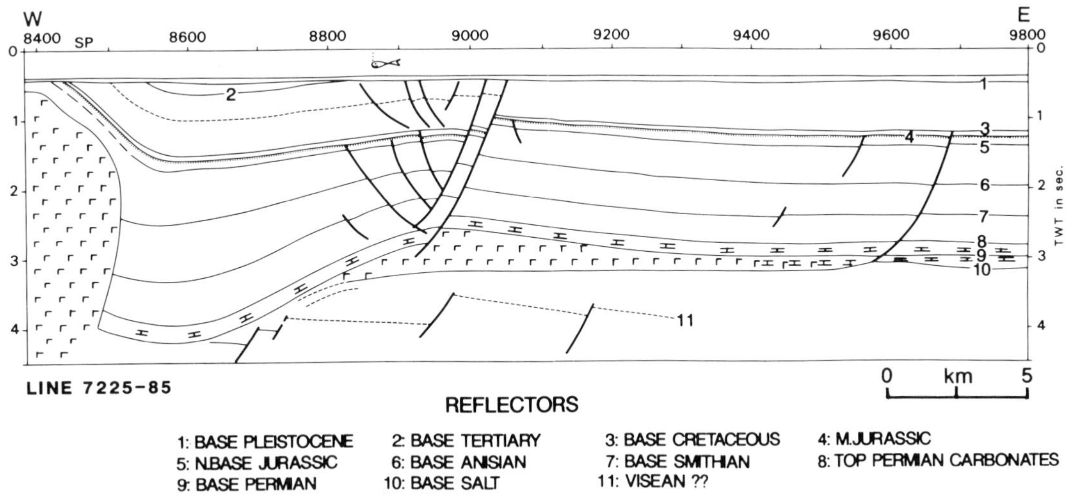

Fig. 5. Line drawing of a seismic section across the eastern margin of the Nordkapp Basin (line 7225-85, Fig. 2). From the known age of the overlying carbonate sequence, a Late Carboniferous to Asselian age is inferred for the salt. A pre-salt tectonic episode is clearly seen on this line.

Post-Carboniferous development of the Nordkapp Basin

The Post-Carboniferous development of the Nordkapp Basin will be illustrated using a reconstruction of a typical seismic line (GNB-87-108) from the central part of the basin (Fig. 6). We use a reconstruction in two-way time instead of depth, to allow tie and sequence identification from our line.

Before halokinesis started in Late Scythian time the salt in the Nordkapp Basin was covered by a regional Early Permian carbonate sequence, Late Permian shale and chert sequences and Early Scythian shales (Fig. 6). The even thicknesses of these sequences on the platforms and in the basin indicate that no differential subsidence or salt movement took place at that time.

In the nearest wells on the Bjarmeland Platform, the Permian carbonate sequence is 540–800 m thick, the Late Permian sequence is 220–410 m and the Early Scythian sequence is 820–1010 m. These are present-day non-decompacted thicknesses. It is, therefore, reasonable to assume that before halokinesis started, the salt in the Nordkapp Basin was covered by 1600–2200 m of sediments, comprising a rigid "lid" of carbonates and a rapidly deposited shaly sequence. When 2000–4000 m of salt were covered by such a sequence, a strong gravitational instability must have existed immediately prior to the onset of diapirism.

Triassic diapirism

Diapirism commenced in Late Scythian time, and the whole Nordkapp Basin developed as one major rim syncline (Fig. 6). At the same time salt pillows and growth faults formed at the basin margin (Figs. 5 and 6). The detailed development of these salt pillows are explained by Gabrielsen et al. (this volume). Near the salt diapir in Fig. 6, minor erosion of the Early Scythian sequence is seen, indicating that a salt pillow may have existed. All diapirs in the basin seems to have been initiated at the same time. The diapirism was strongest from Late Scythian to Late Triassic time, and enormous secondary rim synclines formed. Stratigraphic traps may have formed in these rim synclines (Fig. 6). The reflectors in the secondary rim synclines diverge towards the diapirs and there is no evidence of syn-diapiric sediment progradation into the basin.

In Late Triassic time, the diapirism stopped, and the diapirs were covered by sediments (Fig. 6). There was still some growth faulting at the basin margin, and the salt pillow also grew a little (Fig. 6). The diapirism probably stopped because the gravitational instability had disappeared and the salt source layer had been exhausted.

Post-Triassic history

The Jurassic sequences covered the diapirs and there was only minor differential subsidence of the Nordkapp Basin at that time (Fig. 6). A regional tectonic episode affected the Barents Sea in Late Jurassic to Early Cretaceous time and evidence of this episode is seen in and near the Nordkapp Basin. But the faults are small, and the episode had little influence in the Nordkapp Basin area.

A 1000–1500 m thick Cretaceous sequence covered the area. There are no indications of differential subsidence or salt movement in the Nordkapp Basin at that time (Fig. 6). From well information, we know that 1000–1500 m of Early Tertiary sediments covered the area (Fig. 6) and the weight of these post-Ladinian sequences probably recreated a gravi-

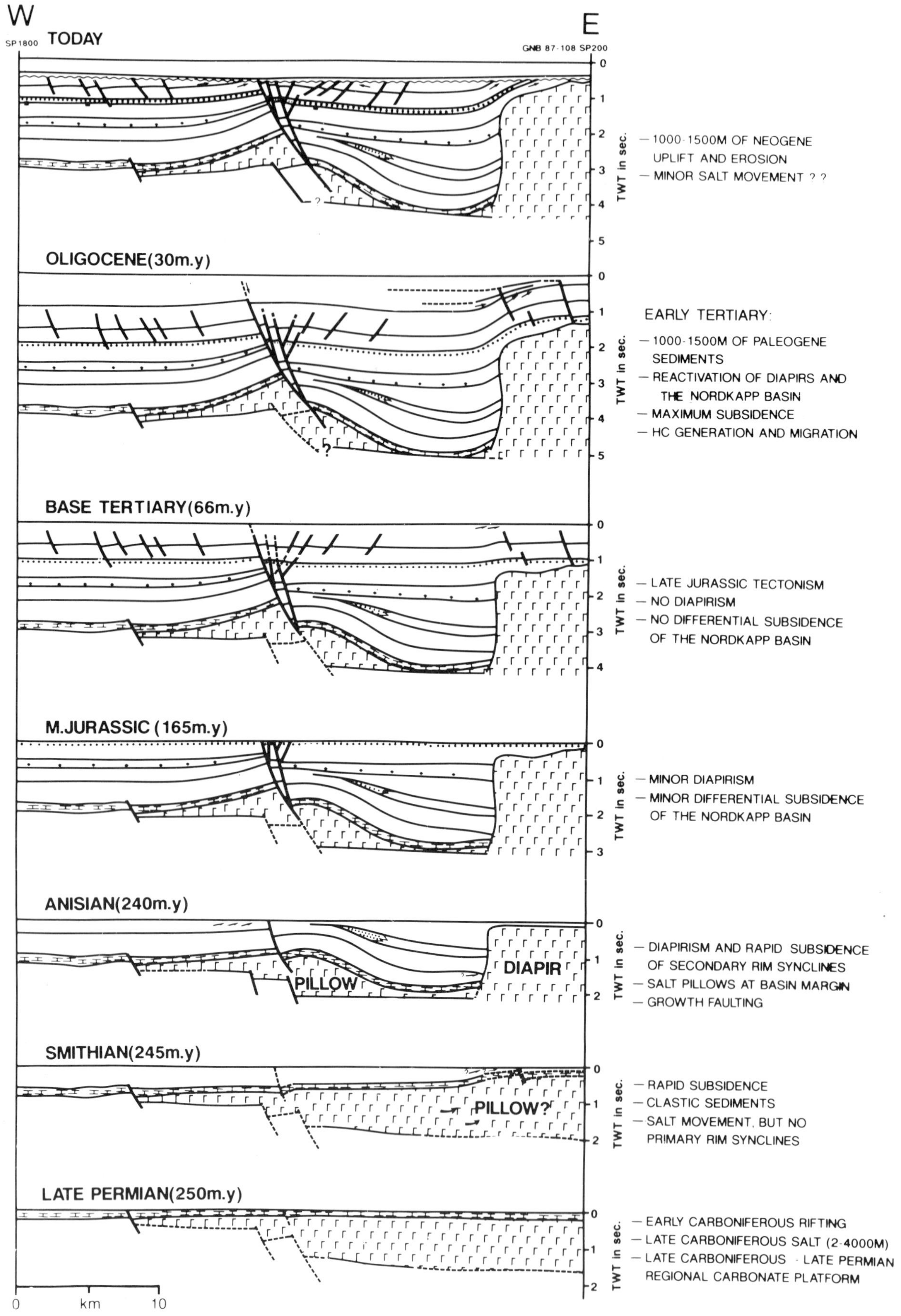

Fig. 6. The geological development and halokinesis of the Nordkapp Basin illustrated by a reconstruction of seismic line GNB-87-108. The location of the line is shown in Fig. 2. Further details are explained in the text.

tational instability in the Nordkapp Basin. A regional compressive episode affected the area in Early Tertiary time, and the combined effect of gravitational instability and compression caused some reactivation of the diapirs in the Nordkapp Basin. Clear evidence of this reactivation is seen in the southwestern part of the basin, where Early Tertiary sediments are preserved.

In Late Tertiary time a regional episode of uplift and erosion affected the Barents Sea (Nyland et al., this volume). At the location of the reconstructed line, the erosion was of the order of 1200–1500 m, and all the Early Tertiary sediments were removed (Fig. 6). The Neogene erosion probably removed some of the salt. The area then subsided and was covered by 100–200 m of Pleistocene sediments and 300–400 m of water (Fig. 6).

A halokinetic model for the Nordkapp Basin

Trusheim (1957) introduced the term halokinesis and presented a general model for gravity-driven salt tectonics (Fig. 7). His model is now widely accepted and seems to apply to many of the worlds salt basins (Halbouty, 1979; Seni and Jackson, 1983; Jackson and Talbot, 1986; Jaritz, 1987). Trusheim's model comprises a salt pillow phase with related primary rim synclines, followed by a diapiric phase with formation of secondary rim synclines (Fig. 7).

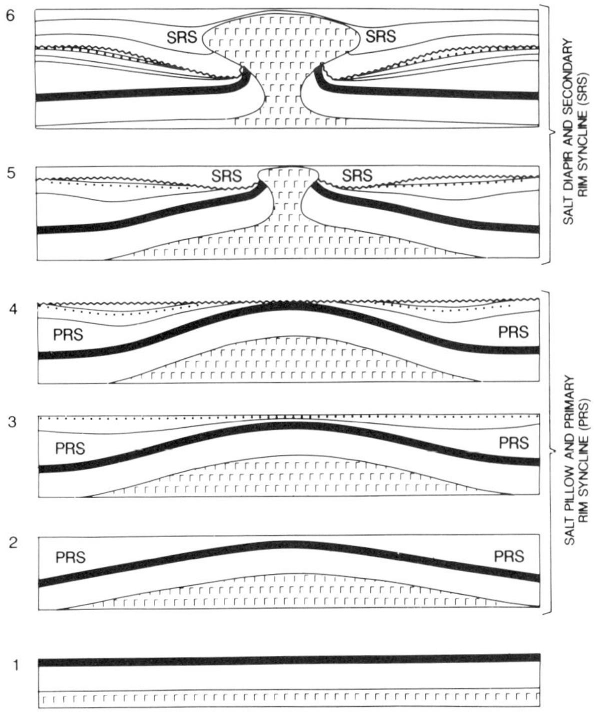

Fig. 7. The Trusheim (1957) model for halokinesis. The salt-tectonic development of the Nordkapp Basin clearly deviates from this model (see Figs. 6 and 8). *PRS* = primary rim syncline; *SRS* = secondary rim syncline.

Bishop (1978) later criticized this buoyancy model and contributed to the understanding of piercement diapirism. Jenyon (1988, p. 316) mentioned a possible situation in which salt diapirism may occur without a pillow phase and the formation of primary rim synclines. Such a situation would be an exception to the general Trusheim model (Fig. 7). We believe that the whole Nordkapp Basin developed in this way in Triassic times (Fig. 8).

The driving force for the diapirism in the Nordkapp Basin must have been a strong gravitational instability, but the triggering mechanism is less clear. Near the diapir in Fig. 6, there is a minor erosion of the Early Scythian sequence, and a short-wavelength pillow or "bulge" may have existed (Fig. 6). This model is shown in Fig. 8 (diapir C). But this phenomenon is only seen at a few diapirs in the Nordkapp Basin and there are no indications of related primary rim synclines.

A tectonic episode may have been the triggering mechanism for the diapirs. A minor growth faulting episode of Late Scythian age is known from the Bjarmeland Platform, and it probably also affected the Nordkapp Basin. Because of the strong gravitational instability in the Nordkapp Basin at the onset of this extension, salt may have been injected into the faults and developed directly into a piercement (diapirs A and B, Fig. 8). Jackson and Talbot (1986, their fig. 5f) and Jenyon (1988) both mention such a mechanism.

On the Finnmark Platform, prograding sequences of Early Scythian age are seen from the Finnmark coast to the margin of the Nordkapp Basin. These sequences are 200–500 m thick and prograde towards the north and northwest. The progradational pattern has not been identified in the Nordkapp Basin, but the influx of sediments from the Finnmark Platform may have contributed to the initiation of diapirism. Such a triggering mechanism has been suggested by Bishop (1978, fig. 3), Halbouty (1979) and Woodbury et al. (1980, fig. 3).

Each of the above mechanisms or a combination of them, may have triggered the unusual diapirism in the Nordkapp Basin. Unfortunately the intruding diapirs and later erosion together have destroyed the evidence for these triggering mechanisms, and a diapir "frozen" in the initial phase does not exist in the Nordkapp Basin.

In the absence of clear evidence for the triggering mechanism within the Nordkapp Basin an analogous example can help. We have searched for published analogues from salt basins around the world, but have been able to find only two. The Louann salt in the Gulf of Mexico is an analogue, but only with respect to salt thickness and the closely spaced diapirs

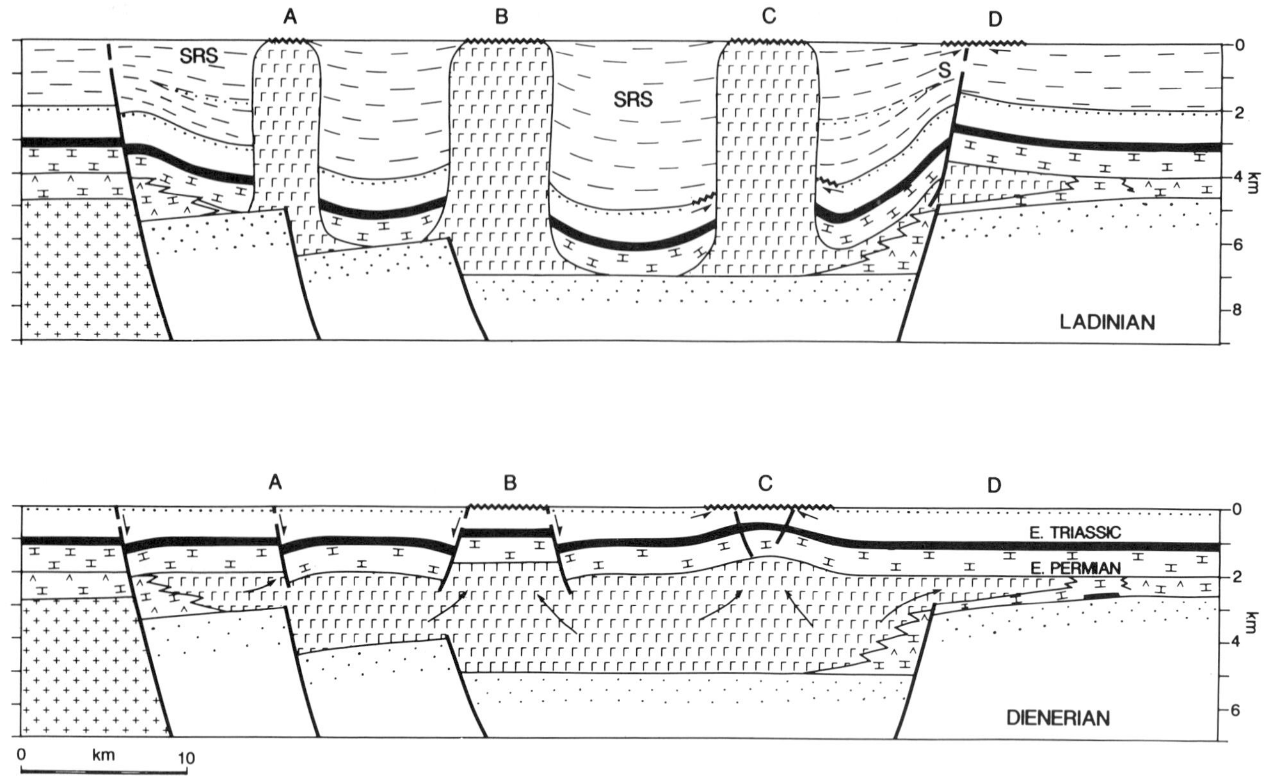

Fig. 8. A halokinetic model for the Nordkapp Basin. Details are explained in the text.

on the outer shelf (Seni and Jackson, 1983, fig. 1). This is because the Louann salt appears to have moved laterally under the weight of the prograding Tertiary sediments (Halbouty, 1979; Woodbury et al., 1980). The salt in the Nordkapp Basin, however, is confined to a rift basin and has not moved laterally as is the case for the Louann salt. The Gulf of Mexico is, therefore, not an exact analogy.

The second analogous example is the North Caspian Depression in the U.S.S.R. Only a few translated publications exist for this basin. According to Kapustin et al. (1981), 2–3.5 km of Kungurian salt were deposited in this rift basin. A profile published by Avrov et al. (1986) may indicate that the closely spaced diapirs in the North Caspian Depression developed in the same way as those in the Nordkapp Basin. The North Caspian Depression is a prolific oil-producing basin and may be the only exact analogy to the Nordkapp Basin.

Conclusions

The tectonic and halokinetic models for the Nordkapp Basin can be summarized as follows:

(1) There may be a genetic relation between the Late Paleozoic rift basins in the Barents Sea and the contemporaneous reactivation of the Trollfjord–Komagelv Fault Zone. The extension needed to create these basins was accommodated as strike-slip on the Trollfjord–Komagelv Fault Zone. This genetic relationship may also explain why the offshore continuation of the fault zone is so difficult to define.

(2) The Trollfjord–Komagelv Fault Zone dies out in an area of regional extension west of the Loppa High.

(3) The salt in the Nordkapp Basin is of Late Carboniferous to Asselian age, and only one salt layer is needed to explain the features seen in the basin.

(4) An original salt volume of 22 500–30 000 km^3 was deposited in the Nordkapp Basin. Based on this volume we estimate an original maximum salt thickness of more than 2 km in the southwestern part and more than 4 km in the northern part of the basin.

(5) During Triassic times the whole Nordkapp Basin developed as a number of secondary rim synclines related to closely spaced diapirs. Evidence of a preceding salt pillow phase with primary rim synclines has not been found. The Nordkapp Basin is, therefore, a rare exception to Trusheim's (1957) halokinetic model.

The halokinetic model for the Nordkapp Basin has implications for the hydrocarbon exploration in the basin. It is important to realize that we cannot expect to find a number of hydrocarbon trap types related to salt pillows and primary rim synclines (cf. Halbouty, 1979, fig. 6-1) because this phase of halokinetic development is absent in the Nordkapp Basin. Moreover, the Late Tertiary erosion has removed the possibility of finding hydrocarbon traps over the crest of the diapirs. We are, therefore, left with traps related to the flanks of the diapirs and the secondary rim synclines. A number of such traps have already been mapped.

Acknowledgements

We thank the management of Statoil Exploration for permission to publish this paper and we acknowledge the helpful comments from Anthony Spencer, Elin Kjemtrup and two anonymous referees.

References

Avrov, V.P., Balgimbayev, M.B., Krylov, N.A. and Sagingaliyev, A.B., 1986. Possibilities for discovery of new oil fields in subsalt carbonate sediments of the North Caspian Depression. Translation in: Pet. Geol., 22(9): 391–399.

Bergendahl, E., 1989. Halokinetisk utvikling av Nordkappbassengets sørvestre segment. Unpublished Cand. Scient. Thesis, University of Oslo, 120 pp.

Bishop, R.S., 1978. Mechanism for emplacement of piercement Diapirs. Am. Assoc. Pet. Geol. Bull., 62: 1561–1583.

Davis, G.A. and Burchfiel, B.C., 1973. Garlock Fault: an intracontinental transform structure, Southern California. Geol. Soc. Am. Bull., 84: 1407–1422.

Faleide, J.I., Gudlaugsson, S.T. and Jacquart, G., 1984. Evolution of the Western Barents Sea. Mar. Pet. Geol., 1: 123–150.

Gabrielsen, R.H., Færseth, R., Hamar, G.P. and Rønnevik, H., 1984. Nomenclature of the main structural features on the Norwegian Continental Shelf north of the 62nd parallel. In: A.M. Spencer et al. (Editors), Petroleum Geology of the North European Margin. Graham and Trotman, London, pp. 41–60.

Gabrielsen, R.H., Færseth, R., Jensen, L.N., Kalheim, J.E. and Riis, F., 1990. Structural elements of Norwegian Continental Shelf. Part I: The Barents Sea region. Norwegian Petroleum Directorate, Bull., 6, 33 pp.

Gabrielsen, R.H., 1984. Long-lived fault zones and their influence on the tectonic development of the Southwestern Barents Sea. J. Geol. Soc. London, 141: 651–662.

Gabrielsen, R.H. and Færseth, R.B., 1989. The inner shelf of North Cape, Norway and its implication for the Barents Shelf-Finnmark Caledonide boundary. A comment. Nor. Geol. Tidsskr., 69: 57–62.

Gabrielsen, R.H., Kløvjan, O.S., Rasmussen, A. and Stølan, T., 1992. Interaction between halokinesis and faulting: structuring of the margins of the Nordkapp Basin, Barents Sea region. In: R.M. Larsen, H. Brekke, B.T. Larsen and E. Talleraas (Editors), Structural and Tectonic Modelling and its Application to Petroleum Geology. Norwegian Petroleum Society (NPF), Special Publication 1. Elsevier, Amsterdam, pp. 121–131 (this volume).

Gjelberg, J.G., 1981. Upper Devonian (Famennian)-Middle Carboniferous succession of Bjørnøya. Nor. Polarinst. Skr., 174, 67 pp.

Halbouty, M.T., 1979. Salt Domes, Gulf Region, United States and Mexico, Gulf Publishing Co., Houston, Texas, 2nd ed., 561 pp.

Harding, T.P, Vierbuchen, R.C. and Christie-Blick, H.N., 1985. Structural styles, plate-tectonic settings, and hydrocarbon traps of divergent (transtensional) wrench faults. In: K.T. Biddle and H.N. Christie-Blick (Editors), Strike-slip Deformation, Basin Formation and Sedimentation. Soc. Econ. Paleontol. Mineral., Spec. Publ., pp. 52–77.

Hazeldine, R.S., 1984. Carboniferous North Atlantic paleogeography: stratigraphic evidence for rifting, not megashear or subduction. Geol. Mag., 121: 443–463.

Jackson, M.P.A. and Talbot, C.J., 1986. External shapes, strain rates and dynamics of salt structures. Geol. Soc. Am. Bull., 97: 305–323.

Jaritz, W., 1987. The origin and development of salt structures in Northwest Germany. In: I. Lerche and J.J. O'Brien (Editors), Dynamical Geology of Salt and Related Structures. Academic Press, New York, N.Y., pp. 479–493.

Jensen, L.N. and Sørensen, K., 1988. Geology and salt tectonics of the Nordkapp Basin, Barents Sea. Abstract, 18. Nord. Geol. Vintermøde, København, 1988, Danm. Geol. Unders., pp. 187–188.

Jensen, L.N. and Broks, T.M., 1988. Late movements on the Trollfjord-Komagelv Fault Zone and rifting in the Nordkapp Basin. VI Annu. TSGS Meet., Inst. Geol., Univ. Oslo, Intern Skriftser., 54, pp. 24–25 (abstract).

Jenyon, M.K., 1988. Some deformation effects in a clastic overburden resulting from salt mobility. J. Pet. Geol., 11: 309–324.

Kapustin, I.N., Kiryukhin, L.G., Payrazyan, V.V. and Rasmyshlyayev, A.A., 1981. Geologic history of the North Caspian Depression and time of formation zones of oil-gas accumulation. Translation in: Pet. Geol., 22(10): 483–485.

Kjøde, J. Storetvedt, K.M. Roberts, D. and Gidskehaug, A., 1978. Paleomagnetic evidence for large-scale dextral movement along the Trollfjord–Komagelv Fault, Finnmark, north Norway. Phys. Earth Planet. Inter., 16: 132–144.

Krokan, B., 1988. Et gravimetrisk studium av Nordkappbassenget. Cand. Sci. Thesis, Univeristy of Oslo, 174 pp (unpublished).

Lind, E., 1987. The Nordkapp Basin, a major salt feature in the SW Barents Sea. 4th Annu. Meet., TSGS Group, Nor. Geol. Tidsskr., 67: 435 (abstract).

Lippard, S.J. and Roberts, D., 1987. Fault systems in Caledonian Finnmark and the southern Barents Sea. Nor. Geol. Unders. Bull., 410: 55–64.

Nyland, B., Jensen, L.N., Skagen, J., Skarpnes, O. and Vorren, T., 1992. Tertiary uplift and erosion in the Barents Sea; magnitude, timing and consequences. In: R.M. Larsen, H. Brekke, B.T. Larsen and E. Talleraas (Editors), Structural and Tectonic Modelling and its Application to Petroleum Geology. Norwegian Petroleum Society (NPF), Special Publication 1. Elsevier, Amsterdam, pp. 153–162 (this volume).

Reading, H.G., 1980. Characteristics and recognition of strike-slip fault systems. Spec. Publ. Int. Assoc. Sedimentol., 4: 7–26.

Rice, A.H.N., Fearn, P. and Townsend, C., 1989. Fault patterns in Northeast Finnmark and its implications for the basement structure of the Southern Barents Sea. Paper presented at Norwegian Petroleum Society (NPF) Workshop, 18–20 October 1989, Stavanger.

Rønnevik, H. and Jacobsen, H.P., 1984. Structural highs and basins in the western Barents Sea. In: A.M. Spencer et al. (Editors), Petroleum Geology of the North European Margin. Graham and Trotman, London, pp. 98–107.

Seni, S.J. and Jackson, P.A., 1983. Evolution of salt structures, east Texas diapir province, Parts 1 and 2. Am. Assoc. Pet. Geol. Bull., 67: 1219–1244; 1245–1274.

Steel, R.J. and Worsley, D., 1984. Svalbard's post-Caledonian strata — an atlas of sedimentational patterns and paleogeographic evolution. In: A.M. Spencer et al. (Editors), Petroleum Geology of the North European Margin. Graham and Trotman, London, pp. 109–135.

Sørensen, K., 1986. Rim syncline volume estimation and salt diapirism. Nature, 319: 23–27.

Worsley, D. and Aga, O., 1986. The Geological History of Svalbard. Statoil, 121 pp.

Trusheim, F., 1957. Über Halokinese und ihre Bedeutung für die strukturelle Entwicklung Norddeutschlands. Z. Deutsch. Geol. Ges., 109: 111–151.

Woodbury, H.O., Murray, I.B. and Osborne, R.E., 1980. Diapirs and their relation to hydrocarbon accumulation. In: A.D. Miall (Editor), Facts and Principles of World Petroleum Occurrence. Canadian Society of Petroleum Geologists, pp. 119–142.

L.N. JENSEN *Statoil A/S, P.O. Box 300, N-4001 Stavanger, Norway*

K. SØRENSEN *Statoil A/S, St. Annæ Plads 13, DK-1298 København K, Denmark*

Interaction between halokinesis and faulting: structuring of the margins of the Nordkapp Basin, Barents Sea region

R.H. Gabrielsen, O.S. Kløvjan, A. Rasmussen and T. Stølan

Salt is found in three alternative structural settings in the southwestern Barents Sea region: (1) as domes and pillows in the platform areas, (2) as pillows, stocks and walls in some basins, and (3) as pillows along some graben margins.

In the Nordkapp Basin there is clear evidence that halokinesis has interacted with faulting during structuring of the graben margins. Furthermore, periods with salt pillow growth are associated with condensed sedimentary sequences in the footwalls and expanded sequences in the hanging walls relative to the deep faults with which the pillows are associated. In general, this suggests flow of salt towards the basin margins.

Some salt pillows are divided into subparallel elongated structures where they overlie deep faults. The crestal parts of the salt pillows are frequently faulted, and in some cases characterized by reverse fault geometries. The following sequence of events is proposed to explain this configuration: (1) salt flow directed towards basin margins; (2) continued deep faulting below the salt pillows associated with continued basin subsidence and basinward rotation of the basin margin areas; and (3) reactivation of crestal faults under differential compaction.

Introduction

Since the early days of exploration in the Barents Sea continental shelf, the presence of large amounts of evaporites in some of the basins of this area has attracted attention (Hinz and Weber, 1975; Rønnevik et al., 1975; Hinz and Schlüter, 1978). From the early works it soon became clear that there is much contrast between the different basins as regards the amount of evaporites and, consequently, tectonic styles. The major salt deposition seems to have been concentrated in the Tromsø and Nordkapp Basin areas (Øvrebø and Talleraas, 1976, 1977; Rønnevik, 1981; Rønnevik et al., 1982; Faleide et al., 1984), although several basins of Late Palaeozoic age in the Barents Sea continental shelf are known to contain evaporites (Jensen and Sørensen, this volume).

Salt is found in three different settings in the southern Barents Sea:

- domes, pillows and sheets in platform areas
- pillows, stocks and walls in the basins
- pillows along the graben margins.

The present study concentrates upon a set of pillow-like salt structures which are found along the northwestern and southeastern margins of the Nordkapp Basin, i.e., along the Nysleppen and the Måsøy Fault Complexes (Figs. 1 and 2).

The tectonic setting and history of the Nordkapp Basin

The Nordkapp Basin is an ENE–WSW-trending structure located between 71°30′N, 25°E and 73°30′N, 34°E. It is more than 100 km long and has a width varying between 30 and 80 km. A kink in the basin margin at about 72°30′N, 28°30′E gives the structure the shape of a dog-leg. The basin is bordered by the Nysleppen, Måsøy and Thor Iversen Fault Complexes separating it from the Finnmark Platform to the southeast, and the Bjarmeland Platform to the northeast. Although the early (i.e., pre-late Carboniferous) history of the basin is poorly known, it is likely that the major fault trends of the area were established by late Devonian to early Carboniferous times (Rønnevik et al., 1982).

The southeasternmost part of the Nordkapp Basin is delineated to the northwest by the Norsel High, which is a part of the Bjarmeland Platform (Gabrielsen et al., 1990). Defined on the pre-Permian levels, the southwestern part of the Nordkapp Basin is a half-graben, whereas sections across its central and northern parts show a more symmetrical structure.

A NW–SE-oriented cross-section through the graben shows that the Nordkapp Basin has a pronounced axial development of large salt stocks,

Structural and Tectonic Modelling and its Application to Petroleum Geology edited by R.M. Larsen, H. Brekke, B.T. Larsen and E. Talleraas. NPF Special Publication 1, pp. 121–131. Elsevier, Amsterdam.

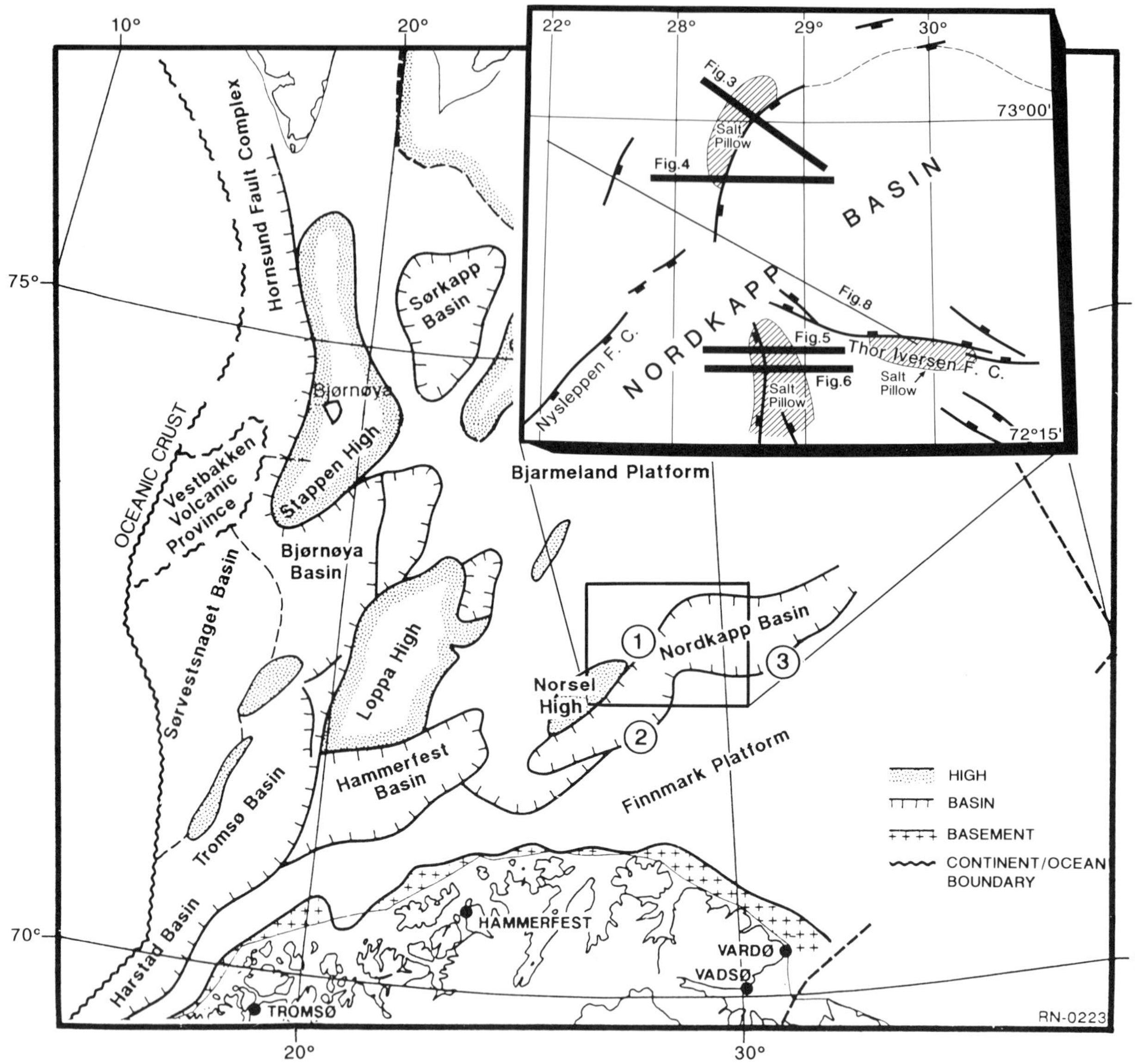

Fig. 1. Structural elements of the Barents Sea. From Gabrielsen et al. (1990). *1* = Nysleppen Fault Complex; *2* = Måsøy Fault Complex; *3* = Thor Iversen Fault Complex. The location of the lines (Figs. 3–6 and 8) are indicated in the inset.

whereas the basin margin zones are characterized by extensional faults associated with salt pillows (Fig. 2). Based upon volume calculations, the original thickness of the salt has been estimated to about 2000 metres in the southern, and up to 5000 metres in the northern part of the basin (Jensen and Sørensen, this volume; Bergendahl, 1989).

The evaporites of the Nordkapp Basin are believed to be of late Carboniferous to early Permian (Asselian) age, being deposited in a broad epeirogenic basin (e.g., Rønnevik and Jacobsen, 1984). Early works suggested that the salt structures were rooted at two different levels (Faleide et al., 1984), but new data do not seem to confirm this (Jensen and Sørensen, this volume).

The relatively uniform thickness of the Permian and lower Triassic sequences demonstrates that no large-scale differential subsidence took place in the Nordkapp Basin at this point in time, but there are indications that salt movements accelerated by mid-Triassic time, followed by decreasing subsidence in the late Triassic. Tectonic episodes in late Jurassic to early Cretaceous and Tertiary times reactivated the basin. The latter episodes also caused reactivation of the salt structures.

Master fault complexes and associated salt pillows

The Nysleppen Fault Complex

The Nysleppen Fault Complex (Gabrielsen et al., 1990) delineates the Nordkapp Basin against the Bjarmeland Platform, and locally against the Norsel High. Regionally, the fault complex trends NE–SW

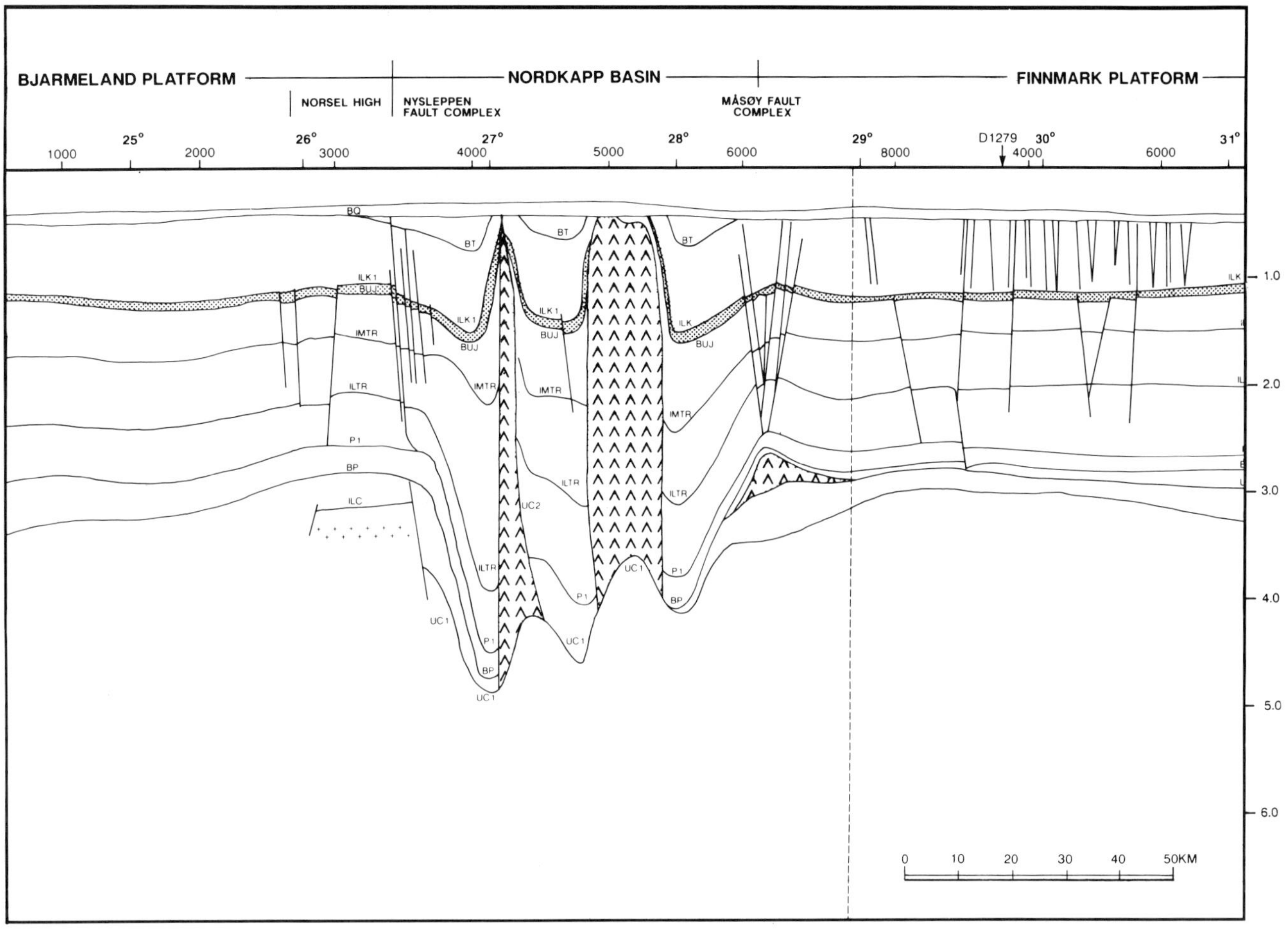

Fig. 2. Regional NW–SE profile through the Nordkapp Basin. Based on Gabrielsen et al. (1990). Same location as Fig. 8, see Fig. 1.

to ENE–WSW, although a N–S to NNE–SSW striking segment northeast of the Norsel High is found (Fig. 1). The structure can be traced northeastwards to approximately 30°E in the form of a flexure defined at intra-lower Cretaceous level. It should be noted that the Nysleppen Fault Complex lies along the continuation of the Troms-Finnmark Fault (Fig. 1), and may represent a segment of a zone of weakness which has wider regional tectonic implications.

As defined at Palaeozoic levels, the Nysleppen Fault Complex consists of one or more subparallel normal faults with an accumulated throw in the order of 500 milliseconds (TWT) or more. At the lower Cretaceous level, the faulted zone is less broad, and the throws are concentrated into one single, or two closely spaced faults. The accumulated throw is considerably less at lower Cretaceous than at Palaeozoic levels. Because of contrasts in deformational style, the different number of faults and the lack of fault plane correlation between the deeper and shallower levels, it is natural to propose that an extensional detachment is present in the lower Cretaceous sequence (Fig. 3).

In the area in which the Nysleppen Fault Complex follows a northerly trend, the fault zone is associated with a major salt pillow. At base salt level, two large faults are identified (Fig. 3), whereas the fault pattern at the shallower levels is dominated by one fault. At the position at which the deep faults transect the base of the salt, the salt may be divided into two or even three almost isolated smaller pillows. Along its strike, the faulted salt pillow associated with the Nysleppen Fault Complex accordingly includes all the characteristic styles described by Jenyon (1986a, b), namely those structures in which the faults affect only the top of the salt, those in which only the base of the salt is faulted and those in which the whole salt sequence is influenced by faulting.

In contrast to a section to the south, which has a much smaller salt structure in the footwall, the footwall above the salt in the northern part of the salt pillow is strongly rotated towards the platform side (compare Figs. 3 and 4). It can be demonstrated that a gradual rotation towards the platform takes place as the salt thickness gradually increases. Furthermore, in areas where the salt is thin or absent,

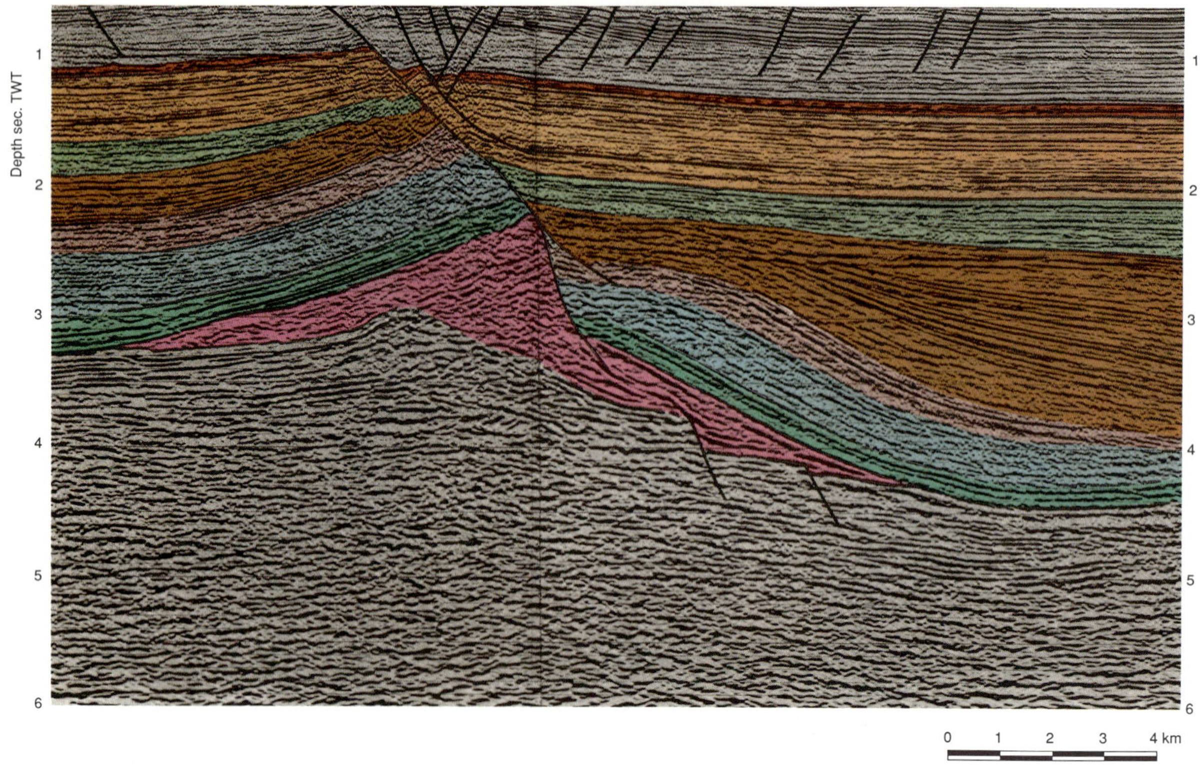

Fig. 3. NW–SE-oriented section through the Nysleppen Fault Complex (see Fig. 1 for location). Note the faults on the base salt level and the rotation of the footwall towards the platform side above the salt. The colour code is as in Fig. 4.

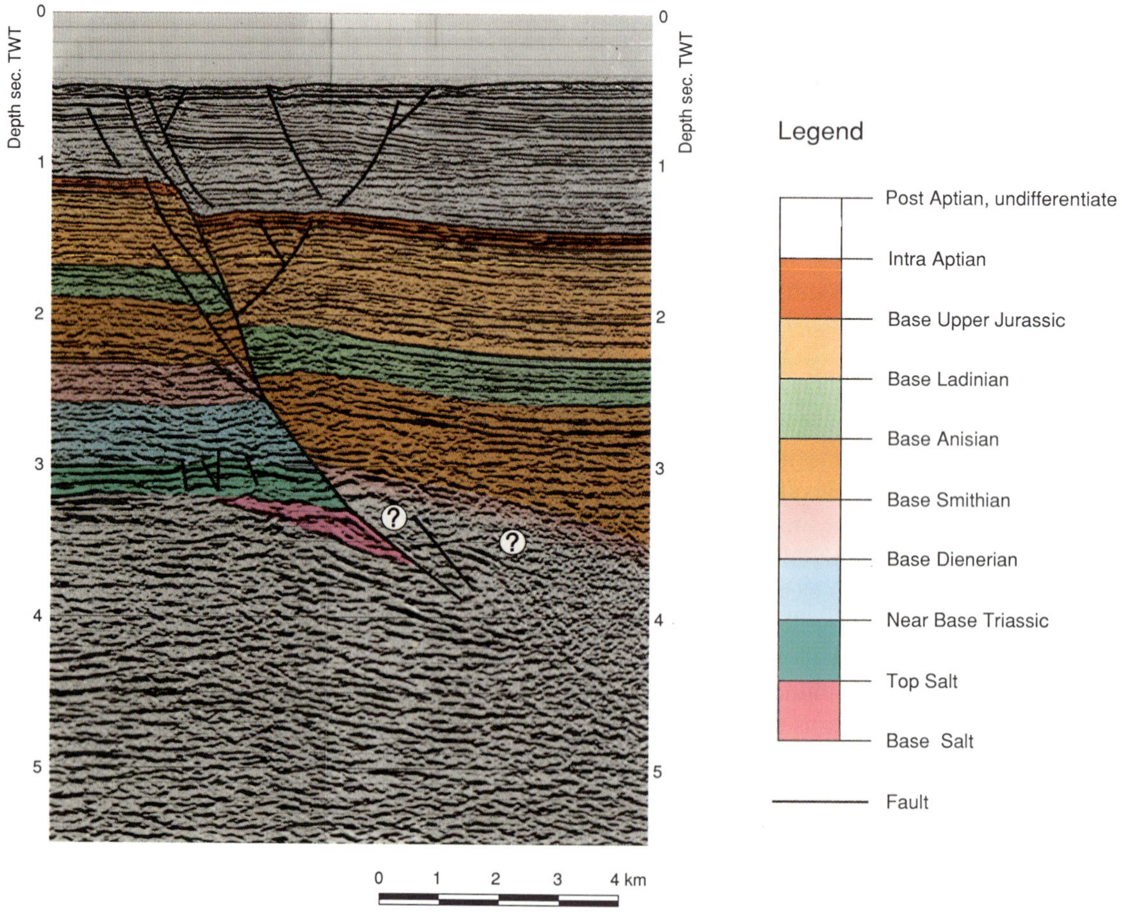

Fig. 4. E–W-oriented section through the Nysleppen Fault Complex (see Fig. 1 for location), in an area where the salt pillow is small. Note the lack of rotation of the sequence above the salt pillow.

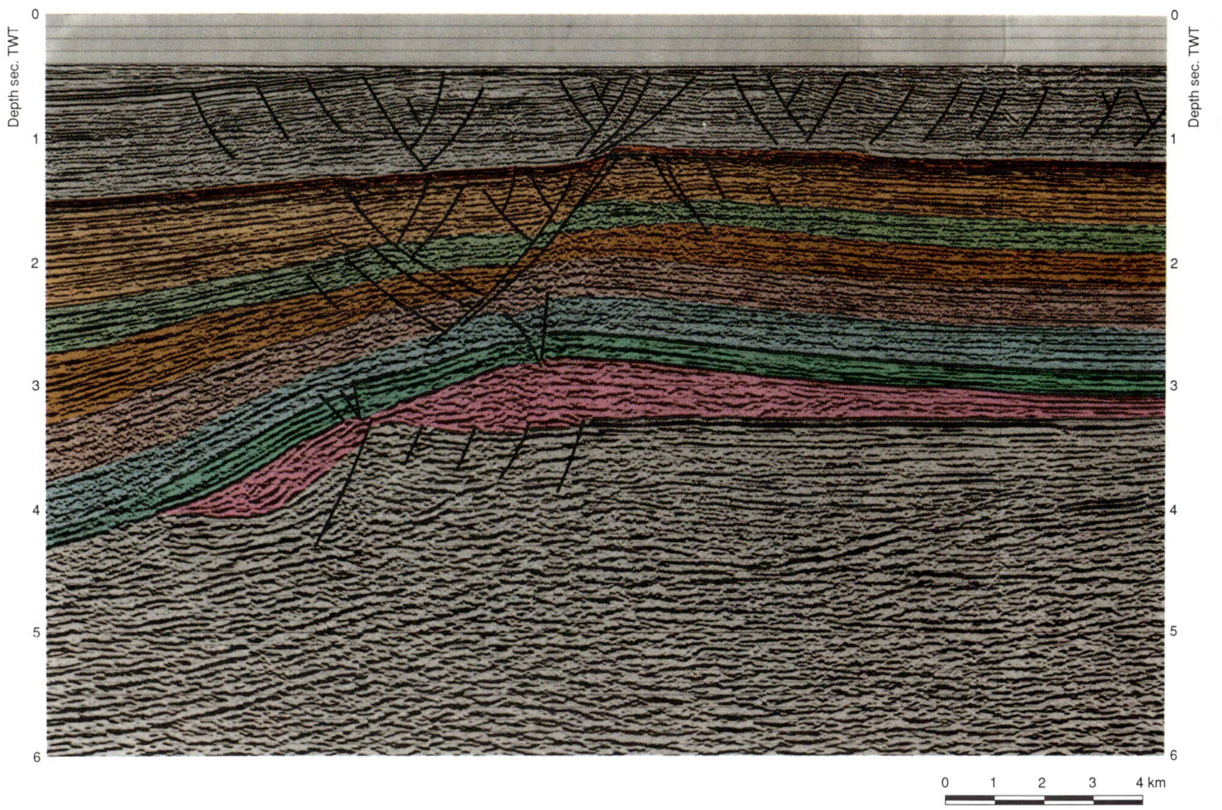

Fig. 5. E–W-oriented section through the Måsøy Fault Complex (see Fig. 1 for location). Note the brittle deformation of the carbonate sequence (green) immediately above the salt layer. The colour code is as in Fig. 4.

Fig. 6. E–W-oriented section through the Måsøy Fault Complex. Note apparent reverse fault at the crest of the salt pillow. The colour code is as in Fig. 4.

there seems to be a tendency for no detachment to occur between the shallower and the deep faults.

These two observations suggest that the strong rotation of the footwall fault block, as well as development of the fault detachment surface in the Cretaceous part of the sequence in the hanging wall, are governed by halokinetic processes.

Locally, the fault planes which are defined at Mesozoic and Cenozoic levels are complex and irregular and are characterized by splays cutting into the footwall as well as into the hanging walls. In such areas apparent folds and locally strongly tilted fault blocks are also observed. This indicates a late, local inversion along the Nysleppen Fault Complex.

The Måsøy Fault Complex

This fault defines the southeastern margin of the Nordkapp Basin between 71°27′N, 24°40′E and 72°35′N, 28°40′E (Gabrielsen et al., 1990). The main trend of the structure is ENE–WSW, but the northernmost branch turns in a more northerly direction. Except for the opposite sense of throw, the geometry of the Måsøy Fault Complex is generally similar to that of the Nysleppen Fault Complex. However, as the southwestern segment of the Nordkapp Basin is developed as a half-graben at pre-Permian levels, with the larger subsidence along its northwestern margin, the throws recorded along the Måsøy Fault Complex are smaller.

Måsøy Fault Complex is also associated with salt pillows with their long axes oriented parallel or subparallel to the basin margin (Figs. 1 and 5). It should be noted that even here a set of deep faults divide the major salt pillow into two separate structures in a manner similar to that of the Nysleppen Fault Complex. Furthermore, the layers situated at the top of the salt are broken into straight segments, some of which are separated by faults with reverse geometries (Fig. 6).

Timing of tectonic events in the Nordkapp Basin

NW–SE and N–S-oriented key lines through the Nordkapp Basin have been depth-converted, backstripped and restored with the use of the program ECHO/PAL (Welldrill 1988). Decompaction was not performed. The assumptions and constraints affiliated with this method are reviewed by Gabrielsen et al. (1986). The general results for the timing of the tectonic events of the basin area are summarized in Fig. 7.

Although the present interpretation gives very little information on the basin history prior to deposition of the salt, it is reasonable to assume that a basin was initiated as a half-graben in Late Devonian to Early Carboniferous times (Rønnevik et al., 1982; Gabrielsen et al., 1990), possibly covering a wider area than the present Nordkapp Basin (Jensen and Sørensen, this volume). Shortly after deposition of the carbonates a slight subsidence in the present Nordkapp Basin area can be recorded (Scythian). The timing of this is very close to the initiation of the halokinesis in the central part of the basin. From this point in time subsidence prevailed in the Nordkapp Basin and lasted well into the Cretaceous (end Albian). However, the subsidence pattern varied considerably within the basin, probably depending upon the local salt movements. According to Jensen and Sørensen (this volume), the salt empiercement peaked in Middle Triassic, and slowed down considerably towards the end of the Triassic. The salt was then reactivated in Tertiary time. This general picture was confirmed in our analysis, even though it may be added that slight, symmetrical subsidence took place in the basin in areas which were less affected by halokinesis, also during the Early and Middle Jurassic.

As seen from slight thinning of the Triassic sequences along the basin margins where salt is present (e.g., in the Måsøy Fault Complex; Fig. 5), salt movements were locally initiated along the graben margins simultaneously with the activation of the major salt stocks in the central part of the Nordkapp Basin. This effect is illustrated in Fig. 8 displaying two depth-converted and restored sections across one of the major salt pillows within the Måsøy Fault Complex (Fig. 1). The sections are flattened on the lower Triassic and base upper Jurassic levels, respectively, and demonstrate that uplift above the crest of the salt pillow was significant in both these time intervals. Since large-scale, secondary horizontal flow in the sediments of the sequences above the salt is considered to be very unlikely, it is implied that the flow of salt towards the basin margin was initiated early in the basin history, and that halokinesis affected the surface topography. The thinning of some units at the crestal points of the pillows relative to that on the platform indicate that salt has not only flowed towards the basin margin from the basin, where it competed with the central stocks and walls, but also from the platform areas.

Thinning of the Scythian sequence associated with salt pillows is not seen everywhere (e.g., in parts of the Nysleppen Fault Complex; Fig. 3). On the contrary, a pronounced thinning of the Ladinian to Middle Jurassic sequences, if recorded above this salt structure, suggests that flow in the pillows took place at different points in time, and that the salt in

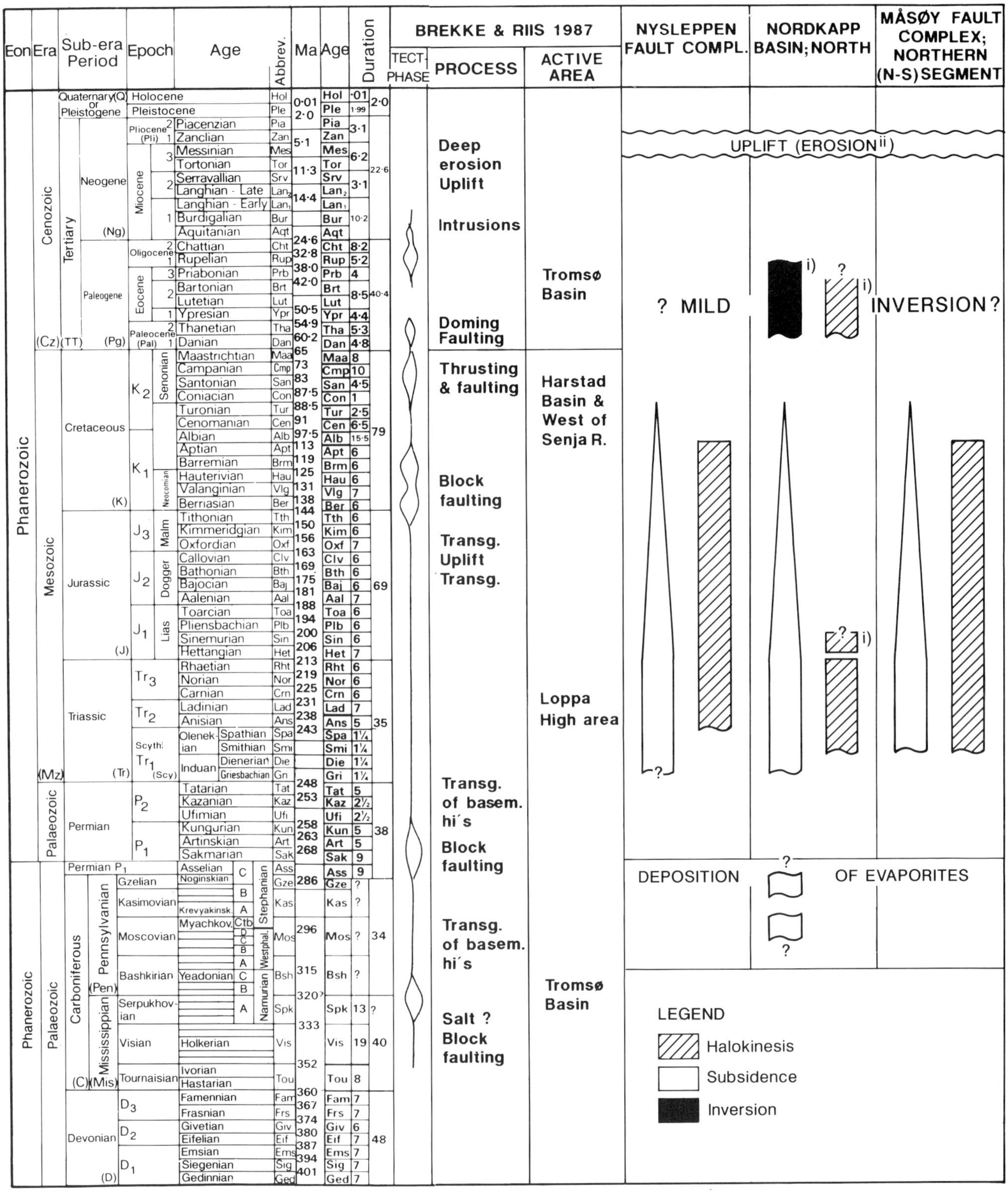

Fig. 7. Timing of the main tectonic events in the Nordkapp Basin, the Nysleppen and Måsøy Fault Complex. The table is based upon results from back-stripping and restoration of one regional seismic line of which parts are shown in Fig. 8 and the lines are shown in Figs. 3–6. For locations see Fig. 1. Data from Brekke and Riis (1987; *ii*) and Jensen and Sørensen (this volume; *i*) have been used in addition to the present data. The time scale is from Harland et al. (1982).

the marginal pillows moved in several stages as basin subsidence proceeded.

As already pointed out in the descriptions of the Nysleppen and Måsøy Fault Complexes, the shoulders of the footwall fault blocks frequently stand out as the most elevated points as defined on the base of the Upper Jurassic and shallower levels. As seen in Fig. 5, the strata on both sides of the high area

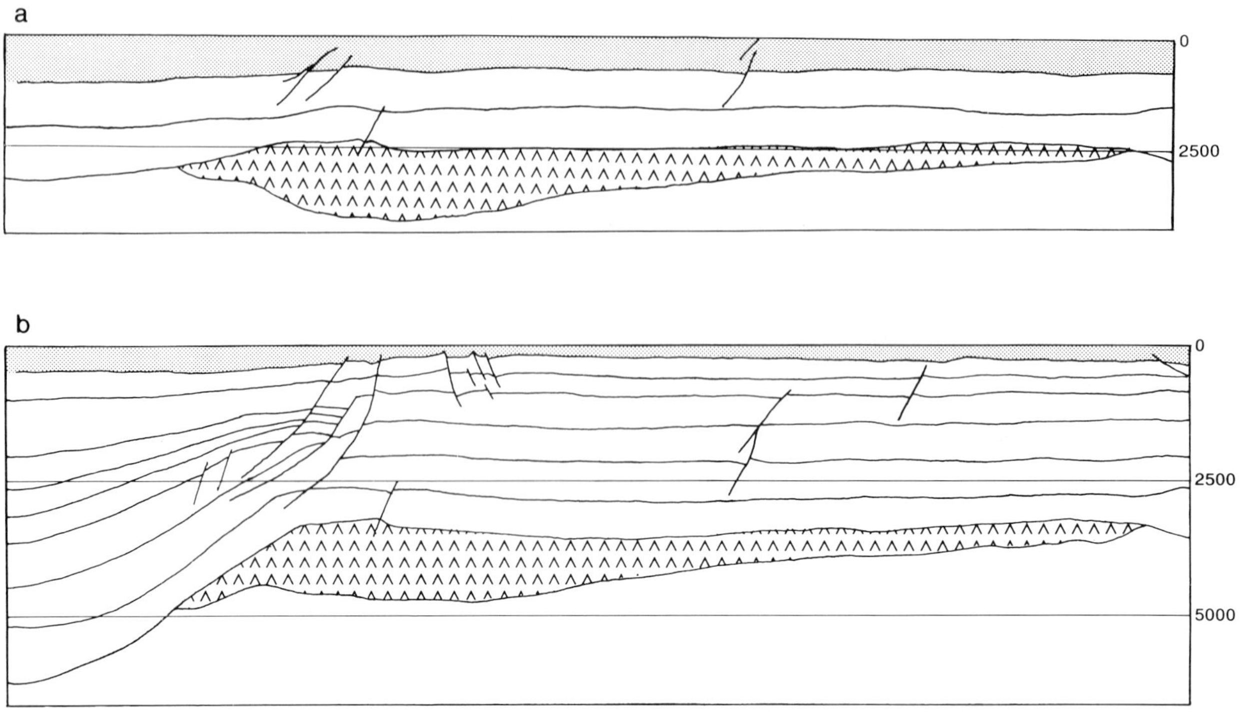

Fig. 8. (a) Restored, depth-converted section at the lower Triassic levels across the southern margin of the Nordkapp Basin. Note slight thinning above salt pillow. Shaded sequence, Base Dienerian–Base Smithian; hatched sequence, salt. The section is not decompacted. The depth scale is in metres. For location see Fig. 1. (b) Restored, depth-converted section at the pre-Late Jurassic levels across the southern margin of the Nordkapp Basin. Note thinning in the footwall above the salt pillow. Base Upper Jurassic–Base Ladinian; hatched sequence, salt. The depth scale is in metres. The section is not decompacted. For location see Fig. 1.

laps onto this reflector, suggesting that this point was elevated also in the Late Jurassic and Early Cretaceous.

Due to the heavy halokinetic effects, to decipher the regional stress situation at any time from the present data from Nordkapp Basin and the adjacent fault complexes would not be trivial. Nevertheless, we do suggest that the development of the Nordkapp Basin took place under mild extension. There are, however, indications of Tertiary regional inversion, as also mentioned by Jensen and Sørensen (this volume).

A model for the structuring of the basin shoulders

The following observations should be taken into account when developing a model of the salt structures along the margins of the Nordkapp Basin:

- the pillows are found along the basin margins and have their long axes are oriented subparallel to the margins;
- the pillows are detached from the salt in the central parts of the basin;
- the pillows are frequently seen to be divided into separate elongated structures where deep faults affect the base of the salt;
- the top of the salt may be faulted and in some cases faults at these levels are characterized by reverse geometries.

It is well documented that salt commonly tends to flow towards fault zones (Jenyon 1986a, b, 1988). As fault zones are associated frequently with differences in sediment thicknesses, e.g., along basin margins, it is reasonable to ascribe this coincidence to primary lateral thickness, or lateral density contrasts. This is indeed the case for the Nysleppen and Måsøy Fault Complexes.

It is noted that halokinesis may be triggered at shallow burial depths and salt has been reported to flow under as little as 150–200 metres of overburden where overlain by dense rocks such as anhydrites and carbonates (Brink, 1974). Furthermore, low-temperature creep in salt is shown to be common under the presence of water (Richter-Bernburg, 1970; Talbot and Jarvis, 1984; Talbot and Jackson, 1987).

With this in mind, and based on the observations of the geometric relations of the salt pillows associated with the Nysleppen and Måsøy Fault Complexes, we propose the following model for the development of the pillows.

Flow of salt towards the margins of the Nordkapp Basin started shortly after deposition of the carbonates (Scythian). A high water content may have facilitated the flow and, most likely, the driving

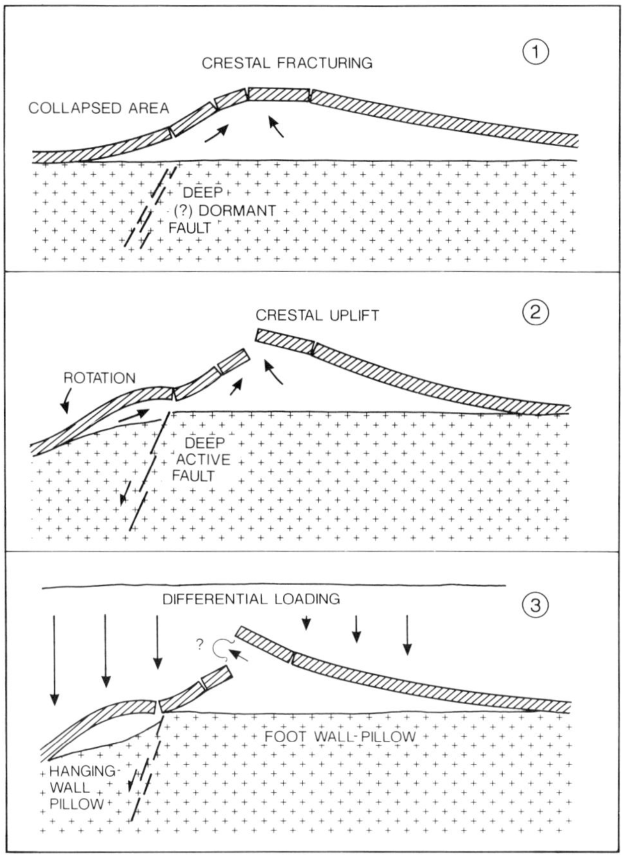

Fig. 9. Schematic model for development of salt pillows along the margins of the Nordkapp Basin. (1) Stiff layer on top of the flowing salt responds with brittle deformation and crestal collapse. Note extensional separation along the faults. (2) Rotation towards the basin axis, thinning above the rotational axis and subdivision of the salt into two separate structures. (3) Break-through of salt in the crestal areas of the salt pillow, following burial and escape of salt through the crestal faults.

mechanism was differential loading along the graben margin. It is, however, not known whether active faulting took place along the master fault systems or whether the basin margin was developed as a flexure above a deeper fault zone.

The development of the salt structures and their associated sedimentary sequences along the basin margins may be illustrated by a three-stage model (Fig. 9).

Stage 1 is the initial pillow stage. Figure 9a shows how the stiff layer of carbonates capping the flowing salt may have responded by brittle deformation and crestal collapse. This process would be accompanied by extension of the brittle top layer and accordingly extensional separation along the faults.

Stage 2. As salt continued to flow towards the graben margins, subsidence in the central parts of the basin (either due to active stretching, retraction of salt or a combination of these) took place. At this stage continued activity was seen along the deep faults below the salt pillows. This process was probably followed by rotation towards the basin. This would cause thinning of the salt pillows directly above the rotational axis and subdivision of the salt into two separate structures (Fig. 9b).

Stage 3. The local accumulation of salt, together with the regional subsidence of the basin floor, would accentuate the contrasts in sediment thickness between the basin and the platform areas and, hence, differential compaction would increase with a compaction maximum in the basin area. This may have triggered the breakthrough in the central areas of the salt pillows, reactivating the pre-existing crestal faults as salt flowed from the highly pressurized basin areas towards the low-pressure platforms. It is expected that this would have lifted the crestal parts of the pillows now situated on the platform margins. During continued burial, salt may have escaped through the crestal faults developing a reverse fault geometry (Fig. 9c).

A plastiline model can be used to illustrate the main stages in this development. The plate of chocolate in Fig. 10a illustrates the break-up of the stiff carbonate lid, simultaneously with separation of the salt pillow into two structures above the fault at the base of the salt. In Fig. 10b plastiline has been squeezed out through the faults at the crest of the pillow during compaction following continued sedimentation.

It might be suggested that the pronounced uplift of the footwall above the salt is associated with fault-block rotation of "domino-type" or "bookshelf-type" fault blocks (Thompson, 1960; Mandl, 1986; Yielding, 1990) decoupling in the salt layer or, alternatively, uplift according to the "cantilever model" (Roberts and Yielding, 1991). However, the bookshelf model in its basic form requires a system of rotating blocks (Wernicke and Burchfiel, 1982; Sclater and Celerier, 1989; Sclater and Shorey, 1989), which is not found in the present area. Even more serious, the uplift is exclusively associated with the salt pillows (Fig. 4), and is not observed in the absence of salt pillows (Fig. 5). It is felt that these two observations rule out both the alternative models for uplift of the basin shoulders.

Conclusions

The salt accumulations along the margins of the Nordkapp Basin demonstrate that salt flow towards the faulted areas has taken place. It is suggested that the margin-directed flow of salt started early

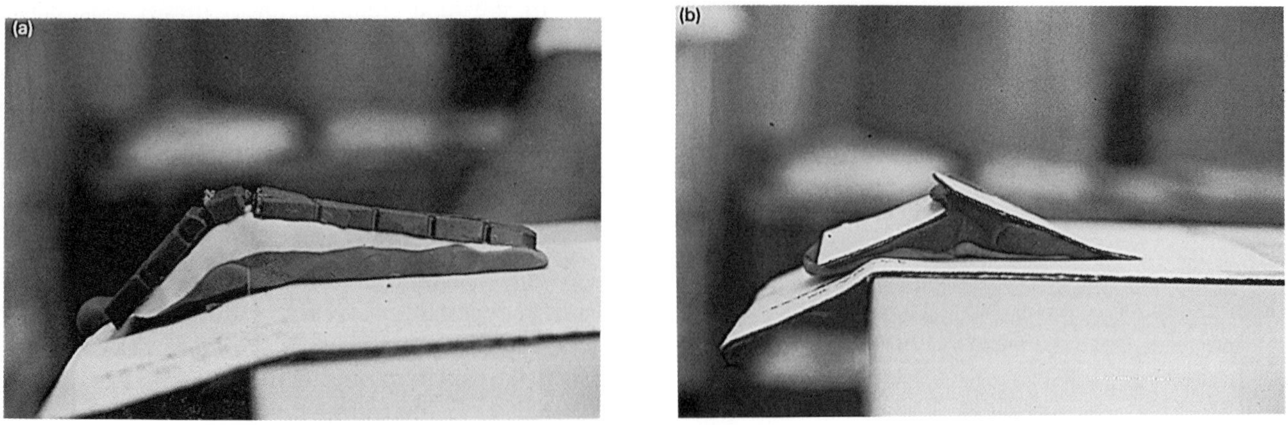

Fig. 10. Plastiline models which illustrate the main stages in the development of the salt pillow associated with the Måsøy Fault Complex. (a) Break-up of the stiff cap lid due to doming. (b) Plastiline is squeezed out through the faults at the crest of the pillow simultaneously with the separation of the pillow into two structures above the "fault" at the base of the plastiline. The effects in (b) happen during vertical compression, which mimics compaction.

in the history of the basin and that it has continued periodically throughout most of its subsequent development.

Generally, the marginal salt accumulations are elongated pillows, locally subdivided into several, subparallel structures, and associated with deep faults. Rotation toward the platform of the sedimentary sequences above the salt pillows in the footwalls is common, and the amount of rotation can be directly correlated with salt thickness. Accordingly, the elevation of the basin margins is not associated with rotational fault blocks nor with "cantilever" effects.

Some pillows (notably those of the southern basin margin) have crestal faults with reverse geometries. These structures can be explained by a simple three-stage model involving: (1) salt flow directed towards faults along the basin margins; (2) continued deep faulting below the salt pillows associated with continued basin subsidence and basinward rotation; and (3) reactivation of crestal faults under differential compaction.

The fault complexes generally show evidence of reactivation and, locally, of inversion. Even though reverse fault geometries are closely associated with deformation of crestal areas of salt pillows, other geometries are not. It is therefore recognized that some inversion structures are difficult to explain by halokinesis, and that these may indicate a late phase of reactivation in the Nordkapp Basin area associated with regional tectonic events.

Acknowledgements

The authors are indebted to Norsk Hydro a.s., Harstad for support and permission to publish this paper. We would also like to thank Dr. Ron J. Steel, Norsk Hydro a.s. who corrected the English text, and Dr. C. Needham, Fina Exploration and an anonymous referee whose comments and suggestions have been of great help.

References

Bergendahl, E., 1989. Halokinetisk utvikling av Nordkappbassengets sørvestre segment. Cand. Scient. Thesis, University of Oslo, 120 pp. (unpublished).

Brekke, H. and Riis, F., 1987. Tectonics and basin evolution of the Norwegian shelf between 62° and 72°. Nor. Geol. Tidsskr., 67: 295–322.

Brink, A.H., 1974. Petroleum geology of the Gabon Basin. Am. Assoc. Pet. Geol. Bull., 58: 216–235.

Faleide, J.I., Gudlaugsson, S.T. and Jacquart, G., 1984. Evolution of the western Barents Sea. Mar. Pet. Geol., 1: 123–150.

Gabrielsen, R.H., Ekern, O.F. and Edvardsen, A., 1986. Structural development of hydrocarbon traps, Block 2/2, Norway. In: A.M. Spencer et al. (Editors), Habitat of Hydrocarbons on the Norwegian Continental Margin. Norw. Petrol. Soc., Graham and Trotman, London, pp. 129–141.

Gabrielsen, R.H., Færseth, R.B., Jensen, L.N., Kalheim, J.E. and Riis, F., 1990. Structural elements, Barents Sea, Norwegian Shelf. Norwegian Petroleum Directorate Bull., 6: 33 pp.

Harland, W.B., Cox, A.V., Llewellyn, P.G., Pickton, C.A.G., Smith, A.G. and Walters, R., 1982. A Geologic Time Scale. Cambridge University Press, Cambridge, 131 pp.

Hinz, K. and Schlüter, H.-U., 1978. The geological structure of the western Barents Sea. Mar. Geol., 26: 199–230.

Hinz, K. and Weber, J., 1975. Zum geologischen Aufbau des Norwegischen Kontinentalrandes und der Barents-See nach reflexionsseismische Messungen. Compendium 75/75, Erganzungsband der Zeitschrift Erdöl und Kohle, Erdgas, Petrochem., 3–29.

Jensen, L.N. and Sørensen, K., 1988. Geology and salt tectonics of the Nordkapp Basin, Barents Sea (abstract). In: K. Binzer, I. Marcussen and P. Konradi (Editors), 18. Nordiske Geologiske Vintermøde, Abstracts. Danm. Geol. Unders., Copenhagen, pp. 187–188.

Jensen, L.N. and Sørensen, K., 1992. Tectonic framework and halokinesis of the Nordkapp Basin, Barents Sea. In: R.M. Larsen, H. Brekke, B.T. Larsen and E. Talleraas (Editors), Structural and Tectonic Modelling and its Application to

Petroleum Geology. Norwegian Petroleum Society (NPF), Special Publication 1. Elsevier, Amsterdam, pp. 109–120 (this volume).

Jenyon, M.K., 1986a. Salt Tectonics. Elsevier, Barking, 191 pp.

Jenyon, M.K., 1986b. Some consequences of faulting in the presence of a salt rock interval. J. Pet. Geol., 9: 29–52.

Jenyon, M.K., 1988. Some deformation effects in a clastic overburden resulting from salt mobility. J. Pet. Geol., 11: 309–324.

Lind, E., 1986. The Nordkapp Basin, a major salt feature in the SW Barents Sea. Abstract Volume, 4th Annu. Meet., Tectonics and Structural Geology Studies Group, Geological Society of Norway, Tectonics and Deposition in Sedimentary Basins, Stavanger, p. 23 (abstract).

Mandl, G., 1986. Tectonic deformation by rotating parallel faults: the "bookshelf" mechanism. Tectonophysics, 141: 277–316.

Øvrebø, O. and Talleraas, E., 1976. The structural geology of the Troms area. In: Exploration Geology and Geophysics. Offshore North Sea Conf., Article G/TV-6, Norwegian Petroleum Society.

Øvrebø, O. and Talleraas, E., 1977. The structural geology of the Troms area, Barents Sea. Geol. J., 1: 47–54. Richter-Bernburg, G., 1970. Post-depositional structures in salt. Proc. Symp. Geol. Technol. Gulf Coast Salt, Louisiana State Univ., Baton Rouge, La., pp. 145–146.

Roberts, A. and Yielding, G., 1991. Deformation around basin-margin faults in the North Sea/mid Norway Rift. In: A. Roberts, G. Yielding and B. Freeman (Editors), The Geometry of Normal Faults. Geol. Soc. London, Spec. Publ., 56: 61–78

Rønnevik, H. and Jacobsen, H.P., 1984. Structural highs and basins in the western Barents Sea. In: A.M. Spencer et al. (Editors), Petroleum Geology of the North European Margin. Norwegian Petroleum Society, Graham and Trotman, London, pp. 98–107.

Rønnevik, H.C., 1981. Geology of the Barents Sea. In: L.V. Illing and G.D. Hobson (Editors), Petroleum Geology of the Continental Shelf of Northwest Europe. Institute of Petroleum, London, pp. 395–406.

Rønnevik, H.C., Bergsaker, E.I., Moe, A., Øvrebø, O., Navrestad, T. and Stangenes, J., 1975. The geology of the Norwegian Continental Shelf. In: A.W. Woodland (Editor), Petroleum and the Continental Shelf of North-West Europe, Vol. 1. Institute of Petroleum, London, pp. 117–129.

Rønnevik, H.C., Beskow, B. and Jacobsen, H.P., 1982. Structural and stratigraphic evolution of the Barents Sea. Can. Soc. Pet. Geol. Mem., 8: 431–440.

Sclater, J.G. and Celerier, B., 1989. Errors in extension measurements from planar faults observed on seismic reflection lines. Basin Res., 1: 217–221.

Sclater, J.G. and Shorey, M.D., 1989. Mid-Jurassic through mid-Cretaceous extension in the Central Graben of the North Sea — part 2: estimates from faulting observed on a seismic reflection line. Basin Res., 1: 201–215.

Talbot, C.J. and Jackson, P.A., 1987. Internal kinematics in salt diapirs. Am. Assoc. Pet. Geol. Bull., 71: 1068–1093.

Talbot, C.J. and Jarvis, R.J., 1984. Age, budget and dynamics of an active salt extrusion in Iran. J. Struct. Geol., 6: 521–533.

Thompson, G.A., 1960. Problem of late Cenozoic structure of the Basin Ranges. Proc. 21st Int. Geol. Congr., Copenhagen, 18: 62–68.

Wernicke, B. and Burchfiel, B.C., 1982. Modes of extensional tectonics. J. Struct. Geol., 4: 105–115.

Yielding, G., 1990. Footwall uplift associated with Late Jurassic normal faulting in the northern North Sea. J. Geol. Soc. London, 147: 219–222.

R.H. GABRIELSEN *Geological Institute, Department A, University of Bergen, Allégaten 41, 5007 Bergen, Norway*
Present address: *Norsk Hydro A/S, Forskningssenteret, N-5020 Bergen, Norway*

O.S. KLØVJAN *Norsk Hydro A/S, Harstad, P.O. Box 31, 9401 Harstad, Norway*

A. RASMUSSEN *Norsk Hydro A/S, Harstad, P.O. Box 31, 9401 Harstad, Norway*

T. STØLAN *Norsk Hydro A/S, Harstad, P.O. Box 31, 9401 Harstad, Norway*

Seismostratigraphic study of a synrift megasequence, from the Barents Sea, northern Norway

K. Gerdes, J. Hurst and R. Jeans

This paper discusses the development and content of a package of late Jurassic–early Cretaceous synrift sediments from the Bjørnøya Basin. The discussion is based on the result of a detailed structural and seismostratigraphic interpretation of seismic data, limited well data from the area and models for synrift sedimentation developed from extensive field studies in well-exposed rift zones.

Structural maps were constructed at two scales to elucidate the first order (regional) and second order (local) structural controls on deposition. The approximate water depths and inclination of depositional surfaces were estimated using backstripped sections tied to well data. These techniques produced an estimate of the basin architecture during deposition of the synrift megasequence.

The synrift megasequence was then subdivided into five seismic sequences. This subdivision depended largely on a high resolution seismic survey of limited areal extent. Each sequence was then subdivided into seismic facies units and isochron and facies maps were constructed for each sequence. A correlation between seismic facies and lithology was made using the only available well in the area. The resultant maps were then interpreted using information on the controls on synrift sedimentation gleaned from extensive fieldwork in East Africa, Greece and the Gulf of Suez and well-constrained subsurface examples.

Depositional models for each sequence were generated and areas most likely to collect reservoir prone sediments were highlighted. The main critical factors concerning the predictive nature of the study are the quality of the tie between lithology and seismic character and the assumption of a coarse grained sediment supply.

Introduction

This paper deals with the development and content of the late Jurassic–early Cretaceous synrift megasequence, found in the Bjørnøya South area of the Barents Sea, offshore northern Norway, (Fig. 1). The aims of the study were twofold.

(1) To develop an integrated analysis of the interplay between extensional faulting and the development of the synrift megasequence sediment wedges.

(2) To document the seismic facies within the synrift megasequence and infer from their extent and distribution the likely presence and variation in reservoir facies.

The study presented in this paper was undertaken as part of the evaluation of the hydrocarbon prospectivity of a Barents Sea licence block (Fig. 1). The primary exploration target in this area was a pre-rift play. The synrift megasequence was recognised as a secondary target. The synrift sediment packages were poorly imaged on the east–west aligned seismic data available prior to the block being licensed. Once the licence was awarded, BP's partners made available dip aligned surveys of higher resolution which they had previously acquired over the block. In addition to these data the site survey shot prior to drilling of the first well was of sufficiently high resolution to clearly image the external geometry and internal character of the synrift megasequence and allowed a seismostratigraphic subdivision of the synrift megasequence. The area of the case study block covered by these dip aligned surveys is termed the 'detailed study area' in this paper (Fig. 2).

A new detailed structural map for the whole case study block was also constructed using these data. The well results, new structural map and seismostratigraphic subdivision of the synrift section form the database of this study. The principles and concepts developed during a research project on synrift sedimentation were used to interpret these data and generate the depositional models presented in this paper.

Regional setting

The regional setting of the Barents Sea and its sequence stratigraphy has been described in detail in the literature (Rønnevik and Jacobsen, 1983; Faleide et al., 1984) and only a brief summary will be given

Structural and Tectonic Modelling and its Application to Petroleum Geology edited by R.M. Larsen, H. Brekke, B.T. Larsen and E. Talleraas. NPF Special Publication 1, pp. 133–152. Elsevier, Amsterdam.

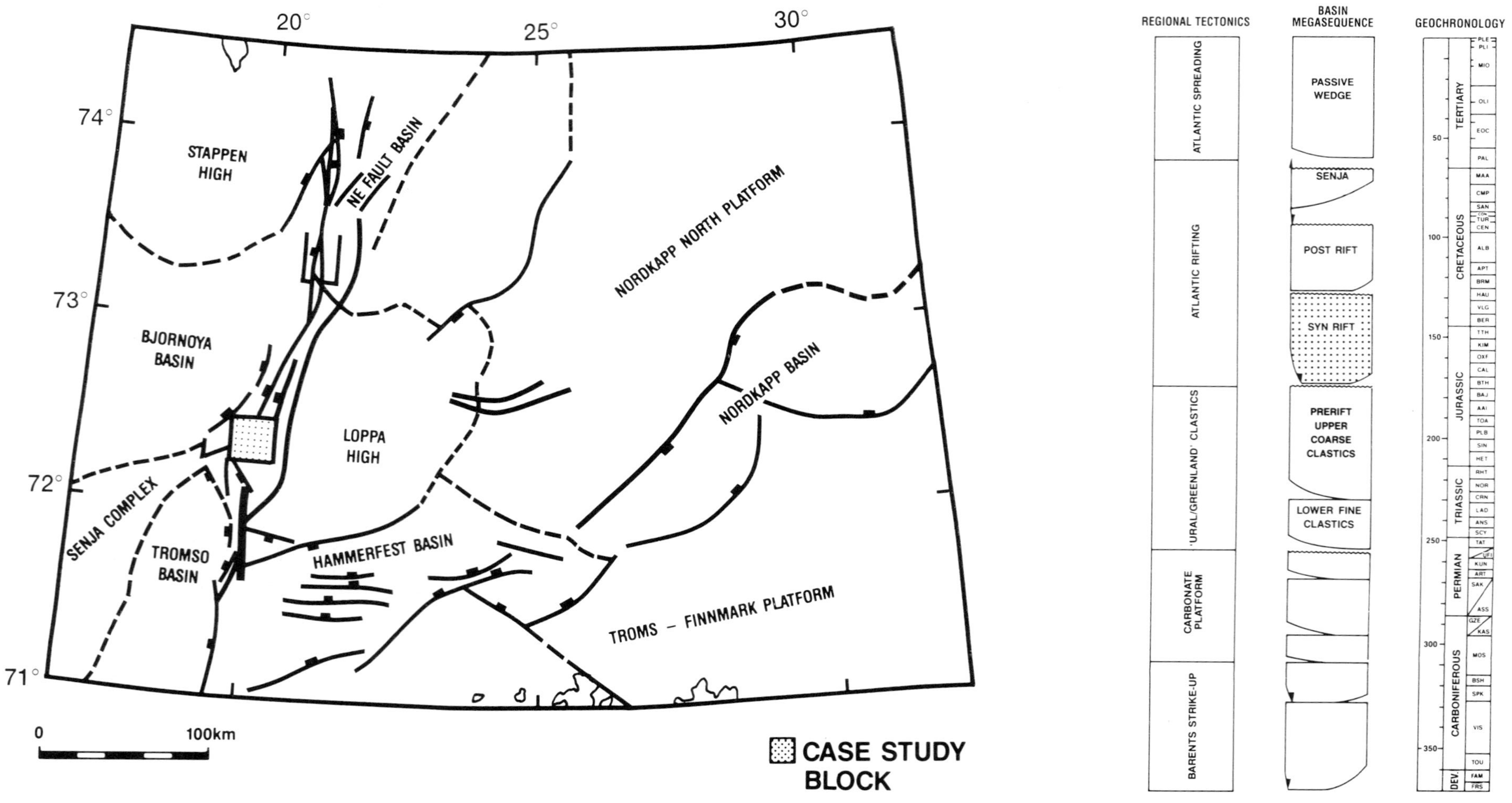

Fig. 1. Regional location map and Barents Sea sequence stratigraphy.

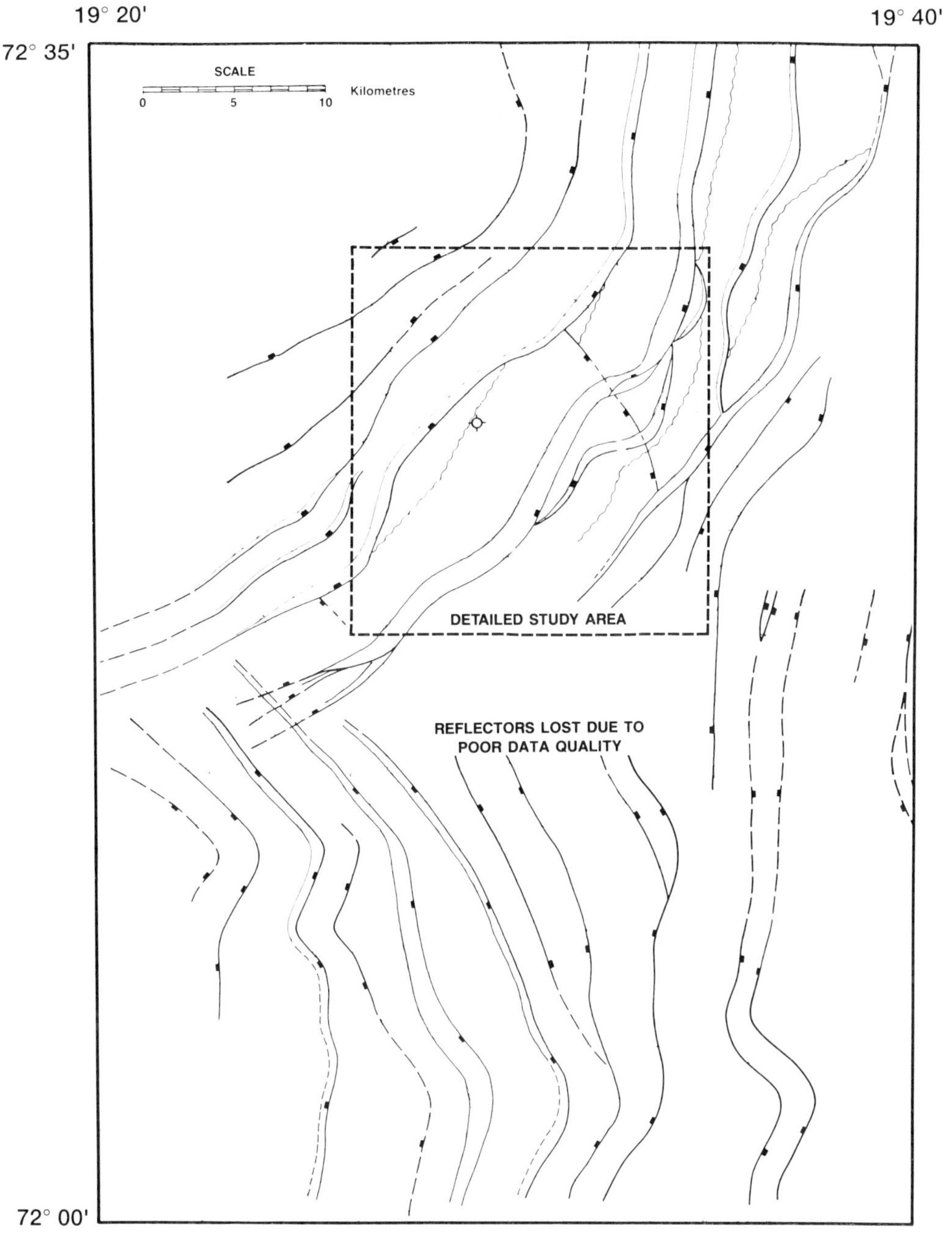

Fig. 2. Case study block — schematic regional structural map with area of detailed study highlighted.

here. During the late Jurassic to early Cretaceous period, an extensional regime developed in the Barents Sea area in response to the initial opening of the Atlantic Ocean south of the Charlie Gibbs Fracture Zone. The resultant multiphase rifting led to the development of the Tromsø and Bjørnøya Basins (Fig. 1). A series of fault block terraces developed on the flanks of the Bjørnøya Basin upon which a synrift megasequence of variable thickness was deposited. These sediments were then buried beneath a great thickness of post-rift deposits, during the mid-Cretaceous to early Tertiary. Recent uplift and erosion has removed between 1.0–2.0 km of post-rift section from parts of the Bjørnøya Basin. This allows the synrift megasequence to be imaged on the flanks of the basin at relatively shallow depths.

The case study block consists of a series of fault block terraces developed during the Mesozoic on the eastern flank of the Bjørnøya Basin. To the southeast of these terraces lies the Mesozoic rift flank of the Loppa Ridge, to the northwest of the terraces the fault blocks are buried beneath large thickness (5 seconds +) of post-rift sediments. The detailed study area covers a segment of one of these fault terraces, within which the synrift megasequence is particularly well developed and imaged. The well drilled in the block was situated on the crest of this tilted fault block terrace.

Database

Approximately 2720 km of seismic data were available for the study. These data were at a horizontal scale of 1:25 000 and a vertical scale of 10 cm:1 s and are summarised in Table 1. The seismic database is shown in Fig. 3.

The well in the study area was drilled to test a deeper, pre-rift target and penetrated only the distal part of the synrift megasequence wedge (Fig. 4). No other wells have been drilled in the licence block. The well was situated on the crest of the tilted fault block terrace and indicated that even at the highest structural point on the fault terraces, the synrift sediments were fully marine. The interpretation of these data, therefore, relies heavily on information gleaned from the research and development project concerned with synrift depositional models. Hence, the final models incorporate the results of observations made at a seismic scale during extensive field studies of well exposed rift systems from around the world.

A number of features intrinsic to the database control the degree of confidence which can be placed on the resultant seismostratigraphic interpretation.

TABLE 1

Summary of seismic database

Survey	Year	Line (km)	Line Spacing (km)	Alignment	Stacked	Migrated
Western [a]	1984	700	4.5	N/S-E/W	√	√
Geco [a]	1984	680	4.5	N/S-E/W		√
Fina	1986	360	2.5 (6)	dip (strike)		√
Mobil	1985	250	4 (8.15)	dip (strike)		√
Geoteam	1987	180	0.3 (0.5)	dip (strike)		√
Regional	pre-1984	550	10	various	√	√

[a] Western and Geco surveys shot to interleave and give an overall line spacing of 2.25.

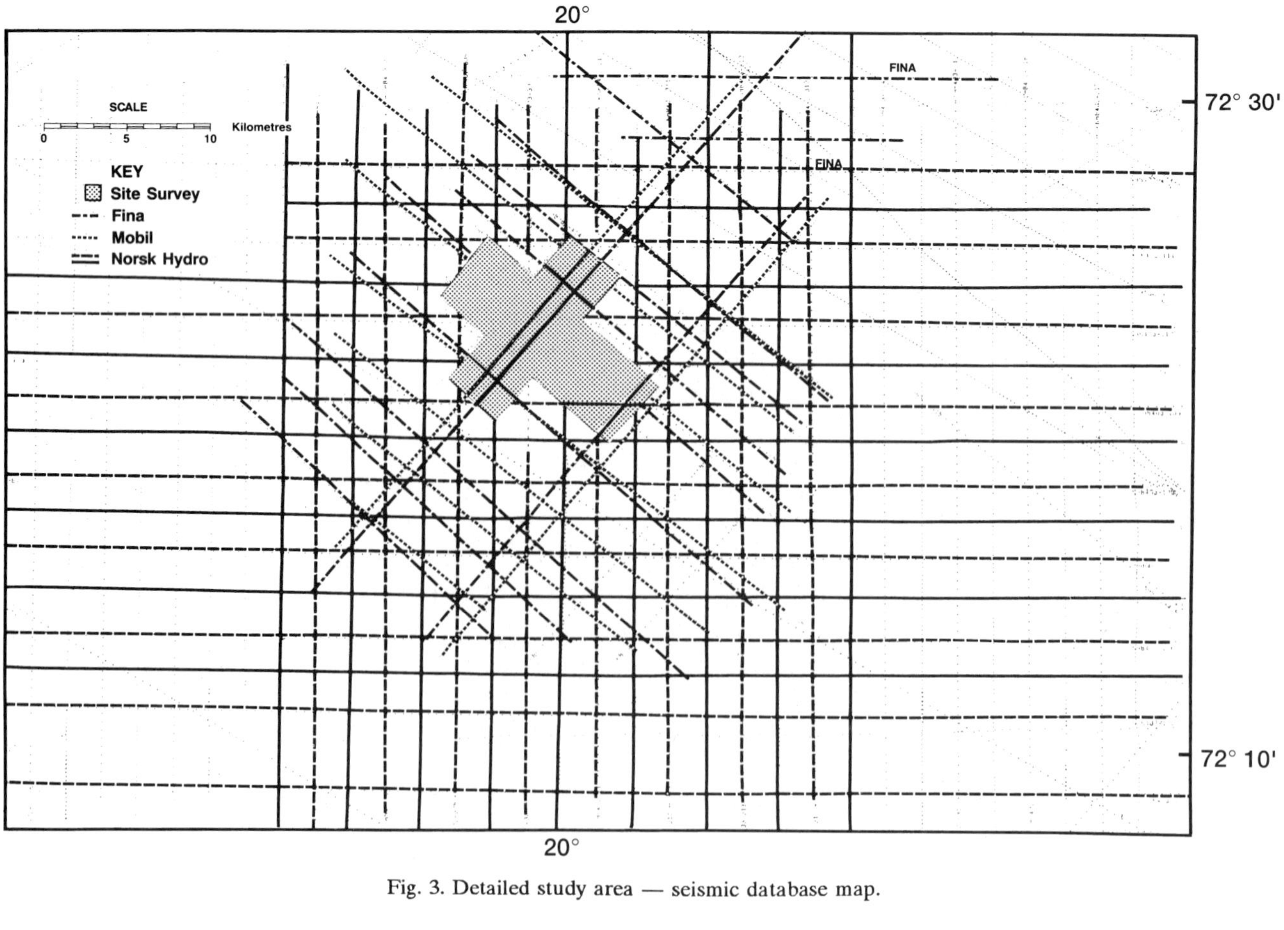

Fig. 3. Detailed study area — seismic database map.

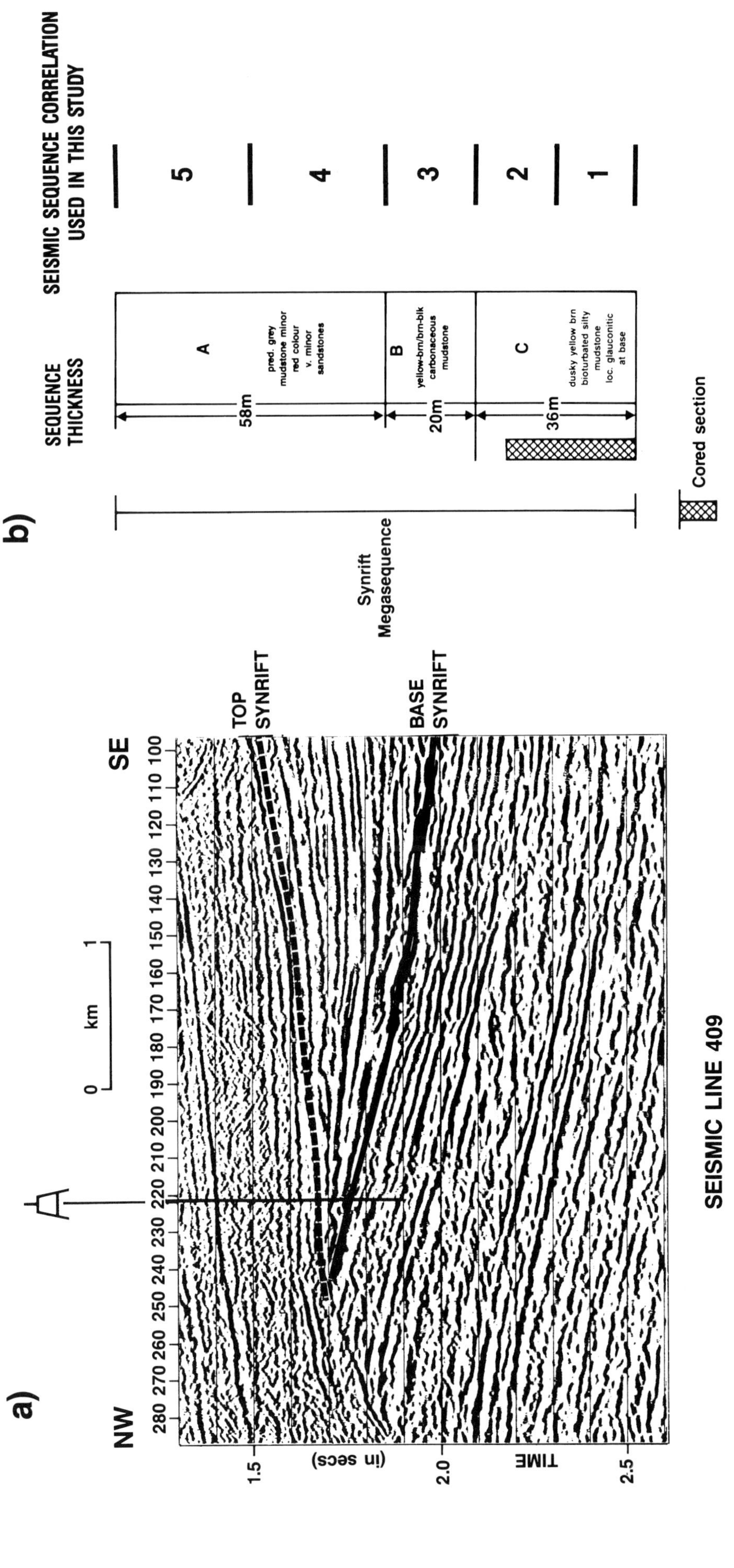

Fig. 4. (a) Seismic/well character tie. (b) Well results with seismic correlation.

A summary of the main limiting features of the database is given below.

(1) Using well data the parallel continuous reflector character observed on the seismic data was correlated with mudstones indicative of a low energy marine environment over the fault block crest (see Fig. 4). The megasequence interval in the well was mud dominated. The presence of coarse grained, sand-prone deposits within the synrift sequences alluded to in this study, is based on the assumption that the change in seismic character observed, from parallel continuous events to more disturbed events, can be correlated with a change in lithofacies from low energy mudstones to coarser grained sand-prone deposits.

(2) Data from six surveys was used for the seismic interpretation (see Fig. 3). Three surveys were aligned approximately along the dip to the main structural trend, one of which was a site survey of high resolution. The other surveys were aligned north–south, east–west. Correlating from survey to survey, in the absence of consistent reprocessing of all the datasets, introduced a variable amount of uncertainty.

A short sensitivity test illustrated that the error in fault positioning between the east–west aligned data and the dip-aligned data was as much as 1 kilometre. The convention used throughout the mapping was that, in the event of a positioning discrepancy, the position on the dip-aligned survey was given the most weight.

(3) The intrinsic variability between the surveys meant that the correlation of seismic character from survey to survey was made with caution. The site survey data offered the highest resolution by far and, hence, these were the data predominantly used for the purpose of seismic sequence definition.

(4) The disposition of the seismic lines and faults in the area prevented the tying of synrift reflectors on the main fault block to the wedge shaped sequences on the higher fault blocks. The same problem meant that these reflectors could not be directly tied across the lateral fault on the main fault block (see Fig. 5). The exact position of this lateral fault is obscured, on the few lines that cross it, by an area of disturbed data probably caused by gas escape. Hence, there is no independent method available for estimating the timing and amount of throw on this fault at particular periods during the development of the synrift sequences.

(5) Although the site survey was of highest resolution and covered the area of main interest, it was not acquired with the present type of study in mind. The strike lines are concentrated in the area where the synrift megasequence is thin or absent and the majority of the lines in this survey only cover the thinnest half of the synrift wedge. Hence, the data density available to tie across the thicker sections of the wedge nearest the major bounding fault is substantially less than that indicated by the average line spacings and survey specifications.

Methodology

The main aim of this study was to investigate whether the expanded seismic database could be used to highlight areas of the synrift megasequence which were likely to be sand-prone. This problem was addressed using two independent lines of investigation: the expanded seismic database was used to construct a more detailed structural map of the entire case study block (Fig. 2) to determine whether the synrift structure could highlight areas likely to focus sand-prone sediments. Secondly, in the detailed study area the synrift megasequence was subdivided into sequences and seismic facies units and these were used in conjunction with seismic character/facies correlation from the nearby well, to infer areas which might be sand-prone.

The seismic interpretation was completed on sections with a horizontal scale of 1:25 000 and a vertical scale of 10 cm:1 s. Structure maps were constructed at a scale of 1:100 000 and 1:50 000 for the case study block (Fig. 2). Two-way time maps to the top and base of the synrift megasequence were then constructed in the area where the two boundaries had been correlated (Fig. 6). The interpretation concentrated on faults which cut the top and basal reflectors of the megasequence sediment wedges. No attempt was made to correlate fault trends beyond the eastern seismic subcrop of the megasequence where the unit is thin or absent.

The synrift wedge on the main fault block of interest was then divided into five seismic sequences bounded by unconformities using seismostratigraphic principles. The seismic picks are shown in Fig. 11. Each unconformity was a prominent onlap surface. The subdivision of the megasequence was restricted to the main fault block.

Isopach maps were then constructed for each seismic sequence (Figs. 12, 13, 15 and 17). The seismic character within each sequence was then annotated along each survey line at a scale of 1:25 000. The amplitude, frequency and continuity of the seismic events within each sequence were described and facies areas were drawn onto the isopach maps. Within each facies area the seismic character was broadly similar, making a qualitative allowance for variations in data quality between surveys.

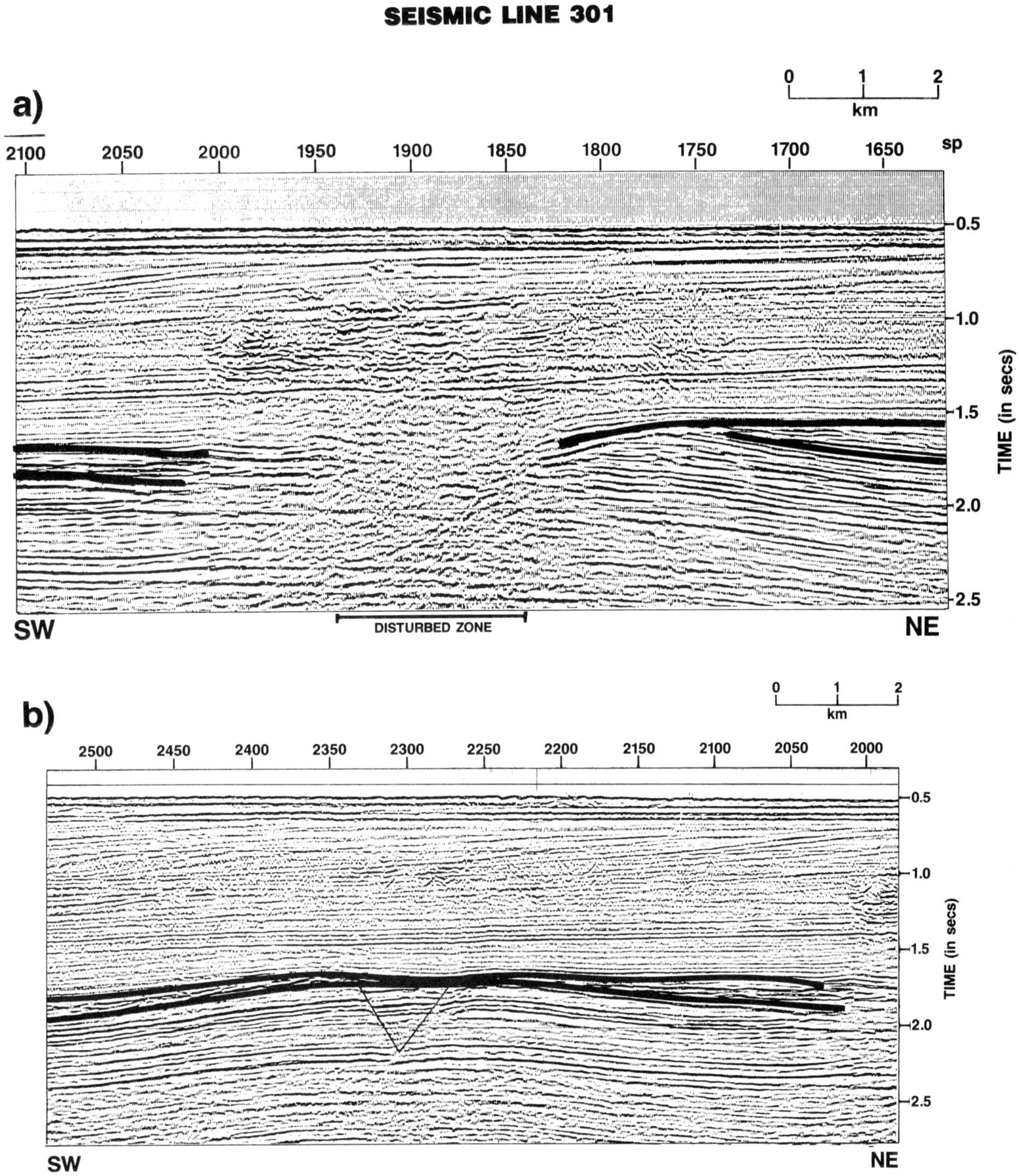

Fig. 5. (a) Main NW–SE lateral fault crossing the main fault block. Noise on section on the fault position thought to be caused by escape of gas into the shallow section. Exact number and location of faults within the disturbed zone unknown. (b) Collapse graben at the apex of broad antiform which is thought to control the area of non-deposition of the synrift megasequence on the main fault block.

Depth maps to top and base synrift and a preliminary isopach map of the megasequence were also constructed, using velocity analyses obtained from the BP Norway office and data from the well. The time and depth to base synrift surfaces were restored to their original orientation by removing the effects of post depositional tilting (see section 4). A composite regional line was then backstripped to aid with the reconstruction of the paleoenvironment during the deposition of the synrift megasequence.

Structural mapping and regional setting

Regional setting

Figure 2 illustrates that the extensional trend changes azimuth sharply in the centre of the Case

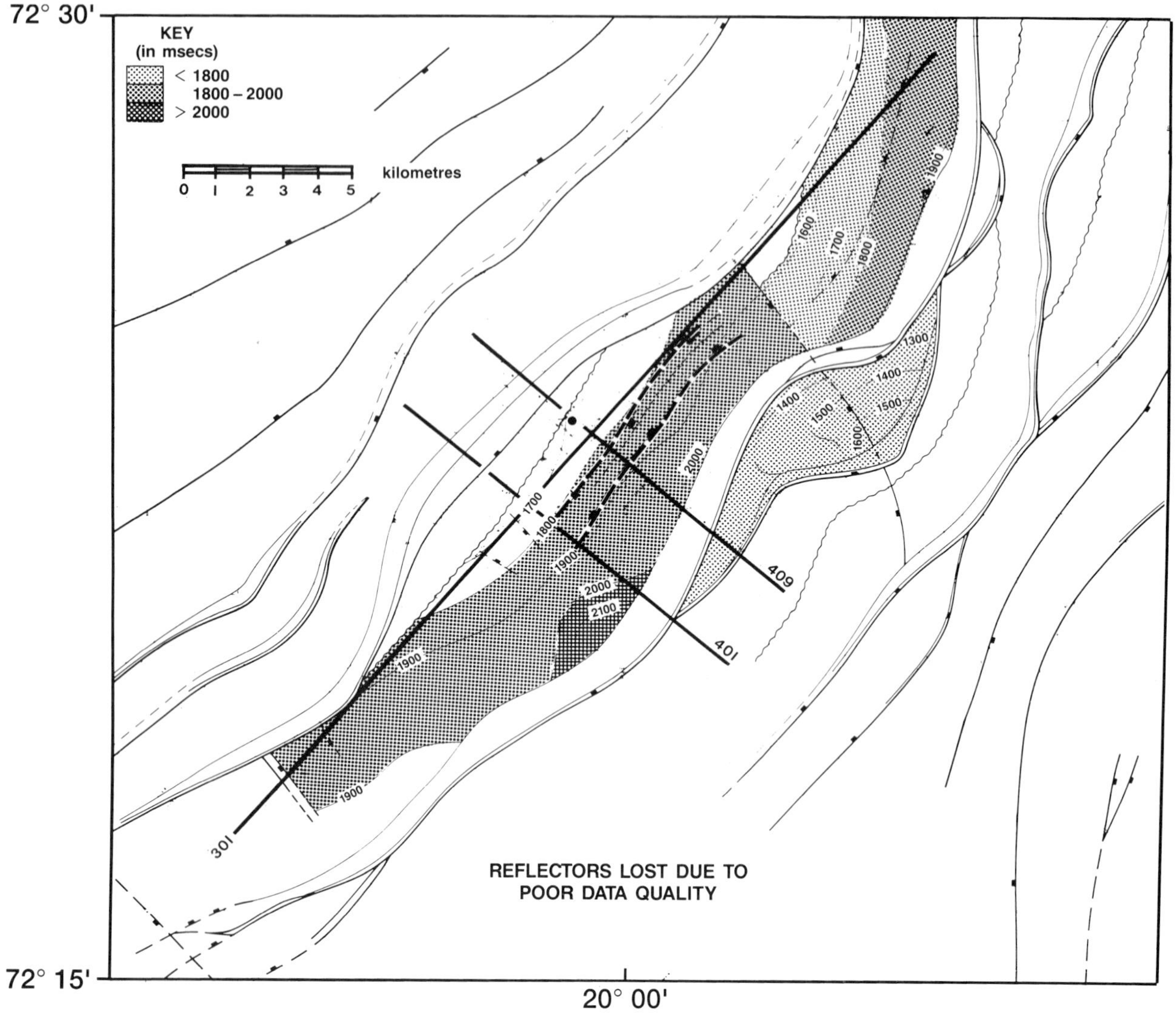

Fig. 6. Two-way time map to base synrift.

Study Block. To the south of this central zone, extension is taken up along faults trending NNW–SSE which turn to a truer NW–SE orientation as they approach this zone. North of the zone the main extensional faults trend NE–SW, becoming closer to NNE–SSW farther north. This latter set includes the main fault bounding the pre-rift target. Both sets of faults lose displacement as they approach this central area and on some lines the displacement on them is dispersed through a system of splays, although this was often difficult to map in detail due to poor data quality. The area in which both sets of extensional faults lose their displacement is interpreted as being the location of a transfer zone. The zone itself is elevated above the fault terraces, is about six kilometres in width and lies at the approximate intersection of the fault terrace and the northwesterly extension of the Trollfjord–Komagelv fault trend from the Hammerfest Basin.

The transfer zone appears to have acted as a conduit for the escape of gas into the shallow section. The presence of the transfer zone produces a morphological ramp which dips downwards along the fault terraces from the transfer zone. Two cusps (or back basins) were mapped in the footwall of the main fault block.

The first order set of faults north of the transfer zone is cross-cut by a few faults which have the same approximate trend as the faults south of the transfer zone and throw down to the southwest. Only two of these faults cut both the top and basal reflectors of the megasequence, but they may have acted as important controls on the deposition of the megasequence in the area of interest. The most important of these is the fault which effectively bisects the fault block. This fault was seen on two seismic lines, whereas the fault of similar trend to the southwest was seen on only one line (Fig. 5a).

Two-way time map to base synrift

The base synrift two-way time map (Fig. 6) shows the main fault block in closer detail and a number of structural features particular to the area around the well location. The fault block crest includes an area of around 12 km in length and up to 2 km in width where the synrift sediment wedge is absent. Uplift of the crest appears to have been accentuated along this zone with respect to the rest of the block crest as evidenced by the increased separation between the fault block crest and the top of the sediment wedge on its hanging wall. The fault block crest shows evidence of what may be minor footwall collapse features on the extreme northwestern edge on some lines.

The fault block surface dips down into the major fault with a relief of up to 400 m from the crest of the block. This trend is interrupted by a lateral fault which bisects both the main fault block and the main footwall cusp (Fig. 6). The throw on this fault is poorly constrained. The interplay of the fault block rotation and this lateral fault define the first order morphology of the base synrift surface. This is further modified by a set of structural features which only affect the base synrift surface.

South of the lateral fault, the morphology of the base synrift is that of the broad antiform with the apex located beneath the area in which the synrift deposits are absent. This is shown in detail in Fig. 5b which shows a broad updoming centred around a small collapse graben.

In addition to this broad doming there is also a minor antiform, which runs subparallel to the main fault trend with an average amplitude of 10 m, on the southern half of the fault block. This is bounded to the southeast by a small fault down to the northwest which runs irregularly along the length of the fault block mapped.

The origin of these minor features is uncertain. They further enhance the relative relief on the base synrift surface and emphasise the morphological depression which runs beneath the main fault scarp southwest of the lateral fault. This area should preferentially have collected the earliest sediments shed along the fault block axis.

The base synrift surface should approximate to the topography upon which the initial synrift sediments were deposited after allowing for the effects of any post-depositional deformation. An approximate correction for the effects of post-depositional tilting was made by tilting the base synrift surface down to the southeast by between 3° and 4° and down to the northeast by approximately 0.9° before flattening onto the top synrift reflector. These rotation angles were obtained from a series of measurements made on the seismic database.

Controls on synrift sedimentation

Introduction

Studies in presently active rift systems have indicated the dominant role which structurally controlled topography plays in the distribution of sand-prone sediments, (Crossley, 1984). Such topography controls the location of sediment entry points into the rift system and also affects erosion and, hence, sediment production potential. Field studies indicate that sediment derived from the flanks of the rift system gains access to the low lying rift axis predominantly via areas such as transfer zones and footwall cusps or back-basins. Examples of this structural control on the access of rivers to the rift system and location of sediment deposition in a rift are given in Figs. 7 and 8 from the Gregory Rift of Kenya and the Gulf of Suez, respectively. A number of different types of transfer zone have been identified. The nature of the associated sediment bodies varies with the scale and style of topography generated within the transfer zone.

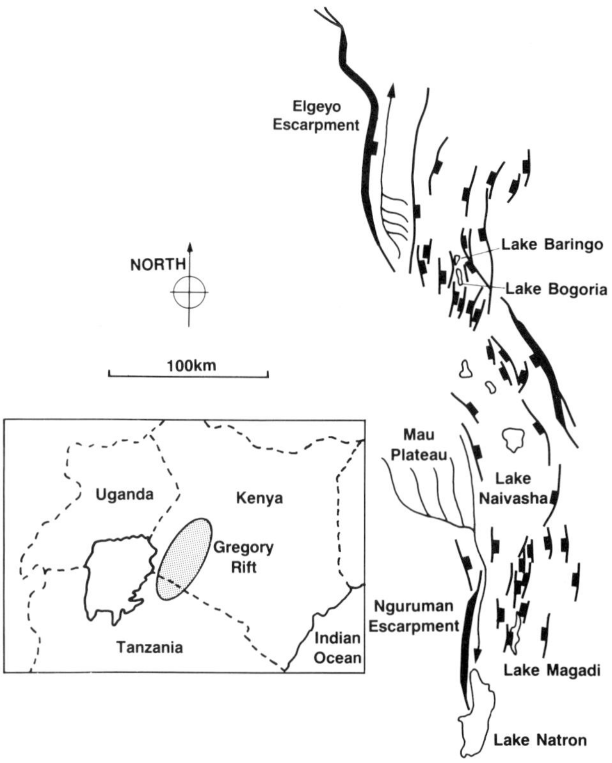

Fig. 7. Role of transfer zones on drainage patterns: Gregory Rift, Kenya.

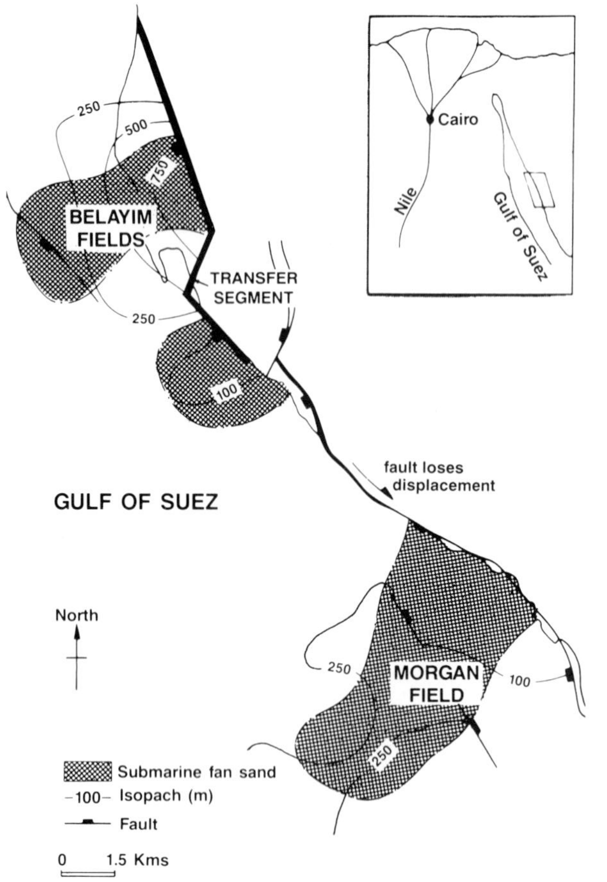

Fig. 8. Role of transfer zones and footwall cusps in distribution of sand prone sediments: Gulf of Suez, Egypt. (1) Lateral sand distribution down the footwall slope enters the graben at transfer points, where the throw on the main fault is lost. (2) Lateral sand input is focused at the intersection of cuspate faults with the main footwall.

Structural and depositional controls of the synrift megasequence

Structural

The large scale schematic structural map (Fig. 2) shows two extensional fault trends, one NW–SE and the other NE–SW. The throws on both sets of faults die out into a complex transfer zone in the centre of the case study block. It was not possible to show during this study whether the two trends separated by the transfer zone were active simultaneously. The true extension direction was probably east–west, and hence both sets of faults would have had an amount of oblique motion along them. The detailed study area is located north of the transfer zone, and the surface morphology generated in the proximity of the transfer zone during the rifting was almost certainly an important first order control on sedimentation. The transfer zone itself would have formed a positive topographic feature during synrift deposition. The faults leading from it increase in throw away from the zone. The fault blocks within the fault terrace north of the transfer zone have a regional dip down to the northeast (see Fig. 6). The cuspate faults southeast of the main fault block may also have had an effect on sediment input. If the sediment transport direction were predominantly axial then these cusps would have controlled any additional lateral input from higher fault blocks.

Superimposed on these extensional geometries are additional transfer elements which consist of NW–SE cross-cutting faults. These are more difficult to identify on the data, but are shown to partition the fault blocks in Fig. 6. These faults laterally offset the main NE–SW extensional trends and the largest throw on the main fault occurs at the point of intersection of the two trends.

In addition to these first order structural controls on sedimentation, more subtle second order effects exist which modify the details of the depositional surface. The doming of the pre-rift deposits accentuates the relative relief on the main fault block and clearly affects the youngest sequence. It apparently controls the position and extent of the area of non-deposition at the fault block crest. This doming may have been caused by diapirism at deeper levels or by minor compression associated with the transfer zone. Either of these processes may have created the slope instability at the fault block crest which led to the slumps down the hanging wall seen to affect seismic sequence 3 (see below).

Provenance

The megasequence is absent in an extremely limited area on the main fault block. Hence internal sediment production within the block itself can be considered to be minimal and the dominant provenance areas for synrift sediments deposited on the fault terrace are the rift flanks lying to the east of the fault terrace.

The subcrop of pre-rift sequences was clearly observed on seismic data from the rift flanks. This truncation appears mainly to affect sequences of mid-Jurassic age. The available well data from the region indicate that these sequences largely comprise highly porous quartz sandstones (Olaussen et al., 1984). Hence, at least the earliest sediments of the synrift megasequence should include the products of reworking of these quartz sandstones.

Depositional surfaces

The base synrift surface approximates to that on which the synrift sediments were deposited, after allowing for the effects of post-depositional deformation. Whether deposition is subaerial or submarine, sediment pathways developed on this surface would

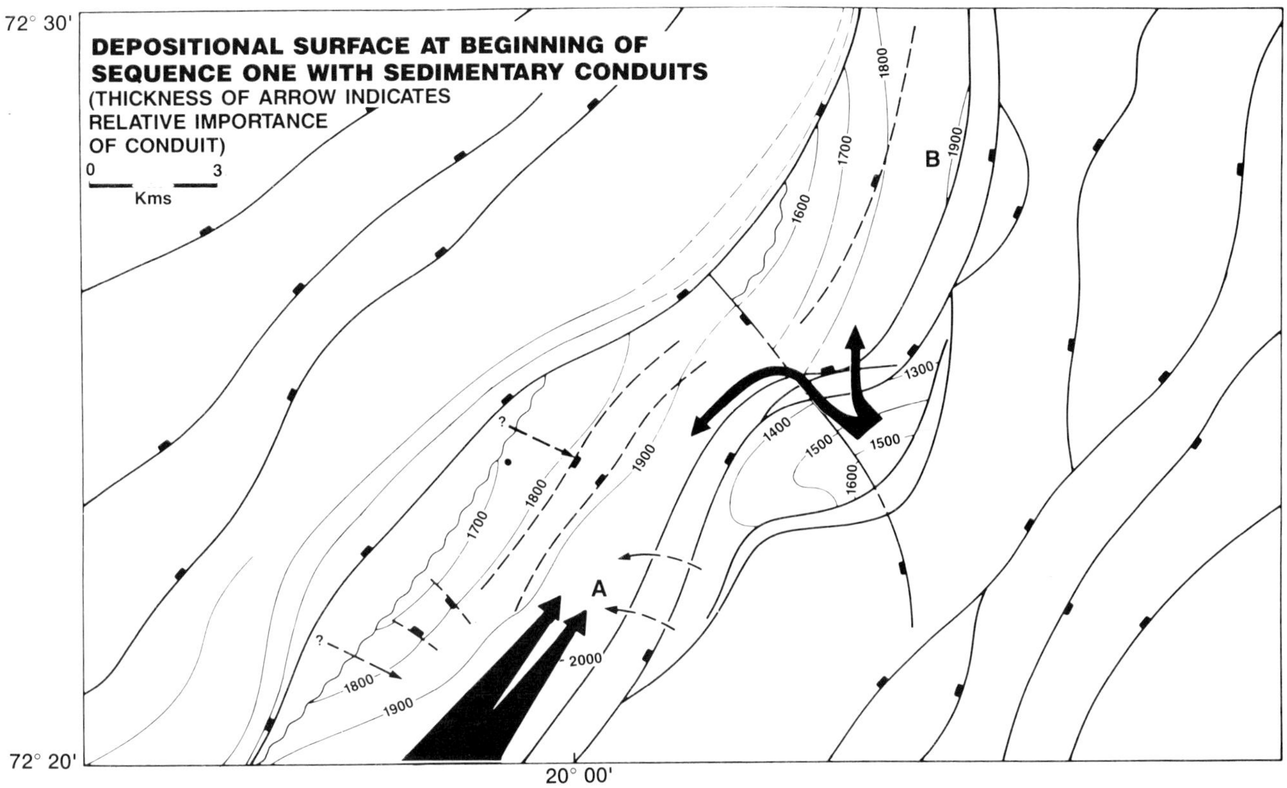

Fig. 9. Sedimentary conduits for seismic sequence 1.

have focused the sediment into the morphological deeps, i.e., areas A and B in Fig. 9. The sediments would have been further ponded into these areas by the second order affect of the small NE–SW oriented structure approximately 4 km from the crest of the fault block (see Fig. 6). The preliminary well results indicated that the lowermost synrift sediments on the fault crest were glauconitic and bioturbated and were interpreted as being submarine. Since this area is the shallowest location on the fault terrace, it follows that the deposition down-dip from this location was also marine, passing further down-dip to deep marine. Any coarse sediments present in this setting would have to have been introduced by mass flow processes given the depths of water calculated from the relative relief between the top and base of each sequence, i.e., between 200 and 400 m.

Sediments would have been introduced axially at point A (Fig. 9) at the onset of rifting, but the timing of sediment input from the upper fault block via the footwall cusp shown in Fig. 9 cannot be determined. Any lateral input of sediment would obviously have postdated the filling of the footwall cusp itself. Figure 10 assumes that the footwall cusp was filled prior to the deposition of seismic sequence 2.

Sequence mapping and depositional models

Synrift megasequence

The base of the megasequence was dated from the well as being of late-Jurassic age. It overlies an angular unconformity which is the boundary between the pre- and syn-rift (Fig. 11). This boundary is marked by a seismic event consisting of a high amplitude, moderate frequency, generally continuous positive peak. It became more difficult to distinguish this event in the vicinity of the main boundary fault, where it was often poorly migrated and partially obscured by migration and other noise from the fault and footwall.

The top of the megasequence was dated as being early Cretaceous. This angular unconformity is recognised as a major onlap surface (Fig. 11). The defining seismic event followed was a strong, continuous, positive peak of moderate amplitude and high frequency.

The two-way-time isopach map for the synrift megasequence is shown in Fig. 12. The fault terraces on all the isopach maps have been divided into zones dependent upon the presence or absence of the megasequence penetrated at the well. In Zone A the megasequence is very thin or absent, in Zone B small wedge shaped bodies exist, but they are isolated from

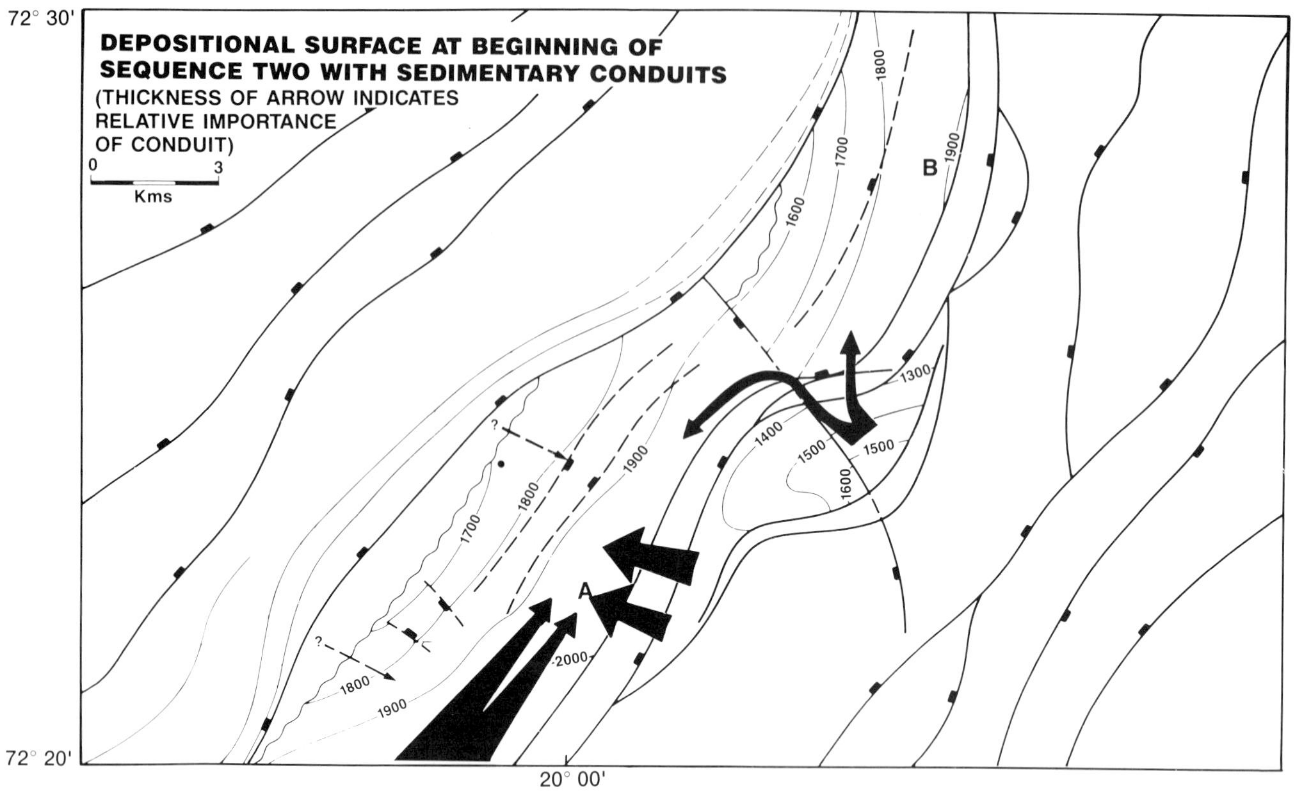

Fig. 10. Sedimentary conduits for seismic sequence 2.

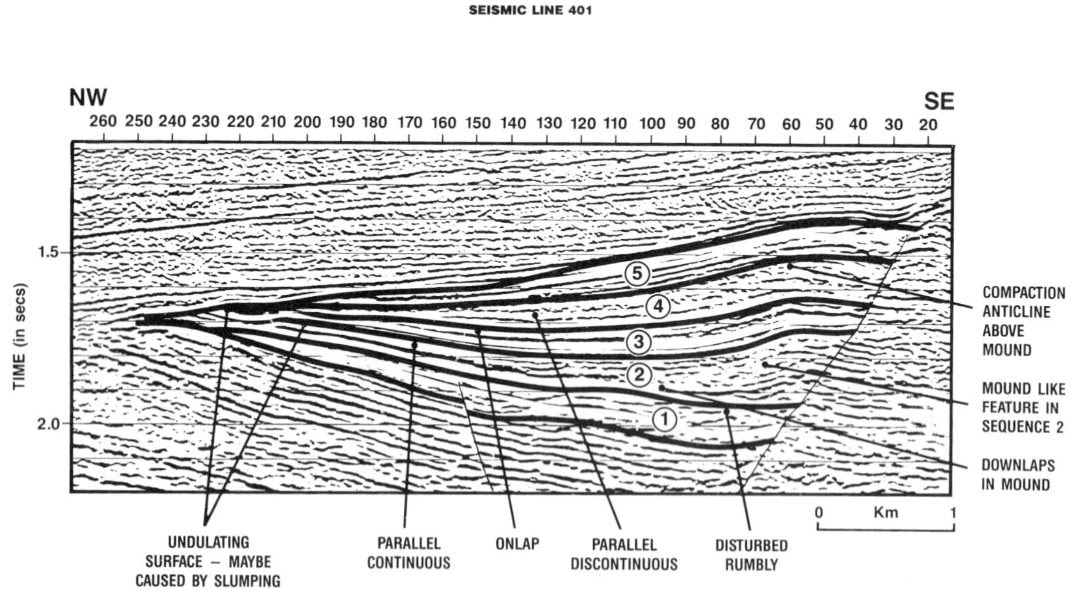

Fig. 11. Synrift megasequence, sequence boundaries and seismic facies used in this study. Reflector characters and morphological features referred to in the text are annotated.

correlatable deposits. It is not yet proven, but it is likely, that all these bodies are contained within the megasequence. Zone C includes the main fault block and footwall cusp within which the sediment bodies can be correlated as being part of the synrift megasequence. Finally, Zone D covers the area to the northwest of the main fault block of interest where any synrift deposits present are obscured by extensive thickness of post-rift sediments. Within Zone C there are two main depocentres on the main fault block and a small depocentre in the footwall cusp. The two depocentres on the main fault block shown in Fig. 12 are separated by the lateral fault.

The megasequence is absent from a triangular area approximately 2 × 3 km on the upthrown side of the lateral fault and also from a lens shaped area about 6

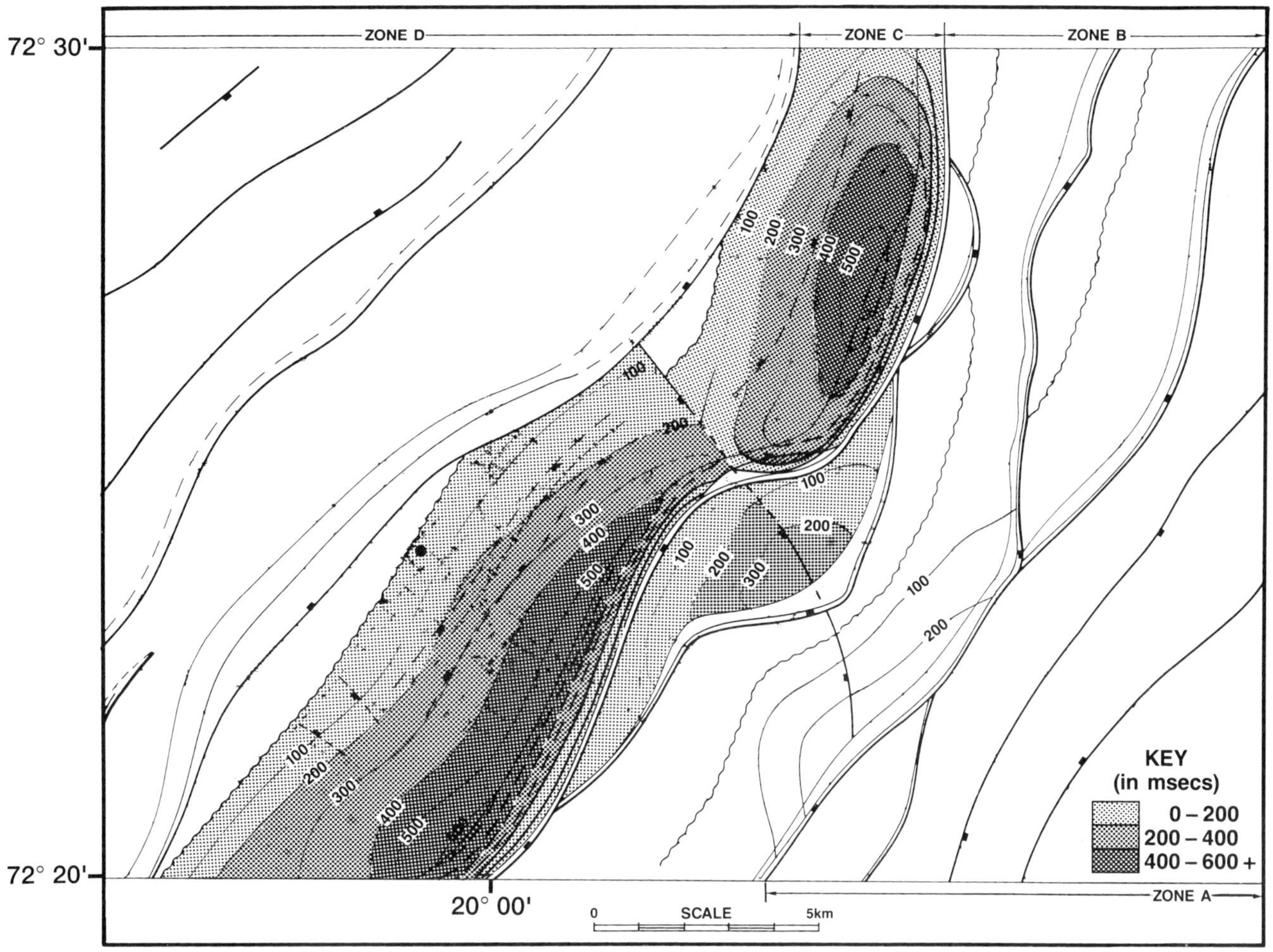

Fig. 12. Isopach map of synrift megasequence.

km south of the lateral fault. This second area has a maximum width of nearly 2 km centred over the apex of an elongate antiform in the pre-rift.

Both depocentres hug the base of the fault scarp. The southern depocentre is thicker and more laterally continuous than the northern one. They are centred over the morphological deeps mapped on the base synrift surface (Fig. 6) and achieve maximum thickness of 500 m in the north and over 600 m in the south.

The distribution of the megasequence is shown in Fig. 12. On the higher terrace east of the main fault block, the megasequence is restricted. West of the main fault block, the megasequence thickness is indeterminate. Well results indicated that the distal part of the megasequence wedge is marine. Backstripping indicated that the coastline was in the vicinity of the fault dividing zones C and D in Fig. 12. It appears, therefore, that the megasequence was deposited in shallow to deep marine environments, with the paleowater depth increasing as shown by the TWT map in Fig. 6.

Seismic sequence 1

The basal unconformity of this sequence is the base synrift reflector in Fig. 11, and was dated from the well as being of late-Jurassic age. Seismic sequence 1 is correlated with the lowermost part of sequence C in the well (Fig. 4).

The isopach map of this unit is shown in Fig. 13. Two depocentres run along the base of the main boundary fault and are separated by the main lateral fault cutting the fault block (Fig. 6). The sequence achieves a maximum thickness of over 160 m in the southern depocentre and 100 m in the north. Ponding of this sequence is further accentuated by the second order features on the basal surface. A small antiformal structure appears to separate the main ponded area from the section of the sequence which gradually thins onto the fault block crest. It also forms the approximate boundary between the undisturbed and disturbed seismic reflector characters. The seismic sequence onlaps onto the fault crest and terminates very close to the pinch-out trace of the megasequence.

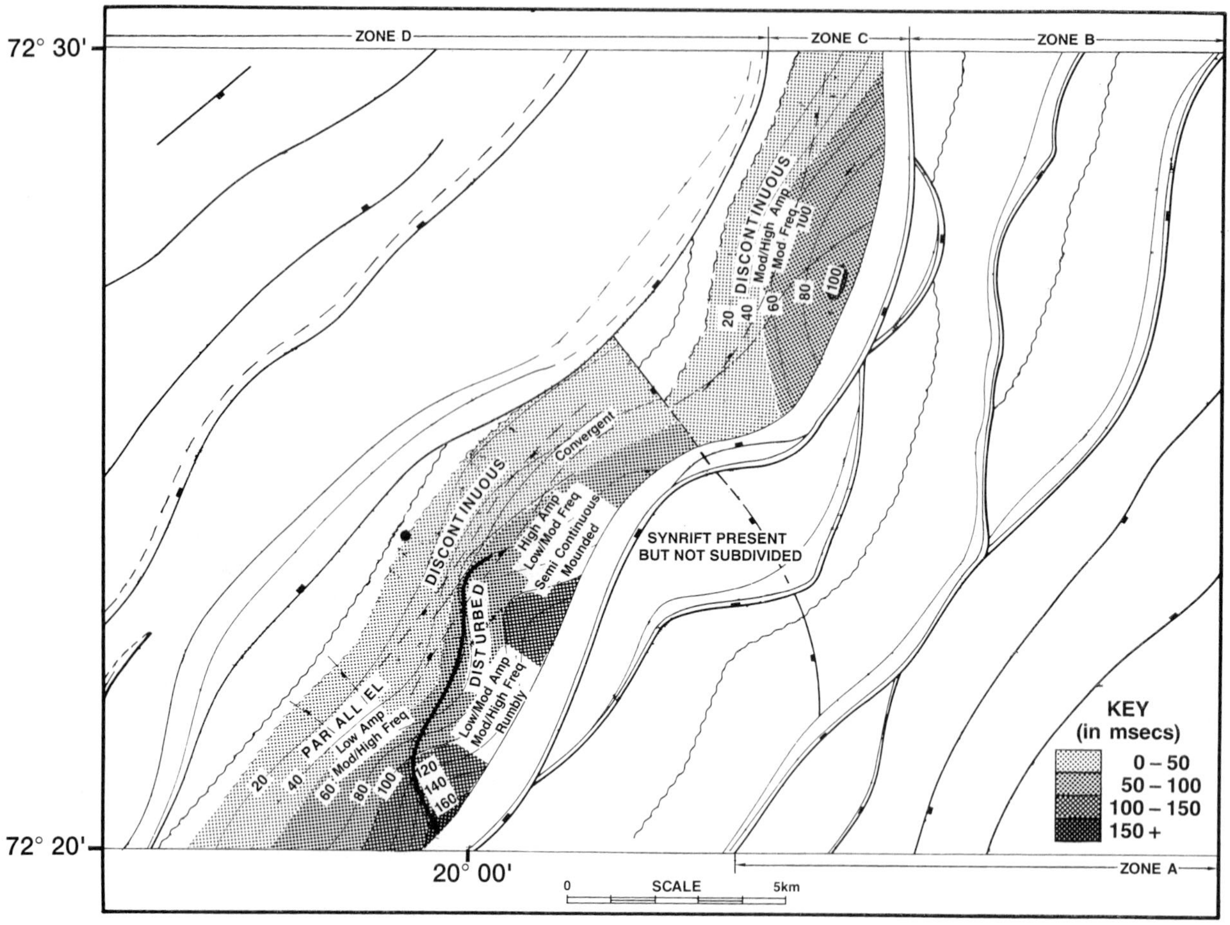

Fig. 13. Isopach map of seismic sequence 1 showing seismic facies.

This seismic sequence was divided into two seismic facies. A zone outlined on the map between the main fault and the small antiformal structure on the base synrift surface enclosed a disturbed seismic character which became rumbly on the high resolution data. This seismic facies illustrates strong, high angle downlaps and the events are generally of moderate amplitude and high frequency with occasional bursts of high amplitude, low frequency energy.

The seismic character of the remainder of the sequence was generally continuous, with events of lower amplitude and higher frequency onlapping onto the fault block crest and disappearing prior to the pinch out of the whole megasequence. The onlap in this part of the sequence is of a much lower angle than in the rest of the sequence.

Sedimentary interpretation and depositional model

The well results indicate that the lowermost units of the megasequence consist of glauconitic, bioturbated siltstones and mudstones, indicative of low energy marine conditions (Fig. 4). The sequence 1 sediments down-dip of this location were deposited in an even deeper marine environment.

Parallel, continuous seismic facies calibrate with low energy marine mudstones in the well. The change in seismic facies in this sequence down-dip to a disturbed character, is used to infer a corresponding change in lithology from mud dominated to sand-prone sediments. Given the paleogeography and structural configuration at this time, the only mechanisms which could transport sand-prone sediments into this deep marine setting would be mass flow processes (Nardin, 1983). The manner in which the disturbed seismic facies is confined to the morphological deeps is consistent with the interpretation of mass flow deposits. The more rumbly seismic character within this facies is also located at the point of closest approach to the two main sedimentary conduits postulated for the sequence, as shown in Fig. 9, which is consistent with the areal distribution of sand-prone sediment being controlled by these structural features.

Correlations between similar seismic facies and lithofacies have been made in a number of other setting, e.g., Nardin (1983), Geoquest manual (Macurda, 1900) and the Brae field of the UKCS.

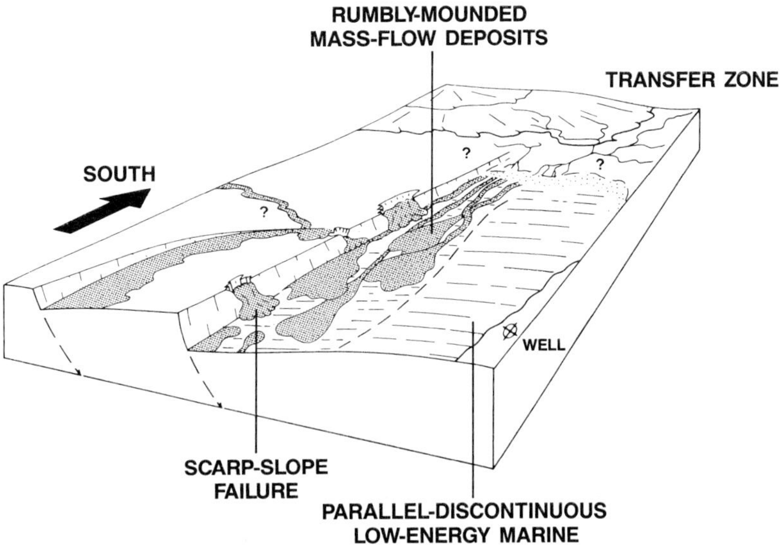

Fig. 14. Depositional model for seismic sequence 1 (seismic character above, sedimentary interpretation below).

In the Magnus area of the UK North Sea, a similar change in seismic facies occurs in a similar structural setting and the disturbed seismic facies relates to high density sand-prone turbidites (M. Shann, pers. commun.).

The geometry and internal configuration of seismic sequence 1 is not consistent with any of the classic radial or elongate submarine fan depositional models (Stow et al., 1981; Mitchum, 1985; Surlyk, 1978). It is most closely comparable with the submarine ramp model proposed recently by Surlyk (1987) for the Upper Jurassic Hareelv Formation of central East Greenland. In this model axially transported sediment passes down a ramp, along the fault terrace with input via the transfer zone and is deposited in deep water under poorly oxygenated conditions during a period of increasing base level. The sand-prone sediments form anastomosing bodies embedded in a shaly matrix. Lateral connectivity between sand bodies is extremely limited (see Fig. 14). This forms the basis for the depositional model developed for seismic sequence 1 shown in Fig. 14. Requejo et al. (1989) have shown that the oil found in seeps within the sandstones of the Hareelv Formation was generated from the Hareelv Formation interbedded shales. Hence, the depositional environment proposed for seismic sequence 1 includes both reservoir and source rocks.

Seismic sequence 2

The basal unconformity of this sequence (Fig. 11) was defined as the first onlap surface above the basal unconformity. This event is a moderately continuous, high frequency, moderate amplitude positive peak. Seismic sequence 2 was correlated with the upper part of sequence C (Fig. 4).

The isopach map of this unit is shown in Fig. 15. This sequence is the most extensive of all sequences mapped, reaching a maximum thickness in both depocentres of over 180 m. Prior to the deposition of this sequence, a substantial amount of the topography on the fault block had been infilled by sequence 1. Hence, none of the second order features on the basal synrift surface appear to affect the distribution of sequence 2. A distinct, laterally impersistent mound runs subparallel and very close to the fault scarp (Fig. 11).

In a zone broadly defined by the main fault and the 100 m contour the events are of a lower amplitude, higher frequency and are less continuous than the rest of the sequence.

The sequence northwest of the 100 m contour consists of very continuous events of higher amplitude and lower frequency (Fig. 11). The character in this area is also affected by the attenuation of the beds as the sequence thins onto the fault block crest.

Sedimentary interpretation and depositional model

The geometry of this seismic sequence most closely resembles that of a "classic" synrift wedge (Mitchum, 1985). The parallel continuous seismic character of the distal part of this sequence (Fig. 11) is correlated with the low energy marine, probably condensed, mudstones and siltstones encountered in the well. The disturbed seismic character down-dip of this facies is inferred to indicate a lithology change to more sand-prone turbidites and other mass flow units. This facies also shows poorly developed internal downlap.

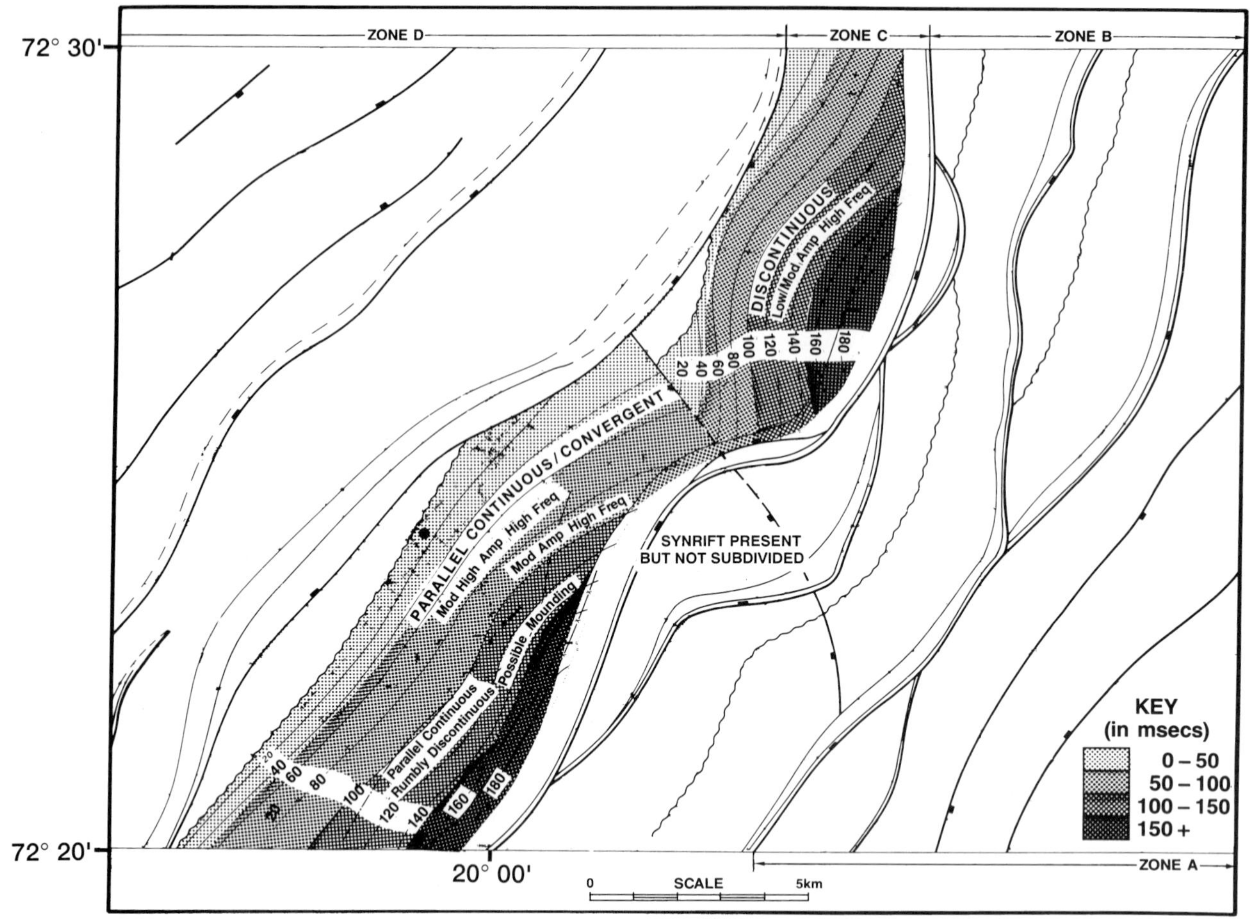

Fig. 15. Isopach map of seismic sequence 2 showing seismic facies.

A further indication of the possible sand-prone nature of the mound feature is the antiform in the overlying sequences (Fig. 11). Examples of such features on seismic data from the Far East have been shown to be caused by differential compaction. Drilled examples from Thailand indicate that the mounds consist of sand-prone alluvial fans encased and overlain by low energy mudstones (C. Sladen, personal communication). Since such an effect is dependent upon the physical properties of sand and shale, one may expect to see it in all mixed depositional environments.

The features of this seismic sequence are taken to indicate a substantial lateral input of sediment via the footwall cusp during deposition. The characteristics of this seismic sequence are more indicative of fans radiating out from the footwall, such as borderland fans associated with small transfer zones (Nardin, 1983).

An important consideration in the development of the depositional model for this sequence is the structural position and external geometry of the disturbed seismic facies unit. This facies unit incorporates a distinct elongated mound, the apex of which parallels the trend of the main fault and which is localised around the footwall cusp (Fig. 15). The location, morphology and presence of internal downlap within this feature is taken to indicate the possibility of a lateral input of sediment from the higher fault terrace via the footwall cusp (Fig. 10). Lateral derivation of sediments into a major graben is commonly associated with overspill from small footwall grabens, e.g., Gulf of Corinth, Greece, Gulf of Suez, East African Rift System (Jackson et al., 1982; Rosendahl et al., 1986; Leeder and Gawthorpe, 1987; Gawthorpe et al., 1990). A direct consequence of such a model would be that the bulk of the sediments within the footwall cusp correlate with the earliest seismic sequence on the main fault block. This depositional model is similar to those proposed for the Brae field (Stow, 1981) and the late Jurassic fans of the Wollaston Foreland of East Greenland (Surlyk, 1977) (Fig. 16). In these models the sand-prone sediments are ponded up against the fault scarp (see Fig. 16). However, the differences in scale should be borne in mind: radial fans in seismic sequence 2 are no more than 2 km across and between 300–400 metres thick which is less than half

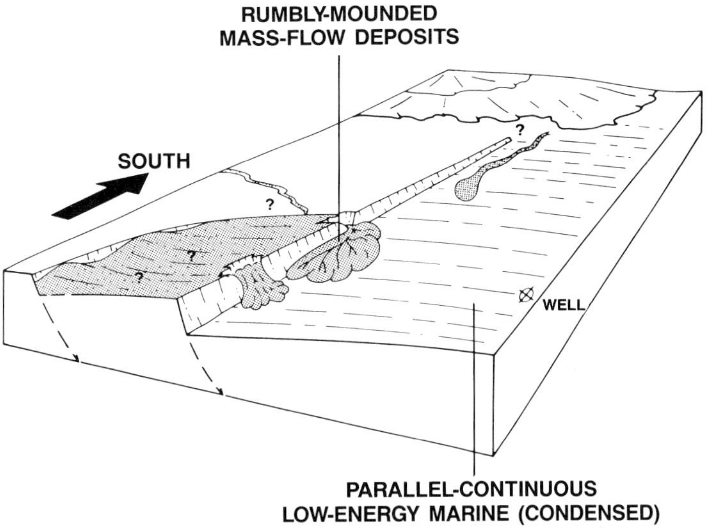

Fig. 16. Depositional model for seismic sequence 2 (seismic character above, sedimentary interpretation below).

the size of those in the Brae and Woolaston Foreland examples.

Seismic sequence 3

The base of this sequence, see (Fig. 11) was the most prominent unconformity and strongest event in the dataset. The event is a very continuous, high amplitude, moderate frequency positive peak with high frequency events onlapping onto it. The event is deformed at the fault block crest with two impersistent cuspate features running sub-parallel to the subcrop trace of the wedge. These features were mapped and are shown in Fig. 17. Seismic sequence 3 was correlated with sequence B from the well results (Fig. 4).

The isopach map of this sequence is shown in Fig. 17. This sequence is the thinnest of all the seismic sequences mapped and shows the least variation in thickness. It attains a maximum thickness of over 80 m in the southern depocentre and around 40 m in the northern area. This "blanket like" morphology indicates that the basal synrift surface morphology has been completely masked prior to the deposition of seismic sequence 3 and that the sequence was deposited on a semi-continuous inclined plane. The sequence extends over the fault block northwestwards to the pinchout trace of the whole megasequence. The higher resolution data to the south show the sequence to consist of very continuous, high amplitude, moderate–high frequency events which onlap at a low angle onto the basal unconformity.

Sedimentary interpretation and depositional model

This seismic sequence is correlated at the well with low energy, deep marine organic rich mudstones (Fig. 3). The truncation of reflectors at the footwall crest, illustrated on Fig. 17, is possibly the result of low angle slumping in unconsolidated sediments. Similar features have been observed on three-dimensional data from analogous structural settings, e.g., Magnus field (Price, personal communication). In the Magnus example, sediments of the same age and lithologies as in Bjørnøya South slide down the hanging wall to form cuspate reflector geometries directly analogous to those observed within seismic sequence 3.

The entire sequence consists of one seismic facies which was calibrated at the well with deep marine mudstones. The external sequence geometry and internal characteristics are consistent with deposition on a low energy slope apron as shown in Fig. 18 (Pickering, 1984).

Seismic sequence 4

The basal unconformity of this sequence (Fig. 11) was defined as an onlap surface. This unconformity is thought to be the base of sequence A from correlation with the well (Fig. 4). The isopach map of this sequence is not given. The sequence thickens regularly into the main fault to a maximum of over 120 m in the south and 160 m in the north. It thins to zero prior to the pinch out trace of the whole wedge along the fault block crest. The thickening of the sequence into the fault probably represents reactivation of motion along this feature after the period of relative quiescence associated with the deposition of seismic sequence 3.

The sequence consists of parallel continuous, high frequency, low amplitude events which onlap at a low angle onto the basal unconformity. The area of

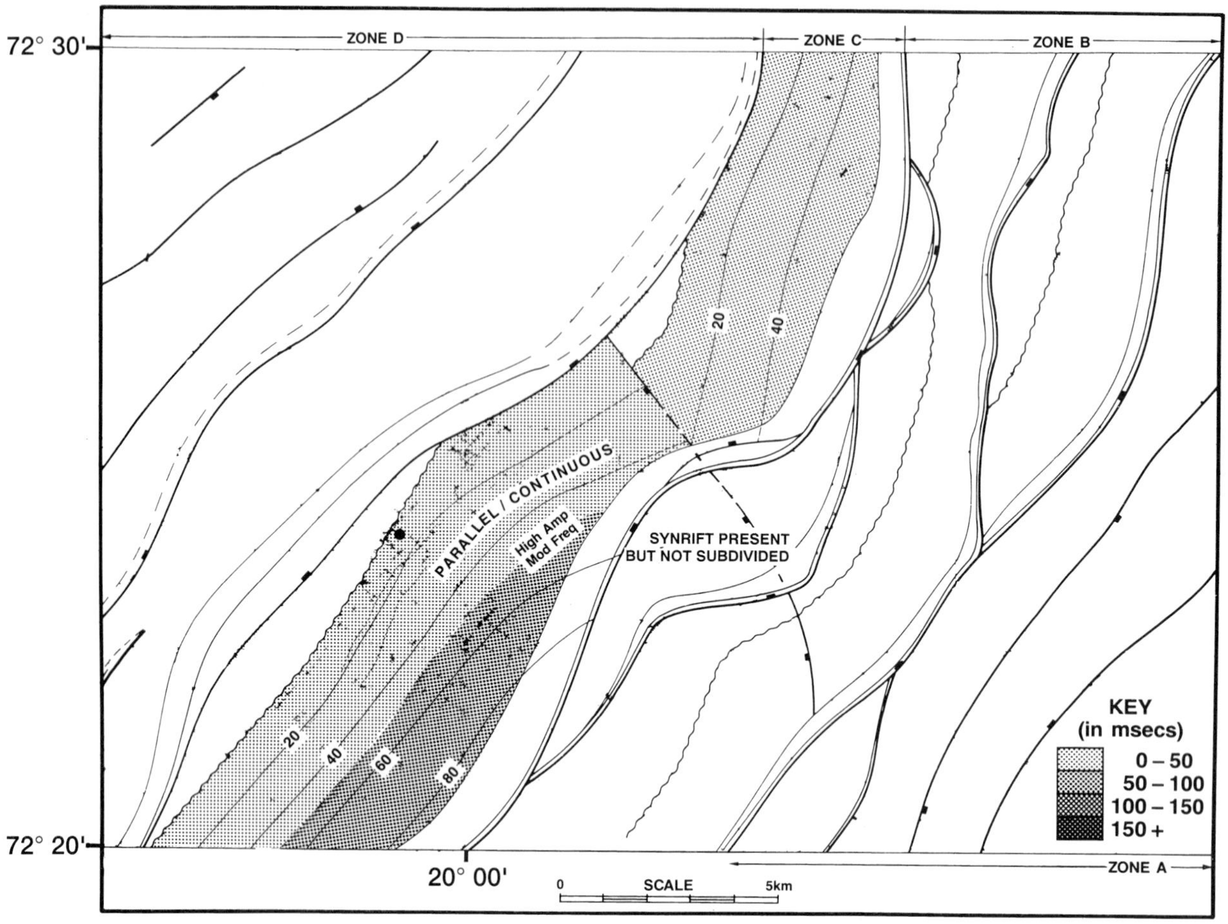

Fig. 17. Isopach map of seismic sequence 3 showing seismic facies.

discontinuous events shown on the map may well be a function of data quality as both sequences 4 and 5 were difficult to pick in this area. The parallel/convergent continuous seismic facies at the fault block crest is correlated with low energy mudstones in the well. Towards the main graben fault a slightly rumbly seismic facies was seen. This could indicate the possibility of coarser grained, more sand-prone material in this area, possibly laterally derived through the footwall.

Seismic sequence 5

The basal unconformity of this sequence is a very continuous positive peak of moderate amplitude and frequency and events of similar character onlap onto it.

The sequence has the same general shape as that of sequence 4, with slightly less thickening into the main fault. Sequence 5 attained a maximum thickness of over 120 m in the southern part of the fault block and 140 m in the more poorly constrained northern part. The sequence consists of very continuous, parallel and sub-parallel events of high amplitude and frequency. These events onlap at a low angle onto the basal reflector (Fig. 11).

Sedimentary interpretation and depositional model for sequences 4 and 5

The wedge shaped geometry of both seismic sequences is indicative of continued (albeit reduced) movement along the main fault during their deposition. The parallel, continuous seismic facies of sequences 4 and 5 are correlated at the well with red mudstones and minor associated sandstones (Fig. 4). Similar deposits of Valanginian age are seen in northern East Greenland at the crests of tilted fault blocks and are interpreted as condensed units deposited in low energy, deep marine environments (Surlyk, 1977, 1978). A similar depositional environment is proposed for seismic sequences 4 and 5.

Closer to the main fault, this gross depositional environment may be modified by the input of laterally derived material via the footwall. This situation may be broadly analogous to that described by Pickering (1984) from the late Jurassic of Scotland along the Great Glen Fault where debris spalled off the footwall and is locally concentrated along the scarp. However, the majority of seismic sequences 4 and 5

Tertiary Uplift and Erosion in the Barents Sea: Magnitude, Timing and Consequences

B. Nyland, L.N. Jensen, J. Skagen, O. Skarpnes and T. Vorren

In 1920 Fridtjof Nansen postulated an uplift and erosion in the order of 500 m in the Barents Sea. Hydrocarbon exploration in the 1980's has since confirmed a significant uplift and erosion in this area.

In 1985 Statoil initiated a multidisciplinary study to evaluate and quantify the amount of uplift and erosion. The methods used were: Shale compaction curves, Dc-exponent from drilling, vitrinite reflectance trends, diagenesis of clay minerals, apatite fission track studies, volumetric mass balance calculations of erosion and deposition, and PVT-modelling of known reservoirs.

Studies of porosity and density trends in Tertiary shales from wells showed an uplift and erosion of the order of 500–1000 m in the Hammerfest Basin, increasing to about 1700–1800 m towards the Fingerdjupet Subbasin area. Vitrinite trends from exploration wells and IKU shallow drillings confirmed these results also showing increasing erosion towards the north (3–4 km erosion at Bjørnøya).

Depending on the paleo-drainage area chosen, volumetric calculation of the Late Tertiary deposition along the western shelf margin gives a magnitude of erosion of about 1–1.2 km in the southern part of the Barents Sea, increasing to approximately 3 km on Svalbard.

Used together, the methods above give a fairly accurate estimate on the magnitude of the erosion but not on timing. Fission track studies indicate one or more phases of uplift from Late Oligocene to Pliocene/Pleistocene, while biostratigraphic redatings of sediments from the wells on the Senja Ridge suggest that a major part of the erosion is related to glacial processes during the last 2–3 Myr.

Extensive residual oil staining found below the present hydrocarbon/water contacts indicates that reservoirs, such as the Snøhvit field in the Hammerfest Basin, once contained larger quantities of oil. PVT-modelling has shown that this oil was lost probably as a consequence of gas expansion during uplift, and can explain in part the lack of commercial oil finds so far in the Barents Sea.

Introduction

Based on bathymetry, seabottom topography and geological information from surrounding areas, Fridtjof Nansen postulated a Late Tertiary uplift and erosion of about 500 m in the south western part of the Barents Sea (Nansen, 1920). Later, coal petrographic studies on Svalbard (Manum and Throndsen, 1978) confirmed a substantial uplift of the area, and about 1.7 km erosion was suggested for the central part of Svalbard.

The first exploration wells drilled in the Hammerfest Basin (Fig. 1) demonstrated a marked unconformity between Plio-Pleistocene and underlying strata. This has later been confirmed by all wells drilled in the area, and this major unconformity is clearly seen on the seismic data. Where Tertiary sequences are present, they consist mainly of Paleocene and Eocene sediments, whereas Oligocene and Miocene sediments are absent.

Another observation from wells in the Hammerfest Basin is the unusually low sandstone porosities at present depth. Olaussen et al. (1984) suggested 1000–1500 m deeper burial of the sediments to explain these low values. Quartz cementation by pressure solution also indicates that the Jurassic sandstones in the Hammerfest Basin have been more deeply buried in the past (Berglund et al., 1986).

Maturity data also indicate a maximum burial depth of at least 1000 m deeper burial, or higher heat flow than observed at present to explain the level of maturation seen for the source rocks penetrated.

Three large fan systems are deposited along the western margin of the Barents Shelf; outside the Bjørnøya Trough, outside the Storfjorden Trough and north of Spitsbergen. These fans are interpreted to be depocentres for the erosional products from the shelf (Vorren et al., 1988, 1990; Nøttvedt et al., 1988).

A multidisciplinary study has been continuing since 1985 to quantify the magnitude and timing of the uplift and erosion. The different methods and

Structural and Tectonic Modelling and its Application to Petroleum Geology edited by R.M. Larsen, H. Brekke, B.T. Larsen and E. Talleraas. NPF Special Publication 1, pp. 153–162. Elsevier, Amsterdam.

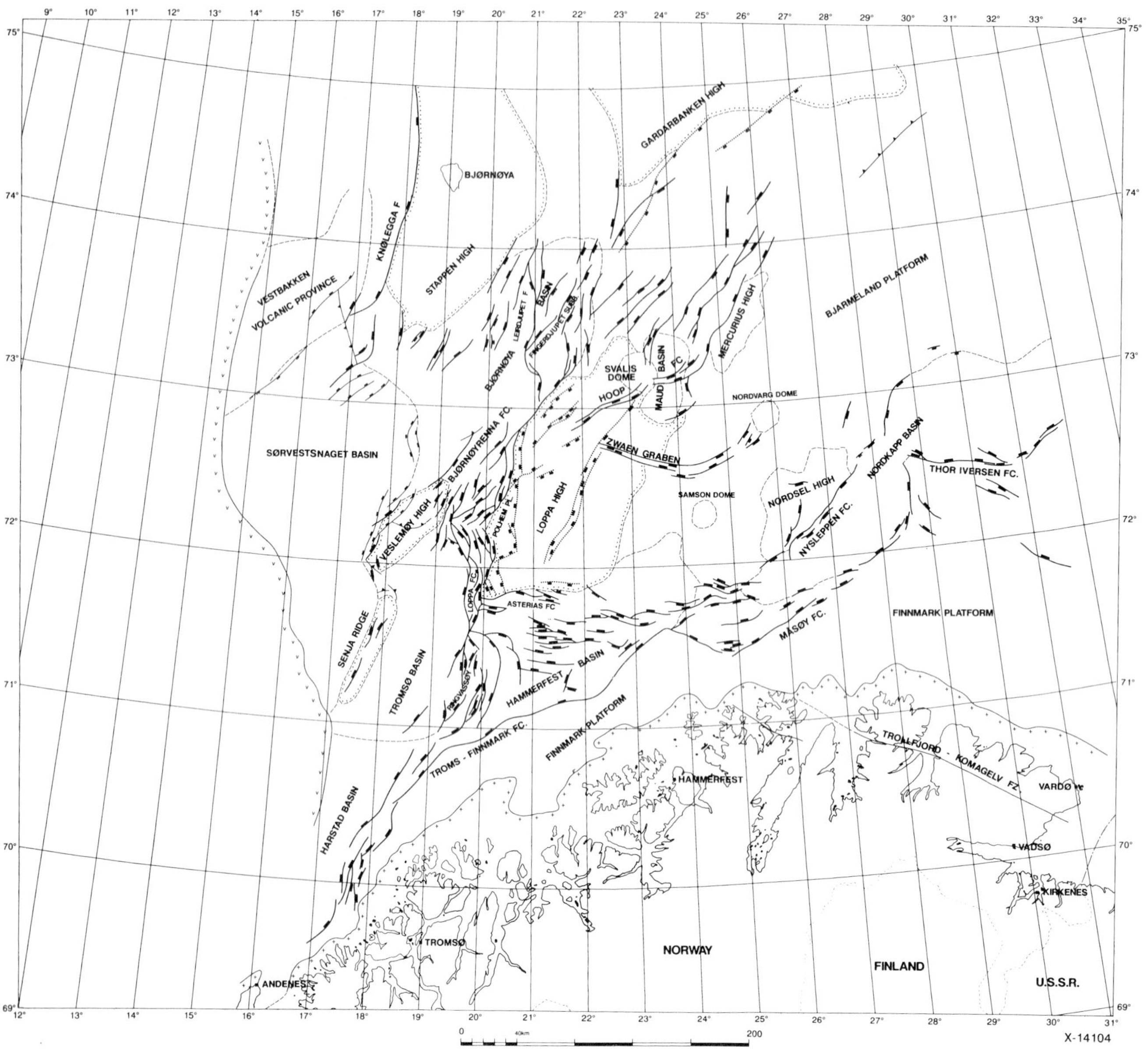

Fig. 1. Barents Sea; main structural elements (after Gabrielsen et al., 1990).

the main results will be summarized in this paper.

The possible effects of the uplift and erosion on a hydrocarbon filled reservoir with respect to PVT-modelling, has also been studied in order to test if this can explain gas discoveries with only traces of residual oil and the presence of residual oil in most of the dry wells.

Methods

Different methods have been used in order to quantify the uplift and erosion (Table 1). The database used consists of data available from exploration wells, IKU (Continental Shelf and Petroleum Technology Research Institute) shallow drillings, seismic data, and results from fieldwork on Svalbard and Bjørnøya. Since most of the well data used are confidential, the results have to be presented in a general way without showing any raw data.

In addition, regional mapping and evaluation of the Cenozoic sequence in the western part of the Barents Sea have been performed at the University of Tromsø.

TABLE 1

Methods used to quantify uplift and erosion

Vitrinite reflectance profiles
Shale compaction curves
Drilling parameters
Fission track analysis
Diagenesis of clay minerals
Volumetric mass balance

Vitrinite reflectance profiles

Vitrinite reflectance analyses are available from all exploration wells and most IKU shallow drillings, and are used as a measure of the level of maturation. A typical vitrinite reflectance profile through a well in the Hammerfest Basin is shown in Fig. 2. The profile can be used to estimate the thickness of a missing sequence by extrapolating the trendline to VR = 0.2%, which is anticipated to be the reflectivity in the sediments deposited on seabottom (Dow, 1977). The differences between the Base Quaternary unconformity and a VR = 0.2% will represent the thickness of the sequence missing, which is about 1100 m for the well shown in this case.

The use of this method is highly debatable (Katz et al., 1988; Feazel and Aram, 1990) since the vitrinite reflectance is controlled by parameters as the heat flow, burial rate, type of organic material, etc., but also on the quality of the sample materials used (cores, cuttings, etc.) and that the measurements are performed on in-situ material. The method is also sensitive for small changes in tracing the gradient. Due to these sources of errors the database was carefully examined and poor samples were excluded. We also compared wells with the same geological setting

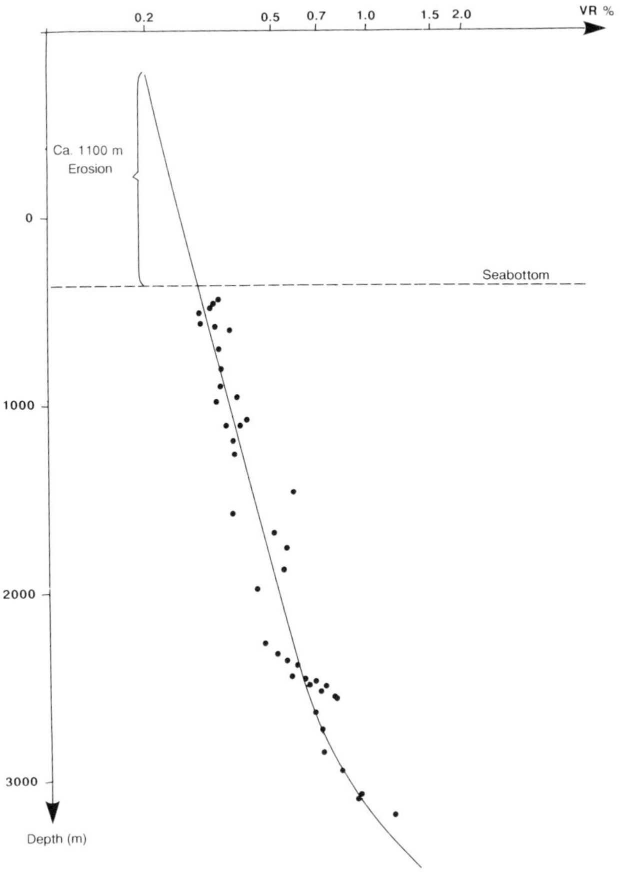

Fig. 2. Vitrinite reflectance profile well A, Hammerfest Basin.

and heat flow history to try to avoid some of the problems mentioned. Based on these considerations, this method gives an error within ±200 m in our experience.

Shale compaction curves

This study is based on the use of published curves for compaction of shales in the central North Sea Basin (Sclater and Christie, 1980). The method uses the irreversible compaction of shale (also sandstone, limestone and shaly sandstone) as a function of depth. Values from density logs define the input into the equation used. Compared with the compaction curves, a higher density than expected at the present depth indicates a deeper burial, and the differences in depth represent the amount of erosion. An advantage of the method is that calibration points from the entire well can be obtained. The great number of data points can then be further analysed statistically and also different lithologies can be used.

A statistical programme has been applied to density values from log data and the results compared with the published curves in order to construct compaction curves for the Barents Sea. Erosion from shale compaction curves are calculated for all wells and the results seem to be consistent. A source of error using this method is that the parameters taken from the functions used by Sclater and Christie (1980) may not be representative for other areas such as the Barents Sea. This problem can to some extent be eliminated by using compaction curves from wells with minimal erosion in the area (as the wells on the Senja Ridge), and then to compare the other wells in the area.

Erosion from drilling parameters

An internal Statoil study has shown that different parameters measured during drilling can be used to estimate the missing sequence in a well (J. Skagen, pers. commun., 1989). Tests have shown that the rate of penetration (ROP) is related to the compaction trend of shales with depth, given that other drilling parameters are constant.

The Dc-exponent (Jordan and Shirley, 1966), used in the prediction of pore pressure, is a function of the ROP (and other drilling parameters) and gives an estimate on the compaction trend in a well. The Dc-exponent is dependent on the depth and the porosity and will, during drilling, indicate anomalous compaction trends caused by, for instance, uplift and erosion if present. The uncertainty of the method is mainly associated with the picking of representative values from the Dc-curve and a deviation of about

±400 m can be expected. Only the wells drilled in the Hammerfest Basin have been examined by this method.

Fission track analysis

The previously mentioned methods used in this study give an estimate on the amount of erosion, but they say nothing about the timing of this event.

Apatite is present as an accessory mineral in sandstones and contains small quantities of uranium which decays radioactively by fission through time. The fission process creates small defects in the crystal structure of the apatite (fission tracks) and the length of these defects can be measured. The density of tracks in an apatite grain is a function of the amount of uranium, the temperature, and the time spent below the annealing temperature. By measuring the density of tracks and the uranium content it is possible to estimate the age of the apatite. Knowing the age of the apatite and the fission track length distribution, it is possible to model the time and temperature history of the rocks containing the apatite (Gleadow et al., 1983).

In this study the method has been used mainly to establish the timing of the uplift. However, together with well temperature information an estimate of the magnitude of uplift can be given.

Diagenesis of clay minerals

The Crystallinity Index (CI) of authigenic illite, calculated from XRD, can be used to evaluate the level of diagenesis reached in sandstone and with that the depth of burial (Foscolos and Kodama, 1974; Foscolos et al., 1976). The CI's measured from Lower to Middle Jurassic sandstones on the Tromsøflaket have been compared with CI's from equivalent formations that have undergone continuous burial in the Haltenbanken area.

A source of error possibly influencing the results is the difference in temperature history between these two areas. In addition, pore water composition and reaction kinetics are other factors that may influence on the illite formation (K. Gibbons, Statoil, pers. commun., 1988).

Amount of uplift

Vitrinite reflectance profiles and shale compaction curves give the most consistent data coverage for estimates of uplift and erosion in the Barents Sea. Profiles, as shown in Fig. 2, have been constructed for all wells, and the thickness of the missing section calculated. Effects caused by higher heat flow in tectonically more active areas, and around salt diapirs have been modelled and taken into consideration in the evaluation. Based on the results from vitrinite reflectance profiles, a regional map showing the amount of erosion has been constructed (Fig. 3).

The least uplift and erosion are seen in the western part of the area; on the Senja Ridge and in the Tromsø Basin (0–500 m). In the Hammerfest Basin and on the Loppa High the values are between 1000–1500 m. On the Finnmark Platform about 1500–2000 m is absent, decreasing towards the south and east. The magnitude of uplift and erosion increase towards the north and northwest, and more than 2000 m of uplift and erosion are observed on the Stappen High and Bjørnøya. Data from Andøya also suggest erosion of the order of 2000 m.

The results from the use of compaction curves seem to follow the results from the vitrinite reflectance profiles. The main difference is that the amount of erosion is lower in the Hammerfest Basin (800–1000 m), but the differences are within the limits of confidence established for this study. As with the vitrinite reflectance profiles, the shale compaction curves show an increase in magnitudes towards north, reaching about 2000 m in the Fingerdjupet Subbasin.

The results from drilling parameters are in accordance with the results achieved using shale compaction curves and vitrinite reflectance profiles, with an order of 800–1500 m erosion in the central part of the Hammerfest Basin and decreasing erosion moving westwards in the basin.

The mean Crystallinity Indices (CI) calculated for wells in the Hammerfest Basin are in the range 12–17. These values correspond to a burial depth of about 900–1100 m deeper than present day compared with the Haltenbanken area. Most uplift and erosion is seen in the central part of the Hammerfest Basin (in the Snøhvit and Albatross fields), decreasing westwards towards block 7119/12 (300–400 m). Least uplift is observed in the wells drilled on the Senja Ridge.

Based on a map of the Upper Regional Unconformity (URU) combined with bathymetric maps (Vorren et al., 1989, 1990), the drainage system sourcing the three depositional areas in the Bjørnøya Trough, Storfjorden Trough and north of Svalbard has been constructed (Fig. 4). Volumetric calculations of the fans indicate about 1.2 km of uplift and erosion in the area draining the southwestern part of the Barents Shelf, while a thickness of about 3.1 km sediments have been removed from the Svalbard drainage area (Vorren et al., 1990). Calculations also indicate that the deposition rate increased from Miocene to Pleis-

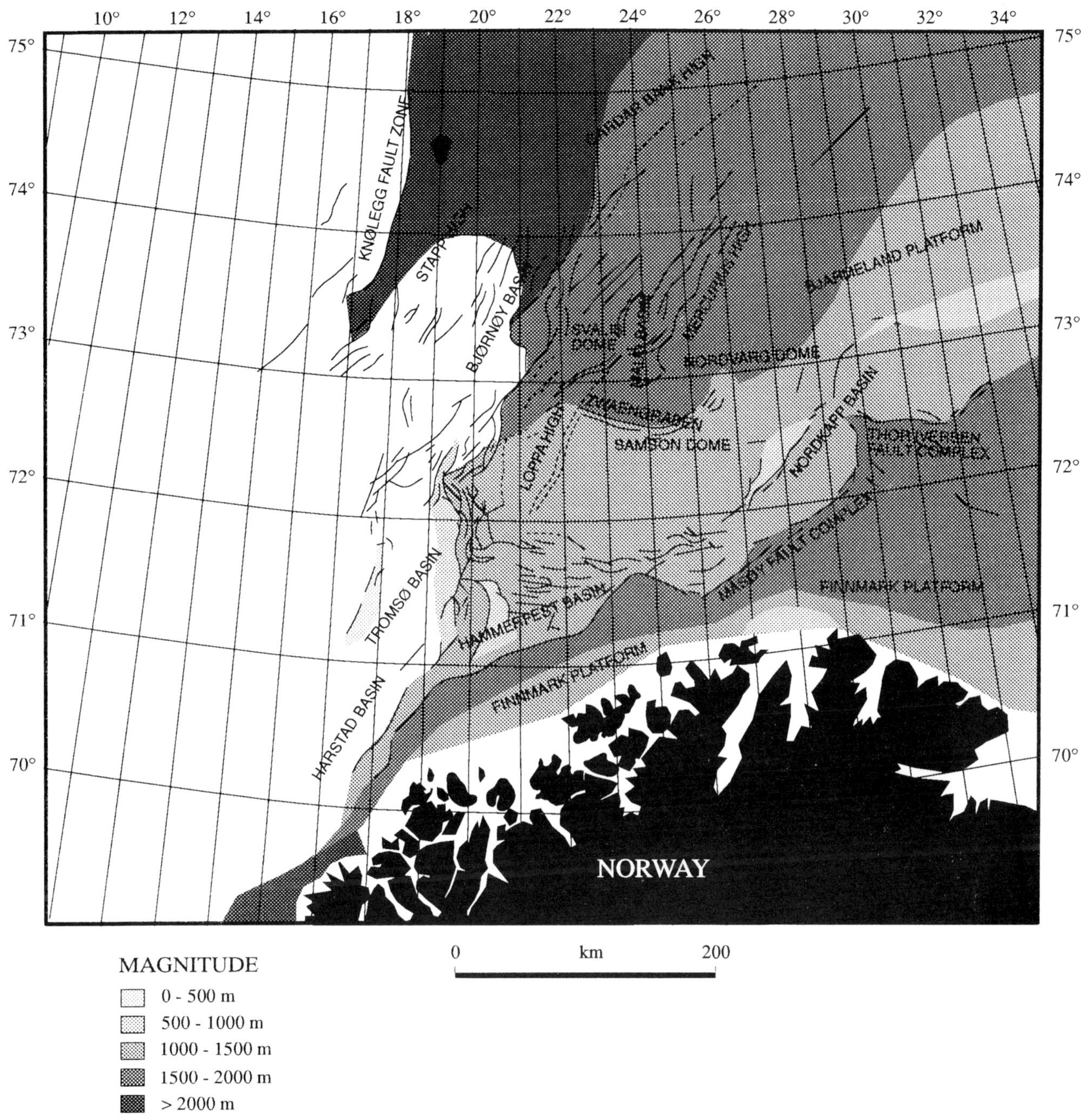

Fig. 3. Postulated uplift and erosion. Based on vitrinite reflectance profiles.

tocene, possibly reaching a maximum in the Late Pliocene.

Fission track analysis can also be used to some extent to model the burial history, since the temperature at maximum burial is one of the parameters given. Generally, the measurements show that most of the samples have been exposed to temperatures of 90–110°C. The difference between the measured formation temperatures in the wells and the paleotemperature given by the fission tracks can be used to predict the uplift. Depending on the temperature gradient used, this gives between 1500–2000 m uplift and erosion in the Fingerdjupet Subbasin, and 1000–1300 m in the Hammerfest Basin.

Timing of uplift and erosion

The timing of uplift and erosion is an important factor controlling the generation and migration of hydrocarbons. The generation of hydrocarbons will cease during uplift and the main phase of generation and migration will be at maximum burial.

The uplift and erosion history can be related to two important regional events. These are the

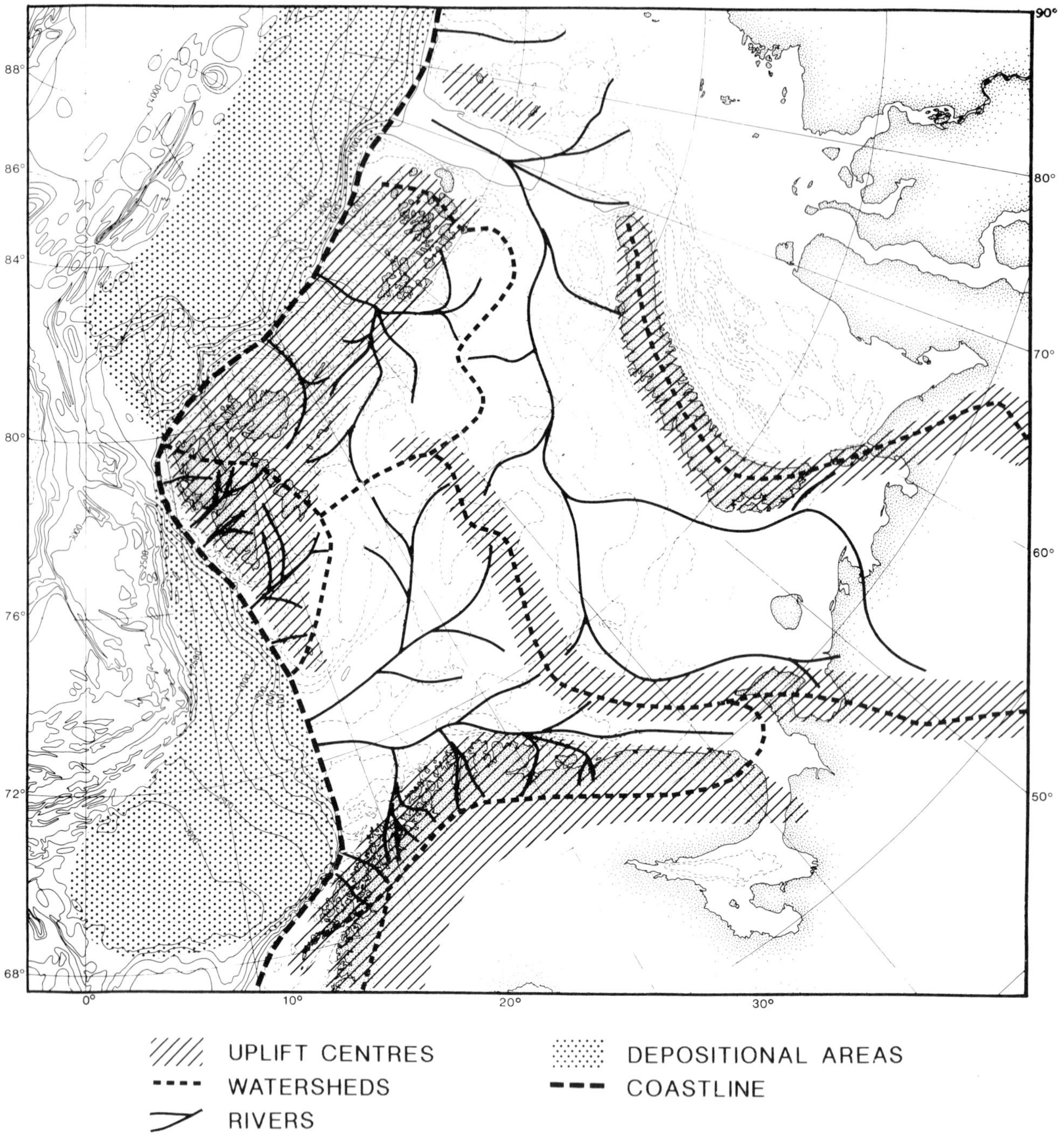

Fig. 4. Reconstruction of the Neogene fluvial drainage pattern in the Barents Sea. Present bathymetry and geography are used as base map.

structural development related to the opening of the Norwegian–Greenland Sea (Eldholm et al., 1984; Faleide et al., 1984; Spencer et al., 1984; Myhre and Eldholm, 1988), and the Late Cenozoic glaciations (Matishov, 1980; Elverhøi and Solheim, 1983; Vorren et al., 1988, 1990; Eidvin and Riis, 1989).

The lack of Late Tertiary sediments in all the wells except for those drilled on the Senja Ridge (Spencer et al., 1984), have for a long time been used to explain the uplift and erosion as a response to the opening of the Norwegian–Greenland Sea in Oligocene time.

The use of fission track analysis indicates two distinct periods of uplift and erosion (Fig. 5). The first phase took place about 40–50 Myr ago, which might have been in response to the early rifting of the Norwegian–Greenland Sea. The second phase occurred about 5–10 Myr ago. This phase might reflect isostatic uplift in response to glaciations.

Fission track modelling in wells also indicates that the magnitude of the first phase of uplift and erosion

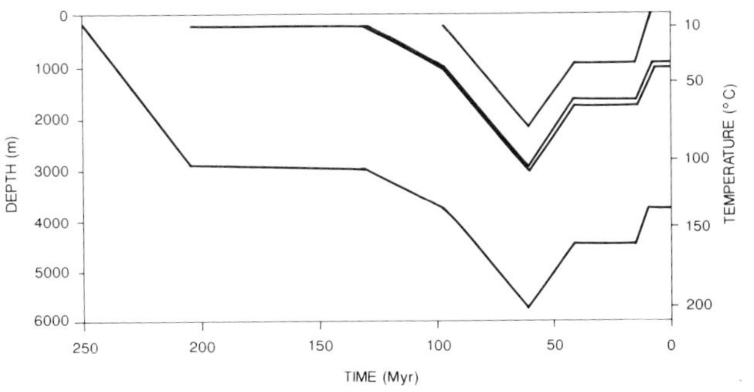

Fig. 5. Apatite Fission Track modelling. Tertiary cooling in two phases of uplift and erosion from the southwestern Barents Sea.

was approximately 500 m greater than the later event (Fig. 5).

However, modelling of fission track data from Svalbard suggests a continuous uplift and erosion since Early Oligocene time, and not two phases as indicated for the south western part of the Barents Shelf.

Most of the works performed on the Cenozoic sequence in the Barents Sea have been based on biostratigraphic datings of sequence boundaries in the wedge penetrated by the wells on the Senja Ridge (Fig. 1) (Spencer et al., 1984; Rønnevik, 1981; Faleide et al., 1984; Henriksen and Richardsen, 1988; Vorren et al., 1989, 1990). Redatings of sediments in these wells (Eidvin and Riis, 1989) indicate that the main part of the wedge (reaching about 4 km in thickness) consists of Pliocene and younger sediments. The redatings of the wedge are supported by Sr-isotope datings giving an age of 2.5–3 Myr (Eidvin and Riis, 1989). This suggests that the main part of the erosion took place during the glacial periods.

Consequences of uplift on hydrocarbon exploration

Observations indicate that the hydrocarbon potential in the area has been significantly affected by uplift and erosion. To date the exploration on the Barents Shelf has been disappointing. But in almost all wells drilled, good shows of residual oils have been found indicating that oil has been generated in the area. In wells where both gas and oil have been tested, high concentrations of extractable hydrocarbons are observed far below and above the present contacts (Fig. 6). We suggest that gas expansion during uplift caused oil to spill out of the structures as a result of pressure reduction, as outlined in Fig. 7.

This model has been tested on the Snøhvit field in the Hammerfest Basin, assuming that the major drop in the C_{15+} concentration (Fig. 6) represents the paleo oil–water contact (PALEO OWC) at maximum

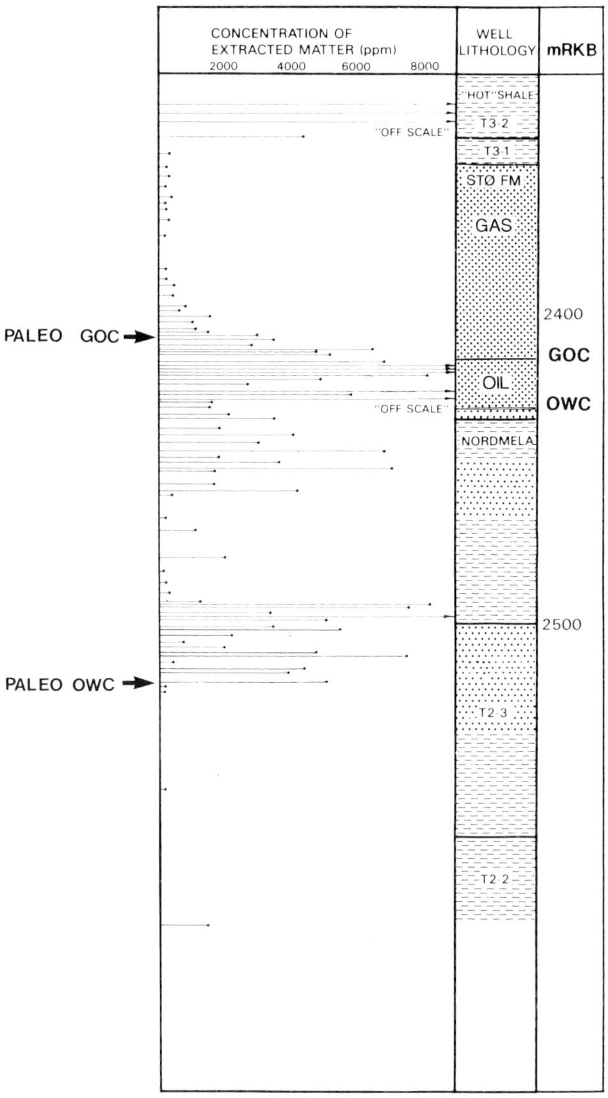

Fig. 6. Paleo oil/water and oil/gas contacts in a Snøhvit well estimated from concentrations of extracted organic matter.

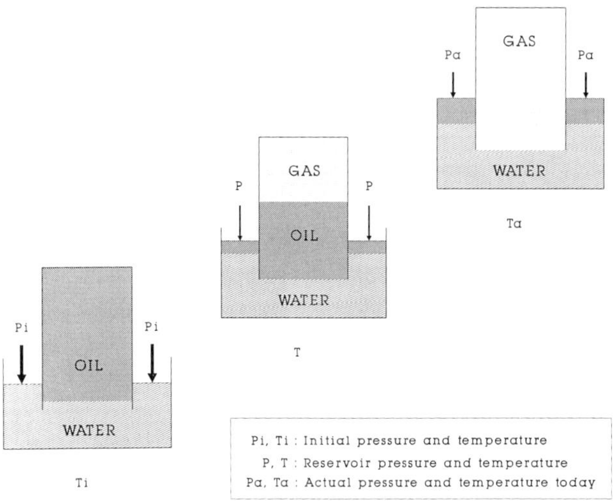

Fig. 7. Phase change caused by pressure and temperature reduction. P_i, T_i = initial pressure and temperature; P, T = reservoir pressure and temperature; P_a, T_a = actual pressure and temperature today.

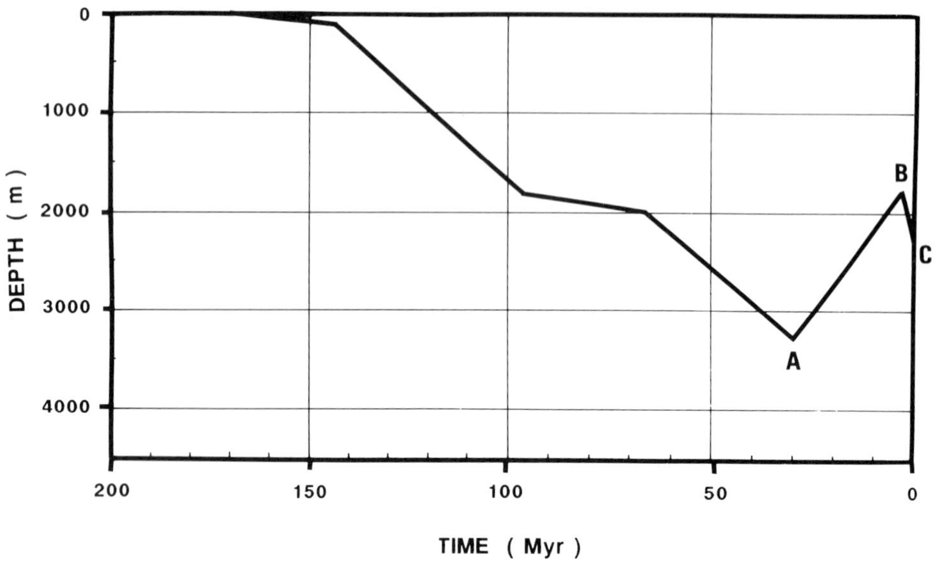

Fig. 8. Burial graph based on well stratigraphy for the Snøhvit field at reservoir level. Only one phase of uplift is used in the modelling.

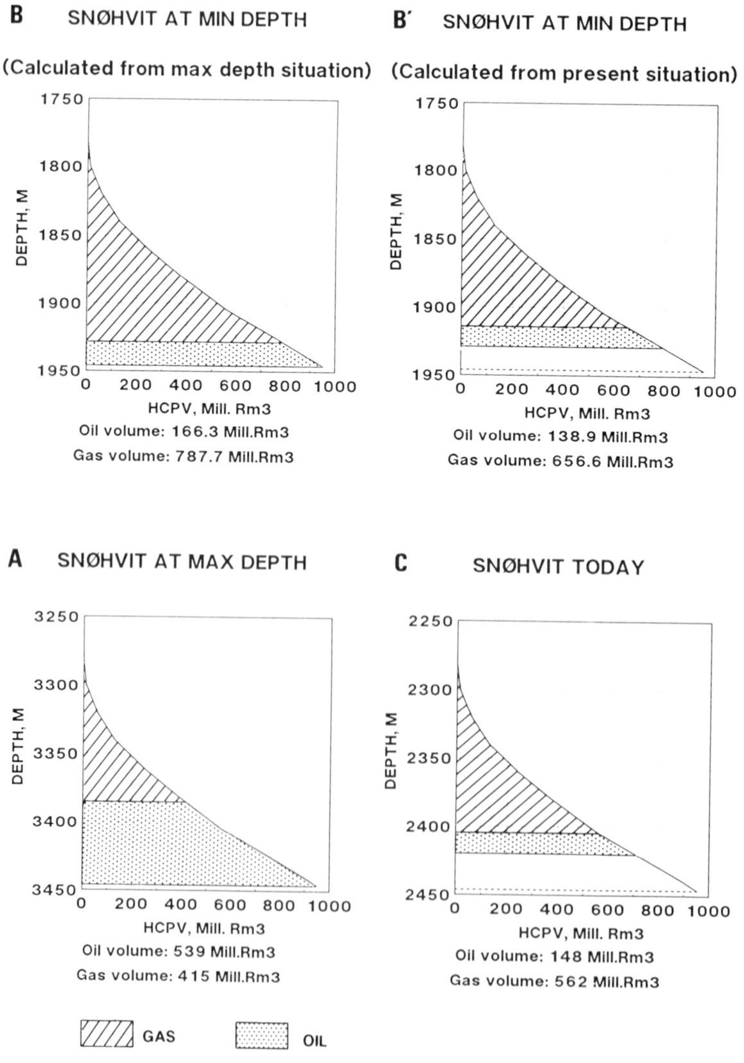

Fig. 9. Volume calculations estimating the effect of uplift and erosion on the Snøhvit field at different burial depths (see text for further explanations).

burial. The low concentration of heavier hydrocarbons ($>C_{15+}$) in the upper part of the reservoir can be an indication of a former paleo gas–oil contact (PALEO GOC) shallower than the contact seen today (GOC).

The burial history at reservoir level in the well used (Fig. 8) is the input for the calculations where *A* is maximum burial, *B* represents the time of maximum uplift and *C* is the depth of the reservoir today.

Knowing the hydrocarbon pore volume (HCPV) at different depths to spillpoint, and the composition of oil and gas in the structure, is it possible to calculate volumes at different depths.

To be able to run this model, some assumptions and simplifications have to be done and the restrictions in the modelling are: (1) The HCPV of the reservoir has not changed during uplift, erosion and renewed burial; (2) the composition of the oil and the gas is constant, and the oil is saturated at "bubble" point at any depth; (3) the pressure in the aquifer is constant (no overpressure); and (4) the temperature gradient has been constant through time.

Figure 9 summarizes the results from the volume calculations from burial at maximum depth (A) to minimum depth (B), and the same calculations performed for Snøhvit today (C) to minimum depth (B′). All calculations are based on the paleocontacts indicated in Fig. 6.

In the first situation (A to B), the modelling shows that about 370 million Rm^3 oil have been expelled from the structure, while the gas has expanded to almost twice the original volume.

In the next case (C to B′) modelling gives slightly lower volumes than the first calculation. However, the difference in volumes between B and B′ represents only 3–4 m difference in depth to spillpoint. The decrease in gas volume from B/B′ to C is within the expected range for gas diffusion loss during the last 2 Myr.

Our model further indicates that the Snøhvit field would have contained gas only if the area had been uplifted just a further 150 m.

Uplift and erosion on the Norwegian shelf

To date the major studies of uplift and erosion of the Norwegian Shelf have focused on the Barents Sea. However, evidence of an uplift in Late Cenozoic time is observed along the entire Norwegian coast, but very little has been published about the amount and consequences of this uplift.

A Late Cenozoic wedge similar to that in the Barents Sea has been mapped in the Haltenbanken area, and volumetric calculations suggest that these deposits originated from about 1000 m of erosion from the Norwegian mainland during the glacial period (Riis and Fjeldskaar, this volume).

In a recent presentation, Jensen and Schmidt (1990) demonstrated 1000–1500 m of Late Cenozoic erosion in the near-shore area along the Skagerrak coast and in Northern Jylland (Denmark) using the same methods as in this paper.

Uplift may further have caused the tilting of the reservoir contacts seen in the Troll field (Gray, 1987). Assuming that Troll was at one time an oil field with a minor gas cap as suggested from the residual oil present, the tilting may have provided more space for the gas to migrate within the structure (Riis, pers. commun., 1989).

Conclusions

The evidence for uplift and erosion with variable magnitude on the Barents Shelf has been presented and quantified by a number of methods. The magnitude is in the order of 1000–1500 m in the southwestern part, increasing to more than 2000 m towards the Stappen High.

The timing of the uplift must be post-Paleocene. Fission track data suggest two distinct periods with uplift and erosion. The first situation might be correlated with the opening of the Norwegian Greenland Sea, the second (and major event) with the post Miocene glaciations. Both the fission track data and evaluation of the sedimentation rate from the wedges indicate that the bulk of the erosion took place in the Late Neogene (Eidvin and Riis, 1989).

Uplifts on the order of magnitude seen on the Barents Shelf will have an effect on hydrocarbon accumulations, especially where reservoirs have a gas cap and a light oil at "bubble" point. In such reservoirs the combined effect of gas expansion and evaporation from the oil will cause oil to be expelled during uplift.

Studies of uplift and erosion have so far been neglected in other places on the Norwegian Shelf, but increasing amounts of new data suggest that there has been a major uplift and erosion, probably as an isostatic response to the glacial overburden in Late Cenozoic time. The effect is best seen in near shore areas, where the sequences are subcropping. Tilting of the Troll field during a late phase of uplift may explain the gas column and residual oil contact seen today.

References

Berglund, L.T., Augustson, J., Færseth, R., Gjelberg, I. and Ramberg-Moe, H., 1986. The evolution of the Hammerfest Basin. In: A.M. Spencer (Editor), Habitat of Hydrocarbons on the Norwegian Continental Shelf. Norw. Pet. Soc., Graham and Trotman, London, pp. 319–338.

Dow, W.G., 1977. Kerogen studies and geological interpretations. J. Geochem. Explor., 7: 79–99.

Eidvin, T. and Riis, F., 1989. Nye dateringer av de tre vestligste borehullene i Barentshavet. Resultater og konsekvenser for den tertiære hevingen. NPD Contrib. (preprint).

Eldholm, O., Sundvor, E., Myhre, A.M. and Faleide, J.I., 1984. Cenozoic evolution of the continental margin off Norway and western Svalbard. In: A.M. Spencer (Editor), Petroleum Geology of the North European Margin. Norw. Pet. Soc., Graham and Trotman, London, pp. 3–18.

Elverhøi, A. and Solheim, A., 1983. The Barents Sea ice sheet, a sedimentological discussion. Polar Res., N.S., 1: 23–42.

Faleide, J.I., Gudlaugson, S.T. and Jacquart, G., 1984. Evolution of the western Barents Sea. Mar. Pet. Geol., 1: 123–150.

Feazel, C.T and Aram, R.B., 1990. Interpretation of discontinuous vitrinite reflectance profiles: Discussion. Am. Assoc. Pet. Geol., Bull., 74: 91–93.

Foscolos, A.E. and Kodama, H., 1974. Diagenesis of clay minerals from Lower Cretaceous shale of northeastern British Columbia. Clays Clay Miner., 22: 319–335.

Foscolos, A.E., Powell, T.G. and Gunther, P.R., 1976. The use of clay minerals and inorganic and organic geochemical indicators for evaluating the degree of diagenesis and oil generating potential of shales. Geochim. Cosmochim. Acta, 40: 953–966.

Gabrielsen, R.H., Færseth, R.B., Jensen, L.N., Kalheim, J.E. and Riis, F., 1990. Structural elements of the Norwegian continental shelf. Part I: The Barents Sea Region. NPD Bull., 6, 33 pp.

Gleadow, A.J.W., Duddy, I.R. and Lovering, J.F., 1983. Fission track analysis; a new tool for the evaluation of thermal histories and hydrocarbon potential. APEA J., 23: 93–102.

Gray, D.I, 1987. Troll. In: A.M. Spencer (Editor), Geology of the Norwegian Oil and Gas Fields. Norw. Pet. Soc., Graham and Trotman, London, pp. 389–401.

Henriksen, E. and Richardsen, G., 1988. Cenozoisk geologisk utvikling av kontinentalmarginen i det sørvestlige Barentshavet. Thesis, University of Tromsø (unpublished).

Jensen, L.N. and Schmidt, B.J., 1990. Late Tertiary uplift and erosion in the Skagerrak area; magnitude and consequences. Nord. Geol. Vintermøte, Stavanger, 19: 61 (abstract).

Jordan, J.R. and Shirley, O.J., 1966. Application of drilling performance data to overpressure detection. J. Pet. Technol., 18: 1387–1399

Katz, B.J., Pheifer, R.N. and Schunk, D.J., 1988. Interpretation of discontinuous vitrinite reflectance profiles. Am. Assoc. Pet. Geol., Bull., 72: 926–931.

Manum, S.B. and Throndsen, T., 1978. Rank of coal and dispersed organic matter and its geological bearing in the Spitsbergen Tertiary. Nor. Polarinst. Årbok 1977, pp. 159–177.

Matishov, G.G., 1980. Geomorphological signs of the action of the Scandinavian, Novaya Zemlya, and Spitsbergen ice sheets on the floor of the Barents Sea. Oceanography, 20: 440–447.

Myhre, A.M. and Eldholm, O., 1988. The western Svalbard margin. Mar. Pet. Geol., 5: 134–156.

Nansen, F., 1920. En ferd til Spitsbergen. J. Dybwad, Oslo, 227 pp.

Nøttvedt, A., Berglund, T., Rasmussen, E. and Steel, R., 1988. Some aspects of Tertiary tectonics and sediments along the western Barents Shelf. In: A.C. Morton and L.M. Parson (Editors), Early Tertiary Volcanism and the Opening of the NE Atlantic. Geol. Soc. London, Spec. Publ., 39: 421–425.

Olaussen, S., Gloppen, T.G., Johannessen, E. and Dalland, A., 1984. Depositional environment and diagenesis of Jurassic reservoir sandstone in the eastern part of Troms I area. In: A.M. Spencer (Editor), Petroleum Geology of the North European Margin. Norw. Pet. Soc., Graham and Trotman, London, pp. 61–80.

Riis, F. and Fjeldskaar, W., 1992. On the magnitude of the Late Tertiary and Quaternary erosion and its significance for the uplift of Scandinavia and the Barents Sea. In: R.M. Larsen, H. Brekke, B.T. Larsen and E. Talleraas (Editors), Structural and Tectonic Modelling and its Application to Petroleum Geology. Norwegian Petroleum Society (NPF), Special Publication 1. Elsevier, Amsterdam, pp. 163–185 (this volume).

Rønnevik, H.C., 1981. Geology of the Barents Sea. In. L.V. Illing and G.D. Hobson (Editors), Petroleum Geology of the Continental Shelf of North-West Europe. Institute of Petroleum, London, pp. 395–406.

Sclater, J.G. and Christie, P.A.F., 1980. Continental stretching: An explanation of the post mid-Cretaceous subsidence of the Central North Sea Basin. J. Geophys. Res., 85: 3711–3739.

Spencer, A.M., Home, P.C. and Berglund, L.T., 1984. Tertiary structural development of the western Barents Shelf: Troms to Svalbard. In: A.M. Spencer (Editor), Petroleum Geology of the North European Margin. Norw. Pet. Soc., Graham and Trotman, London, pp. 199–209.

Vorren, T.O., Lebesbye, E., Henriksen, E., Knutsen, S.-M. and Richardsen, G., 1988. Cenozoic erosion og sedimentasjon i det sørlige Barentshav. XVIII Nord. Geol. Vintermøte, København, 1988, p. 424 (abstract).

Vorren, T.O., Lebesbye, E., Andreassen, K. and Larsen, K.-B., 1989. Glacigenic sediments on a passive continental margin as exemplified by the Barents Sea. Mar. Geol., 85: 251–272.

Vorren, T.O., Richardsen, G., Knutsen, S.-M. and Henriksen, E., 1990. The western Barents Sea during the Cenozoic. In: U. Bleil and J. Thiede (Editors), Geological History of the Polar Oceans: Arctic versus Antarctic. NATO ASI Series C 308. Kluwer Academic Publishers, Dordrecht, pp. 95–118

B. NYLAND *Norwegian Petroleum Directorate, P.O. Box 600, N-4001 Stavanger, Norway*
L.N. JENSEN *Statoil A/S, P.O. Box 300, N-4001 Stavanger, Norway*
J. SKAGEN *Statoil A/S, P.O. Box 40, N-9401 Harstad*
O. SKARPNES *Statoil A/S, P.O. Box 40, N-9401 Harstad*
T. VORREN *University of Tromsø, P.O. Box 3085, N-9001 Tromsø*

On the magnitude of the Late Tertiary and Quaternary erosion and its significance for the uplift of Scandinavia and the Barents Sea

F. Riis and W. Fjeldskaar

Geological evidence related to the Late Tertiary uplift and erosion of the Barents Sea and Scandinavia includes marine seismic data, well data and field observations on land. The amount of erosion has been studied using geochemical and sedimentological data from wells and topographical data and data on the block fields onshore. The amount of erosion has been compared to the amount of deposition in the huge Pliocene-Pleistocene wedges situated offshore Norway and in the Barents Sea.

The data suggest that there is a close relation between the uplift and erosion of Scandinavia and the Barents Sea, and that the uplift is related to two main events: (1) mountain building in Central South Norway and along the trend Lofoten-Bjørnøya-Svalbard due to Palaeocene to Oligocene tectonic phases; and (2) glaciation in the Late Pliocene and Pleistocene, which was related to plateau uplift of large areas.

Based on lithospheric flexural rheology parameters from studies of the post-glacial uplift, modelling shows that the compensating isostatic uplift is close to 70% of the magnitude of erosion where crystalline basement is eroded. The magnitude of the observed uplift in Scandinavia seems, however, to be greater than the erosion in some areas, and therefore mechanisms other than pure isostatic compensation must be sought to explain this. One part of the uplift is thought to date back to Paleogene tectonism, but observations suggest that there is an uplift in excess of these two components. We have investigated the possibility that migration of phase boundaries in the lithosphere may be the mechanism of this additional uplift.

The response of a phase transition in the lithosphere to pressure relief caused by erosion leads to thickening of the upper lighter phase and thus greater uplift than expected from isostasy alone. Migration of the phase boundary is a time-dependent process. Theoretical models show that for load cycles of the order of 1 Ma, the surface deflections associated with the movement of the phase boundary can be large, up to 50% of the isostasy. Therefore, modelling of the phase migration combined with isostasy in response to erosion might lead to an uplift larger than the amount of eroded rocks. Thus, theoretically the phase migration model is of interest, although it is not supported by independent geological evidence.

Introduction

The late phase of erosion in the Barents Sea significantly affected the petroleum accumulated in traps (e.g., Nyland et al., this volume). Also, in large parts of the platform areas, Triassic and Jurassic source rocks situated in the oil window have been uplifted and cooled. Therefore, much effort has been put into the task of quantifying the uplift and erosion in the Barents Sea.

This study was initiated as a consequence of new dating results of the Late Cenozoic sedimentary wedge penetrated by two wells on the Senja Ridge (Fig. 1). Our results indicate the wedge to be of the Late Pliocene and Pleistocene age (Eidvin and Riis, 1989). The size of the wedge corresponds to an erosion of 500–1000 m in the Southern Barents Sea (Vorren et al., 1988; Vorren et al., 1990; Nøttvedt et al., 1988).

A similar wedge is situated on the shelf offshore Mid-Norway (Fig. 2), although it contains a smaller volume of sediments. This wedge has been penetrated by several wells on Haltenbanken, and its age has been determined to be Late Pliocene and Pleistocene. Thus, this age correlates with the wedge outside Bjørnøyrenna.

Results from ODP wells on the Vøring Plateau (Jansen et al., 1988) indicate that the first major glaciations took place at approximately 2.5 Ma. This age correlates well with the age dating of the wedges. Examination of the coarse fraction of the cutting material has yielded large amounts of angular fragments of crystalline basement in the wedges (Eidvin and Riis, 1989).

In both areas, the Miocene and Early Pliocene deposits appear to be thin, on average in the order of a few hundred metres at most. Thus, geological evidence indicates that in the Late Pliocene and

Structural and Tectonic Modelling and its Application to Petroleum Geology edited by R.M. Larsen, H. Brekke, B.T. Larsen and E. Talleraas. NPF Special Publication 1, pp. 163–185. Elsevier, Amsterdam.

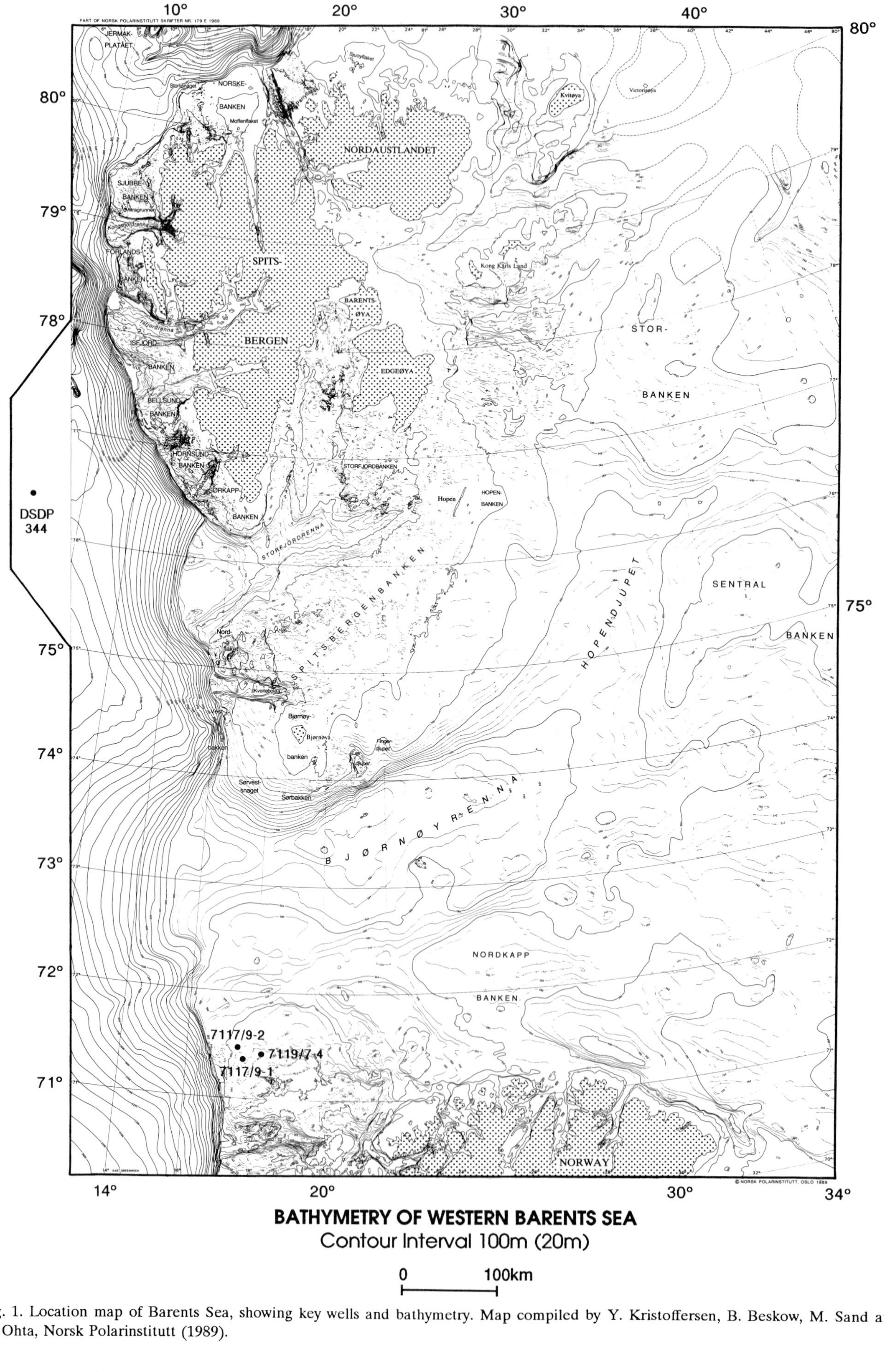

Fig. 1. Location map of Barents Sea, showing key wells and bathymetry. Map compiled by Y. Kristoffersen, B. Beskow, M. Sand and Y. Ohta, Norsk Polarinstitutt (1989).

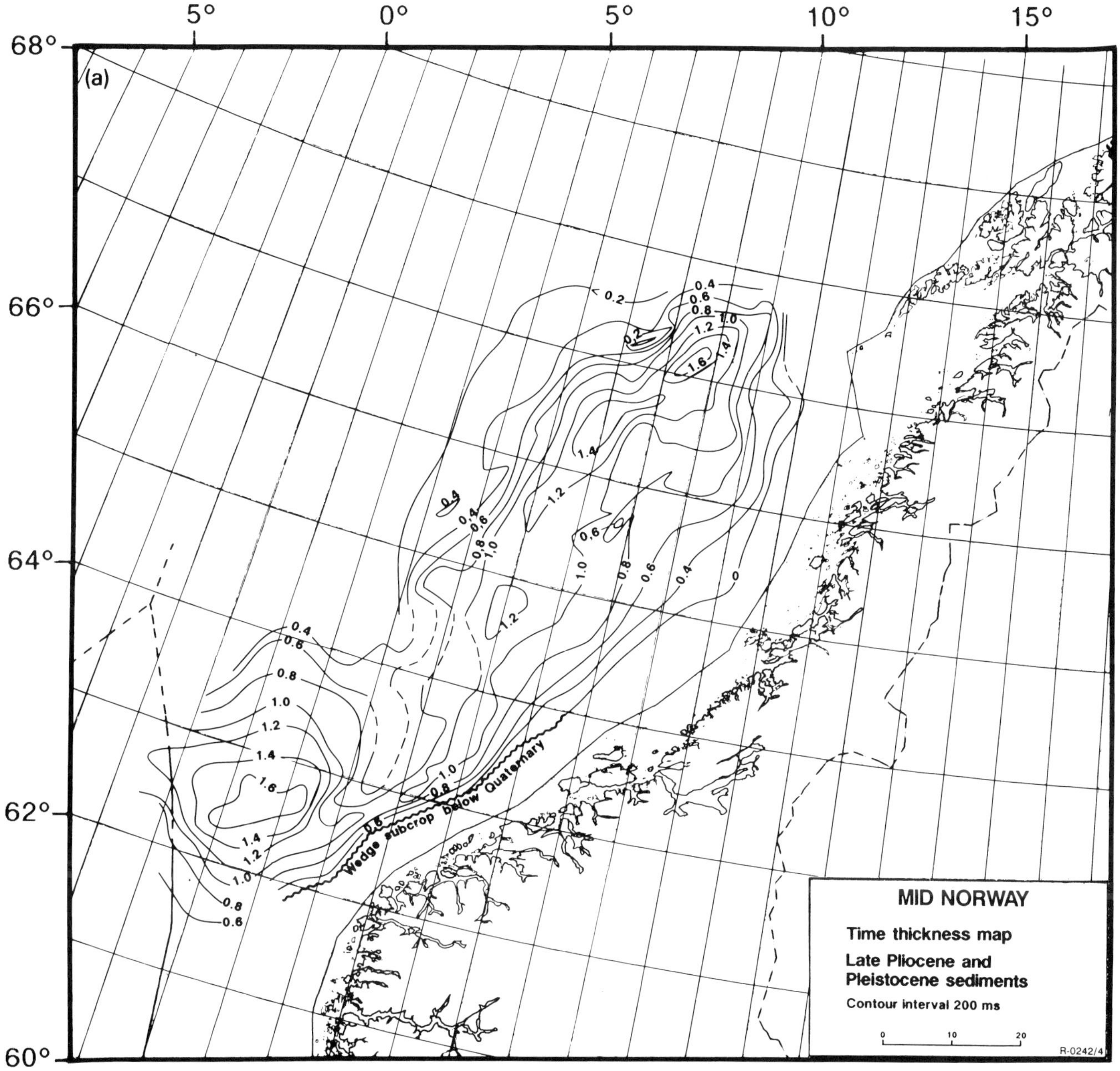

Fig. 2. (a) Time thickness map of Late Pliocene-Pleistocene wedges, Mid-Norway shelf.

Pleistocene period a dramatic increase in deposition rates along the shelf was related to glaciation of the Barents Sea and of Scandinavia. The late phase of erosion in the Barents Sea must, therefore, be seen in a larger context. The observation that large amounts of erosion were related to glaciations led us to investigate the thermal and isostatic effects of the rapid loading and unloading resulting from the erosion process. The study is based on simplified maps showing estimated amounts of erosion and deposition in Scandinavia and in the Barents Sea in glacial times (the last 2.5 million years).

The aim of this study is to determine whether the observed uplift pattern can be modelled using simple assumptions about the effect of erosion and glaciation, and whether the interpreted uplift related to erosion and glaciation can be separated from uplift due to older tectonic phases related to the opening of the North Atlantic.

Late Pliocene-Pleistocene geology

Erosional and depositional pattern

Occurrence and dating of sedimentary wedges

Along the Barents Sea margin, large sedimentary wedges occur at the mouth of bathymetric troughs (Fig. 1). The wedge fed by Bjørnøyrenna is more than 4 km thick (Fig. 3), and the wedge off Storfjordrenna reaches the same thickness, although its areal extent

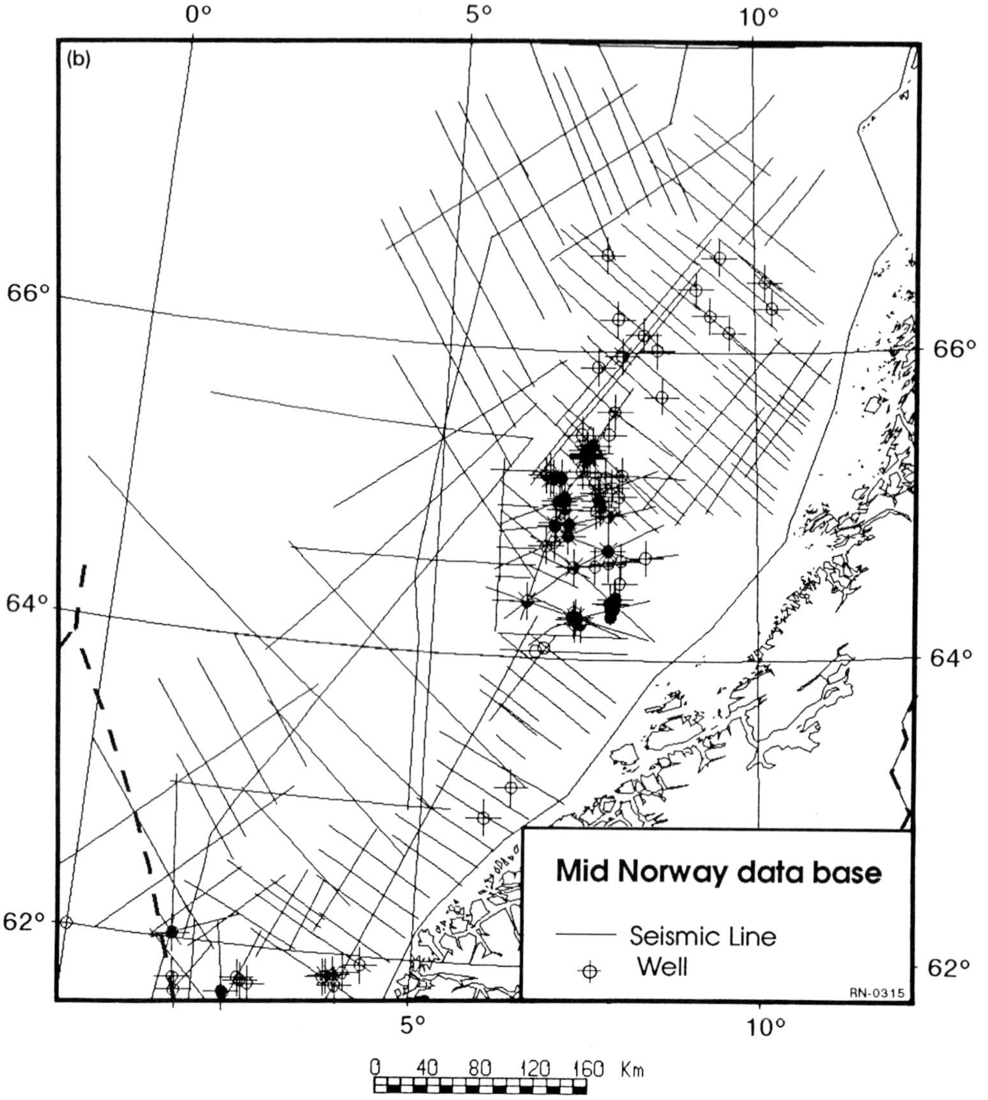

Fig. 2 (continued). Maps (b) and (c) show location of wells and seismic lines and bathymetry.

is smaller. These two wedges are associated with large positive Bouguer gravity anomalies. Fjeldskaar and Riis (1988) relate parts of these anomalies to a lack of isostatic compensation due to rapid recent deposition.

Along the northern Barents Sea margin, a similar pattern appears. Sparse seismic data collected by NPD and by Baturin (1988) and regional bathymetry (Perry and Fleming, 1985) indicate that sedimentary wedges occur in a similar pattern, although probably with smaller sediment volumes.

The wedge off Bjørnøyrenna (Fig. 1) was sampled by the two wells, 7117/9-1 and 2, which penetrate approx. 800 m Upper Pliocene and Pleistocene (Eidvin and Riis, 1989). The dating of this sequence has been much debated due to resedimentation (e.g., Spencer et al., 1984; Nøttvedt et al., 1988). However, the micropalaeontology, Sr isotope analysis and study of the basement fragments all strongly indicate that the depositional age of the sequence is Late Pliocene and Pleistocene.

The wedge off Storfjordrenna was penetrated by one DSDP well in a distal position close to the Knipovich Ridge (Fig. 1). The section was dated as Pliocene (Talwani et al., 1976). The uppermost, mainly Quaternary part of the wedge can be correlated seismically to the wedge off Bjørnøyrenna.

In the Mid-Norwegian shelf, there are two main wedges which are also linked to major bathymetric troughs — the wedge at the mouth of Norskerenna and the "Mid-Norway wedge" at the mouth of troughs extending from Vestfjorden and Ranafjorden (Figs. 2 and 4). These two wedges have a maximum thickness in the order of 2000 m. Both have been penetrated by a number of wells (Fig. 2), and they may be assigned a Late Pliocene–Pleistocene age (Rokoengen and Rønningsland, 1983; Dalland et al., 1988) (Naust Formation).

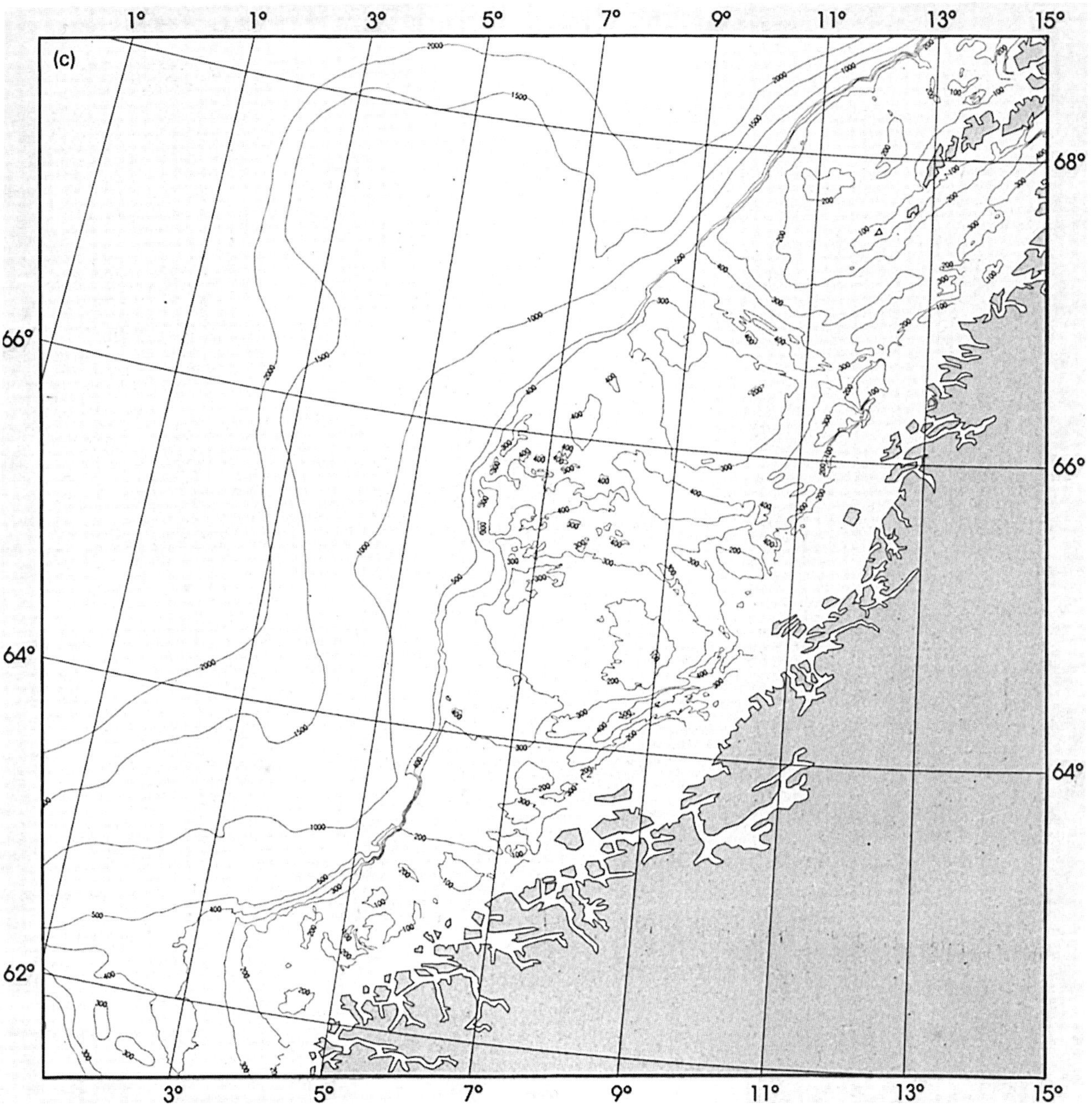

Fig. 2 (continued).

Estimates of erosion

In the Barents Sea and in Scandinavia, the amount of erosion has been estimated by different methods. For the Southern Barents Sea and for Mid-Norway, it has been assumed that all erosion products which were transported to the west have been deposited in the respective offshore wedges. This mass balance gives an independent control on the erosion map, although the drainage divides cannot be established with certainty. We have concentrated on applying the method of mass balance on Mid-Norway-North Sweden, because the uncertainty of the size of the drainage area is not too great. For the southern Barents Sea and Finnmark, the drainage area is less well known, and for South Norway-South Sweden, the area of deposition is not easily mappable. Furthermore, there is no major fault activity along the coast of Nordland post Triassic times. This means that the geometry of the uplift should be simple. However, Lofoten was an active structural element in the Late Cretaceous and Paleogene.

(1) Barents Sea. The magnitude of erosion in the Barents Sea was estimated by geochemical data (vitrinite reflectance and T_{max}) from the deep wells (Fig. 5). In the western parts, the map is based on the occurrence of a seismic reflection which cuts across the seismic stratification in the Late

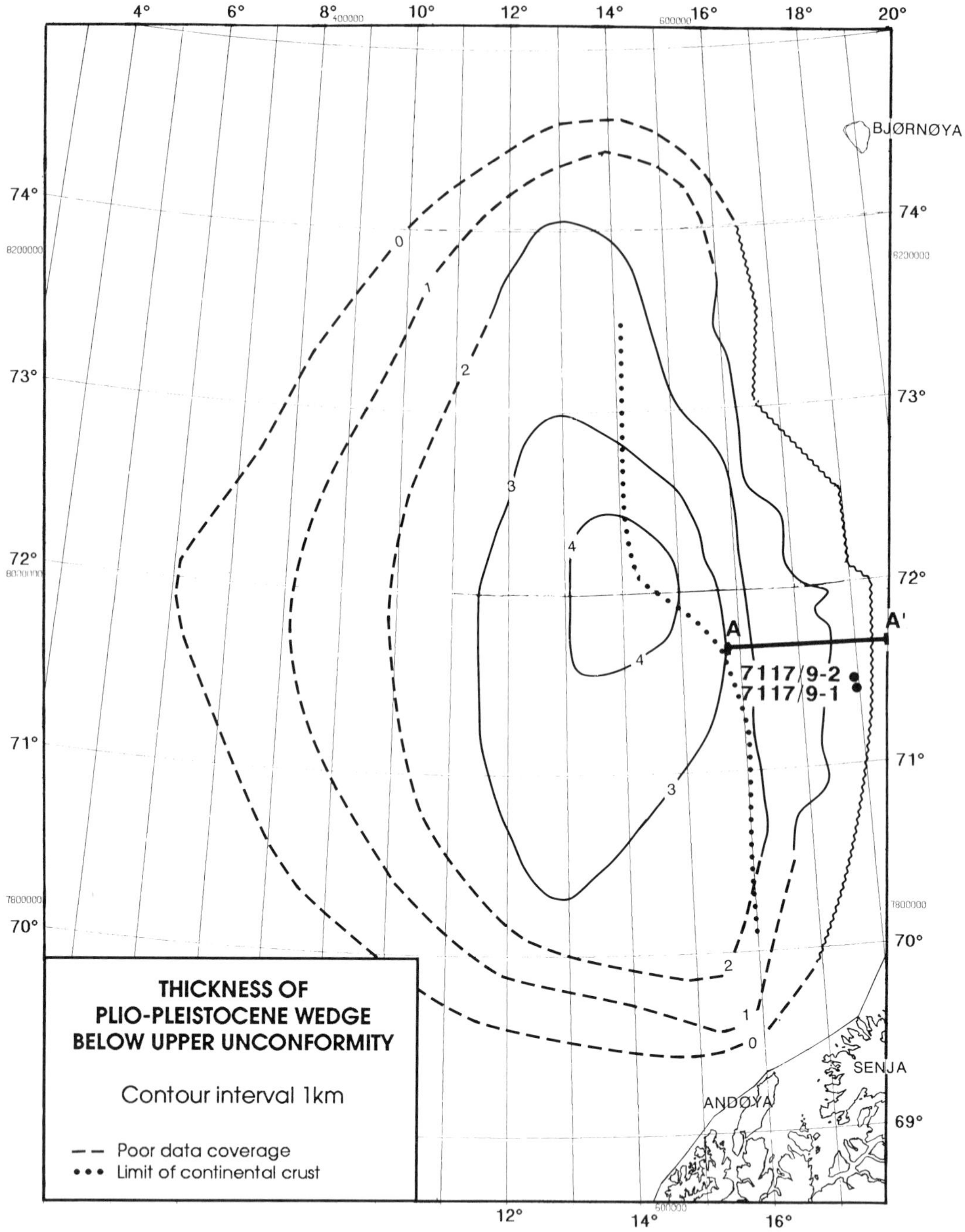

Fig. 3. Thickness map of Late Pliocene and Pleistocene sediments below the Upper Regional Unconformity, southwestern Barents Sea.

Paleocene and Early Eocene layers. This reflection shows a regional dip to the west (Fig. 6), and is correlated to the transition from Opal A to microcrystalline quartz (opal CT) observed in well 7117/9-1 (Ramberg Moe et al., 1988). This transition takes place at approx. 50°C, and the dip of the reflector indicating the diagenetic change may be explained by tilting and erosion after its formation. The deepest level observed for the reflector was assumed to be the level of no erosion, and the erosion in the rest of the area was estimated by the difference in elevation, assuming no lateral change in temperature gradient.

In the eastern and northern parts of the Barents Sea, there are very few data points (Antonsen et al., 1991). Seismic stacking velocities (Norwegian Petroleum Directorate, unpublished work) and geological data from Svalbard (Manum and Throndsen, 1978) suggest a general decrease in erosion from west to east in this area.

It is well known that Spitsbergen was uplifted and eroded as a consequence of Eocene-Oligocene tectonism (e.g., Steel and Worsley, 1984). Also, the Bjørnøya-Stappen High area was uplifted along the Knølegga Fault in different tectonic phases in the

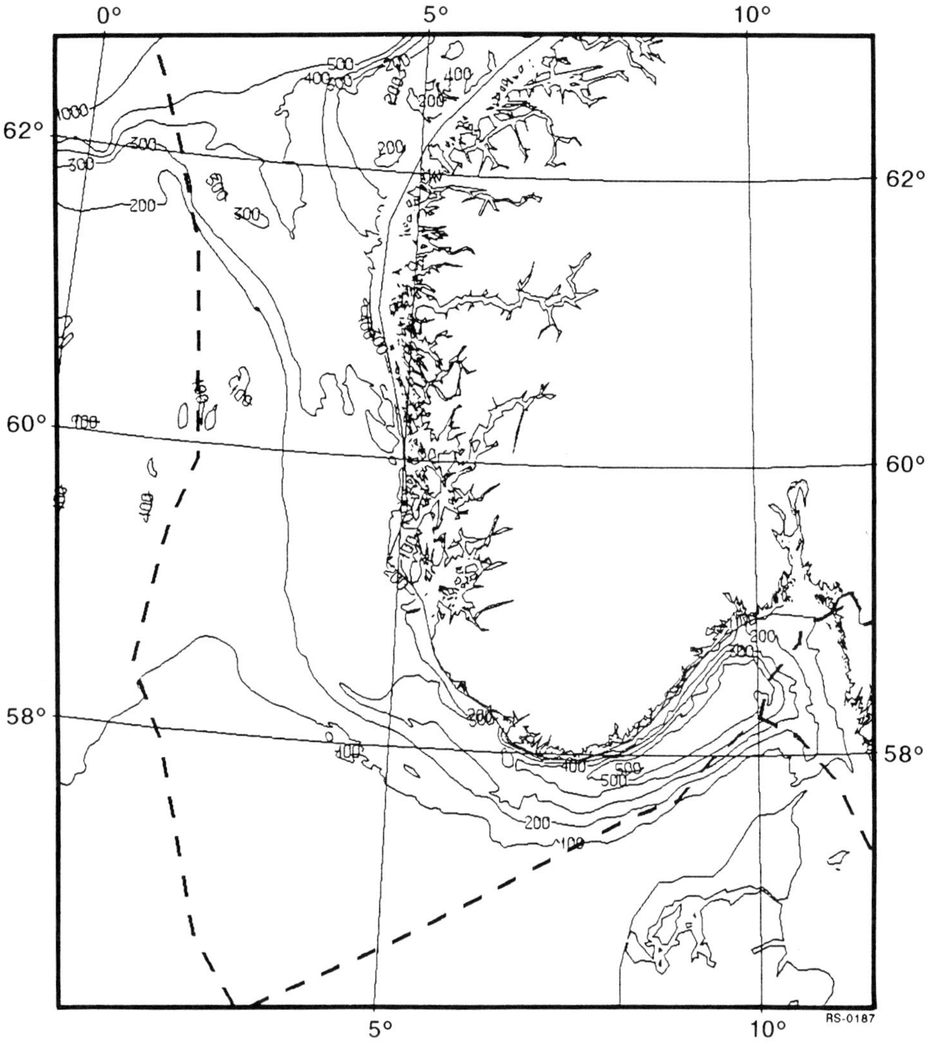

Fig. 4. Bathymetry of the North Sea. Based on data from Statens Kartverk.

Eocene and Oligocene (Fig. 7). This happened also to the Senja Ridge and Lofoten islands farther south (Brekke and Riis, 1987). As a consequence, the large amount of erosion observed in the western and northern parts of the Barents Sea can be related partly to tectonic uplift in the Paleogene and partly to the glacial effects we are concerned with in this study.

Farther east and south in the Barents Sea, the Early Tertiary tectonic effects can be expected to be more subdued. Thus, an estimate of 800–1000 m of Late Pliocene-Pleistocene erosion based on the sediment volume in the wedge off Bjørnøyrenna seems reasonable for the central parts of the Southern Barents Sea.

It should be pointed out that these values are surprisingly high compared to studies of present day erosion rates at Svalbard, estimated at 0.2–0.5 mm/year (Løvø et al., in preparation), and at 0.5–1 mm/year for active glaciers (Elverhøi et al., 1980, 1983). 1000 m of erosion is equivalent to 0.4 mm/year, suggesting that the Barents Sea was not submerged for much of the glaciation period.

(2) Mid Norway. The calculated erosion is based on the assumption that a pre-glacial surface can be mapped onshore in Scandinavia. If the present height and topography of the pre-glacial surface can be established, the amount of erosion can be calculated by subtracting the present mean heights from the height of the pre-glacial level.

Block fields are observed on many of the high mountains in South Norway. Nesje et al. (1988) suggest that the block fields were formed by weathering processes prior to glaciation. It is assumed that the lowest limit of the block fields approximates a pre-

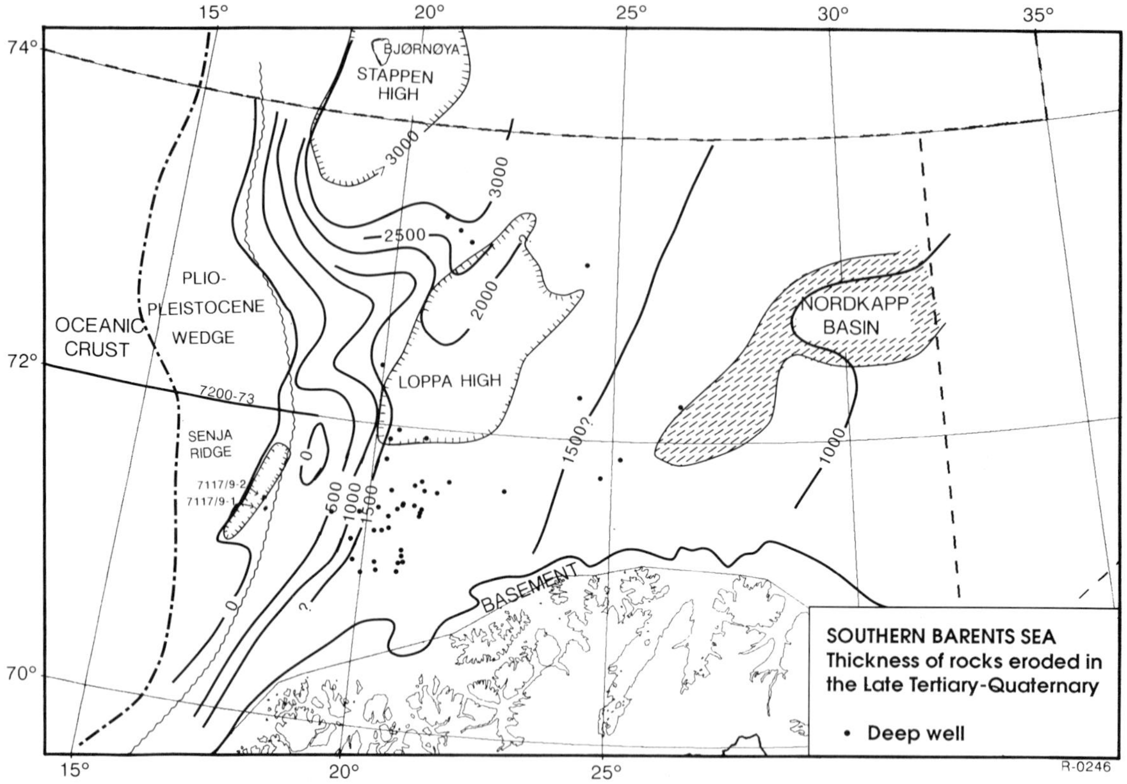

Fig. 5. Map of southern Barents Sea showing interpreted amount of Tertiary erosion (in metres). This interpretation is based on vitrinite and pyrolysis data from deep wells as well as on Figs. 6 and 12, profiles *EE'* and *FF'*.

glacial surface. In Møre, these pre-glacial surfaces may be traced almost to sea level.

Fig. 8 shows a simplified map of mean heights contoured from an 8 × 8 km grid prepared by the Geographical Survey of Norway and the Geographical Survey of Sweden (1963). The summit level map (Fig. 9) is based on smoothed heights of the block fields where they exist, combined with summit levels in areas where there are few data on block fields or where block fields do not occur. This map may be regarded as an approximation of a pre-glacial surface, but as it does not take into account topography in this surface, it could be regarded as a pre-glacial surface of maximum probable elevation.

A. Nesje (University of Bergen, unpublished work) has compiled the paleic surface in Norway (Fig. 10). The interpretation is based on geomorphological field work in Southern and Mid-Norway, and mainly on map studies in Northern Norway. This map may be interpreted as an approximation of the present day level of the pre-glacial topography, and it must be regarded as a better approximation than the summit level map.

Onshore erosion (Fig. 11) was calculated by subtracting the mean height map (Fig. 8) from the summit level map (Fig. 9), and therefore it must be regarded as a maximum case of eroded rock in glacial times. In the offshore areas, erosion has been calculated by extrapolating the eroded part of the sedimentary sequence to the shore. The pre-glacial pinch out line of sedimentary rocks on the crystalline basement was inferred by using the profiles AA' to FF' (Fig. 8), and is shown in Fig. 11. The pinch out line was drawn so that the block field regions were not given a pre-glacial sediment cover.

A comparison was made between the deposited volume of sediments in the wedge off Mid-Norway and the eroded volume in Mid-Norway suggested in Fig. 11, using tentative drainage divides as shown on Fig. 9. The result is given in Table 1. The eroded thicknesses suggested in Fig. 11 are in the order of 1.5–2 times too large to fit into the wedge. This sup-

TABLE 1

Erosion and accumulation Mid-Norway wedge

	Area (km^2)	Volume (km^3)
Glacial wedge (Fig. 2)	139000	100000
Eroded sedimentary rocks (Fig. 11)	88600	64900
Eroded basement (Fig. 11)	108000	89300
Corrected basement (2.7/2.2)	108000	109000
Total erosion: 173900 km^3		
Total deposition: 100000 km^3		

The assumptions are seismic velocity of 2200 m/s and density of basement/density of sediment = 2.7/2.2.

ports the idea that the paleic surface map (Fig. 10) may be a better approximation of the pre-glacial surface than the summit level. If this is the case, the topography before glaciation may have been quite rough even in Nordland, with local relief in the order of 500–700 m. In this case, a "summit level" surface must be interpreted as older than Neogene. Alternative ways of explaining the volume difference would be to change the drainage divides and to reduce the amount of eroded sedimentary rocks.

Our present knowledge of Tertiary tectonic events and thickness of the Oligocene to Early Pliocene section on the shelf, suggests fairly low deposition rates and, thus, a fairly low relief at the onset of glaciation. It seems that more detailed studies are needed to resolve this problem.

In conclusion, Fig. 11 is a maximum case for the amount of material eroded in the glacial period, and the volume could be reduced by 30–40%. Even so, the amount of erosion in the time of glaciation is higher than was expected.

Structural pattern

Six profiles were constructed across Scandinavia in different structural positions (Fig. 12). Profiles AA′ and DD′ describe the situation with a fairly large uplift onshore, no obvious tectonic activity along the margin and, for profile DD′, a thick wedge of Plio-Pleistocene sediments deposited on the shelf. Note that for those profiles the summit level can be projected into the sedimentary sequence, suggesting a pinch out of sediments in a position presently onshore before erosion and uplift took place.

Profile BB′ shows a large uplift onshore, tectonism offshore, and a small Plio-Pleistocene wedge. Onshore, the paleic surface can be traced almost to sea level, fitting the position of the base of the wedge offshore. This supports a pre-glacial, but Tertiary age of the paleic surface.

Profile CC′ traverses the anomalously low Trondheim area, where Jurassic sediments are known in a fault-bounded basin in Beitstadfjorden (Bøe, 1989). Here the summit level projects into the basement-sediment boundary offshore. Accordingly, the sedimentary cover in this profile has been extended farther onshore, to give the onshore dome a shape similar to the other sections. The sediments removed were included in the calculation of erosion (Fig. 11).

Profile EE″ and profile FF′ have been constructed by adding a sedimentary cover in the Barents Sea according to Fig. 5. The profiles suggest that the summit level inferred from Troms and Finnmark correlates fairly well with the erosion interpreted in the Barents Sea, given that there was not a marked topographic break between these areas prior to the onset of major erosion.

The summit level map (Fig. 9) and the cross-sections AA′ to FF′, combined with a regional overview (Fig. 13), show that there are some common structural features related to the uplifted Scandinavia and Barents Sea.

– The sequence of sedimentary rocks surrounds the coast from Skagerrak to the White Sea with a fairly uniform dip away from the mainland. The boundary between sediments and crystalline basement is at a fairly regular distance from the coast line and commonly coincides with a bathymetric low.

– The base of Late Pliocene (glacial) sediments combined with the surface of the summit level together form an asymmetric dome structure with the steep side facing offshore. The structuring of the base of the glacial sediments shows that the dome formation is recent. The dome has two maxima — in North Sweden and in South Norway — and in map view it turns around the coast (Figs. 9 and 13). The boundaries of the dome do not seem to be faulted to a large extent, although they coincide with the boundary faults along some segments of the coast.

– The maximum amount of erosion seems to occur in a zone along the coast. The extremal value is commonly located close to the boundary of sedimentary cover and crystalline basement.

The pattern of a structural dome in Scandinavia may be regarded as part of a geomorphological/structural pattern of concentric highs and lows, as pointed out by White (1972). The following observations may be of importance.

– The exposed crystalline shield is geographically linked to the areas of glaciation.

– A central shallow depression is found in Bottenvika (Fig. 13), (even when allowing for a residual glacial rebound of approx. 100 m, as proposed, e.g., by Balling, 1980).

– Marginal highs occur concentrically, including Scandinavia, Kola and the hills at the Finland-Soviet boundary. The outcrop of the Cretaceous in Denmark and Scania indicates that even in the south there has been a large amount of erosion and uplift.

– In the periphery outside the marginal highs we find shallow marginal morphological depressions. Offshore, they include the bathymetric lows along the coast of Norway and Kola, the White Sea and the Gulf of Finland. Onshore, they include the great lakes of Vänern, Vättern, Ladoga and Onega (Fig. 13). This zone of marginal lows is connected to the boundary between crystalline basement and sedimentary cover.

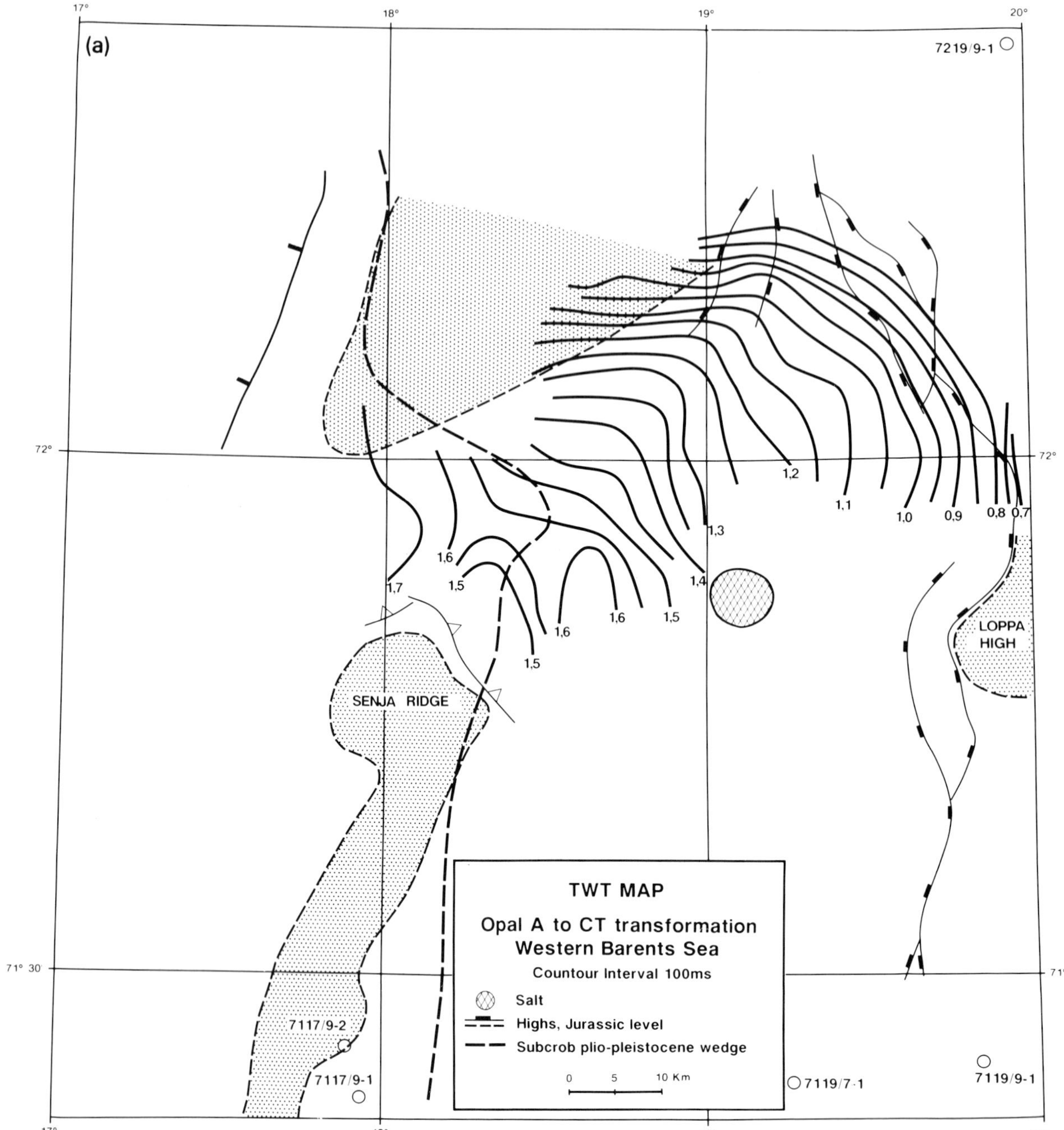

Fig. 6. (a) Map of two-way travel time to seismic anomaly representing the Opal A to CT transition in the Barents Sea.

– The sedimentary cover surrounding the glaciated area in general dips away from the exposed core of basement rocks.

White (1972) noted such a concentric geomorphic pattern in Fennoscandia and compared it with a similar pattern in North America. He connected this pattern to glaciation, but his explanation involved unrealistically high values of erosion (Sugden, 1976).

It may briefly be commented that a similar pattern of a central depression, marginal highs and huge sedimentary wedges occurs even in connection with the Greenland ice sheet (Weidick, 1976). The Barents Sea ice sheets were developed on thick deposits of sedimentary rocks and the present topography differs from the other areas by its low elevation 100–400 m below sea level. However, in this context, Svalbard, Bjørnøya, Franz Josef Land and Novaya Zemlya may be regarded as marginal highs.

It is concluded that the development of the concentric pattern of highs and lows is related to glaciation both in time and space, and hence we suggest that its development should be explained in terms of effects related to glaciation itself.

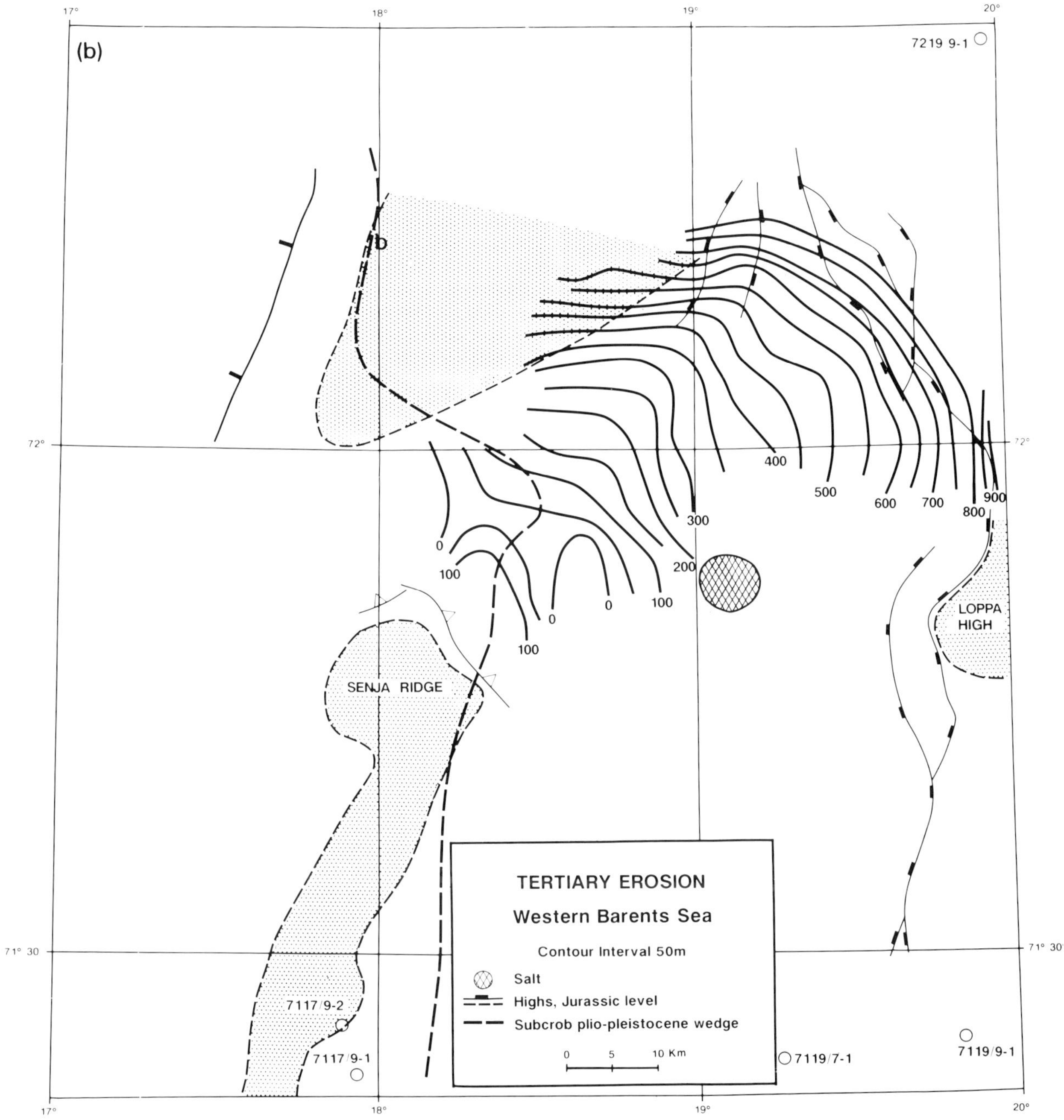

Fig 6. (continued). (b) A conversion of the map to show the eroded thickness of the sediment.

Modelling of crustal response

The removal of load in eroded areas and additional loading in the areas of deposition causes isostatic response of the Earth's lithosphere. The isostatic response is the most obvious mechanism for the uplift/subsidence in the study area. The magnitude of the observed uplift in some areas, however, seems to be greater than the erosion. The isostatic process alone cannot explain an uplift of this magnitude. An additional process is needed. Here we have looked at one possible candidate which may contribute to the uplift. This is the migration of phase boundaries in the lithosphere in response to pressure relief caused by erosion.

In addition, rifting in the North Atlantic causing uplift in Scandinavia (Sales, 1989) is regarded as a part of the story. The modelling presented in this paper is an attempt to calculate the magnitude of the effects expected from glacial erosion, so that these fundamentally different tectonic events can be quantified and separated.

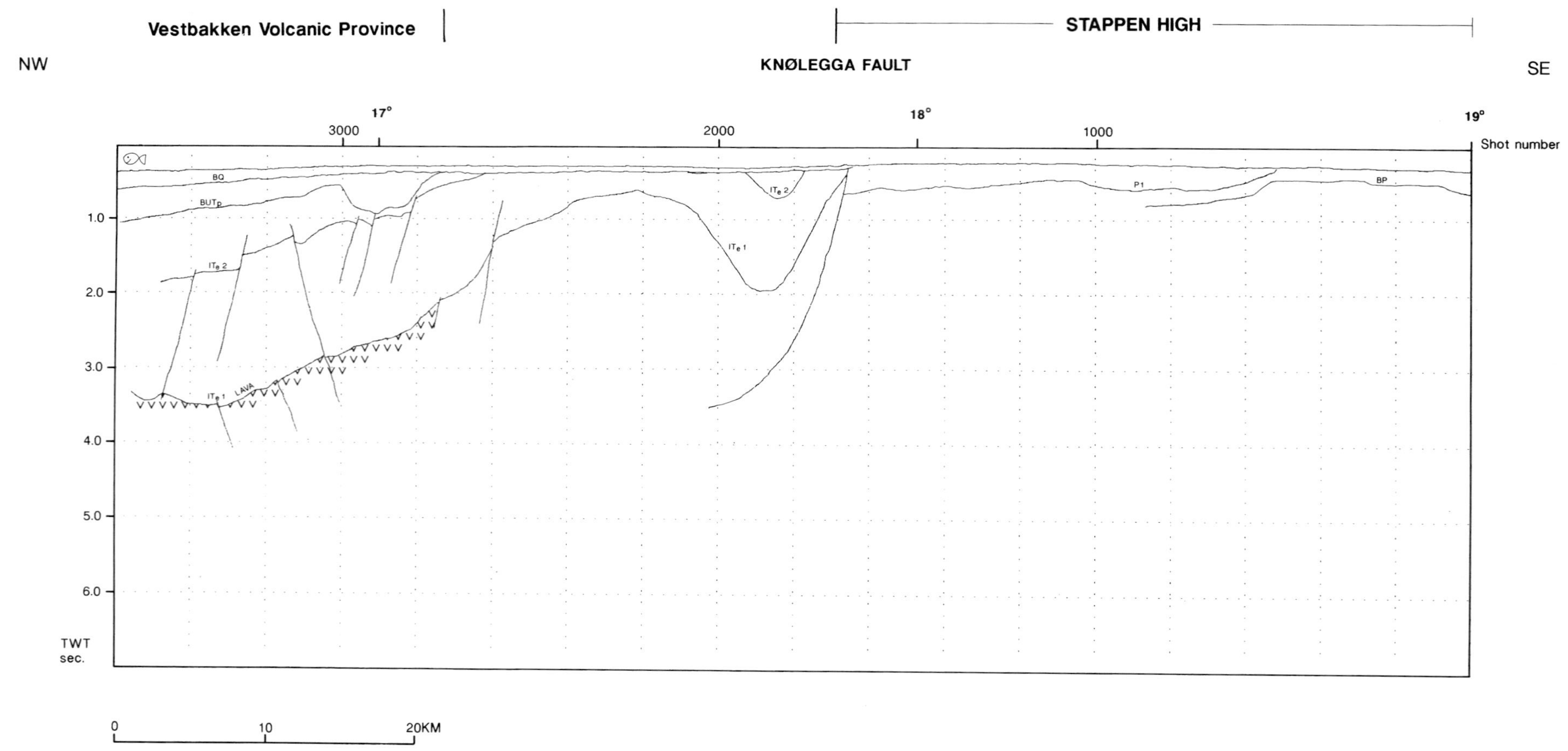

Fig. 7. Cross-section across the Knølegga Fault southwest of Bjørnøya indicating motion in the Eocene and post-Eocene. Line BV-12-86. *BQ* = base of Quaternary; *BUTp* = base of Upper Pliocene; ITe_1 and ITe_2 = intra Eocene; P_1 = Permian; *BP* = base of Permian. From Gabrielsen et al. (1990).

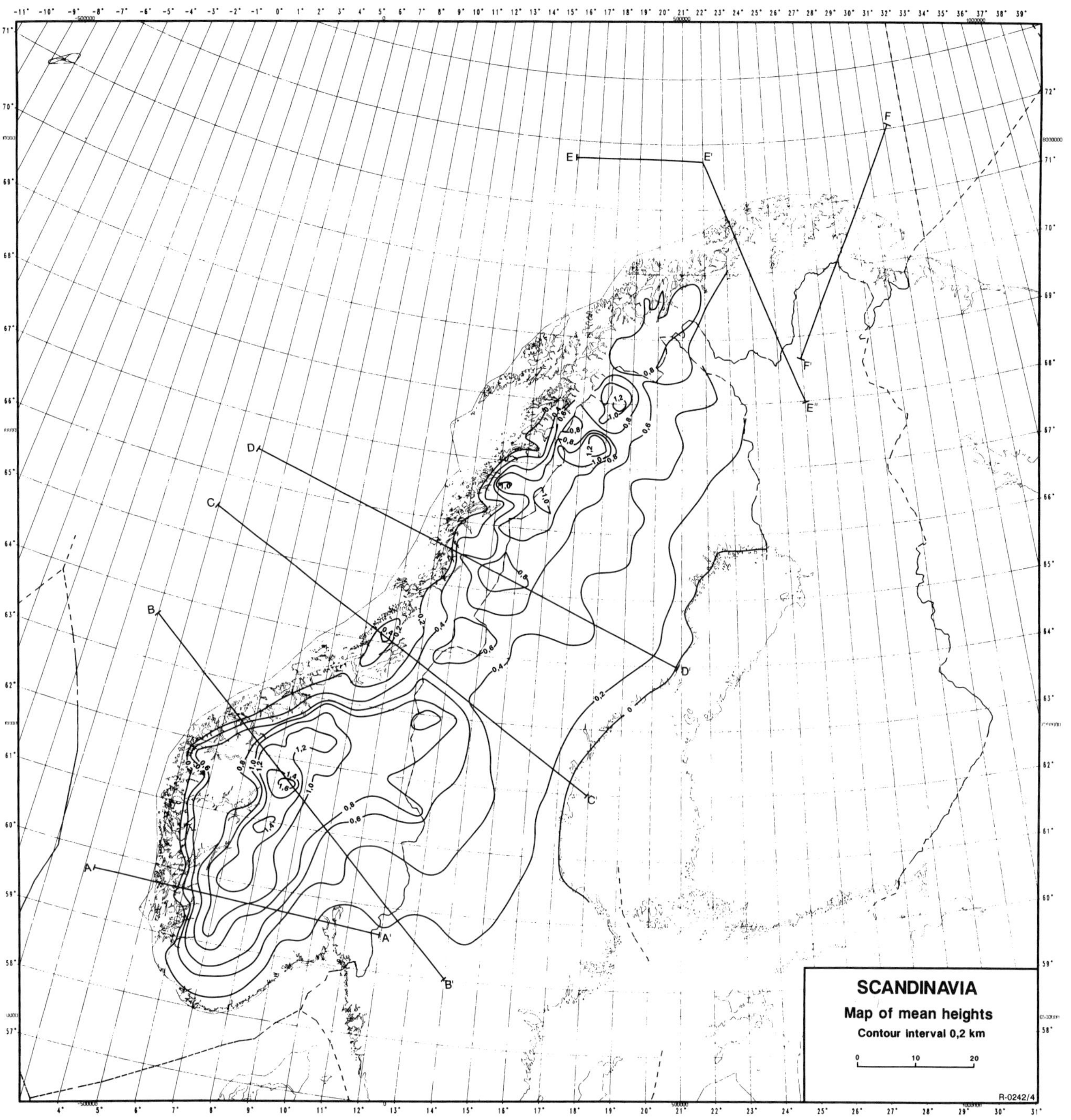

Fig. 8. Mean height map of Scandinavia, based on contouring of 8 x 8 km grid values.

Isostasy

The isostatic process is simulated by a non-spherical three-dimensional model, in which the lithosphere is a uniformly thick elastic layer overlying a viscous asthenosphere. With this model, we are able to treat the isostatic problem analytically using the Fourier transform technique. The mathematical model is described in Appendix A.

A (positive or negative) load applied to the Earth's surface will be partly balanced by the elastic lithosphere and partly by the buoyant forces of the asthenosphere. Loads of short wavelengths will be balanced by the lithosphere, so that the lithosphere acts as a low-pass filter. The characteristics of this filter depend on the flexural rigidity. In this study we used a value for the flexural rigidity of $D = 10^{23}$ Nm. A study based on data on the present rate of uplift and postglacial sea level changes has revealed a value of the flexural rigidity of Scandinavian mainland of less than 10^{24} Nm (Fjeldskaar and Cathles, 1991). The results of the isostatic response reported here

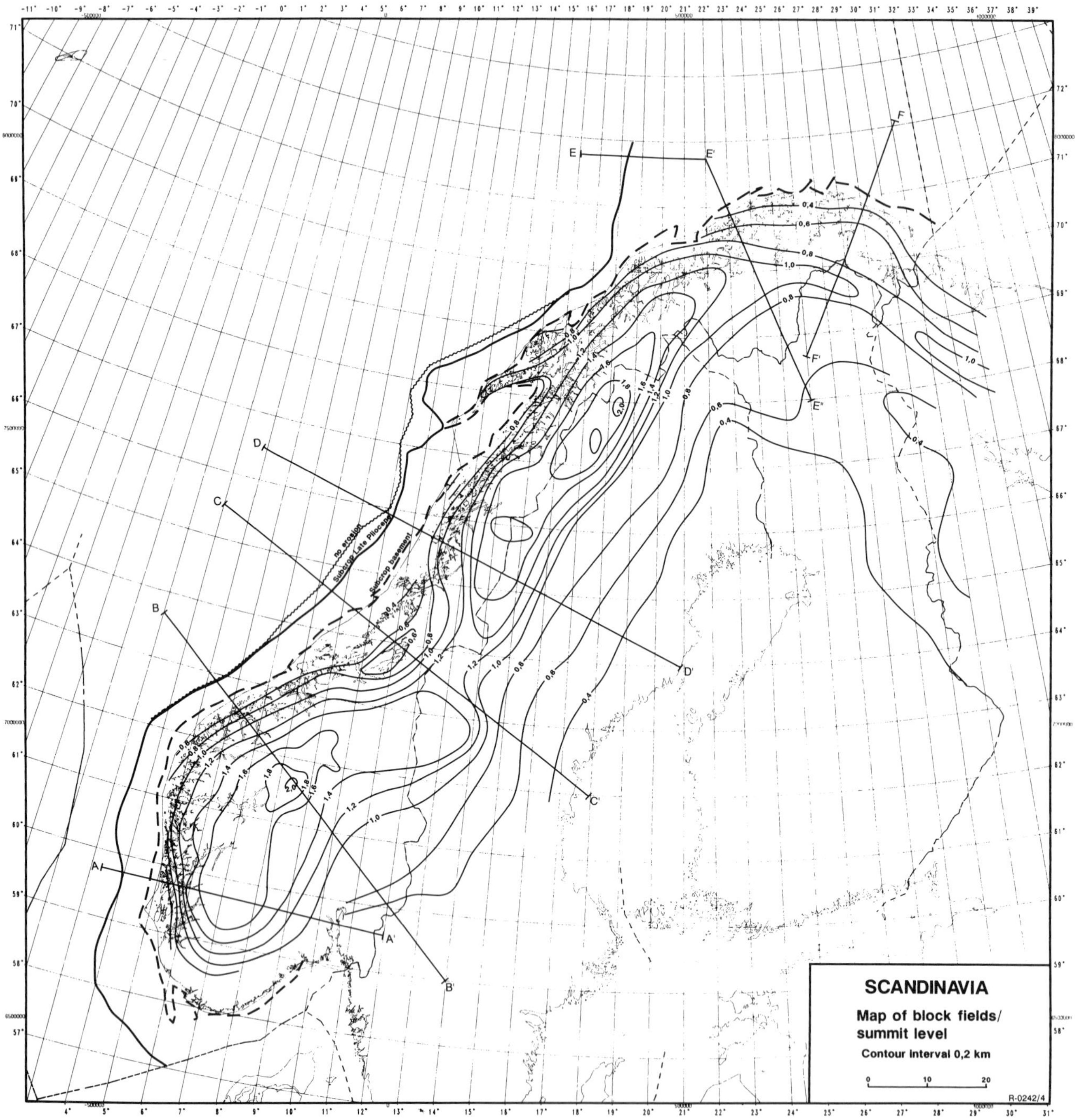

Fig. 9. Summit level map of Scandinavia. Constructed from height of block fields (Nesje et al., 1988) and by drawing isolines through the highest summits.

(Fig. 14) can, therefore, be considered as maximum values.

The relaxation time for the isostatic flexural compensation estimated from the postglacial uplift is only a few thousand years (Fjeldskaar and Cathles, 1991). Thus it is reasonable to assume that the flexural equilibrium was achieved during the erosional process. As the time period considered here is less than 5 million years, the viscous properties of the lithosphere (as described by Fjeldskaar and Pallesen, 1989) are not taken into account.

The resulting isostatic adjustments of the Earth's surface due to Pleistocene/Pliocene erosion/deposition are shown in Fig. 14a. The modelling was based on the erosion model presented in Fig. 13 and the parameter values are given in Table 2.

Calculation based on the amounts of erosion inferred show that the isostatic uplift of Late Pliocene/Pleistocene age amounts to 900–1400 m in the western Barents Sea, and the subsidence of the sediment

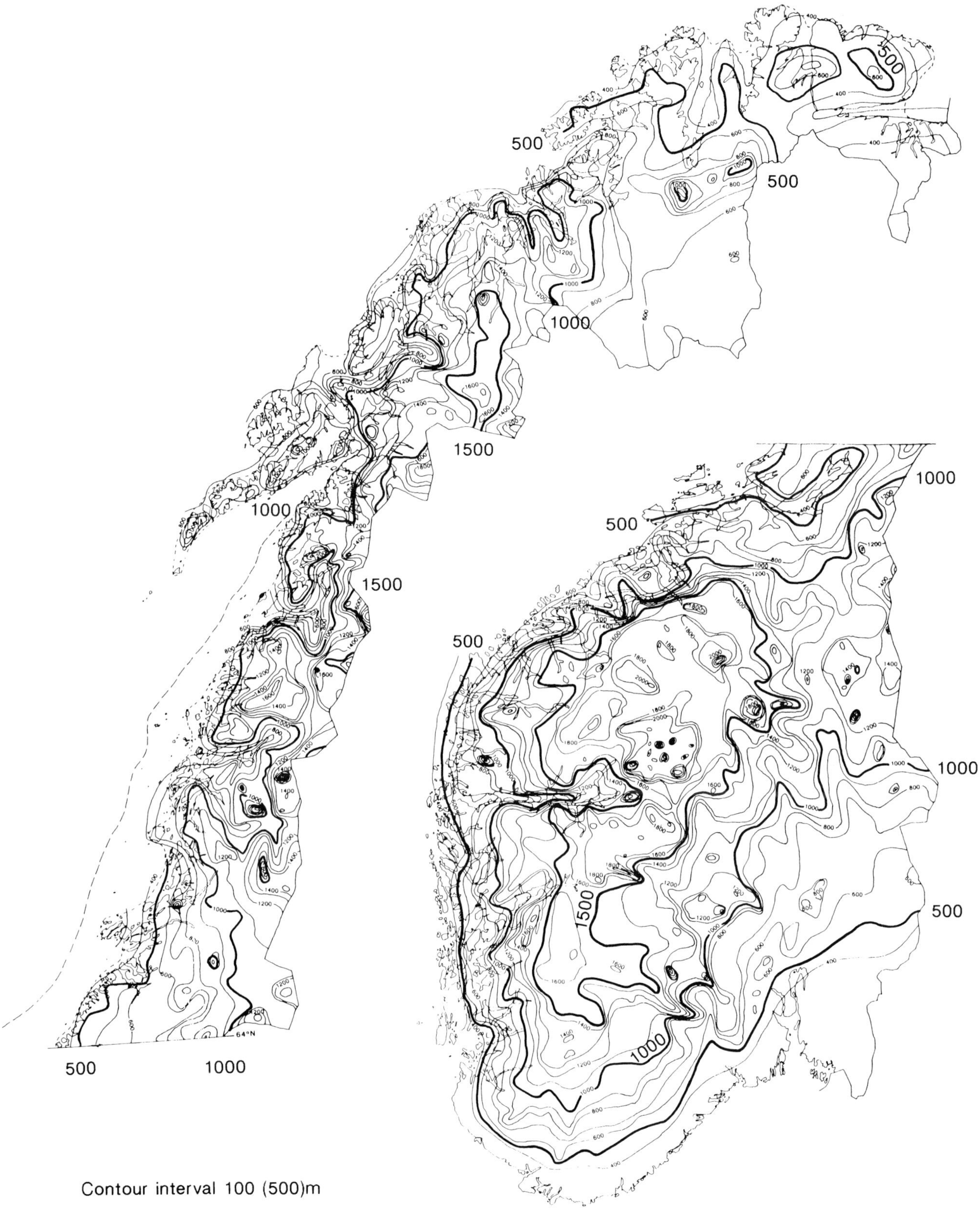

Fig. 10. Map of altitude of paleic surface in Norway. Compiled by Atle Nesje (University of Bergen, unpublished work). 500, 1000 and 1500 m contours are reinforced.

wedge is 700–1800 m. Onshore, the most severe erosion took place in the Nordland area, which caused an uplift of 200–800 m, depending on the lithosphere rigidity of the area.

Phase boundary migration

It has been proposed (e.g., Kennedy, 1959) that phase transitions in the Earth's upper layers provide

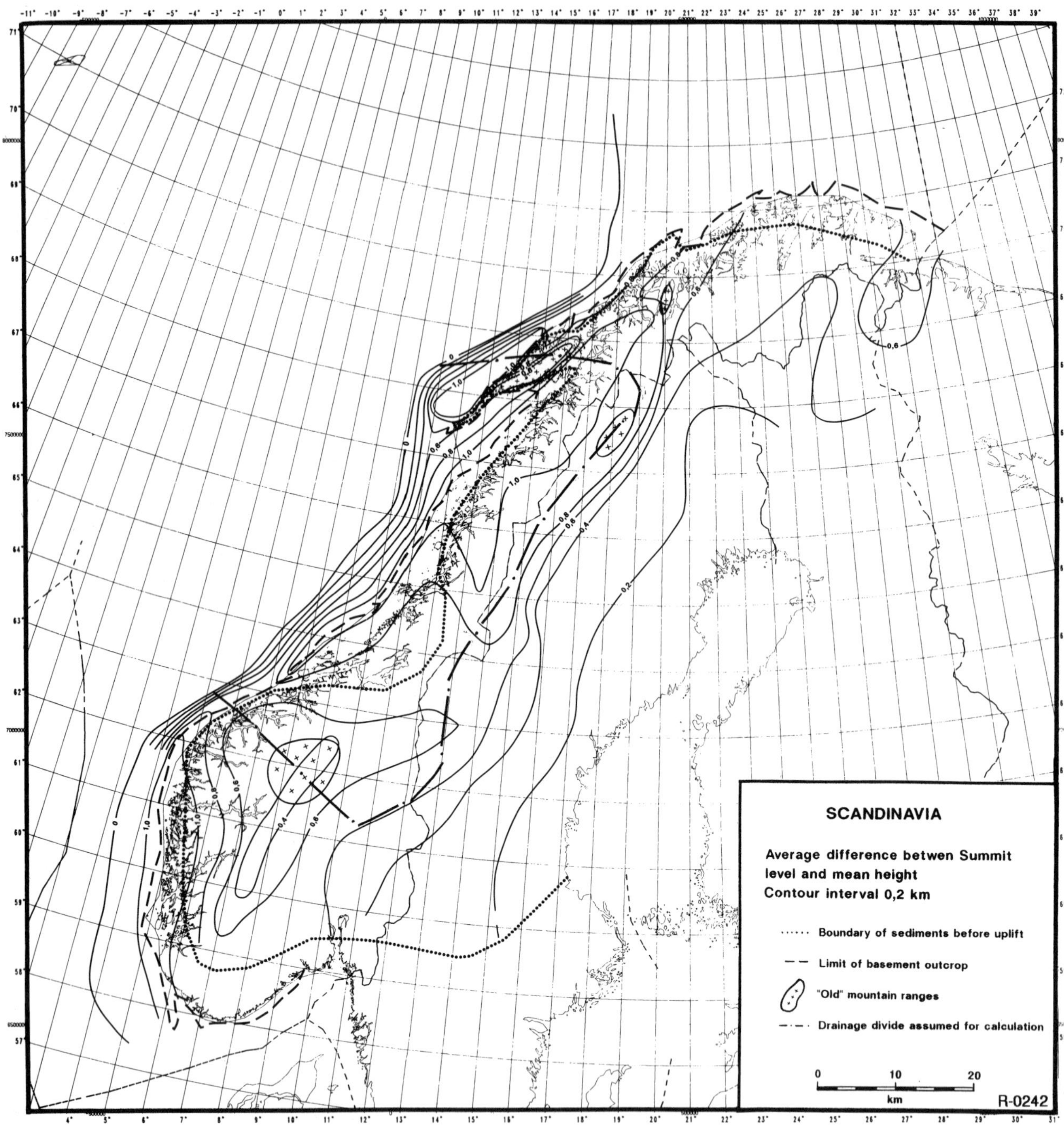

Fig. 11. Smoothed map showing difference between summit levels and mean height.

TABLE 2

Parameter values

Young's modulus	(E)	10.3×10^{10} Nm^{-2}
Poisson's ratio	(υ)	0.25
Lame's parameter	(μ,λ)	3.34×10^{10} Nm
Asthenosphere density	(ρ)	3.3×10^{3} kg/m^3
Inverse slope of Clausius Clapeyron curve	(γ)	75 K/kbar
Geothermal gradient	(β)	12 K/km
Density of upper phase	(ρ_1)	2.8×10^{3} kg/m^3
Density of lower phase	(ρ_2)	3.2×10^{3} kg/m^3
Bulk modulus	(K)	5.56×10^{10} Nm

Partly based on O'Connell and Wasserburg (1967).

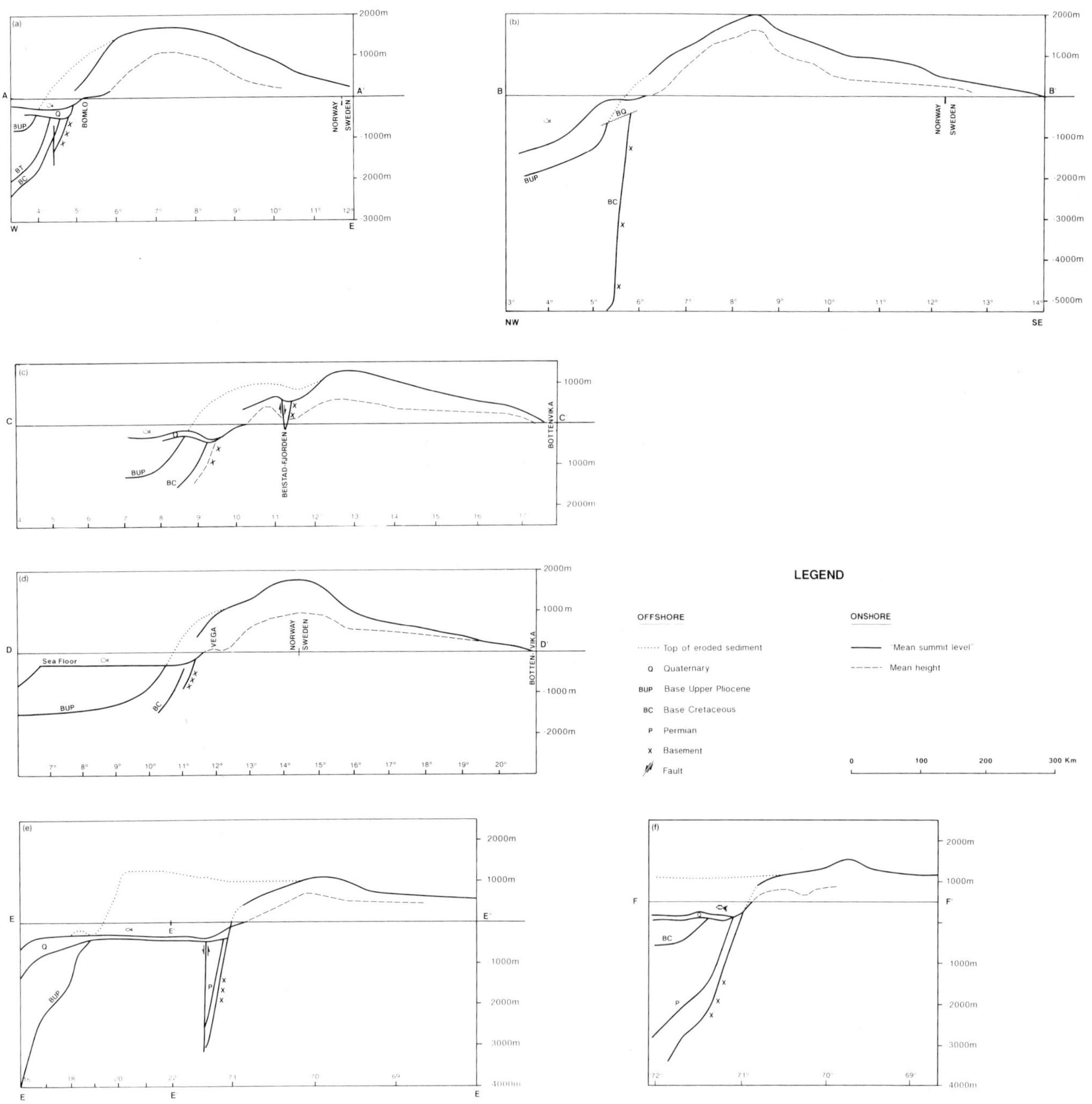

Fig. 12. Structural cross-sections showing summit level (solid line), mean height (dashed line), basement outcrop and Pliocene-Pleistocene deposits. For location of profiles, see Fig. 8.

a mechanism of uplift and subsidence of the Earth's surface. This is based on the assumption that phase transitions at depth respond to pressure changes at the surface. Subsidence at the surface occurs when an increase in pressure (by sedimentation) causes the upward migration of the phase boundary. Uplift would follow a decrease in pressure (by erosion) causing downward motion of the phase boundary. The mechanism has been studied by several authors (e.g., O'Connell and Wasserburg, 1967, 1972; Gjevik, 1972; O'Connell, 1976; Mareschal and Gangi, 1977; Mareschal and Lee, 1983). In these works, analytical and numerical approximations of the movement of the Earth's surface by phase boundary migration have been established.

The phase transition considered here is the transition from gabbro to eclogite at the base of the crust at a depth of 30–50 km. Migration of this phase boundary has been suggested as an explanation for the formation of sedimentary basins (e.g., Mareschal and Lee, 1983).

We have adopted an equilibrium approximation, since it is easily shown, using the analytical approximations of O'Connell (1976), that the phase bound-

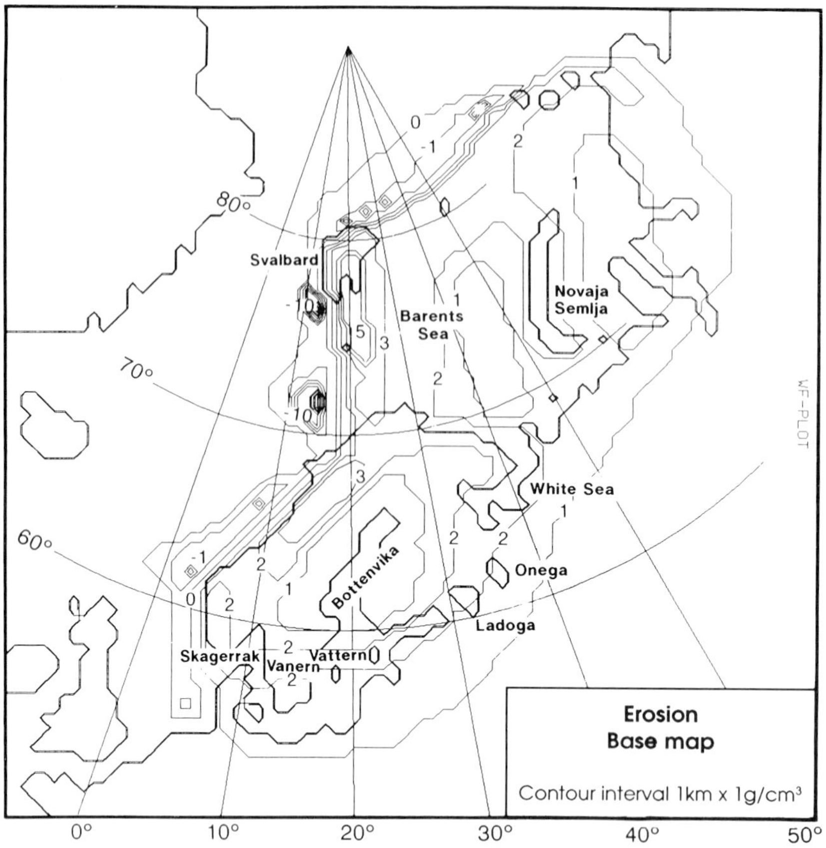

Fig. 13. Contoured grid showing values of erosion in the area covered by the Scandinavian and Barents Sea ice sheets in the glacial period. Erosion is given in terms of thickness (km) multiplied by density. The map is based on computer contouring of grid values. Contour interval: 1 km x 10 kg/m^3. Areas of sedimentation are given with negative values.

ary migration will be 90% compensated in 3 million years, using the parameter values given in Table 2 and a phase transition temperature of 1000 K. The results shown in Fig. 14b are based on an assumption of 50% phase transformation in the area. In spite of this, we consider the resulting surface adjustments to be maximum values. The mathematical model is formulated in Appendix B.

We are aware that field studies of eclogites indicate that granulites may be metastable for a long time, and eclogitization may take place only along shear zones where fluid has moved (Jamtveit et al., 1990).

The modelling of surface adjustments caused by possible changes in phase boundaries shows that this effect may be significant, of the order of 500 m of uplift in the western Barents Sea, and similar amounts of subsidence in the depositional wedge. Approximately 300 m of the uplift in the Nordland area may be explained by phase changes in the crust/mantle boundary.

The resulting adjustments of the Earth's surface due to movements of the phase transition are shown in Fig. 14b. The calculation was based on the erosion/deposition model in Fig. 13, and the parameter values given in Table 2.

Discussion of modelling results

The calculation of crustal response was based on a grid representing the thickness of material eroded in the glacial period multiplied by its density (Fig. 13), as well as the amount of sediment accumulations in the offshore wedges (negative numbers).

For the modelling, it was necessary to input information for the total area of glaciation and erosion and, therefore, very tentative values have been given in the eastern areas on which we have no information. However, errors in the eastern parts will not seriously affect the results in Norway and the Barents Sea.

The contoured values in the western Barents Sea do not coincide with data in Fig. 5, as we have tentatively tried to separate the last (glacial) phase from the Paleogene phase of uplift and erosion.

The three-dimensional modelling results indicate that the uplift of the Scandinavian mountains should be comparable with the amount of erosion. The areas of sediment accumulation subside substantially, allowing space for continued rapid deposition.

In the Barents Sea and other areas where sediments are removed, the net effect of the erosion

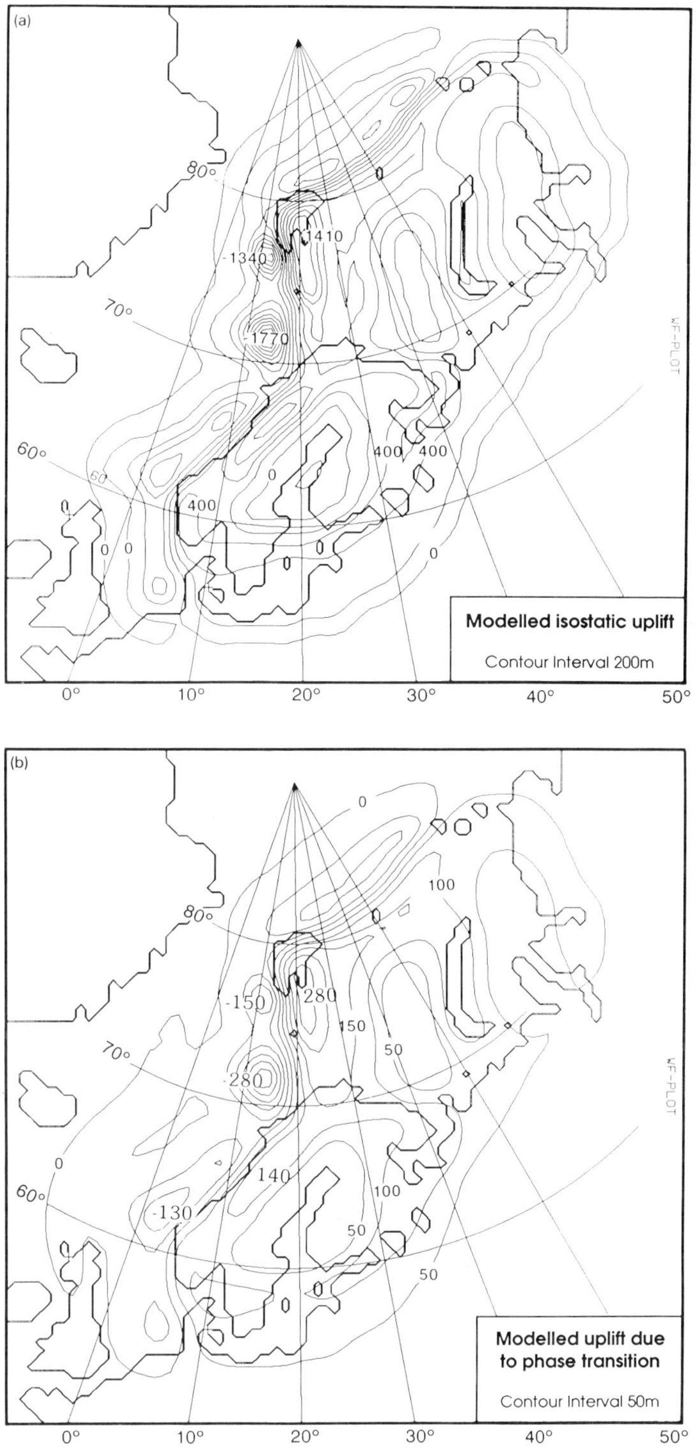

Fig. 14. Three-dimensional modelling of crustal response in the Scandinavia-Barents Sea areas. (a) Uplift and subsidence due to isostatic adjustment, using a flexural rigidity of 1×10^{23} Nm. The contour interval is 200 m. (b) Uplift and subsidence as a result of phase boundary migration, using a crustal thickness of 35 km and a density contrast of 0.4×10^3 kg/m^3. The contour interval is 50 m.

process is lowering of the surface. In the southern Barents Sea, the model predicts that removing 1200–1400 m of sediments (assuming a density of 2.2), results in an uplift of 600–800 m (compare Figs. 13 and 14). Hence, a pre-glacial elevation of 300–400 m above sea level is needed to explain the present elevation. In the western and northern Barents Sea, the elevation must have been higher. Similar results are obtained from the southern part of Sweden and Denmark, where mainly sediments were removed.

The three-dimensional modelling gives results which simulate quite well the observed structural pattern with the marginal highs and peripheral lows. The marginal highs are formed where dense base-

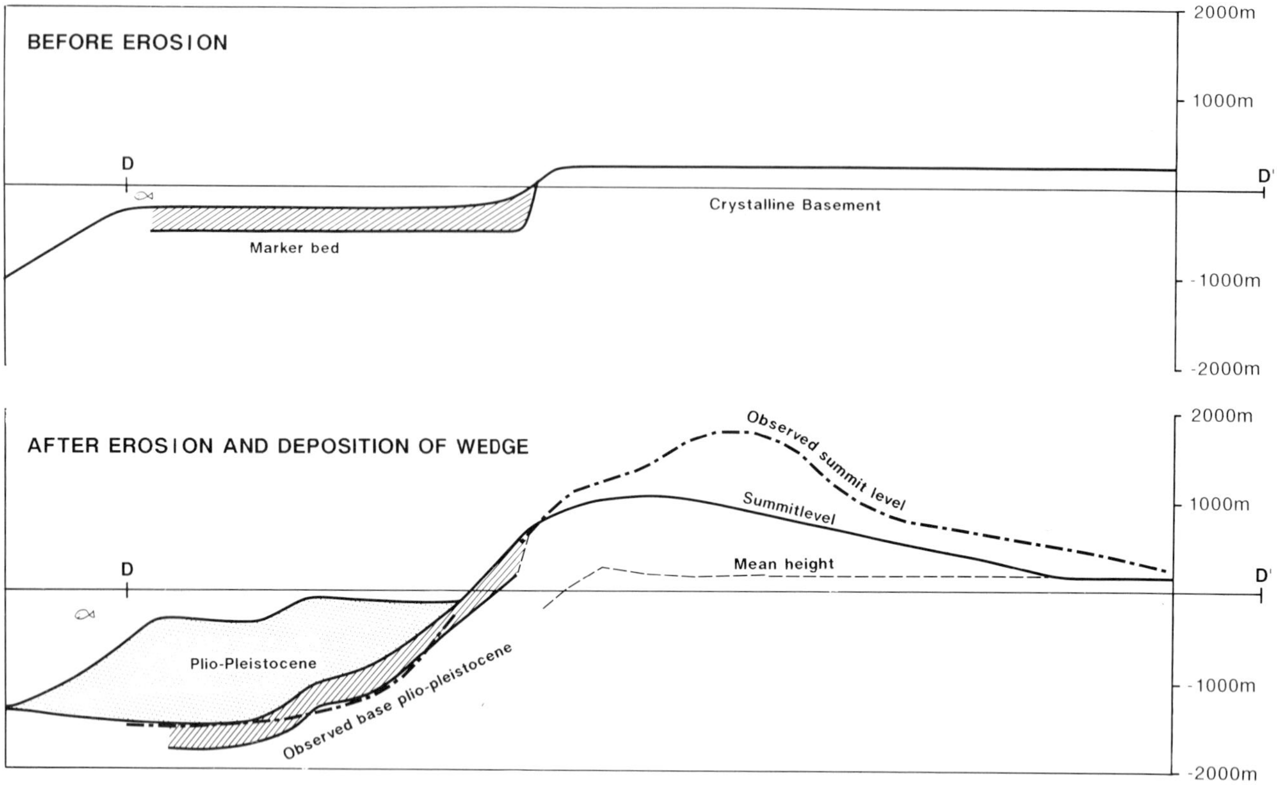

Fig. 15. Two-dimensional modelling of crustal response along profile *DD'*. The shape of a simplified pre-glacial topography is shown after erosion and deposition. The modelled results are compared with the observed profile *DD'*. For location, see Fig. 8.

ment rocks are eroded, and they are slightly displaced towards the central parts of the ice sheet from the area of maximum erosion. The peripheral lows seem to form where the erosion is at a maximum, while the subsidence/uplift is small due to the hinge line position at the boundary between crystalline basement and sedimentary cover. Where sediments only are eroded, the model predicts surface lowering, as is observed. In the central depression of Bottenvika, subsidence is predicted, and observed.

The low elevation of the Trondheim area can be explained by the assumption that it was covered by sedimentary rocks considerably lighter than the basement before onset of erosion.

However, the total theoretical adjustment is not in accordance with observation everywhere. Adjustment due to isostasy and phase change was calculated on profile DD' crossing Vega, Nordland, extending from the shelf edge to Bottenvika (Fig. 12). This profile is regarded as a cross-section in an area which was not affected strongly by Paleogene tectonic events, and the uplifted dome as well as the depositional wedge are well developed.

The results of calculation (Fig. 15) show that the isostatic and phase change effects may comprise almost the amount of erosion in the onshore area. The calculated uplift is considerably smaller than observed, leaving a difference in the order of 500 m. Offshore, the subsidence was calculated by adding the isostatic and phase change effects to the effect of compaction of the underlying sediments. For the calculation of compaction, we used a general porosity trend based on well data from Haltenbanken. Figure 15 demonstrates that the subsidence modelled is slightly smaller than that observed. However, the fit is better than for the onshore part.

There are many possible explanations for the discrepancy observed. One way out of the problem could be to fit the onshore part by assuming that the pre-erosional elevation was much higher than 250 m. This explanation is supported by the paleic surface map, Fig. 10. However, the low sedimentation rates on the shelf from the Oligocene through Early Pliocene suggest that the relief was not high at this time. This points in the opposite direction.

It is thought that some anomalies can be explained by assuming a pre-erosional high relief. The best example is South Norway. Figure 10 indicates that the paleic surface is elevated to more than 1500 m between Sognefjorden and the high peaks of Jotunheimen. These high peaks represent a topography where the summit level is elevated several hundreds of metres above the paleic surface. In the North Sea and Møre Basin, thick Paleogene sedimentary wedges are observed with a progradational direction roughly from the east. Thus, there is independent evidence that the uplift of the highest mountains in South Norway took place largely prior to the Pliocene.

Hence, separation of the Paleogene uplift from the Plio-Pleistocene is problematic. Our calculations indicate that an important part of the concentric morphology can be explained by rapid erosion and deposition of the material in the glacial periods. However, if the calculated uplift is removed from the present topography, we will still be left with a concentric morphology, although more subdued. We consider it unlikely that plate tectonic effects related to the opening of the Atlantic in the Paleogene should create such a topography. Also, the deformation of the base of the Upper Pliocene and even Quaternary layers (Fugelli and Riis, in press) into the regional dome structures is such that it cannot be explained totally using the present model.

Conclusions

– The uplift of Scandinavia and the Barents Sea is related to Paleogene thermal effects due to the North Atlantic rifting and opening, and to isostatic adjustments due to Pliocene/Pleistocene glaciation and erosion.

– The models presented do not explain satisfactorily the total magnitude of the observed uplift of the Scandinavian mountains, although they simulate well the observed structural pattern.

– Observation and modelling suggest that the erosion effects are very important in the study of recent structures in glaciated areas worldwide.

Acknowledgements

The authors are grateful to the Norwegian Petroleum Directorate for supporting this study. We also thank our colleagues in the NPD who have taken part in the discussion on the uplift problem. Further, we acknowledge the important contribution to this paper by Dr. Atle Nesje, Bergen University.

Appendix A

Isostatic model

The lithosphere is modelled as a thin elastic plate overlying an incompressible fluid substratum, expressed in the Fourier domain with the initial topographic load $F(k)$ and the resulting flexural deformation $h_0(k)$ written as a function of wavenumber k. The equilibrium isostatic compensation is achieved by subsidence (Cathles, 1975):

$$h_0(k) = \frac{F(k)}{\rho g \alpha(k)}$$

where g is gravity, ρ is density of the mantle, and $\alpha(k)$ is a parameter characterizing the flexure response of the lithosphere. The lithosphere flexure depends on the elastic strength of the lithosphere and $\alpha(k)$ is given by the equation (Cathles, 1975):

$$\alpha(k) = \frac{\frac{2\mu k}{\rho g}\left[S^2 - (kH)^2\right] + (CS + kH)}{S + kHC}$$

where S is $\sinh kH$, C is $\cosh kH$, and μ is Lamé's parameter.

The mechanical thickness of the lithosphere, H, is given by the equation:

$$H = \left[\frac{12\,(1-\nu^2)\,D}{E}\right]^{1/3}$$

where D is flexural rigidity, E is Young's modulus, and ν is Poisson's ratio.

Appendix B

Theoretical formulation of phase boundary migration

Let us first, for the purpose of illustration, consider a one-dimensional case assuming a phase boundary at depth in the mantle, separating two phases of the same component. Suppose the temperature at which the phase change takes place (T_c) is related to the pressure by the following equation (Clausius-Clapeyron equation):

$$T_c = T_0 + \gamma P_c = T_0 + \gamma g \rho_1 z$$

T_0 is the transition temperature at the Earth's surface, ρ_1 is the density of the upper phase, z is the depth of the phase boundary and γ is the inverse slope of the Clausius-Clapeyron curve.

Consider a uniform erosion of a half-space which undergoes a phase change at depth z_p. The decrease in hydrostatic pressure due to the erosion is ΔP. If the geothermal gradient is initially zero, the phase boundary will migrate downward until the pressure decrease is cancelled. The equilibrium migration of the phase boundary is:

$$\Delta z_p = \frac{\Delta P}{\rho g}$$

This migration will cause an uplift of the Earth's surface of:

$$\Delta z = \Delta z_p \cdot \frac{\Delta \rho}{\Delta \rho + \rho}$$

where $\Delta\rho$ is the density difference between the upper and lower phases.

In this study we use an analytical approximation for the equilibrium position of a phase boundary under the horizontally varying surface loads given by Mareschal and Gangi (1977). Under the assumptions that the Earth behaves as an elastic solid above the phase boundary, that the equilibrium position does not depend on deviatoric stress and that the two phases have the same thermal properties, they found that the vertical displacement of the phase boundary due to the applied load $P(k,a)$ is in the Fourier domain:

$$S(k) = Y(k) \cdot P(k,a)$$

where:

$$Y(k) = \frac{\left[\dfrac{-\gamma}{(\gamma g \rho_1 - \beta)}\right] \exp(-|ka|)}{1 + \left[\dfrac{\mu}{(\lambda + \mu)}\right]\left(\dfrac{\alpha K}{a}\right)\left[\dfrac{ka[\sinh(2ka) - 2ka]}{D(ka)}\right]}$$

and:

$$D(y) = \left[\frac{\mu}{\lambda + \mu}\right]^2 + \left[\frac{\lambda + 3\mu}{\lambda + \mu}\right] \cosh^2 y + y^2$$

$$\frac{\alpha K}{a} = \frac{\rho_2 - \rho_1}{\rho_1} \frac{\left(\dfrac{\gamma K}{a}\right)}{\gamma g \rho_1 - \beta}$$

γ is the inverse slope of Clausius-Clapeyron curve, β is the geothermal gradient, ρ_1 is the density of the upper phase, ρ_2 is the density of the lower phase, a is the depth of the phase change, K is the bulk modulus, and λ and μ are Lamé's parameters.

References

Antonsen, P., Elverhøi, A., Dypvik, H. and Solheim, A., 1991. Shallow bedrock geology in the Olga Basin area, northwestern Barents Sea. Am. Assoc. Pet. Geol., Bull., 75(7): 1178–1194.

Balling, N., 1980. The land uplift in Fennoscandia. In: N.A. Mörner (Editor), The Earth Rheology, Isostasy and Eustasy. J. Wiley and Sons, New York, N.Y., pp. 297–321.

Baturin, G.N., 1988. Dokl. Akad. Nauk SSSR, 299, No. 2.

Brekke, H. and Riis, F., 1987. Tectonics and basin evolution of the Norwegian shelf between 62°N and 72°N. Nor. Geol. Tidsskr., 67: 295–321.

Bøe, R., 1989. The structural development of the Mesozoic rocks in Beitstadfjorden, Edøyfjorden and Frohavet, Central Norway. Terra Abstracts (Eur. Union Geosci.), 1: 22–23.

Cathles, L.M., 1975. The Viscosity of the Earth's Mantle. Princeton University Press, Princeton, N.J, 386 pp.

Dalland, A., Worsley D. and Ofstad, K., 1988. A lithostratigraphic scheme for the Mesozoic and Cenozoic succession offshore mid- and northern Norway. Norw. Pet. Directorate, NPD Bull., 4, 65 pp.

Eidvin, T. and Riis, F., 1989. Nye dateringer av de tre vestligste borehullene i Barentshavet. Resultater og konsekvenser for den tertiære hevingen. NPD Contrib., 27, 43 pp.

Elverhøi, A., Antonsen, P., Flood, S.B., Solheim, A. and Vullstad, A.A., 1988. The physical environment Western Barents Sea 1 : 1 500 000. Shallow bedrock geology. Nor. Polarinst. Skr. 179 D.

Elverhøi, A., Liestøl, O. and Nagy, J., 1980. Glacial erosion, sedimentation and microfauna in the inner part of Kongsfjorden, Spitsbergen. Nor. Polarinst. Skr., 172: 33–61.

Elverhøi, A., Lønne, Ø. and Seland, R., 1983. Glaciomarine sedimentation in a modern fjord environment, Spitsbergen. Polar Res., 1: 127–149.

Fjeldskaar, W. and Cathles, L.M., 1991. Rheology of mantle and lithosphere inferred from postglacial uplift in Fennoscandia. In: R. Sabadini et al. (Editors), Glacial Isostasy, Sea Level and Mantle Rheology. Kluwer Academic Publishers, Dordrecht, pp. 1–19.

Fjeldskaar, W. and Riis, F., 1988. Modellering av tertiærbasseng på den vestlige marginen av Barentshavet. In: K. Binzer (Editor), Abstracts, 18. Nordi. Geol. Vintermøde, København, 1988. Danmarks Geologiske Undersøgelse.

Fjeldskaar, W. and Pallesen, S., 1989. The application of a viscoelastic lithosphere model to isostatic subsidence in backstripping. In: J. Collinson (Editor), Correlation in Hydrocarbon Exploration. Graham and Trotman, London, pp. 53–59.

Fugelli, E.M.G. and Riis, F. in press. Neotectonism in the Jæren area, southwest Norway. In: L.N. Jensen and F. Riis (Editors), Post-Cretaceous Uplift and Sedimentation Along the Fennoscandian Shield. Nor. Geol. Tidsskr.

Gabrielsen, R.H., Færseth, R.B., Jensen, L.N., Kalheim, J.E. and Riis, F., 1990. Structural elements of the Norwegian continental shelf. Part I: The Barents Sea Region. NPD Bull., 6, 33 pp.

Geographical Survey Office of Sweden, 1963. Mean height map of Sweden. Rikets Allmänna Kartverk, Meddelande, Ser. A, No. 30.

Gjessing, J., 1967. Norway's paleic surface. Nor. Geol. Tidsskr., 21: 69–132.

Gjevik, B., 1972. Surface adjustment due to a subcrustal phase transition. Phys. Earth Planet. Inter., 5: 403–408.

Jamtveit, B., Bucher, K. and Austrheim, H., 1990. Fluid controlled eclogitization of granulites in deep-crustal shear zones, Bergen Arcs, Western Norway. Contrib. Mineral. Petrol., 104: 184–193.

Jansen, E., Bleil, U., Henrich, R., Kringstad, L. and Slettemark, B., 1988. Paleoenvironmental changes in the Norwegian Sea and the Northeast Atlantic during the last 2.8 Ma: DSDP/ODP Sites 610, 642, 643 and 644. Paleoceanography, 3: 563–581.

Kennedy, G.C., 1959. The origin of continents, mountain ranges and ocean basins. Am. Sci., 47: 491–504.

Løvø V., Elverhøi, A., Antonsen, H.P., Solheim, A., Butenko, G., Gregersen, O. and Liestøl, O., 1990. Submarine permafrost and gas hydrates in the Northern Barents Sea. Nor. Polarinst., Rapportser., 56, 171 pp.

Manum, S. and Throndsen, T., 1978. Rank of coal and dispersed organic matter and its geological bearing in the Spitsbergen Tertiary. Nor. Polarinst., Årbok 1977, pp. 159–177.

Mareschal, J.C. and Gangi, A.F., 1977. Equilibrium position of a phase boundary under horizontally varying surface loads. Geophys. J.R. Astron. Soc., 49: 757–772.

Mareschal, J.C. and Lee, C.K., 1983. Initiation of subsidence in a sedimentary basin underlain by a phase change. Geophys. J.R. Astron. Soc., 74: 689–712.

Nesje, A., Dahl, S.O., Anda, E. and Rye, N., 1988. Block fields in southern Norway: significance for the Late Weichselian ice sheet. Nor. Geol. Tidsskr., 68: 149–169.

Nyland, B., Jensen, L.N., Skagen, J.I., Skarpnes, O. and Vorren, T., 1992. Tertiary uplift and erosion in the Barents Sea; magnitude, timing and consequences. In: R.M. Larsen, H. Brekke, B.T. Larsen and E. Talleraas (Editors), Structural and Tectonic Modelling and its Application to Petroleum Geology. Norwegian Petroleum Society (NPF), Special Publication 1. Elsevier, Amsterdam, pp. 153–162 (this volume).

Nøttvedt, A., Berglund, T., Rasmussen, E. and Steel, R., 1988. Some aspects of Tertiary tectonics and sediments along the western Barents Shelf. In: A.C. Morton and L.M. Parson, (Editors), Early Tertiary Volcanism and the Opening of the NE Atlantic. Geol. Soc. London, Spec. Publ., 39: 421–425.

O'Connell, R.J., 1976. The effects of mantle phase changes on postglacial rebound. J. Geophys. Res., 81: 971–974.

O'Connell, R.J. and Wasserburg, G.J., 1967. Dynamics of the motion of a phase change boundary to changes in pressure. Rev. Geophys. Space Phys., 3: 329–410.

O'Connell, R.J. and Wasserburg, G.J., 1972. Dynamics of submergence and uplift of a sedimentary basin underlain by a phase change boundary. Rev. Geophys. Space Phys., 10: 335–368.

Perry, R.K. and Fleming, H.S., 1985. Bathymetry of the Arctic Ocean. Map sheet, scale 1:4704073. Naval Research Laboratories, Washington D.C.

Ramberg Moe, H., Roaldset, E. and Gjelsvik N., 1988. Siliceous sediments and seismic response — a correlation tool in exploration on Norwegian shelf. Norwegian Petroleum Society, Correlation in Hydrocarbon Exploration, 3–5 October 1988 (abstract).

Rokoengen, K. and Rønningsland, T.M., 1983. Shallow bedrock geology and Quaternary thickness in the Norwegian sector of the North Sea between 60°30′N and 62°N. Nor. Geol. Tidsskr., 63: 83–102.

Sales, J., 1989. Late Tertiary uplift — West Barents shelf. Character, mechanisms and timing. Paper presented at the Norwegian Petroleum Society Workshop, 18–20 October 1989, Stavanger.

Spencer, A.M., Home, P.C. and Berglund, L.T., 1984. Tertiary structural development of the western Barents Shelf: Troms to Svalbard. In: A.M. Spencer et al. (Editors), Petroleum Geology of the North Atlantic Margin. Graham and Trotman, London, pp. 99–109.

Steel, R.J. and Worsley, D., 1984. Svalbard's post-Caledonian strata — an atlas of sedimentational and palaeogeographical evolution. In: A.M. Spencer (Editor), Petroleum Geology of the North European Margin. Graham and Trotman, London, pp. 98–107.

Strøm, K.M., 1948. The geomorphology of Norway. Geogr. J., 62: 19–27.

Sugden, D.E., 1976. A case against deep erosion of shields by ice sheets. Geology, 4: 580–582.

Talwani, M., Udintsev, G. et al., 1976. Initial reports of the Deep Sea Drilling Project, Vol. 38. U.S. Government Printing Office, Washington, D.C., 1256 pp.

Vorren, T.O., Lebesbye, E., Henriksen, E., Knutsen, S.M. and Richardsen, G., 1988. Cenozoisk erosjon og sedimentasjon i det sørlige Barentshav. Abstr., XVIII Nord. Geol. Vintermøte, København, 1988, 424 pp.

Vorren, T.O., Lebesbye, E., Andereassen, K. and Larsen, K.-B., 1989. Glacigenic sediments on a passive continental margin as exemplified by the Barents Sea. Mar. Geol., 85: 251–272.

Vorren, T.O., Richardsen, G., Knutsen, S.-M. and Henriksen, E., 1990. The western Barents Sea during the Cenozoic. In: U. Bleil and J. Thiede (Editors), Geologic History of the Polar Oceans: Arctic versus Antarctic. NATO ASI Series. Kluwer Academic Publishers, Dordrecht, pp. 95–118.

Weidick, A., 1976. Glaciation and the Quaternary of Greenland. In: A. Escher and W. Stuart Watt (Editors), Geology of Greenland. Geological Survey of Greenland, Copenhagen, pp. 430–458.

White, W.A., 1972. Deep erosion by continental ice sheets. Geol. Soc. Am. Bull., 83: 1037–1056.

F. RIIS — *Norwegian Petroleum Directorate, P.O. Box 600, N-4001 Stavanger, Norway*

W. FJELDSKAAR — *Rogaland Research Institute, P.O. Box 2503, Ullandhaug, N-4004 Stavanger, Norway*

Pleistocene glacial isostasy — implications for petroleum geology

A. Kjemperud and W. Fjeldskaar

The theory of glacial isostasy evolved in the second half of the 19th century and by the beginning of the 20th century it was fully accepted. The onshore data in Fennoscandia were overwhelming with traces of marine influence 200–300 m above the present shoreline. Very few studies have, however, been performed in offshore areas. Only recently, Quaternary research has reported phenomena interpreted as a result of glacial isostatic movements. Among petroleum geologists the effect has been almost neglected, and no public reports on how glacial isostasy may effect hydrocarbon bearing rocks (source or reservoir) exist.

Data from the Quaternary and latest Tertiary research have been compiled to establish a stratigraphic framework for the last 2–3 million years. Realistic ice models have been made which incorporate the observations made on three different sizes of ice sheets in northern Europe. These ice sheet models form the input for the modelling of isostatic deformation of the crust. The earth is modelled by a non-spherical viscous fluid, overlain by a uniform thick elastic lithosphere. The isostatic problem is treated using the Fourier transform technique.

The main product of the modelling is isobase maps describing the equilibrium deformation of the crust during the different glacial events. The model is calibrated using Fennoscandian sea-level curves after the last glaciation. The maps are therefore supposed to be valid also in the offshore parts of northern Europe including shelf areas with considerable oil and gas accumulations.

At least once and perhaps several times during the last 2–3 million years the central North Sea has been downwarped and subsequently uplifted by up to 400 m, on the Haltenbanken up to 250 m and in the Barents Sea as much as 650 m. The vertical movements have been differential in the sense that the areas overlain by thick ice domes have been downwarped more than peripheral areas. The modelled gradients are more than 1.0 m/km in graben areas of the North Sea and parts of Haltenbanken and about 1.3 m/km in the western parts of the Barents Sea.

The differential movements will have affected the entire crust including the sediments with their fluid content. The geometric change during the maximum downwarping may have important implications for the secondary migration of hydrocarbons. Hydrocarbons in traps filled to the spill point will start to spill during the isostatic movements and today the structure may be only partly filled. This effect can be severe in flat and in areally large structures. Our calculations show that as much as 30% of the total hydrocarbon volume can be lost during maximum glacial isostatic downwarping.

Introduction

It is generally accepted that crustal layers have a tendency to approach a position of hydrostatic equilibrium. Dutton (1889) proposed the term isostasy for this process. However, there are different mechanisms by which this may be achieved. Pratt (1855) postulated that mountains are composed of rocks which are less dense than average, while Airy (1855) suggested that mountains have roots, i.e., material of uniform density which extends deeper into the underlying denser material beneath the mountain than beneath the lowlands.

Realizing that the Earth's mantle is not rigid, Jamieson (1865) postulated that the load of an ice cap will deform the surface of the earth, giving rise to glacial isostasy.

The theory of glacial isostasy as an explanation for the marine deposits found high above the present shoreline in Scandinavia was generally accepted in the late 19th century when De Geer (1888, 1890) published his famous work on the uplift of Fennoscandia. A full understanding of the interplay between glacial isostasy and eustasy was, however, not reached before the work of Nansen (1922, 1927) and Ramsey (1924). Nansen (1922) was also the first to point out the possibility of a peripheral forebulge around an ice sheet

Large isostatic adjustments have occurred in Fennoscandia and adjacent areas in the last 2–3 million years. The cause of these adjustments is extensive glaciations with ice caps up to several kilometres thick in some areas. A number of glaciations have taken place during the Quaternary, but the

Structural and Tectonic Modelling and its Application to Petroleum Geology edited by R.M. Larsen, H. Brekke, B.T. Larsen and E. Talleraas. NPF Special Publication 1, pp. 187–195. Elsevier, Amsterdam.

isostatic movements will in principle be the same for all, although the magnitude and duration will be different.

Since the mid-sixties an extensive exploration for oil and gas has taken place in the North Sea, and somewhat later in the Haltenbanken area and in the Barents Sea. It has long been clear that these areas have been completely or partly covered by glaciers at least once and perhaps several times during the Quaternary. However, very little attention has been paid to the effects of glaciations on hydrocarbon generation, migration and trapping. The objective of this study is to examine the implications of glacial isostasy for petroleum geology.

Pleistocene glaciations in Northern Europe

Timing of glaciations

The initiation of northern hemisphere glaciations has been a subject of debate. Based on early DSDP results several authors (Laughton et al., 1970; Berggren, 1972; Berggren and Van Couvering, 1974) defined the onset of glaciations to approximately 3 Ma. Dating of glacial marine deposits from the Amarasian basin in the central Arctic ocean (Clark et al., 1980) gave approximately 5.3 Ma for the first ice-rafted material. The chronostratigraphy of this report has, however, been questioned (Grantz et al., 1982).

Based on results from Pacific cores (Shackleton and Opdyke, 1976; 1977) suggest initiation of northern hemisphere glaciation at 3.2 Ma, with a severe deterioration at 2.5 Ma. A correlation of the oxygen isotope stage zonation of four detailed Pleistocene oxygen isotope records from the Pacific (V28-239, DSDP 504), Caribbean (DSDP 502B) and the North Atlantic (DSDP 552) (Williams 1988) show a good correlation of nine cold events in the last 0.8 million years and in the time interval between 2.2 Ma and 0.8 Ma as much as 17 cold events are interpreted in one or all three areas. We have used the most complete oxygen isotope curve from the Pacific (Fig. 1), which is shown to correlate with the North Atlantic curves (Williams, 1988), to construct a eustatic curve (actually a curve showing the ocean water volume change given in metres of global change, cf., Fjeldskaar, 1989) which may represent the magnitude and importance of the glacial events in the last 2 million years A calibration of 0.11‰ $\delta^{18}O$ per 10 m of sea-level change has been used (Fairbanks and Matthews, 1978).

A recent investigation (Jansen et al., 1988) concluded that a major advance of the Scandinavian ice sheet took place at approximately 2.6 Ma based

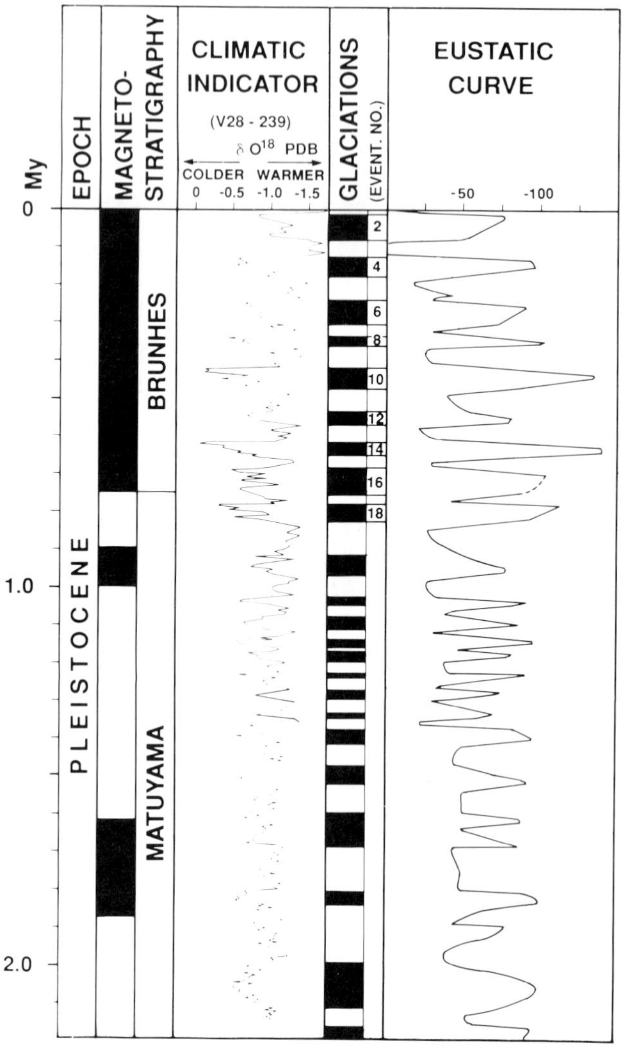

Fig. 1. Isotope records and deduced glacial events and eustatic curve (isotope curve from Schackleton and Opdyke, 1977)

on material from DSDP and ODP cores from the North Atlantic and the Norwegian Sea. This is based on oxygen isotope analyses, magnetostratigraphy and observation of ice-rafted debris. Several severe glaciations took place in the period 2.6–2.0 Ma. Between 2.0 and 1.2 Ma the glaciations in Scandinavia were small, while a transition towards larger glaciations took place during the period 1.2–0.6 Ma. In the last 0.6 million years the climate has been more variable with short warm interglacial periods between period with extensive glaciations seen as pulses of ice-rafted material in the cored material. The oxygen benthic isotope record in the Norwegian Sea (borehole ODP 644 A) shows relative changes of up to 1‰ (occasionally 2 units), which represent approximately 100 m (occasionally 200 m) of eustatic change, using the same method as for the Pacific core.

The extent and thickness of ice sheets

The regional extent and the thickness of ice sheets are the two controlling factors for glacial isostasy. From the stratigraphical record it seems to be certain that the North European area has been influenced by both small and large ice sheets during the last 2–3 million years

Ice sheets can be reconstructed on the basis of analogy with observational data from recent sheets and traces left behind by ancient sheets. The ice sheets used in this study have been reconstructed by the method of Hughes (1981). The most important input for determining the size of the ice-covered area has been the reported observations of dated or age estimated end moraines, while ice sheet gradients from recent sheets are the major input for estimating ice thickness and volume.

As a basis for our calculations of isostatic effects, three different ice sheets have been reconstructed. They are all based on different interpretations of observations from the last glaciation. The three ice models thus represent realistic geometries which most probably closely resemble the real situation in one or several of the previous glacial events.

Ice model I (Fig. 2). The ice sheet is characterized by three large ice domes. One over Novaya Zemlya in the East Barents Sea reaching 3200 m, one in central Fennoscandia of the same thickness and one in northern Great Britain reaching 2000 m. These three domes represent the growth centre of the ice sheets which has coalesced in the saddle areas in the south east Barents Sea and in the central North Sea.

An ice distribution like this has been proposed for the last ice age by Grosswald (1980) and Andersen (1981, maximum alternative). There is, however, an ongoing discussion as to whether the North Sea and the Barents sea were glaciated during the last ice age (see discussion in Boulton, 1979; Grosswald, 1980; Andersen, 1981; Solheim and Kristoffersen, 1984; Sejrup et al., 1987; and Vorren et al. 1989).

It can, however, be concluded that both areas were glaciated once or most probably several times during the Pleistocene. Sejrup et al. (1987) conclude that in the North Sea the central area (Fladen Ground) was overridden by ice several times in the time interval 1–0.7 Ma. The area was also glaciated sometime between 0.2 and 0.13 Ma (the Saalian). In the Barents Sea glacigenic processes have been active at least four times during the Pleistocene and as far west as the shelf break according to Elverhøi and Solheim (1987) and Vorren et al. (1989).

In the Haltenbanken area marginal moraines situated at the shelf edge have been mapped by Andersen (1979). Between 64°N and 66°N the marginal moraines are found approximately west of 6°E, which means that the entire shelf area most probably was covered by ice during the last glacial event. Since the Haltenbanken is situated relatively close to the mountainous area and central Fennoscandia, even minor glacial events could have covered the shelf.

Ice model II (Fig. 3). A large ice dome in central Fennoscandia (thickness 3200 m) is isolated from an ice dome in northern Great Britain (thickness

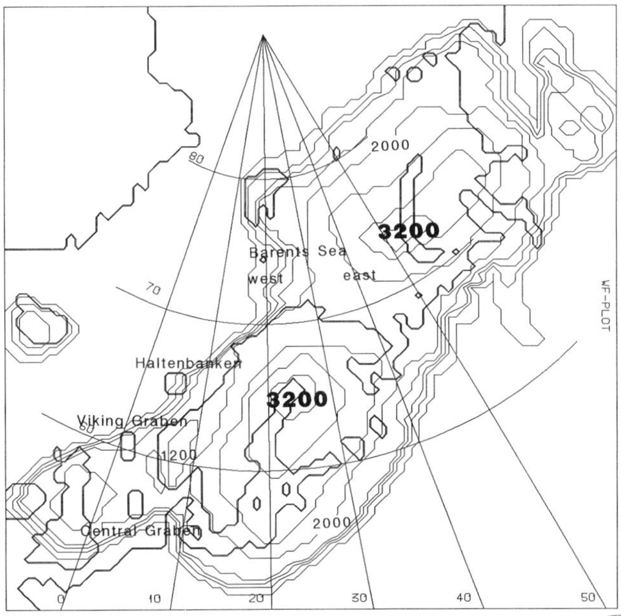

Fig. 2. Ice model I. Based on the maximum model of Andersen (1981). Isopacks are each 400 m. Outer line 400 m.

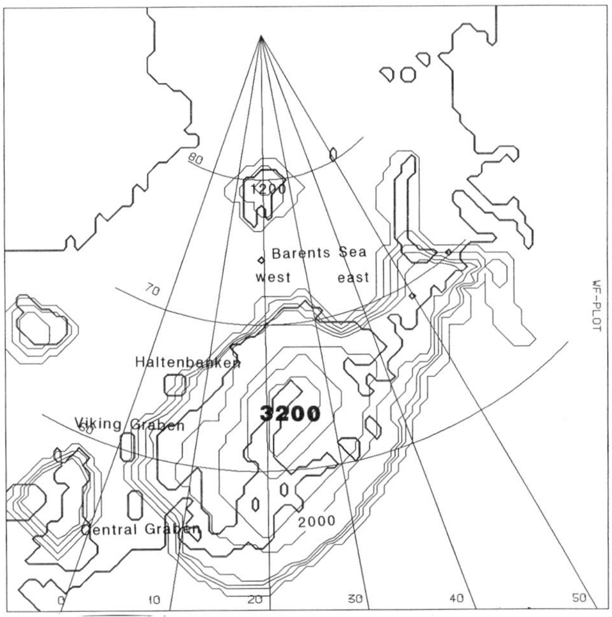

Fig. 3. Ice model II. Based on the minimum model of Andersen (1981) for the Barents Sea and on Sejrup et al. (1987) for the North Sea. Isopacks are each 400 m. Outer line 400 m.

1600 m) and one in Svalbard (thickness 1200 m). Most of the North Sea and the entire Barents Sea are ice free. The ice cover in the Haltenbanken area is the same as in ice model I. The model is based on Andersen's (1981) minimum alternative for the Barents Sea, and that of Sejrup et al. (1987) for the North Sea.

This model can represent the last glaciation (Weichselian) in the North Sea. Data from the western part of the North Sea and from Scotland (Sutherland, 1984) indicate that the British ice sheet did not reach much further east than today's land area. This supports the conclusions made by Sejrup et al. (1987) that the British and the Scandinavian ice did not coalesce in the Fladen Ground area (Sleipner Field).

Ice model III. This model is similar to model II except for a reduced ice thickness in the mountain areas in the eastern part of Norway according to the theory of Nesje et al. (1988). This isostatic modelling, however, gives results very similar to those obtained when using model II (Table 1).

A detailed history of outbuilding and downmelting of the ice sheets has not been applied in any of the models. Both processes are modelled as being instantaneous.

Methods of glacial isostatic modelling

The Earth is modelled by a non-spherical viscoelastic fluid in which the viscosity may vary with depth, overlain by a uniform thick elastic lithosphere. With this flat earth model, we are able to treat the isostatic problem analytically using the Fourier transform technique. The method used is developed by Cathles (1975).

The Fourier transform allows a function such as the ice sheet configuration to be expressed as the sum of harmonic components of different wavelengths.

The asthenosphere is treated as a viscous fluid in which the rate of displacement varies with the wavelength of the harmonic component of the load. The elastic lithosphere is treated as a low-pass filter, because loads of small size tends to be balanced by the lithosphere itself and not by buoyancy.

Mantle viscosity

Data on the present rate of uplift and sea-level changes after the last glaciation show that the mantle is of low viscosity (Fjeldskaar and Cathles, 1991a, b).

In calculations of the time-dependent isostatic deflections, a detailed glacial history must be taken into account. This is not done in the present study. Only the situation at isostatic equilibrium (which is achieved 20000 years after loading/unloading) is modelled. The isostatic model is then rather simple, as the important parameter is only the lithospheric rigidity.

Lithosphere modelling

If an ice load is applied to a fluid, the surface of the fluid will deform until the weight of the fluid displaced from the equilibrium level balances the applied load (local isostatic compensation). If an elastic lithosphere covers the fluid the applied load will be supported partly by the lithosphere and partly by the buoyant forces of the fluid beneath acting through the lithosphere (regional isostatic compensation). If the size of the load is small compared to the lithospheric thickness, the base of the lithosphere is not deformed at all and the entire load is balanced by the lithosphere. If the load is large, the lithosphere supports none of the load, and the deformation at the base of the lithosphere is the full isostatic deformation.

The lithosphere thus acts as a lowpass filter. The characteristics of this filter depend on the elastic strength of the lithosphere. A measure of the elastic strength of the lithosphere is a parameter called the flexural rigidity. The elastic strength of the lithosphere is a function of mechanical thickness and is determined by the following equation:

$$\text{Flexural rigidity } D = \frac{EH^3}{12(1-\nu^2)}$$

where H is elastic thickness, E is Young's modulus, ν is Poisson's ratio.

The regional isostatic equilibrium compensation due to a harmonic load $F(x)$ is achieved by subsidence (Cathles, 1975):

$$h_0 = \frac{F(k)}{\rho g \alpha(k)}$$

For a local compensation model the subsidence is:

$$h_0 = \frac{F(k)}{\rho g}$$

where $F(k)$ is Fourier transformed ice load, ρ is density of the upper mantle, and g is gravity.

The "lithosphere filter" is:

$$\alpha(k) = \frac{\frac{2\mu k}{\rho g}(S^2 - k^2H^2) + (CS + kH)}{S + kHC}$$

where $S = \sinh kH$; $C = \cosh kH$, H is mechanical thickness of the lithosphere, μ is Lame's parameter, and k is wavelength.

In this study we use a mechanical thickness of the lithosphere equal to 70 km (flexural rigidity of 50×10^{23} Nm). This is based on a study by Fjeldskaar and Cathles (1991a) concerning data on the present rate of uplift and sea-level changes after the last glaciation in Scandinavia (Kjemperud 1986).

The model presumes a uniformly thin lithosphere. This is of course a simplification. The spatial variation of the lithosphere thickness is, however, difficult to estimate. Moreover, the effects introduced by this presumption are probably of second-order only.

The flexural rigidity for the shelf areas is not known, the study mentioned above is based on data from the Scandinavian mainland. It is, however, reasonable to assume the same rigidity as for Scandinavia, because the shelf areas also have continental crust. The large sediment thickness on the shelfs does not influence the flexural rigidity; the elastic thickness of the lithosphere is almost 70 km, while the sediment package is only 5–10 km thick.

The geophysical model of the Earth used in this study is shown in Fig. 4. The geological and geophysical processes associated with periods of glaciations/deglaciations are also shown. These processes are isostasy, eustasy, flow in asthenosphere, glacial erosion and deposition.

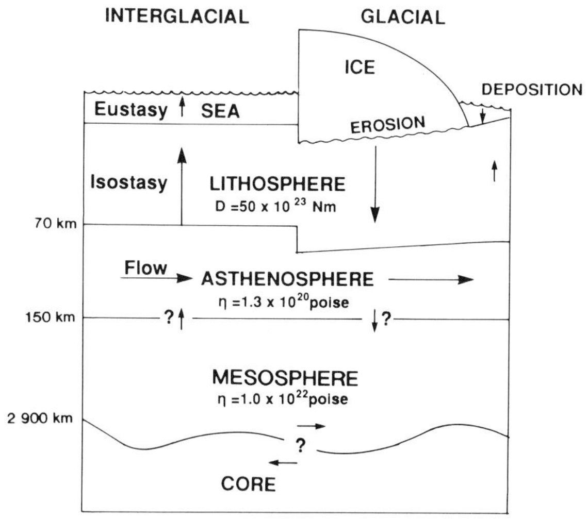

Fig. 4. Geophysical model of the earth as used in this study. Geological and geophysical processes during interglacial periods (left) and glacial periods (right) are shown.

Results from isostatic modelling

The results from the isostatic modelling are given in the isobase maps (Figs. 5 and 6) and in Table 1.

The largest downwarping of the lithosphere seen is in the eastern Barents Sea where values of more than 600 m are reached by applying ice model I. In the North Sea the maximum values exceed 400 m and on Haltenbanken it reaches 250 m.

The largest gradients, 1.3 m/km east–west, are seen in the western Barents Sea. However, the other areas also show values above 1.0 m/km when applying ice model I.

Because of the lack of a deglaciation history, the time-dependent forebulge is not represented in the modelling results (Table 1). A transient model has, however, been applied in the North Sea and in the Haltenbanken; indicating that the forebulge reaches a maximum just after the deglaciation and is somewhat higher than the values shown

TABLE 1

Isostatic effects in key areas of the Norwegian shelf

Ice models	Central graben	North Viking graben	Halten banken	Barents Sea west of 22°E	Barents Sea east
Ice model I					
Ice cover (m)	1200 to 1600	1200 to 1600	400 to 800	400 to 1200	1200 to 2400
Isostasy (max) (m)	400	450	230	400	650
Forebulge	no	no	no	no	no
Gradients N–S	−1.1	1.2	0.6	0.4	0.3
Gradients E–W	−0.5	0.4	1.0	1.3	0.8
Ice model II					
Ice cover (m)	0	0	400 to 800	0	0
Isostasy (max) (m)	−25 to +70	110	250	−10 to +160	−30 to +160
Forebulge	yes	no	no	yes	yes
Gradients N–S	−0.4	0.2	0.8	0.9	0.8
Gradients E–W	−0.3	0.8	1.0	0.5	0.3
Ice model III					
Ice cover (m)	0	0	0 to 800	0	0
Isostasy (max) (m)	−5 to +40	50	230	−10 to +160	−30 to +160
Forebulge	insignificant	no	no	yes	yes
Gradients N–S	−0.3	0.1	0.7	0.9	0.8
Gradients E–W	−0.3	0.5	1.1	0.5	0.3

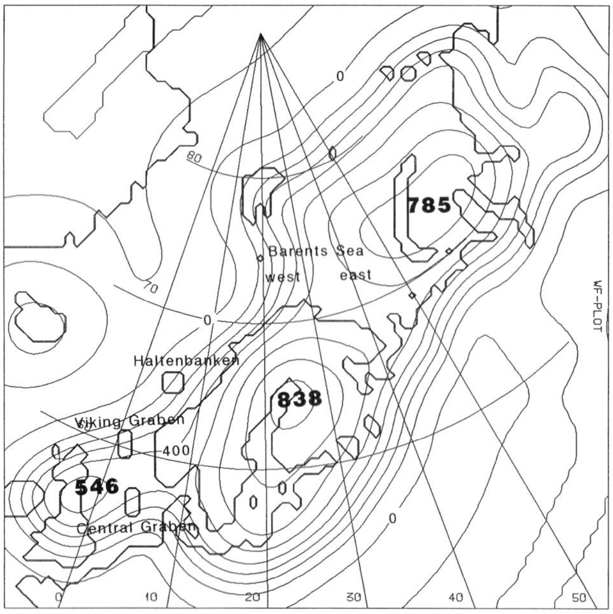

Fig. 5. Modelled isostatic effect of ice model I. Isobases each 100 m.

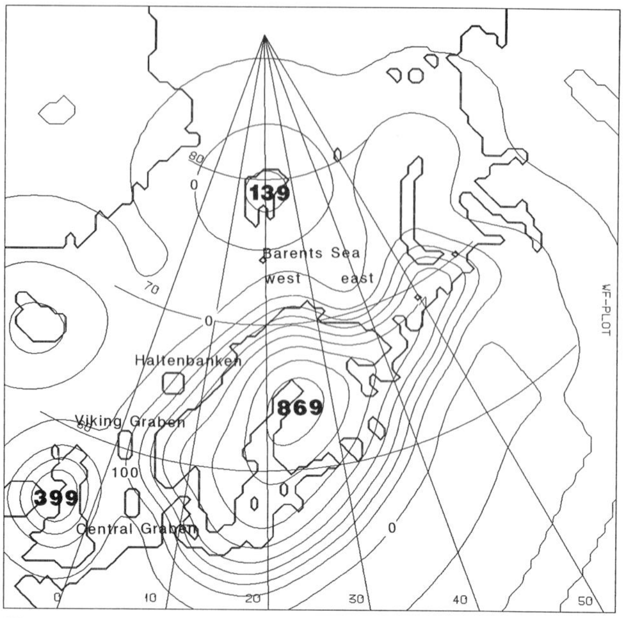

Fig. 6. Modelled isostatic effect of ice model II. Isobases each 100 m.

Table 1. The height of the forebulge does not seem to exceed 60–70 m.

Implications for hydrocarbon migration and trapping

The theory of glacially isostatic induced hydrocarbon migration combines the glacial isostatic effects with the fact that liquid interfaces are equipotential surfaces. In normal pressure areas they are parallel to the geoid. Figure 7 shows how a simple structure behaves during one glacial cycle. It is assumed that the structure is filled with hydrocarbons to the spill point at the onset of vertical movements.

When the lithosphere and the structure are downwarped differentially due to glaciation, hydrocarbons start to spill out of the structure. This goes on until the maximum tilt situation is reached (Fig. 7b). The closure volume is now reduced proportional to the differential movements.

When the lithosphere and the structure are uplifted to the starting position due to deglaciation, a new hydrocarbon-water contact (HCWC) is established (Fig. 7c). The freeboard which is the vertical difference between the HCWC and the spill point will be defined by the magnitude of the differential movement.

Fossil HCWCs can theoretically be found at the spill point level for the two extreme positions. Residual oil can be found in the water zone down to the original spill point level. In a gas-filled reservoir with an oil leg originally, residual oil can also be found in the gas-filled part, limited by the uppermost fossil HCWC.

The effect of the glacially induced hydrocarbon migration is determined by several factors:

– *The gradient of the deformation.* The gradient is dependent on the geometry of the ice sheet. The result observed today is the effect of the ice sheet creating the highest gradients through time.

– *The number of isostatic events.* Tilting will be the result of each isostatic event. A number of isostatic events will create overlapping and chaotic patterns of fossil contacts and residual oil which will be very difficult to interpret.

– *The orientation of the long axis of the hydrocarbon-filled structure.* The direction of the length axis of the field will greatly influence the volume changes during the isostatic events. The differential downwarping will be highest when the longest axis is parallel to the main gradient. A 30 × 5 km field directed along a 1 m/km gradient give a differential tilt value of 30 m, while the perpendicular orientation would give a differential value of only 5 m.

– *The initial fill of the trap.* An incompletely filled trap could possess enough freeboard to accommodate for the differential changes. The *freeboard height* is the difference between the spill points of the trap at maximum uplift and maximum downwarping.

– *The spill point orientation.* If the spill point is situated in the area of least relative downwarping the entrapped volume will be reduced and vice versa. In light of the isobase maps (Figs. 5 and 6) fields in the Barents Sea with spill points in the west and north, and fields on Haltenbanken with spill points towards

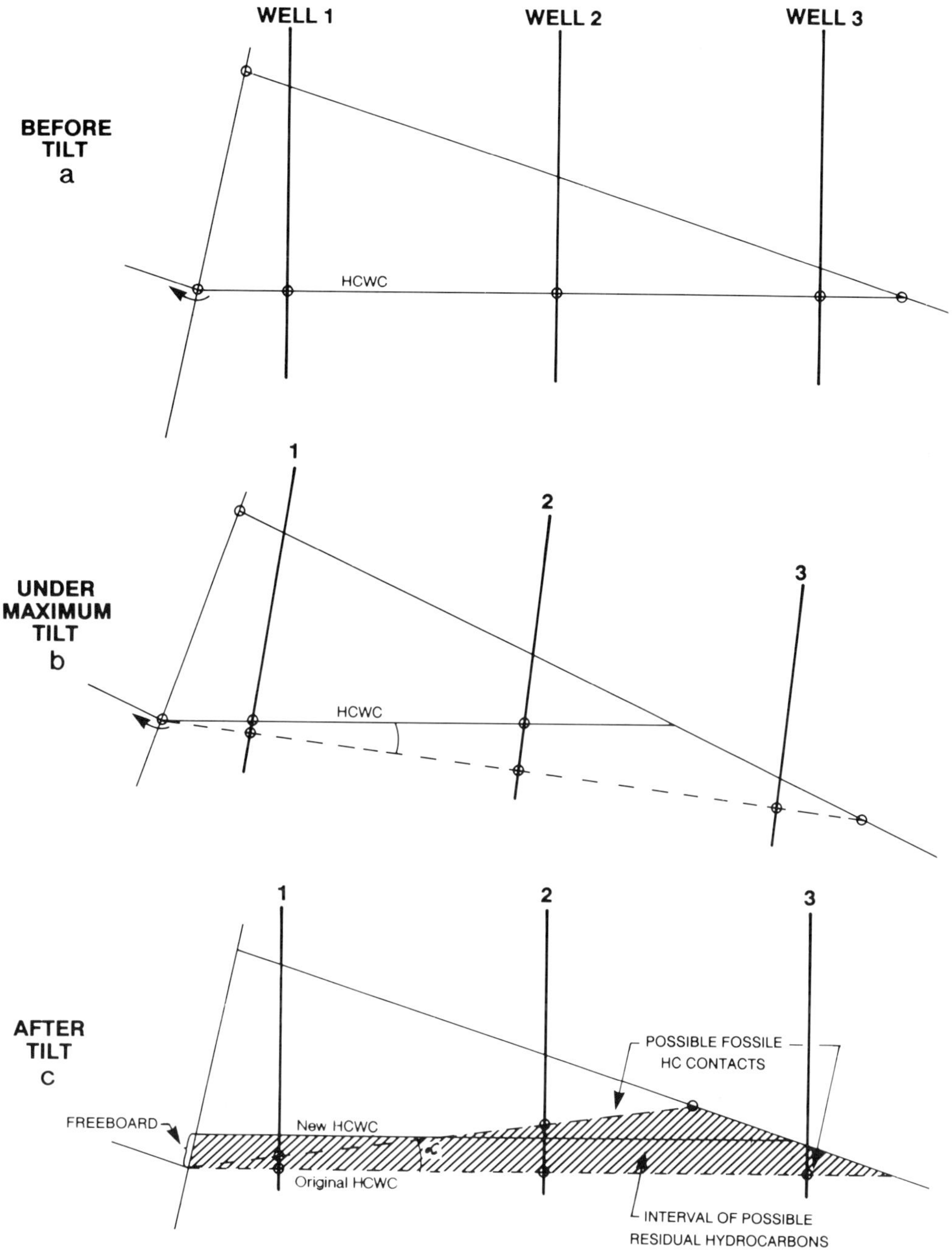

Fig. 7. Geometric changes of schematic reservoir during one cycle of glacial isostatic deformation.

the north/west will get a reduced closure volume during glaciation. In the northern North Sea a spill point in the north and east will give reduced closure volume, while in the southern North Sea a spill point in the east and south will give the same effect.

– *The geometry of the receiving area.* The geometry of the receiving area can give several scenarios for the fluids migrating out of the trap. If a saddle area is opened, the fluids can migrate into a higher lying trap or be lost to the surface. Alternatively, a surrounding flat area can become a part of the trap during the downwarping and fluids will migrate back into the trap under uplift.

A hypothetical example from the Barents Sea

By using a realistic geological model for the last 0.3 million years we have calculated the changes in closure volume in a hypothetical field in the western Barents Sea. Three isostatic events are seen in this period of time, i.e., the Elsterian (event 6 in Fig. 1) the Saalian (event 4) and the Weichselian (event 2).

The field has the following characteristics:

Dimensions:	length, 30 km; width, 5 km
Orientation:	east–west
Spill point:	west
Initial freeboard:	0 m

TABLE 2

Glacial isostatic induced migration in a hypothetic oil/gas field in the western Barents Sea

Glacial event (ice model)	Vertical movement (maximum) (m)	Diff. vertical movement of spillpoint (m)	HC-column (m)	Freeboard (m)	Relative volume
Initial	0	–	100	0	1.0
During 6 (II)	−160	−15	–	0	1.0–0.9
After 6	+160	+15	92	7	0.9
During 4 (I)	−300	−40	–	0	0.9–0.7
After 4	+300	+40	80	20	0.7
During 2 (I)	−300	−40	–	0	0.7
Present	−250	30	85	15	0.7

The calculations are based on isostatic compensation for the two first events, and a remaining uplift of 50 m after the last.

Fluid:	gas and oil
Initial volume:	1
Initial HC column:	100 m
Field geometry:	bar triangular in section

The results of the calculations based on the isobase maps of ice model I (for event 2 and 4) and ice model II (for event 6) are given in Table 2.

During event 6 (Elsterian) the field is tilted to the south and east with gradients of 0.9 and 0.5 m/km, respectively (Table 1). This gives a differential vertical movement of 17 m and a loss in volume of approximately 10%.

During event 4 (Saalian) the change is more drastic. The gradients are 1.3 m/km towards the east and 0.4 m/km towards the south. The differential vertical movement will be 40 m and the total loss of volume will now add up to nearly 30% compared to the initial value. The field will reach a maximum downwarping of 300 m during this event.

During event 2 (Weichselian) there will not be any dramatic effect because a high freeboard is established after event 4, and although the isostatic effect is of similar magnitude during event 2 the implications for the volume are insignificant.

The present situation is not in isostatic equilibrium and a remaining uplift of some 50 m will give 5 m remaining differential uplift at the spill point.

The effects of pressure differences and fluid equilibrium are not taken into account, but are believed to be of minor importance.

This exercise, although hypothetical, is realistic and shows that up to 30% of the initial bulk volume can be affected during the glacial isostatic downwarping. In a more or less homogeneous reservoir this would mean that nearly 30% of the hydrocarbon volume would be lost. In fields where there is an oil leg this lost volume would mainly be oil. The lost volume is controlled by field geometry to a large extent and the 30% is only related to the example given.

Summary and conclusions

Northern Europe including the continental shelf areas has been covered by extensive ice sheets at least once and perhaps several times during the last 2–3 million years

Three ice sheet models have been established based on stratigraphic and geomorphologic data from the North European mainland and shelf areas.

The most extensive ice sheet model covers both the entire North Sea and the Barents Sea with ice thicknesses in excess of 1500 and 2000 m, respectively. The least extensive ice sheet model covers only parts of the shelfs, except for in the Haltenbanken area where the shelf edge is the ice limit. All models fulfil the onland data of glacial uplift after the last glaciation.

The isostatic response to the ice load is modelled by a flat earth model and the Fourier transform technique. Isobase maps for the ice sheet models have been calculated. In the shelf areas the largest isostatic effect, more than 600 m, is found in the eastern Barents Sea when the most extensive ice model is used. The highest deflection gradients are seen in the western Barents Sea reaching 1.3 m/km.

A theory of glacial isostatic induced hydrocarbon migration is established and a hypothetical hydrocarbon accumulation in the western Barents Sea is being modelled using a realistic geological history for the last 0.3 million years The results show that as much as 30% of the closure volume can be lost during the most extensive glacial events.

Acknowledgements

This study was supported by funding from the Norwegian Council for Technology (NTNF).

References

Airy, B.G., 1855. On the computation of the effect of the attraction of mountain masses disturbing the apparent astronomical latitude of stations in geodetic surveys. Philos. Trans. R. Soc. London, 145: 101–104.

Andersen, B.G., 1979. The deglaciation of Norway 15,000–10,000 B.P. Boreas, 8: 79–87.

Andersen, B.G., 1981. Late Weichselian ice sheets in Eurasia and Greenland. In: G.H. Denton, and T. Hughes (Editors), The Last Great Ice Sheets. John Wiley and Sons, New York, N.Y., pp. 1–65.

Berggren, W.A. 1972. Late Pliocene–Pleistocene glaciation. In: A.S. Laughton, W.A. Berggren et al. (Editors), Initial Reports of the Deep Sea Drilling Project, Vol. 12. U.S. Government Printing Office, Washington, D.C., pp. 953–963.

Berggren, W.A. and Van Couvering, J.A., 1974. The Late Neogene: development in paleontology and stratigraphy 2. Elsevier, Amsterdam, 216 pp.

Boulton, G.S., 1979. Glacial history of the Spitsbergen archipelago and the problem of a Barents shelf ice sheet. Boreas, 8: 31–57.

Cathles, L.M., 1975. The Viscosity of the Earth's Mantle. Princeton University Press, Princeton, N.J., 386 pp.

Clark, D.L., Whitman, R.R., Morgan, K.A. and Scudder, D.M. 1980. Stratigraphy and glacial-marine sediments of the Amarasian Basin, central Arctic Ocean. Geol. Soc. Am. Spec. Paper, 181: 1–57.

De Geer, G., 1888. Om Skandinaviens nivåforandringer under Quartarperioden. Geol. Foren. Stockh. Forh., 10: 366–379.

De Geer, G., 1890. Om Skandinaviens nivåforandringer under Quartarperioden. Geol. Foren. Stockh. Forh., 12: 61–110.

Dutton, C.E., 1889. On some of the greater problems of physical geology. Proc. Am. Philos. Soc. Washington, 11: 51–64; 536–537.

Elverhøi, A. and Solheim, A., 1984. The physical environment, Western Barents Sea 1:1 500 000, sheet B; surface sediment distribution. Nor. Polarinst. Skr., 179B: 1–23.

Elverhøi, A. and Solheim, A., 1987. Shallow geology and geophysics of the Barents Sea. Nor. Polarinst. Rapportser., 37: 1–71.

Fairbanks, R.G. and Matthews, R.K., 1978. The marine oxygen isotopicrecord in Pleistocene coral, Barbados, West Indies. Quaternary Res., 10: 181–196.

Fjeldskaar, W. and Cathles, L.M., 1991a. Rheology of mantle and lithosphere inferred from post-glacial uplift in Fennoscandia. In: R. Sabadini et al. (Editors), Glacial Isostasy, Sea-Level and Mantle Rheology. Kluwer Academic Publishers, Dordrecht, pp. 1–19.

Fjeldskaar, W. and Cathles, L.M., 1991b. The present rate of uplift of Fennoscandian implies a low-riscosity asthenosphere. Terra Nova, 3: 393–400.

Grantz, A., Johnson, G.L. and Sweeney, J.F. 1982. The Arctic Region. In: A.R. Palmer (Editor), Perspectives in Regional Geological Synthesis, Planning for the Geology of North America. D-NAG Spec. Publ., 1: 105–115.

Grosswald, M.G., 1980. Late Weichselian Ice Sheet of Northern Eurasia. Quaternary Res., 13: 1–32.

Hughes, T.J., 1981. Numerical reconstruction of Paleo-Ice Sheets. In: G.H. Denton and T.J. Hughes (Editors), The Great Ice Sheets. John Wiley, New York, N.Y., pp. 222–274.

Jamieson, T.F., 1865. On the history of the last geological changes in Scotland. Q. Geol. Soc. London, 1, 21: 161–203.

Jansen, E., Bleil, U., Henrich, R., Kringstad, L. and Slettemark, B., 1988. Paleoenvironmental changes in the Norwegian Sea and the Northeast Atlantic during the last 2.8 m.y. Deep Sea Drilling Project/Ocean Drilling Program sites 610, 642, 643 and 644. Paleoceanography, 3: 563–581.

Kjemperud, A., 1986. Shore line displacement in Trøndelag, Central Norway. Boreas, 15: 61–82.

Laughton, A.S., Berggren, W.A. et al. 1970. Deep Sea Drilling Project, Leg 12. Geotimes, 15(9): 10–14.

Nansen, F., 1922. The strandflat and isostasy. Nor. Vidensk. Akad. Oslo, Mat. Nat., 1927, 11: 1–350.

Nansen, F. 1927. The earth's crust, its surface forms and isostatic adjustments. Nor. Vidensk. Akad. Oslo Mat. Nat., 1928, 12: 1–121.

Nesje, A., Dahl, S.O., Anda, E. and Rye, N. 1988. Block fields in southern Norway: significance for the Late Weichselian ice sheet. Nor. Geol. Tidsskr., 68: 149–169.

Pratt, J.H., 1855: On the attraction of the Himalayan mountains and of elevated regions upon the plumb line in India. Philos. Trans. R. Soc. London, 145: 53–100.

Ramsey, W., 1924. On relation between crustal movements and variations of sea-level during the late Quaternary time, especially in Fennoscandia, Fennia, 44: 1–39.

Sejrup, H.P., Aarseth, I., Ellingsen, K.L., Reither, E. and Jansen, E., 1987. Quaternary stratigraphy of the Fladen area, central North Sea: a multidisciplinary study. J. Quaternary Sci., 2: 35–38.

Shackleton, N.J. and Opdyke, N.D., 1976. Oxygen isotope and paleomagnetic stratigraphy of Pacific core V28-239, late Pliocene to latest Pleistocene. Geol. Soc. Am. Mem., 145: 449–464.

Shackleton, N.J. and Opdyke, N.D., 1977. Oxygen isotope and paleomagnetic evidence for early northern hemisphere glaciation. Nature, 270: 216–219.

Solheim, A. and Kristoffersen, Y., 1984. The physical environment, Western Barents Sea, 1 : 1 500 000, sheet B; Sediments above the upper regional unconformity: thickness, seismic stratigraphy and outline of the glacial history. Nor. Polarinst. Skr., 179 B: 1–26.

Sutherland, D. 1984. The Quaternary deposits and landforms of Scotland and the neighbouring shelves: a review. Quaternary Sci. Rev., 3: 157–254.

Vorren, T. and Kristoffersen, Y. 1986. Late Quaternary glaciation in the south-western Barents Sea. Boreas, 15: 51–59.

Vorren, T. Lebesbye, E., Andreassen, K. and Larsen, K.-B., 1989. Glacigenic sediments on a passive continental margin as exemplified by the Barents Sea. Mar. Geol., 85: 251–272.

Williams, D.F., 1988. Evidence for and against sea-level changes from the stable isotopic record of the cenozoic. In: A. Wilgus et al. (Editors), Sea-Level Changes — An Integrated Approach. Soc. Econ. Paleontol. Mineral., Spec. Publ., 42: 31-36.

W. FJELDSKAAR *Rogaland Research Institute, P.O. Box 2503, Ullandhaug, N-4004 Stavanger, Norway*

A. KJEMPERUD *READ Production Geology Services A/S, P.O. Box 145, N-1300 Sandvika, Norway*

Present address: *PETEC, P.O. Box 63, N-3478 Nærnes, Norway*

Structural style and the habitat of hydrocarbons in the North Sea

I.J. Stewart, R.P. Rattey and I.R. Vann

The structural development of the North Sea Basin is described. While economic interest has largely focussed on structuring at Jurassic and younger levels, the structural framework was inherited from earlier times. The northern North Sea is separated from the central North Sea by a postulated extension of the Tornquist Line, south of which Early Permian Variscan relaxation with associated subsidence accommodated the Late Permian salt basin. The northern North Sea is typified by relaxed Caledonian structure, extended in Devonian and Triassic times. The two structural provinces exhibit significantly different structural styles at Jurassic levels.

In the northern North Sea the earliest significant Jurassic extension is of Bathonian to Early Callovian age and is succeeded by a major phase in the Oxfordian. Patterns of deposition and the amount of footwall erosion can be related to the fault pattern which in general has a broad structural wavelength. A further major faulting episode of latest Jurassic and earliest Cretaceous age is related to Atlantic rifting and is only observed on a significant scale in the extreme north where it has played a fundamental role in trap formation in the Magnus area. In the central North Sea, where the major period of Jurassic extension is Oxfordian to Late Kimmeridgian, the Permian evaporites have had a considerable influence in the control of cover structure. Extension in the basement largely utilised planar Early Permian fault surfaces, cut up-section, and was accommodated in the cover by shallow detaching faults and short wavelength structures. The Early Cretaceous extension of the northernmost North Sea is locally apparent in this area and has accentuated the earlier geometries.

Both areas have suffered post-rift structural modification through episodic Cretaceous and Tertiary inversion, associated salt diapirism, and gravitational sliding. The significant effects of inversion and diapirism are restricted to the central North Sea where they have resulted in trapping geometries in both the rift-fill and the post-rift sections.

Introduction

The North Sea is a major hydrocarbon province that is now entering a mature phase of exploration. Continued technical, and more particularly commercial, success will increasingly depend on our ability to correctly interpret structural geometries and also to recognise the structural controls on sediment entry and dispersal. With this in mind, we have constructed this paper partly as a review and partly as a series of constrained, largely regional observations, and in keeping with this conference volume we have given the paper a slight Norwegian bias. Individual hydrocarbon traps are not described in detail, but rather placed in a more regional structural context.

The paper is divided into four parts, centering on the Jurassic extensional history. The first section outlines the pre-Jurassic structural grain of the central and northern North Sea basins. Secondly, the influence of this grain on Jurassic structuring is described by means of a few simple regional backstripped sections and maps. Thirdly, the structural controls on sediment entry and dispersal are discussed with reference to specific examples, while the final section covers post-Jurassic structural modification.

Figure 1 is a simple regional structure map of the North Sea and adjacent areas. The area described here is restricted to the central and northern North Seas, that is the area to the north of the Mid North Sea High, and to the south of the Møre Basin. The present-day geometry of the basin is shown on three simple regional sections in Fig. 2 and the location of these traverses is given in Fig. 3. The sections are necessarily squashed for display purposes.

Structural grain

It is becoming increasingly clear that many extensional sedimentary basins form on relatively low-angle inherited crustal surfaces (e.g. McGeary et al., 1987), and that the structures within the basin form part of a linked system (Gibbs, 1984). However, most basins cannot be described this simply; various models exist for fault plane geometry and the mechanism and depth of detachment (see for instance Coward et al., 1987; Brun et al., 1989;

Structural and Tectonic Modelling and its Application to Petroleum Geology edited by R.M. Larsen, H. Brekke, B.T. Larsen and E. Talleraas. NPF Special Publication 1, pp. 197–220. Elsevier, Amsterdam.

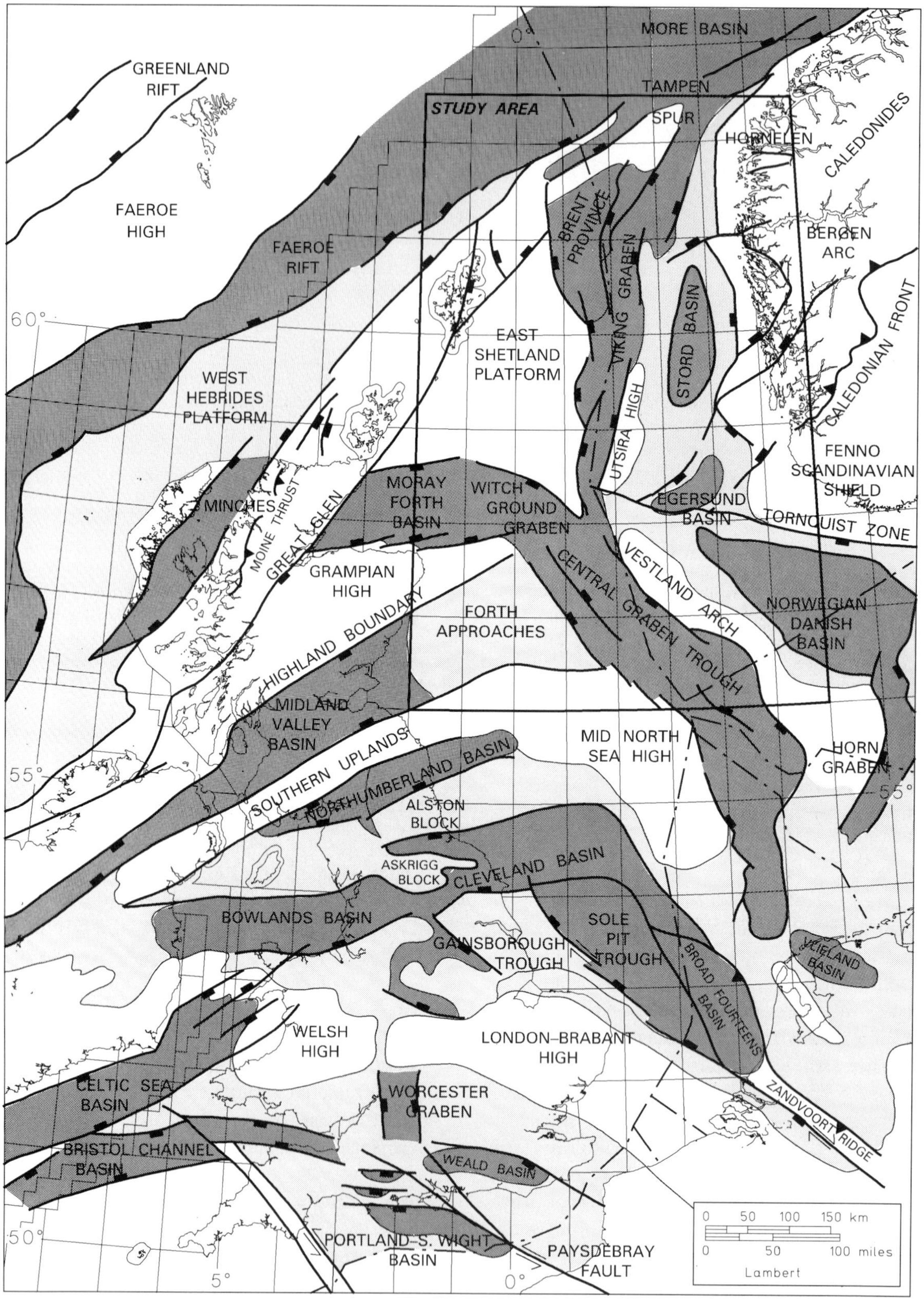

Fig. 1. Simple tectonic elements map of the North Sea area. This paper concentrates on the area north of the Mid North Sea High and south of the Møre Basin.

Marsden et al., 1989; Newall and Kusznir, 1989; and Yielding and Roberts, this volume), and more focussed interpretations are showing the interaction of both dip-slip and strike-slip processes in basin development (e.g., Gibbs, 1987a).

In North Sea hydrocarbon exploration, the seismic data used in structural interpretation are commonly restricted to a 6 or 7 second two-way time window,

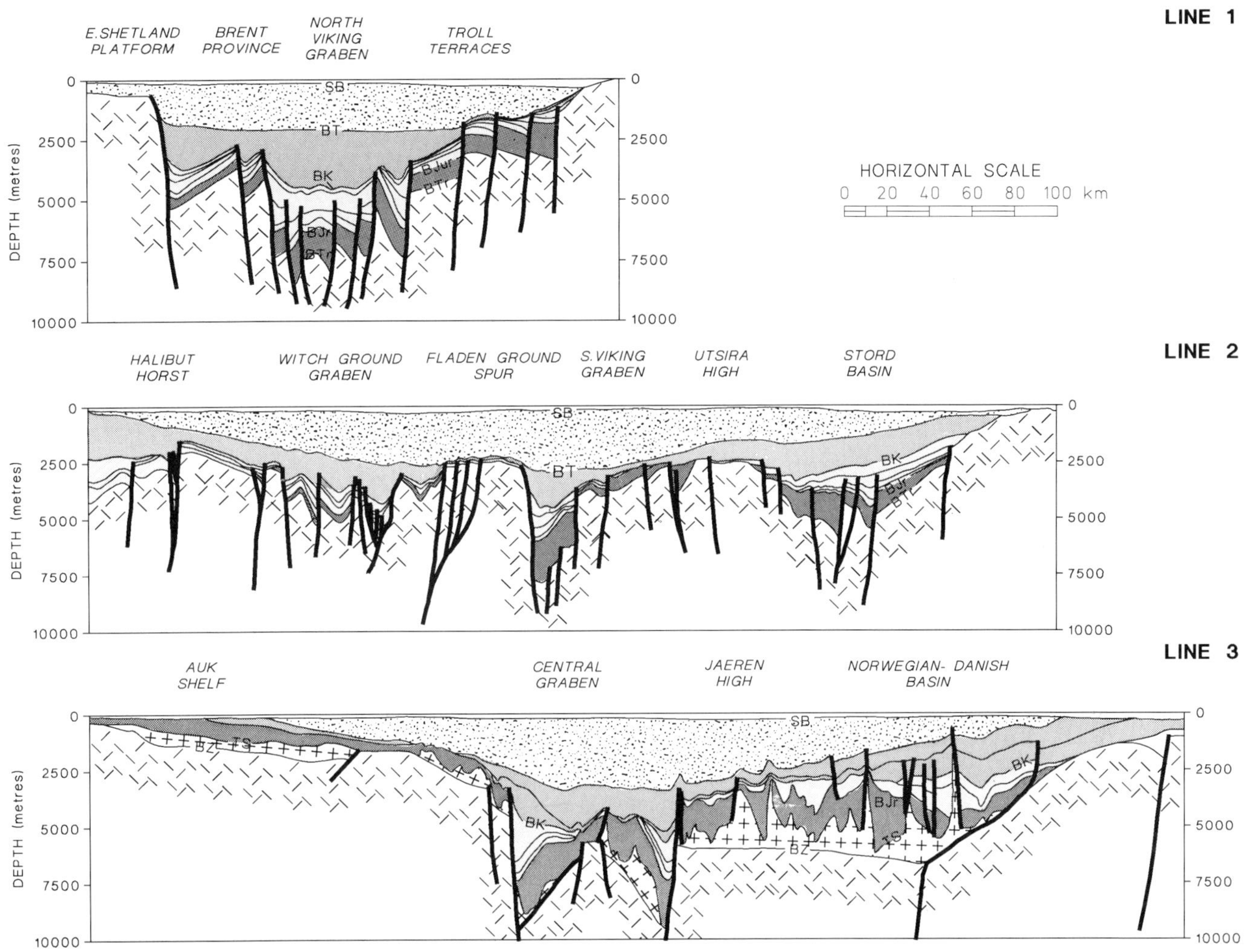

Fig. 2. Present day simple structural cross sections. Locations shown on Fig. 3. The sections are squashed for display purposes; vertical exaggeration ×10. Dashed areas show Early Permian and older rock. *BZ* = Base Zechstein; *TS* = top salt; *BTr* = Base Triassic; *BJr* = Base Jurassic; *BK* = Base Cretaceous; *BT* = Base Tertiary. Unlabelled lines are other sequence boundaries.

and the fundamental inherited crustal control on the basin location and its geometry are rarely observed. Identification of the control then has to be based on the few deep seismic reflection lines over the North Sea and adjacent areas, extrapolation from the shallower, seismically defined, geometries and the regional history of the basin and its exposed margins.

The basic structural grain of the North Sea was established through the Caledonian orogen. In the Scandinavian Caledonides the thrusts verge to the south-east, whilst the Scottish Caledonides verge to the north-west (Fig. 1). The reversal is interpreted to occur beneath the Midland Valley (Needham and Knipe, 1986), although the position offshore in the North Sea is unclear. While it is beyond the scope of this paper to describe in detail the Caledonian grain (see for instance Gee and Sturt, 1985; Johnson and Dingwall, 1981), major oblique-slip motion along a north-east strike is indicated by lateral offsets of both the Caledonian Front and internal terranes (e.g., Bergen Arc, lateral margins of Hornelen and related Devonian basins north of Bergen) in Norway, and the Great Glen — Highland Boundary Fault — in Britain (Fig. 1).

The eastward verging Caledonian Front of Norway is terminated to the south by the Fjerritslev Fault Zone, an interpreted extension of the Tornquist Zone (Pegrum, 1984). Observations of cover geometry clearly support the extension of the Tornquist zone as a crustal feature beyond the Oslo Graben (see also Sørensen et al., this volume) to the southern part of the South Viking Graben, and at least through to an intersection with the offshore extension of the Highland Fault (Doré and Gage, 1987).

In the northern North Sea, deep seismic profiling has been interpreted to indicate that the crustal faults dip eastwards and flatten towards Norway on a deep crustal or top Moho detachment (Beach, 1985; Beach, 1986; Gibbs, 1987b) or were accommodated by pure shear (White, 1989), although the interpre-

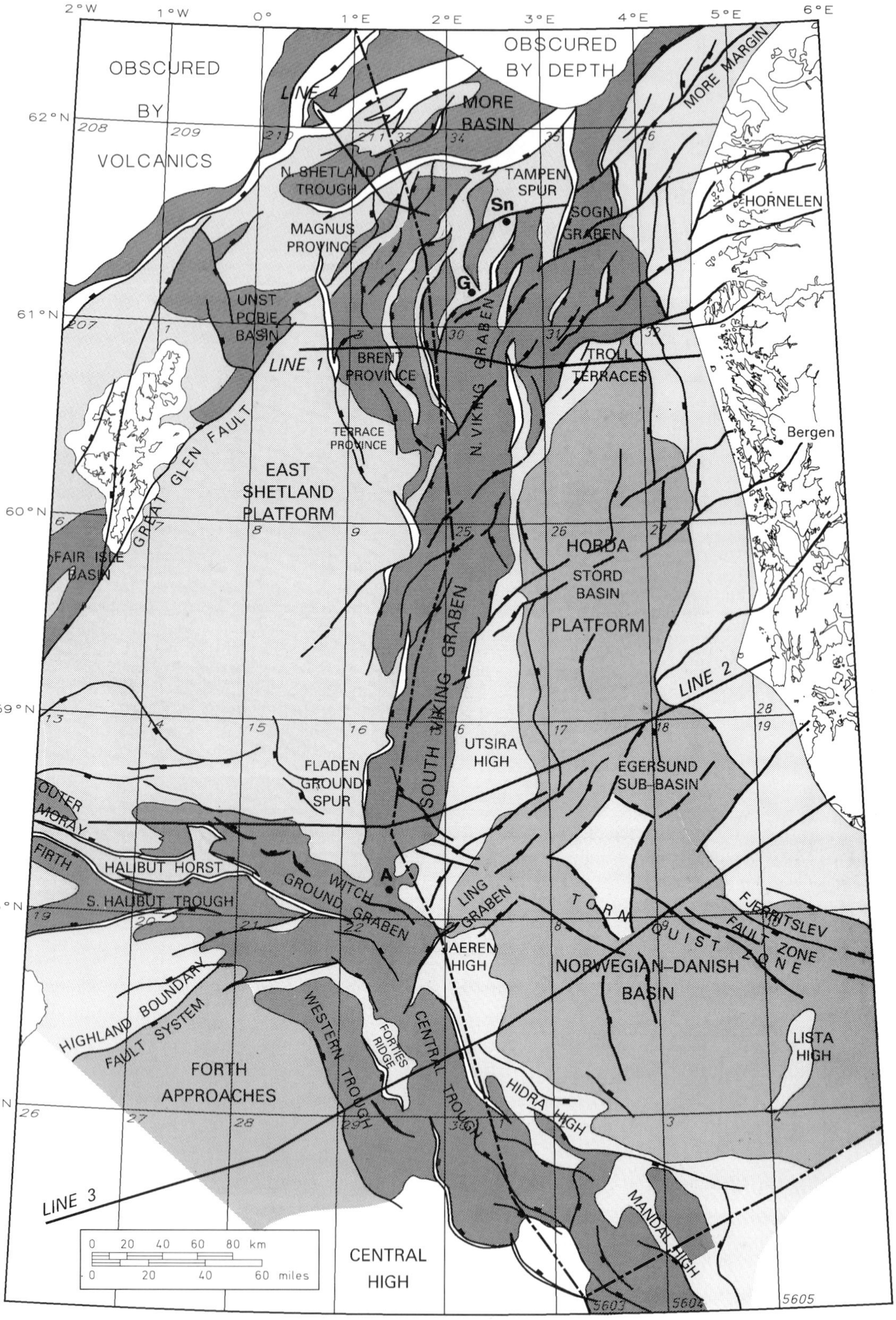

Fig. 3. North Sea structural elements. *G* = Gullfaks Field; *A* = Andrew Field; *Sn* = Snorre Field.

tation of Beach (1985) of similar easterly dipping geometries beneath the Central Graben is more speculative. The argument that the Viking Graben crustal faults are relaxed Caledonian contractional structures, however, is consistent with the observations elsewhere of Brewer and Smythe (1984) in the MOIST data, Cheadle et al. (1987) and Snyder (1990) in the BIRPS data (see also the review by McGeary et al., 1987) and the West Orkney interpretations of Enfield and Coward (1987); the primary

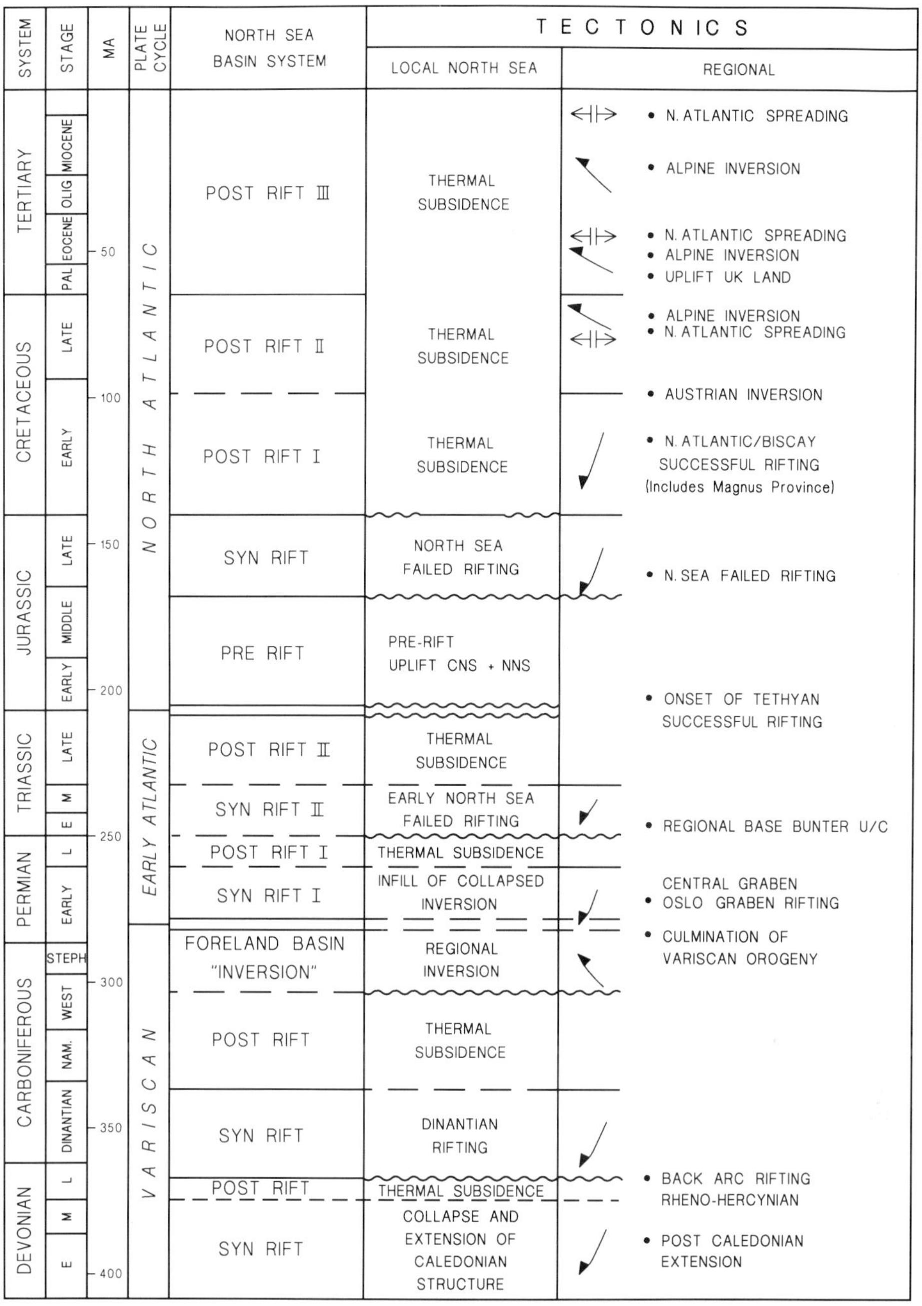

Fig. 4. North Sea. Principal post-Caledonian basin forming processes.

control of the extensional basins to the north and west of Scotland is the relaxation of Caledonian faults. The Middle Devonian age of extension identified by these authors may then be anticipated to be represented in the deep fill of North Viking Graben half-grabens (Fig. 4).

On the Norwegian side, Steel et al. (1985) have summarised the structural models for the Hornelen area basins north of Bergen (Fig. 3) that accommodate Lower to Middle Devonian sediment. The geometries here combine both shallow westerly dipping detachments inherited from a high level in the Caledonian pile (e.g., Hossack, 1984) and north-easterly trending oblique-slip faults; Jurassic fault blocks seen on offshore seismic data, however, imply the major relaxed Caledonian structures continue to dip east and the facing reversal of Caledonian structure in this area apparently runs close to the present-day Norwegian coastline. The pattern is complicated by several changes in younger offshore half-graben polarity across the north-east striking faults.

Early post-Caledonian extension, however, is less well defined in the North Sea area south of the eastward projection of the Tornquist Zone. There are deep fault blocks seen locally on seismic data in the Ling Graben, Egersund sub-Basin, Norwegian

Danish Basin and the Mid North Sea High area, and Devonian and Carboniferous extension is recognised onshore Britain (Fig. 1) from the Midland Valley southwards on relaxed Caledonian trends (Fraser et al., 1990). It should thus be more widely expected in the central North Sea area.

As well as a major Caledonian lineament, the Tornquist Zone also played a major role in post-Caledonian extension and in the subsequent Variscan phase. Here it acted as a dextral strike-slip system, and in the North Sea area as a limit to the Variscan foreland (Pegrum, 1984). The North Sea basin lies north of the "Variscan Front" which lies beneath southern England, although Late Carboniferous inversion (Fig. 4) can be traced through the onshore U.K. basins (Bowland Basin, Northumberland Trough) and into the Midland Valley.

Early Permian extension is well represented in the central North Sea and the Moray Firth, and may be represented in the southern Stord Basin, although for the major part, the extension is limited to the south of the projection of the Tornquist Zone and its intersection with the Highland Boundary Fault Zone (Doré and Gage, 1987). It may be that the extension was limited to the relaxation of Variscan faults, probably on deep through-crustal structures, but nonetheless it has fundamental consequences for the subsequent structural history. The Late Permian (Zechstein) marine evaporites of the north Permian sag basin (Ziegler, 1982) are effectively restricted to the central North Sea area south of the Highland Boundary Fault and the Tornquist Zone intersection (Figs. 1 and 3), and only locally extend northwards through the Ling Graben into the Egersund sub-Basin and the southern South Viking Graben. The Forth Approaches Basin (Fig. 1) at Permian levels is a simple, more or less saucer-shaped depression with few faults, although in both the Central Trough area and the Norwegian–Danish Basin, extensional faults struck west north-west to east south-east and were accommodated by north north-west to south south-east strike-slip surfaces (e.g., Vejbæk and Andersen, 1987; Brown et al., this volume).

Triassic extension is recognised in both the northern and central North Sea areas. The Inner Moray Firth area extended and subsided in an arc eastwards and southwards through the Witch Ground Graben, accommodated by dextral strike-slip on the Great Glen Fault. This extension can also be interpreted to be linked to relaxed Variscan structure with increased subsidence and localised faulting in the Forth Approaches/Central Trough/Norwegian–Danish Basin. This is in contrast to both easterly and westerly directed extension in the Stord Basin (Fig. 3) on detachments that flipped across north-east striking oblique-slip transfer faults, and in the Viking Graben to the east on the easterly dipping relaxed Caledonian trend; the East Shetland Platform remained fixed. Accommodation between these areas may have been achieved by transtensional motion along the major Caledonian oblique-slip trends (such as the Ling Graben and the southern Stord Basin) and the Tornquist Zone as described by Pegrum (1984).

The North Sea as a whole can thus be viewed as having a long extensional history, interrupted by Variscan compression, from Devonian until Early Jurassic times (Fig. 4). The structural grain was established from Caledonian relaxation in the northern North Sea, and by Caledonian and Variscan relaxation in the central North Sea. It is clear that the overall pattern of the subsequent Jurassic extensional system was in place by Late Triassic times.

Jurassic extension

The Jurassic extension of the North Sea was the primary structural control on the majority of known hydrocarbon traps. It has been studied extensively, but in detail it is still a subject of much debate.

The first problem is the scale and extent of the Bajocian (pre-Brent) uplift. A central North Sea domal uplift of this age, centering at the junction of the Central Trough and the South Viking Graben, has long been advocated by Ziegler (e.g., 1975, 1982) to explain the poor preservation of Liassic strata in the central North Sea, the supposed northward progradation of the Brent system, and the progressive onlap of Callovian and younger Jurassic strata onto the "mid-Cimmerian" (rift-onset) unconformity. The onlap would be expected under most current rift models and in itself is not evidence of thermal uplift. On the first point though, stratigraphic evidence of a major Pliensbachian to Early Toarcian sedimentary break in the drilled Liassic sections of the central North Sea suggests a major base-level fall corresponding to the period of Cook Formation deposition on the northern shelf, and that much, if not all, of the missing section may have been removed by sea withdrawal in Late Liassic times. By the Bajocian (Figs. 5 and 6) coastal plain sedimentation was established over much of the central North Sea and, thus, a persistent local uplift seems unlikely. Secondly, the Brent Group represents a 300 km fluvial-deltaic system (Fig. 5) prograding roughly northwards over the site of the Viking Graben; it is difficult to see a single uplift as the principal sediment source and recent work (e.g., Richards et al., 1988) supports the idea that sediment entered the rift basin transversely from regionally elevated margins, and was worked axially.

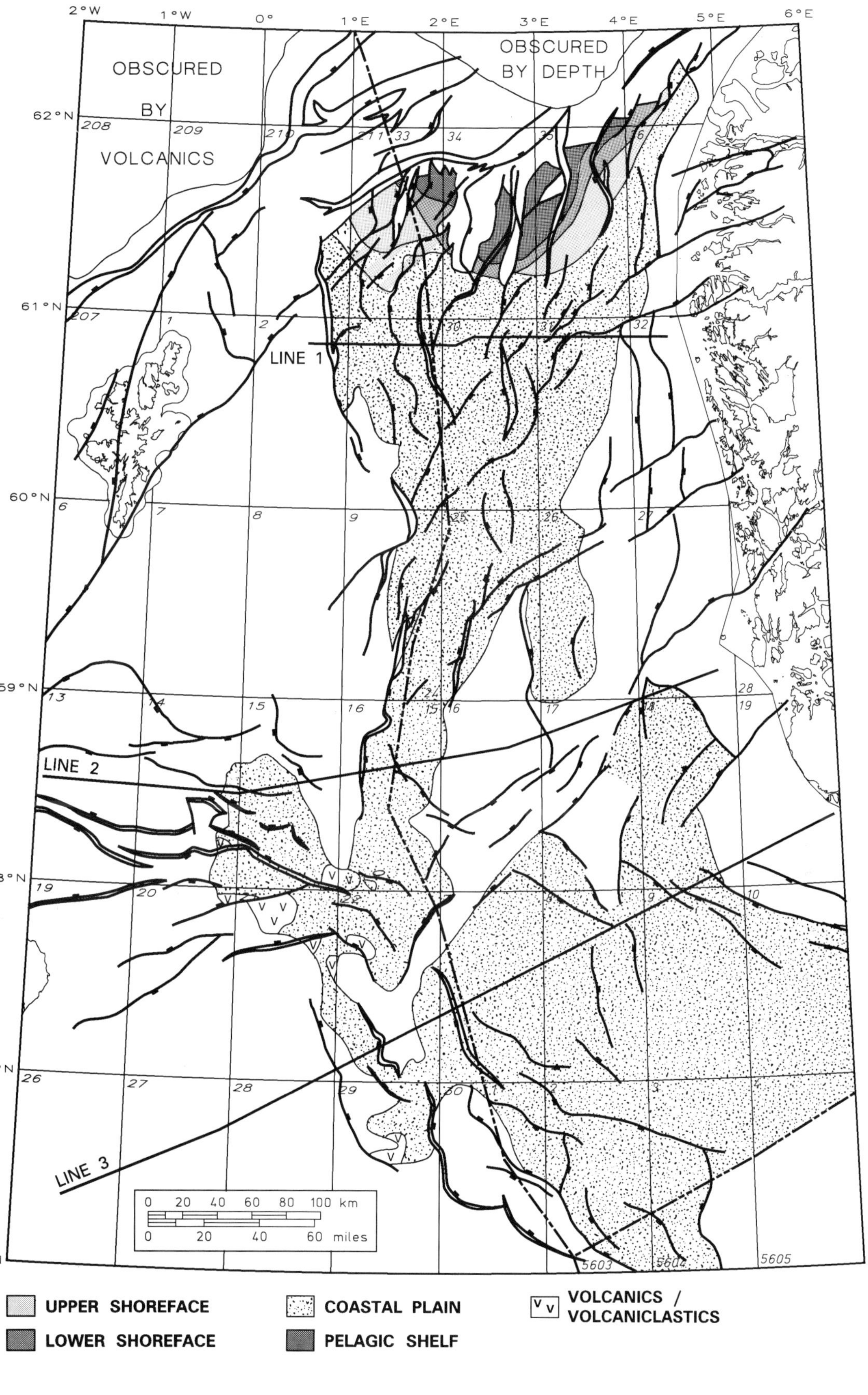

Fig. 5. Bajocian palaeogeography.

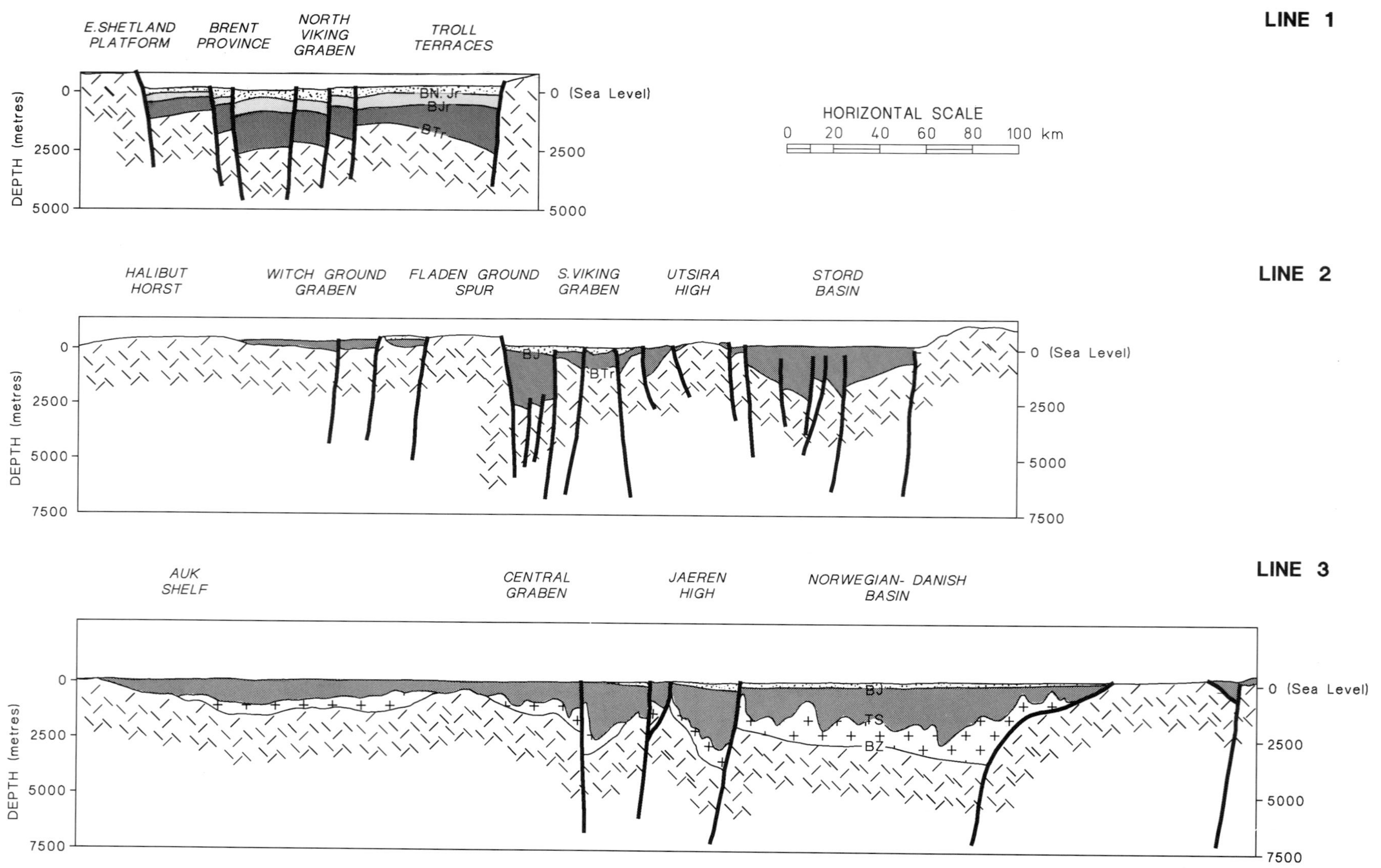

Fig. 6. Simple cross sections of Fig. 2 restored and decompacted to Late Oxfordian. Vertical exaggeration ×10. Labelling as in Fig. 2. Unlabelled horizons are sequence boundaries.

A second problem is the major change in structural style from the northern to central North Sea and the nature of the so-called triple junction. In both areas, the Jurassic extension utilised the existing structural grain, although clearly there are fundamental differences in the structural style. The northern North Sea, particularly the north Viking Graben, is largely characterised by long wavelength (ca. 5–20 km +) tilted fault blocks on easterly dipping, often gently concave-up synthetic faults (the relaxed Caledonian) which have a broad north-south strike (Fig. 3). Onto these link an antithetic westerly dipping set over a large part of the Norwegian sector (Beach, 1986; Gibbs, 1987b). These fault blocks typically contain the pre-rift (pre-Tarbert) Brent Group and older strata and provide a prolific and well described hydrocarbon play (e.g., Campbell and Ormaasen, 1987; Badley et al., 1984).

In the central North Sea, apart from the Moray Firth and Witch Ground graben areas, the structural pattern is complicated by the presence of the Late Permian salt. Jurassic extension in the Central Trough area utilised the Early Permian extensional system (the relaxed Variscan); these apparently planar faults seen on conventional seismic data, cut-up section and were locally controlled by the salt wall and Triassic sediment pod geometries. The result is a series of short wavelength commonly complex structures governed by irregular fault plane and salt wall geometry, superimposed on a longer wavelength structural pattern formed by the underlying pre-Zechstein fault blocks. However, whilst the presence of salt and the associated complexity accounts for the generally smaller Jurassic structural hydrocarbon trap size compared to the Viking Graben, it does not on its own explain the more fundamental observations.

The structural elements map (Fig. 3) shows two clear trends for the north Viking Graben area (see also Halstead, 1975; and Johnson and Dingwall, 1981). The sense of dip-slip extension is more or less east-west, although the extensional system is repeatedly offset by inherited north-east to south-west transfer or strike-slip faults. The interaction of these two trends has created a mosaic of rhomb-shaped elements (a classic example is the Gullfaks Field, Erichsen et al., 1987) which together are bound to the north-west by the Great Glen and related fault systems, and to the south-east by a series of transfer faults related and parallel to the Hitra and Verran fault zone (Møre margin faults) of Mid-Norway, and the major Caledonian oblique-slip trends to the north of the Troll Field area (Badley et al., 1988) and onshore Hornelen. We believe that the map observations strongly support the deep seismic interpretations made by Beach et al. (1987), and the interpretation of Beach (1985) that the North Viking Graben is a major transtensional basin. The oblique-slip is accommodated along the South Viking Graben boundary fault zone and the western Utsira Hight by a series of transfer faults or fault zones with the inherited north-east to south-west strike, and one of these separates the Fladen Ground Spur from the East Shetland Platforms and forms the Bruce–Beryl embayment.

This structural style has to be offset against the Central Trough–Witch Ground Graben–Outer Moray Firth area, where extension was primarily dip-slip in an arc extending eastwards and south-eastwards from the Great Glen fault (Fig. 1). Transfer faults and zones are again prominent with a north-east to south-west trend and the polarity of the basins is switched locally across these. Within the Moray Firth, deep seismic data suggest that extension took place within a system above a fault that dips southwards under north-east Scotland. As the basin is traced south-eastwards towards the Witch Ground Graben the polarity apparently switches (e.g., Beach, 1984). The Central Trough can be similarly interpreted as part of a linked extensional system (e.g., Gibbs, 1984), above a detachment that continues beneath the Forth Approaches and the Central (Mid North Sea) High (Fig. 3).

The critical separation between the Viking Graben and the Central Trough is marked by a zone of complex structure that extends along the southern margin of the Utsira High (Pegrum, 1984) and beneath the Andrew Field in the southern part of UKCS Quadrant 16 (Fig. 3). This area represents the extension of the Tornquist Zone (Pegrum, 1984), which together with the Ling Graben faults and the oblique-slip trends of the southern Stord Basin, apparently accommodated the necessary rotation of Norway relative to Scotland.

The simple cross-sections are shown restored to Late Oxfordian times in Fig. 7, and a palaeogeographic map of this timeslice is given in Fig. 8. One observation is paramount; the Late Oxfordian of the northern North Sea was deposited in a complex, and in fact starved rift basin with great submarine relief, whereas in the central North Sea sedimentation kept up more or less with sea level. The ages of extension in the Northern North Sea were initially Bathonian to Early Callovian, which, coupled with a rise in sea level, saw the drowning of the Brent delta system ("Tarbert transgression"), and Middle Oxfordian. The latter was a period of great change with extremely rapid, probably near instantaneous, hanging-wall subsidence and footwall uplift; the mid-Oxfordian to Kimmeridgian strata show little

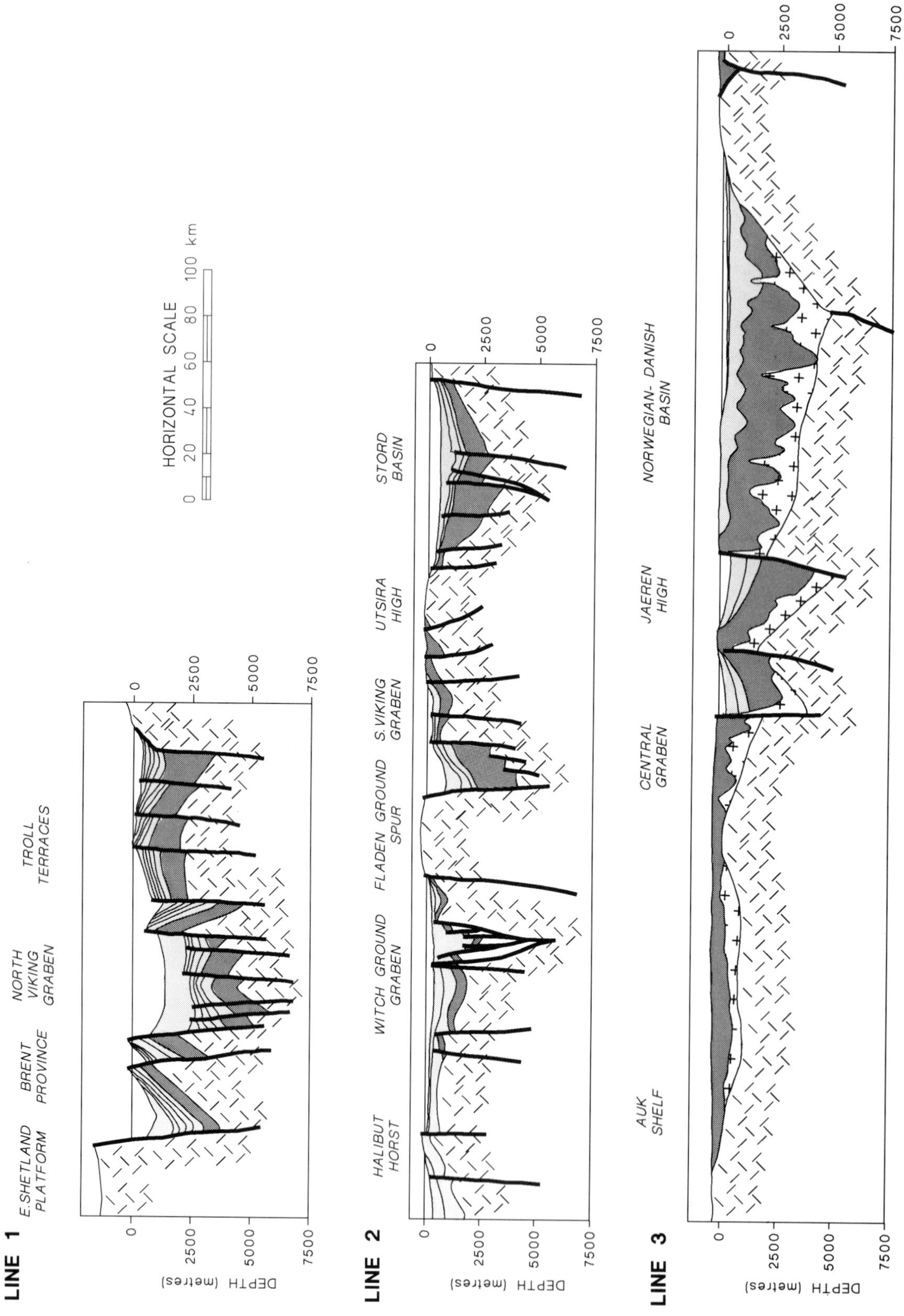

Fig. 7. Simple cross section of Fig. 2 restored and decompacted to Late Oxfordian. Vertical exaggeration ×10. Labelling as in Fig. 2.

Fig. 8. Late Oxfordian Palaeogeography.

rotational growth into the faults and much of the observed hanging-wall geometry seen on seismic data is a consequence of subsequent differential compaction which is removed when sections are restored and decompacted.

A Middle Oxfordian age of extension is similarly recognised in the Inner Moray Firth (Andrews and Brown, 1987), where a complex, relatively deep-water basin was established against the Great Glen Fault system, although eastwards into the Witch Ground Graben and Central Trough areas the basin shallowed and was dominated by shallow-marine clastics (commonly fault accommodated, e.g., Johnson et al., 1986). Minor deeper marine troughs developed over the eastern and western areas of the Central Trough (Fig. 8), although more regional deeper marine conditions as a consequence of extension were not established in the central North Sea area until Late Kimmeridgian to Early Portlandian times. This observation and the map highlight the major stratigraphic difference in the Jurassic hydrocarbon play types between the northern and central sectors; the pre-rift marginal to shallow-marine Jurassic sequences are the main target in the north, and the rift-fill shallow-marine Jurassic sequences provide the major play in central areas.

A further major fault episode of the North Sea is related to the attempted opening of the Atlantic in earliest Cretaceous times. The major effects of this are noted in the extreme north of the northern North Sea where east north-east striking faults displace the Oxfordian faults and their depositional systems (Fig. 9). The scale of faulting is much greater than in the North Sea failed rift and resulted in the development of the Møre Basin and the Shetland Trough, but again is believed to have developed almost instantaneously. The episode was accommodated to variable extents throughout the North Sea basin by dip-slip and oblique-slip motion on the existing fault surfaces.

Structural controls on reservoir distribution

The Jurassic extension of the North Sea provided a complex framework in which both deep-water and shallow-water rift-fill Jurassic sequences accumulated, and a wide array of play types in both the northern and central sectors is recognised. For the purposes of this paper only a few representative examples are described. In simple terms the plays can be separated by their structural position in the footwalls or hanging walls of extensional faults, or as in the Magnus example (Fig. 9), in areas of cross-cutting faults.

Footwall uplift characterises the basin margin and the majority of the fault blocks of the Viking Graben (see for instance Marsden et al., 1989). The crests of the blocks are typically eroded with varying amounts of pre-rift Jurassic and Triassic section removed, and there appears to be a simple relationship between the width of the individual fault blocks, the throw on the bounding faults and the amount of erosion. Where extension is taken up on a few faults, the fault throw and structural wavelength is increased, as is the footwall uplift. This is particularly apparent in the Snorre area (Fig. 3) of the Norwegian sector where Triassic rock in the footwall subcrops rocks of Early Cretaceous age.

The structural dip and the consequent drainage ensured that the majority of the erosion products could not enter the hanging-wall in front of the individual blocks, and these were transported down the dip slope to accumulate as narrow shelf systems normal to the dip. A cartoon through the Brent Province showing the geometry and the chronostratigraphy (Fig. 10) indicates that this process occurred subsequent to both the Bathonian to Early Callovian extension and the younger Middle Oxfordian extension, and a series of stacked retrogradational shelf systems can be anticipated, becoming narrower as the dip slope was rotated in the extension phases. The pattern would depend on the relative dip and the elevation/erosion level and the rate of subsidence/sediment supply in each of the blocks. The Statfjord Nord accumulation (Fig. 10) in the Munin sand is of this type (Gradijan and Wiik, 1987), where the lithofacies are interpreted as shelfal or storm-beach sands.

Deeper marine sandstones may also occupy a comparable structural position and are related to local (individual fault block) or more regional lowstand events, where the shelf system was eroded or bypassed and the erosional products were deposited on the dip slope or in the immediate hanging-wall area of the controlling fault. The Magnus Field (Fig. 9) is contained within submarine-fan sandstone (De'Ath and Schuyleman, 1981) and lies in the footwall to a cross-cutting Atlantic margin fault, yet on the dip slope of an Oxfordian North Sea fault (Fig. 11; Young, this volume). The field area itself has been elevated above sea level and partly eroded through footwall uplift related to the younger extensional event.

In the hanging-wall position, the pattern of footwall uplift and fault geometry governs the entry of coarse clastics. Only a small amount of footwall erosional products will enter the hanging-wall across the faults scarp, and these are typically poorly ordered apron-fans derived from the mass wastage of the

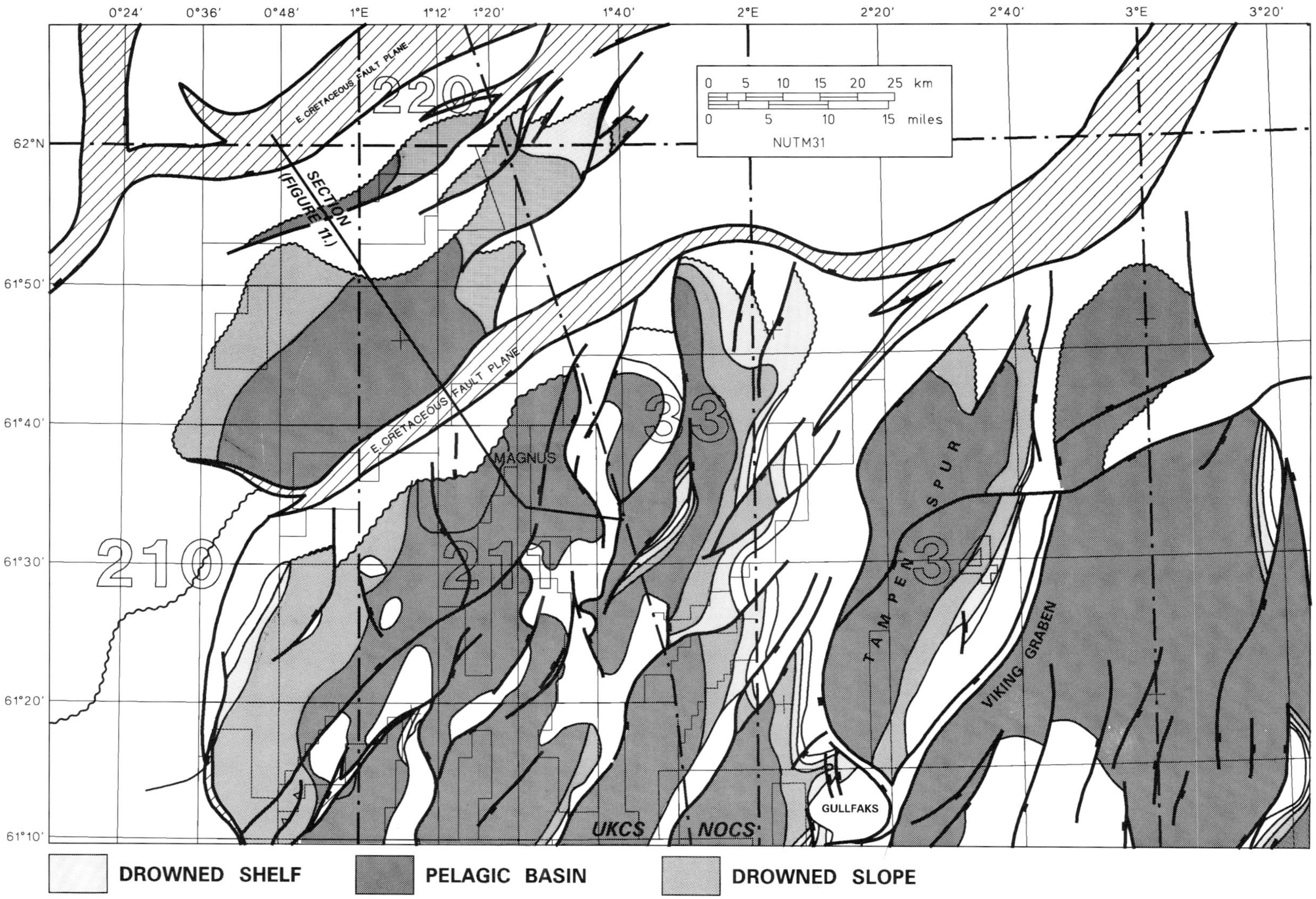

Fig. 9. Northern North Sea: Magnus area. the timeslice shown is Middle Kimmeridgian. The earliest Cretaceous Atlantic Margin faults displace the earlier extensional structure and the associated depositional systems.

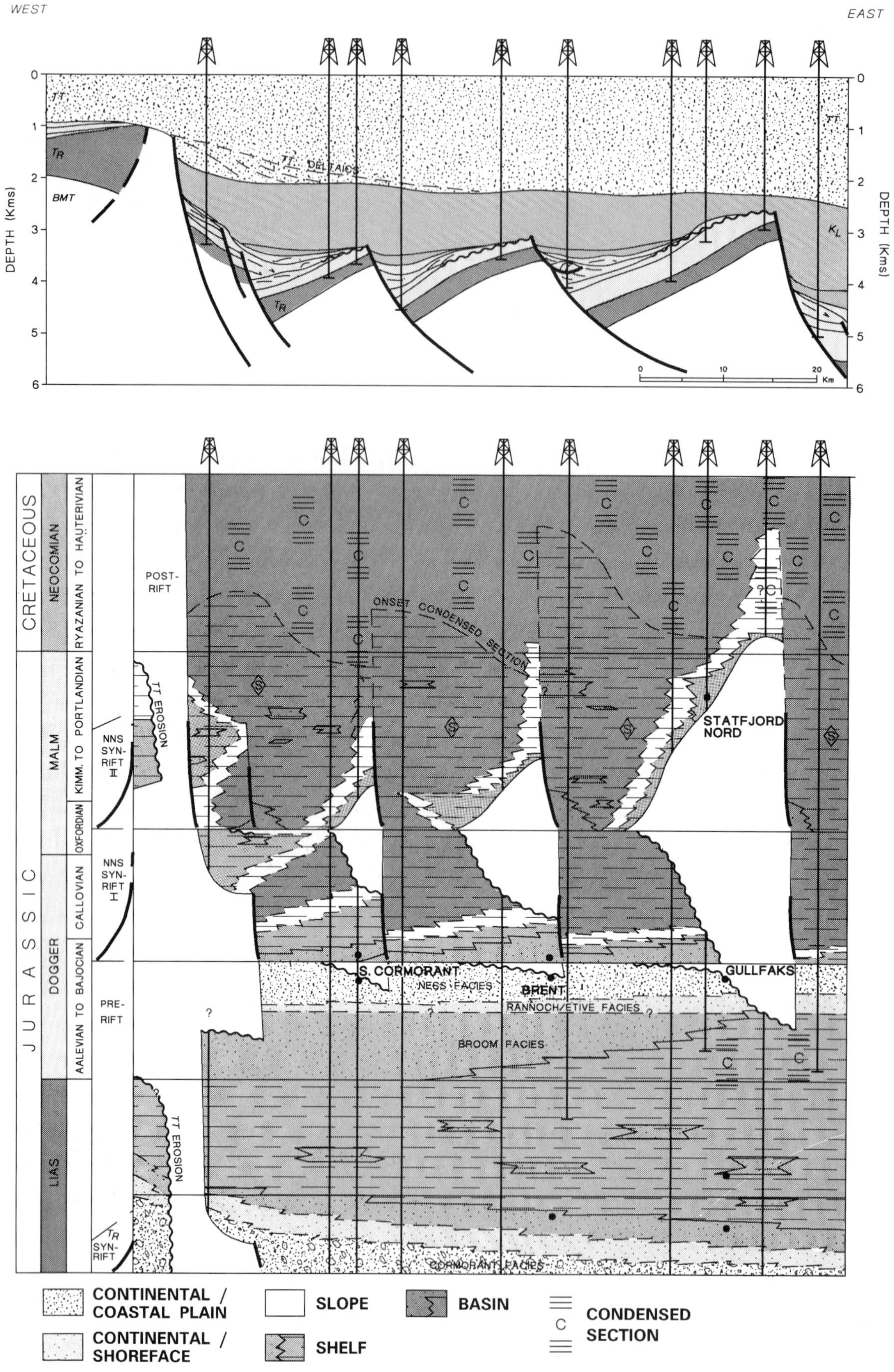

Fig. 10. Summary chronostratigraphy Brent Province/Tampen Spur. The section is a cartoon summarising the Jurassic and Early Cretaceous depositional sequences and their geometry. The lithofacies are shown on the chronostratigraphic diagram.

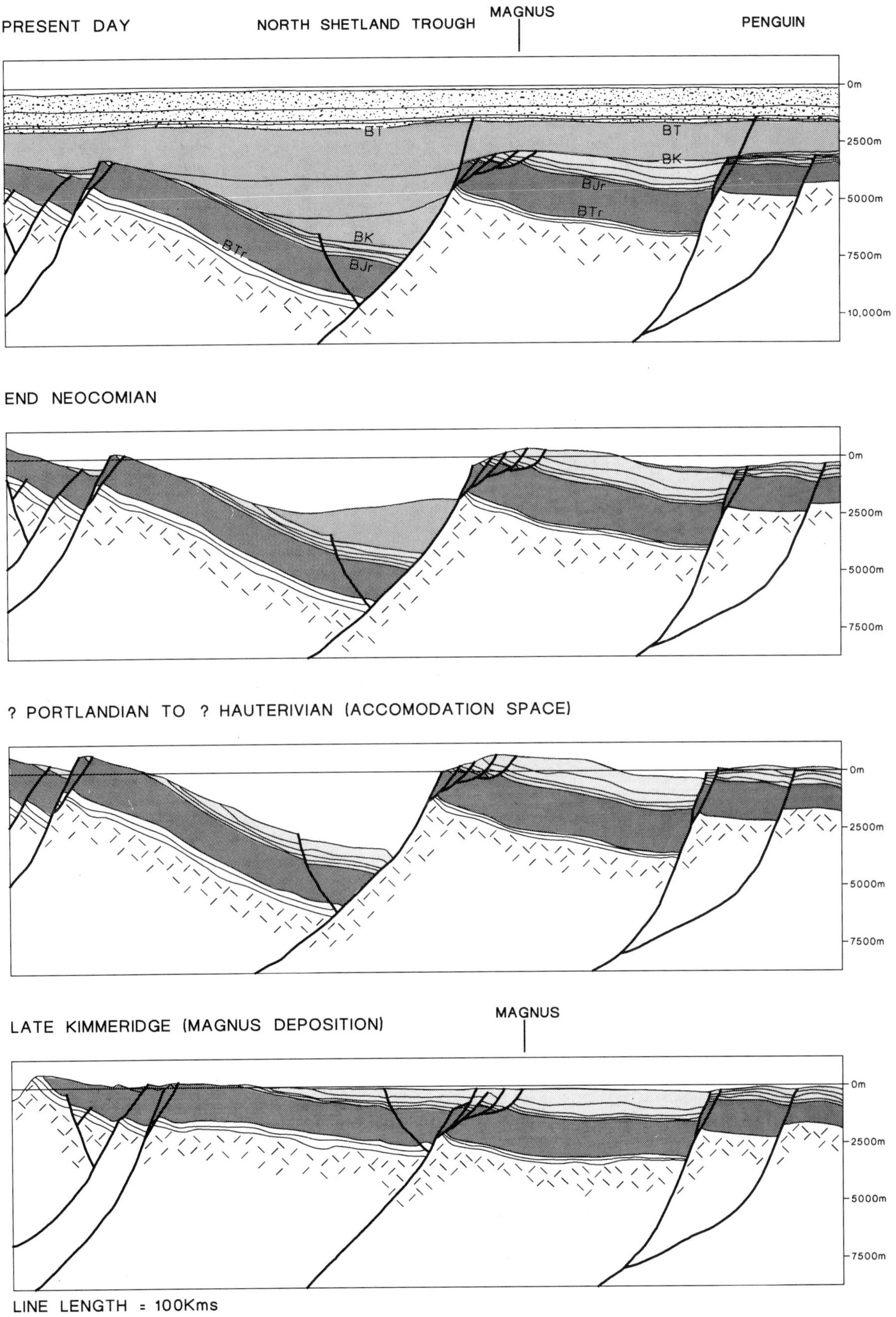

Fig. 11. Magnus Province decompacted and restored section based on Young (this volume). The location of the section is shown as Line 4 on Fig. 3 and on Fig. 9. Labelling of horizons as in Fig. 2. Unlabelled horizons are sequence boundaries.

Fig. 12. The Brae Trend of the South Viking Graben. The timeslice shown is Middle Ryazanian. The organised fans enter the graben from the Fladen Ground Spur through transfer systems or across tip points of the extensional faults.

fault scarp itself during uplift, or slumped lithified blocks. The more ordered systems, both in shallow and deeper marine settings, require a specific entry point and invariably this is provided by cross-cutting or transfer faults, flexures (transfer-zones), or at the tip points of extensional faults. A large-scale example of this is the progradation of the Sognefjord and Fensfjord sequences that provide the reservoir at the Troll Field, from the major oblique-slip system that defines the northern edge of the Horda Platform (Fig. 3).

The South Viking Graben, with dip-slip and transfer segments along its bounding fault related to Jurassic and earlier extension, received large volumes of submarine clastics in the Late Jurassic. Initially these were apron-fan deposits (e.g., Central Brae, South Brae), although by Ryazanian times, the depositional system had evolved with the erosional products of the footwall (Fladen Ground Spur) being worked down the dip slope and following structural topography to enter the graben through the transfer system (Fig. 12). These provide the reservoir for the Miller Field which is closed by subsequent compaction-drape over the fan body.

The structural style of the Central North Sea, particularly the Central Trough area, has been argued to be significantly different to that of the Viking Graben. The obvious difference is the presence of salt, although the major structural difference is the re-utilisation of Permian dip-slip and strike-slip faults during Jurassic extension. Fig. 13 includes a simple geoseismic section across the eastern margin of the Central Trough, through the Gyda Field, and shows the Late Permian salt wall and Triassic pod geometry to have stabilised by at least Late Triassic times. The Jurassic depositional basin margin is controlled by a dip-slip westerly dipping fault, which in Jurassic extension cut-up section through the Triassic and salt wall stratigraphy to involve the cover in a complex hanging-wall structure. The steeper Permian faults to the east of the dip-slip fault on the section map out as near north–south trending strike-slip surfaces, which on a regional scale accommodated the dip-slip extension as well as creating cover pull-aparts (e.g., the Ula Field; Figs. 13 and 15; Brown et al., this volume) and enforcing the inversion of Triassic pods up salt wall flank faults. Thick Jurassic shallow-marine sequences occupy the hanging-wall to the marginal fault system, the local geometry of which governs the position of the shelf and the shelf-break (e.g., fig. 7 of Bergan et al., 1989). Detailed unpublished mapping has shown that sediment entry into the hanging-wall occurred through north-east striking cover transfer faults. The pattern of the Norwegian sector is replicated in, for example, the Fulmar-Clyde area of the U.K. sector, although here more extensive flat detachments at the base of the salt link into the originally Permian faults.

Throughout the central and northern North Sea, the post-rift thermal subsidence led to drowning of the majority of the footwall highs and the basin flanks, and the deposition of the Ryazanian hot shale as a pelagic drape (Bertram and Milton, 1989). The infill of the rift topography ceased, and a substantial marine topography was established that was progressively onlapped by deep marine Cretaceous sediment.

Post-rift structural modification

The structural pattern of the North Sea, the central area in particular, has been modified through post-rift tectonic processes. The most striking is structural inversion which was accompanied in the area of the salt basin by diapirism. The inversion is widespread (Fig. 14) and can be traced from the Central Trough in Denmark (Gowers and Sæbøe, 1985) and through into the U.K. Sector, the Norwegian–Danish Basin and into the Egersund sub-basin, the southern part of the Stord Basin (Biddle and Rudolph, 1988) and the south-eastern Witch Ground Graben. The northern limit of marked inversion corresponds in the south-east to the Tornquist Zone (Pegrum, 1984), while the structuring in the southern Stord Basin appears related to the intersection of the north-east striking transfer zones and the basin margin faults. There is thus a remarkable consistency between the surfaces that accommodated the different extensional patterns of the Viking Graben and the central North Sea during Jurassic and earlier extension, and those utilised in younger inversion.

The inversion was thus governed by the existing structural grain and new faults were only formed at shallow levels where the existing extensional surfaces were unable to accommodate the contraction. Figure 16 shows a section over the Lindesnes Ridge where the inversion enforced the development of short-cut faults through the footwall, and backthrusts in the original hanging wall, the latter utilising originally antithetic faults. The critical aspect of the inversion to hydrocarbon exploration was the development of structural traps at both Jurassic levels and in the post-rift section. In the Ula Trend, the sense of motion of the originally extensional and sinistral strike-slip Early Permian fault surfaces was reversed, such that a series of small wavelength dip-slip related closures was formed over the extensional faults, while the Ula "pull-apart" (Fig. 15) popped-up (Brown et al., this volume).

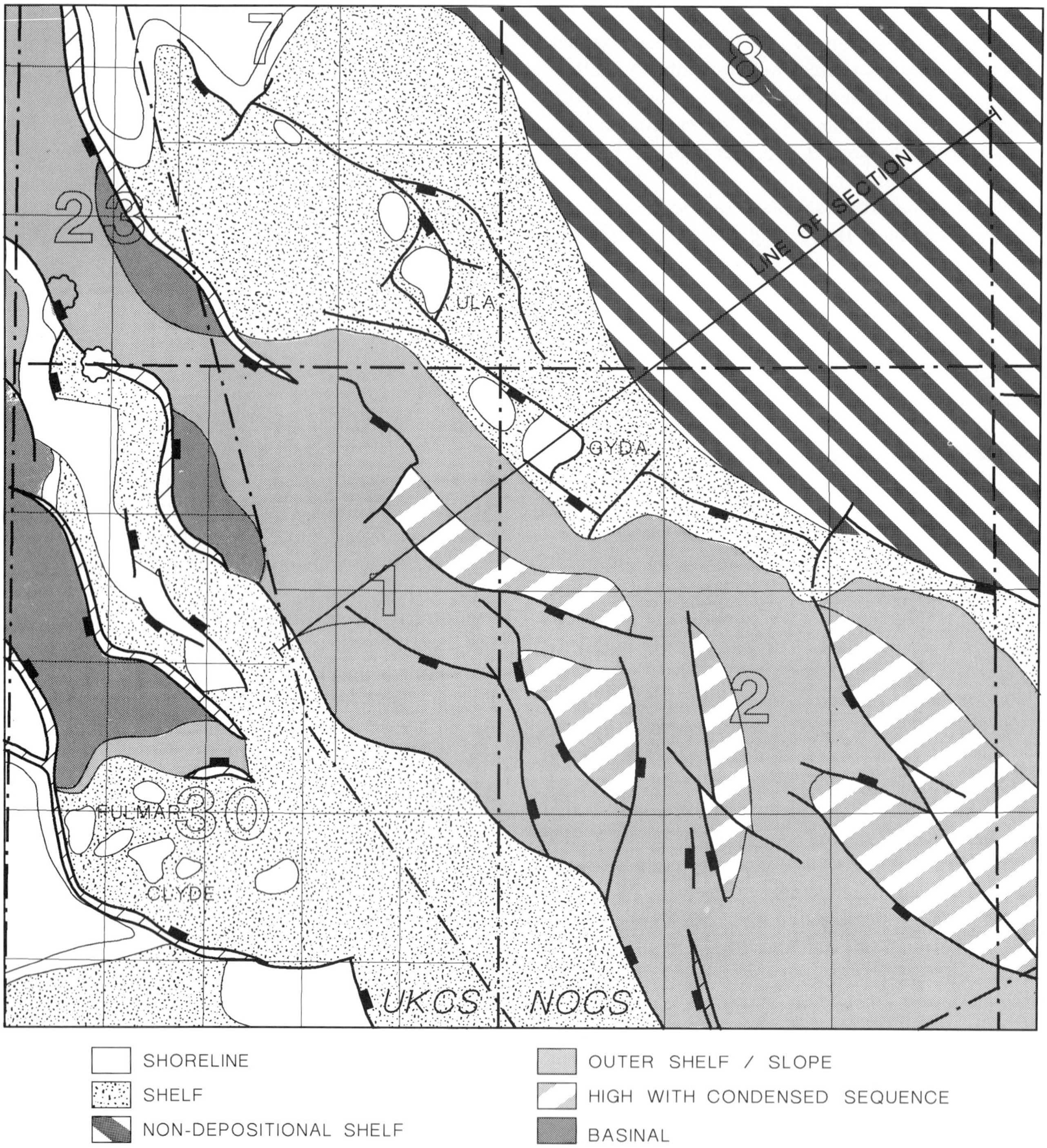
7
8
LINE OF SECTION
23
ULA
GYDA
1
2
30
FULMAR
CLYDE
UKCS
NOCS
SHORELINE
SHELF
NON-DEPOSITIONAL SHELF
OUTER SHELF / SLOPE
HIGH WITH CONDENSED SEQUENCE
BASINAL

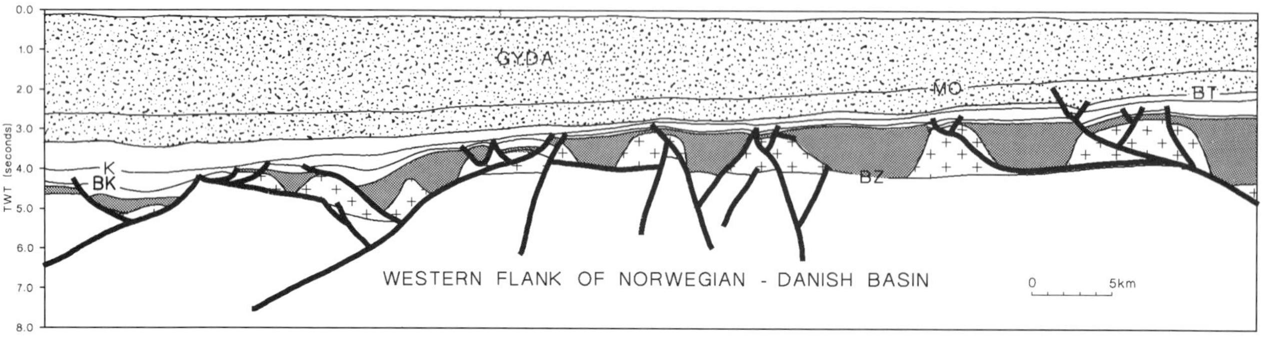
0.0
1.0
2.0
3.0
4.0
5.0
6.0
7.0
8.0
TWT (seconds)
GYDA
MO
BT
K
BK
BZ
WESTERN FLANK OF NORWEGIAN - DANISH BASIN
0
5km

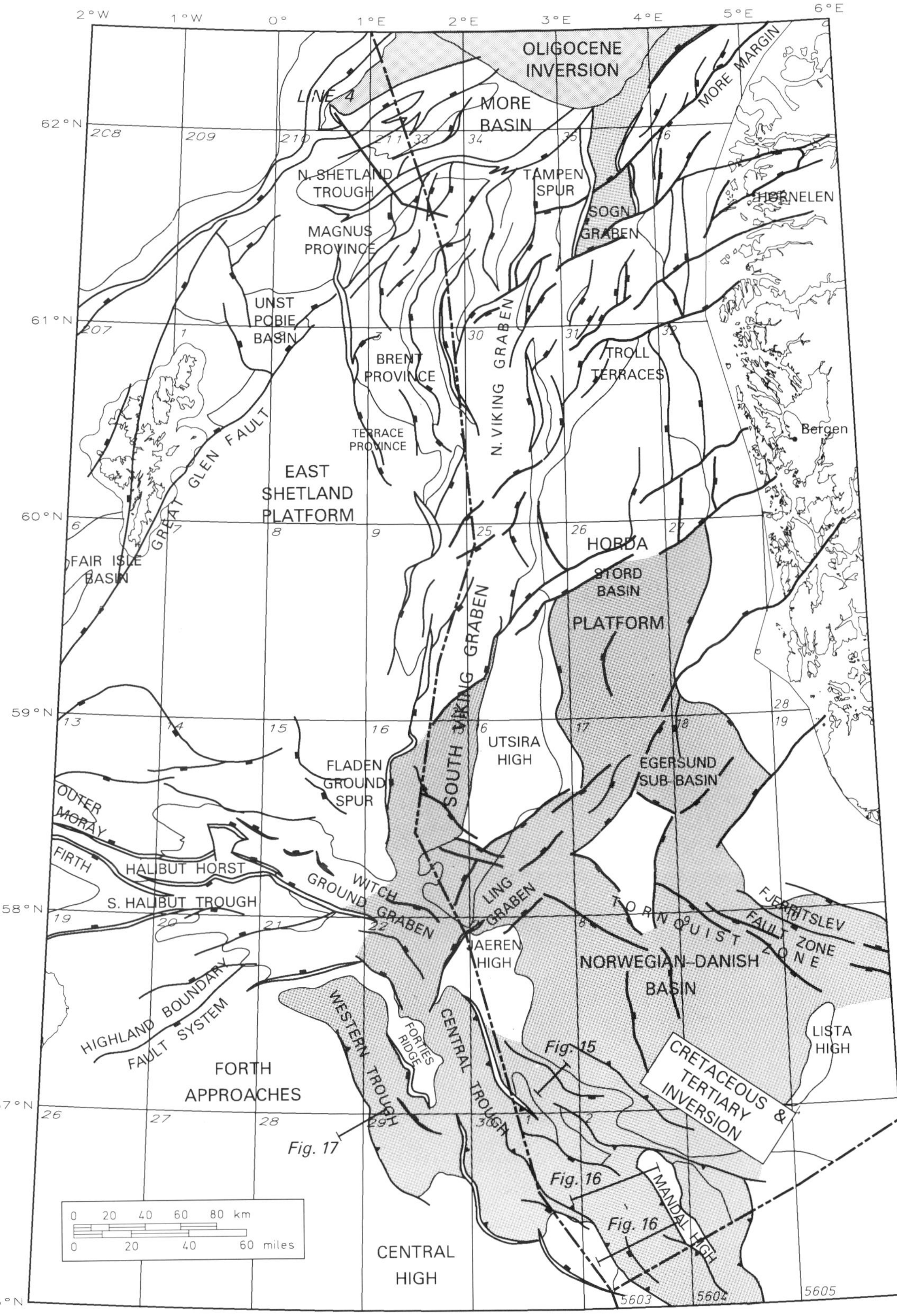

Fig. 14. Inversion structuring in the North Sea. Structural elements map showing areas of Cretaceous and Tertiary structural inversion. Cross sections are shown in Figs. 15 and 16.

Fig. 13. The Ula Trend of the Norwegian central North Sea. The timeslice shown is Middle Kimmeridgian. The underlying section is not depth converted. *MO* = Middle Oligocene; *BT* = Base Tertiary; *K* = Middle Albian; *BK* = Base Cretaceous; *BJ* = Base Jurassic; *BZ* = Base Zechstein.

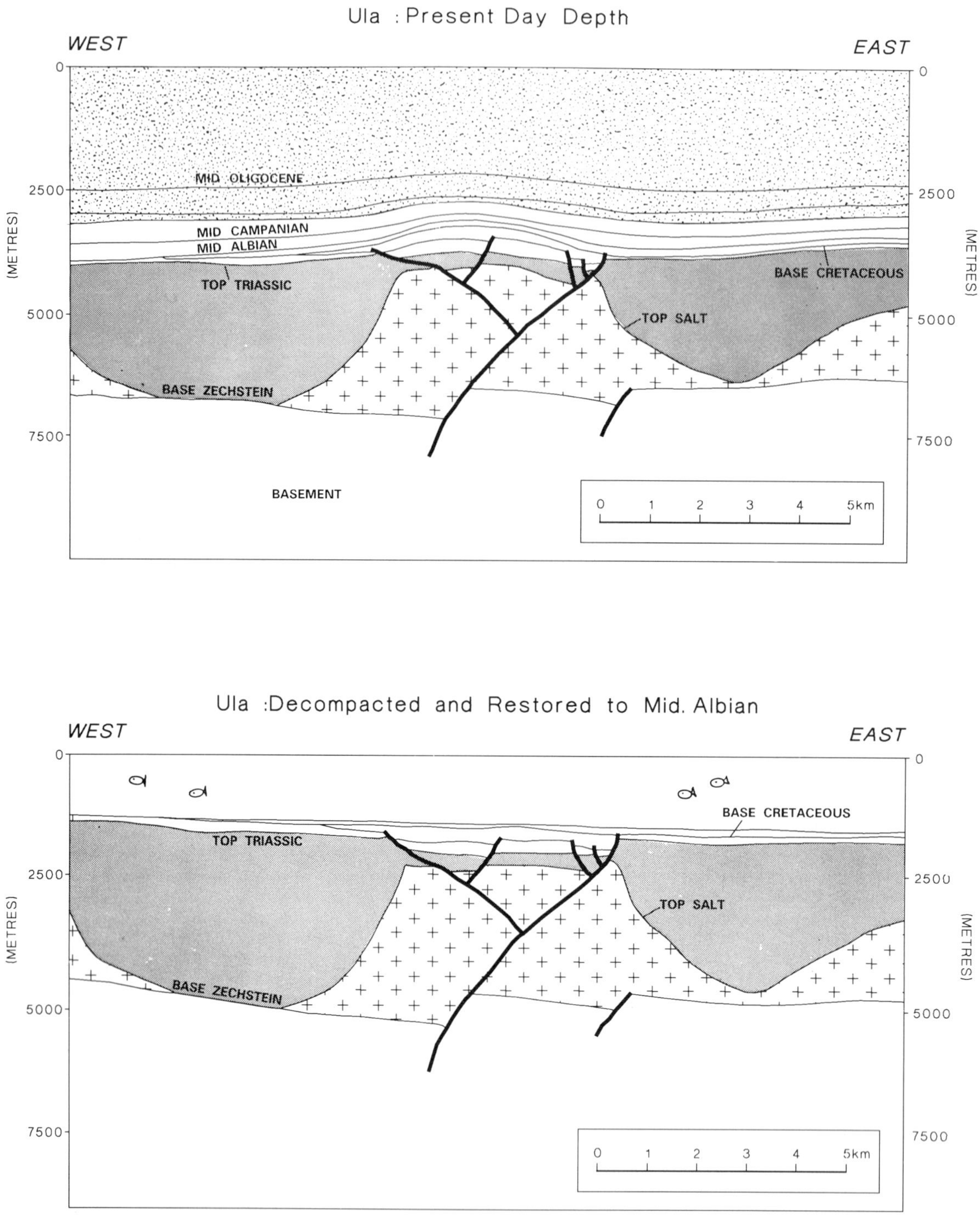

Fig. 15. Ula Field, central North Sea. Present day depth section, and restored and decompacted to Middle Albian. The location of the section is shown in Fig. 14.

Similar dextral strike-slip motion along north–south faults is argued to have governed inversion geometries in the Danish Central Trough (Vejbæk and Andersen, 1987). In the post-rift section, the closures that were formed have a much longer structural wavelength, and can be demonstrated to be related to the fault geometry at Jurassic levels or deeper ramp geometries of the fault surfaces. The chalk fields of the Norwegian sector (e.g., Fig. 16) are all related to the inversion geometries.

The principal ages of inversion are Early Cretaceous (Middle Albian), Late Cretaceous (Middle to Late Campanian) and Tertiary (Ypresian and Middle Oligocene; Fig. 4). Onlap unconformities of these ages have been dated over inversion structures in a large number of exploration wells. The Late Cretaceous and younger inversion episodes have been related to Alpine compression and associated strike-

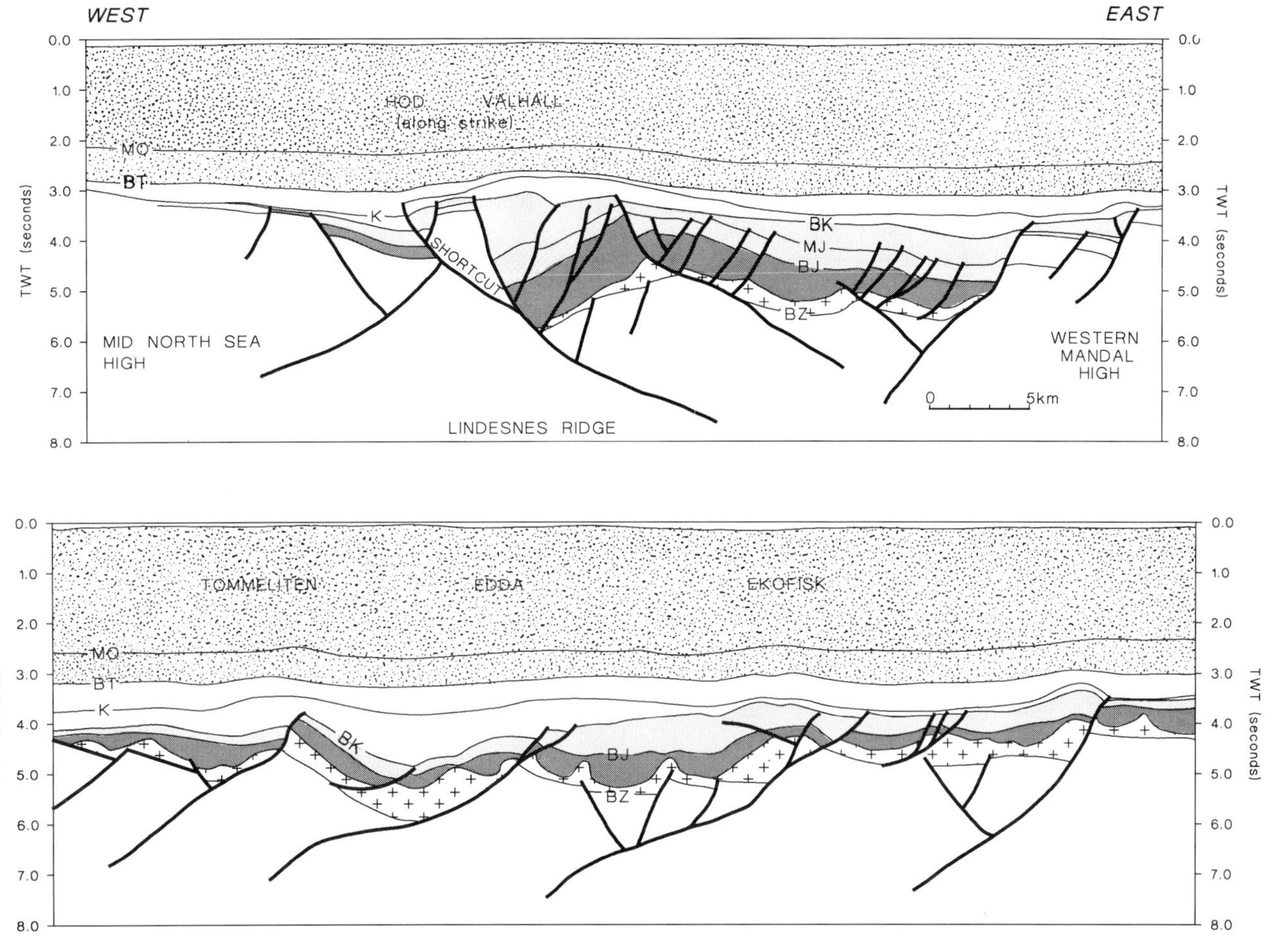

Fig. 16. Inversion geometries in the central North Sea. The upper section covers the Lindesnes Ridge, over which sit the Valhall and Hod fields. The lower section runs through the Ekofisk area. The locations are shown in Fig. 14, upper section is the southern line. *MJ* = top Middle Jurassic; other horizons labelled as Fig. 13.

slip motion along major crustal fault zones (e.g., Pegrum, 1984; Beach, 1987) of which the Tornquist Zone, the offshore extension of the Highland Boundary Fault and the oblique-slip faults of the southern Stord Basin acted in the North Sea area as the northern buttress. There are only minor Cretaceous and Tertiary inversion structures in the northern North Sea, although major Oligocene inversion geometries occur throughout the Cretaceous Atlantic Rift basins of Norway from the Möre Basin to as far north as the western Barents Shelf and may in part be related to ridge-push compression. The Early Cretaceous (approximately Middle Albian) contraction is more difficult to account for. It is recognised over much of the Norwegian sector of the Central Trough (e.g., Ula; Fig. 15), in the Sleipner area (Pegrum and Ljones, 1984) and in the Inner Moray Firth, and in each area appears to be associated with Early Permian and older strike-slip trends. The inversion is apparently slightly younger than the Austrian Tectonic phase (Ziegler, 1982), although it agrees with the initial age of inversion of the Sole Pit Trough (Walker and Cooper, 1987).

In addition to the structural inversion, the central North Sea contains the Zechstein salt which was mobilised in the Cretaceous and Tertiary to form diapiric structures that locally extend up to the sea bed. Over the shallow, relatively unfaulted, basin margins (e.g., Forth Approaches) local pillowing and gravity sliding can be shown to be related to basin tilting and Middle Tertiary and older sediment loading. In the Central Trough area though, the broad salt wall and Triassic sediment pod geometry had stabilised by at least Late Triassic times with the major salt walls developing preferentially from the sites of Early Permian hanging-walls. With the later Jurassic extension re-utilising the Permian system, the salt walls also occupied the hanging-wall areas of the younger faults, although in the Norwegian sector at least, the faults commonly cut-up section through the salt (Figs. 13, 16) leaving the Triassic sediment pods unfaulted. The dating of the main ages of diapiric growth, from the onlap relationships on the diapir flanks, shows that in many examples this was episodic, but rapid, and the ages concur with the inversion unconformities seen in structures which lack salt. It would appear that

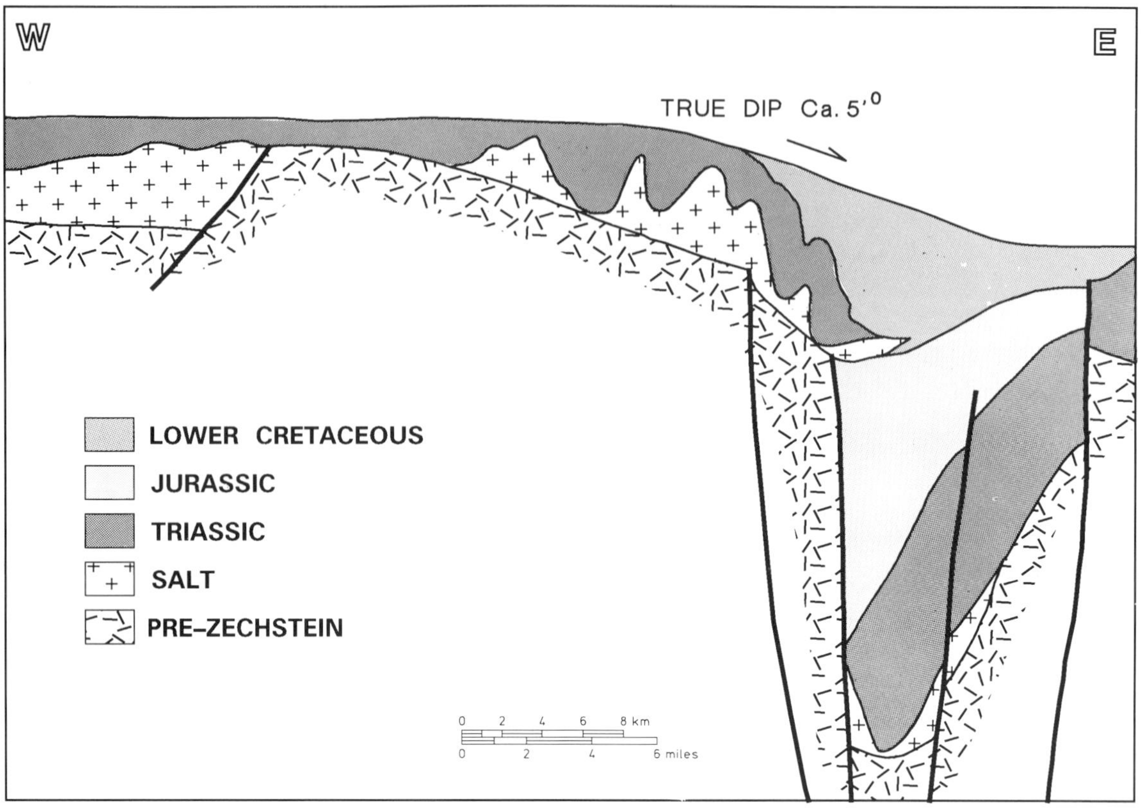

Fig. 17. Section across margin of western Central Trough in the UKCS restored and decompacted to Top Early Cretaceous to reveal palaeo-structural dip. The structure is a gravity slide detached at Base Zechstein and emplaced over the trough margin. The location of the figure is shown in Fig. 14.

the most significant diapiric growth allowing the salt to reach shallow stratigraphic levels occurred when the basin was in contraction. The diapir flank play contains few major discoveries in the central North Sea and success to date has relied on the presence of Early Palaeogene sands and locally fractured chalk, although the former are absent throughout the major part of the Norwegian Central Trough.

In addition to younger growth, the salt in the central North Sea played an important role in the Jurassic extension, providing a major décollement at the base of the salt section. There was also extension utilising the base salt detachment in the post-rift subsidence phase, but not balanced by deeper faults. The most striking example of this is over the Auk Shelf area (e.g., Section 3 on Fig. 2), where the entire section above the base of the salt has detached. The extension was initiated by flexural tilting associated with the earliest Cretaceous thermal subsidence and is balanced regionally by salt-cored contractional folds, or by the emplacement of a toe structure over the edge of the western Central Trough (Fig. 17) which has been proven to contain a few small hydrocarbon pools.

Probably the final, but extremely important, aspect of the post-rift modification of the North Sea Basin is the compaction effect as a result of sediment load. As well as creating simple drape closures (e.g., Montrose Field) over deeper structures or rock, which are less amenable to compaction, the recognition of compaction structure in the post-rift section is commonly critical to structural interpretation. Inversion structures can be distinguished from compaction structures by the vergence of the tip-line fold in the post-rift strata, although deeper compaction structures, particularly on the flanks of incompactible salt walls, or against incompactible basement in the hanging-walls of extensional faults, are frequently misinterpreted. This has to be incorporated into backstripping routines, which are becoming increasingly important in the interpretation of structural geometries (e.g., Young, this volume) and are providing a means to a better understand not only of the structural, but also the stratigraphic habitat of North Sea hydrocarbons.

Conclusions

The central and northern North Sea cover a series of structurally complex sedimentary basins that together form a prolific hydrocarbon province. The observations presented in this paper support the long-held view (e.g., Johnson and Dingwall, 1981) that the basement grain, which was inherited from at least Caledonian times, has played a fundamental role in the younger structuring. The Viking Graben area can be separated from the central North Sea by a projection of the Tornquist Zone as advocated by Pegrum (1984), which together with a series of re-activated north-east striking Caledonian fault zones accommodated different extensional and transtensional patterns between the central and northern North Sea areas. Variscan relaxation in the central North Sea accommodated the Late Permian age salt basin, which has had a major effect on the subsequent structural style and wavelength at Jurassic levels, while the northern North Sea is dominated by longer wavelength relaxed Caledonian trends that provide the classic tilted fault block play. The Magnus area of the northern North Sea provides evidence of younger extension associated with the Cretaceous Atlantic basins. The observations question the application of simple kinematic models such as that proposed by Roberts et al. (1989).

Despite the different geometries in the two areas, the processes of Jurassic rift-fill sedimentation are very similar. The rift-fill sequences provide major hydrocarbon plays throughout the North Sea, although they are particularly productive in the central North Sea (e.g., the Ula-Gyda and Fulmar-Clyde trends) where fault controlled subsidence occurred later on a regional scale. The structures of the central North Sea area in particular have also suffered several phases of structural inversion; this, combined with associated salt diapirism, has complicated the structural pattern at Jurassic levels, but has created longer wavelength structural closures in the post-rift section, notably containing the chalk fields of the Norwegian and Danish sectors.

Acknowledgements

The authors are indebted to many colleagues in Norway and in Britain for useful discussion and their contributions to this paper, in particular Neil Parkinson, Nick Johnson and Tony Hayward. This paper is published with the permission of the Board of BP Exploration.

References

Andrews, J.R. and Brown, S., 1987. Stratigraphic evolution of the Jurassic, Moray Firth. In: J. Brooks and K. Glennie (Editors), Petroleum Geology of North West Europe. Graham and Trotman, London, pp. 785–795.

Badley, M.E., Egeberg, T. and Nipen, O., 1984. Development of rift basins illustrated by the structural evolution of the Oseberg feature, Block 30/6, offshore Norway. J. Geol. Soc. London, 141: 639–649.

Badley, M.E., Price, J.D., Rambech Dahl, C. and Agdestein, T., 1988. The structural evolution of the North Viking Graben and its bearing upon extensional modes of basin formation. J. Geol. Soc. London, 145: 455–478.

Beach, A., 1984. Structural evolution of the Witch Ground Graben. J. Geol. Soc. London, 141: 621–628.

Beach, A., 1985. Some comments on sedimentary basin development in the northern North Sea. Scott. J. Geol., 21(4): 493–512.

Beach, A., 1986. A deep seismic reflection profile across the northern North Sea. Nature, 323: 53–55.

Beach, A., 1987. A regional model for linked tectonics in north-west Europe. In: J. Brooks and K. Glennie (Editors), Petroleum Geology of North West Europe. Graham and Trotman, London, pp. 43–48.

Beach, A., Bird, T. and Gibbs, A.D., 1987. Extensional tectonics and crustal structure: deep seismic reflection data from the northern North Sea Viking Graben. In: M.P. Coward, J.F. Dewey and P.L. Hancock (Editors), Continental Extensional Tectonics. Geol. Soc. London, Spec. Publ., 28: pp. 467–476.

Bergan, M., Torudbakken, J. and Wandaas, B., 1989. Lithostratigraphic correlation of Upper Jurassic sandstones within the Norwegian Central Graben: sedimentological and tectonic implications. In: J.D. Collinson (Editor), Correlation in Hydrocarbon Exploration. Norwegian Petroleum Society, Graham and Trotman, London, pp. 243–251.

Bertram, G.T. and Milton, N.J., 1989. Reconstructing basin evolution from sedimentary thickness; the importance of palaeobathymetric control, with reference to the North Sea. Basin Res., 1: 247–257.

Biddle, K.T. and Rudolph, K.W., 1988. Early Tertiary structural inversion in the Stord Basin, Norwegian North Sea. J. Geol. Soc. London, 145: 603–612.

Brewer, J.A. and Smythe, D.K., 1984. MOIST and continuity of crustal reflector geometry along the Caledonian–Appalachian orogen. J. Geol. Soc. London, 141: 105–120.

Brown, A., Mitchell, A.W., Nilssen, I.R., Stewart, I.J. and Svela, P.T., 1992. Ula Field: relationship between structure and hydrocarbon distribution? In: R.M. Larsen, H. Brekke, B.T. Larsen and E. Talleraas (Editors), Structural and Tectonic Modelling and its Application to Petroleum Geology. Norwegian Petroleum Society (NPF), Special Publication 1. Elsevier, Amsterdam, pp. 409–420 (this volume).

Brun, J.P., Allemand, P. and Ballard, J.F., 1989. Some mechanical aspects of extensional basin development in nature and experiments. Paper presented at Norwegian Petroleum Society Workshop, 18–20 October 1989, Stavanger.

Campbell, C.J. and Ormaasen, E., 1987. The discovery of oil and gas in Norway: a historical synopsis. In: A.M. Spencer et al. (Editors), Geology of the Norwegian Oil and Gas Fields. Graham and Trotman, London, pp. 1–37.

Cheadle, M.J., McGeary, S. Warner, M.R. and Matthews, D.H., 1987. Extensional structures on the western UK continental shelf: a review of evidence from deep seismic profiling. In: M.P. Coward, J.F. Dewey and P.L. Hancock (Editors), Continental Extensional Tectonics. Geol. Soc. London, Spec. Publ., 28: pp. 445–465.

Coward, M.P., Dewey, J.F. and Hancock, P.L., 1987. Continental Extensional Tectonics. Geol. Soc. London, Spec. Publ., 28: 637 pp.

De'Ath, N.G. and Schuyleman, S.F., 1981. The geology of the Magnus oilfield. In: L.V. Illing and G.D. Hobson (Editors), Petroleum Geology of North-West Europe. Institute of Petroleum, London, pp. 342–351.

Doré, A.G. and Gage, M.S., 1987. Crustal alignments and sedimentary domains in the evolution of the North Sea, North-East Atlantic Margin and Barents Shelf. In: J. Brooks, and K. Glennie (Editors), Petroleum Geology of North West Europe. Graham and Trotman, London, pp. 1131–1148.

Enfield, M.A. and Coward, M.P., 1987. The structure of the West Orkney Basin, northern Scotland. J. Geol. Soc. London, 144: 871–884.

Erichsen, T., Helle, M., Henden, J. and Rognebakke, A., 1987. Gullfaks. In: A.M. Spencer et al. (Editors), Geology of the Norwegian Oil and Gas Fields. Graham and Trotman, London, pp. 273–286.

Fraser, A.J., Nash, D.F., Steele, R.P. and Ebdon, C.C., 1990. A regional assessment of the intra-Carboniferous play of Northern England. In: J. Brooks (Editor), Classic Petroleum Provinces. J. Geol. Soc. London, Spec. Publ.

Gee, D.G. and Sturt, B.A. (Editors), 1985. The Caledonide Orogen — Scandinavia and Related Areas. John Wiley and Sons, New York, N.Y., 1266 pp.

Gibbs, A.D., 1984. Structural evolution of extensional basin margins. J. Geol. Soc. London, 141: 609–620.

Gibbs, A.D., 1987a. Development of extension and mixed-mode sedimentary basins. In: M.P. Coward, J.F. Dewey and P.L. Hancock (Editors), Continental Extensional Tectonics. Geol. Soc. London, Spec. Publ., 28: pp. 445–465.

Gibbs, A.D., 1987b. Deep seismic profiles in the northern North Sea. In: J. Brooks and K. Glennie (Editors), Petroleum Geology of North West Europe. Graham and Trotman, London, pp. 1025–1028.

Gowers, M.B. and Sæbøe, A., 1985. On the structural evolution of the Central Trough in the Norwegian and Danish sectors of the North Sea. Mar. Pet. Geol., 2(4): 298–318.

Gradijan, S.J. and Wiik, M., 1987. Statfjord Nord. In: A.M. Spencer et al. (Editors), Geology of the Norwegian Oil and Gas Fields. Graham and Trotman, London, pp. 341–350.

Halstead, P.H., 1975. Northern North Sea faulting. In: K.G. Finstad and R.C. Selley (Coordinators), NPF Jurassic Northern North Sea Symposium Volume, Stavanger, JNNSS/10.

Hossack, J.R., 1984. The geometry of listric growth faults in the Devonian basins of Sunnfjord, W. Norway. J. Geol. Soc. London, 141: 629–638.

Johnson, H.D., MacKay, T.A. and Stewart, D.J., 1986. The Fulmar Oilfield (Central North Sea): geological aspects of its discovery, appraisal and development. Mar. Pet. Geol., 3: 100–125.

Johnson, R.J. and Dingwall, R.G., 1981. The Caledonides: their influence on the stratigraphy of the Northwest European Continental Shelf. In: L.V. Illing and G.D. Hobson (Editors), Petroleum Geology of the Continental Shelf of North-West Europe. Institute of Petroleum, London, pp. 85–97.

Marsden, G., Yielding, G., Roberts, A. and Kusznir, N., 1989. Modelling of extensional basin formation: the northern Viking Graben. Paper presented at the Norwegian Petroleum Society Workshop, 18–20 October 1989, Stavanger.

McGeary, S., Cheadle, M.J., Warner, M.R. and Blundell, D.J., 1987. Crustal structure of the continental shelf around Britain derived from BIRPS deep seismic profiling. In: J. Brooks and K. Glennie (Editors), Petroleum Geology of North West Europe. Graham and Trotman, London, pp. 33–41.

Needham, D.T. and Knipe, R.J. 1986. Accretion- and collision-related deformation in the Southern Uplands accretionary wedge, southwestern Scotland. Geology, 14: 303–306.

Newall, M.J. and Kusznir, N.J., 1989. A coupled quantitative tectonic–sedimentary model of extensional sedimentary basin formation. Paper presented at the Norwegian Petroleum Society Workshop, 18–20 October, 1989, Stavanger.

Pegrum, R.M. 1984 The extension of the Tornquist Zone in the Norwegian North Sea. Nor. Geol. Tidsskr., 64: 39–68.

Pegrum, R.M. and Ljones, T.E., 1984. 15/9 Gamma Gas Field offshore Norway, new trap type for North Sea Basin with regional structural implications. Bull. Am. Assoc. Pet. Geol., 68: 874–902.

Richards, P.C., Brown, S., Dean, J.M. and Anderton, R., 1988. A new palaeogeographic reconstruction for the Middle Jurassic of the northern North Sea. J. Geol. Soc. London, 45: 883–886.

Roberts, A., Yielding, G. and Badley, M., 1989. A kinematic model for the opening of the North Sea/Norwegian Rift, with emphasis on the dip-slip nature of the main graben systems. Paper presented at the Norwegian Petroleum Society Workshop, 18–20 October, 1989, Stavanger.

Snyder, D.B., 1990. The Moine Thrust in the BIRPS data set. J. Geol. Soc. London, 147: 81–86.

Steel, R., Siedlecka, A. and Roberts, D., 1985. The Old Red Sandstone basins of Norway and their deformation: a review. In: D.G. Gee and B.A. Sturt (Editors), The Caledonide Orogen — Scandinavia and related areas. John Wiley and Sons, New York, N.Y., pp. 293–316.

Sørensen, S., Morizot, H. and Skottheim, S., 1992. A tectonostratigraphic analysis of the southeast Norwegian North Sea Basin. In: R.M. Larsen, H. Brekke, B.T. Larsen and E. Talleraas (Editors), Structural and Tectonic Modelling and its Application to Petroleum Geology. Norwegian Petroleum Society (NPF), Special Publication 1. Elsevier, Amsterdam, pp. 19–42 (this volume).

Vejbæk, O.V. and Andersen, C., 1987. Cretaceous–Early Tertiary inversion tectonism in the Danish Central Trough. Tectonophysics, 137: 221–238.

Walker, J.M. and Cooper, W.G., 1987. The structural and stratigraphic evolution of the northeast margin of the Sole Pit Basin. In: J. Brooks and K. Glennie (Editors), Petroleum Geology of North West Europe. Graham and Trotman, London, pp. 263–275.

White, N., 1989. Nature of lithospheric extension in the North Sea. Geology, 17(2): 111–114.

Yielding, G. and Roberts, A., 1992. Footwall uplift during normal faulting — implications for structural geometries in the North Sea. In: R.M. Larsen, H. Brekke, B.T. Larsen and E. Talleraas (Editors), Structural and Tectonic Modelling and its Application to Petroleum Geology. Norwegian Petroleum Society (NPF), Special Publication 1. Elsevier, Amsterdam, pp. 289–304 (this volume).

Young, R., 1992. Restoration of a regional profile across the Magnus Field in the northern North Sea. In: R.M. Larsen, H. Brekke, B.T. Larsen and E. Talleraas (Editors), Structural and Tectonic Modelling and its Application to Petroleum Geology. Norwegian Petroleum Society (NPF), Special Publication 1. Elsevier, Amsterdam, pp. 221–229 (this volume).

Ziegler, P.A., 1975. Geologic evolution of the North Sea and its tectonic framework. Bull. Am. Assoc. Pet. Geol., 59: 1073–1097.

Ziegler, P.A., 1982. Geologic Atlas of Western and Central Europe. Elsevier, Amsterdam, 130 pp.

I.J. STEWART *BP Norway Limited U.A., P.O. Box 197, N-4033 Forus, Norway*
Present address: *BP Exploration Inc., P.O. Box 4587, Houston, TX 77210, U.S.A.*

R.P. RATTEY *BP Exploration, Glasgow, U.K.*

I.R. VANN *BP Exploration, London, U.K.*

Restoration of a regional profile across the Magnus Field in the northern North Sea

R. Young

Confidence in a geological model can be increased by checking it against quantitative reconstructions. The main extensional basin forming mechanisms of faulting, thermal subsidence, loading and compaction are described by simple models and then applied to a regional cross-section over the Magnus Field in the northern North Sea.

The four processes can be approximated by simple numerical models. Upper crustal extension can be modelled using simple geometries. Thermal subsidence can be modelled using a cooling slab model in which the rate of subsidence decreases with increasing time after rifting. Bending of the lithosphere can be modelled using an elastic plate whereby the deflection created by the load is spread over an area larger than the load itself. Compaction can be modelled as an exponential function where porosity varies with depth of burial.

All these models have been used to restore the present day depth section sequentially back through time. The effects of compaction on the underlying sediments has been accounted for, allowing them to expand to their pre-burial thicknesses. The thermal subsidence has been estimated and added back into the section. The load of the overlying units has been removed resulting in lithospheric rebound.

Restoration to Early Cretaceous times indicates that the Magnus Sandstone Member was uplifted above sea-level and could be sub-aerially eroded. This is consistent with both seismic stratigraphic and geochemical evidence.

The regional development of a rifted margin is illustrated by the restorations and shows the relative influence of the basinal processes modelled here.

Introduction

Application of numerical basin modelling to petroleum exploration aids in reducing the risk of finding commercial quantities of hydrocarbons. The confidence in a geological model can be increased if the interpreted sequence of events can be fitted into a quantitative regional setting. This study uses a regional geological cross-section which crosses the Magnus Field in the northern North Sea to show that simple models for the main basin forming mechanisms can be combined to produce quantitative reconstructions.

The main basinal processes that combine to form the stratigraphy of an extensional sedimentary basin such as the North Sea are structure, thermal subsidence, sediment loading and compaction. The upper crust responds to stretching by brittle failure which is expressed as faulting (Hellinger and Sclater, 1983; Leader and Gawthorpe, 1987; and McKenzie, 1978a). The lower crust and lithosphere respond to extension by necking and ductile flow that introduces hotter less dense material which results in thermal uplift (McKenzie, 1978a; Royden and Keen, 1980; Sawyer, 1986; and White and McKenzie, 1988). The thermal anomaly gradually decays as the lithosphere cools down towards its pre-rifted, equilibrium temperature. Figure 1 illustrates the combination of upper crustal and lithospheric responses to stretching that will be modelled.

A basin formed by stretching can accumulate sediments which will represent an external load on the lithosphere. The lithosphere will deform due to this load, thus creating further space for sediments (Watts et al., 1982). As sediments are buried they undergo a loss in porosity due to mechanical compaction (Sclater and Christie, 1980). Loading and compaction can drive basin subsidence.

The basinal processes outlined above can be described using simple numerical models. The models to be used to reconstruct the Magnus regional line are discussed below.

Structure

A simple model for the response of the upper crust to stretching is illustrated in Fig. 2. The dimensions of the model are 100 km long by 10 km deep. The initial unstretched block is shown in Fig. 2a with a listric fault soling out at 12 km depth. The depth

Structural and Tectonic Modelling and its Application to Petroleum Geology edited by R.M. Larsen, H. Brekke, B.T. Larsen and E. Talleraas. NPF Special Publication 1, pp. 221–229. Elsevier, Amsterdam.

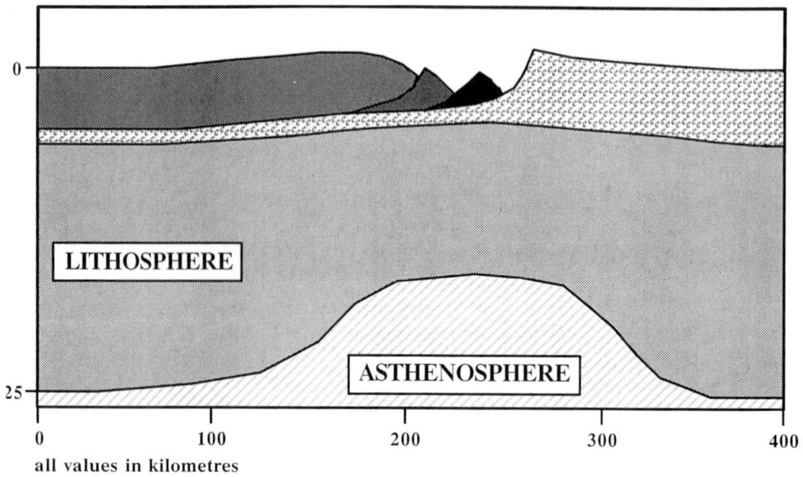

Fig. 1. A simple sketch diagram showing the modelled response of the lithosphere to extension. The upper crust responds by brittle failure resulting the formation of faulted basins. The lower crust and lithosphere responds by ductile flow and necking which results in regional uplift as hot asthenospheric material is brought into the lithosphere.

to detachment used here is within the range of 10–15 km suggested by McKenzie (1978b). Figure 2b shows the result of moving the hanging wall block 4 km. The gap opened up between the hanging wall and footwall can be filled by the vertical shear of the hanging wall (Gibbs, 1983). Figure 2c shows the result of filling the gap from Fig. 2b and the formation of a half-graben structure. The area in Fig. 2c coloured pink represents a load that has been removed from the lithosphere. The effect of regional compensation of the unloading can be calculated and is shown in Fig. 2d. The model used for calculating the deformation is discussed below and its main effect is to spread the deflection over an area larger than the dimensions of the load. This simple model can be used to quantify footwall uplift effects. The magnitude of the footwall uplift in Fig. 2d has a maximum of 700 m and extends 35 km away from the fault.

The deformation modelled in Fig. 2 assumes that the upper crustal extension event occurs relatively quickly and that no sediments fill in the newly formed hole. As the graben fills with sediments the lithosphere will deform under the load and the footwall will be deflected downwards.

Thermal uplift

The lower crust and lithosphere are assumed to respond to extension by necking rather than faulting (Royden and Keen, 1980; and Sawyer, 1986). Stretching raises the lithosphere/asthenosphere boundary above its equilibrium depth and replaces dense mantle material by hotter less dense asthenospheric material. The density contrast between this hot asthenosphere material and the denser mantle material causes a regional uplift across the stretched basin.

The hot asthenosphere gradually cools, losing heat by lateral and vertical conduction and the uplift decays as the material becomes denser. This thermal subsidence can be modelled as a cooling slab resulting in an exponential decay with time (Sleep, 1971; Parsons and Sclater, 1977).

The response of the lithosphere to stretching can be summarised as a combination of structural and thermal effects (Fig. 1). There will be an initial fault controlled subsidence, which occurs relatively rapidly, followed by a gradual longer term subsidence as the lithosphere cools back towards its equilibrium thermal profile (McKenzie, 1978a; Royden and Keen, 1980; and Sawyer, 1986).

Loading

The response of the lithosphere to external loads, such as sediments filling a basin, can be modelled as an elastic plate overlying a fluid substratum (McKenzie and Bowin, 1976; Watts et al., 1980; and McNutt, 1984) and as illustrated in Fig. 3. The load is supported by a combination of the plate rigidity and the buoyancy effect of the density contrasts between the load and the displaced fluid (the mantle in this case). A major effect of including plate rigidity is that deflection is spread over a large area. The rigidity of the lithosphere can be usefully described by a parameter called the effective elastic thickness (EET). The EET is the thickness an elastic plate would have to have in order to respond to loading in the same manner as the lithosphere.

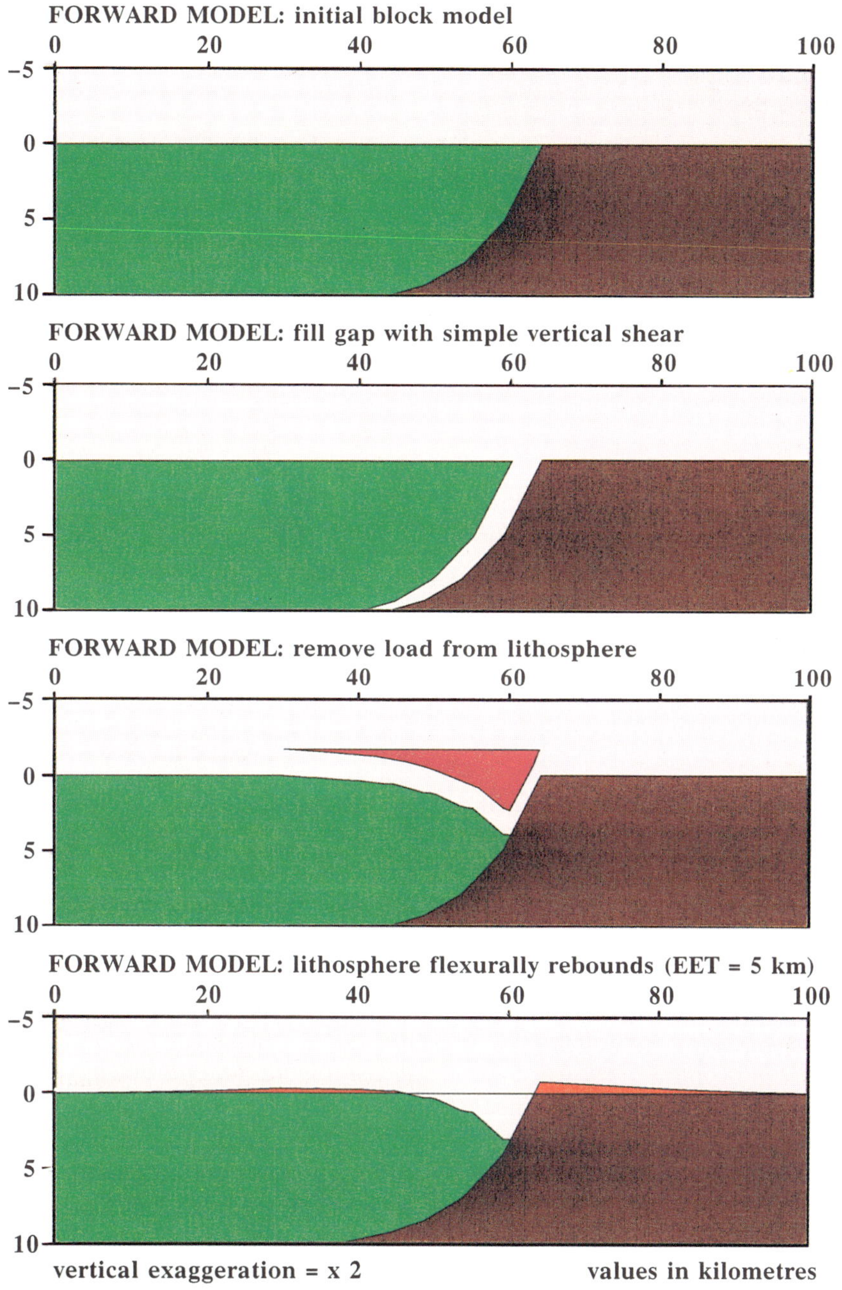

Fig. 2. A simple forward model of upper crustal extension with associated footwall uplift. (a) The initial unstretched block with a listric fault trace; (b) extension causes a gap to open up between the hanging wall and the footwall; (c) this gap can be filled by simple vertical shear with the area of the basin indicated by the pink polygon; (d) deforming the lithosphere by assuming it behaves like an elastic plate the rigidity of which can be expressed as an effective elastic thickness of 5 km. Both the footwall and, to a lesser extent, the hanging wall are deformed. Footwall uplift has a maximum of 700 m.

Sediments infilling a basin represent a load on the lithosphere that will deform the basin and the basin margins. The amount of subsidence due to the load and the distance that the deformation will extend depends on the density of the sediment fill and the rigidity of the lithosphere. Flexural rigidity has been proposed as a simple mechanism to produce a "steer's head" geometry (Dewey, 1982) of a sedimentary basin.

Compaction

The reduction in porosity of sediments as they are buried can be modelled as a mechanical compaction process which depends on the initial or surface porosity and the rate of decrease of porosity with depth. Sclater and Christie (1980) showed that the observed data from wells can be fitted by an exponential decay curve. Different lithologies will have

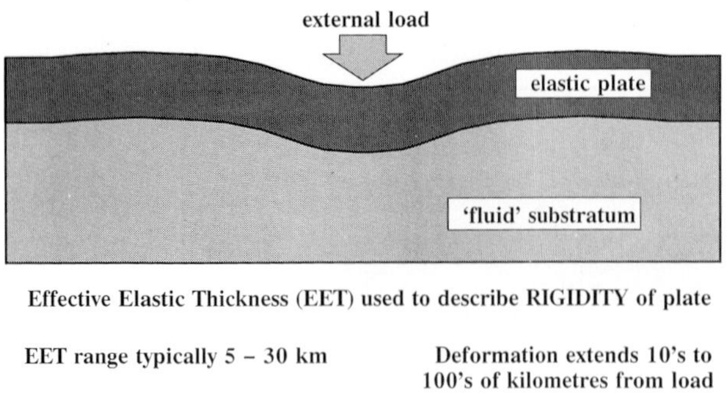

Fig. 3. The elastic plate model for loading/unloading the lithosphere. Loads are supported by a combination of the density contrast between the load and the "fluid" mantle and the rigidity of the plate.

different initial porosities and rates of compaction. Figure 4 shows porosity plotted against depth for shales from four wells across the regional line to be used in this study. The porosities have been derived from both logged sonic and density values.

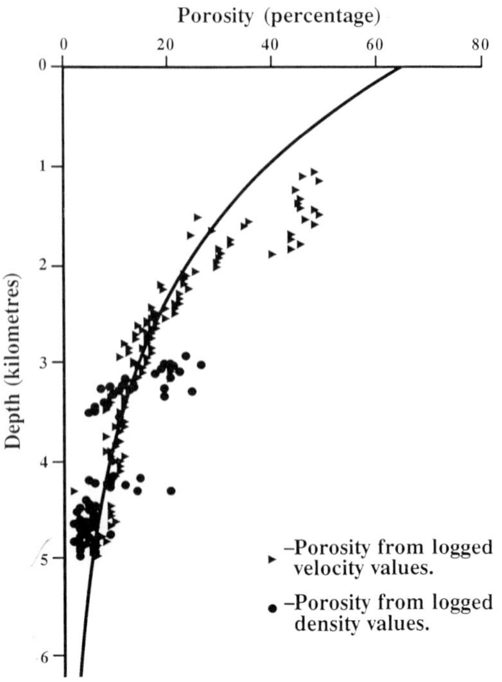

Fig. 4. A plot of porosity, calculated from both logged density and sonic values, against depth for four wells from the vicinity of the Magnus regional line. Values are for shale only and can be reasonably approximated by an exponential decay curve from a surface porosity of 65%.

Application of models to the Magnus regional line

The simple models of basinal processes outlined above can be applied to a regional geological cross-section. The example used here is based on a 100 km long composite seismic line which crosses the Magnus Field from south south east to north north west and extends across the Shetland Trough to the Nordfjord High in the northern North Sea (Fig. 5). Figure 6 shows the present day depth section that will be reconstructed. The Magnus oilfield (De'Ath and Schuyleman, 1981) comprises an east dipping tilted fault block structure, with the reservoir sandstone, the Magnus Sandstone Member (MSM), within the Kimmeridge formation, truncated in the west by the Base Cretaceous unconformity. The Magnus Field lies within the blue coloured sequence in Fig. 6 at a depth of 3 km between 60 and 80 km on the section. All the sediments above the Base Cretaceous unconformity are interpreted as post-rift infill.

The objective of the modelling work is to try and quantify the relative contributions of the basinal processes. In particular the palaeo-bathymetry will be estimated across the section from the present day to the Early Cretaceous, the effects of compaction on the Early Cretaceous sequences will be estimated and the magnitude of the sub-aerial exposure of the Magnus Fault block will be quantified.

Well data were used to constrain the compaction constants for the decompaction calculations. Porosity data were estimated from both density and velocity measurements and plotted against depth for different lithologies. The depth trends, such as that shown in Fig. 4, were fitted with exponential curves, to define surface porosity and rate of decrease of porosity with burial depth (Sclater and Christie, 1980; and Barton and Wood, 1984).

Estimating thermal subsidence

Three of the main basinal processes, loading, compaction and structure, can be removed from the present day depth section to show the relative palaeo-geography across the profile using only the input data of compaction constants, matrix densities and the structural interpretation. The thermal subsidence calculation, however, requires some

Fig. 5. Location sketch map for the regional line used in this study. The cross section (Fig. 6) is 100 km long and crosses the Magnus Field, into the Shetland Trough and up onto the Nordfjord High.

Fig. 6. Regional present day depth section based on a composite seismic line. Cretaceous sequences are red, pink and brown. Tertiary sequences extend from seabed to top brown. The present day bathymetry across the line averages 200 m with deeper water to the leftmost edge (400 m). The Magnus Field is located at the crest of the tilted fault block between 60 and 80 km at a depth of 3 km. The present day relief on the top red sequence (an Early Cretaceous reflector) is 1500 m.

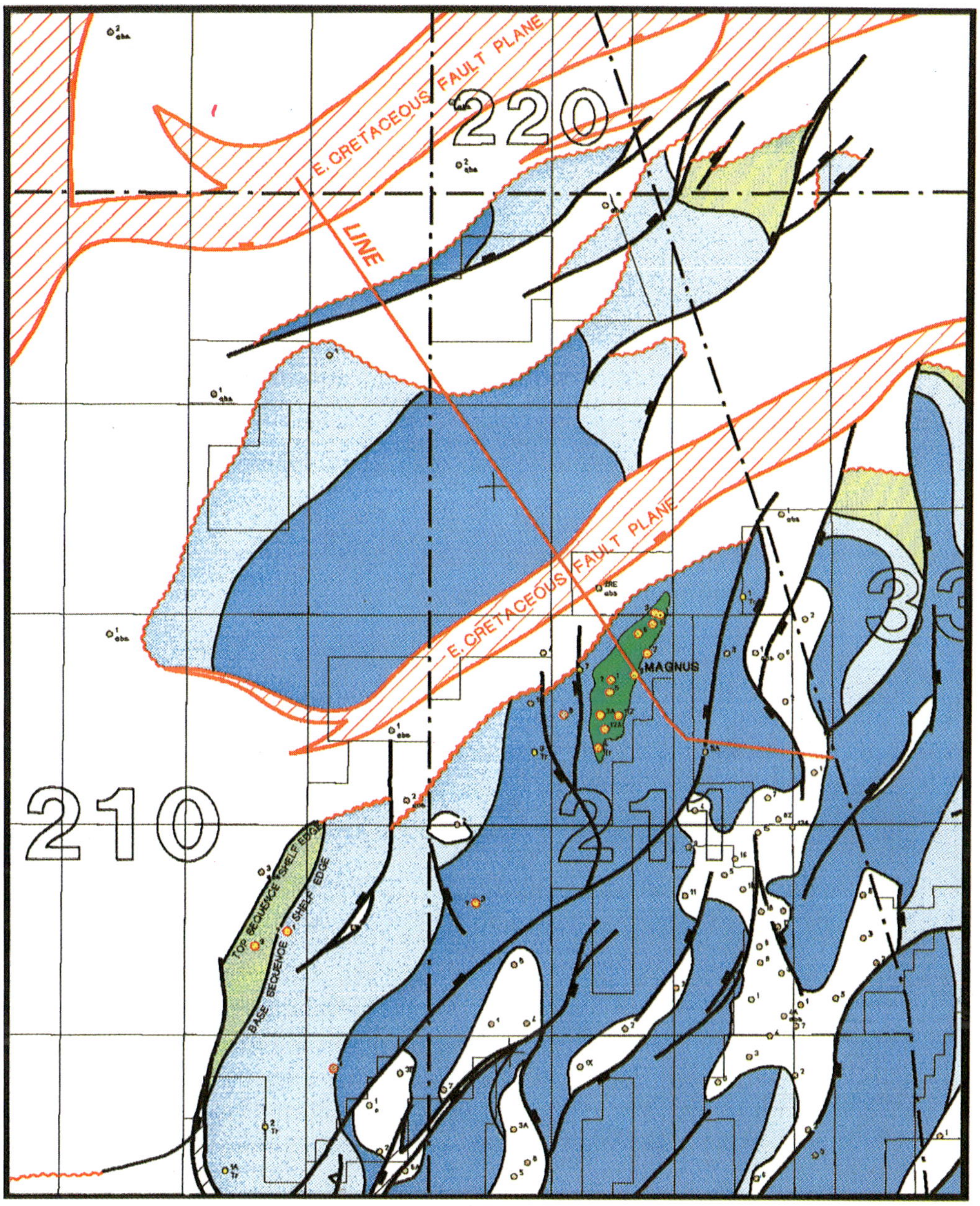
E. CRETACEOUS FAULT PLANE
220
LINE
E. CRETACEOUS FAULT PLANE
MAGNUS
210
211
TOP SEQUENCE SHELF EDGE
BASE SEQUENCE SHELF EDGE

Fig. 5.

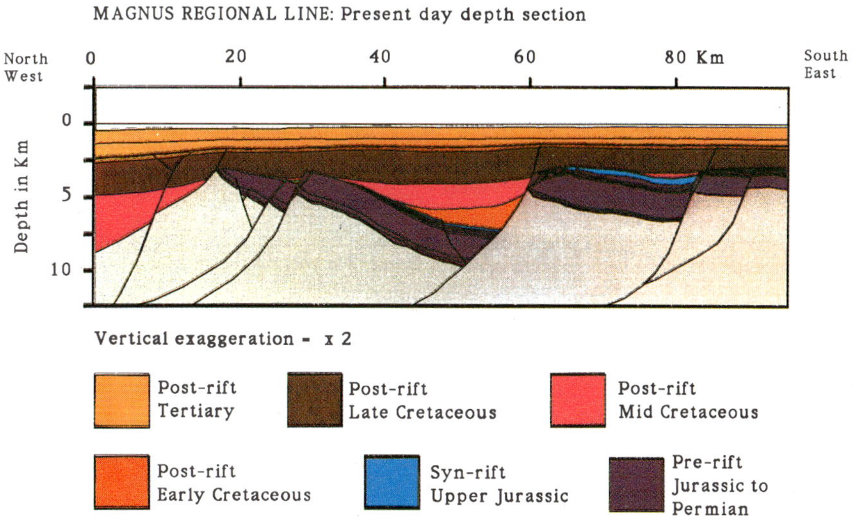
MAGNUS REGIONAL LINE: Present day depth section
North West
South East
0
20
40
60
80 Km
Depth in Km
0
5
10
Vertical exaggeration = x 2
Post-rift Tertiary
Post-rift Late Cretaceous
Post-rift Mid Cretaceous
Post-rift Early Cretaceous
Syn-rift Upper Jurassic
Pre-rift Jurassic to Permian

Fig. 6.

knowledge of the palaeo-bathymetry upon which the exponentially decaying model can be based. Ideally several good markers such as coals would be used at a variety of positions across the model and over a significant time span.

The most useful palaeo-bathymetry marker, or trace as defined by Bertram and Milton (1989), in the present study is the Brent Group which has been interpreted as a shallow marine sandstone (Parry et al., 1981). Applying the models of compaction, flexural loading and structural restoration with accompanying footwall deformation to the present day depth section to obtain a restored section at Top Brent time results in palaeo-bathymetry values of 820 m at the rightmost, least deformed, edge of the section. The difference between the palaeo-bathymetry from modelling three of the basinal processes and the palaeo-bathymetry interpreted for the Brent is assumed to be due to the thermal subsidence from the time of rifting (138 Ma) to the present day.

The structural restoration to Top Brent resulted in a total shortening of the section from 100 km to 89 km which represents a beta factor of only 1.12. If a simple McKenzie model (McKenzie, 1978a) is assumed, with uniform stretching in upper crust and lithosphere, then this gives insufficient thermal uplift to bring the Brent to its interpreted water depth. If a two layer stretching model is used then a lithospheric stretching factor of 1.4 must be input to get 820 m of thermal uplift and restore the MSM to its interpreted palaeo-bathymetry. Figure 7 shows the modelled thermal subsidence for the profile plotted against age in millions of years. The model being used here assumes that thermal subsidence commences at 138 Ma, which is the interpreted age of the Magnus Fault.

The full reconstructions using models for compaction, loading, structure and thermal subsidence are shown and discussed for several time steps in the following sections.

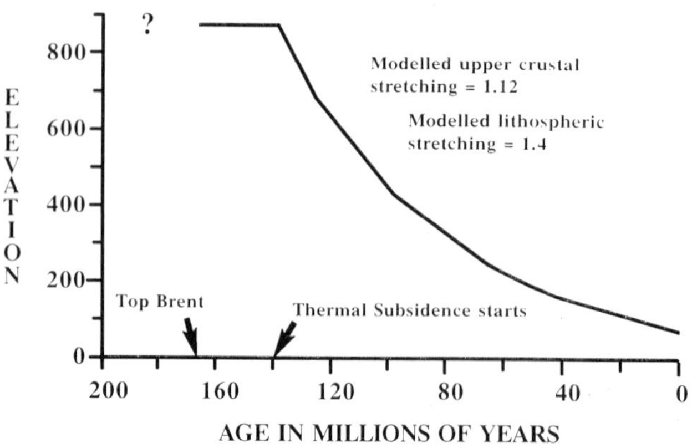

Fig. 7. Thermal subsidence calculated for the cross-section using the Brent Formation as a palaeo-bathymetry marker and starting thermal decay at 138 Ma (the timing of the last phase of faulting). The stretching factors used were 1.12 in the upper crust (as calculated from restoring the section) and 1.4 for the lower crust and lithosphere These were required to obtain 820 m subsidence from Early Cretaceous to the present day.

Reconstruction of the regional line through time

Restoration to Early Tertiary times, Fig. 8, produces a section with palaeo-bathymetry (300 m) only slightly greater than the present day depth section (200 m). The pre-Tertiary sediments have "fluffed out" to their pre-burial thicknesses. The lithosphere has been allowed to rebound by removing the load of the Tertiary to recent sediments and the proportion of the thermal subsidence from 65 Ma to the present day, a total of 200 m, has been added back into the profile. It was not necessary to restore structurally at this time.

The unloading calculation was estimated by assuming a rigidity for the plate characterised by an EET of 15 km. However, this value is not critical, since the load that was removed extends almost uniformly across the model. The calculation is relatively insensitive to EET variations for uniform thickness loads. The inclusion of the elastic plate in the calculation

Fig. 8. Restoration to Base Tertiary times. The section has been allowed to rebound due to removing the load of the tertiary sediments. The underlying Cretaceous and older sediments have been restored to their pre-burial thicknesses and the thermal subsidence from Fig. 7 into the model.

Fig. 9. Restoration to Early compaction, loading and thermal subsidence have been applied and the Magnus Fault the top red sequence has been removed due to decompaction.

Fig. 10. Restoration to Earliest Cretaceous times results in further uplift of the Magnus Fault block and indicates a deep (3.75 km) basin, the Shetland Trough, at this time. Calculated sub-aerial exposure is supported by geochemical evidence of porosity enhancement by meteoric water leaching.

Fig. 11. Reconstruction to Base Cretaceous before movement occurs along the Magnus fault. The Kimmeridge Formation (coloured blue) thickens into a fault towards the righthand (eastern) edge of the section. Submarine fan sandstones (the Magnus Sandstone Member) were deposited during Kimmeridge times before finning upwards into the Upper Kimmeridge organic rich mudstones. The observed uplift on the Magnus fault block (Fig. 10) of 500 m can be modelled as a response to extension (Fig. 1). As the hanging wall block is replaced to its prefaulted position the footwall uplift must be removed.

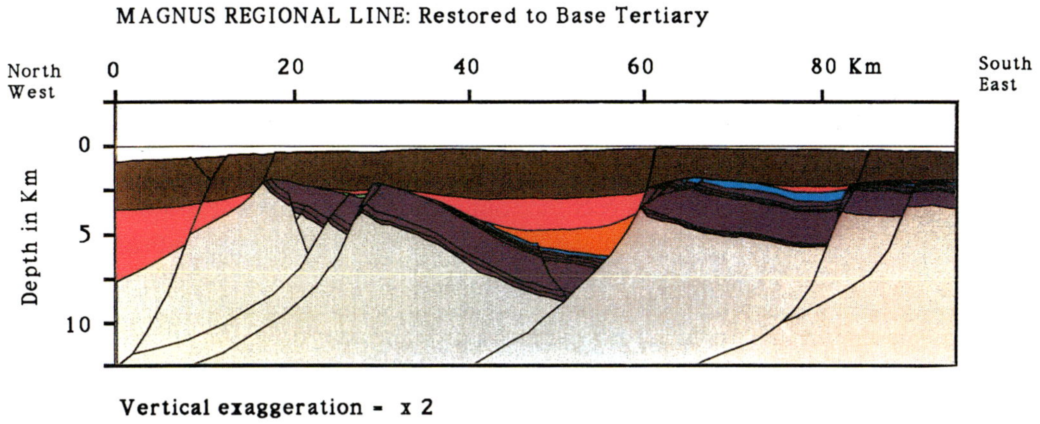

Fig. 8.

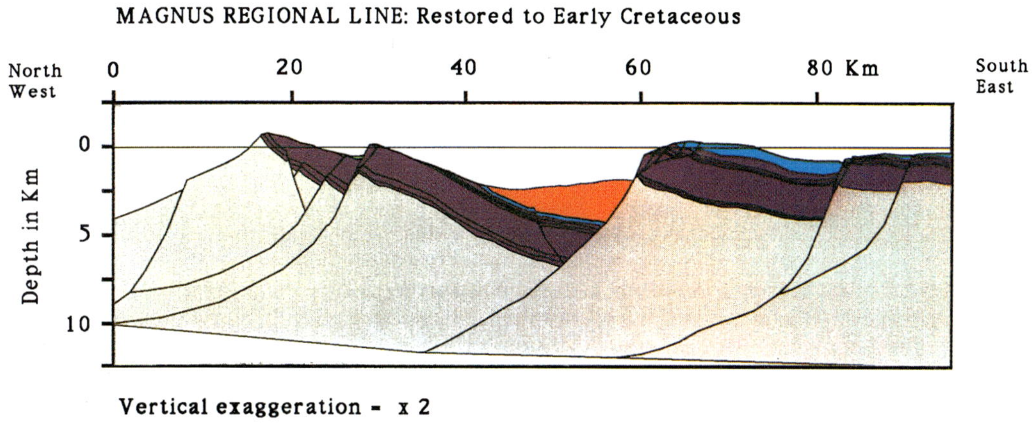

Fig. 9.

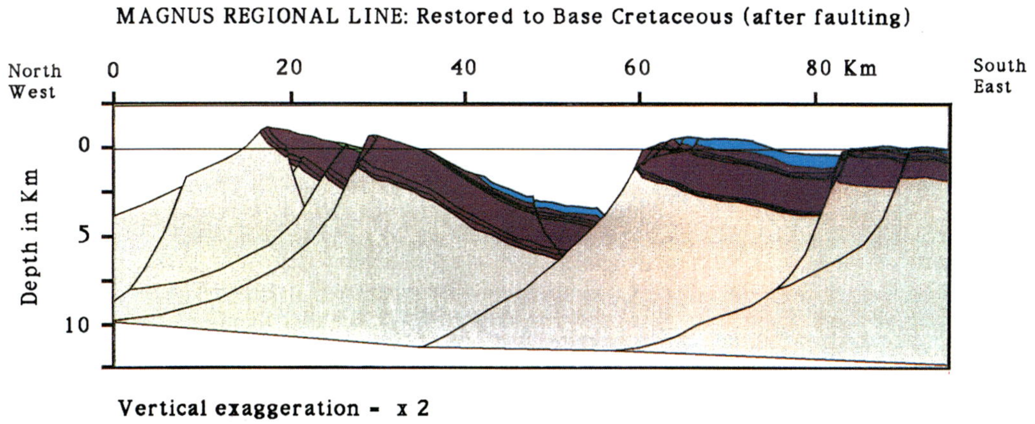

Fig. 10.

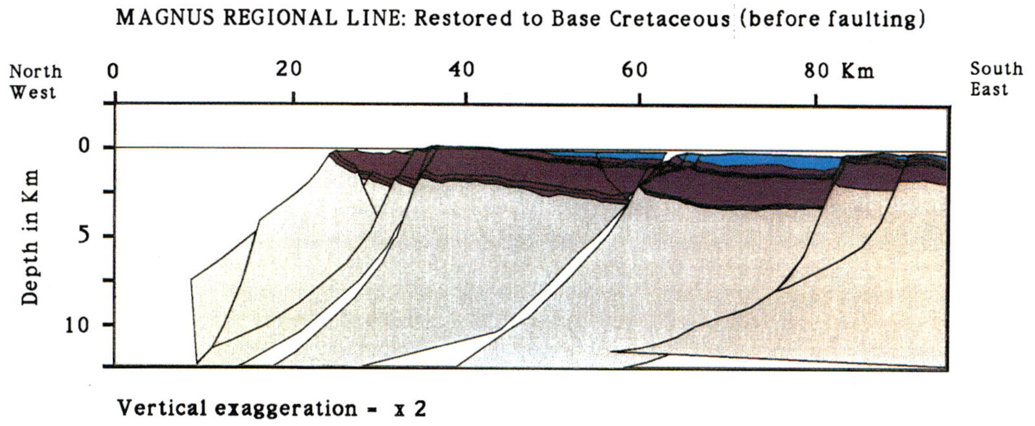

Fig. 11.

serves to iron out the local bumps that would occur if local isostatic response was used.

The implication of the modelling work at this time suggests that a small hole, namely 300 m palaeo-bathymetry, can be filled with a considerable thickness of sediments (in this case 1500 m). The hole in which the sediments are being deposited gets deeper through time due to a combination of thermal subsidence (200 m), lithospheric response to loading and compaction of underlying sediments during burial.

Reconstruction to Early Cretaceous times (Aptian–Barremian, 125 Ma) is shown in Fig. 9. The same three models of the basinal processes have been accounted for as in the previous restoration step. However, there is now a considerable depth of water across the section with shallow regions associated with the tilted fault block structures of Magnus and the Nordfjord High. The Magnus Fault block, between 60 and 80 kilometres in the figures, is sub-aerially exposed with a steep drop into the Shetland Trough, down the Magnus Fault, into palaeo-bathymetries of 2000 m.

Palaeo-bathymetries of the order of 2000 m have been interpreted by Bertram and Milton (1989) from seismic stratigraphy in the North Viking Graben. Nelson and Lamy (1987) also interpret large depths at this time from a depth section close to that used here. However, they interpret sediments at the base of the Shetland Trough as being Upper Jurassic rather than Early Cretaceous as in this study.

There has been a dramatic change in the topography of the top of the red sequence between the present day depth section (relief was 1500 m) and the reconstruction shown in Fig. 9. The top of the red sequence is now flat. The observed topography on the present day depth section can be attributed to differential compaction of the red sequence as it is buried by almost 5 km of sediment. This is consistent with a seismic stratigraphic interpretation that the Early Cretaceous sequences are a deep marine passive infill phase.

Reconstruction to Base Cretaceous is shown in Fig. 10. The loading and compaction effects of the Lower Cretaceous sediments have been removed and more thermal subsidence has been added back into the model (total of 800 m). The result is that the Magnus Fault block is raised approximately 500 m above sea level. Geochemical evidence (Emery et al., submitted to Geology) suggests that the sandstones towards the top of the Magnus Sandstone Member have been flushed by meteoric water resulting in porosity enhancement of the order of 5%. The geochemical evidence is consistent with sub-aerial exposure (Shanmugan, 1987).

A structural restoration has been attempted to replace the major movement on the Magnus Fault and the smaller movements on the other faults. During faulting, footwall uplift will occur as the hanging wall load is removed from the lithosphere as was discussed earlier in the simple structural model outlined in Fig. 2.

To replace the hanging wall of the Magnus Fault we must include deformation of both the hanging wall and the footwall. Figure 11 shows the structural restoration of the hanging wall. Gaps and overlaps are not considered as serious errors during the restoration, but serve to highlight the simplicity of the models used here.

The syn-rift sequences are shown to thicken into a fault to the east of the Magnus Field. There have been two phases of extension in this area. The earlier, Callovian/Oxfordian, phase was controlled by approximately north–south orientated faults (Fig. 5) into which the sequences containing the Magnus Sandstone Member were deposited. A later phase of rifting in Base Cretaceous times orientated more north-east to south-west formed the main Magnus Fault that cuts the earlier syn-rift sequence and uplifts them on the footwall block. Thus, Upper Jurassic submarine fan sandstones are preserved at the top of a fault block tilted during the Early Cretaceous.

Conclusions

The results of this study show that simple models of the main basinal processes can be used to produce quantitative reconstructions starting from a present day geological model.

(1) The palaeo-bathymetry during the Tertiary was little different from that observed in the present day. The 1500 m of Tertiary sediments have been deposited into a 300 m hole due to a combination of thermal subsidence, loading and compaction.

(2) Relief observed on Early Cretaceous sequences can be attributed to compaction effects when buried by up to 5 km of later sediments.

(3) The main result of footwall uplift of the Magnus Fault block could have been interpreted from seismic stratigraphy or geochemistry. The basin modelling carried out in this study increases confidence in the geological model by putting the observations into a quantitative regional setting.

Acknowledgments

BP gave permission for this paper to be published. The author would like to acknowledge the following co-workers Iain Hutchison, Mark Shann, David Jackson and Liz Lasseter with whom the regional study, of which this is only a part, was carried out.

References

Barton, P. and Wood, R., 1984. Tectonic evolution of the North Sea Basin: crustal stretching and subsidence. Geophys. J. R. Astron. Soc., 79: 987–1022.

Bertram, G.T. and Milton, N.J., 1989. Recontructing basin evolution from sedimentary thickness; the importance of palaeobathymetric control, with reference to the North Sea. Basin Res., 1: 247–257.

De'Ath, N.G. and Schuyleman, S.F., 1981. The geology of the Magnus oilfield. In: L.V. Illing and G.D. Hobson (Editors), Petroleum Geology of the Continental Shelf of North West Europe. Institute of Petroleum, London, pp. 342–351.

Dewey, J.F., 1982. Plate tectonics and the evolution of the British Isles. J. Geol. Soc. London, 139: 371–412.

Gibbs, A.D., 1983. Balanced cross-section construction from seismic sections in areas of extensional tectonics. J. Struct. Geol., 5: 153–160.

Hellinger, S.J. and Sclater, J.G., 1983. Some comments on two-layer extensional models for the evolution of sedimentary basins. J. Geophys. Res., 88: 8251–8269.

Leader, M.R. and Gawthorpe, R.L., 1987. Sedimentary models for extensional tilt-block/half-graben basins. In: M.P. Coward, J.F. Dewey and P.L. Hancock (Editors), Continental Extensional Tectonics. Geol. Soc. London, Spec. Publ., 28: 139–152.

McKenzie, D.P., 1978a. Some remarks on the development of sedimentary basins. Earth Planet. Sci. Lett., 40: 25–32.

McKenzie, D.P., 1978b. Active tectonics of the Alpine–Hymalayan belt: the Aegean Sea and surrounding regions. Geophys. J. R. Astron. Soc., 55: 217–254.

McKenzie, D.P. and Bowin, C., 1976. The relationship between bathymetry and gravity in the Atlantic Ocean. J. Geophys. Res., 81: 1903–1915.

McNutt, M.K., 1984. Lithospheric flexure and thermal anomalies. J. Geophys. Res., 89: 11180–11194.

Nelson, P.H.H. and Lamy, J.-M., 1987. The More/West Shetlands area: a review. In: J. Brooks and K.W. Glennie (Editors), Petroleum Geology of North West Europe. Graham and Trotman, London, pp. 775–784.

Parry, C.C., Whitley, P.K.J. and Simpson, R.D.H., 1981. Integration of palynological and sedimentological methods in facies analysis of the Brent Formation. In: L.V. Illing and G.D. Hobson (Editors), Petroleum Geology of the Continental Shelf of North West Europe. Institute of Petroleum, London, pp. 342–351.

Parsons, B. and Sclater, J.G., 1977. An analysis of the variation of ocean floor bathymetry and heat flow with age. J. Geophys. Res., 82: 803–827.

Royden, L. and Keen, C.E., 1980. Rifting process and thermal evolution of the continental margin of Eastern Canada determined from subsidence curves. Earth Planet. Sci. Lett., 51: 363–361.

Sawyer, D.S., 1986. Effects of basement topography on subsidence history analysis. Earth Planet. Sci. Lett., 78: 427–434.

Sclater, J.G. and Christie, P.A.F., 1980. Continental stretching: an explanation of the Post-Mid-Cretaceous subsidence of the Central North Sea. J. Geophys. Res., 85: 3711–3739.

Shanmugan, G., 1987. Origin, recognition, and importance of erosional unconformities in sedimentary basins. In: K. Kleinspehn and C. Paola (Editors), New Perspectives in Basin Analysis. pp. 83–108.

Sleep, N., 1971. Thermal effects of the formation of Atlantic continental margins by continental breakup. Geophys. J. R. Astron. Soc., 24: 325–350.

Watts, A.B., Bodine, J.H. and Steckler, M.S., 1980. Observations of flexure and the state of stress in the oceanic lithosphere. J. Geophys. Res., 85: 6369–6376.

Watts, A.B., Karner, G.D. and Steckler, M.S., 1982. Lithospheric flexure and the evolution of sedimentary basins. Philos. Trans., R. Soc. London Ser. A, 305: 249–281.

White, N.J. and McKenzie, D.P., 1988. Formation of the "steer's head" geometry of sedimentary basins by differential stretching of the crust and mantle. Geology, 16: 250–253.

R. YOUNG *BP Exploration, London, U.K.*
Present address: *BP Norway Limited U.A., Forusbeen 35, P.O. Box 197, N-4033 Forus, Norway*

Patterns in the development of extensional fault-block shapes from comparison of outcrop-scale faults and experimental physical models

S.E. Laubach, B.C. Vendeville and S.J. Reynolds

The geometry and kinematics of normal fault blocks and subsidiary faults within fault blocks are illustrated by centimeter- to meter-scale faults in an exposure of Tertiary sedimentary rocks in Arizona and are interpreted using results from previously published physical models. In map view, the outcrop fault population consists of anastomosing curved and planar faults. In cross-section there are planar faults and curved faults with either smooth or abrupt (ramp-flat) curvature. The fault patterns resemble those of physical models and larger faults in other extended terrains. Ramp-flat faults with planar and curved faults and tilt-block domains are the most prominent structures. Ramp-flat geometry prevails where faults intersect bedding at low angles. Flats are commonly zones of distributed faults rather than a simple throughgoing fault. In tilt-block arrays curved faults that are convex-up are as numerous as concave up (listric) faults and contribute to the overall geometry of the fault array. Subsidiary planar and curved faults prevail in the toes of tilt blocks, whereas horst-and-graben patterns occur in the less faulted block interiors. Internal deformation of rollover structures due to both accommodation and extension of the upper plate was accomplished mainly by slip on arrays of smaller faults in synthetic tilt-block arrays and by crisscrossing horst-and-graben-bounding faults.

Introduction

The structural evolution of normal fault blocks governs the shape, size and distribution of potential reservoir rocks in many extended terrains. Seismic data offer good information on the configuration of major faults, but resolution limits observation of structures that accommodate displacement along major fault planes and distortion within major fault blocks. Restored sections indicate that subsidiary structures must contribute to the structural evolution of fault blocks, but such complexity is difficult to document using seismic data alone. One way to gain a better understanding of the development of the fault-block shape is to study the geometry of small-displacement fault arrays in naturally deformed rocks and the geometry and kinematics of faults in scaled physical models.

This paper describes faults in an outstanding three-dimensional exposure in Tertiary sedimentary rocks from the Basin and Range province in Arizona and compares them with fault geometries in physical models. Because experimental models are created under known conditions of displacement and imposed bulk strain, similarities in geometry between models and the outcrop can be used to refine the interpretation of fault kinematics. The results lend insight into the way in which normal faults interact to produce fault block shapes and how, in larger fault blocks, fault complexity that is below seismic resolution may control reservoir compartmentalization and heterogeneity.

Geologic setting of fault example in Arizona

The study area in west-central Arizona is in the Basin and Range province of the southwestern United States, a region transected by numerous Tertiary detachment faults (e.g., Wernicke, 1981). These faults are large-displacement, low-angle normal faults that have been uplifted and warped into broad, undulating surfaces. The extension and lateral displacement of upper-plate rock was accommodated by the formation of numerous upper-plate fault zones. The Lincoln Ranch basin of the eastern Buckskin Mountains (Fig. 1) is above the Buckskin detachment fault in a domain of northeast-dipping normal faults.

Within the Lincoln Ranch basin there is a canyon exposure (Fig. 2) where the geometry of centimeter- to meter-scale normal faults can be examined locally in 100% outcrop in both map view and in

Structural and Tectonic Modelling and its Application to Petroleum Geology edited by R.M. Larsen, H. Brekke, B.T. Larsen and E. Talleraas. NPF Special Publication 1, pp. 231–241. Elsevier, Amsterdam.

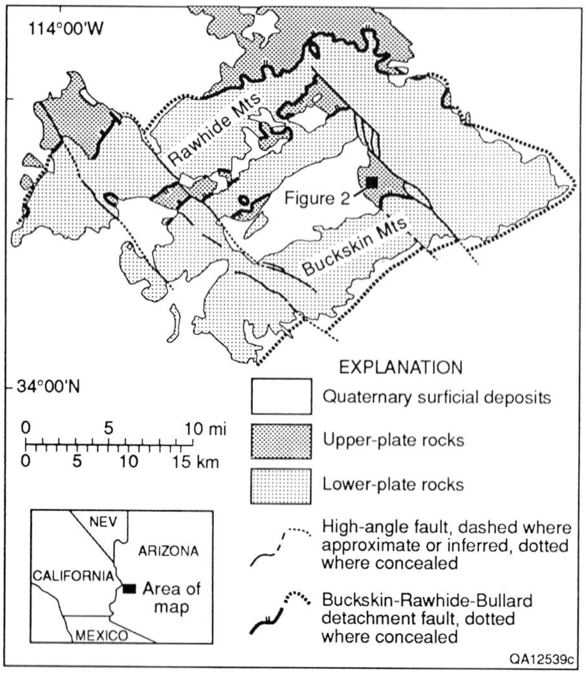

Fig. 1. Structural setting of the Lincoln Ranch basin in the Buckskin Mountains, west-central Arizona (modified from Spencer and Reynolds, 1989).

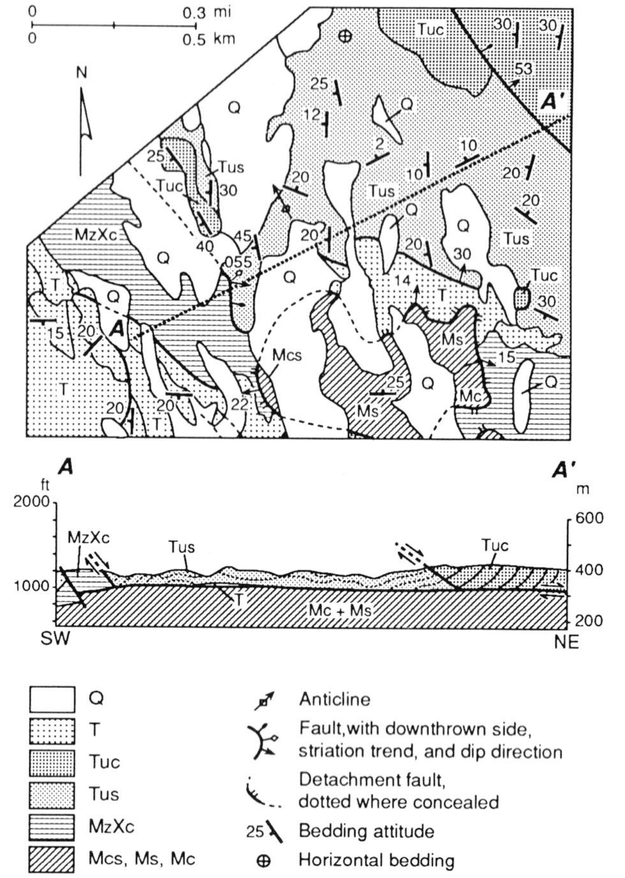

Fig. 2. Geologic map and cross-section of part of the Lincoln Ranch basin showing the setting of the normal fault exposures (modified from Spencer and Reynolds, 1989). *Q* = undifferentiated Quaternary gravel and sand; *T* = undifferentiated Tertiary rocks; *Tuc* = upper conglomerate (Tertiary Chapin Wash Formation); *Tus* = upper sandstone (Tertiary Chapin Wash Formation); *MzXc* = crystalline rocks of upper plate, undivided (Tertiary[?], Mesozoic[?] and Proterozoic[?]). Lower-plate rock: *Mcs* = highly tectonized metasedimentary and crystalline rocks; *Ms* = tectonized metasedimentary rocks; *Mc* = mylonitic crystalline rocks.

cross-section. The faults are in the lower Miocene Chapin Wash Formation, which consists there of fine-grained red to reddish-brown sandstones and siltstones. Sandstone beds range in thickness from centimeters to decimeters, but are typically 10 cm thick or less. Siltstone beds and laminae range from 2 mm to 3 cm thick. Beds do not thicken across faults. Grain fracturing and gouge formation along faults show that the unit was lithified prior to extension. Faulting in this exposure of the Chapin Wash Formation occurred at burial depths of less than 2 km (Spencer and Reynolds, 1989).

General characteristics of the faults

The faults we measured have offsets that range from several millimeters to 50 cm; larger faults have offsets of several meters to tens of meters. Faults range from planar to curved. Individual faults are exposed in cross-section for as much as 35 m and can be traced in map view for tens of centimeters to several meters. Spacing ranges from tens of meters for large faults, meters to tens of centimeters for medium-sized faults, to centimeters for small faults. The mean fault strike is northwest.

Faults have distinct, striated surfaces with no evidence of soft-sediment remobilization or injection of fault gouge into their wall rocks. The indurated fault gouge consists of aligned, scaly, silt-sized particles and is less than 1 mm thick on small-displacement (1 to 10 cm) faults and as much as 2 to 3 cm thick on proportionally larger faults. Slickenline striations on fault surfaces plunge down-dip and consist of straight to slightly curved polished ridges and grooves that match on opposed fracture surfaces.

Fault pattern in map view

In map view, faults range from straight to continuously curved to abruptly curved traces. Abruptly curved traces are composite surfaces comprising interconnected planar and curved parts. Traces are concave and convex toward the hanging wall, with no predominance of either shape. Some faults change from concave to convex along strike. Some faults have irregular or arborescent patterns consisting of short, planar parts and curved parts. Faults are commonly segmented, with segments in subparallel en

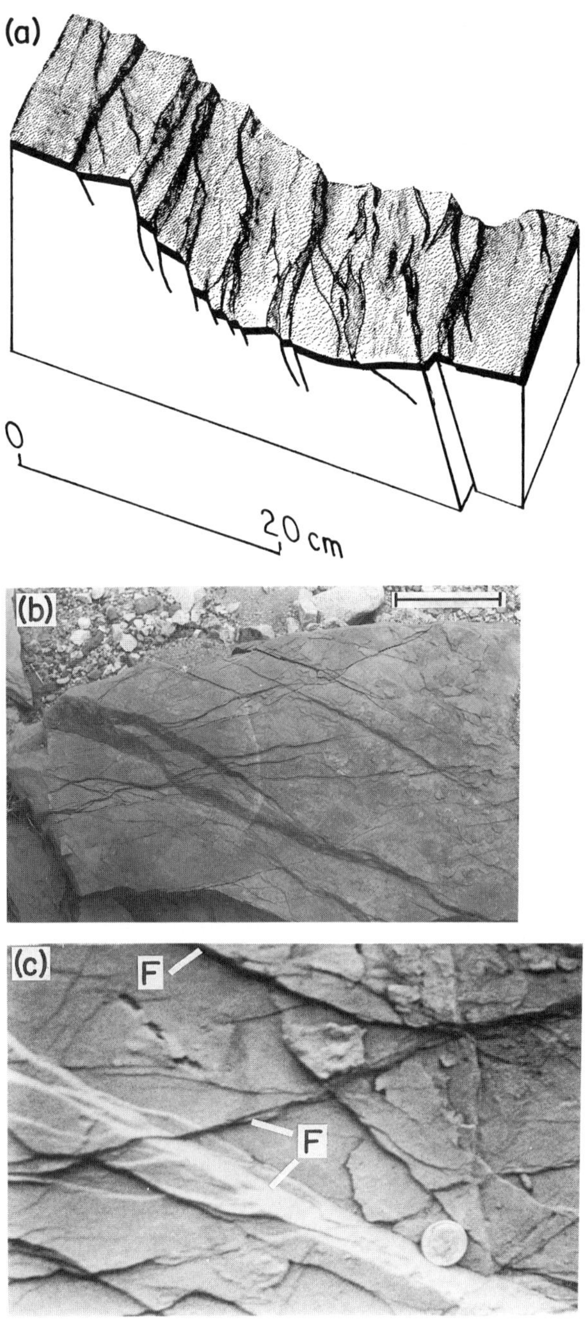

Fig. 3. (a) Map and cross-section view of anastomosing faults in domino-style array. Tracing from photograph. (b) Map view of fault termination. Scale is 30 cm. (c) Map view of fault pattern illustrating branching faults and crosscutting faults. F = main faults. Coin is 1.7 cm wide.

echelon or relay sets. Faults locally intersect and crosscut their neighbors at low and moderate angles, but no faults were observed paralleling the extension direction that could be interpreted as transfer faults. Adjacent faults interact with each other by forming complex systems of lateral ramps. Within lateral ramps, small faults locally have trends that are highly oblique to the main fault trend. Numerous small faults are localized where segments of large faults overlap. Slip sense derived from slickenline trends indicates a gradual change from dip slip to oblique slip along curved fault surfaces.

The faults in map view have a characteristic elongated diamond-shaped, rhomboid, or lenticular pattern created by intersecting planar faults and curved faults (Fig. 3). Aspect ratios of the fault-bounded lozenges are generally 3 : 1 to 5 : 1. The long axes of the fault-bounded lozenges within an outcrop are colinear and perpendicular to the striation trend on the fault surfaces. Diamond patterns in sandstone can result from anastomosing faults that dip in the same direction, consistent with the predominance of domino-style fault sets observed in cross-section, but crisscrossing faults with opposed dips also form these patterns.

Faults terminate along strike by losing slip or by intersecting another fault. The slip decreases abruptly near the fault termination itself. Along-strike loss of slip is locally accompanied by development of splays that branch from the main fault trace at a low angle (Fig. 3b). Anastomosing faults locally abut and terminate against other faults along strike. Some faults also crosscut and offset other faults at a low angle. The mutual crosscutting faults with different strike suggest simultaneous development of multiple faults. This fault pattern resembles those common among large (kilometer-scale) faults in other extended terrains (e.g., Donath, 1962).

Fault pattern in vertical section

The main fault types are curved faults, planar faults and faults with ramp-flat geometry. Curved faults display both downward-shallowing (concave-up) listric and downward-steepening (convex up) "negative listric" shapes (Figs. 4 and 5). Curved faults have curvature radii of 10 cm to several meters. Curved faults and planar faults are approximately equally abundant and concave-up and convex-up curved faults are also equally represented. Concave-up and convex-up faults produce relative stratal rotations of opposite sense, and the equal abundance of both fault types results in approximately constant bed dips across the exposure. Convex-up and concave-up faults that occur in the same area may effectively cancel out each other's stratal rotations even on a local (meter) scale (Fig. 4a). Curved faults exist in both sandstone and siltstone, and many curved faults cut across bedding without changing geometry. This is true for both large- and small-displacement faults.

Ramp-flat geometry occurs where faults cut bedding at a low angle (less than 45°), whereas faults with steep dip (50°–70°) tend to cross bedding with no deflection as they traverse different rock types. Ramps are dominantly in sandstone, and flats are

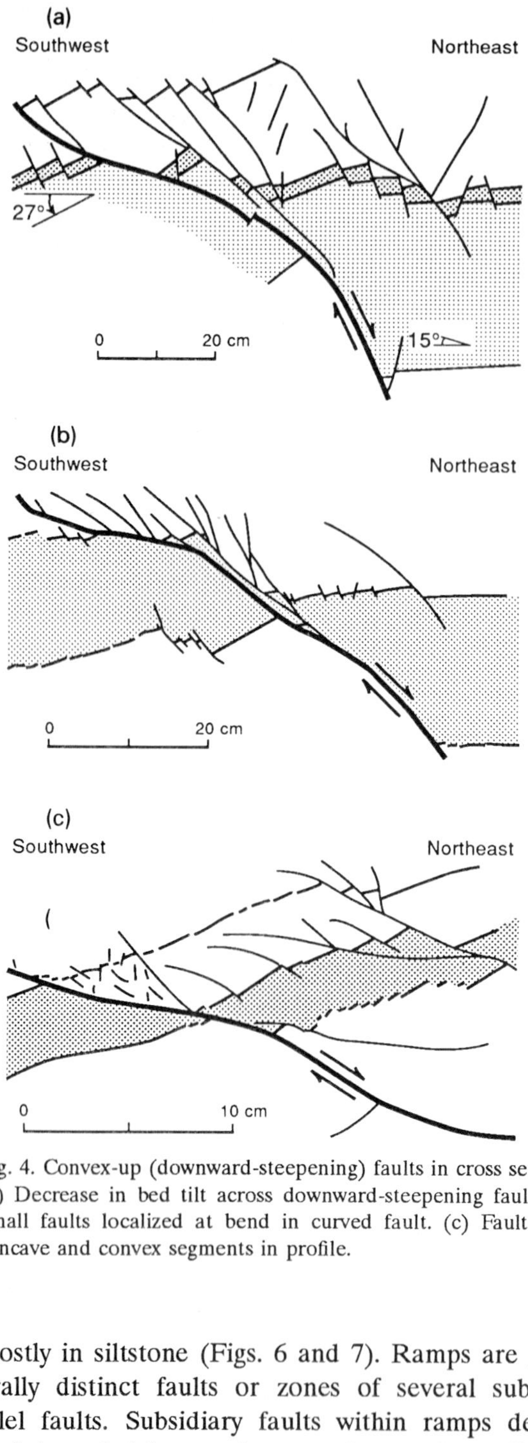

Fig. 4. Convex-up (downward-steepening) faults in cross section. (a) Decrease in bed tilt across downward-steepening fault. (b) Small faults localized at bend in curved fault. (c) Fault with concave and convex segments in profile.

Fig. 5. (a) Curved normal fault in sandstone. Curvature changes from concave up to convex up. Scale is in centimeters. (b) Progressive changes in bed tilt across several curved (steepening-downward) faults. Scale is 10 cm.

mostly in siltstone (Figs. 6 and 7). Ramps are generally distinct faults or zones of several subparallel faults. Subsidiary faults within ramps define fault-bounded lenses that are elongated parallel to the strike of the master fault. Typically, ramps are convex-up (negative listric) in their upper part and concave-up (listric) in their lower part. Ramps with greater displacement are straight and are associated with numerous subparallel fault slivers. Some faults that flatten and become parallel to siltstone layers do not reemerge from siltstone layers within the scale of the outcrop because displacement is accommodated by bedding-plane slip on multiple surfaces.

Flats are complex zones of distributed shear that is manifested in bedding-parallel slip and intersecting, mutually crosscutting arrays of curved and domino-style faults. These faults locally are superimposed on early formed fault-bounded lenses. Older faults in flats, which are crosscut and displaced by younger faults, are much less continuous and more difficult to recognize than younger faults. Subsidiary faults that have both gentle and steep dips are particularly common within and above flats. Steeply dipping faults are truncated by gently dipping faults that define lenses in which bedding is rotated. Mutually crosscutting faults and block rotation can cause abrupt thinning of beds. Closely spaced superposed faults can be so pervasive that the layers within the flat undergo 50% to 60% thinning.

Continuously curved and abruptly curved (ramp-flat) faults have antithetic and synthetic faults in their upper plates that define rollover, tilt-block and horst-and-graben structures. Most common are subsidiary faults that dip in the same direction, but more steeply than the master low-angle slip surface, defining synthetic tilt-block domains. Synthetic and antithetic faults locally crosscut one another. Upper-plate rocks above listric or ramp-flat faults commonly contain rootless graben and horsts bounded by planar or (locally) ramp-flat faults (Fig. 7). These horsts separate fault domains with opposed fault vergence and curvature, and they are less deformed than adjacent parts of the upper plate.

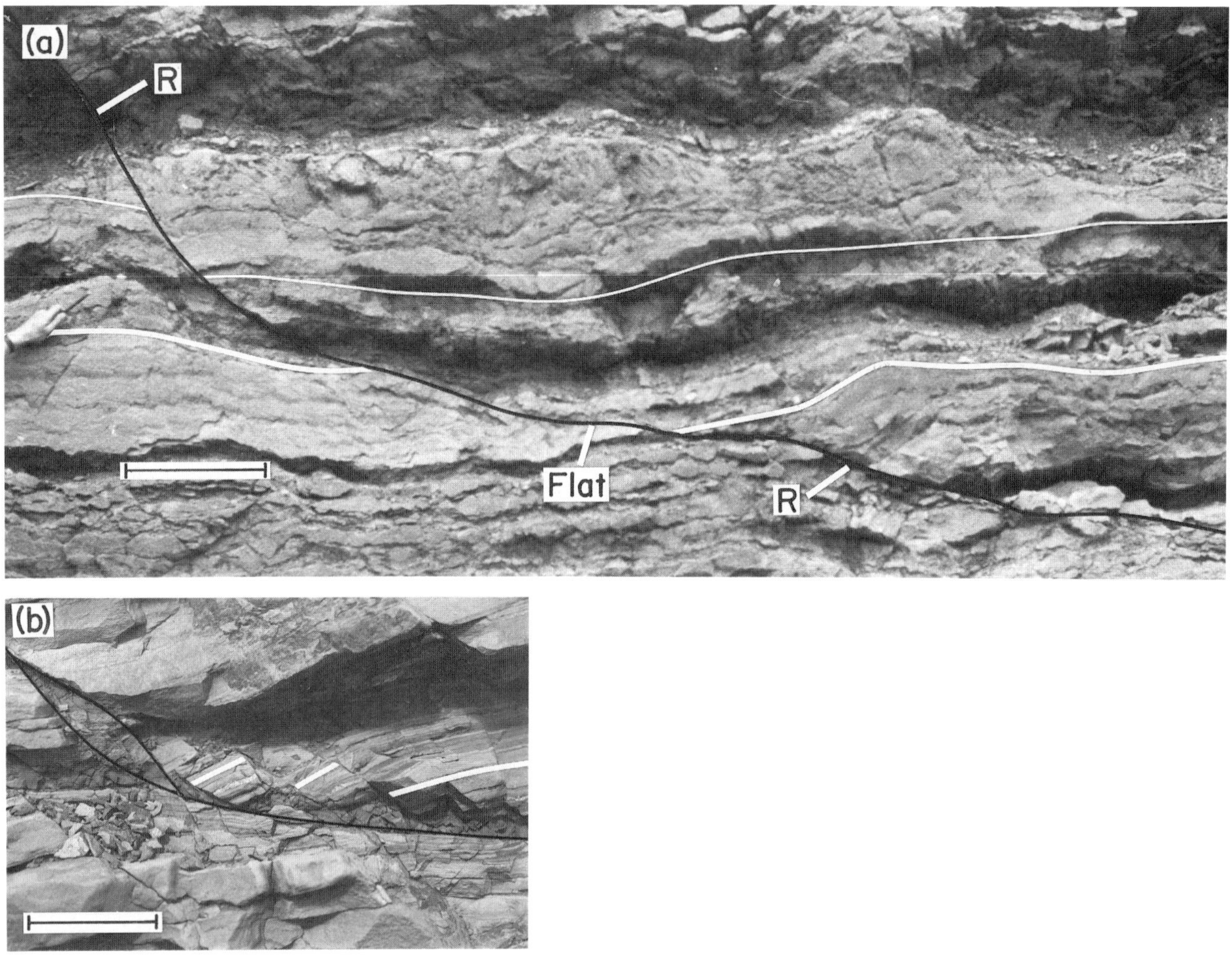

Fig. 6. (a) Fault with ramp-flat geometry resulting in overall fault curvature. *R* = ramp; *F* = flat. Scale is 35 cm. (b) Concave-up fault with subsidiary faults and rollover in the hanging wall. Scale is 10 cm.

Internal deformation of rollover structures due to both accommodation and extension of the upper plate was accomplished mainly by slip on arrays of smaller faults in synthetic tilt-block arrays and by crisscrossing horst-and-graben-bounding faults. Fault sets in upper-plate rocks cannot be interpreted as either vertical (Gibbs, 1983) or nonrotating inclined (White et al., 1986) shear planes, but rather they suggest accommodation along a set of shear planes rotating in domino fashion (Wernicke and Burchfiel, 1982). Moreover, upper-plate deformation is not restricted to the area of rollover accommodation, indicating that the bed-parallel stretch also contributes to upper-plate faulting.

Arrays of planar normal faults with domino (Fig. 8) and horst-and-graben (Fig. 9) patterns are common at all scales we studied. Domino patterns are particularly common in sandstone. The angle between fault plane and bedding is typically large, ranging from 60° to 90°, with 70° to 80° most common. Within an array, domino faults may have either uniform or variable dip. Where dips locally change within a fault domain, fault blocks are internally deformed by smaller faults or by truncation and rotation of domino faults along younger cross-cutting, and locally curved, faults. In some arrays, extensive internal deformation is confined to only a few domino blocks. The tops and bases of domino zones are marked by interstratal slip. Domino and horst-and-graben zones in siltstone are bounded locally above and below by unfaulted sandstone. Some faults within domino arrays of planar faults are curved (concave or convex up) rather than planar.

Faults of a much smaller scale than those of the domino blocks occur within individual dominos and commonly within only a part of the domino block. Subsidiary planar and smoothly curved faults prevail in the toes of domino blocks (Fig. 8), whereas horst-and-graben patterns occur in the less faulted block interiors. Domino-block edges, where weak bedding planes are present, locally display small-offset reverse faults.

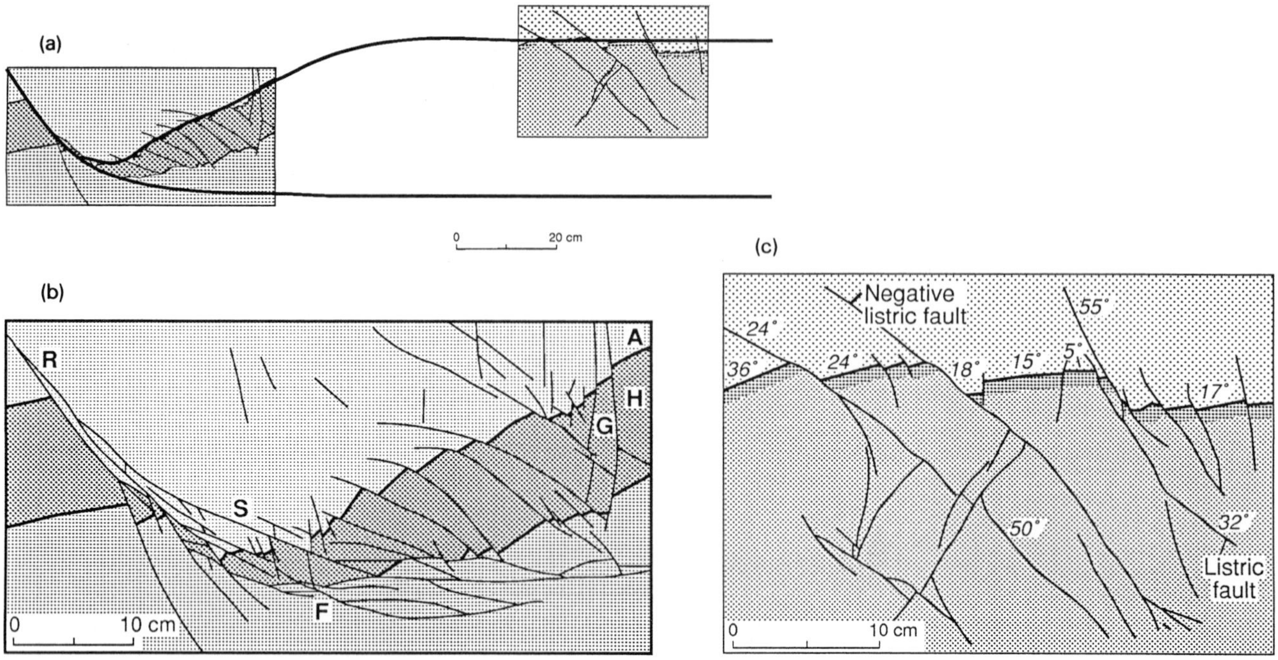

Fig. 7. Ramp-flat fault with rollover anticline. (a) Location overview. (b) R = ramp; F = flat; A = rollover anticline; S = upper-plate fault slivers containing superposed faults; G = upper-plate rootless graben; H = rootless horst. (c) Deformation of hanging wall block by steepening-downward faults in area distant from rollover. Tracings from photographs.

Fig. 8. Vertical section of tilted fault block within domino-style array. Small faults accommodate deformation within the block. S = basal slip surface of array. Width of block is 10 cm.

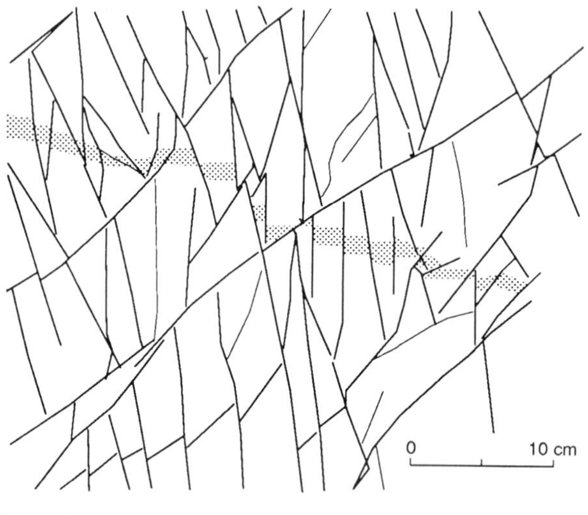

Fig. 9. Criss-crossing faults in outcrop. Tracing from photograph. The dotted areas show bedding plane. For experimental analogs see models presented by Horsfield (1980) and Woods (1988).

Some outcrops contain crisscrossing arrays of planar faults that typically are high-angle faults forming horst-and-graben patterns (Fig. 9). Crisscrossing planar faults offset faults with opposed dip or terminate against them. Arrays of small planar faults with horst-and-graben patterns locally abut larger planar faults. The acute bisectrix of paired faults in horst-and-graben patterns is at an angle of about 30° to the associated master faults and is oblique to layering. Small horst-and-graben patterns are commonly localized adjacent to larger faults; they are common in upper- and lower-plate rocks.

Comparison with physical models

Fault geometries in the Lincoln Ranch basin outcrops are similar to geometries of faults produced in experimental models conducted under controlled boundary conditions (e.g., imposed displacements or bulk strain). Comparison of outcrop examples and experimental results help elucidate the fault kinematics in this area.

Interpretation of map pattern

The outcrop displays three major fault patterns in map view, with the different patterns confined to specific areas or domains. In one type of pattern, lozenge-shaped blocks have fault traces that intersect at small angles; in another, lozenge-shaped blocks

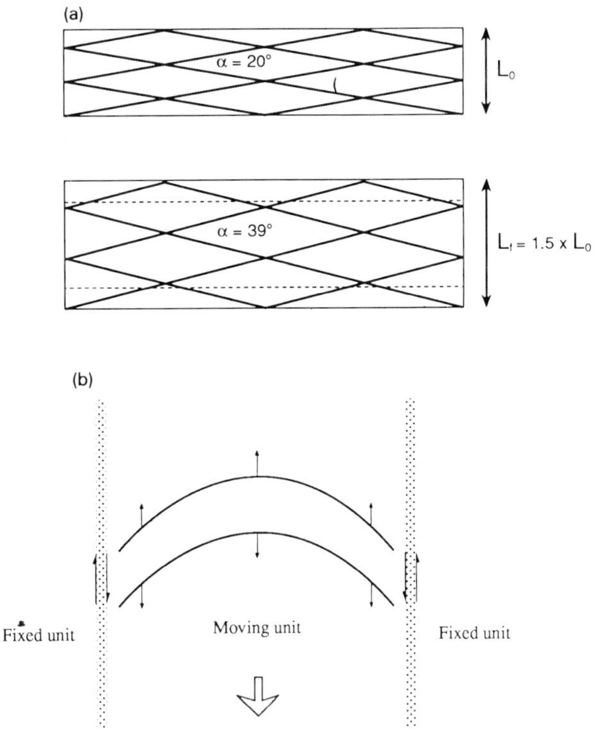

Fig. 10. (a) Increase in obliquity between fault traces induced by uniform uniaxial stretching (map view). Initial obliquity is 20°, and becomes 39° with a uniform stretch of 1.5. (b) Fault trace curvature induced by lateral friction. Ticks on the downthrown side of fault. Based on Vendeville (1987).

have faults that intersect at larger angles (>30°; Fig. 3). A third pattern has parallel faults that locally curve toward the direction of regional extension. The domains of fault traces intersecting at a small angle are similar to patterns created in several clay models subjected to uniaxial stretching (e.g., Cloos, 1955; Oertel, 1965; Freund and Merzer, 1976; Kautz and Sclater, 1988). Both clay experiments and mathematical simulations (Oertel, 1965; Reches, 1978) showed that lozenge-shaped fault blocks defined by four sets of normal faults acting simultaneously can accommodate plane-strain stretching (uniaxial extension). Such faults commonly intersect at small angles (≈15° or less), suggesting that outcrop areas with slightly oblique fault traces indicate plane-strain stretching, not necessarily separate phases of multidirectional stretching. This interpretation is consistent with the unidirectional trend of slip-sense indicators on the fault surfaces.

Locally the angles between crisscrossing fault sets in the exposure are higher (≈30° or greater). Such large angles can be interpreted in two ways. First, high angles may be caused by further uniform uniaxial stretching of initially slightly oblique crisscrossing faults that originally intersected at low angle (Fig. 10a). Second, high angles may reflect biaxial stretching, as in the clay models of Hoeppener et al. (1969), where four sets of faults intersecting at high angles accommodated extension along two perpendicular directions. In these experiments, the extension magnitudes in two stretching directions controlled the attitude and slip along each of the fault sets. Fault traces crosscutting at high angles in the field examples (e.g., Fig. 3b) can indicate areas of local biaxial stretching.

In a third pattern fault traces locally curve toward and merge with the direction of regional extension or have a sinuous pattern. Similar geometries are common in physical models (either sand or clay) where extension varies laterally across the model, causing different units of the model to undergo differential displacements (Cloos, 1968; McGill and Stromquist, 1979; Vendeville, 1987, 1991; Brix et al., 1985; Serra and Nelson, 1988). This effect is also commonly seen near the side walls of models with lateral friction. In these models, normal faults in the center of the model or distant from a unit boundary propagate laterally as oblique- or strike-slip faults with curved traces (Fig. 10b). In the field example this is illustrated by complex splay patterns along and at the lateral terminations of curved faults (Fig. 3c). Fault traces are convex toward the unit with greatest displacement and face either the hanging wall or footwall block, depending on the fault dip within the unit. Slip sense is synthetic to the shear displacement (and shear stress) along the unit lateral boundary (Fig. 10).

Although fault traces that are convex toward the footwall block are common in nature (e.g., Shelton, 1984), fault traces that are concave toward the footwall block have been documented in coal mines (Walsh and Watterson, 1988) and in transfer zones between rift offsets in experimental models (e.g., Faugère et al., 1986; Vendeville, 1988). Our field example shows curved fault traces that are both concave and convex to the hanging wall. Both field examples and experiments suggest that faults with curved traces are not necessarily listric in vertical section.

Fault shapes in vertical section

Both planar and curved faults exist in the outcrop example. We define two main categories of curved faults. The first category comprises faults that curve on a scale of 10 m or more in the outcrop; the entire fault plane is composed of multiple ramp and flat segments, leading to an overall listric geometry. The second category includes smaller faults (meter scale) that are either concave or convex upward; the curvature can be either mechanically controlled (initial) or induced by internal distortion of fault blocks.

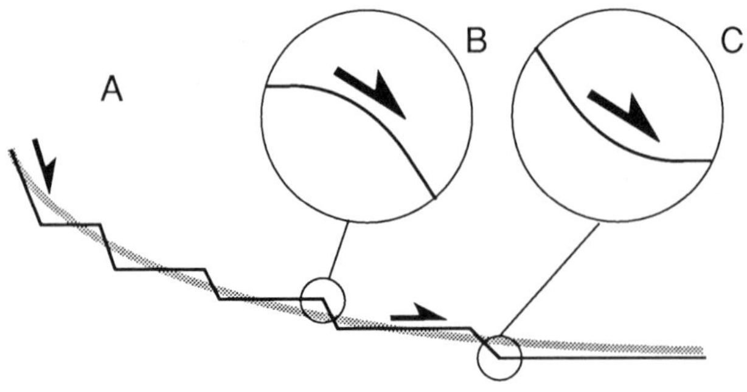

Fig. 11. Fault curvature associated with ramp-flat geometry. (A) Overall fault-plane curvature. Inset shows detail of curved fault segments. (B) Concave-down segment. (C) Concave-up segment.

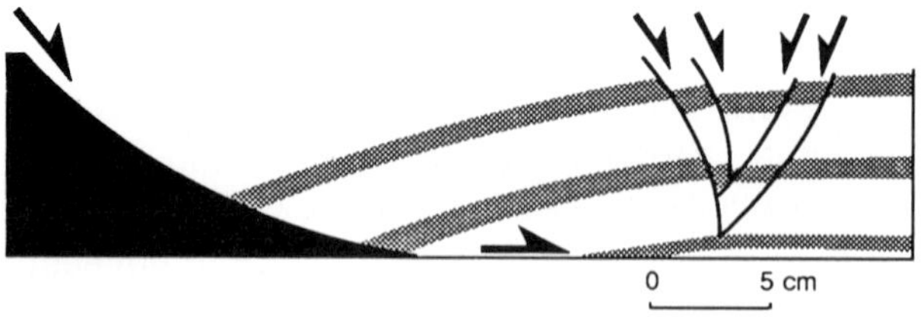

Fig. 12. Curved faults experimentally produced in rollover (in cross-section) (McClay and Ellis, 1987).

Large-scale (decameter) concave-up fault planes are formed by a series of ramp and flat segments. Fault planes climb upsection into sandstone layers, but they can propagate along bedding in shale layers. The proportion of the length of the flat to the total fault length can vary vertically in the outcrop section, leading to overall fault curvature (Figs. 6a, 11a). Initially, when displacement approximately equals bed thickness, beds influence fault geometry, and there are abrupt changes in fault dip between ramp and flat. Faults with ramp-flat geometry have displacement on the same scale as the mechanically significant bed thickness. If fault displacement is large compared with bed thickness, faults are more planar; the fault is smoothed by tectonic erosion. This may result from formation of duplexes and splays adjacent to ramps. Faults that can link ramps into a more planar surface also may develop from flats (Fig. 7). Such faults have not been modelled experimentally in extension so far.

Some small-scale faults bound rigid, unstrained blocks and formed with initial curvature. Examples are faults that link ramps and flats (Fig. 11b, c) and faults that form in sandstone in rollovers above major faults. Such small-scale faults can be either convex-up or concave-up. Experiments of rollover faulting by McClay and Ellis (1987; Fig. 12) show similar geometries and suggest that curved stress trajectories cause initial fault curvature.

Other physical models (W.T. Horsfield, in Mandl, 1988, p. 33; Vendeville, 1991) show that internal block deformation can increase curvature of initially planar or slightly curved faults (Figs. 13 and 14). A vertical change in the magnitude of horizontal extension and associated shear strain within blocks progressively bends the fault plane upward or downward (Fig. 13b, c).

In the outcrop example, convex-up and concave-up faults are equally abundant and affect both highly deformed shale layers and sandstone layers that behave more rigidly. Determining whether fault curvature is initial or acquired is important in interpreting block kinematics. For example, in sandstone, fault curvature controls rotation of adjacent rigid fault blocks. Downward-steepening faults can drastically affect the overall geometry of an extended terrain by counteracting stratal rotations caused by concave-up

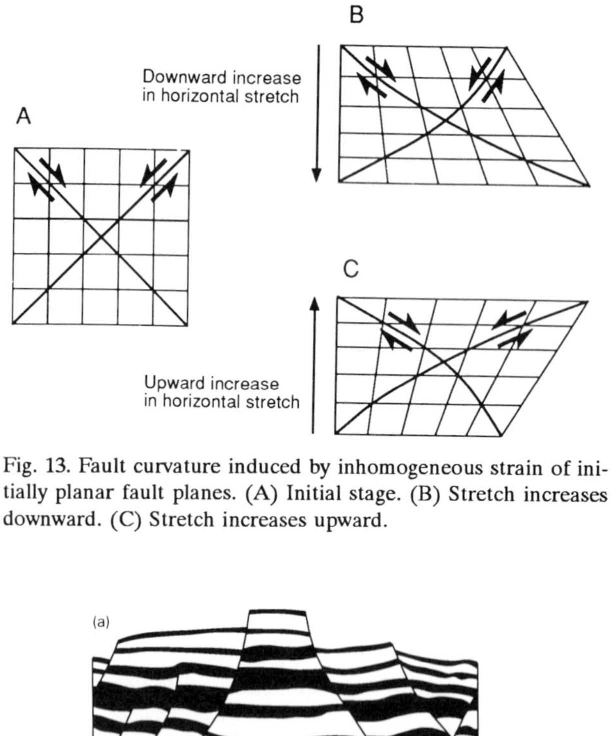

Fig. 13. Fault curvature induced by inhomogeneous strain of initially planar fault planes. (A) Initial stage. (B) Stretch increases downward. (C) Stretch increases upward.

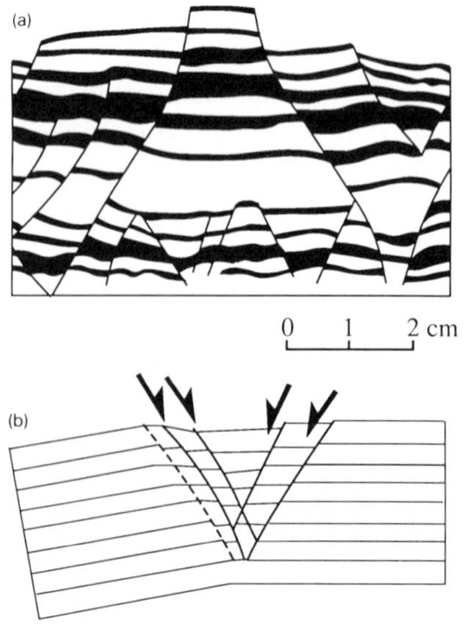

Fig. 14. Physical models of curved faults that were initially planar. (a) Vendeville et al., 1987. (b) Based on W. T. Horsfield in Mandl, 1988.

listric and tilt-block faults and, therefore, they can profoundly affect estimates of extension magnitude based on the dip of rotated planar faults and bedding (Wernicke and Burchfiel, 1982). However, in shale and some siltstone beds, faults are more likely to be distorted by block strain during extension and, therefore, exert little control on block rotation. In contrast to initially curved faults, curved faults in shale may indicate block strain.

Block kinematics and strain

In the outcrop, significant fault-block deformation takes place in three settings: above and below tilt-block domains in penetratively faulted shale layers, locally within tilt blocks, and adjacent to major curved faults in rollover anticlines and zones of distributed faults.

Tilt blocks have been recognized in many extensional terrains (e.g., Morton and Black, 1975; Proffett, 1977), and they have been reproduced in physical models (e.g., Vendeville et al., 1987). In models, domino-style kinematics are generally regarded to indicate the influence of an undeflectable boundary (Mandl, 1987, 1988; Vendeville, 1987) along which some shear stress occurs (Brun et al., 1985). In the outcrop example domino tilt blocks comprise sandstone layers bounded above and below by shales. Zones with tilt blocks are bounded above and below by unfaulted layers (generally sandstone) that act as rigid boundaries (Fig. 8).

Deformation at the base and top of domino blocks is concentrated along the edges of sandstone fault blocks where intervening shale layers are thin. In contrast, thick adjacent shale layers accommodate potential volume problems created by domino rotation by concentrating small-scale faulting. Similarly, models made entirely of brittle layers of sand or clay show penetrative deformation of the base of the block (McClay and Ellis, 1987; Vendeville et al., 1987; Kautz and Sclater, 1988), whereas fault blocks undergo little internal deformation in models underlain by deformable ductile putty (Brun et al., 1985; Vendeville, 1987).

In outcrops, some faults have attitudes that differ either in amount or sense of dip from that which dominates within the array (Fig. 15). Because fault dip, rotation and stretch are related, changes in fault dip within domino domains imply local changes in stretch along the basal detachment and local gaps and overlaps at the base and tops of tilt blocks. In the outcrop, rigid layers above and below tilt blocks prevent any significant gap or overlap between layers. Instead, fault blocks bounded by antithetic or abnormally steep or gently dipping faults are internally deformed by closely spaced small faults.

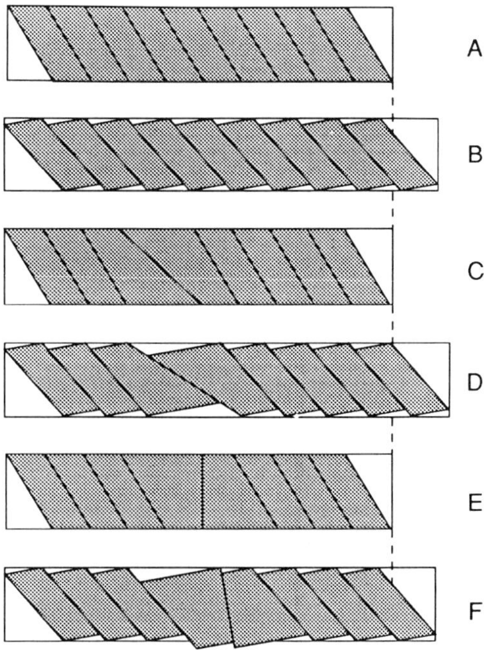

Fig. 15. Kinematics of block rotation with even (A, B) or uneven (C–F) fault dips.

In the hanging wall block of curved faults, subsidiary structures such as crosscutting synthetic and antithetic fault fans, rollover anticlines and crestal collapse graben occur that are closely similar to fault patterns created in analog experiments with sand with a rigid footwall (e.g., McClay and Ellis, 1987) (Figs. 7 and 12). However, the outcrop example differs from McClay and Ellis' model in two respects. First, restoration of cross-sections from the outcrop suggests that the shape of the rollover evolved partly by propagation of synthetic faults into the footwall block. Hanging wall deformation that results from this process has not yet been modelled, since experiments have been carried out with undeformable footwall ramps. Second, in physical models deformation of the hanging wall was solely caused by sliding along the major listric fault plane and did not involve superimposed extension of the hanging wall. Hence, no faults occurred away from the main fault beyond the crestal graben (McClay and Ellis, 1987). Instead, most listric normal faults in the Lincoln Ranch basin display numerous faults that affect the hanging wall away from the main fault and the rollover crest (Fig. 7). This suggests that a significant component of stretching is superimposed on the rollover deformation (Fig. 16).

Implications for interpretation of extended terrains

Excellent exposures that illustrate the details of fault-block shape and their kinematic interpretation

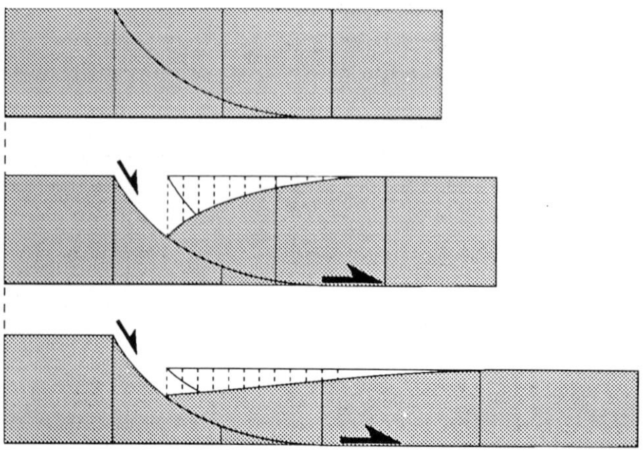

Fig. 16. Rollover profiles without (B) and with (C) superimposed extension of the hanging wall. (A) Shows configuration prior to fault movement. Accommodation is by vertical shear.

based on experimental results lead to the following remarks.

Analysis of three-dimensional fault geometries shows that the upper plate of the detachment fault is compartmentalized into several mesoscale units on the scale of tens of meters that accommodate the bulk extension. Within most units deformation remains plane-strain extension accommodated along four sets of faults that intersect at low angles. Units are separated from each other either by oblique-slip faults with curved traces or by units that have undergone biaxial stretching.

The outcrop example illustrates the significant role of curved faults in the development of overall fault shape and kinematics. Some faults have an overall curvature that in detail consists of straight segments and sharp bends (ramps and flats). Others are smoothly curved and either bound rigid blocks or significantly deformed blocks. Convex-up (steepening downward or negative listric) faults are abundant and play a significant role in the overall geometry of the faulted terrain. Such faults may be more important in extended terrains than is currently appreciated. When they bound rigid blocks, negative listric faults can counteract tilting caused by listric faults or domino-style rotation. Hence they cause underestimation of the amount of extension deduced from block tilt. Underestimation of extension may also result from internal deformation at the base and top of rotating fault blocks as shown in the models of Kautz and Sclater (1988). Similar problems in extension estimation and section restoration arise in stretched hanging wall blocks of major listric faults.

Fault-block shapes reflect crisscrossing and superposed faults having a range of sizes. Complex within-block fault geometry, at medium scale, can result from simple kinematics. Such complex fault geometries are likely to be below resolution of seismic detection (i.e., tens of meters or less) but could influence reservoir compartmentalization and heterogeneity.

Acknowledgments

We thank M.P.A. Jackson, D.D. Schultz-Ela and J.E. Spencer for discussions and reviews, and Bjorn Torudbakken and Rob Young for reviews. This paper is based upon work supported by the Texas Advanced Research Program under Grant No. 2174. Publication authorized by the Director, Bureau of Economic Geology, The University of Texas at Austin.

References

Brix, M., Schwarz, H.-U. and Vollbrecht, A., 1985. Tektonische Experimente als Beitrag zu Strukturanalysen im Ruhrkarbon. Glucksauf-Forschungshefte, 46: 4–11.

Brun, J.P., Choukroune, P. and Faugère, E., 1985. Les discontinuités significatives de l'amincissement crustal: application aux marges passives. Bull. Soc. Géol. Fr., 8: 139–144.

Cloos, E., 1955. Experimental analysis of fracture patterns. Geol. Soc. Am. Bull., 66: 241–256.

Cloos, E., 1968. Experimental analysis of Gulf Coast fracture pattern. Am. Assoc. Pet. Geol. Bull., 52: 420–444.

Donath, F.A., 1962. Analysis of basin-range structure, south-central Oregon. Geol. Soc. Am. Bull., 73: 1–16.

Faugère, E., Brun, J.P. and Van Den Driessche, J., 1986. Bassins asymétriques en extension pure et en décrochement: Modèles expérimentaux. Bull. Centres Recherche Exploration Production Elf-Aquitaine, 10: 13–21.

Freund, R. and Merzer, A.M., 1976. The formation of rift valleys and their zigzag fault patterns. Geol. Mag., 113: 561–568.

Gibbs, A.D., 1983. Balanced cross-section construction from seismic sections in areas of extensional tectonics. J. Struct. Geol., 5: 153–160.

Hoeppener, R., Kalthoff, E., Schrader, P., 1969. Zur physicalischen Tektonik Bruechbildung bei verschieden affinen Deformation im Experiment. Geol. Rundsch., 59: 179–193.

Horsfield, W.T., 1980. Contemporaneous movement along crossing conjugate normal faults. J. Struct. Geol., 2: 305–310.

Kautz, S.A. and Sclater, J.G., 1988. Internal deformation in clay models of extension by block faulting. Tectonics, 7: 823–832.

Mandl, G., 1987. Tectonic deformation by rotating parallel faults — the "bookshelf" mechanism. Tectonophysics, 141: 316–327.

Mandl, G. (Editor), 1988. The Mechanics of Tectonic Faulting. Elsevier, Amsterdam, 407 pp.

McClay, K.R. and Ellis, P.G., 1987. Analogue models of extensional fault geometries. In: M.P. Coward, J.F. Dewey and P.L. Hancock (Editors), Continental Extensional Tectonics, Geol. Soc. London, Spec. Publ., 28: 109–127.

McGill, G.E. and Stromquist, A.W., 1979. The graben of Canyonlands National Park, Utah: geometry, mechanics and kinematics. J. Geophys. Res., 84: 4547–4563.

Morton, W.H. and Black, R., 1975. Crustal attenuation in Afar. In: Pilger, A. and Rosler, A. (Editors), Afar Depression of Ethiopia, Inter-Union Comm. Geodyn. Sci. Rep., No. 14, pp. 55–65.

Oertel, G., 1965. The mechanism of faulting in clay experiments. Tectonophysics, 2: 343–393.

Proffett, J.M., Jr., 1977. Cenozoic geology of the Yerrington district, Nevada and its implications for the nature and origin of Basin and Range faulting. Geol. Soc. Am. Bull., 88: 247–266.

Reches, Z. 1978. Analysis of faulting in three-dimensional strain field. Tectonophysics, 47: 109–129.

Serra, S. and Nelson, R.A., 1988. Clay modelling of rift asymmetry and associated structures. Tectonophysics, 153: 307–312.

Shelton, J.W., 1984. Listric normal faults: an illustrated summary. Am. Assoc. Pet. Geol., Bull., 68: 801–815.

Spencer, J.E. and Reynolds, S.J., 1989. Geology and mineral resources of the Buckskin and Rawhide Mountains, West-Central Arizona. Ariz. Geol. Surv. Bull., 198, 279 pp.

Vendeville, B.C., 1987. Champs de failles et tectonique en extension: modélisation expérimentale. Rennes, France, Mémoires du Centre Armoricain d'Etudes Structurales des Socles, 15, 392 pp.

Vendeville, B.C., 1988. Fault geometries and location in sedimentary cover during basement-controlled deformation: an experimental investigation. Am. Assoc. Pet. Geol., Bull., 72: 1029 (abstract).

Vendeville, B.C., 1991. Mechanisms generating normal fault curvature: a review illustrated by physical models. In: A. Roberts, G. Yielding and B. Freeman (Editors), Geometry of Normal Faults. Geol. Soc. London, Spec. Publ., 56: 241–249.

Vendeville, B.C., Cobbold, P.R., Davy, P., Brun, J.P. and Choukroune, P., 1987. Physical models of extensional tectonics at various scales. In: M.P. Coward, J.F. Dewey. and P.L. Hancock (Editors), Continental Extensional Tectonics. Geol. Soc. London, Spec. Publ., 28: 95–109.

Walsh, J.J. and Watterson, J., 1988. Dips of normal faults in British Coal Measures and other sedimentary sequences. J. Geol. Soc. London, 145: 859–875.

Wernicke, B.P., 1981. Low-angle normal faults in the Basin and Range province: nappe tectonics in an extending orogen. Nature, 291: 645–648.

Wernicke, B.P. and Burchfiel, B.C., 1982. Modes of extensional tectonics. J. Struct. Geol., 4: 105–115.

White, N.J., Jackson, J.A. and McKenzie, D.P., 1986. The relationship between the geometry of normal faults and that of the sedimentary layers in their hanging walls. J. Struct. Geol. 8: 897–910.

Woods, E.P., 1988. Extensional structures of the Jabiru terrace, Vulcan sub- basin. In: P.G. Purcell and R.R. Purcell (Editors), The North West Shelf, Australia. Proceedings, North West Shelf Symposium, Petroleum Exploration Society Australia, Perth, W.A., pp. 311–330.

S.E. LAUBACH *Bureau of Economic Geology, University of Texas at Austin, University Station Box X, Austin, TX 78713-7508, U.S.A.*

B.C. VENDEVILLE *Bureau of Economic Geology, University of Texas at Austin, Austin, TX 78713-7508, U.S.A.*

S.J. REYNOLDS *Arizona Geological Survey, Tucson, AZ 85719, U.S.A.*

Complex hanging-wall deformation above an extensional detachment — example: Gullfaks Field, northern North Sea

A.G. Koestler, A.G. Milnes and A. Storli

Analysis of the geometry and kinematics of the Gullfaks Fault System, based on the Statoil interpretation of the 1985 three-dimensional seismic survey, itself constrained by abundant well data, shows that the overall structure can best be understood in terms of hanging-wall collapse in a pull-apart zone above a sub-Triassic extensional detachment. The "domino"-style faulting in the Western Gullfaks reservoir horizons is part of this pull-apart zone and shows an unusually well-constrained geometry, which, independent of the proposed kinematic model, allows important conclusions to be drawn. In particular, the shape of the irregular fault surfaces and the constancy of bedding can only be understood in terms of bulk "plane strain", with a vertical E–W deformation plane containing the E-plunging slip vector. E–W depth-converted cross-sections can therefore be restored using simple balancing techniques, taking into account the probable original dip of the faults (about 70°) and bedding planes (sub-horizontal) and the probable effect of syn- to post-faulting compaction (shown to be of minor importance). The restored sections show that horizontal extension in the pull-apart zone amounted to about 50%, and that considerable homogeneous shear strain must have affected the reservoir rocks in the fault blocks, a fact which may be significant for reservoir behaviour and modelling.

Introduction

The Gullfaks oil-field lies close to the main axis of the Viking Graben in the northern North Sea (Fig. 1). The geology of the field is well-known because of high-quality three-dimensional seismic surveys and extensive borehole control (Saeland and Simpson, 1981; 1982; Erichsen et al., 1987; Fossen, 1989; Petterson et al., 1990). Structural and tectonic interpretations have been difficult, however, because of the discontinuity of the main reflectors and the uncertainty of fault correlation between different levels. Nevertheless, there is general agreement that the Jurassic reservoir units show complex normal faulting, related to extensional deformation in latest Jurassic and early Cretaceous times. Although there are some indications that the structure may have been affected by later non-extensional

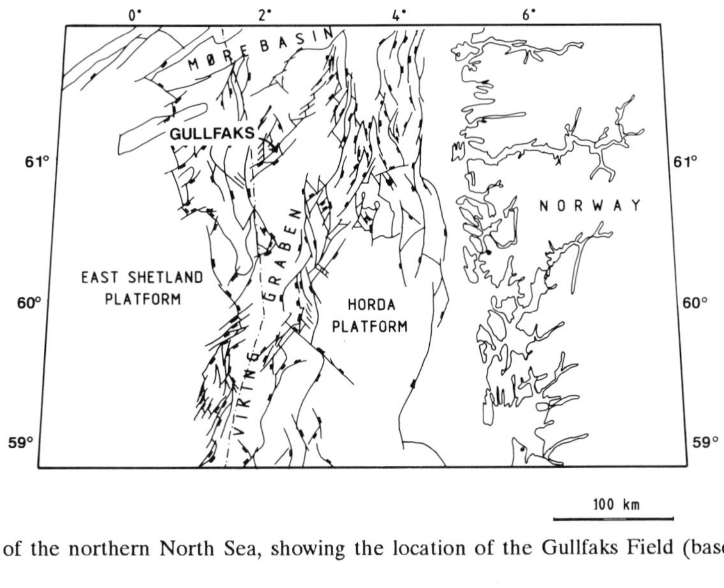

Fig. 1. Tectonic sketch map of the northern North Sea, showing the location of the Gullfaks Field (based on Færseth and Pederstad, 1988).

Structural and Tectonic Modelling and its Application to Petroleum Geology edited by R.M. Larsen, H. Brekke, B.T. Larsen and E. Talleraas. NPF Special Publication 1, pp. 243–251. Elsevier, Amsterdam.

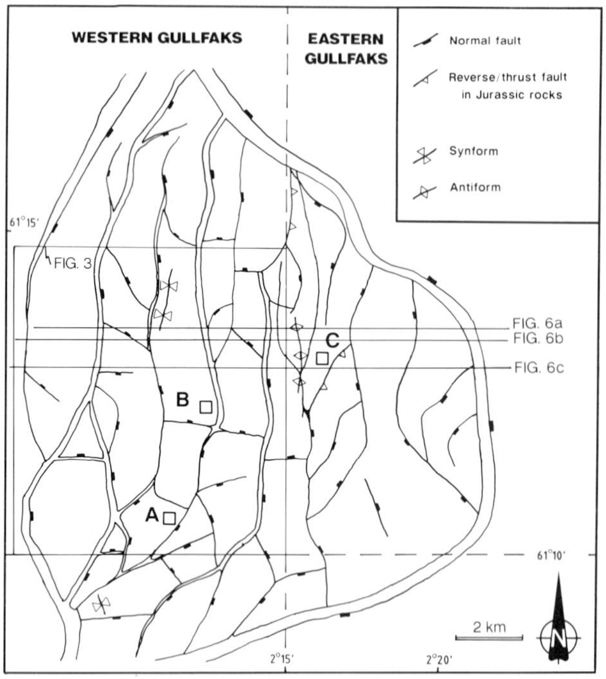

Fig. 2. Structure map of the Gullfaks Field at top Dunlin level, showing the position of Figs. 3 and 6 and of the three production platforms *A*, *B*, *C* (based on Fossen, 1989).

movements (Fossen, 1989), such modifications seem to have been slight compared to the overall extensional strain which characterizes the internal structure of the Gullfaks field. In this paper, we focus on the latter, the main deformation of the reservoir units and, hence, the main determinant of internal reservoir structure. As we shall see, our analysis suggests that it represents a pull-apart and collapse zone above a major sub-horizontal extensional detachment, whose regional significance is at present unknown. In Gullfaks, this complex zone of hanging-wall deformation can be subdivided into two contrasting structural regimes, Western Gullfaks and Eastern Gullfaks, with a gradational boundary approximately along longitude 2°15′ (Fig. 2). In the following, we first summarize the basic seismic structure of these two regimes, before presenting an analysis of the geometrical relationships and a tentative kinematic model.

Seismic structure

The data base for our analysis is the Statoil interpretation of the 1985 three-dimensional seismic survey, in which all the main reflectors have been identified and mapped on the basis of extensive well correlations. The data was available as interpreted seismic time-sections (work station, paper copies) and as depth-converted structure contour maps at several different stratigraphic levels (scale 1 : 25 000, see Petterson et al., 1990; Fig. 6). The quality of the seismic data can be judged by consulting the published lines (uninterpreted and interpreted) in Fossen (1989) and Petterson et al. (1990). In the present work, non-reflective faults are only considered as hard data where they can be fixed in orientation by alignment of three or more reflector terminations. Our aim has not been to reinterpret the seismic data itself, but rather to identify, and to carry out an evaluation of, those structural features of a well-established seismic interpretation which can be regarded as definitive. The quality of the geophysical and well data allows an unusually clear distinction to be made between observation and interpretation, comparable to onshore geological mapping.

Reservoir level, Western Gullfaks

As documented by several prominent Jurassic reflectors (top Statfjord, top Dunlin, top Rannoch, top Brent), the seismic structure of the Gullfaks reservoir horizons west of 2°15′ (e.g., the area surrounding A and B platforms, Fig. 2) is dominated by a prominent system of roughly N–S-trending normal faults, with displacements in the range of a few hundred metres, all with downthrow to the east. In horizontal sections (Fig. 3), the fault traces are sinuous and branching (i.e., the fault surfaces are non-planar/non-parallel). They delineate lozenge-shaped fault blocks with N–S long axes. Geometric analysis of the individual fault surfaces shows them to be strongly "cylindrical" in three-dimensions, i.e., the poles to the fault surfaces measured at various positions along the faults show a great circle distribution in stereographic projection (Fig. 4). This "cylindricity" is repeated throughout the western part of the field, giving a mean "axis of cylindricity" plunging 30–35° due east (Fig. 5). At the same time, the statistical averages of bedding orientation within the various fault lozenges remain constant and tightly confined, with an overall mean dip of 10–15° west (Figs. 4 and 5). This remarkable geometry puts tight constraints on the kinematics of fault movement, which can only have taken place by slip parallel to the mean "axis of cylindricity". Appreciable slip in any other direction would have caused complex and heterogeneously distributed internal adjustments within the lozenges, which the constancy of bedding belies. Hence, the Western Gullfaks fault system at reservoir level corresponds to the "domino" fault model (e.g., Wernicke and Burchfiel, 1982; Axen, 1988; Barr, 1987), although the word "domino" in this context is somewhat misleading, since we are obviously dealing with very irregularly shaped fault blocks. In three-dimensions, the Western Gullfaks structure is an almost ideal example of bulk "plane strain", with a ver-

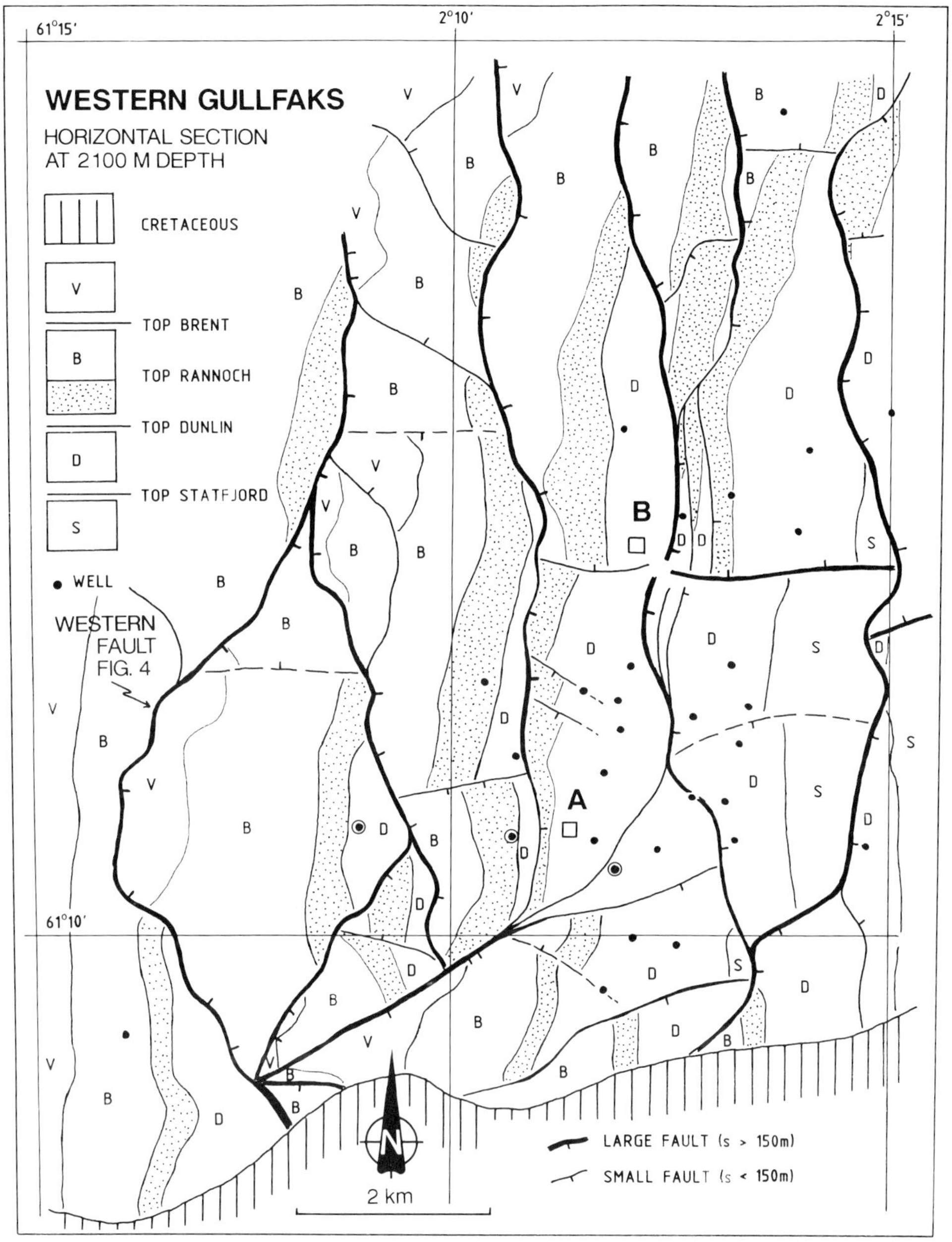

Fig. 3. Horizontal section of the western part of the Gullfaks field at a depth of 2100 m (for location see Fig. 2), based on the Statoil interpretation of the 1985 three-dimensional seismic survey (ST 8511). The section is constructed from 1:25000 depth-converted structure contour maps for top Statfjord, top Dunlin, top Rannoch and top Brent, with stratigraphic control from about 40 wells. Ticks on the fault traces indicate the downthrown side and also mark the positions at which the fault orientation and displacement were measured (for stereographic projection see Fig. 5).

tical E–W-trending deformation plane. This means that all movement has taken place within the deformation plane (fault slip, internal deformation of fault blocks) and there has been no dimensional changes in the N–S direction. Whether by chance or by design, the lines of the 1985 three-dimensional seismic survey parallel precisely the deformation plane, hence the ideally "domino" aspect of the faults in the western part of the Gullfaks Field on seismic sections (Fig. 6).

Reservoir level, Eastern Gullfaks

The seismic structure of the Gullfaks reservoirs east of about 2°15′ (e.g., the area surrounding C platform, Fig. 2) does not show such neat geometrical relationships. Figure 6 shows the reflector reconstructions from three lines of the 1985 three-dimensional survey, taken from the Statoil seismic interpretation as published in Fossen (1989). On such sections, only identified reflector traces and those segments of fault traces which can be "pinned" by lines of reflector terminations are relevant to the present

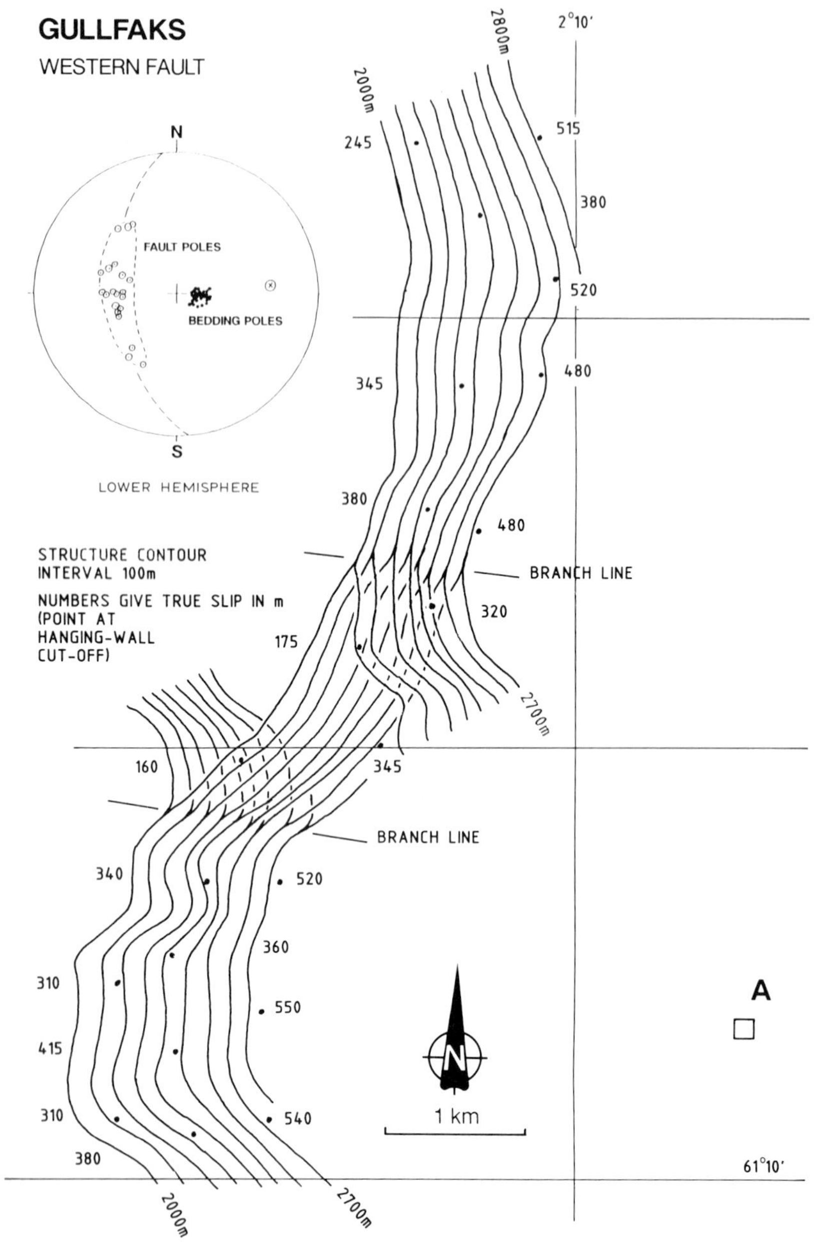

Fig. 4. Example of a structure contour and true slip variation map of a large fault from the Gullfaks field (western fault, see Fig. 2). The inset is a stereogram of the fault surface, showing poles to the fault plane at the points of slip determination. The partial great circle distribution of the fault plane poles results from the "cylindricity" of the fault surface, and the axis of the girdle (X) marks the true slip direction (for explanation see text).

study. The "domino"-style faulting of Western Gullfaks is seen to change eastward into an irregular association of both eastward- and westward-dipping normal faults, possibly with some small reverse faults and flexures. Some of these complications may be due to minor superimposed non-extensional deformation (transpression, cf. Fossen, 1989), but they are also of the type encountered in any area of extensional deformation where the boundary conditions do not allow compatible movements (e.g., Coffield and Schamel, 1989). The most constant structural feature of Eastern Gullfaks is a large subvertical fault with a downthrow of a few hundred metres to the west, i.e., antithetic to the Western Gullfaks faults (Fig. 6). This fault marks the western margin of the so-called Gullfaks "horst" (Saeland and Simpson, 1982; Fossen, 1989), a deeply eroded high along the eastern edge of the field. To the east, this high is bounded by the Main Boundary Fault, a major structure, with a throw of 1–2 km, which can be traced on deep seismic profiles far down into the crust (cf. Beach, 1986; Gibbs, 1987). The seismic structure of the Eastern Gullfaks high differs from the rest of the field in that it shows only small, widely spaced dis-

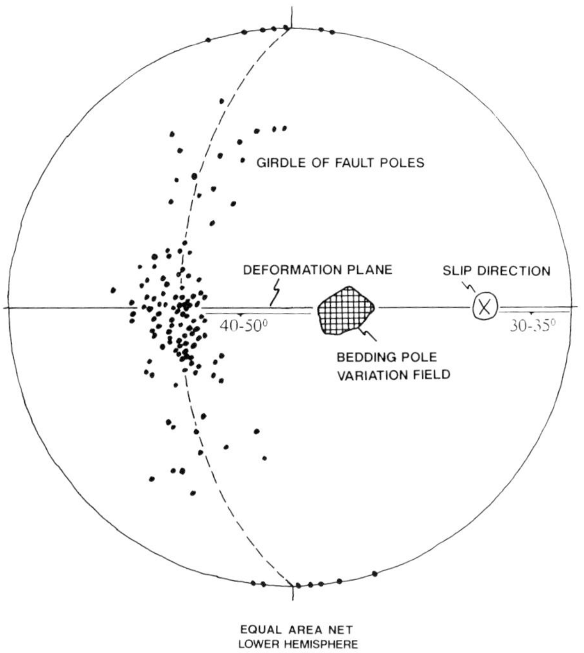

Fig. 5. Compilation stereogram for fault plane measurements in the western part of the Gullfaks field (area of Fig. 3). The great circle distribution of the fault plane poles strongly suggest a plane strain movement picture for the western Gullfaks fault system, with a sub-vertical E–W deformation plane and slip direction marked by the girdle axis (plunge 30–35° towards E).

continuities in the reflector pattern, mostly normal faults both synthetic and antithetic to the fault along its western margin (Fig. 6).

Intra-Triassic level

The prominent array of reflectors at intra-Triassic level (e.g., top Carnian, Fig. 6) are not as well constrained as the Jurassic reservoir horizons described above (no well control), and the constructed fault orientations are less reliable (fewer reflector terminations). Hence, data from this level cannot be analysed so definitively. Nevertheless, it is interesting to note that it can best be interpreted as a similar structure to that documented above, i.e., as a "domino"-style fault system to the west and more complex relations of similar geometry to the east (as drawn in Fig. 6). The apparent lack of continuity between the faults at Jurassic and intra-Triassic levels may be structurally significant (e.g., implying decoupling and ductile deformation in intervening incompetent horizons) or it may be an artifact of processing and interpretation. Be that as it may, the congruence of structural style at the two levels suggests that they both belong to the same tectonic regime. This point leads us to place the main detachment of our kinematic model below the intra-Triassic reflector array (see below).

Kinematic model

The seismic structure described above (see also Petterson et al., 1990) suggests a kinematic interpretation in terms of hanging-wall collapse, in a region of pull-apart above an extensional detachment (Fig. 7). The interpretation is based on the remarkable similarities between different parts of the Gullfaks structure and several published geometrical and experimental models of similar fault systems (e.g., Wernicke and Burchfiel, 1982; Ramsay and Huber, 1987; McClay and Ellis, 1987; Barr, 1987; Vendeville et al., 1987; Axen, 1988; Groshong, 1988; Kautz and Sclater, 1988). However, the single ramp fault of most detachment models, which leads to the asymmetrical structure of the trailing edge of the hanging-wall ("reverse drag", "roll-over" anticline, half-graben, etc.) is replaced in our model by two separate faults (ramp fault and trailing-edge fault) bounding a broad region of pull-apart which collapses gravitationally to fill the space, as the detachment becomes active. In the case of Gullfaks, we suggest that this collapse took the form of extension and thinning of the more competent formations by "domino"-style faulting, possibly accompanied by concomitant ductile deformation in adjacent less competent units. The constant eastward dip of the faults in Western Gullfaks suggests that they were originally a system of small faults within the pull-apart zone, synthetic to the main ramp fault (which lies, on this interpretation, to the west of Gullfaks, presumably between it and the Statfjord Field). The more complex relations in Eastern Gullfaks would be caused by the accommodation of this fault system to the antithetic trailing-edge fault, which is the western marginal fault of the Eastern Gullfaks high, as mentioned above. The high itself represents the trailing edge of the upper "plate" above the detachment. Because of the apparent congruence in tectonic style between the Jurassic reservoir horizons and the intra-Triassic horizons (see Fig. 6), we would place the detachment at a sub-Triassic level, possibly at the basement/cover contact, which in this area is supposed to lie at a depth of 5–6 km (Ediriweera, 1988; see also Swallow, 1986).

A somewhat different interpretation of the Gullfaks structure in terms of detachment tectonics is given by Fossen (1989), although the main extensional deformation of the Gullfaks field was not the subject of his paper. On his interpretation, the detachment ramps downwards into the basement along the Main Boundary Fault (Fossen, 1989; Fig. 7). Our interpretation of the time relations, however, suggests that a large part of the movement on the Main Boundary Fault took place during the deposi-

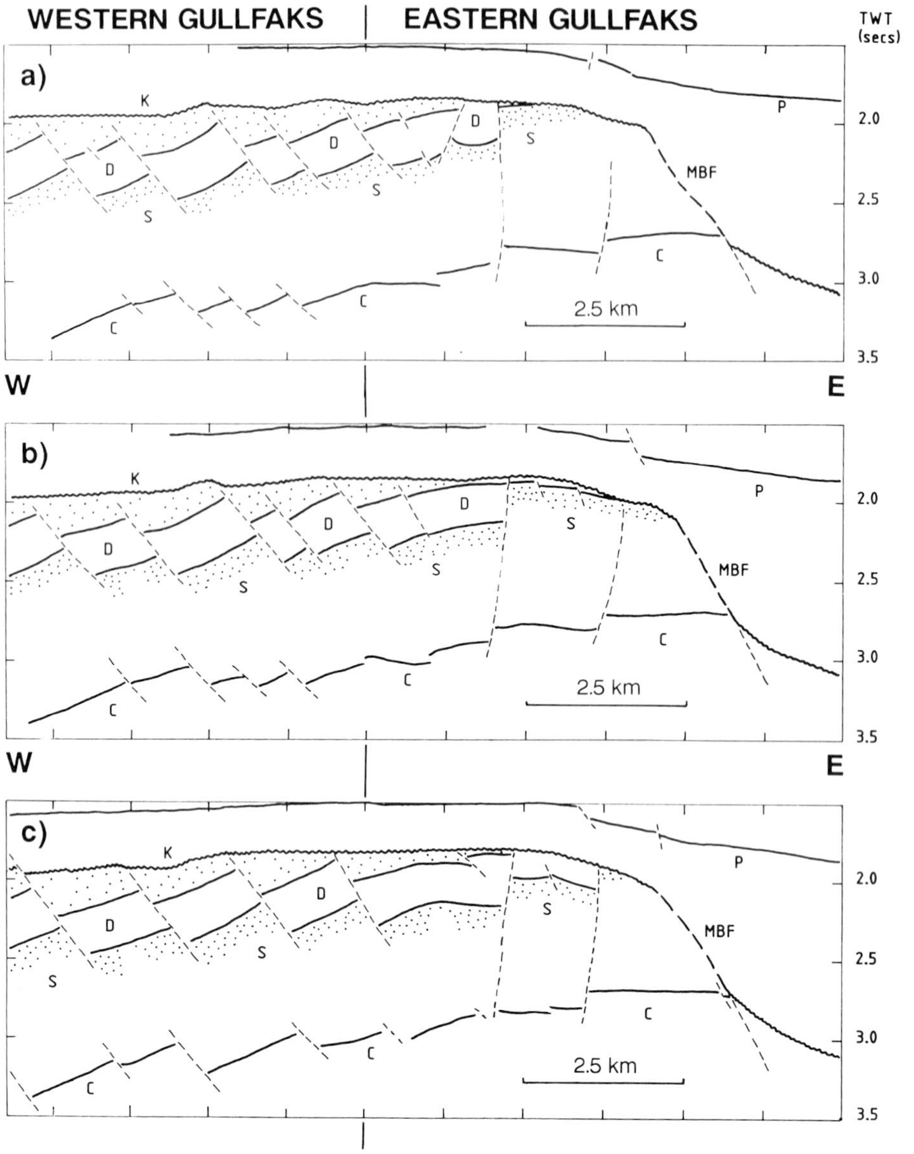

Fig. 6. Reflector reconstructions for three profiles from the 1985 three-dimensional seismic survey, taken from the published seismic sections in Fossen (1989). The indicated faults are not reflectors: they are only indicated where they can be reconstructed definitively from the displacement of the stratigraphic reflectors. The reflectors *S* (top Statfjord) and *D* (top Dunlin) are controlled by abundant well data, whereas reflector *C* (intra-Trias, presumably top Carnian) is uncontrolled. The profiles are not depth-converted and the faults appear steeper than they actually are (cf. Fig. 8). Other symbols: *K* = base Cretaceous, *P* = near top Paleocene, *MBF* = Main Boundary Fault.

tion of the lower Cretaceous formations in the basin to the east, at the time when the Gullfaks high was being uplifted and eroded (creating the base Cretaceous angular unconformity). There is stratigraphic evidence from the Gullfaks South Field, where the post-Brent formations are better preserved, that activity on the Gullfaks Fault System had ceased by uppermost Jurassic times (Heather formation unfaulted). Also, the sharp truncation of several faults of the Gullfaks system by the Main Boundary Fault along the southern boundary of the Gullfaks field is a further indication of temporal separation. Hence, we feel that it is likely that most movement on the Main Boundary Fault took place after the sub-Gullfaks detachment had ceased to be active, truncating it and its hanging-wall east of the trailing edge fault and thus isolating the Eastern Gullfaks high (compare Figs. 6 and 7). That extensional tectonics often take place in distinct phases, producing intersecting fault systems, is well known, both from the northern North Sea (e.g., Badley et al., 1988; Gibbs, 1987) and from other continental margins (e.g., Mutter and Larson, 1989; Williamson et al., 1989). More data from surrounding areas may prove that the Gullfaks structure is of a similar type.

As indicated in the last section, the fault and reflector geometry of Western Gullfaks (Figs. 3–5) shows that almost ideal "plane strain" conditions

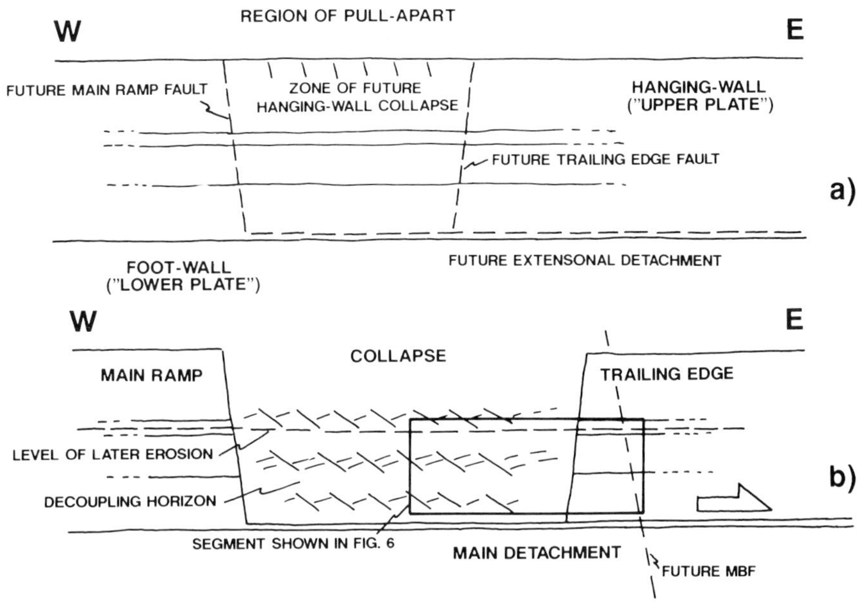

Fig. 7. Schematic illustration of the proposed kinematic model for the Gullfaks Field in terms of detachment tectonics (for explanation see text).

must have reigned during the development of the Gullfaks Fault System. The deformation plane, containing the E-plunging slip vector, was orientated E–W, so that E–W depth-converted sections are ideally suited to applying balancing techniques (cf. Gibbs, 1983). As an example, straightforward area balance has been applied to the Dunlin unit on part of a depth-converted E–W section by restoring the fault planes to an original dip of about 70° and the bedding planes to horizontal (Fig. 8). The original fault dip of 70° is an assumption based on field observations and theoretical considerations (cf. Ramsay and Huber, 1987; Walsh and Watterson, 1988; Xiao and Suppe, 1989). The other assumption in the procedure is that the cut-off trace length (distance between the intersections of the top and bottom of the formation and the fault traces) has remained constant. Physically, this means that most compaction is assumed to have taken place before faulting occurred. The following arguments indicate that this is likely to be a good approximation to the true situation. Porosity reduction in shales, which make up most of the Dunlin Group, takes place mainly during the first 200–300 m of burial. Below this, compaction occurs much more slowly, at about the same rate as in sandstone, and can be shown to alter the dip angles of pre-compaction faults by only about 1°/km (Xiao and Suppe, 1989). Since the Gullfaks faulting is certainly post-Brent in age, possibly even later, and since the Brent formation in Gullfaks is about 270 m thick (Petterson et al., 1990), the Dunlin shales can be assumed to have reached depths of 300 m before faulting started. This is supported by the lack of a discernible angular difference between the fault traces in Dunlin and the underlying and overlying sandy formations (Statfjord, Brent), i.e., faulting took place after the reservoir horizons had reached a depth regime in which differential compaction of sandstone and shale was small, (after a few hundred metres of burial, Xiao and Suppe, 1989). Hence, although compaction clearly continued during and after faulting, it can be assumed to have had only a minor effect on the observed geometrical relations.

Two interesting facts emerge from the restoration shown in Fig. 8. Firstly, it gives a quantitative estimate of the amount of stretching in the pull-apart zone. The horizontal extension in this cross-section (XY to $X'Y'$, Fig. 8) is approximately 50%, a value which is typical for the Western Gullfaks structure as a whole. If the width of the pull-apart zone and the depth to detachment were known, as may be possible with increasing knowledge of surrounding areas, this extension estimate will allow the amount of slip on the detachment fault to be estimated. Secondly, the restoration shows that the fault "blocks" themselves must have undergone a significant amount of internal deformation, since the bedding/fault angle has changed from about 70° to 40–50° in the course of the faulting (Fig. 8). Since compaction, as shown above, is unlikely to have modified the geometry to any significant extent, this indicates that the Gullfaks reservoirs have undergone considerable homogeneous shear strain, which must have been accommodated by adjustments at the level of individual grains and/or by movement on small-scale

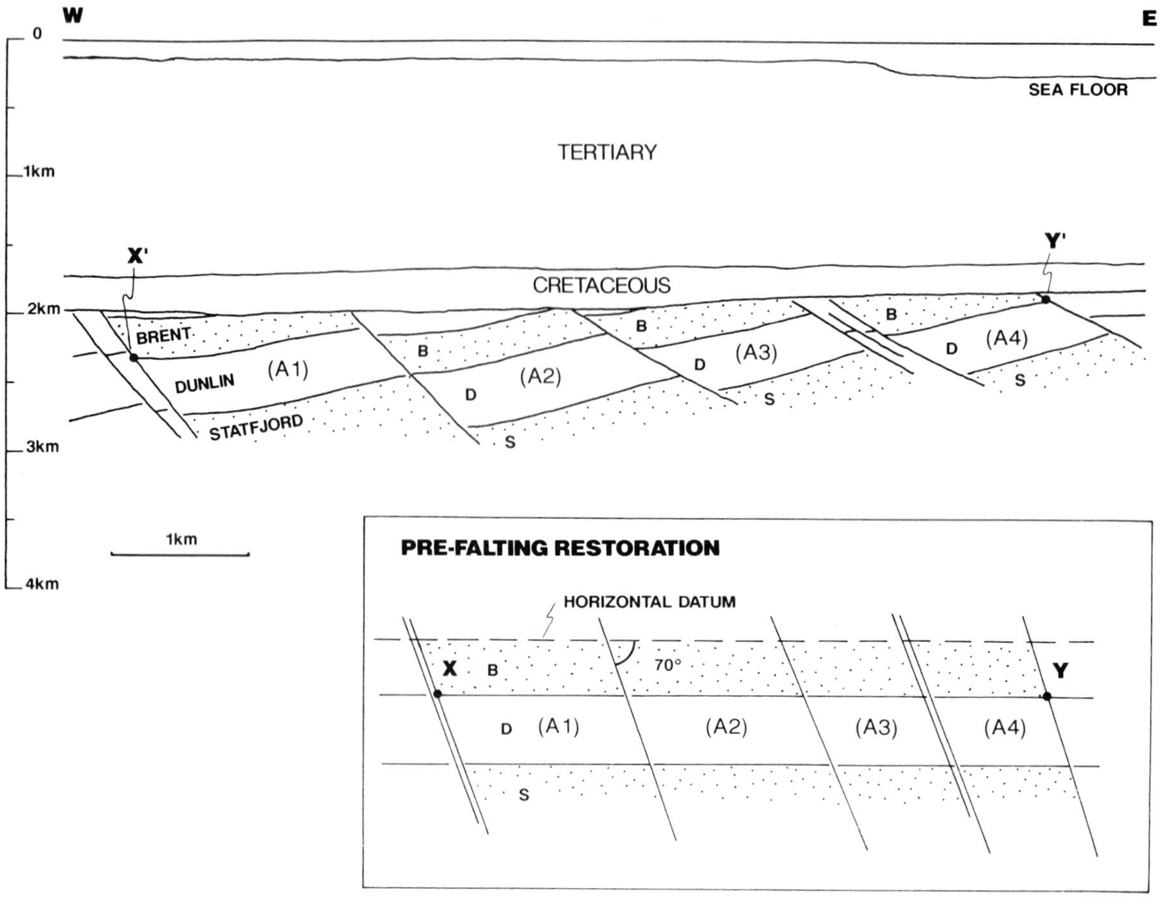

Fig. 8. Depth-converted E–W section through the reservoir horizons in part of Western Gullfaks (approximately along the line of section C, Fig. 6). Inset: pre-faulting restoration assuming that the original fault trace dip was about 70° and the original bedding was horizontal (for explanation see text).

(sub-seismic) distributed fractures. The combination of movement on distinct fault surfaces and homogeneous block-internal "ductile" deformation is a common feature of sand-box and clay models of "domino"-style faulting (McClay and Ellis, 1987; Kautz and Sclater, 1988; Vendeville and Cobbold, 1988), but as far as we are aware, this is the first time it has been proved on a field example, using site-specific data.

Conclusions

Our analysis of the geometry and kinematics of the Gullfaks Fault System is based on the Statoil interpretation of the 1985 three-dimensional seismic survey, itself constrained by abundant well data, and leads to the following conclusions.

(1) Fault surface geometry and the constancy of bedding orientation show that the overall deformation of the "domino"-style faulting in Western Gullfaks is "plane strain", with a vertical E–W deformation plane containing the E-plunging slip vector.

(2) This faulting is part of a fault system which can be interpreted in terms of complex hanging-wall collapse in a pull-apart zone above an extensional detachment. The detachment fault lies below the intra-Triassic reflectors, possibly at the basement/cover contact (here at 5–6 km depth). The Eastern Gullfaks high ("horst") represents the trailing edge of the upper plate: it is a high relative to the faulted units to the west because it did not collapse, and it looks like a "horst" because it was later truncated to the east by the Main Boundary Fault, here interpreted as mainly post-detachment in age.

(3) The tightly constrained geometry of the Western Gullfaks faults allows pre-faulting restoration of depth-converted E–W cross-sections by simple area balancing. We show that most compaction took place before faulting and that compaction during and after faulting is unlikely to have had a major effect on the structural relations. Comparison of the restored and present cross-sections allows a quantitative estimate of horizontal extension in the pull-apart zone (50%). It also shows that the Gullfaks reservoirs must have undergone a considerable amount of internal deformation (homogeneous shear strain, accommodated by grain-scale adjustmen s and/or movement on sub-seismic discontinuities) during faulting. This may be

of significance for reservoir behaviour and production performance and needs to be taken into account in reservoir modelling.

Acknowledgements

The above ideas are based on the interpretation of the Gullfaks ST 8511 survey by J.O. Henden and H. Fossen. We would like to thank them for generously allowing us to use the results of their work, and for many discussions. Our thanks are also due to an anonymous reviewer and Rob Young, for stimulating criticism which resulted in significant improvements to the manuscript, and to Statoil a.s., for publication permission. We are, however, solely responsible for the ideas presented and opinions expressed.

References

Axen, G.J., 1988. The geometry of planar domino-style normal faults above a dipping basal detachment. J. Struct. Geol., 10: 405–411.

Badley, M.E., Price, J.D., Rambech Dahl, C. and Adgestein, T., 1988. The structural evolution of the northern Viking Graben and its bearing upon extensional modes of basin formation. J. Geol. Soc. London, 145: 455–472.

Barr, D., 1987. Structural/stratigraphic models for extensional basins of half-graben type. J. Struct. Geol., 9: 491–500.

Beach, A., 1986. A deep seismic reflection profile across the northern North Sea. Nature, 323: 53–55.

Burchfiel, B.C., Wernicke, B., Willemin, J.H., Axen, G.J. and Cameron, C.S., 1982. A new type of décollement thrusting. Nature, 300: 513–515.

Coffield, D.Q. and Schamel, S., 1989. Surface expression of an accommodation zone within the Gulf of Suez rift, Egypt. Geology, 17: 76–79.

Ediriweera, K.K., 1988. A geological and geophysical investigation of the Viking Graben area, North Sea. Dr. Ing. Thesis, NTH, Trondheim.

Erichsen, T., Helle, M., Henden, J. and Rognebakke, A., 1987. Gullfaks. In: A.M. Spencer et al. (Editors) Geology of the Norwegian Oil and Gas Fields. Graham and Trotman, London, pp. 273–286

Færseth, R.B. and Pederstad,. K., 1988. Regional sedimentology and petroleum geology of marine, late Bathonian-Valanginian sandstone in the North Sea. Mar. Pet. Geol., 5: 17–33.

Fossen, H., 1989. Indication of transpressional tectonics in the Gullfaks oilfield, northern North Sea. Mar. Pet. Geol., 6: 22–30.

Gibbs, A.D., 1983. Balanced cross-section construction from seismic sections in areas of extensional tectonics. J. Struct. Geol., 5: 153–160.

Gibbs, A.D., 1987. Deep seismic profiles in the northern North Sea. In: J. Brooks and K.W. Glennie, (Editors), Petroleum Geology of North-West Europe, Vol. 2. Graham and Trotman, London, pp. 1025–1028.

Groshong, R.H., 1988. Half-graben structures: balanced models of extensional fault-bend folds. Geol. Soc. Am. Bull., 101: 96–105.

Kautz, S.A. and Sclater, J.G., 1988. Internal deformation in clay models of extension by block faulting. Tectonics, 7: 823–832.

McClay, K.R. and Ellis, P.G., 1987. Geometries of extensional fault systems developed in model experiments. Geology, 15: 341–344.

Mutter, J.C. and Larson, R.L., 1989. Extension of the Exmouth Plateau, offshore northwestern Australia: deep seismic reflection/refraction evidence for simple and pure shear mechanisms. Geology, 17: 15–18.

Petterson, O., Storli, A., Ljosland, E. and Massie, I., 1990. The Gullfaks Field-Geology and reservoir development. In: A.T. Butler, E. Berg, O. Hjelmeland, J. Kleppe, O. Torsæter and J.O. Aasen (Editors), North Sea Oil and Gas Reservoirs — II. Graham and Trotman, London, pp. 67–90.

Ramsay, J.G. and Huber, M.I. (Editors), 1987. The Techniques of Modern Structural Geology, Vol. 2. Folds and Fractures. Academic Press, London, 700 pp.

Saeland, G.T. and Simpson, G.S., 1981. 3-D seismic data help to delineate subconformity trap in Norwegian North Sea. Oil Gas J., 79(45): 381–398.

Saeland, G.T. and Simpson, G.S., 1982. Interpretation of 3-D data in delineating a subconformity trap in block 34/10, Norwegian North Sea. Am. Assoc. Pet. Geol., Mem. 32: 207–216.

Swallow, J.L., 1986. The seismic expression of a low angle detachment (sole fault) from the Beryl Embayment, central Viking Graben. Scott. J. Geol., 22: 315–324.

Vendeville, B. and Cobbold, P.R., 1988. How normal faulting and sedimentation interact to produce listric fault profiles and stratigraphic wedges. J. Struct. Geol., 10: 649–659.

Vendeville, B., Cobbold, P.R., Davy, P., Brun, J.P. and Choukroune, P., 1987. Physical models of extensional tectonics at various scales. Geol. Soc. London, Spec. Publ., 28: 95–107.

Walsh, J.J. and Watterson, J., 1988. Dips of normal faults in British Coal Measures and other sedimentary sequences. J. Geol. Soc. London, 145: 859–873.

Wernicke, B. and Burchfiel, B.C., 1982. Modes of extensional tectonics. J. Struct. Geol., 4: 105–115.

Williamson, P.E., Collins, C.D.N. and Falvey, D.A., 1989. Crustal structure and formation of a magnetic quiet zone in the Australian southern margin. Mar. Pet. Geol., 6: 221–231.

Xiao, H.-B. and Suppe, J., 1989. Role of compaction in listric shape of growth normal faults. Am. Assoc. Pet. Geol., Bull., 73(6): 777–786.

Paper submitted October 1989.

A.G. KOESTLER *Geo-Recon A/S, Bernhard Herres vei 3, 0376 Oslo 3, Norway*

A.G. MILNES *Department of Earth Sciences, ETH, Zürich, Switzerland*
Present address: *Geological Institute, Department A, University of Bergen, Allegt 41, N-5007 Bergen, Norway*

A. STORLI *Statoil A/S, Bergen, Norway*

Physical changes in fault blocks caused by rotation and their implications for hydrocarbon accumulations

J.E. Iliffe, I. Lerche and K. Nakayama

Physical changes that occur when a fault block is rotated in the Earth's gravitational field are assessed using a computer model. Parameters such as lithology, height, width, angle of tilt and fault dip are varied. Results describe how and where the physical properties of the block are changed by rotation, e.g., seismic reflection coefficients, geothermal gradients and bulk density of the block.

Fault blocks preferentially compact on their downthrown side, up to and including, the crest. Reflection coefficients in layers in the uppermost 1 km are moderated at particular depths in the compressed field of the block. The footwall area of the block acts as a conductive rod to heat; the juxtaposed low conductive areas making the latter a more attractive hydrocarbon generating area. Dewatering fluids are predicted to flow from the hotter compacted hanging wall to the cooler uncompressed footwall, which may contribute to shale diapirism seen in muddy deltaic sequences and improve the hydrocarbon potential of the footwall area. Also, calculations of extension estimates may be erroneous, due to either shortening of the surface layer or changes in deeper bed lengths by compactional folding. The stress configuration of the rotated compacted block suggests a plane of failure may occur in lieu of compactional strain. The angle this secondary fault makes to the original fault is consistent with field observations.

Introduction

The importance of block faulted extensional basins as hydrocarbon producing areas is exemplified in the North Sea, Gulf of Suez and Renconcavo Basin.

The common element in extensional regimes is the existence of rotated fault blocks. Geohistory models exist to approximate the change in the physical properties of sediments through time in these areas (e.g., Guidish et al., 1984), but analysis of what happens to the intrinsic properties of the fault blocks themselves due to just the rotation — i.e., without syn-sedimentary deposition, needs to be examined.

This paper assesses the physical changes which occur in fault blocks, tilted in the Earth's gravitational field, by means of a forward mathematical model. The model calculates changes in porosity, thermal conductivity, reflection coefficient, thicknesses and lengths, overall block shape and internal stress field caused by the rotation, for the different regions of the block. Forward models have several advantages over field analyses in that the designer knows exactly what all the parameters of the system are and how they are manipulated. A vast array of variability of situations can be tested and the results analyzed to gain a better understanding of the factors influencing a wide range of variations.

Model description

The model consists of three unrotated identical blocks of width, W and vertical height, Th, bounded by faults with an angle, τ, with layers of equal thickness (z_1) (Fig. 1a). All these parameters may be varied in experiments.

The characteristics of different lithology blocks may be broadly recreated by varying the scaling constant (ℓ) and surface porosity (ϕ_o) in the porosity/depth function of Magara (1976) (Fig. 2).

$$\phi(z) = \phi_o \exp\left(\frac{-z}{\ell}\right) \quad (1)$$

where $\phi(z)$ is the calculated porosity at depth z.

Compaction

The most important change in physical properties of the blocks due to rotation is of porosity due to compaction. The compaction effect is determined by solving the differential equations relating the porosity depth function (eqn. 1) to the overburden in terms of the original porosity $\phi(z)$, original frame weight $Wt(x',z')$ and the resultant frame weight $Wt(x,z)$:

$$\Delta z = \frac{\phi(z')\,[Wt(x,z) - Wt(x',z')]}{[1 - \phi(z')^2]\,\rho_m g} \quad (2)$$

Structural and Tectonic Modelling and its Application to Petroleum Geology edited by R.M. Larsen, H. Brekke, B.T. Larsen and E. Talleraas. NPF Special Publication 1, pp. 253–267. Elsevier, Amsterdam.

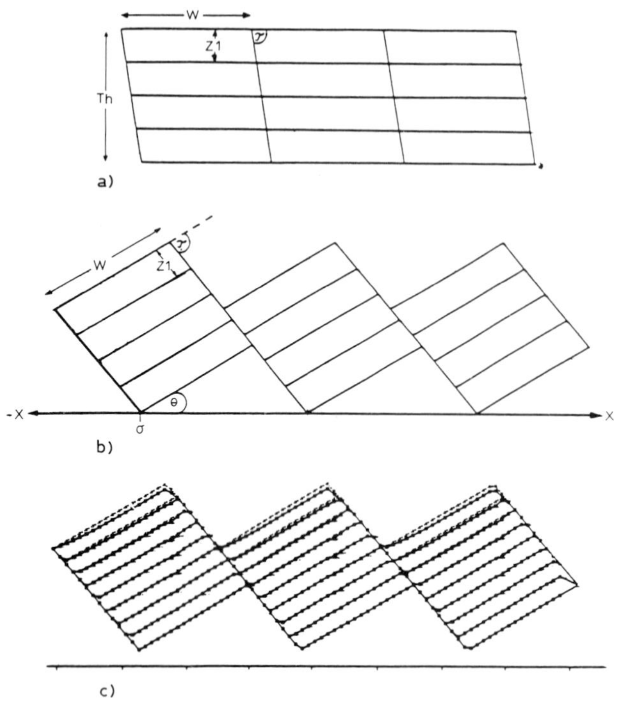

Fig. 1. (a) Diagram showing parameters used to define the pre-rotated blocks. W = block width; Th = block height; $z1$ = strata thickness; τ = fault angle. (b) Diagram illustrating the method of post-rotation sampling of depth (z) in increments of Δx, and parameters used to describe the block in the model. (c) Model output illustrating the results of post-rotation compaction on shale blocks tilted 30°. The dashed line with crosses denotes the pre-compacted block outline, the solid line denotes the resultant compacted block.

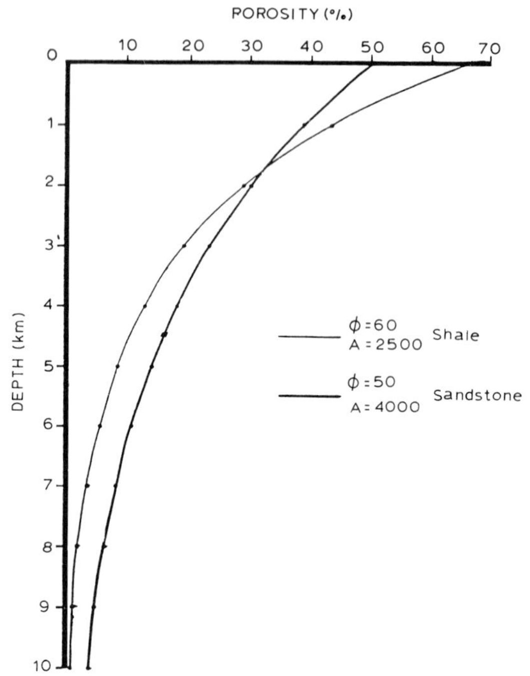

Fig. 2. Diagram comparing porosity depth curves of sandstone and shale using values of ϕ_o and ℓ from Sclater and Christie (1980) in eqn. 1 of Magara (1976).

where Δz is the amount of compaction, ρ_m is the fully compacted rock density and g is the acceleration due to gravity.

The pre-rotation total frame weight $[Wt(x',z')]$ acting along each stratal boundary is calculated by summing the weight between the top of the block and the depth to the stratal boundary (Fig. 1a). The new (post-rotation) total frame weight $[Wt(x,z)]$ is assessed at each point along a stratal boundary by summing the weights calculated for a set of depth values measured perpendicular to the strata over the final vertical depth (Fig. 1b). If the new weight $[Wt(x,z)]$ is less than, or equal to, the old weight $[Wt(x',z')]$, no compaction takes place. Compaction occurs when $Wt(x,z)$ is greater than $Wt(x',z')$.

If the blocks are assumed to have no horizontal strength, compaction is limited to the vertical plane. The Δz values for each z (vertical) point at each x (horizontal) location are summed cumulatively from the base layer upwards. The resulting displacement (Δz) is now subtracted from its original vertical position to yield the compaction corrected section (Fig. 1c).

Geothermal effects

When the porosity of a region within the block is decreased as a result of compaction, the thermal conductivity of that region is increased and may be calculated using the relationship

$$K_R = K_m^{1-\phi(z)} K_f^{\phi(z)} \tag{3}$$

where K_R is the whole rock thermal conductivity at depth z, K_m is the thermal conductivity of the matrix and K_f is the fluid thermal conductivity (Lewis and Rose, 1970). An estimate of the temperature field within the block is then made using Fourier's law of heat conduction:

$$\nabla \cdot (K \nabla T) = 0 \tag{4}$$

with $q = -K(\mathrm{d}T/\mathrm{d}z)$ on the basement and:

$$\frac{\mathrm{d}T}{\mathrm{d}z} = \frac{-q}{K} \tag{5}$$

where q is heat flux through the basement (assumed constant); K is thermal conductivity; and $\mathrm{d}T/\mathrm{d}z$ is the vertical temperature gradient. In order to correct for unrealistic topographic effects, in the one-dimensional case we take the area between the tops of the block and a theoretical horizontal surface (e.g., seafloor with a temperature of 6°C) (*A–B* in Fig. 3a) to be "filled" with a weightless material with the same geothermal properties as the sediments in the blocks. With this boundary condition, the tem-

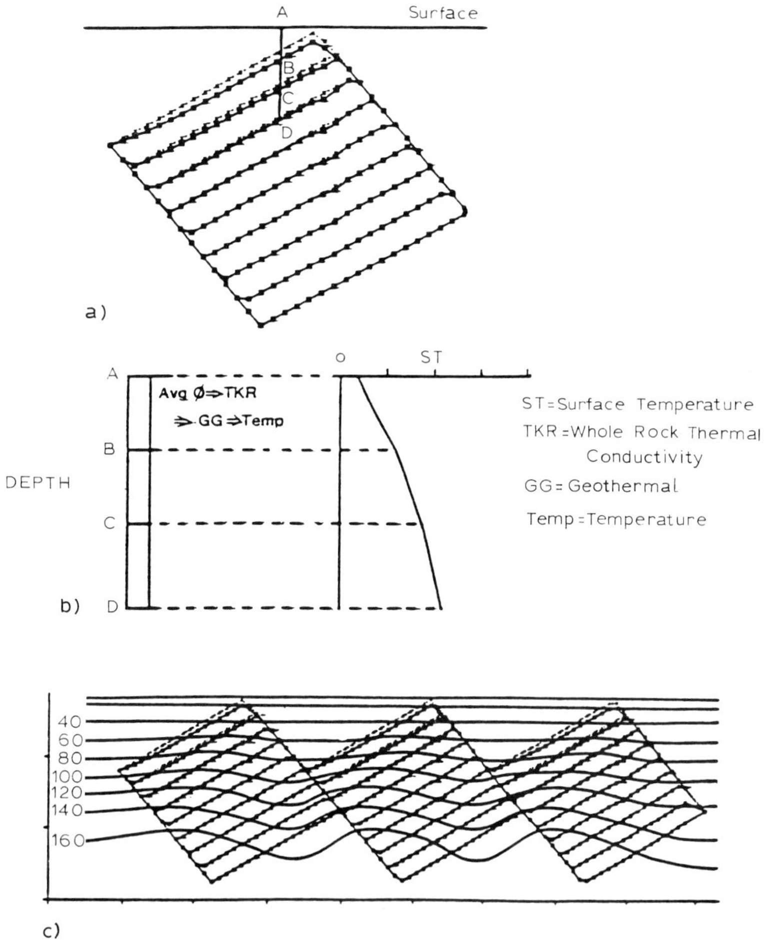

Fig. 3. (a) A column of sediment A, B, C, D is considered as units of rock of homogenous properties. The block is composed of shale, 4 km high, 4 km wide with a fault angle of 80°, rotated 30°. Each layer is 500 m thick. (b) The average porosity of each layer is calculated, yielding an average whole rock thermal conductivity and consequently an average geothermal gradient. The temperature profile with depth is then calculated from an input surface temperature in a cumulative fashion. (c) Resulting pattern of isotherms due to rotation and compaction. Note that regions where isotherms are close together correspond to upper sections of the block being buried, whereas those areas with isotherms more widely spaced correlate with the exhumed deeper parts of blocks with lower porosity. The blocks are the same as those in (a).

perature at each layer surface may be calculated as follows.

The porosities within each layer are calculated from eqn. 1 and the layer's average porosity found. Next, the average thermal conductivity of the layer is calculated from eqn. 3. The average geothermal gradient (dT/dz) of each layer is calculated from eqn. 5, using an assumed value for heatflow. The temperature field is now determined from the horizontal surface down, by adding the product of the thickness of the (compacted) layer and its geothermal gradient (Fig. 3b). Isotherm contours are then drawn by linear interpolation (see Fig. 3c).

Figure 3c shows that isotherms towards the tops of the blocks are closer together than those which are deeper, a result of the relationship of porosity to geothermal gradient. The temperatures in areas of high porosity change with depth more rapidly than those in areas of low porosity.

Reflection coefficients

The calculation of a reflection coefficient is usually undertaken for the boundary between two layers of different density and seismic velocity. In the present form of the model only homogeneous blocks of rock whose density varies as a function of the porosity of the rock are considered (Wyllie et al., 1956), in accord with the porosity-depth function (eqn. 1). We, therefore, calculate the change in reflection coefficients of rotated layers relative to their unrotated state.

To calculate the reflection coefficient, a relationship of velocity with depth is required or, specifically, change of Lame constants with depth. These may be estimated from the formula:

$$\lambda_z + 2\mu_z = (\lambda_o + 2\lambda_o) + [\lambda_{max} + 2\mu_{max} - (\lambda_o + 2\mu_o)] \tanh(z/L) \qquad (6)$$

where λ and μ are the Lame constants at depth,

denoted by subscripts "o" surface, "z" at depth z; and "max", maximum Lame constants, and L is the scaling depth over which the Lame constants vary from surface to deep values.

From an estimate of the maximum (V_{max}) and minimum seismic velocities (V_0) of a shale as, e.g., 6000 and 2000 m/s, respectively, it is possible to construct a formula for the velocity at any depth, z, of the type constructed by Wyllie et al. (1956) (i.e., $V(z)^2 = [\lambda(z) + 2\mu(z)]/\rho(z)$).

The reflection coefficient is calculated from:

$$RC = \frac{1}{2}\frac{\partial \ln V\rho}{\partial z} \tag{7}$$

i.e.,

$$RC = \frac{1}{4}V^{-2} \times \frac{dV^2}{dz} - \frac{1}{2}\frac{\rho_m - \rho_w}{\rho(z)}\frac{d[\phi(z)]}{dz} \tag{8}$$

The only unknown in this equation is dV^2/dz, which may be calculated numerically. After the blocks have tilted, the reflection coefficients calculated along each layer are then compared to those of the unrotated state and the changes are plotted.

Experiments

The effects of variation in lithology, height to width ratio of the block, angle of the faults and degree of tilt, on compaction, geothermal characteristics, reflection coefficients and stress configurations are investigated by measuring changes in bed lengths, bed thicknesses, angles and overall shape of the output relative to the input.

In the tests performed, two lithologic types were considered, those of pure shale and of homogeneous sandstone, the values of ℓ and ϕ_o being taken from Sclater and Christie (1980; Fig. 2).

The size effect of the blocks, specifically the ratio of the height of the block to its width, was tested using three basic types. The long flat "warehouse type" block (Fig. 4) was compared with a perfectly square block (Fig. 5) and with a tall, thin "skyscraper type" block (Fig. 6).

As the lithology and block dimensions were varied the fault angle and degree of rotation were also incrementally changed.

Compaction

The distribution of the compaction effect is shown in Fig. 7. Compaction is restricted to the downthrown hanging wall of the block and to the upper 2500 m at that. The footwall side of the block remains unaffected, except for a few meters down the fault from the surface/fault intersection. When a point at the top of a formation on the hanging wall side of the block is rotated the load acting on it is increased by the secant of the angle of rotation (in this case 30°). A point at the top of the same formation towards

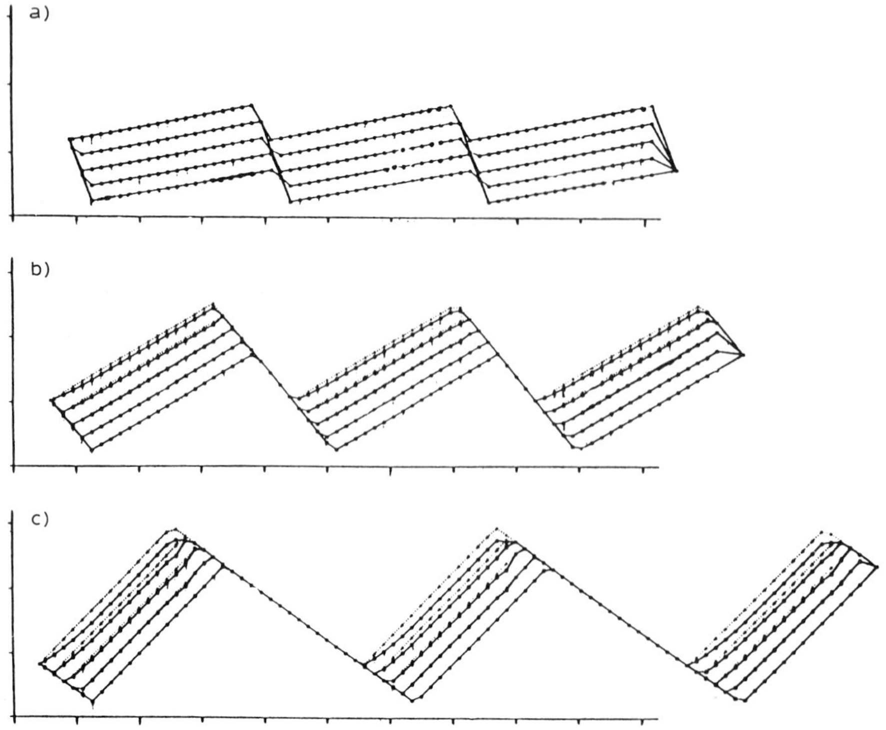

Fig. 4. (a) Fault block of shale, width (W) of 6 km, height (Th) of 2 km and angle of tilt (θ) of 10°; (b) θ = 30°; (c) θ = 45°.

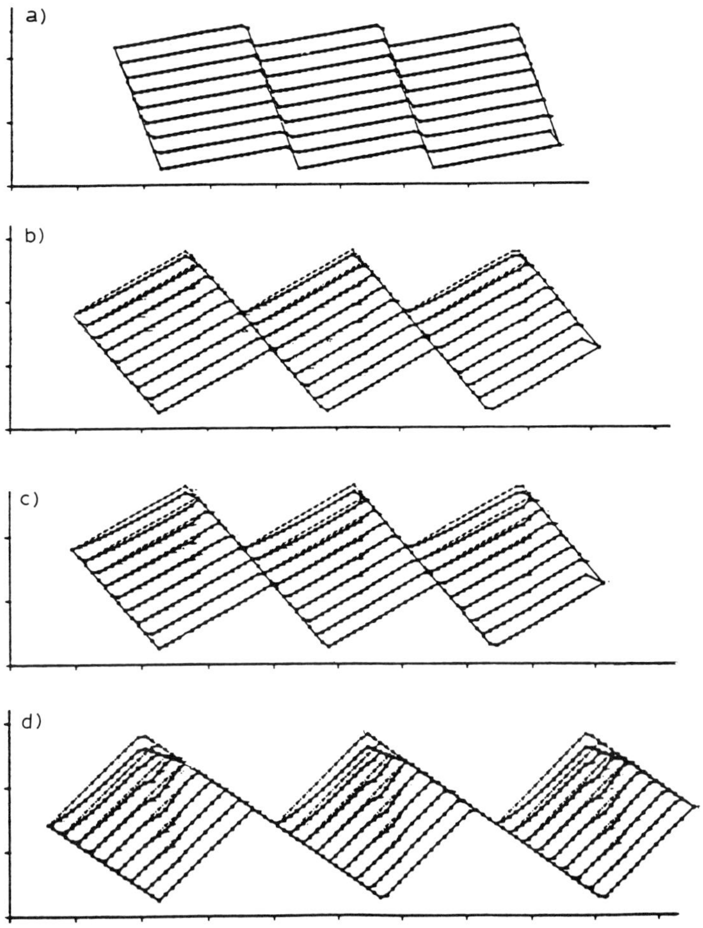

Fig. 5. (a) Fault blocks of shale of width (W) of 4 km, height (Th) of 4 km and angle of rotation (θ) of 10°; (b) θ = 30°; (c) fault blocks of sandstone, θ = 30°; (d) fault block of shale, θ = 45°.

the footwall side of the block has had its load decreased by the rotation and/or faulting, implying no compression.

The compaction effect is not only differently distributed laterally, but it also varies with depth. It is easier to squeeze a very porous surface mud than fully lithified shale at 3 km depth. The optimal depth for compaction is dictated by the scaling factor ℓ in eqn. 1, which is a function of the lithology (Fig. 2) and is related to surface porosity ϕ_o values.

Although the reduction of thickness due to compaction is greater in shale, sandstone has a broader zone of impact (Fig. 8). Increasing the angle of rotation amplifies the compaction effect (Fig. 8). The reduction of the post-rotated (30°) total height of the block by compaction is 4.6% for sandstone and 3.3% for shale. For blocks rotated 45° these values are 9.7% for sandstone and 7.3% for shale.

There is also a point on a formation where greatest compaction occurs. This point is very close to the change-over from compacting to non-compacting sides of the block and is important in the subsequent discussion.

The fault planes themselves are unaffected by compaction, since they are unloaded.

Degree of tilt and height to width ratio

In the skyscraper type blocks (Fig. 6) deformation in the upper formations is intense, but a large part of the block is unaffected. The wide warehouse type blocks deform throughout their length and increase in intensity with increasing tilt (Fig. 4). As the angle of tilt is increased there is always more block to be added to the load (because of its length) in the rough form of $\sec\theta$ times the original weight. The skyscraper type blocks, because of their narrow width, have only a very small maximum potential increase in load (Fig. 6).

The most compaction occurs in blocks rotated 90°. The maximum change in weight takes place below the top left hand corner of the blocks, the excess weight acting on this point can be crudely described as $\rho_{Rav}gW$. Since ρ_{Rav} (bulk density of rock) and g (gravity) are constants, the limiting factor is the width of the block, W. The depth to which compaction is evident is also controlled by the width of the block

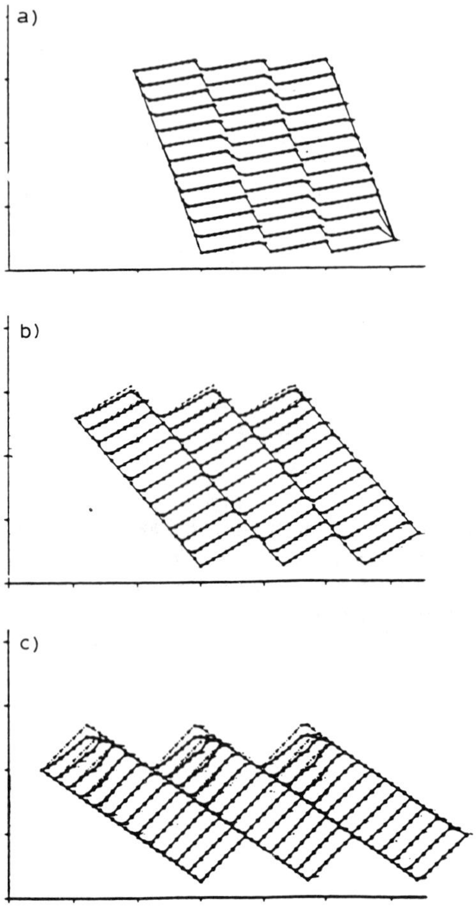

Fig. 6. (a) Fault block of shale, width (W) of 2 km and height (Th) of 6 km tilted by $\theta = 10°$; (b) $\theta = 30°$; (c) $\theta = 45°$.

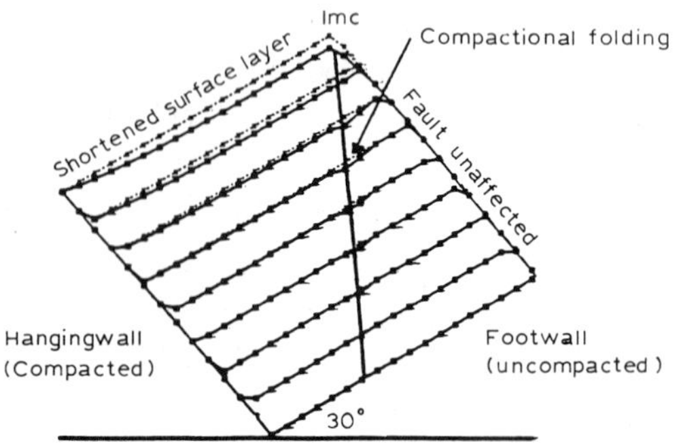

Fig. 7. One post-rotation compacted block exhibiting features of folds, line of maximum compaction, an uncompacted footwall and compacted hanging wall. The block is of shale 4 km high, 4 km wide with a fault angle of 80°, rotated 30°. Layers are 500 m thick.

(compare Fig. 4b and Fig. 6b). The maximum depth at which compaction occurs may be written as:

$$z_{max} = W \sin\theta \tag{9}$$

where z_{max} is the maximum depth of compaction. The compaction distribution of the blocks is related not only to the surface porosity, ϕ_o, and scaling factor, ℓ, i.e., lithologic type, but also, in part, to the

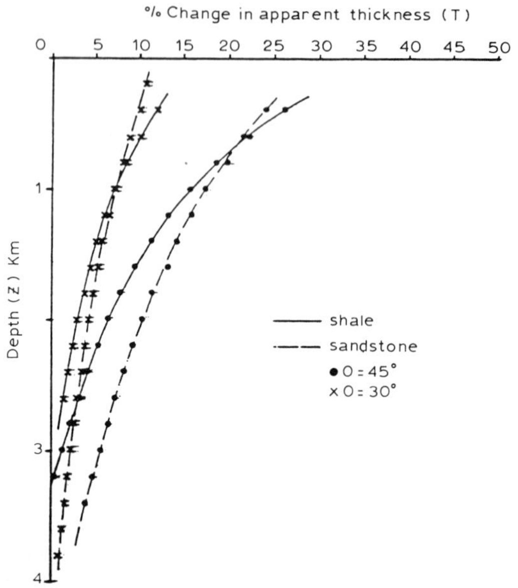

Fig. 8. Graph illustrating the percentage change in apparent thickness of layers 200 m thick, due to compaction following rotation of 30°(•) and 45° (×) of a block of sandstone and shale.

width of the block and angle of rotation.

When considering a block wider than the vertical scaling factor, ℓ, compactional effects are minimal until a tilt of 30° is reached. The rate of change of the compaction effect is roughly proportional to $\sin\theta$.

Fault dip

The distribution and maximum depth of compaction vary with the changing fault angle (τ). As τ is decreased from 90°, the area of the block in the zone of compaction is decreased by its physical rotation movement laterally out of the compacting region. In this case the maximum depth, z_{max}, at which compaction can occur is limited once again by the scaling factor, ℓ, (eqn. 1) and width of the block, with:

$$z_{max} = W \sin\theta + W \cos\theta \tan(\tau - \theta) \tag{10}$$

The results of changing τ, z_{max} and θ are shown in Fig. 9. Consider a block 4 km wide, tilted at an angle of 45°, and of infinite depth. Vary the fault angle by 90° (Fig. 9a), 80° (Fig. 9b) and 60° (Fig. 9c) then z_{max} is 5.6, 4.75 and 3.6 km, respectively. The fault angle, (τ), therefore plays a significant role in determining the amount of compaction taking place and the depth range over which it occurs.

Extension calculations

The surface of the blocks modelled appear shortened, whereas compactional folding occurs at depth (Figs. 4, 5 and 6). This type of deformation is not always obvious in field measurements. Areas

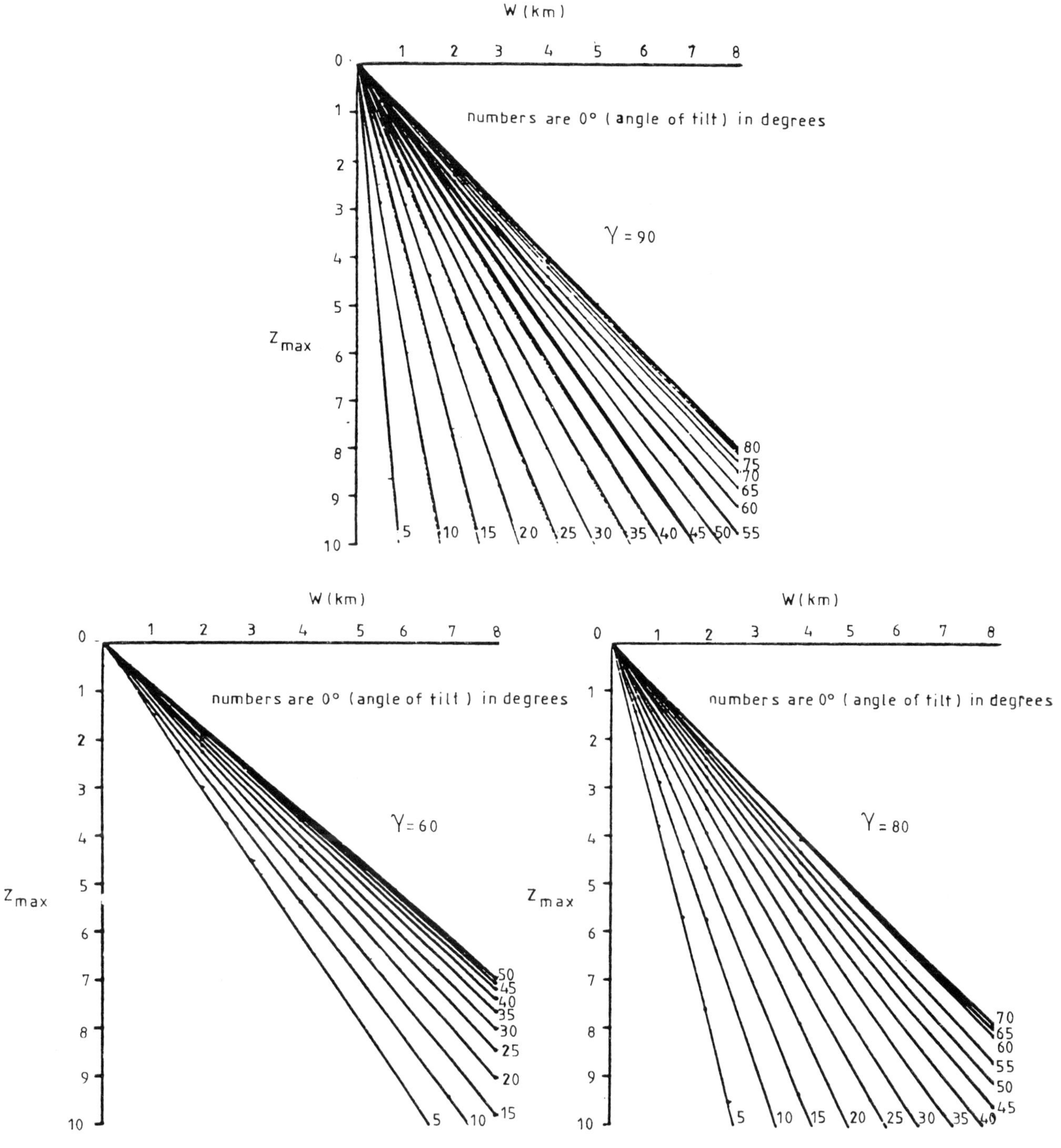

Fig. 9. Diagram showing the variation in maximum depth of compaction ($z_{\max}$) with changing fault angles (τ).

of extension are somewhat quantified by extension estimates and, where subsurface control is limited, this extension is generally obtained by comparing the bed lengths of pre-faulted layers with those of post-faulted layers. Compaction due to rotation will alter the estimate of original bed length because original bed length is now deformed, and differentially so depending on whether this deformation is at the surface or at depth (Fig. 6c).

In this study the error in estimates of extension considering angle of tilt and the depth of the formation in the block were measured and was found to be low in blocks tilted less than 20°, but in blocks tilted 45° the error at the surface was of the order of 20%. However, at depths below 1000 m the error decreased to less than 5%.

It is, therefore, recommended that bed length be measured at a depth of greater than 1000 m if block rotation is greater than 20° in the section being assessed.

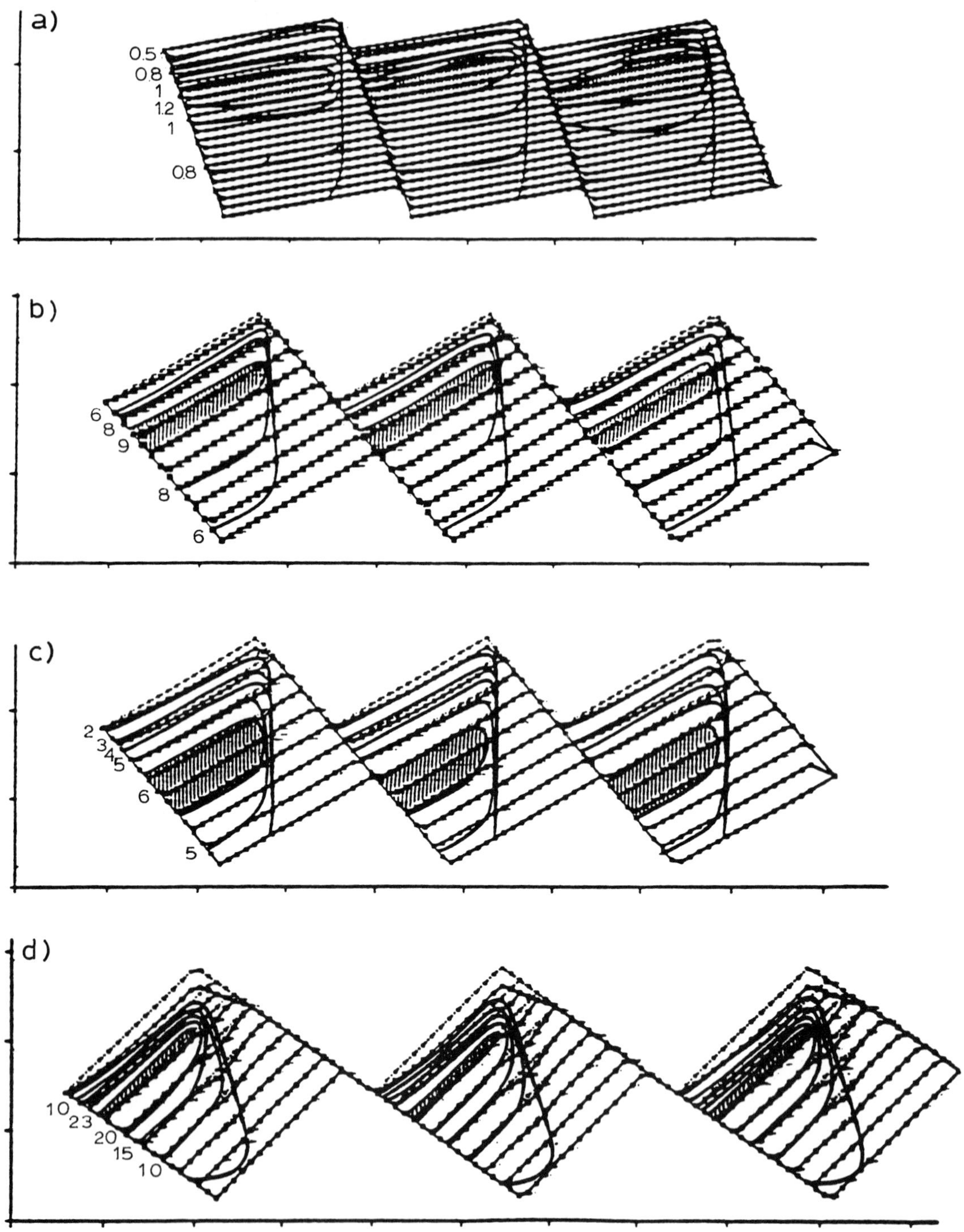

Fig. 10. Diagrams showing the rates of change in reflection coefficients with depth along formation boundaries as a result of block rotation and compaction. All blocks are 4 km wide and 4 km high with a fault angle of 80°. (a) Shale block $\theta = 10°$, (b) shale block $\theta = 30°$, (c) sandstone block, $\theta = 30°$, (d) shale block $\theta = 45°$. Note: values are multiplied 10^5 times.

Reflection coefficients

This model considers blocks of a homogeneous lithology in which seismic waves propagate in a smoothly varying depth dependent medium.

The most important factor affecting reflection coefficients (*RC*) is the change in porosity with depth, $d\phi(z)/dz$, which is dependent on the input lithologic parameters ℓ and ϕ_o which are used through eqn. 1 to give $\phi(z)$ at various depths. For this reason the model was tested in the same way in which compaction was considered, by changing the angle of rotation (θ) and comparing the results of pure shale and homogeneous sandstone blocks (Fig. 10).

Compaction is essentially a porosity decrease (Fig. 2) which causes an increase in bulk density (ρ) and velocity (V) of pressure waves (*P* waves). The values

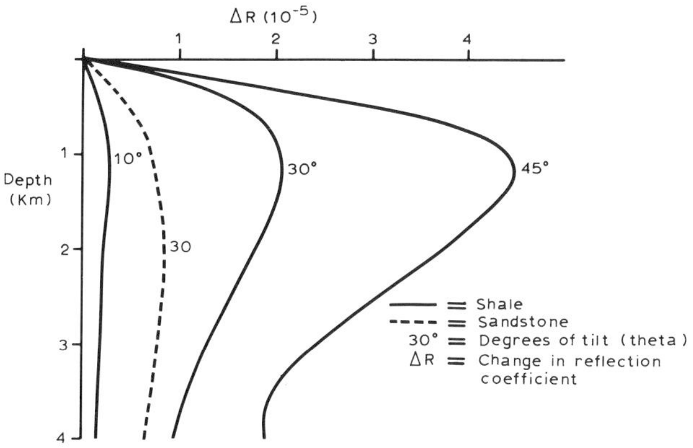

Fig. 11. Graph illustrating the difference between ΔRC before compaction and after compaction at each layer interface for varying angles of tilt and lithology.

of *RC* at each interface are decreased by compaction.

The areas of blocks where the greatest change in ΔRC has occurred are predicted to become more transparent to *P* waves (Fig. 10a–d). But the maximum change in *RC* caused by compaction (Fig. 10d) is only 2–3% of the ΔRC variation over a depth of 4 km.

RC values do not change laterally beyond the line of maximum compression (lmc) and the change occurs rapidly in the upper layers and peters out gradually below the optimum depth (Fig. 11). The values of ΔRC are an order of magnitude greater in blocks tilted 45° than in those tilted 30°. Also, the zone of greatest change in *RC* is much narrower in the vertical dimension in the 45° tilted blocks than in those rotated 10°. These results are implicitly tied to changes in compaction, consequently *RC* values do not significantly change in blocks tilted by less than 30°.

Comparing blocks of pure shale and sandstone, the region of greatest change in *RC* (not to be confused with the optimal region for reflections) is at about 1000 meters for shale (see also Fig. 11) and in all the shale cases (Fig. 10a, b and d) the maximum ΔRC occurs in a relatively narrow band. In the sandstone case (Fig. 16c), the maximum ΔRC is not only deeper (between 1500 and 2500 meters), but is also much wider than the shale cases — a direct result of the lithologic effect, i.e., the ΔRC distribution is responding to the $d\phi(z)/dz$ curves (eqn. 1, Fig. 2).

Geothermal effects

The first assessment of the geothermal effects was to determine whether a two-dimensional calculation was necessary. A horizontal line was drawn across the tops of the "waves" of the 100°C isotherm at 1600 meters depth in a shale block tilted 30° (Fig. 3c). The temperature varied 10°C or 10% of the isotherm over a depth range of 400 meters. The same exercise was carried out for the 150°C isotherm. The maximum departure from the curve using a one-dimensional calculation was of the order of 15°C, again only 10% of the value of the isotherm. This temperature variation occurred over a range of 600 meters down from a depth of 3100 meters. Since bottom hole temperatures measured in wells are accurate only to a confidence level of ±10%, a two-dimensional analysis would not yield any more useful information.

The undulations of the isotherms in Fig. 3c are a direct result of rotation, since topographic effects are removed by assuming the downthrown basin is filled with material with a thermal conductivity appropriate to a depth relationship consistent with the unrotated blocks (eqn. 1). The compaction effect on changing thermal conductivity is minor relative to the rotational effect.

Areas of high thermal conductivity are rotated closer to the surface in the footwall, whereas less compacted sediments are buried on the hanging wall. The hanging wall areas are, therefore, cooler relative to footwall areas. The oil kitchen would therefore be on the downthrown side of the block (hanging wall) at a depth of greater than 1600 meters. The gas kitchen is at a depth of 3100 meters in the lowest part of the footwall and hanging wall.

As the fault angle is increased, the amplitude of the isotherms also increases (compare a and b of Fig. 12). Since sandstone has a higher conductivity than shale (compare b and c of Fig. 12) the isotherms are spaced further apart i.e., the geothermal gradient is less for sandstone. A small change in basement heat flow from 1.2 (Fig. 12b) to 1.5 HFU (Fig. 12d) greatly affects the position of the isotherms, although the general pattern is maintained.

Stress considerations

During the calculations to determine the change in position of a formation due to compaction, the original weight $Wt(x',z')$ and final weight $Wt(x,z)$ are calculated. The difference between these weights is an indication of the change in magnitude of stress along each bedding plane. The compaction of shale and sandstone is therefore strain (in the form of alignment of shale grains and dewatering) as a result of this stress. Beds act as passive strain markers.

Consider a block of identical dimensions, tilted the same amount, but consisting of homogeneous unjointed granite. Although the magnitude of the changes in overburden have changed, the points at which the new weight is greater are in approximately the same positions as in the compressible sedimentary blocks. The pattern of stress increase is

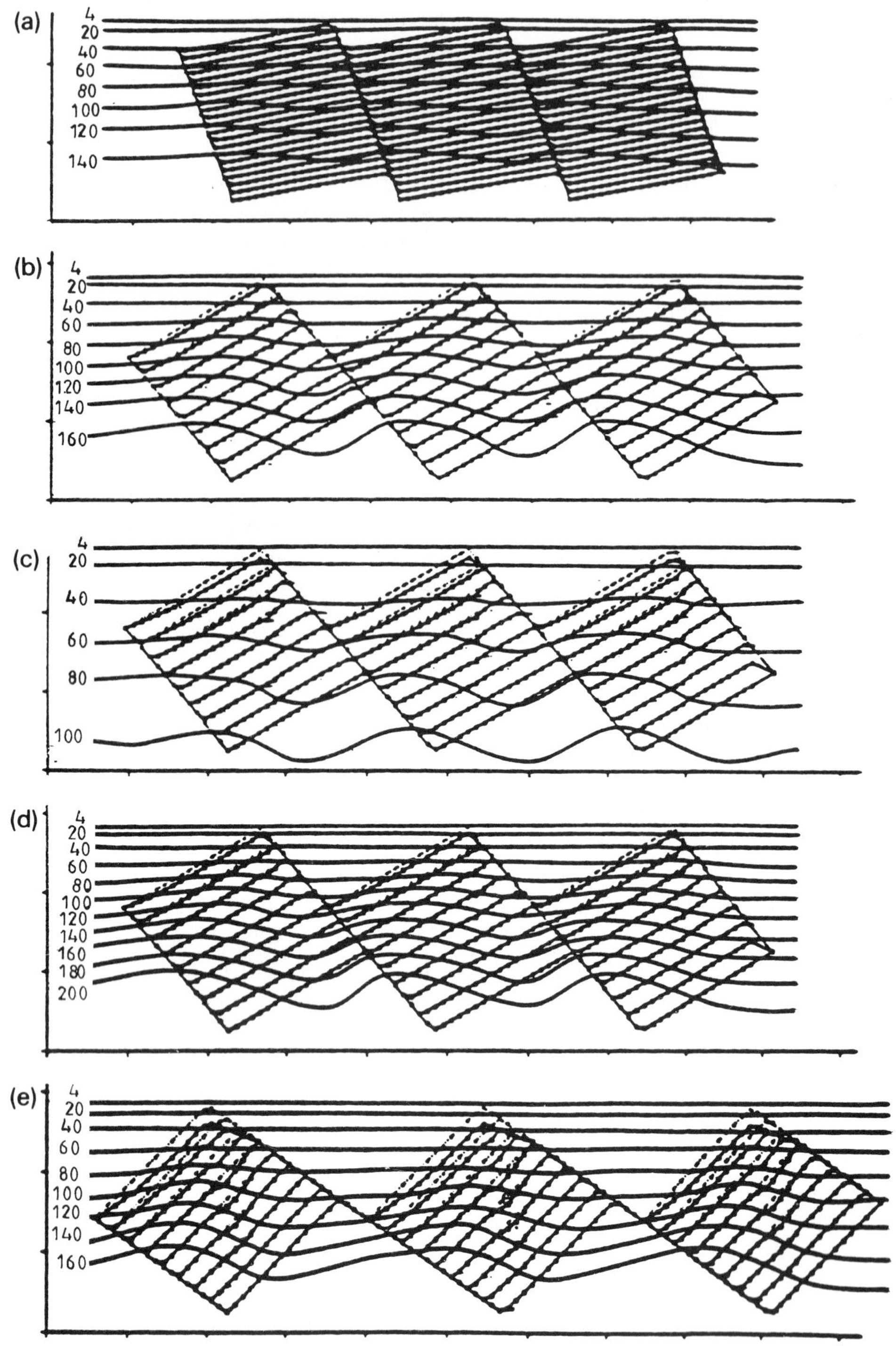

Fig. 12. Temperature distribution caused by differential compaction after rotation calculated using the one-dimensional method. All blocks are 4 km wide and 4 km high. (a) Shale, $\theta = 10°$, $Q = 1.2 HFU$; (b) shale, $\theta = 30°$, $Q = 1.2$ HFU; (c) sandstone, $\theta = 30°$, $Q = 1.2 HFU$; (d) shale, $\theta = 30°$, $Q = 1.5$ HFU; (e) shale, $\theta = 45°$, $Q = 1.2$ HFU.

insensitive to block composition.

A phenomenon inherent in the rotated blocks of sediment is a point along each formation at which compaction is greatest (Fig. 7). These maximum compactional points for each of the formations may be joined to form a line from the top of the block down. This line represents the region in the block where the stress has undergone the greatest change (line of maximum compaction or lmc). If this block were of a brittle rock (such as granite) and the increase in stress was large enough to overcome the shear strength of the material, then a fault may appear in this position. If the granite were indurated with joints in various directions, the pre-existing joint which lies in the most favorable orientation (closest to that predicted here) will release the rotation-created strain by acting as a fault, but requiring less stress than before.

Tests to date suggest that the angle of intersection between this secondary fault and the main fault below is a nearly constant 35° regardless of the degree of tilt, for a fault angle of 80°. If the fault angle is 60°, the angle of the intersection becomes 45°.

This relationship has been independently observed in the field. Miller et al. (1983) working in the Snake Range, Nevada, U.S.A., describe "second generation" faults which intersect older faults at an angle of 40°. They also describe the faults as beginning in a near vertical orientation as predicted by this model.

Similar observations have been made by Angelier and Colletta (1983) in the Gulf of Suez, Egypt; Western Gulf of California, Mexico; and the Southern Basin and Range, U.S.A. They describe the occurrence of near vertical (80–90° dip) tension gashes formed in the early stages of extension, when blocks are tilted 5–10°, which later become active as faults as tilting progresses.

The 80–90° dip of the tension gash with a tilt of 20° in a fault with an original dip of 60° yields an intersection of fault and tension gash at 40°.

Thorough testing of different block dimensions, fault angles and angles of tilt has been carried out, providing firm corroborative support for the ruggedly stable nature of the effect. The model suggests a mechanism for the development of secondary faults that is not limited to compressible sediments and is substantiated by reported field accounts.

Discussion

The problem with all synthetic models lies with their intrinsic assumptions. These may be of a forgivable oversimplification, such as assuming a fault block to be composed of a single lithology or of a more serious nature when the actual physical phenomena are not fully understood. It is therefore important to question the limitations of assumptions assessing the validity of a model.

Interbedded scenarios

Consider the limitations and validity of the assumption of a lithologically homogeneous block. Clearly, situations with interbedded varying lithologies would be more geologically appropriate.

Above a depth of about 1000 m, shale layers compact more than sandstone (Fig. 8), between 1000 m and 1600 m they compact by the same amount and their porosities are quite similar (Fig. 2), but below 1600 m, sandstone layers actually compress more than shale and their porosity is lower. Thus, in an interbedded sandstone and shale block, would compactional folding occur?

Since the amount of compaction of layers differs only slightly between lithologies in the zone between 1000 and 1600 m, folding in the layers in an interlayer situation would be just as prevalent as in the monolithologic blocks.

When considering the change in reflection coefficients in an interbedded scenario it is first appropriate to consider how reflection coefficients will vary with depth in the unrotated state. Following the discussion above of change in thickness and porosity of layers, it is possible to sketch an approximate variation of reflection coefficients with depth (Fig. 13). If these blocks are subsequently buried by rotation, the effect of compaction will change the reflection coefficient by decreasing them in the top of the block (above 1000 m), since the shale compacts towards the porosity and density of the sandstone. The region between 1000 and 1600 m will be largely unchanged, whereas directly below this zone, the reflections will decrease once more. Because the more porous sandstones compact more than the less porous shales, the result is a tendency of layers to be closer together. Reflectors should become more transparent in areas of compaction (i.e., footwall) in the zone above 1000 m and below 1600 m, whereas those between these depths will be largely unchanged, but very transparent (Fig. 14).

Disruption of reflectors in the areas predicted by

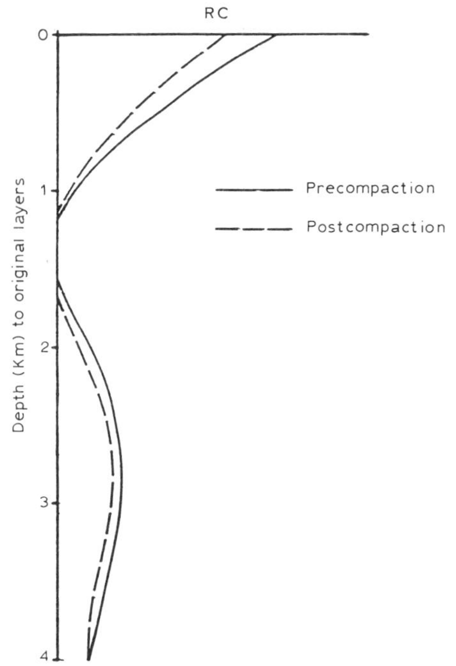

Fig. 13. Sketch illustrating the suggested change in reflection coefficients with depth pre- and post-compaction for a block of interlayered sandstone and shale.

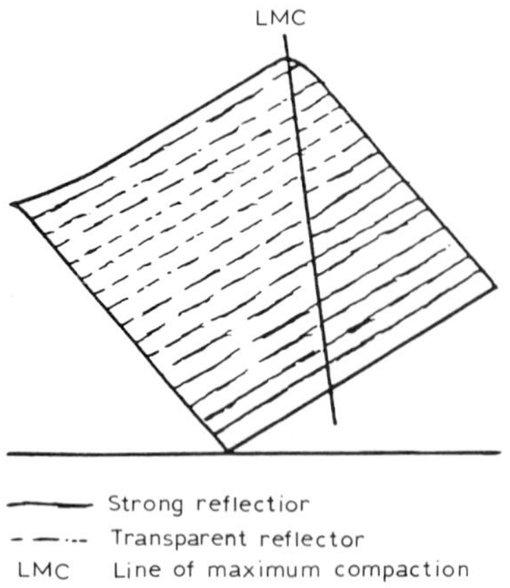

Fig. 14. Sketch illustrating the distribution of reflectors in a compacted block of interlayered sandstone and shale. The intensity of the lines represents the relative visibility of the reflectors.

the model is visible on seismic sections (Fig. 15a and b), although other explanations of the same phenomenon may be possible.

Interbedded layers will not influence the basic trend of isotherms, but because the geothermal gradient is dependent on the thermal conductivity (which varies with lithology), a mixed block will have a lower average geothermal gradient than pure sandstone, but higher than homogenous shale. The result will be intermediate between the sandstone and shale cases (Fig. 12b and c).

The effect of stress change in an interlayered situation would not be very different to that in homogenous blocks, except that the depth at which the stress change occurs would be deeper than in shale.

Fault geometry

Another assumption which requires validation is that of the boundary conditions of planar faults as opposed to listric faults. No compaction occurs on the faults themselves, although the rotation could just as easily be accommodated by listric faults. The simplicity of the mathematical functions required to

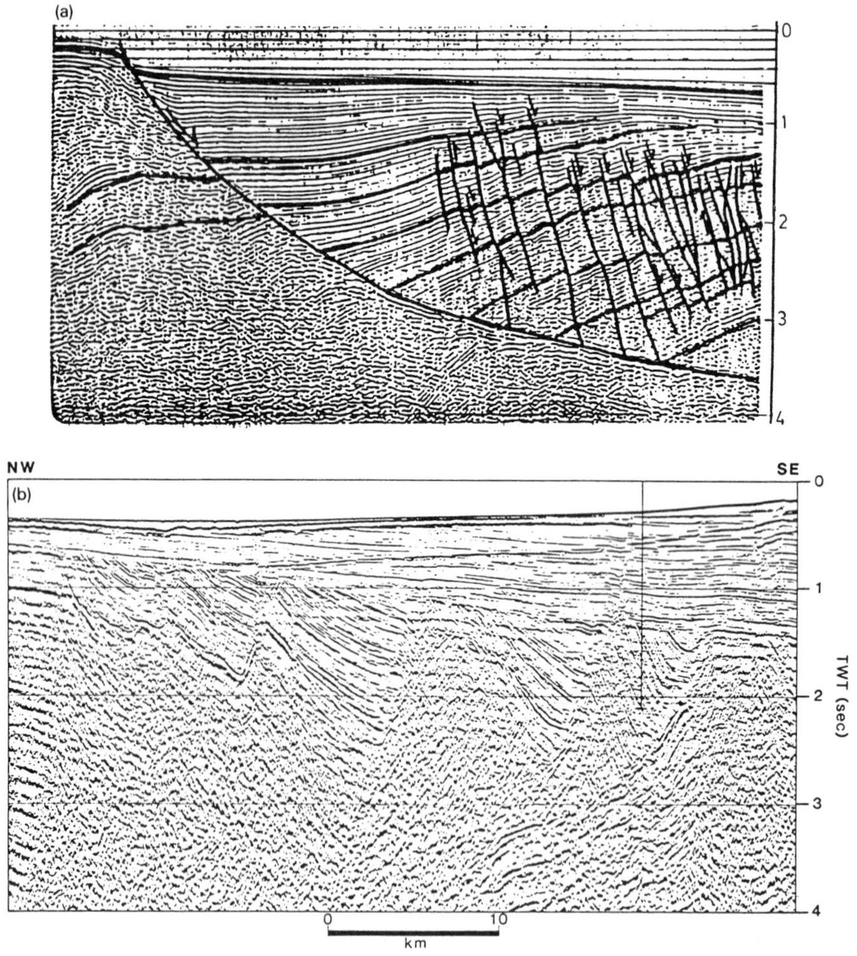

Fig. 15. (a) Seismic section from Wernicke and Burchfiel (1982, their fig. 15), exhibiting folding of upper layers of "skyscraper" type fault blocks, sequential secondary fault development and regions of transparent seismic reflectors in the hanging wall. (b) Seismic line 3630-1/2-85 across the Emerald Basin, Nova Scotia. Note the less contiguous nature of the reflectors in the hanging walls, and the occurrence of secondary faults in the approximate positions predicted by the model (from Tankard and Welsink, 1989). They suggest that diapirism has also occurred in the footwalls of these fault blocks.

rotate blocks on planar faults does not preclude extrapolation of results into listric faulted regions (Fig. 15a and b). Indeed, the problem of the void beneath the rotated blocks would be accommodated by listric faults, whereas vertical foundering of the situations modelled here may transpose the effects of compaction. The speculation of a set of secondary faults may be more applicable to the listric fault setting.

Instantaneous rotation and lack of syn-tectonic deposition

First, let us consider the effect that subsequent deposition might have. Selected sediment deposition on the hanging wall side of the block, as in growth faults, will cause increased compaction in that region, and the changes described above will be accentuated. If, however, the basin as a whole subsided and a "sag fill" (Harding, 1984) type deposition ensued, all the effects would be moderated, since compaction effects are decreased with depth, making natural examples of phenomena described here more difficult to find.

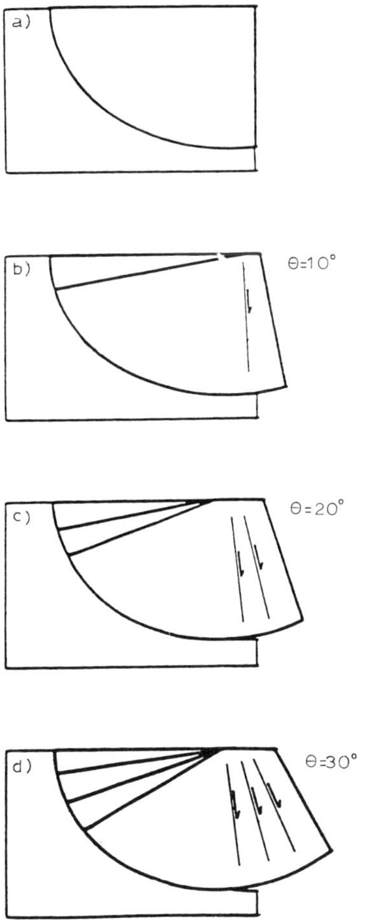

Fig. 16. Sketches suggesting a possible sequential formation of secondary faults towards the footwall during progressive rotation, as suggested in the seismic section of Fig. 15a Wernicke and Burchfiel (1982).

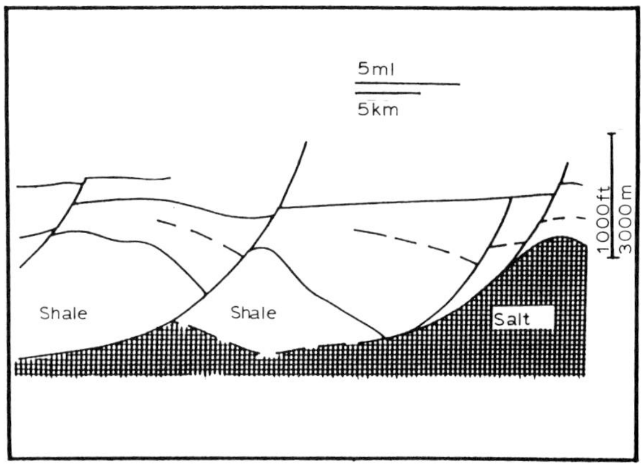

Fig. 17. The observed relationship of shale diapirs, faults and salt movement in offshore Texas and Louisiana based on Woodbury (1973).

If blocks rotated and compacted gradually, the changes in stress applied at certain points in the blocks would also constantly change, affecting the orientation and frequency of secondary faults, perhaps in a sequential fashion not unlike the model of development proposed by Angelier and Colletta (1983) (Figs. 15 and 16).

In general, therefore, we conclude that the observed results of the model are consistent with naturally occurring examples.

Diapirism

A further implication of the model is a possible cause of diapirism. Rotation causes compaction. If the water loss occurs after rotation and if the blocks were composed of interbedded shales, the fluid motion would be curtailed by impermeable shale layers and flow would be more along the bedding to the uncompacted sides of the block, rather than upwards. This concentration of fluids in the hanging wall of the block might be a causative mechanism for mud diapirism and/or areas of accumulation of hydrocarbons. The effect would be amplified by the dehydration of smectite to illite at around the depth at which the temperature attains 100°C (Bruce, 1981). This effect is, of course, concordant with the accepted ideas of diapir formation and shale diapirs in the footwalls of faults in muddy deltaic settings, as was observed in Tertiary of offshore Louisiana, U.S.A. by Woodbury et al. (1973) (Fig. 17) and in the Upper Carboniferous of the West Coast of Ireland (Rider, 1978), and would seem to provide observational support of the conjecture.

Salt diapirism may also be triggered by fault block rotation. If a salt layer acts as an extensional zone of detachment for blocks to tilt on, the changes in variable load beneath different regions of the block may encourage the salt to flow "up gradient" into

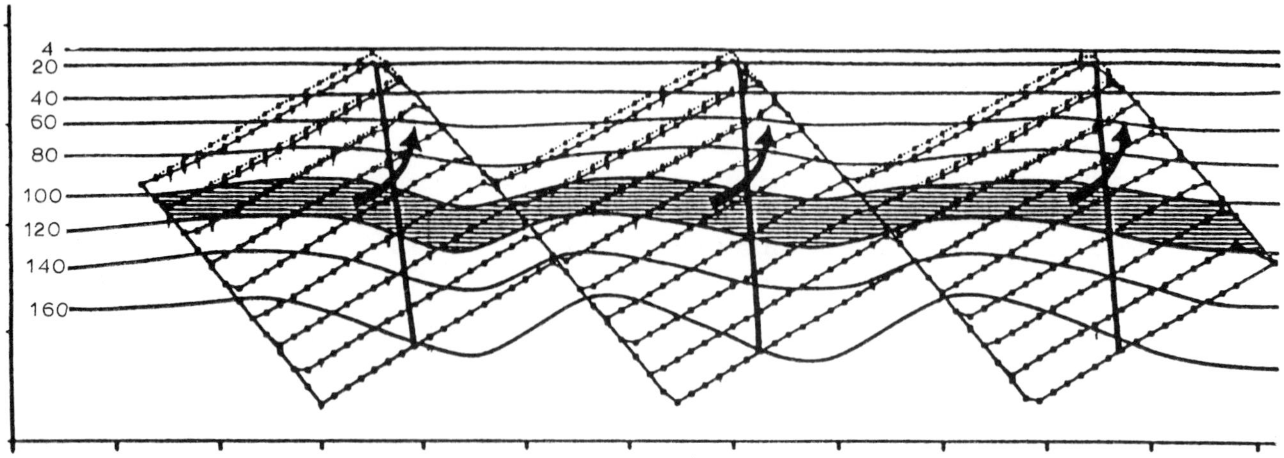

Fig. 18. Diagram summarizing the effects of rotation and compaction on temperature and fluid flow of possible hydrocarbons. Flow is from the hanging wall to the footwall as shown by arrows. Blocks are 4 km high and 4 km wide composed of shale, tilted 30° with a fault angle of 80°.

the uncompacted and unloaded region below the footwall of the block, as also observed by Woodbury et al. (1973).

Implications for hydrocarbon accumulations

The implication that fluid will migrate from a compacted area in the hanging walls of fault blocks to the uncompacted region in the footwall suggests that hydrocarbons reservoirs may be prevalent in these areas. Such reservoirs are widely recognized. What is interesting is that the compacted hanging walls are also the hottest areas, and thus may be kitchens for hydrocarbon maturation, with migration to the footwalls (Fig. 18). It is also possible that the upfaulted high thermally conductive areas may act as conduits for heat to reach upper, less conductive areas of the block and heat them above their expected temperatures, much like salt diapirs do in the Gulf of Mexico, U.S.A. (Vizgirda et al., 1985). This principle can be compared to putting a skewer in a potato to make it bake more quickly.

Conclusions

(1) The maximum depth at which compaction may occur is governed by the scaling factor (ℓ), the width of the block (W), the angle of tilt (θ) and dip of faults (τ). The dominant factors are ℓ and W.

(2) Compaction is confined to the hanging wall side of the blocks and is greater in the upper 1000 meters and is also greater for changes of tilt in excess of 30° and results in an overall rounding of the shape of the block. Layers are reduced in thickness by up to 10% in moderately tilted blocks.

(3) Faults dipping more steeply yield more compaction than those dipping gently, although the faults themselves are not affected by compaction, since they are exhumed and unloaded.

(4) Strata lengths are increased by low amplitude folding due to compaction. Increases in bed length cause underestimation errors in extension estimates of up to 10%. This value decreases with depth and reliable bed lengths can be obtained at depths greater than 1 km. Surface measurements are reduced, suggesting an overestimation of extension.

(5) The change in reflection coefficient with depth due to compaction is a minor effect (<3%) compared with the cumulative change with depth. Reflection coefficients in interlayered sand and shale scenarios are reduced above 1000 m depth and below 1600 m, whereas the zone between remains largely unchanged. These effects are increased with increasing tilt of the blocks.

(6) Seismic evidence is consistent with the prediction of the occurrence of compactional folds, secondary faults and zones of more transparent reflectors in the region between 1000 and 1600 m and more coherent reflectors in the footwall side compared to the hanging wall.

(7) Model results show that the hanging walls of rotated blocks have an elevated temperature compared to the footwall, mainly due to rotation. Isotherm *trends* do not change with lithology although isotherm *values* do change. The act of rotation brings the higher conductive regions of the block to the surface which may act to focus heat from below and dissipate it higher in the lithologic section. This effect can cause local elevation of temperatures, particularly in syn-rift deposits, if present, which may have important connotations on hydrocarbon play designs in such tectonic regions.

(8) Analysis of the change of overburden in the

areas where compaction takes place indicates that the stress configuration is changed due to rotation. A line of maximum compaction can be drawn through points along each formation which have the greatest compaction. This line of change in stress is a likely position for strain release in the form of a secondary fault. This result is consistent with the independent field observations. Secondary faulting will occur even in incompressible rocks provided the increase in stress is sufficient to overcome the shear strength of the material.

(9) Fluid migration will take place from the compacted side of the block up-gradient to the uncompacted side and then vertically upwards. Such flow may manifest itself in nature as observed shale diapirs in muddy deltaic regions or serve to redistribute any hydrocarbons present. Local changes in overburden over an originally horizontal salt layer may also cause halokinesis.

(10) Finally, further burial of the rotated blocks, as commonly seen in extensional basins, will reduce the effects discussed above.

Acknowledgements

Thanks and appreciation go to the Industrial Associates of the University of South Carolina Basin Modelling Group for financial support, and also to Donna Black for typing the manuscript. The paper has benefited greatly by the critical and insightful reviews of Bjorn Tøundbakenin and Rob Young.

References

Angelier, T. and Colletta, B., 1983. Tension fractures and extensional tectonics. Nature, 301: 49–51.

Bruce, C.H., 1981. Smectite dehydration — its relation to structural development and hydrocarbon accumulation in Northern Gulf of Mexico Basin. Am. Assoc. Pet. Geol., Bull., 66: 217–224.

Guidish, T.M., Kendall, C.G.St.C., Lerche, I., Toth, D.J. and Yarzab, R.F., 1984. Basin evaluation using burial history calculations: an overview. Am. Assoc. Pet. Geol., Bull., 69: 92–105.

Harding, T.P., 1984. Graben hydrocarbon occurrences and structural style. Am. Assoc. Pet. Geol., Bull., 68: 333–362.

Lewis, C.R. and Rose, S.C., 1970. A theory relating high temperatures and overpressures. J. Pet. Technol., 22: 11–16.

Magara, K., 1976. Water expulsion from elastic sediments during compaction — direction and volumes. Am. Assoc. Pet. Geol., Bull., 60: 543–553.

Miller, E.L., Grans, P.B. and Garing, T., 1983. The Snake Range decollement: an exhumed mid-Tertiary ductile–brittle transition. Tectonics, 2: 234–263.

Rider, M.H., 1978. Growth faults in Carboniferous of Western Ireland. Am. Assoc. Pet. Geol., Bull., 62: 2191–2213.

Sclater, J.G. and Christie, P.A.F., 1980. Continental stretching: an explanation of the post-mid-Cretaceous subsidence of the central North Sea basin. J. Geophys. Res., 85: 3711–3739.

Tarkard, A.J. and Welsink, H.J., 1989. Mesozoic extension and styles of basin formation in Atlantic Canada, in extensional tectonics and stratigraphy of the North Atlantic Margin. In: A.J. Tarkard and H.R. Balkwill (Editors), Am. Assoc. Pet. Geol., Mem., 46, 12, pp. 175–195.

Vizgirda, J., O'Brien, J. and Lerche, I., 1985. Thermal anomalies on the flanks of a salt dome. Geothermics, 14: 553–565.

Wernicke, B. and Burchfiel, B.C., 1982. Modes of extensional tectonics, J. Struct. Geol., 4: 105–115.

Woodbury, H.O., Murray, I.B., Pickford, P.J. and Akers, W.H., 1973. Pliocene depocenters, outer continental shelf, Louisiana and Texas. Am. Assoc. Pet. Geol., Bull., 57: 2428–2439.

Wyllie, M.R.J., Gregory, A.R. and Gardner, L.W., 1956. Elastic wave velocities in heterogenous and porous media. Geophysics, 21: 41–46.

J.E. ILIFFE *Department of Geological Sciences, University of South Carolina, Columbia, SC 29208, U.S.A.*
I. LERCHE *Department of Geological Sciences University of South Carolina, Columbia, SC 29208, U.S.A.*
K. NAKAYAMA *Department of Geological Sciences University of South Carolina, Columbia, SC 29208, U.S.A.*

Techniques for the geometrical restoration of sections: an example from the Bjørnøya Basin, Barents Sea shelf

Alastair Beach and Mark G. Rowan

Structural analysis and modelling are important aspects of seismic interpretation for both exploration prospect evaluation and field development studies. One of the principal methods for validating structural interpretations of seismic data is by carrying out systematic restoration of sections, and the routine application of these techniques greatly improves the understanding of the geological development of a structure, including the fault geometry, the stratigraphic development and the reservoir geometry. Without the availability of sophisticated software, such methods are time consuming and often inaccurate, but utilising such software allows the techniques to be carried out objectively, with accuracy and in a way that can be repeated. This paper describes the application of these methods to evaluate a specific section from the Bjørnøya Basin on the continental shelf of the Barents Sea.

The principal structures shown in the section are a result of two stages of fault movement, firstly during Jurassic extension and secondly during Cretaceous subsidence. The greatest uncertainty in the original interpretation relates to the geometry of the faults, the correlation of horizons across the faults and the thickness and position of units within each fault block. The restorations enable modifications to be made which improve the consistency and viability of the interpretation and reduce the uncertainty.

Introduction

A seismic line has been selected for analysis from a data set covering the Bjørnøya Basin, to the west of the Loppa High on the continental shelf of the Barents Sea. This analysis illustrates several aspects of section restoration, carried out with the computer software Geosec-20 from Geo-Logic Systems, Inc. It is only since the development of software packages that restoration of sections can be attempted routinely and with precision. The concept of the balanced section was described sometime ago (Dahlstrom, 1969; Hossack, 1979), and more recently developed computer techniques have concentrated largely on methods for testing and verifying that a section is balanced. In compressional terranes, these include checking that bed lengths and cross-sectional area are preserved, predicting the depth to detachment and checking that the geometry is a viable one that can be restored to its unstrained state. In extensional terranes, sections are often balanced by predicting linked listric fault trajectories from the shapes of the hanging walls (Gibbs, 1983; White et al., 1986; Williams and Vann, 1987). Recent geological studies of normal faults, however, raise questions about the assumptions made in employing this geometrical technique in drawing balanced sections. For example, it is recognised that hanging wall rollovers can develop against planar faults, making it difficult to predict fault profiles from hanging wall geometry (Gibson et al., 1989). Similarly, there is usually enough uncertainty in some of the factors that affect the results, such as decompaction, determining the undeformed regional geometry and identifying the appropriate algorithm for fault projection (Rowan and Kligfield, 1989), so much so that application of the technique can be quite misleading.

Two important points emerge from the above considerations. The first is that section balancing in extensional terranes is, at the present time, a poorly understood process. It is more instructive to carry out sequential restorations of interpreted sections, back through geological time, rather than trying to produce unique, balanced sections. The distinction between restoration and balancing is important: in balancing, a set of rules and assumptions, which may or may not be valid, are applied to constrain an interpretation, whereas in restoration, the deformation is removed so that the undeformed and intermediate geometries can be evaluated. The second point is that the computer software used for carrying out the restorations should make no geological assumptions about fault geometry in the section and that it should be equally applicable to listric or planar

Structural and Tectonic Modelling and its Application to Petroleum Geology edited by R.M. Larsen, H. Brekke, B.T. Larsen and E. Talleraas. NPF Special Publication 1, pp. 269–276. Elsevier, Amsterdam.

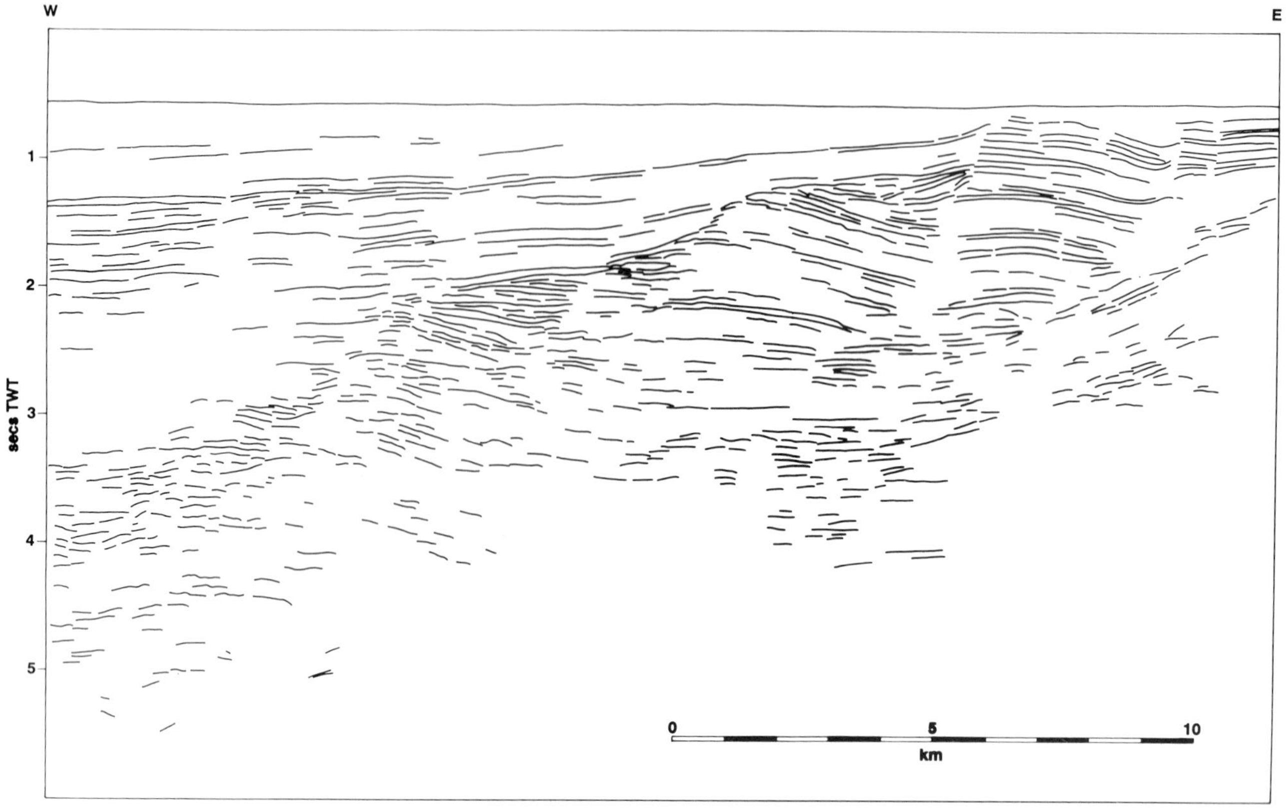

Fig. 1. Line drawing from a seismic section on the flanks of the Loppa High and the Bjørnøya Basin. The line and its exact location are confidential. In this figure, seismic events have been traced from the original section without any interpretation and in an attempt to show objectively the reflections present.

faults. The sequence of restorations then provides a precise record of the geometrical evolution of the section through successive geological stages.

The analysis of the example line from the Bjørnøya Basin (Fig. 1) illustrates the fundamental steps in restoring a section and is typical of many exploration problems where the seismic data are of reasonable quality, but with some ambiguities, and where few well data are available. The starting point is an initial interpretation of the horizons and faults (Fig. 2) and sequential restoration of this section will help answer questions such as:

– is the interpretation of the general structural style acceptable?

– what is the most consistent interpretation of the fault geometry?

– are the horizon correlations across the faults acceptable?

– is the reservoir geometry within each fault block valid?

– what is the deformation history of the interpreted structures?

Geometrical restoration of sections is primarily concerned with checking the validity of interpretations of seismic data and with verifying that these interpretations are based on geological models that are both consistent and acceptable to the geologist. It, thus, provides for a more objective assessment of the structure, stratigraphy and reservoir geometry, although it does not necessarily provide a unique solution.

Prior to carrying out any restorations, it is necessary to establish the most accurate depth geometry for the section. Any errors in the geometry at this stage will be maintained and carried forward in the restored sections, and may give rise to incorrect geological interpretations. The amount of data available for depth conversion will vary enormously according to whether any wells have been drilled and if so, how many. In association with the line studied here, no wells are available, and depth conversion has been based on the use of stacking velocities. Seismic velocities used for depth conversion are: seawater, 1480 m/s; Tertiary, 2600 m/s; Cretaceous, 3000 m/s at the top increasing to 4000 m/s at the base; Jurassic–Triassic, 4000 m/s at the top increasing to 4250 m/s for the deepest layer interpreted. As more data become available, a more accurate set of restored sections becomes possible, as a result of having an improved depth section to start with.

The depth geometry is also affected by compaction of sediments and, therefore, decompaction must be

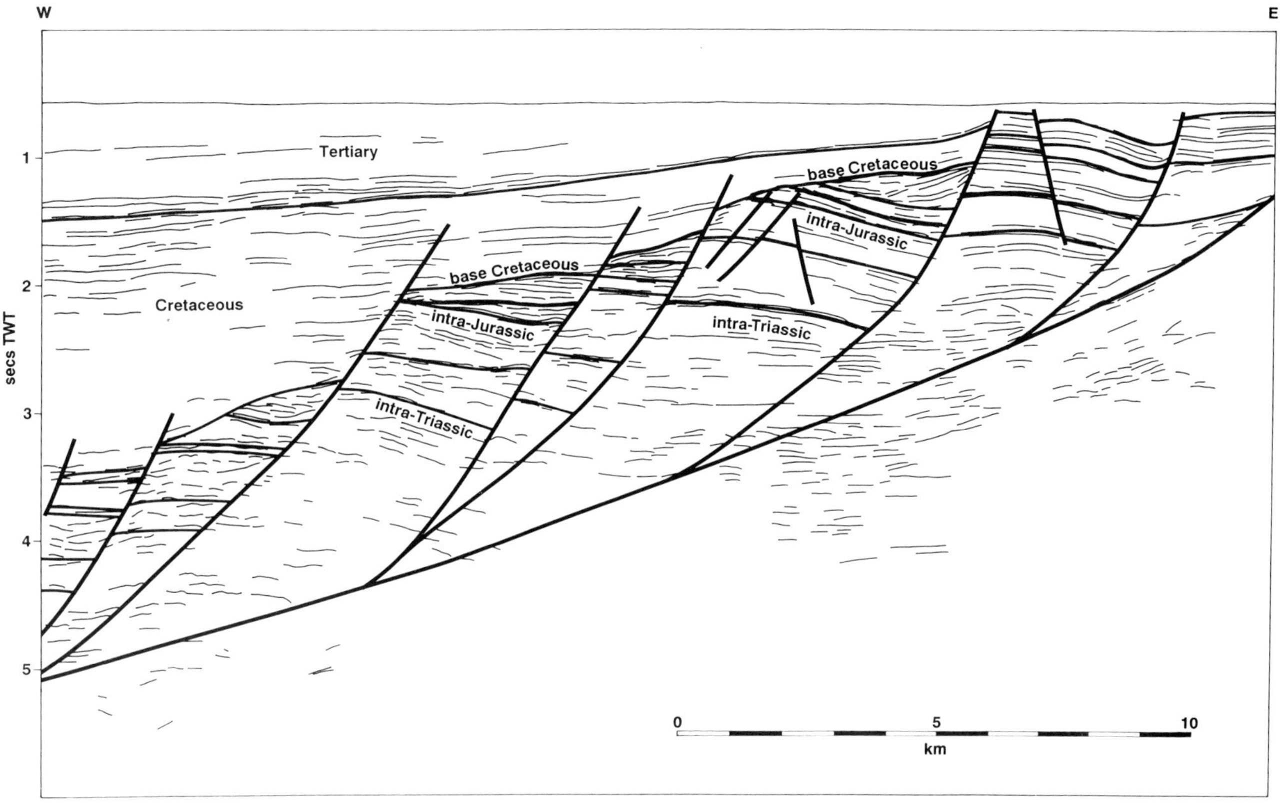

Fig. 2. A preliminary interpretation of the line drawing shown in Fig. 1, based on event correlations and cut-offs. The base Tertiary and base Cretaceous are readily identified, but the stratigraphy in the pre-Cretaceous section is not as clearly understood.

carried out during restoration. Few data are available for the line studied here and decompaction functions from Sclater and Christie (1980) have been used, assuming that the lithologies are claystone in the Tertiary and Cretaceous and a 50/50 mix of interbedded sandstone and claystone in the Jurassic and Triassic.

Section restoration

A single routine restoration of a section is a relatively simple process involving depth conversion, decompaction and then restoration to the chosen geological time horizon. However, if the restoration is to be used to evaluate the interpretation and to make modifications where deemed necessary, the procedure becomes much more complicated and involves interactive manipulation of both the deformed and restored geometries (Fig. 3). After decompaction and restoration to one horizon, the resulting geometry must be evaluated in order to identify any problems arising from a poor interpretation. Any such problems can subsequently be modified in one of three ways. The time interpretation can be adjusted and then depth converted, decompacted and restored again, so that a better interpretation eventually results through trial and error. A variation is to modify the section in depth and then decompact and restore, but any changes must be converted back to time to check for compatibility with the data. Finally, a more powerful method is to make changes to the restored section, then compact and forward model to the deformed geometry. The resulting interpretation

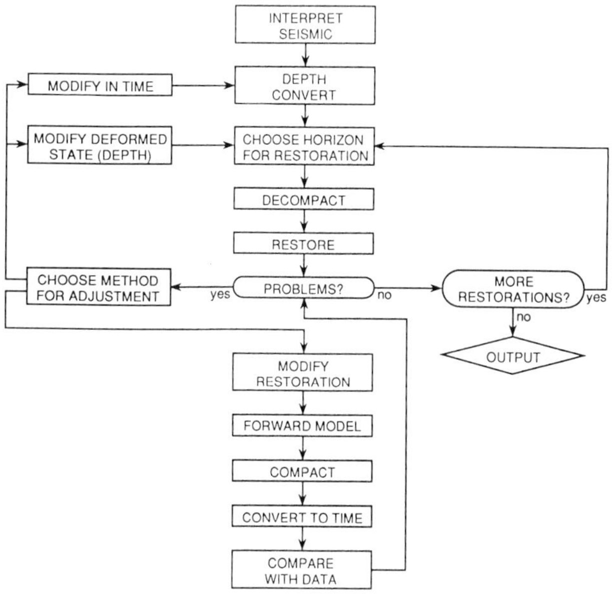

Fig. 3. Flow chart for the simple restoration and modification of seismic interpretations in extensional terranes. See text for discussion.

is one which restores to an acceptable geometry, but it too must be converted to time for comparison with the data. Usually, some combination of all these procedures is employed in modifying and improving an interpretation, and only the use of appropriate software, with its inherent speed and precision, makes it a practical proposition.

A standard simple restoration assumes an original horizontal depositional surface and, thus, uses a flat geometry as the template for the horizon being restored. Alternatively, if the structure is a linked fault system, the basal detachment can be used as the restoration template. Neither of these methods, however, takes into account possible original topography and paleobathymetry at the time of the chosen restoration. For example, whereas a formation deposited regionally at or near sea level can be restored accurately using a flat template, a formation that only partially filled semi-starved basins bounded by fault scarps would be restored incorrectly by the same method. Where the data are available, it is possible to incorporate isostatic effects and thermal subsidence models (e.g., McKenzie, 1978) to calculate the paleobathymetry at any given point and time (e.g., Bertram and Milton, 1989). This can be done for numerous points along a section, thereby recreating the paleotopographic profile which forms the template for a more accurate restoration. In the case of the example from the Barents Sea, an unknown amount of Tertiary uplift and erosion affected the area (Wood et al., 1989), so that such calculations can not be made. Restorations, therefore, were confined to the simple flat regional and basal detachment templates. For both of the horizons restored, however, these form adequate approximations, and potential errors should not affect the conclusions drawn from the restorations.

Figure 1 shows a line drawing of the seismic line, made by tracing reflections from the original data, and Fig. 2 shows an initial interpretation of this line. Figure 4 is the result of depth-converting the interpretation and is simplified in that the minor, intra-block faults are not shown. The depth section shows a westerly dipping, low-angle detachment fault with a sequence of rotated fault blocks in its hanging wall. Although the data could be interpreted with non-linked, more planar faults, other nearby lines support at least the local existence of a dipping detachment at the shallow, eastern end of the line. Whether or not it continues to the west under the deeper fault blocks is unknown, but should not change the main conclusions derived from the restorations. The fault blocks are truncated by the base Cretaceous unconformity and overlain by Cretaceous and Tertiary sequences that thicken to the west. After depth conversion, the basal detachment is seen to have a more gentle dip to the west than is apparent in time (compare Figs. 2 and 4). In addition, the anticlinal folding of events in fault blocks 2 and 3 (Fig. 2) disappears in depth (Fig. 4) and hence these are considered to be velocity anomalies in the time section.

Figure 5 shows the first stage of section restoration where the Cretaceous and Tertiary sequences have been removed and the section restored to a continuous and horizontal base Cretaceous unconformity. This unconformity is interpreted to have been caused by erosion, at or near sea level, of the rotated and uplifted footwalls to the major faults during extension and deformation. Making the unconformity horizontal in the restoration is probably a valid approximation for most of its length, as evidenced by the abundance of truncated events. In reality, however, there may have been paleotopography at the

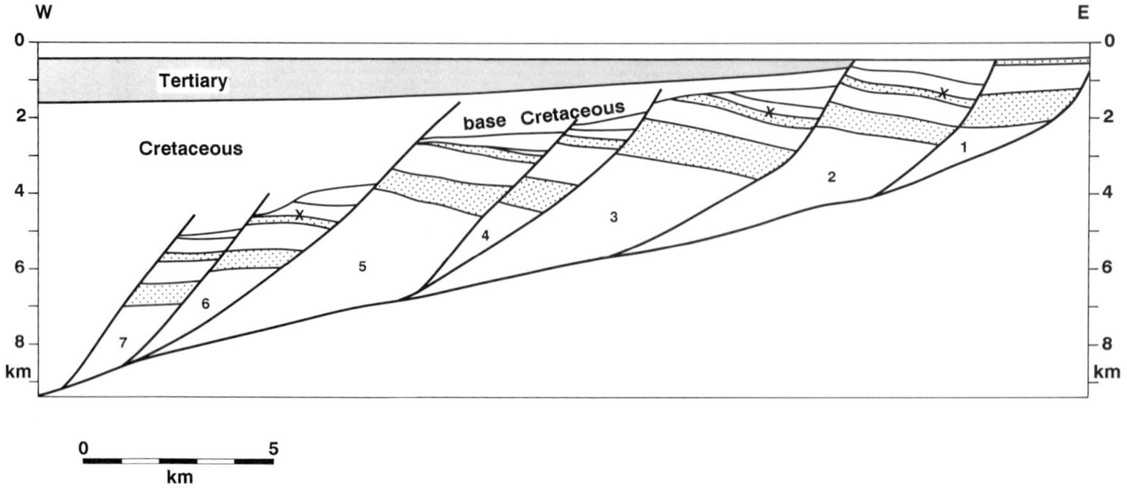

Fig. 4. Depth conversion of the interpretation shown in Fig. 2. Numbers are fault blocks referred to in the text and x is the upper Jurassic horizon to be restored.

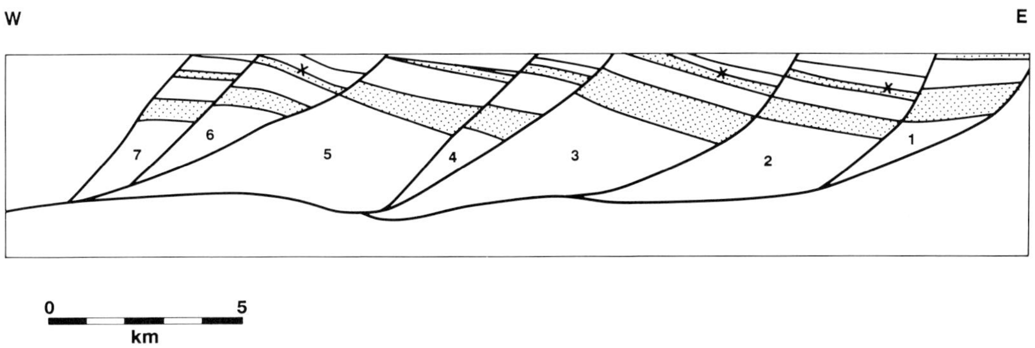

Fig. 5. Restoration of the depth section shown in Fig. 4 to a horizontal base Cretaceous unconformity, thereby removing the overlying differential subsidence of the Tertiary and Cretaceous, the small fault offsets at base Cretaceous and decompacting the underlying sequences. Numbers refer to fault blocks and x marks the next horizon to be restored.

end of the Jurassic in the form of small relict basins in the immediate hanging walls of the major faults.

Three stages are involved in constructing the restoration shown in Fig. 5: (1) the removal of the Cretaceous and Tertiary sequences and the accompanying decompaction of the underlying sediments; (2) the restoration of fault offsets at the base Cretaceous; and (3) the restoration of the unconformity to horizontal. The post-Jurassic regional differential subsidence and associated faulting has been removed and, hence, the fault geometry at the end of the main phase of Jurassic extension has been established. This shows a series of rotated and truncated fault blocks above a roughly horizontal detachment.

The next stage restores the section to one of the upper Jurassic events (marked x in Figs. 4 and 5), using the basal detachment and its footwall as the template onto which successive fault blocks are restored by removing the displacement at the chosen horizon (Fig. 6). Where the restored horizon has been truncated by erosion, its geometry has been projected using underlying events as a guide, so that the missing material has been reconstructed in the restoration.

The bulk deformation has been modelled in both restorations by vertical simple shear, as opposed to the inclined simple shear suggested by White et al. (1986). Since changing the angle of shear used in restoration affects the total amount of extension in the section, one way of evaluating which angle to use is to restore on a fault template and choose the angle which brings the chosen restoration horizon back to regional by reversing the appropriate amount of extension. For example, if removing fault offsets using vertical simple shear on a given fault template leaves the horizon below the level regional, a shear angle inclined antithetic to the fault will raise the horizon and thereby improve the restoration. However, it is crucial that the appropriate fault template be used. In the example presented here, if the upper Jurassic event is restored in one step from the present deformed geometry, using the template provided by the dipping detachment of the original depth section (Fig. 4), a shear angle of 70° antithetic to the fault is found to work best. If, instead, the base Cretaceous unconformity is first restored as described above, so that the Jurassic event is subsequently restored using the level fault template of Fig. 5, vertical simple

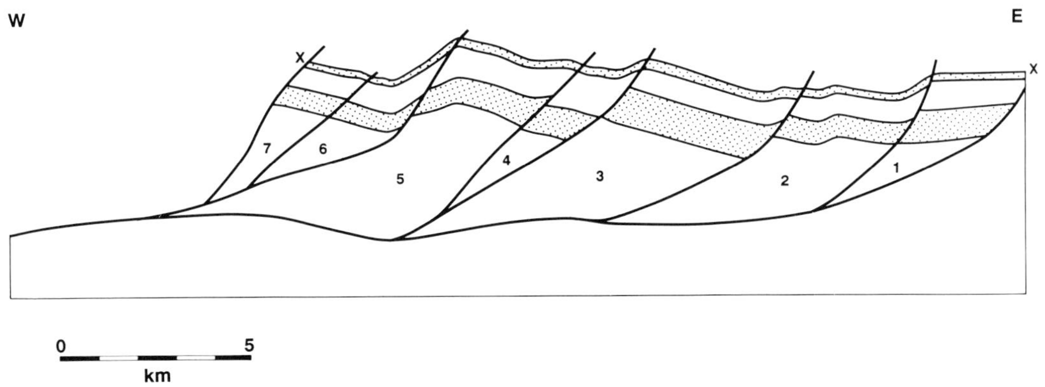

Fig. 6. Restoration of the geometry shown in Fig. 5 to the upper Jurassic horizon (x) using the basal detachment of Fig. 5 as the template. See text for discussion of problems apparent in the restoration. Numbers refer to fault blocks.

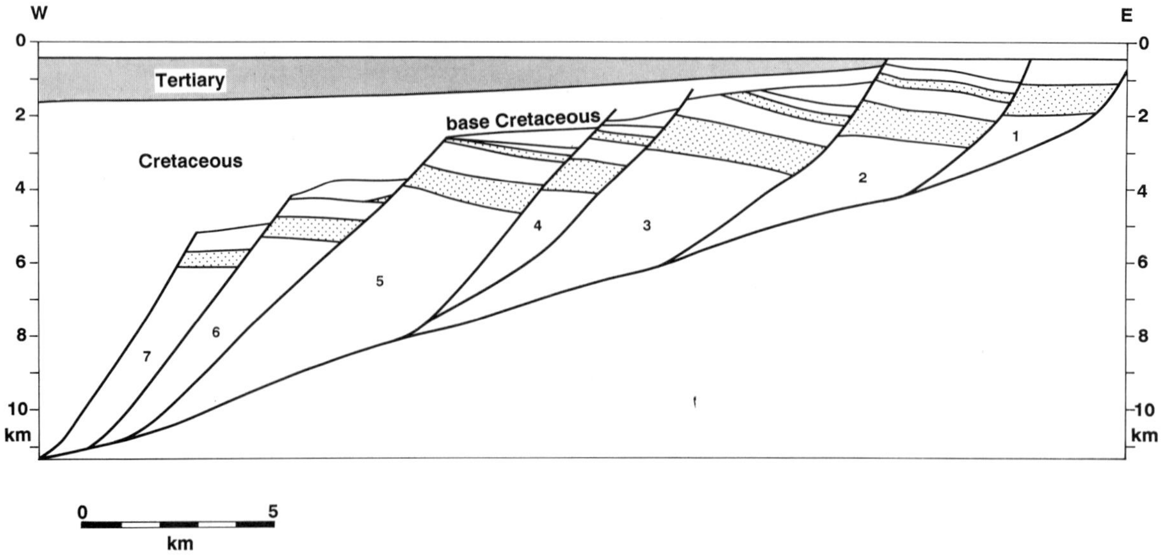

Fig. 7. Final interpretation in depth, generated by multiple cycles of restoration and modification, showing a consistent geometry of faults, horizons and unit thicknesses (compare with Fig. 4). Numbers are fault blocks referred to in the text.

shear works better. The implication is that, at least for this example, if the section is restored properly in a sequential manner, vertical simple shear is a better algorithm for approximating the bulk deformation (see also Rowan and Kligfield, 1989).

Figure 6 may be used as a preliminary check on the validity of the interpretation, for example, by evaluating fault trajectories, the geometry of the restored horizon and the correlation of horizons across faults. The principal ways in which the section does not restore properly are (Fig. 6): firstly, in the irregularity of the uppermost Jurassic event from which fault offsets were removed; secondly, in the reversals of dip along the restored basal detachment; and thirdly, the mismatches between the same horizons across the restored faults. The latter case is especially obvious for fault block 2, which has a thinner sequence at one level than the blocks to either side and for the fault between blocks 4 and 5, which shows reverse offset in the restored geometry.

Iterative modifications can be made to the geometry to produce a more acceptable interpretation and restoration. These can be carried out in the deformed state and their effect tested by restoration, or they may be made to the restored geometry, followed by forward modelling, compaction and conversion to time to check for compatibility to the original data (Fig. 3). The first method has been used to make major changes and the second method for minor adjustments. Numerous such modifications were made, with interactive manipulation of both the deformed and restored geometries. The net effect of all changes is an improved depth interpretation (Fig. 7) and its time equivalent (Fig. 8). The major modifications are (compare Figs. 2 and 8, and Figs. 4 and 7): a steepening of the basal detachment; minor adjustments of the deeper portions of some of the fault trajectories; a reinterpretation of the unconformity in fault block 6; significant adjustments to the interpreted horizons, especially in fault blocks 2 and 4; and a raising of the entire stratigraphic sequence in blocks 6 and 7.

It is, of course, always possible to continue making minor modifications to a section in order to perfect its restoration. However, the limitations of the method are exceeded after a certain point and it is felt that with the data available here, the restorations and modifications made are sufficient to remove the most important inconsistencies in the initial interpretation and adequately produce an acceptable section with horizon correlations that are internally consistent. Restoration of the section provides a clear and objective assessment of the structural geometry and the stratigraphic development of the area. This can be used as the basis for an exploration prospect evaluation and reduces considerably the geological uncertainty in such an evaluation.

Conclusions

Fault movements occurred during two stages, firstly during Jurassic extension and secondly during Cretaceous subsidence. At the eastern end of the section, Jurassic–Triassic sequences occur at seabed and the Cretaceous and Tertiary thicken from zero to over 4 km to the west. This is a period of regional differential subsidence, primarily thermal, of the western flank of the Loppa Ridge. There is no

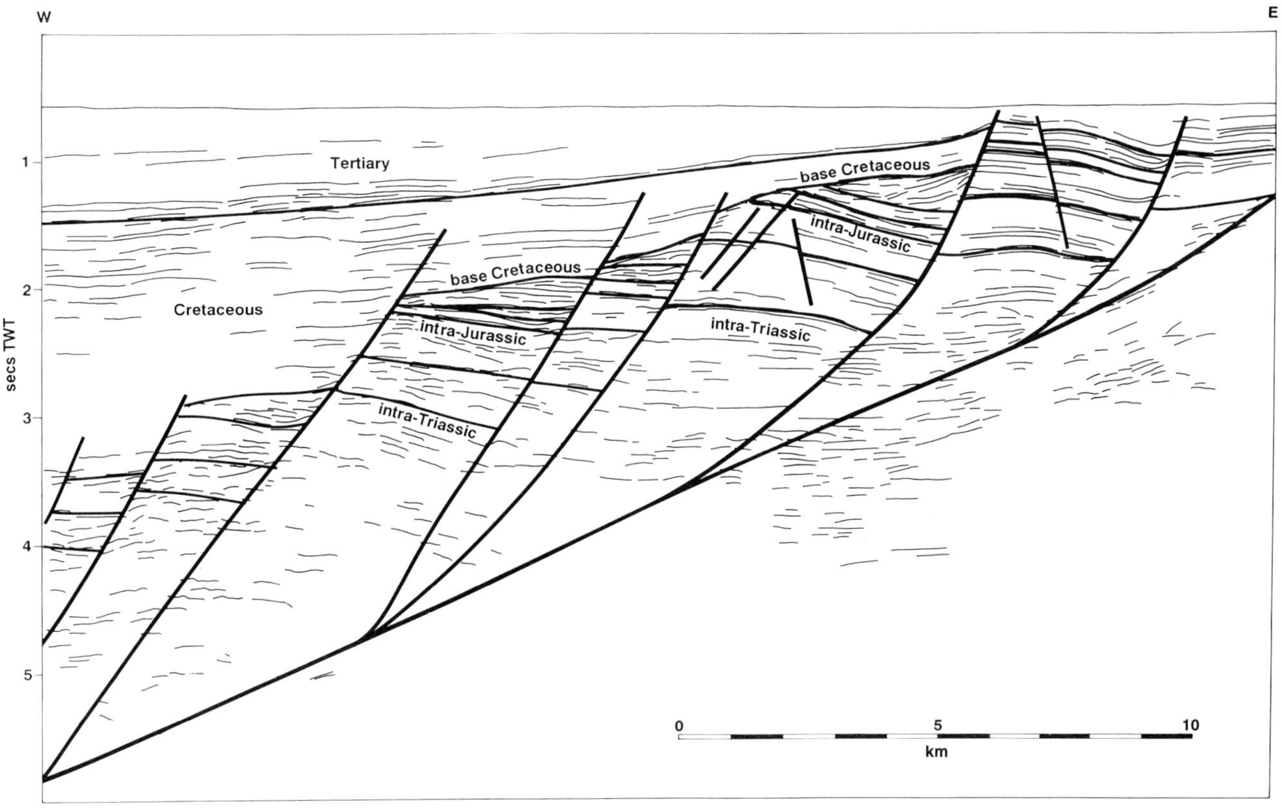

Fig. 8. Final interpretation of the seismic data which corresponds to the depth section of Fig. 7. The differences between this and the original interpretation are discussed in the text.

obvious evidence on the seismic line studied here for compressional structures of either Cretaceous or Jurassic age. However, some slight compressional reactivation along the basal detachment in the Tertiary may explain the present lack of rotation of beds in fault block 1.

The restoration to the base Cretaceous unconformity shows that a pre-Cretaceous geometry of extensional fault blocks was established. These fault blocks lie above a basal detachment fault that probably dipped gently to the west at this time. Rotation of the fault blocks occurred and some footwall crests underwent erosion. The structure restores and reconstructs in a way that suggests that it formed largely by dip-slip extensional movements. Although some strike-slip movement may have occurred, it is not necessary to explain the current geometry.

From east to west across the section studied, there are differences in the structure between adjacent fault blocks. At the eastern end, Triassic–Jurassic sequences are interpreted to be sub-horizontal and there is some element of rotation in fault block 2, though Jurassic is still present at sea floor. The third fault block shows a marked rotation and the strongest angular discordance of Jurassic sequences below the base Cretaceous unconformity. This correlates well with the greater width of the block and represents the locus of maximum footwall uplift during the Jurassic faulting. Successive fault blocks further west all show rotation of Jurassic sequences, but less marked truncation below the unconformity. Analysis of the truncations and the syn-rift sediments indicates that blocks 4 and 5 form a major coherent unit divided by a relatively minor fault and that the same is probably true for blocks 6 and 7.

In conclusion, the geological modelling carried out on the seismic line has resulted in an interpretation that depicts a structural geometry and horizon correlation that is internally consistent. Although it is not the only possible valid interpretation, it is certainly more acceptable than one which does not restore to an acceptable geometry. The restoration procedure provides a clear and objective basis for assessing the structure and stratigraphic sequences in any potential exploration target and the results can easily be extended to adjacent areas.

Acknowledgements

We thank Geo-Logic Systems, Inc., of Boulder, Colorado, for access to the GEOSEC-20 software, and several anonymous reviewers for their comments.

References

Bertram, G.T. and Milton, M.J., 1989. Reconstructing basin evolution from sedimentary thickness: the importance of paleobathymetric control, with reference to the North Sea. Basin Res., 1: 247–257.

Dahlstrom, C.D.A., 1969. Balanced cross sections. Can. J. Earth Sci., 6: 743–757.

Gibbs, A.D., 1983. Balanced cross-section construction from seismic sections in areas of extensional tectonics. J. Struct. Geol., 5: 152–160.

Gibson, J.R., Walsh, J.J. and Watterson, J., 1989. Modelling of bed contours and cross-sections adjacent to planar normal faults. J. Struct. Geol., 11: 317–328.

Hossack, J.R., 1979. The use of balanced cross-sections in the calculation of orogenic contraction: a review. J. Geol. Soc. London, 136: 705–711.

McKenzie, D.P., 1978. Some remarks on the development of sedimentary basins. Earth Planet. Sci. Lett., 40: 25–32.

Rowan, M.G. and Kligfield, R., 1989. Cross-section restoration and balancing as an aid to seismic interpretation in extensional terranes. Am. Assoc. Pet. Geol., Bull., 73: 955–966.

Sclater, J.G. and Christie, P.A.F., 1980. Continental stretching: an explanation of the post mid-Cretaceous subsidence of the Central North Sea Basin. J. Geophys. Res., 85: 3711–3739.

White, N.J., Jackson, J.A. and McKenzie, D.P., 1986. The relation between the geometry of normal faults and that of the sedimentary layers in their hangingwalls. J. Struct. Geol., 8: 897–910.

Williams, G. and Vann, I., 1987. The geometry of listric normal faults and deformation in their hanging-walls. J. Struct. Geol., 9: 789–798.

Wood, R.J., Edrich, S.P. and Hutchinson, I., 1989. Influence of North Atlantic tectonics on the large scale uplift of the Stappen High and Loppa High, Western Barents Shelf. In: A.J. Tankard and H.R. Balkwill (Editors), Extensional Tectonics and Stratigraphy of the North Atlantic Margins, Am. Assoc. Pet. Geol., Mem., 46: pp. 559–566.

A. BEACH *Alastair Beach Associates, 11 Royal Exchange Square, Glasgow G1 3AJ, Scotland, U.K.*

M.G. ROWAN *Alastair Beach Associates, 11 Royal Exchange Square, Glasgow G1 3AJ, Scotland, U.K.*

Numerical simulation of jointed and faulted rock behaviour for petroleum reservoirs

A.G. Koestler, A. Makurat, N. Barton and K. Monsen

Shearing along fractures and faults, and changes in fracture and joint conducting apertures are key elements of the behaviour of petroleum reservoirs and in the evaluation of cap rock integrity. Improved joint characterization, constitutive joint modelling and numerical block interaction codes have recently been developed, which can be utilized for parametric studies in these complex engineering fields. The program UDEC-BB (Universal Distinct Element Code with Barton–Bandis joint behaviour model) can be applied on simulations of reservoir compaction behaviour, in studies of gas storage in lined caverns and in larger-scale studies of tectonic behaviour of faulted sediments under extensional deformation. The paper describes the input data requirements, some special features of the scale-dependent joint model, and some recent results of numerical modelling of faulted rock mass behaviour.

Fault configurations from offshore Norway are still little understood with respect to stress, displacement and rotation distributions within fault blocks and fault zones. An improved understanding would provide a useful tool for the prediction of minor faulting, which often cannot be mapped with traditional seismic surveys. This study concentrates on a large-scale fault configuration similar to parts of the Gullfaks field.

The importance of non-linear joint behaviour in cases involving large changes of fluid pressure are emphasized. The use of linear joint models with constant shear and normal stiffness and linear frictional properties are clearly limited when effective stress changes of several tens of megapascals take place.

Introduction

The attempts in petroleum geology to predict and model tectonic heterogeneities at reservoir scale cover a wide range of approaches, such as studying onshore analogues, analyzing fault patterns at different scales and characterizing them e.g. with fractal geometry, modelling of fault patterns with complex random parameter distributions, incorporating theories from other scientific fields such as percolation theory, cluster distribution theory, chaos theory, etc. The interest of simulation of fault patterns focuses mainly on two points: (1) modelling of fault networks for areas with limited and poor data; and (2) modelling of fault networks below seismic resolution for specific hydrocarbon reservoirs.

The better characterization of reservoirs in terms of horizontal heterogeneities (sedimentological origin) and vertical heterogeneities (tectonic origin) is of vital importance for hydrocarbon production and for enhanced oil and gas recovery.

In this study an attempt was made to simulate stress changes in a fault block configuration using numerical methods. A tectonic setting was chosen with fault blocks above a low-angle detachment and gravitational extension. This type of geological setting is based on analogue situations reported from actual hydrocarbon fields offshore Norway, such as the Gullfaks Field (see Fig. 9; Fossen, 1989; Petterson et al., 1990; Koestler et al., this volume). The large fault blocks above the detachment were designed in a way to give indications on the deformation behaviour of blocks of similar size, the influence of antithetic faults, and the differences in a very narrow fault slice. The hanging wall with a total thickness of about 1600 m consists of homogeneous sandstone fault blocks, while the underlying detachment is designed as a shale layer of about 330 m thickness. The footwall consists of sandstone of the same properties as the hanging-wall sandstone. Additionally the model is covered by 500 m of sediments. The two-dimensional modelling is carried out in the profile section representing the section of plane strain.

The application of a recently developed numerical simulation technique (UDEC-BB: Universal Distinct Element Code with Barton-Bandis joint behaviour model), which allows the realistic modelling of joint deformation behaviour as a function of normal and shear stress, and modelling of the intact block de-

Structural and Tectonic Modelling and its Application to Petroleum Geology edited by R.M. Larsen, H. Brekke, B.T. Larsen and E. Talleraas. NPF Special Publication 1, pp. 277–288. Elsevier, Amsterdam.

formation and translation along joint due to forces acting on these blocks, is supposed to lead to a better understanding of the stress distribution inside the intact blocks and along the faults.

The advent of discontinuum modelling techniques pioneered largely by Cundall (1971, 1980) has opened up a new field of geomechanics. This new field is being actively applied to a variety of geomechanics problems. An especially interesting application has been the modified Universal Distinct Element Code termed FRIP (Fluid–Rock Interaction Program) reported by Pine and Cundall (1985). These authors and other have applied the code to the Cambourne School of Mines hot dry rock geothermal project (CSM-HDR) in Cornwall, England, in attempts to explain fluid–rock interaction in the fractured granite at approximately 2 km depth. Anisotropic in-situ horizontal stresses, and sub-vertical jointing striking at some 30° to the principal stress, causes joint shearing when the effective normal stress is sufficiently reduced by fluid injection. Figure 1 can be used as a simple illustration of this mechanism, where the hydraulic connection between well 1 and 2 will never be achieved.

Pine and Cundall (1985) used an over-simplified linear sub-routine to model joint shearing, dilation, opening and closure, and their coupling of joint aperture with hydraulic conductivity was also over-simplified. They justified their simplified joint geometry and the joint behaviour model with the shortened calculation time achieved.

Harper and Last (1989) utilized Cundall's Universal Distinct Element Code (UDEC) directly in theoretical investigation of the influence of various joint patterns in rock–fluid interaction. They referred to modelling of fractured sandstone petroleum reservoirs but utilized several of their results in attempts to explain some of the phenomena observed or deduced from the CSM-HDR project referred to above. These authors also used over-simplified, linear joint behaviour models, which may have influenced the validity of their conclusions. They also used a direct coupling between joint aperture and hydraulic conductivity which is erroneous for small apertures, due to the areas of rock-to-rock contact required for normal stress transfer.

In studies of reservoir compaction in connection with the subsidence of the North Sea Ekofisk reservoir Barton et al. (1986) utilized both linear an non-linear joint behaviour sub-routines in UDEC. Reservoir fluid pressure draw-down of 20 MPa resulting from 15 years of production were simulated, and were shown to result in shearing along steeply dipping conjugated fractures in the chalk. Figure 2 illustrates the principal set up of the model.

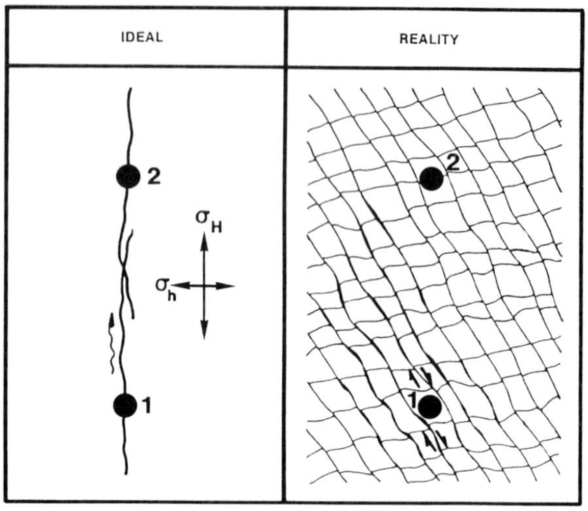

Fig. 1. Classic hydraulic fracturing to make hydraulic connection between two wells may be impossible if pre-existing jointing is non-parallel to an anisotropic principal stress system.

The decrease in reservoir pressure due to depletion and subsequent increase from modelled injections was simulated in a more simplified manner than Pine and Cundall (1985) or Harper and Last (1989). Flow was not modelled since near-well effects were not of direct interest. Therefore it was possible to utilize a more sophisticated non-linear joint subroutine in the calculations. Some important aspects of this BB (Barton and Bandis) joint model will be described here to illustrate differences with respect to commonly used linear models.

Linear and non-linear joint modelling

A linear joint model as utilized by Harper and Last (1989) is illustrated in Fig. 3. A single shear stiffness (K_s) defines the slope of the shear stress–shear displacement lines (see Fig. 2a). No differentiation is made in this value of K_s with increasing normal stress. Similarly, a single dilation angle (d^0) is used to define the slope of the normal displacement–shear displacement lines, independent of normal stress (see Fig. 2b). In normal closure, a single value of normal stiffness (K_n) defines the slope of the aperture–normal stress line up to some critical stress level, at which point a constant residual aperture (a_r) is attained, which does not reduce with increasing normal stress.

Most of the features illustrated in Fig. 3 are fundamentally incorrect. Work conducted by Bandis (1980), Barton and Choubey (1977), Gale (1982) and others has shown that the shear stiffness (K_s) and normal stiffness (K_n) increase significantly with normal stress. A further error is the assumption that hydraulic conductivity (k) is equal to 1/12 of (a_r^2), where a_r is defined in Fig. 3c.

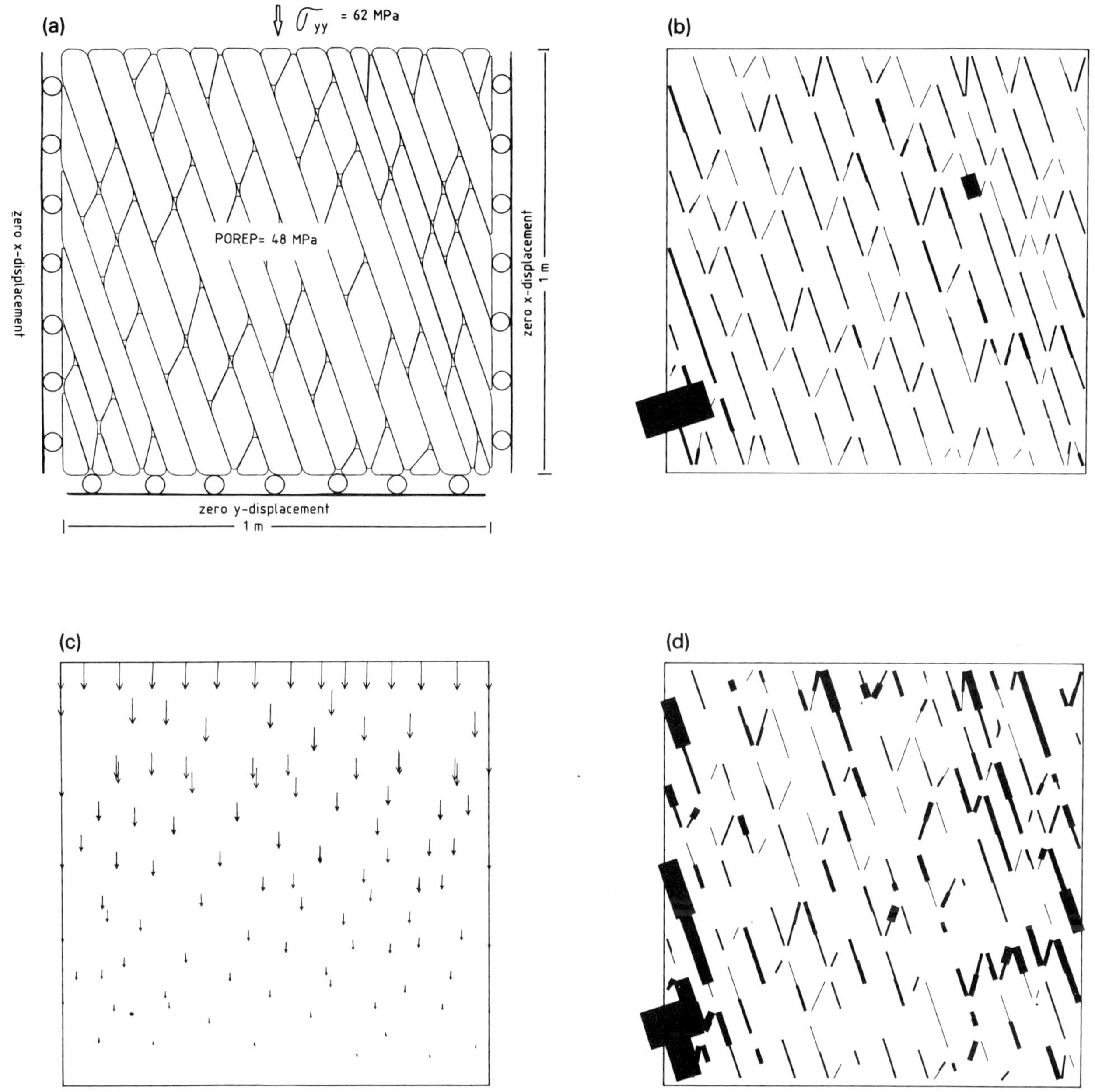

Fig. 2. Results of non-linear distinct element modelling with UDEC-BB in studies of the Ekofisk reservoir compaction (Barton et al., 1986). (a) Model set up. The 1 x 1 m high porosity chalk is surrounded by roller boundaries at the left, right and bottom boundary. 62 MPa vertical stress is applied along the top boundary. (b) Hydraulic fracture aperture after pore pressure reduction from 48 to 18 MPa. Thickness of lines is proportional to fracture aperture. Maximum joint aperture = 31.6 μm. One line thickness = 0.5 μm. (c) Total displacement vectors due to pore pressure reduction (48–18 MPa). Maximum displacement = 4.5 mm. (d) Shear displacement due to pore pressure reduction (48–18 MPa). Thickness of lines is proportional to the amount of shear. Maximum shear displacement = 499 μm. One line thickness = 10 μm.

An example of a more realistic non-linear joint model (based on experiments), as developed by Barton et al. (1985), is shown in Figs. 4 and 5. Shear stress–shear displacement curves show increasing shear stiffness (K_s) with increasing normal stress, and joint size dependence is also modelled. Dilation curves are "delayed" by joint size (or block size) variation, not by normal stresses. Figure 4c illustrates conductivity–displacement relationships from "origins" A, B and C. The source of these initial conductivities is shown in Fig. 5. Physical apertures (E) developed at three different normal stress levels (10, 20 and 30 MPa) are converted to smaller conducting apertures (e) to derive the conductivity (k) given by:

$$k = \frac{e^2}{12} \tag{1}$$

The relationship between the theoretical conducting aperture (e) and the physical aperture (E) is derived from the test data presented in Fig. 6. The

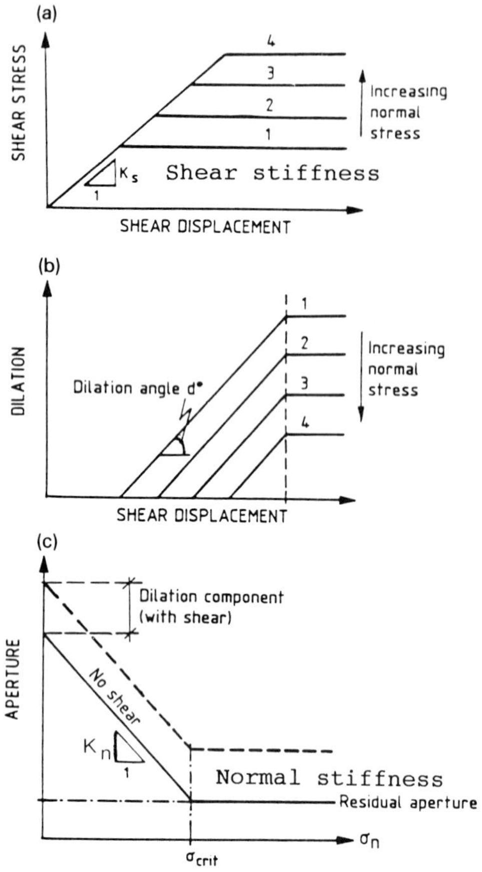

Fig. 3. The linear joint model assumed by Harper and Last (1989). (a) Shear stress vs. shear displacement. Shear stiffness K_s is assumed to be constant until peak. (b) Dilation vs. shear displacement. No dilation occurs until peak shear stress is reached. (c) Joint aperture vs. normal stress. Normal stress is assumed to be independent of normal stress until residual aperture is reached.

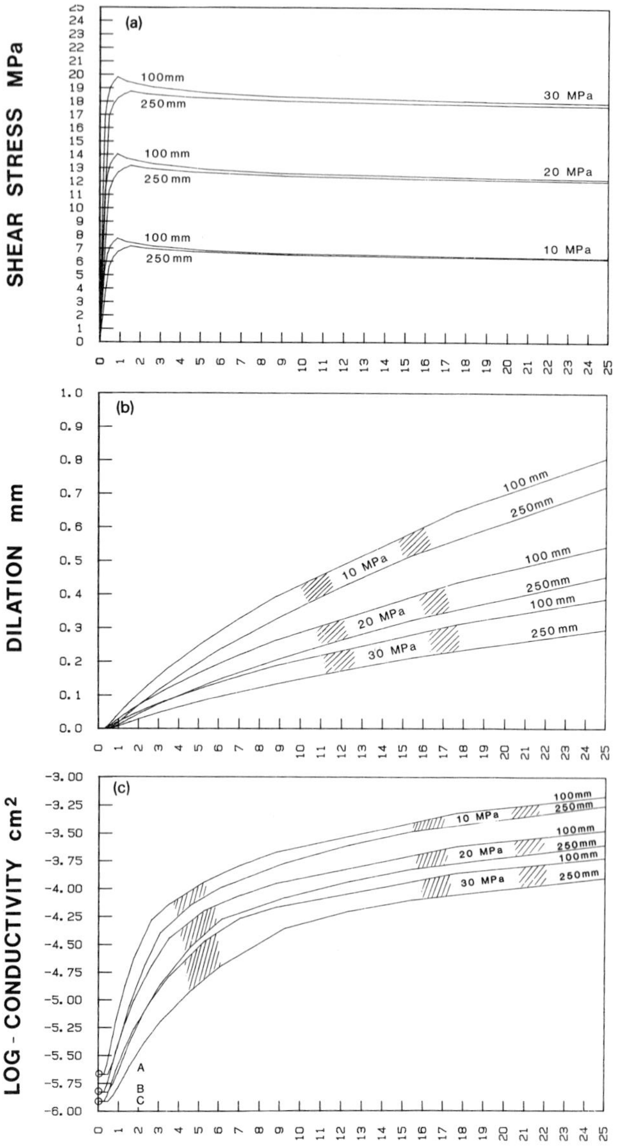

Fig. 4. Non-linear shear–delation–conductivity model, showing stress and joint size dependence (after Barton et al., 1985). The plots show curves for 100 and 250 mm long joint samples, sheared under 10, 20 and 30 MPa normal stress (JCS_0 = 82.5 MPa, JRC_0 = 9.0, ϕ_r = 29.5°). (a) Shear stress vs. shear displacement. (b) Dilation (increase in mechanical aperture) vs. shear displacement. (c) Conductivity vs. shear displacement. *A*, *B*, and *C* correspond to the initial conducting apertures prior to shear for 10, 20 and 30 MPa normal stress taken from Fig. 5.

ratio *E*/*e* increases with increasing normal stress, and also shows roughness dependence.

Input data for non-linear joint modelling

The input data needed to generate the behaviour curves shown in Figs. 4 and 5 are provided by four easily measurable parameters. The first three of these parameters describe the frictional strength of rock joints. The peak friction angle (ϕ) under an effective normal stress (σ_n) is given by the following relationship:

$$\phi = JRC \log \left(\frac{JCS}{\sigma_n} \right) + \phi_r \qquad (2)$$

Simple index tests are used to measure the parameters:

JRC = joint roughness coefficient; self-weight tilt tests on jointed core or on naturally jointed blocks of rock (Barton and Choubey, 1977).

JCS = joint compressive strength; Schmidt hammer tests (for weathered joints) or point load or unconfined compression tests for unweathered rock joints (where $JCS = \sigma_c$) (Barton and Choubey, 1977).

ϕ_r = residual friction angle; tilt tests on flat sawn surfaces of the rock give basic friction angles (ϕ_b) which are converted to ϕ_r via a simple expression involving the ratio *r*/*R*, where *r* = Schmidt hammer rebound on weathered joint, *R* = Schmidt hammer rebound on unweathered rock surface.

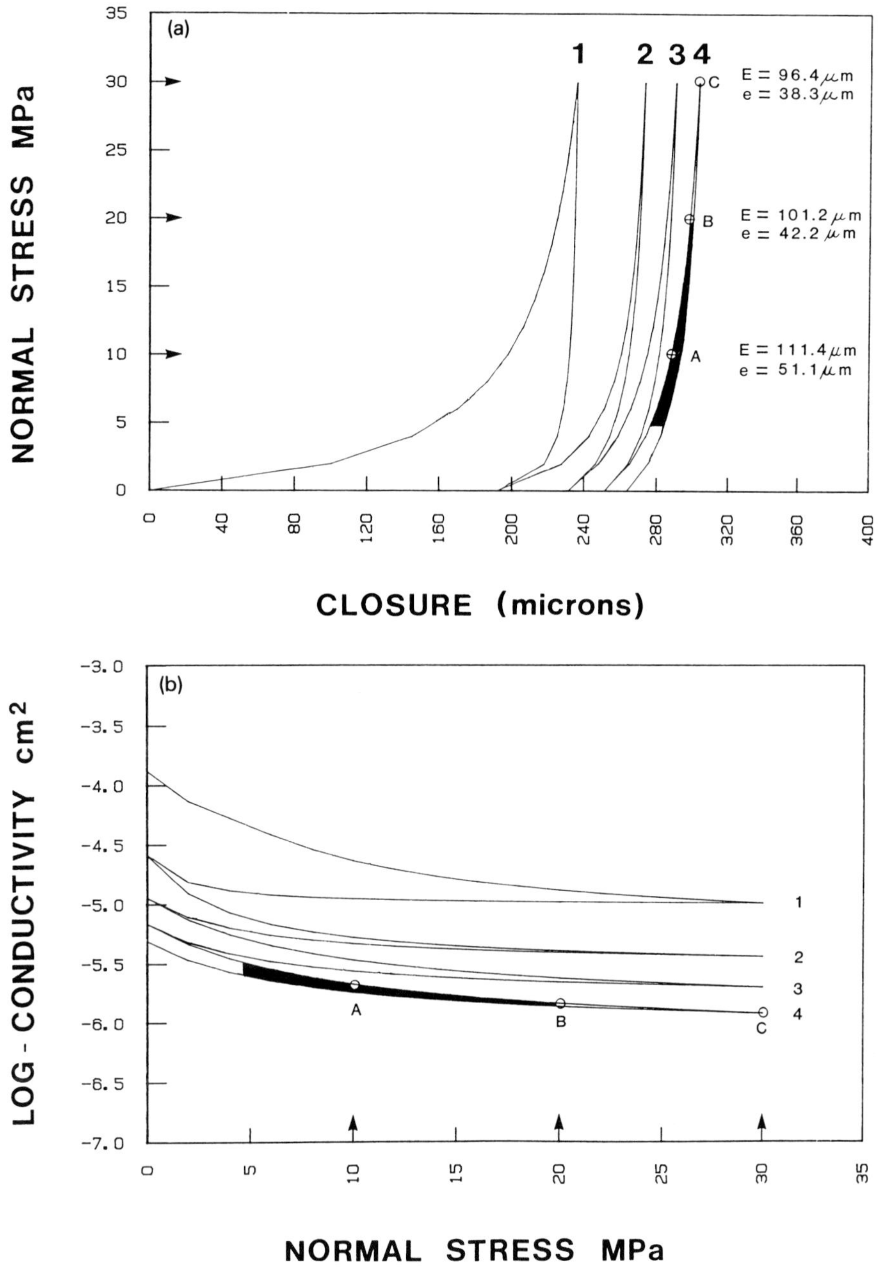

Fig. 5. Non-linear stress–closure–conductivity model (after Barton et al., 1985). The joint is loaded in four cycles to 30 MPa normal stress (JCS_o = 82.5 MPa, JRC_o = 9.0). The fourth cycle is considered to reflect in-situ conditions. (a) Normal stress vs. joint closure. The mechanical (E) and hydraulic (e) joint apertures are given for 10, 20 and 30 MPa normal stress. As input for the shear–dilation–conductivity modelling, e-values are used (see Fig. 4). (b) Joint conductivity vs. normal stress. A, B and C reflect joint conductivities of 10, 20 and 30 MPa normal stress.

L_n = in-situ joint length (m).

These simple tests, which can be performed cheaply on joint samples from drill cores, are described in detail by Barton and Choubey (1977).

Coupled stress flow behaviour

Joint apertures strongly depend on the effective normal stress acting across the joint plane. Due to the non-linear relation between joint conducting and joint mechanical apertures (see Fig. 6) variation in flow (due to stress variation) through rock joints will not be matched by the $e = E$ model of Witherspoon et al. (1980); a model also used by Pine and Cundall (1985) and Harper and Last (1989) in their fractured reservoir codes.

The simple index tests used to obtain JRC, JCS gives only a crude estimate of an initial unstressed aperture (E_o). More sophisticated small-scale tests, in which flow through a natural rough joint is measured under normal and shear stress, have shown that two orders of magnitude increase in conductivity

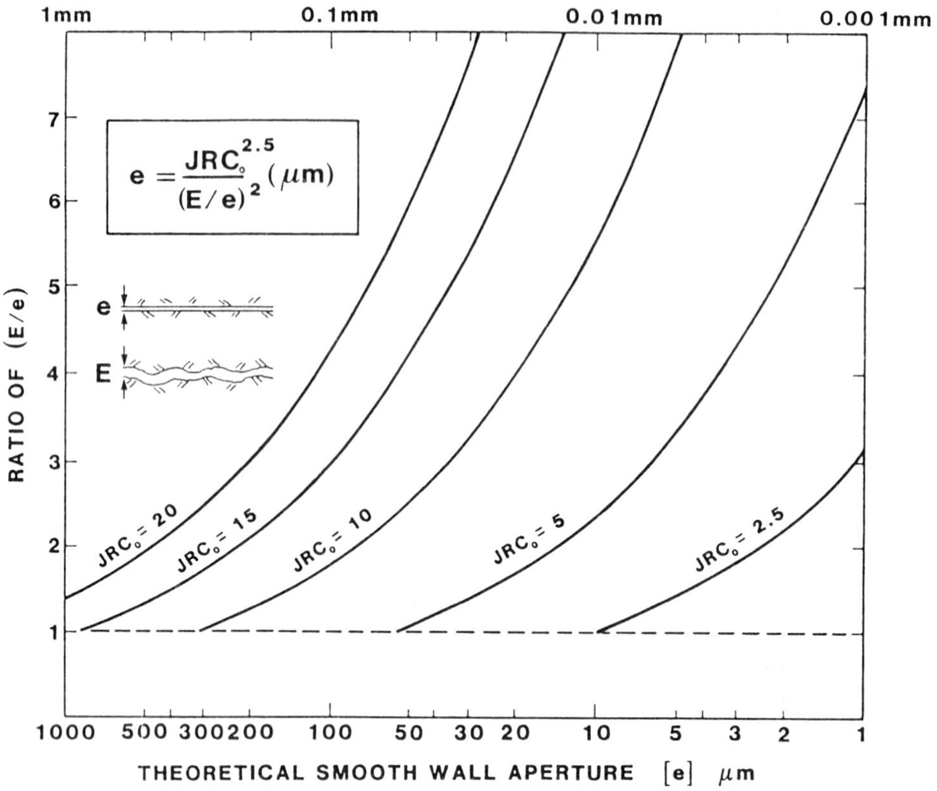

Fig. 6. An empirical relationship incorporating joint roughness (*JRC*) and aperture (E = mechanical aperture; e = hydraulic aperture) (after Barton, 1982).

(for a granitic joint) can result from a modest 1–2 mm of shear displacement (Makurat, 1985), broadly in agreement with our predictive model (Fig. 4c).

Extensive testing of fractures in weaker petroleum reservoir rocks such as sandstones, chalks and cap rock shales indicates that gouge production may partially or wholly block the established flow channels, as shear and asperity degradation occur (Makurat et al., 1990). The results of a coupled shear-flow test (CSFT) on a reservoir sandstone are shown in Fig. 7. Figure 7a shows that the joint dilates over the full range of 5 mm of shear. However, as illustrated in Fig. 7b, only the first 1 mm of shear lead to a major increase in joint conductivity. From 2 to 4 mm shear displacement no further increase of k was observed, and during the last 1 mm even a slight reduction of k was measured. In all CSFT-tests on softer rocks conducted so far, a certain amount of fines could be observed at the outlet of the test apparatus. It is assumed, however, that most of the fines produced due to joint surface degradation stay inside the joint and reduce the effective cross-sectional area available for fluid flow. The key to gouge production and flow channel blockage (with $e \ll E$) is the ratio JCS/σ_n appearing in eqn. (2). Low values imply greater gouge production, as observed in laboratory tests (Makurat et al., 1990). The conductivity curves for $\sigma_n = 30$ MPa in Fig. 4c will probably be lower in practice than those predicted. Some soft cap rock shales have shown reductions in conductivity with shear due to the low values of JCS/σ_n applied (for example $JCS = 10$ MPa, $\sigma_n = 10$ MPa).

Application of UDEC-BB to faulted block behaviour

Geometry and material properties

In a first step a faulted block configuration was chosen which is supposed to represent a border of a graben, where several fault blocks are located above a deep detachment zone (see Fig. 8c). This model could be a highly simplified representation of some areas along the eastern border of the Tampen Spur (see Fig. 8a).

The footwall of the 4 × 3 km model consists of sandstones with elastic properties similar to the Berea sandstone (see Table 1), which is an often used standard sandstone in rock mechanics.

The fault blocks, separated by normal faults with dip angles of 70°, consist of sandstone with identical material properties. They rest on a double layer of shale (for properties see Table 2), of which the lower one contains an additional low angle joint set.

The extra low angle joint set was thought to ease the onset of slip along the shale–sandstone interface,

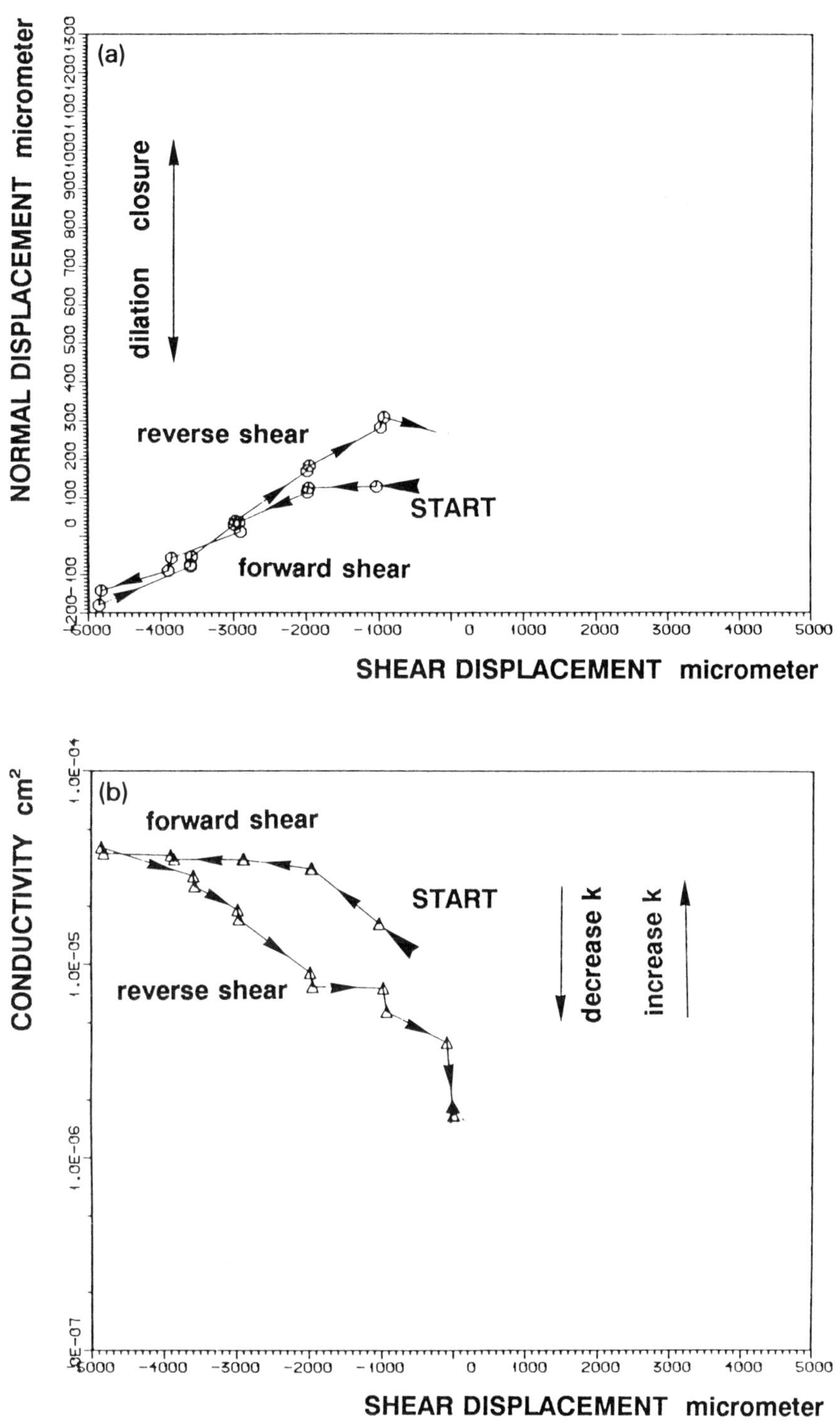

Fig. 7. Coupled shear-flow test on a reservoir sandstone. (a) Normal displacement vs. shear displacement. Forwards and reverse shear resulted in 150 μm reduction of fracture aperture. (b) Conductivity vs. shear displacement. Forward shear starts to increase k due to dilation, but keeps it almost constant for larger shear displacement due to starting gouge production, which is more increased during reverse shear resulting in reduced k when movement is back to initial position.

and to give additional information on the influence of varying dip angles on the direction of the axis of principal stresses.

The gently dipping discontinuities between the sandstone and the shale and all joints inside the shale are modelled as linear joints (joint set 1) with no cohesion and extremely low friction coefficient (see Table 3), in order to create low-friction slip planes.

All faults between the sandstone blocks are modelled as non-linear BB-joints with proper ties as described in Table 4, where a *JRC* value of 4

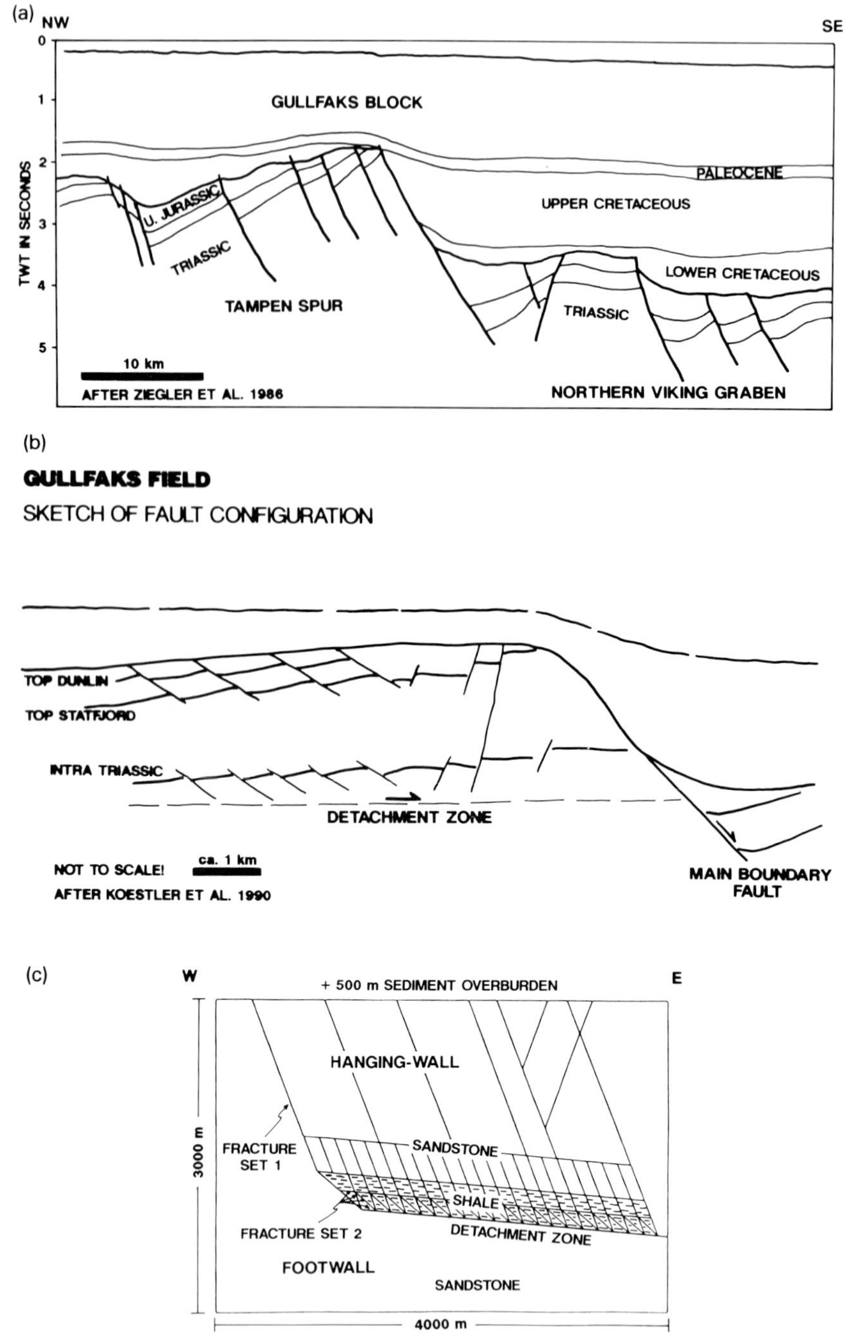

Fig. 8. The Gullfaks block as an analogue to the model set up. (a) Line drawing of a seismic section across the Gullfaks Block (after Ziegler et al., 1986). (b) Configuration sketch of the hanging-wall fault blocks which are separated from the footwall by detachment zone (after Koestler et al., this volume). (c) Model set up for the numerical simulation. The footwall block is surrounded by roller boundaries. Left, bottom and right boundaries are roller boundaries during consolidation. Additional load corresponding to 500 m sediment is applied to the top boundary. During the extension simulation, an outward directed velocity boundary is applied to the right-hand boundary of the hanging wall.

TABLE 1

Material properties of Berea sandstone

Youngs modulus	5.0×10^3 MPa
Poisson's ratio	0.38
Bulk modulus	6.9×10^3 MPa
Shear modulus	1.8×10^3 MPa
Density	2.1×10^3 kg/m^3

TABLE 2

Shale material properties

Young's modulus	5.6×10^4 MPa
Poisson's ratio	0.31
Bulk modulus	4.9×10^4 MPa
Shear modulus	2.1×10^4 MPa
Density	2.6×10^3 kg/m^3

TABLE 3

Linear joint material properties (joint set 1)

Normal stiffness	2.6×10^4 MPa/m
Shear stiffness	1.3×10^3 MPa/m
Cohesion	0.0 MPa/m
Tensile strength	0.0 MPa
Friction coefficient	0.01 tangent
Dilation coefficient	0.0 tangent

TABLE 4

Barton–Bandis joint material properties (joint set 2)

Joint roughness coefficient *JRC*	4.0
Joint compressive strength *JCS*	10.0 MPa
Intact rock uniaxial compressive strength	20.0 MPa
Initial joint aperture	0.35 mm
Residual friction angle	20°

represents a moderate rough joint. A *JRC* of 1 would represent a planar fracture, whereas a *JRC* of 20 would correspond e.g. to an extreme rough tensile fracture.

Boundary conditions and simulation sequence

Under consolidation roller boundaries (zero friction) are applied to the vertical and the lower horizontal boundaries. Gravity is acting on the model and an additional load of 10.5 MPa is applied along the top boundary, corresponding to 500 m overburden (see Fig. 8c).

After consolidation is finished, which means that all blocks are in a steady-state condition, extensional tectonics is simulated by applying a small outward-directed velocity to the upper 2 km of the right model boundary, thus limiting rollers to the basement sandstone (see Fig. 9). Table 5 illustrates the major steps in the simulation sequence.

TABLE 5

Major steps in the simulation sequence

Simulation steps	Figure/Table
Set up geometry	Fig. 8
Define material and joint properties define boundary conditions	Tables 1–4
Consolidation to in-situ stresses	Fig. 9A
Evoke extensional tectonics by applying outward directed velocity to right model boundary	Figs. 9B, 10A
Delete the right side sandstone blocks after they lost contact with the model	Fig. 10C

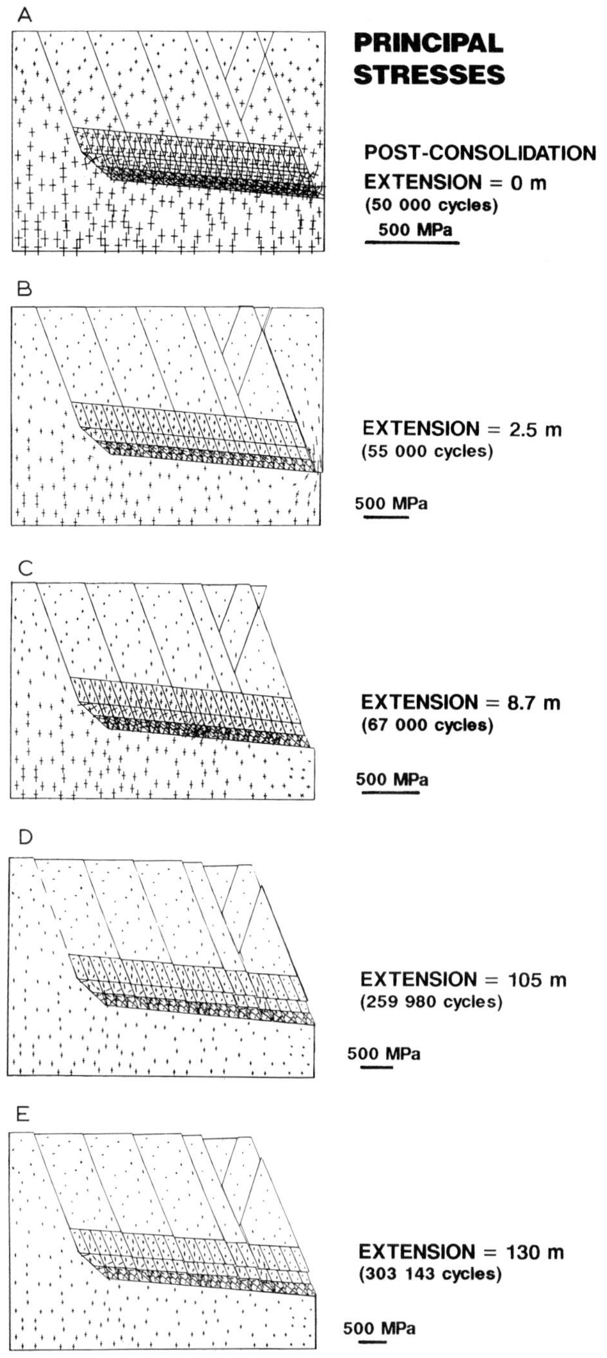

Fig. 9. Development of principal stresses during consolidation (A) and increasing extension (B to E).

Results and discussion

Three areas of interest have been identified in the numerical simulation of a large-scale fault block model:

(1) Stress variation with increasing extension;
block rotation;
distribution of fault slip and block deformation.

(2) Can the simulation be used for the prediction of reservoir structuration?

(3) Is the presented large-scale simulation geological reasonable?

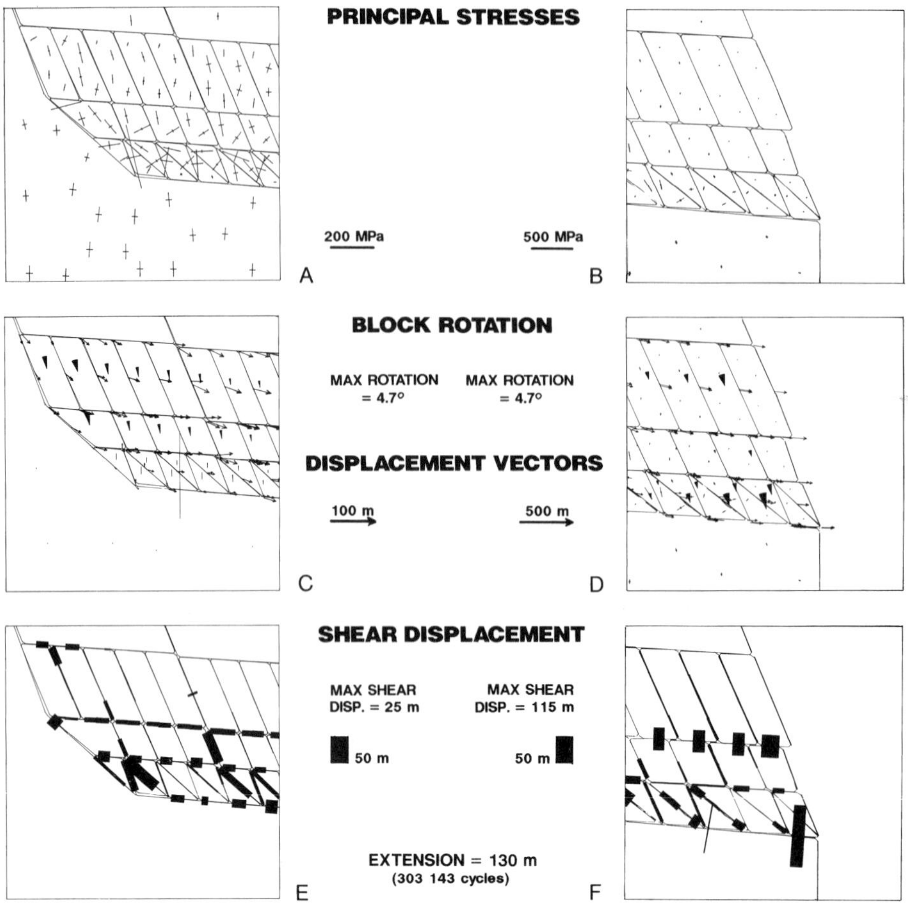

Fig. 10. Principal stresses, block rotation and displacement vectors, and shear displacement after maximum extension close to the detachment zone at the trailing edge (A, C, E) and at the landing edge (B, D, F).

The *principle stresses* after consolidation (before extension) display a sub-regular pattern, with the expected stress increase with depth and the main stress axis oriented vertically and horizontally (Fig. 9A). Minor reorientation due to boundary effects and the influence of shear fracture orientation can specifically be observed in the lower shale layer. Already minor extension (55,000 cycles, E = 2.5 m) produces a more heterogeneous stress distribution. Stress axes are rotated mainly in the right-hand part of the model, where extension results in unloading of the blocks next to the large block, which is pulled towards the right-hand border of the model. During further extension, stress axes are oriented parallel and perpendicular to the fault planes in the vicinity of the major faults (see Fig. 9E). Stress axis orientation in the detachment horizon is highly irregular. Within the shales e.g., stresses are oriented parallel to the chosen detachment orientation (see for detail Fig. 10A). Stress concentrations, with increased V/H ratios, can be observed at the end to the left-hand side of the detachment zone.

The *displacement vectors* indicate how contact points between the blocks move during extension (Fig. 11A–D). Minor extension (E = 33 m) only activates block displacement at the right-hand model side (e.g. 10–20 m for the diamond-shaped block). Increasing extension affects more and more the blocks towards the left-hand model side. The final stage of simulated extension (E = 130 m) displays a regular pattern of curved displacement lines indicated by the orientation of the displacement vectors. The amount of displacement is continuously decreasing from a maximum of about 130 m in the eastern part of the model to about 20 m in the westernmost block of the hanging wall. The lower part of the sandstone blocks and the small shale blocks show sub-horizontal displacements. Large displacements can also be observed in the area where fault blocks are bounded by synthetic and antithetic faults (uppermost part of the right-hand model side). The block rotation indicated by the block displacement vectors is sinistral for all blocks and decreases from east to west for the large sandstone blocks, but increases for the small sandstone blocks and the shale blocks close to the detachment (Fig. 12).

Shear displacements are mainly concentrated on the low-friction interfaces within the shale layers and

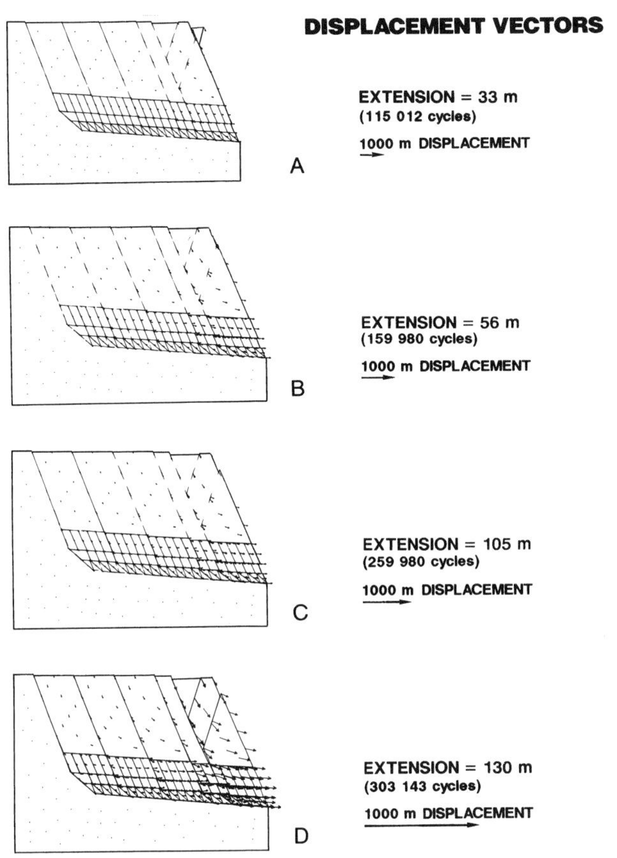

Fig. 11. Displacement vectors at different simulation stages.

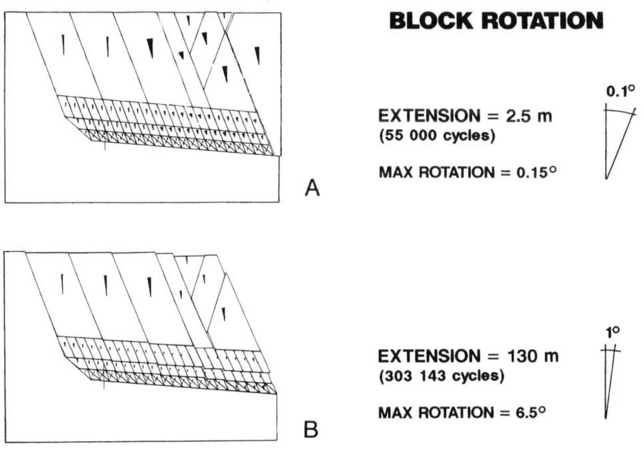

Fig. 12. Block rotation after very little extension (A), and after maximum extension (B).

partly at the bottom of the small sandstone blocks (Fig. 13). Little to no shear displacement happened between the small sandstone blocks, except in the elongation of the major faults. Large displacements are also seen on the major faults mainly in the area where syn- and antithetic faults interact and along the fault which bounds the three equal-sized blocks in the west against the more complex area in the east.

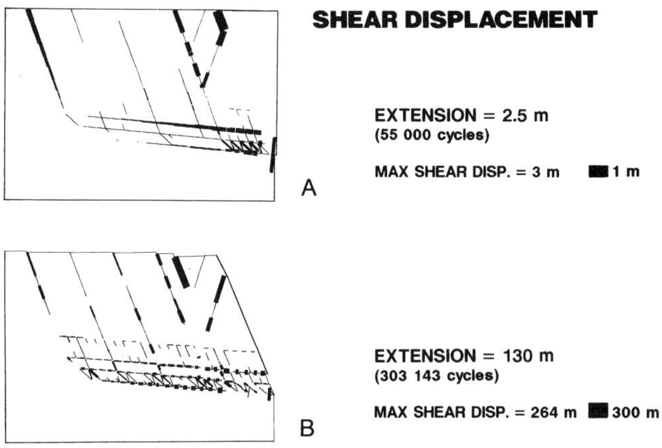

Fig. 13. Slip along shear fractures and faults after very little extension (A), and after maximum extension (B).

Conclusions

(1) Since UDEC-BB is a two-dimensional code, all faults and fractures are assumed to strike perpendicular to the modelled cross section. The Gullfaks Field is an example with a well-defined slip plane and little indications for mass transport normal to it (Koestler et al., this volume) and has therefore been chosen for this pilot study.

(2) The results are considered to be preliminary since different fault geometries have to be simulated and an extension to three-dimensional modelling could drastically improve the applicability of the presented method.

(3) The occurrence of big voids along faults with increasing extension is not expected to happen in nature, but could indicate dilated areas with improved fluid flow potential.

(4) The variation and reorientation of stresses with increasing extension can be investigated in great detail. Stress concentrations and stress ratios can be used to identify possible areas for fault nucleation.

(5) High stress concentrations at the western end of the detachment might favour the generation of new faults, which illustrates the potential for footwall collapse towards the hinterland. High stresses close to the detachment zone imply the chance for small-scale faulting.

(6) The deformation/displacement and rotational behaviour of the fault blocks was shown to be highly dependent on block size, shape and fault block configuration.

(7) Little displacement was observed on faults between blocks of equal size. Antithetic faults favour block rotation and block deformation/displacement.

(8) The applied non-linear joint behaviour model simulates the stress dependent shear displacement–dilation behaviour of faults and fractures in a more

realistic way than linear joint models. The results from laboratory coupled shear flow tests (CSFT) indicate a strong influence of the stress/strength ratio and shear displacement on the fracture surface properties. These results have to be incorporated into similar future studies in order to evaluate the sealing potential of faults as a function of stress and displacement.

(9) The presented method in combination with improved input data (e.g. fault surface properties) can be used as a tool evaluating critical processes for reservoir structuration. Further, the applied method could be used as a stress–displacement/deformation controlled back-stripping tool, which allows to identify critical fault configurations and their reaction on tectonic stresses at different scales.

References

Bandis, S., 1980. Experimental studies of scale effects on shear strength, and deformation of rock joints. Ph.D. Thesis, Dep. Earth Sci., University of Leeds.

Bandis, S., Makurat, A. and Vik, G., 1985. Predicted and measured hydraulic conductivity of rock joints. Int. Symp. Fundamentals of Rock Joints, Bjørkliden, pp. 269–280.

Barton, N., 1982. Modelling rock joint behaviour from in-situ block tests: Implications for nuclear waste repository design. Off. Nucl. Waste Isolation, Columbus, Ohio, ONWI-308, p. 96.

Barton, N. and Choubey, V., 1977. The shear strength of rock joints in theory and practice. Rock Mechanics, Springer, Vienna, No. 1/2, pp. 1–54.

Barton, N., Bandis, S. and Bakhtar, K., 1985. Strength, deformation and conductivity coupling of rock joints, Int. J. Rock Mech. Min. Sci. Geomech. Abstr. 22, No. 3, pp. 121–140.

Barton, N., Hårvik, L., Christianson, M., Bandis, S.C., Makurat, A., Chryssanthakis, P. and Vik, G., 1986. Rock mechanics modelling of the Ekofisk reservoir subsidence. 27th U.S. Symp. Rock Mech., Key to Energy Production, pp. 267–274.

Cundall, P.A., 1971. A computer model for simulating progressive large scale movements in blocky rock systems. Proc. Symp. Int. Soc. Rock Mech., Nancy, Vol. 1.

Cundall, P.A., 1980. A generalized distinct element program for modelling jointed rock. Eur. Res. Off., U.S. Army.

Fossen, H., 1989. Indication of transpressional tectonics in the Gullfaks oil-field, northern North Sea. Mar. Pet. Geol., 6: 22–30.

Gale, J.E., 1982. The effects of fracture type (induced versus natural) on the stress-fracture closure-fracture permeability relationship. Proc. 23rd U.S. Symp. Rock Mech., Univ. Calif., Berkeley, Calif., pp. 290–298.

Harper, T.R. and Last, N.C., 1989. Interpretation by numerical modelling of changes of fracture system hydraulic conductivity induced by fluid injection. Geotechnique, 39(1): 1–11.

Koestler, A.G., Milnes, A.G. and Storli, A., 1992. Complex hanging-wall deformation above an extensional detachment — example: Gullfaks Field, northern North Sea. In: R.M. Larsen, H. Brekke, B.T. Larsen and E. Talleraas (Editors), Structural and Tectonic Modelling and its Application to Petroleum Geology. Norwegian Petroleum Society (NPF), Special Publication 1. Elsevier, Amsterdam, pp. 243–251 (this volume).

Makurat, A., 1985. The effect of shear displacement on the permeability of natural rough joints. Hydrogeology of rocks of low permeability, Proc. 17th Int. Congr. Hydrogeol., Tucson, Ariz., pp. 99–106.

Makurat, A., Barton, N., Rad, N.S. and Bandis, S., 1990. Joint conductivity variation due to normal and shear deformation. In: N. Barton and O. Stephansson (Editors), Rock Joints. Proc. Int. Symp. Rock Joints, Loen. Balkema, Rotterdam, pp. 535–540.

Petterson, O., Storli, A., Ljosland, E. and Massie, I., 1990. The Gullfaks Field — Geology and reservoir development. In: Buller et al. (Editors), North Sea Oil and Gas Reservoirs — II. Proceedings. Graham and Trotman, London, pp. 67–90.

Pine, R.J. and Cundall, P.A., 1985. Applications of the fluid–rock interaction program (FRIP) to the modelling of hot dry rock geothermal energy systems. Int. Symp. Fundamentals Rock Joints, Bjørkliden, pp. 293–301.

Witherspoon, P.A., Wang, J.S.Y., Iwai, K. and Gale, J.E., 1980. Validity of cubic law for fluid flow in a deformable rock fracture. Water Resour. Res., 16: 1016–1024.

Ziegler, W.H., Doery, R. and Scott, J., 1986. Tectonic habitat of Norwegian oil and gas. In: A.M. Spencer et al. (Editors), Habitat of Hydrocarbons on the Norwegian Continental Shelf. Graham and Trotman, London, pp. 3–20.

Paper submitted March 1990

A.G. KOESTLER *Geo-Recon A/S, Bernhard Herres vei 3, 0376 Oslo 3, Norway*

A. MAKURAT *Norges Geoteknisk Institutt, P.O. Box 40 Tåsen, 0801 Oslo 1, Norway*

N. BARTON *Norges Geoteknisk Institutt, P.O. Box 40 Tåsen, 0801 Oslo 1, Norway*

K. MONSEN *Norges Geoteknisk Institutt, P.O. Box 40 Tåsen, 0801 Oslo 1, Norway*

Footwall uplift during normal faulting — implications for structural geometries in the North Sea

G. Yielding and A. Roberts

In recent years, studies of major normal faults in actively extending regions (Aegean, Basin and Range) have documented the vertical motions associated with normal faulting. In addition to the expected subsidence of the hanging wall, it has been found that uplift of the footwall occurs during earthquakes on normal faults. This uplift has been modelled as an elastic effect, treating the fault as a dislocation in an elastic medium. Over geological time scales, however, creep processes below the upper crust act to restore isostatic equilibrium, and so a purely elastic model is not appropriate. Instead, a flexural model is used, in which the upper crust is treated as an elastic layer, overlying a "fluid" substratum. A single fault cutting such a layer gives rise to buoyancy forces that produce reverse drag in both footwall and hanging wall. The wavelength of the reverse drag depends principally on the effective elastic thickness of the lithosphere. The relative amplitude of the footwall and hanging wall vertical movements is dominated by the nature of the fill in the half-graben: loading by sediment pushes the whole structure down, whereas loading by water (sediment-starved) allows more footwall uplift. When a number of faults occur adjacent to one another, their reverse-drag fields overlap to produce the familiar tilted-block profile of normal-faulted terrains. If the major faults are evenly spaced, dips within the fault blocks will be relatively uniform and can be approximated by a "domino" model. Such a fault model can be readily incorporated into a lithospheric stretching model (which causes subsidence of the whole basin), to predict vertical motions relative to a fixed datum such as sea level.

We apply these concepts to a variety of fault systems in the North Sea and the mid-Norway shelf. Normal faulting in the Late Jurassic created sediment-starved half-graben with significant bathymetric relief. The footwalls of major platform boundary faults are typically eroded, often showing post-rift sediments resting on basement (e.g., East Shetland Boundary Fault, Brae Fault). In these cases, no direct evidence remains of the amount of footwall rotation, so that the eroded crest might be assumed rigid and erroneously projected into the basin as a regional datum for "section balancing". However, some boundary faults preserve a condensed footwall sequence, or contain intra-basement reflections, which demonstrate that flexural uplift of the footwall accompanied normal faulting. The Frøya Fault (Haltenbanken) is an example of these. Erosion of marginal footwalls was an important source of clastic material, and is, therefore, a major factor in the distribution of Upper Jurassic reservoir formations (e.g., Frøya Formation, Brae Formation, Claymore Sst. Member, Fulmar Sand Formation). Within the graben, arrays of fault terraces can be described by the domino model, leading to the conclusion that the extent of erosion on intra-basinal footwalls is largely a function of fault spacing. Thus, in the Brent Province, the Snorre fault block suffered severe erosion during the Late Jurassic extension, whereas on the Brent fault-block erosion was less marked, approximate fault spacings being 30 km and 15 km, respectively. Erosion of intra-basinal footwalls has also been a significant source of Late Jurassic sediment for reservoirs, e.g., the Magnus member and Munin member.

Introduction

Deformation in the earth is complex. The rheology of rocks is dependent on many factors, of which temperature, pressure, lithology, strain rate and fluid content are probably the most important. It is, therefore, apparent that no single "model" can describe all aspects of earth deformation. If such a unified model were developed, it would be so complicated as to be of little use on a day-to-day basis in hydrocarbon exploration. The essential features of a useful model are:

(1) It should take into account the significant geological processes operating at the appropriate scale of study. (For example, when studying the gross basin architecture, processes of fault seal can be neglected, but isostasy cannot.)

(2) It should make quantitative predictions. The only reason for performing structural modelling in a commercial context is to increase the understanding of an area and thereby to reduce the risk associated with its exploration or development. This requires that quantitative predictions be made of parameters which cannot be observed directly.

(3) It should be simple to use, so that predictions can be made without spending months performing the calculations.

Clearly, when addressing a particular structural

Structural and Tectonic Modelling and its Application to Petroleum Geology edited by R.M. Larsen, H. Brekke, B.T. Larsen and E. Talleraas. NPF Special Publication 1, pp. 289–304. Elsevier, Amsterdam.

problem, the appropriate model should be chosen carefully, depending on the questions that need answering. A good analogy can be made between models and maps. Both are representations of physical reality, but contain a very limited amount of information about the real world. A road map at a scale of 1:500 000 may be completely adequate to drive from one end of a country to the other. However, it would be of no use in choosing the best climbing route up a particular mountain, for which a large-scale topographic map would be needed. Similarly, a structural model that gives good results for a whole basin may be useless for a single fault zone, and vice versa.

The purpose of this paper is to review briefly a number of selected models concerned with normal faulting and crustal extension. The emphasis is on those models that make predictions about footwall deformation by normal faulting. We specifically exclude listric-fault models that assume a rigid, undeformed footwall. Whilst such models may apply in areas of gravitational collapse of thick sediments (e.g., Gulf Coast, Niger delta), they are of little relevance to areas of crustal extension unless weak decoupling layers such as salt are present.

The models that we consider can be divided into four groups on the basis of scale:

(1) elastic dislocation models, appropriate for smaller faults and for single slip events on larger faults;

(2) flexural/isostatic models, appropriate for larger faults, i.e., those that create significant topography/bathymetry;

(3) domino models, where major faults are approximately evenly spaced;

(4) stretching models, which give the gross basin morphology and ignore the individual faults.

After discussing the elastic modelling in terms of presently active faults, we concentrate on flexural/isostatic and domino models and their application to normal faulting in the North Sea and mid-Norway shelf. We then discuss some aspects of the three-dimensional nature of fault geometry, and the influence this has on sediment transport routes within an extending basin.

Active normal faulting

Most, though not all, fault movement occurs during rapid slip events, i.e., earthquakes. Observations of earthquakes on active faults can, therefore, contribute a great deal to our understanding of how normal faults and their associated structures develop. The largest normal-faulting earthquakes result from a few metres of fault slip, on faults some tens of kilometres long (Jackson and White, 1989). This fault slip produces measurable deformation of the ground surface over an area extending many kilometres from the fault.

The best-documented example of the geodetic changes associated with major normal faulting is the 1983 Borah Peak earthquake (Stein and Barrientos, 1985). This magnitude 7.0 earthquake occurred on the Lost River Fault in the Basin and Range Province of the United States. Geodetic, geological and seismological data are summarised in Fig. 1, taken from Stein et al. (1988). The Lost River Fault is a major normal fault with over 4 km of displacement, separating the Lost River Range from the Thousand Springs Valley (Fig. 1b). The focal mechanism and hypocentre of the earthquake indicate that the fault dip is approximately 45° at a depth of approximately 15 km (Fig. 1c). Since the fault dip at the ground surface is also approximately 45°, the fault is approximately planar through the uppermost 15 km of the crust. This thickness corresponds to the brittle, seismogenic layer, as shown by the distribution of aftershocks (Fig. 1c).

Figure 1a presents geodetic data obtained by resurveying a levelling line that happened to cross the Lost River Fault. The co-seismic elevation changes (i.e., those that accompanied the earthquake) amount to almost 1.4 m of subsidence in the hanging wall adjacent to the fault and almost 0.3 m of uplift in the footwall. Both hanging wall subsidence and footwall uplift display a "reverse drag" profile, i.e., they exponentially decrease in magnitude with distance from the fault. The co-seismic deformation extends approximately 20 km into the hanging wall and approximately 15 km into the footwall. The ratio of subsidence to uplift is approximately 4:1. Also shown in Fig. 1a is a model fit to the geodetic data, derived by assuming the fault to be a dislocation in an elastic medium (Mansinha and Smylie, 1971). The model parameters were 2 m of slip over a planar rectangular fault (23 × 18 km) dipping at 47°. The fit between model and data is excellent.

A number of important conclusions can be drawn from the Borah Peak example:

(1) In normal faulting earthquakes, there is uplift of the footwall as well as subsidence of the hanging wall.

(2) Uplift and subsidence decrease exponentially from the fault.

(3) A hanging wall rollover does not imply a listric fault geometry. A planar dislocation embedded in an elastic medium can produce an excellent match to the data.

Similar observations, though less complete, have been made following other normal-fault earthquakes. Geodetic profiles for the 1928 Bulgaria earthquake

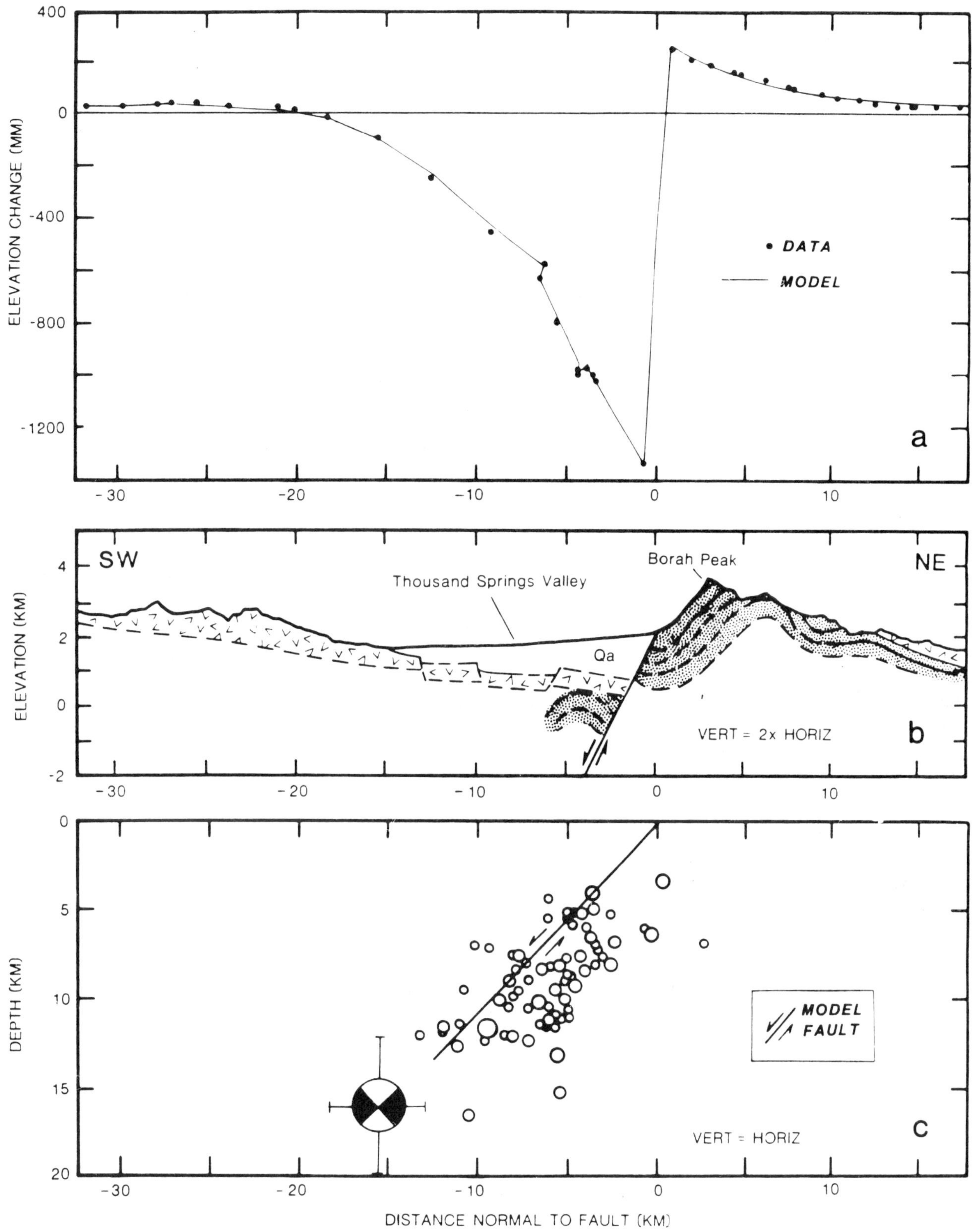

Fig. 1. Geodetic and geological data from the Lost River Fault in Idaho, on which occurred the 1983 Borah Peak earthquake (from Stein et al., 1988). (a) Coseismic deformation associated with the earthquake, measured by resurveying a levelling line across the fault. The model curve was calculated assuming a planar fault in an elastic medium. "Kinks" in the model profile result from changes in azimuth of the levelling route across the section (see Stein et al., 1988, for a detailed map). (b) Geological cross-section across the Lost River Fault. *Qa* = Quaternary alluvial fill; *v* = Tertiary volcanics; dotted area = Palaeozoic and PreCambrian. (c) Seismological data for the 1983 earthquake sequence. Error bars bracket the main-shock location, with the focal mechanism shown at the hypocentre. Small circles indicate aftershock locations. The "model fault" is the fault surface used to generate the geodetic curve in (a).

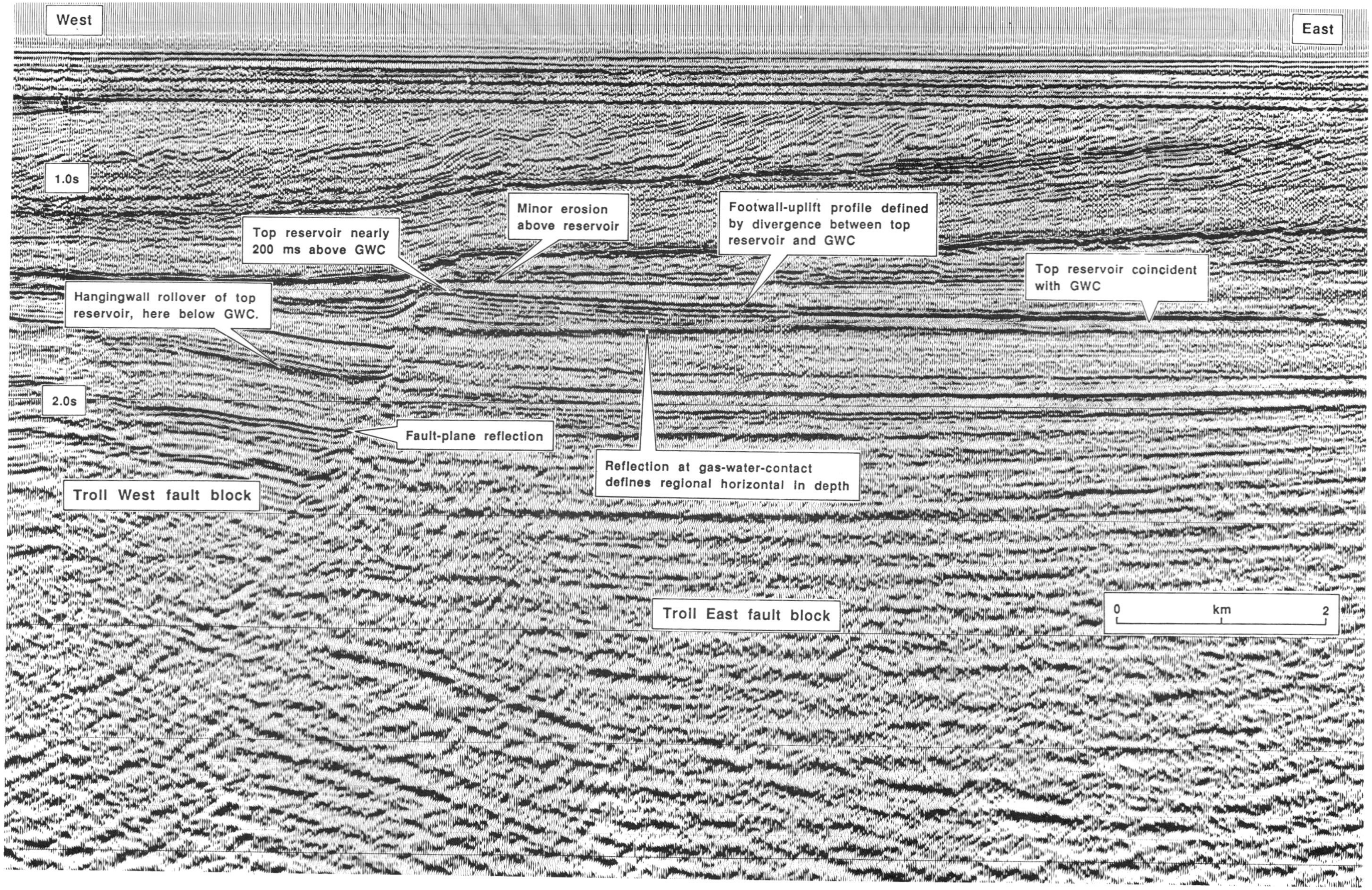

Fig. 2. Seismic line across the Troll East structure (from Roberts and Yielding, 1991). Both the hanging wall and footwall are dominated by reverse drag. Footwall uplift is highlighted by the divergence between the hydrocarbon-water flat-spot and the top reservoir reflection. (The apparent minor normal drag in the hanging wall adjacent to the fault is a post-faulting compaction effect.)

(Jankhof, 1945; Richter, 1958), the 1954 Fairview Peak earthquake (Whitten, 1957) and the 1959 Hebgen Lake earthquake (Myers and Hamilton, 1964) all show a pattern of uplift and subsidence similar to the Borah Peak earthquake. Vita-Finzi and King (1985) report a footwall uplift proportion of approximately 10% (0.2 m in a total throw of 1.7 m) for the 1981 Corinth earthquakes which occurred on near-planar faults dipping at approximately 42°. (No geodetic profile crosses this fault system; uplift and subsidence were measured locally relative to the sea surface.)

The geological structure around faults is built up through many seismic cycles. Each earthquake represents one small increment of slip on the fault surface. Thus, to some extent the geological structure can be modelled by summing together the effects of many successive earthquakes (e.g., King and Vita-Finzi, 1981; Jackson et al., 1982).

Although the co-seismic deformation is most readily measured around large faults, it is believed that a similar deformation field is produced by seismic slip on small faults. As a small fault nucleates and grows, the geological structure that forms around it will be the cumulative result of all the slip events during its growth history. Gibson et al. (1989) used an approximation to the elastic dislocation model to describe bed geometry adjacent to evolving planar normal faults. They used a realistic displacement distribution on the fault surface, and obtained a somewhat tighter "reverse drag profile" than that seen in the earthquake geodetic data.

Such reverse drag, on both sides of a normal fault, is very commonly observed in geological cross-sections. A number of examples are given by Barnett et al. (1987) and Gibson et al. (1989), from both mine plans and seismic reflection data. Another case where the footwall reverse drag is particularly clear is in the eastern part of the Troll Field on the Horda Platform of the northern North Sea (see Fig. 2). Here, a flat internal datum is present within the fault-block in the form of the seismic "flatspot" produced by the hydrocarbon/water contact (Birtles, 1986; Hellem et al., 1986). Reflections from the Viking Group reservoir show the steepest dip to be adjacent to the main fault, decreasing to zero at a distance of approximately 8 km. Footwall uplift above the flatspot datum is approximately 200 m at the fault.

Using an elastic dislocation model to describe the geological deformation around faults is only strictly appropriate if the co-seismic deformation is the dominant structural process. This is probably so for small faults. Here, we are using the distinction between "small" and "large" faults in the same sense as Jackson and White (1989), i.e., a "large" fault is one that cuts right through the seismogenic layer of the upper crust (approximately 0–15 km). A "small" fault is one that is confined within the seismogenic layer. On the time scale of a single earthquake (20 s or less), the whole of the crust and mantle is an elastic solid, which is why elastic models account so well for the co-seismic deformation (Fig. 1). On a longer time scale, however, the mantle and lower crust (beneath the seismogenic layer) deform by thermally activated creep processes (e.g., Brace and Kohlstedt, 1980; Kusznir and Park, 1987). Their response is more akin to a viscoelastic solid, i.e., they show an instantaneous elastic response and then exponential relaxation by viscous flow. Therefore, after an earthquake on a large normal fault, there is a slow rebound to a new equilibrium position. This post-seismic rebound takes the form of a broad up-doming centred on the fault (but with no further fault slip). Modelling of post-seismic movements (e.g., Rundle, 1982; Cohen, 1984) suggests that the rebound is complete within approximately 1000 years of an earthquake, with movement being fastest immediately after the seismic slip. This is less than or comparable to a typical earthquake recurrence time (e.g., Wallace, 1984), and so the rebound following one earthquake is probably completed before the next occurs. Reilinger (1986) documents approximately 30 cm of uplift in the 24 years following the 1959 Hebgen Lake earthquake (approximately 7 m co-seismic throw); post-seismic rebound here is presumably far from complete.

The importance of post-seismic movement is that, for large faults cutting the upper crust (seismogenic layer), the long-term (geological) structure is not simply the sum of all the earthquake slip events. More appropriate models for this scale of faulting are discussed in the next section. However, for small faults, (i.e., those which produce the internal deformation of blocks bounded by large faults), elastic dislocation models appear to work well. This is because small faults do not reach the zone where creep becomes significant. Beds cut by such faults show a local reverse drag profile consistent with the accumulated earthquake slip on the fault surfaces (Gibson et al., 1989).

Flexural/isostatic models

The term flexure is used to describe the bending of the uppermost layer of the Earth (the seismogenic layer, or "elastic lid") in response to loads. A rapidly imposed load will not be in isostatic equilibrium. On a geological time scale thermally activated flow at depth will act to restore isostatic equilibrium. Only the relatively cool seismogenic layer retains sig-

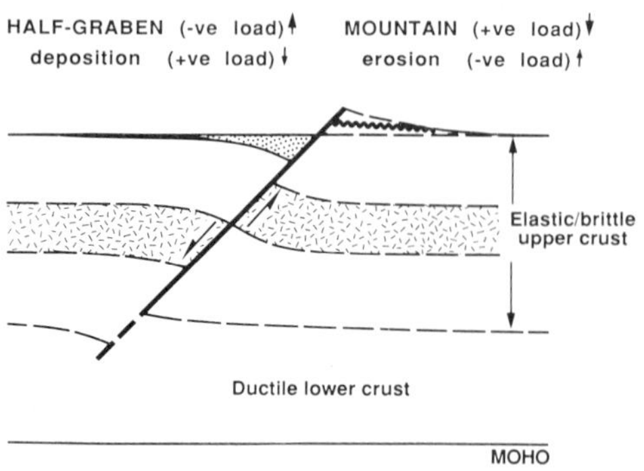

Fig. 3. Schematic diagram illustrating the isostatic loads generated by a large normal fault. Uplift of the footwall and subsidence of the hanging wall generate positive and negative loads, respectively. Erosion of the footwall, and sediment deposition in the hanging wall graben, act to reduce these initial loads. The isostatic loads are supported by the flexural rigidity of the elastic/brittle upper crust (seismogenic layer). Modified from King et al. (1988).

nificant strength. The mathematical model used to investigate flexure is that of an elastic layer (representing the upper crust) overlying a fluid substratum (representing the lower crust and mantle). (See Turcotte and Schubert 1982 for a general discussion of flexural models.)

Figure 3 shows how large normal faults generate loads that in turn cause flexure. Co-seismic uplift of the footwall creates a topographic load at the earth's surface, which in turn will tend to cause post-seismic subsidence. This load may be reduced by erosion of the footwall. Similarly, the half-graben is a "hole", or negative load, at the earth's surface, and will tend to cause post-seismic uplift. Deposition of sediment will tend to reduce this negative load. However, since the half-graben is much bigger than the uplifted footwall (cf. Fig. 1), post-seismic movement around normal faults is dominated by uplift.

The resistance of the elastic layer to bending is described by its flexural rigidity. The flexural rigidity, D, is given by:

$$D = \frac{ET_e^3}{12(1-\nu^2)}$$

where T_e is the thickness of the elastic layer, E is Young's modulus, and ν is Poisson's ratio. However, the earth does not have a well-defined elastic layer with a sharp base, overlying a fluid. The elastic thickness of the lithosphere is, therefore, usually referred to as an apparent or effective elastic thickness, corresponding to a perfect elastic layer with the same response as the earth. Although the seismogenic layer is typically about 10–15 km thick (e.g., Chen and Molnar, 1983), the effective elastic thickness of continental lithosphere is generally estimated to be 5 km or less (Barton and Wood, 1984; Watts, 1988; King et al., 1988; Fowler and McKenzie, 1989). It is not yet clear whether a thin, truly elastic, plate persists in the upper crust, or whether the apparent rigidity is diminished over a broad depth range (so that a thin layer is simulated). In either case, the effective elastic thickness is a parameter that is estimated by comparing observed structural wavelengths with those predicted by modelling. Thus, the wavelengths (approximately 15 km) of the reverse drag deformation around major normal faults are consistent with an effective elastic thickness of less than 5 km (e.g., King et al., 1988; Marsden et al., 1990).

The relative amplitude of the footwall and hanging wall vertical movements is dominated by the sediment loading in the half-graben. A sub-aerial fault system with relatively little sediment input will have approximately equal uplift and subsidence (King et al., 1988). This is the case for the Lost River Fault shown in Fig. 1 (Stein et al., 1988), though the co-seismic movements were unequal. If the footwall is subaerial and the hanging wall covered by sediment, the footwall uplift is only 5–25% of the total throw (cf. "circa 10%" quoted by Jackson and McKenzie, 1983). More footwall uplift occurs if the hanging wall is loaded mainly by water (i.e., a sediment-starved basin), since water is less dense than sediment. This has important implications for Late Jurassic faulting in the North Sea, since sediment input was relatively small and water depths in the half-graben were substantial (Bertram and Milton, 1989).

In the flexural model illustrated by Fig. 3, deformation is measured relative to the undeformed surface at a distance from the fault. In areas near sea level it is often useful to measure (and predict) the deformation relative to sea level, as a fixed datum. It is then necessary to include the effects of basin subsidence in addition to the more local movements adjacent to the faults. The basin subsidence is mainly determined by the distributed extension throughout the lithosphere column (cf. McKenzie, 1978). Kusznir et al. (1988) and Kusznir and Egan (1990) have developed a model of coupled simple

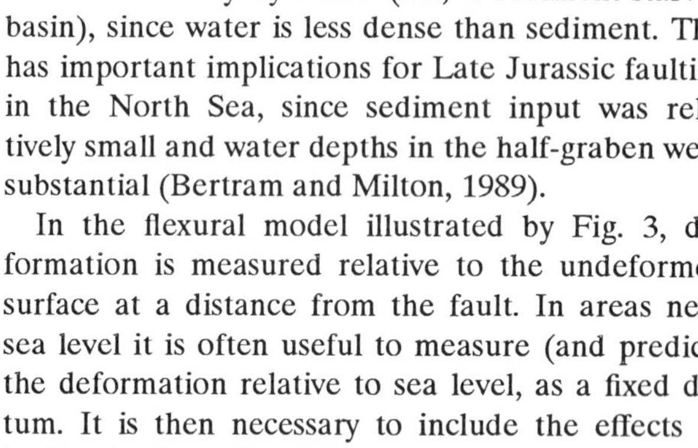

Fig. 4. The "flexural cantilever" model of Kusznir et al. (1988). Extension is accomplished by simple shear on a major fault(s) in the upper crust, and by pure shear spread over a width W in the lower crust and mantle. The isostatic loads generated by the extension are supported by the flexural rigidity of the upper crust.

shear and pure shear, in which lithospheric extension is achieved by major planar faults in the upper crust and by distributed pure shear in the lower crust and mantle. This "flexural cantilever" model is shown schematically in Fig. 4. Distributed extension in the lower crust and mantle is made equal to the fault-related extension in the upper crust. The flexural-isostatic response to upper crustal faulting, pure shear at depth, and sedimentary loading, is calculated in the wavenumber domain.

Application of flexural/isostatic models

We have used the flexural cantilever approach of Kusznir et al. (1988) to model both individual structures and complete basin cross-sections in the North Sea/mid-Norway area. Detailed discussion is given in Roberts and Yielding (1991) and Marsden et al. (1990), and only brief examples are given here.

Figure 5 is a depth section across the western boundary fault of the Trøndelag Platform (mid-Norway shelf), in the vicinity of the Njord and Draugen Fields (Ellenor and Mozetic, 1986). The Halten Terrace lies immediately to the west. Seismic markers in the platform are tied to wells in the Draugen Field and elsewhere. Within 10 km of the boundary fault erosion of the Triassic–mid-Jurassic sequence has occurred. The erosion surface (top of the platform) shows little relief, but the underlying sequence has been progressively uplifted as the boundary fault is approached. This uplift has been accomplished by a series of normal faults, antithetic to the main structure. Using the regional dip of the intra-Triassic salt marker in the "undeformed" eastern part of the section, a "smoothed" footwall uplift profile can be drawn through the small fault blocks. Maximum uplift at the fault is then estimated at approximately 1500 m. The age of the uplift is constrained as between mid-Jurassic and mid-Cretaceous. The main period of faulting was Late Jurassic, and thus we conclude that the footwall to the basin-margin fault zone became uplifted and emergent during extension.

Figure 6 shows two model cross-sections (obtained using the flexural cantilever model) for the fault zone shown in Fig. 5. Three kilometres of extension has been used, on a fault with an initial dip of 50° (the fault rotates to a shallow dip during extension). The basin fill has been given an average density of only 2.0 g cm^{-3}, since the Upper Jurassic sediments do not fill the half-graben and substantial water depth must have been present during rifting. An effective elastic thickness of 3 km was used, to match the observed radius of reverse drag on the depth profile. The upper section in Fig. 6 was modelled with no erosion, whereas the lower section has complete erosion down to sea level. The section with erosion matches the observed profile very well, with 1500 m of erosion adjacent to the fault.

An important aspect of the model profile is the return of the original surface below sea level east of the erosion zone. A potential depocentre is, therefore, predicted for some of the erosion products from the uplifted footwall. On the depth profile of Fig. 5, this depocentre corresponds to the Draugen Field, where Upper Jurassic sands (Frøya Formation) form the reservoir (Ellenor and Mozetic, 1986). These sands are probably the erosion products of the emergent high immediately to the west. Calibration of flexural models against known geology (such as Fig. 5) then enables us to use the models in a predictive way, where the syn-rift sands have yet to be confirmed by drilling.

The largest faults in the northern North Sea, such as the Brae Fault and the boundary fault to the East Shetland Platform, typically have severely eroded footwalls with post-rift sediments resting on basement (see fig. 3 of Roberts and Yielding, 1991). In these cases, no direct evidence remains of the amount of footwall erosion. However, flexural modelling of these fault systems predicts that several kilometres of section may have been eroded from the footwall. It would therefore be wrong to assume that the flat crest of such a footwall represents a stable, rigid high; this crest cannot be projected into the basin as an undeformed regional datum for geometric "section balancing".

Marsden et al. (1990) have applied the flexural cantilever model to a complete basin cross-section across the Viking Graben. A reasonably good fit to the observed structural geometry and sediment thicknesses can be achieved by using observed fault heaves and an effective elastic thickness of 3 km. Uplift and erosion of fault-block crests throughout the basin is predicted, in good agreement with detailed well and seismic data. Similar estimates of footwall uplift on intra-basin fault blocks can also be achieved using a simpler, geometric approximation, as described in the next section.

Domino models

The flexural cantilever model, discussed in the previous section, considers the flexural deformation of arrays of fault blocks within the framework of basin subsidence caused by lithospheric extension. However, even with low values for the effective elastic thickness, the wavelength of the flexural deformation is 15–20 km. Normal faults rarely occur in isolation, and so each fault is generally within the reverse drag radius of adjacent faults. The result of this superpo-

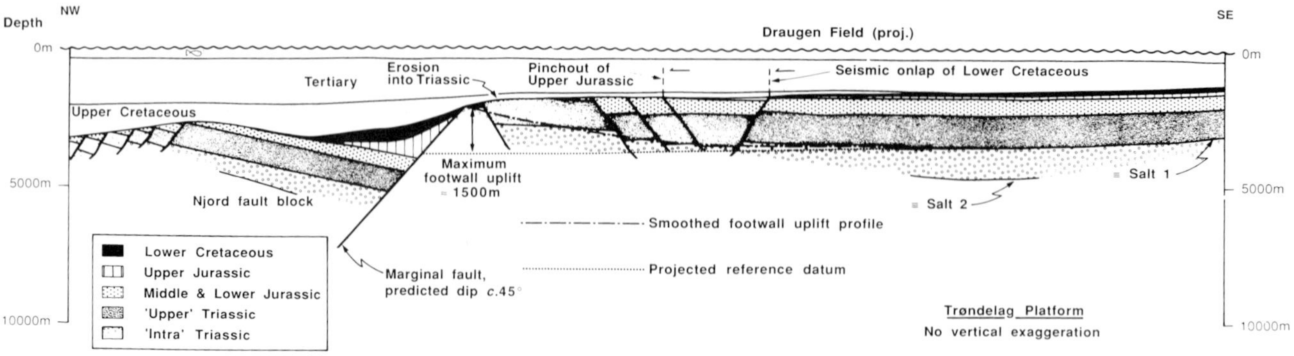

Fig. 5. Depth section across the Trøndelag Platform/Halten Terrace basin margin (mid-Norway Shelf), from Roberts and Yielding (1991). Note the 1500 m of erosion into the footwall of the main fault. The position of the Draugen Field is projected from approximately 10 km off the profile.

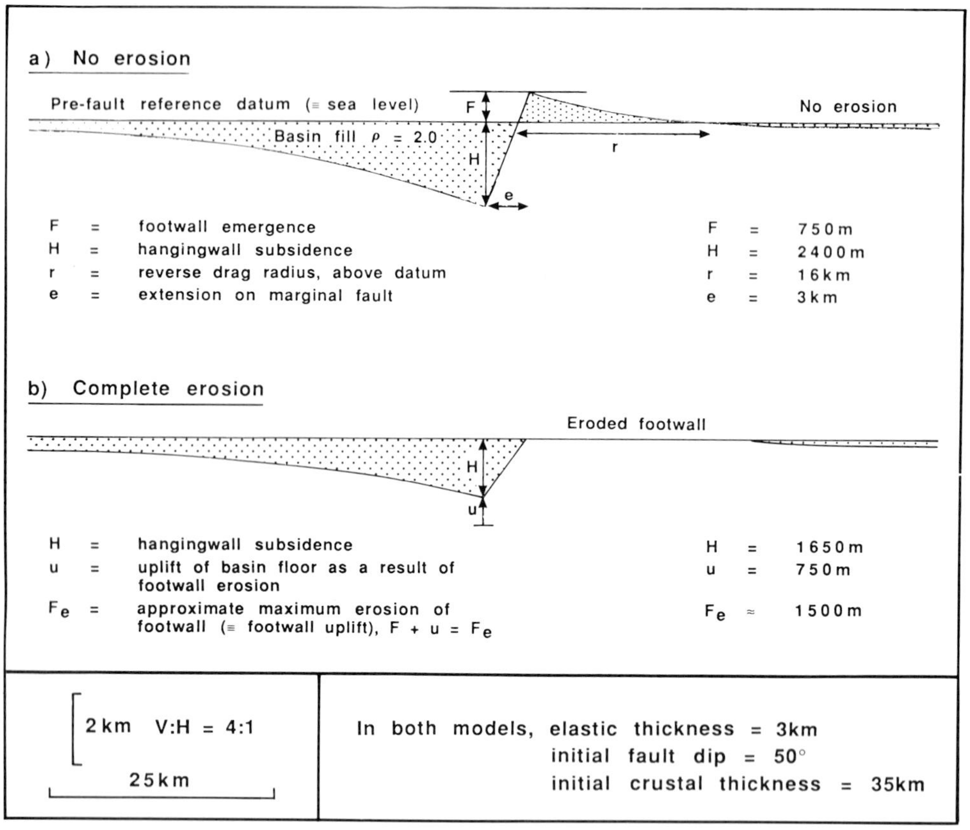

Fig. 6. Flexural-cantilever models of the basin-margin fault of Fig. 5. (a) is without erosion, (b) is with complete erosion to sea level. Note the 1500 m of footwall erosion predicted by model (b). From Roberts and Yielding (1991).

sition of reverse drag fields is that each block has a relatively uniform dip from one fault to the next (Fig. 7). This geometry has long been recognised from observations of normal-faulted terrains, and is generally described by the "rotating domino" model (e.g., Ransome et al., 1910, Wernicke and Burchfiel, 1982, Jackson and McKenzie, 1983). In its simplest form, all the fault blocks are the same size, all the faults have the same dip and all the blocks and faults rotate as the array is extended (see Fig. 8).

The appeal of the domino model is its extremely simple geometry. Calculations of extension are, therefore, very straightforward: the stretching factor is given by:

$$\beta = \frac{\sin\theta}{\sin(\theta - \Psi)}$$

where θ is the original fault dip and Ψ is block tilt (see Fig. 8). Moreover, the fault-block geometry is very similar to the much more complicated approach

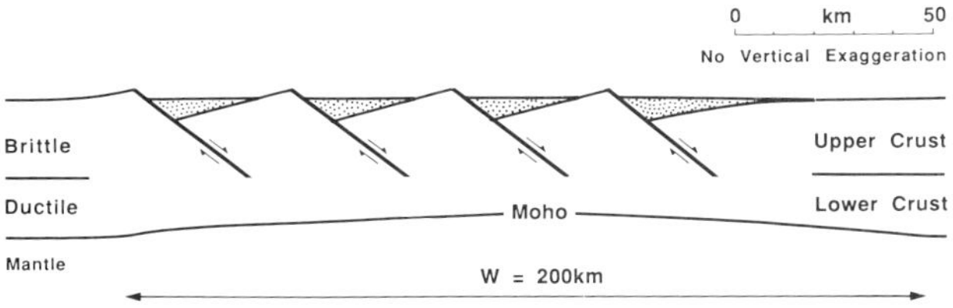

Fig. 7. Flexural-cantilever model of array of planar faults. Note that the fault blocks within the basin adopt the characteristic "domino" form typical of extended terrains. Each fault block is rotated by uplift at one bounding fault and subsidence at the other.

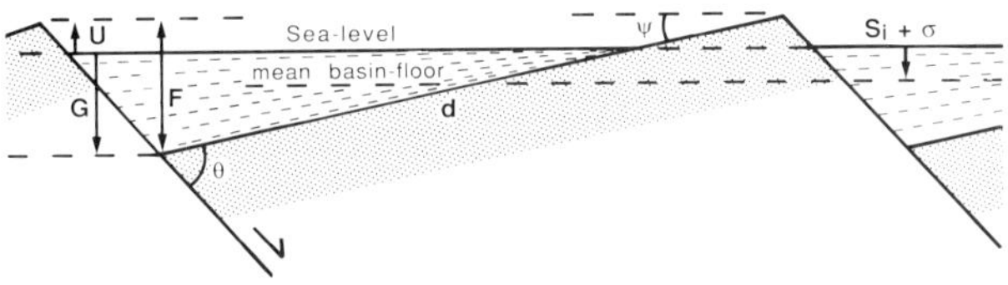

Fig. 8. The "rigid domino" model of continental extension. All fault blocks and faults rotate simultaneously as extension proceeds. The stretching factor $\beta = [\sin\theta]/[\sin(\theta - \Psi)]$, and the fault throw $F = d \sin\Psi$. The average level of the basin floor undergoes a tectonic subsidence S_i that is a function of β, and there is a further subsidence, σ, caused by sedimentary loading. Footwall uplift U is given by $U = F/2 - S_i - \sigma$. See Jackson et al. (1988) for further details.

of modelling the flexure associated with each fault. Provided that an array of major planar faults of roughly equal spacing exists, the domino model is a good first-order approximation to the fault-block topography. The fault blocks are assumed to be perfectly rigid, so that the upper surface of each block is straight and undeformed.

A common criticism of the domino model is that a space problem is created at the keels of the rotating blocks. However, it must be remembered that we are considering large normal faults that cut right through the brittle (seismogenic) upper crust. The transition from the brittle upper crust to the ductile lower crust is probably temperature controlled. If the keels of the blocks protrude into the underlying ductile material, conduction of heat will cause the temperature structure to re-equilibriate very quickly (Jackson et al., 1988). The "brittle-ductile transition" will, therefore, not be significantly perturbed by block rotations.

The domino model describes the geometry of the upper-crustal extension. Barr (1987a, b) was the first to incorporate this fault model into a framework of lithospheric extension. He used the simple McKenzie (1978) model of uniform stretching to define the overall tectonic subsidence of the basin floor. The domino fault blocks are then assumed to be the upper layer of this uniform stretching, i.e., they share the same β factor. Thus, although the upper crust is taken as completely rigid within a fault block, Airy (local) isostasy is assumed on a length scale greater than one fault block. The equations relating fault-block rotation to stretching, and subsidence to stretching, are readily combined, so that the movement of fault-block crests relative to sea level can be calculated. Jackson et al. (1988) derived a similar set of equations, but accounted for sediment loading in a more explicit (and less approximate) way.

Application of domino models

The northern North Sea is a good place to test the domino model because of the abundance of seismic data, and well data at footwall crests. Fault blocks in the East Shetland Basin have been modelled as rotating dominoes by Barr (1987a), Yielding (1990) and White (1990), though with somewhat differing interpretations of the overall amount of Late Jurassic extension. In the East Shetland Basin, it is established that the basin floor was approximately at sea level prior to the Late Jurassic stretching because the Middle Jurassic Brent Group contains coals, and blankets the basin (Brown, 1986, Bertram and Milton, 1989). Well data and depth profiles can be used to calculate the amount of section lost from footwall crests by erosion and degradation, as indicated in Fig. 9. This in turn can be used as an approximate estimate of the amount of footwall uplift above sea level, i.e., relative to the "pre-rift" position of the basin floor.

In cases where the crestal erosion surface is relatively flat (Fig. 9a), it is reasonable to assume that this represents erosion near sea level and, therefore, the eroded section equals the amount of uplift above sea level. The crest of the Snorre Field has this geometry, with Lower Cretaceous limestones covering eroded Triassic sandstones and siltstones (Hollander, 1987). In most cases, however, the crest is more complicated, as shown in Fig. 9b; the Brent Field is a published example with this type of profile (Bowen, 1975; Livera and Gdula, 1990). At Brent there is an east-dipping surface between the present structural crest and the east-dipping main fault. This surface is interpreted by Livera and Gdula (1990) as a surface of submarine degradation of the active fault scarp, with considerable slumping of footwall material along low-angle faults. The present structural crest is generally overlain by a thin cover of marine shales (Late Jurassic Kimmeridge Clay Formation). The structural crest was, therefore, below sea level at the end of fault-block rotation (end Jurassic). A maximum estimate of the amount of footwall uplift is given by the elevation difference between the present crest and the projected footwall cutoff at Top Middle Jurassic level (Fig. 9b). Note that the depth of

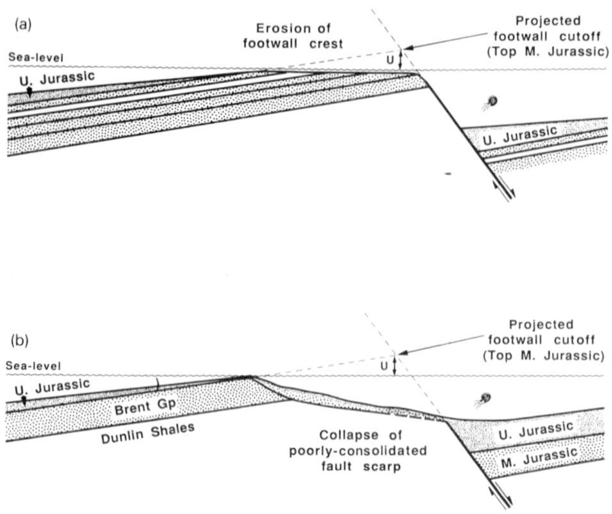

Fig. 9. Schematic erosion profiles across North Sea fault blocks, showing how footwall uplift, U, relative to sea level can be calculated. (The depositional surface of the Middle Jurassic Brent Group was originally very near to sea level.) In (a), the footwall crest represents an erosion surface. In (b), significant degradation of the submarine fault scarp has occurred, increasing the "depth of erosion" into the footwall.

erosion immediately adjacent to the fault will overestimate the amount of uplift relative to sea level, since this degradation surface was always below sea level.

The difference in profile between the Snorre (cf. Fig. 9a) and Brent (Fig. 9b) Fields may be related to the rock types at the block crest. At the Brent Field, submarine degradation and slumping of relatively unconsolidated Middle and Lower Jurassic sands and clays might well be expected at the fault scarp. At Snorre, however, there was much more footwall uplift, so that erosion had cut right down into the Triassic section. It is likely that the Triassic sandstones and siltstones were already more compacted and cemented than the younger Jurassic sediments. Therefore, the main fault scarp would have been more resistant to slumping, and erosion at the wave base was the dominant process for removing footwall material.

Figure 10 compiles footwall erosion estimates from many fields in the northern North Sea, taken mainly from published profiles. Estimates were made as outlined above, so that we are attempting to measure the amount of footwall uplift relative to sea level. These estimates of uplift are plotted against the present size of the fault block (i.e., distance between major faults, measured across strike). There is a clear correlation between amount of footwall erosion and fault-block size. Small blocks such as Thistle and Murchison show negligible crestal erosion, whereas large blocks such as Gullfaks and Snorre are severely eroded (approximately 1 km of section lost).

This relationship corresponds closely to that expected from the domino model. A given extension can be accomplished by a few widely spaced large

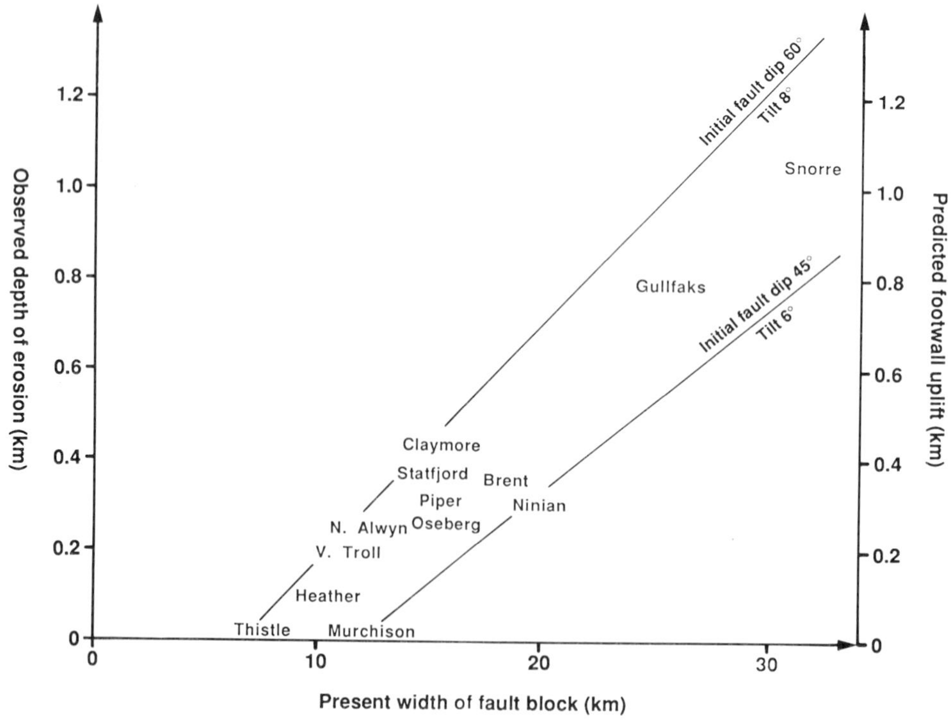

Fig. 10. Graph showing the "observed depth of erosion" at thirteen footwall fields in the North Sea, measured as shown by U in Fig. 9. Based on depth profiles given by Spencer and Larsen (1990), Maher and Harker (1987) and Yielding (1990). The sloping lines show predicted footwall uplift from the domino model, for two different sets of input parameters. Note that all the observed erosion profiles can be accounted for by a small range of variation in initial fault dip and fault-block tilt.

faults, or by a greater number of smaller faults (less throw). The larger fault-blocks are, therefore, expected to be bounded by the larger faults and, therefore, to have the greatest amount of footwall uplift and erosion. The sloping lines on Fig. 10 show two domino-model predictions of the relationship between block size and footwall uplift, given slightly different fault dips and block tilts. The upper line is for an initial fault dip of 60° and a block tilt of 8°, the lower line is for an initial fault dip of 45° and a block tilt of 6° (both correspond to a β value of about 1.1). These values are reasonable bounds on the observed structure for most of the fields shown on Fig. 10. Footwall uplift was calculated using the expressions of Jackson et al. (1988), using a basin-fill density of 1.65 g cm^{-3}. As argued by Yielding (1990), this low value is reasonable, since a significant part of the basin-fill was water (cf. Bertram and Milton, 1989). The two calculated lines bound the spread of observed amounts of erosion. Thus, the domino model appears to work well for predicting the amount of footwall uplift in North Sea fields.

The advantage of the domino model is its simplicity, whilst still making useful quantitative predictions. It is not applicable to isolated normal faults, or basin-margin faults. In its simple form, it specifically excludes any description of the internal deformation of fault blocks, which may locally be significant. Modifications of the domino model have been published which incorporate block deformation by inclined simple shear (Jackson and White, 1989; Faure and Chermette, 1989). However, a more realistic approach will probably require quantitative assessment of the population of smaller-scale faults that must inevitably accompany the large block-bounding faults (Childs et al., 1990). Such models are referred to as "soft-domino" models (e.g., Walsh and Watterson, 1991). Assuming that the dominoes are rigid will always underestimate the amount of extension.

Notwithstanding these limitations, the rigid-domino model can be useful in undrilled areas to assess the prospectivity of intra-basinal fault blocks. Fault-block spacing is generally readily measured on regional seismic profiles, and we can then use the domino model to make predictions about the amount of uplift or erosion on blocks of particular sizes. For example, blocks above a particular threshold size may have no chance of preserving a particular pre-rift reservoir within structural closure at their crests. In more equatorial climates, reef development near footwall crests may be enhanced for particular uplift histories, e.g., those which remain emergent for the longest period (see Fig. 4 of Jackson et al., 1988).

Three-dimensional aspects of uplift and subsidence

In the previous sections uplift and subsidence have been discussed in terms of sections perpendicular to the faulting, i.e., a two-dimensional approach. In reality, fault displacements vary along strike and, therefore, the associated uplift and subsidence will also vary. The topography/bathymetry of the basin floor will be strongly controlled by the interaction of the three-dimensional deformation fields around the faults.

Real faults have a finite length, and often end laterally at a point of zero displacement (e.g., Barnett et al., 1987). The maximum displacement is at or near the centre of the fault, and will be associated with the maximum amounts of footwall uplift and hanging wall subsidence (Jackson and McKenzie, 1983; Gibson et al., 1989). Figure 11 shows schematically the influence of a single large normal fault on a single horizon. The decrease in footwall uplift from the fault centre to the lateral tips gives rise to the characteristic shape of the structural closure that is observed on many footwall fields (see compilation by Spencer and Larsen, 1990).

If the horizon shown in Fig. 11 is considered as the basin floor, the tilt on that horizon will control the local sediment transport direction. For large fault throws, material may be eroded from the area of maximum footwall uplift. Transport of this material may be predominantly down the dip-slope of the footwall, rather than directly over the fault-scarp (Leeder and Gawthorpe, 1987). Such transport directions are shown schematically on Fig. 11. Development of new, syn-rift, reservoir sediments requires the presence of suitable material within the uplifted footwall (e.g., new sands are unlikely to be generated if the footwall rocks are predominantly shaly). The Upper Jurassic sands of the Draugen Field (Fig. 5) were probably derived from the adjacent footwall in this way. Roberts and Jackson (1991) document

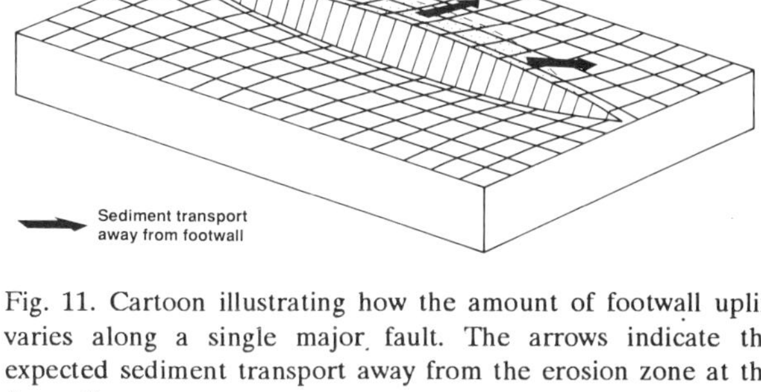

Fig. 11. Cartoon illustrating how the amount of footwall uplift varies along a single major fault. The arrows indicate the expected sediment transport away from the erosion zone at the footwall crest.

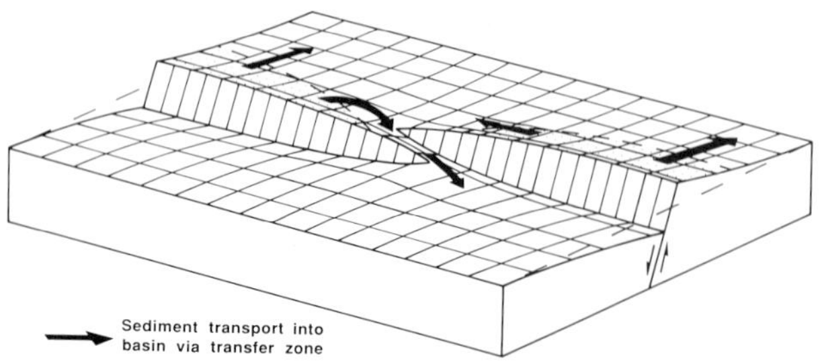

Fig. 12. Cartoon illustrating an en échelon zone between two major normal faults. The "relay ramp" or transfer zone between the two faults necessarily has a lower footwall elevation than the central parts of each fault. Therefore, it acts as a sediment transport route from the footwalls to the hanging wall basin. Syn-rift sands will be preferentially deposited near these transfer zones.

similar controls on sediment transport in the present fluvial systems of central Greece.

However, there are cases in which the majority of the eroded footwall material is transported directly on to the hanging wall rather than down the dip-slope of the footwall as shown in Fig. 11. Where the footwall rocks are only poorly consolidated, failure of the fault scarp may be the main control on the removal of material from the footwall crest (see Fig. 9b). This is the case on the Brent Field (Livera and Gdula, 1990) and many of the other major fields in the East Shetland Basin. In these examples, most of the reworked debris from the footwall is found on the hanging wall side of the main fault.

A common feature of large normal faults is that they are rarely continuous along strike for more than about 20 km (Jackson and White, 1989). This is most readily documented for onshore fault systems for example in the Basin and Range Province (e.g., Schwarz and Coppersmith, 1984; Bruhn et al., 1987). Individual fault segments are often related in an en échelon fashion, with transfer zones or "relay ramps" (Larsen, 1988) between them. Physical linkage of the two faults by a single transfer fault is not particularly common, though is known in some cases. The importance of the en échelon geometry is that it allows sediment input into the hanging wall basin (see Fig. 12). As the footwall uplift on each fault segment decreases towards its tip in the relay zone, this step in the fault pattern must inevitably be one of the lowest points along the footwall ridge. This structural saddle, therefore, attracts the hinterland drainage or sediment transport developed on the dip-slope of the footwalls, and channels it into the hanging wall basin. Steps and discontinuities in fault patterns have long been recognised as sediment entry points (e.g., Leeder and Gawthorpe, 1987), and Roberts and Jackson (1991) demonstrate this in detail in central Greece.

The usefulness of the ideas discussed in this paper can be seen from Fig. 13, which summarises the distribution of known Upper Jurassic ("syn-rift") sands in the northern North Sea. (This map will undoubtedly be proved incomplete by further drilling or release of presently confidential wells.) All of the known sand development is controlled to a greater or lesser degree by the Late Jurassic normal faulting. The Viking Group sands on the Norwegian Horda Platform were probably sourced by regional uplift of Norway, but their route into the basin appears to have been structurally controlled (Badley et al., 1988). In the north part of the East Shetland Basin, both the Munin sand member and the Magnus Sandstone Member have been derived by erosion of uplifted intra-basinal footwalls. The Munin sand member (Gradijan and Wiik, 1987) seen in the Statfjord Nord Field is probably derived at least in part from the eroded Snorre block immediately to the east. The Magnus Sandstone Member was derived from the eroded footwall(s) to the northwest of the Magnus Field (Young, this volume). Minor developments of syn-rift sands occur adjacent to the main boundary fault to the East Shetland Platform (e.g., Wheatley et al., 1987) and are concentrated at discontinuities (relay zones?) in that fault system.

The most prolific depocentres for Upper Jurassic sands have been those adjacent to the Fladen Ground Spur (Fig. 13). The South Viking Graben is essentially a single half-graben with a very large bounding fault at its western side, with at least 7 km of displacement (Harris and Fowler, 1987, fig. 4). The Fladen Ground Spur is the footwall to this major fault. The well-known Brae Formation constitutes a complex set of submarine fans in the hanging wall of the main fault (e.g., Turner et al., 1987, Harris and Fowler, 1987). Jackson and White (1989) have suggested that this fan development in the hanging wall may have been controlled by relay zones, i.e., where major fault segments die out or change strike. Flexural/isostatic modelling of such major faults indicates that several kilometres of erosion may have occurred in the immediate footwall (see earlier section of this paper). The Brae fans are unlikely to account for all of this material, and we would suggest that a large proportion of the sediment was shed southwestwards from the Fladen Ground Spur, into the Witch Ground Graben (see Fig. 13). The deltaic Piper Formation blankets much of the Witch Ground Graben, and like the Brae Formation its development commenced in the Late Oxfordian. Stratigraphic cross-sections through the Piper Formation (Boote and Gustav, 1987, figs. 7–9) indicate sediment transport from the northeast, i.e., from the Fladen Ground Spur. Later sands, e.g., the Volgian

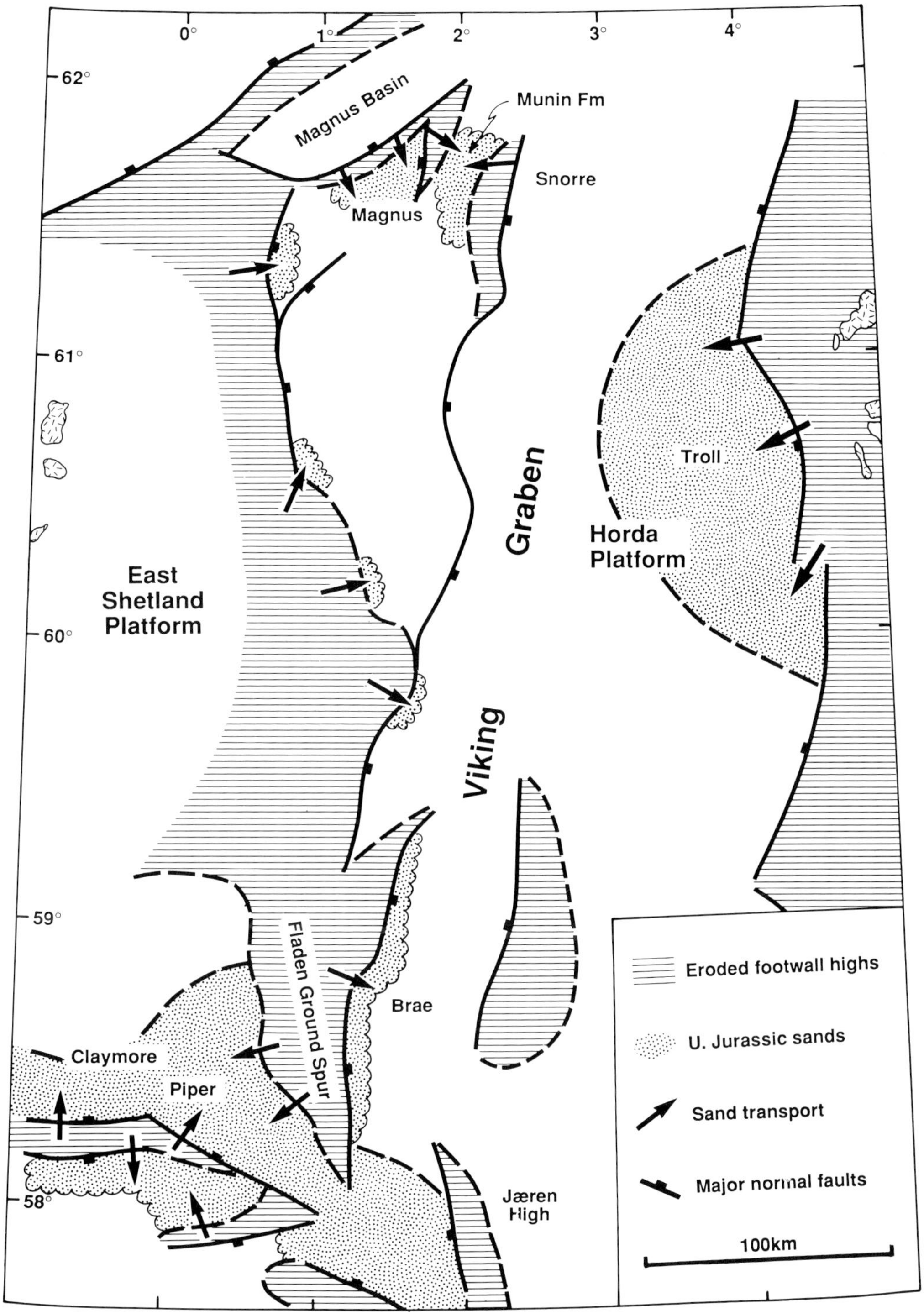

Fig. 13. Map illustrating the main developments of "syn-rift" (Upper Jurassic) sands in the northern North Sea, based mainly on Spencer and Larsen (1990). Note the importance of footwall erosion in providing sediment sources, and the control exerted by fault geometry on the sediment route to the basins.

turbidite systems, show derivation from the southwest (Claymore), as well as the northeast (Galley), as footwall uplift on the Halibut Horst developed.

Further south in the Central Graben, the main "syn-rift" reservoir is the Fulmar Formation. Like the Upper Jurassic sands of the northern North Sea, the depositional pattern of the Fulmar Formation is closely controlled by the major fault systems (see Roberts et al., 1990, for a more detailed discussion).

Conclusions

(1) Uplift of the footwall is the rule during normal faulting associated with crustal extension.

(2) Modelling of faults as dislocations in an elastic medium is appropriate for "small" normal faults (the "production scale"), and for co-seismic (earthquake) slip on "large" normal faults. For geological modelling of the "large" basin-forming faults (the "exploration scale"), flexural/isostatic models are more appropriate. The domino model can be used where large faults are evenly spaced.

(3) Footwall ("pre-rift") reservoirs generally show a reverse drag profile, often modified by erosion or degradation at the crest.

(4) Eroded footwalls, both at basin margins and within the basin, are an important sand source for "syn-rift" reservoirs.

(5) Sand transport routes during rifting are strongly controlled by the basin morphology, which in turn is controlled by footwall uplift and hanging wall subsidence.

Acknowledgements

We are particularly grateful to Terkel Olsen and Charles Jourdan for their interest in the applicability of these ideas. We also thank Dave Barr, James Jackson, Geof King, Nick Kusznir, John Walsh, Juan Watterson and Nicky White for many stimulating discussions over recent years.

References

Badley, M.E., Price, J.D., Rambech Dahl, C. and Agdestein, T., 1988. The structural evolution of the northern Viking Graben and its bearing upon extensional modes of basin formation. J. Geol. Soc. London, 145: 455–72.

Barnett, J.A.M., Mortimer, J., Rippon, J.H., Walsh, J.J. and Watterson, J., 1987. Displacement geometry in the volume containing a single normal fault. Am. Assoc. Pet. Geol., Bull., 71: 925–37.

Barr, D., 1987a. Lithospheric stretching, detached normal faulting and footwall uplift. In: M.P. Coward, J.F. Dewey and P.L. Hancock (Editors), Continental Extensional Tectonics. Geol. Soc. London, Spec. Publ., 28: 75–94.

Barr, D., 1987b. Structural/stratigraphic models for extensional basins of half-graben type. J. Struct. Geol., 9: 491–500.

Barton, P. and Wood, R., 1984. Tectonic evolution of the North Sea basin: crustal stretching and subsidence. Geophys. J. R. Astron. Soc., 79: 987–1022.

Bertram, G.T. and Milton, N., 1989. Reconstructing basin evolution from sedimentary thickness; the importance of palaeobathymetric control, with reference to the North Sea. Basin Res., 1: 247–257.

Birtles, R. 1986. The seismic flatspot and the discovery and delineation of the Troll Field. In: A.M. Spencer et al. (Editors), Habitat of Hydrocarbons on the Norwegian Continental Shelf. Graham and Trotman, London, pp. 207–216.

Boote, D.R.D and Gustav, S.H., 1987. Evolving depositional systems within an active rift, Witch Ground Graben, North Sea. In: J. Brooks and K. Glennie (Editors), Petroleum Geology of North West Europe. Graham and Trotman, London, pp. 819–828.

Bowen, J.M., 1975. The Brent Oil-field. In: A.W. Woodland (Editor), Petroleum and the Continental Shelf of Northwest Europe. Applied Science Publishers, London, pp. 353–360.

Brace, W.F. and Kohlstedt, D.L., 1980. Limits on lithospheric stress imposed by laboratory experiments. J. Geophys. Res., 85: 6248–6252.

Brown, S. 1986. Jurassic. In: K.W. Glennie (Editor). Introduction to the Petroleum Geology of the North Sea. Blackwell Scientific Publications, Oxford, 2nd ed., pp. 133–159.

Bruhn, R.L., Gibler, P.R. and Parry, W.T., 1987. Rupture characteristics of normal faults: an example from the Wasatch fault zone, Utah. In: M.P. Coward, J.F. Dewey and P.L. Hancock (Editors), Continental Extensional Tectonics. Geol. Soc. London, Spec. Publ., 28: pp. 337–353.

Chen, W.P. and Molnar, P., 1983. Focal depths of intracontinental and intraplate earthquakes and their implications for the thermal and mechanical properties of the lithosphere. J. Geophy. Res., 88: 4183–4214.

Childs, C., Walsh, J.J. and Watterson, J., 1990. A method for estimation of the density of fault displacements below the limit of seismic resolution in reservoir formations. In: A.T. Butler et al. (Editors), North Sea Oil and Gas Reservoirs. Graham and Trotman, London, pp. 309–318.

Cohen, S.C., 1984. Postseismic deformation due to subcrustal viscoelastic relaxation following dip-slip earthquakes. J. Geophys. Res., 89: 4538–4544.

Ellenor, D.W. and Mozetic, A., 1986. The Draugen oil discovery. In: A.M. Spencer et al. (Editors), Habitat of Hydrocarbons on the Norwegian Continental Shelf. Graham and Trotman, London, pp. 313–316.

Faure, J.L. and. Chermette, J.C., 1989. Deformation of tilted blocks, consequences on block geometry and extension measurements. Bull. Soc. Géol. Fr., 8: 461–476.

Fowler, S. and McKenzie, D., 1989. Gravity studies of the Rockall and Exmouth Plateaux using SEASAT altimetry. Basin Res., 2: 27–34.

Gibson, J.R., Walsh, J.J. and Watterson, J., 1989. Modelling of bed contours and cross-sections adjacent to planar normal faults. J. Struct. Geol. 11: 317–328.

Gradijan, S.J. and Wiik, M., 1987. Statfjord Nord. In: A.M. Spencer et al. (Editors), Geology of the Norwegian Oil and Gas Fields. Graham and Trotman, London, pp. 341–350.

Harris, J.P. and Fowler, R.M., 1987. Enhanced prospectivity of the Mid-Late Jurassic sediments of the South Viking Graben. In: J. Brooks and K.W. Glennie (Editors), Petroleum Geology of North West Europe. Graham and Trotman, London, pp. 879–898.

Hellem, T., Kjemperud, A. and Øvrebø, O.K., 1986. The Troll Field: a geological/geophysical model established by the PL085 Group. In: A.M. Spencer et al. (Editors), Habitat of Hydrocarbons on the Norwegian Continental Shelf. Graham and Trotman, London, pp. 217–240.

Hollander, N.B., 1987. Snorre. In: A.M. Spencer et al. (Editors), Geology of Norwegian Oil and Gas Fields. Graham and Trotman, London, pp. 307–318.

Jackson, J.A., Gagnepain, J. Houseman, G., King, G.C.P., Papadimitriou, P., Soufleris, C. and Virieux, J., 1982. Seismicity, normal faulting, and the geomorphological development of the Gulf of Corinth (Greece): the Corinth earthquakes of February and March 1981. Earth Planet. Sci. Lett., 57: 377–397.

Jackson, J. and McKenzie, D., 1983. The geometrical evolution of normal fault systems. J. Struct. Geol., 5: 471–482.

Jackson, J.A., White, N.J., Garfunkel, Z. and Anderson, H., 1988. Relations between normal-fault geometry, tilting and vertical motions in extensional terrains, an example from the southern Gulf of Suez. J. Struct. Geol., 10: 155–70.

Jackson, J.A. and White, N.J., 1989. Normal faulting in the upper continental crust: observations from regions of active extension. J. Struct. Geol., 11: 15–36.

Jankhof, K., 1945. Changes in ground level produced by the earthquakes of April 14 to 18, 1928 in southern Bulgaria. In: Tremblements de terre en Bulgarie, Nos. 29–31, Institut Météorologique central de Bulgaria, Sofia, pp. 131–136 (in Bulgarian).

King, G.C.P. and Vita-Finzi, C., 1981. Active folding in the Algerian earthquake of 10 October 1980. Nature, 292: 22–26.

King, G.C.P., Stein, R.S. and Rundle, J.B., 1988. The growth of geological structures by repeated earthquakes, 1. Conceptual framework. J. Geophys. Res., 93: 13307–13319.

Kusznir, N.J. and Park, R.G., 1987. The extensional strength of the continental lithosphere: its dependence on geothermal gradient, and crustal composition and thickness. In: M.P. Coward, J.F. Dewey and P.L. Hancock (Editors), Continental Extensional Tectonics. Geol. Soc. London, Spec. Publ., 28: 35–52.

Kusznir, N.J., Marsden, G. and Egan, S., 1988. Fault block rotation during continental lithosphere extension: a flexural cantilever model. Geophys. J. 92: 546 (abstract).

Kusznir, N.J., and Egan, S.S., 1990. Simple-shear and pure-shear models of extensional sedimentary basin formation: application to the Jeanne d'Arc Basin, Grand Banks of Newfoundland. In: A.J. Tankard and H.R. Balkwill (Editors), Extensional Tectonics and Stratigraphy of the North Atlantic Margins. Am. Assoc. Pet. Geol., Mem., 46: 305–322.

Larsen, P.H., 1988. Relay structures in a Lower Permian basement-involved extension system, East Greenland. J. Struct. Geol., 10: 3–8.

Leeder, M.R. and Gawthorpe, R.L., 1987. Sedimentary models for extensional tilt-block/half-graben basins. In: M.P. Coward, J.F. Dewey and P.L. Hancock (Editors), Continental Extensional Tectonics. Geol. Soc. London, Spec. Publ., 28: 139–152.

Livera, S. and Gdula, J.E., 1990. The Brent Oil Field. In: E.A. Beaumont and N.H. Foster (Editors), Atlas of Oil and Gas Fields, Structural Traps, II. Traps Associated with Tectonic Faulting. American Association of Petroleum Geologists, Tulsa, Okla., pp. 21–63.

Maher, C.E. and Harker, S.D., 1987. Claymore Oil Field. In: Brooks, J. and Glennie, J. (Editors), Petroleum Geology of North West Europe. Graham and Trotman, London, pp. 835–845.

Mansinha, L. and Smylie, D.E., 1971. The displacement fields of inclined faults. Bull. Seismol. Soc. Am., 61: 1433–1440.

Marsden, G., Yielding, G., Roberts, A. and Kusznir, N.J., 1990. Application of a flexural cantilever simple-shear/pure-shear model of continental lithosphere extension to the formation of the northern North Sea Basin. In: D.J. Blundell and A.D. Gibbs (Editors), Tectonic Evolution of the North Sea Rifts. Oxford University Press, Oxford, pp. 240–261.

McKenzie, D. 1978. Some remarks on the development of sedimentary basins. Earth Planet. Sci. Lett., 40: 25–32.

Myers, W.B. and Hamilton, W., 1964. Deformation accompanying the Hebgen Lake earthquake of August 17, 1959. U.S. Geol. Surv. Prof. Paper, 435: 55–98.

Ransome, F.L., Emmons, W.H. and Garrey, G.H., 1910. Geology and ore deposits of the Bullfrog district, Nevada. Bull. U.S. Geol. Surv., 407: 1–130.

Reilinger, R., 1986. Evidence for postseismic viscoelastic relaxation following the 1959 $M = 7.5$ Hebgen Lake, Montana, earthquake. J. Geophys. Res., 91: 9488–9494.

Richter, C.F., 1958. Elementary Seismology. W.H. Freeman, San Francisco, Calif.

Roberts, A.M. and Yielding, G., 1991. Deformation around basin-margin faults in the North Sea/Norwegian rift. In: A.M. Roberts, G. Yielding and B. Freeman (Editors), The Geometry of Normal Faults. Geol. Soc. London, Spec. Publ., 56: 61–78.

Roberts, A.M., Price, J. and Olsen, J.S., 1990. Late Jurassic half-graben control on the siting and structure of hydrocarbon accumulations: UK/Norwegian Central Graben. In: R. Hardman (Editor), Tectonic Events Controlling Britain's Oil and Gas Reserves. Geol. Soc. London, Spec. Publ., 55: 229–257.

Roberts, S. and Jackson, J., 1991. Active normal faulting in Central Greece: an overview. In: A.M. Roberts, G. Yielding and B. Freeman (Editors), The Geometry of Normal Faults. Geol. Soc. London, Spec. Publ., 56: 125–142.

Rundle, J.B., 1982. Viscoelastic-gravitational deformation by a rectangular thrust fault in a layered Earth. J. Geophys. Res., 87: 7787–7796.

Schwartz, D.P. and Coppersmith, K.J., 1984. Fault behavior and characteristic earthquakes: examples from the Wasatch and San Andreas Fault zones. J. Geophys. Res., 89: 5681–5698.

Spencer, A.M. and Larsen, V.B., 1990. Fault traps in the northern North Sea. In: R.P.F. Hardman (Editor), Tectonic Movements Responsible for Britain's Oil and Gas Reserves. Geol. Soc. London, Spec. Publ., 55: pp. 281–298.

Stein, R.S. and Barrientos, S.E., 1985. Planar high-angle faulting in the basin and range: geodetic analysis of the 1983 Borah Peak, Idaho, earthquake. J. Geophys. Res., 90: 11355–11366.

Stein, R.S., King, G.C.P. and Rundle, J.B., 1988. The growth of geological structures by repeated earthquakes, 2. Field examples of continental dip-slip faults. J. Geophys. Res., 9: 13319–31.

Turcotte, D.L. and Schubert, G., 1982. Geodynamics — applications of continuum physics to geological problems. John Wiley and Sons, New York, N.Y.

Turner, C.C., Cohen, J.M., Connell, E.R. and Cooper, D.M., 1987. A depositional model for the South Brae oilfield. In: J. Brooks and K.W. Glennie (Editors), Petroleum Geology of North West Europe. Graham and Trotman, London, pp. 853–64.

Vita-Finzi, C. and King, G.C.P., 1985. The seismicity, geomorphology and structural evolution of the Corinth area of Greece. Philos. Trans. R. Soc. London, Ser. A, 314: 379–407.

Wallace, R.E. 1984. Patterns and timing of late Quaternary faulting in the Great Basin Province and relation to some regional tectonic features. J. Geophys. Res., 89: 5763–5769.

Walsh, J.J. and Watterson, J., 1991. Geometric and kinematic coherence and scale effects in normal fault systems. In: A.M. Roberts, G. Yielding and B. Freeman (Editors), The Geometry of Normal Faults. Geol. Soc. London, Spec. Publ., 56: 193–203.

Watts, A.B., 1988. Gravity anomalies, crustal structure and flexure of the lithosphere at the Baltimore Canyon Trough. Earth Planet. Sci. Lett., 89: 221–238.

Wernicke, B. and Burchfiel, B.C., 1982. Modes of extensional tectonics. J. Struct. Geol., 4: 105–115.

Wheatley, T.J., Biggins, D., Buckingham, J., and Holloway, N.H., 1987. The geology and exploration of the Transitional Shelf, an area to the west of the Viking Graben. In: J. Brooks and K.W. Glennie (Editors), Petroleum Geology of North West Europe. Graham and Trotman, London, pp. 979–990.

White, N., 1990. Does the uniform stretching model work in the North Sea? In: D.J. Blundell and A.D. Gibbs (Editors), Tectonic Evolution of the North Sea Rifts. Oxford University Press, Oxford, pp. 217–239.

Whitten, C.A., 1957. Geodetic measurements in the Dixie Valley area. Bull. Seismol. Soc. Am., 47: 321–325.

Yielding, G., 1990. Footwall uplift associated with Late Jurassic normal faulting in the northern North Sea. J. Geol. Soc. London, 147: 219–222.

Yielding, G., Badley, M.E. and Freeman, B., 1991. Seismic reflections from normal faults in the northern North Sea. In: A.M. Roberts, G. Yielding and B. Freeman (Editors), The Geometry of Normal Faults. Geol. Soc. London, Spec. Publ., 56: 79–89.

Young, R., 1992. Restoration of a regional profile across the Magnus Field in the northern North Sea. In: R.M. Larsen, H. Brekke, B.T. Larsen and E. Talleraas (Editors), Structural and Tectonic Modelling and its Application to Petroleum Geology. Norwegian Petroleum Society (NPF), Special Publication 1. Elsevier, Amsterdam, pp. 221–229 (this volume).

G. YIELDING *Badley Ashton and Associates Limited, Winceby House, Winceby, Horncastle, Lincolnshire, LN9 6PB, U.K.*

A. ROBERTS *Badley Ashton and Associates Limited, Winceby House, Winceby, Horncastle, Lincolnshire, LN9 6PB, U.K.*

Modelling secondary hydrocarbon migration in Haltenbanken, Norway

L. Hermans, A.D. van Kuyk, F.K. Lehner and P.S. Featherstone

This paper reports the results of a map-based modelling study of hydrocarbon migration in the Haltenbanken area, offshore Mid-Norway. The study represents the first field application of a previously published mathematical model that treats secondary migration as a buoyancy-driven, separate-phase flow. Fluid viscosities exceeding that of formation water are required by the theory, thus gas migration is not included. From palaeo-structure maps and maps of hydrocarbon supply rate at different times in the geological history, the model computes the evolution of hydrocarbon migration paths and accumulation patterns throughout a given carrier bed. A general finding from numerous model tests is that system properties such as variations in carrier-bed permeability and fluid mobility can profoundly affect accumulation patterns.

For the Haltenbanken study, hydrocarbon charge was assumed to derive from the Spekk Formation (Kimmeridge Clay equivalent) only, expelling a single hydrocarbon phase — a representative light oil — downward into the carrier bed (the Garn Formation), which was assumed to be homogeneous and continuous over the study area. A series of computer simulations of the model was used to evaluate the response of the migration system to variations in the hydrocarbon supply rate and the carrier-bed properties. Leakage through cap rocks and along faults was not considered and this, combined with the abundance of generating source rock within the study area, resulted in solutions with most of the structures presently filled to spilling point. While the model described in this paper clearly is an oversimplification of the actual situation, even in its present form it has proven its usefulness for assessing the effect of essential migration system parameters such as supply rate, carrier bed and fluid properties, and the accuracy of palaeo-structure maps, on migration and accumulation patterns.

Introduction

Deterministic hydrocarbon-habitat modelling aims at predicting the quantity and distribution of hydrocarbon occurrences through geological time and can, therefore, play a major role in the decision of where to drill for oil or gas. In the sequence of processes leading to the eventual entrapment of hydrocarbons, secondary migration plays a key role in the hydrocarbon distribution.

The literature on the mechanism(s) of secondary oil migration is extensive, and many different processes have been proposed. In recent discussions, an hypothesis of separate-phase flow has generally been favoured (Hobson, 1954; McAuliffe, 1979; Neglia, 1979; Schowalter, 1979; Jones, 1981; Du Rouchet, 1981; Tissot and Welte, 1984; Ungerer et al., 1984; England et al., 1987; Lehner et al., 1987; Dembicki and Anderson, 1989). Other mechanisms, e.g., transport in solution as argued by Price (1976, 1981), could operate in parallel and occasionally even in series, as is evident from the occurrence of diffusive losses of dissolved lighter components through capillary barriers (Krooss et al., 1988). However, these are more difficult to reconcile with constraints imposed by low and skewed hydrocarbon solubilities, limited supply of compaction water and oil/source-rock correlations (McAuliffe, 1979; Jones, 1981; Tissot and Welte, 1984; Ungerer et al., 1984; England et al., 1987).

In contrast with oil, the effect of dissolution might be more important for hydrocarbon gases, notably methane, because of the combination of the much higher water solubility (of the order of 1000 ppm under in-situ conditions: Dodson and Standing, 1944; McAuliffe, 1979) and the increased diffusion coefficient (Krooss and Leythaeuser, 1988). An order-of-magnitude calculation shows that saturation of a carrier bed with gas by diffusion might take place in one million years. However, in view of the much higher transport velocity, separate-phase flow can become dominant even in unsaturated aquifers. Moreover, the usually vast volumes of gas generated far exceed the dissolution capacity of the average aquifer, again suggesting separate-phase flow to be the prevailing transport mechanism.

Based on the assumption of segregated flow, a mathematical model was developed that describes secondary hydrocarbon migration in hydrostatic aquifers. The background and underlying theory of

Structural and Tectonic Modelling and its Application to Petroleum Geology edited by R.M. Larsen, H. Brekke, B.T. Larsen and E. Talleraas. NPF Special Publication 1, pp. 305–323. Elsevier, Amsterdam.

this dynamic flow model were discussed comprehensively in a previous publication (Lehner et al., 1987). A brief outline is given below. The present paper reports a field application of the numerical simulator SECMIG, which is based on the model, to the Haltenbanken area, offshore Mid-Norway.

The Haltenbanken area was chosen because it contains oil and gas accumulations in an area of generally good seismic data quality and coverage, although source-rock richness and distribution are too poorly constrained for the area to serve as a full calibration set. Nevertheless, the structural and lithostratigraphic setting, together with the known hydrocarbon occurrences and the existence of new prospects, make it a good location for exploring the capabilities of the technique.

The migration model — SECMIG

Modelling hydrocarbon migration is essentially a three-dimensional problem, and this, combined with the size and structural complexity of the systems to be modelled and the typically very high migration velocities involved (metres/year) compared with the time interval to be simulated (many millions of years), would make an approach based on the solution of the standard equations governing immiscible two-phase flow in three dimensions computationally unfeasible.

A substantial simplification was found in exploiting the phenomenon of gravity segregation of oil and water, which may be expected to occur in permeable carrier beds (Hobson, 1954). By analogy with the analysis of the motion of the phreatic surface in unconfined groundwater flow (see, for example, Wooding and Chapman, 1966), lateral updip migration of a separate hydrocarbon phase under seals is conceived in our model as a Boussinesq-type, transient, free surface flow (Boussinesq, 1904; Bear, 1972), with source and sink terms representing supply from source rocks and leakage through cap rocks and faults, respectively. Thus, we assume that hydrocarbons, upon entering a carrier bed, will form accumulations below a cap rock that are discharged by a buoyancy-driven flow up-structure. This enables a two-dimensional, map-view mathematical description of a three-dimensional time-dependent secondary migration system, the dependent variable being the height of the hydrocarbon column below the seal. This analysis implies a sharp distinction between vertical migration (i.e., cross-stratigraphy), giving rise to the source and sink terms, and lateral migration. The latter is essentially strata-bound and forms the part which is actually modelled, a segregated flow of a buoyant oil phase in individually hydrostatic aquifers. Vertical migration across carrier beds is assumed to take place instantaneously on the time scale set by the expulsion rates. Vertical migration through cap rocks and seals can be modelled as a quasi-steady flow, governed by the aquifer potential drops between superimposed carrier beds, though this approach has not been included in this study.

The input required by the model's simulator, SECMIG, consists of structure-contour maps at the base of the topseal, rate maps of hydrocarbon expulsion (supply), and the properties of the carrier bed and hydrocarbon fluid for a geologically relevant period of time. The program output consists of calculated hydrocarbon column heights below the seal. These may be plotted in different colour shades on a structure-contour map of the seal at any time during the simulation.

The simulator with which this study was performed has a number of simplifications and restrictions:

– in the derivation of the model the resistant force of the water phase has been disregarded; hence, transient flow will be modelled adequately only if hydrocarbon fluid viscosities significantly exceeds that of the formation water;

– a sharp transition between the oil and water phases is assumed, ignoring capillary pressure effects;

– the current version of the program is restricted to single carrier-bed/impermeable cap rock systems;

– the aquifer is assumed to be hydrostatic;

– carrier-bed thickness is expected to exceed the hydrocarbon fluid column height at every location during the simulation (unconfined flow);

– faults are treated merely as vertical steps at the base of the cap rock and no specific fault-zone properties (i.e., permeability or capillary threshold pressure) are allowed for;

– only a single hydrocarbon phase (oil) of fixed composition is modelled;

– zero residual oil saturation is assumed within the carrier bed;

– fluids are treated as incompressible.

Since any migration system will contain capillary barriers, we have implicitly assumed that these can be specified in advance, based, for example, on lithological criteria. Estimates of residual oil in those carrier-bed regions that have been hydrocarbon saturated at a certain stage in the migration history could easily be incorporated in the present model. In addition, hydrocarbon losses within the remainder of the carrier bed are expected to be generally low because of efficient focussing of the oil flow during vertical migration. This assumption seems reasonable and is supported by the recently published experimental results of Dembicki and Anderson (1989).

Study area-regional setting

The migration simulation study was performed for a rectangular block in the Haltenbanken region, offshore Mid-Norway, covering an area of 82.5 × 72 km situated between 6°30′ and 8°00′E, and 64°08′ and 64°45′N. Figure 1 shows a location map. The geology is well described in the literature (Bukovics and Ziegler, 1985; Cohen and Dunn, 1986). The eastern part of the study area is defined by the west flank of the relatively stable and undeformed Trøndelag Platform, containing the Draugen (undersaturated oil), Mikkel (gas condensate + some oil) and, on the Frøya High, the Njord (oil + gas) accumulations (Pedersen, 1988). A major fault zone tending to SW-NNE separates the Trøndelag Platform from the heavily structured and faulted Halten Terrace which covers the rest of the map and forms the main kitchen area. The major fault activity took place in the Late Jurassic/Early Cretaceous period. Just beyond the northern edge of the study area, oil, gas and condensate have been discovered in South Tyrihans and North Tyrihans (Pedersen, 1988). Within the study area, an uneconomic gas accumulation was found by well 6407/4-1 in a very tight formation. All fields produce from Jurassic strata.

Stratigraphy and assumed model lithology

Figure 2 shows an isometric depth display of the main structurally mapped seismic horizons, and Fig. 3 shows the local stratigraphy in cross-section. The main source rock in the sequence is the Upper Jurassic Spekk Formation (Kimmeridge Clay lithostratigraphic equivalent), the lowermost section having the best-developed source rocks. It is overlain by sealing Cretaceous mudstones and claystones that thicken progressively towards the deeper parts of the basin. The underlying Jurassic sandstones are expected to form the main migration avenue for hydrocarbons generated and expelled from the Spekk Formation. The carrier beds in the Middle/Lower Jurassic (Garn and Tilje Formation) are extensive throughout the region; however, the Upper Jurassic carrier beds (Rogn Formation) are, in general, restricted (Cohen and Dunn, 1986). Downward hydrocarbon expulsion from the Spekk Formation into these carrier beds may be hindered by the locally intervening shaly Melke Formation. The Melke Formation is absent on a number of structural highs but well developed in the lows, particularly towards the northwest within the study area. The very thick Lower Jurassic/Triassic fluvio-deltaic Åre Formation contains numerous coal seams, and is also considered a possible source rock (mainly for gas) in the area.

Since the present version of SECMIG can handle only one caprock/carrier-bed interface at a time, the local stratigraphy had to be simplified in the model as indicated in Fig. 4, with the Spekk Formation acting both as source rock and cap rock for the hydrocarbons expelled downward into the underlying Jurassic carrier bed. For the simulations, the seismically

a b

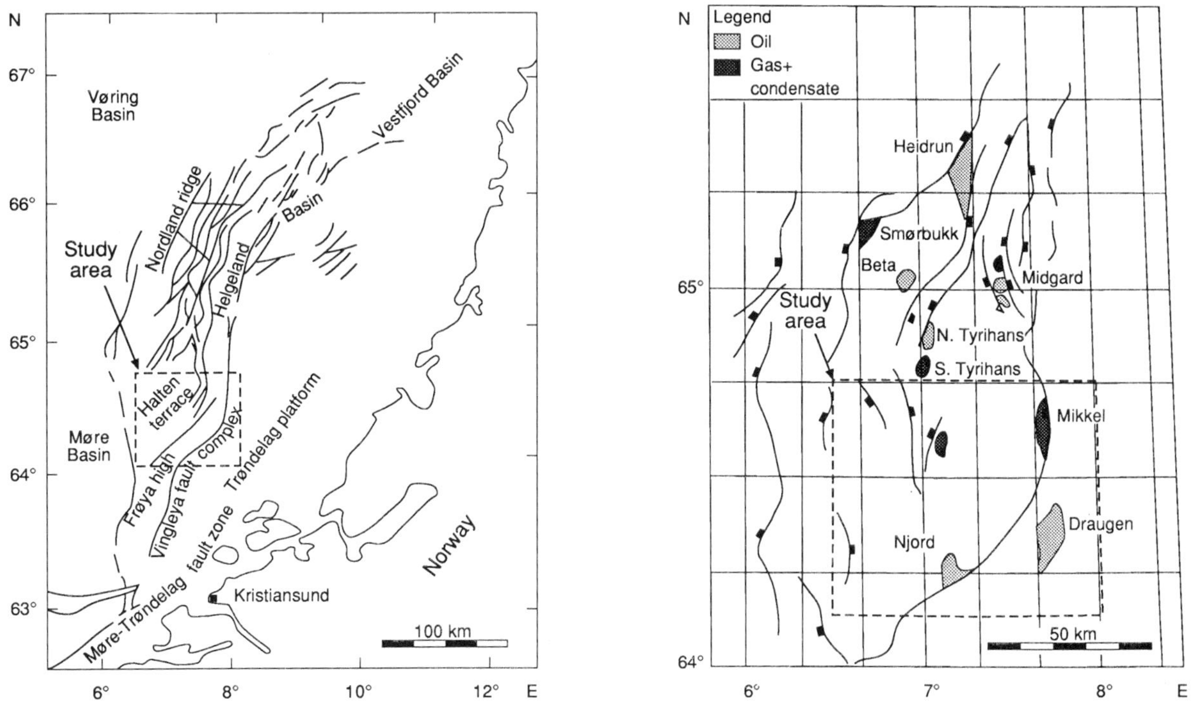

Fig. 1. Location map of the Haltenbanken area, showing (a) the main structural elements and (b) the distribution of the hydrocarbon accumulations. Figure modified from Cohen and Dunn (1986) and Hvoslef et al. (1988).

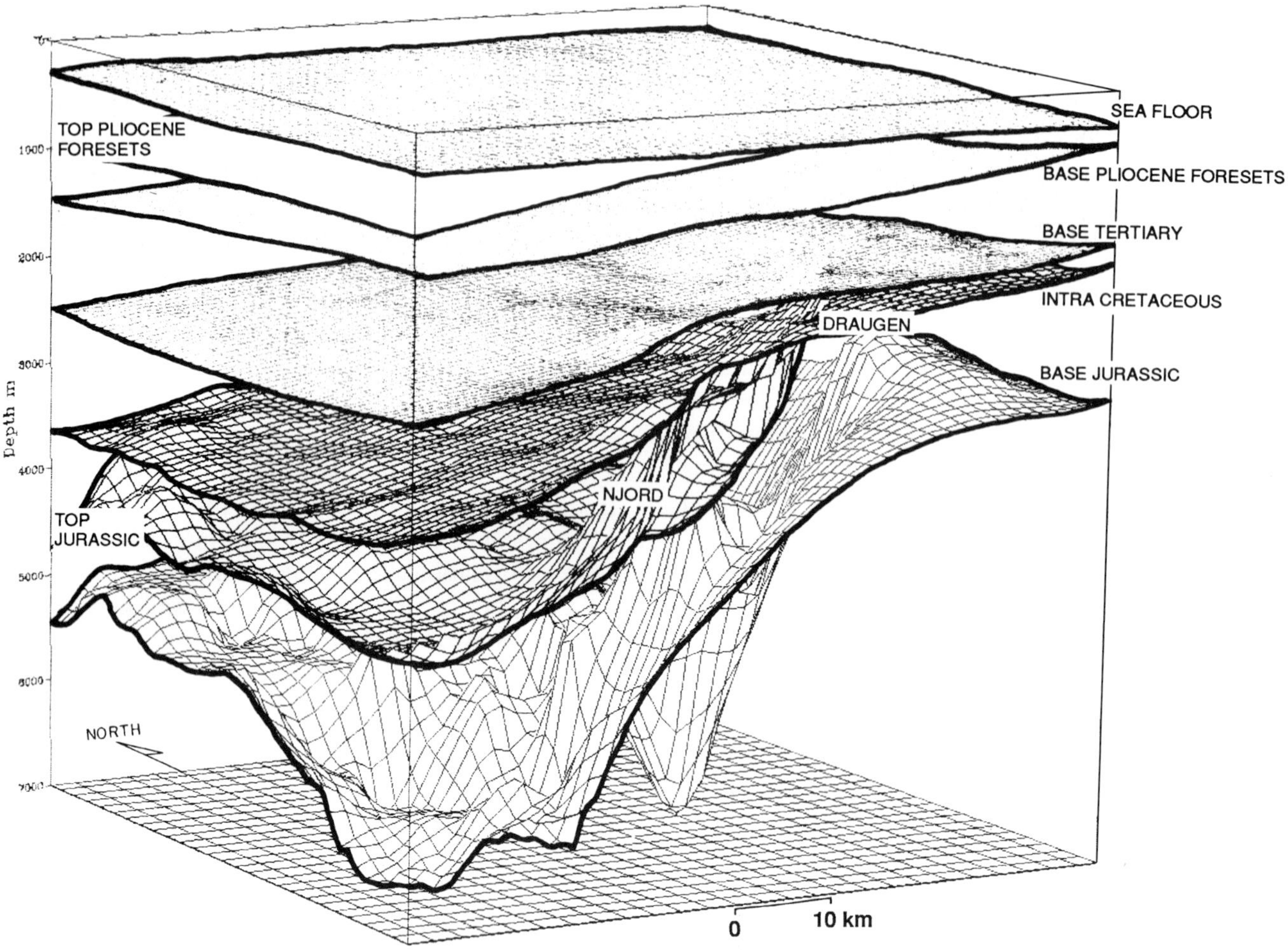

Fig. 2. Isometric display of the main horizons in the Haltenbanken study area.

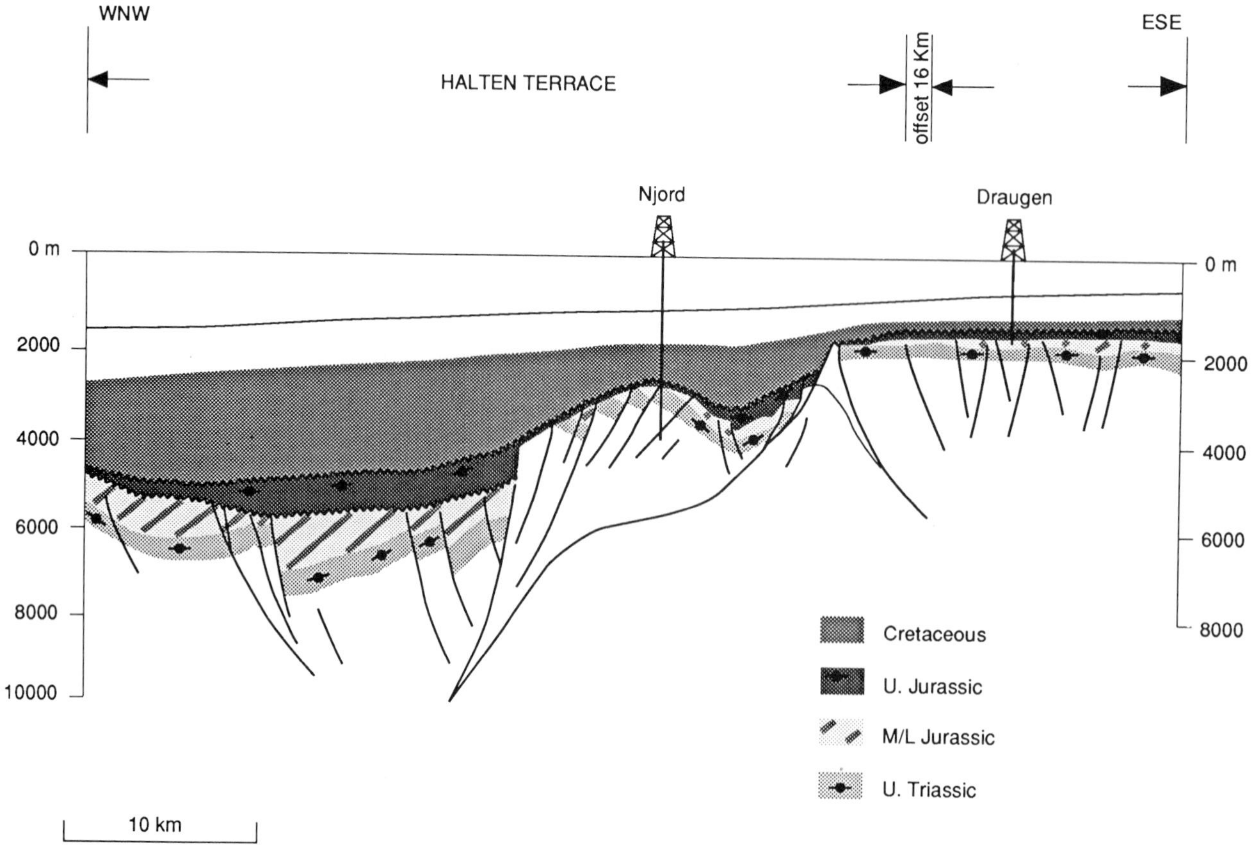

Fig. 3. Stratigraphy of the Haltenbanken study area. The location of the cross-section is indicated in Fig. 5.

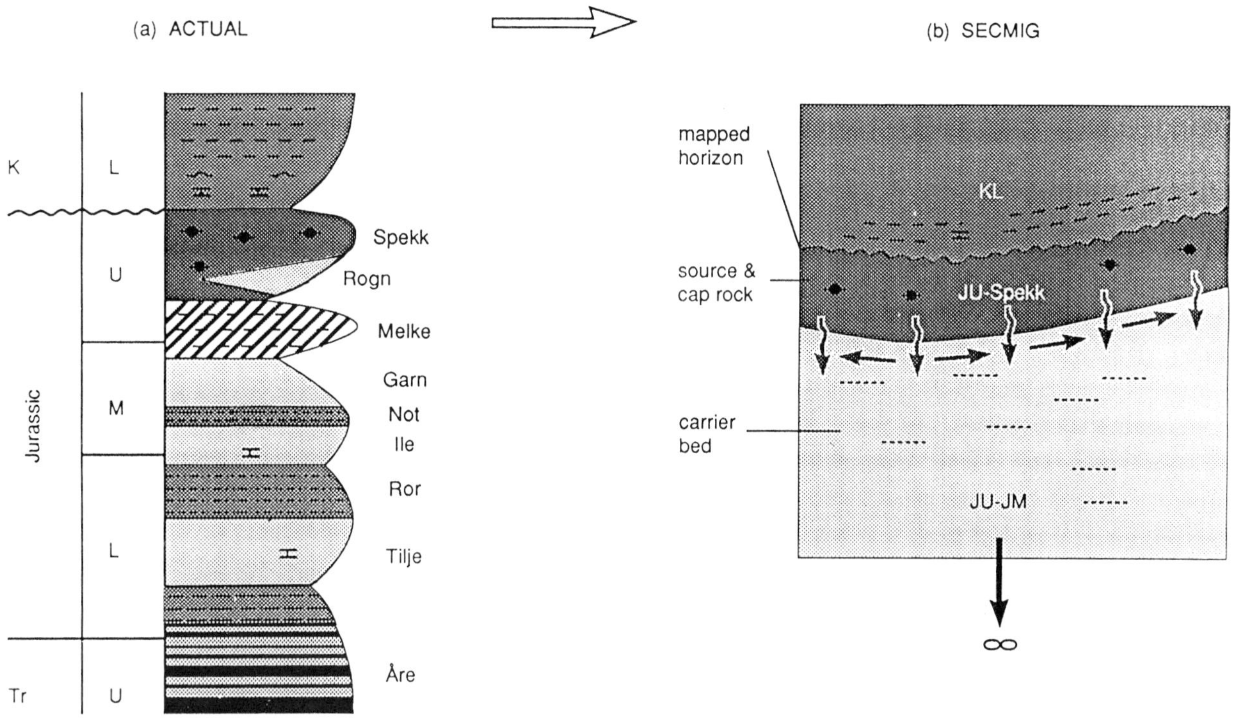

Fig. 4. Stratigraphic column of the Jurassic in the Haltenbanken study area: (a) actual, (b) model stratigraphy used in SECMIG.

mappable base Cretaceous/Upper Jurassic marker was used as the sealing horizon. This was judged sufficiently close to the actual source-rock/carrier-bed interface, in view of the map's accuracy and the near-parallel nature of the base Cretaceous and base source-rock horizons. The present-day contour map of this marker horizon as used in the simulations is shown in Fig. 5. The Melke Formation has been disregarded. Its effect on expelled hydrocarbon volumes is provisionally accounted for by an expulsion factor (see below). Supply from the Åre Formation was not considered, since most of it is expected to have passed the maturity window at the time of trap formation. The cap rock is assumed to be impermeable over the mapped area. Faults are treated merely as steps in the cap rock, hence migration losses by leakage along faults are not considered. The large regional faults on the edge of the Halten Terrace have throws which are most likely greater than the actual carrier-bed thickness. In computer runs I, II and III of this study they were assumed to be permeable, enabling migration to take place from the downthrown into the upthrown block as if the carrier beds were in communication. In run IV they were assumed to be sealing, which was accomplished in the simulation by making the grid points in the fault zone impermeable.

SECMIG runs

The input required for SECMIG consists of base-seal contour maps and hydrocarbon supply rate maps throughout the simulated burial history, together with the physical properties of the carrier bed and the oil. The value of the model's prediction, therefore, relies heavily on the accuracy of the preceding basin reconstruction and the choice of parameters. A series of four computer runs of the model has been made to evaluate the response of the migration system to input variations. The runs differed in the following input characteristics:

- run I: best-fit constant sediment heat flow in space and time;
- run II: same input as in I, but heat flow variable in space and time;
- run III: same input as in II, but with a gentler decline of carrier-bed porosity and permeability with depth;
- run IV: same input as in III, but with a barrier to lateral migration introduced into the map (the fault zone west of Draugen).

In runs III and IV only the final 15 Ma have been calculated with the specified parameter setting. To speed up the calculation, the initial 125 Ma of these runs was modelled with a lower permeability, while careful checks were made to ensure that this procedure did not affect the final results.

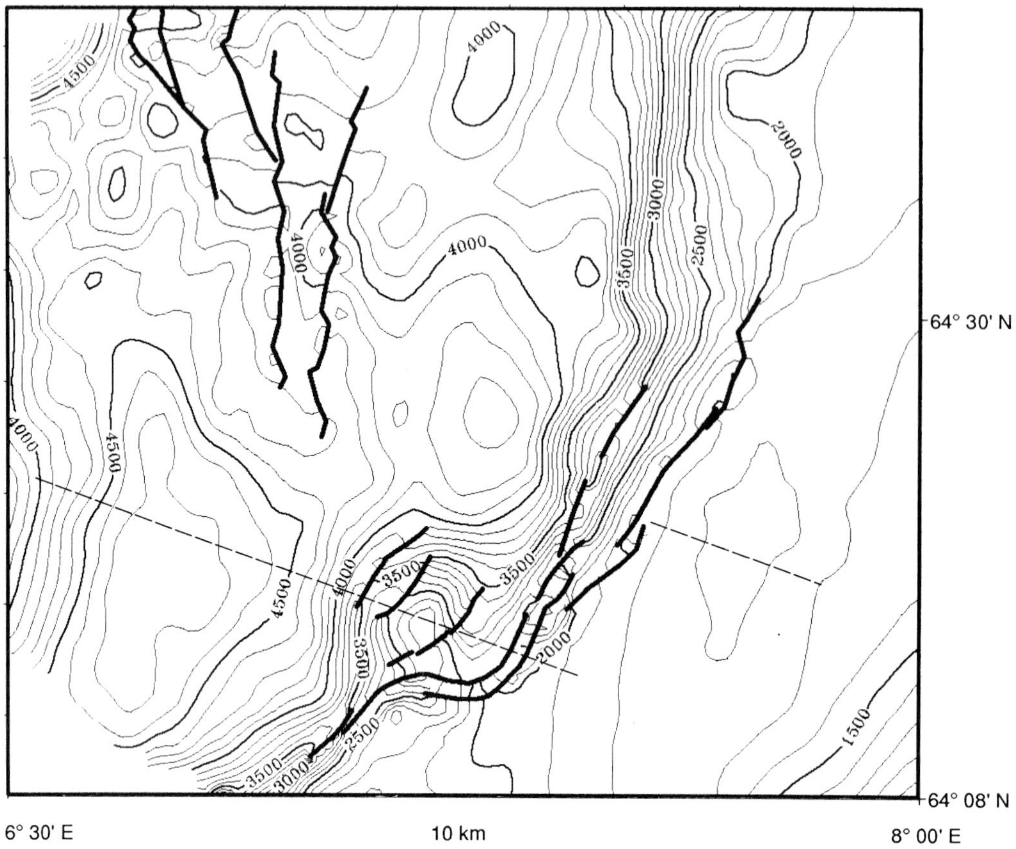

Fig. 5. Present-day depth map of the base Cretaceous/Upper Jurassic marker horizon in the Haltenbanken study area, along which secondary hydrocarbon migration was modelled. The dashed lines indicate the location of the cross-section of Fig. 3.

Calculation procedure

Table 1 and Fig. 6 provide a scheme of the calculation procedure and the chosen parameter settings.

Well data and mapped seismic horizons were used to reconstruct the base-seal/source-rock burial history, thus obtaining palaeo-depth maps. Single-point, palaeo-heat-flow values for the region, obtained from best fits to present-day well bottom-hole temperature data during burial reconstruction, were contoured over the map area and used as input for calculating the palaeo-temperature at source-rock level. Having calculated the source-rock time/temperature history, rate maps of hydrocarbon generation could be derived using the recently published kinetic model of Klomp and Wright (1989) with laboratory-measured kinetic parameters appropriate to the Spekk Formation (Kimmeridge clay type) source rock, and source-rock properties that were averaged over the map area. To obtain the hydrocarbon-supply-rate history, the result of the generation calculation should be input to an expulsion model. Since a suitable model is not available at present, an expulsion/generation ratio was estimated. The carrier-bed and fluid parameters, forming the additional input, were made a function of depth and temperature to simulate in-situ variations in these parameters upon burial. Only the hydrocarbon density was held constant. Although gas is also present in the area, only a single hydrocarbon phase (oil) could be modelled with the current version of the simulator. In the simulations a flux boundary condition was used, thus allowing hydrocarbon outflow from the map area, but excluding inflow.

Input

Base-seal burial and temperature reconstruction

The burial reconstruction over the area was based on present-day depth maps of five horizons: Top and Base Pliocene, Base Tertiary, Intra-Cretaceous and Top Jurassic, shown in the isometric display (Fig. 2). For run I (heat flow constant in space and time) an average sediment heat flow of 55 mW/m^2 was calculated and used as input to the palaeo-temperature calculations. More realistically, in the other runs heat-flow variation in space and time was included. In these runs a crustal-rifting model modified from McKenzie (1978) was used for the heat-flow calculations. The resulting palaeo and present-day heat-flow maps (Fig. 7) show a steady

TABLE 1

Calculation sequence with parameter settings for Haltenbanken study

(1) INPUT FOR BURIAL & TEMP. HISTORY RECONSTRUCTION OF SOURCE ROCK/SEAL HORIZON				
present-day depth maps of relevant (digitised seismic) horizons				(Fig. 2)
well BHT data within and around study area				
map area	82.5x72 km^2 (55x48 grid nodes)			(Fig. 1)
	(RUN 1	75x75 km^2 (50x50 nodes)		
heat flow	RUN 1	constant, 55 mW/m^2, best-fit constant heat flow		
	RUNS 2-4	variable, based on a crustal-rifting model		(Fig. 7)
↓				
burial and temperature history of Upper Jurassic Spekk source rock from onset of deposition (140 Mabp) to present				
(2) INPUT FOR HC GENERATION CALCULATION				
time/temperature result from (1) and				
first-order kinetic model with appropriate type-II source-rock kinetic parameters				
source rock	thickness		100 m	
	init. volume fraction kerogen		0.1	
	init. atomic H/C ratio kerogen		1.2	
↓				
history of hydrocarbon generation rate for Upper Jurassic source rock				
(3) INPUT FOR HC SUPPLY ("EXPULSION") CALCULATION				
result from (2) and (multiplied by)				
source rock expulsion/generation ratio			0.1	
↓				
history of hydrocarbon expulsion rate for Upper Jurassic source rock				
(4) INPUT FOR SECMIG				
burial history seal/carrier bed interface (~source rock) from (1) + result from (3) +				
oil density	(constant)		800 kg/m^3	
oil viscosity	(variable)		$7.67.10^{-7}$. exp (2310/temp) Pa.s	(Fig. 6)
carrier-bed porosity	(variable)	RUNS 1-2	0.59. exp (-z/2340)	(" ")
		RUNS 3-4	0.42. exp (-z/4000)	(" ")
carrier-bed permeability	(variable)	RUNS 1-2	exp ((porosity - 0.162)/0.020) mD	(" ")
		RUNS 3-4	exp ((porosity - 0.100)/0.022) mD	(" ")

decline from a maximum around 120 Ma to the present-day situation.

Source-rock characterisation

The Spekk source-rock properties were taken to be constant over the area. Although lateral variations obviously do occur, these were not mapped in view of the uncertainties involved and biased sampling. The latter is the case particularly for the measured thicknesses. Well thicknesses are in the range of 0 to 100 m, but since all data were obtained from structural highs, where the source rock is most sensitive to weathering and erosion, and locally has been eroded completely, thicknesses greater than 100 m may be present in deeper basin regions. Estimates of the maximum Spekk Formation thickness within the lows based on seismic interpretation vary considerably from "thinly developed" (Bukovics and Ziegler, 1985), a few hundred metres (Ellenor and Mozetic, 1986), to a few kilometres (Heum et al., 1986). In the case of uniformly increasing basin subsidence, formation thickness might relate to present-day burial depth. However, plotting the measured source-rock thicknesses based on well data against present-day depth (Fig. 8) reveals no clear correlation. In view of these uncertainties, the source rock was simply assumed to be continuous over the study area with a uniform thickness of 100 m, which is probably an underestimate. An average richness of 4% wt. organic matter (10% vol.) was taken, based on geochemical analyses of Spekk formation samples from within the study area and further to the north. The initial atomic H/C ratio of the immature source rock was set at 1.2, a value typical of a type II source rock. For the expulsion/generation ratio we have taken the very conservative value of 0.1 to compensate for losses in the Melke Formation and in the Cretaceous Formations on top of the Spekk Formation.

Hydrocarbon fluid (oil) properties

The ≈40°API undersaturated light oil encountered in Draugen was used as the simulation fluid. This was a compromise, since different types of light oil (and gas) are found in the region. The oil density was kept at a constant 800 kg/m^3. The actual change in this parameter is limited, since during the up-

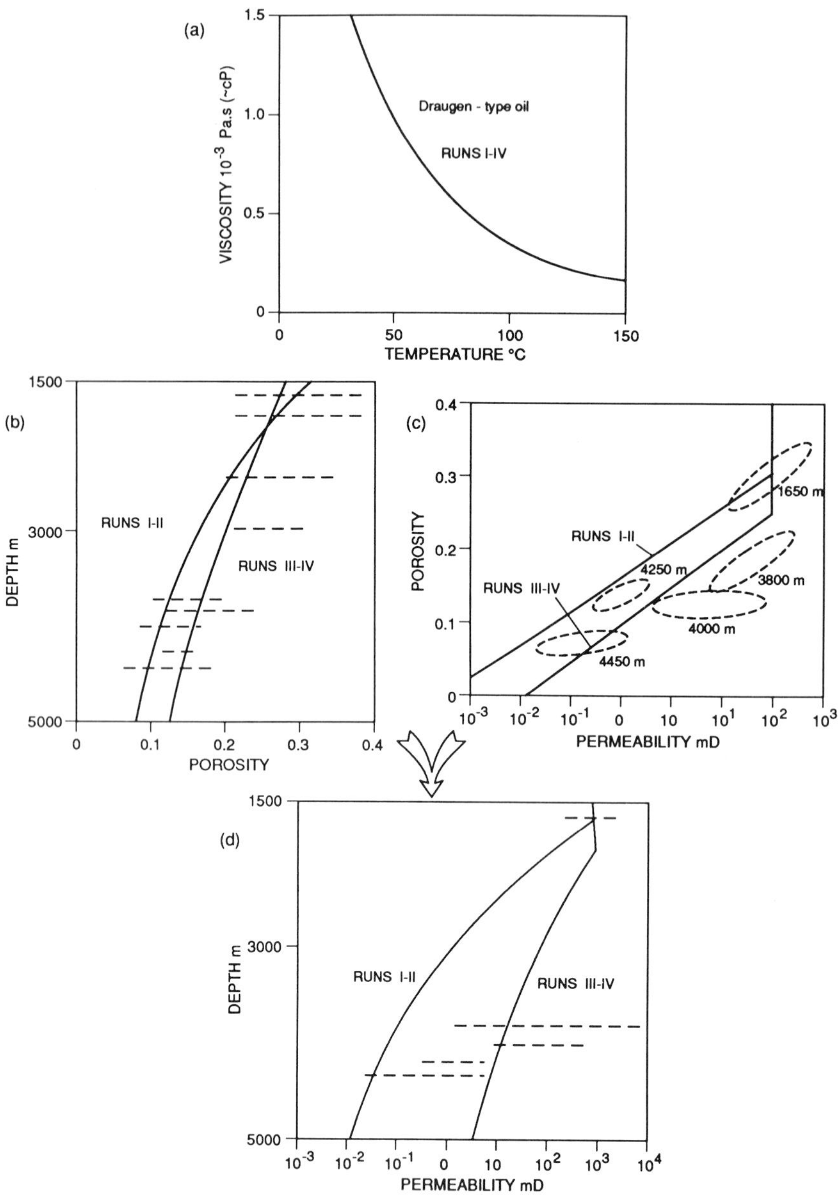

Fig. 6. (a–c) Oil viscosity and carrier-bed porosity and permeability trends used as input for the Haltenbanken SECMIG simulations. Relationship (d) results from a combination of (b) and (c). Measured Garn Formation data are indicated by dashes. See Table 1 for run input characteristics.

ward flow of hydrocarbons the effects of decreasing pressure (giving a density decrease) and temperature (giving a density increase) partly compensate for one another. The oil viscosity was modelled to decrease with increasing temperature according to a standard reservoir engineering relationship, which in SECMIG is approximated by the Andrade exponential formula (Latil, 1980, pp. 146–149), with Draugen oil parameters:

1.3×10^{-3} Pa s (1.3 cP) at 37°C

0.7×10^{-3} Pa s (0.7 cP) at 68°C — Draugen reservoir temperature

0.35×10^{-3}Pa s (0.35 cP) at 104°C.

Carrier-bed properties

The carrier bed was assumed to be homogeneous and continuous over the study area, except for the fault zone at the eastern edge of the Halten Terrace in run IV, which was made impermeable to simulate a lateral migration barrier (Fig. 9).

The present SECMIG version assumes an exponential porosity/depth trend, although, for the Haltenbanken data, linear compaction with depth might have been more appropriate, as is apparent from Fig. 6b and from data shown by Bjørlykke et al. (1986). For the simulations two compaction curves were chosen, both to fit a fractional porosity of around 0.30 at a depth of 1650 m (Draugen): a lower porosity trend

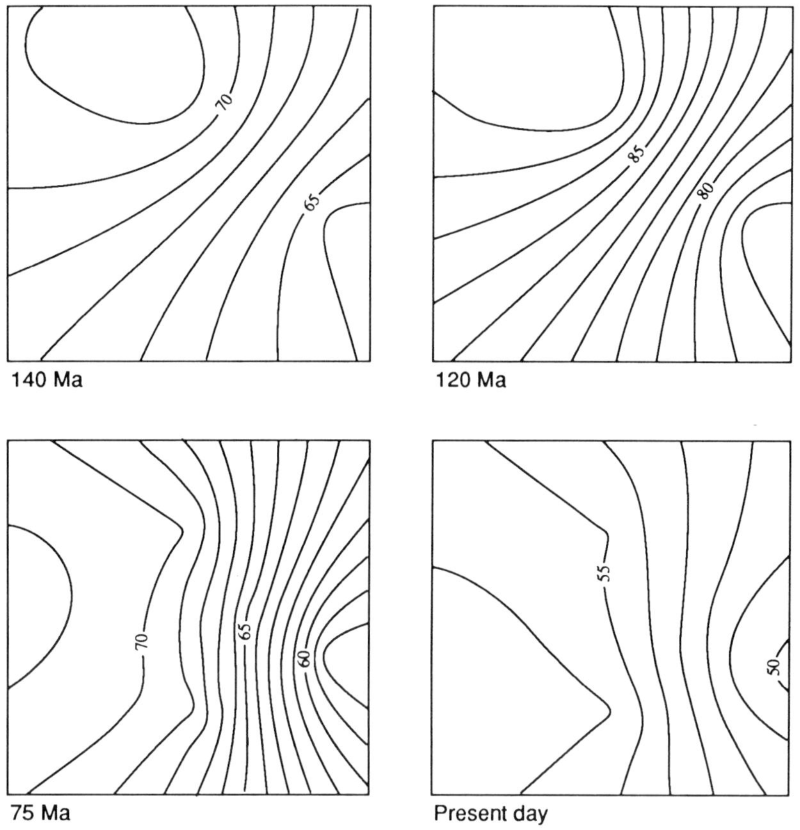

Fig. 7. Palaeo-heat-flow maps of the Haltenbanken study area based on a crustal-rifting model modified from McKenzie (1978), used in SECMIG runs II–IV.

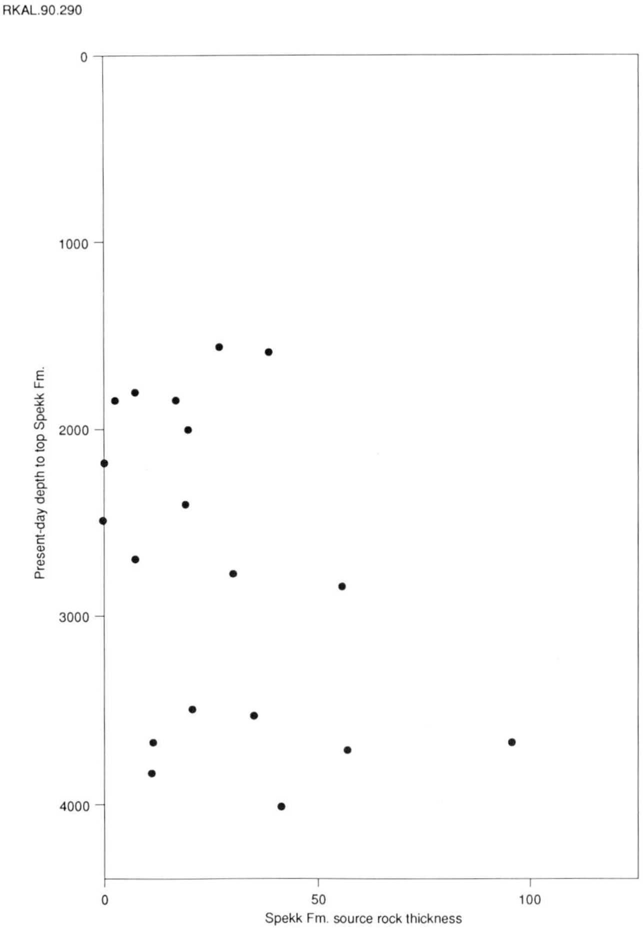

Fig. 8. Measured Spekk Formation source-rock thickness from well data versus present-day depth of source-rock in Haltenbanken.

to fit (0.07, 5000 m) for runs I and II, and a higher porosity trend to fit (0.12, 5000 m) for runs III and IV. In comparison with the local reservoir data, somewhat low porosity estimates were used in the simulations, particularly for the shallower (<3500 m) depths. This choice was based on the idea that since most of the reservoir rock samples analysed were impregnated with hydrocarbons, which may have impeded diagenetic porosity reduction, the "average" carrier bed might indeed have a lower porosity.

A least-squares fit on porosity/permeability core data plotted on a linear-log scale is commonly used as a good first approximation for describing permeability trends. Indeed, there are many examples of reservoirs for which good correlation is observed (Archie, 1950; Füchtbauer, 1967). However, when data over a wide permeability/porosity range have to be taken into account, for example, a carrier bed during its entire range of burial depths, a curved trend or a set of depth-dependent logarithm of permeability-versus-porosity trends (Bjørlykke et al., 1986) might be more appropriate, as suggested by field data

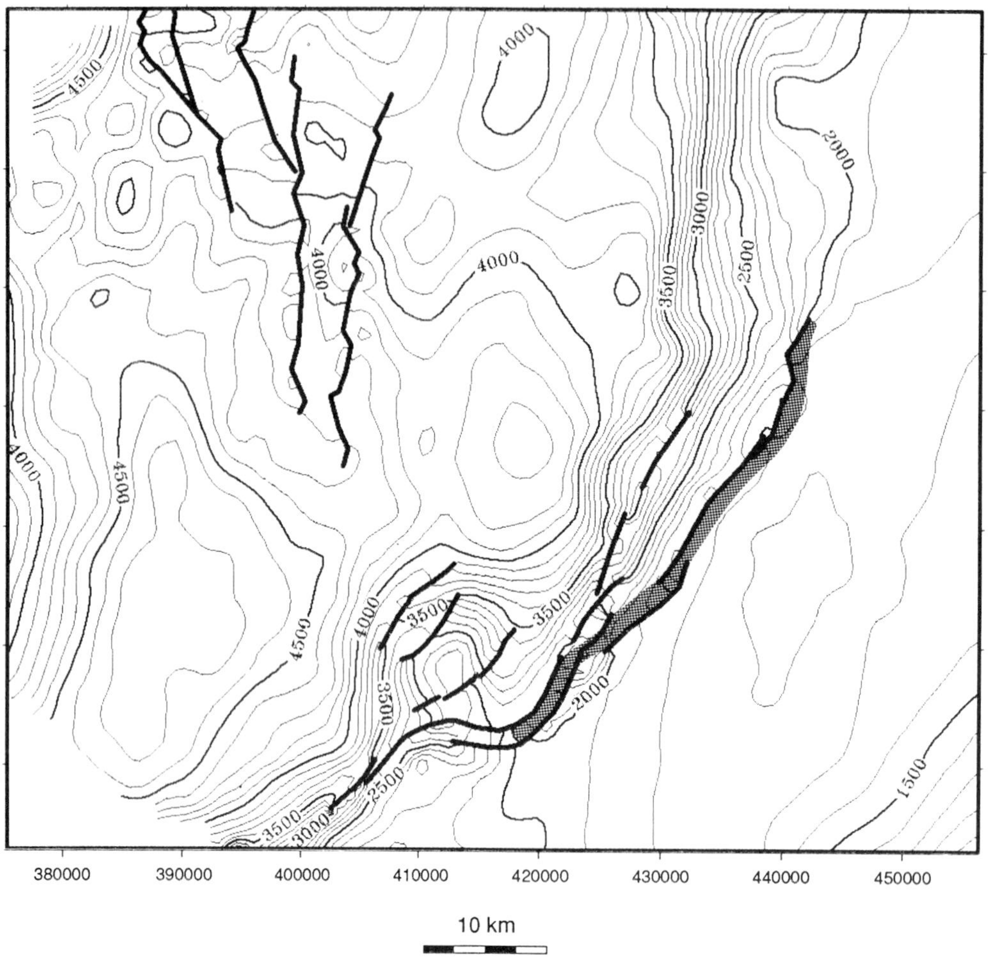

Fig. 9. Present-day depth map used for simulation run IV of the base Cretaceous/Upper Jurassic marker horizon in the Haltenbanken study area, showing a lateral migration barrier (hatched zone) introduced along the Vingleya fault-zone in the southeastern part of the study area.

(Tissot and Welte, 1984; Chilingar, 1963) and predicted by theory (Carman, 1937; Berg, 1975) (Fig. 10). The Garn Formation sample data plotted in Fig. 6c tend to support this view, although combined porosity/permeability measurements were available only at shallow (Draugen; 1650 m) and deep levels (below 3750 m), while according to Fig. 6b the medium-depth samples probably would have tended more toward the centre in the porosity/permeability plot. In view of these uncertainties, the simple linear-log porosity/permeability relationship was still used in the SECMIG runs. For the present simulations it was made to fit the values (0.30, 1000 mD) of Draugen and (0.07, 0.01 mD) in the deepest part of the study area in runs I and II, and the values (0.28, 4000 mD) and (0.12, 2.5 mD) in runs III and IV. To save computer time, permeability was restricted to a maximum of 1 Darcy. Figure 10 offers a closer look at the chosen trend in comparison with other porosity/permeability data sets and correlations predicted by theory.

The permeability estimate remains an uncertain factor in the analysis. Because a range of several orders of magnitude is involved, a sensitivity analysis of this parameter is generally recommended.

Results and discussion

The simulation results are displayed as rate curves and migration maps. Figure 11 shows the hydrocarbon budget for the study area as obtained from the hydrocarbon generation calculation. Figures 12–15 show the SECMIG results of runs I–IV for a number of past and present time steps. Figure 16 is an enlargement of the present-day stage of run III (Fig. 14d) with (re)migration paths indicated. The calculated outflow of oil from the map area as a function of time is sketched in Fig. 17.

In the following sections, five major effects will be discussed.

Oil migration velocity and implications

An appreciation of the hydrocarbon transport rate in the present SECMIG runs may be ob-

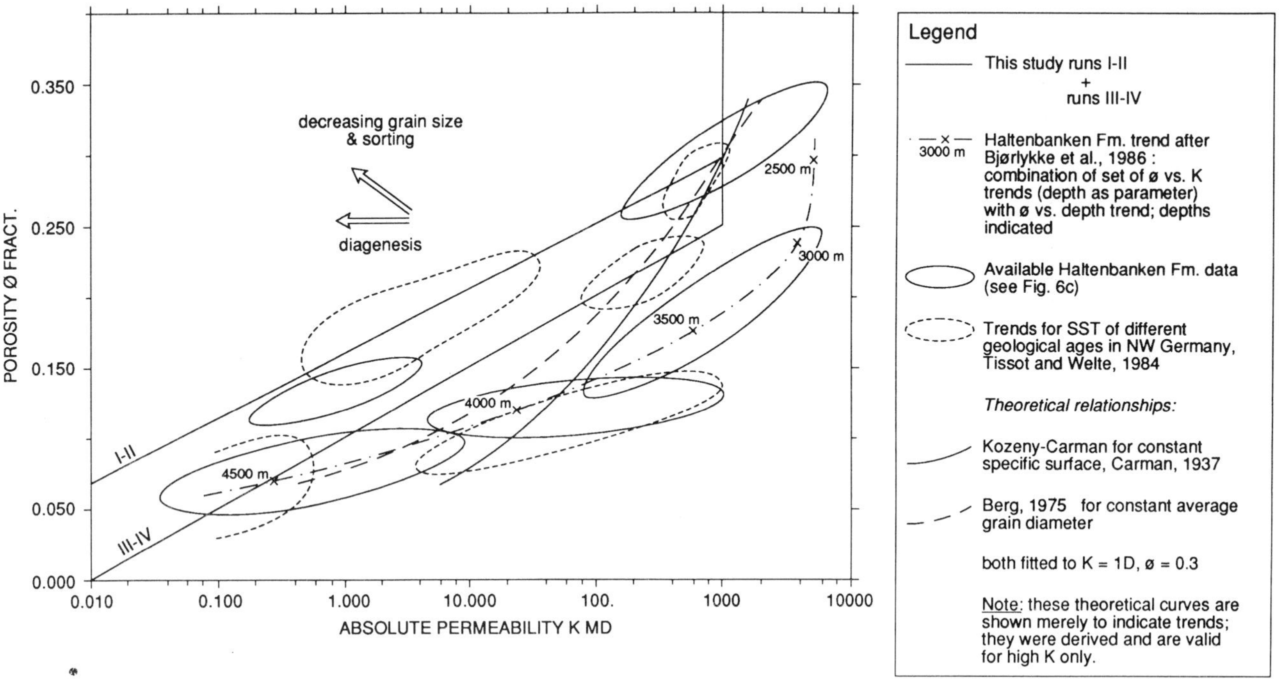

Fig. 10. Carrier-bed permeability/porosity trend used in this study, compared with other data sets and with correlations provided by theory.

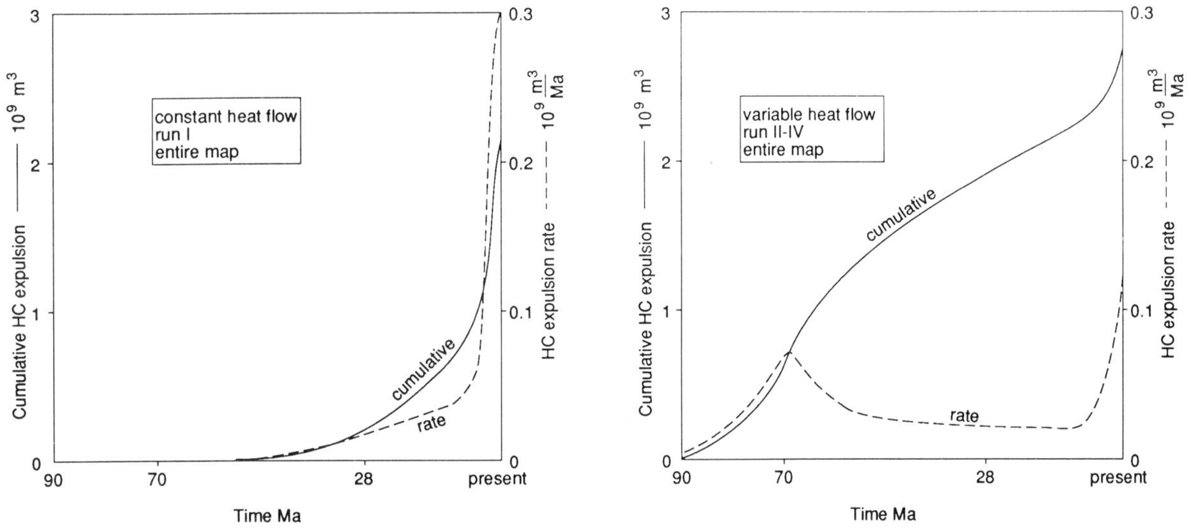

Fig. 11. Hydrocarbon budget of the Haltenbanken study area.

tained by considering the simple example of plane, steady-state up-structure fluid flow, with a stationary hydrocarbon/water interface parallel to the cap-rock boundary. In this case application of Darcy's law, combined with the assumption of a sharp oil/water interface and disregarding the resistant force of the water phase (the water remaining stationary), yields the oil flow rate per unit width (dimension L^3/LT):

$$q_{\mathrm{migr}} = \Delta\rho g \left(\frac{K_{\mathrm{o}}}{\mu_{\mathrm{o}}}\right) h \sin\alpha \cos\alpha \qquad (1)$$

where $\Delta\rho$ = water–hydrocarbon fluid density difference, g = acceleration due to gravity, K_{o} = carrier-bed permeability relative to the hydrocarbon phase, μ_{o} = hydrocarbon fluid viscosity, h = vertical hydrocarbon column height below the cap rock, α = cap-rock dip angle.

In the steady state the hydrocarbon transport rate equals the hydrocarbon supply rate (per unit width). Equation 1, therefore, immediately shows the relationship between supply (expulsion) rate, hydrocarbon column height and the physical parameters for the conditions outlined here.

The value of q_{migr} corresponds to a mean fluid-particle transport velocity (dimension L/T):

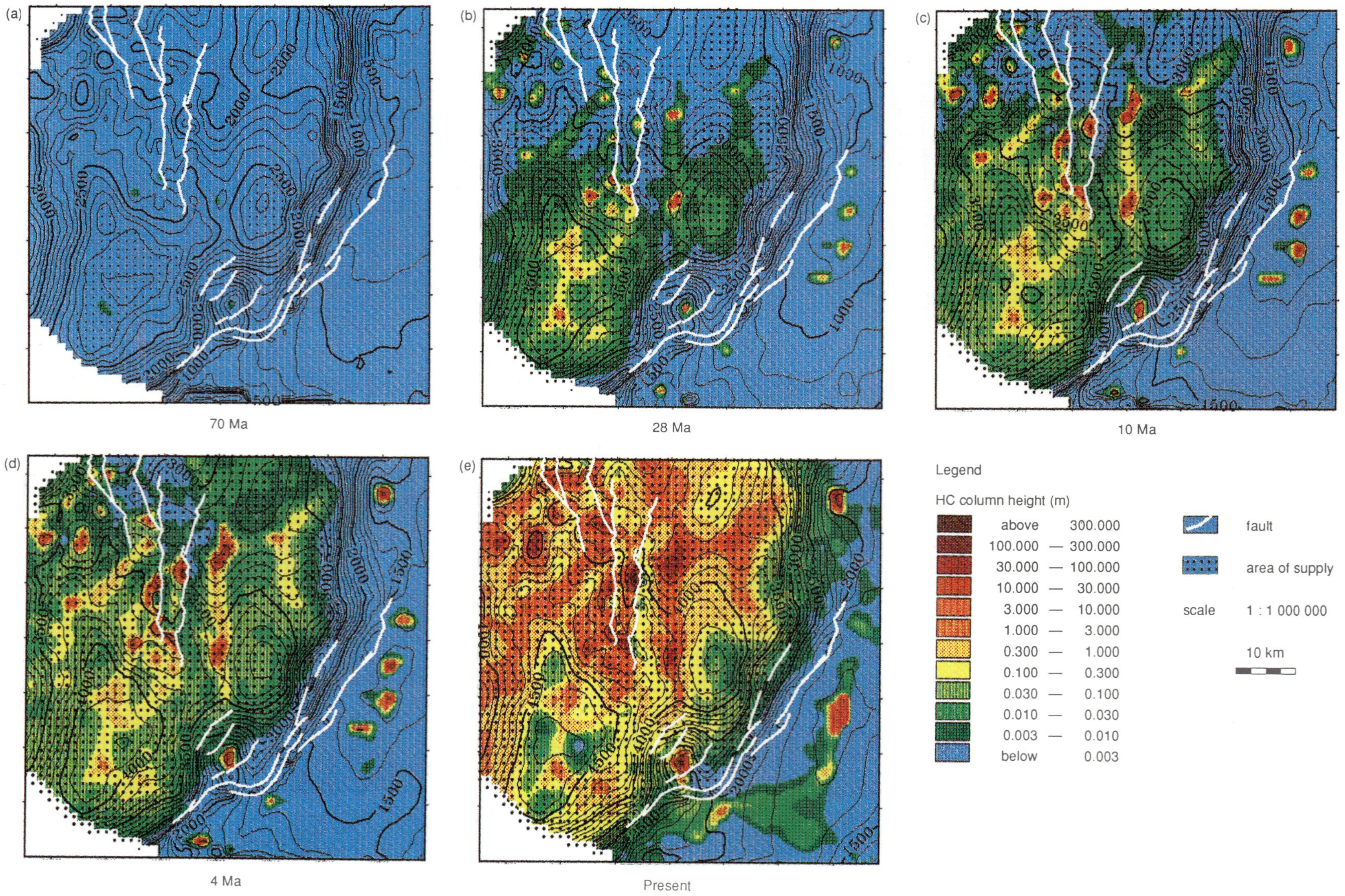

Fig. 12. Modelled secondary hydrocarbon migration in the Upper-Middle Jurassic in the Haltenbanken area, Norway. Run I: constant sediment heat flow. Physical properties are shown in Fig. 6.

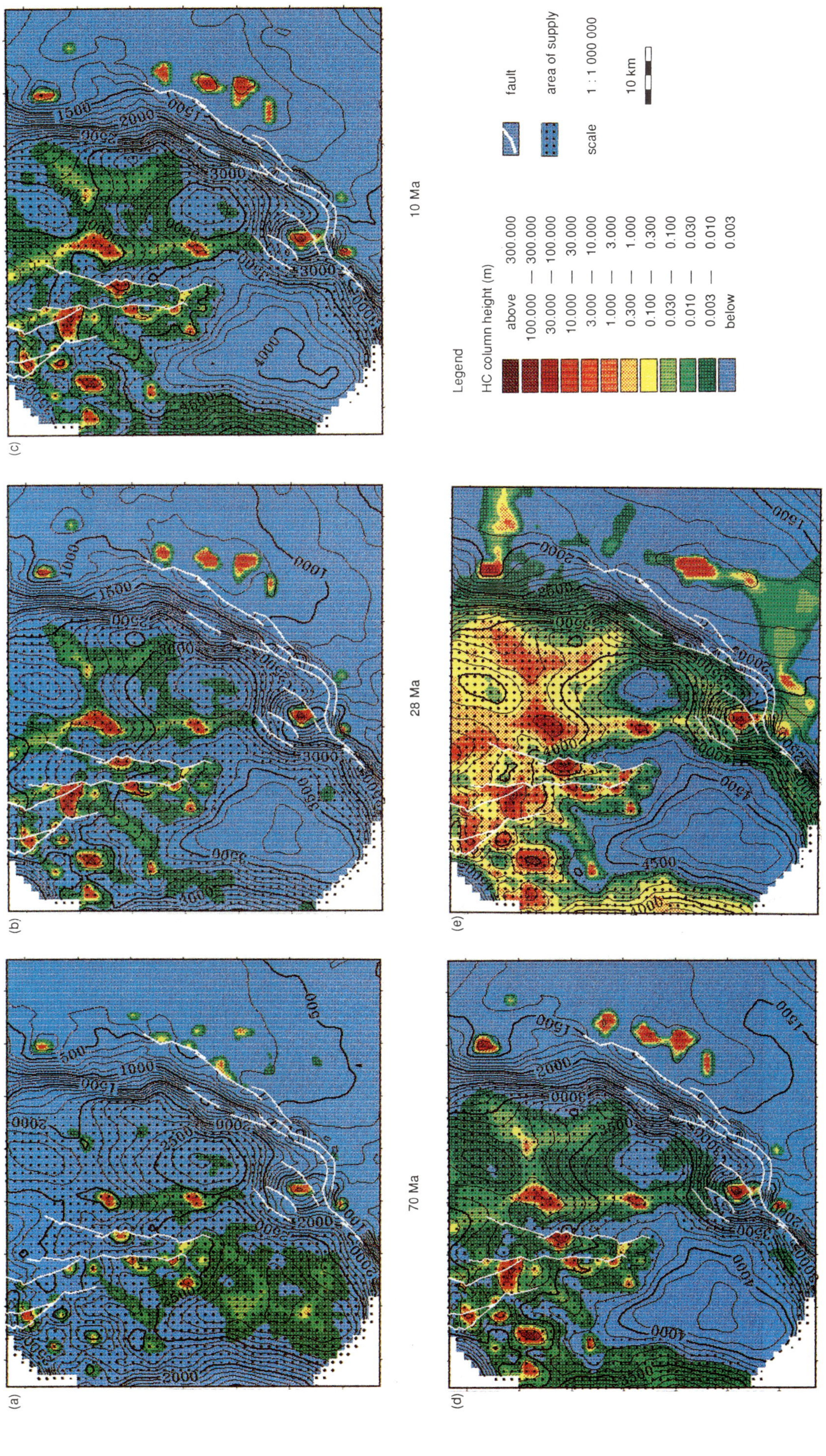

Fig. 13. Modelled secondary hydrocarbon migration in the Upper-Middle Jurassic in the Haltenbanken area, Norway. Run II: same input as in I, except for heat-flow variable in space and time. Physical properties are shown in Fig. 6.

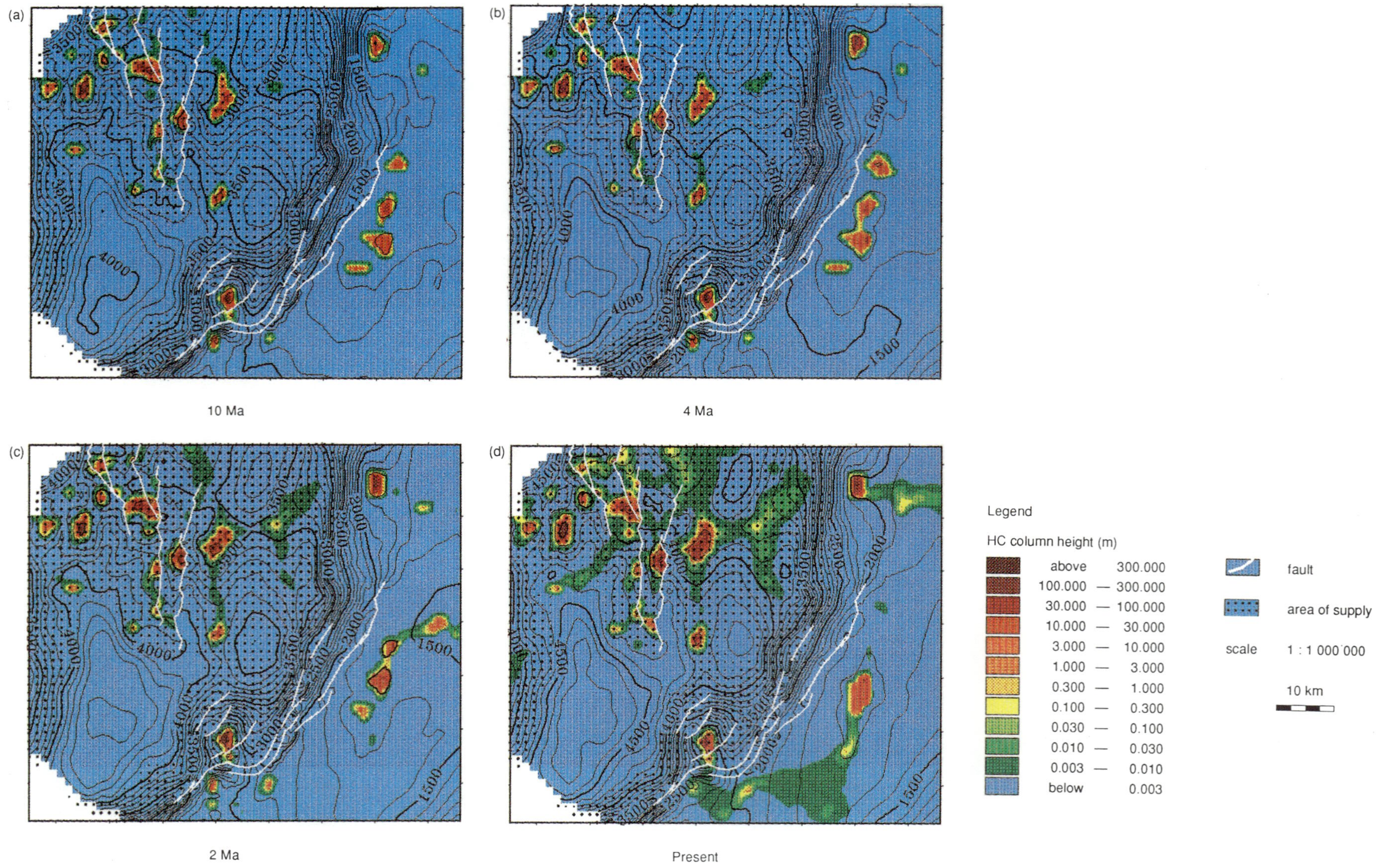

Fig. 14. Modelled secondary hydrocarbon migration in the Upper-Middle Jurassic in the Haltenbanken area, Norway. Run III: same input as in II, but with a gentler decline of carrier-bed porosity and permeability with depth. Physical properties are shown in Fig. 6.

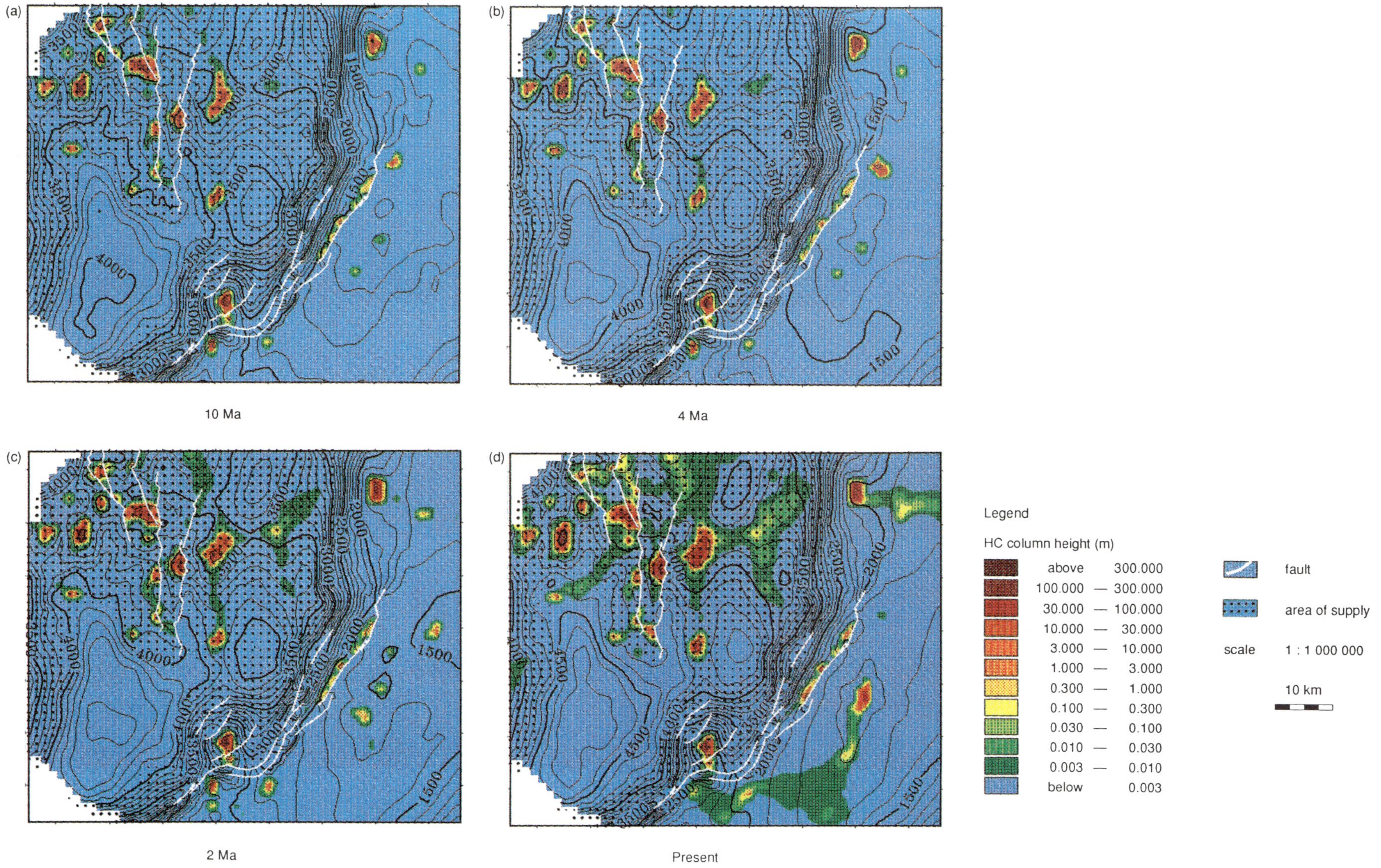

Fig. 15. Modelled secondary hydrocarbon migration in the Upper-Middle Jurassic in the Haltenbanken area, Norway. Run IV: same input as in III, with a lateral migration barrier introduced into the map (fault zone west of Draugen). Physical properties are shown in Fig. 6.

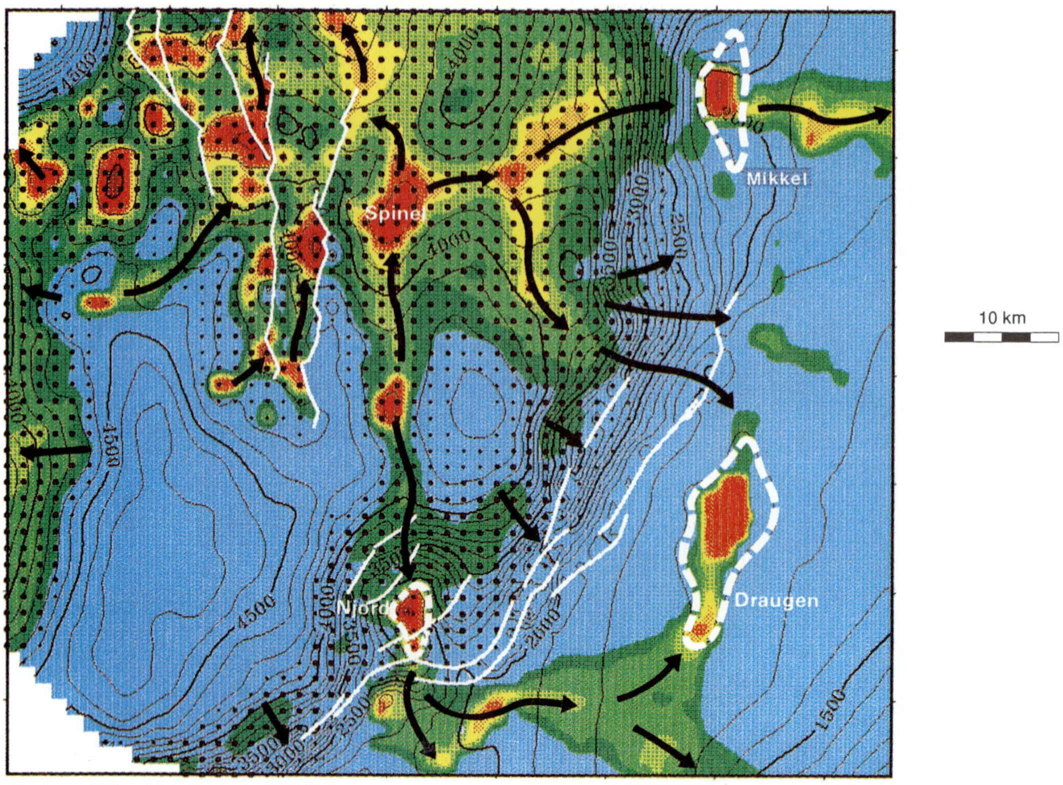

Fig. 16. Enlarged display of Fig. 14d (scale 1 : 700 000), showing the computed present-day hydrocarbon distribution in the Upper-Middle Jurassic carrier system underneath the Spekk Formation source rock in Haltenbanken, Norway. The carrier-bed was assumed to be continuous. Arrows indicate (re)migration pathways.

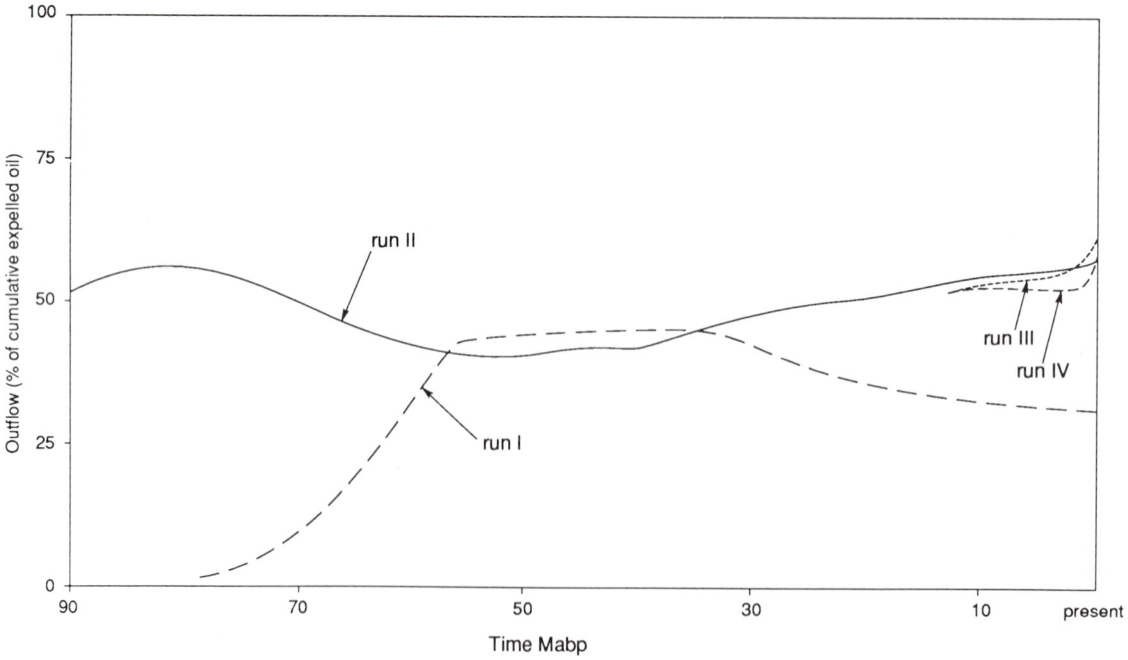

Fig. 17. Hydrocarbon outflow from the Haltenbanken study map area as a function of time. SECMIG runs I–IV.

$$v_{\mathrm{migr}} = \frac{q_{\mathrm{migr}}}{\phi(1 - S_{\mathrm{wc}})h\cos\alpha} = \frac{\Delta\rho g K_{\mathrm{o}} \sin\alpha}{\mu_{\mathrm{o}}\phi(1 - S_{\mathrm{wc}})} \tag{2}$$

where S_{wc} = carrier-bed connate water saturation (assumed 0 in our simulations), and ϕ = fractional carrier-bed porosity.

As a result of the light oil and favourable permeability encountered, the linear oil-migration velocities in the Haltenbanken area are high in comparison with the duration of other geological processes. Substitution of the Haltenbanken parameter values in eqn. 2 reveals lateral migration velocities of the order of 0.1 m/a (runs I and II) and 1 m/a (runs III and IV) at 4000 m depth, increasing to 100 m/a at 1600 m depth for a 4° cap-rock dip angle, the range being caused mainly by the decrease in carrier-bed permeability with depth. Consequently, the hydrocarbon column heights as depicted in the SECMIG output maps will adjust almost instantaneously in response to changes in the input parameters (supply, geometry, physical parameters). That is, transient times will be short. Inherent in the high oil-migration velocity are the typically small hydrocarbon column heights encountered on the migration routes between the accumulations, notably at shallow depth.

Heat-flow history

In the Haltenbanken area with constantly increasing burial with time (for an example burial graph see Fig. 18), present-day depths and temperatures will dominate the present-day cumulative amounts of oil generated. However, the timing of the onset of oil generation will be affected by the heat-flow history. The SECMIG output of run I (Fig. 12) compared with run II (Fig. 13) and the area hydrocarbon budget in Fig. 11 nicely show the impact of heat-flow history on the timing of generation and its effect upon migration: as a result of the elevated palaeo-heat flow in the variable heat-flow case (Fig. 7), both the onset and major phase of generation are shifted backwards in time. This is reflected in earlier trap fill and, where applicable, spill-over and remigration (Bora, Njord, structures in the northwest of the area). The somewhat higher present-day cumulative hydrocarbon generation calculated for the variable heat-flow case (see Fig. 11) will have added to the extent of spill-over and re-migration in the most recent time stages.

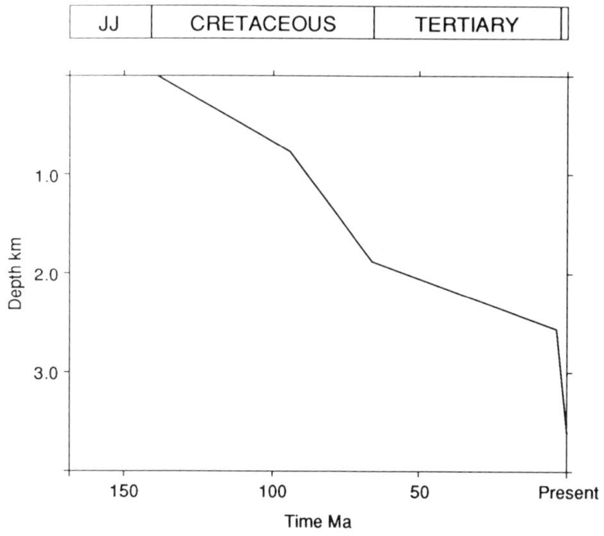

Fig. 18. Burial history at base Cretaceous marker level typical for the Haltenbanken study area.

However, the present-day generation *rate* is much higher in the constant heat-flow run. This results in typically higher (transient) migrating oil columns, particularly in low-permeability (deep) areas in which generation or (re)migration is still taking place, since in these areas the linear migration velocity is low (see eqn. 1). Owing to this transient effect in the constant heat-flow run, less outflow has occurred and more oil is present within the map area in the present-day situation. However, the logarithmic oil column height colour or grey-tone scale tends to exaggerate this difference in oil content visually.

The effect of generation/expulsion timing on the present-day distribution of the hydrocarbons is less significant in these simulations than might be expected, because of the rather uniform subsidence with little palaeo-tilting of the strata.

Permeability/depth trend

In SECMIG the carrier-bed permeability/depth trend is obtained by combining porosity/depth and permeability/porosity relationships, calibrated with measured data (Fig. 6). The carrier-bed porosity range is small and its effect upon migration is restricted. The permeability trend, however, has much more influence, its range covering several orders of magnitude. The drastic permeability reduction with depth translates into decreasing hydrocarbon migration velocities and, hence, increasing migrating oil column heights at a given supply rate (Figs. 12–16). This effect is much too pronounced to be compensated by the decreasing oil viscosity with temperature (depth).

In runs III and IV a different porosity/depth trend is used. The resulting difference in permeability, compared with the other two runs, rises to more than one order of magnitude at a depth of 4000 m, the result of which can be seen by comparing the final stages of run II and run III. In this respect, the relative timing of hydrocarbon generation and of reduction in carrier-bed permeability play key roles in retarding migration.

Map hydrocarbon outflow

The flux boundary condition used in SECMIG will only allow for the outflow of hydrocarbons from the map area and may, consequently, result in a conservative estimate of the amount of hydrocarbons retained, since supply from any kitchens outside the map area is not taken into account. Whenever possible, the boundary should, therefore, be chosen such that it includes the entire kitchen area of the prospects of interest, taking into account the effects of re-migration. In running the program, one may record the outflow across any boundary segment as a function of time.

As shown in Fig. 17, in the Haltenbanken simulations roughly half of the expelled oil has migrated out of the study area by the end of the simulation. Run I shows by far the lowest total outflow (31% of the oil expelled), as a result of the transient effect described in the section on heat-flow history. Clearly, the combination of a low carrier-bed permeability combined with a much higher recent oil supply rate (invoked by the different heat-flow history in run I) is the main factor influencing outflow across the map boundary in this series of simulations. In runs II–IV around 60% of the oil supplied has flowed out, with only minor differences. Not surprisingly, the sealing fault zone in run IV inhibits migration towards the right edge of the map, resulting in a somewhat lower overall oil outflow than that in run III.

Hydrocarbon accumulation and re-migration

The map in Fig. 16 shows most structures filled to spilling point. In practice, added complications of gas, complex stratigraphy and leakage along faults, not included in the model, mean that the results have to be treated with caution. The one structure modelled as being only partially filled is Draugen. This field is known, however, to be filled to spilling point. Three possible explanations for this discrepancy are evident:

(a) the palaeo-structure maps are incorrect; in this context a partial structure opening of Draugen during the simulations from 3 to 1.5 Ma B.P. is indicated (Figs. 14c, 15c);

(b) the recent supply rate from the Draugen kitchen area is underestimated; additional runs not shown in this article have ascertained that a forty-fold increase in the recent supply rate would be required to fill the structure if the palaeo-structure maps were correct;

(c) the migration is longer-range than assumed; the Draugen kitchen extends across our map boundary and there is a hydrocarbon contribution from outside the study area.

Cause (a) definitely has influenced the present-day result, suggesting in view of (b) that the partial opening of Draugen did not actually occur. A higher supply rate might, nevertheless, be realistic, since the present estimate of the expulsion/generation ratio of 0.1 — to compensate for the presence of the Melke Formation and losses in the overlying Cretaceous sequences — is very low, and the potential supply from outside the map area, notably the south, could be substantial.

An additional remark to be made in this respect is that owing to limited interpreted seismic data available at the time the depth maps were prepared for this study, the accuracy in the northeastern and eastern part of the map is indeed low. This is particularly critical for low-relief structures such as Draugen, where a few tens of metres difference in depth can have drastic effects.

Of particular interest are the results of run IV, which suggest that a substantial proportion of the oil present in the Draugen structure originated through re-migration from Njord. Moreover, this re-migration may well be responsible for the observed distributions of lighter (gas) and heavier hydrocarbons over the Draugen and Njord fields, as generally discussed by Gussow in 1954.

Conclusions

The results of a first field-scale application of a secondary hydrocarbon migration simulator to an 82.5 x 72 km^2 block in the Haltenbanken area, offshore Mid-Norway, are reported. Limitations imposed by the current version of the simulator and poor constraints on some of the geological and geochemical input parameters (source-rock data in the deeper-buried regions, fault properties, carrier-bed continuity) inherent in any exploration study compelled the use of a grossly simplified model in this study of the Haltenbanken area. The hydrocarbon charge, a representative light oil, was assumed to derive solely from the Spekk Formation (Kimmeridge Clay equivalent). The oil was assumed to have been expelled downward into a single carrier bed (the Garn Formation), which was assumed to be homogeneous and continuous over the entire area. Faults were modelled as simple steps in the carrier bed. Neither leakage through cap rocks and along faults nor losses due to residual oil were considered and this, combined with the abundance of generating source rock within the study area, resulted in solutions with most of the structures presently filled to spilling point. Although Draugen, the shallowest trap in the migration chain, is actually filled to spilling point, our simulations do not show that this is the case. This may mean that the estimates of expulsion efficiency or migration distance in this study are too conservative.

It is realised that this picture would have been modified substantially if source-rock quality and distribution maps had been available, and if allowance had been made for the complications of leakage, space- and time-dependent carrier-bed parameters and realistic hydrocarbon phase behaviour. Nevertheless, in essential points the modelled present-day distribution of oil column heights and migration routes is found to agree with available field data. While the model clearly is an oversimplification of the actual situation, even in its present form, it has proven its usefulness for assessing the effects of essential migration-system parameters (such as source-rock expulsion rate, carrier-bed permeability, oil viscosity and accuracy of palaeo-structure maps) on migration and accumulation patterns.

References

Archie, G.E., 1950. Introduction to petrophysics of reservoir rocks. Am. Assoc. Pet. Geol., Bull., 34: 943–961.

Bear, J., 1972. Dynamics of Fluids in Porous Media, Elsevier, New York, N.Y., 764 pp.

Berg, R.R., 1975. Capillary pressures in stratigraphic traps. Am. Assoc. Pet. Geol., Bull., 59: 939–956.

Bjørlykke, K., Aagaard, P., Dypvik, H., Hastings, D.S. and Harper, A.S., 1986. Diagenesis and reservoir properties of Jurassic sandstones from the Haltenbanken area, offshore mid Norway. In: A.M. Spencer (Editor), Habitat of Hydrocarbons on the Norwegian Continental Shelf. Norwegian Petroleum Society, Graham and Trotman, London, pp. 275–286.

Boussinesq, J., 1904. Recherches théoretiques sur l'écoulement des nappes d'eau infiltrées dans le sol et sur le débit des sources. J. Math. Pures Appl., 10: 5–78 and 363–394.

Bukovics, C. and Ziegler, P.A., 1985. Tectonic development of the Mid-Norway continental margin. Mar. Pet. Geol., 2: 2–22.

Carman, P.C., 1937. Fluid flow through granular beds. Inst. Chem. Eng. Trans., 15: 150.

Chilingar, G.V., 1963. Relationship between porosity, permeability, and grain size distribution of sands and sandstones. In: L.M.J.U. van Straaten (Editor), Deltaic and Shallow Marine Deposits. Developments in Sedimentology, Vol. 1. Proceedings of the International Sedimentology Congress, Amsterdam–Antwerp. Elsevier, Amsterdam, pp. 71–75.

Cohen, M.J. and Dunn, M.E., 1986. The hydrocarbon habitat of the Haltenbank-Traenabank area, offshore Mid-Norway. In: J. Brooks and K.W. Glennie (Editors), Proceedings of the 3rd Conference on Petroleum Geology of NW Europe, Vol. 2. Graham and Trotman, London, pp. 1091–1104.

Dembicki, H., Jr. and Anderson, M.J., 1989. Secondary migration of oil: experiments supporting efficient movement of separate, buoyant oil phase along limited conduits. Am. Assoc. Pet. Geol., Bull., 73: 1018–1021.

Dodson, C.R. and Standing, M.B., 1944. Pressure volume temperature and solubility relations for natural gas-water mixtures. Drill. Prod. Prac., API, pp. 173–179.

Du Rouchet, J.H., 1981. Stress fields, a key to oil migration. Am. Assoc. Pet. Geol., Bull., 65: 74–85.

Ellenor, D.W. and Mozetic, A., 1986. The Draugen oil discovery, In: A.M. Spencer (Editor), Habitat of Hydrocarbons on the Norwegian Continental Shelf. Norwegian Petroleum Society, Graham and Trotman, London, pp. 313–316.

England, W.A., Mackenzie, A.S., Mann, D.M. and Quigley, T.M., 1987. The movement and entrapment of petroleum fluids in the subsurface. J. Geol. Soc. London, 144: 327–347.

Füchtbauer, H., 1967. Influence of different types of diagenesis on sandstone porosity. Proc. 7th World Pet. Congr., 2: 353–369.

Gussow, W.C., 1954. Differential entrapment of oil and gas: a fundamental principle, Am. Assoc. Pet. Geol., Bull., 38: 816–853.

Heum, O.R., Dalland, A. and Meisingset, K.K., 1986. Habitat of hydrocarbons at Haltenbanken (PVT-modelling as a predictive tool in hydrocarbon exploration). In: A.M. Spencer (Editor), Habitat of Hydrocarbons on the Norwegian Continental Shelf. Norwegian Petroleum Society, Graham and Trotman, London, pp. 259–274.

Hobson, G.D., 1954. Some Fundamentals of Petroleum Geology. Oxford University Press, London, 139 pp.

Hvoslef, S., Larter, S.R. and Leythaeuser, D., 1988. Aspects of generation and migration of hydrocarbons from coal-bearing strata of the Hitra Formation, Haltenbanken area, offshore Norway. In: L. Mattavelli and L. Novelli (Editors), Advances in Organic Geochemistry 1987. Org. Geochem., 13: 525–536.

Jones, R.W., 1981. Some mass balance and geological constraints on migration mechanisms. Am. Assoc. Pet. Geol., Bull., 65: 103–122.

Klomp, U.C. and Wright, P.A., 1989. A new method for the measurement of kinetic parameters of hydrocarbon generation from source rocks. In: B. Durand and F. Behar (Editors), Advances in Organic Geochemistry 1989. Org. Geochem., 16: 49–60.

Krooss, B.M. and Leythaeuser, D., 1988. Experimental measurements of the diffusion parameters of light hydrocarbons in water-saturated sedimentary rocks — II. Results and geochemical significance. Org. Geochem., 12: 91–108.

Krooss, B.M., Leythaeuser, D. and Schaefer, R.G., 1988. Light hydrocarbon diffusion in a caprock. Chem. Geol., 71: 65–76.

Latil, M., 1980. Enhanced Oil Recovery. Editions Technip, Paris, 236 pp.

Lehner, F.K., Marsal, D., Hermans, L. and Van Kuyk, A., 1987. A model of secondary hydrocarbon migration as a buoyancy-driven separate-phase flow. In: B. Doligez (Editor), Migration of Hydrocarbons in Sedimentary Basins. Editions Technip, Paris, pp. 457–471. (Reprinted in: Rev. Inst. Fr. Pétrole, 1988, 43: 155–164.)

McAuliffe, C.D., 1979. Oil and gas migration — chemical and physical constraints. Am. Assoc. Pet. Geol., Bull., 63: 761–781.

McKenzie, D., 1978. Some remarks on the development of sedimentary basins. Earth Planet. Sci. Lett., 40: 25–32.

Neglia, S., 1979. Migration of fluids in sedimentary basins. Am. Assoc. Pet. Geol., Bull., 63: 573–597.

Pedersen, O.E., 1988. Norwegian offshore fields — reserves and production. Scand. Oil Gas Mag., 16: 25–30.

Price, L.C., 1976. Aqueous solubility of petroleum as applied to its origin and primary migration, Am. Assoc. Pet. Geol., Bull., 60: 213–244.

Price, L.C., 1981, Aqueous solubility of crude oil to 400°C and 2000 bars pressure in the presence of gas. J. Pet. Geol., 4: 195–223.

Schowalter, T.T., 1979. Mechanics of secondary hydrocarbon migration and entrapment. Am. Assoc. Pet. Geol., Bull., 63: 723–760.

Tissot, B.P. and Welte, D.H., 1984. Petroleum Formation and Occurrence. Springer-Verlag, Berlin, 2nd ed., 699 pp.

Ungerer, P., Bessis, F., Chenet, P.Y., Durand, B., Nogaret, E., Chiarelli, A., Oudin, J.L. and Perrin, J.F., 1984. Geological and geochemical models in oil exploration. In: G. Demaison and R.J. Murris (Editors), Petroleum Geochemistry and Basin Evaluation. Am. Assoc. Pet. Geol., Mem., 35: 53–77.

Wooding, R.A. and Chapman, T.G., 1966. Ground water flow over a sloping impermeable layer, 1. Application of the Dupuit-Forchheimer assumption. J. Geophys. Res., 71: 2895–2902.

L. HERMANS *Koninklijke/Shell Exploratie en Produktie Laboratorium, Volmerlaan 6, 2288 GD Rijswijk, The Netherlands*
A.D. van KUYK *Koninklijke/Shell Exploratie en Produktie Laboratorium, Volmerlaan 6, 2288 GD Rijswijk, The Netherlands*
F.K. LEHNER *Koninklijke/Shell Exploratie en Produktie Laboratorium, Volmerlaan 6, 2288 GD Rijswijk, The Netherlands*
P.S. FEATHERSTONE *Koninklijke/Shell Exploratie en Produktie Laboratorium, Volmerlaan 6, 2288 GD Rijswijk, The Netherlands*

Faulting processes and fault seal

R.J. Knipe

The understanding of fault seal generation is crucial to the assessment of the migration and trapping of hydrocarbons. The effectiveness of the seal depends upon the porosity and permeability characteristics of the fault zone, which is controlled by the microfabrics present. The range of processes which effect the microfabric evolution and the development of seals in fault zones are reviewed. The examples presented illustrate the use of microstructural analysis for the characterisation of seal properties. The importance of cyclic deformation in fault zones, where different deformation mechanisms operate during different parts of these displacement events, and the role of these mechanisms in producing different types of seal is emphasized. Three broad classes of seal may be recognised: (1) *collapse seals*, where the permeability/porosity of the fault zone is reduced by pore volume decrease achieved by grain reorganisation produced by fracturing, grain deformation and dissolution and by grain boundary sliding; (2) *cement seals*, where the reduced flow across a fault is achieved by the precipitation of phases in and adjacent to the fault; (3) *juxtaposition seals*, generated be the juxtaposition of lithologies with different sealing capacities. All these seal types can contribute to the sealing properties of a single fault and their distribution can be related to the larger fault zone geometry and displacement pattern. Cementation patterns and histories in fault zones can be complex, but analysis of the deformation microstructures and cement sequences in fault zones provides crucial data on the reservoir evolution. This data also helps constrain the conditions, timing and composition of fluid flow events associated with different fault zones. Such analysis also provides important information on the larger scale basin drainage patterns. The sealing capacity of different seals is calculated and the stability of different types of seal during the subsequent geological history is discussed. The examples presented highlight the need for future programmes to combine microstructural studies of fault seal with the analysis of the larger scale fault geometry and the diagenetic evolution in fault zones and reservoirs. Such an integration is needed to understand and quantify the interaction between reservoir development and faulting.

Introduction

Faulting forms one of the main controls on the distribution, size and internal structure of hydrocarbon reservoirs. The role of fault zones in providing both a pathway for secondary hydrocarbon migration and a seal to hydrocarbon accumulations has been recognised for a long time (Hubbert, 1953; Smith, 1966 and 1980). Despite the detailed studies conducted, the assessment of the sealing properties of faults remains an important goal for hydrocarbon exploration and production. Previous studies of fault seal have concentrated on the properties of the reservoir rocks and barrier lithologies juxtaposed by the faulting (Smith, 1966 and 1980; Berg, 1975; Schowalter, 1979; Watts, 1987) or on the structural style or setting (Harding, 1974; Harding and Tuminas, 1988, 1989). Very little attention has been paid to the processes which produce fault rocks and the possible influence of these on sealing properties. This contrasts with the amount of research aimed at assessing the role of fractures in producing high permeability reservoirs (Nelson, 1985; Wilke et al., 1985; Charlaix et al., 1987; Gueguen et al., 1987). Microstructural features such as the pore geometry are known to be crucial to sealing properties, but there are few studies (e.g., Pittman, 1981) which have described and assessed the porosity and permeability development of fault zones in detail.

This paper outlines the microprocesses associated with seal evolution by reviewing the deformation processes which operate in faults and affect the sealing properties of fault zones. This information is used to assess the types of seal which can develop under different conditions, the sealing capacity of faults, the continuity of effective seals and the strength and stability of different seals. The final part of the paper discusses the importance of linking information on the micro-processes of faulting with other aspects of reservoir evolution, particularly the larger scale geometrical development of fault arrays.

Controls on fault seal

Sealing may be defined as the reduction in the flow potential or permeability across a barrier provided by an adjacent medium juxtaposed either stratigraphically or by faulting. For sealing of hydrocarbons

Structural and Tectonic Modelling and its Application to Petroleum Geology edited by R.M. Larsen, H. Brekke, B.T. Larsen and E. Talleraas. NPF Special Publication 1, pp. 325–342. Elsevier, Amsterdam.

to occur, the buoyancy forces associated with the difference between the density of water and hydrocarbons must be matched or exceeded by the capillary pressure or displacement pressure of the seal (Hubbert, 1953; Smith, 1966 and 1980; Berg, 1975; Schowalter, 1979; Watts, 1987). The height of a hydrocarbon column which can be supported at a sealing medium depends upon (1) the difference between the displacement pressure characteristics of the sealing medium and the reservoir, (2) the difference between the density of water and the oil under consideration, and (3) the hydrodynamic or hydrostatic conditions (Fig. 1). A number of authors have provided analyses of the relationships between these variables and the sealing capacity or height of the associated hydrocarbon column (Smith, 1966 and 1980; Berg, 1975; Schowalter, 1979; Watts, 1987). In most cases the approach has been to measure the capillary pressure of core material in the laboratory and to use this information in conjunction with data on the hydrocarbon and water properties to calculate the heights of hydrocarbon columns which may be present at depth. A number of equations have been presented to aid the estimation of the trapped column height (Z). Watts (1987), for example, based on the analysis of Smith (1966) used the following relationship for hydrostatic situations:

$$Z = \frac{Pd_m I_h}{(P_w - P_h)\, g I_m \cos\phi} \qquad (1)$$

where Pd_m is the mercury/air entry (or displacement) pressure, I_h is the hydrocarbon-formation water interfacial tension, P_w and P_h are the densities of the formation water and hydrocarbon phase, g is the acceleration due to gravity, I_m is the mercury-air interfacial tension, and ϕ is the contact angle between mercury and air.

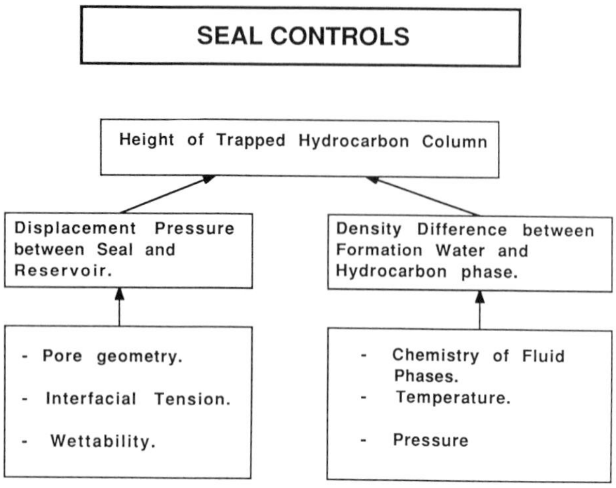

Fig. 1. Summary of controls of fault sealing.

The displacement pressure, d_p, or the mean injection pressure is that which causes displacement of the major part of the wetting fluid and can be defined as (Berg, 1975, eqn. 8):

$$d_p = \frac{(2I\cos\phi)}{r} \qquad (2)$$

where r is a radius of curvature determined by the pore size of the rock.

Because the displacement pressure is dependent upon the pore geometry factor, r, it is also possible to present an equation for the height of trapped hydrocarbons in the hydrostatic situation in terms of the pore geometry, e.g., Berg (1975, eqn. 6):

$$Z = \frac{2I_h(r_t^{-1} - r_p^{-1})}{g(P_w - P_h)} \qquad (3)$$

where r_t is the pore throat radius in the seal and r_p is the average pore radius in the reservoir.

It is clear from these equations that the parameters r_t and r_p are fundamental to the sealing process and are controlled by the texture of the seal whether it is a stratigraphic unit or a fault zone. The factors which influence the evolution of the grain size, shape and sorting together with the characteristics of the cement phases present are fundamental to the pore size, pore throat size and permeability of the sealing medium and, thus, to the sealing properties. Assessment of the diagenetic evolution of these parameters during burial is common and measurements to quantify these parameters are achieved via laboratory measurement of undeformed samples (e.g., Wardlaw and Cassan, 1978; Bjørlykke et al., 1986). Analysis of the porosity evolution during deformation associated with compaction have also been conducted (e.g., Houseknecht, 1988; Wilson and McBride, 1988). There is a need to expand these studies to include characterisation of fault zone material in order to assess the evolution of fault seals and to provide data which will allow estimation of the sealing capacity of fault zones. Only via an integrated analysis of "fault zone diagenesis" and of the operating deformation mechanisms in fault zones can the seal evolution be fully incorporated into assessment of the reservoir development.

Faulting processes and fault seal

This section presents a review of the deformation mechanisms possible in fault zones which are important to sealing. The aim is to provide examples of the range of microfabrics present in fault zones and to discuss the range of processes which contribute to the evolution of these fabrics. The information presented is then used to assess the types of seal developed under different faulting conditions and to estimate the sealing capacity of faults. Particular emphasis is placed on the evolution of the porosity, permeability and pore geometry during fault rock evolution.

The deformation mechanisms which can operate in fault zones depend upon the interaction of the lithological characteristics (mineralogy, texture, porosity, permeability) and the environmental conditions (stress level, confining pressure, fluid pressure, temperature, etc.). Deformation mechanisms have been reviewed by Sibson (1977, 1986b), Groshong (1988), Mitra (1988), and Knipe, (1986a, 1989). The mechanisms can be divided into three categories:

(A) *Diffusive mass transfer (DMT)*, where diffusion of material away from sites of high stress produces deformation (Rutter, 1983; Gratier and Guiguet, 1986). The deformation is a three-stage process involving dissolution, diffusion or transportation of dissolved species, and precipitation. The mechanism is favoured by small grain sizes, small grain contacts and in situations where an efficient diffusion path medium is present. Where water acts as the diffusion path the process is termed pressure solution (see Rutter, 1983; and Gratier and Guiguet, 1986 for reviews) and is important in compaction, cementation and lithification of sediments (e.g., Angevine and Turcotte, 1983; Houseknecht, 1984, 1988; Wilson and McBride, 1988).

(B) *Crystal plasticity* involves the deformation of crystal lattices by intragranular processes such as dislocation motion and twinning. These mechanisms dominate the deformation of minerals under high grade conditions (see Barber, 1985; White, 1985; and Knipe, 1989 for reviews), but are also important to the bending and kinking of phyllosilicates present in sediments and can play an important role in low temperature ductile fracture (see Lloyd and Knipe, 1992).

(C) *Frictional sliding, fracture processes and cataclastic flow*. In this case deformation involves a large group of processes ranging from fracture propagation, frictional sliding on newly created surfaces and independent particulate flow, where deformation is achieved by the frictional sliding of grains without grain breakage. Fracturing involves a large range of processes including brittle fracture and subcritical fracture processes, possibly below the stress levels needed for brittle fracture (see Logan, 1979; Kranz, 1983; Wang, 1986; Tullis, 1986; Atkinson, 1987; Kozak, 1987; and Knipe, 1989 for reviews). Grain boundary sliding or independent particulate flow dominates the flow of unlithified material or material which has been disaggregated by other deformation mechanisms (Borradaile, 1981; Knipe 1986b; Moore et al., 1986; Jones and Preston, 1987; Owen, 1987). Low confining pressures and high fluid pressures favour the operation of these modes of deformation. The permeability and porosity of the aggregate can be transiently increased during the deformation by the dilation which accompanies fracturing and frictional sliding.

Recent research has established the microstructural features which characterise the deformation mechanisms listed above (reviewed in Groshong, 1988; and Knipe, 1989). Although understanding of the details of each process involved in these mechanisms is still incomplete, identification of the type of deformation involved is possible. Because the microstructural changes associated with each mechanism are different, the type of sealing arising from the operation of each also differs. Establishing the deformation mechanisms operating in different faults is therefore critical to the assessment of the development of seals.

The accumulation of displacement in fault zones in the upper crust ranges from the repeated transient, high displacement rate events, i.e., stick-slip behaviour (Frank, 1965; Brace and Byerlee, 1966; Ohnaka, 1987), to continuous displacement or fault creep behaviour (Wesson, 1988). There is a complete spectrum of fault displacement patterns between these two end members. Cyclic changes in deformation mechanisms which accompany faulting are therefore common and evidence for the cyclic alternation of the different mechanisms can be found in the fault rock products (Sibson 1980, 1986a; White and White, 1983; Mawer and Williams, 1985; Knipe, 1986a, b, 1989; Underhill and Woodcock, 1987; Cox and Etheridge, 1989). Fig. 2 presents an example of a deformation mechanism path associated with different fault displacement events. The importance of these cyclic events to the present discussion is that different deformation mechanism paths will produce different types of seal. The stability, strength and effectiveness of the seal depends upon the deformation mechanism paths experienced. In addition the deformation mechanism paths experienced will depend upon the lithological characteristics of the material undergoing faulting, and the conditions of faulting. We can expect different deformation mechanism histories at different depths and at different locations along individual faults because different strain-rate and stress histories are likely in the following situations: (a) where different lithologies are in contact with the fault plane; (b) where different propagation processes operate; (c) where restricted fluid migration pathways cause differences in the cement distribution associated with faults; and (d)

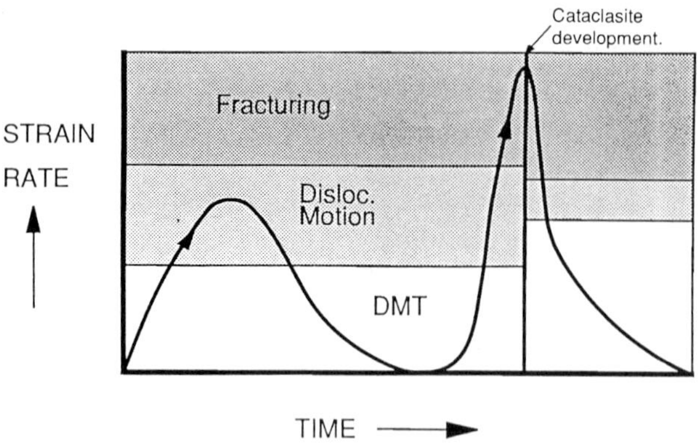

Fig. 2. A deformation mechanism path illustrating the variation in the deformation mechanisms which can dominate different deformation events. Note also the change in the distribution of deformation mechanisms which can arise from changes in the rock fabric which are induced by some deformation events; in this case the DMT (diffusive mass transfer) field is expanded as a result of the grain size reduction which results from cataclasis.

where fault geometry causes different strain histories in and around the fault. Even in situations where the displacement is continuous then important changes in the deformation mechanisms and the fault rock microfabric can be associated with the final stages of displacement or with post-displacement accommodation adjustments in the fault zone. The next sections of this paper outlines the different seals which may be produced by different deformation mechanism histories. The influence of fault geometry is discussed later in the paper.

Seal types

The last section emphasised the cyclic nature of deformation processes which accompany faulting and the likelihood that a combination of processes will contribute to the final sealing properties of the fault zone. This section reviews the different seals which can develop by different deformation mechanism paths by presenting some examples of faults with different microstructures and different sealing characteristics.

Fault seals may be grouped into three broad categories.

(A) *Porosity collapse seals.* Changes in the permeability, porosity and sealing potential within the fault zone are induced by changes in the pore volume, shape and size by grain refinement and rearrangement associated with frictional sliding and/or the cataclasis, dissolution, alteration and plastic deformation of fragments/grains within the fault zone. Smear seals would fall into this category.

(B) *Cement seals.* Changes in the permeability, porosity and sealing potential arising from the growth of new phases within and adjacent to fault zones. Asphalt or tar impregnations are included as cement seals here.

(C) *Juxtaposition seals.* Seals produced by the juxtaposition of two lithologies with different capillary pressures. In this case the seal characteristics are largely independent of the faulting processes and the fault only serves to juxtapose the two lithologies with either different pore geometries and/or fluid pressures (see Hubbert, 1953; Downey, 1984).

Fault seals usually form the lateral seal to a hydrocarbon accumulation where a cap (stratigraphic) seal is present above the reservoir. In addition, fault seals are likely to be dynamic features where there is hydrocarbon flow in and leakage from the reservoir (e.g., Harding and Tuminas, 1988). Juxtaposition seals are the most familiar type of fault seal (Smith, 1966, 1980, Berg, 1975, Downey, 1984; Allen, 1988). However, there is an increasing recognition of the importance of fault seals developed where the juxtaposed lithologies are permeable (Weber et al., 1978; Smith, 1980; Harding and Tuminas, 1989). The sealing properties of a single fault may in part be due to all these factors if the range of conditions and lithologies which are involved in the faulting are large. However, different types of fault rock and therefore characteristic seals are likely to develop in different situations. To illustrate this, the next part of the paper discusses examples of the likely evolution of pore geometry and permeability during faulting in sandstones using examples of the microstructures preserved in such zones. The information obtained from the fault rock microstructure analysis has a twofold usage: (a) information on the grain size, shape and sorting together with the information on the pore geometry can be used to estimate the sealing capacity of the zone; and (b) the texture and mineralogy of the phases present as clasts and cement phases provide important data on the conditions of the fault rock evolution which can be related to the burial history and the diagenetic history of adjacent lithologies. Both these aspects are discussed in more detail after the microstructures and the deformation mechanism histories in the chosen examples have been described.

Figure 3 shows the microstructure of a cataclasite derived from a sandstone. The original grain size (150–175 μm) is reduced by almost two orders of magnitude in the fault zone by cataclasis to fragments with a grain size of 2–7 μm and the final texture retains some (5–10%) porosity. The fault rock is likely to have evolved by repeated deformation (fracturing) events. Transient, high permeability periods are likely to have been separated by low permeability periods between the deformation events (Fig. 4). The

Fig. 3. Secondary electron image of the microstructure of a cataclasite derived from a sandstone. The original grain size (150–175 μm) is reduced in the fault zone by a factor of 20–75 to approximately 2–7 μm. Note the well formed crystal faces present indicating precipitation of overgrowths.

high permeability periods when the fault can act as a fluid migration pathway are considered to represent periods when fluid pressure adjustments take place. The longer the fluid pressure variations or the stress can be maintained at levels which allow large pore throat diameters to persist, the larger the migration potential will be. Further discussion on this aspect is provided in the section on seal continuity below. At the end of each cataclastic deformation event the permeability in the fault zone is reduced by a porosity collapse associated with both the change in size of the pore diameter and pore throat diameter arising from the grain size reduction and from the closing of pores during the drop in fluid pressure. The idealised history show in Fig. 4 shows only a small number of the events which are likely to occur in fault zones during their activity periods (see Watterson, 1986; Sibson, 1989) and also presents an example in which the grain size reduction is progressive. Analysis of the grain size reduction processes in a faulted quartzite by Lloyd and Knipe (1990) demonstrates that grain size reduction can occur during the very early stages of displacement and produce a reduction in >75% of the material in the first few centimetres of displacement. The rate of grain size reduction is important to the sealing potential of the zone, because the porosity of the aggregate will depend upon the grain sorting in the fault zone. However, there are two reasons why the sorting characteristics of the fault zone may be secondary to the pore throat diameter with respect to the sealing potential. Firstly, cataclasites often develop by the progressive increase in the proportion of fine grained cataclastic matrix, where the matrix fragments are often of a restricted size range at least one order of magnitude smaller than the original grains. That is, the grain size evolution in cataclasites is essentially a bimodal one in which the proportion of the fine grained component increases with deformation. Fraser (1953) has illustrated that the permeability in a bimodal mixing of two uniform grain size populations approaches that of the fine grained component when 70–75% of the aggregate is still composed of the coarse grained component. The second reason why sorting may be secondary is the known small dependence of permeability on sorting compared to the influence of sorting on porosity, i.e.,

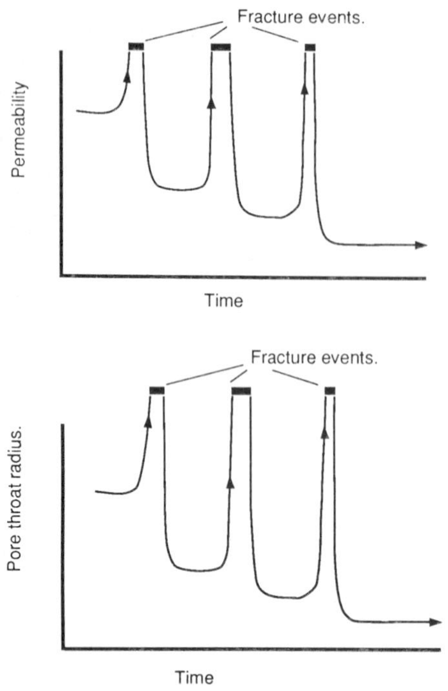

Fig. 4. A possible history of the fault rock shown in Fig. 3 in terms of permeability and pore throat evolution. The fracture events are marked by transient high permeability events which are separated by "rest" periods when the permeability and effective pore throat size is reduced.

poorly sorted material may contain large pores, but these are not interconnected.

The sealing properties in the example shown in Figs. 3 and 4 will be primarily produced by the collapse of the pore volume and of the pore throat size during grain size reduction by cataclasis. The sealing capacity of this type of fault rock will be estimated in a later section.

The second example shown in Fig. 5 illustrates the effect of late stage modification (compaction) of a cataclastic microfabric by diffusive mass transfer (DMT) processes. Such effects emphasise the susceptibility of the fine grained cataclasite to localised grain dissolution. The change to a deformation dominated by DMT may represent the last stages of displacement accommodation in the zone when both the decreasing strain-rate history and the fine grain size favour this deformation process. The result is the production of a compacted microstructure where the porosity is rapidly decreased by both dissolution and re-precipitation of cement. The relative proportion of dissolution and precipitation can vary so that in some cases material is removed but not precipitated in the cataclasite. The decrease in the strain rate and the operation of DMT in a fault zone may be

Fig. 5. Secondary electron image of a compacted cataclasite. The compaction post-dated the grain size reduction in the fault zone and represents the last stages of strain accommodation. Note the interlocking nature of the grains and the reduction in porosity compared with Fig. 3.

Fig. 6. Backscattered electron image of a fault zone where many individual grains were not fractured during the deformation. The deformation is achieved by grain disaggregation and frictional grain boundary sliding within the fault zone (orientated top left to bottom right in the micrograph). Note the reduced porosity in the fault zone which is three or four grains wide.

related to either: (1) a "rest" period between transient faulting events (stick-slip behaviour); (2) the changing pattern of fault activity on a larger scale to one in which during the fault array evolution a new fault takes up the extension and previously formed faults become inactive; or (3) the end of tectonic activity evolution in the area, when activity on all faults decreases. In the first case the compacted cataclasite is likely to be fractured by the subsequent faulting events. The important point in all three cases is that the sealing capacity of a fault can be enhanced by the slow deformation which post-dates the rapid displacement events on the fault and may also be enhanced by the continued burial of an inactive fault.

Figure 6 illustrates the microstructure of a different type of fault zone where many individual grains are not fractured during the deformation. The deformation is achieved primarily by grain boundary fracturing, disaggregation and frictional grain boundary sliding. Such fault zones are likely to occur in partially or weakly lithified material and will be favoured by deformation during high fluid pressures, low confining pressures and at shallow depths of burial. It is interesting to note that lithologies undergoing dissolution of early cements to produce secondary porosity may also be prone to this type of deformation. If this disaggregation takes place, the fault zone may experience a transient increase in porosity and permeability during the faulting, the final post-faulting microstructure may be similar to the pre-faulting structure of grain arrangement. In the simplest case the sealing properties of the rocks should not be affected by the faulting. However, as the example in Fig. 6 illustrates, porosity collapse by some fracturing and grain boundary sliding can accompany the deformation and induce changes in both porosity and permeability. In addition, microstructural analysis of fault zones developed in sandstones which are only partially cemented and have suffered only limited diagenesis reveal that seals can develop in the border zone to the fault. Figure 7 shows an example with collapsed and deformed phyllosilicate aggregates in the pore throats adjacent to the fault zone. It is suggested here that fluid flow pulses associated with faulting may cause collapse and redistribution of the clay phases into the pore throats. This pore throat clogging or "wash-in" seal can occur in the border zones of a fault zone where displacement or strain

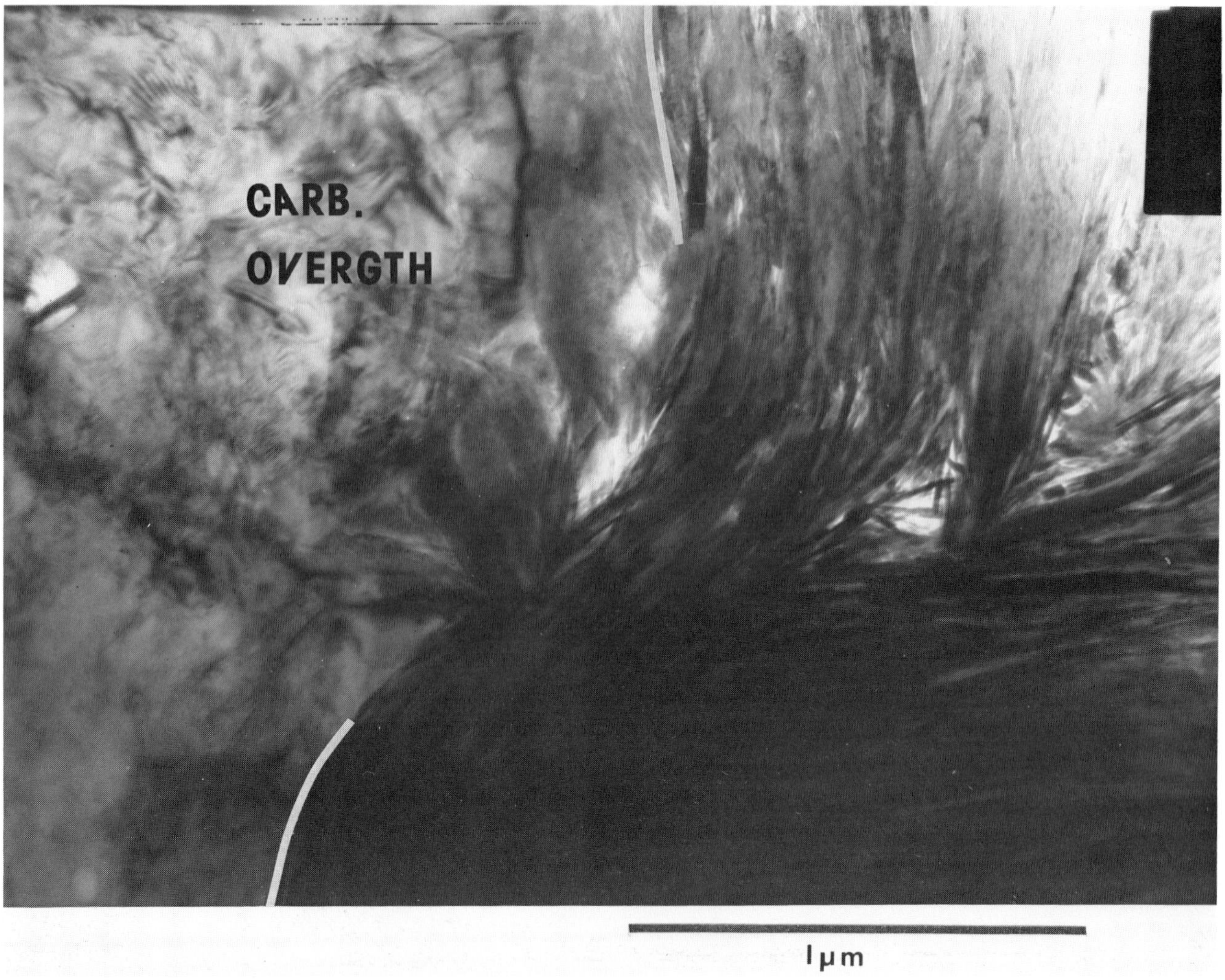

Fig. 7. TEM of aligned and bent phyllosilicates located in a pore throat adjacent to a fault zone. The phyllosilicates are adjacent to a carbonate overgrowth (carb. overgth.) on a detrital grain.

associated with the fault is almost non-existent and is one type of the adjustment to the fault zone borders which can contribute to the sealing properties of the fault zone (Fig. 8). In this case the large changes in permeability can be achieved without significant changes in the porosity; only the pore geometry changes by the redistribution of material.

A more dramatic concentration of phyllosilicates into a fault zone is illustrated in Fig. 9. In this case detrital quartz grains in the host rock are fractured and appear to be corroded in the fault zone. The phyllosilicate concentration is primarily associated with smearing, reorientation and deformation of detrital phyllosilicates. Porosity collapse in fault zones developed in clay rich lithologies has also been discussed by Knipe (1986c) and Moore et al. (1986).

It should be clear from the last two examples discussed that a complete range of "wash-in" seals can develop from those associated with the redistribution of phyllosilicates on the scale of a few grains to the large scale channelling of mud diapirs along fault zones. The disaggregation and redistribution of clay particles during faulting can also lead to more

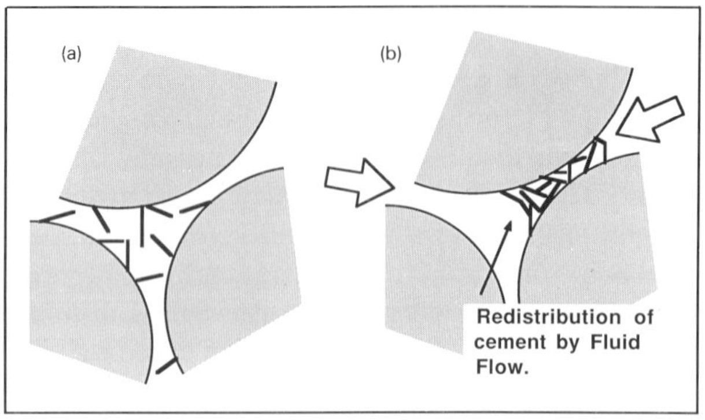

Fig. 8. "Wash-in seal": Sketches illustrating the development of a collapsed phyllosilicate aggregate in a pore throat. (a) The initial configuration of clay plates distributed around a pore between detrital grains. (b) Final fabric of phyllosilicates blocking a pore after a deformation event where a fluid flow pressure wave has redistributed the grains.

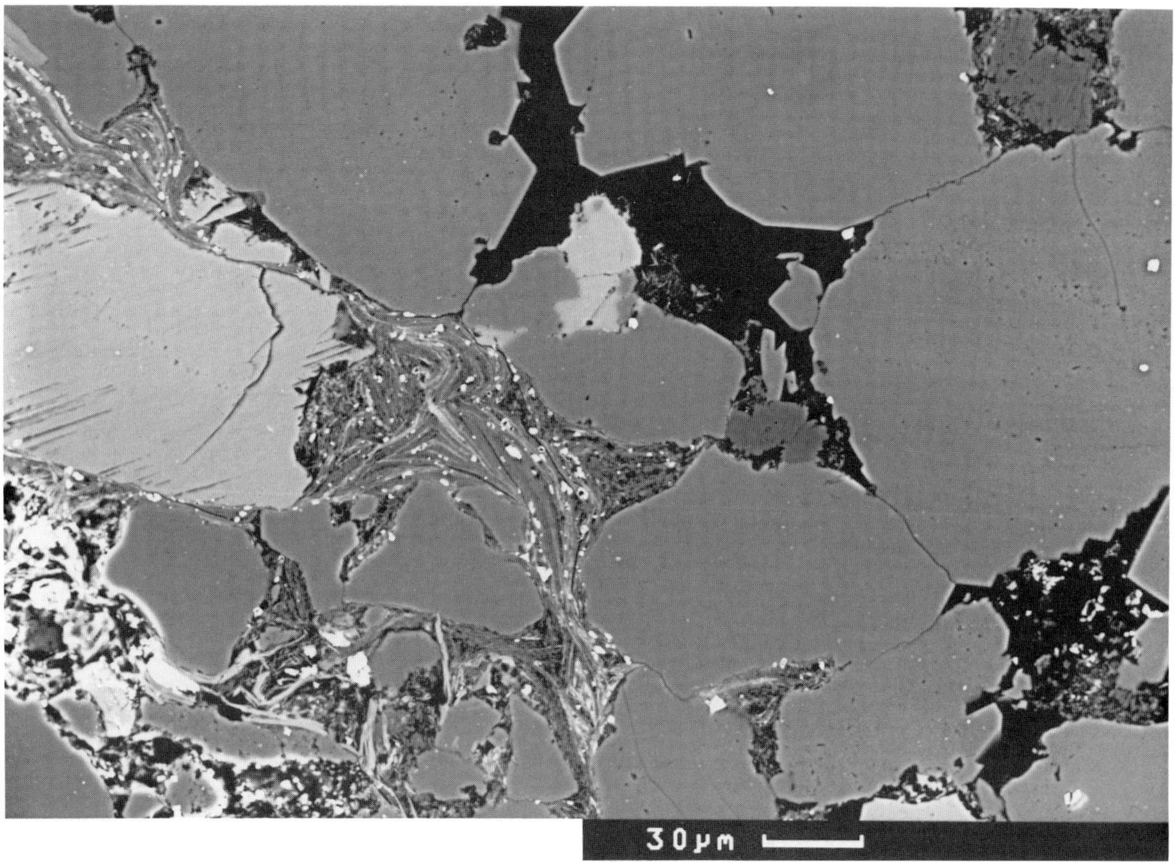

Fig. 9. Backscattered electron image of a fault zone where clays are concentrated in the fault zone (top left to middle bottom of micrograph). Note the high porosity outside the fault and the corroded quartz grains inside the fault zone.

extensive "smear" seals than is predicted by the separation of shale lithologies after displacement. This is achieved by phyllosilicates from the shale unit being carried along the fault zone by flow processes operating during the transient high permeability and disaggregation periods.

Two last examples of fault related seals are presented in Figs. 10 and 11. In these cases the precipitation of cements along and adjacent to the fault zone causes the changes in the pore geometry and permeability which controls the sealing capacity of the fault. The precipitation of cements can be caused by changes in the fluid pressure and temperature experienced by fluids during the faulting process. A detailed analysis of the fluid pressure cycles associated with faulting has been presented by Sibson (1981, 1985, 1989) and the role of such cycles in regional deformation is discussed by Cox and Etheridge (1989). Sibson (1981, 1985, 1989, 1990) presented models for the behaviour of faults as valves which control the channelling of fluids in fault zones and pumps which cause expulsion of the accumulated fluids. Sibson also emphasises the role of dilation sites created along the fault zones during the displacement events (e.g., dilational jogs) as locations where rapid fluid influx and pressure drops can induce preferential precipitation.

The dilation magnitude, together with the fluid pressure and permeability changes during faulting events will influence the volume of cement produced. For example, fault zones where both dilation and fluid pressure changes are large can result in very rapid precipitation and the production of a breccia composed of rock fragments floating in cement. These have been termed implosion breccias by Sibson (1986) and related to the collapse of the wall rock into dilation sites generated by rapid slip events. On the other hand, situations in which the fault zone dilation at the time of precipitation is small and the fragments remain in contact and provide a load-supporting framework, represent the other end of the spectrum, where precipitation of cement creates overgrowths in the pores of the cataclasite. Whatever the detailed mechanism of cement generation the reduction in the porosity and permeability in the fault zone can be large. For example, in sandstones initial porosities of $\approx$20% may be reduced to less than 1% and permeabilities can be decreased by more than an order of magnitude (see also Gabrielsen and Koestler, 1986).

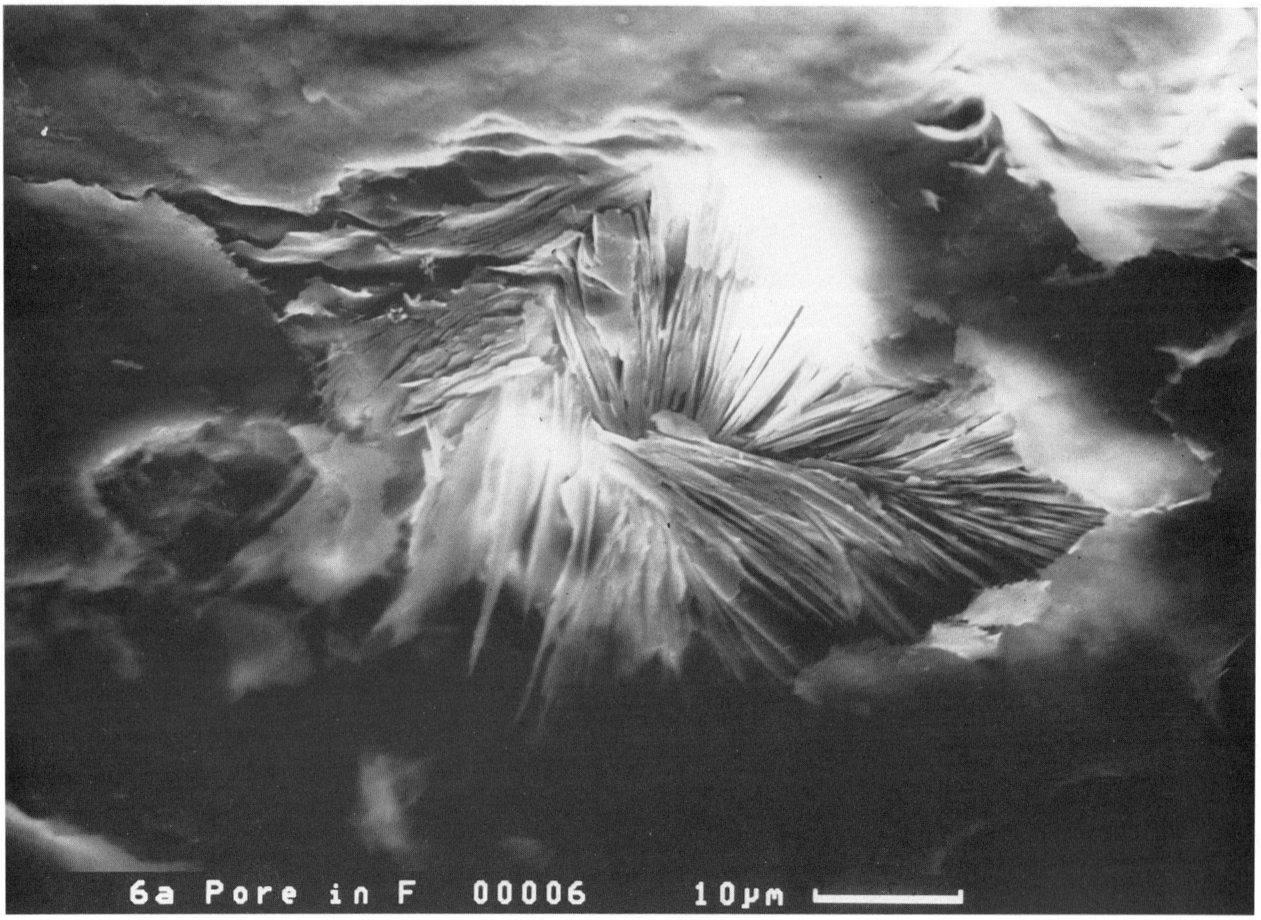

Fig. 10. SEM of kaolinite grains filling pores within a fault zone.

Although the sealing processes in each of the examples presented above is dominated by a small number of factors it should be emphasised that individual faults may contain elements of a number of seal types. For example, the development of sealed fault zone borders by concentrated cementation and/or by pore throat plugging may occur adjacent to a fault zone where cataclasis, dissolution and cement precipitation characterise the internal fault zone structure (Fig. 12). However, it should also be clear that different deformation mechanisms will dominate the deformation under different conditions and produce seals with characteristics which reflect the deformation processes and conditions. Fault sealing at shallow depths, for example, where the lithification is only partially complete and the material strength is low, is more likely to be associated with precipitation of cements adjacent to the fault zone margins or by the injection of material into the fault zone. At deeper levels with increased amounts of diagenesis the possibility of combined cataclasis, grain dissolution and precipitation processes contributing to the seal increases.

Seal timing

The timing of the seal development relative to hydrocarbon maturation and migration timing is crucial in the assessment of reservoir evolution. Integrating the diagenetic history and burial histories of the source and reservoir rocks with an assessment of the faulting mechanisms can provide important constraints on the development of reservoir properties. Figure 13 presents the results of an analysis which links the timing of faulting with burial in a sandstone reservoir. The figure is based upon the work conducted by R.J. Knipe and G.E. Lloyd (1988, unpublished). Each of the faults studied has a distinctive deformation microstructure and cement evolution which can be linked to the burial history of the area. Linking the reservoir diagenesis with the "fault zone diagenesis" by establishing the sequence of cement growth and deformation within and external to the fault zones present has provided information on: (a) the timing of fault activity during burial and the identification of faults active over long periods of reservoir evolution and those active over short time periods; (b) the distribution of faulting (i.e., which

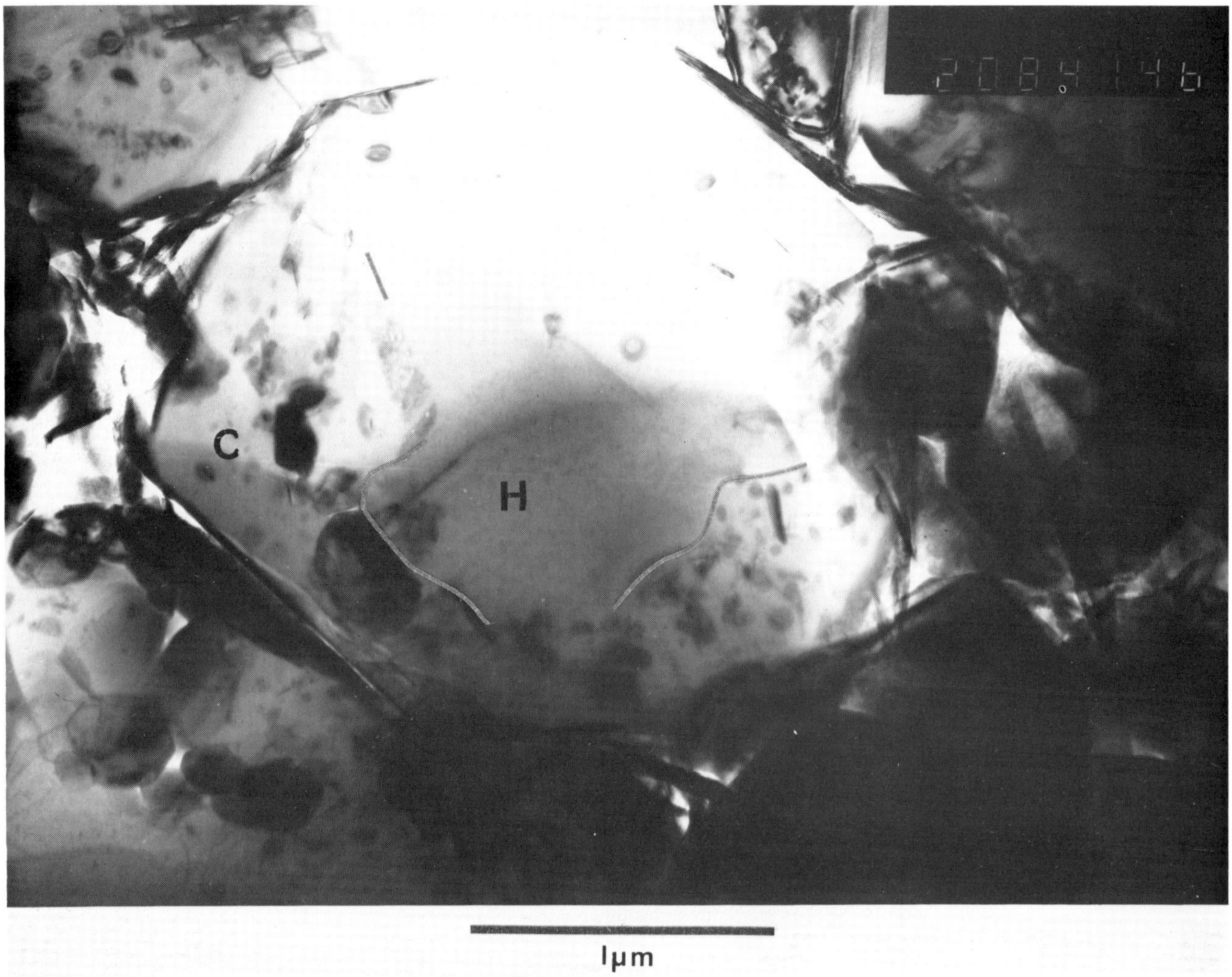

Fig. 11. TEM of a composite cement present in a cataclastic fault zone. The original quartz fragment (*H*) is enclosed in an inclusion rich quartz cement (*C*). Phyllosilicates (illites) form the second cement infill between the quartz overgrowths.

faults were active in different parts of the basin); and (c) a link between the distribution of reservoir cements and fault activity.

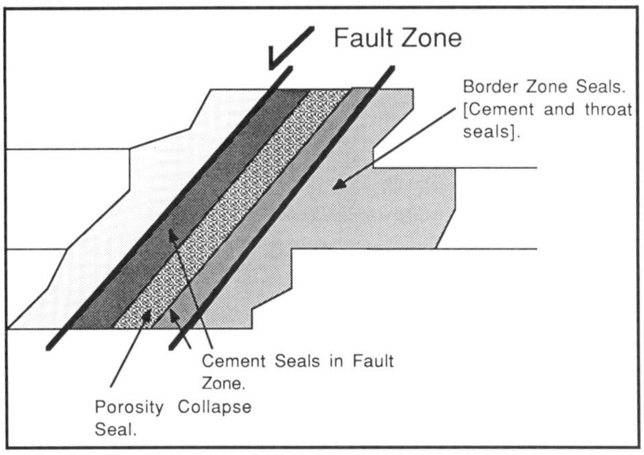

Fig. 12. Fault seal summary illustrating a composite seal where domains within the fault have different characteristics and a border zone seal outside the movement zone is developed to different thicknesses in different layers.

Sealing capacity

The data on the pore microstructure presented in the examples described above are used in this section to estimate the sealing capacity of faults associated with different seal types. The sealing capacity or the height of a hydrocarbon column which these seals could support is calculated using the equation (eqn. 3 above) presented by Berg (1975). Values for the pore size and pore throat diameters used are based on those measured on mercury injection testing, SEM, TEM micrographs and from published data (e.g., Pittman, 1981). The values of the other important variables, which will influence the sealing capacity (i.e., the fluid and hydrocarbon density and the hydrocarbon interfacial tension), used in these estimates enclose the range commonly encountered in reservoirs (and discussed in detail by Schowalter, 1979; and Watts, 1987). The results of the estimates of seal capacity are shown in Fig. 14. The calculations illustrate the differences between the sealing capacities of the different types of seal discussed

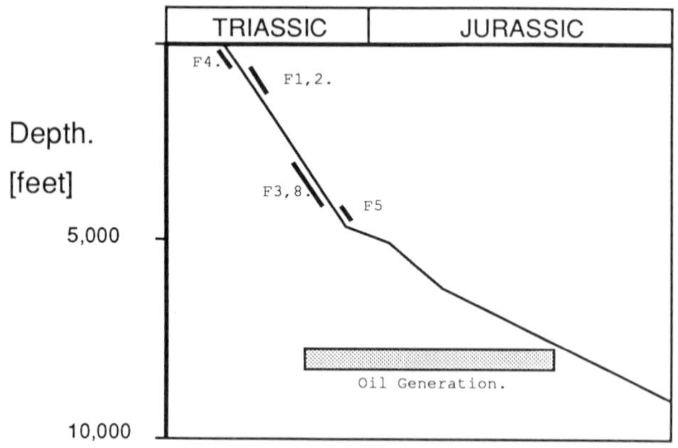

Fig. 13. Summary of the analysis of faulting during burial of a sandstone reservoir. Different faults (*F1–8*) are active for different parts of the burial history. The range of timing of fault activity with respect to burial and overlap with the oil generation period will give rise to different sealing potentials for each type of fault. (Unpublished study by R.J.Knipe and G.E. Lloyd.)

above. The seals which are likely to produce the most effective seals are those with the smallest pore throat.

Seal continuity

The distribution of seals along faults is known to be a fundamental control on the trapping potential of faults. There are a number of factors and processes which will cause the type of seal to vary not only between different faults but also along individual faults. These controls include: (a) the distribution of lithologies intersecting with the fault planes; (b) the propagation and displacement mechanisms of the faults, i.e., the deformation mechanism histories of different faults or fault segments; (c) the nature of the fluid flow along the fault and; (d) the fault plane geometry. These are discussed briefly below and although the factors are discussed separately, the interrelationships and interdependence of these factors are emphasised.

(a) The distribution of lithologies intersecting with the fault planes. Each lithology which intersects the fault plane will have different properties in terms of strength, reaction to the faulting process and, thus, the type of seal likely to develop under the faulting conditions. The type of map described

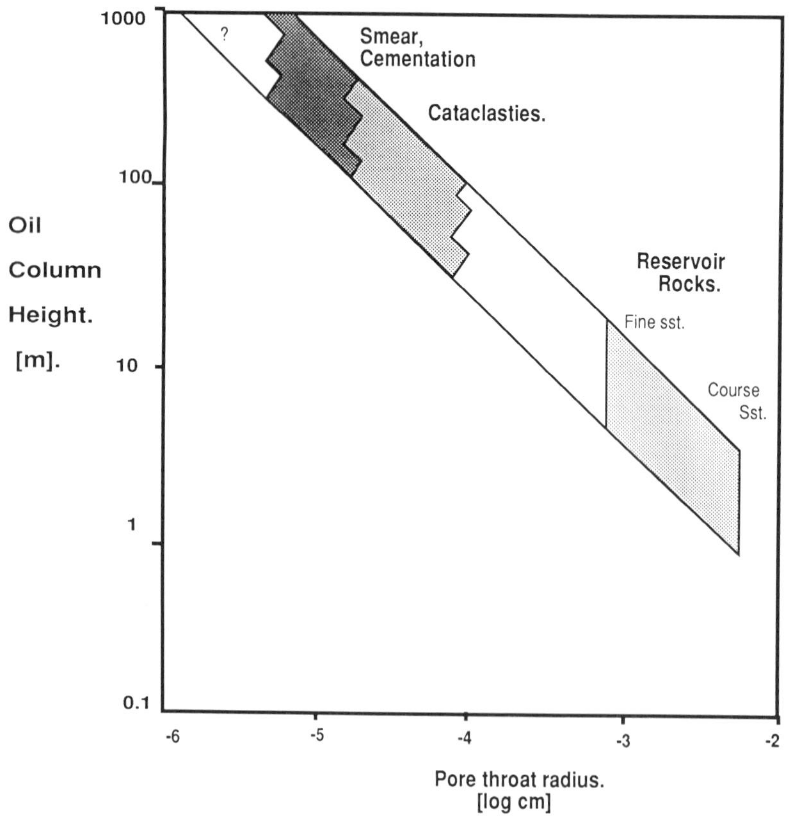

Fig. 14. Plot of sealing capacity of different types of fault seal. The calculations are based on the equation of Berg (1975). The calculations illustrate the oil column height when the values of interfacial tension range from 30 to 35 dynes/cm and the difference between the water and oil densities from 0.1 to 0.385. The pore throat radius values are based upon the microstructural and physical property measurements made by the author and data published in Pittman (1981), Berg (1975) and Fraser (1935).

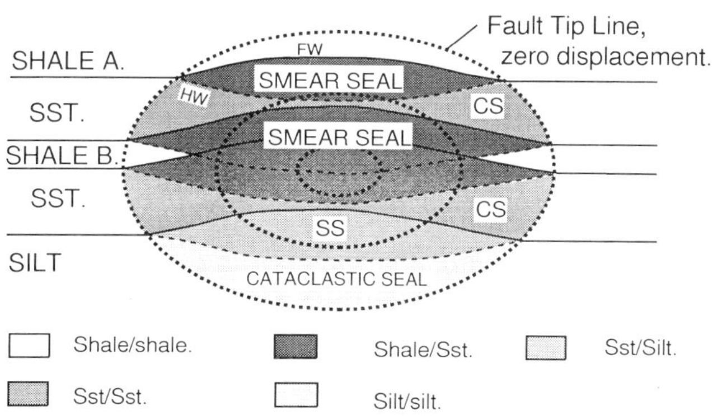

Fig. 15. Fault plane map showing the distribution of seal types on the fault plane which can arise from the intersection of different lithologies with a fault. The intersections of different lithological units from the hanging wall (*HW*) and from the footwall (*FW*) with the fault plane are illustrated. Also shown are displacement magnitude contours, fine dashed line), including the zero displacement, or tip line. Note the distribution of seal types and the zone in which the sandstone cataclasite (*CS*) and the cataclasite derived from the silt overlap (*SS*).

by Allen (1988) to illustrate the separation and juxtaposition of lithologies along fault planes and to asses the distribution of juxtaposition seals can be modified in order to assess also the likely distribution of seal types along the fault planes. An example of a seal-type map is shown in Fig. 15. As illustrated, the fault plane can be divided into areas in which different types of seal will dominate. In addition, there will be transition zones in which the mixing of seal types is likely. For example, fault plane areas above the intersection of a mud/shale with the fault (Fig. 15) may be susceptible to mud injection and the tip zone (loci of zero displacement) of the fault may have a lower seal capacity because of incomplete cataclasis or a different deformation mechanism associated with the lower strain in this zone. These processes will result in a complex distribution of the displacement pressures in the fault zone which will affect the sealing capacity of the fault. The exact pattern of seal types and capacities will depend upon the distribution of lithologies, the fault displacement pattern and the deformation mechanism histories experienced by different parts of the fault. Another important use of the maps is the identification of low displacement pressure locations which will aid assessment of fluid and hydrocarbon flow in and out of the fault zone.

(b) The propagation and displacement mechanisms of the faults. As discussed earlier in the paper the propagation and displacement mechanisms often involve the cyclic accumulation of strain. Different parts of the propagation and displacement cycles may be associated with different deformation mechanism histories. For example, fault tips are zones in which the energy driving the fault propagation is almost dissipated. This can result in a change in the fracture process and the accommodation of strain at a different rate and by different mechanisms. In addition, the tip zones of a main faulting event can also act as a location where aftershocks are concentrated (see Cockerham and Eaton, 1987; Yamashita and Knopoff, 1987; Sibson, 1989). These factors can promote the concentration of fracturing and cementation in tip zones.

Fracturing or cataclastic fault rock development is known to be heterogeneous along faults (Brock and Engelder, 1977; House and Gray, 1982; Hadizadeh and Rutter, 1983; Blenkinsop and Rutter, 1986; Rutter et al., 1986; Sibson, 1986). This may arise from (1) the repeated creation of new tip zones during fault propagation, (2) the deformation of the wall rocks, asperities and perturbations during displacement along the fault zone or (3) different faulting events which are restricted to segments of established faults zones. Sibson (1986) has already recognised the importance of the range of processes operating along fault zones in the production of different types of fault breccia. The result of this variation in the deformation mechanism histories along a fault zone is the potential to develop different seal characteristics along the fault.

(c) The nature of the fluid flow along faults. Fluid flow along fault zones is known to be both episodic and localised, even along individual faults (Sibson, 1981; Etheridge et al., 1984; Sibson et al., 1988). The importance of the fluid pumping to seal evolution is that fluids with different compositions and temperatures can be focussed into the fault zone and the drainage pattern or migration path followed together with the exact PTx history of the fluid in the fault zone will influence the distribution of (1) deformation, (2) the corrosion potential, and (3) the precipitated cement.

It is also possible that the generation of cements adjacent to faults can influence the fluid flow along the fault. Strengthening of the border zones by precipitation will reduce the potential of lateral leakage of migrating fluids from the fault zone. If the cemented zone extends for a distance from the fault zone, which is larger than the width of the strain accommodation zone associated with each increment of slip, then this process will result in the rapid migration of fluid to the tip zone of a propagating fault. This would aid the fault propagation at the tip zone and allow migrating fluids to by-pass sealed horizons. The size of the accommodation strain zone developed at the fault border will depend upon the exact faulting processes, the rock properties and the

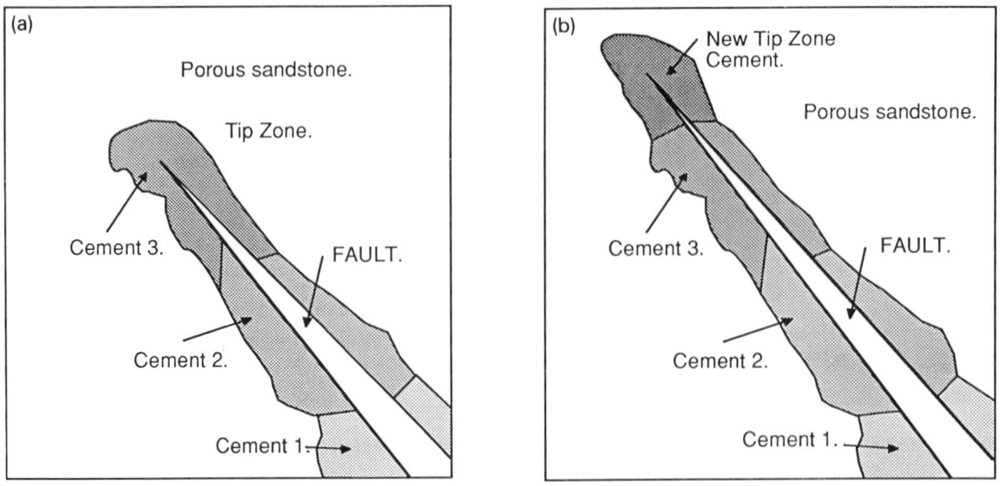

Fig. 16. Development of seal characteristics associated with fault propagation. Cements associated with each propagation event result in a border zone seal. See text for details.

fault zone geometry. The thickness of the cemented zone will depend on the infiltration distance of the fluid into the rock, the amount of solute present in the fluid and the pressure history of the fluid during infiltration. In situations where the rock permeability is high and/or the pressure drop in the fluid associated with infiltration is slow then the thickness of a cemented zone generated during each faulting event may be large. On the other hand thick cement seals may also develop sequentially, where rupture of an earlier border cement seal allows access to the uncemented rock. This latter situation will be favoured by changes in the strain accommodation mechanisms or patterns (see the section on "Fault plane geometry" below). Small scale distributed fracturing in the tip zone (associated either with aftershocks or with the decreasing energy available for fracturing as the propagation rate of the fault dies out) may also result in a concentration of cement growth at the tip zone. These mechanisms provide ways of generating and extending a border zone cement seal sequentially during each increment of fault propagation and extending the fluid flow distance associated with each faulting event (Fig. 16). The continuity of the seal along the fault zone will depend upon the relative extent of the cement zone generated and the propagation distance associated with each fault increment. Another important factor in this context will be that propagation of the fault into a permeable unit may arrest the fault propagation by causing a rapid fluid pressure drop and a focussing of fluid migration along the unit. Sealing of the unit by cementation would then aid continued propagation of the fault. This last statement also emphasises the likelihood of the border cement seals having different characteristics and thicknesses in different lithologies with varying porosity/permeabilities encountered in the same fault (see Fig. 12).

(d) Fault plane geometry. The exact geometry (e.g., curvature) of a fault plane can control the development of fault rock, the characteristics of the strain accommodation zone adjacent to the fault and can, therefore, influence the type of seal produced. Deviations from a planar feature can induce localised dilation or compression during deformations which range in size from grain or fragment sized asperities or perturbations to large scale bends in the fault orientation which occur on the scale of kilometres.

Dilation sites can act as areas of rapid fluid pressure drops during faulting which will lead to precipitation (Sibson, 1989; Sibson et al., 1988) or may act as locations where fluidised gouge generated during displacement is concentrated. Fibrous shear veins are a familiar small scale example of precipitation along faults. These commonly have a patchy distribution on exposed fault planes and are separated by rock contacts where the modification of the rock fabric may be low. Such faults are not effective seals, as leaking may occur between the fibrous patches, unless the displacement or dilation is large enough to allow development of fibres over the entire surface and/or contact areas are marked by fine-grained gouge. Sibson et al. (1988) have presented evidence that dilation can affect large areas associated with dilational jogs on fault planes. Although described in the context of gold mineralisation, the process is also applicable to the present discussion of seals and large cemented areas may develop in extensional zones along faults.

An example of the affect of fault geometry on the distribution of seal types is shown in Fig. 17. The

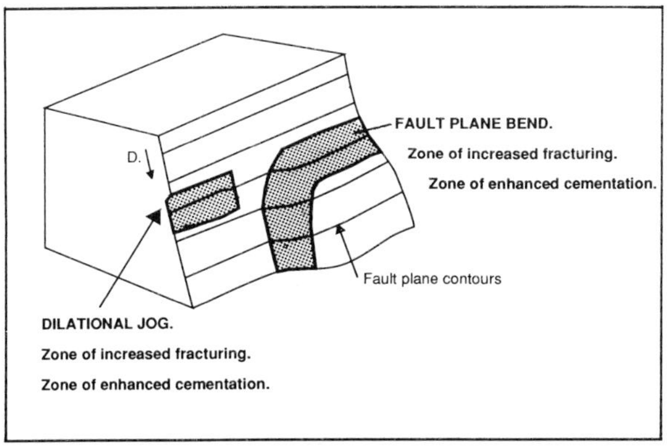

Fig. 17. Fault plane geometry and the distribution of seal types. the sketch illustrates the importance of dilational jogs and fault bends on the distribution of deformation in the adjacent rocks. The shaded areas are zones in which enhanced fracturing and cementation is possible.

size or volume of the cemented zone developed at dilational jogs or fault bends is dependent on the strain accommodation zone developed. The size of such accommodation zones will depend upon the stress levels, strain rates and strain magnitudes and the deformation mechanisms induced in the vicinity of the feature as well as on the size and geometrical characteristics of the bend or jog. The situation is complicated by the likely superposition of repeated cycles of dilation and compression which can be induced in faults with complex geometries. Rocks immediately adjacent to the fault zone can be affected by the large number of small perturbations in the fault plane, while rocks at greater distances from the fault plane may only be affected by the larger scale fault bends encountered during displacement.

The displacement rate and the curvature of a fault bend or perturbation will affect the processes operating in the strain accommodation zone. Deviations in the fault plane of 5°–10° in the movement direction can induce important changes in the deformation mechanisms of the strain accommodation zone (see Knipe, 1985). Fracture arrays may evolve in the strain accommodation zone if the induced strain rates and stress levels are high enough. These conditions will be favoured by perturbations or bends with a small radius of curvature or during rapid displacement events (see Knipe, 1985). Fracture arrays developed to negotiate parts of a fault zone may provide important migration pathways for fluids and may also induce the development of a wide sealed zone with a complex internal structure. In other words, areas of the faults where bends or perturbations exist can be considered as factory zones for fractures which are then displaced along the fault during subsequent movement.

Seal strength and stability

The strength and stability of a fault seal after its formation is crucial to the trapping potential of the fault. The susceptibility of the seal to failure will determine the probability of hydrocarbon leakage or non-trapping. Watts (1987) recognised the importance of the strength of seals and proposed a classification of cap-rock seals based on their mode of failure. Harding and Tuminas (1988) suggested that older faults are less likely to leak and noted the importance of fault rejuvenation in promoting leakage. The decreased leakage associated with older faults noted by Harding and Tuminas may well relate to the reduction in the permeability and porosity of fault zones during the post-movement concentration of diffusion mass transfer in fine grained fault rocks as illustrated in example 2 described above.

The timing of seal evolution relative to the fault activity and faulting processes on an individual fault is important to the stability of the seal. The most stable seals will be those created late during faulting, as these are less likely to be disturbed by subsequent deformation. Further strengthening of the seal would be promoted if continued compaction and cementation by diffusive mass transfer takes place in the fault zone after the main displacement on the fault. The timing of fault seal development with respect to the sequence of fault activity history on a larger (basin) scale will also affect the seal stability. The most stable seals will be those which are unaffected by subsequent displacement of other members of the fault array. The fault array evolution is critical as the distribution of fault activity at different times during the basin or reservoir evolution will affect the seal survival. Analysis of the deformation microstructures and the diagenetic history in and adjacent to different faults can help provide such information on fault activity. In addition, such analyses can also aid the identification of those faults which have been reactivated during inversion.

The different types of seal discussed in this paper will have different strengths and different probabilities of being breached. Seals which border the fault zones and arise from either cementation and/or pore throat plugging are likely to be more stable features because they can be unaffected by later fault displacement events. Fault creep processes are less likely to cause breaching of border zone seals than faulting by transient, higher strain rate events. The more extensive the fault border seals the more likely they are to survive intact. However, if too extensive, they may then seriously restrict the desirable reservoir properties. As noted by Weber et al. (1978) and Smith (1980), shaley smear seals will be more effective if they do not contain mixtures of sand. Cataclastic seals with extensive cements will be stronger and more resilient to later deformation than uncemented cataclastic fault zones.

As described above, the fault plane geometry, i.e., curvature changes along the displacement path followed by the reservoir horizon, will affect the potential for seal breaching. Reservoir rocks which develop early sealing properties at faults are more likely to survive if changes in fault plane curvature during subsequent displacement are small. Stability of the seal will be enhanced if adjacent faults in the fault array, which are active after the fault under consideration, do not induce reactivation or breaching of the previously sealed fault. For example, listric faults are likely to induce more accommodation strains in the adjacent rocks than movement on a set of planar domino faults. It is clear from the discussion that fault seals can develop early during fault array evolution and survive and it is equally clear that assessment of such situations requires integration of fault analysis on a range of scales.

Concluding statement

The examples presented above indicate that the deformation mechanism histories in fault zones can be both varied and complex and that microstructural analysis can aid the assessment of the detailed processes of sealing.

The range of sealing processes which may affect particular faults is similar to the range of diagenetic changes which can affect reservoirs quality with the added complication of the range of deformation mechanisms which may operate. As with the diagenetic history of a reservoir, there is no simple method of predicting the exact sealing properties of a fault zone and each case will depend upon the timing, conditions, duration and processes of faulting.

Future studies of fractures and faults preserved in core together with onshore studies are needed to build up a data-base on the fault behaviour of different reservoir lithologies in different faulting situations. An integration of such fault zone deformation and fault zone diagenesis studies with analysis of reservoir diagenesis, fault geometries and fault array evolution (timing, distribution, etc.) will aid the unravelling of the detailed role of faults in reservoir development.

Acknowledgements

Discussions with G. Cowan, R. Frost, P. Goldsmith, G.E. Lloyd R.H. Sibson and I. Stewart on the processes of faulting are all acknowledged. Hydrocarbons GB are thanked for granting permission to publish Fig. 3.

References

Allen, U.S., 1988. Model for hydrocarbon migration and entrapment within faulted structures. Am. Assoc. Pet. Geol., Bull., 73: 803–811.

Angevine, C.L. and Turcotte, D.L., 1983. Porosity reduction by pressure solution. A theoretical model for quartz arenites. Geol. Soc. Am. Bull., 194: 1129–1134.

Atkinson, B.K. (Editor), 1987. Fracture Mechanics of Rocks. Academic Press, London.

Barber, D.J., 1985. Dislocations and microstructures. In: H.R. Wenk (Editor), Preferred Orientation in Deformed Metals and Rocks — An Introduction to Modern Texture Analysis. Academic Press, London, pp. 149–182.

Berg, R.R., 1975. Capillary pressure in stratigraphic traps. Am. Assoc. Pet. Geol., Bull., 59: 939–956.

Bjørlykke, K., Aagaard, P., Dypvik, H., Hastings, D.S., Harper, A.S., 1986. Diagenesis and reservoir properties of Jurassic sandstones from the Haltenbanken area, offshore Norway. In: A.M. Spencer (Editors), Habitat of Hydrocarbons on the Norwegian Continental Shelf. Norwegian Petroleum Society, Graham and Trotman, London, pp. 275–286.

Blenkinsop, T.G. and Rutter, E.H., 1986. Cataclastic deformation of quartzite in the Moine Thrust Zone. J. Struct. Geol., 8: 669–682

Borradaile, G.J., 1981. Particulate flow and the generation of cleavage. Tectonophysics, 72: 306–321.

Brace, W.F. and Byerlee, J.D., 1966. Stick-slip as a mechanisms for earthquakes. Science, 153: 990–992

Brock, W.G. and Engelder, J.T., 1977. Deformation associated with movement on the Muddy Mountain overthrust in the Buffington window, SE Nevada. Geol. Soc. Am. Bull., 88: 1667–1677.

Charlaix, E., Guyon, E. and Roux, S., 1987. Permeability of a random array of fractures of widely varying apertures. Transport in Porous Media, 2: 31–43.

Cockerham, R.S. and Eaton, J.P., 1987. The earthquake cycle and its aftershocks, April–September 1984. U.S. Geol. Surv., Bull., 1039: 15–28.

Cox, S.F. and Etheridge, M.A., 1989. Coupled grain scale dilatancy and mass transfer during deformation at high fluid pressure: examples from Mount Lyell, Tasmania. J. Struct. Geol., 11: 47–62.

Downey, M.W., 1984. Evaluating seals for hydrocarbon accumulations. Am. Assoc. Pet. Geol., Bull., 68: 1752–1763.

Etheridge, M.A., Wall, V.J., Cox, S.F. and Vernon, R.H., 1984. High fluid pressures during metamorphism and deformation: implications for mass transport and deformation mechanisms. J. Geophys. Res., 89: 4344–4358.

Frank, F.C., 1965. On dilatancy in relation to seismic sources. Rev. Geophys., 3: 485–503.

Fraser, H.J., 1935. Experimental study of the porosity and permeability of clasitic sediments. J. Geol., 43: 912–996.

Gabrielsen, R.H. and Koestler, A.G., 1986. Description and structural implications of fractures in Jurassic sandstones of the Troll Field, nothern North Sea. Nor. Geol. Tidsskr., 67: 371–381.

Gratier, J.P. and Guiguet, R., 1986. Experimental pressure solution-deposition on quartz grains: the crucial effect of the nature of the fluid. J. Struct. Geol., 8: 845–856.

Groshong, R.H., 1988. Low temperature deformation mechanisms and their interpretaion. Geol. Soc. Am. Bull., 100:

1329–1360.

Gueguen, Y., Darot, M. and Reuschle, T., 1987. Permeability evolution under stress. In: Doligez, B. (Editor), Migration of Hydrocarbons in Sedimentary Basins. IFP Explor. Res. Conf., 45: 281–285.

Hadizadeh, J. and Rutter, E.H., 1983. The low temperature brittle–ductile transition in a quartzite and the occurrence of cataclastic flow in nature. Geol. Rundsch., 72: 493–509.

Harding, T.P., 1974. Petroleum traps associated with wrench faults. Am. Assoc. Pet. Geol., Bull., 58: 290–304.

Harding, T.P. and Tuminas, A.C., 1988. Interpretation of footwall (lowside) fault traps sealed by reverse faults and convergent wrench faults. Am. Assoc. Pet. Geol., Bull., 72: 738–757.

Harding, T.P. and Tuminas, A.C., 1989. Structural interpretation of hydrocarbon traps sealed by basement normal blocks and at stable flank of foredeep basins and at rift basins. Am. Assoc. Pet. Geol., Bull., 73: 812–840.

House, W.M. and Gray, D.R., 1982. Cataclasites along the Saltville thrust, U.S.A., and their implications for thrust-sheet emplacement. J. Struct. Geol., 4: 257–269.

Houseknecht, D.W., 1984. Influence of grain size and temperature on intergranular pressure solution, quartz cementation, and porosity in a quartzose sandstone. J. Sediment. Petrol., 54: 348–361.

Houseknecht, D.W., 1988. Intergranular pressure solution in four quartzose sandstones. J. Sediment. Petrol., 58: 228–246.

Hubbert, M.K., 1953. Entrapment of petroleum under hydrodynamic conditions. Am. Assoc. Pet. Geol., Bull., 37: 1954–2026.

Jones, M.E. and Preston, R.M.F., 1987. Deformation of Sediments and Sedimentary Rocks. Geol. Soc. London, Spec. Publ., 29.

Knipe, R.J., 1985. Footwall geometry and the rheology of thrust sheets. J. Struct. Geol., 7: 1–10.

Knipe, R.J., 1986a. Faulting mechanisms in slope sediments: examples from Deep Sea Drilling Project cores. In: J.C. Moore (Editor), Structural Fabric in Deep Sea Drilling Project Cores from Forearcs. Geol. Soc. Am. Mem., 166: 45–54.

Knipe, R.J., 1986b. Microstructural evolution of vein arrays preserved in Deep Sea Drilling Project Cores from the Japan Trench, Leg 57. In: J.C. Moore (Editor), Structural Fabric in Deep Sea Drilling Project Cores from Forearcs. Geol. Soc. Am. Mem., 166: 75–88.

Knipe, R.J., 1986c. Deformation mechanism path diagrams for sediments undergoing lithification. In: J.C. Moore (Editor), Structural Fabric in Deep Sea Drilling Project Cores from Forearcs. Geol. Soc. Am. Mem., 166: 151–160.

Knipe, R.J., 1989. Deformation mechanisms — recognition from natural tectonites. J. Struct. Geol., 11: 127–146.

Kozak, J. (Editor) 1987. Physics of fracturing and siesmic energy release. Pure Appl. Geophys., 124: 973 pp.

Kranz, R.L., 1983. Microcracks in rocks: a review. Tectonophysics, 101: 449–480.

Lloyd, G.E. and Knipe, R.J., 1992. Deformation mechanisms accommodating faulting of quartzite under upper crustal conditions. J. Struct. Geol., 14: 127–143.

Logan, J.M., 1979. Britlle Phenomena. Rev. Reophys. Space Phys., 17. 1121–1132.

Mawer, C.K. and Williams, P.F., 1985. Crystalline rocks as possible palaeoseismicity indicators. Geology, 3: 100–102.

Mitra, S., 1988. Effects of deformation mechanisms on reservoir potential in Central Appalachian overthrust belt. Am. Assoc. Petrol. Geol., Bull., 72: 536–554.

Moore, J.C., Roeske, S., Lundberg, N., Schoonmaker, J., Cowan, D., Gonzales, E. and Lucas, S., 1986. Scaley fabrics from the deep sea drilling project cores from forearcs. In: J.C. Moore (Editor), Structural Fabric in Deep Sea Drilling Project Cores from Forearcs. Geol. Soc. Am. Mem., 166: 55–74.

Nelson, R.A., 1985. Geologic Analysis of Naturally Fractured Reservoirs. Contributions in Petroleum Geology and Engineering, 1. Gulf Publishing Co., Houston, Texas, 320 pp.

Ohnaka, M., 1987. Nucleation and propagation processes of stick-slip failure and normal stress dependence of the physical parameterss of dynamic slip failure. Natural Disaster Sci., 9: 1–21.

Owen, G., 1987. deformation processes in unconsolidated sands. In: M.E. Jones and R.M.F. Preston (Editors), Deformation of Sediments and Sedimentary Rocks. Geol. Soc. London, Spec. Publ., 29: 11–24.

Pittman, E,D., 1981. Effect of fault-related granulation on porosity and permeability of quartz sandstones, Simpson Group (Ordovician), Oklahoma. Am. Assoc. Pet. Geol., Bull., 65: 2381–2387.

Rutter, E.H., 1983. Pressure solution in nature, theory and experiment. J. Geol. Soc. London, 40: 725–740

Rutter, E.H., 1986. On the nomenclature of mode of failure transitions in rocks. Tectonophysics, 122: 381–387.

Rutter, E.H., Maddock, R.H., Hall, S.H. and White, S.H., 1986. Comparative microstructures of natural and experimentally produced clay-bearing fault gouges. Pure Appl. Geophys., 124: 3–30.

Rutter, E.H., Maddock, R.H., Hall, S.H. and White, S.H., 1986. Comparative microstructures of natural and experimentally produced clay-bearing fault gouges. Pure Appl. Geophys., 124: 3–30.

Schowalter, T.T., 1979. Mechanisms of secondary hydrocarbon migration and entrapment. Amer. Assoc. Pet. Geol., Bull., 63: 723–760.

Sibson, R.H., 1977. Fault rocks and fault mechanisms. J. Geol. Soc. London, 133.

Sibson, R.H., 1980. Transient discontinuities in ductile shear zones. J. Struct. Geol., 2: 165–171.

Sibson, R.H., 1981. Fluid flow accompanyiong faulting: field evidence and models. In: D.W. Simpson and P.G. Richards. (Editors), Earthquake Prediction; An International Review. Am. Geophys. Union, M. Ewing Ser., 4: 593–603.

Sibson, R.H., 1985. Stopping of earthquake ruputures at dilation fault jogs. Nature, 316: 248–251.

Sibson, R.H., 1986a. Brecciation processes in fault zones — inferences from earthquake rupturing. Pure Appl. Geophys., 124: 159–175.

Sibson, R.H., 1986b. Earthquakes and rock deformation in crustal fault zones. Annu. Rev. Earth Planet. Sci., 14: 149–175.

Sibson, R.H., 1989. Earthquakes as a structural process. J. Struct. Geol., 11: 1–4

Sibson, R.H., Robert, F. and Poulsen, K.H., 1988. High angle reverse faults, fluid pressure cycling and mesothermal gold-quartz deposits. Geology, 16: 551–555.

Sibson, R.H., 1990 Conditions of fault-valve behaviour. In: Knipe, R.J. and Rutter, E.H. (Editors), Deformation Mechanisms, Rheology and Tectonics. Geol. Soc. London, Spec. Publ., 54: 15–28.

Smith, D.A., 1966. Theoretical consideration of sealing and non-sealinh faults. Am. Assoc. Pet. Geol., Bull., 50: 363–374.

Smith, D.A., 1980. Sealing and non-sealing faults in the Louisiana Gulf Coast Basin. Am. Assoc. Pet. Geol., Bull., 64: 145–172.

Tullis, T.E. (Editor) 1986. Friction and Faulting. Pure Appl. Geophys., 124: 608 pp.

Underhill, J.R. and Woodcock, N.H., 1987. Faulting mechanisms in high porosity sandstones; New Red Sandstones, Arran, Scotland. In: M.E. Jones and R.M.F. Preston (Editors), Deformation of Sediments and Sedimentary Rocks. Geol. Soc. London, Spec. Publ., 29: 91–105.

Wang, C.H., 1986. Internal Structure of Fault Zones. Pure Appl. Geophys., Spec. Issue, 124: 373.

Wardlow, N.C. and Cassan, J.P., 1978. Estimation of recovery efficiency by visual observation of pore systems in reservoir rocks. Can. Pet. Geol., Bull., 26: 572–585.

Wardlow, N.C. and Taylor, R.P., 1976. Merwy capillary pressure curves and the interpretation of pore structure and capillary behaviour in reservoir rocks. Can. Soc. Pet. Geol., 24: 225–262.

Watterson, J., 1986. Fault dimensions, displacements and growth. Pure Appl. Geophys., 124: 365–373.

Watts, N.L., 1987. Theoretical aspects of cap-rock and fault seals for single and two phase hydrocarbon columns. Mar. Pet. Geol., 4: 274–307.

Weber, K.J., Mandl, G., Pilaar, W.F., Lehner, F., Precious, R.G., 1978. The role of faults in hydrocarbon migration and trapping in Nigerian growth fault structures. Offshore Technology Conf., Paper No. 3356. 2643–2652. (Reproduced in: N.H. Foster and E.A. Beaumont (Editors), 1988. Traps and Seals. Treatise Pet. Geol. Reprint Ser., 6: 81–91.)

Wesson, R.L., 1988. Dynamics of fault creep. J. Geophys. Res., 93: 8929–8951.

White, J.C. and White, S.H., 1983. Semi-brittle deformation within the Alpine Fault Zone, New Zealand. J. Struct. Geol., 5: 579–589.

White, S.H., 1985. Defect structures in deformed minerals. In: J.C. White (Editor), Applications of Electron Microscopy in the Earth Sciences. Mineral. Assoc. Can., Short Course, pp. 121–150.

Wilke, S., Guyon, E. and de Marsily, G., 1985. Water penetration through fractured rocks: test of a tridimensional percolation description. Math. Geol., 17: 17–27.

Wilson, J.C. and McBride, E.F., 1988. Compaction and porosity evolution of Pliocene sandstones, Ventura Basin, California. Am. Assoc. Pet. Geol., Bull., 72: 664–681.

Yamashita, T. and Knopoff, L., 1987. Models of aftershock occurrence. Geophys. J.R. Astron. Soc., 91: 13–26.

R.J. KNIPE *Department of Earth Sciences, The University of Leeds, Leeds, LS2 9JT, U.K.*

Controls on the development of fractured reservoirs in the Monterey Formation of central California

R.G. Hickman and J.B. Dunham

The Miocene Monterey Formation of central California hosts important oil reserves in fractured cherts, dolomites and siliceous shales. Commercial accumulations required favorable initial conditions of sedimentation followed by deep burial that produced embrittling diagenetic changes prior to regional folding. Strata of the Monterey Formation are derived from three main components: (1) fine-grained terrigenous sediment; (2) opaline diatom tests; and (3) calcareous tests of nannofossils. Strata consisting of varying proportions of these components are interlayered on a centimeter–decimeter scale. The most favorable reservoir strata accumulated in isolated structural lows, far from sources of terrigenous sediment. The anoxic character of these basins prevented bioturbation that would have homogenized the sediments and eliminated the distinctive mechanical properties of individual layers. Anoxic conditions also maximized preservation of organic material.

With increasing temperature due to burial, nearly pure opaline sediments were converted progressively to brittle opal-CT (opal-CT is a mixture of cryptocrystalline cristobalite and tridymite) and quartz cherts, while opaline-rich sediments were converted to semi-brittle siliceous shales. Clay-rich sediments remained ductile and incompetent. Diagenetic processes also converted some pure calcareous sediments to dolomites.

Chert, dolomite and siliceous shale of the Monterey Formation are prone to fracturing not only because of their brittle character, but also because of (1) strong ductility contrasts between them and interbedded ductile shales, (2) the thinly interbedded nature of the various lithologies and (3) possibly high fluid pressures due to oil generation in interbedded organic shales at the time of deformation. Fractures in the Monterey Formation include planar extension and shear fractures, breccias developed along these fractures and along faults, and breccias associated with intense small-scale tectonic folding.

Extensive, interconnected subvertical extension fractures developed at an early stage of deformation. These fractures are approximately perpendicular to fold axes. Together with normal faults with similar strikes, these fractures provided conduits for hydrocarbons generated from interbedded organic-rich shales to migrate up the flanks of growing folds. As folding progressed, less continuous extension fractures related to bending developed. These fractures strike parallel to the axes of the folds and improved reservoir quality. The zones of highest porosity and permeability are developed in breccias produced by meter-scale tight folding of cherts and dolomites that are interbedded with ductile shales. Because these breccias are stratigraphically controlled, they can have substantial lateral extent and are important reservoirs.

Introduction

The Miocene Monterey Formation hosts important oil reserves in the Santa Maria Basin of coastal central California, the offshore extension of the basin and the nearby Santa Barbara Channel (Fig. 1). Offshore exploration during the 1980's discovered recoverable reserves estimated at 0.34 billion metric tonnes of oil and 60–80 billion cubic meters of natural gas in several closed anticlinal structures (Ogle et al., 1987). Recoverable reserves are mainly in chert and dolomite reservoirs with very low intergranular porosities and permeabilities, and in siliceous shales with moderate intergranular porosities, but very low permeabilities. Commercial hydrocarbon accumulations in all of these strata require extensive, interconnected fracture networks.

An understanding of the fractured reservoirs in these fields is aided by the occurrence of good surface exposures of the reservoir interval in close proximity to major fields. Insights from this region into the controls of fracture development and accumulation of hydrocarbons have applications to exploration and development of fractured reservoirs elsewhere, particularly those occurring in strata of biogenic origin.

This study is based on work along the coastal outcrops between Point Conception and Point Pedernales (Fig. 1) and the study of associated subsurface samples.

Tectonic setting

Coastal California has had a complex Cenozoic tectonic history, mainly related to relative motions of the American lithospheric plate with the Farallon and Pacific oceanic plates. Paleogene and early Neogene basins of coastal California developed in

Structural and Tectonic Modelling and its Application to Petroleum Geology edited by R.M. Larsen, H. Brekke, B.T. Larsen and E. Talleraas. NPF Special Publication 1, pp. 343–353. Elsevier, Amsterdam.

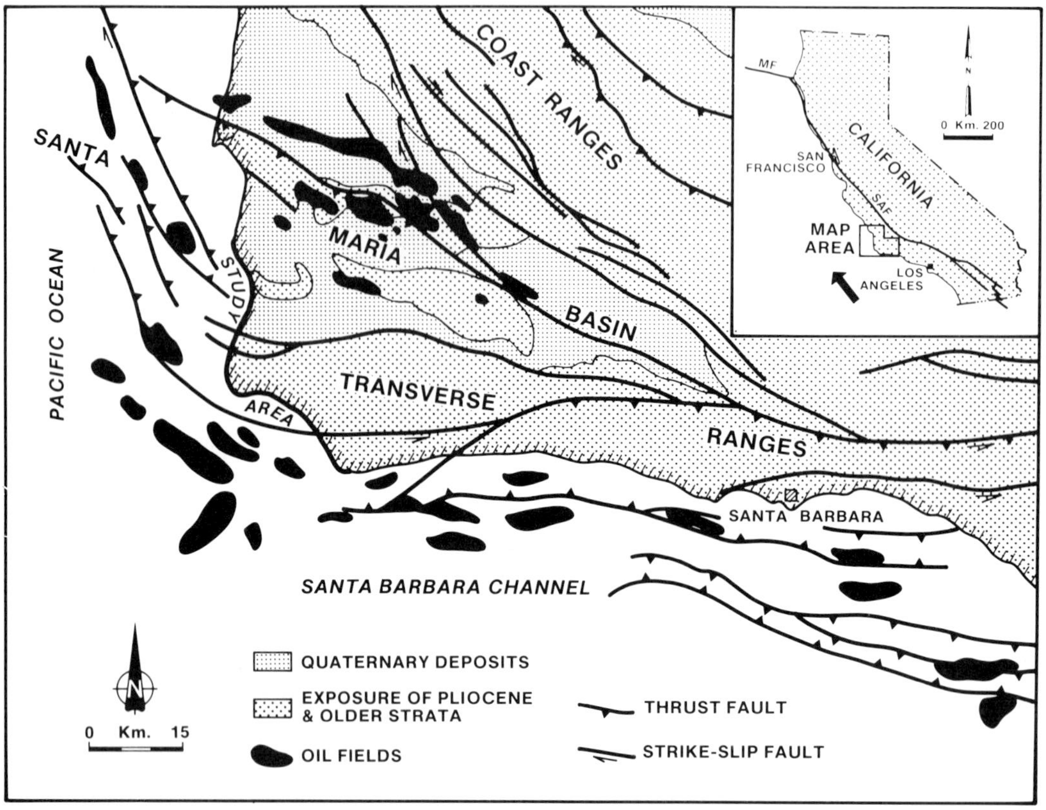

Fig. 1. Map of central coastal California showing oil fields with production from fractured Monterey Formation.

forearc settings inboard of a subduction zone. This setting changed in the early Neogene with the progressive development of a broad transform margin between the Pacific and American plates. The Monterey Formation accumulated in a series of broad basins that developed along this transform margin in Miocene times (Graham, 1987). During the Plio-Pleistocene, the relative Pacific-American plate motion was slightly convergent across this largely transform margin (Crouch et al., 1984). This resulted in segmentation of the Miocene basins and produced compressional folding and thrusting, as well as dextral strike-slip faulting in the basins.

The Point Conception–Point Pedernales area is located at the transition between the W-striking structures of the Transverse Ranges and the NNW-striking structures of the Coast Ranges that formed during the Plio–Pleistocene (Fig. 1). The region is characterized by structures of intermediate strike — NW-striking folds, thrusts and dextral strike-slip faults.

Sedimentary controls on Monterey lithologies

Controls on vertical and lateral variations in sediment composition within the Monterey Formation are related to patterns of detrital and biogenic sedimentation that were present within the basins during Miocene time. Terrigenous detrital sediment entered the basin in turbidity currents which deposited sand, silt and clay in and around submarine fans at the continentward margins of the basins. Pure diatom sediments could not accumulate in or near these areas of high terrigenous influx because the pelagic diatoms raining out from surface waters became diluted by silt and clay. Only clay shale and impure diatomaceous mudstone accumulated at these localities. In contrast, in areas of the basin that were remote from sources of terrigenous influx and thus starved of clastic sediment, the only sediment that did accumulate was deposited by the rainout of siliceous diatom shells and calcareous coccolith shells from surface waters (Ingle, 1981).

The pure diatom sediments were altered to chert during burial diagenesis, while pure coccolith sediments were altered to dolomite during diagenesis. It is important to note that although the chert and dolomite are of diagenetic origin, their distribution was controlled ultimately by original sedimentation patterns of diatoms, coccoliths and clay across the basin. Consequently, as a result of depositional-facies changes, brittle reservoir rocks are not uniformly distributed within the Monterey Formation.

It is critical to understand that silica diagenesis has not in itself resulted in the formation of glassy chert.

Rather, it was a coincidence of special original-sediment composition and sufficient burial heating that brought about the diagenetic generation of chert reservoir rocks at certain areas of the basin.

Dolomite, which is another important reservoir lithology, occurs as 30 cm to 1 meter-thick beds that are interlayered with organic-rich shales and siliceous shales throughout the Monterey section. These dolomite beds have formed as a result of diagenetic replacement of original calcium carbonate sediment which consisted largely of nannofossil-foraminiferal ooze (Pisciotto, 1981b).

Thus in the cases of both dolomite and chert, the depositionally controlled distribution of certain sedimentary facies types within the depositional basin has operated as a primary control on fractured reservoir development. In the cases of both dolomite and chert, in certain areas of the basin where terrigenous influx was low but where biologic productivity in surface waters was high, thick layers of pure-diatom and pure-coccolith oozes were deposited which subsequently altered to brittle rocks. However, in other areas of the basin where terrigenous sediment influx was relatively high, the biogenic sediments became diluted by terrigenous clay, thus resulting in deposition of a mixed sediment that altered during burial diagenesis into soft, plastic calcareous-siliceous shale.

Diagenetic history of siliceous and carbonate rocks

The interlayered cherts and shales of the Monterey are the products of diagenetic alteration of original diatomaceous sediments. Cherts, which consist of greater than 90 wt.% SiO_2, are the product of diagenetic alteration of an originally pure diatomite sediment, while siliceous shales are the result of alteration of less-pure mixed clay and diatom sediment properly termed diatomaceous mudstone. The evolution of biogenic silica from opal-A (original amorphous-silica shells such as diatoms or radiolaria), through an intermediate stage of opal-CT (opal-CT is a mixture of cryptocrystalline cristobalite and tridymite) and eventually into quartz, has been recognized as the method by which Monterey diatomites and diatomaceous mudstones were converted into cherts and siliceous shales (Murata and Larson, 1975; Isaacs, 1980; Pisciotto, 1981a).

The process involves dissolution and reprecipitation of silica at each stage. The principal control on alteration of opal-A into opal-CT and eventually quartz is geothermal heating during burial; the opal-A to opal-CT transition takes place at about 45° to 50°C, and the opal-CT to quartz transition occurs between 75° to 85°C (Isaacs, 1984). Note that since phase changes are controlled by geothermal heating, the depth at which diagenesis occurs is controlled by the geothermal gradient such that depths of phase-change boundaries may vary by hundreds of meters from place to place within a basin due to variations in the geothermal gradient.

Dolomite beds are also the products of diagenesis. Highly variable carbon isotopes within these dolomites reflect the influence of bacterial metabolism on early-diagenetic dolomite formation (Pisciotto, 1981b). Anoxic conditions which persisted on the sea floor during Monterey deposition promoted the growth of sulfate-reducing bacteria which in turn consumed all available sulfate from sediment pore-waters. This led to the alteration of calcium carbonate sediment into lithified dolomite rock (Baker and Kastner, 1981).

Description of fractures

Introduction

"Fractures" within reservoirs of the Monterey Formation are of several general types, each with a characteristic distribution, orientation, continuity and permeability. It is the combined effect of these various fracture types that determines the reservoir properties.

Planar joints

Planar joints are the most widespread fractures. At individual outcrops, two or more joint sets are typically present. The joint sets range from being perpendicular to bedding to making a high angle with bedding. Thus, they divide the bedding surfaces into rectangular or diamond-shaped prisms. Joints range in length from millimeters to meters and, rarely, to tens of meters. Most joints in an outcrop are open, but some are filled with dolomite or chert cement, or tar.

While the number and orientation of joints varies even on a bed-to-bed scale, overall the joints have a strongly preferred orientation. Along the Point Conception–Point Pedernales coastal area the majority of joints have a northeasterly strike (vector mean = N36°E) essentially perpendicular to the trend of mesoscopic and macroscopic fold axial traces (Fig. 2). This orthogonal relationship extends beyond the geographic limits of this study to the Santa Barbara Channel (Belfield et al., 1983). A smaller number of fractures have a northwesterly strike, approximately perpendicular to the northeasterly striking fractures.

Timing of planar joint formation

The clear geometric relationship between the fractures and the folds (Figs. 2 and 3) suggests that

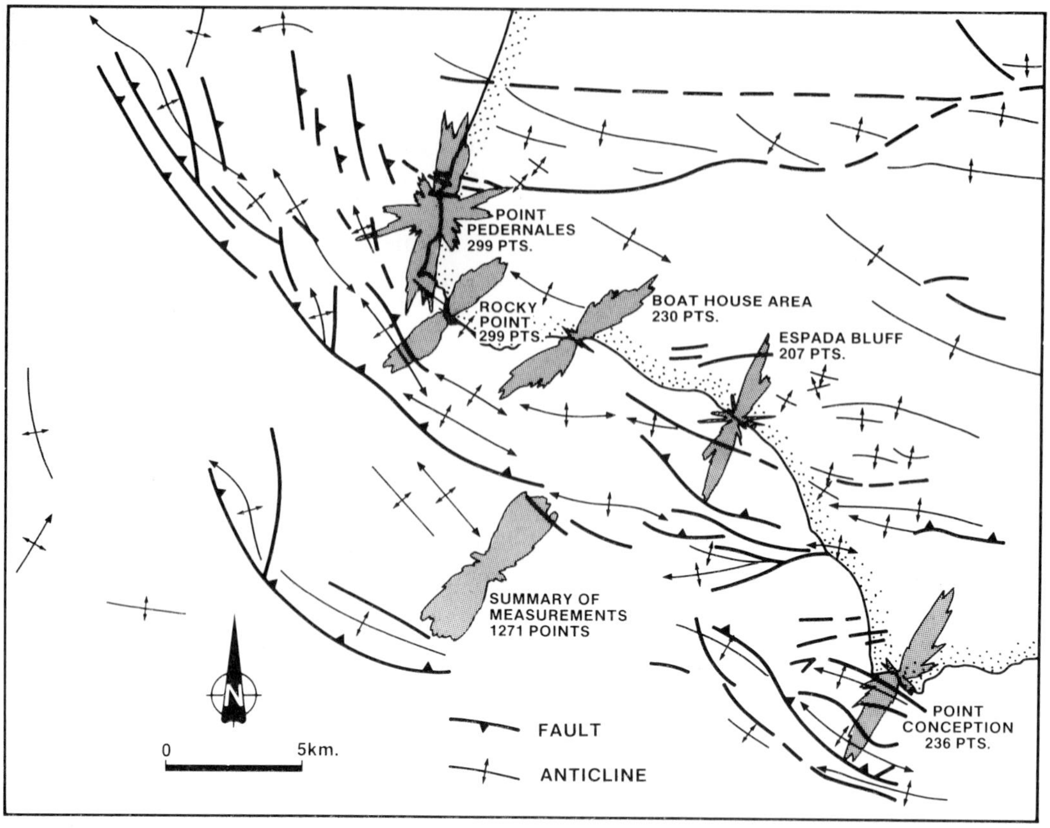

Fig. 2. Relationship of joint orientation to regional structures in the Point Conception area. The strikes of joints are displayed as running-average roses. Each of the roses along the coast summarizes data from several measuring stations. The orientation rose in the offshore area is a summary of all the coastal data (1271 measurements). In the case of steeply dipping beds, calculated joint orientations determined for horizontal bedding were used.

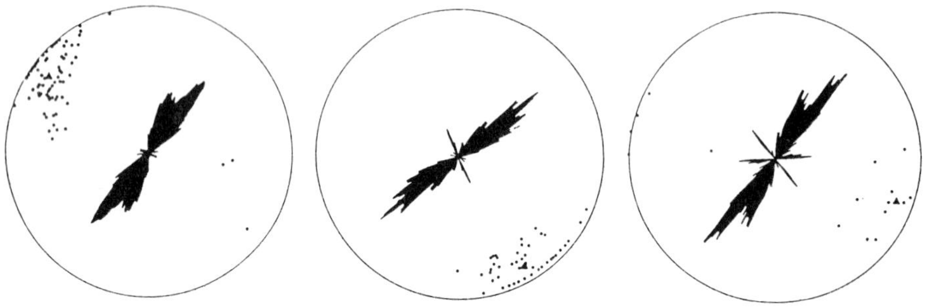

Fig. 3. Relationship of mesoscopic fold axes to fracture orientation for three areas. Running average roses of fracture orientations are superimposed on stereographic projections of fold axes. Each triangle represents the eigenvalue associated with the group of fold axes.

they formed during the Plio–Pleistocene period of folding. This is independently supported by other observations. First, chert beds are brittley deformed by the joints, implying that joint development must have occurred subsequent to the several hundred meters of burial that is needed to convert diatomaceous sediments to chert. This timing is further indicated by the presence of less well developed, but similarly oriented fractures in the Pliocene Sisquoc Formation that overlies the Monterey Formation. The occurrence of identically oriented joints in wells indicates that the joints are not modern features formed by near-surface unloading of the unit. Patterns of joint terminations and offsets indicate that the NW-striking joints formed largely after the NE-striking joints had developed.

Origin of planar joints

The geometric relationship of joints to folds, the timing of their formation, and the brittle manner in which they deform the strata indicate that the joints are of tectonic origin. Tectonically produced fractures may be either shear or extension fractures. In terms of the principal stress directions, extensional

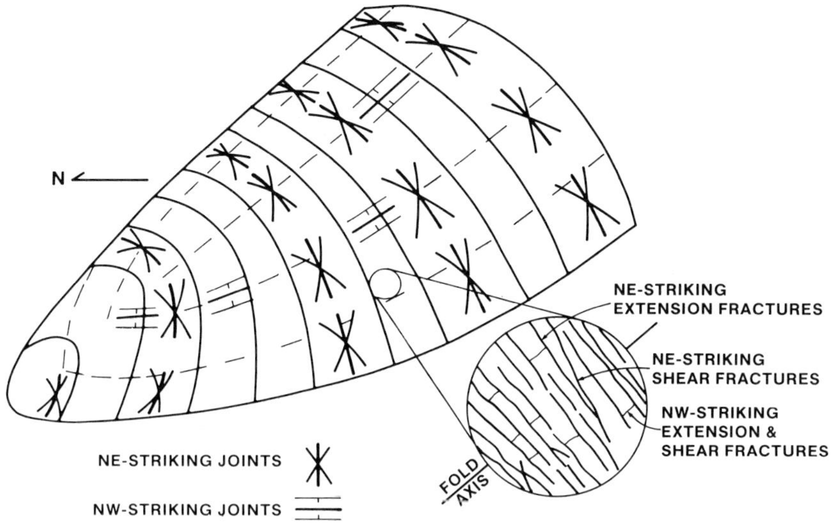

Fig. 4. Schematic drawing of anticline showing the relationship of joints to the fold. The inset shows the predominance of extension fractures approximately perpendicular to the fold axis.

fractures develop in the plane containing the greatest and intermediate principal stresses. This plane bisects the two potential shear planes each of which typically makes an angle of about 30° with the plane.

Shear and extension fractures are produced by the stresses that cause folding, but the pattern can be complex because stress orientation and magnitude vary both spatially and temporally during fold formation. Interpretation is also complicated because not all groups of possible joints may be present in a single outcrop. Additionally, early joints can be rotated by folding. Nevertheless, fractures in the Monterey Formation have a strong preferred orientation that is consistent with patterns associated elsewhere with folded strata (Sterns, 1968).

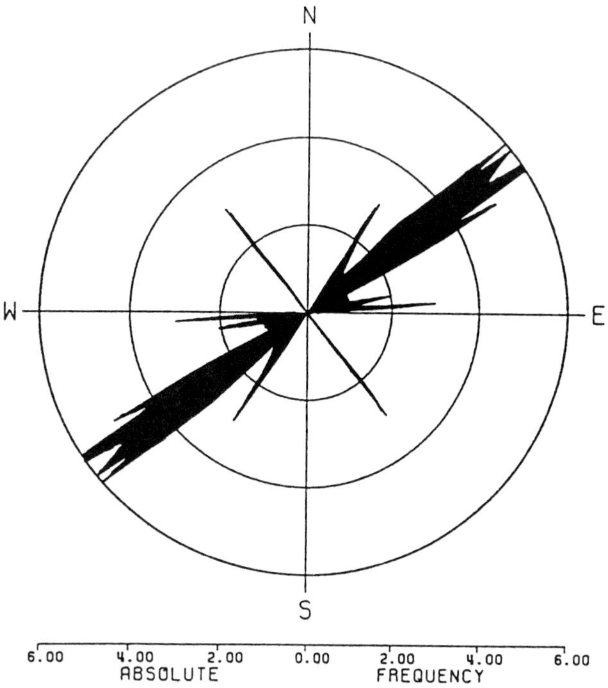

Fig. 5. Running average diagram of the strike of fractures near Point Conception. The diagram includes 25 fractures measured over about 1 square meter of bedding surface. All fractures are essentially vertical. The northeast-striking fractures are interpreted as representing a set of extension fractures flanked by a conjugate set of shear fractures.

The approximately vertical, NE-striking fractures that dominate the region are interpreted to include both extensional and shear fractures formed in response to a regional NE–SW shortening reflected by the folds and thrust faults of the region (Fig. 4). In combining the orientation data from many stations, the extension and shear fractures form a diffuse, NE-striking rose. However, when data from an individual station are examined, it is sometimes possible to distinguish extensional and shear fractures on the basis of differing strike (Fig. 5).

The less well developed NW-striking joints are also interpreted to include vertical extensional fractures and steeply dipping shear fractures. This change in fracture orientation probably developed locally as a result of extension that occurred in the outer arcs of the major folds. The observation that these joints are generally younger than the NE-striking joints is compatible with this interpretation.

Breccias

Breccias in the Monterey Formation of the Point Conception area include at least three types: (1) breccias associated with the NE-striking joints; (2) fault breccias; and (3) breccias associated with mesoscopic folds.

Planar zones of breccia up to 1 meter wide and many meters in horizontal and vertical extent have

Fig. 6. Tightly folded chert bed interbedded with unfolded clay shales. Folds have wavelengths of about 0.5 m.

developed where well developed NE-striking joints are closely spaced. In some cases, brecciation has been aided by very minor displacement across these zones. In surface exposures, pore space in these breccias is always filled entirely with tar. It is not clear whether high fluid pressures helped to develop the breccia zones; however, in one instance, tar was clearly injected from a vertical breccia body 15 meters along a flat-lying bed of brecciated shale. This may be strictly a near-surface phenomenon related to uplift.

Minor breccia zones are also associated with faults. However, in many outcrop cases, breccias formed in this way consist of fragments of brittle rock surrounded by sheared clay shale. Thus, these breccias generally do not have good porosity and permeability.

Volumetrically the most important breccias are associated with tight, meter-scale folds that develop in particular stratigraphic horizons. The folds are best developed in sections of interbedded chert and dolomite (Fig. 6). They typically have wavelengths of 20 centimeters to 3 meters and deform packets of rock 0.5 to 2 meters thick. They are bounded above and below by decollement zones, developed in shales. These zones extend laterally over large distances. Their similar orientation to regional scale folds and brittle deformation associated with the folds indicate that they are of tectonic origin.

The folding increases fracture density by increasing the number of extensional joints in the outer arcs of folds and by the rigid-body rotation of tabular pieces of beds. This latter process creates irregular voids between bed fragments (Fig. 7). Crowding in the cores of folds also produces shearing and brecciation. The net result is that discrete stratigraphic intervals have enhanced fracture porosity.

Causes of fracturing

In the Santa Maria Basin, competent rocks of the Monterey Formation are ubiquitously fractured in both surface outcrops and in the subsurface. "Fractures" include sets of planar joints as well as zones of brecciated rock that occur both discordantly and concordantly with bedding.

There are several reasons for the extensive brittle deformation of the unit. These include: (1) the brittle character of several important Monterey lithologies; (2) the ductility contrast between interbedded ductile and brittle strata; (3) the thinly interbedded nature of the strata; and (4) the possible high pore pressures at the time of deformation. Each of these points is discussed below.

Brittle character of reservoir strata

As outlined above, the formation of brittle rock within the Monterey Formation required a particular combination of the original sediment composition and burial diagenesis. In the parts of the unit that were buried to a sufficient depth, diagenetic changes imparted on the original sedimentary lithologies have produced interbedded sequences of brittle (chert,

Fig. 7. Folded dolomite beds highlighted by light-colored dolomite cement. Void spaces resulting from rigid-body rotation of beds by folding have been filled by dolomite cement. In many cases these voids remain partly unfilled.

dolomites), ductile (organic-rich shales and calcareous shales) and semi-brittle (siliceous shales, impure dolomites) rocks.

These rock types are distinct not only in terms of mineralogy, but also in other properties such as matrix-crystal size and matrix porosity; properties which influence rock ductility. Chert is an extremely brittle lithology because it is a low-porosity rock composed of a tightly interlocked mosaic of micron- to sub-micron-sized quartz crystals. The matrix porosity of the quartz-crystal mosaic is less than 5%, and is limited to submicron-sized sheet pores present between interlocked crystals. The tight interlocked mosaic of quartz crystals is extremely resistant to deformation; it deforms by brittle fracture rather than plastic flow.

Dolomite rock is brittle both because of its composition and because it is a low-porosity rock composed of tightly interlocking crystals. The matrix porosity of pure dolomites of the Monterey Formation is typically less than 5%, and the tight interlocking of crystals within the rock matrix allows little ductile deformation of the rock. Empirical observations, however, indicate that under reservoir conditions, dolomite behaves somewhat less brittley than chert.

The clay shales of the Monterey Formation are much more ductile than the chert or dolomite. Monterey shales are typical shales in that they have a high matrix porosity and a significant detrital clay content. Both the high porosity and the clay content contribute to the ductile nature of these shales; the high-porosity matrix can deform through porosity reduction, and the shale matrix is a heterogeneous mixture of detrital clay plates and quartz or opal-CT crystals which, as a consequence of high porosity, are not tightly interlocked in a crystal mosaic. When the shale is stressed, these grains are free to slip within the rock matrix, thus resulting in a ductile rock. Siliceous shales with very high silica contents have high matrix porosities, but exhibit semi-brittle behavior due to their low clay content.

The brittle character of certain lithologies in the formation is of fundamental importance. While fracture density varies widely at any one location the relative fracture densities between various lithologies are very predictable and increase from shale to dolomite to siliceous shale to chert (Figs. 8 and 9). Commonly, fractures in brittle lithologies terminate at the contact with ductile beds (Fig. 9). Additionally, fractures in shales can be seen to have closed as a result of plastic flow of the shale (Dunham and Blake, 1987). These fractures are very tight, even in surface outcrops, whereas well developed fractures in competent dolomites and cherts contain visible voids and vugs.

Ductility contrast

Ductility contrast, or the juxtaposition of brittle and ductile rocks, is important in two ways. Under early lateral compression, weak ductile layers deformed in part by some combination of layer-parallel shortening, vertical thickening and compaction. This

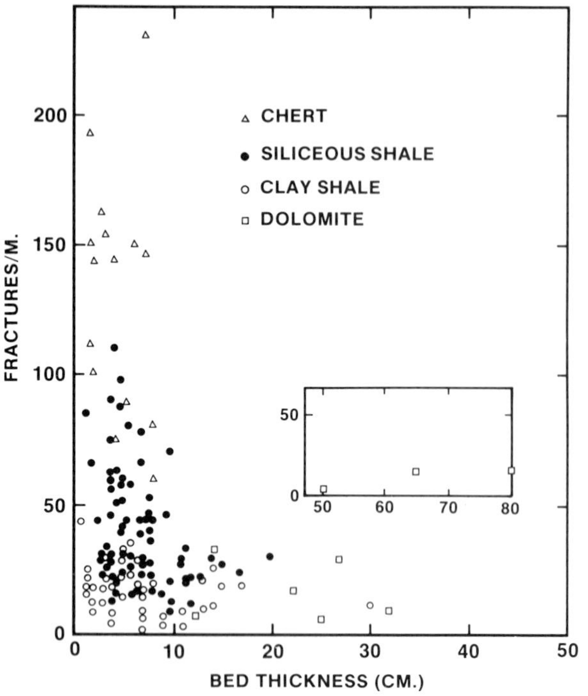

Fig. 8. Fracture density as a function of bed thickness for chert, siliceous shale, clay shale and dolomite of the Monterey Formation.

deformation concentrated the stresses on the competent layers that could only deform brittley. Moreover, large ductility contrasts promote the development of folds (Ramsay, 1967). In the case of the Monterey Formation, this has resulted in the extensive development of mesoscopic intraformational folds. Bending and local extension produced by these folds resulted in enhanced fracturing.

Thin beds

The thin-bedded nature of the strata make it mechanically easier for folds to form (Ramsay, 1967). Thin beds also may be more prone to extensional fracturing. This is because the bedding surface area of thin beds is large relative to their cross sectional areas. This retards bedding plane slip which tends to keep early fractures from widening and, instead, promotes the development of new fractures (Price, 1966).

High fluid pressures

It has been suggested that high fluid pressures may have been partly responsible for the brittle deformation in the Monterey Formation. One possibility is that high fluid pressures produced natural hydraulic fracturing (Redwine, 1981; Roehl, 1981; Snyder, 1987). Less extreme fluid pressures also could have had the effect of reducing the effective confining pressure on the strata, thereby promoting brittle behavior. However, presently the strata are not overpressured.

Timing of sediment diagenesis relative to timing of oil generation

Chert, the best of the fractured reservoir rocks, forms during burial diagenesis as described earlier. Thus, there is a considerable time lag between

Fig. 9. Highly jointed chert bed overlying unjointed clay shale bed. Prominent joints in chert strike to the northeast.

deposition of pure silica sediment and evolution of that sediment into brittle rock. In some areas of the Santa Maria Basin where Miocene sediments have not been deeply buried, pure silica sediments are still preserved as opal-A diatomites. These diatomites are very high porosity rocks which are soft and ductile and not susceptible to extensive fracture development.

Siliceous rocks must undergo at least opal-CT grade diagenesis before they are brittle enough to fracture. Consequently, for maximum development of reservoir-quality fractures, tectonic folding should follow and not precede silica diagenesis. In order for an oil-filled fractured reservoir to be formed in the Monterey Formation, a definite sequence of events must occur. Firstly, silica-rich sediments must be deposited. Secondly, these sediments must be buried at least deep enough to undergo opal-CT diagenesis before application of tectonic stress. Thirdly, the now-brittle siliceous rocks must then undergo tectonism in order to form fractures (Fig. 10). Only then do the rocks contain sufficient fracture porosity to receive an oil charge. Oil generation and migration must take place only after all these other steps.

There are many areas of the Santa Maria Basin where these events have not occurred in the proper order to produce fractured reservoirs. Pure silica, non-fractured diatomites are present where tectonic uplift took place prior to silica diagenesis. Additionally, oil-saturated diatomites attest to oil migration taking place prior to fractured reservoir development. Vast volumes of heavy oil are reservoired within the high matrix porosity of diatomites, but the extremely low permeability of these non-fractured rocks makes it very difficult to recover this oil. In other cases, maximum silica diagenesis occurred after folding. Consequently, fracturing is not well developed in these strata.

Hydrocarbon generation within the Santa Maria Basin

The Monterey Formation contains unusually organic-rich shales with total organic carbon (TOC) contents commonly exceeding 5%, with peaks of over 18 wt.% TOC. The organic richness derives from low terrigenous influx to dilute organic matter within the sediment, as well as anoxic conditions at the sediment-water interface during deposition which permitted preservation of the organic matter. Thus, the Monterey Formation contains its own source rocks and reservoir rocks. In areas of the Santa Maria Basin where tectonic subsidence has carried

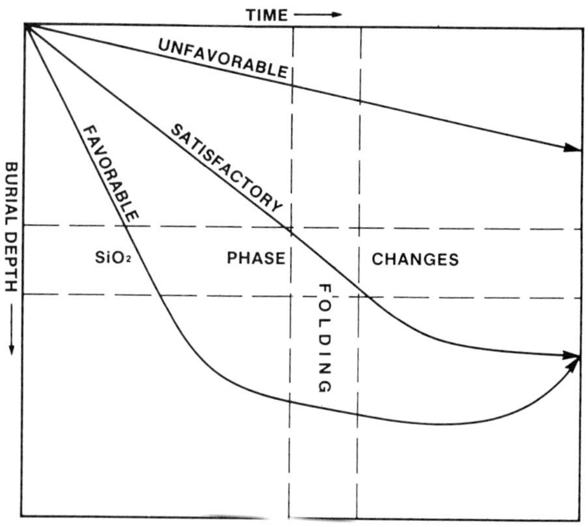

Fig. 10. Time versus burial depth diagram for siliceous sediments. The most favorable burial path for development of fractured reservoirs in the Monterey Formation was the situation where silica phase changes occurred prior to folding. Shallowly buried sediments were never converted to brittle chert.

the organic-rich shales to depths where temperatures exceed about 120°C, oil generation has taken place.

Hydrocarbon migration

Under favorable conditions where diagenetic changes occurred prior to deformation, and deformation occurred prior to or during oil generation, oil migration was controlled by the orientation of fracture networks. Since brittle, fracturable rocks are interbedded with source rocks, oil expelled from organic shales immediately entered the fracture network in the brittle rocks. The most interconnected and permeable of these are the extensional joints that are oriented perpendicular to fold axes (Fig. 11). This joint network encouraged oil generated in synclinal axes to migrate updip, perpendicular to the fold axes into anticlinal crests (Fig. 12). If a trap is present, the oil accumulates in a fractured reservoir. The trap may be a stratigraphic pinchout of reservoir facies, but more likely the trap is an anticline. If, on the other hand, a trap is not present, the oil may become absorbed by high porosity diatomites, or it may leak to the surface to form one of the numerous oil seeps that occur in the area.

Summary

Tectonics and sedimentation have combined to localize major hydrocarbon accumulations within fractured reservoirs of the Santa Maria Basin. *Economic oil accumulations are present only at localities where a particular depositional facies, at an appropriate diagenetic grade, have been fractured during folding and*

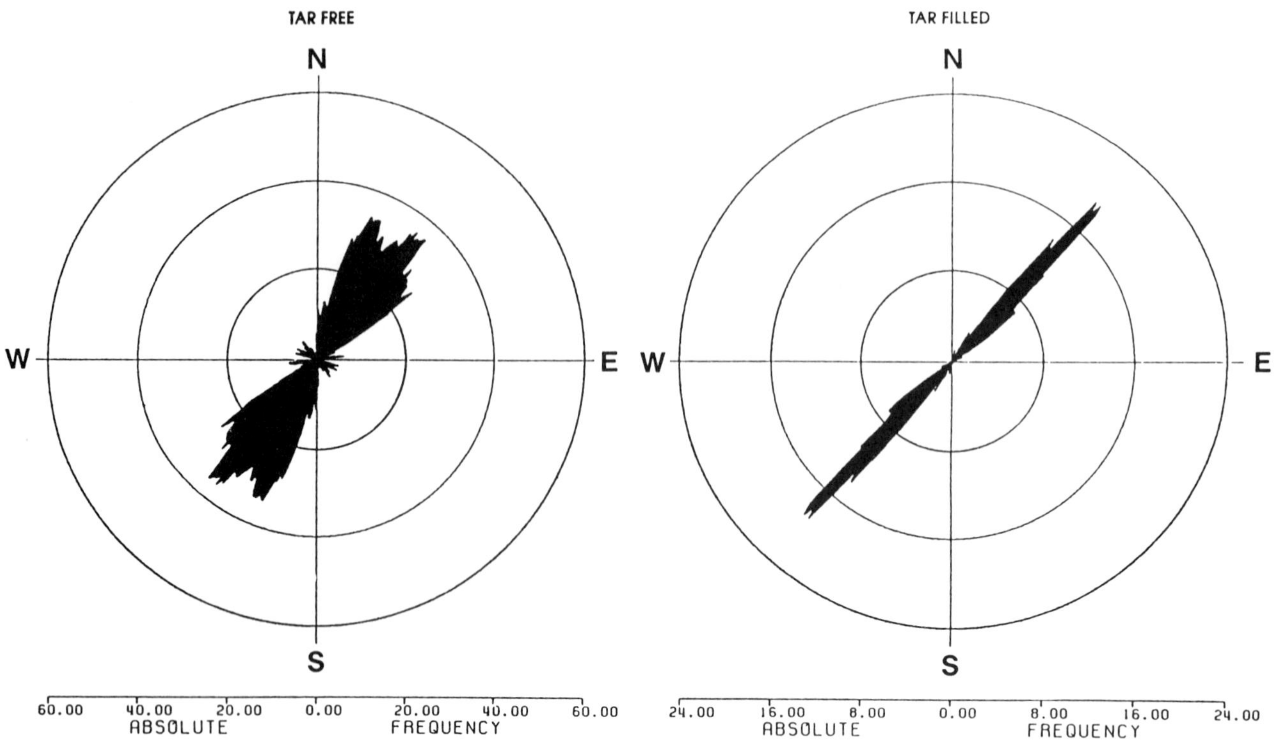

Fig. 11. Running average plots of the strike of tar-filled and tar-free joints at Point Arguello. The region is characterized by generally NE-striking joints and some NW-striking joints. Tar occurs only in NE-striking joints with a limited range of strikes. These are interpreted to extension fractures that allowed oil migration because of their degree of interconnectivity and their high permeabilities.

growth of anticlinal traps prior to peak oil generation from interbedded organic-rich shales. This coincidence of sedimentology, diagenesis and structural growth, combined in the proper order, has led to the accumulation of several billion metric tonnes of oil within fractured reservoirs. Ultimate recovery from known fields, and future discoveries may approach a billion metric tonnes.

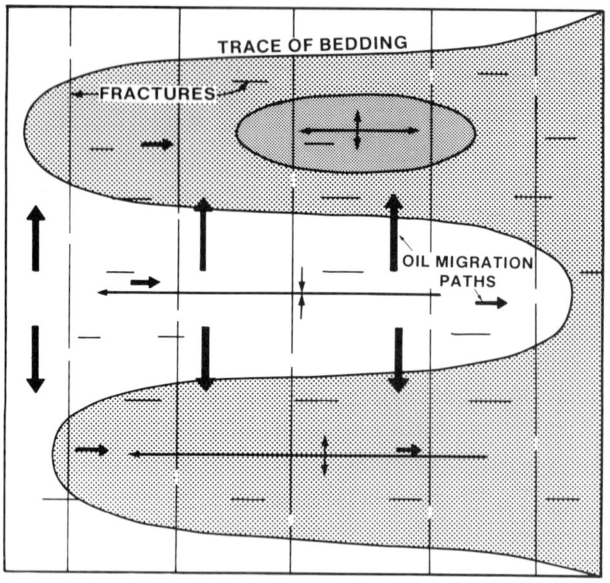

Fig. 12. Diagrammatic view of oil migration in the Monterey Formation. Oil generated from Monterey Formation shales in synclinal areas migrates toward anticlines via the fracture network in Monterey Formation cherts and dolomites. The more continuous and open network of extensional fractures orthogonal to the folds encourages oil migration directly up the flanks of the anticlines.

Acknowledgements

This paper benefited from the thoughtful reviews of R. Knipe, Alan Roberts and Bob Varga. We thank Unocal for granting permission to publish this paper.

References

Baker, P.A. and Kastner, M., 1981. Constraints on the formation of sedimentary dolomite. Science, 213: 214–216.

Belfield, W.C., Helwig, J., LaPoint, P. and Dahleen, W.K., 1983. South Elwood oil field, Santa Barbara Channel, CA. A Monterey fractured reservoir. In: C.M. Isaacs and R.E. Garrison (Editors), Petroleum Generation and Occurrence in Miocene Monterey Formation, California. Pacific Section, Society of Economic Paleontologists and Mineralogists, pp. 213–221.

Crouch, J.K., Bachman, S.B. and Shay, J.T., 1984. Post-Miocene compressional tectonics along the central California margin. In: J.K. Crouch and S.B. Bachman (Editors), Tectonics and Sedimentation Along the California Margin. Pacific Section, Society of Economic Paleontologists and Mineralogists, 38: 37–54.

Dunham, J.B. and Blake, G.H., 1987. Guide to coastal outcrops of the Monterey Formation of western Santa Barbara County California. Pacific Section, Society of Economic Paleontologists and Mineralogists, 53, 37 pp.

Graham, S.A., 1987. Tectonic controls on petroleum occurrence in central California. In: R.V. Ingersoll and W.G. Ernst (Editors), Cenozoic Basin Development of Coastal California, Rubey Volume VI. Prentice-Hall, Englewood Cliffs, N.J., pp. 47–63.

Ingle, J.C. Jr., 1981. Cenozoic depositional history of the northern continental borderland of southern California and the origin of associated Miocene diatomites. In: C.M. Isaacs (Editor), Guide to the Monterey Formation in the California Coastal Area, Ventura to San Luis Obispo. Pacific Section, American Association of Petroleum Geologists, 52: 1–8.

Isaacs, C.M., 1980. Diagenesis in the Monterey Formation examined laterally along the coast near Santa Barbara, California. Ph.D. Dissertation, Stanford University, Stanford, Calif., 329 pp.

Isaacs, C.M., 1984. The Monterey-key to offshore California boom. Oil Gas J., 82: 75–81.

Murata, K.J. and Larson, R.R., 1975. Diagenesis of Miocene siliceous shales, Temblor Range, California. U.S. Geol. Surv., J. Res., 3: 553–566.

Ogle, B.A., Wallis, W.S., Heck, R.G. and Edwards, E.B., 1987. Petroleum geology of the Monterey Formation in the offshore Santa Maria/Santa Barbara areas. In: R.V. Ingersoll and W.G. Ernst (Editors), Cenozoic Basin Development of Coastal California, Rubey Volume VI. Prentice-Hall, Englewood Cliffs, N.J., pp. 382–406.

Pisciotto, K.A., 1981a. Diagenetic trends in the siliceous facies of the Monterey Shale in the Santa Maria region, California. Sedimentology, 28: 547–571.

Pisciotto, K.A., 1981b. Review of secondary carbonates in the Monterey Formation, California. In: R.E. Garrison et al. (Editors), The Monterey Formation and Related Siliceous Rocks of California. Pacific Section, Society of Economic Paleontologists and Mineralogists, Spec. Publ., pp. 273–284.

Price, N.J., 1966. Fault and Joint Development in Brittle and Semi-Brittle Rocks. Pergamon Press, Oxford, 176 pp.

Ramsay, J.G., 1967. Folding and Fracturing of Rock. McGraw-Hill, New York, N.Y., 568 pp.

Redwine, L.E., 1981. Hypothesis combining dilation, natural hydraulic fracturing, and dolomitization to explain petroleum reservoirs in Monterey Shale, Santa Maria area, California. In: R.E. Garrison, R.G. Douglas, K.E. Pisciotto, C.M. Isaacs and J.C. Ingle (Editors), The Monterey Formation and Related Siliceous Rocks of California. Pacific Section, Society of Economic Paleontologists and Mineralogists, pp. 221–248.

Roehl, P.O., 1981. Dilation brecciation — a proposed mechanism of fracturing, petroleum expulsion and dolomitization in the Monterey Formation, California. In: R.E. Garrison, R.G. Douglas, K.E. Pisciotto, C.M. Isaacs and J.C. Ingle (Editors), The Monterey Formation and Related Siliceous Rocks of California. Pacific Section, Society of Economic Paleontologists and Mineralogists, pp. 285–315.

Snyder, W.S., 1987. Structure of the Monterey Formation: stratigraphic, diagenetic and tectonic influences on style and timing. In: R.V. Ingersoll and W.G. Ernst (Editors), Cenozoic Basin Development of Coastal California, Rubey Volume VI. Prentice-Hall, Englewood Cliffs, N.J., pp. 321–347.

Sterns, D.W., 1968. Certain aspects of fractures in naturally deformed rocks. In: R.E. Riecker (Editor), National Science Foundation Advanced Science Seminar in Rock Mechanics for College Teachers of Structural Geology. Air Force Cambridge Research Laboratories, Bedford, Mass., pp. 97–118.

R.G. HICKMAN *Unocal Science & Technology Division, 5460 East La Palma Avenue, P.O. Box 68076, Anaheim, CA 92817-8076, U.S.A.*

J.B. DUNHAM *Unocal International Division, P.O. Box 7600, Los Angeles, CA 90051, U.S.A.*

Reservoir compartmentation by faults in Cormorant Block IV, U.K. northern North Sea

W. Sassi, S.E. Livera and B.P.R. Caline

An important petroleum engineering challenge relates to the effect of partially sealing faults on reservoir compartmentation. This problem has been examined in the Cormorant Block IV, U.K. northern North Sea.

Sedimentological review of all the Brent Group core in this block has led to a refined classification of the reservoir units and more detailed models of facies architecture. Structural descriptions of the core-scale fractures and faults were compared with the large-scale fault pattern obtained from the interpreted top Brent three-dimensional seismic map to investigate the impact of several scales of faults on potential compartmentalization of the reservoir.

Large-scale faults are thought to be sealing by means of juxtaposition and clay smearing, for which some evidence was found in the cores. In the small scale, minor shears present some degree of grain size reduction and cementation, the combination of which leads to appreciable reduction in permeability. However, their effect as barriers to the reservoir flow behavior is found to be negligible. Orientations of small-scale faults in cores could be measured and were found to be consistent with the large-scale fault-strike distributions, suggesting a genetic link between small- and large-scale features. This observation has helped to validate a statistical model for a hierarchic fault throw distribution that can be used to quantify the occurrence of medium-scale sub-seismic features. When the statistical model is linked to a reservoir subdivision describing variations in sand thickness and shale content, approximate hydrocarbon volumes in individual fault-block compartments can be calculated. These predictions agree with well test analyses.

The approach used suggests that certain reservoir units or sequences are more susceptible to fault-block compartmentation than others. It can be used for heavily faulted reservoirs to help in assigning different recovery factors to each sequence.

Introduction

The concept of fault self-similar relationships has been highlighted by many authors (e.g., King, 1983; Haldorsen et al., 1988; Villemin and Sunwoo, 1987; Chiles, 1988; Van der Molen, 1989; Childs et al., 1990) who have observed that fault lengths and fault displacement magnitudes tend to be distributed in a scale-independent manner. Such empirical relationships can be illustrated using hypothetical hierarchic fault construction models honoring a given homogeneous bulk strain (King, 1983). Although the problem of the creation of a self-similar network of faults is not yet fully understood, the use of these empirical relationships is found to have an important practical application in predicting fault distribution below the level of seismic resolution (e.g., fault throw magnitudes below 50–100 feet).

This study presents an analysis of the fault pattern characterizing the Cormorant Block IV accumulation, using core and seismic data-sets. A self-similar relationship for the distribution of fault throw is used to characterize the implications of fault-block compartmentation on reservoir quality and production behavior.

The research project was intended to provide a reservoir geological model of the Block IV accumulation using all core, log, pressure and field performance data. An integrated approach was adopted in which the sedimentology was reviewed in detail on the extensive core data. The outcome of this part of the project is only summarized below and more detail can be found in Livera and Caline (1990). This review was then linked with the structural analysis discussed below, which is based on data obtained from cores and the 1987 seismic interpretation map of the top Brent Group reservoir level (see Fig. 2).

The Cormorant field (Fig. 1) is situated 150 km north-east of the Shetland Islands (U.K., North Sea), and is divided into four separated structures situated in four main fault blocks (I to IV, see Fig. 2), of which Block IV contains the most oil. Block IV consists of an area of approximately 30 km^2 and overlies a large, partly syndepositional, normal fault (the Block IV western boundary fault).

The fault downthrows to the east and has a total displacement of about 1000 ft (300 m) at the top reservoir (Brent Group) level. Block IV is dome-shaped and is broken up into a number of smaller fault-bounded compartments (Fig. 2). This compos-

Structural and Tectonic Modelling and its Application to Petroleum Geology edited by R.M. Larsen, H. Brekke, B.T. Larsen and E. Talleraas. NPF Special Publication 1, pp. 355–364. Elsevier, Amsterdam.

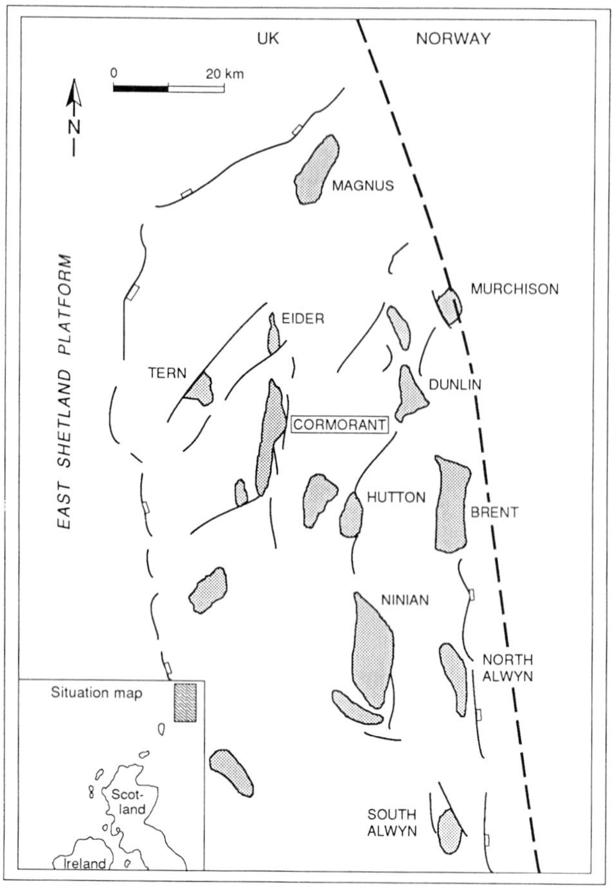

Fig. 1. Location of the Cormorant field and adjacent oil fields in the northern North Sea, U.K. sector.

ite fault block has quite distinctive sedimentological and structural features, the oil/water contacts are deeper than elsewhere in the field and it is being developed as a separate accumulation. Moreover, several different oil/water contacts within Block IV have been recorded in the accumulation and many faults appear to be at least partially sealing during production (i.e., only slow recharge occurs across faults when wells are closed in). Shell Expro are producing the low volatile oil by waterflooding within individual major fault blocks.

About 43% of the penetrated Brent Group section in Block IV has been cored (amounting to over 4100 ft or 1200 m) and a total of 28 wells had penetrated the structure at the time of the study. This is an exceptional amount of core and it has enabled a more thorough understanding of the sedimentology and structural geology than could be achieved from wireline log data alone.

Structural setting, sedimentology and reservoir framework

The Cormorant Block IV structures result from an overall E–W extension of the Northern Viking graben, inducing mainly N–S trending normal faulting in the cover formations above the Caledonian basement. The 1987 top Brent Group reservoir map, derived from a three-dimensional seismic survey,

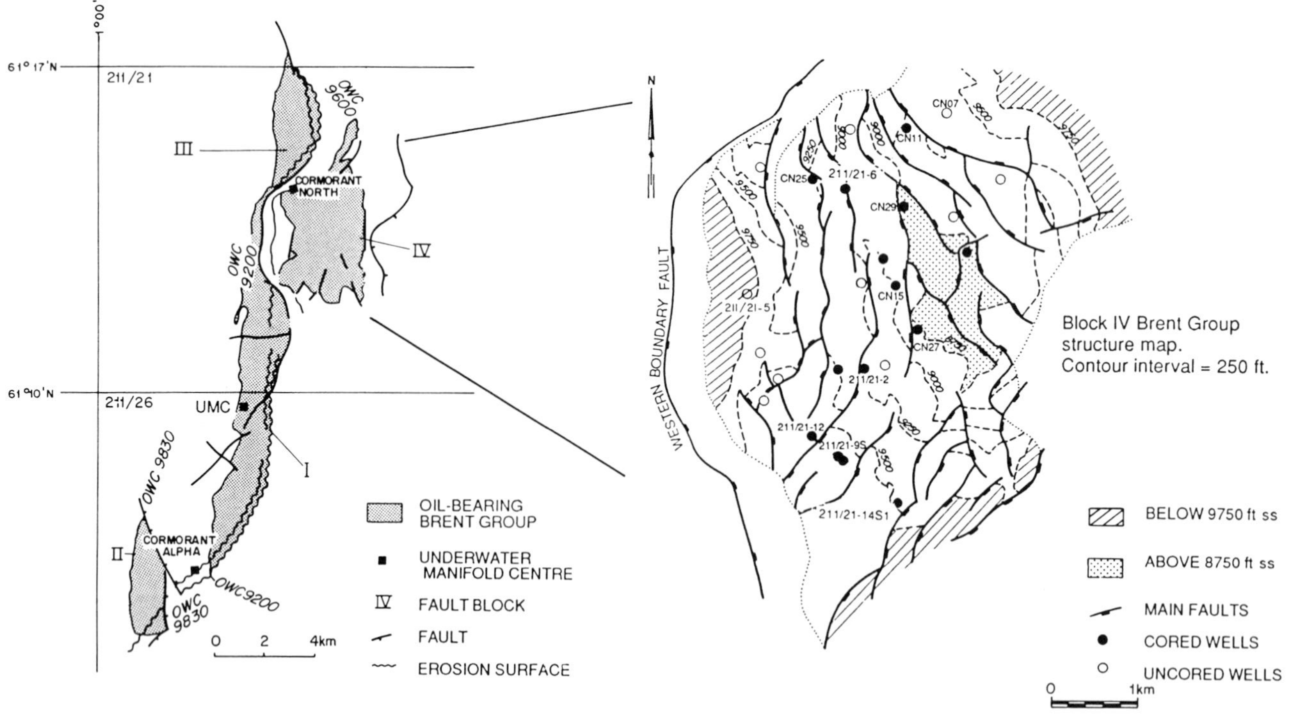

Fig. 2. Map of the Cormorant field showing the four main hydrocarbon accumulations, termed fault blocks I–IV. The enlargement of the downthrown Block IV shows the curved outline of the Block IV western boundary fault (separating it from Blocks I and II), the domed nature of the accumulation and the numerous faults that occur within the structure. (Only the wells referred to in the text or later figures are numbered.)

reveals that the dominant fault directions change within Block IV. N–S faulting dominates in the central part of the block, while NW–SE and NE–SW are the major trends in the northern and southern parts of Block IV, respectively. The role of the main basement structures and the large-scale structural setting, have been discussed by Speksnijder (1987). He suggested that E–W extension has involved some strike-slip movement along pre-existing NW–SE and NE–SW basement boundary faults, and an accompanying eastward displacement of the central block. A decollement horizon was suggested from the western boundary fault separating Block IV from the rest of the Cormorant field, the depth and shape of which remains undetermined.

The western boundary fault of Block IV (Fig. 2) had a complex history of reactivation of normal movements (e.g., indicated by growth throughout the Middle and Upper Jurassic sequences across this major boundary fault) from the Triassic to the Tertiary. This fault was active during some periods of Brent Group deposition, as deduced from correlations between Block IV and Blocks I and III to the west. Significant increases in reservoir thickness occur across the fault in the Broom, Upper Ness and Tarbert horizons.

The lower reservoir (Broom, Rannoch, Etive and Lower Ness Formations, Fig. 3)

The Broom Formation thickens abruptly across the western boundary fault from 7 to 37 ft in Blocks I and III to 40 to 61 ft in Block IV. It forms a more significant reservoir sequence in Block IV than elsewhere in the Cormorant field. Although the reservoir quality of the package is generally good, intense diagenetic modification (including pervasive carbonate cementation) makes its prediction to uncored wells difficult.

The 100-ft to 160-ft thick upward-coarsing sequence of the Rannoch and Etive Formations is the most important reservoir interval in Block IV and displays considerable lateral variability. Although the block is heavily faulted, this sequence is sufficiently thick and sandy that only a few of the largest, seismically resolvable faults are sealing. Sedimentological aspects, therefore, have an impact on the waterflood behavior of these horizons.

The variations in the lower part of the Rannoch Formation can be tied into time lines established by correlations based on grain size, degree of bioturbation and wireline response. These are illustrated in Fig. 4. These time lines dip to the NNE at approximately 0.3° and confirm the NNE direction of progradation established for southern Cormorant (Budding and Inglin, 1981). The time lines determine the correlation of better reservoir quality in the Lower Rannoch Formation. Data from a survey with the repeat formation tester (RFT) indicate that a correlatable carbonate-cemented horizon at the top of one of the upward-coarsing profiles in unit 4.4 (Fig. 4) is a vertical pressure barrier, across which a differential pressure of 300 psi has been measured.

The Lower part of Ness Formation can be demonstrated to be chronostratigraphically equivalent with the Etive Formation in Block IV (Fig. 4). Thus a layered reservoir zone of thin (<10 ft) sands separated by vertically sealing shales (holding 50–100 psi differential pressures) is laterally equivalent to a blocky sand of excellent reservoir quality to the north. Dark anoxic lagoonal mudstones near the base of the Lower Ness Formation in the southern part of the block hold over 2000 psi in severely depleted wells.

The upper reservoir (Upper Ness and Tarbert Formations)

Compared with the underlying sequences, the Upper Ness Formation is laterally more consistent, and most reservoir units (Fig. 3) can be correlated uniformly across the whole of Block IV. Two sand-rich packages (units 2.2–2.3 and 2.6, Fig. 3) are separated by vertically sealing 20–30 ft thick lagoonal shales. Internal variation within these packages is marked, and a number of different types of sand body are encountered.

Because wireline log response is ambiguous, we rely on core data for much of the interpretation of the Tarbert Formation. The sediments are thickest in the north and west (107 ft in 211/21-5, Fig. 2), and thin to the south and east approximately onto the crest of the structure. There is an erosive contact between the Tarbert and Upper Ness Formations in some crestal wells, complementing similar observations in Blocks I and III. The facies associations recognized in the core appear to onlap successively against the domed area in the center of the block, suggesting that internal unconformities are present within the Tarbert Formation and that is was deposited during continued structuring of the Block. As a result there is not a simple relationship between the present top Brent structure map and the thickness of the Tarbert Formation. Further modelling of the structural development of Block IV will be required to understand the distribution of the many Late Jurassic and Early Cretaceous unconformities.

Structural analysis

Faults were analyzed both at field scale from the seismic map and at core scale.

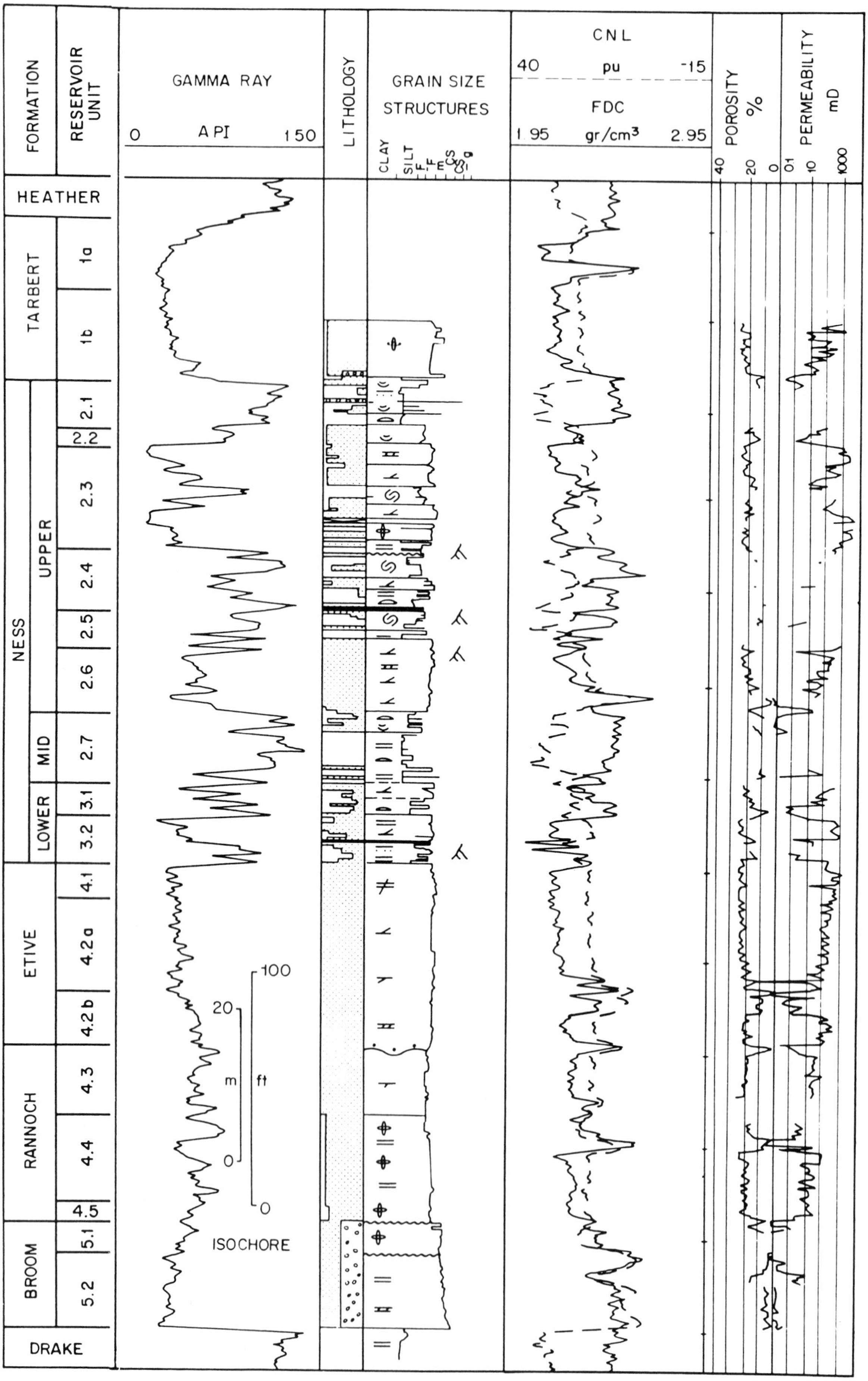

Fig. 3. Type log of Cormorant Block IV (from well CN 25, Fig. 2).

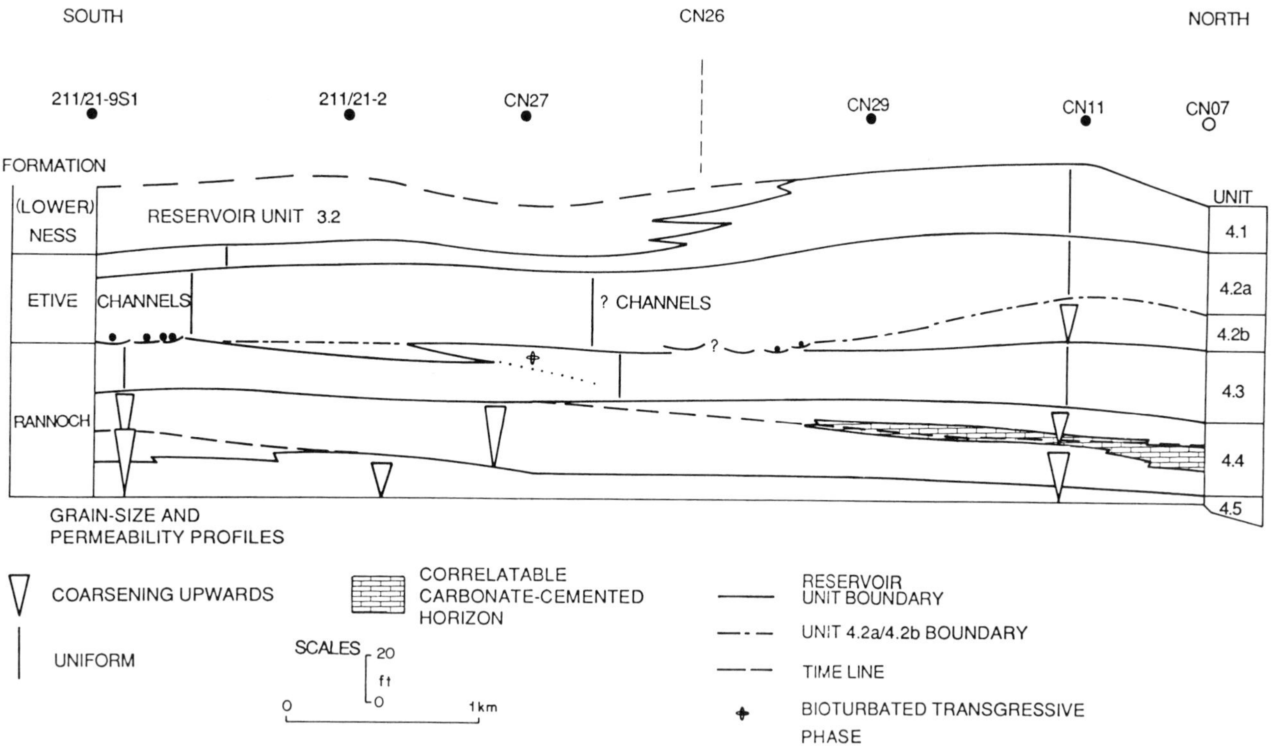

Fig. 4. South-to-north cross-section illustrating the lateral variation in the Rannoch, Etive and Lower Ness Formations of Cormorant Block IV (for location of wells, see Fig. 2).

Field-scale fault pattern

The 1987 top Brent map shows an improved resolution compared to older two-dimensional data. This map contains 126 faults (not all reported in Fig. 2) from which we have examined both the areal fault strike distributions (Fig. 5) and the cumulative fault throw frequency distributions (Fig. 6). In Fig. 5, the largest rose diagrams show that the dominant fault direction is N–S in the central part of Block IV, NW–SE in the north and NE–SW in the south.

The log-log diagram in Fig. 6 shows the cumulative number of faults per square kilometer in the mapped area that have a throw greater or equal to a given value. The throw values used are the maximum fault displacement mapped for each fault. Although the fault displacements span over two orders of magnitude, the data-set seems to approach a reasonably linear trend for faults with a displacement above 50 feet. Below this level, it is expected that the number of faults is underestimated owing to the limited seismic resolution (King, 1983; Villemin and Sunwoo, 1987; Childs et al., 1988). It is therefore proposed that the fault throw distribution be characterized by a self-similar relationship with a slope factor of -1.5 obtained by linear regression using the most representative data points. The significance of such a high value is somewhat subjective, since there is no clear understanding of its origin. Such diagrams will have a slope factor of -1.0, if one assumes, for example, that an homogeneous deformation is accommodated by a simple hierarchic fault construction in which the cumulative number of faults doubles each time the throw is reduced by half (King, 1983; Van der Molen, 1989). This discrepancy between a theoretical value of -1 and an observed one of -1.5, could suggest a look back at the interpretation map, but this would put too much weight on such a conceptual model of discontinuous deformation. However, considering the fact that the deformation in the Cormorant Block IV is not homogeneous, we assume the observed slope value to be a kind of property parameter, as suggested by others (Villemin and Sunwoo; 1987). These authors obtained high slope values (-1.3) from measurements taken from outcrops in coal mines in northern France.

Small-scale faults from core observation

Small-scale normal faults and fractures visible on cores were recorded on 1:200 scale core-log summary sheets. From these data and observations of fault cut-outs from preliminary correlations, certain cores were selected for more detailed description and measurement.

Fractures are found in many sandstones, particularly in the Etive and Rannoch Formations. These features are thin (1–5 mm wide) cemented shears, with displacements in the order of centimeters. Shear gouges are characterized by a light-colored cement,

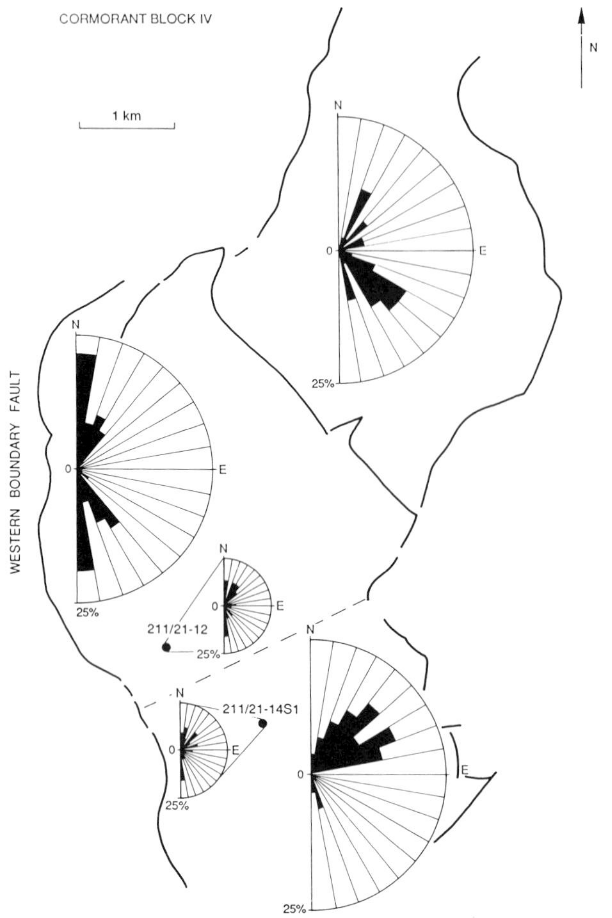

Fig. 5. Fault azimuth-frequency diagrams derived from the top Brent map (large diagrams) compared with fracture measurements in well 211/21-12 and 14S1 (small diagrams).

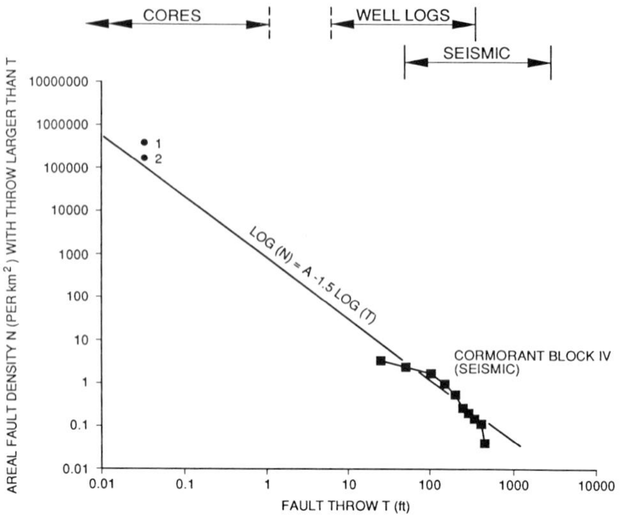

Fig. 6. Cumulative fault throw size-frequency diagram for Cormorant Block IV. ● = Estimates derived from core data and probabilistic methods; *1:* from Etive Rannoch well 211/21-12; *2:* from Etive Rannoch well 211/21-14S1.

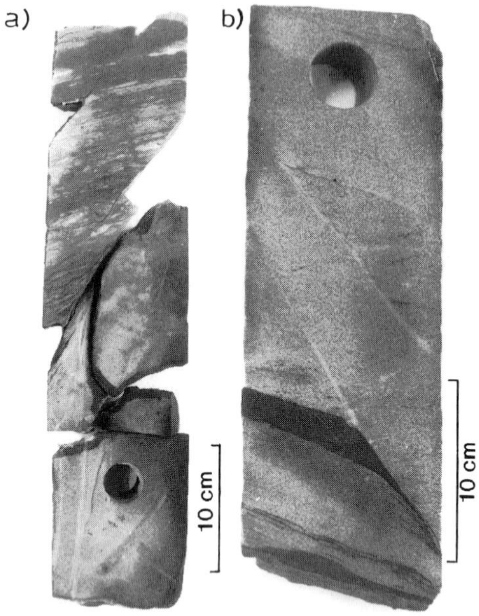

Fig. 7. (a) Photograph of slabbed core showing small fault in the Ness Formation. Note the soft sediment deformation in the downthrown block and the associated fractures in the sandstones (possibly of hydraulic origin). Well 211/21-6 (core depth 9040 ft). (b) Photograph of slabbed core showing a fault and clay smear in the Etive Formation. Displacement is greater than 10 cm. Note the associated fracture above. Well 211/21-14S1 (core depth 10342 ft).

and sometimes include dark micaceous streaks. These fractures are considered as small-scale normal faults. Some degree of cataclasis, reorientation of micas and cement overgrowths were observed in the thin sections. They are recognized as barriers of the flow from permeability reduction measurements made across them using standard core plugs from selected samples. A reduction factor of one-tenth of the matrix permeability was measured across these sections.

Over 80 of these minor normal faults (with over 1 cm offset) in cores of the Etive and Rannoch Formations of wells 211/21-12 and 211/21-14S1 were reoriented using bedding planes and well deviation parameters to establish their true azimuth and dip. Measurements are plotted on azimuth-frequency diagrams in Fig. 5.

Larger-scale normal faults, with offsets measured in decimeters, are also present in some cores, these show evidence of clay smear and ductile behavior of shales (Figs. 7a, b). Both types of feature indicate that the faults were formed before the sediments were consolidated, thus confirming earlier suggestions that clay smearing can be an effective sealing mechanism (Speksnijder, 1987) in this field.

Relationship between small- and large-scale structures

There is a good agreement between the fault azimuth-frequency diagrams measured from the top

Brent map and the oriented core measurements (Fig. 5). This suggests that the large- and the small-scale structures could have been formed under a similar extensional stress regime, as shallow or syn-sedimentary features associated with the Block IV western boundary fault.

It was assumed that the two scales of structures are genetically linked and a comparison between their horizontal (mappable) areal densities was attempted. It was necessary to derive equivalent areal densities for the small-scale normal faults observed on the cores. In order to do so, the approach used in this work requires an estimate of the linear (vertical) fault density and the frequency distribution of fault dips together with the volumetric fault density and the mean fault length. The first two parameters are measured on the cores, while the latter parameters had to be estimated numerically using a stochastic approach in which fault termination observations within the cores are taken into account (Tiller, 1986). The ratio of intersected and terminating small-scale faults, and an assumed distribution law for the fault length spectrum (i.e., normal, log-normal or exponential) are the most important input parameters.

Horizontal areal densities were derived using a recurrent relationship (see Matheron, 1967) relating linear and areal or areal and volumetric fault densities. These relationships can be expressed as follows:

$$l_s = l_v \cdot E(L \cdot \sin\theta) \qquad (1)$$

$$l = l_v \cdot E(L^2 \cdot \cos\theta) \qquad (2)$$

where l is the linear (vertical) fracture density, l_s is the areal fracture density, l_v is the volumetric fracture density, θ is the fracture dip angle, and L is the fracture length; $E(\,)$ stands for the mean value.

Rearranging eqns. 1 and 2, we obtain:

$$l_s = \frac{l \cdot E(L \cdot \sin\theta)}{E(L^2 \cdot \cos\theta)} \qquad (3)$$

The results, plotted in Fig. 6, show good agreement between the density of small faults as derived from cores and as predicted from seismic data. This further supports the use of the fault-throw distribution relationship.

Effect of fault sealing on the reservoir framework

Effect of small scale faults

The calculated distribution of length data for the small scale normal faults, the vertical spacing as

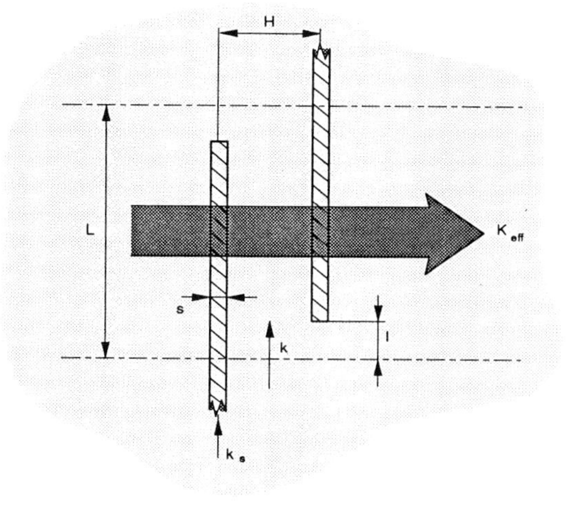

$$K_{eff} = k\left(\frac{H-s}{L}\right)\left(\frac{H}{T}\right)\left(\frac{L-2l}{L}\right)\left(\frac{k\,k_s\,H}{sk+(H-s)\,k_s}\right)+\left(\frac{2l}{L}\right)\left(\frac{k\,k_s\,H}{0.5\,sk+(H-0.5\,s)\,k_s}\right)$$

EXAMPLE FROM ETIVE FORMATION

INPUT l = 0.8 m L - l = 2.5 m s = 0.5 cm (core observation) H = 50 cm (core observation) T = L + 2H - 2l - s	
MATRIX PERMEABILITY K = 304 mD	K_{eff} (2 - D effective matrix permeability including fractures)
PERMEABILITY MEASURED ACROSS FRACTURE k_s = 23 mD	K_{eff} = 290 mD
ASSUMED PERMEABILITY k_s = 2.3 mD	K_{eff} = 169 mD

Fig. 8. Estimation of the impact of small-scale faults and fractures on effective horizontal reservoir permeability. The calculation is derived from the approach of Martin and Cooper (1984) assuming a regular "brick" type of arrangement for fault distribution within the reservoir.

measured from the cores and the core plug permeability measurements across the cemented minor faults were combined to estimate their effects on bulk permeabilities in the reservoir, using the approach of Martin and Cooper (1985). For the data from the core of well 211/21-12, the minor normal faults in the Rannoch and Etive Formations have on average a 50 cm spacing and a mean horizontal length of 5 m. However, given the small difference between matrix and bulk permeability indicated in Fig. 8, these minor faults are unlikely to influence the permeability of the reservoir unless they are completely tight.

Effect of medium-scale faults

The Rannoch, Etive and Broom Formations are too sand-rich and thick to be compartmentalized by sub-seismic faults. Medium-scale (10–50 ft throw)

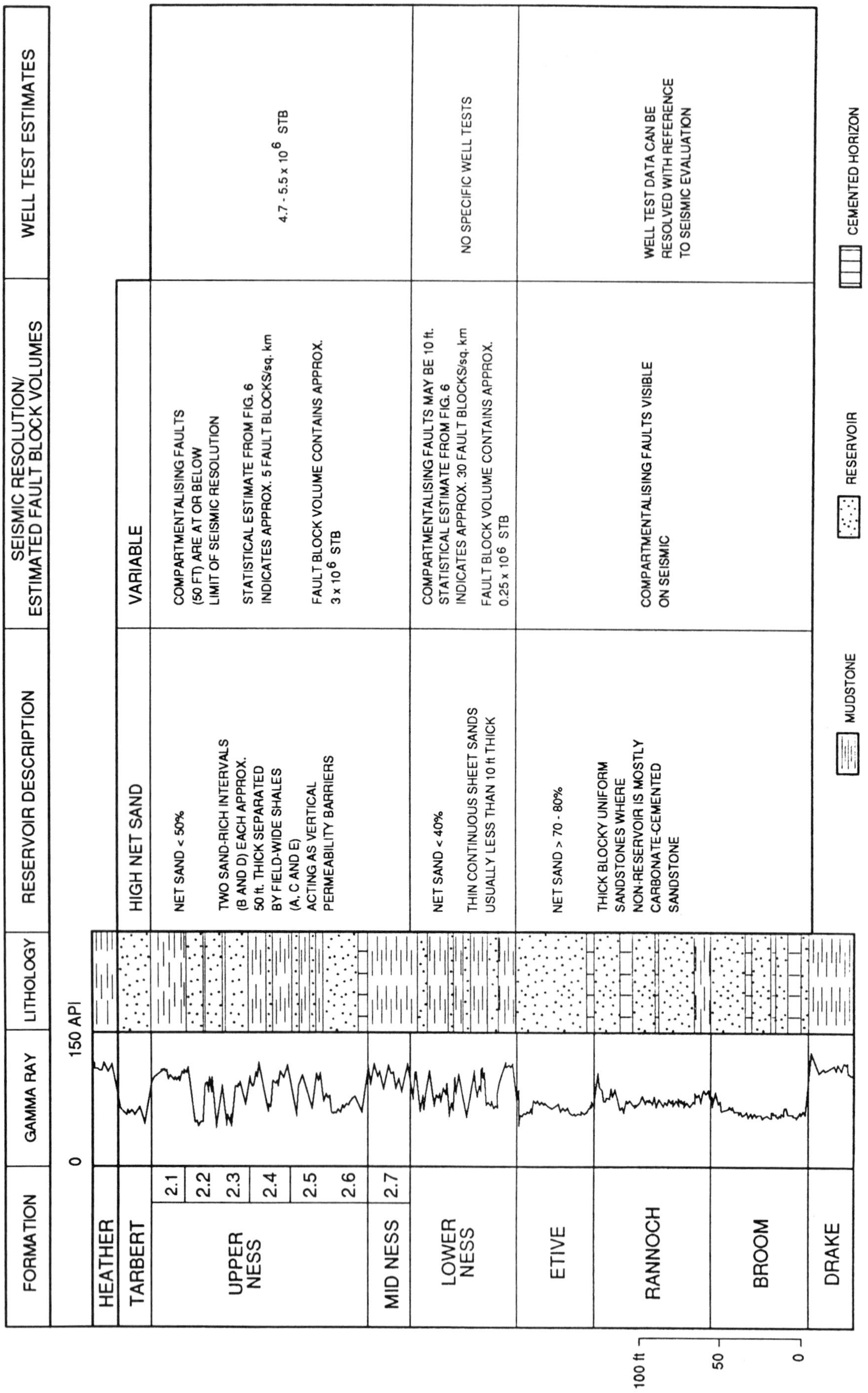

Fig. 9. The effect of a 50 ft fault in Cormorant Block IV and implications for reservoir compartmentalization.

faults would, however, be effective in compartmentalizing the thinner bedded and shaly sequences of the overlying Ness Formation by means of juxtaposition and clay-smearing effects. Although localized later, cement occurs on some faults, there is no evidence of completely cemented (i.e., tight) faults.

In this way, a 50-ft fault (the thickness of the Upper Ness sandy reservoir intervals in well 211/21-9, Fig. 9) could provide a complete seal across the Upper Ness Formation. From Fig. 6 it can be estimated that some four or five faults per square km have a throw greater than 50 ft. They could subdivide the reservoir into about five fault blocks per square km. Volumetric calculations indicate that such fault blocks contain about 3×10^6 STB of oil. Well test data for the Upper Ness indicate similar volumes (Fig. 9), giving added weight to this statistical analysis.

The analysis can be carried further and applied to more sensitive sequences such as the Lower Ness Formation (which consists of thin (<10 ft) sheet sands in a moderate net: gross sequence) for which a 10 ft fault might be completely sealing. About 30 faults per square km have a throw of 10 ft or more according to Fig. 6, giving a compartmentalized Lower Ness fault block volume of 0.25×10^6 STB. This amount of oil in individual fault blocks could only be produced by depletion unless a paired water injector and producer were placed in each fault compartment.

The volumes calculated above are only approximate values that should be investigated with more, specific, well tests. The relationship between fault throw and fault length was found not to be systematic and, indeed, one of the characteristics of the fault pattern in the mapped area is that the faults show a rather constant displacement along most of their length. Thus, the assumption that the faults are always long enough to compartmentalize blocks may be incorrect and requires further investigation. However, the present analysis already suggests that recovery efficiencies deduced for each subdivision should reflect their susceptibility to fault compartmentalization.

Conclusions

In Cormorant Block IV, production behavior in the Brent Group reservoir is strongly controlled by sealing or partially sealing faults. This problem has been examined using an integration of different scales of fault measurements, from core to seismic scale. The gap in resolution between the two distinct sources of information has been tentatively linked using fault self-similar relationships to characterize the fault-throw distribution within the accumulation. The relationship for the fault-throw distribution from the seismic map was estimated to have a characteristic slope factor of −1.5. This slope value could be compared successfully with areal small-scale fault frequencies derived from core observations.

The approach described in this study is most effective if integrated with the sedimentological characteristics of the sequence, by which the reservoir is subdivided into its component units. The amount of compartmentalization varies with reservoir thickness, and recovery factors for each reservoir unit should reflect this. Specific well tests of particular flow units are, however, required to reduce the uncertainty related to the statistical analysis of the structural data set.

Acknowledgements

The authors would like to thank Shell U.K. Exploration and Production, Esso Exploration and Production U.K., and Shell International Research Maatschappij for granting permission to publish this paper. We thank A. Steward and R.H. Gabrielsen for a helpful review.

References

Budding, M.C. and Inglin, H.F., 1981. A reservoir geological model of the Brent sands in southern Cormorant. In: L.V. Illing and G.D. Hobson (Editors), Petroleum Geology of the Continental Shelf of North-West Europe, Institute of Petroleum, London, pp. 326–334.

Childs, C., Walsh, J.J. and Waterson, J., 1990. A method for estimation of the density of fault displacements below the limit of seismic resolution in reservoir formations. In: A.T. Buller, E.W. Berg et al. (Editors), Proceedings of the 2nd International Conference on North Sea Oil and Gas Reservoirs, Trondheim, 8–11 May 1989. Graham and Trotman, London, pp. 309–318.

Chiles, J.P., 1988. Fractal and geostatistical methods for modelling of a fracture network. Math. Geol., 20: 631–654.

Haldorsen, H.H., Brand, P.J. and Macdonald, C.J., 1988. Review of the stochastic nature of reservoirs. In: S. Edwards and P.R. King (Editors), Proceedings of The Mathematics of Oil Production, Cambridge, U.K., 6–7 July 1987. Clarendon Press, Oxford, pp. 109–209.

King, G.C.P., 1983. The accomodation of large strains in the upper lithosphere of the earth and other solids by self-similar fault systems: the geometrical origin of B-value. Pure Appl. Geophys., 121: 761–1093.

Livera, S.E. and Caline, B.P.R., 1990. The middle Jurassic Brent Group in Cormorant Block IV. J. Pet. Geol., 13: 367–396.

Martin, J.H. and Cooper, J.A., 1985. An integrated approach to the modelling of permeability barrier distribution in a sedimentologically complex reservoir. Paper SPE-13052, pp. 1–9.

Matheron, G., 1967. Éléments pour une théorie des milieux poreux. Masson, Paris, 166 pp.

Speksnijder, A., 1987. The structural configuration of Cormorant Block IV in context of the northern Viking Graben structural framework. Geol. Mijnbouw, 65: 357–379.

Tiller, G., 1986. Bellaire Research Center, Shell Oil, Houston, Texas (unpublished report).

Van der Molen, I., 1989. The size-frequency relationship of faults. Abstract Vol., Meeting on Geometry of Normal Faults, as Interpreted from Seismic Reflection Profiles and Other Data Sources, Geol. Soc. London Tectonic Study Group, London, 14–15 June, 1989.

Villemin, T.F and Sunwoo C., 1987. Distribution logarithmique self-similaire des rejets et longueurs de failles: exemple du bassin houiller Lorrain. C.R. Acad. Sci. Paris, 305: 1309–1312

W. SASSI *Koninklijke/Shell Exploratie en Produktie Laboratorium, Volmerlaan 6, 2288 GD Rijswijk ZH, The Netherlands*
Present address: *Institut Géologie-Géochimie, Institut Français du Pétrole, 14 Avenue de Bois Préau, B.P. 311, 92506 Rueil-Malmaison Cedex, France*

S.E. LIVERA *Koninklijke/Shell Exploratie en Produktie Laboratorium, Volmerlaan 6, 2288 GD Rijswijk ZH, The Netherlands*

B.P.R. CALINE *Koninklijke/Shell Exploratie en Produktie Laboratorium, Volmerlaan 6, 2288 GD Rijswijk ZH, The Netherlands*

Permeability of natural and simulated fracture patterns

Noelle E. Odling

A model based on the resistor network method is used to determine the permeability of two examples of natural, two dimensional tectonic fracture patterns. This type of model has, in the past, been most frequently applied to problems within material science. In the model, individual fractures are discretized along a grid containing up to 400 x 400 nodes. For each grid size, checks are performed to ensure that the intersection properties of the fracture pattern are preserved. The permeability of the whole pattern in the direction of the pressure gradient is determined from the total flow through the network. By applying the model to the fracture patterns in varying orientations, a picture of the directional permeability is constructed. Using the model, the effects of varying fracture aperture and matrix permeability on the permeability of the fracture patterns as a whole are investigated. The two patterns were found to show very different permeability characteristics and this is discussed in terms of the properties of their geometry such as connectivity and the nature of fracture intersections. It is concluded that permeability is a function of the number of fracture sets and the nature of fracture intersections as well as fracture density.

Three methods of simulating the natural patterns are investigated and the resulting patterns are analysed for permeability using the model. The simulations are constructed to have the same fracture density (defined as the total fracture length per unit area) as the natural pattern. The results are compared with those obtained from the natural fracture patterns as a measure of the simulation goodness. Good fits were found only at low fracture to matrix permeability ratios and become progressively worse as this ratio increases, most markedly in the direction of maximum permeability. Both over- and underestimation of the permeability occurs. The discrepancies are discussed in terms of the differing geometries of the natural and simulated patterns.

Introduction

Research into fluid flow through fractured media has increased markedly in recent years, motivated, in part, by an interest in the movement of water, petroleum, chemical wastes, etc. through bedrock. Approaches to the problem have been based on either analytical solutions to an equivalent porous medium or numerical studies of fluid flow through fracture networks. In numerical studies, finite element models have been used to determine flow through two-dimensional simulations of crack patterns (Long et al., 1982; Long and Witherspoon, 1985; Andersson and Thunvik, 1986; Long and Billaux, 1987; Khaleel, 1989). In three-dimensions both finite element (Long et al., 1985) and boundary element (Elsworth, 1986; Piggot and Elsworth, 1989) models have been used. A different approach was employed by Hudson (1982) and La pointe and Hudson (1985) who measured the electrical conductance of a simulated fracture and a natural fracture pattern printed on photographic film, drawing an analogy between conductance and permeability. All these models assume an impermeable matrix and the majority have examined only artificial realizations of fracture patterns.

While the permeability of fractured impermeable rock is of importance for predicting the movement of fluid in rocks such as granite, the permeability of fractured permeable rocks has special relevance to the petroleum industry. In a fractured petroleum reservoir, the rock stores the fluid, whereas the fractures, whether more or less permeable than their matrix, can significantly control the flow when a pressure gradient is applied. In this paper, fluid flow through permeable fractured media is studied using a model based on the "resistor network" method. This method has been used over the last 30 years, largely within material science, for calculating transport properties of inhomogeneous systems, e.g., metal semiconductor transitions, fluid flow through static random media and conductance of compacted mixtures of conducting and nonconducting materials (Kirkpatrick, 1973). In the study presented in this paper, two-dimensional fracture patterns have been mapped from the Devonian sandstones of the Hornelen district in Norway and a model based on the resistor network method has been used to determine the permeability properties of these patterns under different matrix and fracture conditions. The method is also applied to various types of simulation which have been used in past studies of flow in frac-

Structural and Tectonic Modelling and its Application to Petroleum Geology edited by R.M. Larsen, H. Brekke, B.T. Larsen and E. Talleraas. NPF Special Publication 1, pp. 365–380. Elsevier, Amsterdam.

tured rock, and the results are compared with those for the natural patterns. This comparison provides an objective assessment of the goodness of these simulations.

The resistor network method

In the resistor network method, the fracture pattern is discretized to lie along the elements of a square grid (Fig. 1). Elements corresponding to fractures are given a permeability value representative of the fractures, while the remainder are assigned values of the matrix. A pressure gradient is applied in the X direction, by setting the pressure at all grid nodes along the left hand margin to zero and all those at the right hand margin to a value equal to the dimension of the grid. Periodic boundary conditions in the Y direction are imposed by connecting the grid nodes along the top of the model to those at the base which is equivalent to wrapping the pattern around a cylinder. This results in a pressure profile at the top and bottom of the model which is determined by the fracture pattern modelled and ensures that no fluid is lost or gained from the top and bottom of the model. As a result, permeabilities for equal but opposite directions are identical. Although this condition is rather artificial, it is considered better than imposing nonpermeable boundaries at top and bottom, which could restrict the flow, or in assigning linearly increasing pressure profiles regardless of the fracture pattern modelled which can result in fluid loss or gain at the top and bottom of the model and dissimilar permeabilities for equal but opposite directions (Long et al., 1982).

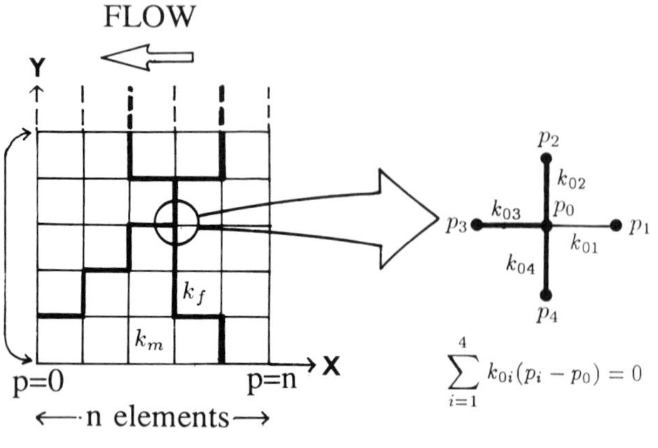

Fig. 1. The fracture pattern is discretized on a square grid (left) in which elements are assigned permeabilities representative of fracture (k_f) or matrix (k_m). Periodic boundary conditions are assumed for the top and bottom of the model (see text for further details). A pressure gradient is applied from right to left and the resulting pressures at each node are computed by assuming conservation of mass, i.e., the sum of flows towards each node equals zero (right).

At each node in the grid, it is assumed that the sum of flows in the four connecting grid elements is zero (Fig. 1). That is:

$$(p_0 - p_1)k_1 + (p_0 - p_2)k_2 + (p_0 - p_3)k_3 + (p_0 - p_4)k_4 = 0$$

where p_0 to p_4 are the pressures at grid nodes and k_1 to k_4 are the permeabilities of the grid elements.

This results in a series of linear equations which is solved for the pressure field from which the flow in each grid element can be obtained. The total flow through the model in the X direction then gives the permeability of the fracture pattern in this direction (direction of the pressure gradient).

A fracture is treated as a one-dimensional, smooth-walled channel in which flow is assumed to be laminar and for which Darcy's Law for fluid flow can be applied. The fracture aperture is, thus, not the physical width of the fracture but the "hydraulic" aperture. In general, the hydraulic aperture is much smaller than the physical aperture (up to one eighth), due to fracture wall roughness (Bandis et al., 1985). In the grid, each element can be thought of as representing a small block of permeable medium with a size dependent on the grid size. Since fracture apertures (around 1–10^3 μm) are five or more orders of magnitude smaller than the fracture lengths (around 10^2–10^3 m) and, therefore, the pattern as a whole, grid elements designated to a fracture represent not only the fracture but also some of the surrounding porous medium. Thus, the permeability assigned to fracture grid elements in the model is an average permeability of this block, rather than the permeability of the fracture alone.

In the model, the fracture permeability is normalized with respect to the matrix and the result from the model is, therefore, the global permeability with respect to the matrix, in the direction of the pressure gradient. The permeability assigned to fracture grid elements in the model is given by:

$$k_{gr} = Nb(r - 1) + 1$$

where N is the number of grid elements in the direction of the applied pressure gradient, b is the hydraulic fracture aperture and r is the permeability ratio between fracture and matrix.

One of the challenges of the griding procedure is ensuring that the connectivity properties of the fracture pattern are preserved. At smaller grid sizes, fracture traces that are very close together can become joined when discretized. This situation is avoided by checking all intersections after griding against a list of intersections from the original pattern and eliminating unwanted intersections by converting appropriate fracture grid elements to matrix. In this way, the connection properties of the pattern are preserved regardless of the grid size.

In order to obtain a reasonable representation of the fracture pattern, large grids are required. The model is designed to compute the pressure field for grids up to 400 × 400. To solve the resulting large system of equations as expediently as possible, a conjugant gradient method (NAG subroutine) is used and the model is run on an IBM 3090 using the extended architecture facility with 70 megabytes of memory. The time taken to solve the problem ranges up to 5 minutes of C.P.U. time for a grid size of 400 × 400 nodes and depends on the fracture pattern as well as the grid size. Higher fracture to matrix permeability ratios and patterns containing many unconnected fracture segments increase the computation time. However, for the fracture patterns presented here, a grid size of 200 × 200 nodes was found to be adequate and individual runs took 1 minute of C.P.U. time or less.

The resistor network approach represents a versatile method of investigating fluid flow in fractured, permeable media. Large grid sizes allow the actual geometry of fracture patterns to be faithfully reproduced and a wide range of fracture to matrix permeability ratios can be simulated. Since the pressures at each grid node are derived, the flow in individual bonds can be computed and the global flow patterns examined. In addition, it is also possible to model inhomogeneities in the matrix and variations in fracture aperture. Finally, this model could also be extended to examine three dimensional flow in fractured, permeable media.

The fracture patterns

The fracture patterns to which the model was applied were mapped from the Devonian sandstones and shales of the Hornelen district in Western Norway. The sediments were deposited under probably a dominantly tensional tectonic regime (Steel, 1976; Hossack, 1984; McClay et al., 1986; Norton, 1986), but were later compressed resulting in gentle to moderate upright folds with east–west trending fold axes (Steel, 1976). From the relationship between the fractures and the folds, the fracture pattern in the rocks is interpreted to be related to this folding and the relationship between fracture sets suggests a single age for all sets.

Fracture pattern A (Fig. 2) was mapped from a 12 × 12 m area on the bedding surface of a sandstone bed which is approximately 2 m thick and dips 16° towards 89° (UTM coordinates KP965439). All fractures down to 3 cm long were mapped and the total length of all fractures which extended out of the mapped region was noted. The sample contains 554 fractures and the fracture density (defined as fracture length per square metre) is 2.75 m/m^2. The trends of all fractures were analysed to find the best fit mixed von Mises distribution to the data (Table 1). Application of the chi-square test to the observed and theoretical distribution showed the fit to be statistically valid. Four fracture sets were identified whose mean directions trend was 42°, 71°,

Fig. 2. Fracture patterns A (top) and B (bottom) were mapped from the bedding surfaces of Devonian sandstone and shale layers, respectively, in the Hornelen district of Norway. Pattern A is 12 m across and pattern B is 2 m.

TABLE 1

Analysis of trend and length distributions for pattern A and B.

(a) Von Mises analysis of trend distributions			(b) Analysis of natural logarithms of fracture trace lengths.			
Amplitude	Concentration	Mean	Mean	Stand. dev.	*KS*	D_{crit}
Pattern A (5° class interval) [a]			*Pattern A* (0.2 m class interval)			
1.41	18.88	175	−0.821	0.981	0.027	0.530
21.42	21.78	42				
62.40	1.42	71				
21.81	4.62	138				
Pattern B (5° class interval) [b]			*Pattern B* (0.1 m class interval)			
27.0	40.81	75	−1.343	1.061	0.063	0.111
4.3	6.41	151	−2.225	0.599	0.075	0.224
6.5	0.75	38	−2.400	0.562	0.155	0.392

[a] Chi-square test statistic = 28.0; critical value at 10% point = 33.2. [b] Kolmogorov-Smirnov test statistic = 0.034; critical value = 0.158. *KS* = Kolmogorov-Smirnov test statistic for goodness of fit to lognormal distribution; D_{crit} = Kolmogorov-Smirnov critical value

138° and 175°. No significant difference between the fracture trace length distributions of each fracture set was found and the length data for the whole sample showed a statistically valid fit to a lognormal distribution (Kolmogorov-Smirnov test, Table 1).

Fracture pattern B (Fig. 2) was mapped from the surface of a thin (approximately 1 cm) shale layer (UTM coordinates KP968433), dipping 17° towards 100° which occurs between two beds of sandstone, located approximately 1 km south of fracture pattern A. All fractures down to 2 cm were mapped and, as in fracture pattern A, the lengths of fractures extending out of the mapped area were noted. The sample contains 155 fractures and the fracture density is 9.34 m/m^2. The best fit mixed von Mises distribution indicates three sets of fractures whose means trend was 75°, 151° and 21°, of which 75° trend is dominant. In this case, the chi-square test was found to be inappropriate due to low frequencies in some class intervals. The Kolmogorov-Smirnov test was applied instead and the fit between observed and theoretical distributions was found to be valid. A significant difference in the trace length distributions of the different fracture set was found with those not belonging to the dominant set being shorter. All sets showed a valid fit to a lognormal distribution (Kolmogorov-Smirnov test, Table 1).

The fracture set mean orientations (Table 1) show that patterns A and B have three sets in common. From the amplitude parameters (Table 1), the dominant sets in both patterns have similar orientations (71°, pattern A; and 75°, pattern B) and closely coincide with the orientation of the southern bounding fault and fold trends of the Hornelen basin. These may be interpreted as shear fractures associated with movement along the bounding fault. The other two sets shared by patterns A and B (42° and 138°, pattern A; 38° and 115°, pattern B) are symmetrically disposed about the fold trend and can be interpreted as conjugate shears associated with folding. The remaining set (175°, pattern A) is perpendicular to fold trends and could represent tension fractures.

Despite the similarities in fracture set orientations, the two fracture pattern A and B are otherwise markedly different, three sets showing approximately equal amplitude in pattern A and one dominating in pattern B. This probably relates to the differing mechanical properties of the shale and sandstone.

In the following sections, these two patterns are used as examples of natural fracture patterns and their permeability properties are investigated using the resistor network method.

Analysis of model results from patterns A and B

The permeability of a fractured medium is commonly characterized by determining the equivalent porous medium for which a permeability tensor can be written. If this can be done, the effects of individual fractures and the fracture network can be ignored and the whole medium may be treated as an homogeneous continuum.

The existence of such equivalent porous media for fractured media and the conditions that must be satisfied for such an approximation to be valid have been discussed by several authors (e.g., Long et al., 1982; Khaleel, 1989). In the case of an anisotropic homogeneous porous medium, where the permeability (k) is measured in the direction of the pressure gradient, $1/\sqrt{k}$ describes an ellipse (Scheidegger, 1974). The degree to which a fractured medium approximates an homogeneous porous medium can, therefore, be judged by the fit of $1/\sqrt{k}$ to an ellipse (Long et al., 1982; Khaleel, 1989). In addition to the above condition, the sample region must be large enough for all characteristics of the fracture pattern to be represented, so that no significant changes in the permeability occur for small changes in area or location. This aspect has been investigated by Long et al., (1982) and Khaleel (1989) for simulated fracture patterns by examining the change in the permeability pattern as the size of the area is increased. In this study, the values of $1/\sqrt{k}$ determined for patterns A and B were fitted to an ellipse using a least-squares routine and the chi-square test applied to the measured and best fit ellipse values to determine the validity of the fit using the 10% point of the chi-square distribution as the critical value. The stability of the results with respect to varying area size is then tested.

Permeability patterns with varying sample size

The stability of the permeability patterns with change in sample size and the validity of an equivalent porous medium has been investigated for fracture patterns A and B for different fracture to matrix permeability ratios. A grid size of 200 × 200 nodes was found to give a reasonable representation of the fracture patterns and no significant change in the results occurred using larger grids. Initially, the largest rotatable square that lies inside each sample was determined. Then square areas with side lengths of 25, 50, 75, 100, 125, 150, 175 and 200, all centered on the midpoint of the original 200 × 200 grid were selected for analysis. Permeability was determined at 10° intervals as these regions were rotated through 180°.

Fig. 3 shows the maximum and minimum permeability results for the case of a fracture hydraulic aperture of 80 μm (fracture permeability 6×10^{-10} m^2) and matrix permeabilities of 6×10^{-12}, 6×10^{-14} and 6×10^{-16} m^2, which represent good to poor quality reservoirs. The resulting best fit homogeneous anisotropic porous medium results are plotted where the fit of $1/\sqrt{k}$ to an ellipse was found to be statistically valid and where this was not the case, maxima and minima of the original data was used.

As the matrix permeability and sample size decreases, the fit to an anisotropic homogeneous porous medium becomes less good and is only invalid at the lowest matrix permeability (6×10^{-16} m^2). In all cases, the maxima show much greater fluctuations than the minima which die out as the sample size increases. From Fig. 3, it can be seen that the fitting of the data to an equivalent anisotropic medium for pattern B can result in minimum values less than one. This is due to the ellipse fitting procedure and, since the model cannot produce values less than the matrix permeability (1.0), these values indicate that the minimum permeability is indistinguishable from that of the matrix.

The effect on increasing area size is most marked in pattern B. High maximum permeabilities for small sample sizes are due to the presence of fractures which transverse the whole sample area thus allowing high flow in specific directions, whereas at larger area sizes, no single fracture spans the entire sample area. This illustrates the need to establish a minimum representative size for the fracture pattern. The plots in Fig. 3 suggest that fracture patterns A and B are large enough to be representative of the permeability ratios tested.

Orientations of the maximum permeability direction for the areas sizes are plotted with respect to that of the largest area in Fig. 4. Where the fit to an anisotropic homogeneous porous medium was invalid, the maximum value has a possible range of ±10° as measurements were taken at 10° intervals. Pattern A shows large fluctuations for smaller sample sizes, while Pattern B shows a relatively stable direction for all sizes. Both patterns show a convergence to the direction of that of the largest area which suggests that both areas are representative of the fracture patterns with respect to the maximum permeability orientation.

Effects of varying matrix and fracture permeability

The permeability was determined for fracture patterns A and B for a fracture aperture of 80 μm (fracture permeability 6×10^{-10} m^2) and matrix permeabilities ranging from 6×10^{-12} m^2 to 6×10^{-20} m^2 (Fig. 5). The permeabilities show a valid fit to an anisotropic, homogeneous medium for all matrix permeabilities except the lowest (6×10^{-17} m^2) for pattern A. For both patterns, the maximum and minimum permeabilities are indistinguishable from that of the matrix at high matrix permeabilities. However, the two patterns show rather different behaviour as the matrix permeability decreases. For pattern A, the minimum tends to a value about one order of magnitude greater that the matrix, whereas for pattern B, the minimum closely follows the value of the matrix. The maximum of pattern A tends to a constant value about four orders of magnitude less that of the fractures themselves. The maximum of pattern B tends to a value approximately one order of magnitude higher than the matrix, showing behaviour similar to that of the minimum for pattern A.

The difference in behaviour of the patterns can be linked to geometrical properties of the patterns themselves. Pattern A has a "backbone" of linked fractures in the direction of maximum permeability,

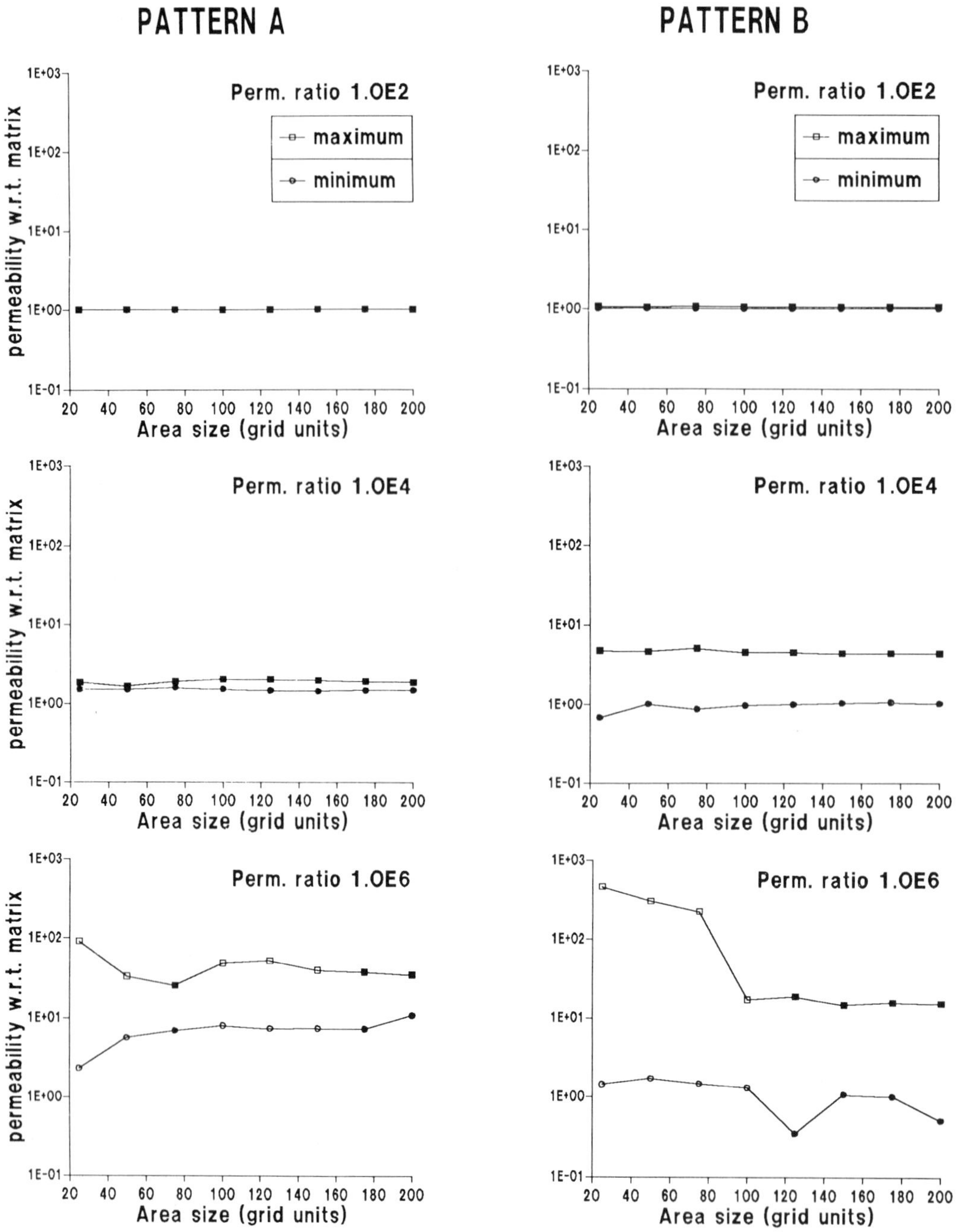

Fig. 3. Plots showing the variation in the maximum and minimum permeabilities of patterns A and B with varying sample size, for three values of matrix permeability (6.0 x 10^{-14}, 6.0 x 10^{-15} and 6.0 x 10^{-16} m^2). The fracture aperture was held constant at 80 μm in all cases. The samples occupy parts of the 200 x 200 grid with a common centre. Filled symbols indicate a valid fit to a homogeneous, anisotropic porous medium. In the case of an invalid fit (open symbols) the largest and smallest values are plotted.

i.e., fluid can transverse the pattern without entering the matrix. As the matrix permeability tends to zero the maximum permeability of the pattern tends to a value of the backbone alone, which is dependent only on the fracture permeability. The backbone, a concept used in the percolation theory (Stauffer, 1985), is an expression of the connectivity of the fracture pattern and is illustrated in Fig. 6 for the direction of maximum permeability of pattern A. In the direction of the minimum permeability, pattern A has no backbone, but fractures with some component of their orientation parallel to the minimum direction contribute to the global permeability. Thus, the minimum permeability is controlled by the matrix permeability, but enhanced by the presence of fractures.

Pattern B has no backbone in any direction. In the maximum direction, which is parallel to the dominant fracture set, the permeability is enhanced by the presence of fractures and the behaviour is similar to

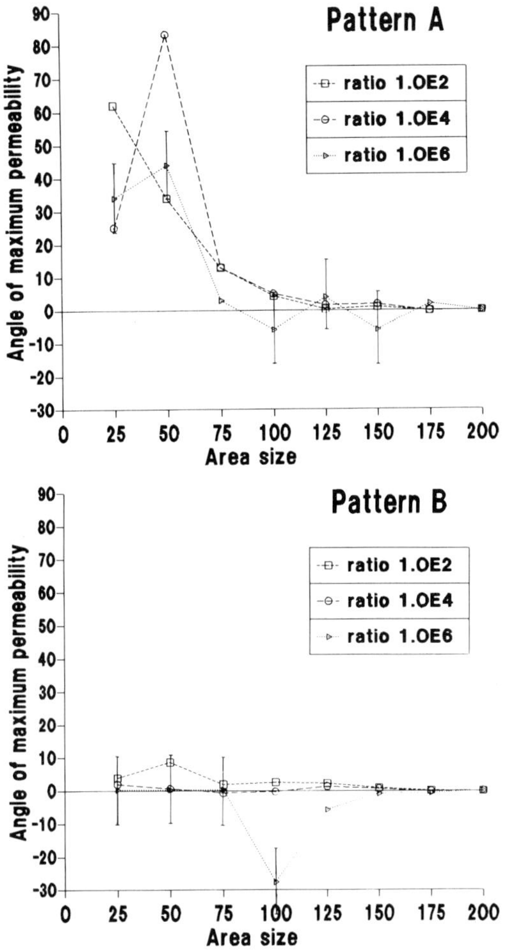

Fig. 4. Variation in orientation of the maximum permeability for patterns A and B for varying sample size. Filled and open symbols as for Fig. 3.

that of the minimum for pattern A. Pattern B has few fractures with a component of their orientation parallel to the direction of the minimum permeability and, thus, in this direction, the global permeability remains close to that of the matrix.

The effect of varying the hydraulic fracture aperture from 102 μm down to 4.25 μm with a constant matrix permeability of 6.0×10^{-16} m^2 is also shown in Fig. 5. All except the largest fracture aperture (102 μm, permeability ratio of 1.96×10^{6} m^2), showed a valid fit to an anisotropic homogeneous porous medium. In pattern A, both the maximum and minimum decrease and approach the matrix permeability as the fracture aperture decreases and at a fracture aperture of 8.5 μm (permeability ratio of 1.0×10^{-4} m^2) the permeability is almost indistinguishable from that of the matrix. Pattern B shows a similar pattern at small apertures, but at wider apertures the maximum permeability tends towards a constant value of approximately 1.0×10^{-14} m^2. This suggests that the matrix has a limiting effect on the permeability so that increasing the fracture aperture beyond 2.16×10^{-10} m^2 has little effect on the total permeability.

Fig. 7 shows the change in orientation of the maximum permeability direction as the matrix permeability and fracture aperture vary. As matrix permeability decreases or as fracture aperture increases, the maximum direction of pattern A rotates from 057° to 79°. This rotation can be attributed to the relative importance of the fracture sets and the backbone in determining the permeability anisotropy at differing fracture/matrix permeability ratios. At low ratios (high matrix permeability or small fracture apertures), flow is influenced most by the orientations of the dominant fractures sets (042° and 071°) resulting in an intermediate maximum permeability orientation of 057°. As the ratio increases (decreasing matrix permeability or increasing fracture aperture), the backbone structure, which has an mean orientation close to that of the dominant fracture set (071°), increasingly controls the maximum permeability direction. Pattern B has one dominant fracture set (075°) and no backbone in any direction with the direction of maximum permeability being thus controlled by the direction of this set throughout.

Simulations and their permeability

Initial attempts to model flow through fractured rock used very simple geometries composed of usually orthogonal sets of parallel, infinitely long fractures for which flow could be analytically determined (e.g., Warren and Root, 1963). Realization that these geometries did not adequately represent natural fracture patterns and the development of numerical techniques for studying flow in fractured media has since led to attempts to create simulations that incorporate known characteristics of real fracture patterns. In these simulations, various orientation distributions (uniform, wrapped normal, von Mises) and length distributions (uniform, negative exponential, lognormal) have been employed (Hudson, 1982; Long et al., 1982; La Pointe and Hudson, 1985; Long et al., 1985; Andersson and Thunvik, 1986; Balberg, 1986; Long and Billaux, 1987). Most of these simulations have assumed uniform, random fracture spatial distributions although La Pointe and Hudson (1985) investigated periodic fracture zones and linear increase in fracture frequency and Long and Billaux (1987) applied differing fracture densities to subregions of the total simulation area. In addition to these more general approaches, special cases have also been investigated, such as polygonal jointing in basalts (Khaleel, 1989).

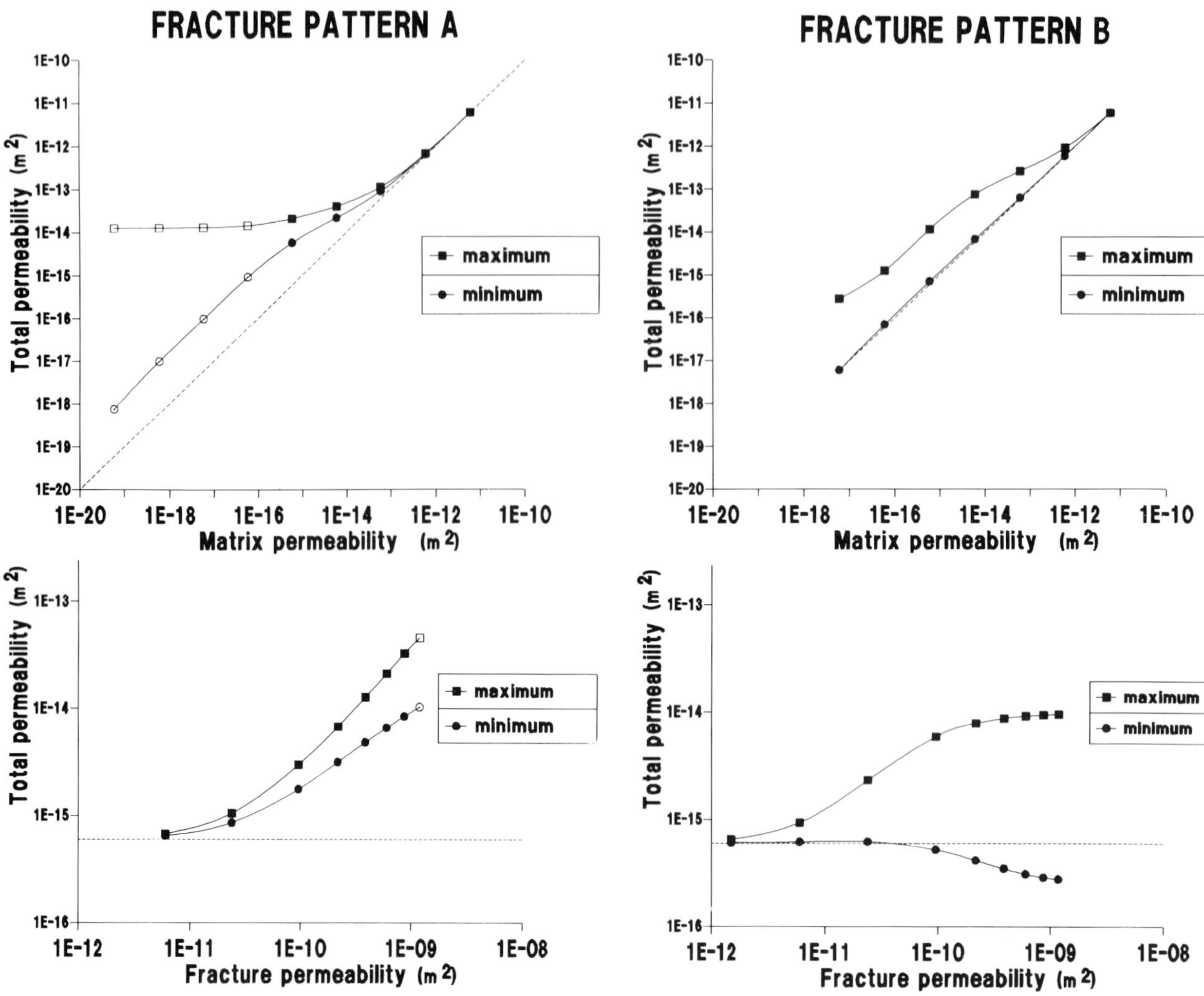

Fig. 5. Plots of maximum and minimum permeability of patterns A and B for varying matrix permeability and fracture permeability. Filled and open symbols indicate valid and invalid fits to a homogeneous, anisotropic porous medium.

Several authors have applied numerical models of fluid flow to simulated fracture patterns and used the results to draw conclusions about the nature of fractured rock permeability. Although attempts have been made to include specific characteristics of the fracture network, few have questioned how well these simulations represent natural fracture patterns. Because the permeability of fractured rock is usually of interest in areas where the natural fracture patterns can seldom be examined in detail, there is a need for reliable simulation techniques which can be employed where there is limited data. A first step is to test existing simulation techniques against real fracture patterns. Here, three types of simulation which have been used in previous studies are tested for their permeability properties and compared to the results described above for naturally occurring patterns A and B. All three simulations were constrained to have the same fracture density (defined as fracture length per unit area) as the natural patterns and in all, fracture centres (defined as fracture trace midpoints) are randomly distributed spatially. The simulations increase in complexity from random orientations with constant fracture lengths in the first, to constant orientations and length for each fracture set in the second and finally to orientation and length distributions taken from those determined for the natural fracture patterns A and B. An example of each simulation type for each of patterns A and B is shown in Fig. 8.

Simulation 1

In this simple model, fracture centres are located randomly, orientations are randomly assigned from a random, uniform distribution and fracture lengths are held constant at the average length calculated from fracture patterns A and B. As there is no preferred orientation in this model, permeability is theoretically isotropic. This model obviously cannot reproduce the permeability anisotropy of patterns A and B but, in the absence of good orientation data, it could represent the best available simulation. Such

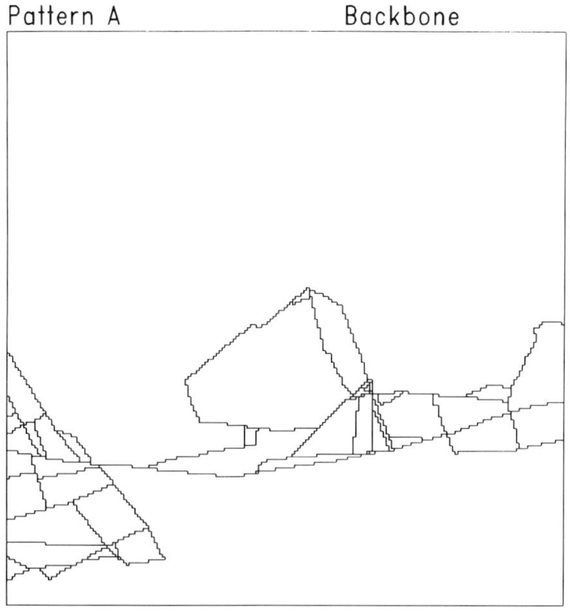

Fig. 6. The "backbone" of pattern A in the direction of maximum permeability. The backbone is composed of all possible paths a fluid particle may take along the fracture system in order to traverse the pattern from left to right (or vice versa) without duplicating any part of its journey.

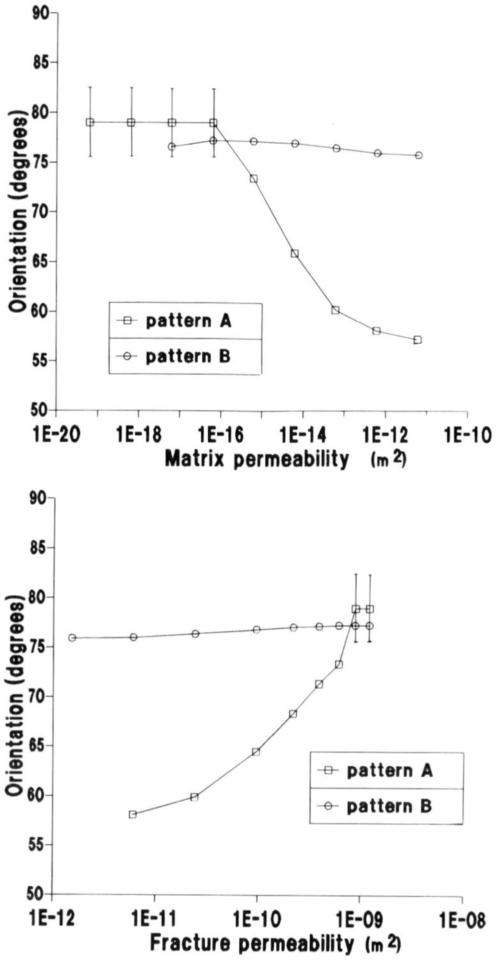

Fig. 7. Variation in orientation of the direction of maximum permeability for patterns A and B for variable matrix and fracture permeabilities. Where the fit to an homogeneous anisotropic porous medium is invalid (open symbols), the direction of the permeability maximum is known to within $\pm 10°$ (measurement interval).

a model has been used to look at cluster statistics of fracture patterns by Balberg (1986) and been used to investigate permeability of fractured rock by Andersson and Thunvik (1986).

Four simulations for each of patterns A and B were produced from different samples of random numbers. The permeability of each simulation in 18 directions, 10° apart, was determined for a fracture aperture of 8 μm and matrix permeabilities of 6.0×10^{-14} m^2, 6.0×10^{-15} m^2 and 6.0×10^{-16} m^2. The fit to an homogeneous anisotropic porous medium becomes worse as the matrix permeability decreases, but was found to be statistically valid in all except one case.

Simulations of pattern A showed maximum to minimum permeability ratios up to 1.27 with most less than 1.1, close to isotropic permeability as predicted. The maximum and minimum permeability values, plotted together with those for pattern A in Fig. 9 show that the simulations agree quite well with pattern A for the greatest matrix permeability, but increasingly underestimate the permeability as the matrix permeability decreases. None of the simulations possessed a backbone (i.e., were connected) in any direction, in contrast to pattern A. Thus, at high fracture to matrix permeability ratios where the backbone dominates the maximum permeability in pattern A, the simulations underestimate pattern A. Although pattern A is not connected in the direction of the minimum, a similar, though reduced, effect is seen. This is due to the presence of a fracture set parallel to the minimum direction in pattern A which increases the component of total fracture length in this direction compared to the simulations. The results of the permeability analysis of four simulations of pattern B (Fig. 9) show maximum to minimum permeability ratios greater than 1.1 in all but one case, the highest being 12.4. Thus, although permeability is theoretically isotropic, the simulations show anisotropic results. One of the simulations patterns was found to possess a backbone (i.e., was connected) and in others connection in some directions was very close to complete. The existence of a backbone in one of the simulations of pattern B indicates that these are close to the "percolation threshold" i.e., the density at which a through-going network forms. Close to the this threshold in such systems, behaviour is known to be erratic (Stauffer, 1985) and this accounts for the

Fig. 8. Examples of simulations 1, 2 and 3 for pattern A (top row) and pattern B (bottom row). Simulation 1 has random orientations and average fracture lengths from patterns A and B. Simulation 2 has mean orientations and average lengths of each fracture set from patterns A and B. Simulation 3 uses the orientation and length distributions determined from patterns A and B. All patterns have random fracture locations and fracture densities equal to those of patterns A and B.

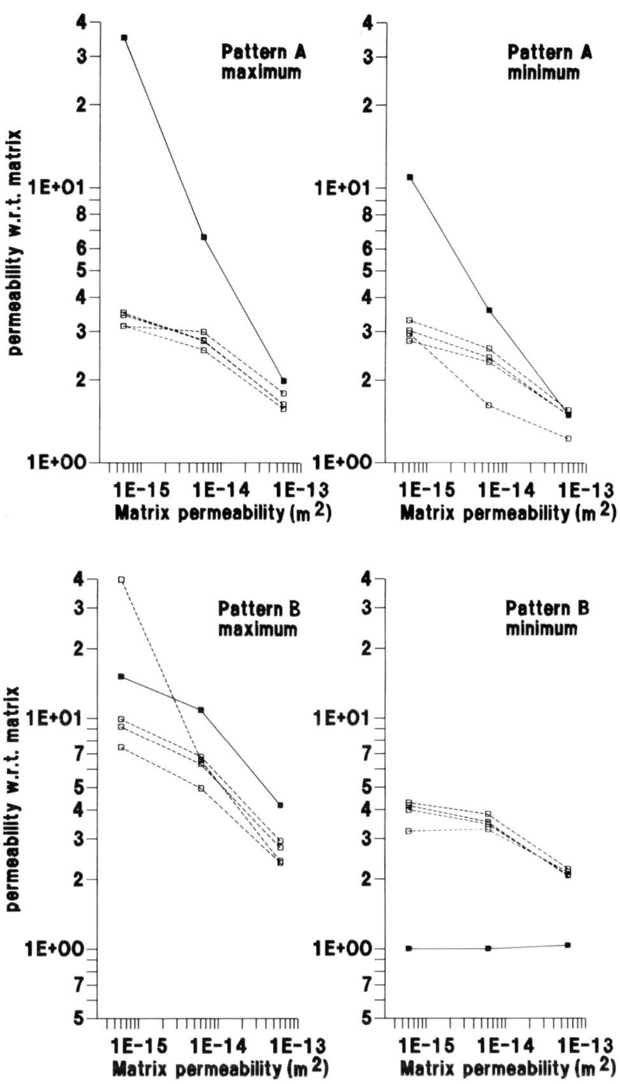

Fig. 9. Permeability results for type 1 simulations of patterns A and B (open symbols, dashed lines). The fracture aperture is 80 μm and the three matrix permeabilities are 6.0×10^{-14}, 6.0×10^{-15} and 6.0×10^{-16} m^2. These are compared to the results for patterns A and B (filled symbols, solid line).

anisotropy and wide range of permeability results between simulations. The results for simulations of pattern B (Fig. 9) show that the maximum is largely underestimated and the minimum overestimated. This reflects the difference between the orientation distributions of pattern B, where the fractures are strongly concentrated in the direction of maximum permeability, and the simulations, where they are uniformly distributed. The effect of connectivity, discussed above, gives rise to a wider spread of results in the simulations especially noticeable in the maximum direction.

Simulation 2

In this simulation, fracture length and orientation are kept constant at the average values for each fracture set as determined in section 2. As in simulation 1, fracture centres are located randomly and the fracture densities were constrained to be equal to that of the natural patterns. Since the preferred direction of fracture sets are now included, something of the permeability anisotropy of the natural fracture patterns can be expected. This model was used by Long et al. (1982) to examine the effects of varying fracture density on the permeability.

A fracture aperture of 80 μm was used and for each of three matrix permeabilities (6.0×10^{-14} m^2, 6.0×10^{-15} m^2 and 6.0×10^{-16} m^2) four different simulations were tested for permeability. The results are plotted on polar graphs of permeability versus direction together with the results for patterns A and B in Fig. 10. All simulations gave good fits to an homogeneous anisotropic porous medium.

Simulation results for pattern A show noticeably anisotropic results with major permeability directions that agree with that for pattern A to within 17° for the greatest matrix permeability and to within 8° for the smallest. For the greatest matrix permeability (6.0×10^{-4} m^2) the maximum shows a good fit to pattern A. However, all minima, and both maxima and minima at lower matrix permeabilities, increasingly underestimate the values for pattern A, as the matrix permeability decreases, reaching discrepancies of an order of magnitude. None of the simulations were found to be connected in any direction, unlike pattern A which is connected in the direction of the maximum permeability. In pattern A some 55% of fracture intersections are T-junctions, whereas in the simulations most are cross-junctions. This means that there are less "dead ends" in pattern A than in the simulations. Thus the fractures in pattern A are arranged in a fashion that favours connectivity in comparison to the simulations, resulting in greater permeability in all directions.

The simulations of pattern B also resulted in highly anisotropic permeability with a maximum direction in good agreement with the natural pattern (within 6°) at all matrix permeabilities. In all cases the minima show good fits to that of pattern B (close to that of the matrix) but there is an increasing discrepancy in the direction of the maximum from 25% at matrix permeability of 6.0×10^{-4} m^2 to 75% at 6.0×10^{-6} m^2. Neither pattern B nor any of the simulations were connected in any direction and, since one fracture set dominates, fracture intersections are scarce in both pattern B and the simulations. However, in the natural pattern, fractures often show an en échelon arrangement such that very small gaps between fractures ends frequently occur. It was found from plots of fluid flow rate, derived from the model, that flow is greatly enhanced in the matrix between such closely spaced fractures and that this

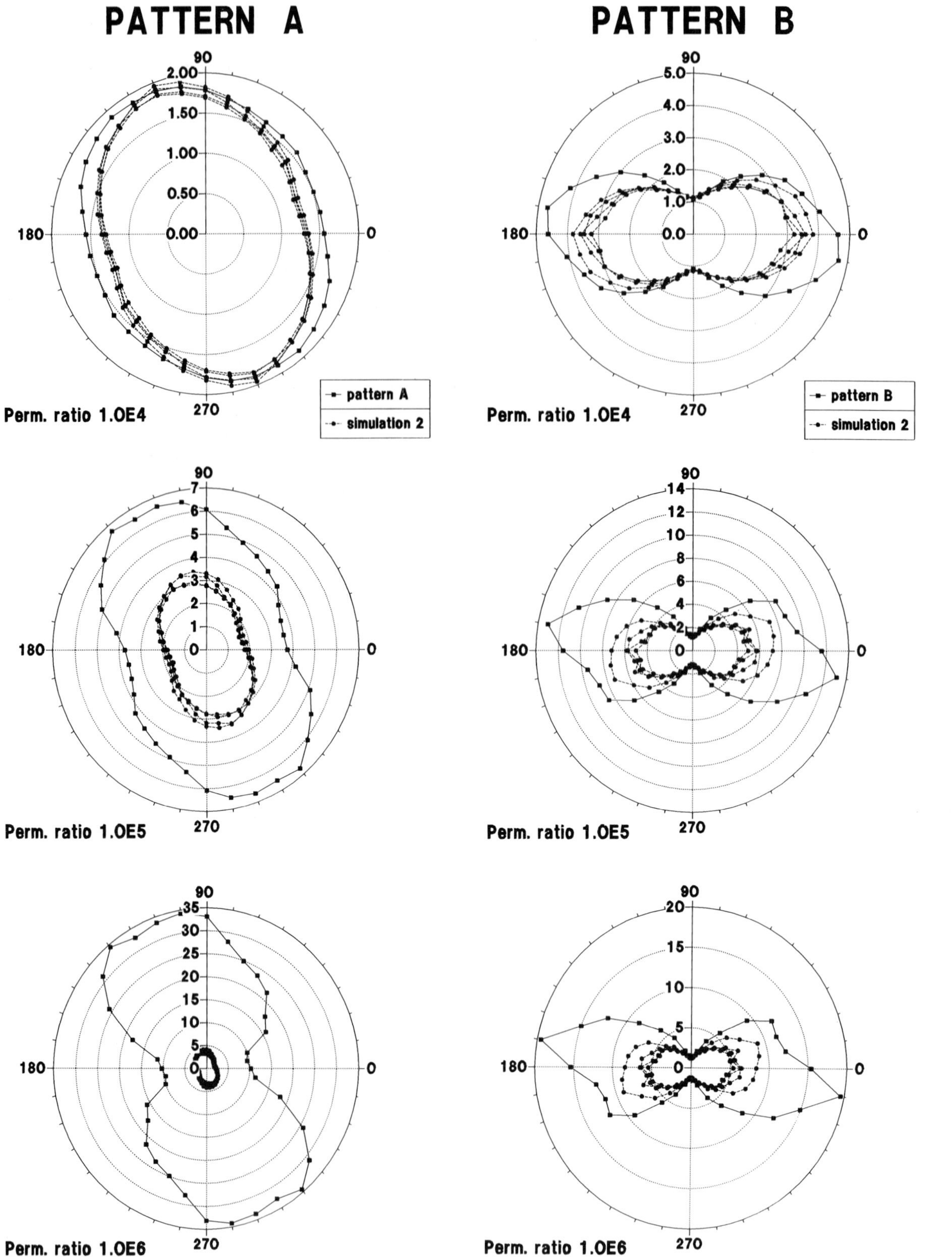

Fig. 10. Polar graphs showing the directional permeability of type 2 simulations (open symbols, dashed lines) and patterns A and B (filled symbols, solid lines). Fracture aperture and matrix permeabilities are as for Fig. 9.

effect increases as the permeability contrast between fracture and matrix increases. Thus, although this geometry does not result in connection within the fracture network, it does serve to enhance flow. In the simulations, fracture centres are randomly distributed and this feature is not common. This results in an underestimation of the maximum permeability in the simulations.

Simulation 3

In this simulation, the same random centre locations as in simulations 1 and 2 were used, but fracture orientations and lengths were taken from the distributions determined for fracture patterns A and B. This is the kind of simulation that has been most used in past studies of permeability of fractured rock (Long et al., 1982; Long and Billaux, 1987; Long et al., 1985; Hudson, 1982). The results of four simulations for a fracture aperture of 80 μm and the three matrix permeabilities used in simulations 1 and 2 are compared to the results from patterns A and B in Fig. 11.

The simulations of pattern A all show acceptable fits to homogeneous anisotropic porous media although there is a noticeable increase in the spread of the simulation results as the matrix permeability decreases. The simulations result in a permeability anisotropy with a maximum direction within 25° of that of pattern A. The minimum permeability is well reproduced by the simulation with only a small underestimation for the lower matrix permeabilities. However, the maximum permeability is underestimated by an increasing amount as the matrix permeability decreases, from about 35% at 6.0×10^{-14} m^2, to 80% at 6.0×10^{-16} m^2, although there is a considerable improvement on the results from type 2 simulations. All four simulations were found to be connected, in contrast to simulations of type 2, but to a lesser extent than in pattern A (i.e., the backbone is less extensive). This smaller degree of connectivity and the larger number of dead ends arising from the nature of intersections, as for type 2 simulations, accounts for the underestimation of the permeability anisotropy.

The simulations of pattern B showed valid fits to homogeneous, anisotropic porous media in only the greatest matrix permeability case (6.0×10^{-14} m^2) and one case at 6.0×10^{-15} m^2. The results (Fig. 11) showed anisotropy with a maximum permeability direction in generally good agreement with pattern B (within 15°). The minimum permeability is well reproduced by the simulations. However, the maximum values range from a good fit for the highest matrix permeability case to an overestimation of over an order of magnitude at the lowest matrix permeability. As a result, the anisotropy is overestimated. All four simulations were found to be connected in contrast to pattern B. This is a result of the introduction of variable orientations which has allowed fractures of the dominant set to intersect one another, a situation which does not occur in pattern B due to the regular spacing of fractures. Thus permeability anisotropy is enhanced in this simulation type, in contrast to simulations of type 2 which resulted in an underestimation.

Conclusions and discussion

The permeability results show that patterns A and B have significantly different properties, reflected by the maximum and minimum permeabilities and their directions. As matrix permeability decreases, the maximum permeability of pattern A tends to a constant value, whereas that of pattern B tends to a constant multiple of the matrix value. This implies that as the matrix permeability tends to zero, pattern A will conduct fluid in selected directions, while pattern B will be nonpermeable. As the fracture aperture increases, the maximum permeability of pattern A tends to a constant multiple of the fracture permeability, while that of pattern B tends to a constant value. This suggests that above a certain fracture aperture, further increase in the aperture has little effect on the permeability of pattern B.

Investigation of the permeability properties of patterns A and B has highlighted the importance of the degree to which the fracture patterns are connected. Pattern A was found to be connected, while pattern B is not, despite its higher fracture density (9.34 m/m^2 pattern B and 2.75 m/m^2 pattern A). In general, fractures of a single set are rarely observed to intersect, even when en échelon geometries are common and very small gaps exist between fractures, as in pattern B. This implies that fracture patterns where one fracture set is markedly dominant (as in pattern B) are unlikely to be connected in any direction regardless of the fracture density. Where two or more sets of more equal concentration are present, the existence of a through-going network (backbone) is more likely and could be present even at low fracture densities. The degree of connectivity is therefore a function of the number of fracture sets and their relative importance as well as the fracture density.

The existence of a backbone also influences the direction of maximum permeability as matrix permeability and fracture aperture vary. In pattern B, which does not possess a backbone, the direction of the maximum permeability varies little, but in pattern A, which does possess a backbone, the maximum permeability direction rotates through about 20°. This has been related to the relative importance of direction of maximum net fracture length (dominant at low fracture to matrix permeability ratios) and the direction of the most extensive backbone (dominant at high permeability ratios). Since, in a reservoir rock, fracture apertures can be affected by fluid pressure, it is possible, therefore, not only for the degree of permeability anisotropy but also the orientation to vary as the fluid pressure varies.

In the models presented here, all fractures are

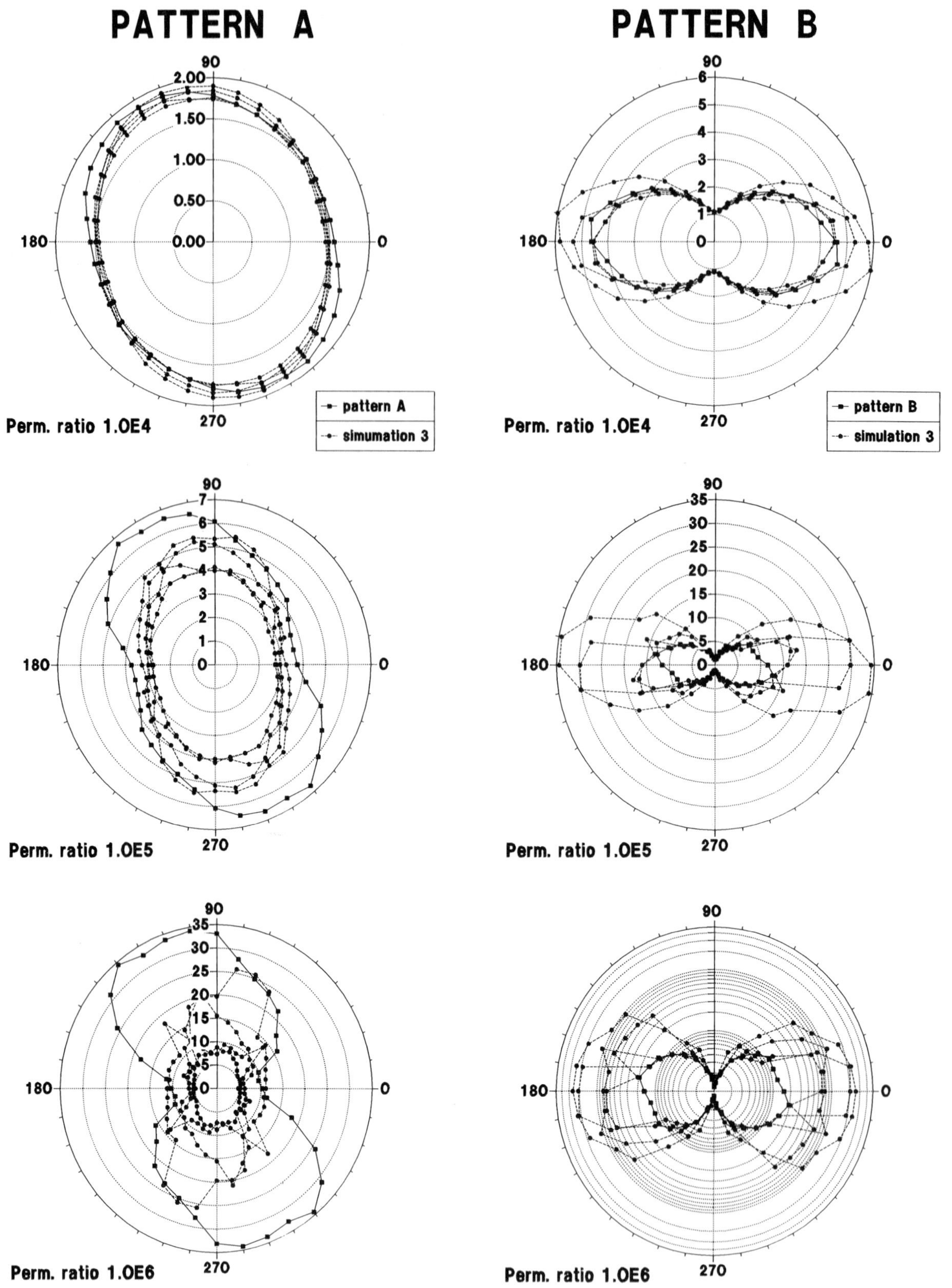

Fig. 11. Polar graphs showing the directional permeability of type 3 simulations (open symbols, dashed lines) and patterns A and B (filled symbols, solid lines). Fracture aperture and matrix permeabilities are as for Fig. 9.

assumed to have equal apertures. In reality, fracture apertures are variable and dependent on fracture length and effective normal stress. Fracture apertures decrease with increasing normal stress to a minimum where asperities on the fracture walls are in contact, although further reduction in aperture can take place if significant shear displacement or cyclical loading has occurred (Bandis et al., 1985).

At maximum closure, the fracture aperture is of the order of the root mean square height of asperities and where displacements (correlation distances) are of the order of 1 mm, was found to vary by a factor of approximately 1.5 (Brown and Scholtz, 1985). Thus a large range in apertures for joints, where displacements are small, is not expected. The model presented here, therefore, most closely simulates the case of isotropic compressive stress (so that fracture closure is at a maximum) and fractures which have displacements of less than 1 mm (joints). Since the natural and simulated fracture patterns are treated alike, the comparison of results remains valid despite the simplifications in the model. It is planned to build variable fracture apertures into the model at a future date so that the effects of fracture length and anisotropic stress on permeability can also be assessed.

Of the three types of simulation investigated, none provided a satisfactory fit to the permeability of the natural fracture patterns, for the range of matrix permeabilities tested. Although reasonably good fits were sometimes obtained at the lowest fracture matrix permeability ratio (simulations 1 and 3 for pattern A and simulation 3 for pattern B), errors increased (up to one order of magnitude) as the permeability ratio increased. Both underestimation and overestimation of the permeabilities of the natural patterns occurred. The main cause for the discrepancies was found to be the differences in the degree of connectivity between the simulations and the natural patterns. In the case of pattern A, which possesses a backbone in specific directions, the simulations were either found to be unconnected (1 and 2) or less well connected (3). In the case of pattern B, which is not connected in any direction, the simulations 1 and 3 were either connected or very nearly so. A secondary cause for discrepancy, which affects the simulations of pattern A, is thought to be the nature of fracture intersections. Pattern A contains a large proportion of T-junctions, whereas in the simulations almost all junctions are of a cross type. This results in a larger number of dead-ends, i.e., fracture segments which are not part of the backbone, in the simulations, which has an increasing effect on the permeability as the contrast between matrix and fractures increases.

The main difference between the natural patterns and the most sophisticated simulation (type 3) is the spatial distribution of the fractures. A random distribution of fracture centres, as used in all simulations above, allows fractures of the same set to intersect and produces dominantly cross-junctions at intersections, in contrast to the natural patterns. Such a distribution also fails to reproduce the regular spacing which is a common feature of fracture patterns and illustrated well by pattern B. Attempts have been made to reproduce the periodicity of fracture zones (La Pointe and Hudson, 1985) and variable fracture density (Long and Billaux 1987). Other simulations have been designed to incorporate spacing (Lemos et al., 1985) and intersection characteristics (Dershowitz, 1979; Khaleel, 1989). However, to the author's knowledge no fracture simulations that attempt to combine the length and orientations distribution with the spatial and intersection characteristics of natural patterns exists. A possible approach to the problem may be to "grow" fractures numerically, enabling different intersection characteristic to be produced, in a fashion similar to that suggested by Einstein and Herda (1986). Inspection of the spatial distribution of fracture centres in patterns A and B, suggests that they are clustered and that the clustering intensity is related to fracture length. In order to produce better simulations, more information must be obtained about the spatial distribution of fractures. To this end, a study of the spatial distribution of fractures in natural examples is planned.

The resistor network method has been applied here to two-dimensional natural and simulated fracture patterns. In reality, fracture networks are three-dimensional and fractures unconnected in the data plane may be connected in the third dimension. This implies that the global permeabilities determined here are minimum values. However, under certain circumstances such two-dimensional analyses can also be representative of the three-dimensional picture. In the case where the rock sequence is composed of alternating layers of different lithology, where one is of low permeability, e.g., sandstone and shale sequences, flow will be largely parallel to the bedding. In such circumstances, fractures are commonly perpendicular to the bedding, and the fracture pattern as displayed on the surface of the bed can be representative of the layer as a whole. The fracture patterns A and B, which come from a sandstone-shale sequence, are markedly different in geometry. The intersection of the two patterns as they are juxtaposed in space will, therefore, be a set of points, except for widely spaced large scale fractures which transverse several sandstone-shale units. Thus, is seems probable that, despite the presence of fractures in both lithologies, flow perpendicular to the bedding will be inhibited except along larger scale fractures.

Acknowledgements

I would like to thank Itzak Webman of the computational physics group at the Bergen Scientific Centre for his advice and help in constructing the model and on the subject of percolation theory. I would also like to thank Øistein Bøe for discussions on the subject of fluid flow in fractured permeable media and Johnny Petersen for his help with computational matters.

References

Andersson, J. and Thunvik, R., 1986. Predicting mass transport in discrete fracture networks with the aid of geometrical field data. Water Resour. Res., 22: 1941–1950.

Balberg, I., 1986. Connectivity and conductivity in 2D and 3D fracture systems. In: R. Englman and Z. Jaeger (Editors), Fragmentation, Form and Flow in Fractured Media. Ann. Israel Phys. Soc., 8: 89–101.

Bandis, S.C., Makurat, A. and Vik, G., 1985. Predicted and measured hydraulic conductivity of rock joints. In: Proceedings of the International Symposium Fundamentals of Rock Joints, Bjørkliden, pp. 269–279.

Brown, S.R. and Scholz, C.H., 1985. Broad bandwidth study of topography of natural rock surfaces. J. Geophys. Res., 90: 12575–12582.

Dershowitz, W.S., 1979. A probabilistic model for the deformability of jointed rock masses. M.S. Thesis, Massachusetts Institute of Technology, Cambridge, Mass.

Einstein, H.H. and Herda, H.H., 1986. Discrete modelling of the geometry of rock joint systems. In: R. Englman and Z. Jaeger (Editors), Fragmentation, Form and Flow in Fractured Media. Ann. Israel Phys. Soc., 8: 52–71.

Elsworth, D.A., 1986. A hybrid boundary element-finite element analysis procedure for fluid flow simulation in fractured rock masses. Int. J. Numer. Anal. Methods Geomech., 10: 569–584.

Hossack, J.R., 1984. The geometry of listric growth faults in the Devonian basins of Sunnfjord, W. Norway. J. Geol. Soc. London, 141: 629–637.

Hudson, J.A., 1982. Computer graphics and printed circuits for studying rock jointing patterns. Geol. Soc. London, Misc. Paper, 15, 69–80.

Khaleel, R., 1989. Scale dependence of continuum models for fractured basalts. Water Resour. Res., 25: 1847–1855.

Kirkpatrick, S., 1973. Percolation and conduction. Modern Phys., 45: 574–588.

La Pointe, P.R. and Hudson, J.A., 1985. Characterization and interpretation of rock mass joint patterns. Geol. Soc. Am., Spec. Paper, 199, 37 pp.

Lemos, J.V., Hart, R.D. and Cundall, P.A., 1985. A generalized distinct element program for modelling jointed rock mass. In: Proceedings of the International Symposium Fundamentals of Rock Joints. Bjørkliden, pp. 335–343.

Long, J.C.S. and Billaux, D.M., 1987. From field data to fracture network modelling: an example incorporating spatial structure. Water Resour. Res., 7: 1201–1216.

Long, J.C.S. and Witherspoon, P.A., 1985. The relationship of the degree of interconnection to permeability in fracture networks. J. Geophys. Res., B90: 3087–3098.

Long, J.C.S., Remer, J.S., Wilson, C.R. and Witherspoon, P.A., 1982. Porous media equivalents for networks of discontinuous fractures. Water Resour. Res., 18: 645–658.

Long, J.C.S., Gilmour, P. and Witherspoon, P.A. 1985. A model for steady state flow in random three-dimensional networks of disc-shaped fractures. Water Resour. Res., 21: 1105–1115.

McClay, K.R., Norton, M.G., Coney, P. and Davis, G.H., 1986. Collapse of the Caledonian orogen and the Old Red Sandstone. Nature, 232: 147–149.

Norton, M.G., 1986. Late Caledonian extension in western Norway: a response to extreme crustal thickening. Tectonics, 5: 195–204.

Piggot, A.R. and Elsworth, D., 1989. Physical and numerical studies of a fracture system model. Water Resour. Res., 25: 457–462.

Scheidegger, A.E., 1974. The Physics of Flow Through Porous Media. University Press of Toronto, Toronto, Ont., 236 pp.

Stauffer, D., 1985. Introduction to Percolation Theory. Taylor and Francis, London, 124 pp.

Steel, R.J., 1976. Devonian basins of western Norway — sedimentary response to tectonism and to varying tectonic context. Tectonophysics, 36: 207–224.

Steel, R. and Gloppen, T.G., 1980. Late Caledonian (Devonian) basin formation, western Norway: signs of strike slip tectonics during infilling. Spec. Publ. Int. Assoc. Sedimentol., 4: 79–103.

Warren and Root, 1963. The behaviour of naturally fractured reservoirs. J. Soc. Pet. Eng., 3: 245–251.

N.E. ODLING *Continental Shelf and Petroleum Technology Research Institute (IKU), NHåkon Magnusson's gt. 1B, P.O. Box 1883, Jarlesletta, 7001 Trondheim*

Present address: *Bergen Environmental Centre, IBM, Thormøhlensgt. 55, Bergen, N-5008, Norway*

Structure of the Mid-Norway Heidrun Field and its regional implications

W.J. Schmidt

Heidrun Field is formed by a trap-door structure located at the transition between the Nordland Ridge and Halten Terrace. Ample evidence exists for two major extensional events at Heidrun Field. Based upon seismic interpretation and regional constraints, a major extensional event occurred in mid-Permian(?) time. This event, which followed the NE trend already established in the Caledonian basement, created NE-striking tilted fault blocks. Displacement, which exceeds 500 m on some faults, occurs on east-dipping listric normal faults that may coincide with reactivated Caledonian thrusts.

During latest Jurassic to Early Cretaceous time the second major extensional event (the late Kimmerian phase) occurred. Four coeval fault sets developed at this time: two NE-trending sets which in places involve reactivated mid-Permian(?) faults and two N-trending sets which are often antithetic to the NE-trending fault sets. The multiple fault sets most likely developed in response to three-dimensional strain caused by basement-controlled rifting. An ESE–WNW extensional direction for the event can be inferred from the orthorhombic symmetry of the four coeval fault sets. Apparently strain associated with late Kimmerian extension could not be accommodated entirely by reactivation of older NE-trending faults, thus resulting in the formation of the N-trending fault sets. Along the Nordland Ridge a minor component of sinistral displacement most likely accompanied late Kimmerian deformation based on the inferred extension direction, the pull-apart basin geometry of the Halten Terrace and observed minor strike-slip displacement.

Introduction

A variety of models have been proposed to explain the tectonic development of the Mid-Norway continental margin (Bukovics et al., 1984; Brekke and Riis, 1987; Caselli, 1987). All of the models are poorly constrained by the lack of detailed information concerning Paleozoic stratigraphy and structure. Few wells in the area have penetrated rocks older than Triassic and available two-dimensional seismic data often lacks the resolution necessary to map Paleozoic features. In contrast, a three-dimensional seismic survey over Heidrun Field clearly images structures developed in the Paleozoic sequence, including metamorphic basement. The aim of this paper is to describe the structure of Heidrun Field and discuss the relevance of Heidrun tectonics to the tectonic development of the continental margin of Mid-Norway.

Regional setting

The major structural features of the Mid-Norway continental shelf are shown in Fig. 1. For the most part these features formed during latest Jurassic to earliest Cretaceous deformation, although pre-Jurassic deformation undoubtedly played a role in the overall structural configuration of the area (Aanstad et al., 1981; Gabrielsen and Robinson, 1984).

The structurally complex Halten Terrace lies in a transition zone between the Trøndelag Platform to the east and the Vøring and Møre Basins to the west. In essence the Halten Terrace occurs at a N-trending dogleg bend in the overall NE-trending Kristiansund–Bodø Fault Complex. Heidrun Field is located at the transition between the Halten Terrace and the Nordland Ridge. This transition zone is an ideal area in which to study the effect that Paleozoic structures have had on the tectonic development of the overlying Mesozoic sequence. This is due to the relatively shallow depths of basement in this area caused by uplift of the Nordland Ridge. At Heidrun Field, basement is shallow enough to be imaged on high-quality three-dimensional seismic data.

Although uplift associated with the Nordland Ridge appears to affect Heidrun Field, the detailed structure is similar to that observed elsewhere in the Halten Terrace. Bukovics et al. (1984) note that fault patterns on the Halten Terrace are characterized by interference of NE- and N-trending fault segments (Fig. 1). A similar fault pattern is observed at Heidrun Field (Fig. 2).

Structural and Tectonic Modelling and its Application to Petroleum Geology edited by R.M. Larsen, H. Brekke, B.T. Larsen and E. Talleraas. NPF Special Publication 1, pp. 381–395. Elsevier, Amsterdam.

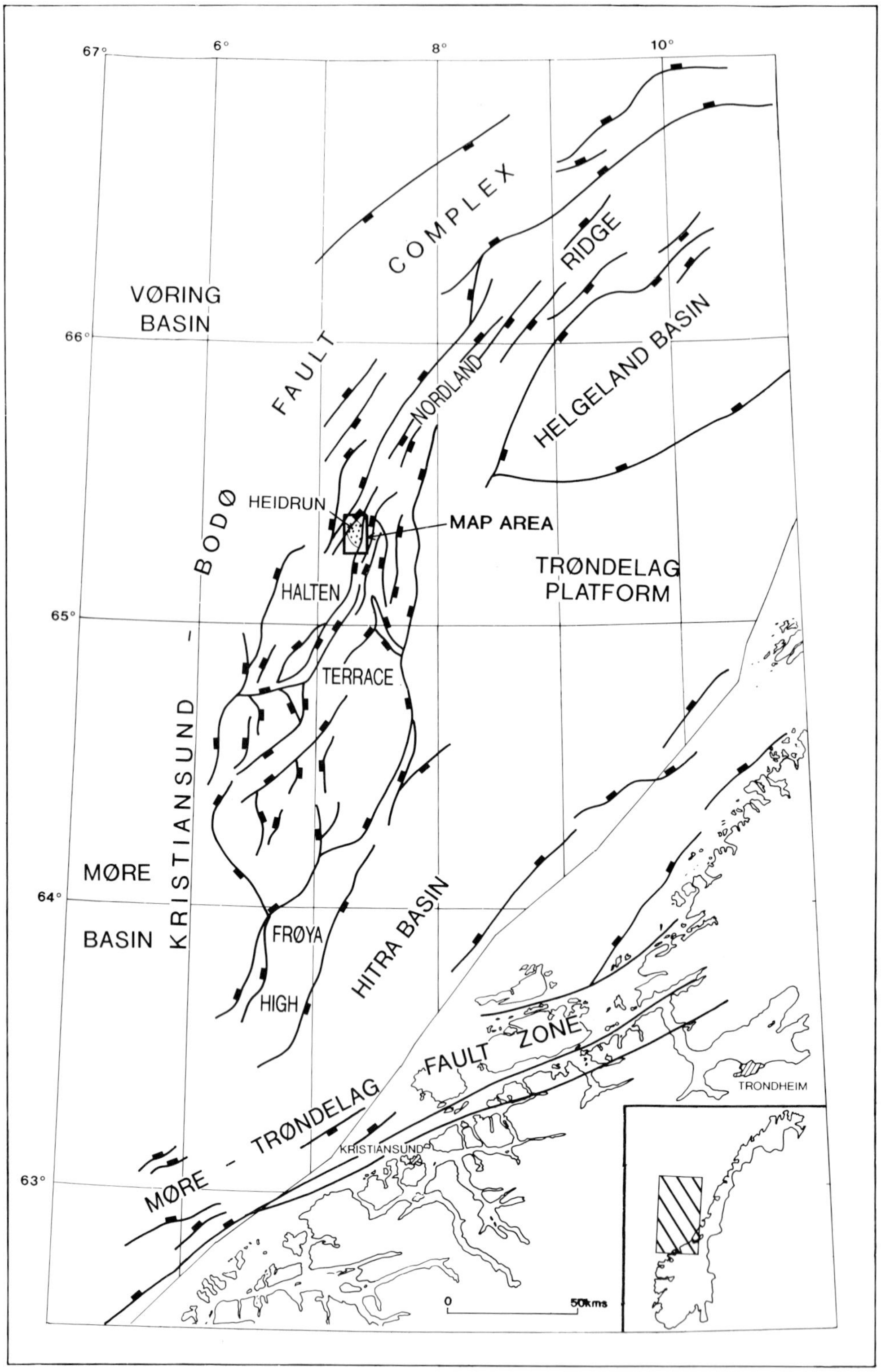

Fig. 1. Tectonic features of the Mid-Norway continental shelf. The map is compiled from Conoco Norway Inc. regional mapping with nomenclature after Gabrielsen et al. (1984). The map area at Heidrun Field for Figs. 2, 5 and 6 is also shown.

Fig. 2. Heidrun Field time structure map (TWT in seconds) on the top Middle Jurassic sand (Fangst Group) based on Conoco's three-dimensional seismic interpretation. Large normal faults at the field's eastern, northern and western flanks bound a trap-door structure, similar to those described by Harding (1984). Inset shows rose diagram of fault trends based on cumulative length. See Fig. 1 for map area.

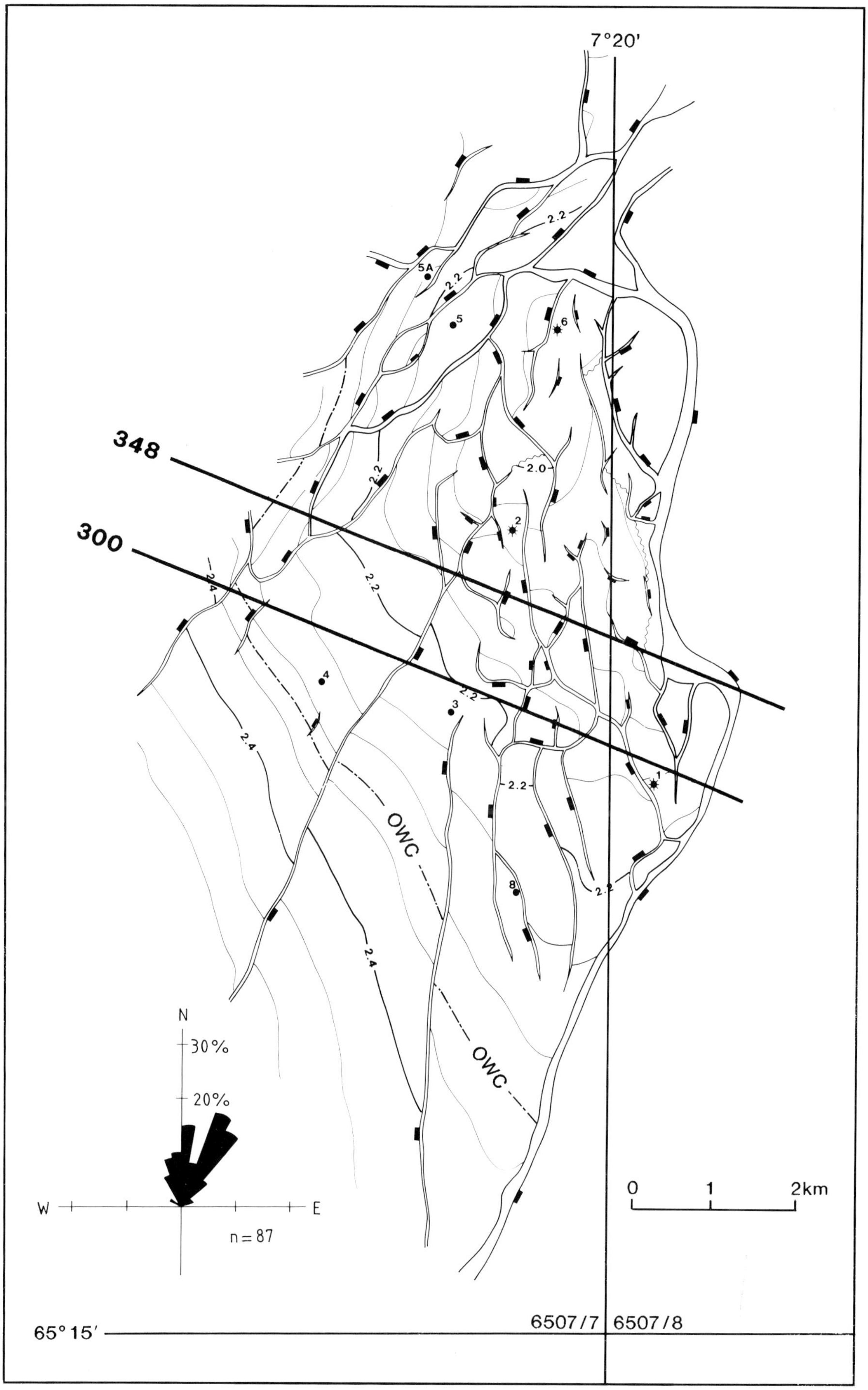
7°20'
348
300
2.2
2.0
2.4
5A
5
6
2
4
3
1
8
OWC
N
30%
20%
W
E
n = 87
0
1
2km
65° 15'
6507/7
6507/8

Structure and tectonic evolution

A generalized stratigraphic column with associated tectonism for Heidrun Field is shown in Fig. 3. It should be noted that a significant amount of uncertainty exists for the Paleozoic due to lack of well control. The eight wells drilled to date on the Heidrun structure have not penetrated rocks older than Lower Jurassic to Upper Triassic (i.e. the Åre Formation).

The crystalline basement underlying Heidrun Field is most likely comprised of metamorphic and igneous rocks that formed as a result of the collision between the Laurentian and Fennoscandian plates during the Late Silurian Caledonian Orogeny (Dewey, 1969; Ziegler, 1981). Metamorphic basement rocks have

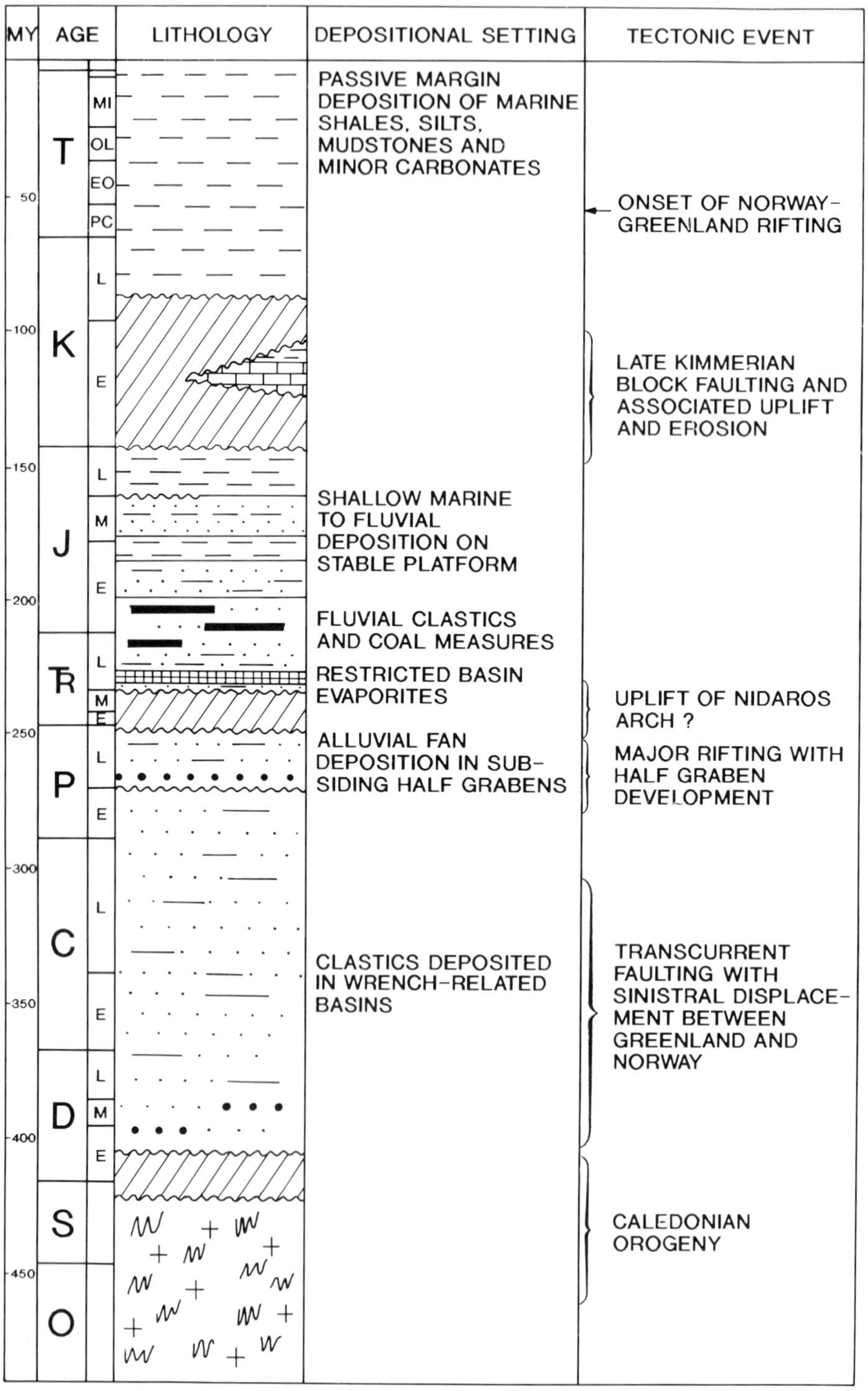

Fig. 3. Simplified stratigraphic column and tectonic events for Heidrun Field based on well control and three-dimensional seismic interpretation. Lithostratigraphy is based on nomenclature of Dalland et al. (1988).

been penetrated in a well on the Nordland Ridge, northeast of Heidrun. The Caledonian metamorphics underlying Heidrun are assumed to have a NE-striking fabric based upon the trend of Caledonian structures in East Greenland (Surlyk, 1977), Mid-Norway (Aanstad et al., 1981) and the overall NE trend of the Caledonian structures in the North Sea (Johnson and Dingwall, 1981).

A deep and very strong seismic event observed beneath Heidrun Field is thought to represent the top of crystalline basement (Fig. 4). This event corresponds to acoustic basement and occurs as deep as 3000 m (based upon seismic stacking velocities) below the Upper Triassic salt succession. All basement faults beneath Heidrun Field trend northeasterly (Fig. 5). Basement appears to be offset in part by listric normal faults. The age of the basement faulting is probably Permo-Triassic and is discussed in more detail below.

The presence of thick Devonian, Carboniferous and Permian strata offshore Mid-Norway can be inferred from rocks exposed in East Greenland (Bukovics and Zielger, 1985). Sedimentation during this period was related in part to Devono-Carboniferous rift and wrench systems which developed as a result of sinistral displacement between Greenland and Norway (Ziegler, 1981; Bukovics and Ziegler, 1985). In the Heidrun area seismic stacking velocities indicate that the thickness of the inferred Devonian to mid-Permian sequence (Fig. 4) is in excess of 2500 m.

A major rifting episode most likely occurred during the mid-Permian (Brekke and Riis, 1987), based upon analogy with East Greenland, where Upper Permian basal conglomerates rest with angular discordance upon underlying rocks that were tilted during mid-Permian block faulting (Surlyk et al., 1984). The mid-Permian half grabens of East Greenland look very similar to those developed beneath the Upper Triassic sequence at Heidrun Field (Fig. 4). In both areas it appears that the trend of the half grabens follows the older Caledonian structural grain. Furthermore, as in East Greenland, the top of the block-faulted sequence coincides with an angular unconformity.

There is extensive evidence of Permo-Triassic rifting and associated half-graben development in the northern North Sea. On the Shetlands–Orkneys Platform, Permo-Triassic half grabens dissect shallow Caledonian metamorphics (Duindam and van Hoorn, 1987). Kirton and Hitchen (1987) note that half-graben formation in the West Shetland Basin most likely initiated in the Permo-Triassic based on the age of the graben fill. In the North Viking Graben, Badley et al. (1988) note that there is unequivocal evidence for a pre-Jurassic rifting event, probably Permo-Triassic in age.

Throughout the northern North Sea a very distinctive NE–SW trending, post-Caledonian half-graben system follows trends of the Caledonian basement (Johnson and Dingwall, 1981). Based on deep seismic profiles from the West Shetland Basin, the Permo-Triassic half-grabens are bounded by SE-dipping listric faults which are interpreted as being reactivated Caledonian thrusts (Brewer and Smythe, 1984; Duindam and van Hoorn, 1987; Nelson and Lamy, 1987). A similar listric geometry is observed at Heidrun where pre-Late Triassic SE-dipping faults offset basement and mid-Permian reflectors (Fig. 4). Faults cutting these horizons trend NE (Figs. 5 and 6), presumably following the NE trend in Caledonian metamorphic basement, and may coincide with reactivated Caledonian thrusts. If this is the case, it implies that Caledonian thrusts at Heidrun have a dip similar to the inferred dip of Caledonian thrusts from the West Shetland Basin.

At Heidrun Field, strata of inferred Late Triassic age rest with angular discordance on the older block-faulted sequence. The strong seismic event which occurs about 100 ms above the Permo-Triassic angular unconformity (Fig. 4) correlates well with the top of the Upper Triassic (Late Carnian) salt succession observed on the Trøndelag Platform to the east (Jacobsen and van Veen, 1984). The Lower to Middle Triassic section appears to be absent at Heidrun. This may be due to uplift of the Nidaros Arch, as described by Bukovics et al. (1984).

Well and seismic data from Heidrun Field indicate that the Late Triassic to Late Jurassic was a period of relative tectonic quiescence. During this time, fluvial and shallow marine sediments were deposited on a relatively stable platform.

The effects of latest Jurassic to earliest Cretaceous deformation, i.e. the late Kimmerian phase (Rawson and Riley, 1982), are evident throughout the Mid-Norway continental margin (Bukovics et al., 1984). At Heidrun faulting associated with this event occurred in latest Jurassic to Early Cretaceous time based on the age of the youngest faulted strata (Oxfordian) and the oldest overlying rocks (Coniacian) unaffected by faulting.

During late Kimmerian deformation, NE- and N-trending fault sets developed at Heidrun (Fig. 2). The NE-trending faults are related, in part, to reactivation of the NE-striking faults which formed during the Permo-Triassic rifting episode, although most of the Permo-Triassic faults were not reactivated during the late Kimmerian event. The observed offset on the reactivated faults is relatively small (less than 100 m)

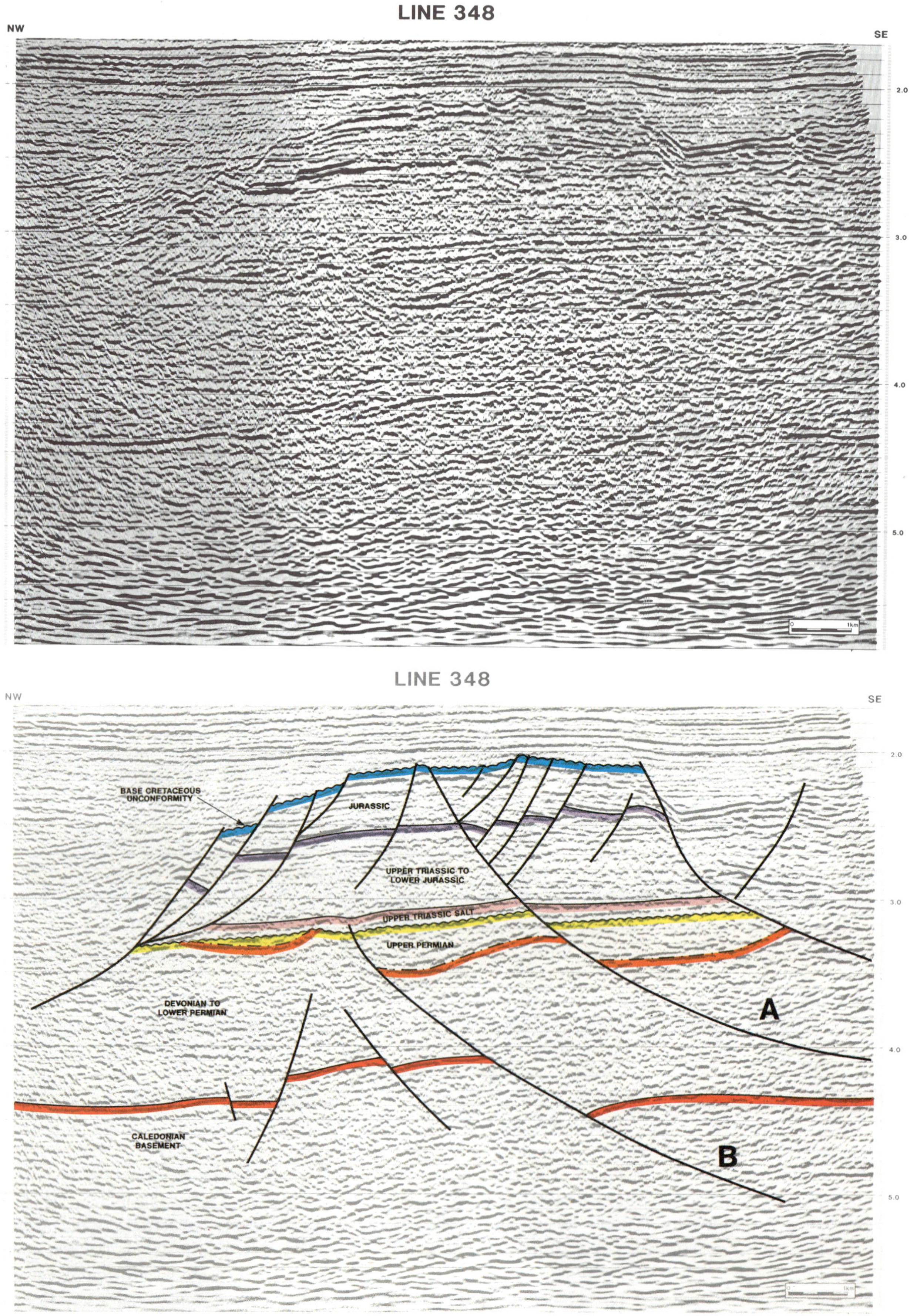

Fig. 4. Two migrated three-dimensional seismic sections over Heidrun Field. See Figs. 2, 5 and 6 for line locations. Fault *A* is a Permo-Triassic age fault partially reactivated during late Kimmerian deformation. Fault *B* is also Permo-Triassic in age but was not reactivated during the late Kimmerian event. Note that there are numerous antithetic, westerly-dipping faults associated with fault *A*, which trends NE.

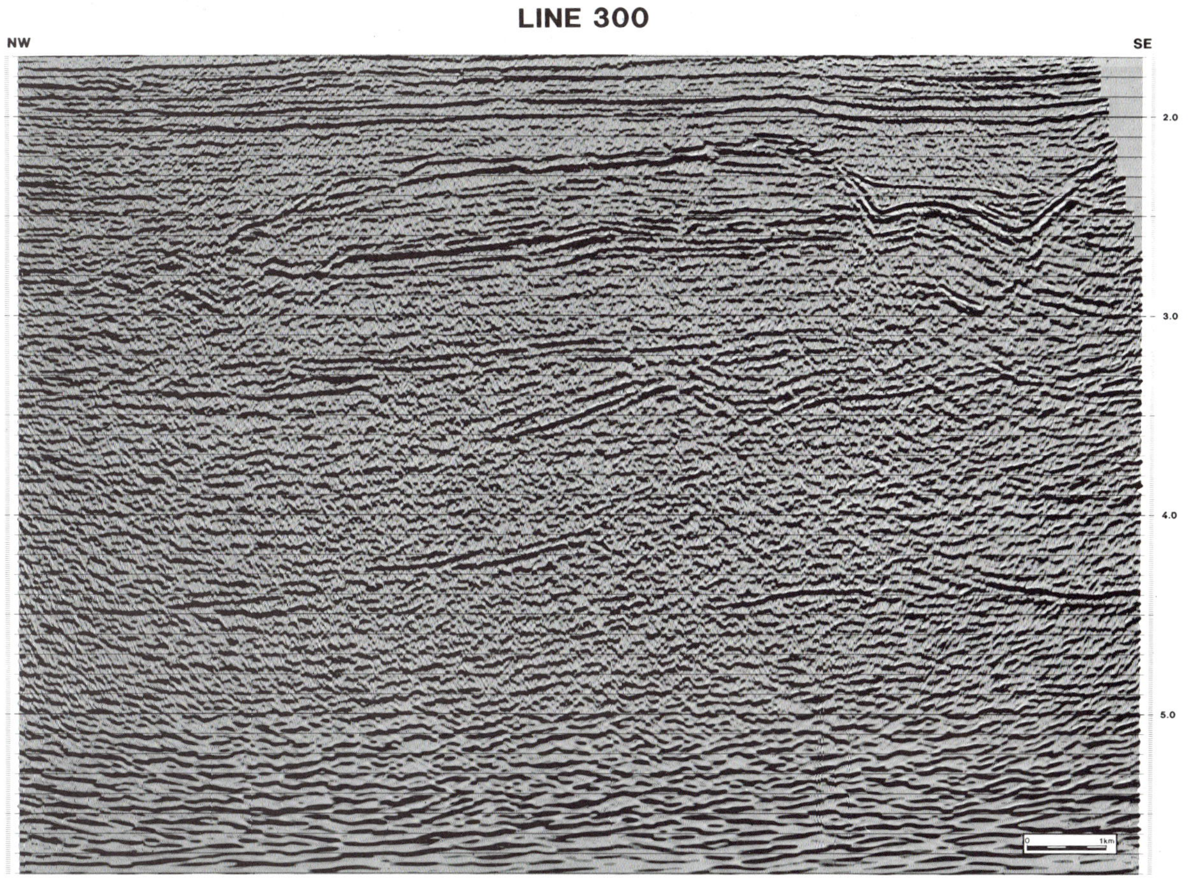

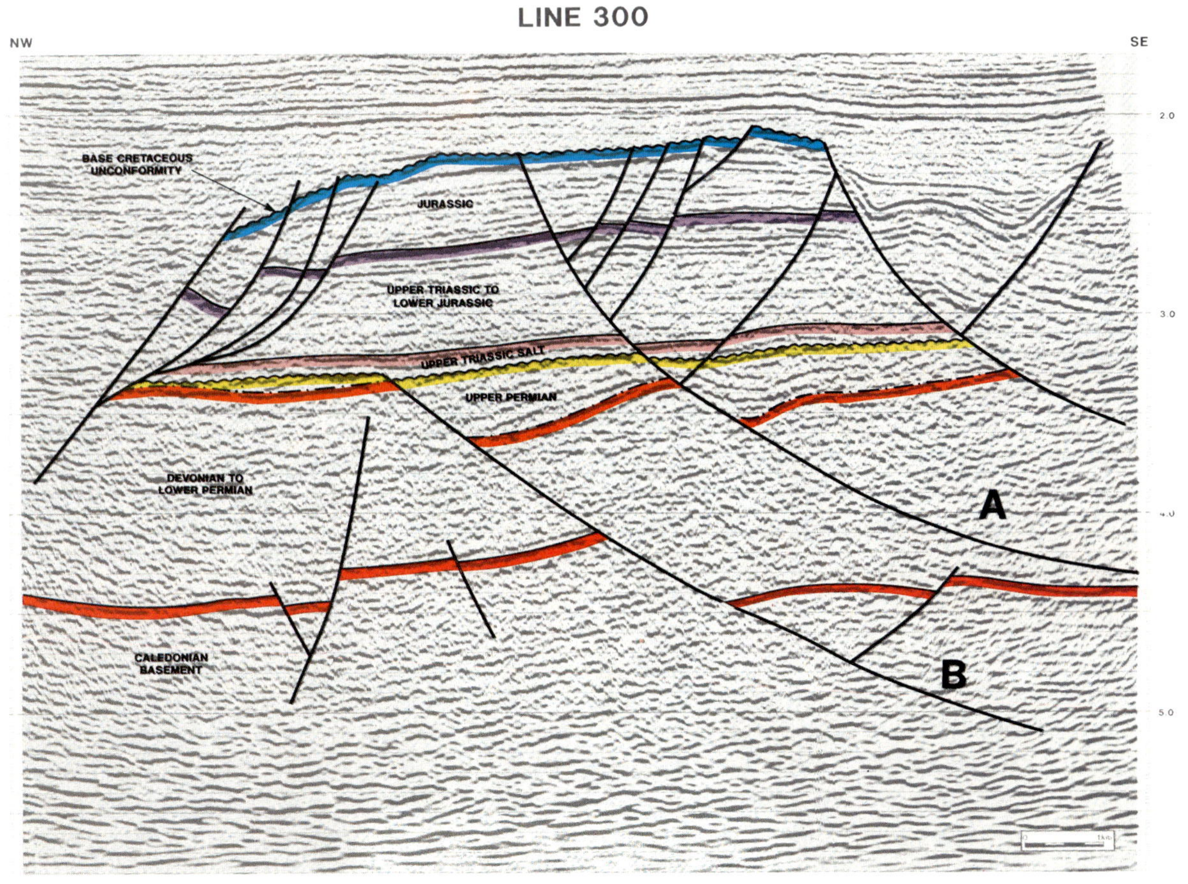

Fig. 4 (continued).

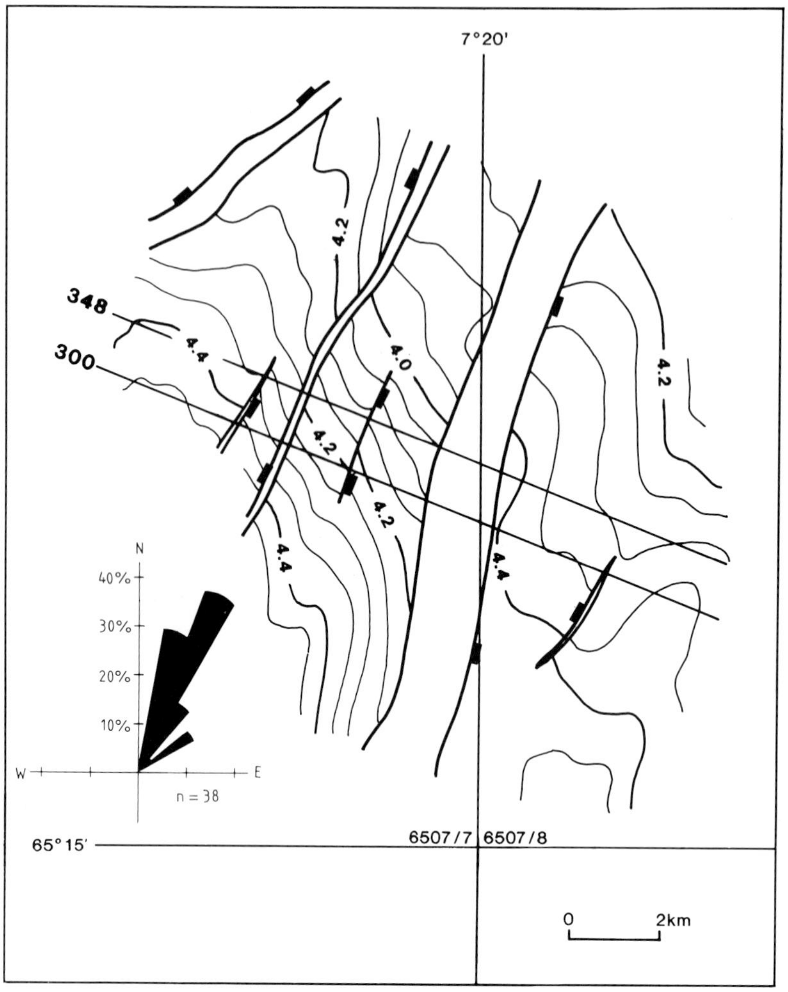

Fig. 5. Heidrun Field time structure map (TWT in seconds) on the top Caledonian basement(?). Rose diagram shows the fault trends based on cumulative length. See Fig. 1 for map area.

but may be considerably larger on the flanks of the structure (Fig. 4).

At Heidrun Field the observed offset on the late Kimmerian faults is always dip-slip, based on detailed mapping of Triassic- and Jurassic-age horizons. No evidence of strike-slip or reverse faulting has been observed, barring one possible exception. The large graben bounding the eastern flank of the field (Fig. 4) may be related to a component of strike-slip faulting based on the negative flower structure geometry (Harding, 1983). The rather complex geometry of the graben may also be related strictly to normal faults which sole out at depth, as shown in the interpretation in Fig. 4. There is clear evidence of faults soling out into the Triassic salt sequence on the western flank of the Heidrun structure (Fig. 4).

The N- and NE-trending fault sets are mutually offsetting (Fig. 2) indicating that they formed simultaneously. The N-trending faults are often antithetic to NE-trending faults (Fig. 4). The dip of late Kimmerian faults in the Triassic and Jurassic sequence averages about 55–60°.

Evidence is lacking for any significant tectonic activity in the Heidrun area following the late Kimmerian event. Late Cretaceous sediments onlap the Heidrun structure, which was uplifted and eroded at its crest during latest Jurassic to Early Cretaceous time.

The effect of the separation of Norway from Greenland in the Late Paleocene to Early Eocene (Eldholm et al., 1984) is probably reflected in the observed increase in sedimentation during the Eocene to Pliocene.

Late Kimmerian deformational model

The N- and NE-trending normal fault sets observed at Heidrun are characteristic of late Kimmerian faulting in the Halten Terrace (Fig. 1). Bukovics et al. (1984) note that the fault pattern of the Halten Terrace, where mappable on detailed seismic grids, is characterized by the interference of N- and NE-striking fault segments.

A similar N- and NE-trending, late Kimmerian

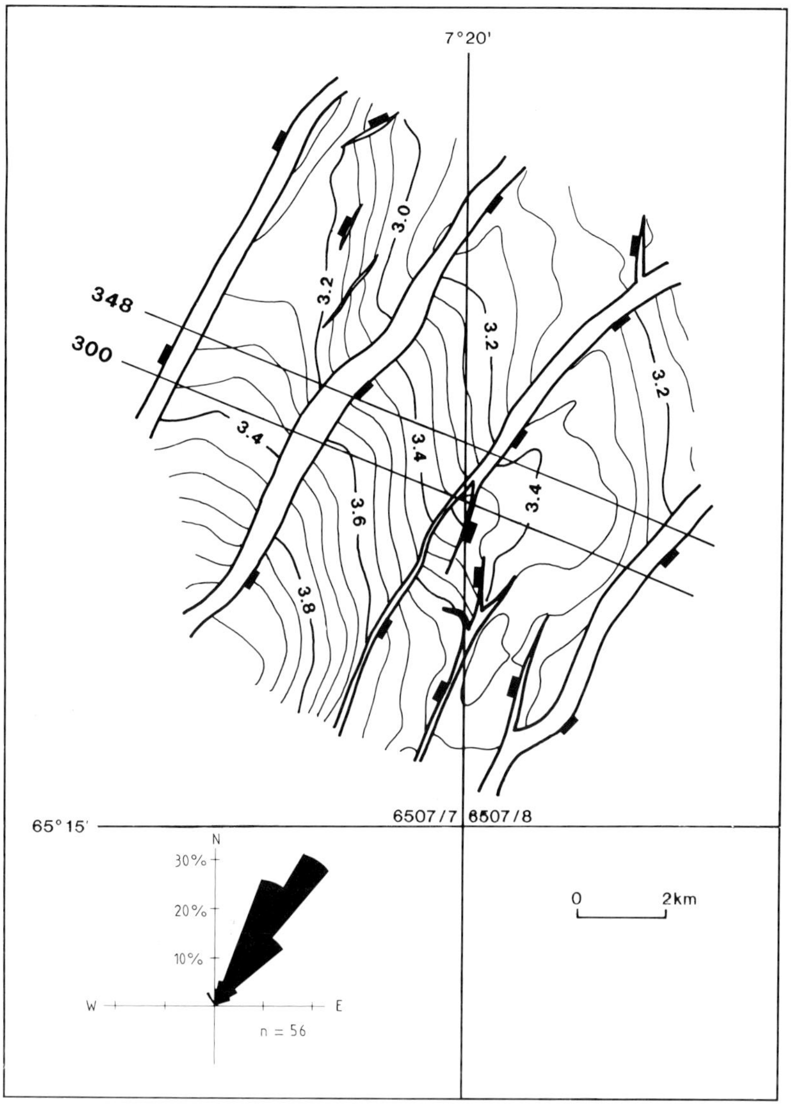

Fig. 6. Heidrun Field time structure map (TWT in seconds) on the top mid-Permian(?). Rose diagram shows the fault trends based on cumulative length. See Fig. 1 for map area.

faulting pattern has been observed by Hay (1978) in the North Viking Graben (Fig. 7). As on the Halten Terrace, late Kimmerian multiple fault sets on the prolific western flank of the North Viking Graben appear to have formed synchronously (Hay, 1978; Harding, 1984) and apparently have a dominant dip-slip displacement. Faulting in both the Halten Terrace and the North Viking Graben occurs on four fault sets: two N-trending sets which dip E or W, and two NE-trending sets which dip NW or SE (Figs. 1 and 7). A similar gridded fault pattern comprised of four fault sets is observed at numerous other localities (Harding, 1984; Ramberg and Smithson, 1975) including the Rhine Graben (Illies, 1981), the Rio Grande Rift (Ramberg et al., 1978) and the San Rafael Swell (Aydin and Reches, 1982).

The widely accepted theory of failure proposed by Anderson (1951) cannot explain the origin of three or more coeval fault sets (Reches, 1978; Krantz, 1988). Anderson's theory describes failure as a result of two-dimensional strain (i.e. plane strain) where a maximum of two conjugate fault sets develop and intersect in the intermediate principal strain and stress directions (Fig. 8a). Reches (1978; 1983) and Krantz (1988) have derived models for faulting in a general, three-dimensional strain field (i.e. where strain in the intermediate principal direction is not necessarily equal to zero). Both models indicate that four sets of faults are required to accommodate three-dimensional, coaxial (i.e. non-rotational) strain. The four fault sets exhibit orthorhombic symmetry with respect to the principal axes of stress and strain (Fig. 8b).

The orthorhombic symmetry of late Kimmerian faulting at Heidrun is remarkably similar to that predicted by the three-dimensional strain models

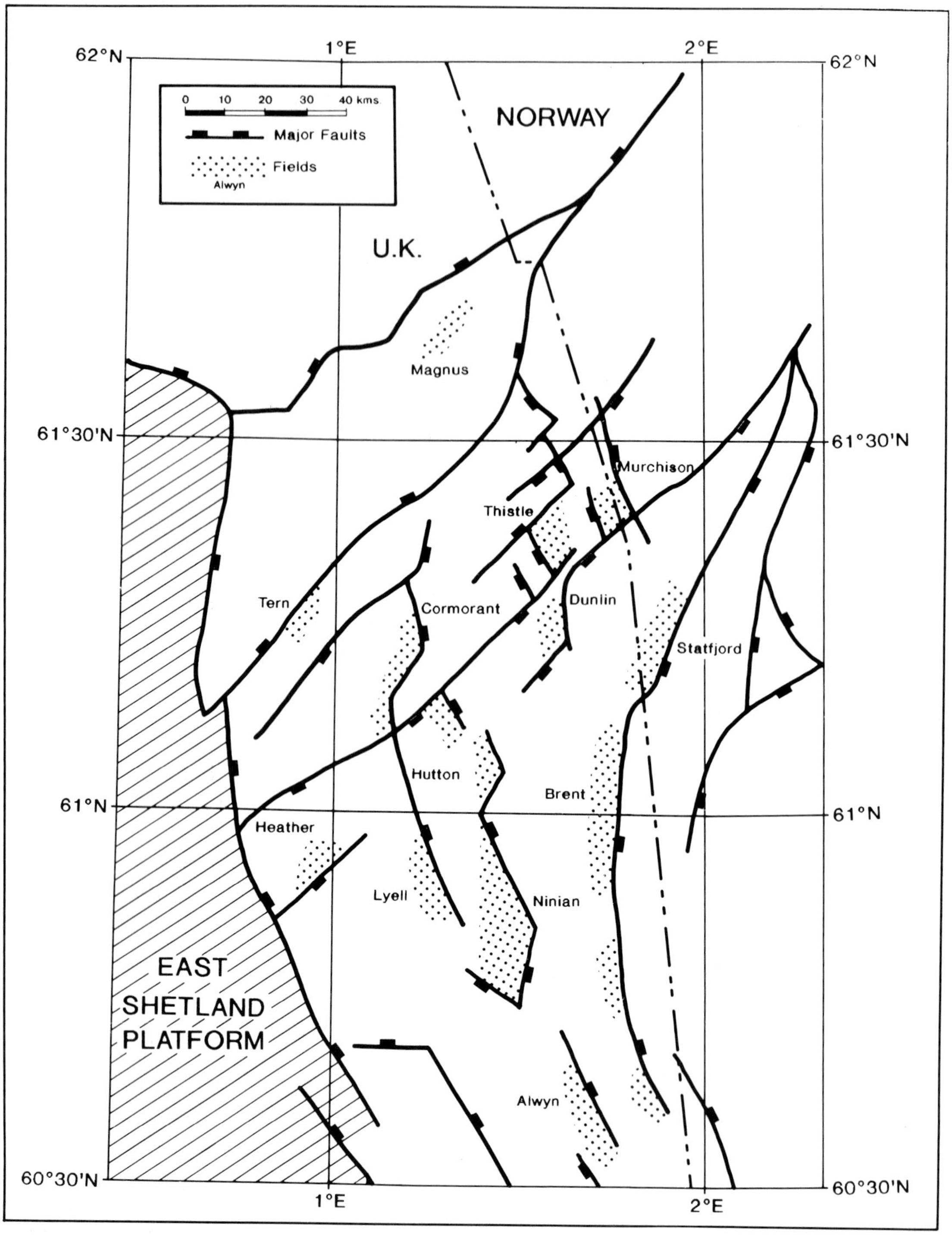

Fig. 7. Fault pattern and fields of the NW-flank of the North Viking Graben (after Hay, 1978).

(Fig. 2). Due to the fact that Heidrun deformation involved coeval faulting on four fault sets, which exhibit orthorhombic symmetry, it appears likely that the late Kimmerian extensional phase at Heidrun involved three-dimensional strain. If this is the case, then an ESE–WNW extension direction can be inferred from the orthorhombic late Kimmerian fault pattern (Fig. 9). Due to the similarities in style and nature of deformation throughout the Halten Terrace (Fig. 1) and North Viking Graben (Fig. 7), it is suggested that late Kimmerian structures in these areas also resulted from three-dimensional strain.

Other explanations for the gridded fault pattern observed in the northern North Sea have been proposed. Halstead (1975) postulates that the gridded fault pattern is related to superposition of Permian to Cenozoic normal faulting upon a pre-existing strike-slip fault system developed in the Caledonian base-

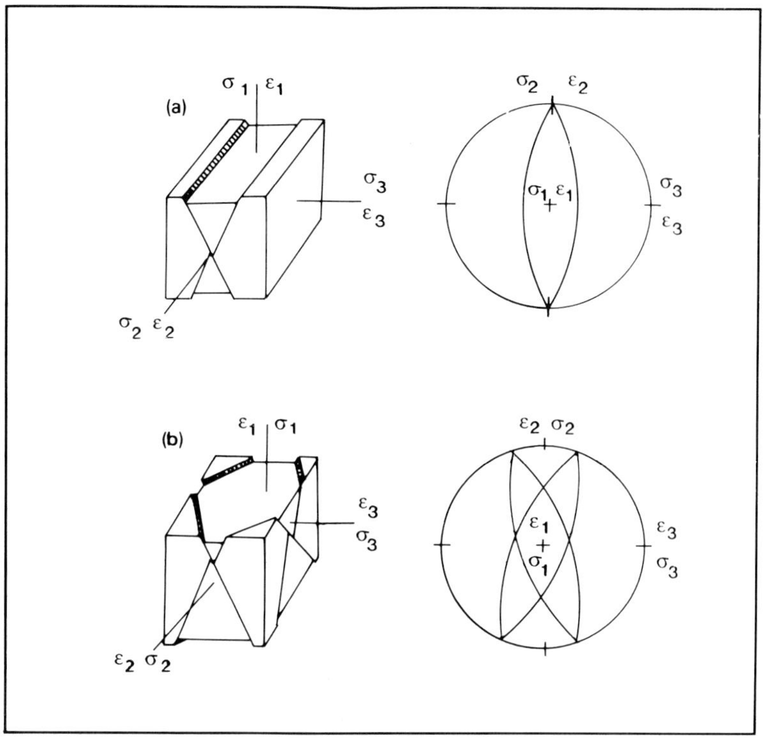

Fig. 8. (a) Conjugate set of normal faults associated with coaxial, plane strain according to Anderson's (1951) theory. Principal axes of stress and strain are indicated. (b) Four sets of normal faults capable of accommodating coaxial, three-dimensional strain. Note the orthorhombic symmetry of the four fault sets in stereographic projection (after Reches, 1983).

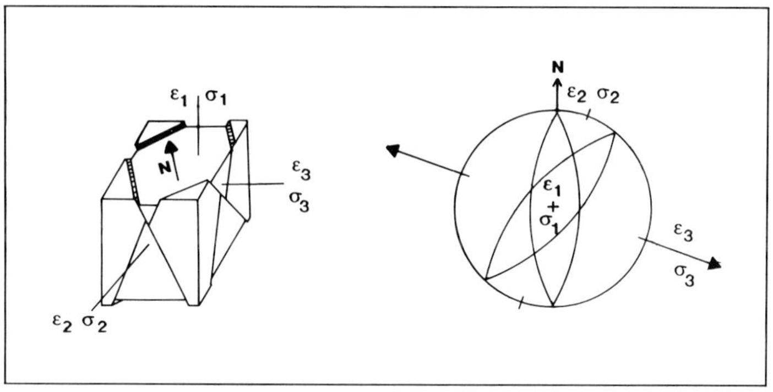

Fig. 9. Block diagram and stereographic projection illustrating the inferred ESE–WNW extension direction for Heidrun Field based on the trend of the four coeval late Kimmerian fault sets which exhibit orthorhombic symmetry. Fault trends are derived from the rose diagram shown in Fig. 2.

ment. Hay (1978) relates the gridded fault pattern to reactivation of strike-slip faulting along old NE Caledonian trends during E–W Jurassic extension.

It is important to note that gridded fault patterns can apparently form without strike-slip or basement-controlled faulting. Harding (1984) notes that gridded fault patterns, like those observed in the northern North Sea, often form at dogleg bends in graben systems where faults of the regional and oblique directions intersect. Based upon data gathered from a number of different grabens, Harding (1984) observes that gridded faults are generally synchronous, exhibit dip-slip displacement and are not dependent upon the basement structural grain for their orientation. If these observations are applicable to the northern North Sea, then the models to describe the gridded fault pattern in terms of basement-controlled strike-slip faulting (Hay, 1978) or reactivation of multidirectional faulting in the basement (Halstead, 1975) may not apply.

The gridded fault pattern of the Halten Terrace fits well with the general observations Harding (1984) has made with respect to the occurrence of multidirectional faulting. The Halten Terrace occurs in a N-trending dogleg of an overall NE-trending extensional system (Fig. 1). Late Kimmerian faulting in the Halten Terrace reflects the interference of the N-oblique and NE-regional trends. Both trends of faults formed synchronously and exhibit predominantly dip-slip displacement.

It is worthwhile noting that the basic constraint on models developed for three-dimensional strain (Reches, 1983; Krantz 1988), namely that coaxial three-dimensional strain involves synchronous displacement on four independent fault sets exhibiting orthorhombic symmetry, is compatible with Harding's (1984) general observations on gridded fault patterns.

As noted above, an ESE–WNW extension direction can be inferred for Heidrun Field from the orthorhombic late Kimmerian fault pattern, assuming the deformation resulted from three-dimensional, coaxial strain. The same multidirectional faulting is observed throughout the Halten Terrace, where the N- and NE-trending coeval faults may also be used to infer an ESE–WNW extensional direction for the late Kimmerian event.

An ESE–WNW extensional direction places certain kinematic constraints upon late Kimmerian deformation along the Kristiansund–Bodø Fault Complex. Figure 10 shows the inferred extension direction for Heidrun Field superimposed upon the Kristiansund–Bodø Fault Complex. Figure 11 shows the same extension direction superimposed on the Nordland Ridge and Halten Terrace. Unlike the Halten Terrace, the Nordland Ridge is comprised of only NE-trending faults; N-trending fault sets are not observed. Therefore, the inferred oblique extension direction should result in a component of sinistral displacement, as shown in Fig. 11. On Halten Terrace the ESE–WNW extension can be accommodated on the N- and NE-trending fault sets without any strike-slip component to strain (i.e. coaxial deformation). Thus, the Halten Terrace may have experienced a larger amount of extension than the Nordland Ridge. This may be one factor involved in explaining why the Halten Terrace has subsided more than the Nordland Ridge.

As noted above, the Halten Terrace lies in a N-trending dogleg of the overall NE-trending fault complex comprised of the Kristiansund-Bodø Fault Complex and the Møre–Trøndelag Fault Zone (Fig. 1). Crowell (1974) describes the kinds of structures observed at dogleg bends in strike-slip settings. The rhombohedral depression associated with the Halten Terrace is very similar in geometry to pull-apart basins described by Crowell (Fig. 12). Brekke and Riis (1987) note that the rhombic shape of the Halten Terrace probably formed as a pull-apart basin in a sinistral fault system during late Kimmerian deformation.

Caselli (1987) postulates that the pull-apart basin associated with the Halten Terrace is the result of right-lateral slip movements. This interpretation appears to be incompatible with the observed rhomb-shaped depression that is associated with the Halten Terrace. With dextral offset a component of compressional uplift should be observed (Fig. 12). There is some uplift observed along the western boundary of the Halten Terrace, i.e. the West Haltenbanken High (Bukovics et al., 1984; Caselli, 1987), but this high is most likely related to footwall uplift in response to the deep subsidence of the Møre and Vøring Basins to the west. Based upon arguments developed by Barr (1987), it is possible to have over 1500 m of footwall uplift at the margins of deeply subsided basins in purely extensional settings.

Price and Rattey (1984) discuss the evidence supporting dextral displacement in Mid-Norway. The dextral displacement inferred by Price and Rattey (1984) occurred in the mid-Cretaceous along ENE-trending faults. The effects of a mid-Cretaceous event are not observed at Heidrun Field.

The model for late Kimmerian deformation proposed here, i.e. three-dimensional strain associated with ESE–WNW extension, may be used to describe the major features comprising the Kristiansund–Bodø Fault Complex. In this model (Fig. 11) a component of sinistral strike-slip deformation should be observed along the Nordland Ridge but the majority of the strain is the result of the extensional component. Evidence for late Kimmerian strike-slip movement along the Nordland Ridge is discussed by Caselli (1987).

Within the Halten Terrace the coeval, multiple fault sets with orthorhombic symmetry are best explained by faulting in a three-dimensional strain field. The NE-trending fault sets are related, in part, to reactivation of Permo-Triassic structures with the same NE trend.

The three-dimensional state of strain in Halten Terrace may have been caused by ESE–WNW extension that could not be entirely accommodated by reactivation of NE-trending faults. Unfortunately, the models for three-dimensional strain are not far enough advanced to predict the impact of basement-controlled faulting or non-coaxial deformation; both may have played a role in late Kimmerian deformation although the nature of the faulting in the Halten Terrace appears to favor, as a whole, coaxial strain.

Conclusions

The main conclusions from this investigation are as follows:

(1) The multiple, coeval fault sets associated with late Kimmerian deformation at Heidrun Field, and

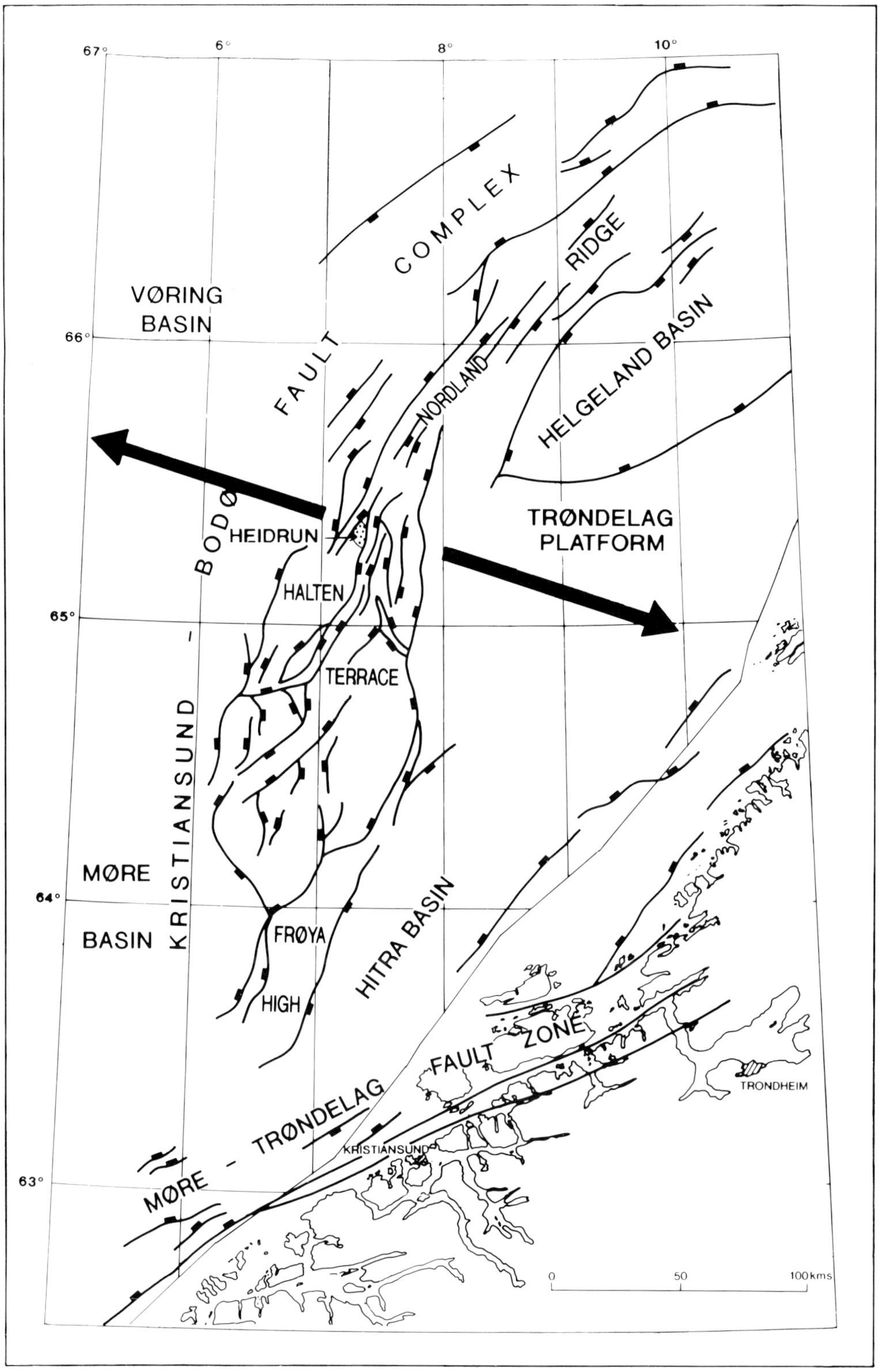

Fig. 10. Inferred ESE–WNW extension direction for Heidrun late Kimmerian deformation superimposed on the Kristiansund–Bodø Fault Complex.

the Halten Terrace as a whole, appear to be related to three-dimensional strain which developed at a N-trending dogleg of the overall NE-trending Kristiansund–Bodø Fault Complex. An ESE–WNW extension direction for late Kimmerian deformation can be inferred from the four coeval, normal fault sets.

(2) Based upon the inferred ESE–WNW extension direction, kinematic modelling (Fig. 11) implies that a minor component of sinistral displacement oc-

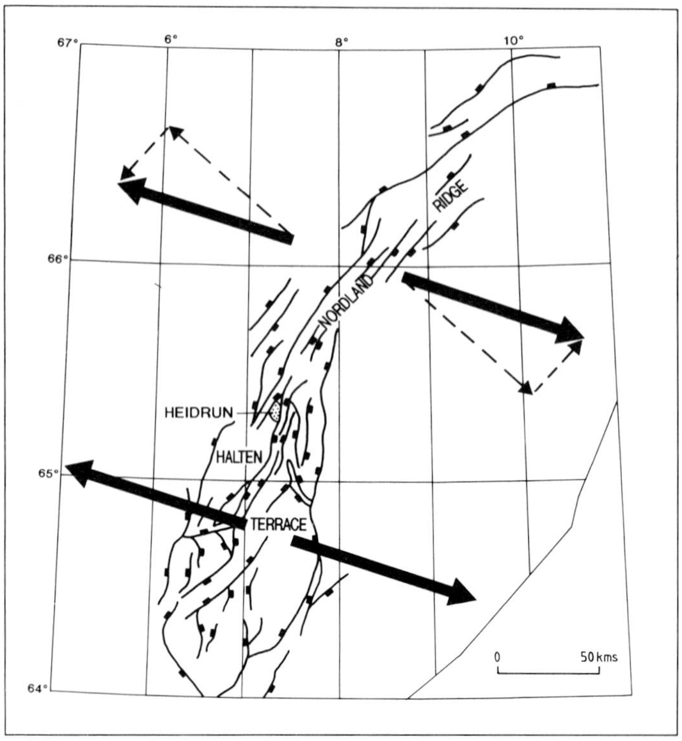

Fig. 11. The inferred ESE–WNW extension direction for Heidrun Field superimposed on the Nordland Ridge and Halten Terrace. Note that the extension can be divided into sinistral strike-slip and normal dip-slip components with respect to the NE-trending Nordland Ridge.

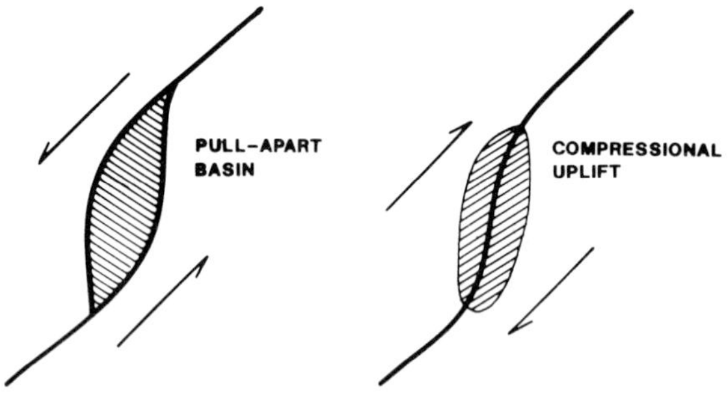

Fig. 12. Sketch showing the formation of a pull-apart basin and compressional uplift based on sense of shear along a strike-slip fault with a dogleg bend (based upon Crowell, 1974).

curred along the NE-trending Nordland Ridge during the late Kimmerian phase. This may be due to the fact that, unlike Halten Terrace, the Nordland Ridge lacks development of four coeval, normal fault sets required to accommodate three-dimensional, coaxial strain.

(3) Data from Heidrun Field implies that a major rifting event and associated half-graben formation occurred in Permo-Triassic time, probably in the mid-Permian. Trends of faults follow the old NE-Caledonian grain. Displacement occurs on SE-dipping listric normal faults which may coincide with reactivated Caledonian thrusts. If this is the case, it implies that Caledonian thrusts dip to the southeast in the Heidrun area, similar to the inferred dip of Caledonian thrusts from the West Shetland Basin.

Acknowledgements

The author is indebted to Conoco Norway Inc. and the Heidrun Unit Owners (Statoil, Arco Norway, Norsk Hydro and DNO) for permission to publish this investigation. P. van den Heuvel, B. Reid and A.G. Doré provided helpful suggestions for the manuscript. K.A. Pedersen coordinated the drafting of the figures and I. Thorsnæs and L. Moen typed the manuscript.

References

Aanstad, M.K., Gabrielsen, R.H., Hagevang, T., Ramberg, I.B. and Torvanger, O., 1981. Correlation of offshore and onshore structural features between 62°N and 68°N, Norway. In: Norwegian Symposium on Exploration. Norw. Pet. Soc., Bergen, pp. 1–24.

Anderson, E.M., 1951. The Dynamics of Faulting. Oliver and Boyd, London, 206 pp.

Aydin, A. and Reches, Z., 1982. Number and orientations of fault sets in the field and in experiments. Geology, 10: 110–112.

Badley, M.E., Price, J.D., Rambech Dahl, C. and Agdestein, T., 1988. The structural evolution of the northern Viking Graben and its bearing upon extensional modes of basin formation. J. Geol. Soc. London, 145: 455–472.

Barr, D., 1987. Lithospheric stretching, detached normal faulting and footwall uplift. In: M.P. Coward, J.F. Dewey and P.L. Hancock, (Editors), Continental Extensional Tectonics. Geol. Soc. London, Spec. Publ., 28: 75–94.

Brekke, H. and Riis, F., 1987. Tectonics and basin evolution of the Norwegian shelf between 62°N and 72°N. Nor. Geol. Tidsskr., 67: 295–322.

Brewer, J.A. and Smythe, D.K., 1984. MOIST and the continuity of crustal reflector geometry along the Caledonian–Appalachian orogen. J. Geol. Soc. London, 141: 105–120.

Bukovics, C. and Ziegler, P.A., 1985. Tectonic development of the Mid-Norway continental margin. Mar. Pet. Geol., 2: 2–22.

Bukovics, C., Shaw, N.D., Cartier, E.G. and Ziegler, P.A., 1984. Structure and development of the Mid-Norway continental margin. In: A.M. Spencer et al. (Editors), Petroleum Geology of the North European Margin. Norw. Pet. Soc., Graham and Trotman, London, pp. 407–423.

Caselli, F., 1987. Oblique-slip tectonics Mid-Norway shelf. In: J. Brooks and K. Glennie (Editors), Petroleum Geology of North West Europe. Graham and Trotman, London, pp. 1049–1063.

Crowell, J.C., 1974. Origin of Late Cenozoic basins in Southern California. In: W.R. Dickinson (Editor), Tectonics and Sedimentation. Soc. Econ. Paleontol. Mineral., Spec. Publ., 22: 190–204.

Dewey, J.F., 1969. Evolution of the Appalachian/Caledonian Orogen. Nature, 222: 124–129.

Dalland, A., Worsley, D. and Ofstad, K., 1988. A lithostratigraphical scheme for the Mesozoic and Cenozoic succession, offshore mid- and northern Norway. Norw. Pet. Dir. Bull., 65 pp.

Duindam, P. and Van Hoorn, B., 1987. Structural evolution of the west Shetland continental margin. In: J. Brooks and K. Glennie (Editors), Petroleum Geology of North West Europe. Graham and Trotman, London, pp. 765–773.

Eldholm, O., Sundvor, E., Myhre, A.M. and Faleide, J.I., 1984. Cenozoic evolution of the continental margin off Norway and western Svalbard. In: A.M. Spencer et al. (Editors), Petroleum Geology of the North European Margin. Norw. Pet. Soc., Graham and Trotman, London, pp. 3–18.

Gabrielsen, R.H. and Robinson, C., 1984. Tectonic inhomogeneities of the Kristiansund–Bodø Fault Complex, offshore Mid-Norway. In: A.M. Spencer et al. (Editors), Petroleum Geology of the North European Margin. Norw. Pet. Soc., Graham and Trotman, London, pp. 397–406.

Gabrielsen, R.H., Foerseth, R., Hamar, G. and Rønnevik, H., 1984. Nomenclature of the main structural features on the Norwegian and Continental Shelf north of the 62nd parallel. In: A.M. Spencer et al. (Editors), Petroleum Geology of the North European Margin. Norw. Pet. Soc., Graham and Trotman, London, pp. 41–60

Halstead, P., 1975. Northern North Sea faulting. In: Jurassic Northern North Sea Symposium, Article 10. Norsk Petroleumsforening, Oslo.

Harding, T.P., 1983. Divergent wrench fault and negative flower structure, Andaman Sea. In: A.W. Bally (Editor), Seismic Expression of Structural Styles. Vol. 3, Am. Assoc. Pet. Geol., Stud. Geol. Ser., 15, pp. 4.2-1–4.2-8.

Harding, T.P., 1984. Graben hydrocarbon occurrences and structural style. Am. Assoc. Pet. Geol., Bull., 3: 333–362.

Hay, J.T.C., 1978. Structural development in the Northern North Sea. J. Pet. Geol., 1: 65–77.

Illies, J.H., 1981. Mechanism of graben formation. Tectonophysics, 73: 249–266.

Jacobsen, V.W. and Van Veen, P., 1984. The Triassic offshore Norway north of 62°N. In: A.M. Spencer et al. (Editors), Petroleum Geology of the North European Margin. Norw. Pet. Soc., Graham and Trotman, London, pp. 317–327.

Johnson, R.L. and Dingwall, R.G., 1981. The Caledonides: their influence on the stratigraphy of the northwest European continental shelf. In: L.W. Illing and G.D. Hobson (Editors), Petroleum Geology of the Continental Shelf of North-West Europe. Institute of Petroleum, London, pp. 85–97.

Kirton, S.R. and Hitchen, K., 1987. Timing and style of crustal extension N of the Scottish mainland. In: M.P. Coward, J.F. Dewey and P.L. Hancock (Editors), Continental Extensional Tectonics. Geol. Soc. London, Spec. Publ., 28: 501–510.

Krantz, R.W., 1988. Multiple fault sets and three-dimensional strain: theory and application. J. Struct. Geol., 10(3): 225–237.

Nelson, P.H.H. and Lamy, J.M. 1987. The Møre/West Shetlands area: a review. In: J. Brooks and K. Glennie (Editors), Petroleum Geology of North-West Europe. Graham and Trotman, London, pp. 775–784.

Price, I. and Rattey, R.P., 1984. Cretaceous tectonics off Mid-Norway: Implications for the Rockall and Færoe Shetland troughs. J. Geol. Soc. London, 141: 985–992.

Ramberg, I.B. and Smithson, S.B., 1975. Gridded fault pattern in a late Cenozoic and Paleozoic continental rift. Geology, 3: 201–205.

Ramberg, I.B., Cook, F.A. and Smithson, S.B., 1978. Structure of the Rio Grande rift in southern New Mexico and West Texas based on gravity interpretation. Geol. Soc. Am. Bull., 89: 107–123.

Rawson, P.F. and Riley, L.A., 1982. Latest Jurassic–early Cretaceous events and the late Cimmerian unconformity in the North Sea. Am. Assoc. Pet. Geol., Bull., 66: 2628–2648.

Reches, Z., 1978. Analysis of faulting in a three-dimensional strain field. Tectonophysics, 47: 109–129.

Reches, Z., 1983. Faulting of rocks in three-dimensional strain fields II. Theoretical analysis. Tectonophysics, 95: 133–156.

Surlyk, F., 1977. Mesozoic faulting in East Greenland. Geol. Mijnbouw, 56: 311–327.

Surlyk, F., Piasecki, S., Rolle, F., Stemmerik, L., Thomsen, E. and Wrang, P., 1984. The Permian base of east Greenland. In: A.M. Spencer et al. (Editors), Petroleum Geology of the North European Margin. Norw. Pet. Soc., Graham and Trotman, London, pp. 303–315.

Ziegler, P.A., 1981. Evolution of sedimentary basins in North-West Europe. In: L.V. Illing and G.D. Hobson (Editors), Petroleum Geology of the Continental Shelf of North-West Europe. Institute of Petroleum, London, pp. 3–39.

W.J. SCHMIDT *Conoco Norway Inc., Norway*
Present address: *Conoco, 600 N. Dairy Ashford, Houston, TX 77079, U.S.A.*

Characterization of faulting and fracturing in Ekofisk Field from seismic, core and log data

J. Dangerfield, I. Knight and H. Farrell

Early well test results and core descriptions indicated that the chalks of Ekofisk field are highly fractured and that the fracturing controls reservoir permeability. The initial conventional seismic was obscured by the presence of overpressured gas in the overlying shales and was considered to be of no use in interpreting faults. Improved seismic methods have enabled significant faulting in the chalks to be mapped. The orientations of these faults correspond closely to the orientations of natural fractures interpreted from oriented cores and formation microscanner borehole images. Regional faults and natural fractures are oriented NNE/SSW in the northwestern quadrant of the field. Elsewhere faulting is less frequent and natural fractures are predominantly oriented radially around the elliptical dome that forms the field. The direction of the resulting permeability anisotropy is anticipated to be a function of the natural fracture orientation, which varies with location on the structure.

Introduction

Ekofisk is a mature oil and gas field, currently (August 1990) with 40 producing wells, 17 water injectors and 5 gas injectors. It has been producing from chalks of the Tor and Ekofisk Formations since 1972 (Brewster et al., 1986), yet offers frequent surprises. The field is undergoing further development with a waterflood injection program. The need to predict the waterflood behaviour has driven activity to define the fracture system.

Seismic data over the crest of Ekofisk have been poor because the presence of gas in the overlying section distorts the view of the reservoir. The gas has leaked from the reservoir over geological time and is now concentrated in the Tertiary sediments above the crest of the field. The gas reduces the velocity of sound to less than that of water, resulting in seismic sections showing very large time depressions, multipathing and apparent faults. The field appears to be a collapsed anticline in seismic sections (Fig. 1). The field had been producing for two and a half years before a well was drilled into the collapsed area. The reservoir came in about 800 ft higher than predicted, showing the presence of a very large, domal structure (Fig. 2). By 1983, with 55 wells drilled on the field, all the available data indicated that the structure was a simple, effectively unfaulted dome.

Borehole profiles and carefully positioned surface seismic acquired since 1984 have changed our interpretation of the field; it is now believed to be heavily faulted in a predominantly NNE–SSW direction. Other methods showing faults and fractures and their orientations all fit with this interpretation. The Formation Microscanner logging technique showed that strong fracturing was present over much of the field with this orientation. Core data indicated the presence of both this NNE–SSW trend and a radial fracture trend. Anelastic strain data indicated that the present day orientation of the stress system closely reflects the core data.

The results of a three-dimensional seismic program shot in spring 1989, and of a three-dimensional borehole profile shot in summer 1989, plus further core and log data, will shortly be available to help clarify the importance of the fracture sets as the waterflood is extended throughout the field.

Seismic data

The problems created by the gas are shown by ray tracing. Using a simple gas model and considering shots from outside the gas, the problems of ray bending, multi-pathing and apparent fault creation become obvious (Fig. 3). Placing a geophone downhole, and much closer to the reservoir, constrains the acoustic wavefront to a local and largely undistorted image (Fig. 4). Borehole profiling therefore provides significant advantages.

Surface seismic line 100 is a typical seismic line, running NE–SW over the crest of the field, showing time delays and loss of reflectors (Fig. 5). The borehole profile line from the same place shows

Structural and Tectonic Modelling and its Application to Petroleum Geology edited by R.M. Larsen, H. Brekke, B.T. Larsen and E. Talleraas. NPF Special Publication 1, pp. 397–407. Elsevier, Amsterdam.

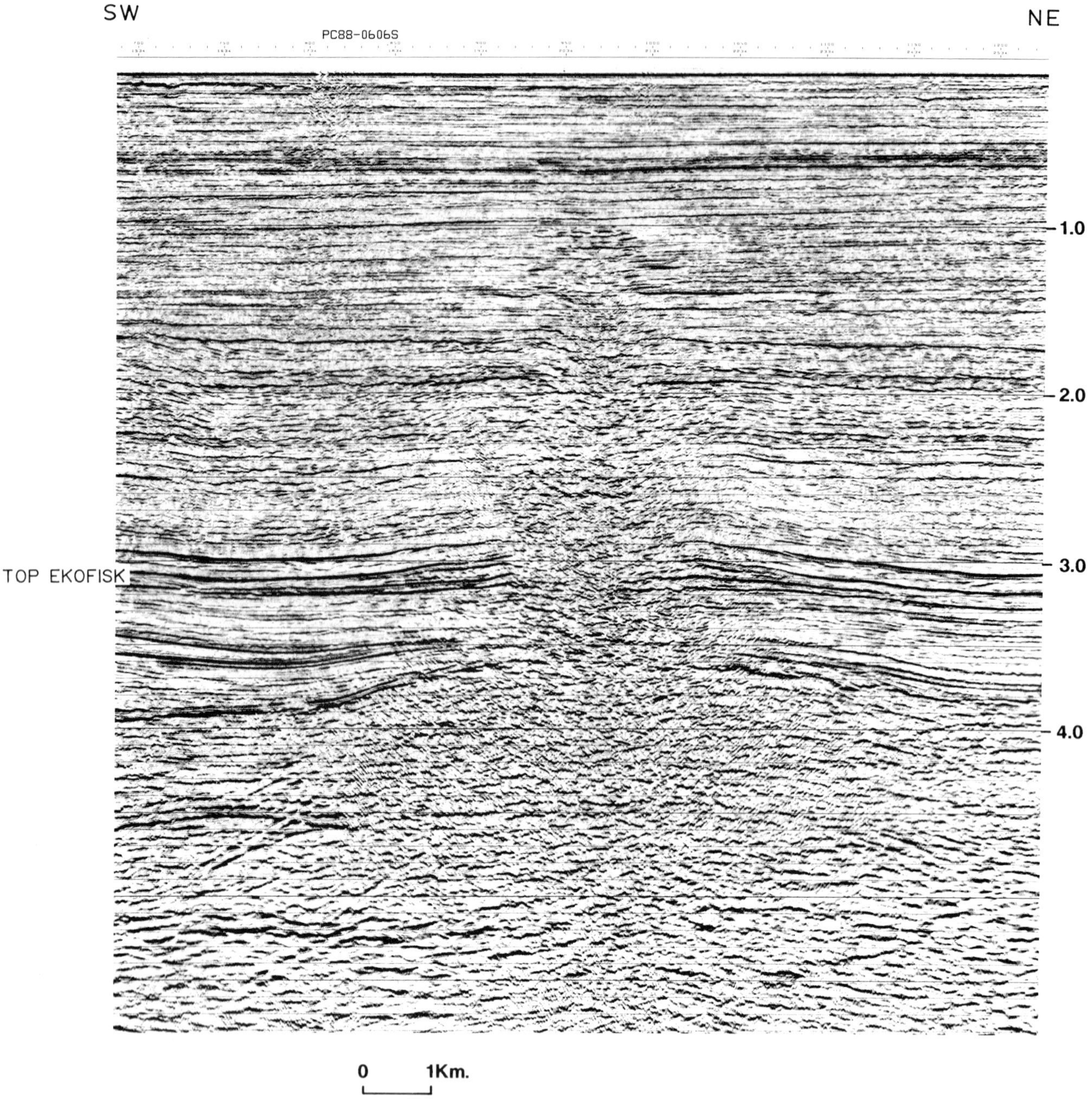

Fig. 1. Seismic line PC880606.

the major improvement possible with downhole geophones (Fig. 6). Although the borehole profile image was also depressed in time by gas effects, it revealed a graben with 45 m of throw (Christie and Dangerfield, 1984). The image location is marked by the line *C–D* in Fig. 7. The fault on the west side of the graben was later encountered in the reservoir during drilling. It is believed to be part of a major crestal graben, such as commonly found above salt domes. A total of eight wells were subsequently used to obtain the seventeen borehole profile lines shown in Fig. 7. The profiles showed major faults and rapid variation in character between profiles. They confirmed that the field has some large faults, particularly in the lower reservoir, the Tor Formation, but there were not enough lines for a definitive interpretation (Fig. 8). The crestal graben appeared to be the dominant structure, with a NNW–SSE orientation.

Borehole profiling is roughly twenty times as expensive as surface seismic per kilometre of line; another method of obtaining surface seismic data was therefore tested in 1987. Instead of shooting lines in a normal star pattern across the field, the seismic lines and the hydrophone cable were kept tangential to the edge of the gas cloud and just outside it (Fig. 9). This minimized sound travel through

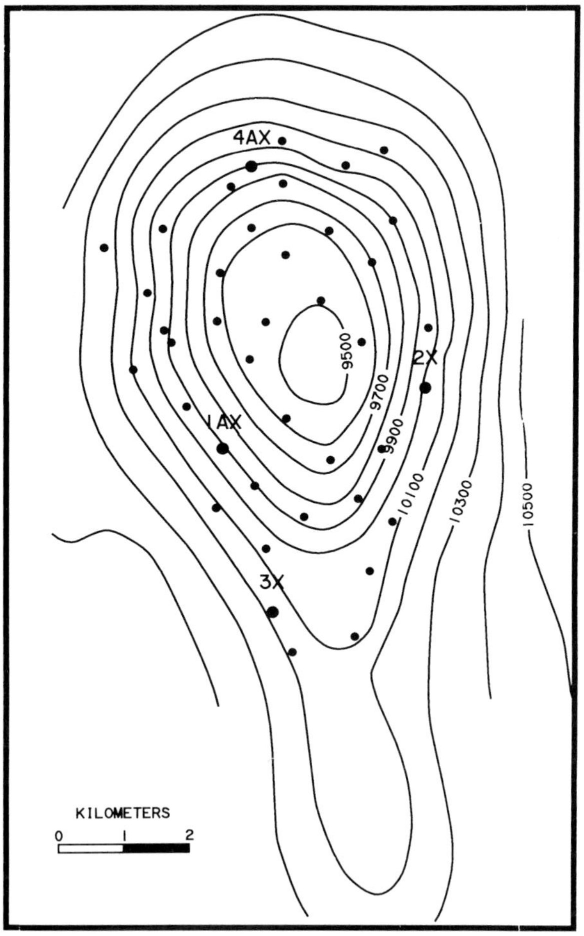

Fig. 2. Top Ekofisk depth map from 1979.

gas affected layers (Fig. 10). The lines imaged more of the crest than any previous data and were of excellent quality. Faults in the reservoir were considered real, as opposed to due to gas, when the Paleocene reflector was unbroken (Fig. 11). The results made a considerable difference to our interpretation: unexpected faults were imaged with up to 70 m of throw. The data set was still too sparse for a definitive interpretation but a strong NNE–SSW orientation seemed likely (Fig. 12).

The new map provided immediate benefit to the reservoir simulation model. The presence of major faults explained why pressures in the upper reservoir (Ekofisk Formation) and lower reservoir (Tor Formation) are equal within certain restricted areas (Fig. 12). Elsewhere, the Tight Zone separating the reservoirs keeps them in pressure isolation. The reservoir simulation model had been unable to the match the observed pressures without "windows" at the locations of the faults.

The new seismic lines challenged our view of the probable origin of the field as a salt dome. The large faults appear to extend up from the basement where they may have originated as strike slip faults.

Three-dimensional seismic methods shot to minimize gas effects should permit the interpretation of much smaller faults. A three-dimensional survey was shot in the spring of 1989 using two separate shooting directions to obtain data as far as possible up the flanks of the field (Fig. 13). To attempt to obtain data from the gas-affected crest of the field a three-dimensional walkaway borehole profile was shot in the autumn of 1989 (Fig. 14) with 41 walkaway lines, each recorded by an 8-geophone string downhole, to produce 328 lines. The results are described in AAPG Memoir 42 (Dangerfield, 1991).

Core data

Reservoir permeability is thought to be controlled by fracturing because the permeabilities inferred from reservoir performance and the analyses of well tests are as high as 150 mD (Brown, 1987), which is at least two orders of magnitude greater than matrix permeabilities measured in cores (Thomas et al., 1984). To confirm the influence of fracturing, as well as to acquire data for reservoir management, a core-based study of fracturing was undertaken.

Natural fracturing recognized in Ekofisk cores has been classified according to four major fracture types (Brewster et al., 1986). Tectonic, stylolite-associated, and irregular fractures all enhance reservoir permeability. Healed fractures may, at least locally, reduce matrix permeability.

Tectonic fractures form the bulk of the fracturing in the Ekofisk Formation. They are through-going, planar features, dipping from 65 to 80°, that form well developed parallel sets. Where the fracture intensity is high, sets of fractures conjugate to the dominant set are also seen (Fig. 15). Fracture spacing, measured perpendicular to the fracture surface, is highly variable throughout the formations but spacing rarely exceeds 100 cm at any point in the field. It is believed that the fractures form an interconnected network throughout most of the productive reservoir.

Stylolite-associated fractures (Nelson, 1981) occur primarily in the Tor Formation; they are only rarely encountered in the Ekofisk Formation. Most large stylolites have fractures associated with them. The fractures appear to have been initiated at the stylolite seam and extend away from the stylolite into the chalk matrix. The stylolite-associated fractures are typically parallel to the stylolite columns and within 5–10° of vertical (Fig. 16). The fractures usually extend up to 10 cm from the stylolite column, but may be as much as 20 cm in length. They usually form sub-parallel, anastomosing networks; within each network the fractures are interconnected and

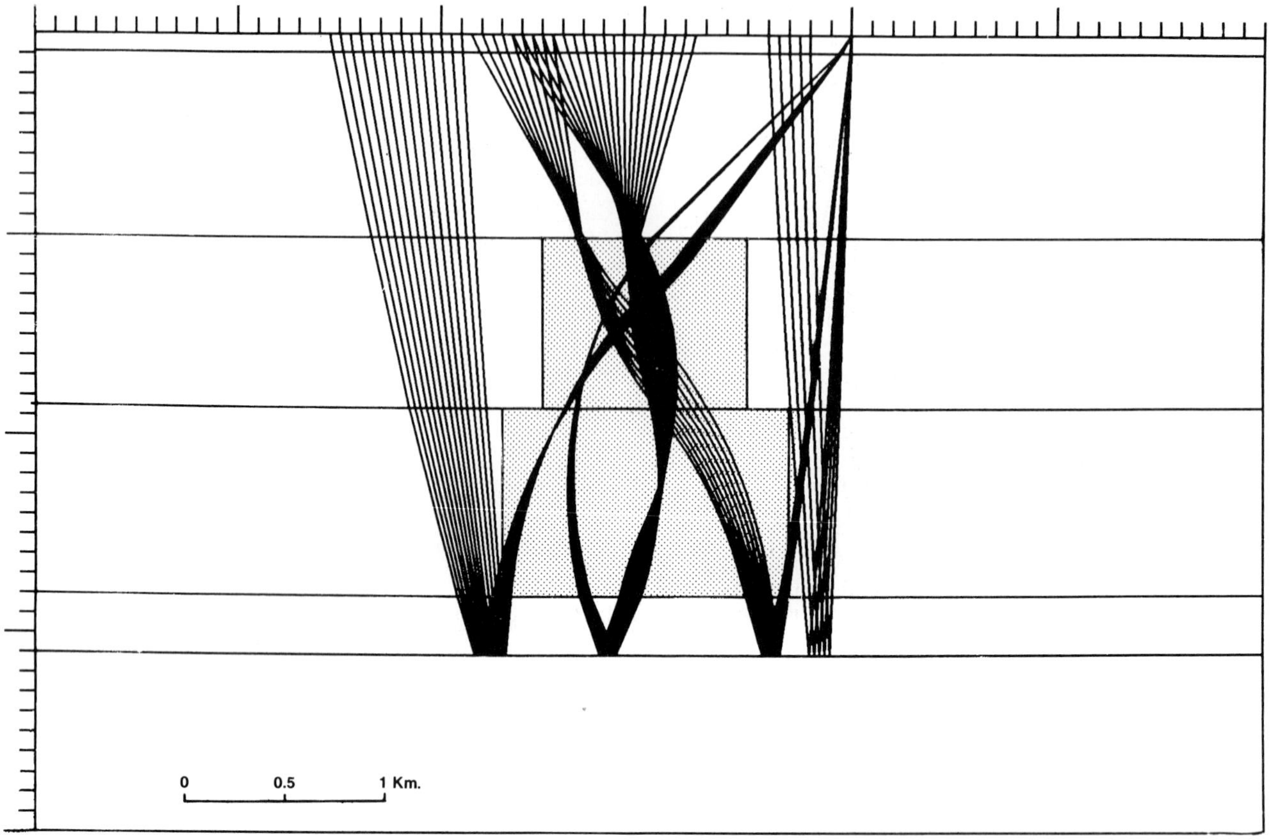

Fig. 3. Raypaths through the gas-affected area from a shot outside the gas.

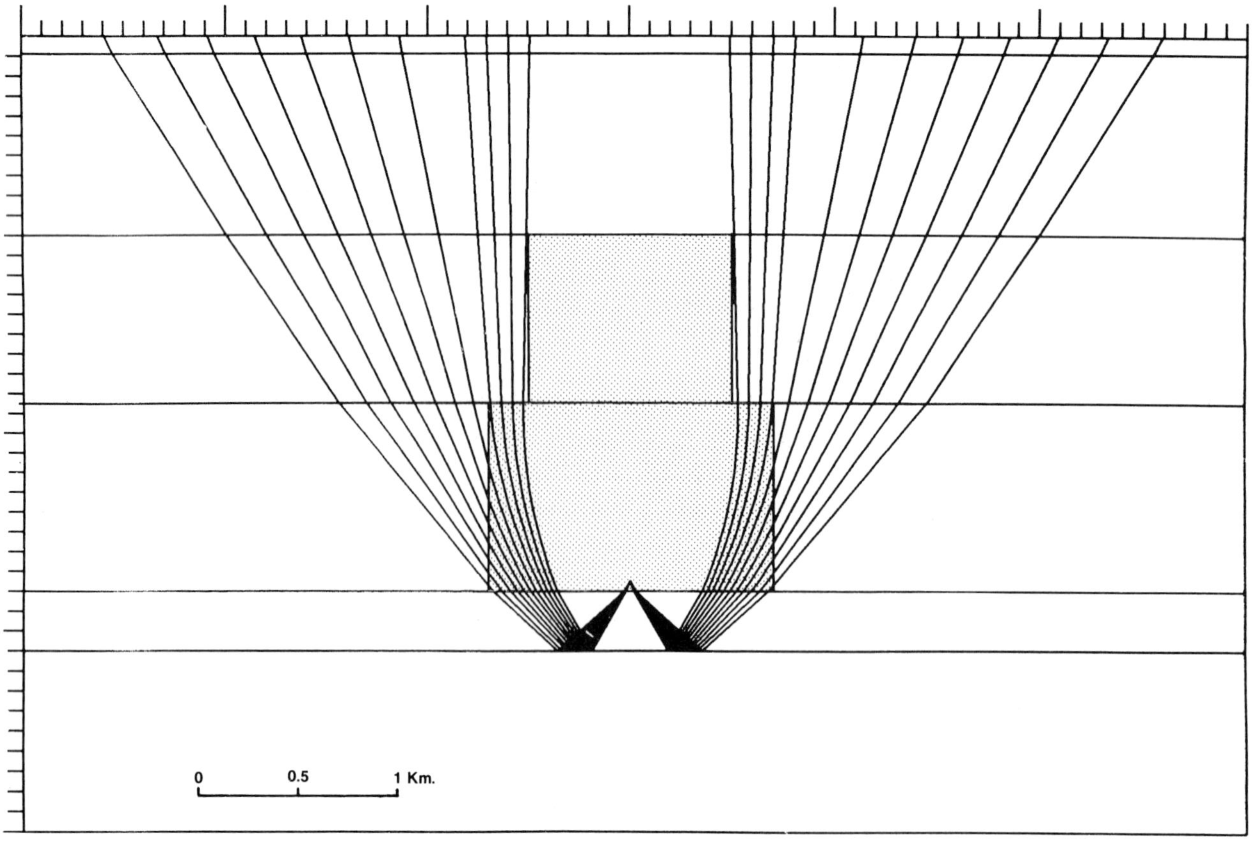

Fig. 4. Raypaths for a borehole seismic survey.

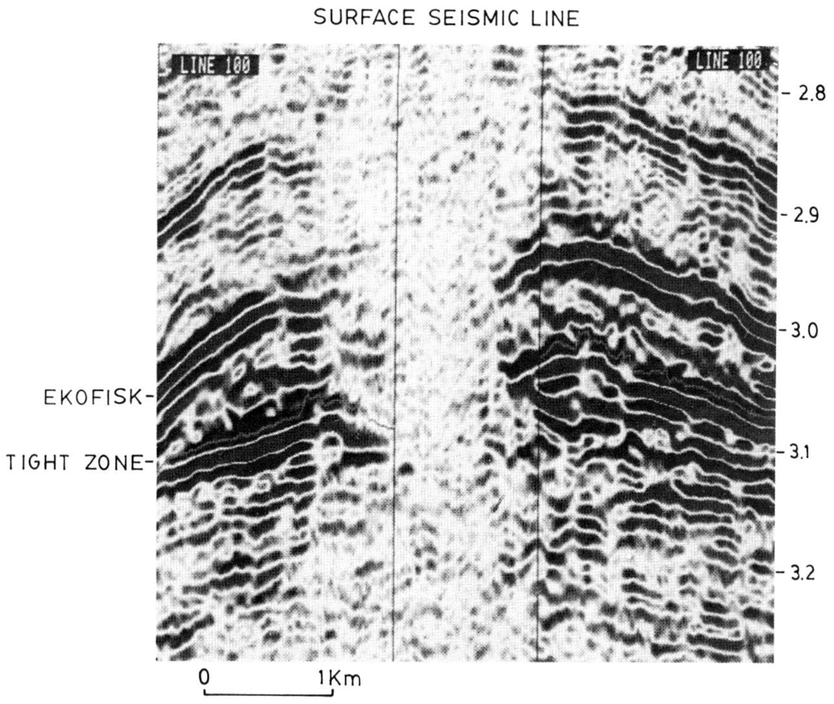

Fig. 5. Surface seismic line 100.

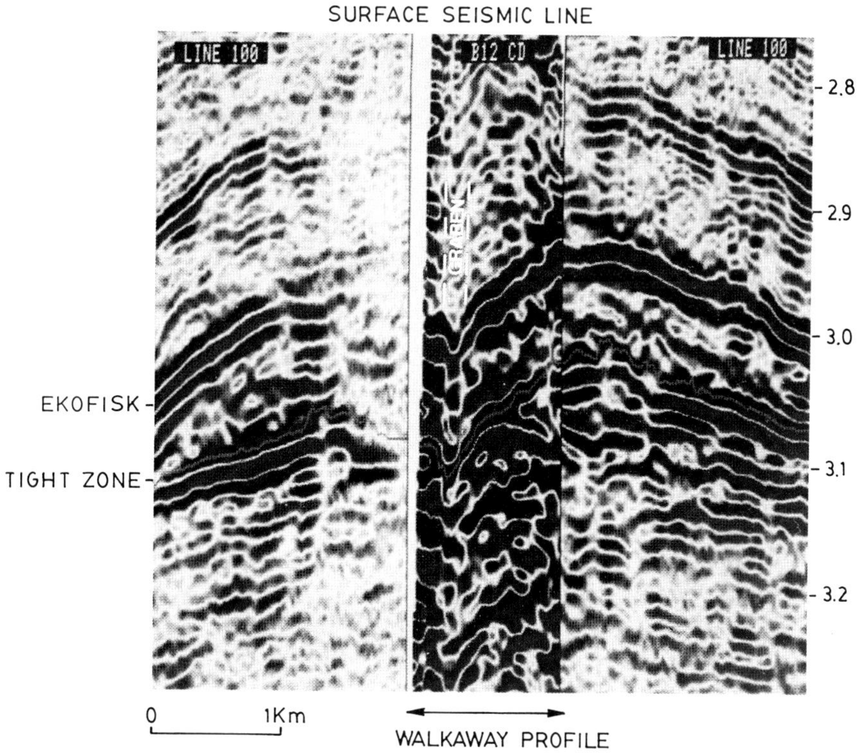

Fig. 6. Surface seismic line 100 with the walkaway borehole profile inserted.

form bedding parallel zones of enhanced permeability. They do not form a pervasive network throughout the formation unless they occur in combination with the tectonic fractures.

Irregular fractures and healed fractures do not appear to enhance reservoir permeability. Any of the fracture types may be partially or completely sealed by secondary minerals. Calcite is the most common fracture filling mineral, although quartz, pyrite and clay fillings are not uncommon.

Tectonic fracture trends for the Ekofisk Formation are shown on Fig. 17. In most of the wells (B19a,

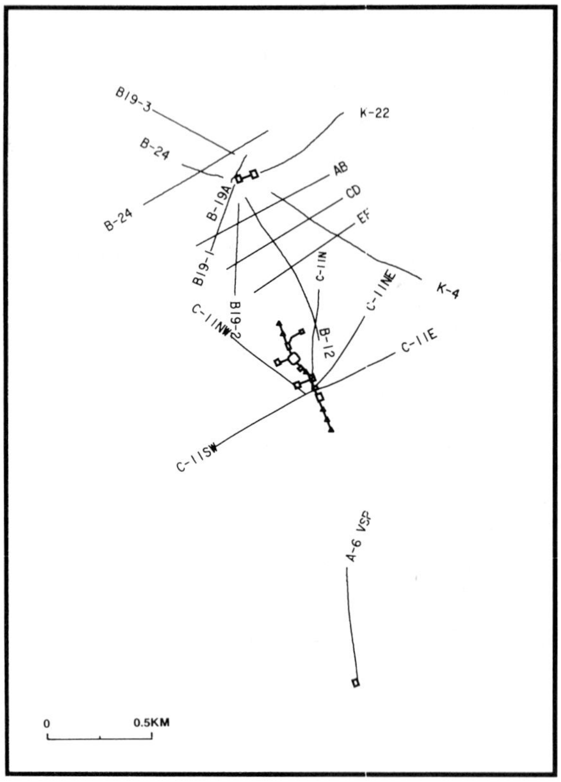

Fig. 7. Borehole profiles.

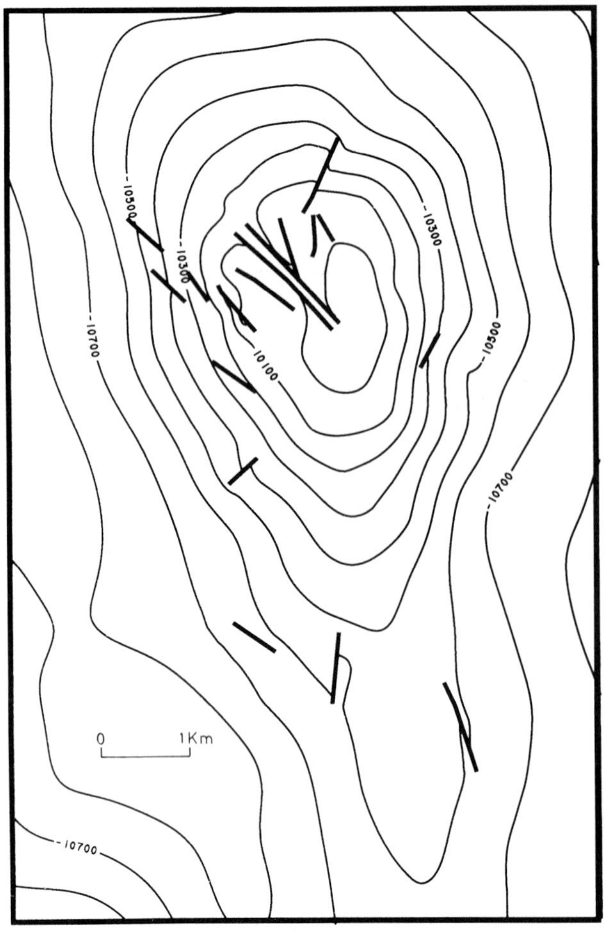

Fig. 8. Top Tor 1988 interpretation.

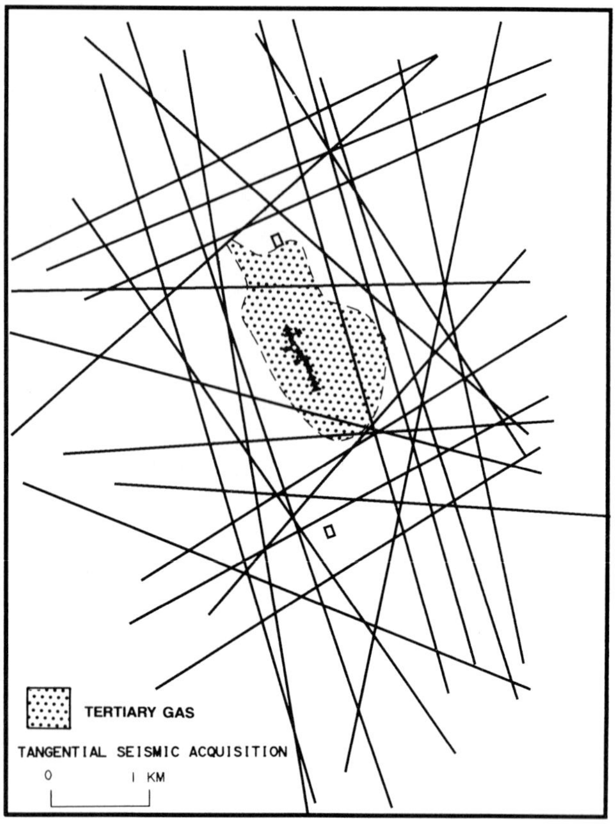

Fig. 9. Tangentially oriented surface seismic.

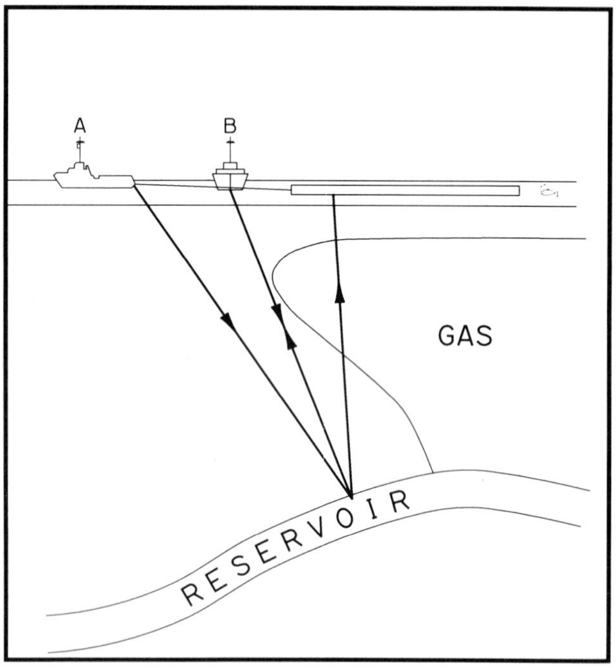

Fig. 10. Advantages of shooting with the cable oriented tangential to the gas.

K13b, C11, A15a and A6) a strong NNE–SSW fracture trend is present. This NNE–SSW fracture trend is parallel to the trend of the major faults that have been mapped from seismic data in the northern area of the field. It is assumed that the fractures with this

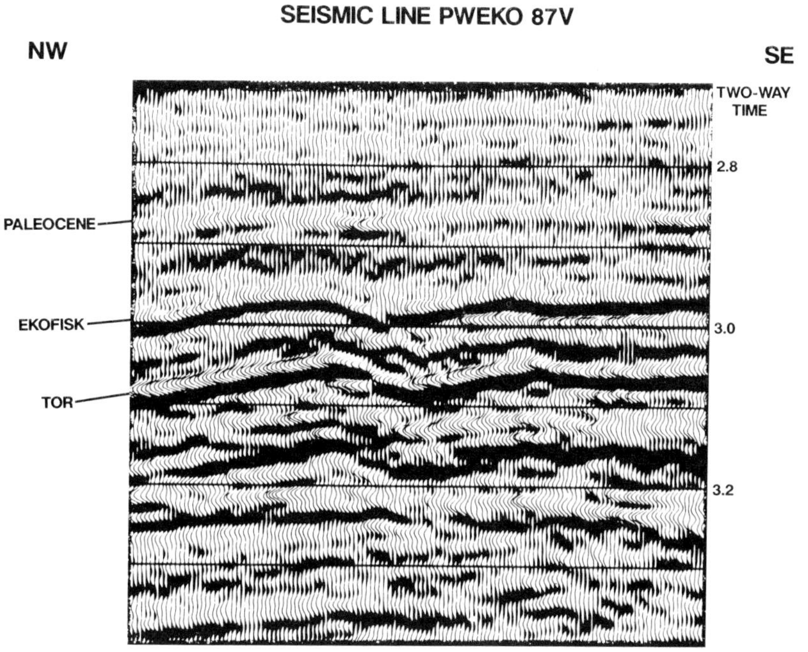

Fig. 11. Surface seismic line PWEKO87V showing good data and clear faults beneath an unbroken Paleocene reflector.

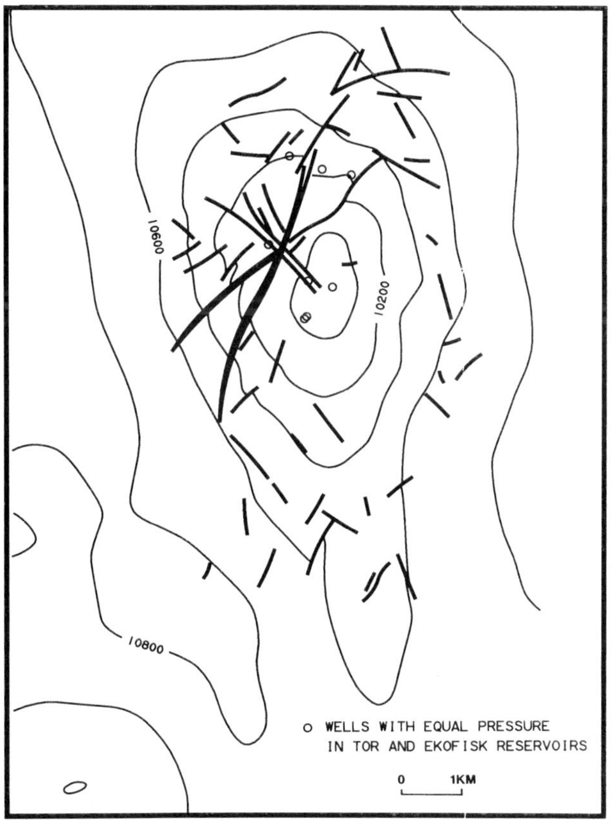

Fig. 12. Top Tor seismic interpretation in 1989, also showing wells with equal pressure in both upper and lower reservoir formations.

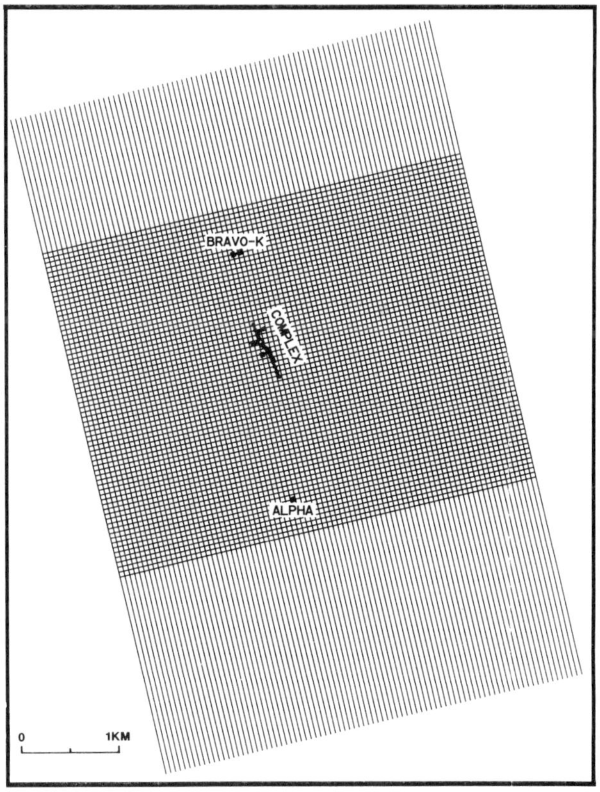

Fig. 13. Three-dimensional surface seismic grid.

orientation are genetically related to this faulting, i.e. they developed at the same time, and in response to the same stresses as the faults. Fractures and faults with this trend, which are geometrically similar features differing only in scale, do not seem to be genetically associated with the Ekofisk structure itself; their location and orientation are not affected by structural position. In general the tectonic fracture trends in the Tor Formation reflect those in the Ekofisk Formation.

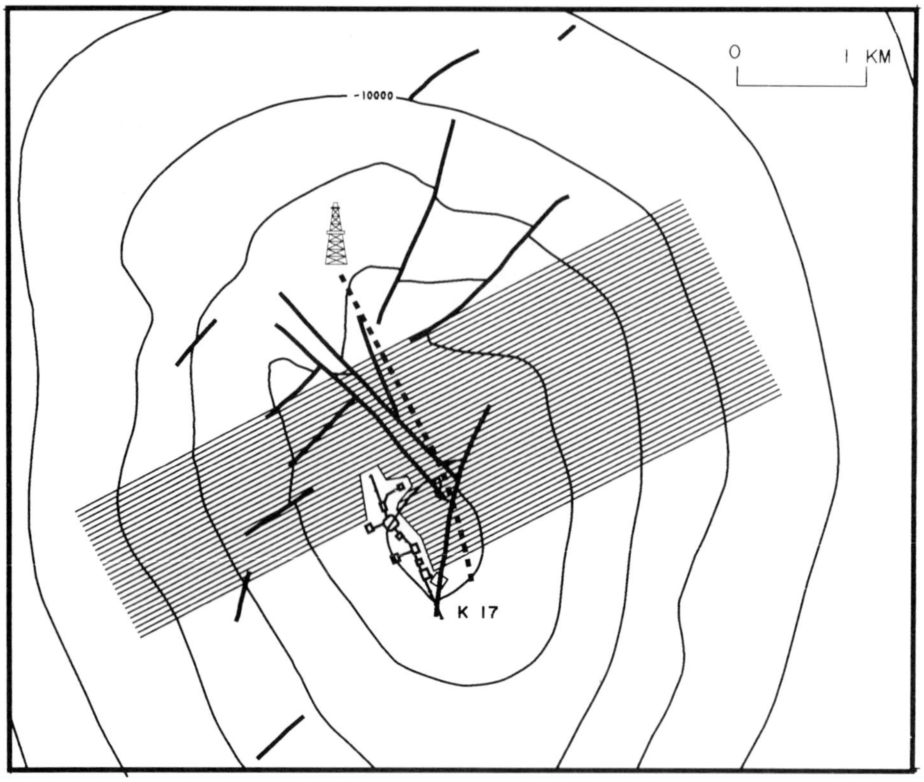

Fig. 14. Three-dimensional borehole profile acquisition of 41 walkaway lines.

Fig. 15. Tectonic fractures.

On the east flank in K4 and C4, the tectonic fractures trend approximately east–west. This trend, which differs by a few degrees in each well, is interpreted to be part of a radial fracture system that is genetically related to the fold. These radial fractures are seen as a subordinate trend in the B19a well and possibly weakly in the K13b well. Several wells (B20b, A15a and b, and C11) are in a structural position that would make radial fractures parallel the NNE–SSW regional trend, therefore the fractures in these wells could have either, or both, origins.

On the east flank of the field radial fractures appear to be dominant. In this area, no major faults have been identified from seismic interpretation. In contrast, in the northwest part of the field, which is heavily faulted, the NNE–SSW trending regional fractures predominate, although radial fractures are also present in some wells.

Core observations indicate that where they occur in the same intervals of core, the average strike of stylolite-associated fractures parallels the strike of the tectonic fractures. The stylolite-associated fractures are therefore the extension fractures and the tectonic fractures are the shear fractures associated with the same stress system.

Fig. 16. Stylolite-associated fractures.

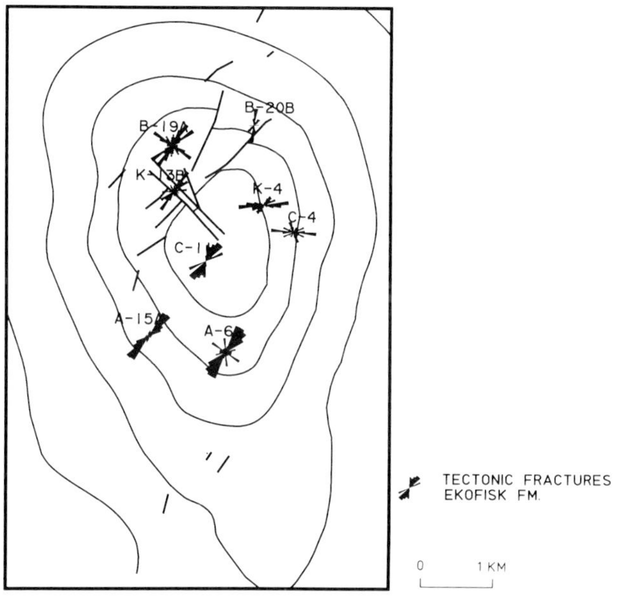

Fig. 17. Orientation of the tectonic fractures in the Ekofisk Formation.

Formation microscanner data

One of the earliest observations made from cores was that fracture type and distribution showed some association with chalk lithofacies. While tectonic fractures are ubiquitous throughout the reservoir interval, the stylolite associated fractures were almost exclusively confined to the lower reservoir section, the Tor Formation. Fracture intensity also suggested a correlation with the stratigraphy; the reworked chalks which form the best reservoir intervals have a higher fracture intensity than the poorer quality, pelagically deposited chalks, of which the Tight Zone is the best example.

Attempts were made to use the extensive conventional log data base by examining various combinations of logs to see if they could identify particular chalk facies and corresponding fracture characteristics. It was found to be impossible to define a statistically reliable method of identifying chalk lithofacies from suites of conventional wireline logs. This essentially prevented the use of a potentially huge data base for any form of quantitative fracture modelling. Fracture identification and description in uncored wells was restricted to recently developed logging tools such as the Formation Microscanner Tool (FMS).

The FMS, developed in 1985, generates an electrical resistivity image of the borehole. It is a four-pad conductivity tool with two rows of eight electrodes on each pad. Data from the older wells in Ekofisk was obtained using an earlier version of this tool that had resistivity electrodes on only two of the four pads. In the final borehole image, variations in conductivity of 35–50% of the borehole wall are displayed. Lighter tones represent low conductivity and darker tones high conductivity. Knowing the relative bearing of the tool pads, it is possible to compute the direction and magnitude of dip for any planar feature, such as fractures or bedding, crossing the wellbore.

The FMS was first run in Ekofisk in 1986 and has now been run in thirteen wells with hole deviations of up to 45°. An excellent example of a fractured chalk section seen in an FMS log is shown in

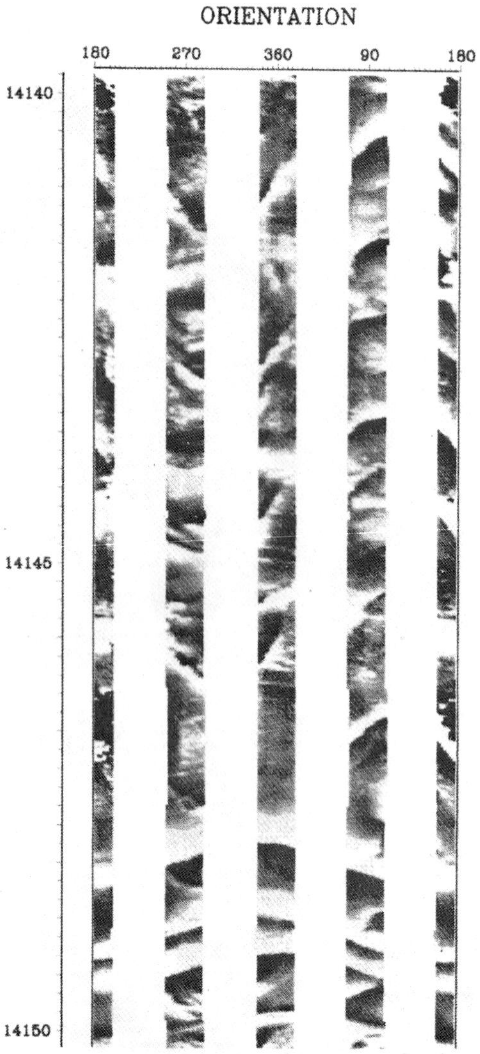

Fig. 18. An FMS log showing fractures.

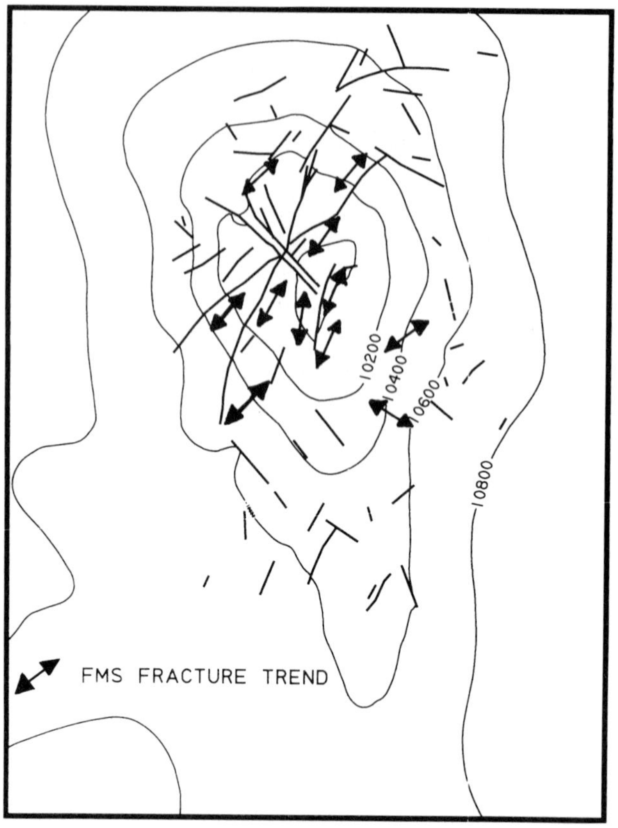

Fig. 19. Top Ekofisk depth map showing FMS fracture trends.

Fig. 18. This image shows the sinusoidal patterns of contrasting matrix and fracture conductivities which correspond to high-angle, open, tectonic fractures. Sealed fractures may be identified by the white (low-conductivity) sinusoids.

Figure 19 shows the fracture orientations recorded from FMS data for the Ekofisk Field. Fracture orientations are highly consistent with orientations recorded from cores, and reflect the strong NNE–SSW fracture strike seen in both seismic and cores over the northern half of the field. In the 2/4A-4A well drilled on the southeast flank of the structure, fracture strike was seen to be northwest–southeast, agreeing with core and anelastic strain data, and interpreted as part of the radial fracture pattern.

The FMS has provided a valuable insight into fracture orientations, but the tool is selective in the type of fractures it can identify. High-angle, open, tectonic fractures are generally easily recognized as a consequence of the relative geometries of the fracture plane to the wellbore. The short, sub-vertical stylolite associated fractures are practically unidentifiable with the processing presently used on the FMS. Consequently, some wells interpreted to be poorly fractured from FMS data have proven to be good producers (with high fracture intensity) on well test analysis.

Other applications for the FMS are also under investigation. The four-arm caliper on the tool provides useful data to assess borehole ellipticity, wellbore break-out and hence present-day stress orientation. This information may have application to the future design of well stimulation programs. Further tool applications should develop with further development of image processing.

Anelastic strain data

For reservoir management it is important to know the relative apertures of fractures of different sets and whether these apertures will change as the stress state within the reservoir is altered by production and associated compaction of the reservoir. In order to examine the distribution of stress in the reservoir *in situ* stress orientations and magnitudes have been determined from measurements of anelastic strain recovery of oriented core (Teufel and Warpinski, 1984).

Data from seven wells indicate that the orientation of the maximum horizontal stress is radial (Teufel, 1991). Thus the present day maximum horizontal stress parallels the radial tectonic fractures. Fracture width, or aperture, is one of the most important parameters controlling fracture permeability. Fracture width is a function of fracture surface roughness and the normal stress across the fracture. Fractures that are parallel to the maximum *in situ* stress will have larger apertures and should be more permeable than those at a high angle to the maximum stress. This indicates that the permeability of individual fractures in the NNE–SSW and the radial fracture systems may be dissimilar.

Conclusion

Seismic imaging of Ekofisk Field, by tangentially oriented surface seismic and borehole profiling, indicates that there are many large-scale faults in the field and that they are predominantly oriented NNE–SSW.

Core and FMS Log studies indicate that small-scale fractures with the same orientation are also present, but can be subdivided into different groups based upon their characteristics. The core-based studies indicate that small-scale radial fractures are also present and do not appear to be related to large-scale faulting. The radial orientation of the maximum horizontal *in situ* stress indicates that the permeability of fractures may be a function of their orientation and location. The fracture network at Ekofisk is therefore extremely complex and difficult to interpret when dealing with problems of reservoir management, such as production anomalies, waterflood performance, and sidetrack locations. Three-dimensional seismic and a strong multidisciplinary team approach are required in our future work.

Acknowledgements

The authors would like to thank the management of Phillips Petroleum Company and their co-venturers on Ekofisk Field for their permissions to publish this paper. The co-venturers are: Petrofina A.S., Norsk Agip A/S, Elf Aquitaine Norge A/S, Norsk Hydro A/S, Total Marine Norsk A/S, Statoil, and Phillips Petroleum Company. All opinions expressed in this paper are those of the authors and do not necessarily reflect the opinion of Phillips Petroleum Company or any of the co-venturers. *In situ* stress measurements were performed and interpreted by L.W. Teufel of Sandia National Laboratories through a USDOE/industry cooperative research program. Discussions with Larry Teufel, John Hensley, Tom Moore, David Ellen, Rod Boade, Michel D'Heur and Guy Fraselle contributed to development of many of the ideas discussed in this paper.

References

Brewster, J., Dangerfield, J.A. and Farrell, H.E., 1986. The geology and geophysics of the Ekofisk Field waterflood. Mar. Pet. Geol., 3: 139–170.

Brown, D.A., 1987. The flow of water and displacement of hydrocarbons in fractured chalk reservoirs. In: J.C. Goff and B.P.J. Williams (Editors), Fluid Flow in Sedimentary Basins and Aquifers. Geol. Soc. London, Spec. Publ., 34: 201–218.

Christie, P.A.F. and Dangerfield, J.A., 1987. Borehole seismic profiles in the Ekofisk Field. Geophysics 52: 1328–1345.

Dangerfield, J.A., 1991. In: A.R. Brown (Editor), Interpretation of 3D Seismic Data. Am. Assoc. Pet. Geol., Mem., 42, 3rd ed.

Dangerfield, J.A. and Brown, D.A., 1987. The Ekofisk Field. In: J. Kleppe et al. (Editors), Proceedings of North Sea Oil and Gas Reservoirs Seminar, NTH, 1985. Graham and Trotman, London.

Nelson, R.A., 1981. Significance of fracture sets associated with stylolite zones. Am. Assoc. Pet. Geol., Bull., 65: 2417–2425.

Teufel, L.W., 1991, Influence of lithology and geologic structure in "in situ" stress: examples of stress heterogeneity in reservoirs. In: L.W. Lake, H.B. Carroll and T.C. Wesson (Editors), Reservoir Characterization, II. Academic Press, New York, N.Y., pp. 565–576.

Teufel, L.W. and Warpinski, N.R., 1984. Determination of "in situ" stress from anelastic strain recovery measurements of oriented core: comparison to hydraulic fracture stress measurements. Proc. 25th U.S. Nat. Rock Mech. Symp., Evanston, Ill., pp. 176–185.

Thomas, L.K., Dixon, T.N., Evans, C.E. and Vienot, M.E., 1984. Ekofisk Waterflood Pilot. SPE Paper 13120.

J. DANGERFIELD — *Phillips Petroleum Company Norway, Utenlandsk Aksjeselskap, P.O. Box 220, N-4056 Tananger, Norway*

I. KNIGHT — *Statoil A/S, P.O. Box 300, 4001 Stavanger, Norway*
Present address: *Phillips Petroleum Company Norway, Utenlandsk Aksjeselskap, P.O. Box 220, N-4056 Tananger, Norway*

H. FARRELL — *Phillips Petroleum Company Norway, Utenlandsk Aksjeselskap, P.O. Box 220, N-4056 Tananger, Norway*

Ula Field: relationship between structure and hydrocarbon distribution?

A. Brown, A.W. Mitchell, I.R. Nilssen, I.J. Stewart and P.T. Svela

The Ula Field is located on the Norwegian Offshore Continental Shelf, approximately 260 km SW of Stavanger. During the course of field development, technical perception of the reservoir has evolved from a simple, partially filled anticlinal structure with a single oil–water contact (OWC), to a complex accumulation with a difference in OWC between the west and the east flanks of the field of more than 280 m.

Data from development drilling indicates that the oil columns on both sides of the field are in excellent pressure communication, thus indicating no effective compartmentalisation. The real petroleum distribution in the Ula Field was not discovered until 2 years after production start-up. As a result, the well data collected to date are insufficient to allow the precise definition of the mechanisms which controlled the original fluid distribution in the field.

Three models are presented to explain the different OWC values. In one model, the crestal NW–SE striking fault system which bisects the field acts as a permeability barrier for a compaction driven aquifer influx from the west. This results in differing aquifer pressures causing displacement of contacts across a fault permeability membrane.

An alternative model is that the same crestal fault system acted as a seal through geological time and the eastern and western reservoirs of the field were charged with hydrocarbons independently. The fault seal has subsequently failed as a result of the large pressure differential that has developed through oil production which has only been from the east flank.

The third model, requires the reservoir to be discontinuous beyond the west flank of the field; that the crestal fault system is not sealing; and that petroleum charged the structure from the east, progressively filling the field crest. This process would continue only until the bottom sealing limits of the field crest were reached. At this point water contained within the west flank would no longer be displaced leaving a trapped west flank aquifer.

Presently, there is insufficient data to identify which of these three models offers the most likely explanation for the phenomena observed on the field to date.

Introduction

This paper reviews the impact of development well drilling upon the structural and reservoir model of the Ula Field. In the development planning stage, Ula Field was considered a simple, four-way dip-closed salt-cored anticline with a single field-wide oil–water contact (Home, 1987). Two years after production start-up, however, drilling of the first downflank injection wells on the east flank of the field demonstrated that the reservoir configuration was much more complicated than had at first been thought. It is now evident that structuring has played an important role in the migration and trapping of hydrocarbons in the Ula structure although the exact relationship is still poorly understood.

Regional setting

The Ula Field lies approximately 260 km SW of Stavanger and 65 km NW of the Ekofisk complex (Fig. 1). It is situated within the Ula Trend, "a Jurassic hydrocarbon province located along the faulted north-eastern margin of the oil-rich Central Graben" (Spencer et al., 1986). The reservoir of the Ula Field is the Upper Jurassic sandstone of the Ula Formation. Mudstones of the overlying Tyne Group formed the cap-rock (Deegan and Scull, 1977) with petroleum sourced from the "hot shale" of the Upper Jurassic Mandal Formation (Spencer et al., 1986). Details of the lithostratigraphy (Fig. 2) are described within that proposed by Deegan and Scull (1977) and modified by Vollset and Doré (1984).

Discovery

BP farmed in to the 019A Licence in May 1976 in return for drilling one well, located and operated by Conoco. Using this well, 7/12-2 (Fig. 3), the Ula field was discovered. The well penetrated 154 m true vertical thickness (TVT) of Upper Jurassic sandstone and reached total depth (TD) in the Triassic. The

Structural and Tectonic Modelling and its Application to Petroleum Geology edited by R.M. Larsen, H. Brekke, B.T. Larsen and E. Talleraas. NPF Special Publication 1, pp. 409–420. Elsevier, Amsterdam.

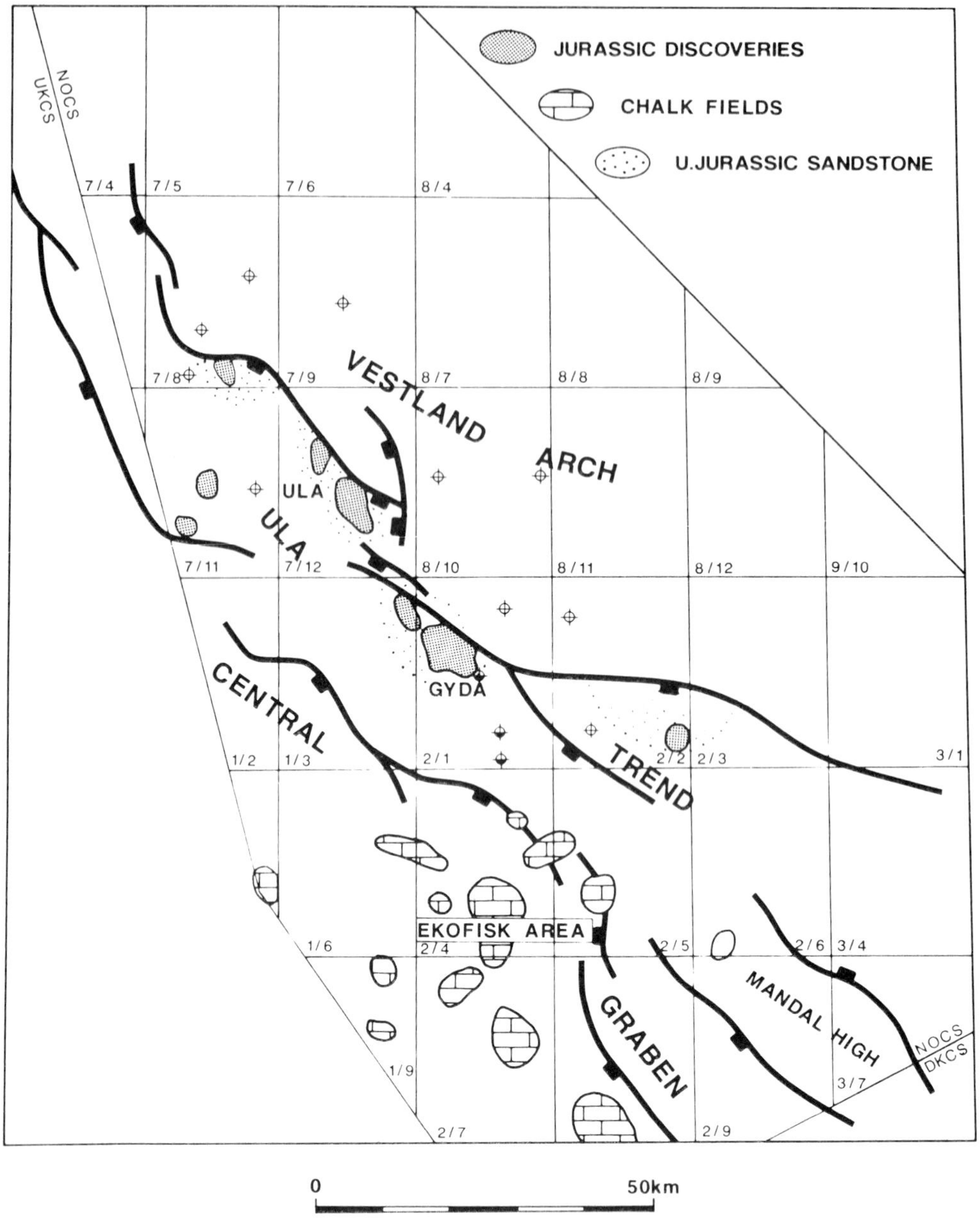

Fig. 1. Ula trend. Fields and discoveries.

Upper Jurassic sand had a net thickness of 148 m with porosities ranging between 14 and 28% and permeabilities ranging from a few milliDarcies (mD) to 2.6 Darcies (D). On testing the Jurassic sands, 40.2 API oil flowed at rates up to 7100 stb/d and the well proved an Oil-Down-To (ODT) of 3507 m TVD s.s.

Appraisal, early reservoir delineation and development scheme

Following the success of 7/12-2, BP took over the operatorship of the PL019A Licence. A suite of two-dimensional seismic data was acquired and in 1977, appraisal wells 7/12-3, 7/12-3A and 7/12-4 were drilled (Fig. 3). Wells 7/12-3 and 7/12-4 were to test possible fault and dip-controlled spill points as well as the quality and thickness variation of the Upper Jurassic sand. Well 7/12-3 encountered an entirely water-bearing reservoir and it was decided to deviate the well up-dip to establish oil-water contact. The sidetrack, 7/12-3A, proved a Water-Up-To (WUT) of 3509 m s.s. Thus, the data from 7/12-2 and 7/12-3A when combined with structure mapping indicated a field-wide oil–water contact at 3508 m s.s. Well 7/12-4 was then targeted to confirm the position of the OWC and to test the reservoir sandstone quality and petroleum distribution on the south-east flank of the field. The well proved only 84 m of Upper Jurassic reservoir and tested oil down to a depth of

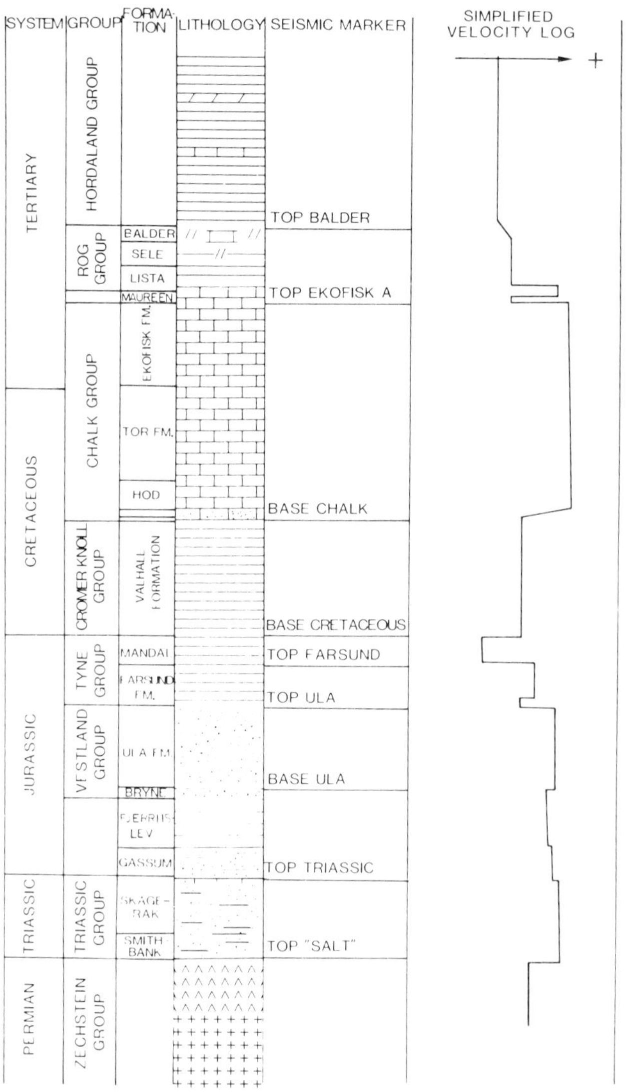

Fig. 2. Ula Field. Generalised stratigraphic section.

3505 m TVD s.s. supporting the interpreted "OWC" at 3508 m TVD s.s.

The remaining unproven reserves lay in the north of the field and were appraised by the well 7/12-6. The well encountered the reservoir shallower than predicted and consequently the sand was fully oil bearing.

The field development scheme was thus based upon a four-way dip-closed reservoir partially filled with a field-wide OWC at a depth of 3508 m s.s. The development plan comprised crestal production wells supported by down-dip water injection.

To date, a three-dimensional seismic survey has been acquired and seven appraisal wells together with thirteen development wells have been drilled on the field. These data have led to a complete revision of our understanding of the reservoir, and the elements of this are discussed below.

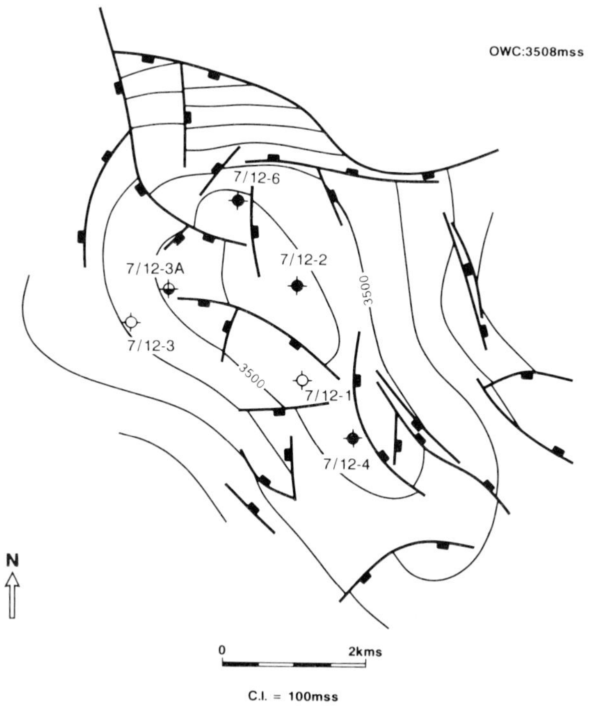

Fig. 3. Top Ula reservoir. 1982 two-dimensional map.

Structure

Both the stratigraphic and three-dimensional seismic data now indicate that the Ula structure is the consequence of structural inversion of an originally complex, extensional system. The structural grain of the area was established by Late Permian time, and this grain has had a fundamental role, being re-utilised in younger structuring.

Permian extension

The main Permian structural element at Ula is a north–south orientated westerly heading fault over which there is significant thickening of the post-rift Late Permian evaporite sequence (Fig. 4). The fault is a single surface over the major part of the three-dimensional grid, although it splays to the north and the south. This surface is interpreted as transtensional; the fault mosaic revealed by the three-dimensional data at Base Zechstein shows no simple pattern that may relate to simple dip-slip extension and the other minor faults at this level are interpreted as oblique conjugate sets between the main fault and a further important north–south fault that lies to the east of the three-dimensional grid.

Superimposed on the Base Zechstein structural grain is a salt wall and sediment pod geometry with associated faults which evolved in Early Triassic times. The broad trend of the salt wall is north-west to south-east.

The salt wall effectively became stable in Early Triassic times leaving elongate grounded basins which

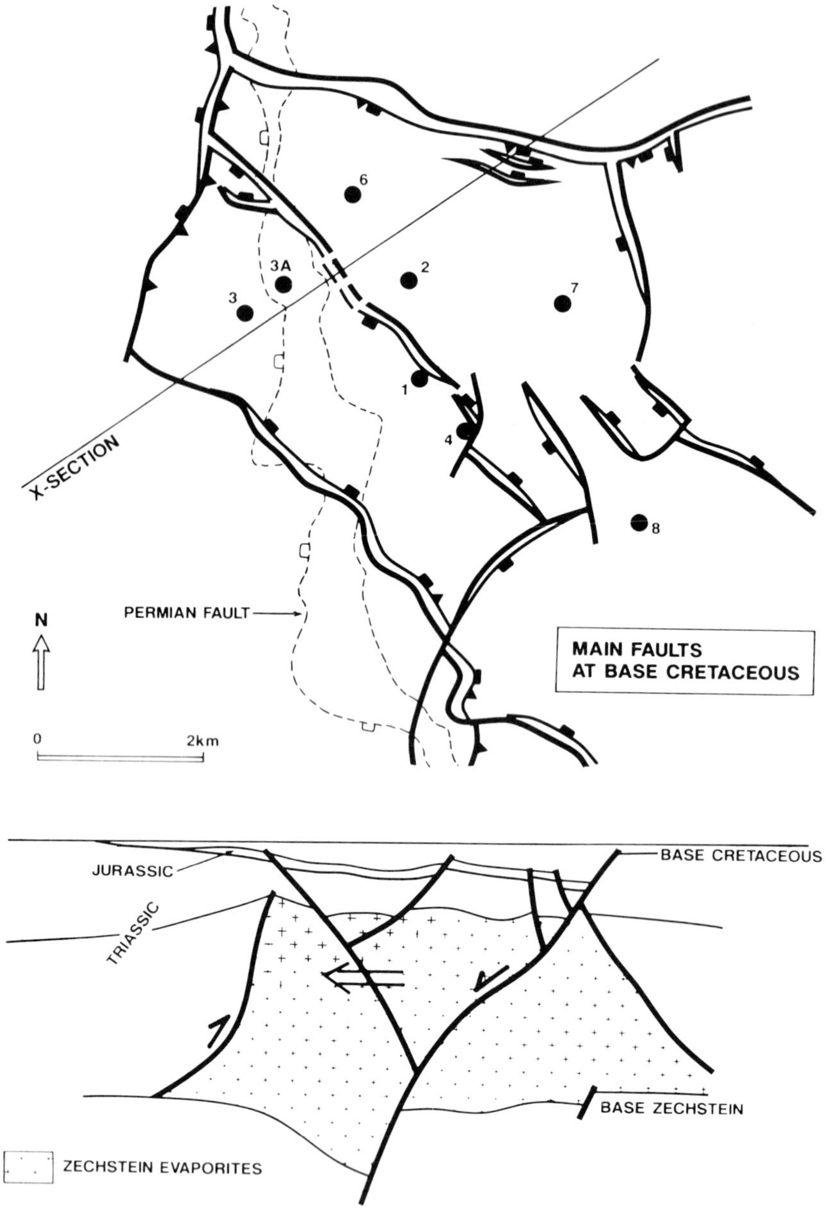

Fig. 4. The Jurassic extensional system.

were largely infilled by largely onlapping Triassic sediments. Accommodation faults developed along the margins of the salt walls. In the Late Triassic, regional extension utilised the Permian fault system, and the faults cut up-section through the salt wall. This structural style was further developed in Late Jurassic extension.

Jurassic extension

Jurassic extension is thought to have been controlled by transtension on the NNW–SSE Base Zechstein Fault, in response to regional dip-slip extension occurring on northwest–southeast orientated Base Zechstein fault sets, south and west of the Ula area. The sense of motion at Ula was sinistral and the effect was to "pull-off" Late Permian and younger cover east of the Base Zechstein Fault towards the south along the shallow faults (Fig. 4). An antithetic fault to the main fault was established and was linked by transfer systems to the east and the west. The overall style is that of a pull-apart above the deeper transtensional fault. The main ages of extension were Early Kimmeridgian to Mid-Portlandian, with a subsequent phase in the Late Ryazanian, after the deposition of the organic rich claystones of the Mandal Formation. The crestal fault system over the Ula structure is considered to have originated in the Early Kimmeridgian in response to the oblique pull-off, although the major throw on these faults is of Late Ryazanian age.

It is the displacement associated with this crestal fault system which has subsequently controlled the distribution of hydrocarbons encountered in the field today.

Cretaceous and Tertiary inversion

The present-day Ula reservoir trap is the result of structural inversion which developed in response to dextral strike-slip or transpressional motion along the base Zechstein Fault. As with many of the inversion structures in the Central North Sea, contraction occurred at several discrete times. The principal ages of inversion are Early Cretaceous, Late Cretaceous (Middle to Late Campanian) and Tertiary (Late Ypresian and Middle Oligocene). The geometries of the inversion structure are directly opposite to the extensional geometries and the same lineaments have been used in contraction as in extension.

The structural history as it can be interpreted from the three-dimensional data set is, therefore, much more complex than originally thought. The trap developed not through halokinesis, but through cover extension and contraction in response to lateral movement along the deeper Base Zechstein Fault surfaces into which the cover faults linked. Initially the sense is transtensional in the Mesozoic, although the system is reversed and acted in periodic transpression after Early Valanginian times.

Reservoir characteristics of the Upper Jurassic

Extensive core coverage from both development and appraisal wells drilled after production start-up has permitted an accurate description of the Upper Jurassic reservoir. It comprises marine sandstones (Bailey et al., 1981 and Home, 1987) which vary in thickness from 84 m in well 7/12-4 to 166 m in well 7/12-3A (Fig. 5). Reservoir thickness is greatest in the north-western area of the field, gradually decreasing towards the flanks.

The sandstones are predominantly fine and medium grained and are relatively homogeneous. Mudstones are absent from the producing reservoir section (Fig. 6). There is, however, variation in reservoir quality across the Ula field and this variation is controlled by an interplay between detrital clay and quartz cement (Gluyas, 1990). Fine-grained argillaceous sandstones have a high proportion of microporosity and relatively low quartz cement content, but also low permeabilities. Clean sandstones on the other hand contain high proportions of quartz cement but nonetheless moderately good porosities and permeabilities are preserved. The proportion of quartz cement in the reservoir also increases significantly with depth.

Well averaged, overburden corrected core porosities vary between 10% and 23% with an average porosity of 16.5%. Permeability ranges between less than 1 mD to a maximum 3.1 Darcies in well 7/12-A-18. The field average core arithmetic permeability is 200 mD.

The observed reservoir homogeneity, combined with the good reservoir quality is not considered to be the principal control of the petroleum distribution.

Petroleum distribution in the Upper Jurassic

Combined appraisal and development well drilling have now demonstrated that Ula Field does not have a single field-wide oil–water contact at 3508 m s.s. In fact, the hydrocarbon column on the east flank of the field extends at least 280 m deeper than the hydrocarbon levels penetrated on the west flank.

Appraisal well 7/12-7 encountered an "effective OWC" at a depth of 3788 m s.s. (Fig. 7). This is the deepest productive oil encountered in the field, to date, and residual oil was logged to the TD of 3822 m s.s. It is considered possible that this "effective OWC" may possibly represent an ODT or producing OWC and that the field may originally have been full to a structural spill point which is currently mapped in the east at a depth of 3820 m s.s.

The oil limits on the west flank of the field were also revised after the drilling of development well 7/12-A-12 (Fig. 7). This well proved an effective OWC at 3462 m s.s. Additionally, the reservoir was found to be depleted to the same pressure as the crestal-producing wells indicating that, at the time of drilling, the crestal fault system was not sealing. No production or injection wells are currently located on the west flank of the field.

The relationship between structure and petroleum distribution

Since reservoir heterogeneity cannot explain such a fluid distribution, a complex relationship must exist between the structure of the Ula Field and the petroleum distribution. Consequently, any explanation of the proven oil distribution relies on an accurate knowledge of the influence of faulting and compartmentalisation that was not observed when the original field development plan was formulated.

The current perception of the structural evolution of the field, indicates that the trap as we see it today did not begin to develop until the onset of structural inversion in Cretaceous times. The fault system which is currently recognised cutting the

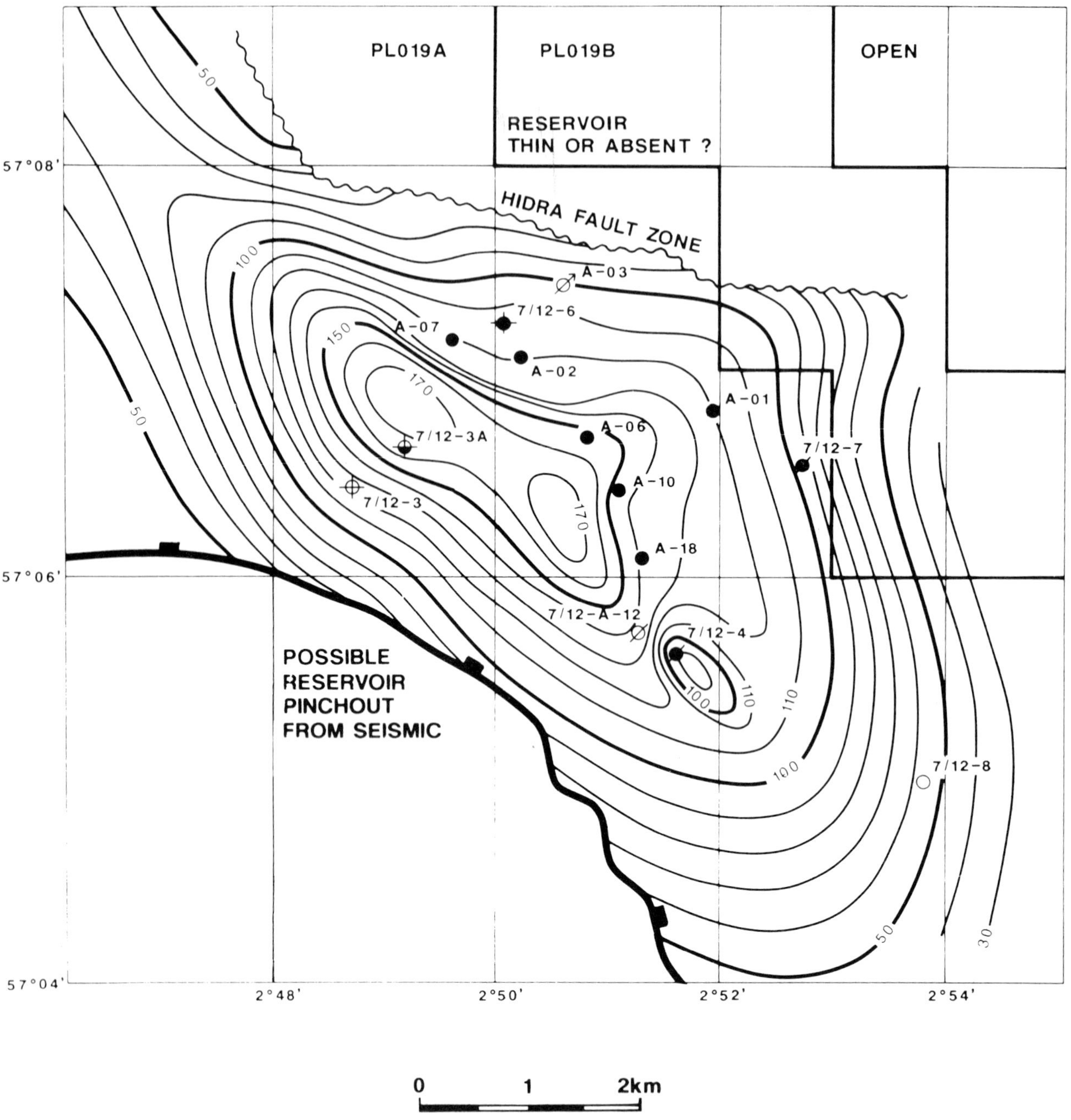

Fig. 5. Ula Field. Reservoir isopach. The map is based on combined well and seismic data.

reservoir has been inherited from earlier Jurassic extension. The majority of faults are discontinuous, but there are three (Hidra, Crestal and Transfer) which are thought to control the hydrocarbon distribution proven by drilling (Fig. 8). These are described in turn below.

(A) The Hidra Fault Zone. This a major structural feature in the Ula trend. This complex fault zone defines the North eastern limits of the Ula field where it strikes in a northwest–southeast direction and downthrows to the southwest. The presently observed throw along the Ula Field section of the fault is highly variable, but difficult to quantify accurately because of a marked deterioration of seismic resolution in the fault zone. The Ula reservoir is estimated to be 70–90 m thick on the hanging wall of the fault. Reservoir presence on the footwall is unconfirmed, but is expected to be very thin or absent. The north-eastern limits of the Ula Field petroleum are controlled by this fault.

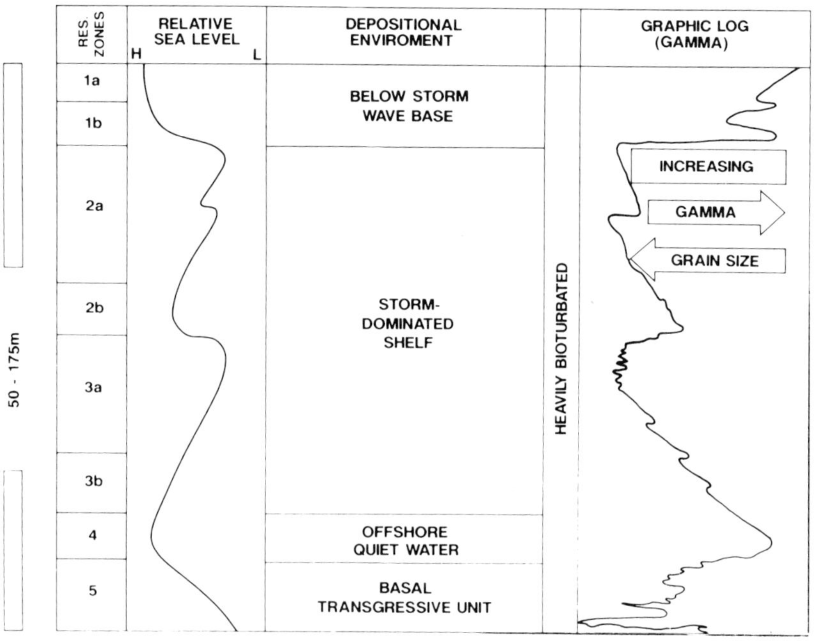

Fig. 6. Ula Field. Stratigraphy and reservoir zonation.

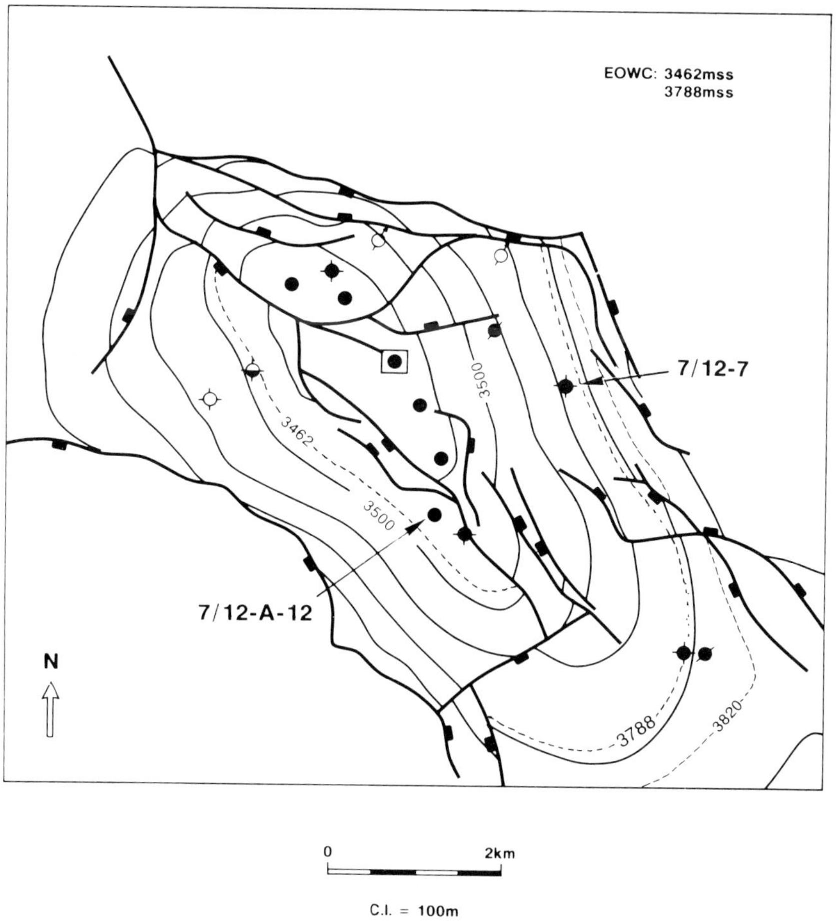

Fig. 7. Top Ula reservoir. 1988 depth map.

(B) The NW–SE trending Crestal Fault System. The throw across this system is variable. Along the field crest, very little or no throw is seen. Furthermore, the south-eastern end of this fault has very little throw and heads in the opposite direction.

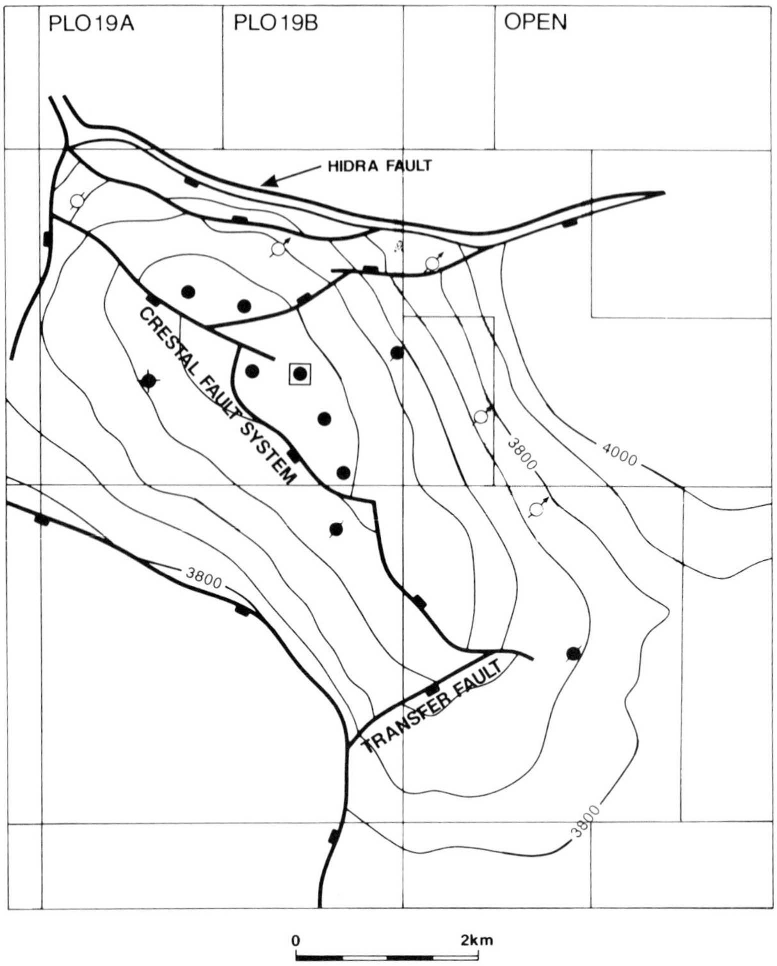

Fig. 8. Ula Field. Top reservoir depth structure.

(C) The NE–SW oriented Transfer Fault. This is located at the southern end of the Crestal Fault and has 30–50 metres of displacement. This is insufficient to displace the reservoir sands yet, it separates water-bearing Ula formation in the footwall from oil-bearing oil in the hanging wall. This fault is regarded as sealing.

Pressure data obtained prior to production start-up are shown in Fig. 9. These data can be used to infer a common OWC on the field, but the only real aquifer data available prior to production start-up are from the west flank well 7/12-3A. It was, therefore, not possible to determine the true relationship between the east and the west flank aquifers prior to the onset of field production. One additional piece of information comes from well 7/12-A-12 completed on the west flank of the field. This well encountered depleted reservoir pressures similar to those on the east flank indicating that, 2 years after production start-up, the east and west flanks of the field were in good communication.

Three models are proposed using the available structural and well data. Each model is partially valid although none fully explains the observed data set. The three models are discussed in turn below and in each the role of the Jurassic crestal fault system is of paramount importance.

Sealing faults model

This model assumes that the Central Fault System was, in this case, sealing during the course of reservoir filling. The reservoir would require charging separately from east and west, with the east flank filled to structural spill point and the west flank only partially charged (Fig. 10).

The main problems associated with this model are that the Crestal Fault System, which has no throw over the central section, must have been completely sealing during the period of reservoir charging and remain so up until the onset of production.

Pressure data from the exploration and appraisal wells do not show oil gradients which differ between the east and west flanks (Fig. 9). An oil-leg pressure difference of approximately 200 psi would be anticipated between east and west if the central fault system had been sealing and the aquifers were in

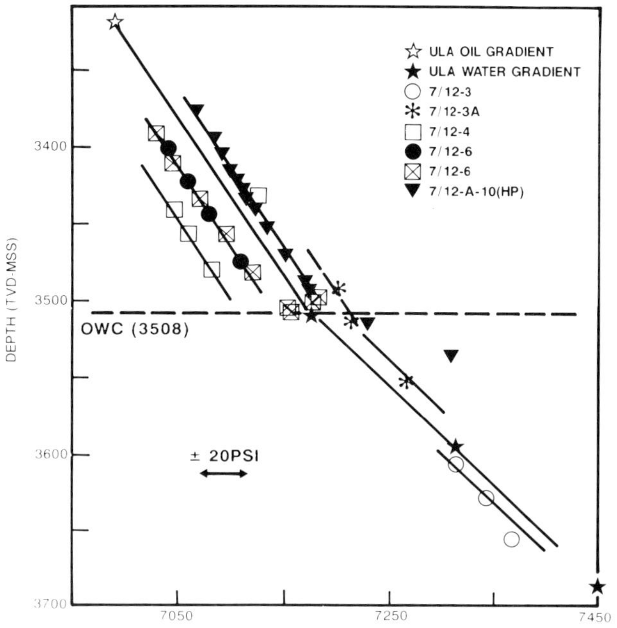

Fig. 9. Ula composite RFT/FMT pressures.

contact. To obtain similar oil gradients with such greatly differing oil columns across a sealing fault requires either that the oils have to be of a radically different composition, which geochemical analysis proves that they are not, or that the aquifers are not in communication.

With respect to the latter point, while an aquifer gradient has been obtained for the wells on the west flank of the field, no such gradient is available for pre-production wells drilled on the east flank. Hence, the original relationship between the east and west flank aquifers will remain unknown.

However, this model fails to explain simply the data obtained from development well 7/12-A-12. This well was drilled on the west flank of the structure, 2 years after production commenced, and proved that the west flank of the field was in direct communication with the east. This indicates that if the Crestal Fault System were sealing through geological time, then the fault seal must have broken down after production start-up. However, there is no obvious change in the pressure history of the field to indicate that the fault seal might have broken down.

Dynamic aquifer model

This model assumes that the difference in fluid levels from west to east is a result of compaction driven active aquifer influx from the west of the field (Fig. 11). Here it is assumed that the Ula Field lies on an active pressure gradient where formation waters are being driven through the Ula structure as a result of compaction of sediments in the basin to the west of the field. In this model, the Crestal Fault System could be either a partial permeability barrier or completely non-sealing. In the former case, this would allow fluids to be transferred from the west to the east when a critical pressure had been reached. This hypothesis would also permit the field to be hydrocarbon charged either from east or west. The fault and stratigraphic configuration of the Ula Field is not consistent with a simple, reservoir-borne aquifer flow from the west to the east of the field, since structural and stratigraphic mapping suggests that the reservoir sand limits to the west of the field are restricted.

Thus, aquifer drive through compaction of shales lying in the basin to the west of the field is seen as the only alternative mechanism to support the dynamic aquifer hypothesis. However, published data suggest that fluid flow rates from compaction of shales in the basin to the west of Ula are inadequate (Giles, 1987) to produce the observed contacts, flowing at rates of less than 1 mm/year (Fig. 12).

Confined water tank model

In this model, the reservoir sand is regarded as being continuous over the Ula Field, but the sand is discontinuous beyond the west flank of the field either through shaling out of the sand (Fig. 13) or through disconnection by down-dip faulting.

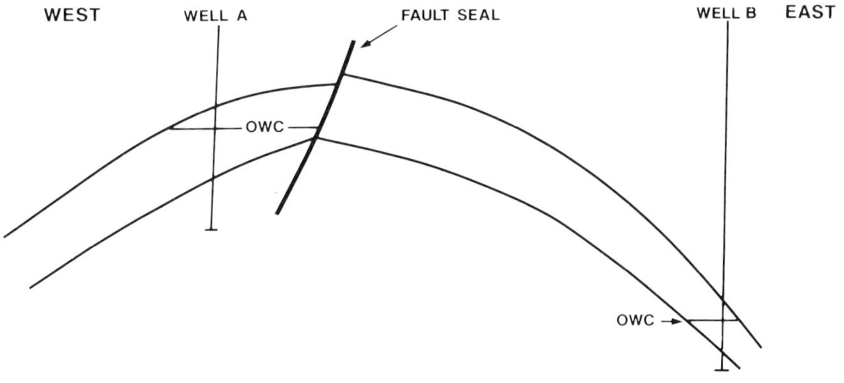

Fig. 10. Fault seal breakdown.

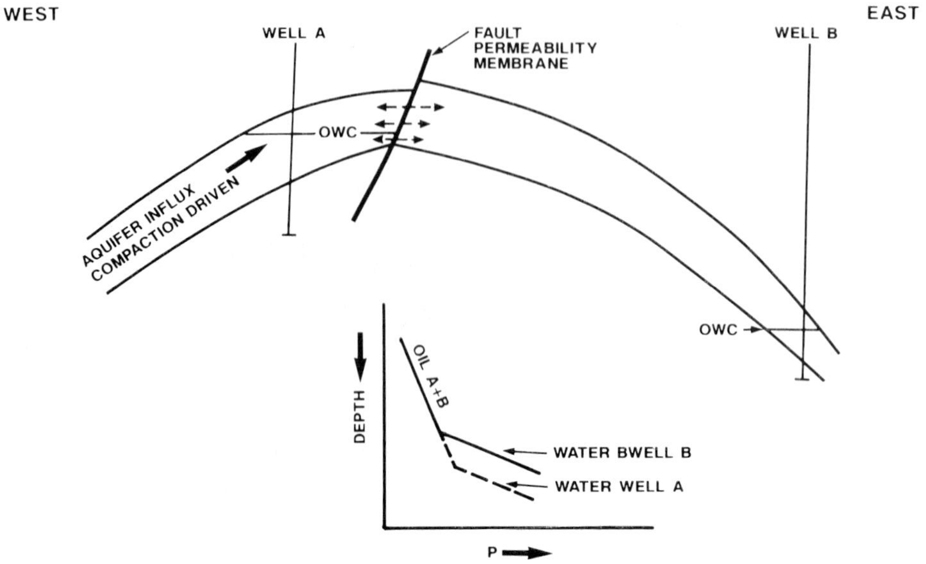

Fig. 11. Active aquifer.

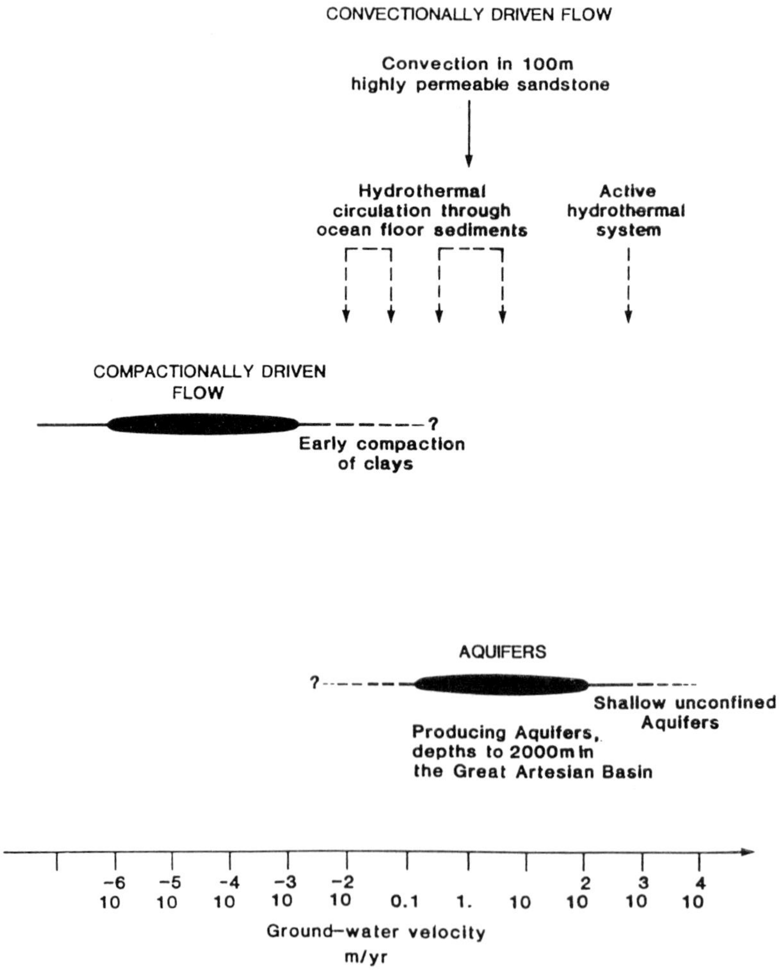

Fig. 12. Rates of water flow in sedimentary basins.

It is assumed that the crestal fault system is not sealing and that petroleum charged the structure from the east, progressively filling the field crest. If the hypothesis were correct, then the process would continue until the bottom sealing limits of the field crest were reached. At this point, water contained within the west flank would no longer be displaced, leaving a trapped aquifer to the west. In

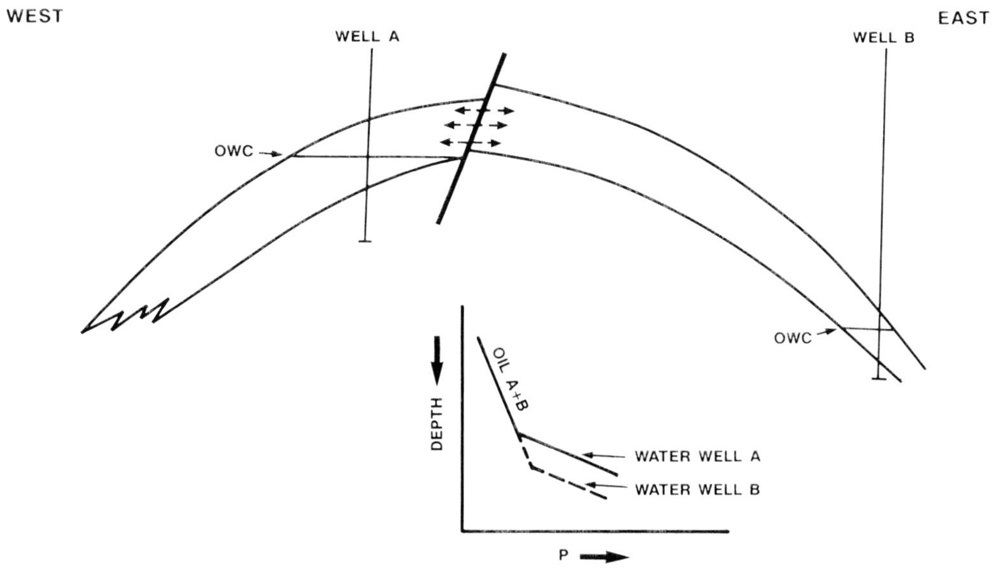

Fig. 13. Confined "water tank".

this case, the pressure gradients of the oil column would be similar and the aquifer pressures would lie on different gradients. Seismic data suggest that the reservoir sand is not laterally extensive to the west, but provides insufficient resolution to indicate unequivocal sand pinchout. Thus, a true limit to the west of the reservoir is not seen. The Crestal Fault System appears to be open both at the crest and also at the down-dip. For this model to work, the crestal sand-against-sand fault contact is required to be non-sealing and the down-dip sand-against-sand contact to be sealing. This model also requires the south-eastern Transfer Fault to be sealing even with very limited throw.

Conclusions

Despite the three-dimensional survey and extensive well data, the controls on the hydrocarbon distribution in the Ula Field are not well understood. Accurate knowledge of the sedimentology of the homogeneous reservoir sands together with the factors controlling their post-depositional reservoir quality indicate that the gross petroleum distribution is not controlled by either primary or secondary reservoir facies. The three-dimensional dataset has revealed that Ula Field has undergone a much more complex structural history than was previously thought and it suggests that the structuring is primarily responsible for the observed fluid phenomena. Although a complex structure, Ula field does not appear to be excessively compartmentalised, yet there is clearly a major variation in the OWC from the western to the eastern flanks.

Because of the late discovery of the "real" fluid distribution, the data set available for interpretation is insufficient to enable identification of the precise controlling mechanism. In fact, three models have been proposed as possible explanations, although the lack of confirmatory data precludes the identification of the most likely.

The preferred model is that of sealing faults which have broken down as a result of the rapid development of a large pressure differential through field production. Only this model has the potential of explaining all the facts, but if correct, we must expect to encounter some new oil–water contacts in future development drilling.

Acknowledgements

The authors wish to thank the British Petroleum Company plc, Svenska Petroleum A/S., Statoil, Norske Conoco A/S and K/S A/S Pelican and Co., for granting permission to publish this paper.

References

Bailey, C.C., Price, I. and Spencer, A.M., 1981. The Ula Oil Field, Block 7/12, Norway. In: Norwegian Symposium on Exploration, Norsk Petroleum-forening, Article 18, pp. 1–26.

Deegan, C.E. and Scull, B.J., 1977. A Standard Lithostratigraphic nomenclature for the Central and Northern North Sea. Institute of Geological Sciences, London, Rep. No. 77/25.

Giles, M.R., 1987. Mass transfer and problems of secondary porosity creation in deeply buried hydrocarbon reservoirs. Mar. Pet. Geol., 4: 188–204.

Gluyas, J.G., Leonard, A.J., Oxtoby, N.H., 1990. Diagenesis and petroleum emplacement: the race for space — Ula Trend, North Sea. In: International Association of Sedimentologists,

Utrecht; 13th Int. Congr., p. 193.
Home, P.C., 1987. Ula. In: A.M. Spencer et al. (Editors), Geology of the Norwegian Oil and Gas Fields. Graham and Trotman, London, pp. 143–151.
Spencer, A.M., Home, P.C. and Wiik, V., 1986. Habitat of hydrocarbons in the Jurassic Ula Trend, Central Graben, Norway, In: A.M. Spencer et al. (Editors), The Habitat of Hydrocarbons in the Norwegian Continental Shelf. Graham and Trotman, London, pp. 111–127.
Vollset, J. and Doré, A.G., 1984. A revised Triassic and Jurassic Lithostratigraphic Nomenclature for the Norwegian North Sea. Norw. Pet. Dir., Bull., 3.

A. BROWN *BP Norway Limited U.A., P.O. Box 197, N-4033 Forus, Norway*
A.W. MITCHELL *BP Norway Limited U.A., P.O. Box 197, N-4033 Forus, Norway*
I.R. NILSSEN *BP Norway Limited U.A., P.O. Box 197, N-4033 Forus, Norway*
I.J. STEWART *BP Norway Limited U.A., P.O. Box 197, N-4033 Forus, Norway*
Present address: *BP Exploration Inc., P.O. Box 4587, Houston, TX 77210, U.S.A.*
P.T. SVELA *BP Norway Limited U.A., P.O. Box 197, N-4033 Forus, Norway*

A frontier basin analysis study based on limited data: the South Mozambique Graben

J.E. Iliffe, I. Lerche and M. DeBuyl

The South Mozambique Graben is assessed for hydrocarbon potential by reconstructing its structural, burial and thermal histories using a seismic line and an array of basin modelling techniques.

A depth converted cross-section is simultaneously restored palinspastically and backstripped at designated pseudowells for each chronostratigraphic package.

The thermal history is predicted from extensional models which relate paleoheat flow to basement subsidence. Maturation potential cross-sections are derived by applying a generation model to each of the pseudowells, assuming each layer to be a potential source rock. The basin is then analyzed by considering the timing of hydrocarbon generation relative to the deposition of potential source and seal rocks and the formation of trapping structures and migration pathways.

When a range of uncertainty is allowed for in the development of these models, the exploration geologist can make a better judgement of the dynamic range of hydrocarbon potential of this frontier basin. Also, this type of study determines the type of additional information required to reduce uncertainties significantly.

In this paper we illustrate how mathematical modelling can enhance the understanding of the hydrocarbon potential of a basin given only limited data. Formerly, one-dimensional burial history models (e.g., Guidish et al., 1984; Cao, 1985) have assumed that sediments must have been deposited, not deposited (hiatus) or eroded during each time phase; meanwhile, the structural development has been largely ignored.

The primary goal is to demonstrate a basin analysis technique which unites the structural, depositional and thermal histories of an area. Such an integrated approach may greatly contribute to an exploration geologist's understanding and assessment of the hydrocarbon potential of a speculative basin.

The method consists of simultaneously removing depositional layers and reconstructing structural cross-sections in sequential chronostratigraphic steps back in time.

Before the technique was applied to real field data from the South Mozambique Graben, the problems of backstripping wells which penetrate faults were investigated by means of a series of tests on synthetic examples and this aspect is discussed first.

Backstripping wells which intersect faults: tests on synthetic examples

The synthetic scenario designed for the initial experiments consisted of a simple fault block which had been rotated on planar growth faults in domino style (Fig. 1). The burial history of this system began with the deposition of three pre-rift layers (*1*, *2* and *3*, Fig. 1) each of which represents 5 million years. These layers were followed by the deposition of two syn-rift layers (A and B, Fig. 1), each one also equivalent to 5 million years. These secondary layers were of a different lithology from the first three.

Pseudowells positioned across the structure were backstripped using a one-dimensional isostatic burial history program similar to that of Guidish et al. (1984). The decompaction of each chronostratigraphic layer was accommodated by means of the porosity/depth functions of Magara (1976) taking lithologic parameters from Sclater and Christie (1980).

After the first unit (A) was backstripped off each well, the underlying layers were decompacted and a new cross-section was drawn by correlating the units in each well (Fig. 1b). This process was repeated for unit B (Fig. 1c). A one-dimensional vertical backstripping restoration of a two-dimensional fault block scenario has shortcomings, mainly in that the throw of the faults cannot be restored, since this would require lateral movement of strata. Other anomalous features of one-dimensional restoration are that the fault angles are steepened and originally planar faults become curved. Also, faults are backstripped to the surface and so appear to migrate towards the downthrown side. Strata beneath the fault are also incorrectly restored.

This latter problem was solved by treating the fault as if it were an unconformity at depth. Sediments are

Structural and Tectonic Modelling and its Application to Petroleum Geology edited by R.M. Larsen, H. Brekke, B.T. Larsen and E. Talleraas. NPF Special Publication 1, pp. 421–437. Elsevier, Amsterdam.

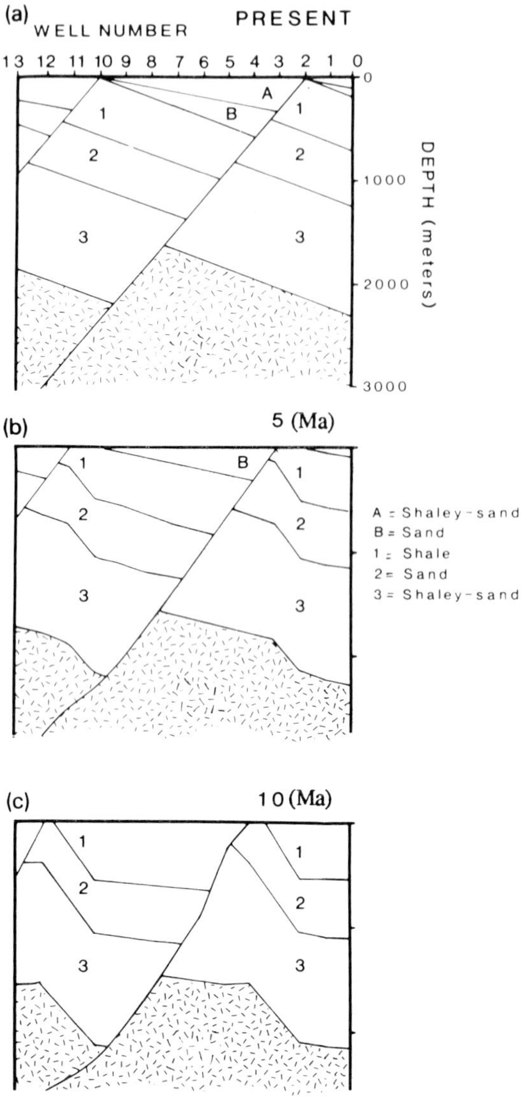

Fig. 1. (a) Basic structural cross-section showing the location of pseudowells used in all of the experiments. Note that in some experiments the fault angle and scale of the system vary. (b) Situation after removing layer "*A*" corresponding to the scenario 5 Ma. (c) After removal of both the syn-faulting layers *A* and *B* (10 Ma). Note layers *1*, *2* and *3* are considerably discordant. Note also the curvature of the fault trace and that sediments beneath the fault are not properly restored.

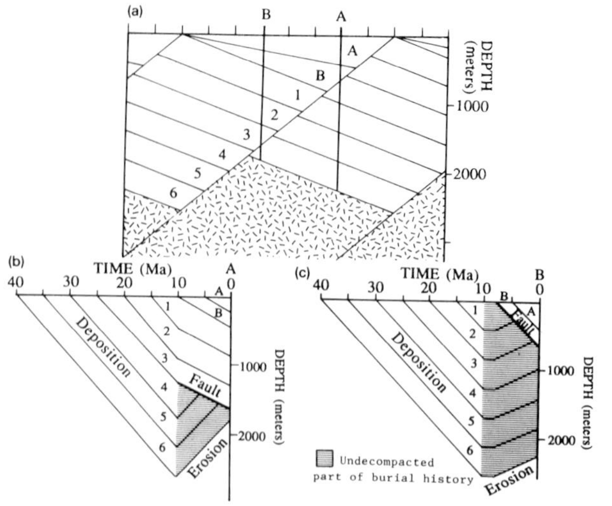

Fig. 2. (a) Input system for the detailed analysis. Note positions of wells *A* and *B*. Layers *1–6* are pre-rift, *A* and *B* are syn-rift. (b) Burial history of well *A* incorporating the "erosion" below the fault solution to compensate for disturbances of bed seen in Fig. 1c. (c) Burial history of well *B*, also using the subfaulting "erosion" solution.

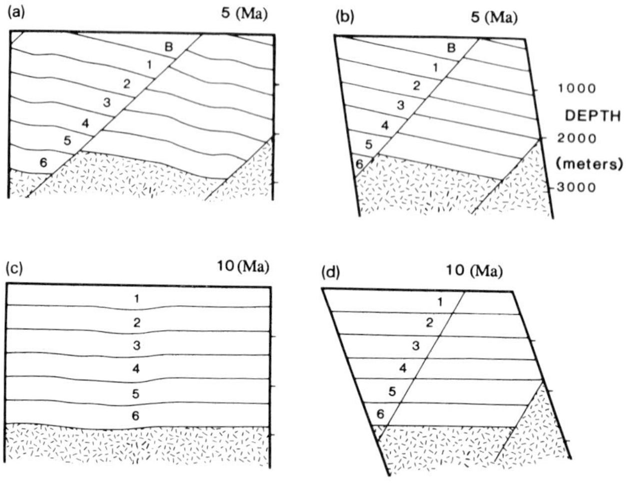

Fig. 3. (a) Backstripped and reconstructed situation of Fig. 3a at 5 Ma. (b) Simple geometric solution (structural, not backstripped) of Fig. 3a at 5 Ma. (c) Backstripped and reconstructed situation at 10 Ma. Note waviness of strata compared with Fig. 3d. (d) Geometrical solution at 10 Ma.

essentially considered to be "eroded" from beneath the fault, while deposition is occurs above the fault (Fig. 2). The values for depths to the strata are read from the burial history plots. The cross-sections at time periods 5 and 10 Ma resulting from this procedure (Fig. 3) were contrasted against those derived from a simple geometrical restoration. Comparison of the throws of the faults at different stratal levels for each solution (i.e., decompacted and undecompacted faults) at the 5 Ma time step (Fig. 4) shows that compaction reduces the throw of the faults and this effect decreases with depth.

A sensitivity study was undertaken for a sequence of experiments similar to the initial case (Fig. 1) where the individual effects of lithologies, spatial scale and initial fault angles were investigated. Two extreme lithologic cases (homogeneous shale or homogenous sandstone) along with two variations in the present day fault angle (40° and 60°) and two different system scales (1:50000 and 1:25000) were backstripped and reconstructed. Results showed that shallow faults are more affected by compaction than steeper faults, the angle of the fault has more variation up the fault in shales than in sandstone, and the variation up the fault is greater in the larger scale model (Fig. 5).

Another pitfall of using one-dimensional models in

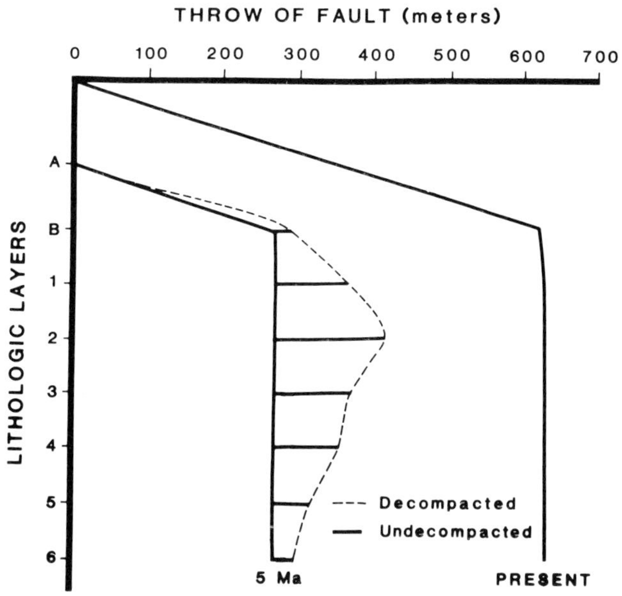

Fig. 4. Comparison of the variation in throw of the fault in the non-decompacted and the decompacted solution. Note that the throw is generally increased by decompaction although this effect decreases with depth.

extensional regimes is related to "contraction" of the basin during backstripping. After each reconstruction step a well's burial history may be radically altered (e.g., well 1, Fig. 6). This problem is alleviated by keeping the position of the pseudowells fixed and "redrilling" each well after each layer of sediment has been removed, and then rerunning the burial history. The composite burial history of the well in that one position is a better geologic representation than a simple "one run" case. These constraint procedures may now be implemented on real geologic situations.

The South Mozambique Graben: a case study

One of the problems often encountered by the exploration geologist is the assessment of the hydrocarbon potential of a speculative basin with limited data which is also often of poor quality. This study exemplifies such a situation and shows how basin modelling may improve evaluation based on the limited resources available.

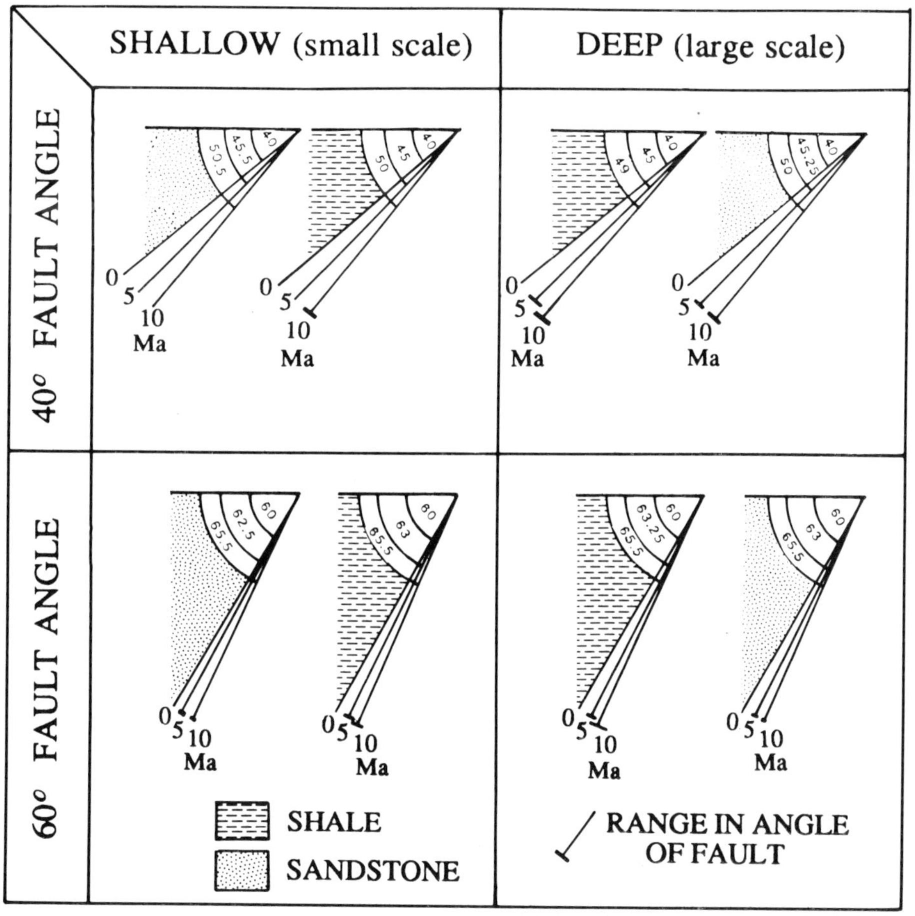

Fig. 5. Summary of results of experiments considering the effects of lithology, fault angle and depth on the backstripping of the input system shown in Fig. 1.

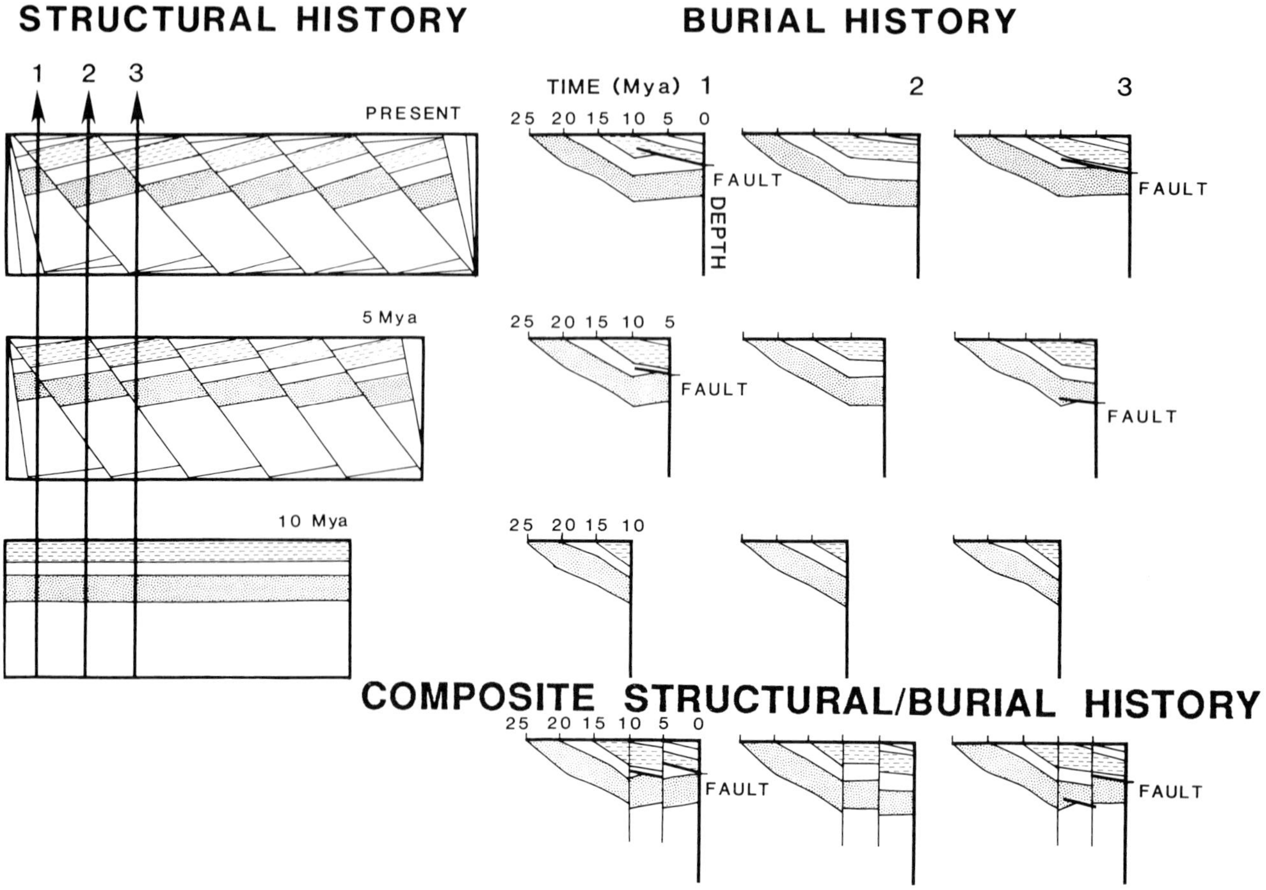

Fig. 6. Variation in burial history at a location due to contraction during reconstruction. Note the changes in burial history between 5 Ma and 10 Ma for each well. Compare with the composite burial history at the bottom.

The available data for offshore South Mozambique consisted of a few seismic lines, two shallow wells and the general literature. In this study the tools applied to these data are a one-dimensional Burial History Program (Guidish et al., 1984), a Tissot and Welte (1978) type generation model (Tissot and Espitalie, 1975), models of basin subsidence (Royden et al., 1980; McKenzie, 1978) and some basic geological knowledge.

Tectonic setting

The seismic data revealed a "V" shaped rift basin trending N–S to NNW–SSE (Fig. 7) bordered to the east by a fault trending NNW–SSE, and to the west by a fault oriented more N–S. Series of tilted fault blocks dipping to the east on westward verging faults are visible within the rift.

The tectonic setting of offshore Mozambique has not been presented coherently to date, mainly due to lack of data. Indeed, there is some debate as to the age of rifting and direction of stress fields (cf. Segoufin, 1978; Rabinowitz et al., 1983; Mougenot et al., 1986; Martin and Hartnady, 1986; Coffin and Rabinowitz, 1987; Lawver and Scotese, 1987), although Darracott (1974) recognized that the area lies between the continental crust of Mozambique and the oceanic crust of the Mozambique channel which separates Madagascar from the African continent. Martin and Hartnady (1986) postulate a Mesozoic plate boundary parallel to the Lombobo Mountains of the Kaapvaal Craton, suggesting that the South Mozambique Graben basement is thinned continental crust. The general consensus is that Madagascar has been transported southwards along a N–S trending transcurrent fault (Coffin and Rabinowitz, 1987) (Fig. 8). This movement has been dated on the basis of magnetic anomalies as being from 150 to 110 Ma (Segoufin, 1978) or 160 to 125 Ma (Rabinowitz et al., 1983). It is also possible that initial rifting began as early as Karoo time (Permo-Triassic to Jurassic) or, more probably, is associated with the major break-up of Gondwana in Late Triassic to Early Jurassic times (Mougenot et al., 1986).

The NNW–SSE to N–S fault trends observed within the South Mozambique Graben originally may have been part of this transcurrent fault system, since flower structures are visible in deeper parts of E–W seismic profiles. The Graben has been described as a pull-apart basin by Chatellier and Slevin (1988). The transcurrent faults may have been reactivated later

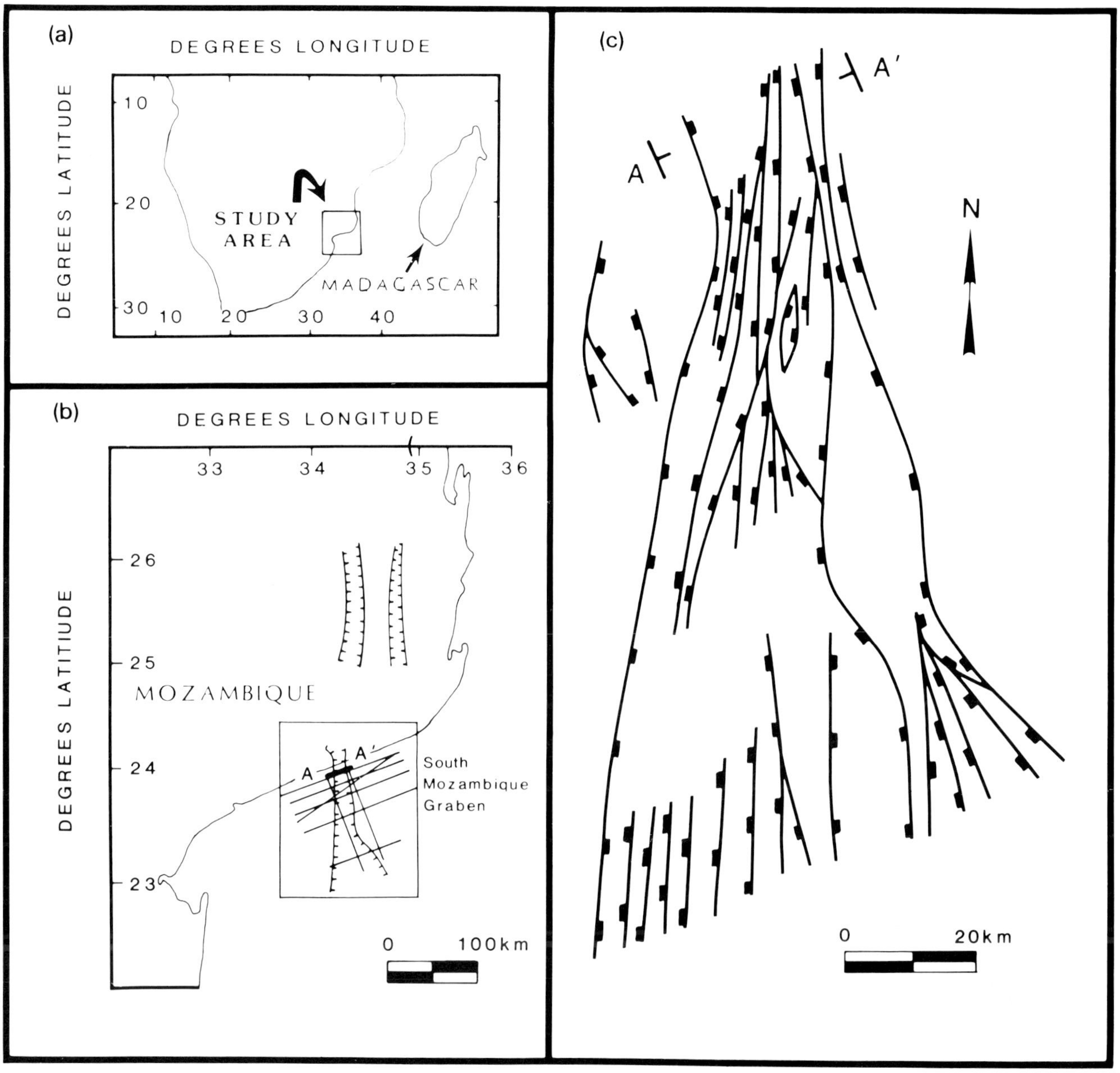

Fig. 7. Location and structural trends which cut Lower Cretaceous strata in the South Mozambique Graben.

in a purely extensional regional stress field in Late Cretaceous times (100 Ma).

Seismic interpretation and depositional history

Interpretation of the seismic data was based on the recognition of seismic units rather than following one or two reflectors regionally, which proved impossible in the rift itself. Basic isopach and structural maps were produced and a seismic section perpendicular to tectonic strike was then selected for subsequent analysis (*A–A'*, Fig. 7).

While describing the general problem of sedimentation for the south Mozambique Graben it is useful to refer to the seismic character of the six chronostratigraphic units recognized (Lafourcade, 1984).

The oldest unit presumed present is a basement of volcanic rocks which make up part of the Kaapvaal Craton (DeBuyl and Flores, 1986). This unit is probably overlain by the Karoo (whose existence offshore has not been proven beyond doubt). These rocks consist of a sequence of continental deposits ranging in age from Permo-Triassic to Upper Jurassic. At the base, the Ecca series exhibit coal measures, shale, siltstone and sands. The overlying Beaufort series contains sandstones and shales and, at the top the Stormberg series, is represented by volcanoclastics, lava flows and sandstones (Kamen-Kaye, 1978; 1982).

The Karoo unit may be present in the South Mozambique Graben, making up part of unit 6 (Fig. 9a). The reflectors in this unit are structurally disturbed and stratigraphically discontinuous typi-

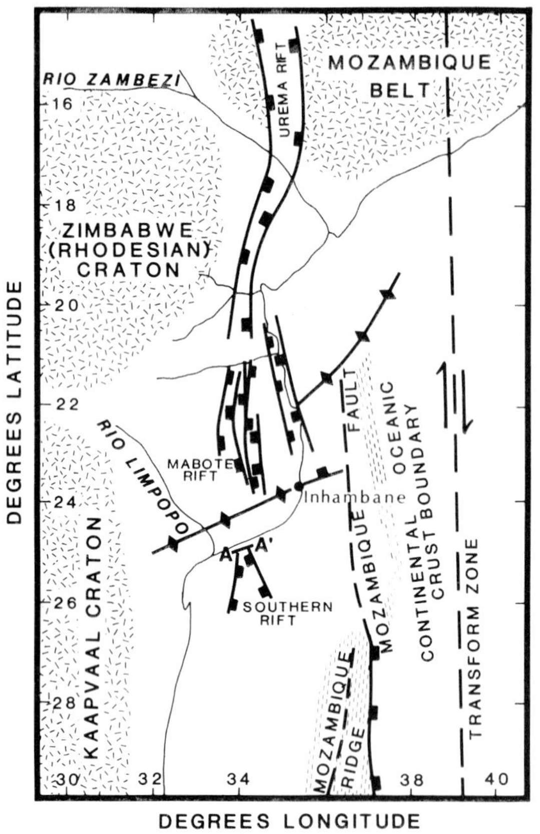

Fig. 8. Main structural trends in the Mozambique Region. Note the transcurrent fault in the area.

fying them as a syn-rift deposit. The age of this unit is speculative, since it has never been drilled. Reports from nearby countries with better data suggest that the Karoo was deposited during Gondwana break-up, in Late Triassic to Early Jurassic times (Mougenot et al., 1986). Accordingly this Karoo unit is taken to be 195 Ma at its base and 140 Ma at its top. This Jurassic syn-rift deposition is not reported onshore, probably because most of the Gondwana break-up in the region now lies offshore, and the onshore areas of this time were probably erosional highs manifested in the rock record as unconformities, as suggested by Forster (1975).

Following deposition of unit 6, a marine transgression occurred, beginning in Early Cretaceous times (Albian?) as recorded by the sands and shales of the Maputo Formation. Reflectors in unit 5 are discontinuous and structurally disturbed.

The Maputo Formation is overlain by the Lower Domo Shales Formation from Cenomanian to Turonian in age. There is a bright acoustic doublet visible at the top and bottom of the corresponding unit (4). Internally the unit is generally seismically transparent. Where visible, reflectors are discontinuous.

The doublet at the top of unit 4 may be the mid-Turonian Domo sands which are truncated by an Upper Cretaceous unconformity overlain by the clayey silts, sands, sandy limestones and marls of the Upper Domo or Grudja Formation. This unit (3) has parallel, sub-horizontal and continuous reflectors which may indicate a basin type depositional setting (Flores, 1973). It is apparent from the seismic section (Fig. 9a) that this unit was either very thin, eroded or not deposited on the flanks of the rift.

Above the Grudja Formation lie an oolitic algal limestone, marls and sands of the Eocene-age platform, and a ramp sequence of the Cheringoma Formation (unit 2).

The ramp, visible on the northeast side of the seismic section (Fig. 9a) contains slump structures. The reflectors are slightly inclined, discontinuous and non-parallel on the ramp and faint or absent in the basinal areas.

Overlying these sediments are the deltaic sediments of Miocene to Recent age (unit 1). Reflectors are gently dipping, thin and sub-parallel.

Backstripping and reconstruction

The seismic profile (Fig. 9a) was converted to a true depth section (Fig. 9b) using pseudowells 1 to 12 which were located at shot points where stacked, interval velocities had been calculated. These velocities were then used to calculate the thickness of each layer. The accuracy of these thickness estimates depends on the accuracy of the velocities, which was not provided.

After obtaining a "true" section, (Fig. 9a) the lengths of each pick were compared and the actual amount of extension as a ratio of the total extension was calculated for each time line, to produce an extension rate history (Fig. 10a).

The structural interpretation was constrained by length balancing an imaginary reflector below and parallel to the deepest seismic reflector at the base of unit 6 (Dahlstrom, 1969).

The data needed for the one-dimensional isostatic burial history calculations are: depths to formation tops, absolute ages at these tops, layer lithologies, paleobathymetry of each layer, density of rock matrix and porosity/depth relationships for various lithologies (Guidish et al., 1984).

Several assumptions are therefore required to develop the analysis. The lithology of unit 6 is presumed to be sands related to Jurassic syn-rift deposition. Owing to the lack of downhole sonic and density information, the lithological porosity/depth parameters were defaulted to those available in Sclater and Christie (1980). Layers of thickness 1000 meters or more were taken to be monolithological. This is obviously in error; in general, however, such a working hypothesis provides an evaluation in keeping with

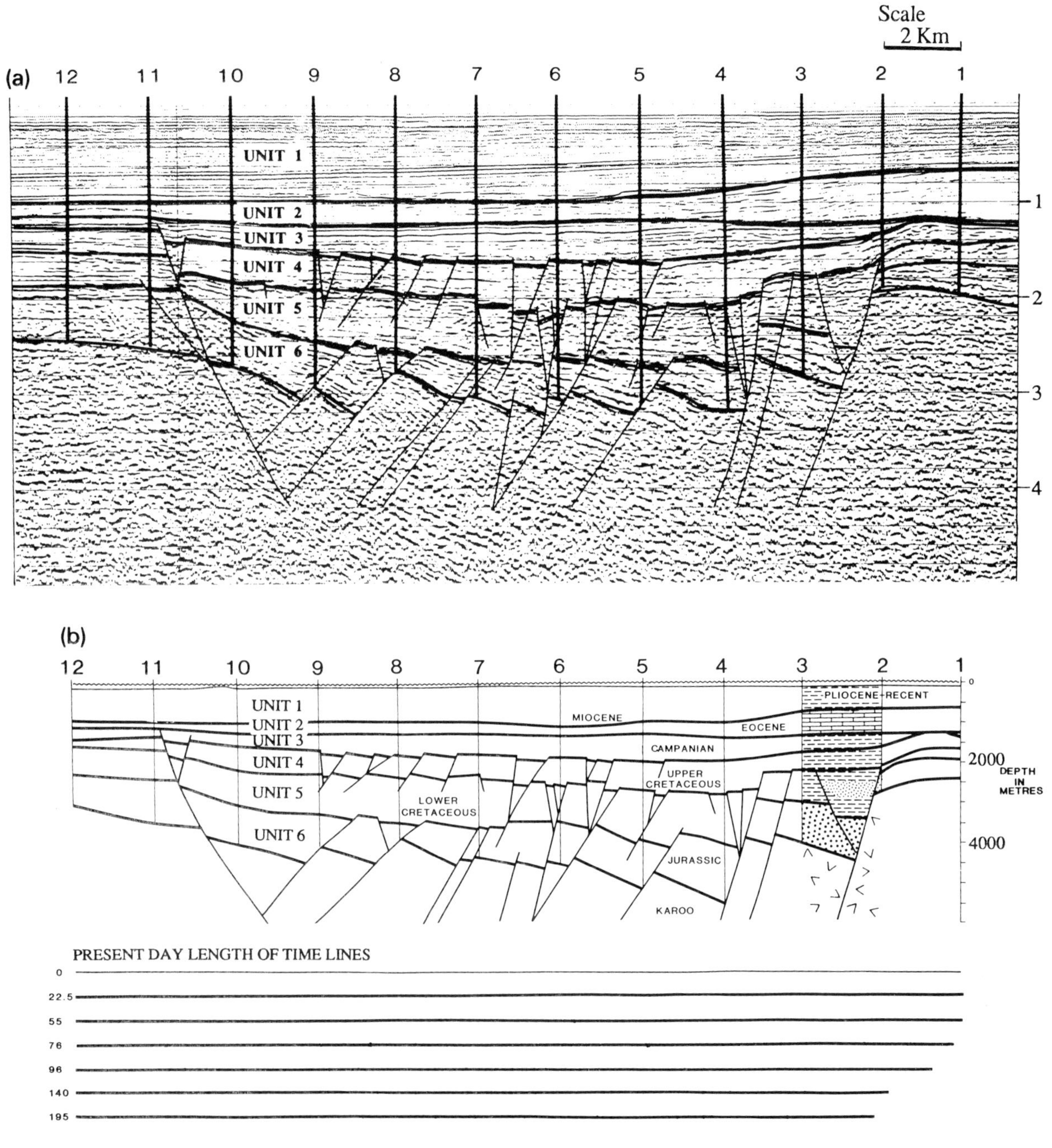

Fig. 9. (a) Interpreted seismic section A–A' trending east–west. (b) Depth converted geologic cross-section with a generalized stratigraphic column.

the resolution of the bulk of the data (Cao, 1988). More critical to the model is the estimation of absolute ages. The paleobathymetry estimation is based mainly on the geologic interpretation and sparse literature.

All twelve pseudowells were backstripped one unit, after which the cross-section was redrawn (Fig. 11a). Since no faulting was evident after 55 Ma, no rebalancing was necessary after unit 1 was removed.

When the second unit was backstripped (76 Ma to 55 Ma), the throw on some faults had to be adjusted according to the procedure described initially in this paper (Fig. 11c). The removal of throw on these faults slightly contracted the section such that those wells that crossed a fault had to be "redrilled". This process of backstripping, restoring, rebalancing, redrawing and redrilling was repeated until the top of the Karoo reflector appeared as a horizontal line at the top of the model.

Each cross-section (Fig. 11a–e) was then analyzed for depth to detachment and length of the uppermost bed using the method described by Gibbs (1983). The final lengths used in the depth to detachment calculations were between the border faults in the

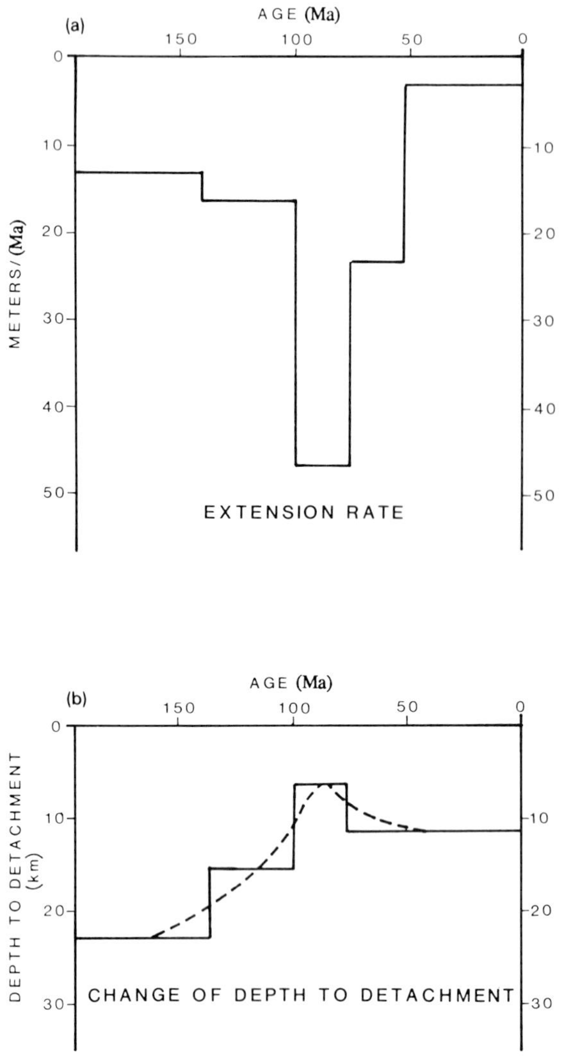

Fig. 10. (a) The extension rate shown for each time period of the backstripping. (b) Variation of the depth to detachment for each time period of the backstripping.

present day true section, i.e., within the zone of deformation.

Results and discussion of structural and depositional reconstructions

When considering the variation of extension rate with time it was found that there was a slow acceleration of extension from 195 Ma to 76 Ma, the acme of rifting being between 96 and 76 Ma (Fig. 10a). It was also noted that the depth to detachment seems to have shallowed from 195 Ma up to the period between 96 and 76 Ma and dropped gently since then (Fig. 10b). It is significant that, even though the data are sparse and of poor quality, events of high accumulation, rapid subsidence and quickest extension are almost coincident at 96 to 76 Ma.

The slow extension rates from 195 to 96 Ma may be attributed to early, subtle, probably strike-slip faulting, sympathetic to the southern emplacement of Madagascar. The occurrence of flower structures in the deeper parts of the rift, the resemblance the "V" shaped graben has to a shear zone (Tchalenko, 1970) and wrench fault settings (Harding, 1974), and the geologic data presented by Martin and Hartnady (1986) are suggestive of early strike-slip faulting manifestation of the north–south separation of the African and Antarctic plates (Simpson et al., 1979; Martin and Hartnady, 1986). Strain was therefore out of the plane of the East–West cross-sections (Fig. 11a–e). It is unclear what the Stromberg volcanics onshore might represent tectonically. The vigorous extension of the rift in Late Cretaceous times resulted in the blocks being tilted in a more systematic normal domino fashion than in the previous event. This event broadly coincides with the docking of Madagascar in its present position relative to Africa, to the onset of rifting between Madagascar and the Antarctic (Segoufin and Patriat, 1980), to the initial separation of Antarctica and Australia (Veevers, 1986) and to postulated changes in poles of rotation of South America relative to Africa (Rabinowitz and LaBreque, 1979; Martin et al., 1982). It is, thus, apparent that a major plate reorganization occurred. Mougenot et al. (1986), working independently to the north, also concluded that the region underwent an initial period of strike-slip faulting followed by a Late Cretaceous phase of extension and volcanism. Such a two-phase history was first proposed by Flores (1970). It may be significant that Upper Cretaceous volcanics also occur at this time (DeBuyl and Flores, 1986).

The depth to detachment in this area may represent the brittle/ductile transition zone. If so, then measured fluctuations in the depth to detachment may be an indication of the heat flow history of the region. Continuing this reasoning, the region may have experienced its highest heat flow during late Cretaceous times. This deduction is somewhat substantiated by the occurrence of Cretaceous sills and feeder dikes in the area (DeBuyl and Flores, 1986).

The fastest extension rate measured in the basin, some 5 cm per 1000 years, is slow when compared to spreading rates of between 1 and 17 cm per year for present day opening oceans. This basin, therefore, does not appear to have been very active and an extensional model to describe the heat flow and subsidence history must be applied with some caution.

Thermal history

Having assessed the structural and burial histories of the graben from the seismic line, some approx-

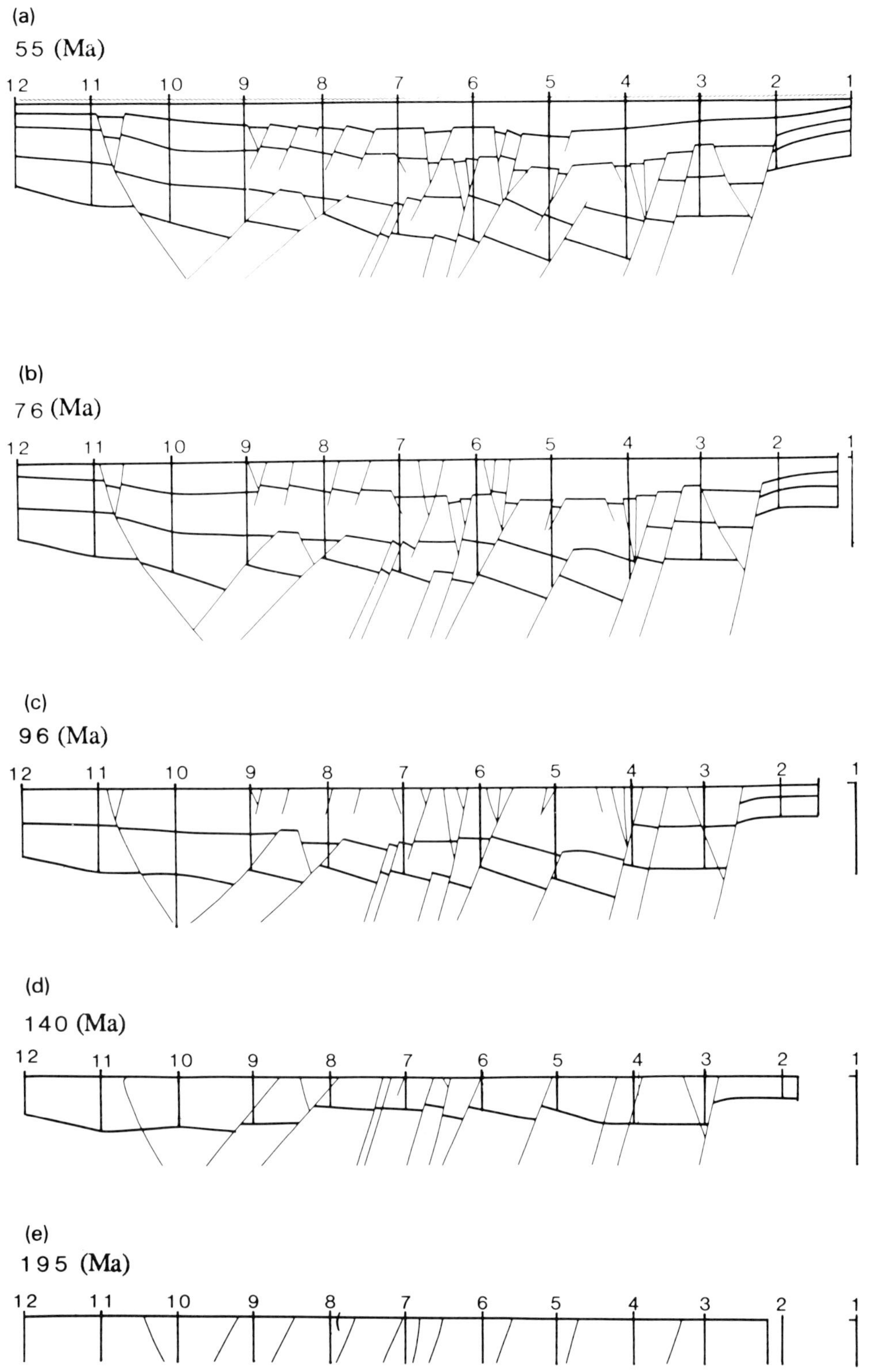

Fig. 11. (a) Cross-section just prior to Eocene deposition 55 Ma. (b) Cross-section at 76 Ma. (c) Cross-section at 96 Ma. (d) Cross-section at 140 Ma. (e) Cross-section at 195 Ma.

imation to the thermal history of the basin must be made in order to apply hydrocarbon generation models in the area to complete the basin analysis. In regions with data that include downhole measurements of thermal indicators, it is possible to invert the information to obtain estimates of paleoheat flow (Lerche et al., 1984). In lieu of any available thermal information — the nearest heat flow values of 1.2 and 1.29 HFU come from the Mozambique channel (Anderson et al., 1977) — the heat flow history of the South Mozambique Graben is estimated from the extensional model of McKenzie (1978) and the dike intrusion model of Royden et al. (1980).

The geologic observations of several dikes or sills in the region implies volcanic activity, which favors the use of the Royden et al. (1980) dike intrusion model. The McKenzie model does not involve any intrusive effects and is, therefore, probably less appropriate. However, in order to bracket the dynamic range of possible thermal histories, both the Royden et al. (1980) and McKenzie (1978) models were used. Another constraint to consider when refining an extension model is the heat flow history derived from depth to detachment calculations. Where the depth to detachment is known, and with the detachment assumed to be the brittle/ductile mid-crustal phase boundary of greenschist to amphibolite metamorphic facies, a crude estimate of heat flow through time can be given using the equation:

$$\frac{Q}{K} = Gtz$$

where Q is heat flow at surface, K is thermal conductivity and Gtz is geothermal gradient. The variation of paleoheat flow will essentially match the shape of the depth to detachment, e.g., the shallowest depth to detachment of 7 km yields the highest heat flow value of between 2.8 and 3.6 heat flux units (HFU), using a thermal conductivity of 5 mcal/cm/°C/s, and a temperature at the depth which the phase change occurs of between 400 and 500°C. This implies a present day heat flow value of around 1.6 to 2.1 HFU which is higher than that observed in the Mozambique Channel (Anderson et al., 1977).

The McKenzie (1978) and Royden et al. (1980) models essentially predict the heat flow at a particular time after rifting in a basin, given certain parameters of that basin. Both models relate basement subsidence (tectonic subsidence) to the amount of extension (McKenzie, 1978) or dike intrusion (Royden et al., 1980) and to the heat flow. It is, therefore, possible to take the basement subsidence curves for each of the pseudowells, which are output from the burial history program (Fig. 12a), and find the best fit predicted basement subsidence curve of the model that corresponds to a specific factor (Fig. 12b). Inspection of the basement subsidence curves warranted the division of the structural history into an initial rifting stage (195–109 Ma) and a later event at 96 Ma (Fig. 12). These two events correspond to known tectonic episodes of a strike-slip, pull-apart basin setting, coincident with the southward emplacement of Madagascar in the early Jurassic, and extension and volcanism in the Late Cretaceous, respectively, as described from seismic and well data in

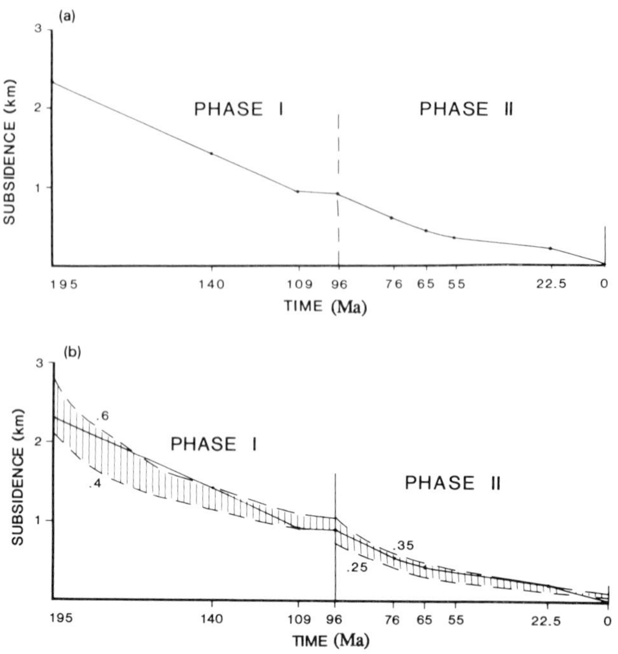

Fig. 12. (a) Observed basement subsidence curve from well 6 showing two subsidence events. (b) Determining the range of extension factors from inspection of best fit between observed curves and those predicted by the extensional models.

this study and reported independently by Mougenot et al. (1986).

In practice, the fitting of the observed to predicted basement subsidence curves has a degree of subjectivity. Therefore, the fit was bracketed to provide some idea of minimum and maximum error (Fig. 12b). For each well, and for each of the two models, there was an upper and lower limit of the possible factors. Each of these upper and lower factors was then used in the paleoheat flow equation of the model under consideration. The observed pattern of basement subsidence shows two events (Fig. 12a), but only the second has any direct supporting field evidence. In order to accommodate all possible variations, three general heat flow histories were considered (Fig. 13): (1) a single thermal heat pulse event at the first rifting (195 Ma); (2) a constant heat flow of 1 HFU until 96 Ma at which time there was instantaneous rifting and an associated thermal event; and (3) an initial thermal event at 195 Ma followed by a later event at 96 Ma (instantaneous rifting).

The Tissot and Welte (1978) type generation model was applied to each well for all three possible cases and a range of extension factors for each of the events. The input of a heat flow curve $Q(t)$ at time t was in the form of the equation:

$$Q(t) = Q_o \exp(\alpha t)$$

where α is a variable which determines the slope and

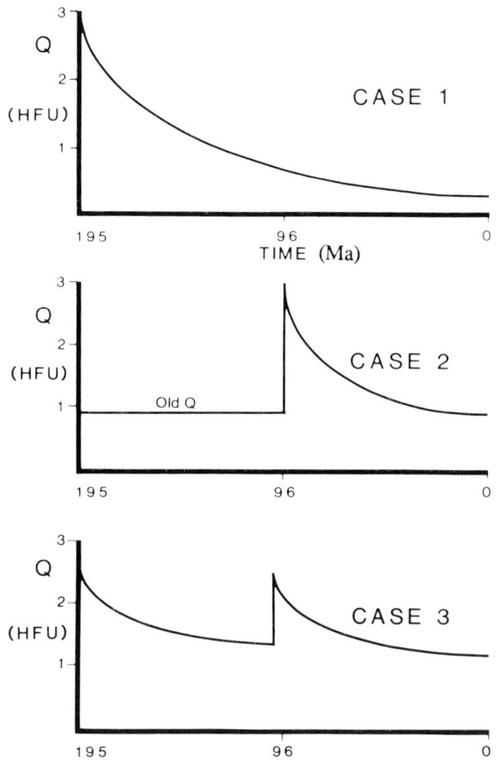

Fig. 13. Three case histories: (a) initial rifting and thermal cooling at 195 Ma; (b) a single pulse of rifting and heating at 96 Ma; (c) two rifting and thermal events at 195 Ma and at 96 Ma.

curvature of the line, and Q_o is the present day heat flow. The extension factors taken from the models were used to formulate an α value for the heat flow behavior with time in the model.

For the first two cases this technique posed no problems, as long as the present day heat flow was assumed known. A conservative estimate of 1.0 HFU was used for the present day heat flow when data was lacking.

In the third case, of two thermal events, it was necessary to calculate what the maximum heat flow would be at the inception of the second rifting, given the α value taken from the extension factors predicted from the observed basement subsidence curves for the first event. The maximum heat flow of the second event can be pinned by backtracking the equation from the present day to 96 Ma with a known α value as shown below:

$$Q_{R2} = Q_o \exp[+\alpha(T_{pd} - T_{R2})]$$

where Q_{R2} is maximum heat flow at the second rifting event, Q_o is present day heat flow (1.0 HFU), T_{pd} is time since initial rifting till present day and T_{R2} is time since initial rifting till second rifting event.

The Tissot and Welte (1978) hydrocarbon generation model is applied to all the strata. From the results of this approach we can ascertain which layers are within the oil window. Those layers may then be studied more closely for their actual source potential. The generation model predicts how many milligrams of oil are produced per gram kerogen for each layer at each well site. This value was then plotted in the center of the formation for each well and contoured (Figs. 14–17) to give maturation potential cross-sections.

Such maturation potential sections were generated for each of the three cases of heat flow and for the upper and lower limits of each extension factor for both the McKenzie and Royden et al. models of extension. In this way, possible variations are taken into account and some degree of confidence may be placed on the results.

If there was only one rifting event at 195 Ma, the Royden et al. (1980) model predicts that the basin is extremely oil prone (Fig. 14c, d) (if a source rock exists!). On the other hand, if there was only one thermal event of minimal magnitude at 96 Ma, the McKenzie (1978) model suggests that most of the basin is undermature, although still with some potential in the east, near wells 4 and 5 (Fig. 15a, b).

The results of the two rifting events case implies that the later smaller heating event at 96 Ma has more effect on maturation of the sediment than the rifting event of greater magnitude but occurring earlier at 195 Ma (Fig. 16a–d). During the initial thermal event the sediments had either not yet been deposited or were still relatively near the surface with lower temperatures. During the later event the same sediments were at a greater depth and so experienced higher temperatures.

The timing of generation shown for each case using well 5 (Fig. 17) also suggests that the late thermal event is more important for the hydrocarbon potential of this basin (Fig. 17b, c), since a later generation facilitates sealing and trapping to a greater extent. It is, therefore, of importance that this event coincides with a raising of the brittle/ductile boundary and the documented lava flows in the region.

The key is not how high the heat flow occurred, but rather how hot the sediments became and how long they remained hot.

Basin prognosis

To evaluate a basin for hydrocarbon potential we should answer five questions:

(1) Does the basin have potential trapping structures? In the case of the South Mozambique Graben the answer is strongly affirmative.

(2) Does the basin have reservoirs? Most likely the syn-rift unit, 6, may provide reservoir quality sands and gravels in the form of fans at the sides of blocks.

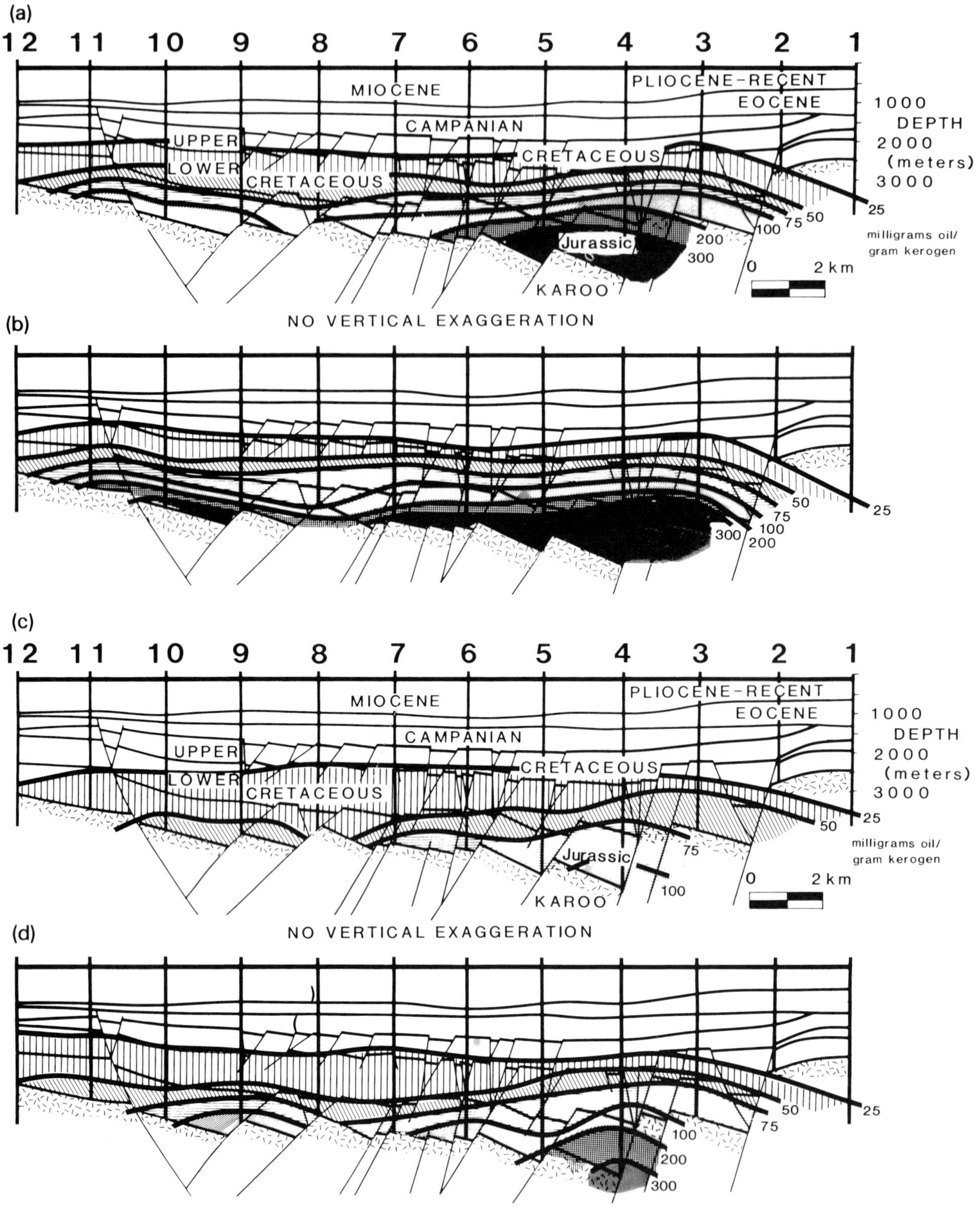

Fig. 14. (a) Royden et al. (1980) model, lower limit, case 1. (b) Royden et al. model, upper limit, case 1. (c) McKenzie model lower limit case 1. (d) McKenzie model, upper limit, case 1.

Kihle (1983) suggests that reservoirs may also be present in the Maputo, Domo sands, Grudja sands and Cherinyoma limestones.

(3) Are sealing conditions present? The Lower Cretaceous and certainly the Upper Cretaceous shales are potential seals. In most cases the point in time at which generation from the lowermost potential sources occurs is after the deposition of these shales (Fig. 17a, b), which would further enhance potential.

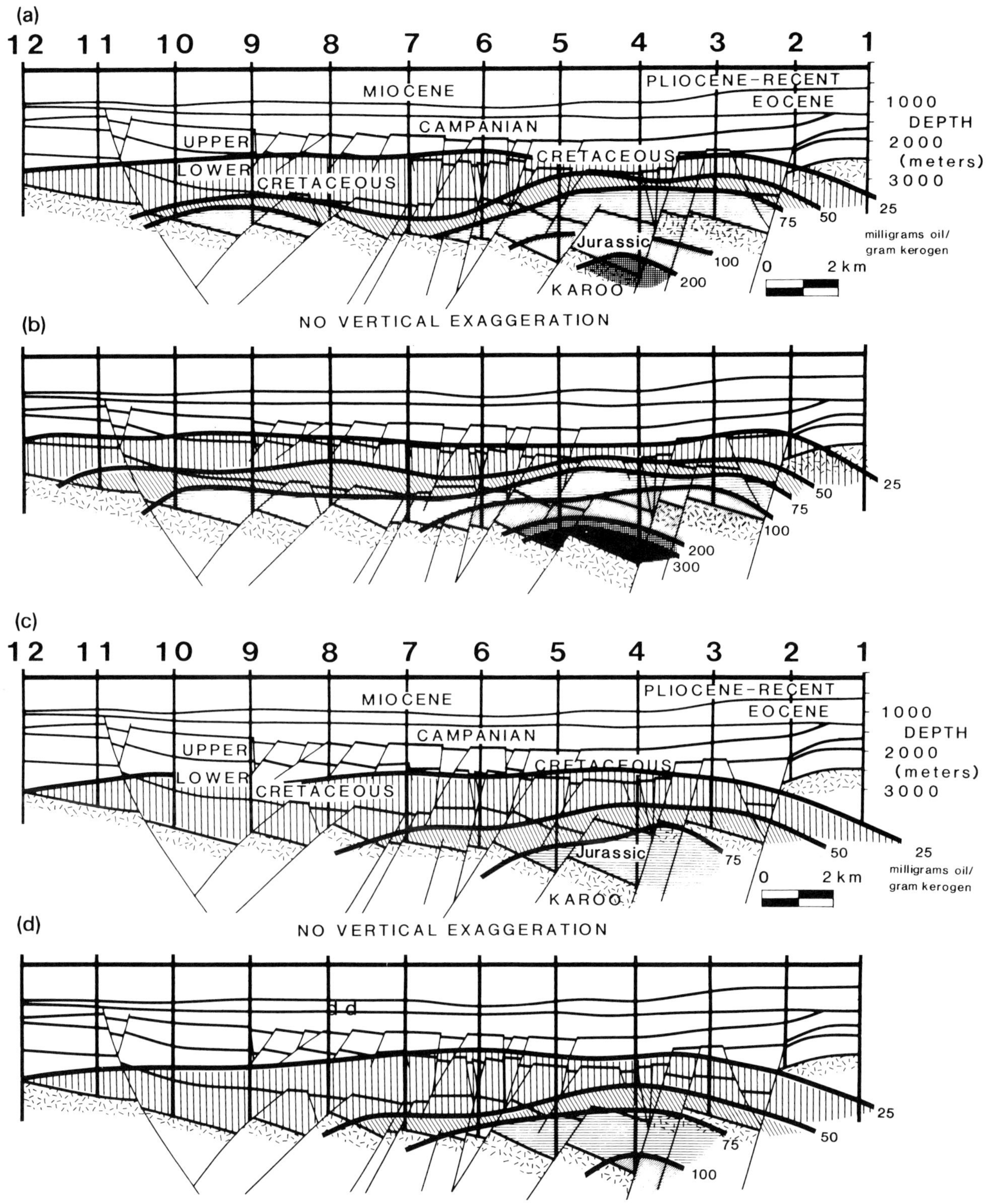

Fig. 15. (a) Royden et al. model, lower limit, case 2. (b) Royden et al. model, upper limit, case 2. (c) McKenzie model, lower limit, case 2. (d) McKenzie model, upper limit, case 2.

(4) Is there a source rock? This is a question which may only be answered definitively with a well. However, the fact that copious shales are present (of both terrigenous and marine provenances) enhances the possibility of a high quality organic source shale being present. Flores (1973) refers to the lower Domo shales as being euxinic in nature, which would increase source potential, and Hartnady (1990) suggests that the Limpopo drainage system is part of a large prograding delta, which also improves the hydrocarbon potential of the region. Marine Jurassic rocks have been reported further north and have

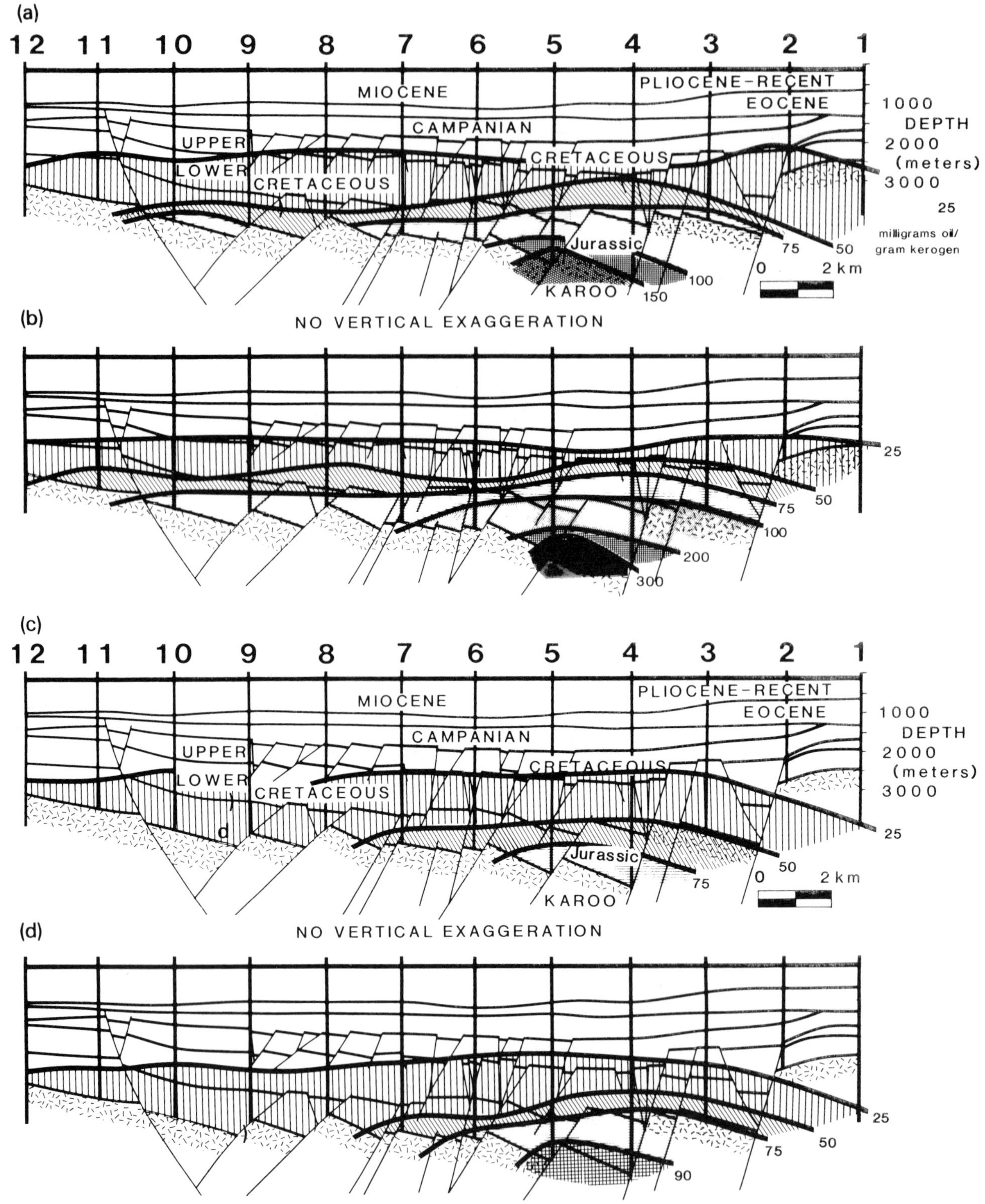

Fig. 16. (a) Royden et al. model, lower limit, case 3. (b) Royden et al. model, upper limit, case 3. (c) McKenzie model, lower limit, case 3. (d) McKenzie model, upper limit, case 3.

been suggested by Kent et al. (1971) to be present in the offshore rifts.

(5) Has the source rock adequately matured? Based on the limits of these analyses, which likely encompass the real situation, hydrocarbons will have been produced from Lower Cretaceous or Jurassic potential source rocks.

The advantage of the technique is in the assimilation of all aspects of the basin to tie down the relative timing of structural, depositional and thermal events.

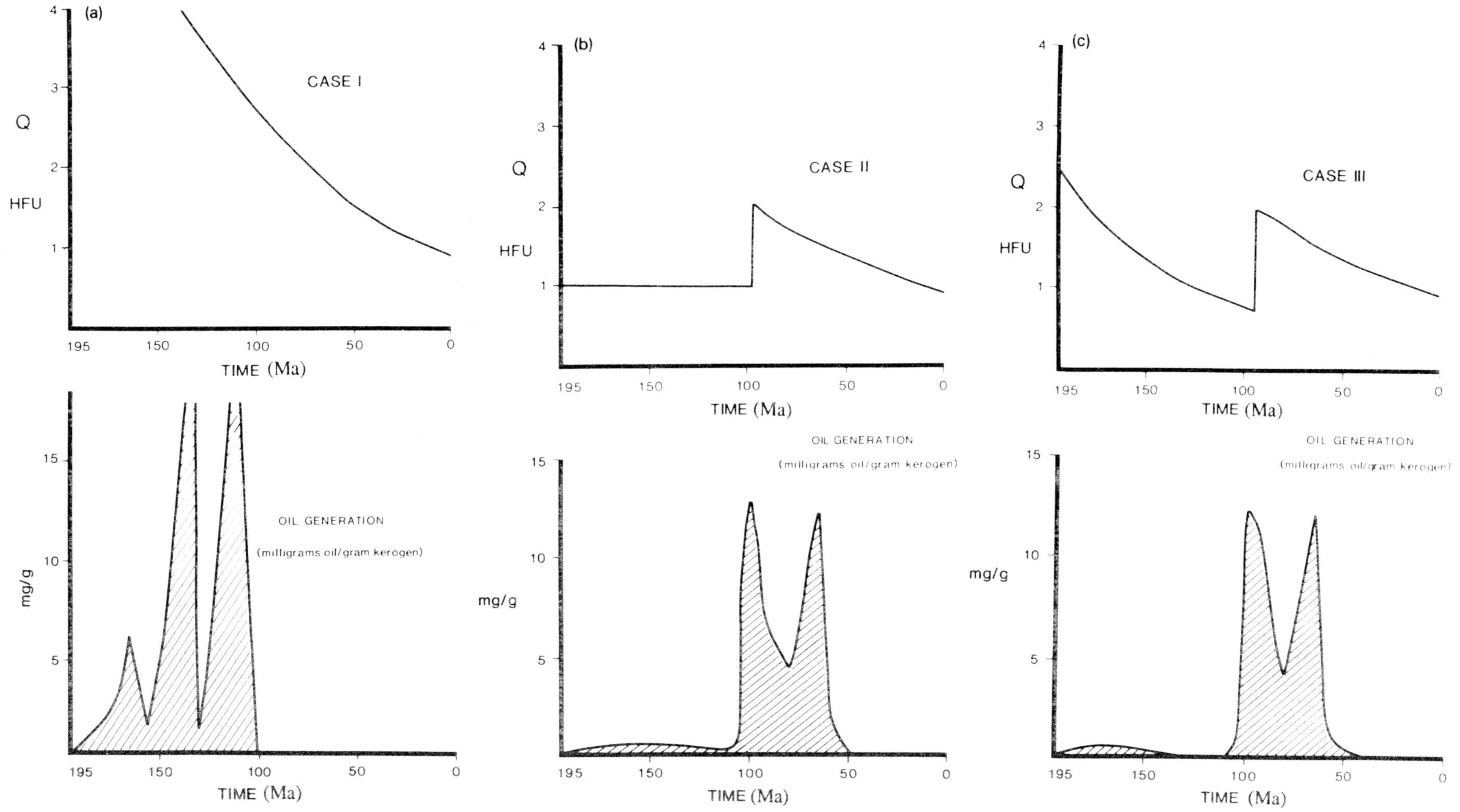

Fig. 17. (a) Oil generation rate of the Royden et al. model, upper limit, case 1, well 5. (b) Oil generation rate of the Royden et al. model, upper limit, case 2, well 5. (c) Oil generation rate of the Royden et al. model, upper limit, case 3, well 5.

It is important to determine whether maturation of source occurred before or after deposition of a sealing layer, or whether traps were formed and sealed before maturation. The South Mozambique Graben has potential source rocks most likely maturing after sealing rock (Upper Cretaceous shales) and traps are in place (Fig. 17a, b).

The biggest problem encountered with an oil-probable region, such as the South Mozambique Graben, is potential recovery. The porosity at depths of between 3000 and 4000 meters is not too promising unless overpressured conditions are experienced. Up dip stratigraphic traps may be more prolific.

The questions which cannot be answered are those of hydrocarbon migration and accumulation, although we can show that generation occurred mostly around the time of deposition of the Upper Cretaceous shale and generally after faults had become inactive. Perhaps the migration potential could be assessed using a two-dimensional fluid flow/compaction model such as that of Nakayama (1987), but the lack of more refined data renders such an endeavor less valuable at the present stage of understanding of the South Mozambique Graben.

Conclusions

Theoretical testing

(1) When wells which intersect faults are one-dimensionally backstripped, strata below the faults are improperly restored. This was solved by assuming that structural "erosion" takes place below the fault.

(2) The changing burial history in a backstripped well, due to contraction of the cross-section during reconstruction, was accommodated by "redrilling", and recalculating the burial history for each time step.

(3) Sensitivity analysis on backstripping planar faulted extensional scenarios revealed that shallow faults are more affected by compaction than steeper faults, the angle of the fault has more variation up the fault in shales than in sandstones, and as the scale of the system gets larger so does the apparent variation.

Tectonic and thermal implications

(1) The structural history of the South Mozambique Graben has two tectonic episodes recognizable on the seismic reflection profiles and subsidence histories of the pseudowells. The first event is of strike-slip nature, possibly sympathetic to the emplacement of Madagascar. The second event is a more normal extension manifested on the seismic section as a sequence of domino-type fault blocks in the Cretaceous section. This type of structural history has recently been reported by Mougenot et al. (1986) from the area just to the north.

(2) Extension rates estimated from variation in strata lengths show an acceleration of extension at about 96 Ma, which also coincides with the point at which the depth to detachment was calculated to be at its shallowest.

(3) Of the three different heat flow histories calculated from the subsidence models of Royden et al. (1980) and McKenzie (1978), the Royden et al. model appears to be more favorable from the maturation standpoint. Also, the later of the two rifting events modelled has more effect on source rock maturation.

Hydrocarbon potential

(1) Structural and stratigraphic traps are present due to the extensive faulting.

(2) Reservoirs may be present in syn-rift sands or Cretaceous rocks.

(3) The Upper Cretaceous shales would suffice as an effective seal.

(4) A source rock which can reach maturity for hydrocarbons likely lies somewhere within the lower Cretaceous or Jurassic strata. The probability of such a source rock being present is enhanced by the marine/terrigenous shale depositional setting for these strata.

(5) Regardless of the model used or heat flow history configuration, the basin is expected to have mature hydrocarbons in the eastern part at a depth of 3000–4000 meters.

(6) The ideal situation of: relative timing of source deposition followed by faulting, then seal deposition followed by maturation of the source, seems to be the case in the South Mozambique Graben.

Acknowledgements

The work reported here was supported by the Industrial Associates of the Basin Analysis Group at USC. Data were generously provided by the Western Geophysical Co. of America. Many thanks to Donna Black for typing and to Elain Hadaddin, Ann Watkins and Mike Garbee for drafting services. The paper benefited considerably from the comments of the anonymous reviewers, to whom we extend our thanks.

References

Anderson, R.N., Langseth, M. and Sclater, J.G., 1977. The mechanisms of heat transfer through the floor of the Indian Ocean. J. Geophys. Res., 82: 3391–3409.

Cao, S., 1985. A quantitative dynamic model for basin analysis and its application to the Northern North Sea Basin. M.S. Thesis, University of South Carolina, Columbia, S.C.

Cao, S., 1988. Sensitivity analysis of 1-D dynamical model for basin analysis. Ph.D. Dissertation, University of South Carolina, Columnia, S.C.

Chatellier, J.-Y. and Slevin, A., 1988. Review of African petroleum and gas deposits. J. Afr. Earth Sci., 7: 561–578.

Coffin, M.F. and Rabinowitz, P.D., 1987. Reconstruction of Madagascar and Africa: evidence from the Davie Fracture Zone and Western Somali Basin. J. Geophys. Res., 92: 9385–9406.

Dahlstrom, C.D.A., 1969. Balanced cross sections. Can. J. Earth Sci., 6: 743–757.

Darracott, B., 1974. On the crustal structure and evolution of southeastern Africa and the adjacent Indian Ocean. Earth Planet. Sci. Lett., 24: 282–290.

DeBuyl, M. and Flores, G., 1986. The Southern Mozambique Basin: the most promising hydrocarbon province offshore East Africa. In: M.T. Halbouty (Editor), Future Petroleum Provinces of the World. Am. Assoc. Pet. Geol., Mem., 40: 399–425.

Flores, G., 1970. Suggested origin of the Mozambique Channel. Trans. Geol. Soc. S. Afr., 73: 1–16.

Flores, G., 1973. The Cretaceous and Tertiary sedimentary basins of Mozambique and Zululand: In: G. Glant (Editor), Sedimentary Basins of the African Coasts, Part 2: South and East Coasts. Association of the African Geological Survey, Paris, pp. 81–111.

Forster, R., 1975. The geological history of the sedimentary basin of southern Mozambique, and some aspects of the origin of the Mozambique Channel. Palaeogeogr., Palaeoclimatol. Palaeoecol., 17: 267–287.

Gibbs, A.D., 1983. Balanced cross-sections construction from the seismic sections in areas of extensional tectonics. J. Struct. Geol., 5: 153–160.

Guidish, T.M., Kendall, C.G.St.C., Lerche, I., Toth, D.J. and Yarzab, R.F. 1984. Basin evaluation using burial history calculations: an overview. Am. Assoc. Pet. Geol., Bull., 69: 92–105.

Harding, T.P., 1974. Petroleum traps associated with wrench faults. Am. Assoc. Pet. Geol., Bull., 58: 1290–1304.

Hartnady, C.J.H., 1990. Gondwana break-up, Limpopo drainage-system evolution and the Kalahari Basin–Mozambique Delta connection. J. Afr. Sci., 7.

Kamen-Kaye, M., 1978. Permian to Tertiary faunas and palaeogeography: Somalia, Kenya, Tanzania, Mozambique, Madagascar, South Africa. J. Pet. Geol., 1: 79–101.

Kamen-Kaye, M., 1982. Mozambique-Madagascar Geosyncline, I: deposition and architecture. J. Pet. Geol., 5: 3–30.

Kent, P.E., Hunt, J.A. and Johnstone, D.W., 1971. The geology and geophysics of coastal Tanzania. Inst. Geol. Sci., London, Geophys. Paper, 6: 101.

Kihle, R., 1983. Recent surveys outline new potential for offshore Mozambique. Oil Gas J., Feb. 28: 126–134.

Lafourcade, P., 1984. Étude géologique et géophysique de la Marge Continentale du sud Mozambique (17°S à 28°S). Ph.D. Dissertation, Laboratoire de Géodynamique Sous-Marine, Villefranche-sur-Mer, Université Pierre et Marie Curie, Paris.

Lawver, L.A. and Scotese, C.R., 1987. A revised reconstruction of Gondwana land. In: G.D. McKenzie (Editor), Gondwana Six: Structure, Tectonics and Geophysics. Am. Geophys. Union, Geophys. Monogr., 40: 17–83.

Lerche, I., Yarzab, R.F. and Kendall, C.G.St.C., 1984. Determination of paleoheatflux from vitrinite reflectance data. Am. Assoc. Pet. Geol., Bull., 68: 1704–1717.

Magara, K., 1976. Water expulsion from elastic sediments during compaction-direction and volumes. Am. Assoc. Pet. Geol., Bull., 60: 543–553.

Martin, A.K., Goodland, S.W., Hartnady, C.J.H. and Du Plessis, A., 1982. Cretaceous palaeopositions of the Falkland Plateau relative to southern Africa using Mesozoic seafloor spreading anomalies. Geophys. J. R. Astron. Soc., 71: 567–579.

Martin, A.K. and Hartnady, C.J.H., 1986. Plate tectonic development of the South West Indian Ocean: a revised reconstruction of East Antarctica and Africa. J. Geophys. Res., 91: 4767–4786.

McKenzie, D., 1978. Some remarks on the development of sedimentary basins. Earth Planet. Sci. Lett., 40: 25–32.

Mougenot, D., Virlogeux, P., Vanney, J.-R. and Malod, J., 1986. La Marge continentale au Nord du Mozambique: résultats préliminaires de la campagne md40/macamo. Bull. Soc. Géol. Fr., 8: 419–422.

Nakayama, K., 1987. A dynamic, two dimensional, fluid flow model for basin analysis. Ph.D. Dissertation, University of South Carolina, Columbia, S.C.

Rabinowitz, P.D. and LaBreque, J.L., 1979. The Mesozoic South Atlantic Ocean and evolution of its continental margins. J. Geophys. Res., 84: 5973–6002.

Rabinowitz, P.D., Coffin, M.F. and Falvey, D., 1983. The separation of Madagascar and Africa. Science, 220: 67–69.

Royden, L., Sclater, J.G. and Von Herzen, R.P., 1980. Continental Margin subsidence and heat flow: important parameters in formation of petroleum hydrocarbons. Am. Assoc. Pet. Geol., Bull., 64: 173–187.

Sclater, J.G. and Christie, P.A.F., 1980. Continental stretching: an explanation of the post-mid-Cretaceous subsidence of the central North Sea basin. J. Geophys. Res., 85: 3711–3739.

Segoufin, J., 1978. Anomalies mésozoiques dans le bassin de Mozambique. C.R. Acad. Sci. Paris, 287: 109–112.

Segoufin, J. and Patriat, P., 1980. Existence d'anomalies Mésozoiques dans le bassin de Sanalie: implications pour les relations Afrique–Antarctique–Madagascar. C. R. Acad. Sci. Paris, Sér. B, 291: 85–88.

Simpson, E.S.W., Sclater, J.G., Parsons, B., Norton, I. and Meinke, L., 1979. Mesozoic Magnetic Lineations in the Mozambique Basin. Earth Planet. Sci. Lett., 43: 260–264.

Tchalenko, S., 1970. Similarities between shear zones of different magnitudes. Geol. Soc. Am. Bull. 81: 1625–1640.

Tissot, B. and Espitalie, J., 1975. L'évolution thermique de la matière organique des sédiments: applications d'une simulation mathématique. Rev. Inst. Fr. Pétrole, 30: 743–777.

Tissot, B. and Welte, D.M., 1978. Petroleum Formation and Occurrence. Springer-Verlag, New York, N.Y., pp. 500–521.

Veevers, J.J., 1986. Breakup of Australia and Antarctica estimated as mid-Cretaceous (95 ± 5 Ma) from magnetic and seismic data at the continental margin. Earth Planet. Sci. Lett., 77: 91–99.

J.E. ILIFFE *Department of Geological Sciences, University of South Carolina, Columbia, SC 29208, U.S.A.*

I. LERCHE *Department of Geological Sciences University of South Carolina, Columbia, SC 29208, U.S.A.*

M. DeBUYL *Western Geophysical, 10001 Richmond Avenue, Houston, TX 77252, U.S.A.*

The Danish North Sea: a basin prognosis

R.O. Thomsen, I. Lerche and J.A. Korstgård

Applying one- and two-dimensional fluid-flow/compaction models to the NW part of the Danish Central Graben, the geohistory, thermal history, source capacity and oil generation were examined. The burial history shows that the NW Danish Central Graben developed through several stages of subsidence — a Late Jurassic differential stage, a Late Cretaceous/Early Tertiary uniform stage and a Late Miocene/Quaternary uniform stage. The thermal history is assessed by inversion of measured vitrinite reflectance with depth and all wells give a consistent cooler past; resolution of the available data on the thermal history is discussed. Pseudo-wells have been "drilled" on seismic sections in areas without well data. Synthetic vitrinite reflectance data have been generated for the pseudo-wells using the modelled thermal history. The sensitivity of the maturity and thermal history modelling to changes in parameters such as erosion events, errors in the present day heat flow and paleo-overpressure evolution is discussed. The modelled maturation history of the Late Jurassic shale shows that the shale reached maturity some 5–50 Ma. Modelling of the hydrocarbon generation indicates that the peak generation took place some time between 10 Ma and the present day. Possible migration paths are determined from modelled excess fluid pressure and areas of possible accumulation of hydrocarbons are indicated. The potentials of the areas are evaluated and an area along the eastern boundary fault between the Tail End Graben and the Ringkøbing-Fyn High is suggested as a prospective new target for further exploration.

Introduction

The task of petroleum exploration is becoming progressively more difficult in a time where the more obvious structural reservoirs have been discovered. Exploration targets will progressively become either more subtle stratigraphic reservoirs in well explored basins or reservoirs in frontier basins in areas where exploration is both costly and logistically difficult. Within the range of uncertainties in the basic geologic assumptions computer aided models provide us with the ability to perform an integrated quantitative evaluation of the hydrocarbon potential of an area.

Although dynamical fluid-flow models for use in quantitative basin analysis have been available for a number of years (Welte and Yukler, 1981; Ungerer et al., 1984; Cao, 1985; Cao et al., 1986; Cao and Lerche, 1987; Nakayama and Lerche, 1987; Wei and Lerche, 1988) these models are only slowly being integrated in the exploration process.

Many advantages lie, however, in the use of dynamical fluid-flow models. In particular such models provide a tool not only for predicting the hydrocarbon potential of a basin, but also for testing different hypotheses concerning the development of the basin based on implications from the available data. In this way a confidence range can be put on the assessed thermal history and fluid-flow conditions, on the predicted hydrocarbon generation timing and locations, and also on hydrocarbon migration and on possible accumulation sites.

Both a one-dimensional and a two-dimensional model have been applied to the Danish North Sea. The one-dimensional model was used as a first approximation, with the two-dimensional model applied to an area which appeared promising through the one-dimensional study. For details about the one-dimensional model the reader is referred to Cao (1985), Cao et al. (1986) and Cao and Lerche (1987). For details of the two-dimensional model the reader is referred to Nakayama and Lerche (1987) and Wei and Lerche (1988).

Geologic setting

The Central Graben developed as part of the North Sea system of rifts and is in itself a complex multi-stage rift system comprised of a series of highs, grabens and half grabens. The division of the area into structural elements has been adapted from Vejbæk and Andersen (1987) (Fig. 1). Evolution of an early fracture system is believed to have originated in the Early Permian (Andersen et al., 1982; Glennie, 1984). The area, however, lacks both well and seismic control of the pre-Permian basinal regime. Thus, it is not possible at present to determine whether or not the system was active at

Structural and Tectonic Modelling and its Application to Petroleum Geology edited by R.M. Larsen, H. Brekke, B.T. Larsen and E. Talleraas. NPF Special Publication 1, pp. 439–455. Elsevier, Amsterdam.

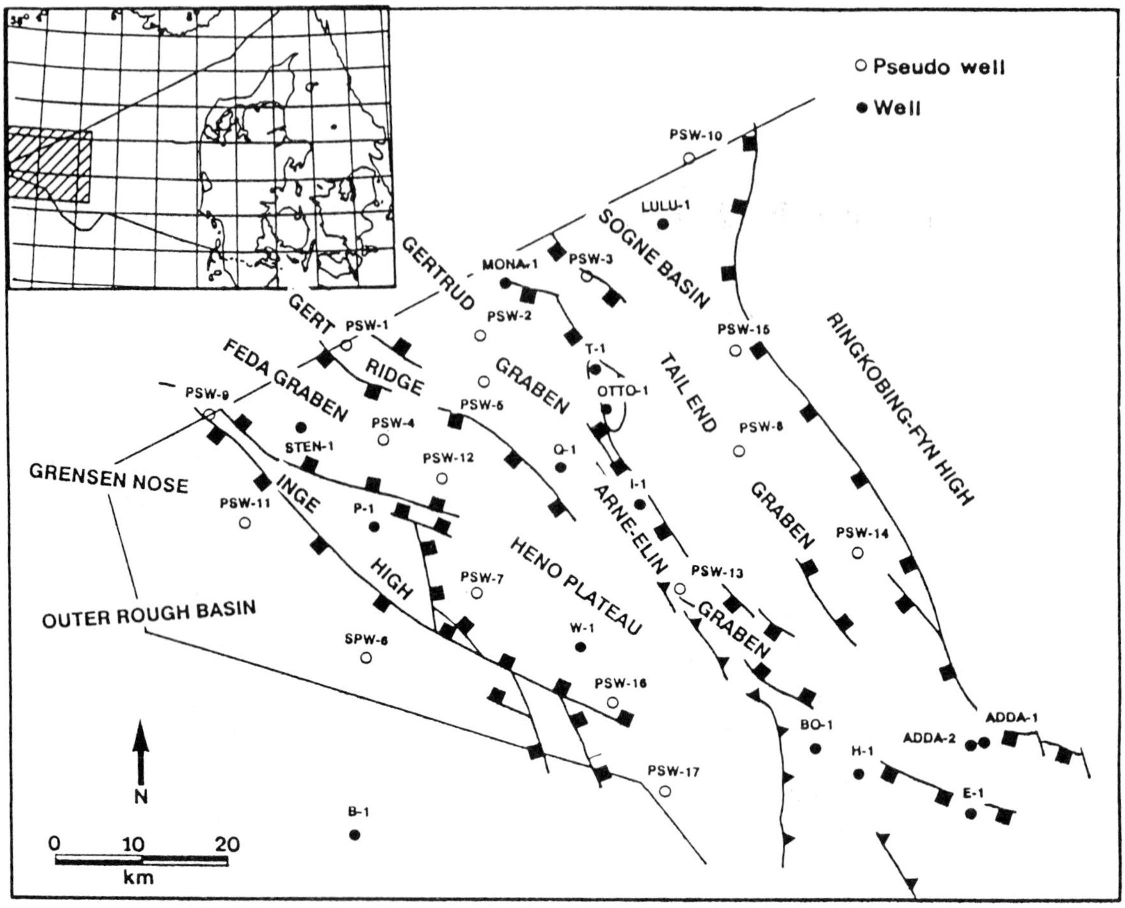

Fig. 1. The main structures and the commonly used subdivision into structural elements in the northern part of the Danish Central Graben. Also shown is the distribution of wells (solid dots) and pseudo-wells (open circles) used in this study. The structural outline is based on Vejbæk and Andersen (1987).

this time (Gowers and Sæbøe, 1985). It would seem that the opening of the North Sea Grabens did not take place any sooner than late Permian or Early Triassic (Ziegler, 1982). The presence of Rotliegendes volcanics in the Danish wells B-1, P-1, Q-1 and W-1, however, could indicate an early rifting stage prior to deposition of the Zechstein evaporites. The present day outline of the Zechstein sequence reveals deposition in a narrow graben trending NNW through the Tail End Graben-Søgne Basin and in the Feda Graben (Gowers and Sæbøe, 1985). The dominant lithologies in the Rotliegendes are volcanics and redbeds, while the Zechstein transgression resulted in marginal shallow water carbonates and thick evaporite sequences (Andersen et al., 1982).

The beginning of the break-up of the Atlantic 200–180 Ma ago resulted in a new structural framework in northwestern Europe, modifying the Permian structural pattern in the North Sea, with the Viking and Central Graben as new major structural elements (Fisher, 1984). The Triassic was marked by a regional regression and return to a continental depositional regime. Due to scarcity of both well data and good seismic resolution, combined with erosion during the Early Cimmerian and mid-Cimmerian tectonic events, very little can be said about the original sediment distribution in large parts of the Central Graben (Andersen et al., 1982).

At the end of the Early Jurassic the mid-Cimmerian tectonic phase induced renewed block faulting in the Northwest European graben systems (Ziegler, 1982). Tectonic uplift in the central North Sea area caused erosion of Early Jurassic sediments and these are now found only in local basins (Andersen et al., 1982). During the Middle and Late Jurassic the structural framework was dominated by N–S trending basement-attached faults and by NW–SE trending right lateral strike-slip faults (Møller, 1986). This new trend left the Søgne Basin inactive, while the major part of the Tail End Graben subsided as one large rotational fault block along its boundary to the Ringkøbing-Fyn High. The Late Cimmerian tectonic event caused a widespread diachronous unconformity which in large parts of the area constitutes the boundary between the Late Jurassic and Early Cretaceous sediments.

The Early Cretaceous setting is a continuation of that of the Jurassic and the depocenters appear to be related to those of the Late Jurassic (Gowers and Sæbøe, 1985). Transgressional sedimentation is evident from on-lap sequences on seismic sections, although internal unconformities occur close to the fault zone at the western margin of the Tail End Graben, suggesting tectonic activity in this area during the Early Cretaceous (Andersen et al., 1982). Tensional and left-lateral trans-tensional movements occurred along the boundary faults of the Gertrud and Arne-Elin Grabens giving rise to normal faulting (Vejbæk and Andersen, 1987).

The Late Cretaceous ushered in a marked change from predominantly siliciclastic sedimentation to deposition of chalk. Tectonically this change marks the transition from rift-dominated subsidence to passive regional subsidence (Vejbæk and Andersen, 1987). Compressional and wrench-induced inversion of previously tensional sub-basins dominated the Late Cretaceous–Early Tertiary with the main inversion occurring during Turonian–Santonian, mid-Maastrichtian and post-Danian–Early Tertiary times (Vejbæk and Andersen, 1987).

Regional subsidence continued and, during the Tertiary and Quaternary, the basin subsided evenly and rapidly. This subsidence pattern was disturbed only by minor tectonic events and, locally, by halokinetic movements (Andersen et al., 1982). The Early Tertiary sedimentation was mainly clay dominated, while sand and silt intercalations increase towards the Late Tertiary (Andersen et al., 1982). A general stratigraphic column with formation names is shown in Fig. 2.

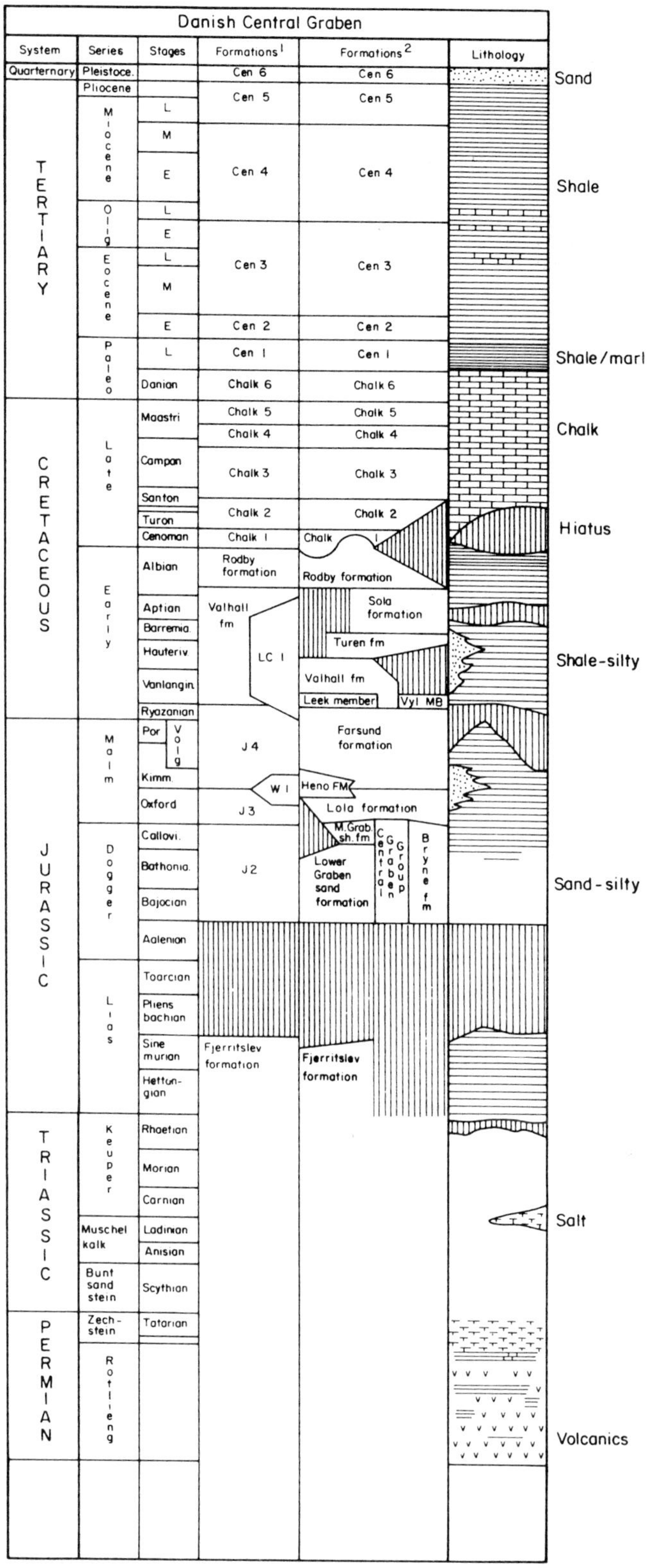

Fig. 2. Generalized stratigraphic column (after Michelsen, 1982 and Michelsen et al., 1987).

The data base

A large amount of data is available in the open domain from publicly released wells and the types of data used in this study are summarized in Table 1. Although there is an uneven spatial distribution of wells with data, most of the structural elements are represented by well data (Fig. 1). To improve the data distribution and to include all structural elements, the seismic survey RTD-81 (GECO, 1981) has been used for "drilling" pseudo-wells with synthetic data. The distribution of wells and pseudo-wells used is shown in Fig. 1. In addition, line RTD-81-23 has been selected for a detailed two-dimensional simulation (Fig. 3). Dating of the individual reflectors is based on a correlation of biostratigraphy and seismic markers for selected wells in the northwestern part of the Danish Central Graben (Stouge, 1988; Stouge and Toxwenius, 1988). Furthermore, an effort has been made to make the seismic interpretation of the pre-Tertiary consistent with the works of Vejbæk and Andersen (1987) and Møller (1986).

For simplicity, an average interval velocity for each sequence was chosen for depth conversion of the seismic sections. Reflector ages, lithologies derived

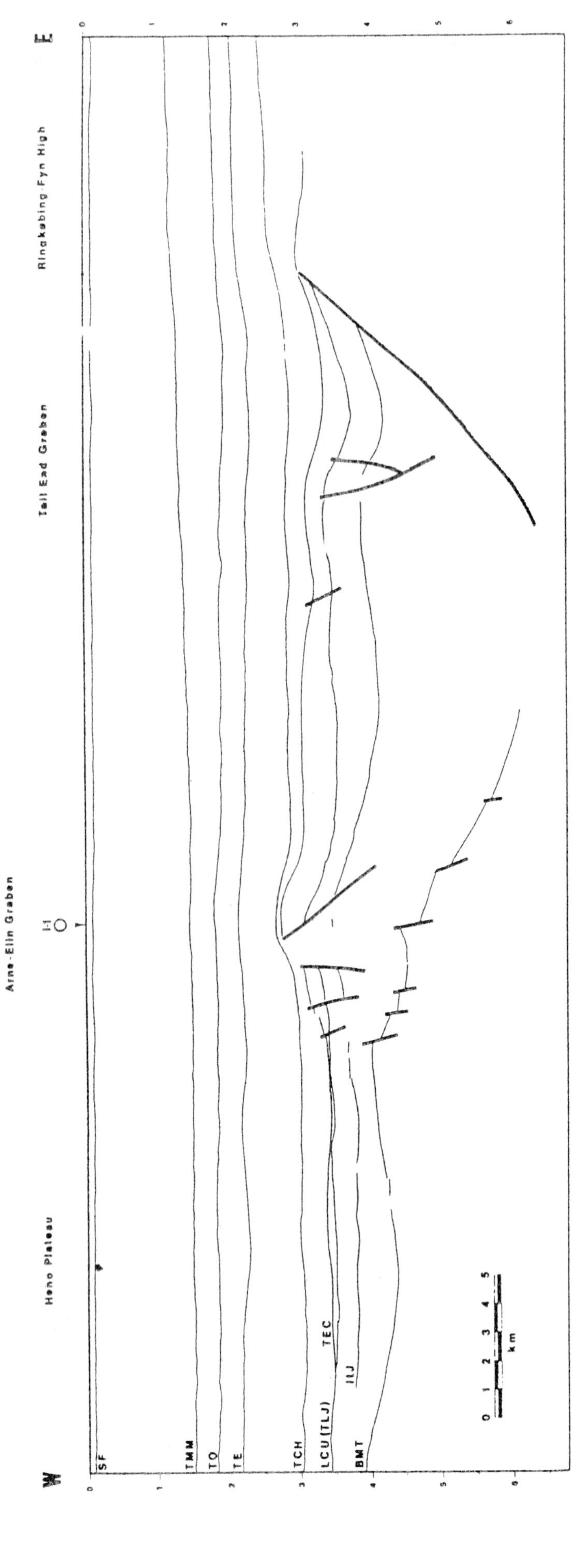

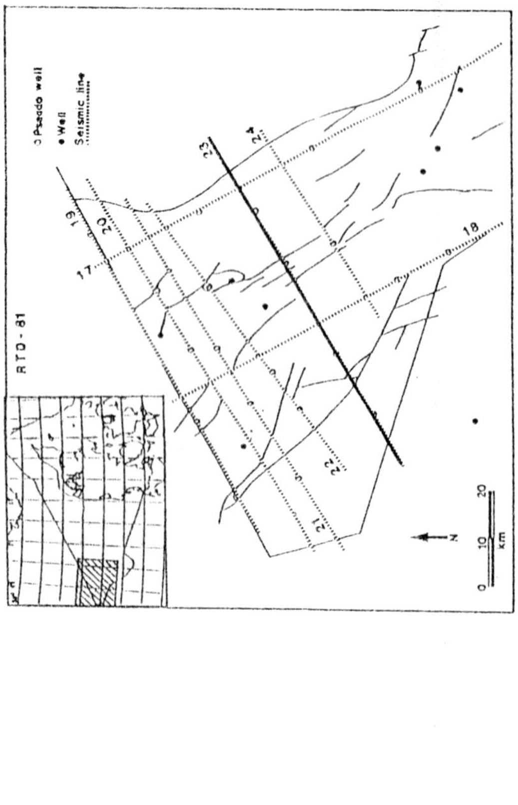

Fig. 3. Interpreted seismic section (RTD-81-23). *SF* = sea floor; *TMM* = Top Middle Miocene; *TO* = Top Oligocene; *TE* = Top Eocene; *TCH* = Top Chalk; *LCU* = Late Cimmerian Unconformity; *TLJ* = Top Late Jurassic; *TEC* = Top Early Cretaceous; *ILJ* = Internal Late Jurassic; *BMT* = Basement.

TABLE 1

Well data used in this study (for location of wells see Fig. 1)

	Stratigraphical log	Wireline log	Composite log	Well data summary	Temperature	Porosity	Vitrinite reflect.	Permeability	Geochemistry
Lulu-1	–	×	×	×	×	–	×	–	×
Mona-1	–	–	×	–	–	–	×	–	×
Sten-1	–	×	×	–	–	–	×	–	×
P-1	×	×	–	×	×	×	–	–	–
Q-1	×	×	–	×	×	×	×	×	×
Otto-1	–	×	×	×	–	–	–	–	–
T-1	×	×	–	×	×	×	–	×	–
I-1	×	×	–	×	×	×	×	×	×
B-1	×	×	–	×	×	×	×	–	×
Bo-1	–	×	–	×	×	–	×	–	×
H-1	–	–	–	–	×	×	–	–	–
Adda-1	–	–	–	–	×	×	–	–	–
Adda-2	–	–	–	–	–	–	–	–	–
W-1	×	×	×	–	×	×	×	–	×
E-1	×	×	–	–	–	–	×	–	×

from well ties and the interval velocities used in the depth conversion are shown in Table 2. The present day temperature-depth relations at the pseudo-wells have been derived from a previous temperature mapping of the Danish Central Trough (Klint Jensen, 1983).

The geochemical character of the Late Jurassic shales was examined from pyrolysis data (E. Thomsen, 1988) with the objective of discriminating the proportions of kerogen types I, II and III in the shales. Consistent with earlier studies (Lindgreen et al., 1982; E. Thomsen, 1989; R. Thomsen, 1989), the Kimmeridge clay equivalent (Farsund Formation) shows good to excellent source capacity, and data favor type II kerogen with some type III kerogen (Fig. 4) (Lindgreen et al., 1982; Goff, 1983; E. Thomsen et al., 1983; Conford, 1984; Waples, 1985; E. Thomsen, 1989; R. Thomsen, 1989). The underlying Oxfordian clay has less source capacity than the Kimmeridge clay equivalent and is gas prone (Conford, 1984). As input for the hydrocarbon generation model we used the proportions of different kerogen types assessed by the geochemical study (R. Thomsen, 1989) for the Upper Jurassic shales (Table 3). Elsewhere, values of 50/50 type II/III kerogen were assigned.

TABLE 2

Reflector ages, lithologies derived from well ties and interval velocities used in the depth conversion

Reflector	Age (Ma)	Lithology	Interval velocity (m/s)
1	0	water	1480
2	1.65	sand	2127
3	10.2	shale	1920
4	26.5	shale	2032
5	36	shale	2173
6	60.2	shale	2560
7	92–96	chalk	3891
8	128–131	shale	2525
9	136–145	shale	2908
10	?–175	sandy shale	3385

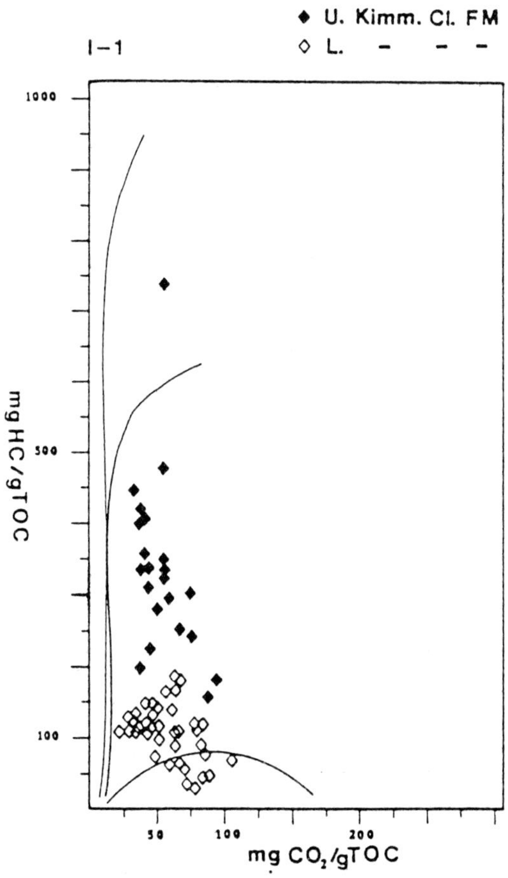

Fig. 4. Modified van Krevelen diagram for the Upper Jurassic in well I-1 showing mainly type II kerogen.

TABLE 3

Proportions of different kerogen types for the Upper Jurassic shales

Well	Type I	Type II	Type III
Lulu-1	0.0	0.4	0.6
Mona-1	0.0	0.3	0.7
Sten-1	0.1	0.3	0.6
Q-1	0.0	0.5	0.5
I-1	0.0	1.0	0.0
W-1	0.0	0.2	0.8
B-1	0.0	1.0	0.0
Bo-1	0.0	1.0	0.0
E-1	0.0	0.8	0.2

Source: R.Thomsen (1989).

Burial history-geohistory

Using the one-dimensional model, the basin subsidence was modelled and expressed in terms of basement subsidence rates in m/m.y. The results are presented as contour maps of the basement subsidence rate at different times in the past. This kind of mapping enables us to interpret the modelled subsidence in terms of tectonic activity of the basement.

The contour outline of Late Jurassic basement subsidence rates (Fig. 5a, b) is characteristic of strong differential subsidence with closed contours confined within distinct areas. Rates exceeding 30 m/m.y. occurred in the Tail End and Arne-Elin Grabens; in the northern Gertrud Graben a total basement subsidence of about 1100 m occurred at rates of over 40 m/m.y. The Søgne Basin and the eastern part of the Gertrud Graben both subsided at lower rates than the Tail End Graben. The Inge High was at this stage already established as a structural high, while the adjacent Feda Graben, Outer Rough Basin and the Heno plateau underwent subsidence at intermediate rates (15 m/m.y.). The Gert Ridge subsided only slowly (4–10 m/m.y.). The Late Jurassic structural pattern was mainly NNW–SSE with minor NW–SE striking faults. Movement was along the boundary fault between the Søgne Basin-Tail End Graben and the area east of Gertrud Graben, normal faulting occurred along the Inge High boundary faults, and movement along the faults marking the outline of the Gert Ridge (Fig. 5a, b).

A significant change in the subsidence pattern is seen during the Early Cretaceous stage. Low to intermediate basement subsidence rates (5–15 m/m.y.) are dominant throughout most of the area. Maximum subsidence rates of 25 m/m.y. are seen in the Gertrud Graben, and 20 m/m.y. in the central part of the Tail End Graben and in the southern part of the Arne-Elin Graben (Fig. 5a, b). The Gertrud Graben developed as a half graben with decreasing subsidence towards the southern part of the graben. The overall structural pattern changed to a more NW–SE trend.

During the early Late Cretaceous the subsidence pattern became more uniform in character. The major subsidence occurred south of the Inge High across the Heno Plateau and the Outer Rough Basin where subsidence rates exceeding 35 m/m.y. are typical. In the Heno Plateau, Gertrud and Feda Grabens, subsidence rates reached 17–24 m/m.y., with the main subsidence where these structural elements meet. Most activity along the Jurassic faults had ceased at this stage, but renewed inverted movements along the boundary faults of the Arne-Elin Graben are inferred by a difference between subsidence rates in the Arne-Elin Graben and the surrounding areas (Fig. 5a, b).

This subsidence pattern continued with only minor changes throughout the Maastrichtian and Danian. Rapid subsidence with rates of 35–38 m/m.y. continued across the Outer Rough Basin, the Inge High, and the Heno Plateau, and in the former Gertrud Graben. In the former Arne-Elin Graben the rate of 15 m/m.y. outlines the NNW–SSE structural trend of inversion tectonics, dividing the area into eastern and western areas of major subsidence (Fig. 5a, b).

The post-Danian Tertiary is characterized by rapid regional subsidence with rates of 30–45 m/m.y. totalling 900–1300 m, interrupted only by a minor quiet period from Early Miocene to Late Miocene. The most rapid subsidence took place from post-Middle Miocene to Present, where the basement dropped more than 1100 m. During the Pleistocene subsidence rates extended from 55 m/m.y to near 300 m/m.y.

Thermal history

The thermal history is assessed differently in the two models, in the one-dimensional model by inversion (Lerche, 1988a, 1988b) of measured vitrinite reflectance with depth for several wells (Table 1) using an exponential heat flux relation:

$$Q(t) = Q_o \cdot \exp(\beta t) \tag{1}$$

where $Q(t)$ is the basement heat flow and Q_o is the present day heat flow, t is time and β is a constant determined by minimizing the difference between predicted and measured vitrinite reflectance with depth (Lerche et al., 1984). In the two-dimensional model the heat flow history is fixed as an input parameter.

The present day heat flow, Q_o, is derived from downhole temperature logs (real wells), and by contour interpolation (pseudo-wells). A contour map of

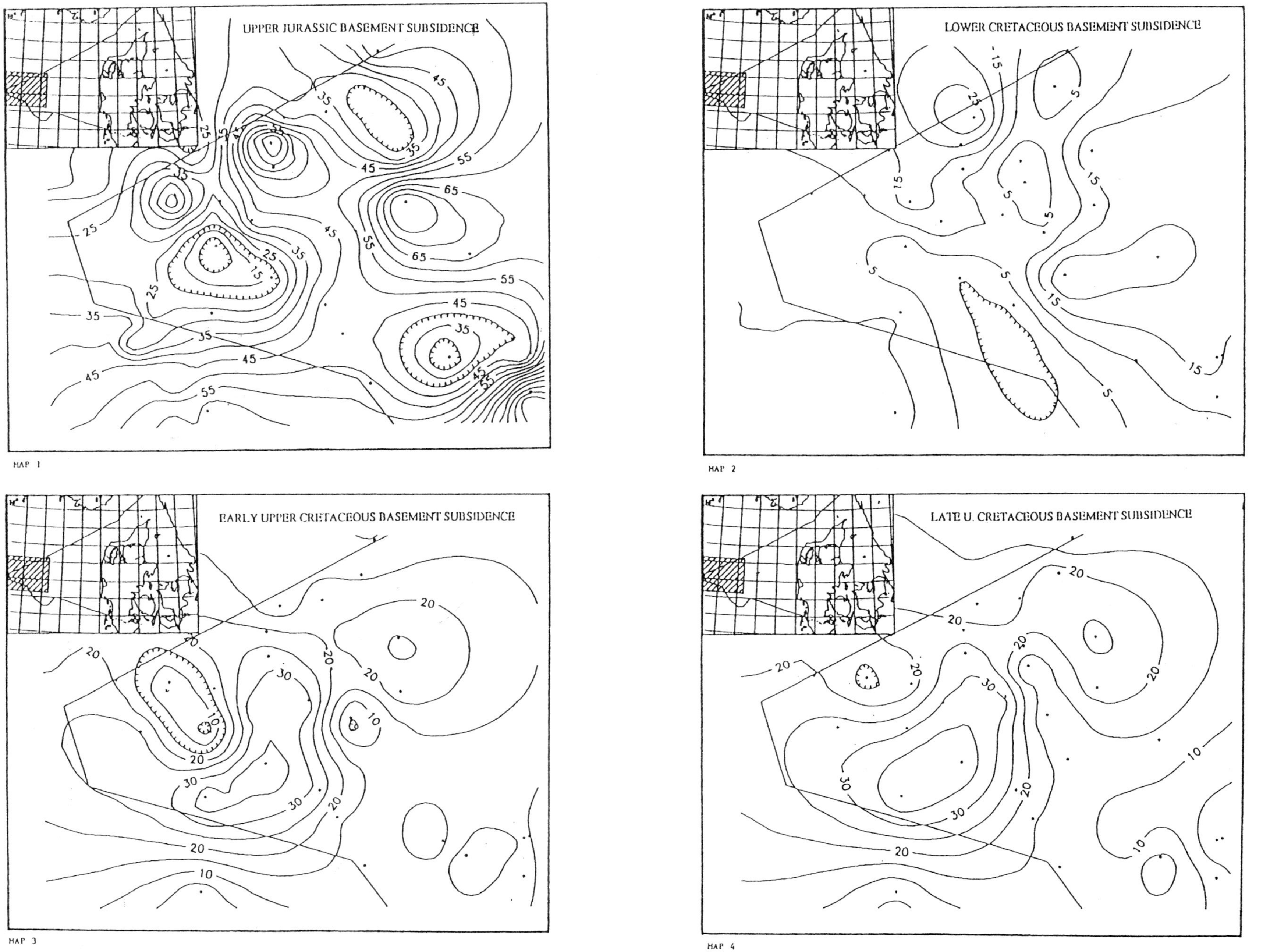

Fig. 5. (a) Subsidence patterns based on basement subsidence rates (in m/m.y.) for different time periods: Late Jurassic, Early Cretaceous, Early Late Cretaceous and Late Cretaceous-Danian.

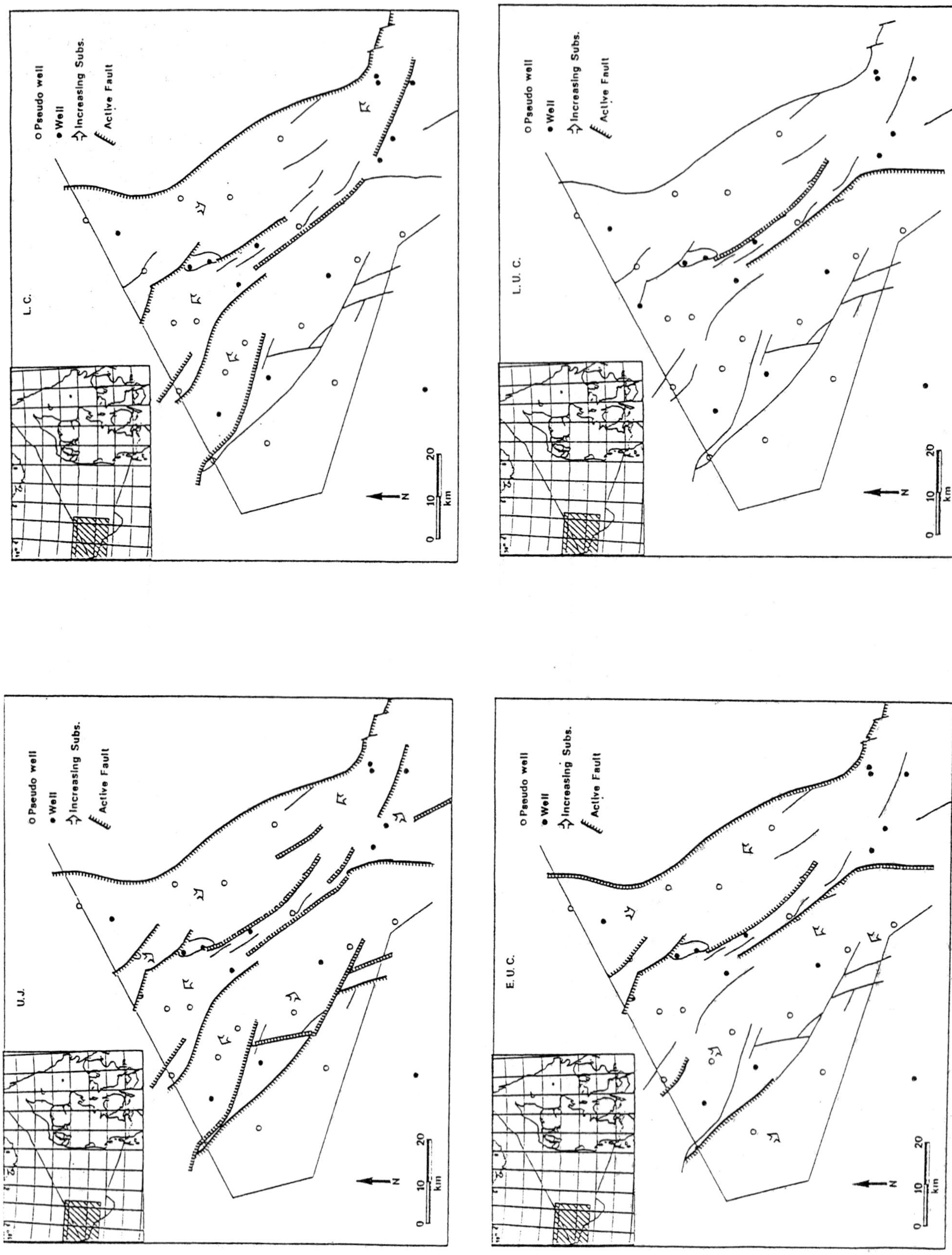

Fig. 5 (continued). (b) Fault activity for the same time periods as in (a).

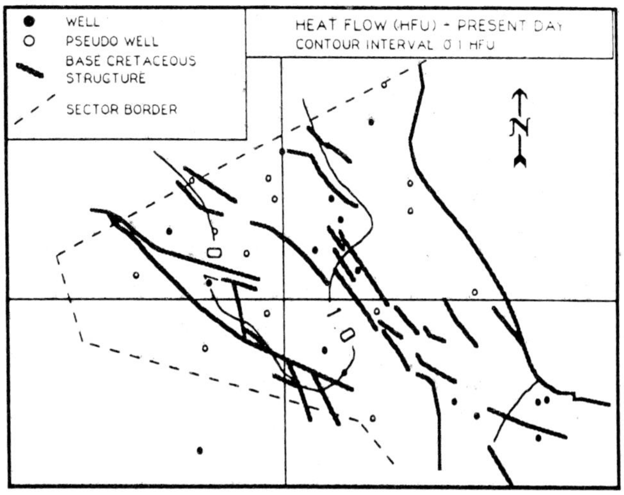

Fig. 6. Contour map of the present day heat flow.

the present day heat flow (Fig. 6) shows an almost uniform heat flow throughout the basin with a lowest heat flow in a N–S trending zone in the middle of the study area. The assessed low heat flow is due to overpressure in the shales causing a decreased thermal conductivity.

The modelled heat flow histories for the nine wells with R_o data show consistently cooler pasts with heat flows approaching zero between 35 and 80 Ma ago (Fig. 7). This fact implies that vitrinite reflectance has no resolution of the early heat flow history. The reason is the dramatic Tertiary drop of the basin causing rapid deep burial of the pre-Tertiary sediments. A consequence is rapid heating of the sediments, which is so dominant that the vitrinite reflectance data has lost resolution of any earlier heating. This behavior also indicates that any significant maturation must be a result of the very recent burial and heating.

Since the inversion of vitrinite reflectance data failed to resolve any early thermal history and the present day heat flow is close to 1 HFU (heat-flow units, 1 HFU = 10^{-6} cal/cm/s), a constant heat flow of 1 HFU was chosen as thermal history input in the two-dimensional model.

Maturity

The maturation history of a given source rock can be expressed in various ways such as a thermal

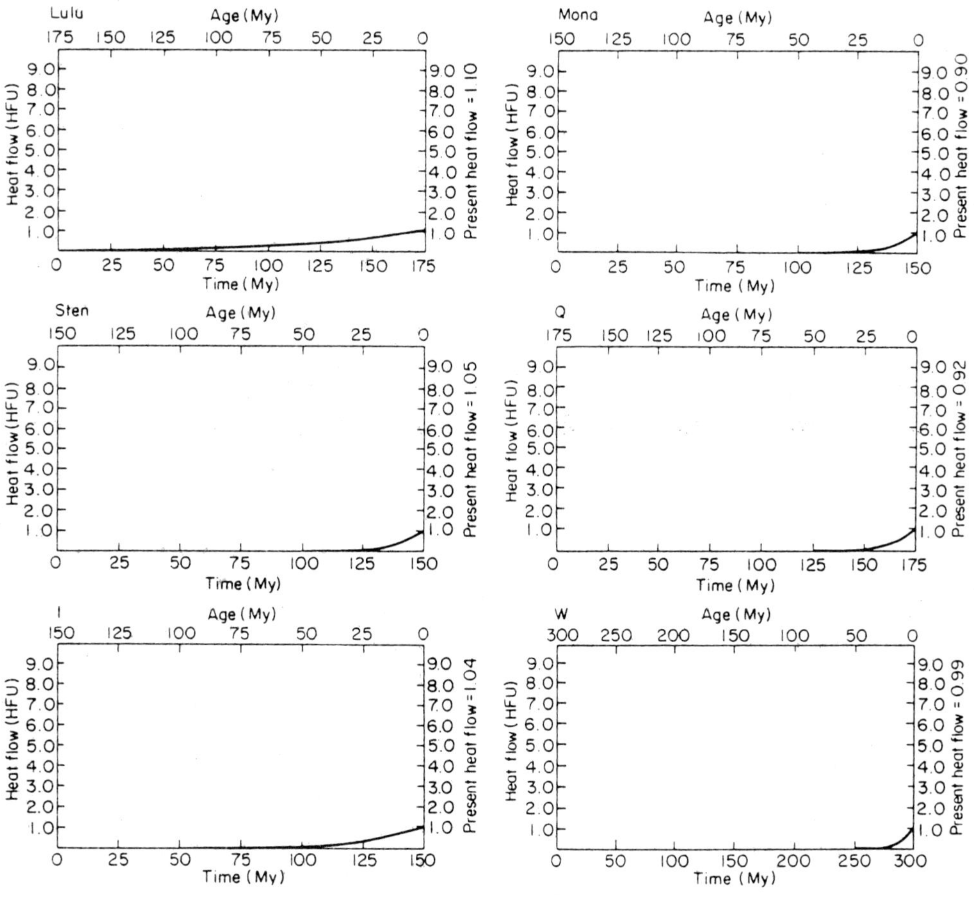

Fig. 7. Heat flow variation with time assessed from inversion of measured vitrinite reflectance.

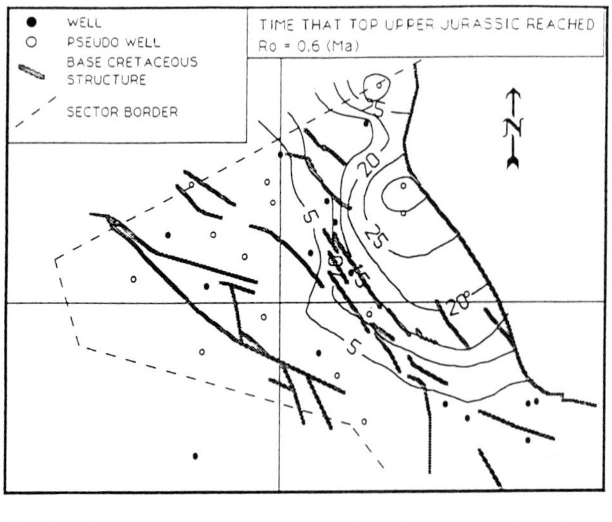

Fig. 8. Contours of the time at which the top of the Upper Jurassic reached a vitrinite reflectance of 0.6.

indicator value correlated to oil generation. Since vitrinite reflectance is the only thermal indicator available as a measure of basinal maturity, vitrinite reflectance variations with time and depth have been modelled for wells and pseudo-wells, using the one-dimensional model, with a focus on the Upper Jurassic shales. A vitrinite reflectance of 0.6 has been chosen as the onset of oil generation, and 1.35 as the end of oil generation and onset of wet gas generation (Waples, 1985).

A time contour map showing the time at which the Late Jurassic shales reached full maturity (R_o = 0.6) has been constructed (Fig. 8). Full maturity (R_o = 0.6) was reached first in the N and E 30 Ma, while the W, SE and SW reached full maturity only 5 Ma ago, indicating that maturation moved through the basin from north and east towards west, southwest and south. In the two-dimensional model the maturation of the area was expressed in terms of temperature evolution with the "oil window" chosen between 100°C and 150°C. This approach gives a consistent result with the Late Jurassic entering the "oil window" between 10 Ma and 1.7 Ma in the Tail End Graben, and between 1.7 Ma and present day in the western part of the modelled section (Fig. 9).

Oil generation

The hydrocarbon potential is assessed by modelling the oil generation through time using the simple scheme of Tissot and Welte (1978). The oil generation history is expressed as oil generation rate in mg/g/m.y. (Fig. 10) Contours of this measure for the Upper Jurassic sediments at different times show that no significant oil generation took place in the Jurassic sequence until between 10 Ma and present day. Present day oil generation rates show the highest generation rates in the western half of the area. At 10 Ma the highest oil generation rates are seen in the eastern part of the area, implying that oil generation had peaked in the eastern part sometime between 10 Ma and the present day, whereas the western parts of the area are approaching peak generation at present. This pattern is consistent with the maturation trend seen in the time contour map of the Late Jurassic reaching a vitrinite reflectance of 0.6 (Fig. 8). Compared to the implications obtained from vitrinite reflectance, with 0.6 chosen as the lower limit for onset of oil generation, a difference in timing is observed, with the oil generation model giving more recent maturation. Both results are model dependent, the difference giving a range of uncertainty in assessing the maturation history of the Upper Jurassic source rock.

In both cases, however, the modelled maturation stresses the significant impact of the Tertiary deep burial and resulting heating on the maturation of the Upper Jurassic source rock.

Fluid-flow pattern and possible migration paths

Once hydrocarbons are being generated in the source rock they tend to migrate out of the source rock towards possible reservoirs. It was long being believed that the buoyancy effect induced by the difference in specific gravity between oil (or gas) and water (known as the "gravitational" theory (Hubbert, 1953)) was the sole driving force for hydrocarbon migration. This belief is, however, only valid in the special case where the pore water is in hydrostatic equilibrium (Hubbert, 1953). The deposition, and later compaction, of sediments in a basin provides a major water flow throughout the basin's geological evolution. Thus, as hydrocarbons are generated in a basin they are subject to both the intrinsic buoyancy drive as well as to a significant hydrodynamic drive.

Consideration and evaluation of possible migration paths from the generating parts of the basin towards possible reservoirs should therefore include an evaluation of the fluid-flow pattern in the basin throughout the time of migration. The change in porosity, permeability, fluid pressure and excess fluid pressure through time was simulated with the purpose of monitoring temporal changes in the above parameters in the Upper Jurassic shales. Because the timing of hydrocarbon generation is between 30 Ma and the present day, these changes were only modelled through the Tertiary. Simulation of the porosity variation gave 47% at the top of the Jurassic shale and 27% at the bottom at 60 Ma, with a decreasing

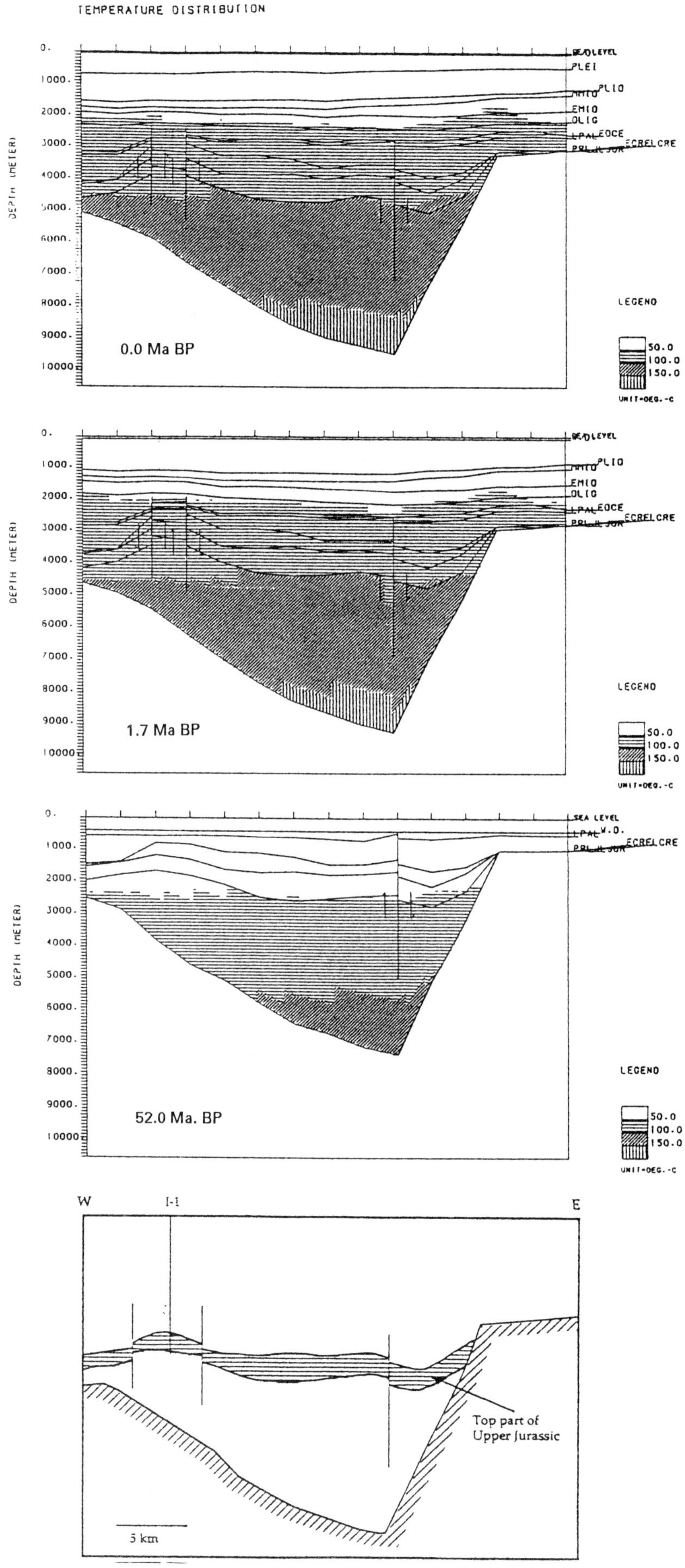

Fig. 9. Two-dimensional simulation of line RTD-81-23 showing the temperature distribution at 52 Ma, 1.7 Ma and present day. The Upper Jurassic reached a temperature of 100°C some 2 Ma ago.

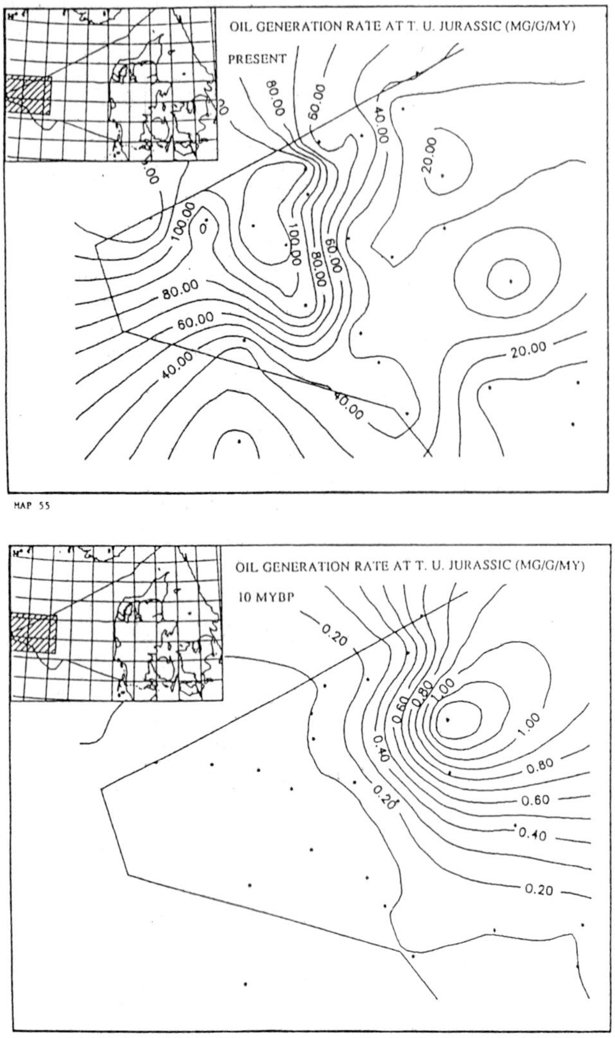

Fig. 10. Contour maps of oil generation rates (in mg/g/m.y.) for the Upper Jurassic shales at 10 Ma and the present day. The maps show that oil generation peaked in the eastern part between 10 Ma and the present day, while not reaching peak generation in the western part until the present day.

porosity with time until the present day yielding values of 32% at the top and 17% at the bottom of the shale. The modelled present day porosities lie within the range of porosities assessed from wireline logs in different wells (3%–30%) (Michelsen, 1982).

The modelled permeability gave a range from 2.10 mD (top) to 0.20 mD (bottom) at 60 Ma, decreasing from 0.35 mD (top) to 0.10 mD (bottom) at the present day. A permeability of 3 mD is reported from the J-2 Middle Jurassic sandy shale in the well U-1 southeast of the study area (Michelsen, 1982).

The excess fluid pressure is the driving force for fluid flow, with fluids flowing from high excess pressure towards low excess pressure. A significant build-up of excess fluid pressure in the Late Jurassic shales took place between 60 Ma and the present day. Maps of excess fluid pressure allow for an interpretation of the fluid-flow pattern in terms of a lateral component acting as a hydrodynamic drive on migrating hydrocarbons (Fig. 11) and reveal potential migration paths for hydrocarbons. The low excess pressure areas are in the Søgne Basin around the Lulu-1 well, across the southern Tail End Graben from the Adda wells to the wells H-1 and Bo-1, and in the Arne-Elin Graben around well I-1. High excess pressures are found in the deep parts of the central Tail End Graben and in the Gertrud and Feda Grabens, indicating a flow of fluids out of the central Tail End Graben, Gertrud Graben and Feda Graben towards the areas of low excess pressure.

The two-dimensional simulation illustrates this behavior slightly differently in that the highest excess fluid pressure gradient is seen out of the deepest part of the Tail End Graben towards the Ringkøbing-Fyn High, indicating a significant lateral component of flow in this area.

Evaluation

By examining the excess fluid pressure map and fluid-flow pattern in conjunction with the oil generation maps, it is seen that oil generation occurred in the areas of high excess pressure. Thus, the generated hydrocarbons flow towards the areas of low excess pressure following migration paths outlined by the fluid-flow pattern. The low excess pressure areas with converging flow act as accumulation sites and are shown in Fig. 12. Hydrocarbons generated in the Gertrud Graben migrate towards the Otto salt structure where the excess pressure favors possible accumulation only in the Upper stratigraphic levels (area *2* on Fig. 12). Hydrocarbons generated in the central Tail End Graben migrate towards the Lulu-1 well (area *1* on Fig. 12), the inverted structure of the Arne-Elin Graben (area *2* on Fig. 12), the eastern boundary fault along the Ringkøbing-Fyn High (area *4* on Fig. 12), and an area between the Adda wells, well H-1 and well Bo-1 (area *3* on Fig. 12).

A potential target

The most prospective areas, as indicated by the model, are shaded in Fig. 12. Three of these areas have already been drilled: (1) Lulu (Lulu-1); (2) Arne (I-1, Q-1, T-1, Otto-1); (3) Roar and Adda (H-1, Bo-1 and Adda-1, Adda-2) with hydrocarbons found. Hydrocarbon migration towards the eastern boundary fault of the Tail End Graben along the Ringkøbing-Fyn High indicates possible accumulation sites in the area marked *4* in Fig. 12. Any hydrocarbon accumulations in this area would likely be sealed either by the tight chalk or, alternatively,

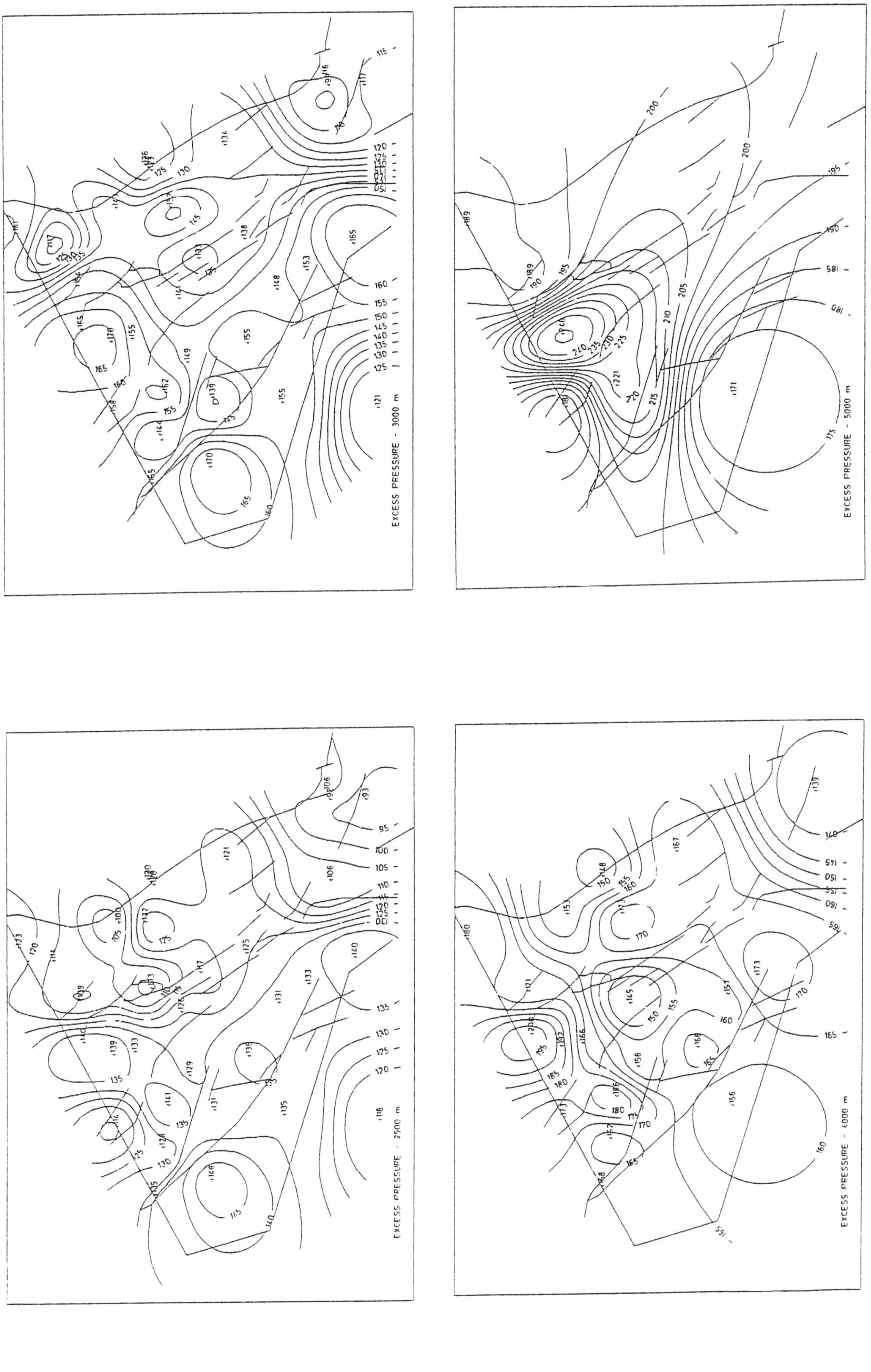

Fig. 11. (a) Contour maps of the excess fluid pressure for the depths 2500 m, 3000 m, 4000 m and 5000 m.

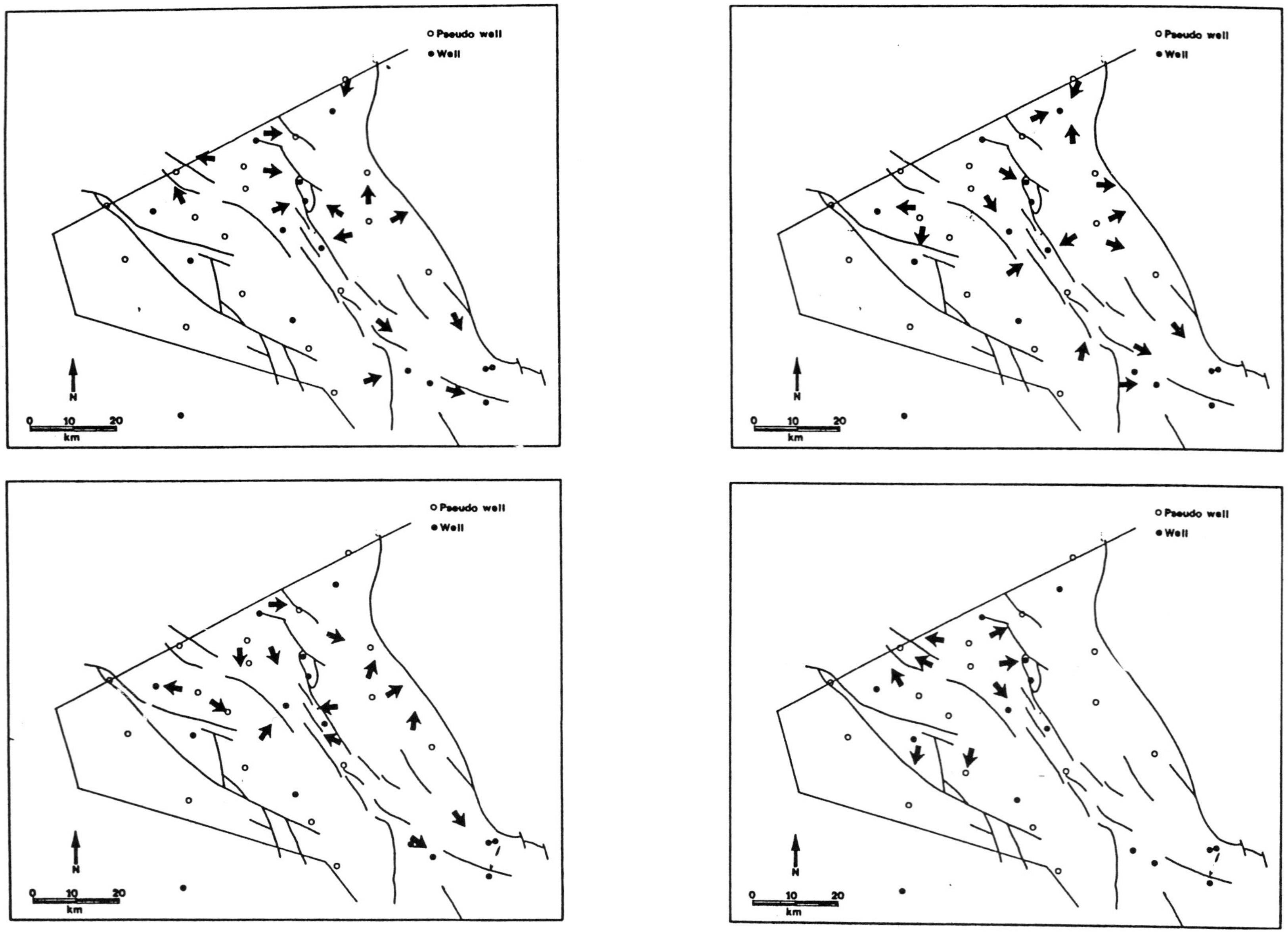

Fig. 11 (continued). (b) Interpreted pressure drive for lateral fluid flow assessed from the contoured excess fluid pressure.

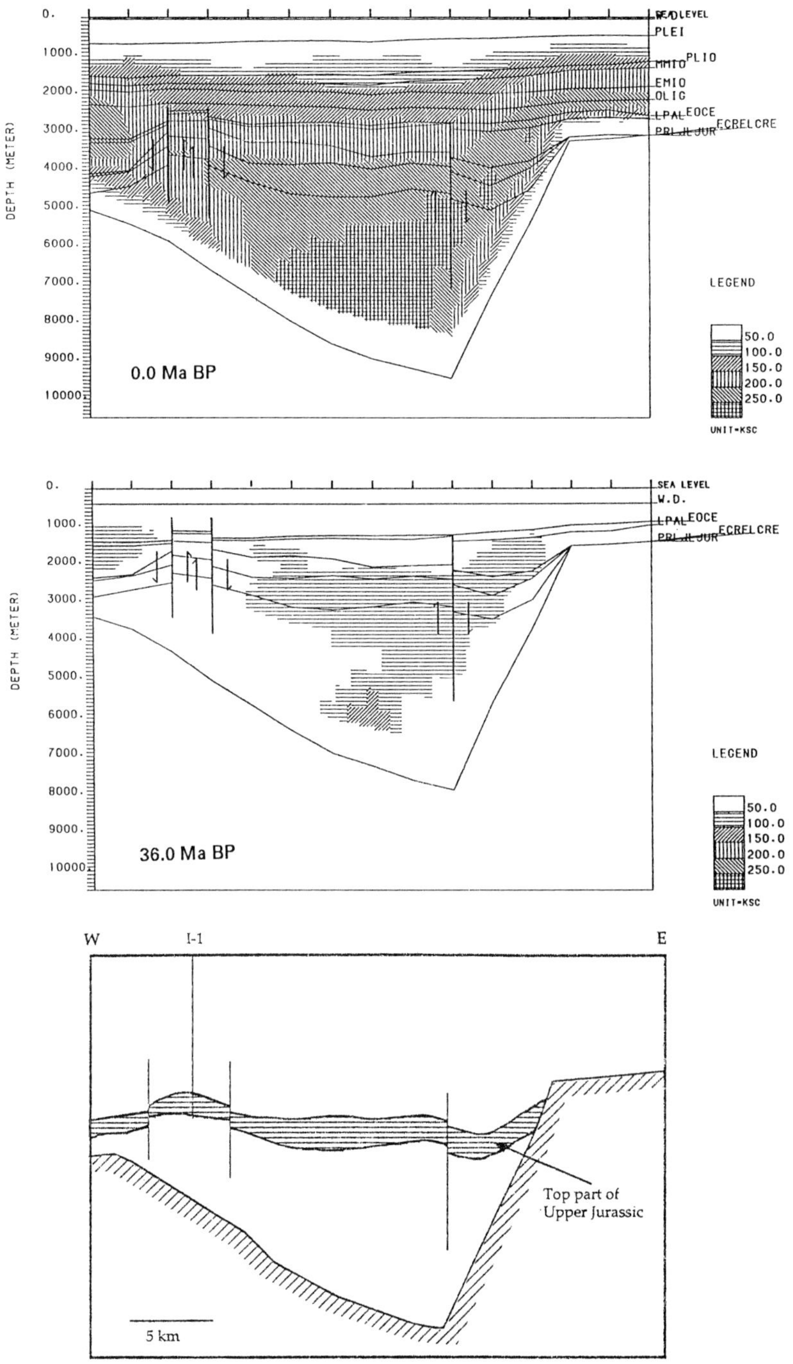

Fig.11 (continued). (c) Excess pressure from the two-dimensional simulation of line RTD-81-23 showing commencement of a build-up of excess fluid pressure around 36 Ma.

the slightly flexured chalk could leak hydrocarbons through to Early Tertiary sand tongues, sealed by overlying shales. The consistency between the observations and the model predictions for areas *1*, *2* and *3* in Fig. 12, lends confidence to the suggestion that the fourth area — along the Tail End Graben-Ringkøbing-Fyn High boundary fault — is a high potential, low risk exploration target.

Conclusions

The subsidence pattern though time from the mid-Jurassic to the present day indicates that the northern Danish Central Graben developed through three stages of subsidence: a Late Jurassic differential stage, a late Cretaceous/Early Tertiary uniform stage and a Late Miocene/Quaternary uniform stage.

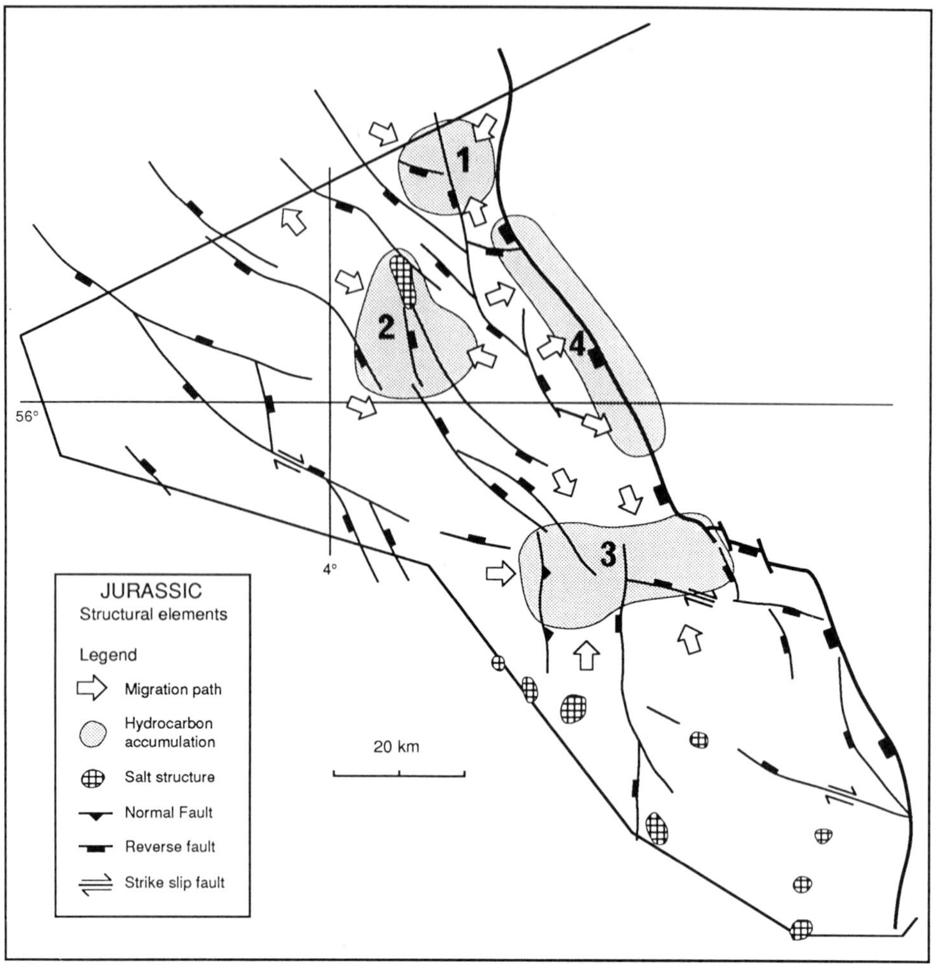

Fig. 12. Map showing the most prospective areas, as indicated by the model (shaded areas). *1* = Lulu; *2* = Arne; *3* = Adda (east) and Roar (west), *4* = Ringkøbing-Fyn High–Tail End Graben boundary.

The porosities, permeabilities and excess pressures within the Upper Jurassic shales indicate that fluid flow apparently did not change considerably since Early Tertiary. Due to a significant pressure driven lateral component, fluid flow is directed out of the Gertrud Graben and central Tail End Graben towards areas of low excess pressure located in the Søgne Basin, the Arne-Elin Graben, the Heno Plateau and the southern Tail End Graben. The thermal history was determined by inversion of vitrinite reflectance data from several wells. Resolution of thermal history degrades beyond 30–50 Ma due to the recent deep burial and heating. Maturation of pre-Cretaceous rocks took place within the Tertiary.

Modelled maturation in terms of vitrinite reflectance indicates that the Upper Jurassic shales entered the R_o = 0.6 "oil window" between 50 Ma and 5 Ma.

Pyrolysis data from nine wells showed that the Upper Jurassic shales contain kerogens of both types II and III. Type II kerogen predominates in some stratigraphic levels in the Farsund Formation (Kimmeridge clay), while type III kerogen predominates in the Late Oxfordian.

A Tissot-Welte oil generation kinetic model showed peaking in the eastern half of the basin between 10 Ma ago and the present day, the western half not reaching significant generation until the present day.

The integrated evaluation showed that hydrocarbon migration was from the Gertrud Graben and central Tail End Graben towards the Lulu, Arne, Roar and Adda structures, and towards the eastern boundary fault of the Tail End Graben.

Acknowledgements

We wish to thank The Statoil Group, Denmark: Total CFP, Denerco K/S, LD Energi A/S, EAC Energi A/S, BHP Petroleum (Denmark) Inc., Dansk Olie-OG Gasproduktion A/S, Statoil Efterforskning OG Produktion A/S, The Danish Research Academy and Aarhus Universitet for financial support; the Basin Modelling Group, University of South Carolina, Columbia, SC, U.S.A., for help and advice and Microlith, Denmark for providing the ISOPAC 3.0 software for constructing the contour maps.

References

Andersen, C., Olsen, J.C., Michelsen, O. and Nygaard, E., 1982. Structural outline and development. In: O. Michelsen (Editor), Geology of the Danish Central Graben. Danm. Geol. Unders., Ser. B, 8: 9–26.

Cao, S., 1985. A quantitative dynamic model for basin analysis. Master's Thesis, University of South Carolina, Columbia, S.C. (unpublished).

Cao, S., Glezen, W.H. and Lerche, I., 1986. Fluid flow, hydrocarbon generation, and migration: a quantitative model of dynamical evolution in sedimentary basins. 18th Annu. Offshore Technology Conf., Houston, Texas, OTC Paper 5182, pp. 2267–2276.

Cao, S. and Lerche, I., 1987. Geohistory, thermal history and hydrocarbon generation history of the northern North Sea Basin. Energy Explor. Exploit., 5: 315–355.

Conford, C., 1984. Source rocks and hydrocarbons of the North Sea. In: K.W. Glennie (Editor), Introduction to the Petroleum Geology of the North Sea. Blackwell, Oxford, pp. 171–200.

Fisher, M.J. 1984. Triassic. In: K.W. Glennie (Editor) Introduction to the Petroleum Geology of the North Sea. Blackwell, Oxford, pp. 85–103.

GECO, 1981. Geophysical Company of Norway A/S, Line RTD-81-20.

Glennie, K.W. (Editor), 1984. Introduction to the Petroleum Geology of the North Sea. Blackwell, Oxford, 236 pp.

Goff, J.C., 1983. Hydrocarbon generation and migration from Jurassic source rocks in E Shetland Basin and Viking Graben of the northern North Sea. J. Geol. Soc. London, 140: 445–474.

Gowers, M.B. and Sæbøe, A., 1985. On the structural evolution of the Central Trough in the Norwegian and Danish sectors of the North Sea. Mar. Pet. Geol., 2: 298–318.

Hubbert, M.K., 1953. Entrapment of petroleum under hydrodynamic conditions. Am. Assoc. Pet. Geol., Bull., 37: 1954–2026.

Klint Jensen, P., 1983. Formation temperatures in the Danish Central Graben. Danm. Geol. Unders., Årbog 1982, pp. 91–106.

Lerche, I., 1988a. Inversion of multiple thermal indicators: quantitative methods of determining paleoheat flux and geological parameters, I. Theoretical development for paleoheat flux. Math. Geol., 20: 1–36.

Lerche, I., 1988b. Inversion of multiple thermal indicators: quantitative methods of determining paleoheat flux and geological parameters, II. Theoretical development for chemical, physical, and geological parameters. Math. Geol., 20: 73–96.

Lerche, I., Yarzab, R.F. and Kendall, C.G.St.C., 1984. Determination of paleoheat flux from vitrinite reflectance data. Am. Assoc. Pet. Geol., Bull., 68: 1704–1717.

Lindgreen, H., Thomsen, E. and Wrang, P., 1982. Source rocks. In: O. Michelsen (Editor), Geology of the Danish Central Graben. Danm. Geol. Unders., Ser. B, No. 8.

Michelsen, O. (Editor), 1982. Geology of the Central Graben. Danm. Geol. Unders., Ser. B, No. 8, C.A. Reitzels Forlag, Copenhagen.

Michelsen, O., Frandsen, N., Holm, L., Jensen, T.F., Møller, J.J. and Vejbæk, O.V., 1987. Jurassic-lower Cretaceous of the Danish Central Trough — depositional environments, tectonism, and reservoirs. Danm. Geol. Unders., Ser. A, 16, 45 pp.

Møller, J.J., 1986. Seismic structural mapping of the Middle and Upper Jurassic in the Danish Central Trough. Danm. Geol. Unders., Ser. A, 13: 1–42.

Nakayama, K. and Lerche, I., 1987. Basin analysis by model simulation: effects of geological parameters on 1-D and 2-D fluid flow systems with application to an oil field. Trans. GCAGS, XXXDVIII: 175–184.

Stouge, S., 1988. Biostratigraphy of selected wells from the northwestern part of the Danish Central Trough. Rep., CENBAS/Danm. Geol. Unders. (unpublished).

Stouge, S. and Toxwenius, B., 1988. Chronostratigraphic position of Cenozoic seismic markers of DGU and some log markers. Rep., CENBAS/Danm. Geol. Unders. (unpublished).

Thomsen, E., 1988. Source Rock Analysis of Jurassic–Lower Cretaceous Deposits in the Wells B-1, Bo-1, E-1, I-1, Q-1, W-1, Lulu-1, Mona-1, Sten-1. Results of Leco-Rock-Eval and Optical Maturity Determinations — Tabulated Data. Rep., CENBAS/Danm. Geol. Unders. (unpublished).

Thomsen, E., 1989, Source rock analysis of Jurassic–Lower Cretaceous deposits in the Danish Central Trough. Results of optical and organic geochemical analysis of the wells B-1, Bo-1, E-1, I-1, Q-1, W-1, Lulu-1, Mona-1, Sten-1. Rep., CENBAS/Danm. Geol. Unders. (unpublished).

Thomsen, E., Lindgreen, H. and Wrang, P., 1983. Investigation on the source rock potential of Denmark. Geol. Mijnbouw, 62: 221–239.

Thomsen, R.O., 1989. The Danish Central Trough: burial history and 1-D modelling — Vol. I: An Integrated Quantitative Analysis. Rep., CENBAS, 66 pp. (unpublished).

Tissot, B. and Welte, D.H., 1978. Petroleum Formation and Occurrence. Springer-Verlag, New York, N.Y.

Ungerer, P., Bessis, F., Chenet, P.Y., Durand, B., Nogaret, E., Chiarelli, A., Oudin, J.L. and Perrin, J.F., 1984. Geological and geochemical models in oil exploration, principles and practical examples. In: G. Demaison and R.J. Murris (Editors), Petroleum Geochemistry and Basin Evaluation. Am. Assoc. Pet. Geol., Mem., 35: 53–77.

Vejbæk, O.V. and Andersen, C., 1987. Cretaceous--Early Tertiary inversion tectonism in the Danish Central Trough. Tectonophysics, 137: 221–238.

Waples, D.W., 1985. Geochemistry in petroleum exploration. IHRDC, Boston, Mass., 232 pp.

Wei, Z. and Lerche, I., 1988. Quantitative dynamic geology of the Pinedale anticline Wyoming, U.S.A.: an application of a two-dimensional simulation model. Appl. Geochem., 3: 423–440.

Welte, D.H. and Yukler, M.A., 1981. Petroleum origin and accumulation in basin evolution — a quantitative model. Am. Assoc. Pet. Geol., Bull., 65: 1387–1396.

Ziegler, P.A., 1982. Geological Atlas of Western and Central Europe. Elsevier, Amsterdam, 130 pp.

R.O. THOMSEN *Geologisk Institut, Aarhus Universitet, DK-8000 Århus C, Denmark; and Department of Geological Sciences, University of South Carolina, Columbia, SC 29208, U.S.A.*
Present address: *Saga Petroleum A/S, P.O. Box 490, N-1301 Sandvika, Norway*

I. LERCHE *Department of Geological Sciences University of South Carolina, Columbia, SC 29208, U.S.A.*

J.A. KORSTGÅRD *Geologisk Institut, Aarhus Universitet, DK-8000 Århus C, Denmark*

RIFT, a model of sedimentary basin evolution by finite rate, non-uniform, pure shear extension of the lithosphere

Noelle E. Odling

A model of sedimentary basin formation by finite rate, non-uniform pure shear extension of the lithosphere is described. This model can be applied to problems of superimposed rifting events and generally non-uniform extension. Existing models of instantaneous pure shear extension (McKenzie, 1978) and subsequent modifications (Jarvis and Mackenzie, 1980, Royden and Keen, 1980) have been applied, with success, to many extensional basins. All these models assume a linear, equilibrium temperature profile throughout the lithosphere prior to rifting. However, where several rifting phases have occurred throughout the history of a basin, the assumption of an initially linear, equilibrium temperature profile is not necessarily valid. The model described here is developed from the models of McKenzie (1978) and Jarvis and McKenzie (1980) to calculate the effects of any extension on any initial temperature profile.

In the present model, the lithosphere is divided into any number of layers (N) of equal thickness, each of which is assigned an extension factor. In this way, any non-uniform stretching event can be modelled. Initial and thermal subsidences and heat flow variation are calculated using the equations of McKenzie (1978) and McKenzie and Jarvis (1980) adapted for the multilayer model. Extension of the lithosphere with a non-equilibrium temperature profile is modelled by adjusting the extension factors to those necessary to obtain the final temperature profile from the equilibrium state. This extension will in general vary with depth, a situation that can be simulated using the multilayer model.

Program RIFT (in standard Fortran) has been written to compute the model. The lithosphere is divided into 100 layers, each of which can be extended by a different amount. Finite rate rifting is modelled as a series of small rifting events, each of which is followed by a short cooling period. Subsidence values are computed for the water filled basin (tectonic subsidence). However, prior to the second and subsequent rifting events, the basin is filled to a specified water depth with sediments composed of 50% shale and 50% sandstone, taking compaction into account. This gives a more realistic estimate for crustal thickness immediately prior to the next rifting event. The program outputs a sequence of tectonic subsidence and heat flow values at time intervals specified by the user. The model is used to estimate the probable effects of Triassic rifting on subsidence and heat flow during and after Jurassic extension in the northern North Sea. Effects are greatest on thermal subsidence and remnant Triassic thermal subsidence can contribute as much as 30% to Jurassic thermal subsidence when Triassic extension factors are in the region of 1.5 and Jurassic factors around 1.25. The advantages and limitations of the model are discussed.

Introduction

The presence of large, normally tilted fault block structures underlying thick sequences of largely undeformed sediments, observed on seismic sections from some sedimentary basins, indicated the importance of extensional tectonics in the formation of these basins. Subsequently, a quantitative, physical model of basin formation based on instantaneous pure shear extension of the lithosphere was developed by McKenzie (1978). Applications of this model and subsequent modifications (Jarvis and McKenzie, 1980, Royden and Keen, 1980) have been successful in modelling subsidence for various basins, e.g., the North Sea (e.g., Barton and Wood; 1984, Giltner, 1987), the East Labrador coast (Royden and Keen, 1980) and the Pannonian Basin (Sclater et al., 1980).

Interpretations of deep seismic data for the North Sea basin have led to different structural models for the North Sea basin that involve either pure shear extension (e.g., Barton and Wood, 1984; Giltner, 1987) or low angle shear zones and asymmetric rifting (Beach et al., 1987). However, evidence from the geometry of the sedimentary basin fill has led several authors to favour extension involving dominantly pure shear (Badley et al., 1988; Klemperer, 1988; White, 1988). Thus it seems probable that the pure shear extension model of McKenzie is a valid approach to the problem.

As the resolution and depth of penetration of seismic data has improved, it has become apparent that basins such as the North Sea are the result of several periods of extension each of which is characterized by major normal faulting. In the case

Structural and Tectonic Modelling and its Application to Petroleum Geology edited by R.M. Larsen, H. Brekke, B.T. Larsen and E. Talleraas. NPF Special Publication 1, pp. 457–467. Elsevier, Amsterdam.

of the North Sea, it is probable that extension started very soon after the end of the Caledonian Orogeny in the Devonian and went through several stages of extensional faulting ending with rifting events in the Triassic and mid-Jurassic (McClay et al., 1986).

In the pure shear extension models of McKenzie (1978) and others, the lithosphere is assumed to be in an initially equilibrium state (linear temperature profile). Extension of the lithosphere produces a passive upwelling of hot asthenosphere and a temperature anomaly. Isostatic adjustments in response to instantaneous extension lead to an initial subsidence when the crust is of normal thickness or thicker and uplift in the case of thin crust. Cooling, accompanied by contraction and density increase, results in subsidence and basin formation. Since many basins are the result of more than one period of rifting (e.g., Tankard and Welsink, 1987, Badley et al., 1988) the assumption of an equilibrium temperature profile prior to later events may not be valid.

In the present paper, a general model of basin formation resulting from non-uniform, pure shear extension of the lithosphere, based on the models of McKenzie (1978), Jarvis and McKenzie (1980) and Royden and Keen (1980), is developed so that any variation of extension with depth may be modelled. As in the case of these previous models, the present model is one dimensional in the sense that there is no lateral variation in extension. As a test case, the model is used to estimate the possible effects of Triassic rifting on subsidence during and post-dating the mid-Jurassic extensional event of the northern North Sea.

Existing models of extensional basin formation

The basic model of sedimentary basin formation resulting from thinning of the lithosphere was developed by McKenzie (1978). Prior to extension, the lithosphere is assumed to be in equilibrium, with a linear decrease in temperature from that of the asthenosphere at the base of the lithosphere to 0°C at the surface of the crust. The lithosphere is instantaneously extended by uniform, pure shear by a factor β (= initial/final lithospheric thickness) resulting in a temperature anomaly (Fig. 1A and B). This deformation is associated with a change in the elevation of the surface of the crust which is a balance between subsidence caused by replacement of light crust by dense asthenosphere and uplift due to thermal expansion. This change in elevation (termed "initial subsidence") is, therefore, sensitive to the initial crustal thickness and the model predicts uplift if the crust is initially thinner than approximately 20 km. Following extension, the temperature anomaly cools and subsidence due to thermal contraction forms the basin. The cooling history is quantified by solving the one-dimensional heat flow equation:

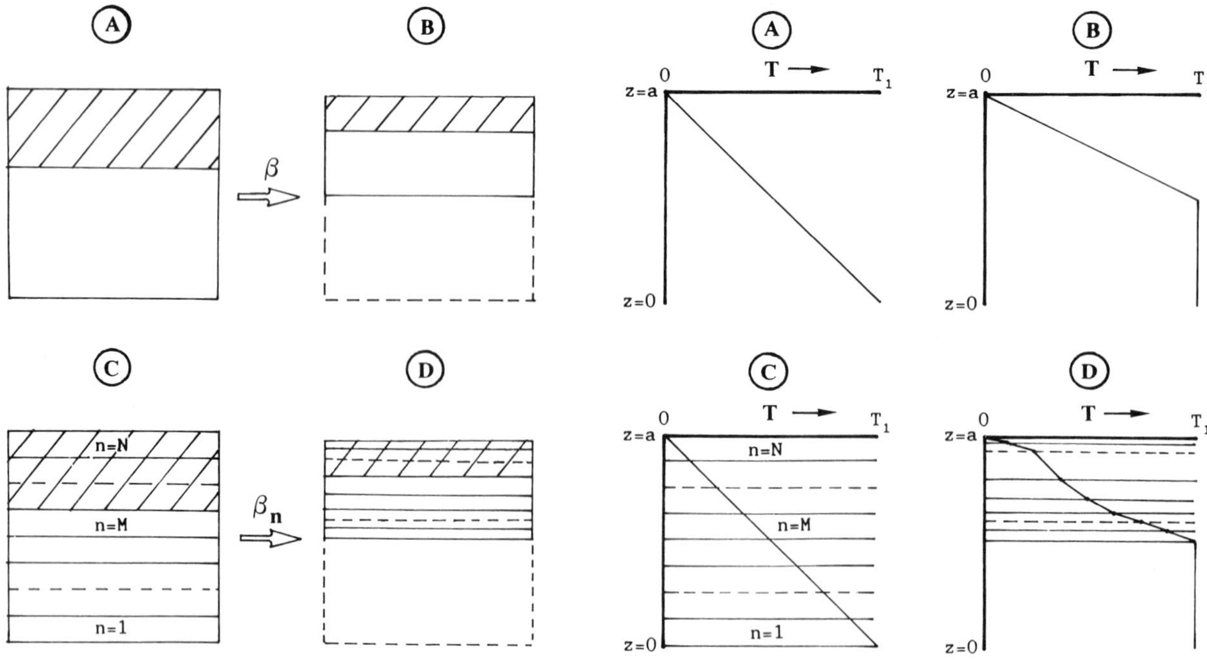

Fig. 1. A, B: Instantaneous uniform, pure shear extension of the lithosphere by a factor β (McKenzie 1978). The resulting temperature profile (B) is linear to the base of the thinned lithosphere below which temperature is constant at T_1, the temperature of the asthenosphere. C, D: instantaneous, non-uniform, pure shear extension of the lithosphere by factors β_n which, in general, vary with depth. The resulting temperature profile within the thinned lithosphere is non-linear (D).

TABLE 1

Symbol definitions and parameter values (mostly after McKenzie 1978)

Symbol	Definition	Parameter value
t_c	crustal thickness	32 km
a	lithospheric thickness	125 km
ρ_m	mantle density at 0°C	3.33 g cm^{-3}
ρ_c	crustal density at 0°C	2.8 g cm^{-3}
ρ_w	density of water at 0°C	1.0 g cm^{-3}
α	thermal expansion coefficient for crust and mantle	3.28 x 10^{-5}°C
κ	thermal diffusivity coefficient for crust and mantle	0.00804 cm^2 s^{-1}
K	thermal conductivity of crust and mantle	0.0075 cal^{-1} cm^{-1} s^{-1}
T	temperature (°C)	
t	time (Ma)	
z	depth (km)	
β	extension factor	

$$\frac{\partial T}{\partial t} = \kappa \frac{\partial^2 T}{\partial z^2} \tag{1}$$

where T is temperature, t is time, z is depth and κ is the thermal diffusivity of the lithosphere (Table 1). McKenzie (1978) solved this equation by Fourier expansion and developed expressions for the variation in heat flow and subsidence with time after instantaneous extension for the case of a water filled basin (tectonic subsidence).

This model assumes instantaneous rifting whereas, in reality, extension takes place over a finite time interval. When rifting times are short, this assumption is reasonable but if extension rates are sufficiently slow, significant heat is lost during the rifting event resulting in significantly different heat flow and subsidence patterns. Jarvis and McKenzie (1980) have modified McKenzie's initial model to take finite rifting time into account by assuming a constant rate of extension and solving the resulting one-dimensional heat flow equation. The result of heat loss during rifting is to increase the subsidence during rifting at the expense of that during cooling (the total subsidence remaining the same). The results of this model show that subsidence and heat flow patterns differ significantly from those derived from the instantaneous extension model when the rifting times are greater than approximately 10 Ma.

Further modifications of the instantaneous pure shear extension model has been developed by Royden and Keen (1980) and Rowley and Sahagian (1986) to deal with non-uniform extension throughout the lithosphere. As the rheological properties of the lithosphere vary with pressure, temperature and chemical composition, it is possible that the degree of thinning due to extension may vary with depth, resulting in a change in the width of the deformation zone. A model in which β varies continuously with depth has been used by Rowley and Sahagian (1986) to provide a mechanism for the uplift of rift flanks during extension. For this, they used an analytical expression for the vertical variation in β, assuming a linear increase in the width of the extension zone with depth, and used modifications of the equations of McKenzie (1978) to calculate uplift and subsidence.

The observation of low angle shear zones on land (Wernicke, 1984) and possible examples on seismic sections (de Charpal et al., 1978; Beach et al., 1987; Tankard and Welsink, 1987) also suggests that decoupling zones which separate domains with differing magnitudes of extension, can exist within the lithosphere. In the model of Royden and Keen (1980), the lithosphere is divided into two layers, separated by a "decoupling" zone placed, in their case, at the base of the crust. They then developed expressions for the Fourier coefficients of the initial temperature distribution for the two layer case and then applied the expressions of McKenzie (1980) to derive heat flow and subsidence with time. By simulating different magnitudes of instantaneous extension in these two layers, preferential crustal and subcrustal stretching can be modelled. Preferential subcrustal stretching was found to reduce initial subsidence and has been used by Royden and Keen (1980) to explain subsidence patterns from the Labrador Shelf. This model has also been employed by White and McKenzie (1988) to explain the "steer's head" geometry of basin fill. They used two different Gaussian distributions to describe the lateral variation of β in the crust and upper mantle, while keeping the total extension of the two layers equal. By stretching the upper mantle by lesser amounts over a wider area, uplift of the rift flanks is generated and subsequent subsidence results in a "steer's head" geometry of the infilling sediments.

The present model

A multilayer lithospheric model for instantaneous extension

The theories of McKenzie (1978) and Royden and Keen (1980) have been applied to a lithosphere divided into any number of equally thick layers. In this way, any variation in extension with depth, which need not necessarily be described as an analytical expression (as in Rowley and Sahagian, 1986), can be modelled. As in previous models, the initial temperature distribution of the lithosphere is assumed to be that of the equilibrium state with the temperature increasing linearly with depth from 0°C at the surface of the crust to T_1, the temperature of the asthenosphere, at the base of the lithosphere. On rifting, each layer within the lithosphere is extended by a factor β_n, resulting in a general non-linear temperature profile (Fig. 1C and D).

This extension is associated with a change in elevation which is a balance between uplift due to thermal expansion and subsidence due to the replacement of light crust by denser mantle material. It is assumed that isostatic equilibrium is maintained and the change in elevation is then calculated by equating the mass of a unit cross-sectional area column to the base of the lithosphere before and after extension. In the case of uniform pure shear, the change in elevation, S_i, is given by (McKenzie, 1978: his eqn. 1):

$$S_\mathrm{i} = \frac{a\left[\dfrac{(\rho_\mathrm{m}-\rho_\mathrm{c})t_\mathrm{c}}{a(1-\alpha T_1 t_\mathrm{c}/2a)} - \dfrac{\alpha T_1 \rho_\mathrm{m}}{2}\right]\left(1-\dfrac{1}{\beta}\right)}{\rho_\mathrm{m}(1-\alpha T_1)-\rho_\mathrm{w}} \tag{2}$$

A list of definitions of the variables in the above and following expressions is given in Table 1. In the multilayer model, the contribution of each layer must be calculated separately and the results summed. The change in elevation is then given by:

$$\begin{aligned} S_\mathrm{i} = & \left[\sum_{n=M}^{N} \rho_\mathrm{c}\left(1-\frac{\alpha}{2}(T_{n+1}+T_n)\right)\left(1-\frac{1}{\beta_n}\right)\right. \\ & + \sum_{n=1}^{M-1} \rho_\mathrm{m}\left(1-\frac{\alpha}{2}(T_{n+1}+T_n)\right)\left(1-\frac{1}{\beta_n}\right) \\ & \left. - \rho_\mathrm{m}(1-\alpha T_1)\left(N-\sum_{n=1}^{N}\frac{1}{\beta_n}\right)\right] \\ & \times \frac{-a}{N[\rho_m(1-\alpha T_1)-\rho_\mathrm{w}]} \end{aligned} \tag{3}$$

Equation 3 gives the elevation change when subsidence occurs and the resulting basin is water filled (tectonic subsidence). In the case of uplift, when the basin is not water loaded, the denominator in the above expression becomes $N[\rho_\mathrm{m}(1-\alpha T_1)]$.

The instantaneous extension event causes an upwelling of hot asthenosphere producing a temperature anomaly, the decay of which results in thermal contraction and decreasing heat flow, thereby creating a basin. The solution of the one-dimensional heat flow equation (eqn. 1) gives the following expression for the temperature distribution with depth after time, t (adapted from Jarvis and McKenzie, 1980: their eqn. 18):

$$\begin{aligned} T(z,t) = & \, T_1\left(1-\frac{z}{a}\right) \\ & + T_1\sum_{k=1}^{\infty} b_k \exp\left[-k^2\pi^2 t\left(\frac{\kappa}{a^2}\right)\right]\sin\frac{\kappa\pi z}{a} \end{aligned} \tag{4}$$

where b_k represents the Fourier coefficients of the temperature distribution immediately after instantaneous extension. The kth Fourier coefficient is given by (Jarvis and McKenzie, 1980: their eqn. 19):

$$b_k = \frac{2}{aT_1}\int_0^a \left[T_{(z)} + T_1\left(\frac{z}{a}-1\right)\right]\sin\frac{k\pi z}{a}\,\mathrm{d}z \tag{5}$$

where $T_{(z)}$ is the temperature distribution at time, $t = 0$. For the multilayer model the Fourier coefficients are determined by adding the results of the above integral applied to each layer, the derivation of which is given in Appendix A:

$$b_k = \frac{2}{k\pi}\sum_{n=1}^{N}\left[\frac{\beta_n-\beta_{n-1}}{k\pi}\sin k\pi\left(1-\frac{1}{N}\sum_{m=n}^{N}\frac{1}{\beta_m}\right)\right] \tag{6}$$

Subsidence and heat flow variation with time for the multilayer model can then be derived using the equations of Jarvis and McKenzie (1980): their eqns. 22, 23, 24 and 27. The elevation above the final level to which the surface of the crust sinks is (Jarvis and McKenzie, 1980: their eqns. 23 and 24):

$$\begin{aligned} e(t) = & \frac{-a\alpha T_1}{\pi[\rho_\mathrm{m}(1-\alpha T_1)-\rho_\mathrm{w}+Bt_\mathrm{c}/a)]} \\ & \times \sum_{k=1}^{\infty}\frac{b_k}{k}\exp\left[-\frac{k^2\pi^2\kappa t}{a^2}\right]\left\{\left[1-\cos k\pi\left(1-\frac{t_\mathrm{c}}{a}\right)\right]\right. \\ & \left. \times(\rho_\mathrm{m}+\rho_\mathrm{c})+\rho_m-(-1)^n\rho_\mathrm{c}\right\} \end{aligned} \tag{7}$$

where:

$$B = \rho_\mathrm{w}\left[1+\frac{\alpha T_1}{(1-\alpha T_1)}\right]\left(1-\frac{\rho_\mathrm{c}}{\rho_\mathrm{m}}\right)+(\rho_\mathrm{c}-\rho_\mathrm{m})$$

As $e(t)$ approaches zero as t approaches infinity, the maximum post-extension subsidence (S_G) is obtained by setting $t = 0$ in eqn. 7 (Jarvis and McKenzie, 1980: their eqn. 27):

$$\begin{aligned} S_\mathrm{G} = & \frac{-a\alpha T_1}{\pi[\rho_\mathrm{m}(1-\alpha T_1)-\rho_\mathrm{w}+Bt_\mathrm{c}/a]} \\ & \times\sum_{k=1}^{\infty}\frac{b_k}{k}\left\{\rho_m\left[1-\cos k\pi\left(1-\frac{t_\mathrm{c}}{a}\right)\right]\right. \\ & \left.+\rho_\mathrm{c}\left[\cos k\pi\left(1-\frac{t_\mathrm{c}}{a}\right)-(-1)^k\right]\right\} \end{aligned} \tag{8}$$

The subsidence at time, t, is then given by:

$$S_t = e(t) - S_\mathrm{G} \tag{9}$$

where subsidence is positive and uplift negative.

Heat flow after time, t, is given by (Jarvis and McKenzie, 1980: their eqn. 22):

$$F(t) = \frac{KT_1}{a}\left[1+\pi\sum_{k=1}^{\infty}kb_k(-1)^{k+1}\exp\left(\frac{-k^2\pi^2 t\kappa}{a^2}\right)\right] \tag{10}$$

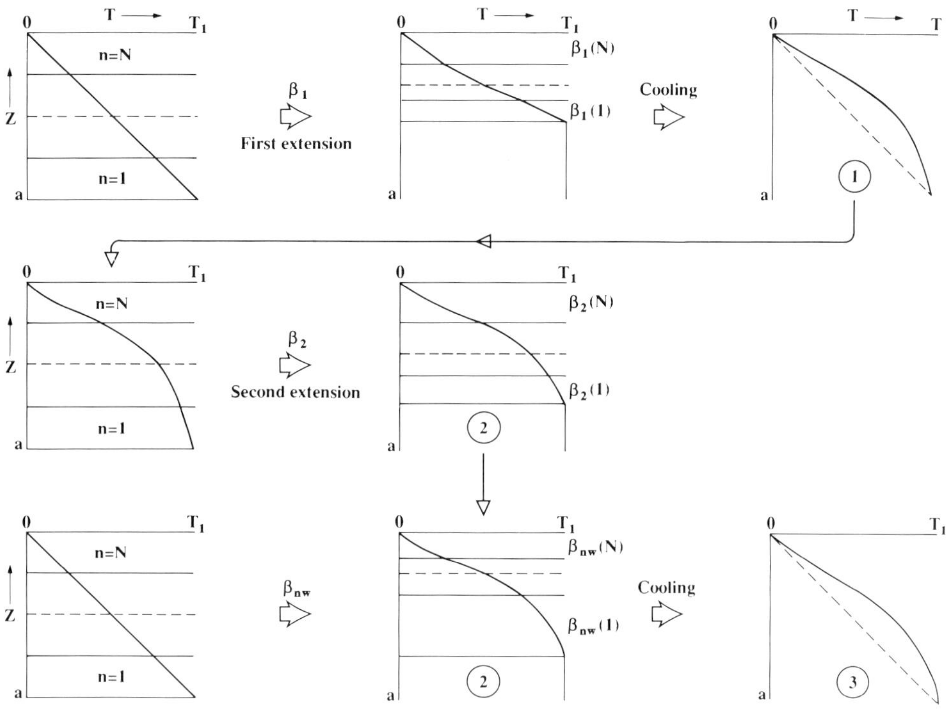

Fig. 2. The superposition of two rifting events. Before the temperature anomaly created by non-uniform rifting phase β_1 decays, the temperature profile of the lithosphere is non-linear. This profile is then extended by a second rifting phase, β_2, to give a new temperature profile (*2*). To determine subsidence and heat flow variation during the decay of this anomaly, a second set of extension factors β_{nw} are calculated. These are the extension factors required to produce temperature profile (*2*) from the equilibrium linear profile. The second set of *beta* factors do not represent physical thinning of the lithosphere, but are the values required by the model to calculate changes in the temperature profile with time.

Superimposing two rifting events and finite rate rifting

The temperature profile after an instantaneous rifting event is the result of the initial profile and the magnitude of the extension. The above expressions for heat flow and subsidence after instantaneous rifting assume an initial equilibrium temperature profile for the lithosphere. However, if extension follows an earlier rifting event before the temperature anomaly caused by that event has completely decayed, the initial temperature distribution may differ significantly from the equilibrium state. This will result in a different temperature perturbation after rifting and thus different subsidence and heat flow patterns. Every temperature profile can be described as an extension superimposed on the equilibrium temperature distribution. The magnitude of this required extension will, in general, vary with depth, a situation that can be modelled using the multilayer model described in the previous section. In the model, the effects of rifting on a non-equilibrium temperature profile are simulated by calculating a second set of β extension values which are those required to produce the final temperature profile from the equilibrium temperature state (Fig. 2). These recalculated β values do not represent physical thinning of the lithosphere but the values required by the model to calculate cooling and resulting thermal subsidence. Using these modified extension factors, eqns. 3–10 described above may then be used to determine subsidence and heat flow after rifting.

To determine the extension values required, the lithosphere, with any temperature profile, is divided into N layers, each of which is assigned an extension factor (β) appropriate for the rifting event. There is no theoretical limit to the difference in β values of adjacent layers. However, the aim of the model is to simulate smooth changes in β with depth and thus large jumps in β between adjacent layers should normally be avoided. Knowing the initial temperature at the base of each layer and the extension factors, the depth of each temperature value after instantaneous extension can be calculated. Temperature is assumed to vary linearly with depth between these values which results in a smooth temperature profile if enough layers are used. In the equilibrium, linear temperature profile, the temperatures at the base of each of N layers of equal thickness are calculated and the location of each of these temperatures in the final temperature profile is determined. The vertical distance between these points with the original thickness of each layer can then be used to calculate the modified extension β values required to produce the final temperature profile from the equilibrium state (Fig. 2).

Using this method, the above equations can be used to determine subsidence and heat flow variation with time for superimposed rifting events where the second event follows the first before the temperature perturbation has completely decayed. Finite rate rifting can be modelled as a series of small rifting events. The method could also be used to model the effects of rifting on other non-equilibrium temperature profiles caused, e.g., by heating events originating in the mantle.

Program RIFT

The program RIFT (standard fortran) has been constructed to compute subsidence and heat flow during one to five rifting events comprising finite rate, pure shear extension of the lithosphere and subsequent cooling of the temperature anomaly. The lithosphere is assumed to be in an equilibrium state prior to the first rifting event, i.e., that temperature decreases linearly from temperature T_1 at the base of the lithosphere, to 0°C at the surface of the crust. Finite rate rifting is modelled as a series of small, instantaneous rifting events each of which is followed by a short cooling period. The lithosphere is divided into 100 layers for the purposes of computation and each layer is assigned an independent extension factor, β (user determined), for each rifting phase. At the end of each finite rate rifting increment and rifting phase, the temperature profile and values of crustal and lithospheric thickness are used as input for the following rifting event. During each rifting phase the basin is assumed to be water filled for the case of subsidence, giving the tectonic subsidence of the basin. However, at the end of a rifting phase, any basin that has formed is likely to be filled with sediment, resulting in a modified crustal thickness. In order to model the most realistic situation, the effect of filling the basin formed by a rifting phase with sediment composed of 50% sand, 50% shale to a user specified water depth are included in the model (method after Sclater and Christie, 1980). The loading effects of this quantity of sediment are calculated using density, initial porosity and lithological compaction factor quoted by Sclater and Christie (1980) and compaction of the pre-rift sequence is also taken into account. The crustal thickness is then adjusted to include this sedimentation prior to the next rifting event. Other physical input parameters used in the model follow values given in McKenzie (1978) and are listed in Table 1.

The results of RIFT for a single rifting phase were compared to those from the finite-rate, uniform pure shear rifting model of Jarvis and McKenzie (1980). Increasingly better fits were obtained as the number of increments used to model finite rifting increased. RIFT was found to overestimate subsidence by a maximum of 4% for an extension factor of 2.0 when 10 or more increments were used. Errors in the heat flow were 1.5% or less.

The model applied to superimposed Triassic and Jurassic rifting in the northern North Sea

The existence of a Triassic phase of rifting in the northern North Sea region is well known although the timing of this event is, as yet, poorly constrained (Badley et al., 1988). Late Permian to Early Triassic faulting is described by Badley et al. (1984) in the Oseberg region, while Late Triassic faulting has been recorded from the western part of the Viking Graben (Davies and Watts, 1977; De'Ath and Schuyleman, 1981; Beach et al., 1987; Karlsson, 1986; Gabrielsen et al., 1989). Estimates for the thickness of Triassic sediments in the North Sea region range up to 3000 m (Ziegler, 1982) which, if attributed to subsidence following rifting, indicates that a major extension event took place in Triassic times. Triassic rifting was followed by a second and better documented rifting phase in Late Jurassic times described by many authors and dated as Bathonian to Ryazanian by Badley et al. (1988).

The thinning of the crust in the northern North Sea since Early Triassic times has been estimated from the depths of the Base Triassic and the Moho, determined from reflection seismic and gravity data by Klemperer (1988) and Holliger and Klemperer (1989). The thickness of this basement ranges from 32 km on the Shetland platform and western Norway (assumed to be the original crustal thickness) to 15 km in the deepest parts of the Viking Graben. This gives a maximum extension factor, β, since the Early Triassic as 2.1. However, discrepancies between the Moho derived from reflection seismic data and gravity data have led Holliger and Klemperer (1989) to speculate that the lower crust may have been thickened by basaltic intrusions in the deepest parts of the Viking Graben, in which case the basement thickness may be as low as 10 km, implying a β factor of up to 3.2.

Giltner (1987) modelled subsidence due to Triassic and Jurassic rifting events in the northern North Sea and obtained extension factors of up to 1.8 for the Triassic event and 1.1 to 1.2 for the Jurassic event (total β of 2.16). Beach et al. (1987) obtained a β value of 1.5 for Triassic rifting and 1.8 to 2.2 for Jurassic rifting from a subsidence curve from the north Viking Graben (total β of up to 3.3). Vially (1988), in a regional study of subsidence in the northern North Sea obtained maximum Triassic β values of 1.5 (Horda platform) and a maximum Jurassic value of 1.5 (Viking Graben) (maximum total β of about 2.0). Badley et al. (1988) estimated maximum extension factors for Jurassic rifting at 1.15 from initial subsidence and 1.49 from thermal subsidence, as average values for the Viking Graben. All of the above estimates assumed that the effects of residual Triassic thermal subsidence were negligible prior to Jurassic extension. Lippard and Liu (this volume) have modelled Triassic and Jurassic rifting in the northern North Sea, using model RIFT and arrived at Triassic β values of up to 1.5 and Jurassic values up to 1.8 (total β of 2.7).

An attempt has been made by Hellinger et al. (1989) to determine the maximum effect of Carboniferous to Triassic rifting on subsidence during and post-dating Jurassic extension in the Central Graben of the North Sea, by modelling combinations of the various rifting events as single phases. For a Carboniferous to Triassic β value of 1.69 and a Jurassic β value of 1.26, they estimated that up to 40% of Jurassic subsidence may be attributable to remnant thermal subsidence of earlier events. To obtain an idea of the possible effects that Triassic rifting may have had on subsidence during and after Jurassic rifting in the northern North Sea, program RIFT was used to calculate theoretical tectonic subsidence and heat flow curves for the superimposed Triassic and Jurassic rifting phases. For simplicity, uniform pure shear is assumed for both rifting phases. The curves obtained for subsidence and heat flow after the start of Jurassic extension are compared to curves of equivalent β value for a single Jurassic rifting event, as a measure of the effects of the Triassic event. This treatment is not intended as a definitive study of the basin development in the northern North Sea, but rather as an indicator of the probable extent of the effects of Triassic rifting on subsidence and heat flow after the start of Jurassic extension. A fuller treatment of subsidence in the northern North Sea from observed and theoretical subsidence curves derived from program RIFT, including depth dependent stretching, is given by Lippard and Liu (this volume).

A Jurassic extension of 25 Ma duration, from 156 to 131 Ma (Bathonian to end Jurassic), (Badley et al., 1988) was used. As the duration of Triassic rifting is unknown, a value of 25 Ma, as for Jurassic rifting was assumed. Two "end member" timings of Triassic rifting, Early Triassic (250–225 Ma) and Late Triassic–Early Jurassic (225–200 Ma), were modelled and a crustal thickness of 32 km (Holliger and Klemperer, 1989) prior to Triassic rifting was assumed. Jurassic extensions of 1.25 and 1.5 were superimposed on two Triassic extension events of 1.25 and 1.5 and the results are plotted in Fig. 3. These results are compared to those for a single Jurassic event (extension 15–131 Ma) using an initial crustal thickness of 32 km (Fig. 3).

Figure 3 shows that Triassic rifting in general causes a decrease in initial subsidence and an increase in thermal subsidence in agreement with observations by Barton and Wood (1984) and Hellinger et al. (1989) on data from the Central Graben of the North Sea. The effect on initial subsidence is a combination of a decrease caused by a thinner crust and an increase induced by remnant Triassic thermal subsidence. The net result is a decrease of up to 6% for the β values modelled here, which is greatest when Triassic extension factors are high and the influence of a thinned crust is most dominant. Jurassic thermal subsidence is increased by the addition of remnant Triassic thermal subsidence in all cases. The greatest effects occur when Triassic β factors are high (around 1.5) and Jurassic β factors are low (around 1.25) when Triassic thermal subsidence contributes up to 23% (Early Triassic rifting phase) and 30% (Late Triassic rifting phase) to subsidence post-dating Jurassic extension. In the case of total subsidence, the effects on initial and thermal subsidence tend to counteract each other but leave an overall positive influence with Triassic rifting contributing up to 10% (Early Triassic rifting) to 16% (Late Triassic rifting).

The effects on initial subsidence of up to 6% probably lie within the errors on subsidence curves derived from stratigraphic data and are, therefore, negligible in the determination of β values from initial subsidence. However, initial subsidence patterns are often complicated by footwall uplift and hanging wall collapse and may be unsuitable for determination of β factors. Because of this, more reliance is frequently placed on the shape of the thermal subsidence curve (e.g., Barton and Wood, 1984). Here the increased subsidence caused by remnant Triassic subsidence will result in an overestimation of β if observed subsidence curves are compared to those for a theoretical single event. Where Triassic extension is low compared with Jurassic extension, the discrepancies are of the order of 10% or less and probably largely within the data uncertainties. However, where Triassic extension are high compared with Jurassic extensions, Triassic thermal subsidence may compose up to 23% (Early Triassic rifting) to 30% (Late Triassic rifting), resulting in an overesti-

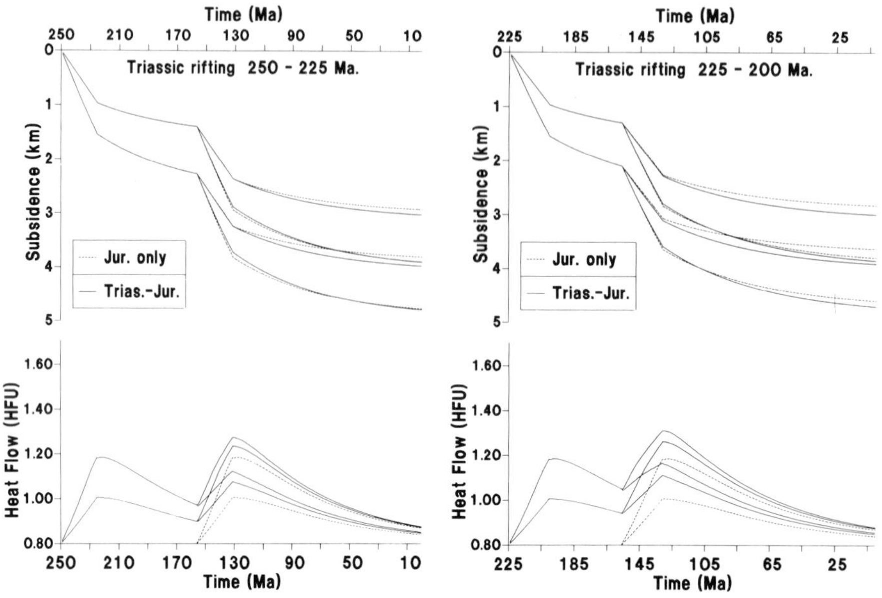

Fig. 3. Subsidence and heat flow histories for superimposed Triassic–Jurassic rifting (solid curves). Two timings of Triassic rifting are shown: Early Triassic, 250–225 Ma (left); and Late Triassic to Early Jurassic, 225–220 Ma (right). Jurassic rifting with extension factors, β, of 1.25 and 1.5 are superimposed on Triassic rifting with extensions of 1.25 and 1.5. The two-phase model curves are compared to those for a single Jurassic event of equivalent β values calculated with an initial crustal thickness of 32 km (dashed curves).

mation of Jurassic β factors of the order of 0.1. The effects of remnant Triassic thermal subsidence on heat flow after the start of Jurassic extension is also shown in Fig. 3, together with the heat flow curves for a single rifting event. The greatest contribution of Triassic rifting occurs at the start of Jurassic extension where it composes 11% (β of 1.25) to 18% (β of 1.5) of the total. In all cases, this decreases to 2% or less 100 Ma after Jurassic extension.

The above examples predict a present day, pre-Triassic basement thickness of 14.2 km in the central parts of the Viking Graben and 20.5 km on its flanks. This is in good agreement with the values determined from seismic reflection data (Klemperer, 1988) of less than 16 km in the central parts of the graben and around 21 km on the flanks. Present day total crustal thicknesses predicted by the model range from 26.5 km to 28.2 km. This agrees quite well with the thicknesses presented by Klemperer (1988) on the flanks of the Viking Graben (around 26 km), but is larger than Klemperer's estimate (less than 21 km) in the graben centre. A possible reason for this discrepancy could be that the crust is more extended than the underlying mantle in this region, resulting in less thermal subsidence.

Discussion and conclusions

The model presented can be used to predict subsidence and heat flow patterns arising from any non-uniform, finite rate pure shear extension of the lithosphere. The program that computes the model is specifically designed to model superimposed rifting events where the temperature anomaly of an event has not completely decayed before the onset of the next.

As a test case, the model has been used to assess the possible influence of Triassic rifting on subsidence and heat flow during and after Jurassic rifting, in the northern North Sea. Two timings of Triassic rifting were modelled, Early Triassic (250–225 Ma) and Late Triassic (225–200 Ma). As would be expected, the effects of Late Triassic rifting were the larger, but only small differences were seen in the results, up to 4% for initial subsidence and up to 9% for thermal subsidence. The effects of Triassic rifting on initial Jurassic subsidence, when compared to a single rifting event with an initial crust of 32 km, were very small (6% or less). This is due to the effects of the crust being thinner than initially (reducing subsidence) which counteracts the effects of remnant thermal subsidence (enhancing subsidence). Remnant Triassic thermal subsidence generally increases thermal subsidence post-dating Jurassic extension and makes the largest relative contribution when Triassic extension factors are high compared to Jurassic values, as has been suggested for the Viking Graben (Giltner, 1987) and the Horda Platform (Giltner, 1987; Vially, 1988). For a Triassic β value of 1.5 and a Jurassic value of 1.25, Triassic remnant thermal subsidence contributes 23–30% of Jurassic thermal subsidence. This could cause an overestimation in β factors estimated using a single phase model of about 0.1.

Thus, the influence of Triassic rifting results in an increase in the total subsidence and a change in the relative proportions of initial and final subsidence. This may go some way to explaining the discrepancies between β factors determined from initial and thermal subsidences, where a single rifting model is employed (Barton and Wood, 1984; Badley et al. 1988). A method that has been used frequently to improve the fit of observed and single phase theoretical subsidence curves is to reduce the initial thickness of the crust, either using the subsidence curves as a guide (Barton and Wood, 1984) or using estimates of Triassic extension (Giltner, 1987). This will tend to lead to an overestimation of β factors as it ignores the effects of remnant Triassic thermal subsidence.

A uniform pure shear model was used in the test cases for the northern North Sea. Most authors modelling subsidence in the North Sea region have used a uniform pure shear model (e.g., Barton and Wood, 1984; Giltner, 1987; Hellinger et al., 1989). Badley et al. (1988) concluded from the mismatch between β factors predicted from initial subsidence and thermal subsidence using a single phase, uniform pure shear model (0.1 to 0.3) in the northern North Sea, that a depth dependent stretching model involving preferential mantle stretching, is more appropriate. However, such a model of preferential mantle stretching everywhere in the basin results in space problems or, if mass is conserved, in a narrowing of the deformation zone with depth, an improbable geometry. For the Horda Platform to the east of the Viking Graben, Lippard and Liu (this volume) have used a two layer model with preferential mantle stretching, in agreement with Badley et al. (1988), to explain subsidence curves.

Models which predict a range of preferential crustal and mantle stretching across the basin have been produced by Rowley and Sahagian (1986) and White and McKenzie (1988). In these models, the crust is preferentially stretched in the graben centre and the mantle preferentially stretched on the graben flanks. In the model of White and McKenzie (1988), the zones between these two regions have approximately uniform pure shear. Space problems are avoided as total extension is the same at all depths. The vertical and horizontal distributions of β imply a widening of the deformation zone with depth. Deep reflectors in the upper mantle were identified on deep seismic reflection profiles by Klemperer (1988) and interpreted by him as probably representing shear zones. Their location at the margins of the major Mezozoic extensional faulting in the northern North Sea and sense of dip outwards from the faulted zone, although implying dominantly uniform pure shear in the crust, are also consistent with a widening of the deformation zone in the mantle with depth. Modelling presented here tentatively suggests that the crust may be preferentially stretched in the deeper parts of the Viking Graben which would also be consistent with this model. Thus, the models of Rowley and Sahagian (1986) and White and McKenzie (1988) may go some way to explaining some of the observations and modelling results of the above authors. However, it is clear that more data and detailed subsidence modelling will be required before the complex question of uniform versus depth dependent stretching in the North Sea region can be resolved.

The model RIFT assumes Airy isostacy, i.e., that isostatic equilibrium is maintained throughout the history of the basin and that any flexural loading effects of sediments are negligible. The question of the importance of flexural loading has been addressed by Barton and Wood (1984) and Hellinger et al. (1989) for the central North Sea and Fowler and McKenzie (1989) for the Rockall Plateau. They estimated the elastic thickness of the crust in both areas to be around 5 km, a very small value, and concluded that there is therefore little error in assuming Airy isostacy. In the northern North Sea, Holliger and Klemperer (1989) found the sediments of the Viking Graben to be isostatically compensated implying that the strength of the lithosphere in this region is drastically reduced. Thus it seems probable that, as in the central North Sea, there is little error in assuming Airy isostacy in the northern North Sea region. Model RIFT is not applicable if the effects of flexural loading are considered important or in areas that are known to be isostatically uncompensated, such as the Inner Moray Firth basin (Barr, 1985).

The model RIFT is one dimensional in that it can be used to model the subsidence of only one point at a time within the basin, as is the case with the model of McKenzie (1978) and the other modifications described above. Two dimensional sections and maps must be constructed from one-dimensional results as in Barton and Wood (1984), Giltner (1987), Hellinger et al. (1989) and others. Questions of area/mass balance and the implied shape of the extending zone must, therefore, be addressed separately.

As with the models of McKenzie (1978), Jarvis and McKenzie (1980) and Royden and Keen (1980), only vertical heat conduction is considered. Thus, effects from lateral heat conduction are not included. Such effects could be expected to increase heat loss (and therefore the subsidence rates) at a rift centre and to retard heat loss (and therefore decrease subsidence rates) from the rift flanks. Extension factors derived from RIFT would then be overestimated in the graben centre and underestimated on the flanks.

However, in the case of the northern North Sea, the generally good agreement between the range of estimated β factors for Triassic and Jurassic extension events (Giltner, 1987; Beach et al., 1987; Vially, 1988; Lippard and Liu this volume) and estimates of the pre-Triassic basement thickness (Klemperer, 1988) would suggest that this is not an important factor. Model RIFT is not applicable if large magma bodies in the lower crust are associated with rifting as this would produce different heat flow and therefore subsidence patterns (Royden and Keen, 1980).

Acknowledgements

The author would like to thank I.K.U., owners of program RIFT, for permission to publish the model. I would also like to thank Alastair Jenkins for his mathematical help during the model construction and Øivind Sylte and Guojiang Liu for checking the model algorithms. I thank S. Lippard for discussions on North Sea geology.

Appendix A: Derivation of the Fourier coefficient for the initial temperature distribution in the multilayer model case

The temperature of the lithosphere in the equilibrium state is assumed to increase linearly from $t = 0°C$ at the surface of the crust ($z = a$) to $t = T_1$ at the base of the lithosphere ($z = 0$). The lithosphere is divided into N number of layers of equal thickness a/n with layer 1 at the base of the lithosphere (Fig. A-1). In the equilibrium state, the base of the nth layer lies at depth:

$$z = \frac{a}{N}(n-1) \qquad \text{(A-1)}$$

The temperature at depth, z, in the nth layer is:

$$T = T_1\left(\frac{n}{N} - \frac{z}{a}\right) + T_1\left(1 - \frac{n}{N}\right)$$

After instantaneous extension of the lithosphere, each layer is extended by factor β_n, where β is the ratio of initial to final layer thickness. The new distance to the base of the nth layer is:

$$z = a\left(1 - \frac{1}{N}\sum_{m=n}^{N}\frac{1}{\beta_m}\right) \qquad \text{(A-2)}$$

and the temperature at depth, z, in the nth layer is:

$$T_{(z)} = T_1\beta_n\left(1 - \frac{1}{N}\sum_{m=n+1}^{N}\frac{1}{\beta_m} - \frac{z}{a}\right) + T_1\left(1 - \frac{n}{N}\right) \qquad \text{(A-3)}$$

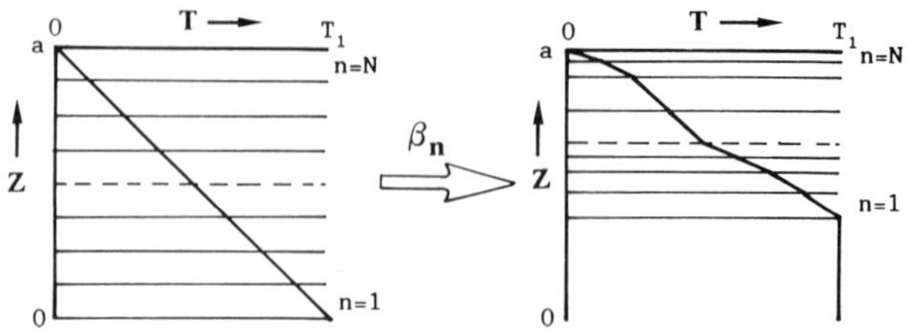

Fig. A-1. Temperature profile of the lithosphere before and after non-uniform, pure shear extension. The equilibrium lithosphere (linear temperature profile) is divided into N layers, each of which is independently assigned an extension factor β_n. Since the extension factor can vary with depth, this results in a generally non-linear temperature profile through the thinned lithosphere.

The general expression for the kth Fourier coefficient, b_k, of this temperature distribution is (Jarvis and McKenzie, 1980, their eqn. 19):

$$b_k = \frac{2}{aT_1}\int_0^a \left[T_{(z)} + T_1\left(\frac{z}{a} - 1\right)\right]\sin\frac{k\pi z}{a}\,dz \qquad \text{(A-4)}$$

In the multilayer model, the contribution that the nth layer makes to this integral is:

$$\frac{2}{a}\int_{aX}^{aY}\left[\beta_n\left(1 - \frac{1}{N}\sum_{m=n+1}^{N}\frac{1}{\beta_m}\right) + \frac{z}{a}(1-\beta_n) - \frac{n}{N}\right]\sin\frac{k\pi z}{a}\,dz \qquad \text{(A-5)}$$

where X and Y are, respectively, the depths to the base and top of the nth layer:

$$X = 1 - \frac{1}{N}\sum_{m=n}^{N}\frac{1}{\beta_m}$$

and

$$Y = 1 - \frac{1}{N}\sum_{m=n+1}^{N}\frac{1}{\beta_m}$$

Evaluating the integral in eqn. 5 gives an expression for the nth contribution to the kth Fourier coefficient:

$$\frac{2}{k\pi}\left(\beta_n Y - \frac{n}{N}\right)[\cos k\pi X - \cos k\pi Y] + \frac{2}{k\pi}(1-\beta_n)\left[\frac{1}{k\pi}\sin k\pi Y - \frac{1}{k\pi}\sin k\pi X + X\cos k\pi X - Y\cos k\pi Y\right] \qquad \text{(A-6)}$$

The final $(n+1)$th contribution comes from the portion of the asthenosphere lying between the original base of the lithosphere ($z = 0$) and the base of layer 1 ($z = z_1$):

$$\frac{2}{a^2}\int_0^{z_1} z\sin\frac{k\pi z}{a}\,dz$$

where:

$$z_1 = a\left(1 - \frac{1}{N}\sum_{m=1}^{N}\frac{1}{\beta_m}\right)$$

Evaluating this integral gives:

$$\frac{2}{k\pi}\left[\frac{1}{k\pi}\sin k\pi\left(1 - \frac{1}{N}\sum_{m=1}^{N}\frac{1}{\beta_m}\right) - \left(1 - \frac{1}{N}\sum_{m=1}^{N}\frac{1}{\beta_m}\right) \times \cos k\pi\left(1 - \frac{1}{N}\sum_{m=1}^{N}\frac{1}{\beta_m}\right)\right] \quad \text{(A-7)}$$

Adding all contributions from the stretched lithosphere (eqn. 5) and that of the remaining thickness to the base of the original lithosphere (eqn. 7) and simplifying gives the final expression for the kth Fourier coefficient, b_k:

$$b_k = \frac{2}{k\pi}\sum_{n=1}^{N}\left[\frac{1}{k\pi}(\beta_n - \beta_{n-1})\sin k\pi\left(1 - \frac{1}{N}\sum_{m=n}^{N}\frac{1}{\beta_m}\right)\right] \quad \text{(A-8)}$$

References

Badley, M.E., Egeberg, T. and Nipen, O., 1984. Development of rift basins illustrated by the structural evolution of the Oseberg feature, Block 30/6, offshore Norway. J. Geol. Soc. London, 141: 639–649.

Badley, M.E., Price, J.D., Rambech Dahl, C. and Agdestein, T., 1988. The structural evolution of the northern Viking Graben and its bearing upon extensional modes of basin formation. J. Geol. Soc. London, 145: 455–472.

Barr, D., 1985. Three dimensional restoration of normal faults in the Inner Moray Firth: implications for extensional basin development. Earth Planet. Sci. Lett., 75: 191–203.

Barton, P. and Wood, R., 1984. Tectonic evolution of the North Sea basin: crustal stretching and subsidence. Geophys. J. R. Astron. Soc., 79: 987–1022.

Beach, A., Bird, T. and Gibbs, A.D., 1987. Extensional tectonics and crustal structure: deep seismic reflection data from the northern North Sea Viking Graben. In: M.P. Coward, J.F. Dewey and P.L. Hancock (Editors), Continental Extensional Tectonics. Geol. Soc. London, Spec. Publ., 28: 467–476.

Davies, E.J. and Watts, T.R., 1977. The Murchison oil field. Norwegian Petroleum Society (NPF), Mesozoic Northern North Sea Symposium, MNNSS/15: 1–24.

De'Ath, N.G. and Schuyleman, S.F., 1981. The Geology of the Magnus oil field. In: L.V. Illing and G.D. Hobson (Editors), Petroleum Geology of the Continental Shelf of Northwest Europe, Institute of Petroleum, London, pp. 342–351.

de Charpal, O., Montadert, L., Guennoc, P. and Roberts, D.G., 1978. Rifting, crustal attenuation and subsidence in the bay of Biscay. Nature, 275: 706–711.

Fowler, S. and McKenzie, D., 1989. Gravity studies of the Rockall and Exmouth Plateaux using SEASAT altimetry. Basin Res., 2: 27–34.

Gabrielsen, R.H., Færseth, R.B., Idil, S. and Steel, R., 1989. Architectural styles of basin fill in the North Viking Graben. Terra Abstracts, 1: 33.

Giltner, J.P., 1987. Application of extensional models to the Northern Viking Graben. Nor. Geol. Tidsskrift, 67: 339–352.

Hellinger, S.J., Sclater, J.G. and Giltner, J.P., 1989. Mid-Jurassic through mid-Cretaceous extension in the Central Graben of the North Sea, Part 1. Estimates from subsidence. Basin Res., 1: 191–200.

Holliger, K. and Klemperer, S.L., 1989. A comparison of the Moho interpreted from gravity data and from deep seismic reflection data in the northern North Sea. Geophys. J., 97: 247–250.

Jarvis, G.T. and McKenzie, D., 1980. Sedimentary basin formation with finite extension rates. Earth Planet. Sci. Lett., 48: 42–52.

Karlsson, W., 1986. The Snorre, Statfjord and Gullfaks oilfields and the habitat of hydrocarbons on the Tampen Spur, offshore Norway. In: A.M. Spencer (Editor), Habitat of Hydrocarbons on the Norwegian Continental Shelf. Graham and Trotman, London, pp. 181–197.

Klemperer, S.L., 1988. Crustal thinning and nature of extension in the northern North Sea from deep seismic reflection profiling. Tectonics, 7: 803–821.

Lippard, S. and Liu, G., 1992. Tectonic modelling of the northern North Sea using program RIFT. In: R.M. Larsen, H. Brekke, B.T. Larsen and E. Talleraas (Editors), Structural and Tectonic Modelling and its Application to Petroleum Geology. Norwegian Petroleum Society (NPF), Special Publication 1. Elsevier, Amsterdam, pp. 43–54 (this volume).

McClay, K.R., Norton, M.G., Coney, P. and Davis, G.H., 1986. Collapse of the Caledonian orogen and the Old Red Sandstone. Nature, 323: 147–149.

McKenzie, D., 1978. Some remarks on the developments of sedimentary basins. Earth Planet. Sci. Lett., 40: 25–32.

Royden, L. and Keen, C.E., 1980. Rifting process and thermal evolution of the continental margin of Eastern Canada determined from subsidence curves. Earth Planet. Sci. Lett., 51: 343–361.

Rowley, D.B. and Sahagian, D., 1986. Depth-dependent stretching: a different approach. Geology, 14: 32–35.

Sclater, J.G. and Christie, P.A.F., 1980. Continental stretching: an exploration of the post-mid-Cretaceous subsidence of the North Sea basin. J. Geophys. Res., 85: 3711–3739.

Sclater, J.G., Royden, L., Horvath, F., Burchfiel, B.C., Semken, S. and Stegena, L., 1980. The formation of the intra-Carpathian basins as determined from subsidence data. Earth Planet. Sci. Lett., 51: 139–162.

Tankard, A.J. and Welsink, H.J., 1987. Extensional tectonics and stratigraphy of Hibernia oil field, Grand Banks, Newfoundland. Am. Assoc. Pet. Geol., Bull., 71: 1210–1232.

Vially, R., 1988. La subsidence dans le Viking Graben (mer du nord septentrionale). Rev. Inst. Fr. Pétrole, 43: 629–646.

Wernicke, B., 1984. Uniform-sense normal simple shear of the continental lithosphere. Can. J. Earth Sci., 22: 108–125.

White, N., 1988. Nature of lithospheric extension in the North Sea. Geology, 17: 111–114.

White, N. and McKenzie, D., 1988. Formation of "steer's head" geometry of sedimentary basins by differential stretching of the crust and mantle. Geology, 16: 250–253.

Ziegler, P.A., 1982. Geological Atlas of Western and Central Europe. Shell/Elsevier, Amsterdam.

N.E. ODLING *Continental Shelf and Petroleum Technology Research Institute (IKU), NHåkon Magnusson's gt. 1B, P.O. Box 1883, Jarlesletta, 7001 Trondheim*

Present address: *Bergen Environmental Centre, IBM, Thormøhlensgt. 55, Bergen, N-5008, Norway*

Structural styles in thrust belts developed through rift basins: a view from the western Alps

A.I. Welbon and R.W.H. Butler

In the external French Alps Mesozoic passive margin development on the flanks of the Tethyan Ocean resulted in repeated phases of extension with the formation of faults and associated stratigraphic variations. Extensional faults are found with great lateral extent throughout the French subalpine chains, often roughly parallel but with antithetic orientations and a dip either towards the foreland or hinterland. Oligocene and Miocene shortening in the area resulted in fold and thrust geometries which show a strong orientational relationship to the original normal faults. These early structures and the related stratigraphy also control the propensity for folding as opposed to simple thrust breaks.

We review published structural sections from the western Alps, together with unpublished data from the French-Swiss border southwards to illustrate the structural style of these stacked rift basins. This recognition of the interaction of basin geometry and thrusts is important in the modelling of the thrust belt structures and their geometric evolution.

The recognition of the structural styles in this tectonic setting presents an opportunity to improve our understanding of structural geometry of hydrocarbon traps for thrust belts developed across rift basins. For example, over-ridden fault blocks, normal faults reactivated as thrusts and folding against normal faults can provide important new exploration plays, but the complexities of structural styles may make these elusive targets.

Introduction

Research into the foreland fold and thrust belt of the Rocky Mountains has led to the acceptance of certain geometrical styles in thrust belts (Dahlstrom, 1970; Boyer, 1986; Mitra, 1986). These have been based upon the recognition of a series of duplexes and thrust structures which deformed a layer-cake or wedge-shaped stratigraphic template (for more examples see Boyer and Elliott, 1982; Suppe, 1983). Similar geometries are also present in other thrust belts such as the Appalachians and Scandinavian Caledonides (see Woodward and Gray, 1985; Morley, 1986; Townsend et al., 1986).

However, in many parts of some thrust belts, a layer-cake or wedge-shaped stratigraphic template was not the pre-existing basin geometry through which thrusting occurred. Often thrust belts have developed within areas of recognised passive margin stratigraphy, consisting of pre-rift basement, tilt-blocks and a clastic syn-rift cover, and an overlying post-rift sequence. Alternatively, continental margin development could have related to a complicated overprinting of rifting events. In both cases thrusting took place through complex stratigraphy and normal faults and resulted in the formation of complicated thrust geometries (McMechan and Price, 1982; Bally, 1984; Dewey et al., 1986; McClay et al., 1989).

Recently, the effect of variations in basin geometry on the development of thrust belts has been a focus of interest, not just in foreland areas, but also in the mechanisms of crustal stacking at depth during collision (Jackson, 1980). The relationship between basin geometry and thrust belt style is of importance if we are to be able to model thrust belt structures and understand their geometrical evolution.

In the European Alps, the stratigraphic template is known to be complicated by normal faults and stratigraphic variations related to the Mesozoic passive margin of the Tethyan Ocean, (Tricart, 1984; Lemoine et al., 1986; Tricart and Lemoine, 1986; Lemoine and Trümpy, 1987; Lemoine and Graciansky, 1988). The influence of this Alpine passive margin configuration on compressional deformation has been known for a long time. The presence of basement highs and lows had a control on nappe geometry, with bounding thrusts often defined at the margins of the pre-existing blocks and basins (Argand, 1916). On a smaller scale, within these major basins, pre-thrusting normal faults periodically controlled cover sequence development. These faults and related stratigraphic variations also had an influence on thrust geometry (Welbon, 1988a; Butler, 1989a). In this paper, we review geometry of the rift basins of the Tethyan margin in the foreland of the French Alps and ideas for the nature of in-

Structural and Tectonic Modelling and its Application to Petroleum Geology edited by R.M. Larsen, H. Brekke, B.T. Larsen and E. Talleraas. NPF Special Publication 1, pp. 469–479. Elsevier, Amsterdam.

teractions between pre-existing basin geometry and Alpine compressional deformation. Much data on this topic is dispersed in Alpine literature or in theses, so here we collate it together with new data to provide an overview of the nature of the structural style in this thrust belt.

Finally, we go on to look at the implications of this style of tectonics for the geometry of structural hydrocarbon traps, and discuss the possibilities of extrapolation of these styles to hydrocarbon provinces such as the Rockies (Schmidt and Perry, 1988).

The French Alps

Geological setting

In the external French Alps repeated phases of extension related to Mesozoic passive margin development of the Tethyan Ocean were followed by limited Eocene then Oligo-Miocene shortening leading to the formation of the thrust belt (Trümpy, 1973; Lemoine et al., 1986; Lemoine and Trümpy, 1987). Excellent exposure makes the western Alps an ideal locality to document structural styles in a thrust belt and their control by inherited basin geometry.

Many review articles exist on the stratigraphy of the western Alps and its tectonic framework (e.g., Lemoine et al., 1986; Tricart and Lemoine, 1986). General agreement exists upon the main characteristics in the region. Hercynian crystalline basement is overlain by Triassic pre-rift sandstones, shales and evaporites. The advent of Tethyan margin extension resulted in Jurassic argillite deposition superseded by development of a major limestone unit, the Tithonian. More uniform regional subsidence in the Cretaceous began with deposition of interbedded sandy limestones and marls followed by carbonate platform development (the Urgonian limestone). Tertiary sedimentation began with syn-rift deposition of limestones and marls and became progressively flysch-like in nature reflecting the onset of compressional tectonics in the developing Alpine orogen.

The structural style in the external French Alpine Thrust Belt varies along strike from a series of fold dominated nappes at the French-Swiss border (the Helvetic Nappes, e.g., Ramsay et al., 1983) to a more thrust dominated sequence to the southwest (Doudoux et al., 1982; Butler, 1989a, b). A consistent feature along the whole External French Alps, and the Swiss Alps to the north, is the presence of laterally continuous intrabasinal faults (Lemoine et al., 1986). These were active at various times from Triassic (Froitzheim, 1988) through Jurassic (Lemoine et al., 1986) to Oligocene (Butler, 1989a). Some normal faults show long-lived tectonic histories, with reac-

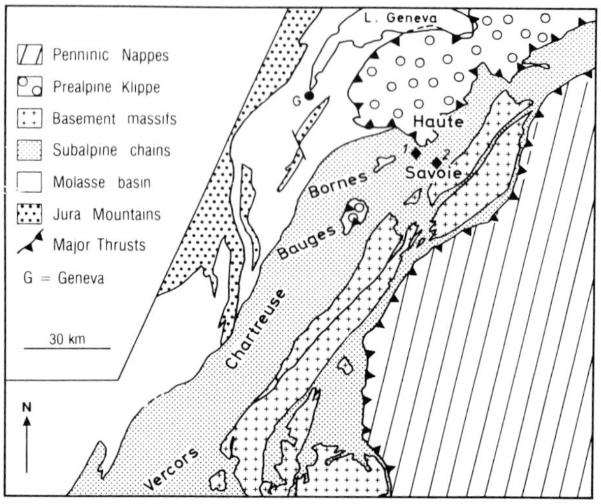

Fig. 1. A geological map of the External French Alps showing the location of the subalpine chains and basement massifs containing the normal and thrust faults described in the text.

tivation taking place, whilst others were only active during one extensional episode. Normal faults may extend from Tertiary cover rocks (Welbon, 1988a, b) as far down as the Hercynian crystalline basement in the more central Alps (Lemoine et al., 1986; Gillcrist et al., 1987). During shortening the variation in basin geometry resulted in the formation of many complicated thrust structures (Welbon, 1988a; Butler, 1989a).

Examples

In Haute Savoie (Fig. 1), in addition to thrust geometries related to thrust development in layer-cake cover sequences, several spectacular examples of normal fault/thrust fault interactions are exposed. Often associated with these outcrops is a pronounced stratigraphic variation between hanging-wall and footwall rocks across thrusts. Examples of these structures can be found in Welbon (1988a, b) and in Butler (1989a, 1992).

In the Sales–Pointe de Platé area (Fig. 1) a thrust fault exhibits a clear variation in stratigraphy from hanging-wall to footwall, although the sequences are of the same age (Fig. 2) (Welbon, 1988a, b). Kinematic indicators (small scale folds, sheared cleavage, striae) show the latest movements on the fault to have been thrust sense. The fault at Sales is well exposed from the level of the Early Cretaceous (Urgonian) limestone facies up to Oligocene sandstones. The Urgonian facies are similar on either side of the fault, but the footwall Late Cretaceous units are 150 m thicker than rocks of the same age in the hanging-wall. Similarly, the footwall Tertiary units exhibit a greater thickness, 70 m as compared to 30

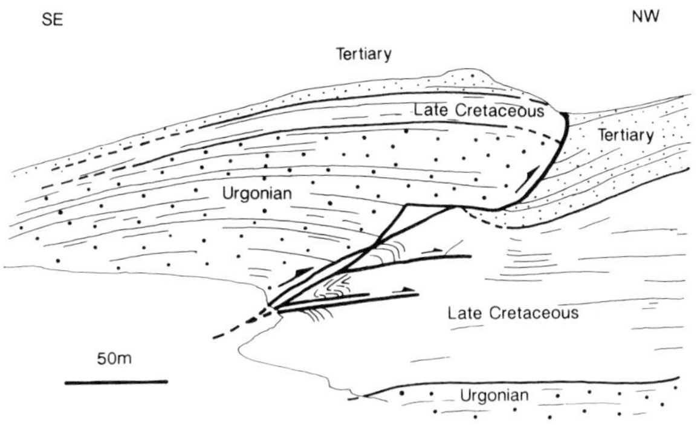

Fig. 2. A sketch of the Sales fault (after Welbon 1988a, b). An increased stratigraphic thickness of footwall rocks relative to those in the hanging-wall is evident. Footwall deformation in the Late Cretaceous limestones consists of folds and thrusts which lose displacement away from the fault.

m in the hanging-wall. Intense deformation in the form of folds and thrusts characterises the footwall, especially the Late Cretaceous units, but restoration of these structures reveals that a considerable stratigraphic thickness difference remains between the rocks on either side of the fault. Shortening increases towards the top of the Late Cretaceous units, which, when restored, indicates that the fault was originally near vertical or dipping in the opposite direction (i.e., NW) to the present SE dip. This is supported by the presence of a NW-dipping Tertiary syn-sedimentary normal fault carried in the hanging-wall which was presumably synthetic to the original major normal fault. Slip surfaces between the Tertiary and Late Cretaceous rocks in the footwall (Welbon, 1988b) would also have accommodated increased shortening in the upper stratigraphic units. From this data, it is concluded that the fault had an early history of extension, prior to and during Eocene deposition, and was then overturned and reactivated as a thrust during Oligo-Miocene shortening.

Along strike at the Pointe de Platé, the fault has a similar variation in stratigraphic thickness as at Sales, but the fault has a splay in its footwall that carries Urgonian limestone (Fig. 3A). The fault has a thrust-sense offset and is interpreted as a reactivated normal fault (Fig. 3B). Only the top part of the normal fault has been used by a thrust after it was overturned during folding. In this case the earlier folding is related to the thrust which carries Urgonian facies lithologies in the footwall to the reactivated fault.

At Balme, in the Arve valley (Fig. 1, Fig. 4A) the structures are more complicated than those at Sales. Here, a graben structure with similar thickening to that found at Sales (i.e., within Late Cretaceous and Tertiary rocks) is exposed. The graben has two oppositely dipping faults that have a strike similar

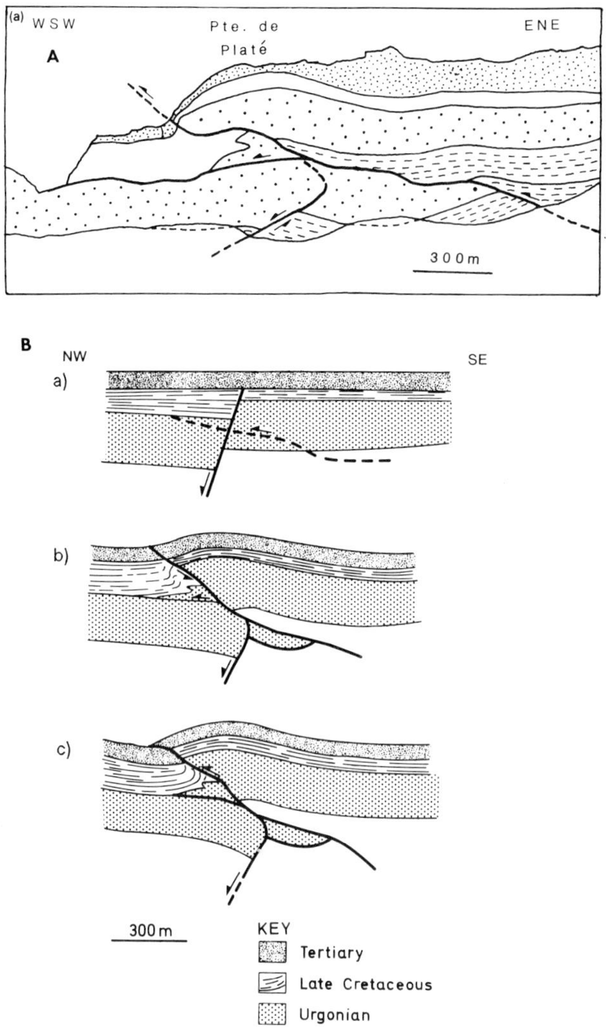

Fig. 3. (A) A line drawing of a photograph of the Pointe de Platé Fault which is highly oblique to its strike. This is an along-strike expression of the Sales Fault 3 km SW of the exposures in the Sales valley. A similar stratigraphic variation to that shown in Fig. 2 occurs. The fault structure has a splay in its footwall which cuts the Urgonian rocks. Key: densely dotted area = Tertiary; blank area = Late Cretaceous; sparsely dotted area = Urgonian limestone; dashed areas = Hauterivian-Berriasian. (B) An evolutionary diagram of the Pointe de Platé Fault. (a) Potential fault trajectory. (b) A fault cuts through the Urgonian rocks and overturns the normal fault carried in its hanging-wall. (c) The overturned part of the normal fault is reactivated as a thrust.

to that of the later thrust faults. Two major NW verging thrusts cut the graben (Fig. 4A, B). The upper one cuts through and overturns a normal fault dipping away from it and ramps ahead of the SE-dipping normal fault. The lower NW verging thrust cuts through both sets of faults, carrying the graben and is exposed 400 m NW of the graben. Again, like the Sales example where the early fault dipped away from the advancing thrust, folding has been an important part of the shortening process, resulting here in partial overturning of the early fault.

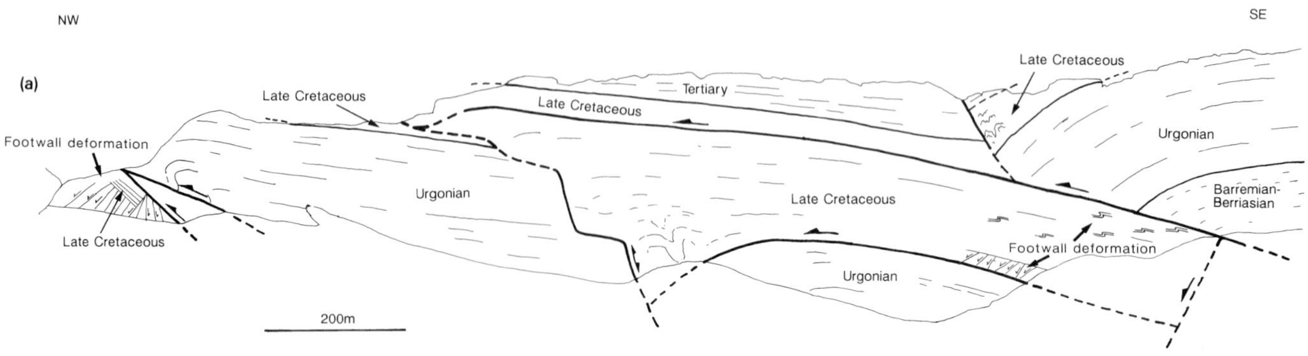

In addition to the formation of major displacement thrusts, minor thrusts (decollement) develop within thrust sheets at contacts between less competent, well bedded, Late Cretaceous limestones and the more competent, massive Urgonian limestones (Fig. 4A, B). The origin of the internal deformation of thrust sheets is the ramp areas of the overlying major thrusts, and transference of displacement is evidenced by footwall deformation linking the overlying thrust plane with the decollement (Welbon, 1988b). The minor thrusts pin at normal faults and shortening above the decollement is accommodated by folding against the normal fault (buttressing). These structures are folded by hanging-wall anticlines developed above underlying major thrusts and are, therefore, formed prior to the formation of the major thrusts.

Further south in the Chartreuse (Fig. 1) stratigraphic thickness variations suggest the presence of a near vertical, SE dipping, normal fault cut by the Corbel Thrust Zone (Fig. 5). The Thrust Zone (Butler, 1992) consists of a complicated zone of branching thrusts developed up dip from the normal fault carried in the hanging-wall. Folding above and below the thrust is minor and, therefore, a simple offset of the normal fault is observed. This is in contrast to the geometries in the Balme and Sales localities where folding within thrust sheets is sometimes sufficient to overturn early faults.

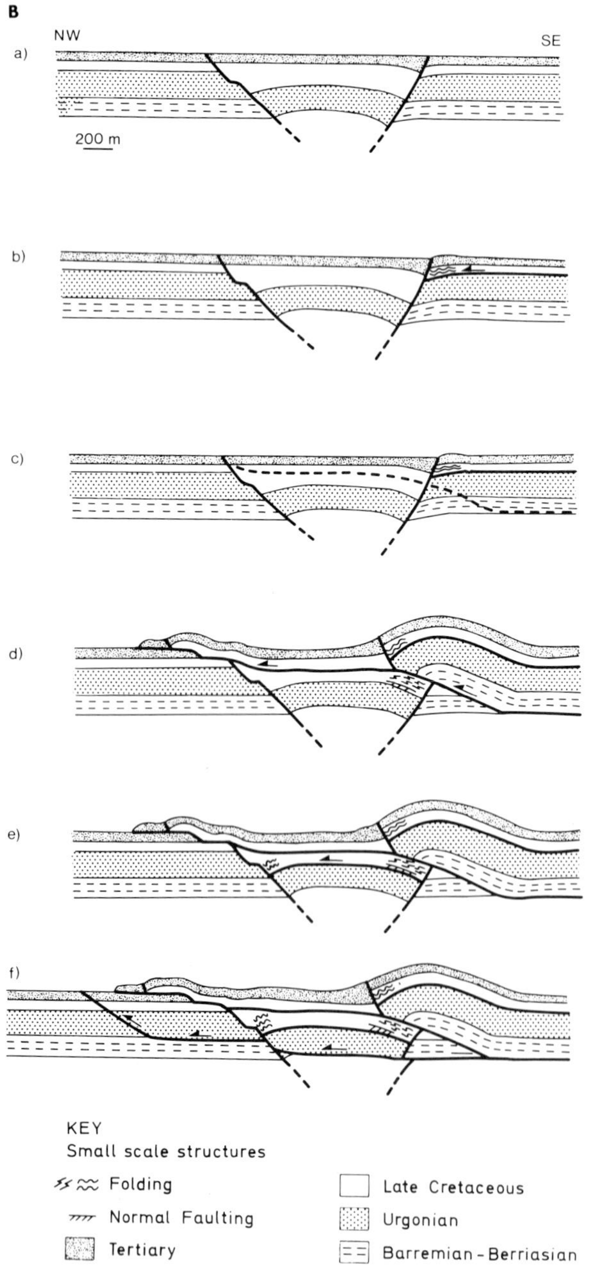

Fig. 4. (A) A diagram and (B) evolutionary model of the fault geometries in Balme (after Welbon 1988a, b). In B: A graben with two oppositely dipping faults is cut by two NW verging thrusts. The graben thickens Late Cretaceous limestones and, in the case of the NW dipping normal fault, Tertiary rocks are thickened. (b) A minor thrust develops at the Urgonian-Late Cretaceous contact and pins at the NW dipping normal fault, causing folding above the thrust. (c, d) The upper major thrust offsets the NW dipping normal fault and the upper part of it is overturned in the hanging-wall anticline developed above the thrust. Note the upper thrust ramps ahead of the SE dipping normal fault and links into less competent units (thinly bedded Late Cretaceous limestones) above the competent Urgonian facies massive limestones. (e) A minor thrust forms at the Urgonian-Late Cretaceous boundary, pins at the SE dipping normal fault and continued shortening results in buttressing (folding) against the normal fault. (f) The lower major thrust transports both normal faults and the upper thrust in its hanging-wall. It forms a ramp approximately 400 m NW of the graben.

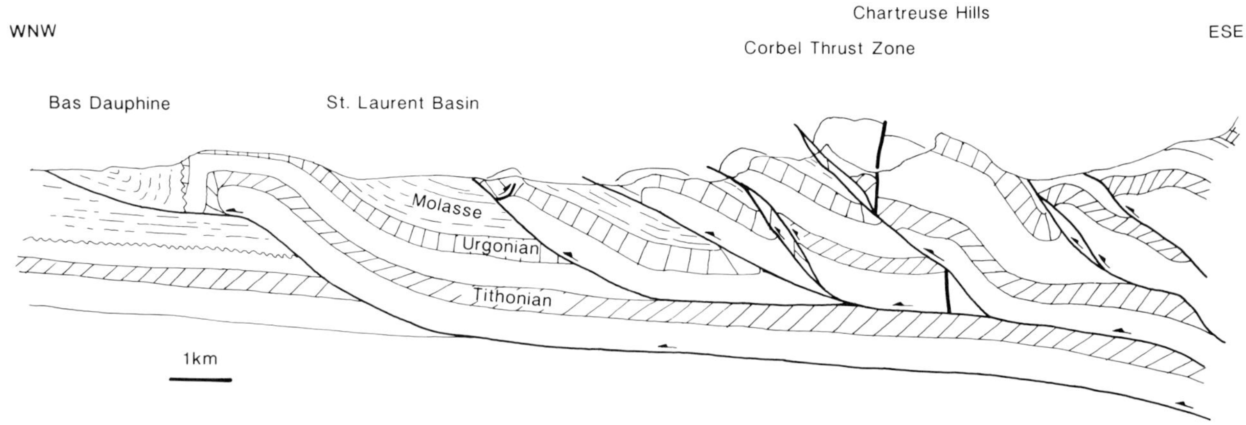

Fig. 5. A section through the Chartreuse region showing a SE dipping normal fault offset by the Corbel Thrust Zone (after Butler, 1992).

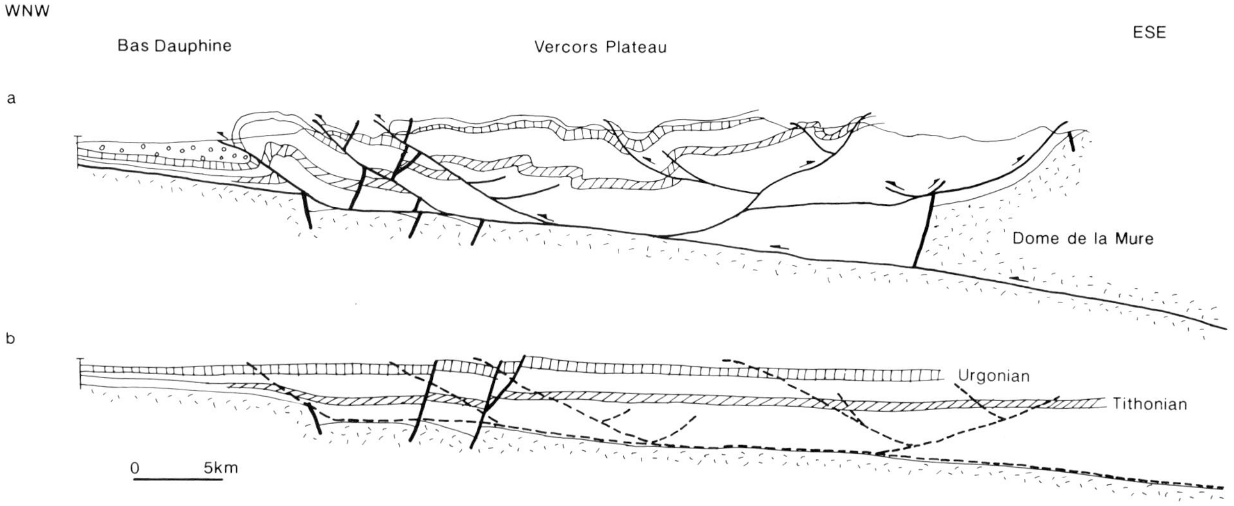

Fig. 6. A section through the Vercors (after Butler, 1989a). Note the backthrusting of cover over the Dome de la Mure massif and the buried NW dipping normal fault at the front of the massif. Within the cover at Royans, near the thrust front, thrusts cut through a pair of WNW dipping normal faults. At the thrust front an ESE dipping normal fault in the basement is over-ridden by a thrust.

In the Vercors region SW of the Chartreuse (Fig. 1) there are more examples of the original rift geometry in this part of the thrust belt (Fig. 6). In the most internal part of the section at the Dome de la Mure basement massif, a WNW dipping early normal fault is predicted from thickness variations in the pre-Aalenian cover rocks (Butler, 1989a). This early normal fault is predicted to cut basement rocks and thicken Liassic sediments. The footwall to this fault, the Dome de la Mure basement massif, was emplaced as a thrust sheet beneath the cover rocks resulting in the formation of a backthrust between the basement and cover. This backthrust carries Jurassic and younger rocks across the thin cover sequences on the crest of the Dome de la Mure massif.

More externally in the Vercors, at Royans, thrusts cut through a pair of WNW dipping normal faults. Again, in a similar fashion to the structural styles in the Chartreuse, there is little overturning of the faults, they are simply cut through. Finally, at the thrust front, an ESE dipping normal fault in the basement is overridden by a thrust fault ramping ahead of it (Fig. 6)

At the front of the Jura mountains (Fig. 7) a series of NW dipping normal faults bounding the Bresse graben are truncated by WNW verging thrust structures (Butler, 1989b). The dip of these normal faults is antithetic to the thrust faults, and, like the Sales example, illustrates that when normal faults dip away from an advancing thrust, they often present no barrier to thrusting and are simply cut through.

In summary, from the geometry of the structures in the western Alps a progressive change in response to shortening of the early faults and their related stratigraphy is suggested. More folding occurs in the northern part of the French Alps at the Swiss border than to the southwest towards the Vercors. It is also evident that a primary influence on thrusting is the original orientation of early faults. Early faults dipping towards the thrust tend to cause ramping of the thrust ahead of the early fault, those dipping

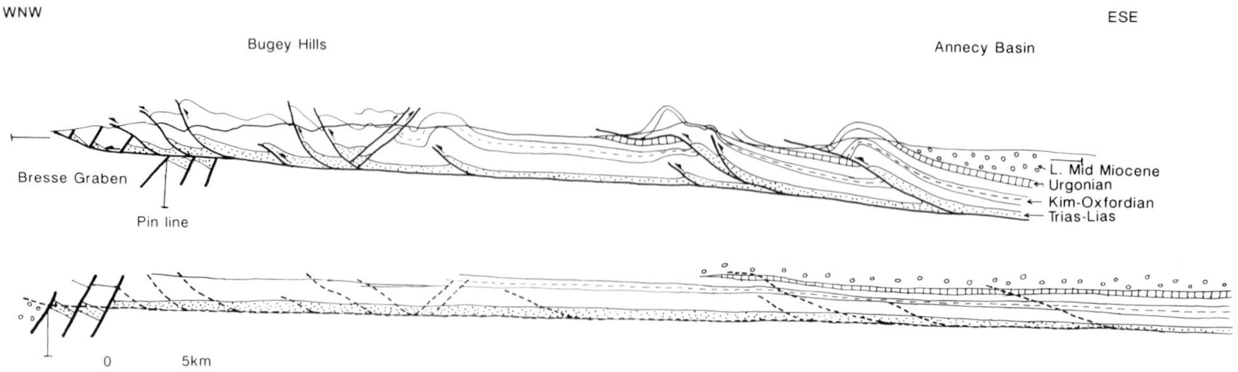

Fig. 7. A section through the Jura Mountains after Butler (1989b). Note WNW verging thrusts cut from the Jura mountains through the NW dipping normal faults bounding the Bresse graben.

away from the thrust are usually overturned and partially reactivated, or cut through. The variation in geometry of the stratigraphic units adjacent to the fault and the differences in competence of these stratigraphic units also plays an important role in determining thrust geometry.

Although we have illustrated many examples of early normal fault-late thrust fault interactions, it is important to note that many thrusts are demonstrably unrelated to normal faults or stratigraphic variation (Doudoux et al., 1982; Ramsay et al., 1983; Butler, 1989b). Thrust spacing is smaller than the spacing between significant basin heterogeneities. Thus, many thrusts within the thrust belt propagate through parts of the stratigraphic sequence where there was no apparent influence from pre-existing structures. The resulting geometries are those such as thrusts, duplexes, with fault-bend and fault propagation folds (Boyer and Elliott, 1982; Boyer, 1986; Mitra, 1986; Welbon, 1988b).

Discussion

The examples of Alpine thrusts discussed in this paper consistently show a series of geometries between several end members. These are given below.

(1) Thrusts cut through early faults (e.g., Balme, Chartreuse, Vercors).

(2) Thrusts ramp ahead of normal faults dipping towards the advancing thrusts (e.g., Balme, Jura, Vercors).

(3) Thrusts cut through and overturn normal faults dipping away from the advancing thrust (e.g., Balme).

(4) Thrusts overturn and reactivate the top part of normal faults originally dipping away from the advancing thrust (e.g., Sales, Pointe de Platé).

(5) Thrusts pin at the early fault and shortening continues above the thrust in the form of folding (e.g., Balme).

Intermediate geometries are possible between these end members dependent on the amount of shortening and, like all rocks, other factors determining strain will also include lithology, pressure and temperature conditions and strain rate (e.g., Knipe, 1985).

However, from comparing the Sales and Balme examples where similar lithological units are involved in thrusting, but with different original geometries due to different extensional fault orientations and throws, it is evident that the degree of folding as opposed to simple thrust breaks and any style of thrusting is greatly dependent on the original orientation of the early fault and the geometry of the stratigraphic units developed around the fault. Often in the Alpine examples the fault itself is not a barrier to thrusting, but the juxtaposition by the fault of mechanically strong lithologies against weak lithologies is common (e.g., Balme) preventing the continuation of any decollement developed in the weak layer and thereby resulting in folding or ramping of thrusts.

Normal faults dipping towards the thrust are likely to cause ramping of the thrust ahead of the early fault. This is likely to be the case where intrabasinal (normal) faults are involved, as in their hanging-walls they tend to thicken younger, less competent stratigraphic units. Thus, in the footwall to the early fault a junction between competent and less competent units will be at a higher level and will act as a stress riser, the thrust will be localised here and folding will be limited (Wiltschko and Eastman, 1983; Schedl and Wiltschko, 1987). Early faults dipping away from an advancing thrust with competent, mechanically stronger units in the footwall produce a stress shadow across potential decollement surfaces in the hanging-wall. Hence these early faults are a barrier to thrusting and are much more likely to be overturned before a thrust will propagate forwards, the thrust often using the overturned fault as an easy slip surface.

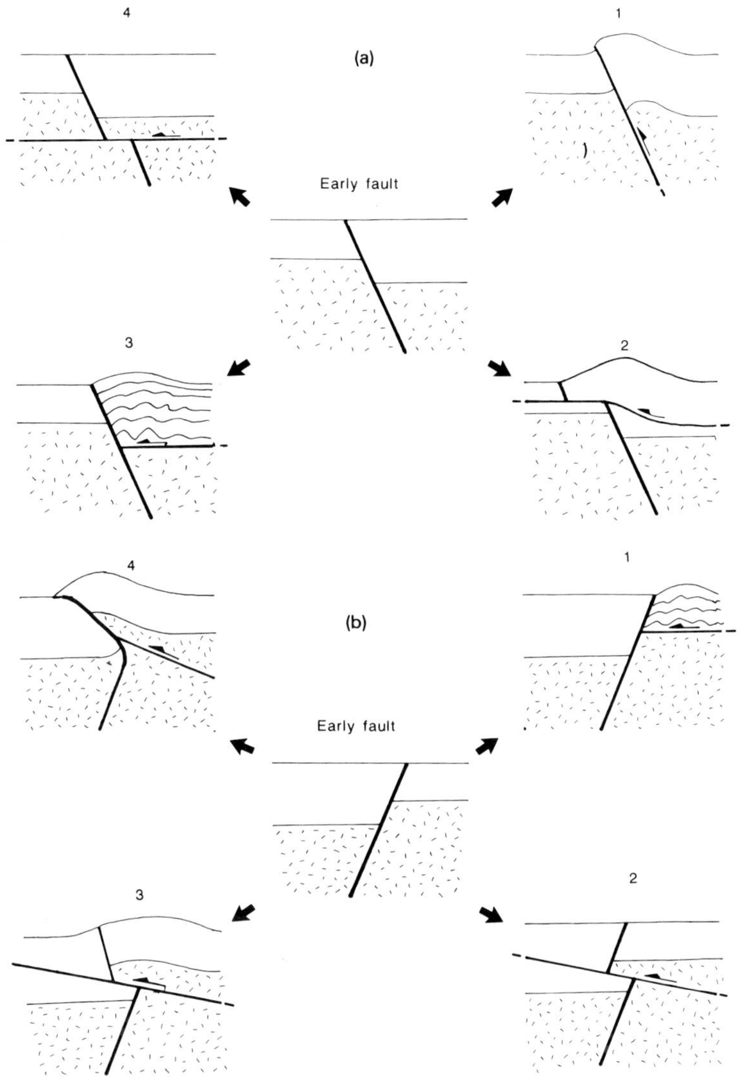

Fig. 8. A catalogue of possible thrust geometries found in areas where pre-existing faults and stratigraphic variations exist. These structures may form hydrocarbon traps. In the following description "basement" (dotted area on the figure) refers to a competent package of lithologies and "cover" (blank area on the figure) refers to a less competent package of lithologies. (a) Upper diagram. Structures formed when a pre-existing fault dips towards an advancing thrust. (1) Simple reactivation of the early fault. (2) The advancing thrust ramps ahead of the early fault and links into the cover above the competent basement. (3) The thrust forms along a basement-cover contact and pins at the early fault. Continued shortening results in folding above the detachment. (4) The early fault is cut through by the thrust. (b) Lower diagram. Structures formed when a pre-existing fault dips away from an advancing thrust. (1) The thrust forms a detachment between basement and cover and continued shortening results in the formation of folds. (2) The early fault is cut through by the thrust. (3) The early fault is cut through by the thrust and is overturned. (4) Folding by a results in overturning into an orientation appropriate for reactivation of the top part of the early fault. N.B., where folding takes place against a fault (a, example 3 and b, example 1), an assumption is made that the fault or the rock types across the fault exhibit relative competence compared to the folded strata.

On the basis of the examples discussed, we have made a catalogue of possible thrust geometries developed through areas of inherited structures and stratigraphic variations (Fig. 8). The figure shows possible structures that can form from shortening across the two most common types of inherited structures found in the Alps. The first set (Fig. 8a) is where an early fault dips towards the advancing thrust and the second set (Fig. 8b) is where a pre-existing fault dips away from the advancing thrust. Note the original early fault geometry and the distribution of mechanically different stratigraphic units shown in the figure could be formed by oblique slip or dip slip faults.

In a hydrocarbon province the resulting geometry after shortening in a thrust belt may provide the setting for structural traps for hydrocarbons. Although Fig. 8 shows only two-dimensional representations of the structures possible in a thrust belt through interaction with inherited structures, the figure can be used to discuss the potential geometries of hydrocarbon traps assuming structural closure in the third dimension, at right angles to the page. Below is an outline of the structural styles of the structures and a brief outline of the possible hydrocarbon traps.

Figure 8a shows the alternative structures formed in the case of an early fault dipping towards oncoming thrusts. Alternative 1 shows simple reactivation of a normal fault which may produce hanging-wall and footwall traps. The hanging-wall traps consist of anticlinal structures, whereas the footwall traps can be produced by combined footwall syncline formation and sealing along the fault. Alternative 2 forms when shortening takes place above a more competent unit which has been previously downfaulted. The thrust ramps ahead of the normal fault and links into less competent units in the footwall. Over-ridden fault blocks then become potential traps. Alternative 3 is similar to the previous model, but illustrates a situation where a thrust forms at a level between two units of different competence and then pins at the early fault or more competent footwall rocks. Continued shortening results in fold formation above the detachment. Depending on the scale of these structures, a potential hydrocarbon trap may form, but examples documented at Balme in the Alps are small and show a very complex geometry unsuitable for structural traps. Alternative 4 shows cutting through of the early fault and the carrying of the fault passively in the hanging-wall to the thrust. In many external fold and thrust belts this is likely to form where the lithologies on either side of the fault and the fault itself are mechanically weak. This can create two possible traps in the over-ridden block on either side of the early fault.

Figure 8b illustrates the alternatives in a scenario where an early fault dips away from the advancing thrust. Again the geometry of the early fault and the distribution of the mechanically different stratigraphic units can be produced by a normal fault or oblique slip fault. Alternative 1 shows a thrust which forms a decollement between competent and less competent units and pins at the early fault. Continued shortening causes folding above the detachment. The early fault (or the hanging-wall rocks in this case) are mechanically strong and act as a buttress to the footwall shortening. This occurred at the SE end of the Balme example, in the original footwall to the early normal fault (now overturned by a major thrust) at the level of the Late Cretaceous limestones. Significant hydrocarbon traps are unlikely to form if, as in Fig. 8a alternative 4, the structure is complex and small scale. Alternative 2 shows the most common structural style encountered in the study area of those in Fig. 8b. This is where the early fault is cut through by the thrust and carried passively. Again there is the possibility of creating structural traps in the over-ridden fault block. Alternative 3 illustrates a situation similar to that in 2, but where folding is more dominant and as a result the early fault carried in the hanging-wall to the thrust is overturned. Folding may also occur in the footwall to the thrust. In addition to traps formed in the over-ridden block, traps may form in the folded areas of the overturned normal fault. Alternative 4 shows overturning of the early fault being sufficient to allow reactivation of the top part of the normal fault as a thrust. Similar traps to those in alternative 3 are possible.

In addition to the possible traps indicated in Fig. 8, early faults may not interact with later thrusts and can be passively tilted or folded within thrust sheets to form hydrocarbon traps. This is particularly likely if the early faults are transported passively in one thrust sheet and are then folded by a culmination related to the emplacement of a lower thrust sheet. Reactivation of these faults may then occur if they are in an appropriate orientation, and this can conceivably result in reactivation of an early fault with its original sense of motion. The Mont Charvin area of Haute Savoie (for stratigraphic setting see Villars et al., 1988) may be an area where this type of reactivation has taken place (Fig. 9). Here there is an early Tertiary normal fault with a throw of more than 500 m which has been overturned by uplift of the underlying external Belledonne massif. Kinematic indicators on part of the normal fault, including folds, striae and an extensional crenulation cleavage in the hanging-wall, indicate that it was reactivated as a backthrust with the same type of offset as the original normal fault, although displacement was small.

It is important to note, especially when considering the hydrocarbon potential of these structures, that although the initial and final geometries of the structures described here are known, it is not always clear what the sequence of formation of the intermediate structures was. Folds may have formed prior to thrusts or vice versa. In many cases it is impossible to determine from geometry alone what the sequence was, although overprinting of varying kinematic indicators on a meso- or microscale may provide the answer. A particular intermediate structural style may result in the loss of hydrocarbons before a trap is created.

Published data from other thrust belts has illustrated some thrust belt styles similar to those in the western Alps. In the Rocky Mountains thrust belt thrust structures and hydrocarbon traps developed within duplexes and imbricate thrust stacks where shortening took place through a layer-cake or wedge-shaped stratigraphic template (Boyer and Elliott, 1982; Mitra, 1986). In other regions of the Rockies,

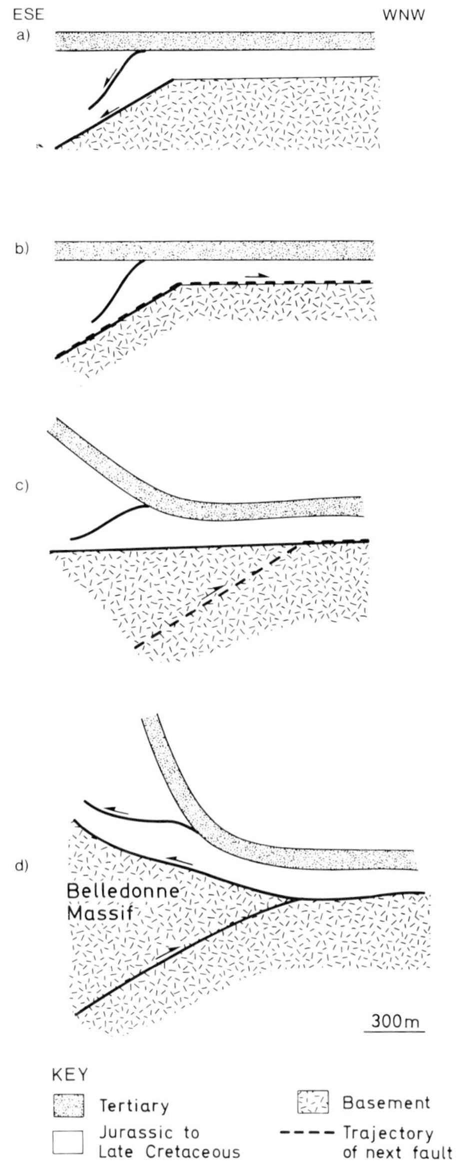

Fig. 9. A diagram and model of the Mt. Charvin area. The intrabasinal extension is well documented and results in thickening of the Late Cretaceous and Tertiary lithologies (Villars et al., 1988). (a) Original pre-Thrusting geometry, with extensional faults within the cover and at the basement cover contact. (b–d) Tilting of this fault by the uplift of the Belledonne massif results in it being in an appropriate orientation for reactivation as a backthrust with the same type of offset as the original normal fault.

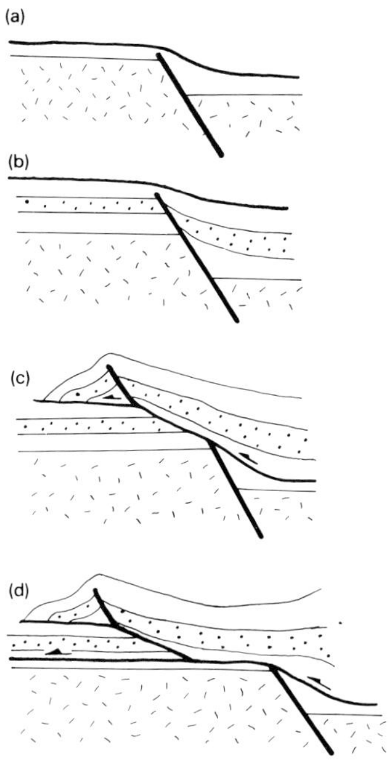

Fig. 10. A schematic history of the Birmingham Anticlinorium after Schedl and Wiltschko (1987) based on Thomas (1982). (a, b) Initiation of the basement normal fault in the Eocambrian, followed by drape folding after Mid-Cambrian, Mid-Silurian and Early Mississipian movement. (c, d) Initiation of a thrust ramp in the early Post-Pennsylvanian followed by transport forward via later faults. The step in the basement is interpreted as acting as a stress riser resulting in the ramping of the advancing thrust ahead of the basement normal fault.

normal faults and stratigraphic variations are more common and have resulted in a control of structural style in the thrust belt. Normal fault reactivation as thrusts, ramping ahead of structures at basin margins or buttressing (folding) against normal fault footwalls are known to have occurred in the foreland of the Rockies and much literature has been published on the nature of this style of tectonics (see Benvenuto and Price, 1979; Schmidt and Perry, 1988).

Appalachian thrust structures within cover rocks, where associated with early faults, are consistently found to ramp up above basement topography (e.g., Thomas, 1982, 1983) (Fig. 10). Again the thrusts link less competent units on either side of the normal faults/basement topography, the normal fault footwall rocks having acted as a stress riser as predicted from finite element modelling of Schedl and Wiltschko (1987) (Fig. 11).

In the Bass Straits/Gippsland basin of Australia the reactivation of deeply buried extensional blocks in a basin is common (Etheridge, 1986). Many styles of reverse and strike-slip reactivation are documented, largely depending on the original orientation of the early fault. This type of reactivation may be the case at a passive margin during the early stages of collision or where a basin has been weakly affected by a compressional orogenic event (see also Ziegler, 1983, for examples of North Sea Alpine inversion).

Conclusions

Accepted geometrical styles of thrusting from foreland areas with layer-cake or near so basin geometry include classic thrust duplexes and related folding. The examples of thrust belt geometry given above show that inherited tectonic structures and stratigraphic variations are a common feature of the

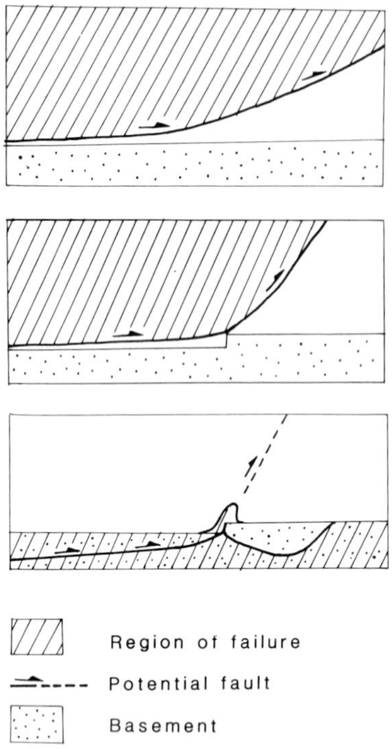

Fig. 11. A diagram of failure regions and potential faults for various geometries of basement faults and basement cover contacts, after Schedl and Wiltschko (1987). The potential geometry depends upon the basement topography and the degree of basement-cover coupling.

western Alps and other thrust belts and that their presence has had an influence on thrust evolution. The variation of early fault geometry and deformation conditions within a thrust belt means that later thrust fault geometries can show a wide variety of forms. More complex structures develop with progressive shortening. Depending on early fault orientation, reactivation, ramping of thrusts ahead of buried faults and folding against the early fault (buttressing) are common. Alternatively, early faults can simply be cut through by later faults. The recognition of these thrust belt styles is important if we are to model them to understand their geometrical evolution.

Structural complexity of this type in a hydrocarbon province can result in structural traps being formed. The type of hydrocarbon trap, like structural style, will vary greatly depending on the orientation of early faults and their related stratigraphic variations relative to thrusts and the level of decollement. The position of buried normal faults in over-ridden blocks may increase or decrease exploration potential, depending on the nature of the lithologies buried in the early fault beneath the thrust. Analysis employing a correlation between surface and seismic data may help constrain subsurface geometries.

The consideration of the presence of these types of compressional structure in any thrust belt or shortened sedimentary basin with non-layer-cake stratigraphy is necessary for a successful assessment of potential hydrocarbon traps during exploration. It may be insufficient to rely on techniques such as the projection of surface data to depth, (e.g., Suppe, 1983), with an assumption of lateral continuity of rock type or to make structural predictions without reference to stratigraphy and the influence rift structures have had on its development.

Acknowledgements

A.W. acknowledges a Natural Environment Research Council (NERC) Ph.D. studentship No. GT4/85/GS/44. A.W. also acknowledges a Royal Society European Science Exchange Programme Postdoctoral Fellowship and thanks Professor Arild Andresen for providing research facilities at the Department of Geology, University of Oslo. Fieldwork (R.W.H.B.) in the French Alps was funded by a Natural Environment Research Council (U.K.) grant No. (GR3/6172). We thank Rob Knipe, Daniel Muller and Francois Villars for discussions on Alpine Geology.

References

Argand, E., 1916. Sur l'arc des Alpes Occidentales. Eclogae Geol. Helv., 14: 145–191.

Bally, A.W., 1984. Tectonogenèse et sismique reflexion. Bull. Soc. Géol. Fr., 26: 279–285.

Benvenuto, G.L. and Price, R.A., 1979. Structural evolution of the Hosmer Thrust Sheet, southeastern British Colombia. Bull. Can. Pet. Geol., 27: 360–394.

Boyer, S.E., 1986. Styles of folding within thrust sheets: examples from the Appalachian and Rocky Mountains of the U.S.A. and Canada. J. Struct. Geol., 8: 325–339.

Boyer, S.E. and Elliott, D., 1982. Thrust systems. Am. Assoc. Pet. Geol., Bull., 66: 1196–1230.

Butler, R.W.H., 1989a. The influence of pre-existing basin structure on thrust system evolution in the Western Alps. In: M.A. Cooper and G.D. Williams (Editors), Inversion Tectonics. Geol. Soc. London, Spec. Publ., 44: 105–122.

Butler, R.W.H., 1989b. The geometry of crustal shortening in the Western Alps. In: A.M.C. Segnor (Editor), Tectonic Evolution of the Tethyan Region. Proceedings of the NATO Advanced Study Institute, C239, pp. 43–76.

Butler, R.W.H., 1992. Structural evolution of the western Chartreuse fold and thrust system, NW French Subalpine Chains. In: K.R. McClay (Editor), Thrust Tectonics. Chapman & Hall, London, pp. 287–298.

Dahlstrom, C.D.A., 1970. Structural geology in the eastern margin of the Canadian Rocky Mountains. Bull. Can. Pet. Geol., 18: 332–402.

Dewey, J.F., Hempton, M.R., Kidd, W.S.F., Saroglu, F. and Segnor, A.M.C., 1986. Shortening of continental lithosphere: the neotectonics of Eastern Anatolia — a young collision zone. In: M.P. Coward and A.C. Ries (Editors), Collision Tectonics. Geol. Soc. London, Spec. Publ., pp. 3–36.

Doudoux, B., Mercier de Lepinay, B. and Tardy, M., 1982. Une interpretation nouvelle de la structure des massifs sub alpins (Alpes occidentales): nappe de charriage oligocènes et déformations superposées. C.R. Acad. Sci. Paris, Sér. II, 295: 63–68.

Etheridge, M.A., 1986. On the reactivation of extensional fault systems. Philos. Trans. R. Soc. London, Ser. A, 317: 179–194.

Froitzheim, N., 1988. Synsedimentary and synorogenic normal faults within a thrust sheet of the eastern Alps (Ortler zone, Graubunden, Switzerland). Eclogae Geol. Helv., 81: 593–610.

Gillcrist, R., Coward, M.P. and Mungnier, J-L., 1987. Structural inversion and its controls. Examples from the Alpine Foreland and the French Alps. Geodyn. Acta (Paris), 1: 5–34.

Jackson, J.A., 1980. Reactivation of basement faults and crustal shortening in orogenic belts and plate tectonics. Nature, 283: 343–346.

Knipe, R.J., 1985. Footwall geometry and the rheology of thrust sheets. J. Struct. Geol., 7: 1–10.

Lemoine, M., Bas, T., Arnaud, H., Gidon, M., De Graciansky, P.C., Rudkiewicz, J-L., Megard-Galli, J. and Tricart, P., 1986. The continental margin of the Mesozoic Tethys in the Western Alps. Mar. Pet. Geol., 3: 179–199.

Lemoine, M. and Graciansky, P.C., 1988. Histoire d'une marge continental passive: les Alpes occidentales au Mésozoique. Introduction. Bull. Soc. Géol. Fr., 8: 597–600.

Lemoine, M. and Trümpy, R., 1987. Pre-oceanic rifting in the Alps. Tectonophysics, 133: 305–320.

McClay, K.R., Insley, M.W. and Anderton, R., 1989. Inversion of the Kechika trough, Northeastern British Columbia, Canada. In: M.A. Cooper and G.D. Williams (Editors), Inversion Tectonics. Geol. Soc. London, Spec. Publ., 44: 235–257.

McMechan, M.E. and Price, R. A., 1982. Transverse folding and superposed deformation, Mt. Fisher area, Southern Canadian Rocky Mountain thrust and fold belt. Can. J. Earth Sci., 19: 1011–1024.

Mitra, S., 1986. Duplex structures and imbricate thrust systems: geometry, structural position and hydrocarbon potential. Bull. Am. Assoc. Pet. Geol., Bull., 70: 1087–1112.

Morley, C.K., 1986 The Caledonian thrust front and palinspastic restorations in the southern Norwegain Caledonides. J. Struct. Geol., 8: 753–765.

Ramsay, J.G., Casey, M. and Kligfield, R., 1983. Role of shear in the development of the Helvetic fold-thrust belt of Switzerland. Geology, 11: 439–442.

Schedl, A.H. and Wiltschko, D.V., 1987. Possible effects of pre-existing basement topography on thrust fault ramping. J. Struct. Geol., 9: 1029–1037.

Schmidt, C.J. and Perry, J., Jr., 1988. Interaction of the Rocky Mountain foreland and the Cordilleran Thrust Belt. Geol. Soc. Am., Mem., 171.

Suppe, J., 1983. Geometry and kinematics of fault bend folding. Am. J. Sci., 283: 684–721.

Thomas, W.A., 1982. Stratigraphy and structure of the Appalachian fold and thrust belt in Alabama. In: W. A. Thomas and T. L. Neathery (Editors), Appalachian Thrust Belt in Alabama; Tectonics and Sedimentation. 95th Annu. Meet. Geol. Soc. Am. Guidebook for Field Trip, Geology Society of America, pp. 55–78.

Thomas, W.A., 1983. Basement-cover relations in the Appalachian fold and thrust belt. Geol. J., 18: 267–276.

Townsend, C., Roberts, D., Rice, A.H.N. and Gayer, R.A., 1986. The Gaissa Nappe, Finnmark, North Norway: an example of a deeply eroded external imbricate zone within the Scandinavian Caledonides. J. Struct. Geol., 8: 431–440.

Tricart, P., 1984. From passive margin to continental collision: a tectonic scenario for the Western Alps. Am. J. Sci., 284: 97–120.

Tricart, P. and Lemoine, M., 1986. From faulted blocks to megamullions and megaboudins: Tethyan heritage in the structure of the western Tethys. Tectonics, 5: 267–276.

Trümpy, R., 1973. The timing of orogenic events in the Central Alps. In: K.A. DeJong and R. Scholte (Editors), Gravity Tectonics. Wiley, New York, N.Y.

Villars, F., Muller, D. and Lateltin, O., 1988. Analyse de la structure du Mont Charvin (Haute Savoie) en termes de tectonique synsedimentaire paleogene. Consequences pour L'interpretation structurale des chaines subalpines septentrionales. C.R. Acad. Sci. Paris, Sér. II, 307: 1087–1090.

Welbon, A.I., 1988a. The influence of intrabasinal faults on the development of a linked thrust system. Geol. Rundsch., 77: 11–24.

Welbon, A.I., 1988b. Deformation styles and localisation of thrust faults in the external French Alps. Ph.D. Thesis, University of Leeds, Leeds (unpublished).

Wiltschko, D. and Eastman, D., 1983. Role of basement warps and faults in localising thrust fault ramps. In: R.D. Hatcher, H. Williams and I. Zeitz (Editors), Contributions to the Tectonics and Geophysics of Mountain Chains. Geol. Soc. Am., Mem., 158.

Woodward, N.B. and Gray, D.R., 1985. Southwest Virginia, Tennessee and Northern Georgia sections. In: N.B. Woodward (Editor), Valley and Ridge Thrust Belt: Balanced Structural Sections, Pennsylvania to Alabama. Univ. Tennessee, Stud. Geol., 12: 40–53.

Ziegler, P.A., 1983. Inverted basins in the Alpine foreland. In: A. W. Bally (Editors), Seismic Expression of Structural Styles — A Picture and Work Atlas. Am. Assoc. Pet. Geol., Stud. Geol. Ser., pp. 3.3.3–3.3.12.

A.I. WELBON *Department of Geology, University of Oslo, P.O. Box 1047 Blindern, 0316 Oslo 3, Norway*

R.W.H. BUTLER *Department of Earth Sciences, The University, Leeds, LS2 9JT, U.K.*

Present address: *Department of Geological Sciences, Queens University, Union Street, Kingston, Ont. K7L 3N6, Canada*

Tertiary decollement thrusting and inversion structures along Billefjorden and Lomfjorden Fault Zones, East Central Spitsbergen

P. Haremo and A. Andresen

Structural investigations on East Central Spitsbergen along the Billefjorden (BFZ) and Lomfjorden Fault Zones (LFZ), indicate that most of the stratigraphic thickness variations within the Mesozoic strata in this area are due to Tertiary compressional tectonics related to the Eocene transpressive West Spitsbergen Orogeny. No convincing evidence of a distinct Mesozoic deformational event, as suggested by previous workers, has been found. The Tertiary compressional tectonics is characterized by a combined thin-skinned/thick-skinned (basement involved) structural style. Two regionally extensive decollement zones are recognized in the Mesozoic strata; a lower one in the Lower Triassic Sassendalen Group, and an upper one in the Upper Jurassic/ Lower Cretaceous Janusfjellet Subgroup. Both zones can be traced from the West Spitsbergen Foldbelt eastwards to Storfjorden. Numerous reverse faults with variable eastward displacement are associated with both zones. Reverse faults and east-verging folds are particularly frequent as the decollement zones cross BFZ and LFZ. Preliminary calculations indicate an eastward displacement of minimum 3–4 km, possibly as much as 10 km, of Lower Cretaceous and younger rocks across BFZ. Along-strike changes in fold and fault geometry and the appearance of thrust faults above the Festningen Sandstone (Lower Cretaceous) towards the south, suggest the existence of transfer zones or lateral ramps across which the upper decollement zone climbs up-section eastward as well as southward. The southward climbing of the thrusts is inferred to be controlled by pre-deformational uplift and southward tilting of Spitsbergen during Upper Cretaceous time. Both BFZ and LFZ represent reactivated steep, opposite dipping, basement involved reverse faults. This has resulted in an uplifting (inversion) of the area between BFZ and LFZ. Interference structures along BFZ suggest that thick-skinned faulting preceded thin-skinned thrusting, whereas the opposite age relationship is inferred for the thick- and thin-skinned deformation observed along LFZ. Although the thick- and thin-skinned deformations appear to represent two separate deformational events, they are both considered to be of Eocene age and associated with large scale transpression west of Spitsbergen.

Introduction

Important hydrocarbon source and reservoir rocks occur in the Upper Palaeozoic and Mesozoic sequences on Spitsbergen. These rocks have been extensively studied and their depositional environment is relatively well known. Considerably less is known about their deformational history, particularly the extent and nature of the Tertiary West Spitsbergen Orogeny. Another important problem concerning hydrocarbon accumulations on Spitsbergen, as for the entire western Barents Sea shelf (e.g., Nyland et al. and Riis and Fjeldskaar, this volume), is the Cenozoic uplift and erosion which on Spitsbergen have removed as much as 1.7 km of Late Tertiary and younger strata (Manum and Trondsen, 1978). In such a setting, an understanding of the tectonic history of major fault zones and how related structural traps are generated, is of great importance in detecting possible hydrocarbon accumulations in this region of the Barents Sea shelf.

The Billefjorden (BFZ) and the Lomfjorden Fault Zones (LFZ) represent two of the most distinct and extensive tectonic lineaments on Spitsbergen (Fig. 1). Both lineaments have a complex tectonic history, probably going back into the Precambrian, and it is well documented that BFZ played an important role during the structural and sedimentological evolution of Spitsbergen in the Late Palaeozoic (Orvin, 1940; Harland et al., 1974; Kellogg, 1975; Steel and Worsley, 1984). Both fault zones were reactivated during the Tertiary (De Geer, 1919; Hagermann, 1925; Orvin, 1940; Livshits, 1965; Parker, 1966; Kellogg, 1975). However, neither the geometry nor the kinematics of these Tertiary structures nor the way in which they relate to the narrow belt of intense deformation along western Spitsbergen have been thoroughly described. In this paper the complex deformation structures found along BFZ and LFZ as a result of interacting thick- and thin-skinned Tertiary deformation are discussed.

Structural and Tectonic Modelling and its Application to Petroleum Geology edited by R.M. Larsen, H. Brekke, B.T. Larsen and E. Talleraas. NPF Special Publication 1, pp. 481–494. Elsevier, Amsterdam.

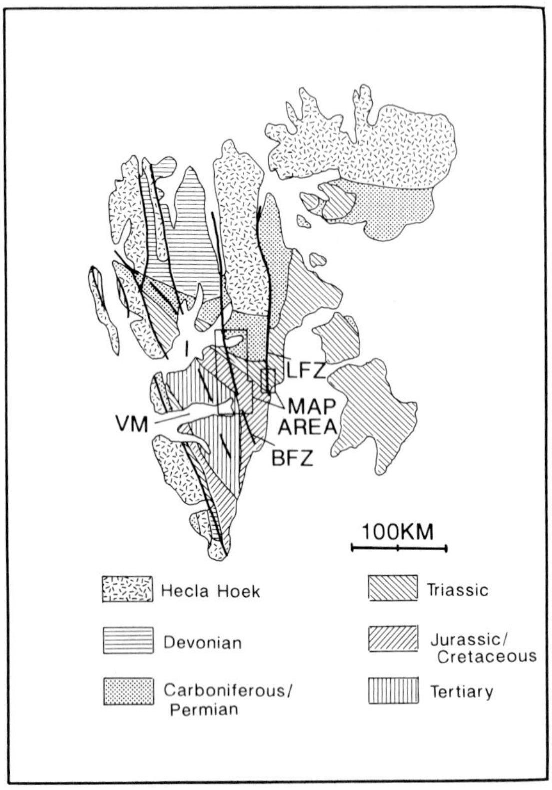

Fig. 1. Simplified geological map of Svalbard. *BFZ* = Billefjorden Fault Zone; *LFZ* = Lomfjorden Fault Zone; *I* = Isfjorden; *VM* = Van Mijenfjorden.

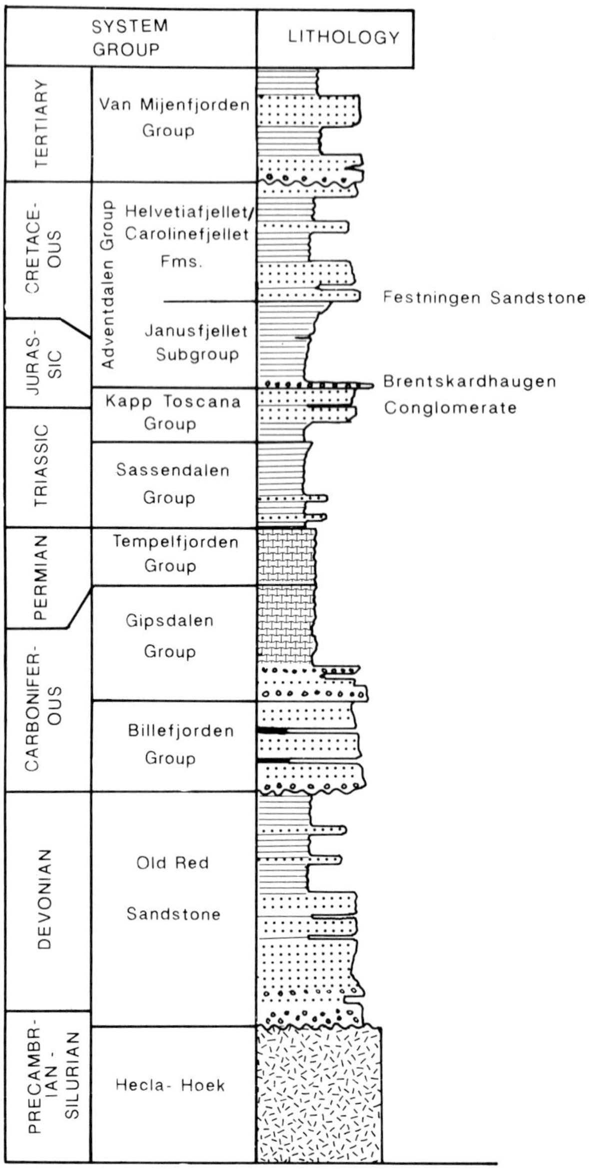

Fig. 2. Generalized stratigraphy of Spitsbergen.

Geological setting and stratigraphy

Spitsbergen displays an almost complete stratigraphical succession from Late Precambrian to Palaeogene. The rocks are generally grouped into five main units separated by major unconformities (Figs. 1 and 2).

Precambrian–Late Silurian

The oldest sequence on Spitsbergen is the Caledonian basement complex (Hecla Hoek) of Late Riphean to Silurian age. Hecla Hoek rocks are dominated by sediments, with minor volumes of igneous rocks, deformed and metamorphosed during the Ny Friesland orogenic phase, but there is also evidence of an older (Greenvillian?) orogenic event (Orvin, 1940; Harland, 1965; Birkenmajer, 1975). Harland and Gayer (1972) argue that BFZ and LFZ were already established and acted as major fault zones during the Precambrian.

Late Silurian–Devonian

Up to 8 km of late/post-orogenic molasse deposits (Old Red Sandstone) of Late Silurian to Middle Devonian age are unconformably overlying the Hecla Hoek sequence (Holtedahl, 1914; Friend and Moody-Stuart, 1972). The Devonian strata were deposited in north–south trending fault-bounded grabens (Fig. 1) and are locally strongly deformed. This deformation is thought to be of Late Devonian age (Vogt, 1929; Birkenmajer, 1981). Both compressional and transcurrent block movements have been proposed to have taken place along BFZ during this "Svalbardian Orogeny" (McWhae, 1953; Harland, 1965; Harland et al., 1974; Harland et al., 1984; Lamar et al., 1986; Reed et al., 1987).

Carboniferous–Permian

Earliest Carboniferous represented a period of regional crustal extension on Spitsbergen, with formation of graben structures controlled by the old, Caledonian, north–south trending lineaments, including BFZ and LFZ (Harland, 1969; Steel and Worsley, 1984). The Lower Carboniferous deposits

(Billefjorden Group, Fig. 2) are characterized by interbedded fluviatile sandstones and shales showing dramatic facies and thickness variations (Cutbill and Challinor, 1965; Gjelberg and Steel, 1981). Resumed tectonic activity along boundary faults in Middle Carboniferous time as evidenced by erosion, renewed graben formation and local compression (Cutbill and Challinor, 1965; Birkenmajer, 1975) has been related to continued strike slip movements along the old lineaments (Birkenmajer, 1975; Steel and Worsley, 1984). Middle Carboniferous to early Permian strata are characterized by continental red beds, carbonates and evaporite sequences (Gipsdalen Group, Fig. 2) reflecting more stable tectonic conditions. During Permian time a transition to a stable platform environment with carbonate deposition developed (Tempelfjorden Group, Fig. 2).

Mesozoic

The platform conditions established during Permian, persisted on Spitsbergen throughout the Mesozoic, although resumed extensional movements along BFZ and LFZ during the Mesozoic have been postulated (Parker, 1966, 1967; Harland et al., 1974; Mörk et al., 1982; Larsen, 1987). However, Haremo and Andresen (1988) and Haremo et al. (1990) argue, based on structural, sedimentological and stratigraphical evidences, that stable tectonic conditions prevailed along the lineaments also during the Mesozoic.

The Mesozoic succession starts with the Sassendalen Group (Early–Middle Triassic, Fig. 2), a 300–400 m thick unit consisting mainly of black shales and siltstones (Buchan et al., 1965; Mörk et al., 1982). The overlying Kapp Toscana Group (Late Triassic–Early Jurassic, Fig. 2) consists of a 300–400 m thick sequence of dark grey and greenish shales interbedded with light brown sandstones. Sandstones are the dominant lithology towards the top of the group.

The Adventdalen Group (Middle Jurassic–Lower Cretaceous, Fig. 2) has a thickness of 800–1600 m. The base of the group is marked by the phosphatic Brentskardhaugen conglomerate which represents a large hiatus comprising parts of the Lower Jurassic and much of the Middle Jurassic (Bäckström and Nagy, 1985). The lower part of the Adventdalen Group comprises the Janusfjellet Subgroup (Dypvik et al., 1991). This subgroup is composed of a 300–400 m thick sequence dominated by black shales. The upper part of the Adventdalen Group consists of a 300–800 m thick sequence of shales and sandstones (Helvetiafjellet and Carolinefjellet Formations). The base is marked by the 5–15 m thick fine to coarse grained Festningen Sandstone (Fig. 2). The boundary between the Adventdalen Group and the Tertiary deposits is marked by a slight angular unconformity formed as a result of the pre-Tertiary uplift of the entire Spitsbergen area (Major and Nagy, 1972).

Tertiary

Dextral strike-slip movement between Svalbard and Greenland, related to opening of the North Atlantic-Arctic Ocean Basins (Eldholm et al., 1987), resulted on Spitsbergen in the Tertiary West Spitsbergen Orogeny and the formation of the Central Spitsbergen Basin (Lowell, 1972; Harland and Horsfield, 1974; Steel et al., 1985).

The up to 2.3 km thick Van Mijenfjorden Group (Fig. 2), making up the Central Spitsbergen Basin, consists of alternating sandstones and shales of mixed marine and continental origin, and is considered Palaeocene–Eocene in age (Ravn, 1922; Major and Nagy, 1972; Manum and Trondsen, 1978, 1986; Steel et al., 1985). Steel et al. (1985) argue that the Central Spitsbergen Basin (the younger part) represents a foreland basin generated as a result of Eocene transpression along the Hornsund Fault Zone located west of Spitsbergen.

Billefjorden fault zone

Permian through Tertiary rocks are exposed along the BFZ in the study area, extending from Gipshuken in the north to Kjellströmdalen in the south (Figs. 1 and 3). Due to the regionally southwesterly dip of strata, progressively younger rocks are exposed towards the south, giving east–west cross-sections at different stratigraphical and structural levels along the fault zone (Figs. 3 and 4).

The structural development along BFZ in the area between Adventdalen/Sassendalen and Kjellströmdalen has been described in detail by Haremo et al. (1990). The most characteristic structural features found along BFZ are two, regionally extensive, sub-parallel, east facing anticlines, the Skolten Anticline to the west and the Tronfjellet Anticline to the east (Figs. 3 and 4). These anticlines, and the Drönbreen Syncline separating them, have been mapped from Adventdalen in the north to Kjellströmdalen in the south (Fig. 3). The Skolten Anticline is particularly well developed between Adventdalen and Reindalen (Fig. 3). In this area it is developed as a rather tight anticline with a wavelength of 600–700 meters and an amplitude of approximately 600 m (Fig. 4d and 5). The folded Festningen Sandstone can be traced southwards to Kjellströmdalen where it plunges southward below gently folded Cretaceous and Tertiary strata. The

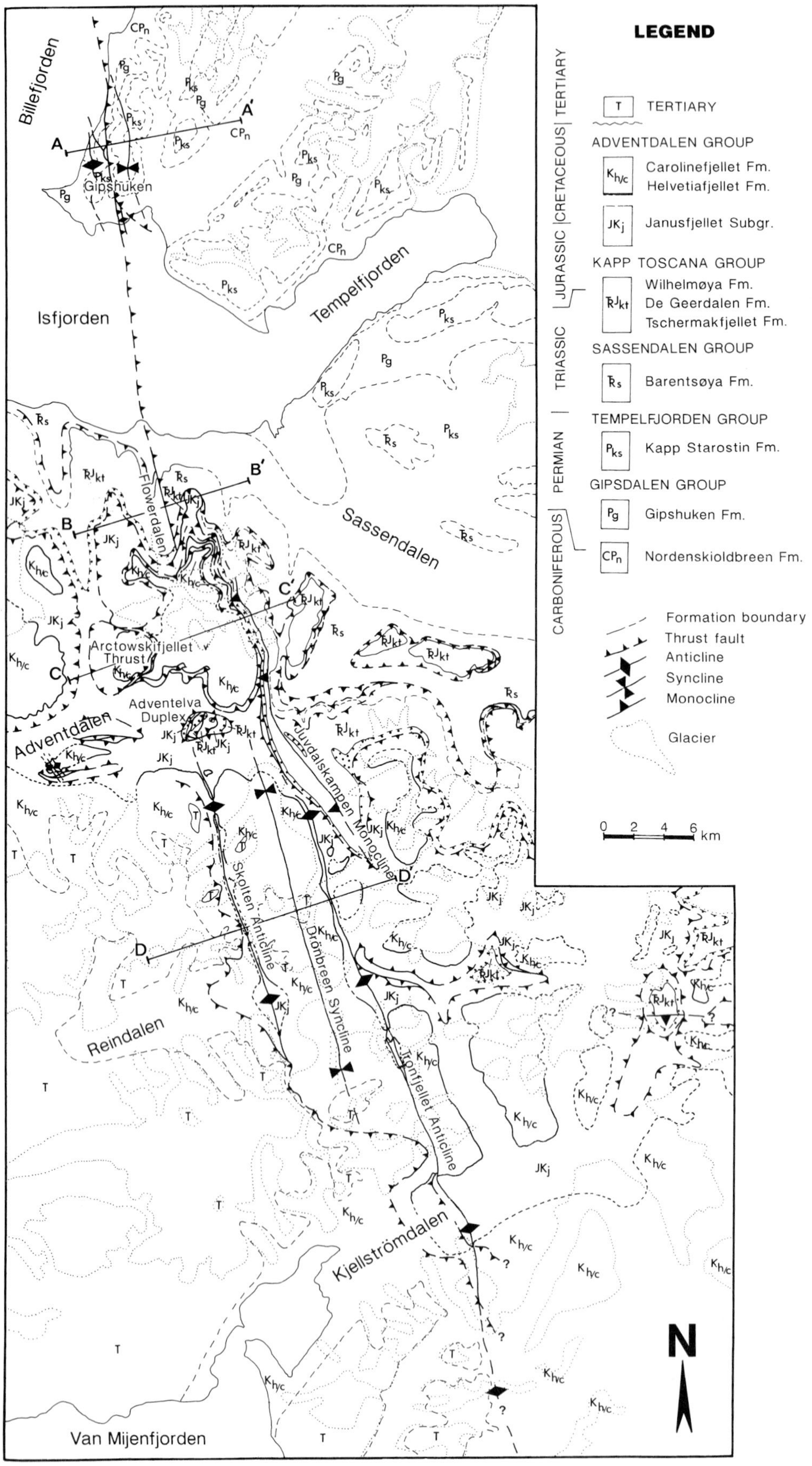

Fig. 3. Geological map along the southern part of the Billefjorden Fault Zone, East Central Spitsbergen.

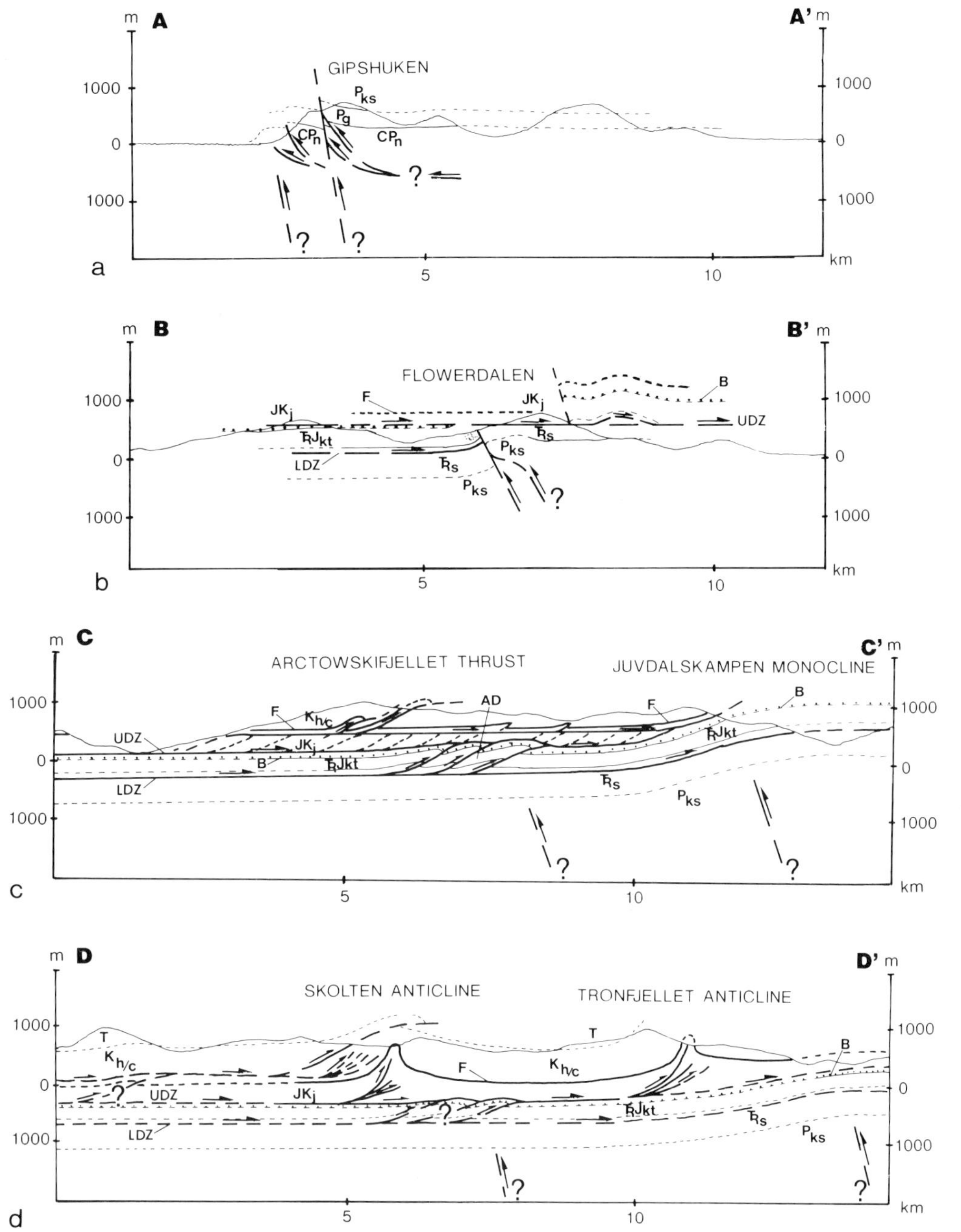

Fig. 4. Profiles across the Billefjorden Fault Zone; (a) Gipshuken, (b) Flowerdalen, (c) Adventdalen, (d) Reindalen. *LDZ* = Lower Decollement Zone; *UDZ* = Upper Decollement Zone; *B* = Brentskardhaugen conglomerate; *F* = Festningen Sandstone. For legend see Fig. 3.

Skolten Anticline cannot be traced north of Adventdalen (Fig. 3). Instead, a major west-dipping reverse fault (Arctowskifjellet Thrust, Fig. 4c and 6) accomplishes the shortening seen south of the valley. Minimum shortening across the Arctowskifjellet Thrust is 1.5 km.

The Tronfjellet Anticline occurs as an asymmetrical anticline at the Festningen Sandstone level, located approximately 5 km east of the Skolten Anticline (Figs. 3 and 4d). The anticline is particularly well developed north and south of Reindalen. Its amplitude is slightly less — and it is not as tight as the Skolten Anticline. To the south, in Kjellströmdalen, the Tronfjellet Anticline is developed as an asymmetrical, kink-like, more open fold with an amplitude of less than 200 meters (Fig. 3).

The two fold structures, developed in Upper Jurassic and younger sequences, are not recognized below the Lower Jurassic Janusfjellet Subgroup (Figs. 3 and 4c, d; Haremo et al., 1990). This structural discontinuity and the existence of numerous imbricate as well as minor duplex structures within the Janusfjellet Subgroup itself, resulting in local thickening, indicate the existence of a decollement zone (Up-

Fig. 5. The Skolten Anticline at Rudiaksla north of Reindalen, viewed from the south. The Festningen Sandstone is developed as a tight, slightly east-verging anticline.

Fig. 6. The Arctowskifjellet Thrust viewed from the south (Adventdalen). This fault ramps up-section through the upper part of the Adventdalen Group with a throw of at least 300 m. Shortening of the Festningen Sandstone across this fault is at least 1.5 km. *F* = Festningen Sandstone.

per Decollement Zone) at this stratigraphical level (Figs. 3 and 4; Haremo et al., 1990).

Compressional structures including folds and reverse faults of variable size and geometry, unrelated to the Skolten and Tronfjellet Anticlines, are also found below the Upper Decollement Zone, in the Sassendalen and Kapp Toscana Groups (Figs. 3 and 4c, d; Haremo et al., 1990). A particularly well developed example of the latter is displayed in Adventdalen, where the upper part of the Kapp Toscana Group and the lower part of the Janusfjellet Subgroup, including the Brentskardhaugen Conglomerate, are repeated three times in a duplex structure (Adventelva Duplex, Figs. 3 and 4c; Haremo et al., 1990). In addition, data from Flowerdalen and Sassendalen (Fig. 3) demonstrate that the upper part of the Sassendalen Group is locally intensely imbricated and thickened (Fig. 7), indicating the existence of another decollement zone (Lower Decollement Zone) located at this stratigraphical level (Figs. 3 and 4).

A major west facing monocline (Juvdalskampen Monocline) located slightly east of the Tronfjellet Anticline (Figs. 3 and 4d), is another character-

Fig. 7. Local thickening of the upper, dark, cliff-forming part of the Sassendalen Group by intense reverse faulting and imbrication. Viewed from the north (Sassendalen).

istic structural feature found along BFZ in the Adventdalen–Reindalen area. This monocline, affecting the entire Mesozoic sequence, is not in any simple way related to any of the previously described fold and thrust structures (Haremo et al., 1990). The Juvdalskampen Monocline is interpreted to be controlled by an east-dipping reverse fault at depth (Fig. 4c). We believe that this fault is exposed further north in Flowerdalen and at Gipshuken (Figs. 3 and 4a, b). The Juvdalskampen Monocline trends north–northwest and intersects the Tronfjellet Anticline with an acute angle (Fig. 3). The monocline apparently dies out southwards, although it may link up with a similar east–west trending monocline in the inner Kjellströmdalen area.

A dramatic thinning of the Mesozoic sequence in the Adventdalen–Flowerdalen area, has previously been interpreted as the result of Mesozoic syndepositional extensional tectonics (Parker, 1966, 1967; Harland et al., 1974). Haremo and Andresen (1988) and Haremo et al. (1990) argue against such a model based on structural observations. Alternatively, it is suggested that an out-of-sequence thrust, associated with the Upper Decollement Zone, cutting down-section in the transport direction, is responsible for the thinning (Figs. 3 and 4b, c; Haremo et al., 1990). Such an explanation is supported by numerous small scale out-of-sequence thrusts within the Janusfjellet Subgroup. These thrusts are observed to cut across previously imbricated, steep, west-dipping reverse faults. Our tectonic model for the reduced thickness of the Janusfjellet Subgroup, considers the out-of-sequence thrusting and thinning to post-date development of the Juvdalskampen Monocline.

Reverse faults, located in the upper part of the Adventdalen Group above the Festningen Sandstone level, are recognized south of Adventdalen (Figs. 3 and 4d). This indicates that lateral ramps may exist, along which reverse faults climb up-section towards south (Haremo et al., 1990). Lateral ramps may also be responsible for the observed local transverse folding of the Skolten and the Tronfjellet Anticlines (Fig. 3).

Lomfjorden fault zone

Triassic through Cretaceous rocks are exposed along the LFZ in the Agardhdalen area (Figs. 1 and 8). The structural development along LFZ in this area is described in detail by Andresen et al. (1992). The dominant structure is a regionally extensive, east-facing, highly asymmetric anticline, the Eistraryggen Anticline (Figs. 8 and 9) affecting the entire Mesozoic sequence including the Festningen Sandstone. The difference in elevation at Eistraryggen between the almost flatlying beds east of the fold structure compared with similar beds west of the anticlinal hinge zone is approximately 500 m. The bedding orientation along Eistraryggen is very steep and locally inverted (Fig. 9a).

The anticline plunges gently to the south and reappears at Klementievfjellet south of Agardhdalen (Fig. 8). Here, several large parasitic fold structures are developed in the core of the anticline. Generally, these folds are asymmetric, kink or chevron-like structures, all facing east (Fig. 9b). The eastern limb

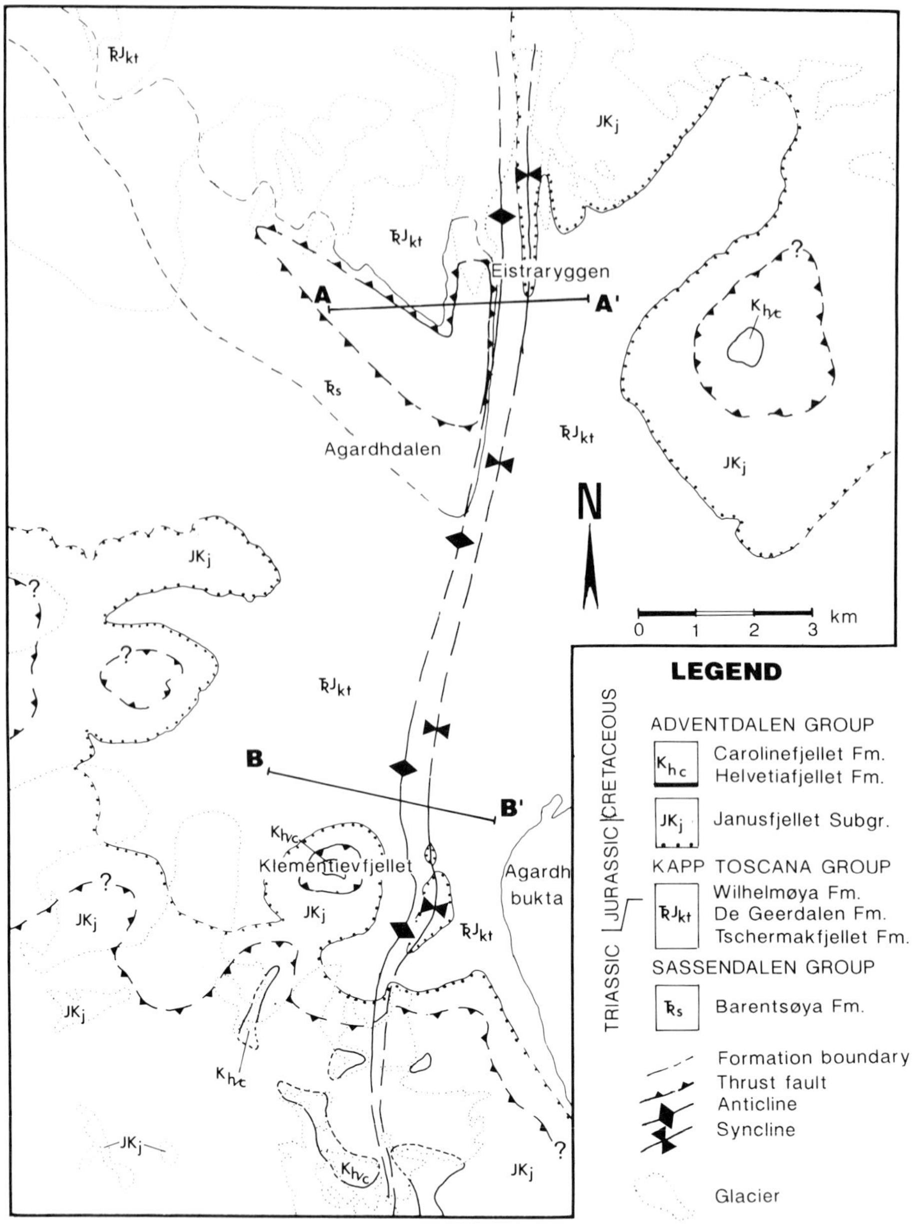

Fig. 8. Geological map along the southern part of the Lomfjorden Fault Zone, East Central Spitsbergen.

of some of the parasitic folds is locally inverted. The fold geometry suggests that considerable shortening has taken place, and there can be no doubt about the compressional origin of the Eistraryggen Anticline.

We prefer to relate the development of the Eistraryggen Anticline to a west-dipping reverse fault at depth (Fig. 9; Andresen et al., 1988, 1992). This is supported by observations along LFZ toward its type area to the north, were pre-Mesozoic rocks including Hecla Hoek rocks are uplifted and exposed west of the fault (Flood et al., 1971; Hjelle and Lauritzen, 1982).

A characteristic structural feature within the upper part of the Sassendalen Group north of Agardhdalen, is the development of numerous, small and medium scale reverse faults, locally forming duplex structures and imbricate fans (Fig. 10). These reverse faults are rotated into a sub-horizontal orientation when traced into the steep, eastern limb of the Eistraryggen Anticline, suggesting that decollement thrusting along LFZ predates the generation of the anticline (Fig. 9). Isolated reverse faults are also observed in many places within the Sassendalen Group west of the Eistraryggen Anticline, indicating that the Lower Decollement Zone found along BFZ continues eastward to LFZ (Andresen et al., 1992).

Small scale imbrications and duplexes are also observed in the Janusfjellet Subgroup, at Klementiewfjellet suggesting the existence of a decollement zone (Upper Decollement Zone) also at this stratigraph-

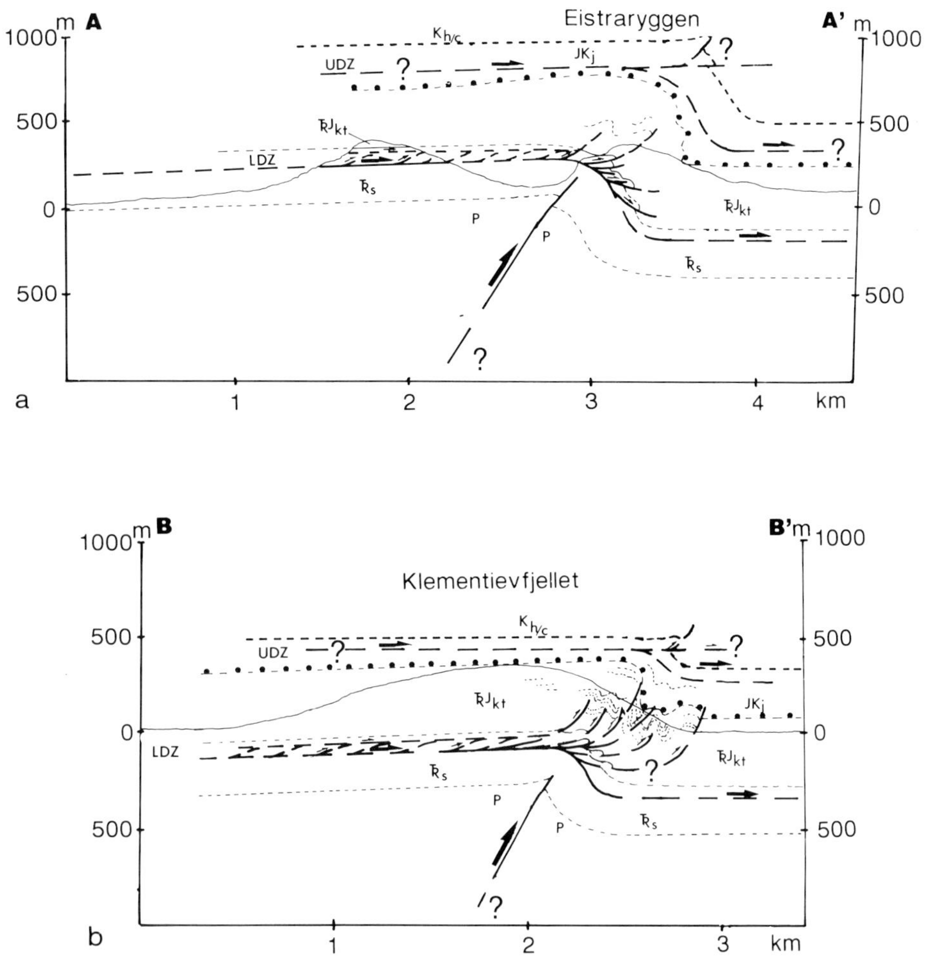

Fig. 9. Profiles across the Lomfjorden Fault Zone. (a) Eistraryggen, (b) Klementievfjellet. *LDZ* = Lower Decollement Zone; *UDZ* = Upper Decollement Zone. For legend see Fig. 8.

ical level. The Janusfjellet Subgroup thins across Klementievfjellet (Fig. 9b), a feature interpreted to be the result of out-of-sequence thrusting, also confirming the presence of an Upper Decollement Zone in the Janusfjellet Subgroup this far east (Andresen et al., 1992).

Discussion

Tertiary tectonic activity along BFZ and LFZ has been described by several authors (De Geer, 1919; Hagermann, 1925; Orvin, 1940; Parker, 1966; Harland et al., 1974). The importance of the Tertiary deformation has, however, been disputed (Andresen et al., 1986; Haremo and Andresen, 1988; Dypvik et al., 1990; Haremo et al., 1990). The structural development described herein, demonstrating that Tertiary deformation along BFZ and LFZ has been underestimated, is also supported by other recent investigations (e.g., Bergh et al., 1988; Faleide et al., 1988; Maher, 1988; Nöttvedt et al., 1988).

The decollement zones recognized in the Mesozoic strata are not restricted to BFZ and LFZ, although deformational structures are particularly frequent along these two fault zones. Local thickening of both the Sassendalen Group and the Janusfjellet Subgroup due to small scale imbrication and duplex development, is recognized in the area between the two fault zones, especially in Sassendalen (Lower Decollement Zone, Figs. 3 and 7) and in the inner parts of Reindalen and Kjellströmdalen (Upper Decollement Zone, Fig. 3). Similar structures are also developed at the same stratigraphical levels west of BFZ (Haremo and Andresen, 1988; Nöttvedt et al., 1988; Haremo et al., 1990), as well as along the western coast of Spitsbergen (Dallmann, 1988; Maher, 1988; Ringseth, 1988). It therefore seems to be well documented that the decollement zones recognized in the Mesozoic strata along BFZ and LFZ have a regional extent.

The data presented above, on the Tertiary deformation on Spitsbergen, suggest that the traditional view of the West Spitsbergen Orogeny representing a transpressive orogen with structures typical of wrench tectonics (Lowell, 1972; Kellogg, 1975; Harland and Horsfield, 1974; Steel et al., 1985), has to be

Fig. 10. Duplex structures developed in the upper part of the Sassendalen Group at Eistraryggen. Viewed toward the south.

modified. The Tertiary deformation structures recognized in the foldbelt, as well as along BFZ and LFZ indicate that Tertiary deformation is purely compressional and similar to structures recognized in typical fold and thrust belts (e.g., Dahlström, 1970; Boyer and Elliot, 1982; Mitra, 1986; Baker et al., 1988).

Placed in this context an evolutionary model, illustrating the Tertiary deformational history along BFZ and LFZ, is presented in Fig. 11. Figure 11, stage 1, illustrates a pre-deformational (pre-Tertiary) stage, in which structural discontinuity is restricted to the existence of basement-rooted extensional faults (BFZ and LFZ) of pre-Mesozoic age. In this model, these faults are thought to act as "structure generators" during later deformation.

In stage 2, Tertiary compression commences, resulting in the initiation of decollement zones in the Sassendalen Group (Lower Decollement Zone) as well as in the Janusfjellet Subgroup (Upper Decollement Zone). Decollement thrusting is, however, thought to be limited during stage 2. Inversion along BFZ is the principal event during this stage. Reactivation of the steep, east-dipping, pre-Mesozoic fault lines results in the generation of the Juvdalskampen Monocline and possibly also in some reactivation along the inferred western fault line (Fig. 11). Inversion along BFZ during this stage, is probably not only a result of simple reactivation along old fault lines. Data from Gipshuken (Ringseth and Andresen, 1988) and the inner part of Billefjorden suggest that a detachment level existed in the Carboniferous evaporite deposits (Fig. 11). Movement along this detachment zone may have resulted in reactivation along LFZ.

During stage 3 (Fig. 11) deformation along BFZ and LFZ is restricted mainly to decollement thrusting along the Upper and Lower Decollement Zones. Along BFZ, movement along the Lower Decollement Zone results in the development of the Adventelva Duplex, the location of which is most likely controlled by an underlying fault. More or less simultaneous movement along the Upper Decollement Zone creates the Tronfjellet Anticline, above the Adventelva Duplex. Decollement thrusting extends eastwards, all the way to LFZ where imbricate fans and duplex structures are developed during stage 3.

The next stage of Tertiary deformation along BFZ (Fig. 11, stage 4) is characterized by out-of-sequence thrusting, related to movements along the Upper Decollement Zone. The Tronfjellet Anticline is decapitated and displaced eastwards. As it reaches the Juvdalskampen Monocline, movement ceases and the Arctowskifjellet Thrust (or the Skolten Anticline to the south) is developed above the Adventdalen Duplex. Along LFZ this late stage deformation is characterized by the development of the Eistraryggen Anticline as a response to reverse movement along the steep, west-dipping, basement-rooted fault. Out-of-sequence thrusting resulting in tectonic thinning of the Janusfjellet Subgroup at Klementievfjellet, is related to movement on the Upper Decollement Zone, and is thought to post-date inversion and to represent the final deformational event along LFZ (Andresen et al., 1992).

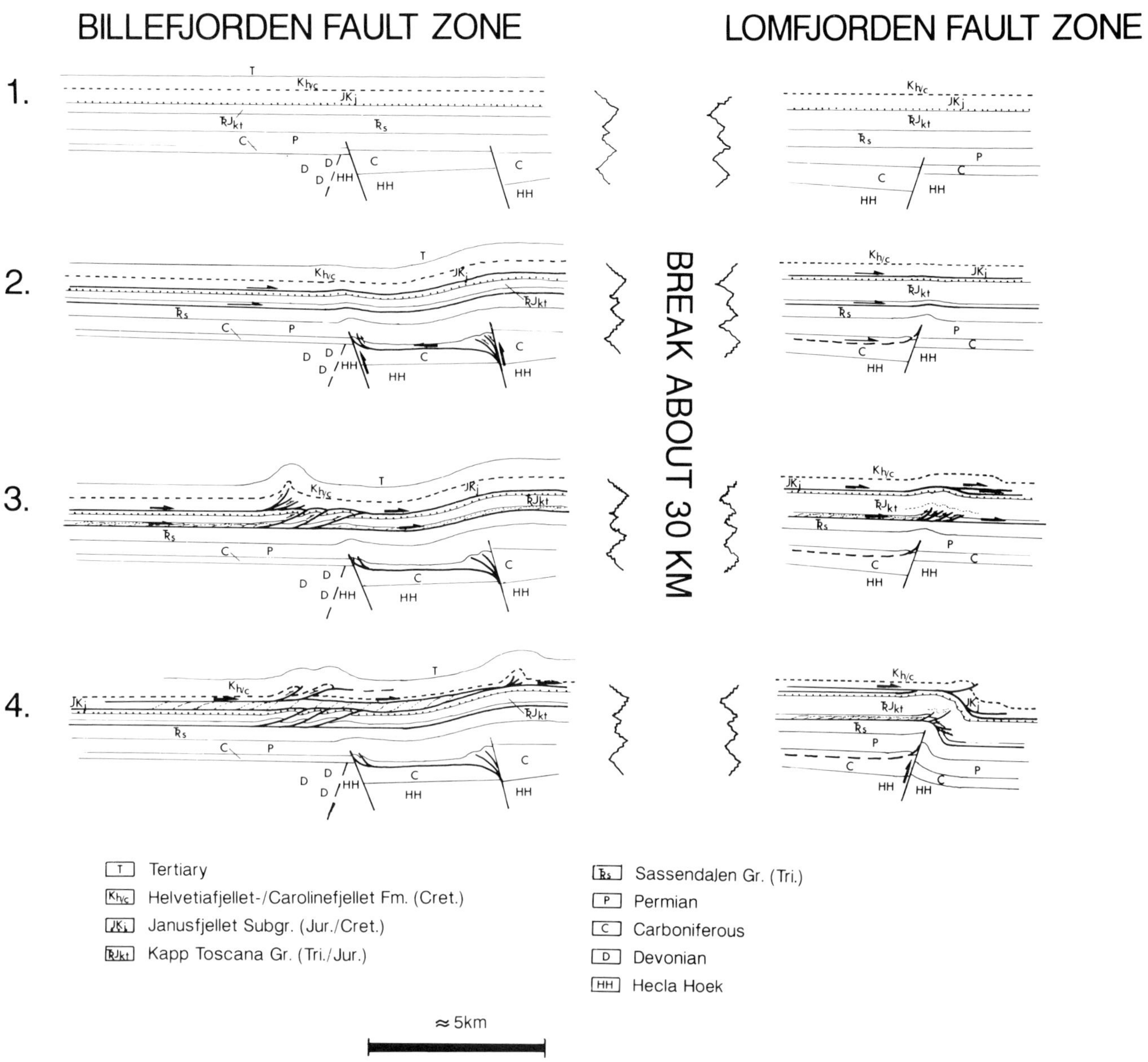

Fig. 11. Evolutionary model illustrating Tertiary deformational history along Billefjorden and Lomfjorden Fault Zone.

The Tertiary deformational history presented above suggest an eastward displacement of Lower Cretaceous and younger rocks across BFZ of minimally 3–4 km, and locally as much as 10 km. A feature not included in the kinematic model presented above is the along-strike variation in geometries observed along BFZ (Figs. 3 and 4). The most interesting observation in this respect, is the appearance of thrust faults at higher stratigraphical levels (above the Festningen Sandstone) to the south. We prefer to view these as caused by the existence of lateral or oblique ramps, probably controlled by the pre-deformational Late Cretaceous/Early Tertiary uplift and southward tilting of Spitsbergen.

The model presented in Fig. 11 proposing that inversion along BFZ preceded most of the movement along the two decollement zones, is mainly based on the tectonic thinning of Mesozoic strata across the Juvdalskampen Monocline, indicating that inversion pre-dates out-of-sequence thrusting (Fig. 4; Haremo et al., 1990). The opposite age relationship is inferred for the thick- and thin-skinned deformation observed along LFZ (Fig. 11). Here, inversion post-dates decollement thrusting as reverse faults generated through decollement thrusting during the development of the Eistraryggen Anticline are rotated into a sub-horizontal or even east-dipping orientation (Fig. 9; Andresen et al., 1992).

The interaction between thick- and thin-skinned deformation described above, of which the latter can be linked to the West Spitsbergen Orogeny, suggests that they are formed more or less contemporaneously. As the thin-skinned folds affect strata of Palaeocene age (Manum and Trondsen, 1986), the deformation is considered post-Palaeocene in age. Since transpression along the Hornsund Fault Zone west of Spitsbergen is thought to be of Late Palaeocene–Middle Eocene age (Steel et al., 1985; Eldholm et al., 1987) an Eocene age is suggested for the deformational structures developed along BFZ and LFZ.

Conclusion

Most lateral thickness variations within the Jurassic and Triassic strata on Central Spitsbergen are due to Tertiary compressional tectonics, linked to the West Spitsbergen Orogeny. The Tertiary deformation is characterized by an interacting thin- and thick- skinned structural style. Two regionally extensive decollement zones are recognized in the Mesozoic strata; a lower one in the Lower Triassic Sassendalen Group, and an upper one in the Upper Jurassic/Lower Cretaceous Janusfjellet Subgroup. Compressional structures (folds and reverse faults) are particularly frequent along the basement-rooted Billefjorden and Lomfjorden Fault Zones. Preliminary calculations indicate an eastward displacement of minimally 3–4 km, possibly as much as 10 km, of Lower Cretaceous and younger rocks across BFZ. The complex Tertiary deformation recognized along BFZ and LFZ are caused by reactivation of steep, opposite-dipping basement-rooted reverse faults during the West Spitsbergen Orogeny. This caused inversion of the Carboniferous-Permian Billefjorden Basin. Thick-skinned faulting preceded thin-skinned thrusting along BFZ, whereas the opposite relationship is inferred along LFZ.

Acknowledgements

The present study is part of the BP R&D project 1068 "Study of the sedimentological and structural evolution of the area between Kjellströmdalen and Adventdalen/Sassendalen, Central Spitsbergen". We thank BP Petroleum Development (Norway), Ltd. for financial support to the project and for permission to publish the results. We also thank the Norwegian Polar Research Institute, Norsk Hydro and Statoil for valuable logistical support during fieldwork. Einar Sverdrup, Eivind Swensson and two unknown referees are thanked for valuable comments on a preliminary draft of the manuscript.

References

Andresen, A., Dypvik, H., Nagy, J., Elverhøi, A., Eikeland, T.A., Haremo, P., Nesteby, H. and Strand, K., 1986. Mesozoisk kontra tertiær tektonisk aktivitet langs Billefjord-lineamentet syd for Isfjorden, Spitsbergen. 10. Nor. Geol. Vintermöte Trondheim 1987. Geolognytt, 21: 14 (abstract).

Andresen, A., Haremo, P. and Bergh, S., 1988. The southern termination of the Lomfjorden Fault Zone; evidences for Tertiary compression on east Spitsbergen. Symp. Tertiary Tectonics of Svalbard, Oslo, 26–27 April 1988, Norsk Polarinst. Rap., 46: 75–78 (extended abstract).

Andresen, A., Haremo, P., Swensson, E. and Bergh, S.G., 1992. Structural geology around the southern termination of the Lomfjorden Fault Complex, Agardhdalen, East Spitsbergen. Nor. Geol. Tidskr., 72 (in press).

Baker, D.M., Lillie, R.J., Yeats, R.S., Johnson, G.D., Yousuf, M. and Zamin, A.S.H., 1988. Development of the Himalayan frontal thrust zone: Salt Range, Pakistan. Geology, 16: 3–7.

Bergh, S.G., Andresen, A., Bergvik, A. and Hansen, A.I., 1988. Tertiary thin-skinned compressional deformation on Oskar II Land, Central Vest-Spitsbergen. Symp. Tertiary Tectonics of Svalbard, Oslo, 26–27 April 1988, Nor. Polarinst. Rap., 46: 51–54 (extended abstract).

Birkenmajer, K., 1975. Caledonides of Svalbard and plate tectonics. Geol. Soc. Denm. Bull., 24: 1–19.

Birkenmajer, K., 1981. The geology of Svalbard, the western part of the Barents Sea, and the continental margin of Scandinavia. In: A.E.M. Nairn, M. Churkin Jr. and F.G. Stehli (Editors), The Ocean Basins and Margins, Vol. 5. The Arctic Ocean. Plenum Press, New York, N.Y., pp. 265–329.

Boyer, S.E. and Elliot, D., 1982. Thrust systems. Am. Assoc. Pet. Geol., Bull., 66: 1196–1230.

Buchan, S.H., Challinor, A., Harland, W.B. and Parker, J.R., 1965. The Triassic stratigraphy of Svalbard. Nor. Polarinst. Skr., 135: 1–72.

Bäckström, S.A. and Nagy, J., 1985. Depositional history and fauna of a Jurassic phosphorite conglomerate (the Brentskardhaugen bed) in Spitsbergen. Nor. Polarinst. Skr., 183: 1–61.

Cutbill, J.L. and Challinor, A., 1965. Revision of the stratigraphical scheme for the Carboniferous and Permian rocks of Spitsbergen and Björnöya. Geol. Mag., 102: 418–439.

Dallmann, W.K., 1988. Thrust tectonics south of Van Keulenfjorden. Symp. Tertiary Tectonics of Svalbard, Oslo, 26–27 April 1988, Nor. Polarinst. Rap., 46: 43–46 (extended abstract).

Dahlström, C.D.A., 1970. Structural geology in the eastern margin of the Canadian Rocky Mountains. Can. Pet. Geol. Bull., 18: 332–406.

De Geer, G., 1919. Om Spetsbergens natur i Sveagruvans omnejd. Ymer, H4: 240–277.

Dypvik, H., Andresen, A., Haremo, P., Nagy, J., Elverhøi, A., Eikeland, T.A., Johansen, H. and Strand, K., 1990. Sedimentation and tectonics in the Jurassic/Cretaceous sequences of Central Spitsbergen. In: V.M. Kotlyakov and V.E. Sokolov (Editors), Arctic Research. Advances and Prospects. Proc. Conf. of Arctic and Nordic Countries on Coordination of Research in the Arctic, Leningrad, December 1988, Part 1. Nauka, Moscow, pp. 348–354.

Dypvik, H., Nagy, J. Eikeland, T.A., Backer-Owe, K., Andresen, A., Haremo, P., Bjærke, T., Johansen, H. and Elverhøi, A., 1991. The Janusfjellet Subgroup (Bathonian to Hauterivian) on central Spitsbergen: a revised lithostratigraphy. Polar Res., 9(1): 21–44.

Eldholm, O., Faleide, J.I. and Myhre, A.M., 1987. Continental-ocean transition at the western Barents Sea/Svalbard continental margin. Geology, 15: 1118–1122.

Faleide, J.E., Gudlaugsson, S.T., Eiken, O. and Hanken, N.M., 1988. Seismic structure of Spitsbergen: Implications for Tertiary deformation. Symp. Tertiary Tectonics of Svalbard, Oslo,

26–27 April 1988, Nor. Polarinst. Rap., 46: 47–50 (extended abstract).

Flood, B., Nagy, J. and Winsnes, T.S., 1971. Geological map of Svalbard 1:500,000, Sheet 1G, Spitsbergen southern part. Nor. Polarinst. Skr., 154A.

Friend, P.F. and Moody-Stuart, M., 1972. Sedimentation of the Wood Bay Formation (Devonian) of Spitsbergen: regional analysis of a late orogenic basin. Nor. Polarinst. Skr., 157: 1–77.

Gjelberg, J.G. and Steel, R.J., 1981. An outline of lower–middle Carboniferous sedimentation on Svalbard: effects of tectonic, climatic and sea level changes in rift basin sequences. Can. Soc. Pet. Geol., Mem., 7: 543–561.

Hagermann, T.H., 1925. Stratigraphic and structural investigations within south-western Spitzbergen. Geogr. Ann., 7: 195–221.

Haremo, P. and Andresen, A., 1988. Tertiary movements along the Billefjorden FaultZone and its relation to the Vest-Spitsbergen Orogenic Belt. Symp. Tertiary Tectonics of Svalbard, Oslo, 26–27 April 1988, Nor. Polarinst. Rap., 46: 71–74 (extended abstract).

Haremo, P., Andresen, A., and Dypvik, H., in prep. Mesozoic extension versus Tertiary compressionalong the Billefjorden Fault Zone south of Isfjorden, central Spitsbergen.

Haremo, P., Andresen, A., Dypvik, H., Nagy, J., Elverhøi, A., Eikeland, T.A. and Johansen, H., 1990. Structural development along the Billefjorden Fault Zone in the area between Kjellströmdalen and Adventdalen/Sassendalen, Central Spitsbergen. Polar Res., 8(2): 195–216.

Harland, W.B., 1965. Discussion of the tectonic evolution of the Arctic-North Atlantic Region. Philos. Trans. R. Soc. London, Ser. A, 258: 59–75.

Harland, W.B., 1969. Contribution of Spitsbergen to understanding of the tectonic evolution of North Atlantic Region. Am. Assoc. Pet. Geol., Mem., 12: 817–851.

Harland, W.B. and Gayer, R.A., 1972. The Arctic Caledonides and earlier Oceans. Geol. Mag., 109: 289–314.

Harland, W.B. and Horsfield, W.T., 1974. The West Spitsbergen Orogen. Geol. Soc. London, Spec. Publ., 4: 747–755.

Harland, W.B., Cutbill, J.L., Friend, P.F., Gobett, D.S., Holliday, D.W., Maton, P.I.,Parker, J.R. and Wallis, R.H., 1974. The Billefjorden Fault Zone, Spitsbergen. Nor. Polarinst. Skr., 161: 1–72.

Harland, W.B., Gaskell, B.A., Heafford, A.P., Lind, E.K. and Perkins, P.J., 1984. Outline of Arctic post-Silurian continental displacements. In: A.M. Spencer (Editor), Petroleum Geology of the North European Margin. Proceedings of the North European Margin Symposium. Norwegian Petroleum Society, Graham and Trotman, London, pp. 137–148.

Hjelle, A. and Lauritzen, Ö., 1982. Geological map of Svalbard, 1:500,000, Sheet 3G, Spitsbergen, Northern part. Nor. Polarinst. Skr., 154C: 1–15.

Holtedahl, O., 1914. On the Old Red Sandstone Series of North Western Spitsbergen. Int. Geol. Congr. CRXIIth Sess., Canada 1913: 707–712.

Kellogg, H.E., 1975. Tertiary Stratigraphy and Tectonism in Svalbard and Continental Drift. Am. Assoc. Pet. Geol., Bull., 59: 465–485.

Lamar, D.L., Reed, W.E. and Douglass, D.N., 1986. Billefjorden fault zone, Spitsbergen: is it part of a major Late Devonian transform? Geol. Soc. Am. Bull., 97: 1083–1088.

Larsen, V.B., 1987. A synthesis of tectonically related stratigraphy in the North Atlantic–Arctic region from Aalenian to Cenomanian time. Nor. Geol. Tidsskr., 67: 281–293.

Livshits, Y.Y., 1965. Tectonics of central Vestspitsbergen. In: V.N. Sokolov (Editor), Geology of Spitsbergen, 1. National Lending Library for Science and Technology, U.K., pp. 59–75.

Lowell, D.L., 1972. Spitsbergen Tertiary Orogenic Belt and the Spitsbergen Fracture Zone. Geol. Soc. Am. Bull., 83: 3091–3102.

Maher, H.D., Jr., 1988. Minimum estimate of Tertiary shortening suggested by surface structures exposed on Midterhuken, Bellsund, Spitsbergen. Symp. Tertiary Tectonics of Svalbard, Oslo, 26–27 April 1988, Nor. Polarinst. Rap., 46: 35–38 (extended abstract).

Major, H. and Nagy, J., 1972. Geology of the Adventdalen map area. Nor. Polarinst. Skr., 138: 1–58.

Manum, S.B. and Trondsen, T., 1978. Rank of coal and dispersed organic matter and its geological bearing in the Spitsbergen Tertiary. Nor. Polarinst., Årbok, 1977: 159–177.

Manum, S.B. and Trondsen, T., 1986. Age of Tertiary formations on Spitsbergen. Polar Res., 4: 103–131.

McWhae, J.R.H., 1953. The Major Fault Zone of Central Spitsbergen. Q. J. Geol. Soc. London, 108: 209–232.

Mitra, S., 1986. Duplex structures and imbricate thrust systems: geometry, structural position, and hydrocarbon potential. Am. Assoc. Pet. Geol., Bull., 70: 1087–1112.

Mörk, A., Knarud, R. and Worsley, D., 1982. Depositional and diagenetic environments of the Triassic and lower Jurassic succession of Svalbard. Can. Soc. Pet. Geol., Mem., 8: 371–398.

Nyland, B., Jensen, L.N., Skagen, J.I., Skarpnes, O. and Vorren, T., 1992. Tertiary uplift and erosion in the Barents Sea; magnitude, timing and consequences. In: R.M. Larsen, H. Brekke, B.T. Larsen and E. Talleraas (Editors), Structural and Tectonic Modelling and its Application to Petroleum Geology. Norwegian Petroleum Society (NPF), Special Publication 1. Elsevier, Amsterdam, pp. 153–162 (this volume).

Nöttvedt, A., Livbjerg, F. and Midböe, P.S., 1988. Tertiary deformation of Svalbard — various models and recent advances. Nor. Polarinst. Rap., 46: 79–84.

Orvin, A.K., 1940. Outline of the geological history of Spitsbergen. Skr. Svalbard Oglshavet, 78: 1–57.

Parker, J.R., 1966. Folding, faulting and dolerite intrusions in the Mesozoic rocks of the fault zone of central Spitsbergen. Nor. Polarinst., Årbok, 1964: 47–55.

Parker, J.R., 1967. The Jurassic and Cretaceous sequence in Spitsbergen. Geol. Mag., 104: 487–505.

Ravn, J.P.J., 1922. On the Mollusca of the Tertiary of Spitsbergen. Result. Nor. Spitsbergeneksp. 1(2): 1–28.

Reed, W.E., Douglass, D.N. and Lamar, D.L., 1987. Devonian Old Red Sandstone sedimentation and tectonic history of Billefjorden fault zone, Spitsbergen. Geol. Mijnbouw, 66: 191–199.

Riis, F. and Fjeldskaar, W., 1992. On the magnitude of the Late Tertiary and Quaternary erosion and its significance for the uplift of Scandinavia and the Barents Sea. In: R.M. Larsen, H. Brekke, B.T. Larsen and E. Talleraas (Editors), Structural and Tectonic Modelling and its Application to Petroleum Geology. Norwegian Petroleum Society (NPF), Special Publication 1. Elsevier, Amsterdam, pp. 163–185 (this volume).

Ringseth, N., 1988. The fold and thrust system of Midterhukfjellet, Bellsund. Symp. Tertiary Tectonics of Svalbard, Oslo, 26–27 April 1988, Nor. Polarinst. Rap., 46: 39–42 (extended abstract).

Ringseth, N. and Andresen, A., 1988. The Gipshuken Fault System — evidence for Tertiary thrusting along the Billefjorden Fault Zone. Symp. Tertiary Tectonics of Svalbard, Oslo, 26–27 April 1988, Nor. Polarinst. Rap., 46: 67–70 (extended abstract).

Steel, R.J. and Worsley, D., 1984. Svalbard's post-Caledonian strata — An atlas of sedimentational patterns and palaeogeographic evolution. In: A. Spencer (Editor), Petroleum Geology of the North European Margin. Proceedings of the North European Margin Symposium. Norwegian Petroleum Society. Graham and Trotman, London, pp. 109–135.

Steel, R.J., Gjelberg, J., Helland-Hansen, W., Kleinspehn, K., Nöttvedt, A. and Rye-Larsen, M., 1985. The Tertiary strike-slip basins and orogenic belt of Spitsbergen. Soc. Econ. Paleontol. Mineral., Spec. Publ., 37: 339–359.

Vogt, T., 1929. Den norske fjellkjedes revolusjonshistorie. Nor. Geol. Tidsskr., 10: 97–115.

P. HAREMO *University of Oslo, Department of Geology, P.O. Box 1047, Blindern, 0316 Oslo 3, Norway*
Present address: *Norsk Hydro A/S, Drammensveiem 264, P.O. Box 200, N-1321 Stabekk, Norway*

A. ANDRESEN *University of Oslo, Department of Geology, P.O. Box 1047, Blindern, 0316 Oslo 3, Norway*

Pre-Cretaceous structural development of the Danish Central Trough and its implications for the distribution of Jurassic sands

O. Michelsen, T.E. Mogensen and J.A. Korstgård

The structural development of the Danish Central Trough has been studied to determine the age of graben development and to reconstruct the paleogeographic distribution of Jurassic sands.

The Late Hercynian compression seems to have influenced the area. The presence of Rotliegende volcanics may be related to transitional movements along a right lateral, NW–SE strike-slip fault system. The Zechstein, Triassic and Early Jurassic times are regarded as stable tectonic periods during which the Central Trough area may have acted as an inter-basinal barrier.

Graben development started along the Coffee Soil Fault in the Middle Jurassic; subdivision into the smaller half grabens took place in the Late Jurassic.

Fluvio-deltaic, Middle Jurassic sands were deposited along the Coffee Soil Fault, indicating a transgression from the north. Oxfordian to early Kimmeridgian, lower delta plain to offshore marine sands were deposited in the area of the later Heno Plateau, Feda and Gertrud Grabens, indicating a progressive westward transgression from the eastern region of the trough.

Introduction

A number of earlier regional papers have dealt with the subject, presenting a simple picture of graben development from the Permian to the Jurassic. Gowers and Sæbøe (1985) presented a more differentiated outline of the structural development, which we regard as more realistic. The Jurassic and Early Cretaceous structural outline described by Møller (1986) and Vejbæk (1986), respectively, will be followed in this paper (see Fig. 1). However, the major bordering fault zone of the Ringkøbing-Fyn High will here be referred to as the Coffee Soil Fault, following Gowers and Sæbøe (1985). The sedimentary facies and depositional environments of the Jurassic sands are described by Frandsen (1986), Jensen et al. (1986) and Johannessen (pers. commun., 1989). The biostratigraphic correlation used in the paleogeographic reconstruction is based on data published in the aforementioned papers or derived from internal reports of the Geological Survey of Denmark. Within the last few years the amount of released data has increased, creating the opportunity to test earlier models. Our primary aims were firstly to investigate how far back in geological time graben development can be justified or proved by the data and secondly to reconstruct the paleogeographic distribution of Jurassic sand.

Our study is based on released seismic and well data (Fig. 1). North of profile 6 (Fig. 1) the seismic database consists of surveys: CGT81, RTD81, ELF8083, SP82, NH82. South of profile 6, published lines from Møller (1986), Vejbæk (1986) and Cartwright (1987) have been used.

Structural and depositional development

Pre-Permian

Precambrian rocks were drilled on the Mandal High (3/7-1 well) and, with a Caledonian overprint, on the edge of the Ringkøbing-Fyn High (Per-1 well); Caledonian metamorphosed rocks were encountered by the P-1 well on the Inge High (Fig. 1). These wells are situated on the footwall part of major basement fault blocks (Fig. 1). These footwalls have been relatively uplifted and suffered deep erosion. The uplift of the footwalls and the configuration of the deep reflectors (see Fig. 2, *A* and *B*) are in accordance with recent models for the elastic and flexural behaviour of the upper crust during faulting. These models, e.g., the "flexural cantilever model" (Kusznir et al., 1988), predict both footwall uplift and a concave upwards configuration of bed geometries in the footwall. Application of these models explains the locally high positions of Caledonian and

Structural and Tectonic Modelling and its Application to Petroleum Geology edited by R.M. Larsen, H. Brekke, B.T. Larsen and E. Talleraas. NPF Special Publication 1, pp. 495–506. Elsevier, Amsterdam.

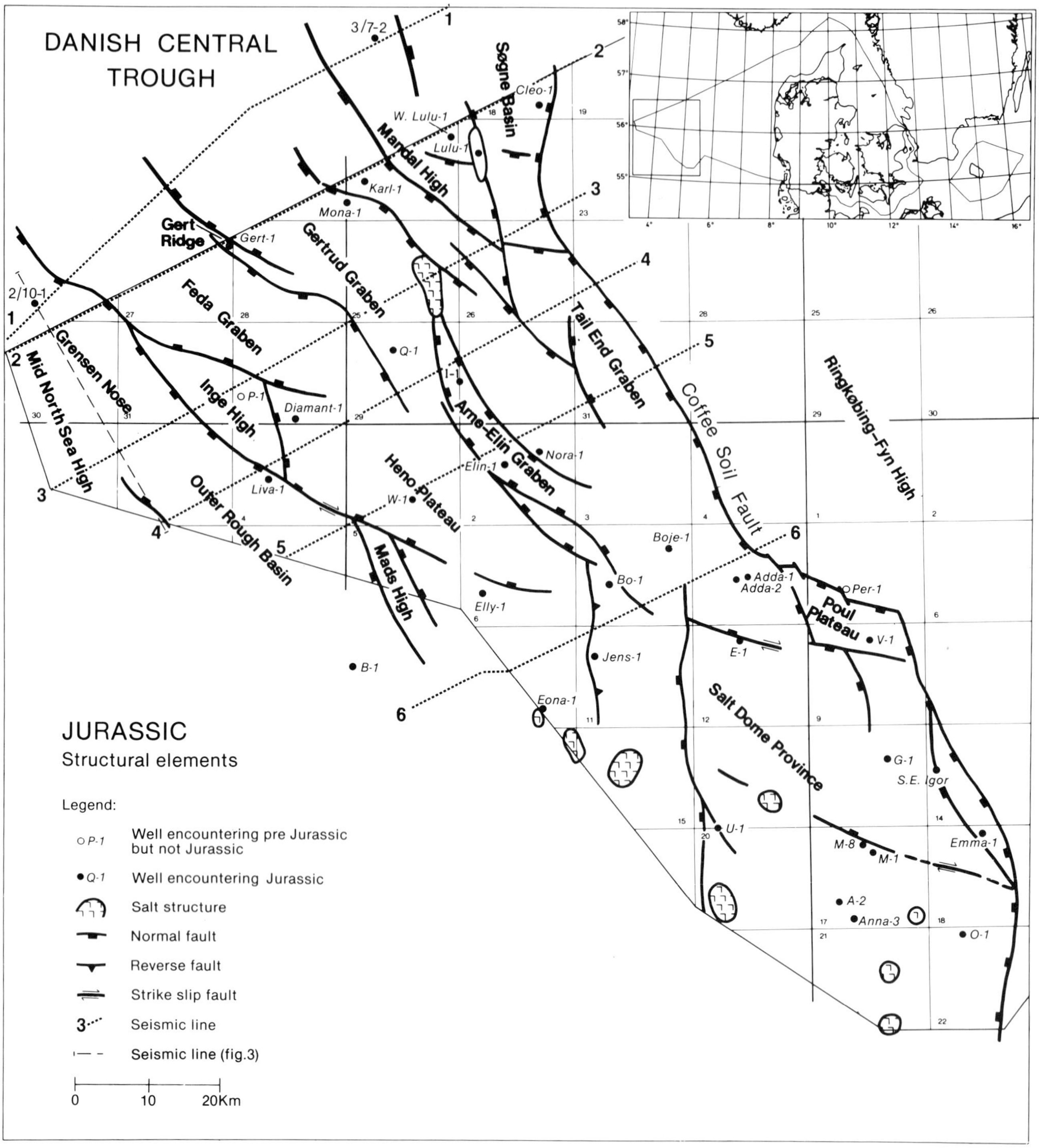

Fig. 1. Jurassic structural elements in the Danish Central Trough. Location of wells available for the study and location of the seismic lines presented in Fig. 2, modified from Møller (1986).

Precambrian basements covered by Late Cretaceous sediments.

Carboniferous sediments are encountered in the P-1 well (at the Inge High) and the 2/10-1 well (in the adjacent Norwegian sector). The Carboniferous sequence can be identified in the same area on the seismic data and is probably present through most of the Danish Central Trough. Seismic reflectors below Top Rotliegende may well delineate the Carboniferous sequence.

At the Mid North Sea High, the uniformly thick Carboniferous sequence forms W–E striking anticlines (Fig. 3) which may indicate a late Hercynian compressive phase, a compression that might have formed the major W–E striking Mid North Sea High–Ringkøbing-Fyn High system (Pegrum, 1984).

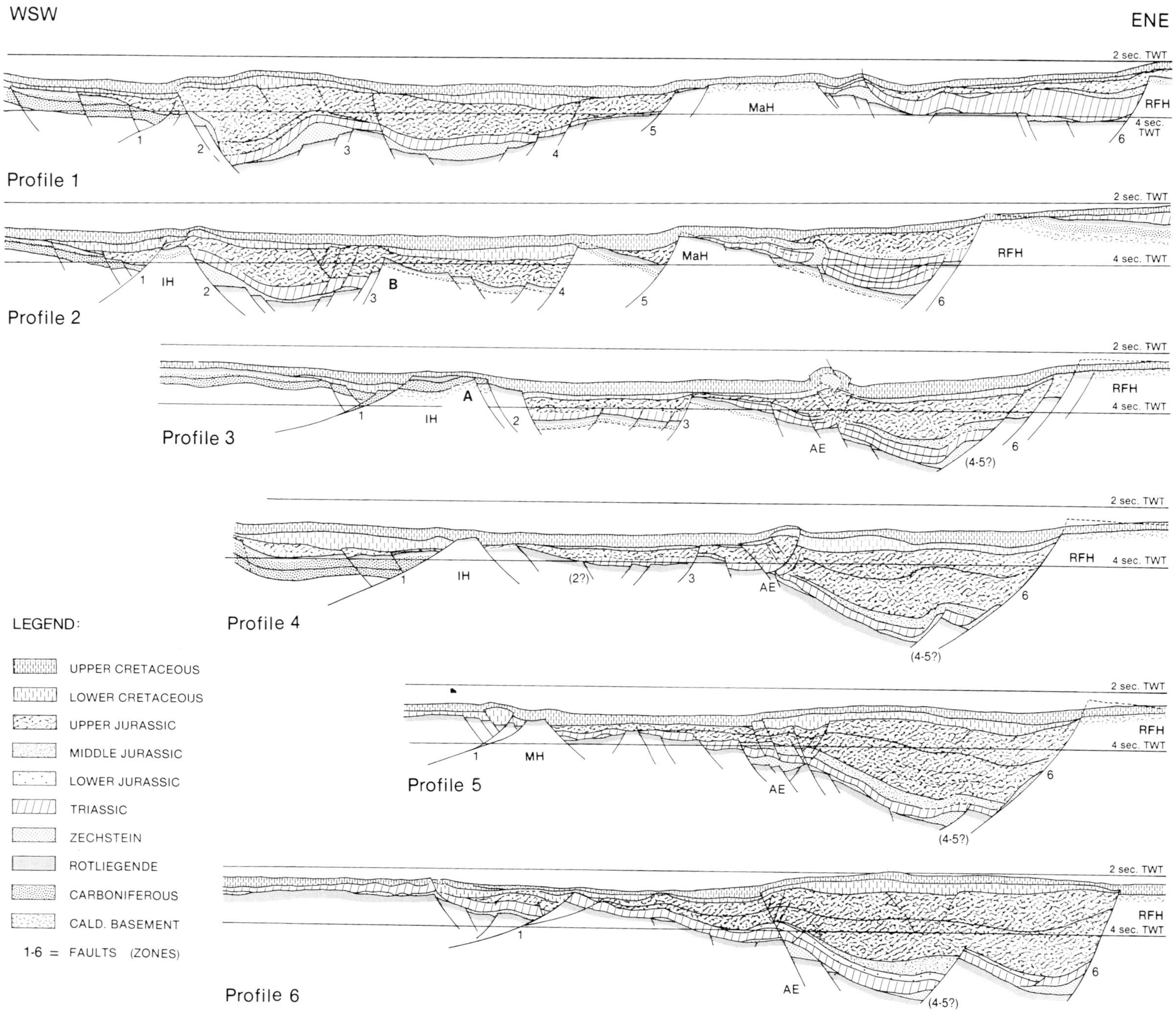

Fig. 2. Profiles based on seismic interpretation across the Central Trough (*A* and *B*, see text). *1–6* indicate connected faults. *AE* = Arne Elin Trend; *IH* = Inge High; *MH* = Mads High; *MaH* = Mandal High; *RFH* = Ringkøbing Fyn High. For locations see Fig. 1.

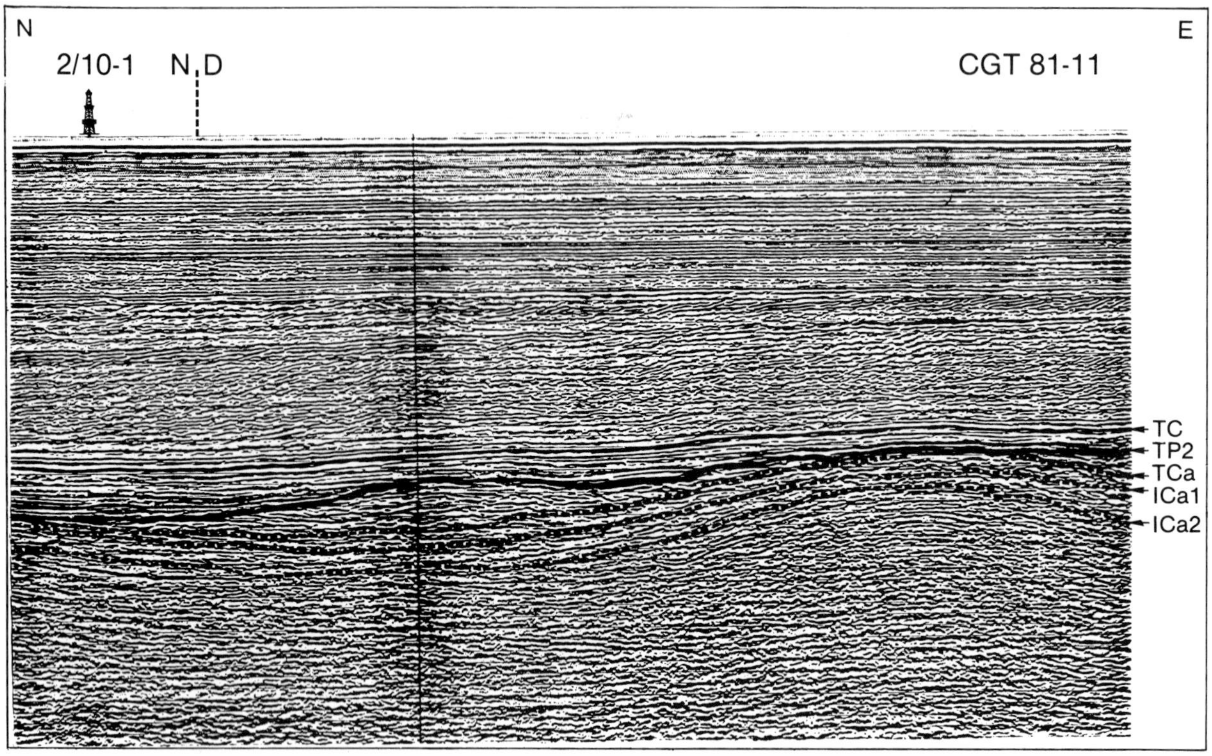

Fig. 3. Seismic section showing a Carboniferous fold on the Mid North Sea High. *TC* = Top Chalk; *TPZ* = Top Pre Zechstein; *TCa* = Top Carboniferous; *ICa 1–2* = Intra Carboniferous 1–2. For locations see Fig. 1.

Rotliegende

According to Glennie (1984, 1986) and Ziegler (1982, 1987), the Hercynian fold belt collapsed in Stephanian-Autunian times, due to the development of a NW–SE orientated right lateral transform fault system. This system caused a complex pattern of conjugate shear faults and related pull-apart structures and volcanism over large areas of Northwest Europe (Fig. 4).

Rotliegende volcanics are encountered in a number of wells drilled in the northern part of Danish Central Trough and adjacent areas, indicating tectonic activity (Fig. 4). Ziegler (1982, 1987) refers the volcanism to evolution of the Mid North Sea High–Ringkøbing-Fyn High system, while Glennie (1984, 1986) relates the volcanism to a postulated opening of a graben structure in the Central Trough area. It is not possible from available data to establish whether the faults, e.g., the Coffee Soil Fault, were part of the fault systems proposed by Ziegler and Glennie (op. cit.). However, a few minor NNW trending Rotliegende faults are recognized in the Søgne Basin, which could be a part of a NW–N trending fault system.

Our suggestion is that the apparent NNW trending distribution of the volcanics (Fig. 4) and the observed N–NW fault orientations, may be related to transtensional movements, through the area of the later Danish Central Trough. Pegrum (1984) and Cartwright (1987) also proposed transtensional movements in the Danish Central Trough area, but they related these to movements along a broad Tornquist Zone.

Zechstein

Zechstein evaporites have been drilled in the Søgne Basin to the north and in the Salt Dome Province to the south (Michelsen and Andersen, 1983) and marginal carbonate facies were encountered in the Diamant-1 well on the Heno Plateau. Seismic mapping shows a distribution of the Zechstein sediments which is in accordance with that described by Gowers and Sæbøe (1985) (see Fig. 5).

The evaporites are found in trends that correlate with the proposed NW–N trending right lateral fault system. However, the present day distribution of the Zechstein sediments is controlled by faults that were primarily active in the Jurassic and the distribution may, therefore, be largely a result of Mesozoic erosion.

Deeply buried Zechstein evaporites have been interpreted from seismic data to be present in the Tail End Graben (Fig. 2), thus indicating a connection between the Northern and Southern Permian Basins through the Søgne Basin–Tail End Graben–Salt Dome Province trend. However, the marginal

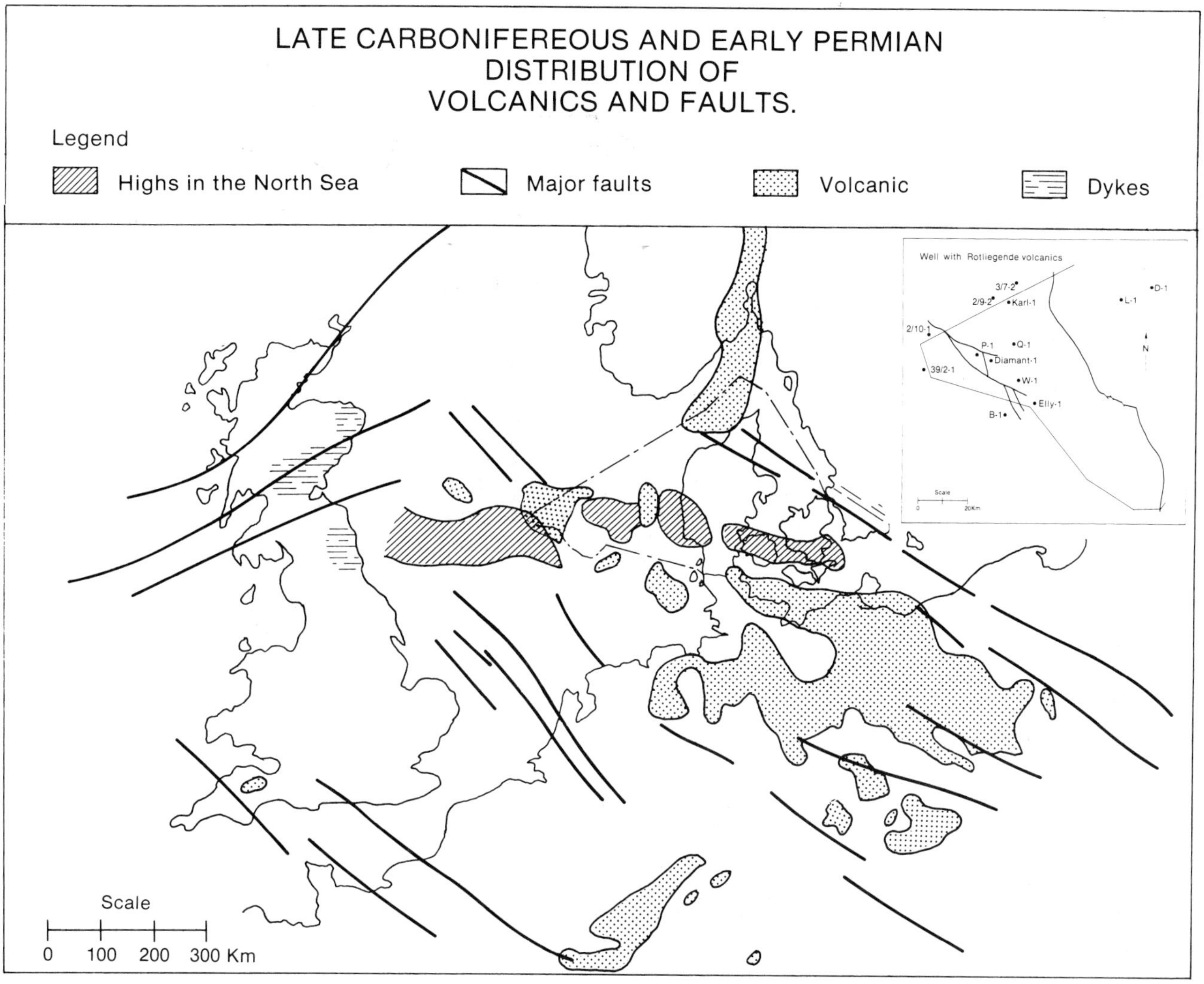

Fig. 4. Stephanian-Autunian distribution of volcanics and major faults. Modified from Glennie (1986). The inset map shows locations for the wells penetrating Rotliegende volcanics in the investigated area.

carbonate facies on the Heno Plateau and the thin evaporites and/or carbonates in the middle of the Tail End Graben point towards an inter-basinal barrier in the Central Trough.

The Zechstein evaporites were possibly deposited along trends subsiding due to thermal effects, caused by the Rotliegende heat event.

Triassic

In the southern part of the Danish Central Trough more than 1500 m of Triassic sediments are present, representing the entire Triassic time interval. The facies succession shows that the sequence is a part of the depositional pattern known from the southern North Sea, which is characterized by repeated marine transgression (Jacobsen, 1982; Michelsen and Andersen, 1983). Farther to the north, in the Q-1, W.Lulu-1 and Cleo-1 wells, 2–400 m of Lower to lowermost Middle Triassic sediments have been penetrated (Fig. 6). These sediments are referred to the

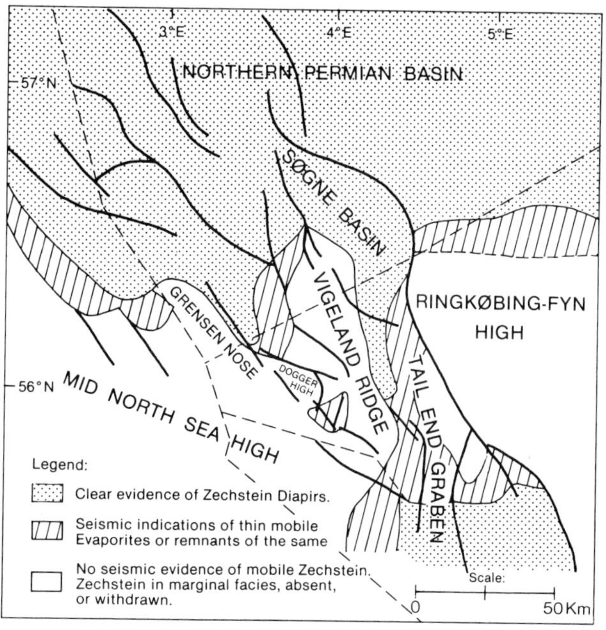

Fig. 5. Distribution of different seismic expressions of the Zechstein sequence. From Gowers and Sæbøe (1986).

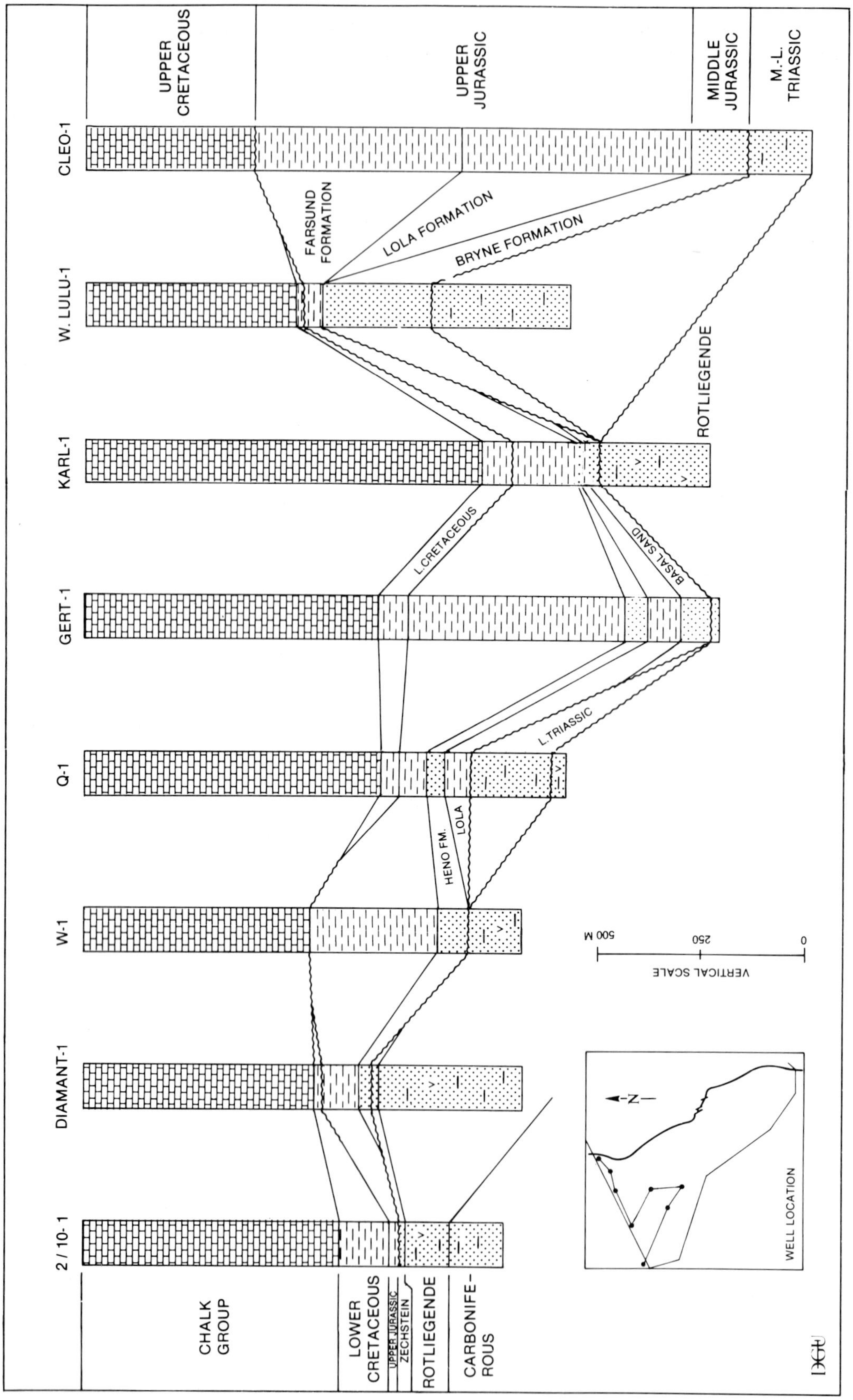

Fig. 6. Lithostratigraphic subdivision of the Mesozoic and pre-Mesozoic in well sections from the northern Danish Central Trough.

northern facies province, characterized by continental plain deposits. However, seismic and well data indicate that Triassic sediments were deposited in the entire trough area, which is in agreement with Gowers and Sæbøe (1985). Triassic sediments are encountered in the V-1 well on the later down-faulted Poul Plateau, indicating that deposition in the Triassic also occurred on the Ringkøbing-Fyn High.

The seismic data show no clear asymmetric sedimentation patterns in the Danish Central Trough, except for halokinetically induced sedimentation, profiles 1-2 (Fig. 2). This is in contrast to the rapid and asymmetric subsidence in the nearby, but differently orientated, Horn Graben. The thinning of the Triassic sequence towards the north and northwest in the Danish Trough area is probably the result of Middle–Late Jurassic erosion.

The halokinetically induced rim-synclines are clearly seen in the Søgne Basin and also identified in the Tail End Graben (Fig. 2 and Mogensen, 1988). Minor fault activity in the Triassic probably triggered the salt movements.

The thickness and facies distribution of the Triassic sediments indicate that the Late Permian structural framework seems to have continued through the Triassic. The trough area probably acted as an interbasinal barrier together with the Mid North Sea High–Ringkøbing-Fyn High trend, separating the northern and southern Triassic facies provinces.

Lower Jurassic

The Lower Jurassic deposits are only drilled in the Salt Dome Province, in the Edna-1, Jens-1, M-8, O-1 and U-1 wells (Fig. 1), but it is generally accepted that the Early Jurassic sea covered most of the North Sea area (see also Gowers and Sæbøe, 1985).

The drilled sequence is referred to the lower part of the Fjerritslev Formation, as known from North Jylland (Michelsen, 1978). Lithostratigraphic and biostratigraphic correlations show that sedimentation rates may have been the same over broad areas.

Seismic data indicate that the Lower Jurassic sediments are present in the southern part of the Tail End Graben, while they are absent in the Søgne Basin.

The drilled Lower Jurassic sequence is truncated so that the Upper Sinemurian beds are the youngest remaining (Michelsen, 1978). The erosion was probably caused by the central North Sea uplift, as described by Ziegler (1982). The overlying sediments are dated to the Bajocian to Callovian stages (Jensen et al., 1986). The erosional event possibly also affected the Triassic sequence in the northern and northwestern Danish Central Trough.

Middle Jurassic

The Middle Jurassic sediments are known from the Edna-1, Jens-1, Lulu-1, M-8, Nora-1, O-1, U-1 and W.Lulu-1 wells in the Søgne Basin, Tail End Graben and Salt Dome Province (Fig. 1). The sequence consists of fluvial-deltaic sediments that show evidence of increasing marine influence upwards (Jensen et al., 1986).

Both seismic and well data show a pronounced increase in thickness towards the east, suggesting that sedimentation was controlled by fault activity along the Coffee Soil Fault. This feature has also been observed in the northern Søgne Basin (Gowers and Sæbøe, 1985). The Middle Jurassic sequence is, thus, the oldest clear syn-rift, wedge-shaped deposit that is recognized in the Danish Central Trough.

Late Jurassic

During the Late Jurassic, marine conditions prevailed in the Central Trough region, caused by the eustatic sea-level rise and increased subsidence. Up to 4000 m of clay-dominated sediments were deposited to the east, in the Tail End Graben. In the Feda and Gertrud Grabens more than 2000 m were deposited and approximately 600 m seems to be present in the Grensen Nose and Outer Rough Basin (Fig. 7; Møller, 1986; Michelsen et al., 1987).

In the Salt Dome Province–Tail End Graben–Søgne Basin trend there seems to have been continuous sedimentation from the Callovian to the earliest Early Cretaceous. The age distribution of the formations indicates a transgression from the north through this basinal trend during the late Middle Jurassic (Michelsen et al., 1987; fig. 3). The present study, furthermore, indicates that during the Late Jurassic the transgression continued westwards onto the Heno Plateau and Feda Graben areas.

The age distribution of sediments subcropping the Middle Jurassic stratigraphic level supports this model (Figs. 8 and 9).

(1) To the south, in the Salt Dome Province, the Middle Jurassic is separated from the Lower Jurassic sediments by a hiatus corresponding to the Pliensbachian and Toarcian stages. The Lower Jurassic overlies Upper Triassic sediments.

(2) In the Søgne Basin the Middle Jurassic rests on lower Middle to Lower Triassic sediments. According to the seismic data this also seems to be the case in parts of the Feda Graben and Gertrud Graben.

(3) At the eastern margin of the Heno Plateau, the Oxfordian Lola Formation rests on lower Triassic sediments.

(4) Farther westwards on the Heno Plateau, the late Oxfordian to early Kimmeridgian Heno Formation overlies Zechstein or Rotliegende sediments.

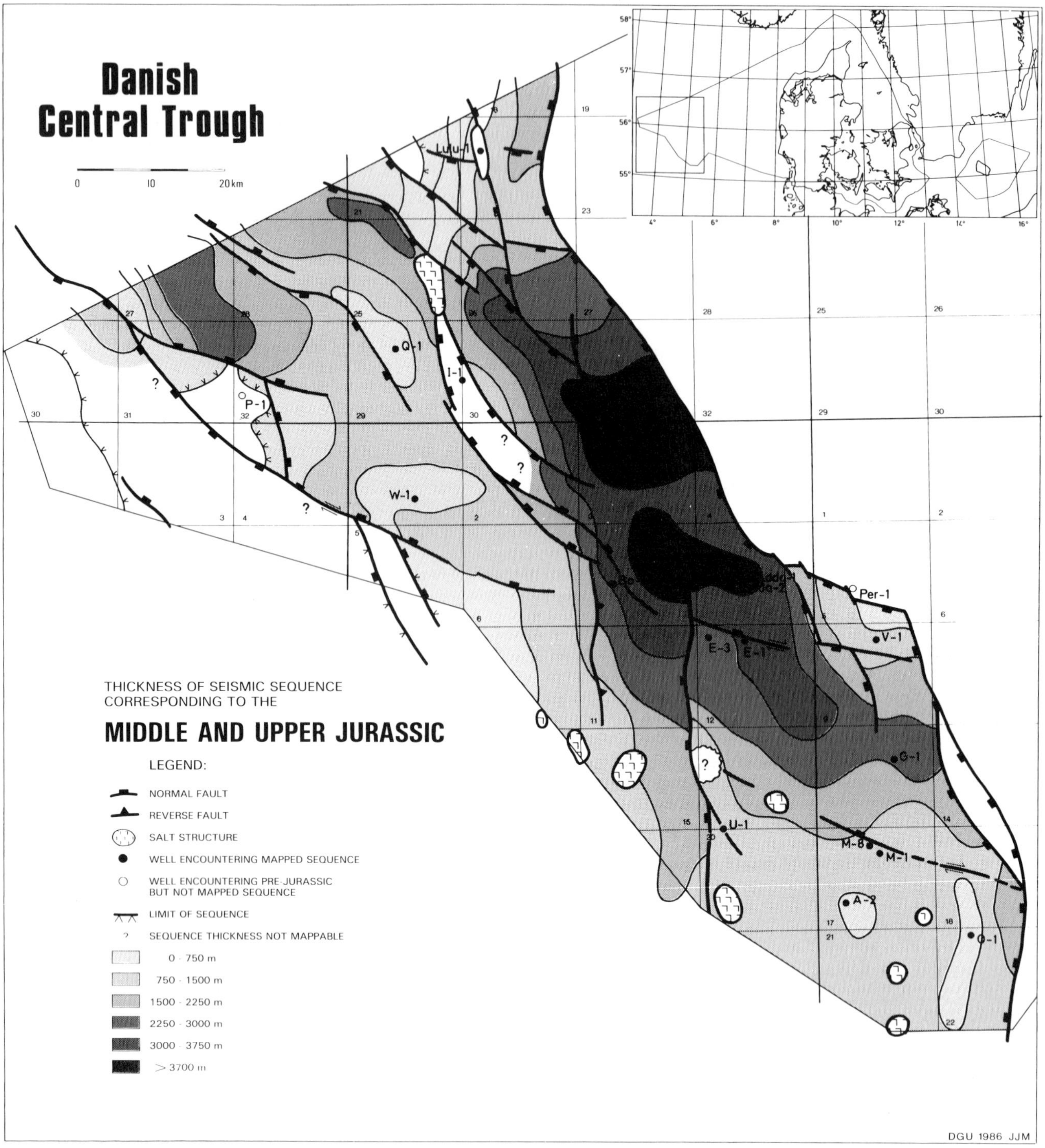

Fig. 7. Thickness and distribution of the seismic sequence corresponding to the Middle and Upper Jurassic section. From Møller (1986).

(5) In the Gert Ridge area, north of the Heno Plateau, the Oxfordian basal sand rests on Rotliegende sediments.

(6) To the north, in the Norwegian sector, the Upper Jurassic rests on Rotliegende at the Mandal High and the uppermost Jurassic Farsund Formation overlies Zechstein carbonates at the Mid North Sea High.

This subcrop pattern may be explained by progressively deeper erosion from east to west in the Danish Central Trough area during the early Middle Jurassic uplift and the initial rifting phase (Fig. 9). Continued rifting, causing subdivision of the trough area into smaller halfgrabens, must have taken place after this erosion phase. Subdivision of the trough area into the Feda Graben and Gertrud Graben occurred in earliest Kimmeridgian, after deposition of Oxfordian basal sand, which is drilled in the Gert-1 well (Fig. 6). Seismic profiles (Fig. 2) show that the

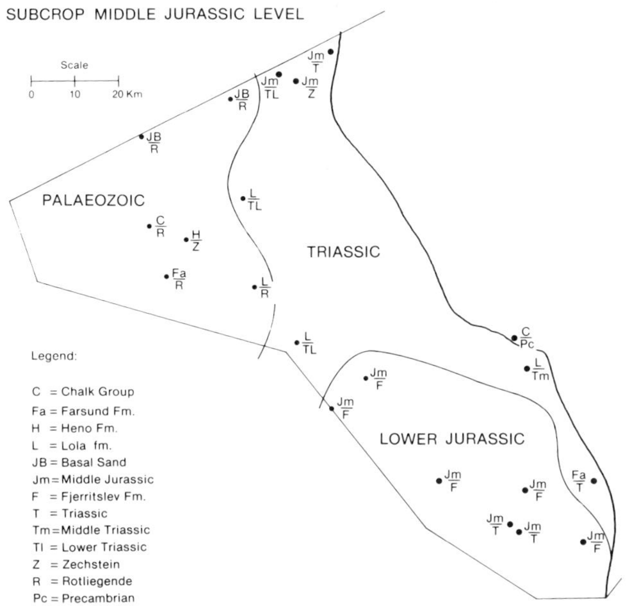

Fig. 8. Age and distribution of sediments subcropping the base Middle Jurassic stratigraphic level, indicating the well locations.

Inge and Mads Highs were part of the Mid North Sea High until Late Jurassic times. The broadening of the Central Trough through the time-span from the late Middle Jurassic to the Early Cretaceous, indicates slow stretching according to the stretching model of McKenzie (1978). Slow stretching causes thermal recovery in the lithosphere and subsequent strain hardening, widening of the rift system and finally abortion of the rift.

The WNW–NW orientation of Middle to Late Jurassic faults cuts to a certain degree the older NW–N Rotliegende grain. This caused a separation of the Søgne Basin and the Tail End Graben and footwall uplift of the Mandal High, the southern Søgne Basin and the Ringkøbing-Fyn High, inducing complex structures at the triple point between the Søgne Basin, the Ringkøbing-Fyn High and the Tail End Graben (Fig. 10). The separation of the Søgne Basin and the Tail End Graben and the subsequent starvation of the Søgne Basin, is clearly seen from the distribution of Cretaceous deposits covering the Late Cimmerian Unconformity surface (Fig. 11).

Middle Late Jurassic paleogeography and related sand deposits

The structural preconditions for reconstructing the Middle and Late Jurassic paleogeography are described above. The early Middle Jurassic erosional surface is documented by well and seismic data. In the southeastern Danish Central Trough the erosion cut down to the middle of the Lower Jurassic sequence, whereas in the northwest the erosion truncates Paleozoic rocks.

Middle Jurassic

Following the early Middle Jurassic uplift and the subsequent erosion fault activity is recognized in the eastern Danish Central Trough, especially along the Coffee Soil Fault. Fluvio-deltaic sediments were deposited in the eastern part of this major halfgraben system (Fig. 12). The environment became increasingly marine in the course of time. The age distribution recognized in the Norwegian, Danish and

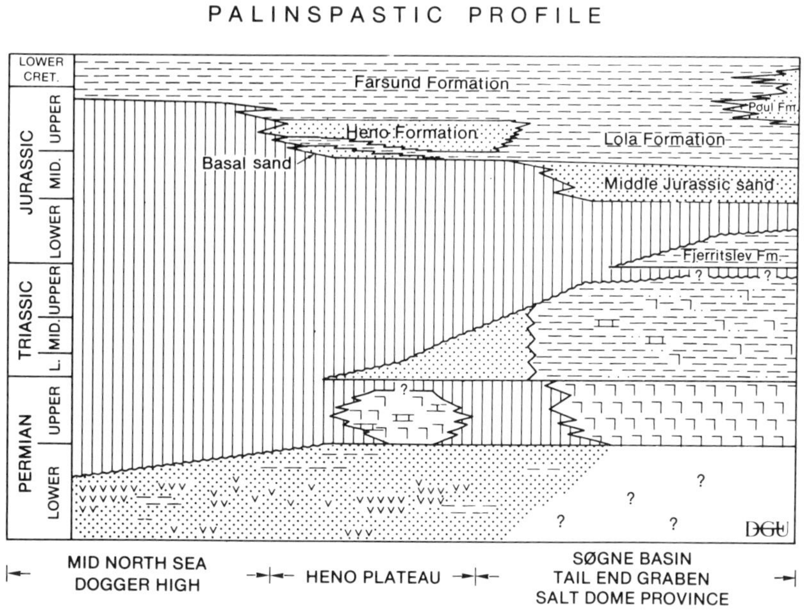

Fig. 9. An approximately east–west palinspastic profile of the Danish Central Trough, including the Permian to lowermost Lower Cretaceous sequences. Note the major hiatus at the base Middle Jurassic stratigraphic level.

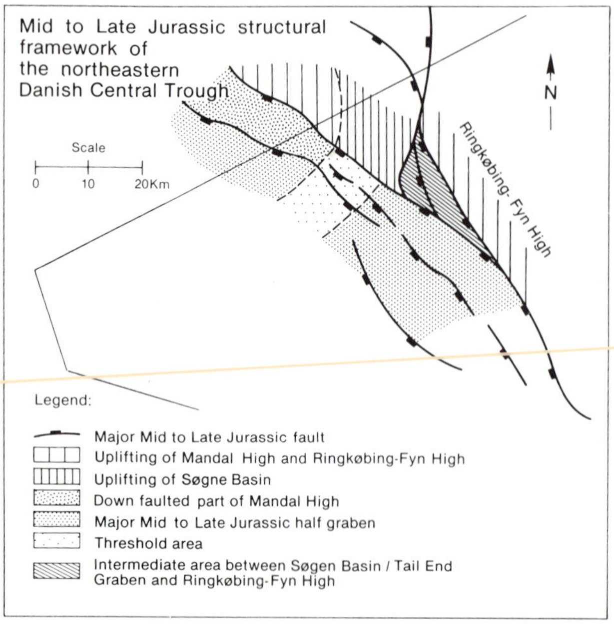

Fig. 10. The mid to Late Jurassic structural framework of the northeastern Danish Central Trough. From Mogensen (1988).

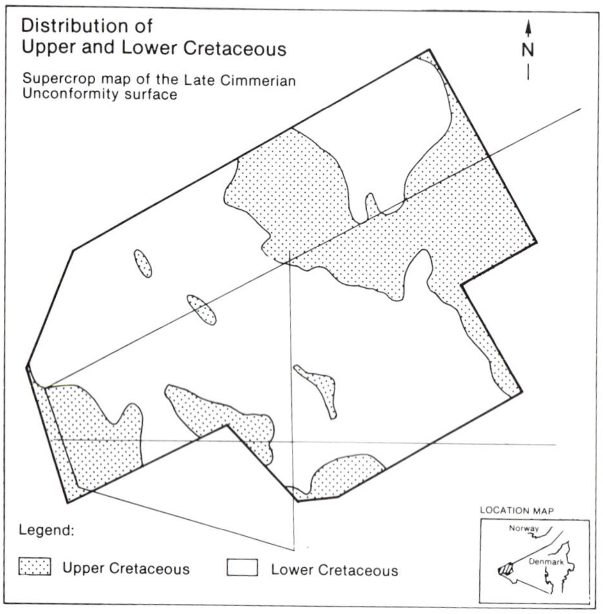

Fig. 11. Distribution of Lower and Upper Cretaceous deposits overlying the late Cimmerian unconformity. From Mogensen (1988).

Dutch sectors indicates that the transgression came from the north during the Callovian (Michelsen et al., 1987).

Oxfordian

The fault activity along the Coffee Soil Fault system continued into the Oxfordian. Farther westwards, the faults delimiting the younger Feda Graben and Gertrud Graben were possibly initiated at the end of this period. In the eastern Central Trough marine conditions prevailed with deposition of shelf muds (Lola Formation), which are also found to the west on the Heno Plateau (Fig. 9). Farther to the northwest, in the Gert Ridge area, coarse-grained beach and lower delta plain deposits are found (the basal sand, Fig. 6). The lower part of the basal sand seems to be time-equivalent with the shelf muds (Lola Formation) on the Heno Plateau, whereas the upper part is time-equivalent with the Heno Formation mentioned below, which consists of shoreface and offshore bar sediments (Fig. 12). This distribution of environments indicates that the sea transgressed westwards from the eastern part of the trough area during the earliest Oxfordian, whereas an apparent regression caused the coast line to migrate eastward again during the latest Oxfordian.

Kimmeridgian

The subdivision of the northwestern trough area into the Feda and Gertrud Grabens occurred in the earliest Kimmeridgian. Shelf muds (Lola Formation) were deposited in this area as well as in the eastern and southern parts of the Central Trough region. On the Heno Plateau, shoreface and offshore coarse-grained sediments (Heno Formation) were deposited in the early Kimmeridgian (Fig. 12). Over a short period, these latter environments prograded into the newly established Feda and Gertrud Grabens (see the Gert-1 and Karl-1 wells, Fig. 6) and farther into the Norwegian sector (Bergan et al., 1989). During the latest Kimmeridgian, rapid subsidence took place in the Tail End, Gertrud and Feda Grabens and shelf muds (Farsund Formation) were deposited in the entire trough region.

Volgian

The rate of subsidence in the grabens increased in the Volgian and a thick sequence of shelf mud (Farsund Formation) was deposited in the entire trough region (Fig. 12). Locally, coarse-grained sediments deposited by density flows (Poul Formation) interfinger with the Farsund Formation. At the end of the period the Inge and Mads Highs were probably separated from the Mid North Sea High and the Outer Rough Basin and the Grensen Nose were established. Elevation of the Mandal High and starvation of the Søgne Basin seems to have occurred at the same time.

Summary

This study has shown that the present Danish Trough has a more differentiated structural development, than is generally accepted in the literature. The main conclusion is that graben development, as

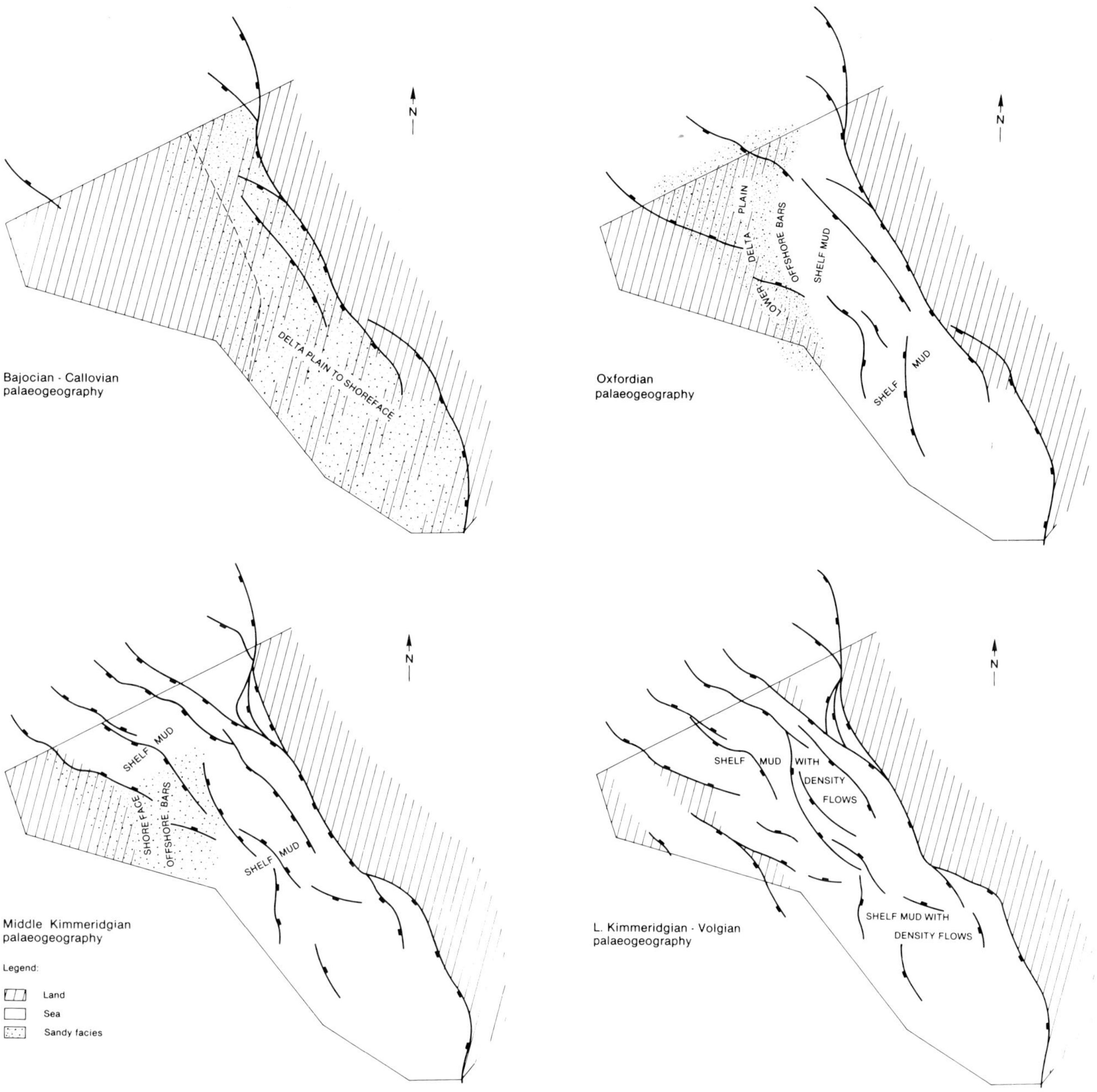

Fig. 12. Middle to Late Jurassic paleogeographic maps of the Danish Central Trough.

reflected by fault controlled, asymmetric sedimentation, started in the Middle Jurassic and continued throughout the Late Jurassic. The dynamic basin development is summarized below.

(1) The trough region and adjacent areas, were uplifted in early Middle Jurassic times. A major halfgraben, the Tail End Graben and the adjacent areas to the west, developed along the Coffee Soil Fault, bordering the Ringkøbing-Fyn High. This represents the oldest documented phase of graben development. The area to the west was deeply eroded during the Middle Jurassic.

(2) Simultaneously with the halfgraben development fluvio-deltaic, sand-dominated sediments (the Bryne Formation) were deposited in the eastern part of this halfgraben. Caused by a transgression coming from the north, the environment became increasingly marine in the course of time.

(3) During the Oxfordian, continued transgression led to deposition of the clay-dominated Lola Formation, which also covers the Heno Plateau to the west. Farther northwestwards, updip the halfgraben structure, coarse-grained beach and lower delta plain sediments (the basal sand) were deposited.

(4) During the latest Oxfordian, a short-lived regression caused an eastward migration of the near shore sedimentation, and the Heno Formation covered the entire Heno Plateau.

(5) High subsidence rates prevailed in the Tail End Graben throughout the Late Jurassic. The Feda Graben and Gertrud Graben started to develop in earliest Kimmeridgian and shelf muds (Lola Formation) were deposited in these areas. Deposition of more coarse-grained sediments continued on the Heno Plateau. Over a short period these latter shallow marine facies prograded into the Feda Graben and the Gertrud Graben and farther into the Norwegian sector.

(6) During the latest Kimmeridgian and the Volgian, rapid subsidence took place in the entire Central Trough region and shelf muds (the Farsund Formation) were deposited. Locally, coarse-grained sediments deposited by density flows intermix the Farsund Formation. The Grensen Nose and Outer Rough Basin developed late in the Late Jurassic. The broadening of the Central Trough was probably the result of slow rifting. Starvation of the Søgne Basin and footwall uplift of the Mandal High took place. The Inge High and Mads High seem to have been parts of the Mid North Sea High until this time.

References

Bergan, M., Tørudbakken, B. and Wandås, B., 1989. Stratigraphic correlation of Upper Jurassic sandstones within the Norwegian Central Graben: sedimentological and tectonic implications. In: J.D. Collinson (Editor), Correlation in Hydrocarbon Exploration. Graham and Trotman, London, pp. 243–251.

Cartwright, J.A., 1987. Transverse structural zones in continental rifts — an example from the Danish Sector of the North Sea. In: J. Brooks and K.W. Glennie (Editors), Petroleum Geology of North West Europe, Vol. 1. Graham and Trotman, London, pp. 441–452.

Frandsen, N., 1986. Middle Jurassic deltaic and coastal deposits in the Lulu-1 well of the Danish Central Trough. Danm. Geol. Unders., Ser. A, 9, 23 pp.

Glennie, K.W., 1984. Introduction to the Petroleum Geology of the North Sea. Blackwell Scientific Publications, Oxford, 236 pp.

Glennie, K.W., 1986. Development of N.W. Europe's Southern Permian Gas Basin. In: J. Brooks, J.C. Goof and B. van Hoorn (Editors), Habitat of Paleozoic Gas in N.W. Europe. Geol. Soc. London, Spec. Publ., 23: 3–22.

Gowers, M.B. and Sæbøe, A., 1985. On the structural evolution of the Central Trough in the Norwegian and Danish sectors of the North Sea. Mar. Pet. Geol., 2: 298–318.

Jacobsen, F., 1982. Triassic. In: O. Michelsen (Editor), Geology of the Danish Central Graben. Danm. Geol. Unders., Ser. B, 8: 32–36.

Jensen, T.F., Holm, L., Frandsen, N. and Michelsen, O., 1986. Jurassic–Lower Cretaceous lithostratigraphic nomenclature for the Danish Central Trough. Danm. Geol. Unders., Ser. A, 12, 65 pp.

Kusznir, N.J., Marsden, G. and Egan, S., 1988. Fault block rotation during continental lithosphere extension: a flexural cantilever model. Geophys. J., 92: 546 (abstract).

McKenzie, D., 1978. Some remarks on the development of sedimentary basins. Earth Planet. Sci. Lett., 40: 25–32.

Michelsen, O., 1978. The Lower Jurassic of the Danish Nordsø O-1 boring, Central Trough, Danm. Geol. Unders., Årbog 1977: 77–87.

Michelsen, O. and Andersen, C., 1983. Mesozoic structural and sedimentary development of the Danish Central Graben. Geol. Mijnbouw, 62: 93–102.

Michelsen, O., Frandsen, N., Holm, L., Jensen, T.F., Møller, J.J. and Vejbæk, O.V., 1987. Jurassic–Lower Cretaceous of the Danish Central Trough — depositional environments, tectonism, and reservoirs. Danm. Geol. Unders., Ser. A, 16, 45 pp.

Mogensen, T.E., 1988. Permo-Jurassisk strukturel udvikling af Central Truget — i Grænseområdet mellem Danmark og Norge. Masters Thesis, Department of Geology, Aarhus University, Aarhus.

Møller, J.J., 1986. Seismic structural mapping of the Middle and Upper Jurassic in the Danish Central Trough. Danm. Geol. Unders., Ser. A, 13, 40 pp.

Pegrum, R.M., 1984. The extension of the Tornquist Zone in the Norwegian North Sea. Nor. Geol. Tidsskr., 64: 39–68.

Vejbæk, O.V., 1986. Seismic stratigraphy and tectonic evolution of the Lower Cretaceous in the Danish Central Trough. Danm. Geol. Unders., Ser. A, 11, 46 pp.

Ziegler, P.A., 1982. Geological Atlas of Western and Central Europe. Elsevier, Amsterdam, 107 pp.

Ziegler, P.A., 1987. Late Cretaceous and cenozoic Intraplate compressional deformations in the Alpine foreland — a geodynamic model. Tectonophysics, 137: 389–420.

O. MICHELSEN *Geological Survey of Denmark, Copenhagen, Denmark*
Present address: *Department of Geology, University of Aarhus, DK-8000 Århus C, Denmark*
T.E. MOGENSEN *Aarhus University, Århus, Denmark*
J.A. KORSTGÅRD *Geologisk Institut, Aarhus Universitet, DK-8000 Århus C, Denmark*

Small faults in sandstones from Spitsbergen and Haltenbanken. A study of diagenetic and deformational structures and their relation to fluid flow

E. Sverdrup and K. Bjørlykke

A study of medium to small scale faults has been carried out on outcrops of Cretaceous sandstones from eastern Spitsbergen (Kvalvågen) and on Jurassic reservoir sandstones from Haltenbanken, offshore mid-Norway. Structural and textural evidences from the faults indicate that they formed at shallow depth when the sand was uncemented. At Kvalvågen the faulting can be demonstrated to be syn-sedimentary, and formed during the progradation of a deltaic sequence. Rapid fluid escape along developing faults gave rise to deformation zones with fault parallel laminae. The high frequency of fault structures observed in cores from the reservoir sandstones of the Njord Field, Haltenbanken, formed at shallow burial depth. This is inferred by the fact that feldspar dissolution and precipitation of authigenic kaolinite, which probably relates to meteoric water flushing, post-date movements on the faults.

Enhanced pressure solution and stylolitization have developed on the fault structures at both localities at greater burial depths, due to enrichment of clay minerals and micas along the fault planes when they formed. Steeply dipping stylolites, such as those observed in some Haltenbanken reservoir sandstones, can therefore be explained by selective pressure solution along fault planes due to overburden stress, and need not imply horizontal tectonic stress.

Permeability measurements of the fault planes show that they are important barriers to horizontal fluid flow and may significantly influence the reservoir properties.

Introduction

Permeability reduction in reservoir sandstones by faults has been observed by several workers (e.g., Pittman, 1981; Nelson, 1985; Gabrielsen and Koestler, 1987), and is generally closely related to deformational textures such as dense physical packing of grains, clay enrichment, grain size reduction by grain breakage, brecciation and/or fault related mineralization. In addition, the primary texture and mineralogy of the faults might be diagenetically altered during deep burial. To predict the effect of faults in sandstones, whether they act as barriers or not, as well as their geometrical distribution within a reservoir, it is important to distinguish between early (pre-lithification) faults and tectonic deformation structures generated at deeper burial. The picture is, however, generally diffuse where syn-sedimentary and shallow generated tectonic faults are to be distinguished, as such deformational depths tend to produce similar textures.

In outcropping sedimentary sequences the relative timing of fault displacement can be inferred from field data. In buried sedimentary basins large scale faults can be interpreted, and partly dated, from seismic sections. Small-scale faults found in cores, however, are more difficult to assess. By relating movements along such faults to a sequence of diagenetic reactions, a relative timing of fault movements and later deformation along fault planes can be obtained. The aims of this paper are to exemplify how fault structures formed at shallow burial depth might become diagenetically altered during late burial, to demonstrate how small-scale faults in sandstones affect permeability, and to explain the development of steeply inclined (fault-parallel) stylolites.

Syn-sedimentary faulting in lower Cretaceous deltaic sandstones at Kvalvågen, Eastern Spitsbergen

The section at Kvalvågen, Eastern Spitsbergen, Arctic Norway (Fig. 1) displays syn-sedimentary faults of Lower Cretaceous age. Detailed descriptions of the geological setting and deformational history are given in Nemec et al. (1988) and Sverdrup and Prestholm (1990).

The cliff exposure shows a strike section through an upper Jurassic deltaic system, which represents the distal development of the Helvetiafjellet and

Structural and Tectonic Modelling and its Application to Petroleum Geology edited by R.M. Larsen, H. Brekke, B.T. Larsen and E. Talleraas. NPF Special Publication 1, pp. 507–517. Elsevier, Amsterdam.

Fig. 1. Location map Kvalvågen, Spitsbergen, Norway.

Janusfjellet Formations (Fig. 2). Open marine shale and thin sandstone beds (the Janusfjellet Formation) are overlain by a fluviodeltaic succession (the Helvetiafjellet Formation). The lower part of the Helvetiafjellet Formation consists of a sandstone unit some 20 m thick, which represents a distributary channel fill sequence. The channel sandstone is medium- to coarse-grained and finely laminated. Subordinates are thin siltstone layers and beds of coarser, locally pebbly material. The channel sandstone is overlain by mouth bar, lagoon and barrier bar deposits (Fig. 2).

Syn-sedimentary faulting in the distributary channel sandstones and the upper part of the Janusfjellet Formation is interpreted to be due to the collapse of the prograding delta front (Nemec et al., 1988). The faults cutting through channel sandstones do not display discrete slip-planes but appear as sandstone-dominated fault zones (Sverdrup and Prestholm, 1990). Vertical throws on the faults fall in the range of 3–35 m, their dips vary between 40 and 60° and widths of the fault zones are between 5 and 40 cm (Fig. 3). The fault zone rims are sub-parallel and the boundary to the adjacent channel sandstones locally exhibits a lithological or structural contrast, or a dark, clay/micaceous seam (Fig. 4). The fault zones are characterized by fault-parallel laminae (Fig. 5).

Small-scale normal faults occur within 2 m of the major fault zones (Fig. 3). One set of such small-scale faults is located in the hanging wall block, is synthetic to the major faults, shows considerable continuity and throws in the order of 1–2 cm. Another set of small-scale faults, located in the footwall blocks, is both syn- and antithetic to the major faults (conjugated) with induced millimetric to centimetric offsets. The small-scale fault structures show no systematic increase of density towards the major fault zones.

The channel sandstone has a very low porosity

Fig. 2. Collapsed channel sandstones of the Helvetiafjellet Formation are draped by mouth bar, lagune and barrier bar sediments.

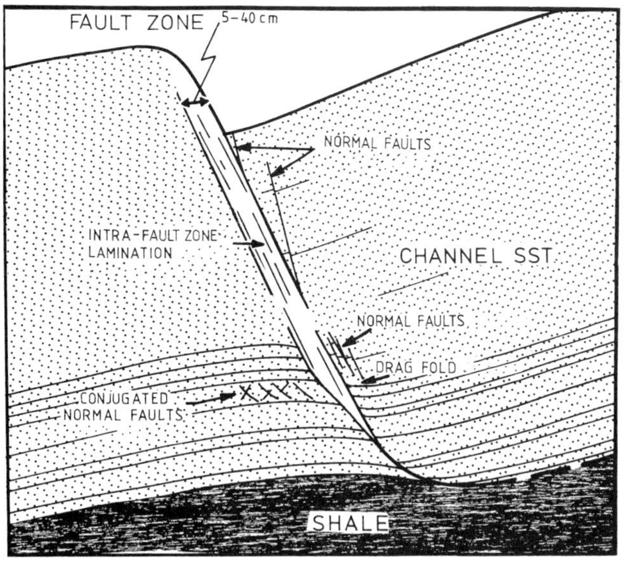

Fig. 3. Idealized sketch of the fault structures at Kvalvågen.

measured from thin sections (0–5%). Quartz-grains and cement comprise 70–90% of the rock. Feldspars (1–5%) and lamina of clay minerals and micas make up the main compositional variations in the sandstones.

By examining grain size and mineralogical composition of the fault zones it has become clear that the material herein is derived from the channel sandstones. No porosity difference exists between the undeformed sandstones and the fault zones. In the microscopic scale the fault zone rim locally appears as a distinct planar feature of a few grain-diameters enriched in clay and micaceous material (Fig. 6). Variations in grain-size and matrix content define the fault-parallel lamination within the fault zones.

The fault zones probably generated from channel sand which got fluidized and remobilized due to water flushing through developing faults during the delta collapse (Sverdrup and Prestholm, 1990). Variable grain size and mineralogical composition of the fluidized material caused separation of particles which constituted the fault zone laminae. As no second generation faults are observed to cut through the described fault zones, the process is suggested to have occurred as a rapid, single-phase injection. It has not been determined whether the small-scale faults represent water-escape structures or ordinary faults.

Thin section studies of the fault zones show no clear evidence of a relative denser physical packing or preferred alignments of grains here compared to the adjacent, undeformed sandstone, which usually characterizes syn-sedimentary faulting (e.g., Allen, 1982). Furthermore, as expected, there is no cataclasis of either quartz or feldspars within the fault zones. The fault structures could, therefore, easily be overlooked in core and/or thin section studies alone. The post-faulting textural development of the fault zones is explained on page 515.

Early diagenetic faulting in reservoir sandstones at Haltenbanken

The rocks from the Njord Field at Haltenbanken are of Middle Jurassic age and have been interpreted as shallow marine deposits (Gjelberg et al.,

Fig. 4. Photograph of a typical fault zone at Kvalvågen. Notice the sharp rim and the discordant character of the fault zone.

Fig. 5. Intra fault zone laminations appear as dark seams. A few small-scale normal faults which offset the primary sedimentary laminae in the channel sandstone are also displayed (scale in cm.).

1987). Seismically defined faults, which probably relate to movement along the western boundary of the Halten Terrace (Kristiansund–Bodø Fault Complex) during the Late Jurassic to Early Cretaceous time (Gabrielsen and Robinson, 1984) cut through the sedimentary sequence.

Sandstones are from depths around 3000 m below sea level, which correspond to their maximum depth of burial. They are homogeneous or faintly laminated, very-fine to fine-grained. Porosities of the sandstones are in the 15–25% range. K-feldspar (5–15%) and muscovite (0–15%) are abundant in the rocks. Matrix, when present, is composed of fine-grained clay minerals (kaolinite, illite) and organic material.

Faults displayed in cores appear as numerous small-scale structures concentrated within certain intervals (Fig. 7). Dark color contrasts commonly make the traces of individual faults visible. The fault structures are single or multiple and the individual fault planes are linear or curved (Fig. 7). Multiple fault planes are either conjugated or sub-parallel. Interweaving fault planes make up anastomosing zones. The displacements of individual faults are on the order of 0.5–2.0 cm, and on the microscopic scale they appear as micro-fault zones with widths between 0.1 and 5.0 mm. Porosity is considerably lower within these micro-fault zones (2–7%) compared to the adjacent, undeformed sandstones (Fig. 8a). The faults are characterized by dense physical packing and preferred orientation of grains and enrichment of matrix material. Locally, grain size reduction, due to mechanically deformed micas and pressure solution of quartz, is present (Fig. 8b).

The absence of mechanically deformed quartz and feldspar grains suggests that the faults formed rel-

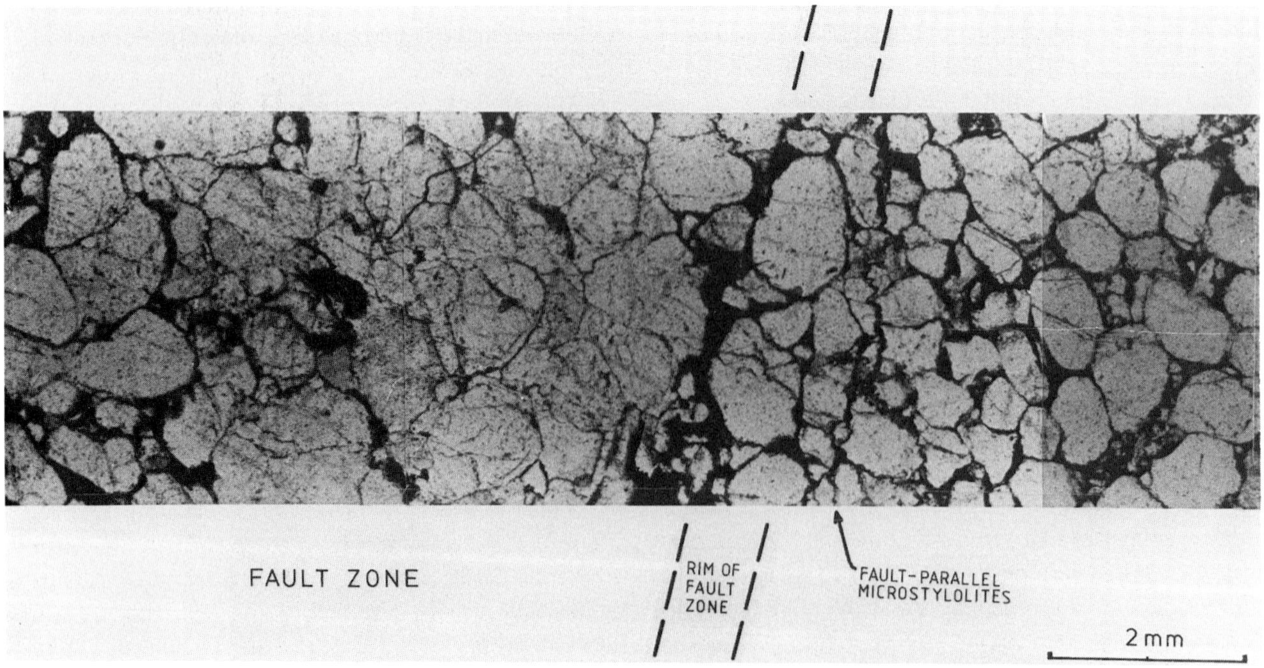

Fig. 6. Photograph (microscale) of the fault zone rim. The rim is defined by a clay seam. The sandstone within the fault zone at this particular position has a lower clay content than the undeformed channel sandstone immediately outside. Notice the fault parallel microstylolites along the rim.

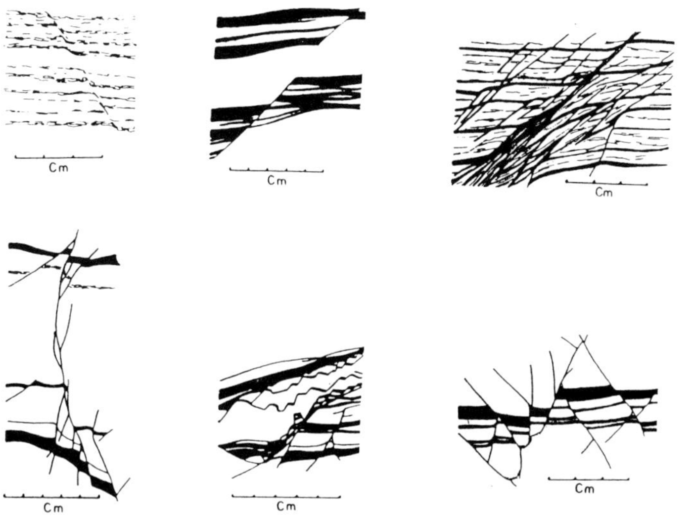

Fig. 7. Small-scale fault structures as they appear in the cores from the Njord Field, Haltenbanken, Offshore Mid-Norway (see text for further descriptions).

atively shallow, prior to any extensive cementation of the sandstone. This is in accordance with Maltman (1984) who pointed out that intragranular deformation dominates in a lithified material due to compaction and cementation, and which also has been visualized through Knipe's (1986) "Deformation Mechanism Path Diagrams". The grain deformation during faulting was characterized by independent particulate flow (Borradaile, 1981) where individual grains, except from micas, remained undeformed. Relative "enrichment" of clay minerals and micas in the micro-fault zone is explained by volume loss caused by the collapse of the primary grain framework and by clay smear of primary sandstone laminae. The deformation also produced preferential alignment of flaky and elongated grains.

The movement on the faults can be dated relatively to two phases of carbonate cementation, feldspar dissolution and pressure solution of quartz. Two generations of carbonate cement (pre- and post-faulting) have been distinguished in the sandstones. Early, microsparry, ferroan calcite occurs within nodules (concretions) and thin layers in fine-grained argillaceous intervals in the cored sequence. The concretions are not cut by faults. The calcite cemented layers, however, exhibit brittle deformation where cut by faults (Fig. 9). These relations strongly suggest that this calcite cement precipitated prior to the faulting process. In certain zones of the sandstones, poikilotopic non-ferroan calcite cement is present. Petrographical relations suggest that this cementation took place after the faulting. The cement shows no brittle deformation (fracturing, etc.) adjacent to the faults and the cement has not precipitated within the densely packed, low-porosity micro-fault zones (Fig. 10).

The two different generations of calcite cement have been analysed with respect to oxygen and carbon isotopes (Fig. 11). The early, "pre-faulting" calcite cement generally displays heavier $\delta^{18}O_{(PDB)}$ values than the late "post-faulting" calcite cement which is rather negative. Early cement with $\delta^{18}O_{(PDB)} > -3‰$ and $\delta^{12}C_{(PDB)} \leq -8‰$ probably reflects precipitation from marine water in the oxydized zone ($\delta^{18}O_{(SMOW)} = -2‰$) at low temperatures ($T$ = 15–20°C). Early cements with more negative values ($-4‰ > \delta^{18}O_{(PDB)} > -8‰$) and $\delta^{13}C$ values between -10 and -20 could have precipitated from meteoric water in the sulphate reducing zone (Irwin et al., 1977). Assuming an average oxygene isotopic composition for this cement at $-6‰$, the $\delta^{18}O_{(SMOW)}$ value for the water at the time of precipitation (Jurassic time) was in the range between -5 and $-6‰$ (T = 12–15°C). The oxygene isotope values of the late calcite cement are generally more negative ($\delta^{18}O_{(PDB)} = -10.5–14.0‰$). By assuming an isotopic composition for the pore water at $\delta^{18}O_{(SMOW)} = -3‰$, which is close to the present day formation water composition, the corresponding crystallization temperature becomes 50–77°C. This is consistent with values obtained (and discussed further) by Bjørlykke et al. (1986) and Saigal and Bjørlykke (1987).

Moulds of feldspar partly filled with kaolinite are common in the sandstones and are also found within the micro-fault zones (Fig. 12). Since this texture has not been deformed, the displacement on the faults must pre-date the feldspar dissolution, which in the Haltenbanken area is interpreted to be due to meteoric water flushing at a relatively shallow burial depth (Bjørlykke et al., 1986).

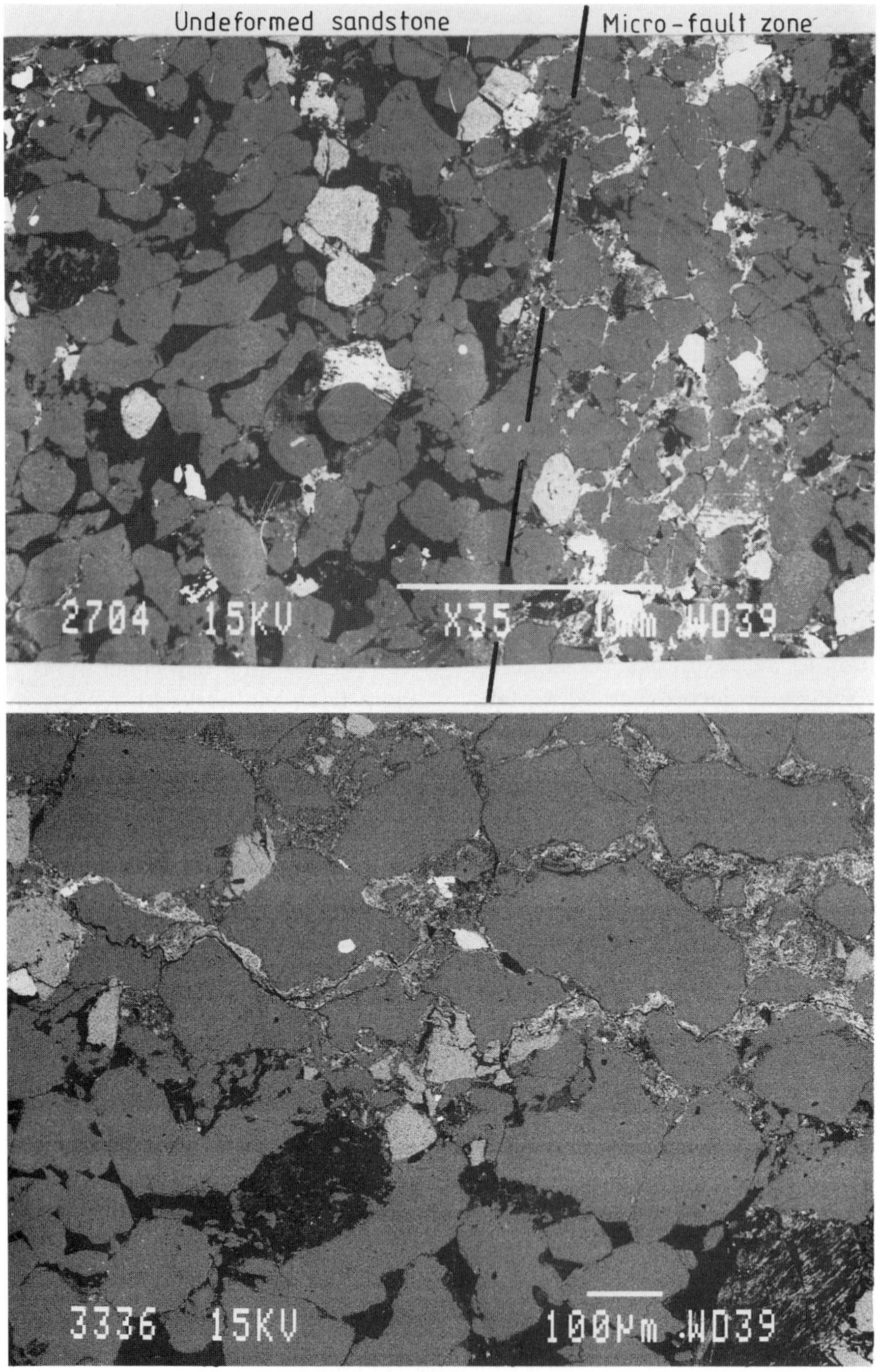

Fig. 8. Top: SEM picture showing the grain arrangements, the porosity loss and the enrichment of clay minerals and fine-grained micas within the micro-fault zone (right). The left side of the picture shows the undeformed sandstone texture. Bottom: SEM picture of a micro-fault zone where pressure solutions of quartz and microstylolites are well developed. Note also the oversized, secondary pores outside the fault generated from dissolved feldspars.

Quartz overgrowth occurs at a great burial depth, and sandstones at Haltenbanken buried less than 2.5–3 km deep generally display a low quartz cement content (Bjørlykke et al., 1986). Furthermore, stylolites are reported to be present in Haltenbanken sandstones at burial depths greater than 2.5 km, which supports quartz dissolution being a late diagenetic phenomenon in this area.

The textural development of the fault structures in relation to the burial and diagenetic history of the sandstones is schematically illustrated in Fig. 13. The two processes: faulting and diagenesis, are both demonstrated as important factors in the textural development of the fault structures.

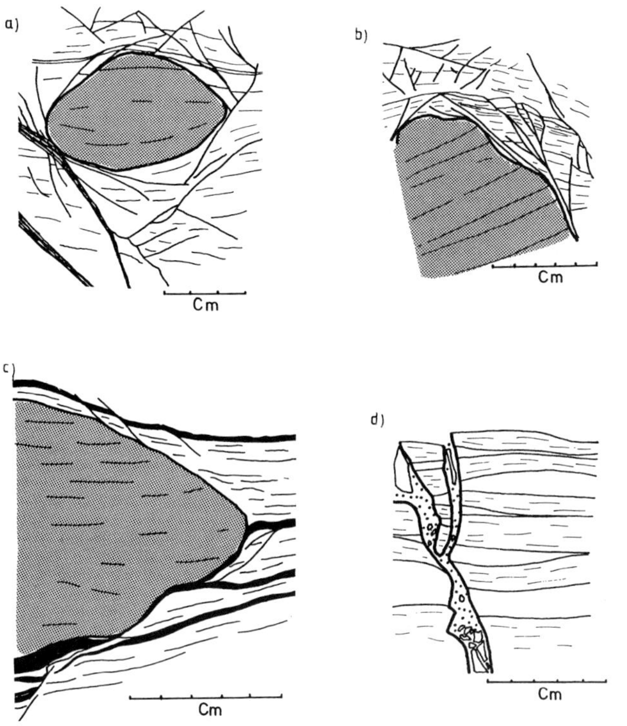

Fig. 9. Structural relationships between the early calcite cemented nodules (a–c) and layers (d).

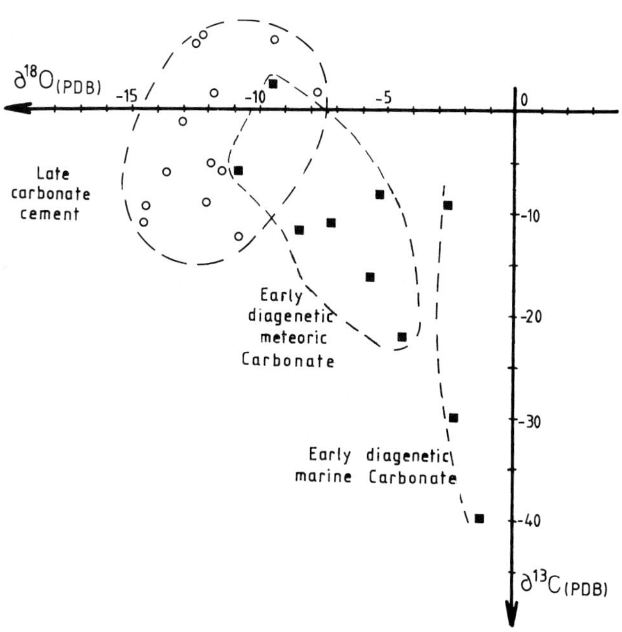

Fig. 11. Isotope data for different generations of carbonate cements from the Njord Field. Early cement which predates faulting (black squares) is characterized by low $\delta^{18}O$ values. The marine cement is formed at the sea floor in the sulphate reducing zone. Late cement which postdate faulting (open circles) is formed at higher temperatures and greater burial depths as indicated by the more negative $\delta^{18}O$ values.

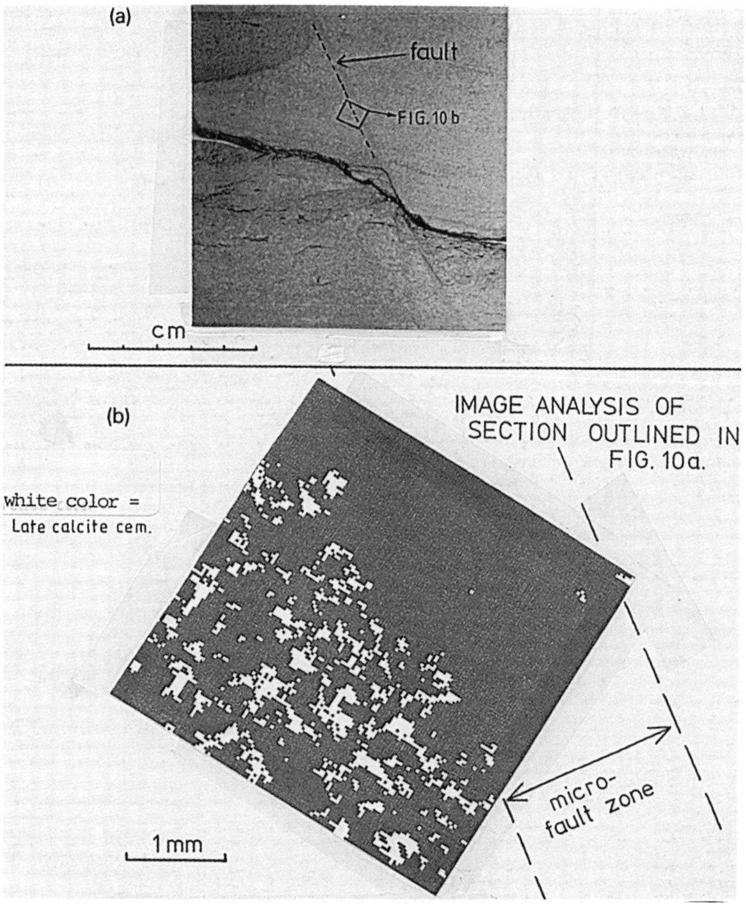

Fig. 10. Textural relationship between the late calcite cement and a fault plane.

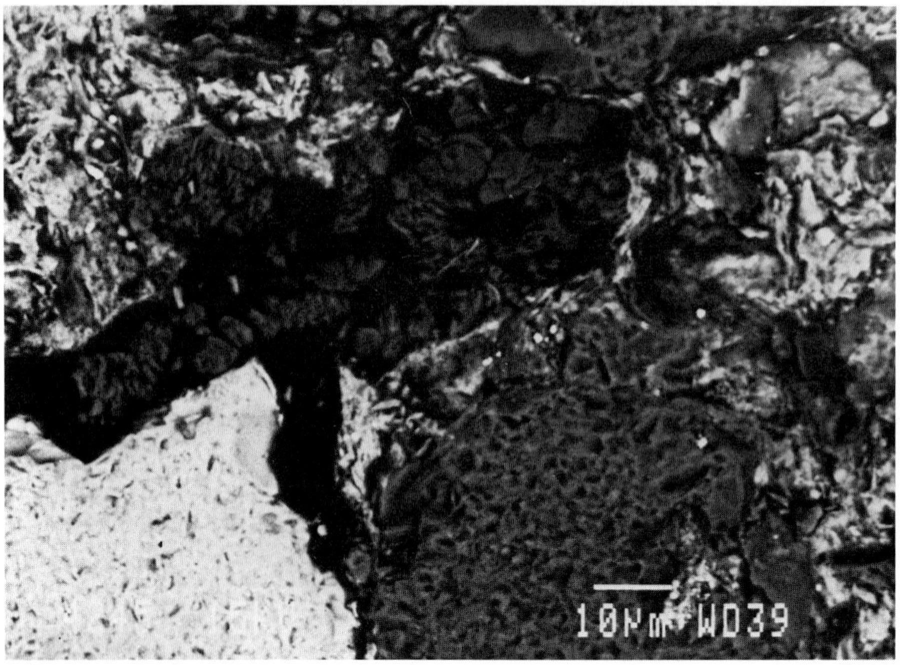

Fig. 12. Secondary porosity within a micro-fault zone due to dissolved feldspar. The mould is partly filled with authigenic kaolinite which indicates that fault movement pre-dates feldspar dissolution and kaolinite precipitation.

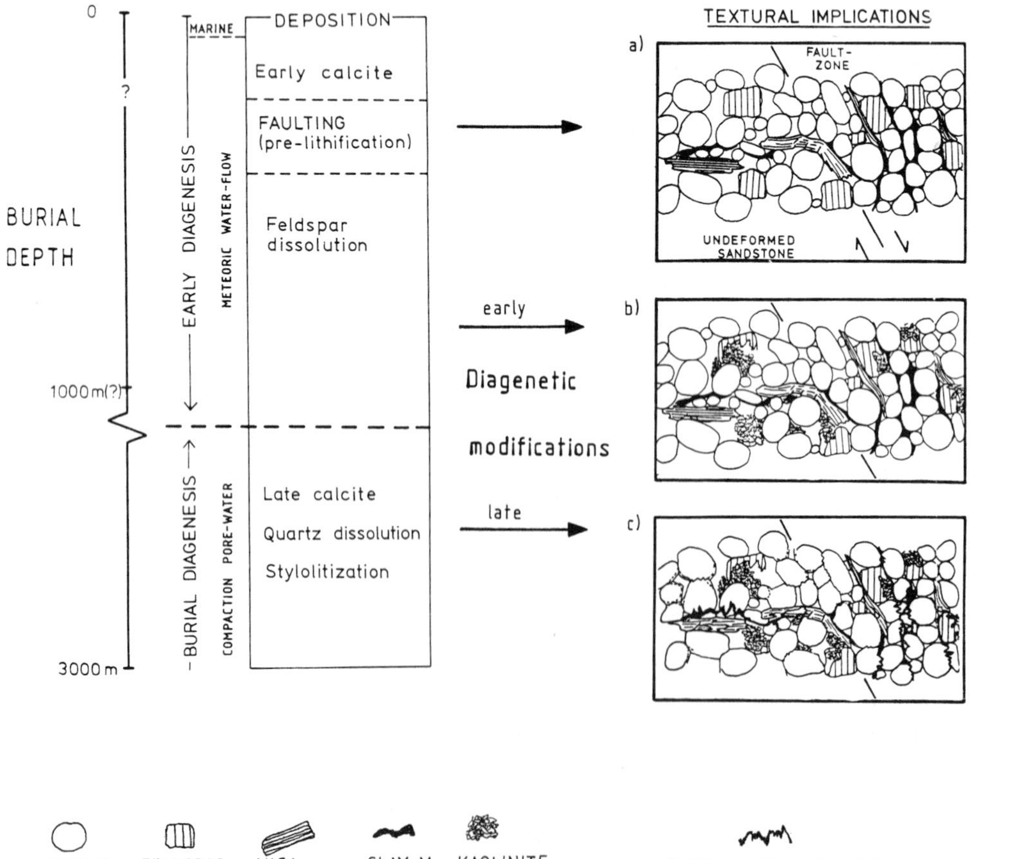

Fig. 13. Composite diagram showing the textural development of a sandstone containing a fault structure during the burial (and diagenetic) history.

Diagenesis superimposed on fault textures

In spite of a shallow deformational depths, the fault structures at Kvalvågen and Haltenbanken show great dissimilarity with respect to fault textures at present. This is partly due to the different deformational processes, but is mainly explained by differences in the post-faulting diagenetic history in the two areas. In the Njord Field, the fault zones are relatively tight with porosities in the lower edge of 5%, compared to an average of about 20% in the surrounding sandstone. This is caused by dense packing of grains and subsequent pressure solution. At Kvalvågen, the sediments have experienced deeper burial (approximately 3500 m; Sverdrup and Prestholm, 1990), and extensive quartz cementation has reduced the porosities to generally less than 5% both in the fault zones as well as in the undeformed sandstone. Consequently, any differences in the early, post-deformational packing arrangements within the fault zones and the channel sands have disappeared and are not observed. A common feature at both localities is the presence of stylolites along the fault structures.

Pressure solution and stylolitization

The most prominent diagenetic processes which have affected the fault structures at both localities, are pressure solution and stylolitization. Stylolites are developed along primary laminae as well as fault surfaces (Figs. 6 and 8b).

Stylolites are generally considered to be formed perpendicular to the maximum principal stress component (e.g., Buchner, 1981; Nelson, 1981). In a subsiding basin, chemical compaction is induced by the vertical lithostatic stress which may produce horizontal (layer-parallel) stylolites. Vertical or steeply dipping stylolites are commonly interpreted in terms of horizontal (tectonically induced) compression in a basin (e.g., Bjørlykke et al., 1986; Mitra, 1988).

Sverdrup and Prestholm (1990) have shown that the stress component of the vertical, lithological stress acting perpendicular to fault surfaces, may be sufficient for stylolites to generate when temperatures are favourable. A calculation performed for the Kvalvågen rocks concluded that the stress component perpendicular to a fault dipping 70° equals 18 MPa. Similar calculations can be performed for the rocks at Haltenbanken.

In the Njord Field at Haltenbanken the vertical effective stress at 2800 m is close to 25 MPa. In sandstones, the ratio between horizontal and vertical stress (σ_z/σ_h) varies between 0.2 and 0.4 (Howard and Fast, 1970), but in general the horizontal component may be taken to be one-third of the vertical

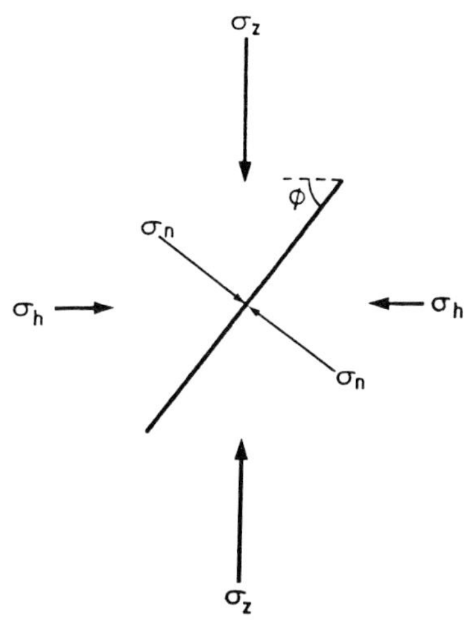

Fig. 14. Stress relationships on a dipping plane. σ_z = vertical stress component; σ_h = horizontal stress component; σ_n = stress component perpendicular to plane; ϕ = dip of plane.

component. The normal stress component on any plane is given by the relation:

$$\sigma_n = \frac{\sigma_z + \sigma_h}{2} + \frac{\sigma_z - \sigma_h}{2} \cdot \cos\phi$$

where ϕ represents the dip of the plane (Fig. 14). From this equation σ_n can be calculated for any dipping surfaces. For $\phi = 60°$ the normal stress component (σ_n) equals 12.5 MPa. This value corresponds to a vertical normal stress at a depth close to 1000 m, with normal pore pressure. As intergranular pressure solution has been observed in sandstones at depths of as little as 900 m (Wilson and Sibley, 1978), and reports from depth of 1000 to 2000 m are common (e.g., Thompson, 1959; Nagtegaal, 1978), the stress value of 12.5 MPa should be sufficiently high for the pressure solution to develop taking into account the temperature at 2800 m. The situation described above, as for the Kvalvågen example, might be visualized in a "Deformational diagram" (Rutter, 1976) (Fig. 15). Here, pressure solution can be demonstrated to be an active process along the fault planes in both localities. The stylolitization observed, taking place along the inclined, clay/mica-rich fault surfaces, has probably evolved as such inhomogeneities in sandstones tend to act as precursors for dissolution of quartz (e.g., Heald, 1959; Weyl, 1959; Boles, 1984).

Permeability of faults

To investigate the permeability effects of small-scale faults, such as those described from Haltenbanken, the permeability perpendicular to small-scale faults was measured in core plugs and compared with the permeability of the adjacent, undeformed sandstones. As also was recognized by

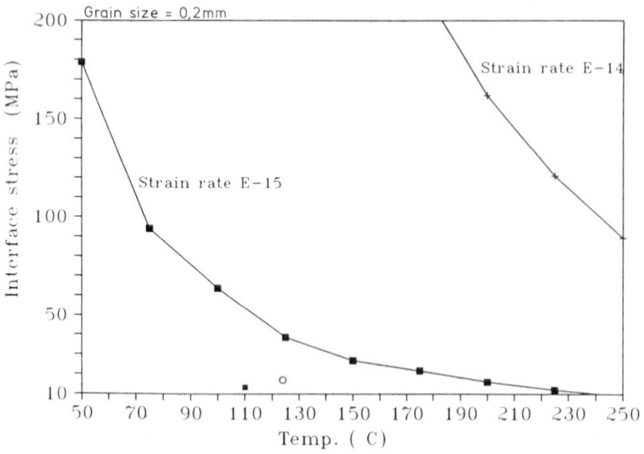

Fig. 15. Deformational diagram constructed from Rutter (1976). The black square describes the stress and temperature situation for a surface dipping 60° at approximately 2800 m depth in the Njord Field. The open circle represents the situation for a surface dipping 70° at approximately 3500 m depth (Kvalvågen faults). The plots are both within the pressure solution field of the diagram. The pressure solution is indicated as an active process taking place at a slow strain rate ($\dot{e} \leq 10^{-15}\ s^{-1}$).

Gabrielsen and Koestler (1987) during their study on small-scale faults from the Troll Field, a relative permeability reduction commonly characterizes the samples containing faults. The method used for permeability measurements, however, also admits the permeability of the fault planes themselves to be calculated by the following relationship:

$$K_{(\mathrm{fault})} = \frac{x \cdot K_{(\mathrm{sst})} \cdot K_{(\mathrm{sy})}}{K_{(\mathrm{sst})} - (1-x)K_{(\mathrm{sy})}}$$

where $K_{(\mathrm{fault})}$, $K_{(\mathrm{sst})}$ and $K_{(\mathrm{sy})}$ are the permeability of the fault, the undeformed sandstone plug and the plug containing both fault and undeformed sandstone, respectively. x is the net thickness of the "tight" zone (fault) as a fraction of the plug length.

By this relationship very low permeability values (in the range between 0.06 mD and 5.28 mD) were obtained for the fault planes. By plotting the permeability values against matrix content and grain size in the parent, undeformed sandstones, a log-normal trend appears (Fig. 16). A few values did, however, not fit into this trend. These values relate to faults where pressure solution and stylolitization have been most extensive, and the faults show lower permeabilities than should be expected from their parent sandstone matrix content and grain size.

Conclusions

(1) The faults at Kvalvågen and the Njord Field formed at a shallow burial depths. Fault textures in both localities have later been modified by diagenesis, resulting in loss of or modifications of earlier soft-sediment deformational textures.

(2) Diagenetic reactions like feldspar dissolution, precipitation of kaolinite, carbonate and quartz cementation, can, both texturally and by the use of geochemical methods, help to date movements on faults.

(3) Syn-sedimentary or early tectonic faults are frequently enriched in clay minerals and micas along their slip surfaces relative to the surrounding sandstones. Stylolites can evolve along the fault planes at greater burial depth, purely due to the overburden stress.

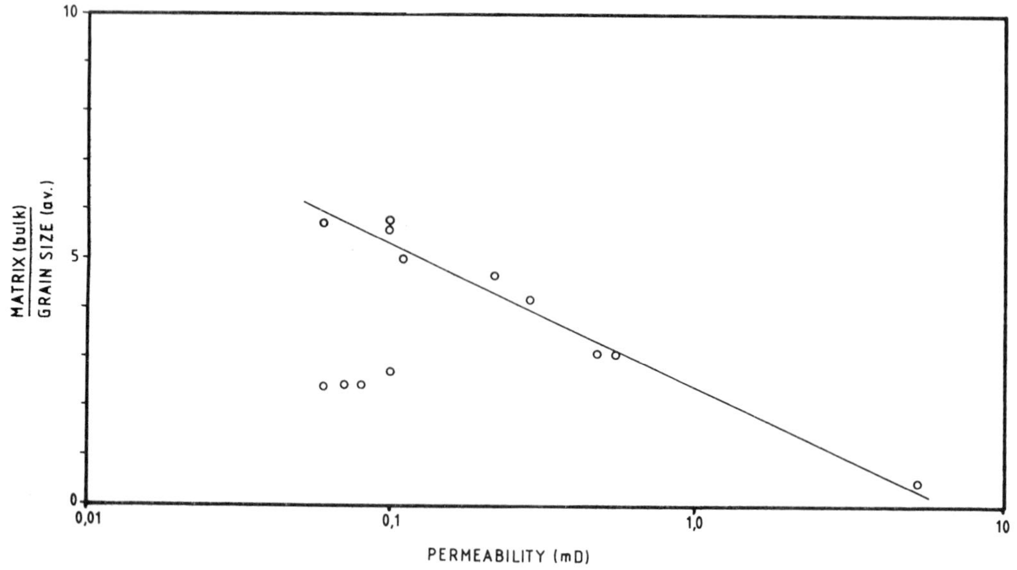

Fig. 16. Log-normal plot of fault permeability versus matrix content (percentage value) above grain size (sorting) of the parent, undeformed sandstone. There is a positive correlation between increased grain size and increased permeability. High matrix content in the parent sandstone tend to lower the permeability of the fault. The plots which deviate from the trend belong to faults with the most extensive pressure solution and stylolitization.

(4) Stylolitization along low-porosity, early fault surfaces may cause further reduction in permeability across the low-permeable structures, because clay minerals and micas are concentrated as quartz dissolves. This is important in moderate to deep buried reservoirs affected by faulting, where pressure solution and stylolitization are common.

Acknowledgements

Financial support for this study was provided by the Norwegian Research Council for Scientific and Applied Research (N.T.N.F.). Norsk Hydro a.s. are acknowledged for their co-operation and for providing rocks samples from the Njord Field. H. Johansen of the Institutt for Energiteknikk (I.F.E., Kjeller) is thanked for discussions and assistance in connection with the Laser sampling procedure for isotope analysis.

References

Allen, J.R.L., 1982. Sedimentary Structures: Their Character and Physical Basis. Elsevier, Amsterdam, 663 pp.

Bjørlykke, K., Aagaard, P., Dypvik, H., Hastings, D.S. and Harper, A.S., 1986. Diagenesis and reservoir properties of Jurassic sandstones from the Haltenbanken area, offshore Mid-Norway. In: A.M. Spencer et al. (Editors), Habitat of Hydrocarbons on the Norwegian Continental Shelf. Graham and Trotman, London, pp. 275–286.

Boles, J.R., 1984. Influence of mica surface on pore water pH. Chem. Geol., 43: 303–317.

Borradaile, G.J., 1981. Particulate flow of rock and the formation of cleavage. Tectonophysics, 72: 305–321.

Buchner, F., 1981. Rhinegraben: horizontal stylolites indicating stress regimes of earlier stages of rifting. Tectonophysics, 73: 113–118.

Gabrielsen, R.H. and Robinson, C., 1984. Tectonic inhomogeneities of the Kristiansund–Bodø Fault Complex off-shore Norway. In: A.M. Spencer (Editor), Petroleum Geology of the North European Continental Margin. Norwegian Petroleum Society, Graham and Trotman, London, pp. 397–406.

Gabrielsen, R.H. and Koestler, A.G., 1987. Description and structural implications of fractures in late Jurassic sandstones of the Troll Field, northern North Sea. In: A.G. Koestler and S. Olaussen (Editors), Tectonics and Deposition in Sedimentary Basins. N.G.T., 67, 4: 371–381.

Gjelberg, J., Dreyer, T., Høie, A., Tjelland, T. and Lilleng, T., 1987. Late Triassic to Mid-Jurassic sandbody development on the Barents and Mid-Norwegian shelf. In: J. Brooks and K. Glennie (Editors), Petroleum Geology of North-West Europe. Graham and Trotman, London, pp. 1105–1129.

Heald, M.T., 1959. Significance of stylolites in permeable sandstones. J. Sediment. Petrol., 29: 251–253.

Howard, G.C. and Fast, C.R., 1970. Hydraulic fracturing. Soc. Pet. Eng., AIME Monogr., 2: 210 pp.

Irwin, H., Curtis, C. and Coleman, M., 1977. Isotopic evidence for source of diagenetic carbonates formed during burial of organic rich sediments. Nature, 269: 209–213.

Knipe, R.J., 1986. Deformation mechanism path diagrams for sediments undergoing lithification. Geol. Soc. Am., Mem., 166: 151–160.

Maltman, A.J., 1984. On the term "soft-sediment deformation". J. Struct. Geol., 6: 589–593.

Mitra, S., 1988. Effects of deformation mechanisms on reservoir potential in Central Appalachian Overthrust Belt. Bull. Geol. Soc. Am., 72(5): 536–554.

Nagtegaal, P.J.C., 1978. Sandstone-framework instability as a function of burial diagenesis. J. Geol. Soc. London, 135: 101–105.

Nelson, R.A., 1981. Significance of fracture sets associated with stylolite zones. Am. Assoc. Pet. Geol., Bull., 65(11): 2417–2425.

Nelson, R.A., 1985. Geologic analysis of naturally fractured reservoirs. Contrib. Pet. Geol. Eng., 1, 279 pp.

Nemec, W., Steel, R.J., Gjelberg, J., Collinson, J.D., Prestholm, E. and Øxnevad, I.E., 1988. Anatomy of a collapsed and re-established delta front in the lower Cretaceous of Eastern Spitsbergen: gravitational sliding and sedimentation processes. Am. Assoc. Pet. Geol., Bull., 72(4): 454–476.

Pittman, E.D., 1981. Effect of fault-related granulation on porosity and permeability of quartz sandstones, Simpson Group (Ordovician), Oklahoma. Am. Assoc. Pet. Geol., Bull., 65: 2381–2387.

Simpson Group (Ordovician), Oklahoma. Am. Assoc. Pet. Geol., Bull., 65: 2381–2387.

Rutter, E.H., 1976. The kinetics of rock deformation by pressure solution. Philos. Trans. R. Soc. London, Ser. A, 283: 203–219.

Saigal, G.C. and Bjørlykke, K., 1987. Carbonate cements in clastic reservoir rocks from offshore Norway — relationships between isotopic composition, textural development and burial depth. In: J.D. Marshall (Editor), Diagenesis of Sedimentary Sequences. Geol. Soc. London, Spec. Publ. 36: 313–324.

Sverdrup, E. and Prestholm, E., 1990. Synsedimentary deformation structures and their implications for stylolitization during deeper burial. Sediment. Geol., 68: 201–210.

Thompson, A.F., 1959. Pressure solution and porosity. Soc. Econ. Paleontol. Mineral., Spec. Publ., 7: 92–110.

Weyl, P.K., 1959. Pressure solution and the force of crystallization — a phenomenological theory. J. Geophys. Res., 64: 2001–2025.

Wilson, T.V. and Sibley, D.F., 1978. Pressure solution and porosity reduction in shallow buried quartz arenite. Am. Assoc. Pet. Geol., Bull., 62: 2329–2334.

E. SVERDRUP *Saga Petroleum A/S, Kjørbovn. 16, P.O. Box 490, N-1301 Sandvika, Norway*

K. BJØRLYKKE *Department of Geology, University of Oslo, P.O. Box 1047 Blindern, N-0316 Oslo 3, Norway*

Subparallel faults and horizontal-stress orientations: an evaluation of in-situ stresses inferred from elliptical wellbore enlargements

R.L. Dart and H.S. Swolfs

Stress-induced wellbore-enlargement (breakout) data from the Marietta basin in south-central Oklahoma and the Bravo dome area of the central Texas Panhandle show a possible variation in the orientation of horizontal stress trajectories across areas bounded by subparallel faults. Regional studies of breakout and hydraulic fracturing-stress data indicate a northeast–southwest-oriented maximum principal horizontal stress (SH_{max}) oriented oblique to westnorthwest-striking subparallel faults in these two study areas. Subtle variation in long-axis orientations of elliptically shaped wellbore enlargements from wells located between and near the ends of mapped subparallel faults in the Marietta basin and in the Bravo dome area indicates a possible local rotation of SH_{max}. From the observed variation in the mean orientations of wellbore enlargements from selected wells, it is inferred that as much as 52° and 22° of stress trajectory rotation occurs between subparallel faults in the Marietta basin and the Bravo dome area, respectively. Assuming the areas bounded by these subparallel faults are functioning as stress field discontinuities, factors affecting local stress trajectory reorientation may include the amount of fault overlap, fault separation and differences in Young's modulus of the faulted rock units. Another factor contributing to or controlling suspected stress trajectory reorientation might be the influence of anisotropic rock-fabric properties.

Introduction

Borehole field studies (Cox, 1970; Brown, 1978; Bell and Gough, 1979, 1982; Gough and Bell, 1981, 1982, Hickman et al., 1985; Plumb and Hickman, 1985; Teufel, 1985; Zoback et al., 1985) and theoretical and laboratory studies simulating borehole stress conditions (Mastin, 1984; Haimson and Herrick, 1985; Zoback et al., 1985) show that wellbore enlargements known as "breakouts" are stress-induced spall zones that typically elongate vertically within the wellbore and are the result of compressional shear failure associated with horizontal compressive stresses about the wellbore. Breakouts are observed to cause the wellbore to be elliptically enlarged in cross section and are typically aligned in the direction of minimum horizontal principal stress (SH_{min}). The idea that breakouts reflect local in situ stress conditions is supported by the agreement between horizontal stress directions inferred from breakouts and stress orientations obtained using other types of stress data such as earthquake focal mechanisms, hydraulic-fracturing measurements and geologic indicators of stress (Bell and Gough, 1982; Gough and Bell, 1981, 1982; Hickman et al., 1985; Plumb and Hickman, 1985; Dart and Zoback, 1989).

Breakout-data analysis has become an important tool in evaluating present-day stress conditions in areas where data from more traditional stress-measurement techniques (earthquake focal mechanisms, hydraulic-fracturing, etc.) may not be available. An understanding of the in-situ state of stress is essential in assessing seismic hazards, recent crustal deformation and the potential for slip on existing faults. In the petroleum industry, knowledge of local stress-field conditions can assist in the successful engineering of borehole drilling and the design of wellbore hydraulic-fracturing treatment (Dart and Zoback, 1989).

This study is intended to illustrate that routine analysis of stress-induced or stress-related borehole features may yield useful information concerning subtle variations in the local stress field, especially where structural features like the pattern of fracturing locally influence the orientation of regional horizontal stresses.

Structural setting

The Texas Panhandle–southern Oklahoma study region (Figs. 1 and 2) is an area of deep sedimentary basins filled with thick sequences of Cambrian and younger Paleozoic marine rocks separated by

Structural and Tectonic Modelling and its Application to Petroleum Geology edited by R.M. Larsen, H. Brekke, B.T. Larsen and E. Talleraas. NPF Special Publication 1, pp. 519–529. Elsevier, Amsterdam.

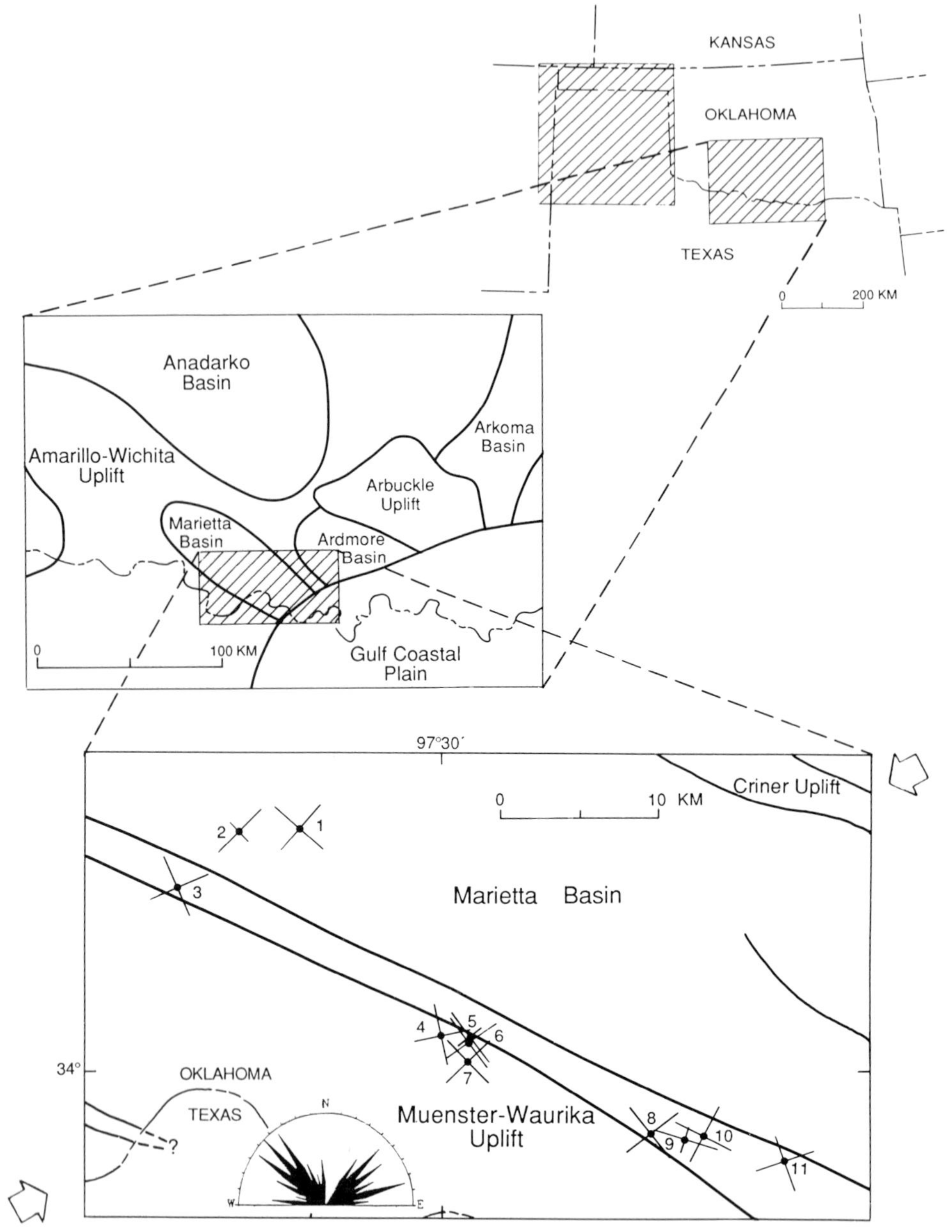

Fig. 1. Marietta basin study area. Well locations are plotted as solid dots with the mean orientations of wellbore data distributions (bars). Wellbore data distributions are shown as either orthogonally oriented trends (bars of equal length) or a unimodal trend (bars of unequal length). Faults are solid lines. Regional state of stress (SH_{max}) is the orientation of the large open arrows. The distribution of wellbore enlargement data is shown in the rose diagram (inset).

uplifted basement structures (Jordan, 1967; Johnson et al., 1972, 1984; Johnson and Denison, 1973). The primary west–northwest structural trend of basins and basement highs, extending across southern Oklahoma into the Texas Panhandle, developed throughout the Paleozoic Era during periods of basement faulting, igneous intrusion, basin subsidence and orogenic deformation (Ham et al., 1964; Hoffman et al., 1974; Wickham, 1978). The stratigraphic history of the region appears to be dominated by episodes of carbonate and shale deposition attributed to eustatic fluctuations and intervals of clastic deposition associated with subareal erosion of uplifted areas.

The Marietta basin in Oklahoma is located at the eastern end of the Amarillo-Wichita uplift (Fig. 1). The Marietta basin, a northwest-trending synclinal structure, is bounded on the northeast by the Criner uplift and on the southwest by the Waurika-Muenster uplift. During the early depositional history of this area, predating the formation of the Marietta syncline, a thick sequence of Cambrian to Ordovician carbonate sediment was deposited. This early depositional sequence is overlain by Upper Ordovician through Lower Mississippian limestone, sandstone and shale strata (Henry, 1968).

The Wichita orogeny in Late Mississippian to

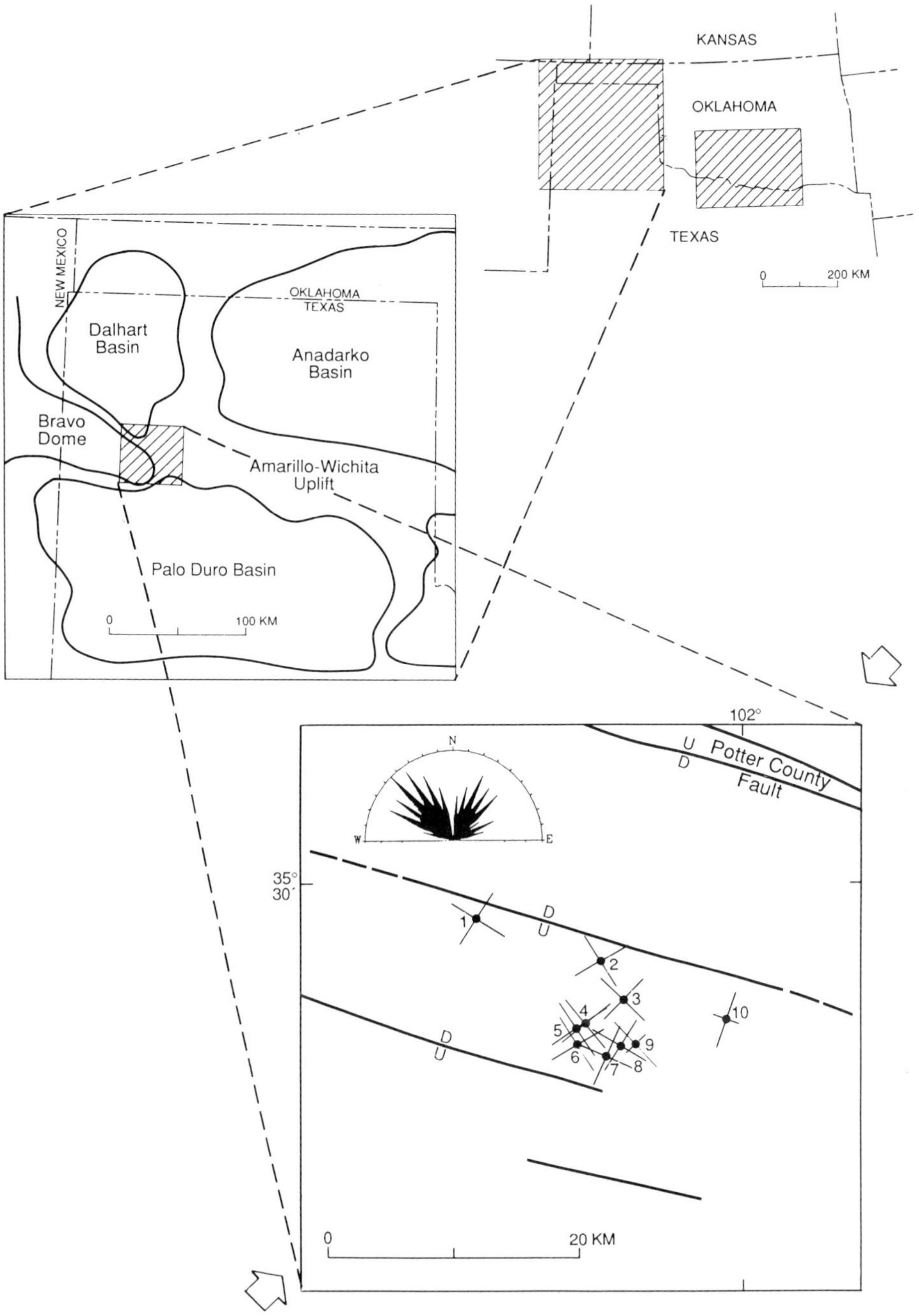

Fig. 2. Bravo dome study area. Well locations are plotted as solid dots with the mean orientations of wellbore data distributions (bars). Wellbore data distributions are shown as either orthogonally oriented trends (bars of equal length) or a unimodal trend (bars of unequal length). Faults are solid lines. Regional state of stress (SH_{max}) is the orientation of the large open arrows. The distribution of wellbore data is shown in the rose diagram (inset).

Middle Pennsylvanian (early Atokan) time and the Arbuckle orogeny in Middle Pennsylvanian (early Desmoinesian) to Late Pennsylvanian (early Virgilian) time were significant events in the structural and depositional history of the area (Tomlinson and McBee, 1959). Formation of the Criner and Muenster-Waurika uplifts and the faulting and folding that occurred during the initial development of the Marietta syncline accompanied the Wichita orogeny (Tomlinson and McBee, 1959). Significant deposition within the syncline began in late Middle Pennsylvanian (Desmoinesian) time (Frederickson and Redman, 1965). Basin subsidence associated with sediment loading continued into post-Permian time (Frederickson and Redman, 1965). During the structural and tectonic history of southern Oklahoma, significant amounts of dip-slip and left-lateral strike-slip displacement have occurred on faults within the Amarillo-Wichita uplift (Harlton, 1963; Butler, 1980; Donovan, et al. 1982; McLean and

Sterns, 1983). The Frontal fault system, a regional fault system associated with the Amarillo-Wichita uplift, strikes west–northwest/east–southeast across southern Oklahoma into the Marietta basin.

The Bravo dome (Fig. 2) is a Late Mississippian to Early Pennsylvanian age tectonic feature positioned between the Dalhart and Palo Duro basins (Budnik and Smith, 1982). Formation of the Bravo dome coincides with development of the Amarillo-Wichita uplift in Early Pennsylvanian time. By mid-Permian time, carbonate and clastic (arkosic) sediment deposited in response to Late Mississippian to Early Pennsylvanian tectonism completely covered structural features like the Bravo dome and filled the smaller basins in the region (McGookey and Goldstein, 1982). A sequence of Permian age halite-bearing formations interbedded with marine mudstone, siltstone, anhydrite and dolomite rock units covers the Bravo dome structure (Presley, 1980). This overlying sedimentary rock accumulation is cut by a series of northwest-striking high-angle faults that define basement structures (Dutton et al., 1982). The most prominent of these faults is the west–northwest-striking Potter County fault separating the Bravo dome from the Dalhart basin (Fig. 2). Basement faults associated with Paleozoic tectonism of the Amarillo-Wichita uplift are probably reverse faults (Dutton et al., 1982). Deformation of Miocene Ogallala Formation strata associated with these faults is indicative of Cenozoic fault movement in the Texas Panhandle (Budnik, 1987).

Wellbore enlargement data

Wellbore enlargement data consist of the measured long-axis orientations of certain elliptically shaped features observed on standard petroleum industry four-arm, high-resolution dipmeter and fracture-identification well logs. These features are inferred to be in-situ stress related. Stress-related wellbore enlargement features are intervals within the wellbore where the borehole diameter has suddenly increased in one direction beyond the normal bit size. Within this interval, normal logging tool rotation is interrupted and the elliptically shaped enlargement interval is tracked vertically. In order for a wellbore enlargement feature of this kind to be considered stress related, its long-axis azimuth must not coincide with the bearing of the vertical deviation of the hole or show an obvious correlation to mechanical "drilling-induced" enlargement of the wellbore.

The two types of wellbore enlargement features important to the stress study are: (1) stress-induced breakouts associated with shear failure at the wellbore, and (2) fracture borehole enlargements inferred to be associated with either drilling-induced hydraulic fracturing, or with favorably oriented, preexisting fractures or closely spaced point sets intersecting the wellbore (Dart and Zoback, 1989). It is assumed that the orientation of borehole enlargements, due to preexisting fractures, reflects the general trend of these fractures, and that fractures intersecting the wellbore, whether hydraulically induced or preexisting, can contribute to a structural failure of the wellbore when aided by the in-situ state of stress. Fracture wellbore enlargements are inferred, therefore, to be stress related when they orient in the direction of SH_{max}, orthogonal to the trend of breakouts. Thus, an azimuthal data distribution containing breakout and fracture wellbore enlargement orientations should have a bimodal appearance with bimodal trends being orthogonal (Dart and Zoback, 1989; Morin et al., 1989). Wellbore enlargement data distributions are shown in Figs. 1 and 2 as either unimodal or bimodal.

Well data sets used in this study were selected from a larger set of available data. The criteria used for data selection were that each data set consist of four or more in-situ stress-associated elliptical wellbore-enlargement intervals and that each azimuthal distribution, either unimodal or bimodal, must have an angular deviation (standard deviation) of $\leq 20°$. In the case of bimodal distributions, each orthogonal trend had to have an angular deviation of $\leq 20°$. The data used are give in Tables 1A and 2A. Additional data tables (Tables 1B–D, 2B–D) show a comparison of the Marietta basin and Bravo dome area data sets. The data presented in these tables illustrate similarities in stratigraphy and in the distribution of wellbore-enlargement azimuthal data.

Stress trajectory rotation

In the Marietta basin and Bravo dome areas, the direction of SH_{max} (Figs. 1 and 2) is oriented oblique (east–northeast; von Schonfeldt et al., 1973; Hooker and Johnson, 1969; Borjeson and Lamb, 1987) to west–northwest/east–southeast-striking subparallel faults. We assume that reorientation or rotation of the stress trajectories will occur in the vicinity of these fault-bounded areas, and that due to fault geometry and the mechanical properties of stress-field discontinuities, the sense of rotation and the amount of rotation will vary from point to point across the fault-bounded area.

It has been shown (Segall and Pollard, 1980; Rodgers, 1984; Hempton and Neher, 1986) that en echelon faults having a left-stepping or right-stepping left-lateral strike-slip sense of displacement

TABLE 1

A. Marietta basin drill hole and wellbore enlargement data

Map location	1	2	3	4	5	6	7	8	9	10	11
# of drill* holes	3	1	1	1	1	1	2	1	1	2	1
Surface elevation	893**	866	793	855	840	876	926**	854	845	785**	918
Total depth	7905**	6871	9934	8028	7708	7760	7725**	8399	8799	9042**	10547
Logged interval	6602-7904**	4471-6464	5407-6527	6028-9771	5590-7799	5708-7745	6174-7710**	6333-8383	6739-8798	6951-9061**	8911-10516
Data interval	7133-7545**	5987-6446	5726-6446	7214-7894	5818-7619	6508-7698	7151-7619**	6833-8189	8337-8635	7794-8729**	8939-10211
Vertical ft. of enlargement (NE/NW)	90/257	216/-	132/20	54/52	-/421	48/125	149/14	92/59	-/162	26/260	28/88
# of enlargement intervals (NE/NW)	2/4	4/-	4/1	4/2	-/10	3/2	5/1	2/2	-/4	1/6	2/2
Distribution	bi	uni	bi	bi	uni	bi	bi	bi	uni	bi	bi
Mode orientation	NE	NE	NE	ENE	-	NE	NE	NE	-	NE	ENE
Mean	43	54	67	77	-	56	46	61	-	58	84
Angular deviation	5	20	11	7	-	12	9	12	-	-	6
Mode orientation	NW	-	NW	NW	NW	NW	NNW	NW	WNW	WNW	NNW
Mean	312	-	318	328	323	323	343	232	289	298	341
Angular deviation	4	-	-	10	18	5	-	5	11	11	12

* Two or more drill holes may be located in close proximity, at the same map location. ** Average value.
Depths are measured from ground surface. A slash (/) separates data from bimodal distributions, modes are oriented generally NE (0 to 90 degrees) and NW (270 to 360 degrees). Abbreviations: bimodal (bi), unimodal (uni).

B. Marrietta basin borehole stratigraphic data

	Sandstone	Limestone	Shale	Clastic Rock	Ss-Ls-Sh	Ss-Ls-Sh-Cong	Ss-Sh	Total
Upper Pennsylvanian	832(.032)							832(.032)
Middle Pennsylvanian	239(.009)		2476(.094)	806(.031)	6709(.25)	1401(.053)	1840(.070)	13471(.51)
Lower Pennsylvanian					2397(.091)			2397(.091)
Upper Devonian			492(.019)					492(.019)
Upper Ordovician		6054(.23)	188(.007)					6242(.24)
Middle Ordovician		346(.014)	2042(.077)		581(.22)			2969(.11)
Total	1071(.041)	6400(.24)	5198(.20)	806(.031)	9687(.37)	1401(.053)	1840(.070)	26403(1.0)

The data are the vertical borehole feet logged and its precentage of the total footage logged, in parenthesis. Abbreviations: sandstone (Ss), limestone (Lm), shale (Sh) and conglomerate (Cong). A hyphen (–) indicates units with mixed lithologies.

C. Marrietta basin wellbore enlargement data

	Sandstone	Limestone	Shale	Ss-Ls-Sh	Ss-Ls-Sh-Cong	Ss-Sh	Total
Middle Pennsylvanian	16(.007)/ 1(.016)	43(.019)/ 1(.016)	180(.078)/ 4(.065)	408(.18)/ 7(.12)		18(.008)/ 1(.016)	665(.29)/ 14(.23)
Lower Pennsylvanian		46(.020)/ 1(.016)	21(.009)/ 1(.016)		134(.058)/ 6(.10)	70(.031)/ 1(.016)	271(.12)/ 9(.15)
Upper Devonian			17(.007)/ 1(.016)				17(.007)/ 1(.016)
Upper Ordovician	76(.033)/ 1(.016)	1096(.48)/ 25(.61)					1172(.51)/ 26(.43)
Middle Ordovician		35(.015)/ 3(.049)	24(.011)/ 2(.033)	91(.039)/ 5(.082)	18(.008)/ 1(.016)		168(.073)/ 11(.18)
Total	92(.040)/ 2(.033)	1220(.53)/ 30(.49)	242(.11)/ 8(.13)	499(.22) 12(.20)	152(.07)/ 7(.12)	88(.038) 2(.033)	2293(1.0)/ 61(1.0)

The data are the vertical feet of borehole enlargement (% of the total vertical feet of enlargement)/number of enlargement intervals (% of total the number). Abbreviations: sandstone (Ss), limestone (Lm), shale (Sh) and conglomerate (Cong). A hyphen (–) indicates units with mixed lithologies.

D. Marrietta basin wellbore enlargement data

	Azimuthal orentation (degrees)									
	W-E (270-281)	WNW (281-304)	NW (304-326)	NNW (326-349)	N-S (349-360/0-11)	NNE (11-34)	NE (34-56)	ENE (56-79)	E-W (79-90)	Total
Middle Pennsylvanian		52(.023)/ 1(.016)	288(.13)/ 6(.10)	126(.055)/ 4(.066)	90(.039)/ 1(.016)	83(.036)/ 1(.016)		26(.011)/ 1(.016)		665(.29/ 14(.23)
Lower Pennsylvanian		39(.017)/ 1(.016)	8(.003)/ 1(.016)	84(.036)/ 2(.033)	70(.031)/ 1(.016)		59(.026)/ 3(.049)	11(.005)/ 1(.016)		271(.11)/ 9(.15)
Upper Devonian								17(.007)/ 1(.016)		17(.007)/ 1(.016)
Upper Ordovician	225(.10)/ 3(.049)	57(.025)/ 2(.033)	330(.14)/ 8(.13)	48(.021)/ 2(.033)		96(.042)/ 1(.016)	134(.058)/ 5(.082)	282(.12)/ 5(.082)	9(.004)/ 1(.016)	1172(.51)/ 26(.18)
Middle Ordovician			44(.019)/ 1(.016)	15(.007)/ 2(.033)			42(.018)/ 3(.049)	36(.016)/ 3(.049)	31(.014)/ 2(.033)	168(.073) 11(.18)
Total	225(.10)/ 3(.049)	148(.065)/ 4(.066)	661(.29)/ 15(.25)	273(.12)/ 10(.16)	160(.070)/ 2(.033)	179(.078)/ 2(.033)	235(.10)/ 11(.18)	372(.16)/ 11(.18)	40(.017)/ 3(.049)	2293(1.0)/ 61(1.0)

The data are the vertical feet of borehole enlargement (% of the total vertical feet of enlargement)/number of enlargement intervals (% of total number).

TABLE 2

A. Bravo dome drill hole and wellbore enlargement data

Map location	1	2	3	4	5	6	7	8	9	10
# of drill holes	1	1	1	1	1	1	1	1	1	1
Surface elevation	3500	3451	3615	4020	3715	3706	3542	3590	3539	3257
Total depth	4757	6881	4475	4202	4076	3733	4064	3596	3750	2781
Logged interval	2740-4747	837-6839	2469-4469	2192-4198	2075-4075	2883-3583	1965-4093	400-3618	1650-3739	1438-2779
Data interval	3110-3747	1227-5691	2788-3655	2322-3244	2340-3157	3031-3505	2428-4019	826-2786	2453-3100	2291-2611
Vertical ft. of Enlargement (NE/NW)	69/16	143/96	142/87	91/60	-/55	21/75	53/75	116/13	-/53	166/-
# of enlargement intervals (NE/NW)	2/2	5/5	3/4	2/6	-/4	1/3	3/3	3/1	-/4	4/-
Distribution	bi	bi	bi	bi	uni	bi	bi	bi	uni	uni
Mode orientation	NNE	NE	NE	NE	-	NE	NE	NNE	-	NNE
Mean	33	59	45	30	-	35	25	33	-	19
Angular deviation	17	14	8	20	-	-	11	9	-	19
Mode orientation	WNW	NW	NW	NW	NW	NW	WNW	WNW	NW	-
Mean	293	336	314	322	326	328	302	302	317	-
Angular deviation	2	9	4	4	12	8	6	-	19	-

A slash (/) separates data bimodal distributions, with modes oriented generally NE (0 to 90 degrees) and NW (270 to 360 degrees. Abbreviations: bimodal (bi), unimodal (uni).

B. Bravo dome area borehole stratigraphic data

	Dolomite	Limestone	Shale	Clastic Rock	Silts-Ls-Anhy-Cong	Igneous rock	Total
Upper Permian	8379(.36)	1482(.063)			274(.012)		10135(.43)
Middle Pennsylvanian		1297(.055)	7518(.32)	3556(.15)			12371(.53)
Lower Pennsylvanian		109(.005)		206(.009)			315(.013)
Precambrian						670(.029)	670(.029)
Total	8379(.36)	2888(.12)	7518(.32)	3762(.16)	274(.012)	670(.029)	23491(1.0)

The data are vertical borehole feet logged and its percentage of the total footage logged, in parenthesis. Abbreviations: sandstone (Ss), limestone (Lm), shale (Sh) and conglomerate (Cong). A hyphen (–) indicates units with mixed lithologies.

C. Bravo dome area wellbore enlargement data

	Dolomite	Limestone	Shale	Clastic Rock	Igneous Rock	Total
Lower Permian	682(.51)/ 33(.60)	32(.024)/ 1(.018)				714(.54)/ 34(.62)
Middle Pennsylvanian			515(.39)/ 16(.29)	82(.062)/ 3(.055)		597(.49)/ 19(.35)
Precambrian					19(.014)/ 2(.036)	19(.0140/ 2(.036)
Total	682(.51)/ 33(.60)	32(.024)/ 1(.018)	515(.39)/ 16(.29)	82(.062)/ 3(.056)	19(.014)/ 2(.036)	1330(1.0)/ 55(1.0)

The data are the vertical feet of borehole enlargement (% of the total vertical feet of enlargement)/number of enlargement intervals (% of the total numer).

D. Bravo dome area wellbore enlargement data

	Azimuthal orentation (degrees)									
	W-E (270-281)	WNW (281-304)	NW (304-326)	NNW (326-349)	N-S (349-360/0-11)	NNE (11-34)	NE (34-56)	ENE (56-79)	E-W (79-90)	Total
Lower Permian		45(.034)/ 5(.091)	152(.11)/ 12(.22)	86(.065)/ 6(.11)	39(.029)/ 2(.036)	152(.11)/ 3(.055)	156(.12)/ 5(.091)	96(.072)/ 2(.036)		726(.55)/ 35(.64)
Middle Pennsylvanian		61(.046)/ 2()	152(.11)/ 5()		12(.009)/ 1(.018)	14(.011)/ 1(.018)	338(.25)/ 8(.15)	8(.006)/ 1(.018)		585(.44)/ 18(.33)
Precambrian					10(.008)/ 1(.018)	9(.007)/ 1(.018)				19(.014)/ 2(.036)
Total		106(.080)/ 7(.13	304(.23)/ 17(.31)	86(.065)/ 6(.11)	61(.046)/ 4(.073)	175(.13)/ 5(.091)	494(.37)/ 13(.24)	104(.078)/ 3(.055)		1330(1.0)/ 55(1.0)

The data are the vertical feet of borehole enlargement (% of the total vertical feet of enlargement)/number of enlargement intervals (% of total number).

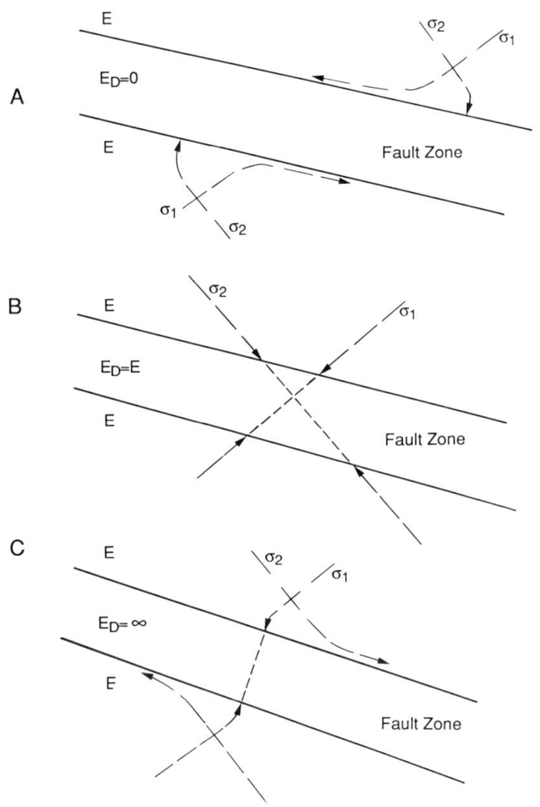

Fig. 3. Schematic diagram illustrating the influence of Young's modulus (E) of displacement discontinuities (E_D) on stress trajectory orientation (σ_1 and σ_2). Shown are displacement discontinuities of fault-bounded areas with a Young's modulus less than (A), equal to (B) and greater (C) than that of the adjacent rock (modified after Hudson and Cooling, 1988).

can initiate stress trajectory rotation in the vicinity of these en echelon faults. The distance separating en echelon faults and the amount of offset or overlap of the ends of these faults will influence the extent of stress trajectory rotation.

Rotation of stress trajectories will also occur when the subparallel faults function as displacement discontinuities. If differences do exist in mechanical properties of the rock units involved, the fault zone or fault-bounded area (Fig. 3) will function as a discontinuity within the regional stress field. If a fault-bounded area functions as a stress-field discontinuity within the regional stress field having a Young's modulus greater than that of the adjacent rock (Fig. 3C), it is predicted (Hudson and Cooling, 1988) that the stress trajectory pattern will rotate either counterclockwise or clockwise when crossing boundary faults.

Stress trajectory rotation may also be predicted if an anisotropic rock fabric, such as strongly oriented jointing, is present. In this case the rock fabric property can act as a stress guide to redirect the stress trajectory orientation (Amadei et al., 1988).

Discussion of results

We make the assumption that reorientation of horizontal stress trajectories across an area bounded by subparallel faults can be indicated by a corresponding change in the orientations of stress-related wellbore-enlargement features in wells located across the study area.

The two areas of interest, the southwestern edge of the Marietta basin and the eastern flank of the Bravo dome, are structurally similar in that they both have a thick sequence of relatively flat-lying sedimentary rock offset by west–northwest/east–southeast-striking subparallel basement faults. The distribution of wellbore enlargement orientations from wells located adjacent to and between these subparallel faults exhibits bimodal trends that are orthogonal in character (Figs. 1 and 2). Because $SH_{\max}$ is regionally oriented northeast (Hooker and Johnson, 1969; von Schonfeldt et al., 1973; Borjeson and Lamb, 1987), these bimodal-orthogonal distributions are assumed composed of northwest-oriented wellbore breakouts and northeast-oriented fracture wellbore enlargements (Dart and Zoback, 1989).

Observed variation in the calculated azimuthal means of wellbore-enlargement data distributions from wells located between subparallel faults in the two study areas has the appearance of a systematic rotation. This apparent rotation is inferred to be indicative of stress-trajectory reorientation between subparallel faults and may best be seen among wells 8 through 11 in the Marietta basin (Fig. 1) and wells 7 through 9 in the Bravo dome study area (Fig. 2). The maximum amount of apparent stress-trajectory rotation is 52° between wells 9 and 11 (Fig. 1) in the Marietta basin and 22° between wells 7 and 9 (Fig. 2) in the Bravo dome area. A bimodal-orthogonal distribution of wellbore-enlargement orientations from the Marietta basin strongly contrasts with the unimodal distribution of wellbore enlargement orientations, inferred to be breakouts, from the eastern Anadarko basin adjacent to and north of the Marietta basin (Dart, 1989; and Dart and Zoback, 1989). In the eastern Anadarko basin, drill holes with stress associated wellbore enlargements were not or were only rarely located between or in the vicinity of mapped subparallel faults.

Drilling-induced hydraulic fracturing is more likely to develop in a normal-faulting stress regime (Healy et al., 1984; Stock et al., 1985). Correspondingly, formation of hydraulic fracture-related wellbore enlargements in drill holes located in the Marietta basin and Bravo dome study areas are common and may indicate local extensional (normal) faulting stress conditions. Rodgers (1984) and Hempton and Neher (1986) described structural settings where extensional and compressional stresses can develop between and near the ends of en echelon strike-slip faults in response to regional horizontal compression oblique to their strikes. Subparallel faults in the Marietta basin and Bravo dome area are not en echelon by definition; however, these two areas appear to structurally resemble the areas described by Rodgers (1984) and Hempton and Neher (1986). Extensional stress conditions (normal-faulting stress regime) might exist within the Marietta basin and the Bravo dome study area where SH_{max} is oriented oblique (northeast) to the strike of subparallel faults.

Values for the Young's modulus of the stratigraphic section of the two study areas are not known. To our knowledge, the necessary field and laboratory studies have not been made to determine the relative similarity or difference in Young's modulus of the rock units between or beyond the subparallel faults in the Marietta basin or the Bravo dome area. Because the rock units on opposite sides of these subparallel faults are offset vertically (Harlton, 1963; Butler, 1980; Dutton et al., 1982; Donovan et al., 1982; McLean and Sterns, 1983), a Young's modulus difference may indeed exist. However, any variation would probably be small because the rock units on either side of the offset appear to be similar

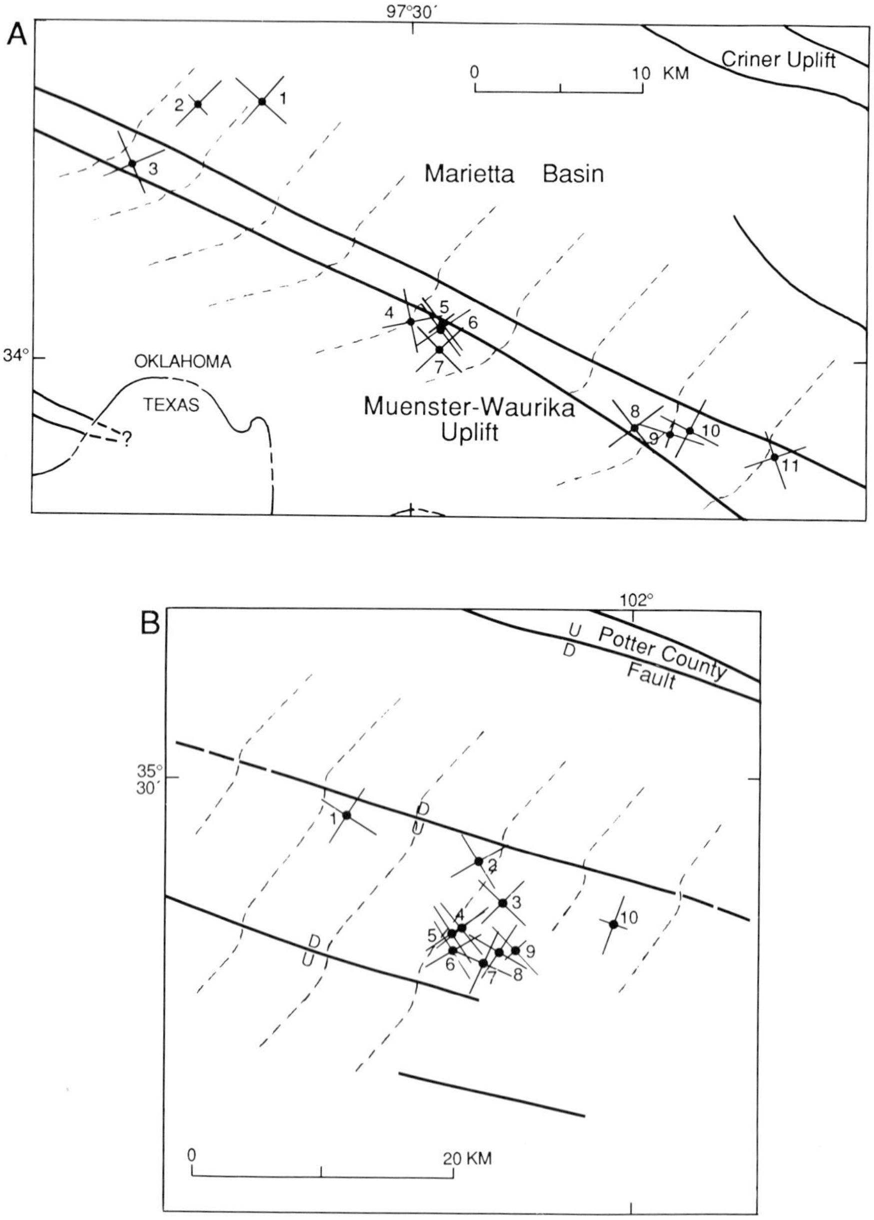

Fig. 4. Schematic diagram illustrating the inferred local rotational pattern of stress trajectories (dashed lines) across the Marietta basin (A) and Bravo dome (B) study areas.

lithologically. Therefore, the influence of Young's modulus on stress trajectory rotation in the two study areas is thought to be, at best, of minor importance.

Preexisting fractures in an array of orientations may exist within blocks bounded by subparallel faults. Well-established fracture trends or joint set patterns could function as anisotropic rock fabric elements to exert an influence on stress trajectory orientation. Studies of fracture and lineament patterns exist for the Bravo dome area and the Slick Hills area of southern Oklahoma (Wilhelm and Morgan, 1986), an area adjacent to and northwest of the Marietta basin. Surface and subsurface fractures in the Bravo dome area form bimodal patterns oriented west–northwest and northeast (Gustavson and Budnik, 1985; Gustavson and Finley, 1985). Mapped lineaments in the Slick Hills area also formed bimodal trends and were oriented northwest and northeast (Wilhelm and Morgan, 1986). From the limited data available, the degree of influence fractures or joint sets may have on the reorientation or rotation of stress trajectories within the Marietta basin or Bravo dome study areas is not known. However, the area between subparallel basement faults in these two areas may have experienced intense fracturing histories resulting in a variety of fracture orientations. Fractures or joint sets between subparallel faults having orientations aligned with the local orientation of $SH_{\max}$ could aid in the development of elliptical enlargements.

From the above discussion, it is not possible to state conclusively that the observed apparent variation in the calculated mean stress orientations of any individual well is a direct indication of stress trajectory rotation within the two study areas. However, in light of the apparent observed shift in the elliptical long-axis orientation of wellbore enlargements, it does seem reasonable to speculate on the possibility of stress trajectory rotation given the location of drill holes with respect to mapped subparallel faults (Fig. 4). A bimodal-orthogonal distribution of wellbore enlargements would be expected in the Bravo dome area where the maximum principal stress (σ_1) is vertical (Stone and Webster Engineering Corp., 1983) and thus, where wellbore enlargements associated with drilling-induced hydraulic fracturing (Healy et al., 1984; Stock et al., 1985) would be more likely to form. A similar bimodal-orthogonal distribution of wellbore enlargement orientations would not necessarily be expected for data from the Marietta basin. The Marietta basin is located within the Midcontinent stress province, a region dominated by east west compressional tectonics (Zoback and Zoback, 1980). However, suspected strain accumulation of east–northeast-striking subparallel faults extending into the Marietta basin, in response to the east–northeast/west–southwest regional $SH_{\max}$, may have generated a left-lateral sense of displacement on these faults. In this scenario, a left-lateral sense of displacement could generate a local extensional state of stress in the vicinity of these faults where stress trajectories would be locally rotated and where the orientations of stress sensitive features like elliptical wellbore enlargements would reflect this local state of stress anomaly.

Conclusions

The appearance of a systematic change in the orientation of stress-induced or stress-related elliptical wellbore enlargements among some closely spaced drill holes located within a structurally well-defined area suggests a local variation in the regional horizontal stress pattern. Detailed knowledge of local variations in the regional state of stress is useful when evaluating the seismic potential of an area or when local stress conditions influence the success of a borehole drilling or a wellbore fracturing program. Questions of this nature can be addressed through a detailed and systematic analysis of stress-related wellbore features.

References

Amadei, B., Swolfs, H.S. and Savage, W.Z., 1988. Gravity-induced stresses in stratified rock masses. Rock Mech. Rock Eng., 21: 1–20.

Bell, J.S. and Gough, D.I., 1979. Northeast–southwest compressive stress in Alberta — evidence from oil wells. Earth Planet. Sci. Lett., 45: 475–482.

Bell, J.S. and Gough, D.I., 1982. The use of borehole breakouts in the study of crustal stress. In: M.D. Zoback and B.C. Haimson (Editors), Proceedings of Workshop XVII — Workshop on Hydraulic Fracturing Stress Measurements. U.S. Geol. Surv., Open-File Rep., 82-1575: 530–557.

Borjeson, R.W. and Lamb, T.J., 1987. Geotechnical borehole testing report Holtzclaw No. 1 well (PD-10), Palo Duro basin. Stones and Webster Engineering Corporation, No. 312.

Brown, R.O., 1978. Application of fracture identification logs in the Cretaceous of north Louisiana and Mississippi. Gulf Coast Assoc. Geol. Soc., Trans., 28: 75–91.

Budnik, R.T., 1987. Late Miocene reactivation of Ancestral Rocky Mountain structures in the Texas Panhandle — a response to basin and range extension. Geology, 15: 163–166.

Budnik, R.T. and Smith, D., 1987. Late Miocene reactivation of Ancestral Rocky Mountain structures in the Texas Panhandle — a response to basin and range extension. Geology, 15: 163–166.

Butler, K.R., 1980. A structural analysis of the Cambro–Ordovician strata on the north flank of the Wichita Mountains, Oklahoma (abstract). Geol. Soc. Am., Abstracts with Programs, 12: 1–2.

Cox, J.W., 1970. The high resolution dipmeter reveals dip-related borehole and formation characteristics. Soc. Prof. Well Log

Analysis, 11th Annu. Logging Symp., May 3–6, 1970, pp. 1–25.

Dart, R.L., 1989. Horizontal stresses from wellbore breakouts and lithologies associated with their formation, Oklahoma and Texas Panhandle. Okla. Geol. Surv. Circ., 90: 97–120.

Dart, R.L. and Zoback, M.L., 1989. Well bore-breakout stress analysis within the Continental United States. Log Anal. J., 30: 12–25.

Donovan, R.N., Sanderson, D.J., Marchini, D., 1982. An analysis of structures resulting from left-lateral strike-slip movement between the Wichita Mountains and Anadarko Basin, southwestern Oklahoma. Geol. Soc. Am., Abstracts with Programs, 14: 476.

Dutton, S.P., Goldstein, A.G. and Ruppel, S.C., 1982. Petroleum potential of the Palo Duro Basin, Texas Panhandle. Texas Bur. Econ. Geol., Rep. Invest., 123: 80.

Frederickson, E.A. and Redman, R.H., 1965. Geology of Love County, in Geology and petroleum of Love County, Oklahoma. Okla. Geol. Surv. Circ., 63: 7–47.

Gough, D.I. and Bell, J.S., 1981. Stress orientations from oil-well fractures in Alberta and Texas. Can. J. Earth Sci., 18: 638–645.

Gough, D.I. and Bell, J.S., 1982. Stress orientation from borehole wall fractures with examples from Colorado, east Texas and northern Canada. Can. J. Earth Sci., 19: 1358–1370.

Gustavson, T.C. and Budnik, R.T., 1985. Structural influences on geomorphic processes and physiographic features, Texas Panhandle — technical issues in siting a nuclear-waste repository. Geology, 13: 173–176.

Gustavson, T.C. and Finley, R.J., 1985. Late Cenozoic geomorphic evolution of the Texas Panhandle and northeastern New Mexico — case studies of structural control on regional drainage development. Texas Bur. Econ. Geol., Rep. Invest., 148: 42.

Haimson, B.C. and Herrick, C.G., 1985. In situ stress evaluation from borehole breakouts, experimental studies. Proc. 26th U.S. Symp. on Rock Mechanics, Rapid City, S.D., 1985, 2: 1207–1218.

Ham, W.E., Denison, R.E. and Merritt, C.A., 1964. Basement rocks and structural evolution of southern Oklahoma. Okla. Geol. Surv. Bull., 95: 302.

Harlton, B.H., 1963. Frontal Wichita Fault system of southwestern Oklahoma. Am. Assoc. Pet. Geol., Bull., 47: 1552–1580.

Healy, J.H., Hickman, S.H., Zoback, M.D. and Ellis, W.L., 1984. Report on televiewer log and stress measurements in core hole USW-G1, Nevada Test Site, December 13–22, 1981. U.S. Geol. Surv., Open-File Rep., 84-15: 47.

Hempton, M.R. and Neher, K., 1986. Experimental fracture, strain and subsidence patterns over en echelon strike-slip faults — implications for the structural evolution of pull-apart basins. J. Struct. Geol., 8: 597–605.

Henry, G.E., 1968. Recent development in the Marietta basin. In: J.S. Wendell (Editor), Basins of the Southwest, 1. American Association of Petroleum Geologists, Tulsa, Okla., pp. 71–78.

Hickman, S.H., Healy, J.H. and Zoback, M.D., 1985. In situ stress, natural fracture distribution and borehole enlargement in the Auburn geothermal well — Auburn, New York. J. Geophys. Res., 90: 5497–5512.

Hooker, V.B. and Johnson, C.F., 1969. Near-surface horizontal stresses, including the effects of rock anisotrophy. U.S. Bur. Mines, Rep. Invest., 7654: 18.

Hoffman, P., Dewey, J.F. and Burke, K., 1974. Aulacogens and their genetic relation to geosynclines and a Proterozoic example from Great Slave Lake, Canada. In: R.H. Dott, Jr. and R.H. Shaver (Editors), Modern and Ancient Geosynclinal Sedimentation. Soc. Econ. Paleontol. Mineral., Spec. Publ., 19: 38–55.

Hudson, J.A. and Cooling, C.M., 1988. In situ rock studies and then measurement. In the U.K. — Part 1, the current states of knowledge. Int. J. Rock Mech., Mining Sci. Geomechan., Abstr., 25(6): 363–370.

Johnson, K.S., Branson, C.C., Curtis, N.M., Jr., Ham, W.E., Marcher, M.V. and Roberts, J.F., 1972. Geology and earth resources of Oklahoma, an atlas of maps and cross sections. Okla. Geol. Surv. Educ. Publ., 1: 8.

Johnson, K.S. and Denison, R.E., 1973. Igneous geology of the Wichita Mountains and economic geology of Permian rocks in southwest Oklahoma. Geol. Soc. Am., Guidebook for Field Trip, 6: 35.

Johnson, K.S., Bruchfield, M.R. and Harrison, W.E., 1984. Guidebook for Arbuckle Formation field trip, southern Oklahoma. Okla. Geol. Surv. Spec. Publ., 84-1: 23.

Jordan, L., 1967. Geology of Oklahoma — a summary. Okla. Geol. Surv., Geol. Notes, 27: 215–229.

Mastin, L.G., 1984. The development of borehole breakouts in sandstone. M.S. Thesis, Stanford University, Stanford, Calif., 101 pp.

McLean, R. and Sterns, D.W., 1983. Fault analysis in Wichita Mountains. Am. Assoc. Pet. Geol., Bull., 67: 511–512 (abstract).

McGookey, D.A. and Goldstein, A.G., 1982. Structural influence on deposition and deformation at the northwest margin of the Palo Duro Basin, in Geology and geohydrology of the Palo Duro Basin, Texas Panhandle — a report on the progress of nuclear waste isolation feasibility studies. Texas Bur. Econ. Geol. Circ., 82-7: 28–37.

Morin, R.H., Anderson, R.N. and Barton, C.A., 1989. Analysis and interpretation of the borehole televiewer log — information on the state of stress and the lithostratigraphy at hole 504B. Proc. Ocean Drilling Program, Scientific Results, Vol. III, College Station, Texas.

Plumb, R.A. and Hickman, S.H., 1985. Stress-induced borehole elongation; a comparison between the four-arm dipmeter and the borehole televiewer in the Auburn geothermal well. J. Geophys. Res., 90(B7): 5513–5521.

Presley, M.W., 1980. Upper Permian salt-bearing stratigraphic units, in geology and geohydrology of the Palo Duro Basin, Texas Panhandle — a report on the progress of nuclear waste isolation feasibility studies. Texas Bur. Econ. Geol. Circ., 80-7: 33–40.

Rodgers, D.A., 1984. Analysis of pull-apart basin development produced by en echelon strike-slip faults In: A.G. Sylvester (compiler), Wrench Fault Tectonics. Am. Assoc. Pet. Geol., Reprint Ser., 28: 345–360.

Segall, P. and Pollard, D.D., 1980. Mechanics of discontinuous faults. J. Geophys. Res., 85(B8): 4337–4350.

Stock, J.M., Healy, J.H., Hickman, S.H. and Zoback, M.D., 1985. Hydraulic fracturing stress measurements at Yucca Mountain, Nevada and relationship to the regional stress field. J. Geophys. Res., 90(B10): 8691–8706.

Stone and Webster Engineering Corporation, 1983. Area geological characterization report for the Palo Duro and Dalhart basins, Texas. Tech. Rep. DOE/CH/10140-1, 429 pp.

Teufel, L.W., 1985. Insights into the relationship between well bore breakouts, natural fractures and in situ stress. Proc. 26th U.S. Symp.on Rock Mechanics, Rapid City, S.D., 1985, 2: 1199–1206.

Tomlinson, L.W. and McBee, W., Jr., 1959. Pennsylvania sediments and orogenies of Ardmore district, Oklahoma. In: Petroleum Geology of Southern Oklahoma, 2. American Association of Petroleum Geologists, Tulsa, Okla., pp. 3–52.

von Schonfeldt, H.A., Kehle, R.O. and Gray, K.E., 1973. Mapping of stress field in the upper Earth's crust of the U.S.. U.S. Geol. Surv., Final Tech. Rep., Grant 14-08-0001-1222, Reston, Va., 78 pp.

Wickham, J.S., 1978. The southern Oklahoma aulacogen In: J.S. Wickham and R.E. Denison (Editors), Structural Style of the Arbuckle Region. Univ. Okla. Guidebook for Geol. Soc. Am. South-Central Annu. Meet. Field Trip, 3: 9–41.

Wilhelm, S.J. and Morgan, K.M., 1986. Utilization of Lansat thermatic-mapper data for lineament analysis of the Slick Hills area. In: R.N. Donovan (Editor), The Slick Hills of Southwestern Oklahoma — Fragments of an Aulacogen. Okla. Geol. Surv., Guidebook, 24: 35–39.

Zoback, M.D., Moss, Daniel, Mastin, R.L. and Anderson, R.N., 1985. Wellbore breakouts and in-situ stress. J. Geophys. Res., 90(B7): 5523–5530.

Zoback, M.L. and Zoback, M., 1980. State of stress in the conterminous United States. J. Geophys. Res., 85(B121): 6113–6156.

R.L. DART *Branch of Geologic Risk Assessment, U.S. Geological Survey, Box 25046, MS 966, Denver Federal Center, Denver, CO 80225, U.S.A.*

H.S. SWOLFS *U.S. Geological Survey, Box 25046, MS 966, Denver Federal Center, Denver, CO 80225, U.S.A.*

References index

Subject index